Grundlehren der mathematischen Wissenschaften 323

A Series of Comprehensive Studies in Mathematics

For further volumes:
www.springer.com/series/138

Jürgen Neukirch
Alexander Schmidt
Kay Wingberg

Cohomology of Number Fields

Second Edition
Corrected Second Printing

Springer

Jürgen Neukirch †

Alexander Schmidt
Mathematical Institute
Heidelberg University
Heidelberg, Germany

Kay Wingberg
Mathematical Institute
Heidelberg University
Heidelberg, Germany

ISSN 0072-7830 Grundlehren der mathematischen Wissenschaften
ISBN 978-3-662-51745-1 ISBN 978-3-540-37889-1 (eBook)
DOI 10.1007/978-3-540-37889-1
Springer Heidelberg New York Dordrecht London

Mathematics Subject Classification: 11Gxx, 11Rxx, 11Sxx, 12Gxx, 14Hxx, 20Jxx

Printed on acid-free paper

Springer is part of Springer Science+Business Media (www.springer.com)

Vorwort

Als unser Freund und Lehrer Jürgen Neukirch Anfang 1997 starb, hinterließ er den Entwurf zu einem Buch über die Kohomologie der Zahlkörper, welches als zweiter Band zu seiner Monographie *Algebraische Zahlentheorie* gedacht war. Für die Kohomologie proendlicher Gruppen, sowie für Teile der Kohomologie lokaler und globaler Körper lag bereits eine Rohfassung vor, die schon zu einer regen Korrespondenz zwischen Jürgen Neukirch und uns geführt hatte.

In den letzten zwei Jahren ist, ausgehend von seinem Entwurf, das hier vorliegende Buch entstanden. Allerdings wussten wir nur teilweise, was Jürgen Neukirch geplant hatte. So mag es sein, dass wir Themen ausgelassen haben, welche er berücksichtigen wollte, und anderes, nicht Geplantes, aufgenommen haben.

Jürgen Neukirchs inspirierte und pointierte Art, Mathematik auf hohem sprachlichen Niveau darzustellen, ist für uns stets Vorbild gewesen. Leider erreichen wir nicht seine Meisterschaft, aber wir haben uns alle Mühe gegeben und hoffen, ein Buch in seinem Sinne und nicht zuletzt auch zum Nutzen seiner Leser fertig gestellt zu haben.

Heidelberg, im September 1999

Alexander Schmidt
Kay Wingberg

Introduction

Number theory, one of the most beautiful and fascinating areas of mathematics, has made major progress over the last decades, and is still developing rapidly. In the beginning of the foreword to his book *Algebraic Number Theory*, J. Neukirch wrote

> „Die Zahlentheorie nimmt unter den mathematischen Disziplinen eine ähnlich idealisierte Stellung ein wie die Mathematik selbst unter den anderen Wissenschaften." *)

Although the joint authors of the present book wish to reiterate this statement, we wish to stress also that number theory owes much of its current strong development to its interaction with almost all other mathematical fields. In particular, the geometric (and consequent functorial) point of view of arithmetic geometry uses techniques from, and is inspired by, analysis, geometry, group theory and algebraic topology. This interaction had already started in the 1950s with the introduction of group cohomology to local and global class field theory, which led to a substantial simplification and unification of this area.

The aim of the present volume is to provide a textbook for students, as well as a reference book for the working mathematician on cohomological topics in number theory. Its main subject is Galois modules over local and global fields, objects which are typically associated to arithmetic schemes. In view of the enormous quantity of material, we were forced to restrict the subject matter in some way. In order to keep the book at a reasonable length, we have therefore decided to restrict attention to the case of dimension less than or equal to one, i.e. to the global fields themselves, and the various subrings contained in them. Central and frequently used theorems such as the global duality theorem of *G. Poitou* and *J. Tate*, as well as results such as the theorem of *I. R. Šafarevič* on the realization of solvable groups as Galois groups over global fields, had been part of algebraic number theory for a long time. But the proofs of statements like these were spread over many original articles, some of which contained serious mistakes, and some even remained unpublished. It was the initial motivation of the authors to fill these gaps and we hope that the result of our efforts will be useful for the reader.

In the course of the years since the 1950s, the point of view of class field theory has slightly changed. The classical approach describes the Galois groups

*) "Number theory, among the mathematical disciplines, occupies a similar idealized position to that held by mathematics itself among the sciences."

of *finite* extensions using arithmetic invariants of the local or global ground field. An essential feature of the modern point of view is to consider infinite Galois groups instead, i.e. one investigates the set of all finite extensions of the field k at once, via the *absolute* Galois group G_k. These groups intrinsically come equipped with a topology, the Krull topology, under which they are Hausdorff, compact and totally disconnected topological groups. It proves to be useful to ignore, for the moment, their number theoretical motivation and to investigate topological groups of this type, the profinite groups, as objects of interest in their own right. For this reason, an extensive "algebra of profinite groups" has been developed by number theorists, not as an end in itself, but always with concrete number theoretical applications in mind. Nevertheless, many results can be formulated solely in terms of profinite groups and their modules, without reference to the number theoretical background.

The first part of this book deals with this "profinite algebra", while the arithmetic applications are contained in the second part. This division should not be seen as strict; sometimes, however, it is useful to get an idea of how much algebra and how much number theory is contained in a given result.

A significant feature of the arithmetic applications is that classical reciprocity laws are reflected in duality properties of the associated infinite Galois groups. For example, the reciprocity law for local fields corresponds to Tate's duality theorem for local cohomology. This duality property is in fact so strong that it becomes possible to describe, for an arbitrary prime p, the Galois groups of the maximal p-extensions of local fields. These are either free groups or groups with a very special structure, which are now known as Demuškin groups. This result then became the basis for the description of the full absolute Galois group of a p-adic local field by *U. Jannsen* and the third author.

The global case is rather different. As was already noticed by *J. Tate*, the absolute Galois group of a global field is *not* a duality group. It is the geometric point of view, which offers an explanation of this phenomenon: the duality comes from the *curve* rather than from its generic point. It is therefore natural to consider the étale fundamental groups $\pi_1^{et}(\mathrm{Spec}(\mathcal{O}_{k,S}))$, where S is a finite set of places of k. Translated to the language of Galois groups, the fundamental group of $\mathrm{Spec}(\mathcal{O}_{k,S})$ is a quotient of the full group G_k, namely, the Galois group $G_{k,S}$ of the maximal extension of k which is unramified outside S. If S contains all places that divide the order of the torsion of a module M, the central *Poitou-Tate duality theorem* provides a duality between the localization kernels in dimensions one and two. In conjunction with Tate local duality, this can also be expressed in the form of a long 9-term sequence. The duality theorem of Poitou-Tate remains true for infinite sets of places S and, using topologically restricted products of local cohomology groups, the long exact sequence can be generalized to this case. The question of whether the group $G_{k,S}$ is a duality group when S is finite was positively answered by the second author.

As might already be clear from the above considerations, the basic technique used in this book is Galois cohomology, which is essential for class field theory. For a more geometric point of view, it would have been desirable to have also formulated the results throughout in the language of étale cohomology. However, we decided to leave this to the reader. Firstly, the technique of sheaf cohomology associated to a Grothendieck topos is sufficiently covered in the literature (see [5], [139], [228]) and, in any case, it is an easy exercise (at least in dimension ≤ 1) to translate between the Galois and the étale languages. A further reason is that results which involve infinite sets of places (necessary when using Dirichlet density arguments) or infinite extension fields, can be much better expressed in terms of Galois cohomology than of étale cohomology of pro-schemes. When the geometric point of view seemed to bring a better insight or intuition, however, we have added corresponding remarks or footnotes. A more serious gap, due to the absence of Grothendieck topologies, is that we cannot use *flat* cohomology and the global flat duality theorem of Artin-Mazur. In chapter VIII, we therefore use an ad hoc construction, the group Б_S, which measures the size of the localization kernel for the first flat cohomology group with the roots of unity as coefficients.

Let us now examine the contents of the individual chapters more closely. The first part covers the algebraic background for the number theoretical applications. Chapter I contains well-known basic definitions and results, which may be found in several monographs. This is only partly true for chapter II: the explicit description of the edge morphisms of the Hochschild-Serre spectral sequence in §2 is certainly well-known to specialists, but is not to be found in the literature. In addition, the material of §3 is well-known, but contained only in original articles.

Chapter III considers abstract duality properties of profinite groups. Among the existing monographs which also cover large parts of the material, we should mention the famous *Cohomologie Galoisienne* by *J.-P. Serre* and *H. Koch*'s book *Galoissche Theorie der p-Erweiterungen*. Many details, however, have been available until now only in the original articles.

In chapter IV, free products of profinite groups are considered. These are important for a possible non-abelian decomposition of global Galois groups into local ones. This happens only in rather rare, degenerate situations for Galois groups of global fields, but it is quite a frequent phenomenon for subgroups of infinite index. In order to formulate such statements (like the arithmetic form of Riemann's existence theorem in chapter X), we develop the concept of the free product of a *bundle* of profinite groups in §3.

Chapter V deals with the algebraic foundations of Iwasawa theory. We will not prove the structure theorem for Iwasawa modules in the usual way by using matrix calculations (even though it may be more acceptable to some mathematicians, as it is more concrete), but we will follow mostly the presentation found in Bourbaki, Commutative Algebra, with a view to more general situations. Moreover, we present results concerning the structure of these modules up to isomorphism, which are obtained using the homotopy theory of modules over group rings, as presented by *U. JANNSEN*.

The central technical result of the arithmetic part is the famous global duality theorem of Poitou-Tate. We start, in chapter VI, with general facts about Galois cohomology. Chapter VII deals with local fields. Its first three sections largely follow the presentation of *J.-P. SERRE* in *Cohomologie Galoisienne*. The next two sections are devoted to the explicit determination of the structure of local Galois groups. In chapter VIII, the central chapter of this book, we give a complete proof of the Poitou-Tate theorem, including its generalization to finitely generated modules. We begin by collecting basic results on the topological structure, universal norms and the cohomology of the S-idèle class group, before moving on to the proof itself, given in sections 4 and 6. In the proof, we apply the group theoretical theorems of Nakayama-Tate and of Poitou, proven already in chapter III.

In chapter IX, we reap the rewards of our efforts in the previous chapters. We prove several classical number theoretical results, such as the Hasse principle and the Grunwald-Wang theorem. In §4, we consider embedding problems and we present the theorem of *K. IWASAWA* to the effect that the maximal prosolvable factor of the absolute Galois group of $\mathbb{Q}^{ab}$ is free. In §5, we give a complete proof of Šafarevič's theorem on the realization of finite solvable groups as Galois groups over global fields.

The main concern of chapter X is to consider restricted ramification. Geometrically speaking, we are considering the *curves* Spec($\mathcal{O}_{k,S}$), in contrast to chapter IX, where our main interest was in the *point* Spec(k). Needless to say, things now become much harder. Invariants like the S-ideal class group or the p-adic regulator enter the game and establish new arithmetic obstructions. Our investigations are guided by the analogy between number fields and function fields. We know a lot about the latter from algebraic geometry, and we try to establish analogous results for number fields. For example, using the group theoretical techniques of chapter IV, we can prove the number theoretical analogue of Riemann's existence theorem. The fundamental group of Spec($\mathcal{O}_k$), i.e. the Galois group of the maximal unramified extension of the number field k, was the subject of the long-standing *class field tower* problem in number theory, which was finally answered negatively by *E. S. GOLOD* and

I. R. Šafarevič. We present their proof, which demonstrates the power of the group theoretical and cohomological methods, in §8.

Chapter XI deals with Iwasawa theory, which is the consequent conceptual continuation of the analogy between number fields and function fields. We concentrate on the algebraic aspects of Iwasawa theory of p-adic local fields and of number fields, first presenting the classical statements which one can usually find in the standard literature. Then we prove more far-reaching results on the structure of certain Iwasawa modules attached to p-adic local fields and to number fields, using the homotopy theory of Iwasawa modules. The analytic aspects of Iwasawa theory will merely be described, since this topic is covered by several monographs, for example, the book [246] of *L. Washington*. Finally, the Main Conjecture of Iwasawa theory will be formulated and discussed; for a proof, we refer the reader to the original work of *B. Mazur* and *A. Wiles* ([134], [249]).

In the last chapter, we give a survey of so-called *anabelian geometry*, a program initiated by *A. Grothendieck*. Perhaps the first result of this theory, obtained even before this program existed, is a theorem of *J. Neukirch* and *K. Uchida* which asserts that the absolute Galois group of a global field, as a profinite group, characterizes the field up to isomorphism. We give a proof of this theorem for number fields in the first two sections. The final section gives an overview of the conjectures and their current status.

The reader will recognize very quickly that this book is not a basic textbook in the sense that it is completely self-contained. We use freely basic algebraic, topological and arithmetic facts which are commonly known and contained in the standard textbooks. In particular, the reader should be familiar with basic number theory. While assuming a certain minimal level of knowledge, we have tried to be as complete and as self-contained as possible at the next stage. We give full proofs of almost all of the main results, and we have tried not to use references which are only available in original papers. This makes it possible for the interested student to use this book as a textbook and to find large parts of the theory coherently ordered and gently accessible in one place. On the other hand, this book is intended for the working mathematician as a reference on cohomology of local and global fields.

Finally, a remark on the exercises at the end of the sections. A few of them are not so much exercises as additional remarks which did not fit well into the main text. Most of them, however, are intended to be solved by the interested reader. However, there might be occasional mistakes in the way they are posed. If such a case arises, it is an additional task for the reader to give the correct formulation.

We would like to thank many friends and colleagues for their mathematical examination of parts of this book, and particularly, *Anton Deitmar*, *Torsten Fimmel*, *Dan Haran*, *Uwe Jannsen*, *Hiroaki Nakamura* and *Otmar Venjakob*. We are indebted to Mrs. *Inge Meier* who TEXed a large part of the manuscript, and *Eva-Maria Strobel* receives our special gratitude for her careful proofreading. Hearty thanks go to *Frazer Jarvis* for going through the entire manuscript, correcting our English.

Heidelberg, September 1999

Alexander Schmidt
Kay Wingberg

Preface to the Second Edition

The present second edition is a corrected and extended version of the first. We have tried to improve the exposition and reorganize the content to some extent; furthermore, we have included some new material. As an unfortunate result, the numbering of the first edition is not compatible with the second.

In the algebraic part you will find new sections on filtered cochain complexes, on the degeneration of spectral sequences and on Tate cohomology of profinite groups. Amongst other topics, the arithmetic part contains a new section on duality theorems for unramified and tamely ramified extensions, a careful analysis of 2-extensions of real number fields and a complete proof of Neukirch's theorem on solvable Galois groups with given local conditions.

Since the publication of the first edition, many people have sent us lists of corrections and suggestions or have contributed in other ways to this edition. We would like to thank them all. In particular, we would like to thank *Jakob Stix* and *Denis Vogel* for their comments on the new parts of this second edition and *Frazer Jarvis*, who again did a great job correcting our English.

Regensburg and Heidelberg, November 2007

Alexander Schmidt
Kay Wingberg

Contents

Algebraic Theory

Chapter I

Cohomology of Profinite Groups

Profinite groups are topological groups which naturally occur in algebraic number theory as Galois groups of infinite field extensions or more generally as étale fundamental groups of schemes. Their cohomology groups often contain important arithmetic information.

In the first chapter we will study profinite groups as objects of interest in themselves, independently of arithmetic applications, which will be treated in the second part of this book.

§1. Profinite Spaces and Profinite Groups

The underlying topological spaces of profinite groups are of a very specific type, which will be described now. To do this, we make use of the concept of inverse (or projective) limits. We refer the reader to the standard literature (e.g. [160], [79], [139]) for the definition and basic properties of limits. All index sets will be assumed to be filtered.

(1.1.1) Lemma. *For a Hausdorff topological space T the following conditions are equivalent.*

(i) *T is the (topological) inverse limit of finite discrete spaces.*

(ii) *T is compact and every point of T has a basis of neighbourhoods consisting of subsets which are both closed and open.*

(iii) *T is compact and totally disconnected.*

Proof: In order to show the implication (i) $\Rightarrow$ (ii), we first recall that the inverse limit of compact spaces is compact (see [15] chap.I, §9, no.6, prop.8). Therefore T is compact. By the definition of the inverse limit topology and by (i), every point of T has a basis of neighbourhoods consisting of sets of the

J. Neukirch et al., *Cohomology of Number Fields*,
Grundlehren der mathematischen Wissenschaften 323,
DOI 10.1007/978-3-540-37889-1_1, © Springer-Verlag Berlin Heidelberg 2008

form $f^{-1}(W)$, where W is a subset of a finite discrete space V and $f : T \to V$ is a continuous map. These sets are both open and closed.

For the implication (ii) $\Rightarrow$ (iii) we have to show that the connected component C_t of every point $t \in T$ equals $\{t\}$. Since T is compact, C_t is the intersection of all closed and open subsets containing t (see [15] chap.II, §4, no.4, prop.6). Since T is Hausdorff, we obtain $C_t = \{t\}$.

It remains to show the implication (iii) $\Rightarrow$ (i). Let I be the set of equivalence relations $R \subseteq T \times T$ on T, such that the quotient space T/R is finite and discrete in the quotient topology. The set I is partially ordered by inclusion and is directed, because $R_1 \cap R_2$ is in I if R_1 and R_2 are. We claim that the canonical map $\phi : T \to \varprojlim_{R \in I} T/R$ is a homeomorphism.

First we see that the map ϕ is surjective, because for an element $\{t_R\}_{R\in I} \in \varprojlim_{R\in I} T/R$, the sets $(p_R \circ \phi)^{-1}(t_R)$ are nonempty and compact. Since I is directed, finite intersections of these sets are also nonempty and compactness then implies that $\phi^{-1}(\{t_R\}_{R\in I}) = \bigcap_{R\in I}(p_R \circ \phi)^{-1}(t_R)$ is nonempty.

For the injectivity it suffices to show that for $t, s \in T$, $t \neq s$, there exists an $R \in I$ such that $(t, s) \notin R$. But since s is not in the connected component of t, there exists a closed and open subset $U \subseteq T$ with $t \in U$, $s \notin U$ (see [15] chap.II, §4, no.4, prop.6). Then the equivalence relation R defined by "$(x, y) \in R$ if x and y are both in U or both not in U" is of the required type. The proof is completed by the remark that a continuous bijection between compact spaces is a homeomorphism. □

In fact one immediately verifies that we could have chosen the inverse system in (i) in such a way that all transition maps are surjective.

(1.1.2) Definition. *A space T is called a* **profinite space** *if it satisfies the equivalent conditions of lemma* (1.1.1).

A compactness argument shows that a subset $V \subseteq \varprojlim X_i$ of a profinite space is both closed and open if and only if V is the pre-image under the canonical projection $p_i : X \to X_i$ of a (necessarily closed and open) subset in X_i for some i. Every continuous map between profinite spaces can be realized as a projective limit of maps between finite discrete spaces. Without giving an exact definition, we want to note that the category of profinite spaces with continuous maps is the *pro-category of the category of finite discrete spaces.*

Recall that a topological group is a group G endowed with the structure of a topological space, such that the group operations $G \to G$, $g \mapsto g^{-1}$, and

$G \times G \to G$, $(g, h) \mapsto gh$, are continuous. The reader will immediately verify that the inverse limit of an inverse system of topological groups is just the inverse limit of the groups together with the inverse limit topology on the underlying topological space.

(1.1.3) Proposition. *For a Hausdorff topological group G the following conditions are equivalent.*

(i) *G is the (topological) inverse limit of finite discrete groups.*

(ii) *G is compact and the unit element has a basis of neighbourhoods consisting of open and closed normal subgroups.*

(iii) *G is compact and totally disconnected.*

Proof: (i) ⇒ (iii): The inverse limit of compact and totally disconnected spaces is compact and totally disconnected.
(ii) ⇒ (i): Assume that U runs through a system of neighbourhoods of the unit element $e \in G$, which consists of open normal subgroups. Then the canonical homomorphism $\phi : G \to \varprojlim_U G/U$ is an isomorphism:

To begin with, ϕ is injective, because G is Hausdorff. In order to show the surjectivity, let $x = \{x_U\}_U \in \varprojlim_U G/U$. Denoting the canonical projection by $\phi_U : G \to G/U$, we have the equality

$$\phi^{-1}(x) = \bigcap_U \phi_U^{-1}(x_U).$$

The intersection on the right side is taken over nonempty compact spaces and finite intersections of these are nonempty. Hence $\phi^{-1}(x)$ is nonempty, and therefore ϕ is surjective. Furthermore, ϕ is open, hence a homeomorphism. Finally, for every such U, the group G/U is discrete and compact, hence finite.
(iii) ⇒ (ii): By (1.1.1), the underlying topological space of G is profinite, hence every point has a basis of neighbourhoods consisting of open and closed subsets. Note that an open subgroup is automatically closed, because it is the complement of the union of its (open) nontrivial cosets. Let U be an arbitrary chosen, closed and open neighbourhood of the unit element $e \in G$. Set

$$V := \{v \in U \mid Uv \subseteq U\}, \qquad H := \{h \in V \mid h^{-1} \in V\}.$$

We claim that $H \subseteq U$ is an open (and closed) subgroup in G. We first show that V is open. Fix a point $v \in V$. Then $uv \in U$ for every $u \in U$ and therefore there exist neighbourhoods U_u of u and V_u of v, such that $U_u V_u \subseteq U$. The open sets U_u cover the compact space U and therefore there exists a finite subcover, $U_{u_1}, \dots, U_{u_n}$, say. Then

$$V_v := V_{u_1} \cap \dots \cap V_{u_n}$$

is an open neighbourhood of v contained in V. Hence V is open and also $H := V \cap V^{-1}$, since the inversion map is a homeomorphism. It remains to show that H is a subgroup. Trivially $e \in H$ and $H^{-1} = H$ by construction. We now check that $xy \in H$ if $x, y \in H$. First we have $Uxy \subseteq Uy \subseteq U$, and so $xy \in V$. In the same way we obtain $y^{-1}x^{-1} \in V$, hence $xy \in H$. This proves that H is an open subgroup of G contained in U. In particular, H has finite index in G and there are only finitely many different conjugates of H. The intersection of these finitely many conjugates is an open, closed and normal subgroup of G contained in U. □

(1.1.4) Definition. *A Hausdorff topological group G satisfying the equivalent conditions of* (1.1.3) *is called a* **profinite group**.

Without further mention, homomorphisms between profinite groups are always assumed to be continuous and subgroups are assumed to be closed. Since a subgroup is the complement of its nontrivial cosets and by the compactness of G, we see that open subgroups are closed and a closed subgroup is open if and only if it has finite index. If H is a (closed) subgroup of the profinite group G, then the set G/H of coset classes with the quotient topology is a profinite space. If H is normal, then the quotient G/H is a profinite group in a natural way.

In principle, all objects and statements of the theory of finite groups have their topological analogue in the theory of profinite groups. For example, the profinite analogues of the Sylow theorems are true (see §6). We make the following

(1.1.5) Definition. *A* **supernatural number** *is a formal product*

$$\prod_p p^{n_p},$$

where p runs through all prime numbers and, for each p, the exponent n_p is a non-negative integer or the symbol ∞.

Using the unique decomposition into prime powers, we can view any natural number as a supernatural number. We multiply supernatural numbers (even infinitely many of them) by adding the exponents. By convention, the sum of the exponents is ∞ if infinitely many summands are non-zero or if one of the summands is ∞. We also have the notions of l.c.m. and g.c.d. of an arbitrary family of supernatural numbers. In particular, any family of natural numbers has an l.c.m., which is, in general, a supernatural number.

(1.1.6) Definition. *Let G be a profinite group and let A be an abelian torsion group.*

(i) *The* **index** *of a closed subgroup H in G is the supernatural number*

$$(G:H) = l.c.m.(G/U : H/H \cap U),$$

where U ranges over all open normal subgroups of G.

(ii) *The* **order** *of G is defined by*

$$\#G = (G:1) = \underset{U}{l.c.m.}\,\#(G/U).$$

(iii) *The* **order** *of A is defined by*

$$\#A = l.c.m.\,\#B\,,$$

where B ranges over all finite subgroups of A.

Given closed subgroups $N \subseteq H \subseteq G$, we have

$$(G:N) = (G:H)(H:N).$$

Furthermore, the order $\#A$ of an abelian torsion group A is just the order of the profinite group $\mathrm{Hom}(A, \mathbb{Q}/\mathbb{Z})$.

(1.1.7) Definition. *Let G be a profinite group. An* **abstract G-module** *M is an abelian group M together with an action*

$$G \times M \to M,\ (g, m) \mapsto g(m)$$

such that $1(m) = m$, $(gh)(m) = g(h(m))$ and $g(m+n) = g(m) + g(n)$ *for all* $g, h \in G,\ m, n \in M$.

A **topological G-module** *M is an abelian Hausdorff topological group M which is endowed with the structure of an abstract G-module such that the action $G \times M \to M$ is continuous.*

For a closed subgroup $H \subseteq G$ we denote the subgroup of H-invariant elements in M by M^H, i.e.

$$M^H = \{m \in M \mid h(m) = m \text{ for all } h \in H\}.$$

(1.1.8) Proposition. *Let G be a profinite group and let M be an abstract G-module. Then the following conditions are equivalent:*

(i) *M is a discrete G-module, i.e. the action $G \times M \to M$ is continuous for the discrete topology on M.*

(ii) *For every $m \in M$ the subgroup $G_m := \{g \in G \mid g(m) = m\}$ is open.*

(iii) *$M = \bigcup M^U$, where U runs through the open subgroups of G.*

Proof: If we restrict the map $G \times M \to M$ to $G \times \{m\}$, then $m \in M$ has pre-image $G_m \times \{m\}$. This shows (i)⇒(ii). The assertion (ii)⇒(iii) is trivial because $m \in M^{G_m}$. Finally, assume that (iii) holds. Let $(g, m) \in G \times M$. There exists an open subgroup U such that $m \in M^U$. Therefore $gU \times \{m\}$ is an open neighbourhood of $(g, m) \in G \times M$ mapping to $g(m)$. This shows (i). □

In this book we are mainly concerned with discrete modules and so the term G-module, without the word "topological" or "abstract", will always mean a discrete module.

If $(A_i)_{i \in I}$ is a family of discrete G-modules, then their direct sum $\bigoplus_{i \in I} A_i$, endowed with the componentwise G-action $g((a_i)_{i \in I}) = (g(a_i))_{i \in I}$, is again a discrete G-module, but this is not necessarily true for the product. The tensor product

$$A \otimes B = A \otimes_{\mathbb{Z}} B$$

of two discrete modules endowed with the diagonal action $g(a \otimes b) = g(a) \otimes g(b)$ is a discrete module. The set $\mathrm{Hom}(A, B) = \mathrm{Hom}_{\mathbb{Z}}(A, B)$ becomes an abstract G-module by setting $g(\phi)(a) = g(\phi(g^{-1}(a)))$. Its subgroup of invariants

$$\mathrm{Hom}_G(A, B) = \mathrm{Hom}(A, B)^G$$

is the set of G-homomorphisms from A to B. If $A = A^U$ for some open subgroup $U \subseteq G$, then $\mathrm{Hom}(A, B)$ is a discrete G-module. This is the case, for example, if G is finite or if A is finitely generated as a $\mathbb{Z}$-module.

The groups $\mathbb{Z}, \mathbb{Q}, \mathbb{Z}/n\mathbb{Z}, \mathbb{F}_q$ are always viewed as *trivial* discrete G-modules, i.e. G-modules with trivial action of G.

So far we have considered totally disconnected compact groups. If A is any topological group, then the *connected component* A_0 (of the identity) of A is a closed subgroup. We have the following general facts for which we refer to [170] sec. 22, and [15] chap. III, §4.6.

(1.1.9) Proposition. *Let A be a locally compact group. Then*

(i) *A_0 is the intersection of all open normal subgroups of A, and A/A_0 is the largest totally disconnected quotient.*

(ii) *A_0 is generated by every open neighbourhood of 1 in A_0.*

(iii) *If $A \to B$ is a continuous surjective homomorphism onto the locally compact group B, then the closure of the image of A_0 is the connected component of 1 in B.*

An essential tool for working with locally compact abelian groups is a duality theorem due to *L. S. Pontryagin*. We consider the group $\mathbb{R}/\mathbb{Z}$ as a topological group with the quotient topology inherited from $\mathbb{R}$.

(1.1.10) Definition. *Let A be a Hausdorff, abelian and locally compact topological group. We call the group*

$$A^\vee := \mathrm{Hom}_{cts}(A, \mathbb{R}/\mathbb{Z})$$

the **Pontryagin dual** *of A.*

Given locally compact topological spaces X, Y, the set of continuous maps $\mathrm{Map}_{cts}(X,Y)$ carries a natural topology, the *compact-open topology*. A subbasis of this topology is given by the sets

$$\mathcal{U}_{K,U} = \{f \in \mathrm{Map}_{cts}(X,Y) \mid f(K) \subseteq U\},$$

where K runs through the compact subsets of X and U runs through the open subsets of Y. For the proof of the following theorem we refer to [170], th. 5.3 or [146], th. 23, [186], th. 1.7.2.

(1.1.11) Theorem (Pontryagin Duality). *If A is a Hausdorff abelian locally compact topological group, then the same is true for $A^\vee$ endowed with the compact-open topology. The canonical homomorphism*

$$A \longrightarrow (A^\vee)^\vee,$$

given by $a \longmapsto \tau_a : A^\vee \to \mathbb{R}/\mathbb{Z},\ \phi \mapsto \phi(a)$, is an isomorphism of topological groups. Thus $\vee$ defines an involutory contravariant autofunctor on the category of Hausdorff abelian locally compact topological groups which moreover commutes with limits. Furthermore, $\vee$ induces equivalences of categories

$$\begin{array}{ccc} (\textit{abelian compact groups}) & \overset{\vee}{\Longleftrightarrow} & (\textit{discrete abelian groups}) \\ (\textit{abelian profinite groups}) & \overset{\vee}{\Longleftrightarrow} & (\textit{discrete abelian torsion groups}). \end{array}$$

For an (abstract) abelian group A we use the notation

$$A^* = \mathrm{Hom}(A, \mathbb{Q}/\mathbb{Z}).$$

Clearly, if A is a discrete torsion group, then $A^\vee \cong A^*$ and we will frequently also write A^* instead of $A^\vee$, at least if we are not interested in the topology of the dual. If A is abelian and profinite, then it is an easy exercise to see that every continuous homomorphism $\phi : A \to \mathbb{R}/\mathbb{Z}$ has finite image. If, moreover, A is topologically finitely generated, then every subgroup of finite index is open in A, and hence also in this case $A^\vee \cong A^*$.

Now assume that we are given a family $(X_i)_{i\in I}$ of Hausdorff, abelian topological groups and let an open subgroup $Y_i \subseteq X_i$ be given for almost all $i \in I$ (i.e. for all but finitely many indices). For consistency of notation, we put $Y_i = X_i$ for the remaining indices.

(1.1.12) Definition. *The* **restricted product**

$$\prod_{i\in I}(X_i, Y_i)$$

is the subgroup of $\prod X_i$ *consisting of all* $(x_i)_{i\in I}$ *such that* $x_i \in Y_i$ *for almost all* i. *The restricted product is a topological group, and a basis of neighbourhoods of the identity is given by the products*

$$\prod_{j\in J} U_j \times \prod_{i\in I\setminus J} Y_i,$$

where J *runs over the finite subsets of* I *and* U_j *runs over a basis of neighbourhoods of the identity of* X_j.

Basic examples of restricted products are the product of groups ($Y_i = X_i$ for all i) and the direct sum of discrete groups ($Y_i = 0$ for all i). We will write

$$\prod_{i\in I} X_i$$

for short if it is clear from the context what the Y_i are. The restricted product is again a Hausdorff, abelian topological group.

(1.1.13) Proposition. *If all* X_i *are locally compact and almost all* Y_i *are compact, then the restricted product is again an abelian locally compact group.*

For the Pontryagin dual of the restricted product, there is a canonical isomorphism

$$\Big(\prod_{i\in I}(X_i, Y_i)\Big)^\vee \cong \prod_{i\in I}(X_i^\vee, (X_i/Y_i)^\vee).$$

Proof: The product of compact topological spaces is compact, therefore the restricted product is locally compact under the given conditions. Furthermore, since Y_i is compact and open in X_i, the same is true for $(X_i/Y_i)^\vee$ in $X_i^\vee$. Hence also $\prod_{i\in I}(X_i^\vee, (X_i/Y_i)^\vee)$ exists and is locally compact.

A sufficiently small open neighbourhood of $0 \in \mathbb{R}/\mathbb{Z}$ contains no nontrivial subgroups. Therefore a continuous homomorphism $\phi : \coprod(X_i, Y_i) \to \mathbb{R}/\mathbb{Z}$ annihilates Y_i for almost all i. In other words, the restriction of ϕ to X_i lies in the subgroup $(X_i/Y_i)^\vee \subseteq X_i^\vee$ for almost all i. This yields a bijection between $(\coprod_{i\in I}(X_i, Y_i))^\vee$ and $\prod_{i\in I}(X_i^\vee, (X_i/Y_i)^\vee)$, which easily can be seen to be a homeomorphism. □

Exercise 1. Show that an injective (resp. surjective) continuous map between profinite spaces may be represented as an inverse limit over a system of injective (resp. surjective) maps between finite discrete spaces.

Exercise 2. Let X be a profinite space and let $X_0 \subseteq X$ be a closed subspace. Show that every continuous map $f : X_0 \to Y$ from X_0 to a finite discrete space Y has a continuous extension $F : X \to Y$ (i.e. $F|_{X_0} = f$) and that any two such extensions coincide on an open neighbourhood of X_0 in X.

Exercise 3. Let G, H be profinite groups. Show that

$$\mathrm{Hom}(G, H) = \varprojlim_{V \subseteq H} \varinjlim_{U \subseteq G} \mathrm{Hom}(G/U, H/V),$$

where the limits are taken over all open normal subgroups V of H and U of G.

Exercise 4. If $K \subseteq H$ are closed subgroups of the profinite group G, then the projection $\pi : G/K \to G/H$ has a continuous section $s : G/H \to G/K$.

Hint: Let X be the set of pairs (S, s), where S is a closed subgroup such that $K \subseteq S \subseteq H$ and s is a continuous section $s : G/H \to G/S$. Write $(S, s) \le (S', s')$ if $S' \subseteq S$ and if s is the composite of s' and the projection $G/S' \to G/S$. Then X is inductively ordered. By Zorn's lemma, there exists a maximal element (S, s) of X. Show that $S = K$.

Exercise 5. A morphism $\phi : X \to Y$ in a category $\mathcal{C}$ is called a *monomorphism* if for every object Z of $\mathcal{C}$ and for every pair of morphisms $f, g : Z \to X$ the implication "$\phi \circ f = \phi \circ g \Rightarrow f = g$" is true. The morphism ϕ is called an *epimorphism* if it is a monomorphism in the opposite category $\mathcal{C}^{op}$ (the category obtained from $\mathcal{C}$ by reversing all arrows).

(i) Show that the monomorphisms in the category of profinite groups are the injective homomorphisms.

(ii) Show that the epimorphisms in the category of profinite groups are the surjective homomorphisms.

Hint for (ii): First reduce the problem to the case of finite groups. Assume that there is an epimorphism $\phi : G \to H$ of finite groups which is not surjective. Assume that $(H : \phi(G)) \ge 3$ (otherwise $\phi(G)$ is normal in H) and choose two elements $a, b \in H$ having different nontrivial residue classes modulo $\phi(G)$. Let S be the (finite) group of set theoretic automorphisms of H. Let $s \in S$ be the map $H \to H$ which interchanges the cosets $a\phi(G)$ and $b\phi(G)$ and which is the identity on the other left cosets modulo $\phi(G)$. Then consider the maps f and g defined by $f(h_1)(h_2) = h_2 h_1^{-1}$ and by $g(h) = s^{-1} f(h) s$.

§2. Definition of the Cohomology Groups

The cohomology of a profinite group G arises from the diagram

$$\cdots \ \overrightarrow{\overrightarrow{\overrightarrow{\rightarrow}}} \ G\times G\times G \ \overrightarrow{\overrightarrow{\rightarrow}} \ G\times G \ \overrightarrow{\rightarrow} \ G,$$

the arrows being the projections

$$d_i : G^{n+1} \longrightarrow G^n, \qquad i = 0, 1, \ldots, n,$$

given by

$$d_i(\sigma_0, \ldots, \sigma_n) = (\sigma_0, \ldots, \hat{\sigma}_i, \ldots, \sigma_n),$$

where by $\hat{\sigma}_i$ we indicate that we have omitted σ_i from the $(n+1)$-tuple $(\sigma_0, \ldots, \sigma_n)$. G acts on G^n by left multiplication.

From now on, we assume all G-modules to be discrete. For every G-module A we form the abelian group

$$X^n = X^n(G, A) = \mathrm{Map}\,(G^{n+1}, A)$$

of all continuous maps $x : G^{n+1} \longrightarrow A$, i.e. of all continuous functions $x(\sigma_0, \ldots, \sigma_n)$ with values in A. X^n is in a natural way a G-module by

$$(\sigma x)(\sigma_0, \ldots, \sigma_n) = \sigma x(\sigma^{-1}\sigma_0, \ldots, \sigma^{-1}\sigma_n).$$

The maps $d_i : G^{n+1} \longrightarrow G^n$ induce G-homomorphisms $d_i^* : X^{n-1} \longrightarrow X^n$ and we form the alternating sum

$$\partial^n = \sum_{i=0}^{n} (-1)^i d_i^* : X^{n-1} \longrightarrow X^n.$$

We usually write ∂ in place of ∂^n. Thus for $x \in X^{n-1}$, ∂x is the function

$$(*) \qquad (\partial x)(\sigma_0, \ldots, \sigma_n) = \sum_{i=0}^{n} (-1)^i \, x(\sigma_0, \ldots, \hat{\sigma}_i, \ldots, \sigma_n).$$

Moreover, we have the G-homomorphism $\partial^0 : A \to X^0$, which associates to $a \in A$ the constant function $x(\sigma_0) = a$.

(1.2.1) Proposition. *The sequence*

$$0 \longrightarrow A \xrightarrow{\partial^0} X^0 \xrightarrow{\partial^1} X^1 \xrightarrow{\partial^2} X^2 \longrightarrow \ldots$$

is exact.

Proof: We first show that it is a *complex*, i.e. $\partial\partial = 0$. $\partial^1 \circ \partial^0 = 0$ is clear. Let $x \in X^{n-1}$. Applying ∂ to $(*)$, we obtain summands of the form $x(\sigma_0, \ldots, \hat{\sigma}_i, \ldots, \hat{\sigma}_j, \ldots, \sigma_n)$ with certain signs. Each of these summands arises twice, once where first σ_j and then σ_i is omitted, and again where first

σ_i and then σ_j is omitted. The first time the sign is $(-1)^i(-1)^j$ and the second time $(-1)^i(-1)^{j-1}$. Hence the summands cancel to give zero.

For the exactness, we consider the map $D^{-1} : X^0 \to A$, $D^{-1}x = x(1)$, and for $n \geq 0$ the maps

$$D^n : X^{n+1} \longrightarrow X^n, \quad (D^n x)(\sigma_0, \dots, \sigma_n) = x(1, \sigma_0, \dots, \sigma_n).$$

These are homomorphisms of $\mathbb{Z}$-modules, not of G-modules. An easy calculation shows that for $n \geq 0$

$$(*) \qquad D^n \circ \partial^{n+1} + \partial^n \circ D^{n-1} = id.$$

If $x \in \ker(\partial^{n+1})$ then $x = \partial^n(D^{n-1}x)$, i.e. $\ker(\partial^{n+1}) \subseteq \operatorname{im}(\partial^n)$ and thus $\ker(\partial^{n+1}) = \operatorname{im}(\partial^n)$ because $\partial^{n+1} \circ \partial^n = 0$. □

An exact sequence of G-modules $0 \to A \to X^0 \to X^1 \to X^2 \to \dots$ is called a **resolution** of A and a family $(D^n)_{n \geq -1}$ as in the proof with the property $(*)$ is called a **contracting homotopy** of it. The above resolution is called the **standard resolution**.

We now apply the functor "fixed module". We set for $n \geq 0$

$$C^n(G, A) = X^n(G, A)^G.$$

$C^n(G, A)$ consists of the continuous functions $x : G^{n+1} \to A$ such that

$$x(\sigma\sigma_0, \dots, \sigma\sigma_n) = \sigma x(\sigma_0, \dots, \sigma_n)$$

for all $\sigma \in G$. These functions are called the (homogeneous) **n-cochains** of G with coefficients in A. From the standard resolution (1.2.1) we obtain a sequence

$$C^0(G, A) \xrightarrow{\partial^1} C^1(G, A) \xrightarrow{\partial^2} C^2(G, A) \longrightarrow \dots,$$

which in general is no longer exact. But it is still a complex, i.e. $\partial\partial = 0$, and is called the **homogeneous cochain complex** of G with coefficients in A.

We now set

$$\begin{aligned} Z^n(G, A) &= \ker(C^n(G, A) \xrightarrow{\partial^{n+1}} C^{n+1}(G, A)), \\ B^n(G, A) &= \operatorname{im}(C^{n-1}(G, A) \xrightarrow{\partial^n} C^n(G, A)) \end{aligned}$$

and $B^0(G, A) = 0$. The elements of $Z^n(G, A)$ and $B^n(G, A)$ are called the (homogeneous) **n-cocycles** and **n-coboundaries** respectively. As $\partial\partial = 0$, we have $B^n(G, A) \subseteq Z^n(G, A)$.

(1.2.2) Definition. *For $n \geq 0$ the factor group*

$$H^n(G, A) = Z^n(G, A)/B^n(G, A)$$

is called the **n-dimensional cohomology group** *of G with coefficients in A.*

For computational purposes, and for many applications, it is convenient to pass to a modified definition of the cohomology groups, which reduces the number of variables in the homogeneous cochains $x(\sigma_0, \ldots, \sigma_n)$ by one. Let $\mathscr{C}^0(G, A) = A$ and $\mathscr{C}^n(G, A)$, $n \geq 1$, be the abelian group of all continuous functions $y : G^n \longrightarrow A$. We then have the isomorphism

$$C^0(G, A) \longrightarrow \mathscr{C}^0(G, A), \quad x(\sigma) \longmapsto x(1),$$

and for $n \geq 1$ the isomorphism

$$C^n(G, A) \longrightarrow \mathscr{C}^n(G, A),$$

$$x(\sigma_0, \ldots, \sigma_n) \mapsto y(\sigma_1, \ldots, \sigma_n) = x(1, \sigma_1, \sigma_1\sigma_2, \ldots, \sigma_1 \cdots \sigma_n),$$

whose inverse is given by

$$y(\sigma_1, \ldots, \sigma_n) \mapsto x(\sigma_0, \ldots, \sigma_n) = \sigma_0 y(\sigma_0^{-1}\sigma_1, \sigma_1^{-1}\sigma_2, \ldots, \sigma_{n-1}^{-1}\sigma_n).$$

With these isomorphisms the coboundary operators $\partial^{n+1} : C^n(G, A) \longrightarrow C^{n+1}(G, A)$ are transformed into the homomorphisms

$$\partial^{n+1} : \mathscr{C}^n(G, A) \longrightarrow \mathscr{C}^{n+1}(G, A)$$

given by

$$\begin{aligned}
(\partial^1 a)(\sigma) &= \sigma a - a && \text{for } a \in A = \mathscr{C}^0(G, A),\\
(\partial^2 y)(\sigma, \tau) &= \sigma y(\tau) - y(\sigma\tau) + y(\sigma) && \text{for } y \in \mathscr{C}^1(G, A),\\
(\partial^{n+1} y)(\sigma_1, \ldots, \sigma_{n+1}) &= \sigma_1 y(\sigma_2, \ldots, \sigma_{n+1})\\
&\quad + \sum_{i=1}^{n} (-1)^i y(\sigma_1, \ldots, \sigma_{i-1}, \sigma_i\sigma_{i+1}, \sigma_{i+2}, \ldots, \sigma_{n+1})\\
&\quad + (-1)^{n+1} y(\sigma_1, \ldots, \sigma_n) && \text{for } y \in \mathscr{C}^n(G, A).
\end{aligned}$$

Setting

$$\begin{aligned}
\mathscr{Z}^n(G, A) &= \ker(\mathscr{C}^n(G, A) \xrightarrow{\partial^{n+1}} \mathscr{C}^{n+1}(G, A))\\
\mathscr{B}^n(G, A) &= \operatorname{im}(\mathscr{C}^{n-1}(G, A) \xrightarrow{\partial^n} \mathscr{C}^n(G, A)),
\end{aligned}$$

the isomorphisms $C^n(G, A) \xrightarrow{\sim} \mathscr{C}^n(G, A)$ induce isomorphisms

$$H^n(G, A) \cong \mathscr{Z}^n(G, A)/\mathscr{B}^n(G, A).$$

The functions in $\mathscr{C}^n(G, A)$, $\mathscr{Z}^n(G, A)$, $\mathscr{B}^n(G, A)$ are called the **inhomogeneous n-cochains, n-cocycles** and **n-coboundaries.** The inhomogeneous coboundary operators ∂^{n+1} are more complicated than the homogeneous ones, but they have the advantage of dealing with only n variables instead of $n + 1$.

For $n = 0, 1, 2$ the groups $H^n(G, A)$ admit the following interpretations.

The group $H^0(G,A)$: We have a natural isomorphism $C^0(G,A) \to A$, $x \mapsto x(1)$, by which we identify $C^0(G,A)$ with A. Then, for $a \in A$, $(\partial^1 a)(\sigma_0,\sigma_1) = \sigma_1 a - \sigma_0 a$, or $(\partial^1 a)(\sigma) = \sigma a - a$ in the inhomogeneous setting, so that

$$H^0(G,A) = A^G.$$

The group $H^1(G,A)$: The inhomogeneous 1-cocycles are the continuous functions $x : G \longrightarrow A$ such that

$$x(\sigma\tau) = x(\sigma) + \sigma x(\tau) \qquad \text{for all } \sigma, \tau \in G.$$

They are also called **crossed homomorphisms.** The inhomogeneous 1-coboundaries are the functions

$$x(\sigma) = \sigma a - a$$

with a fixed $a \in A$. If G acts trivially on A, then

$$H^1(G,A) = \mathrm{Hom}_{cts}(G,A). \quad ^{*)}$$

The group $H^1(G,A)$ occurs in a natural way if we pass from an exact sequence

$$0 \longrightarrow A \xrightarrow{\ i\ } B \xrightarrow{\ j\ } C \longrightarrow 0$$

of G-modules to the sequence of fixed modules. Then we lose the exactness and are left only with the exactness of the sequence

$$0 \longrightarrow A^G \longrightarrow B^G \longrightarrow C^G.$$

The group $H^1(G,A)$ now gives information about the deviation from exactness. In fact we have a canonical homomorphism

$$\delta : C^G \longrightarrow H^1(G,A)$$

extending the above exact sequence to a longer one. Namely, for $c \in C^G$ we may choose an element $b \in B$ such that $jb = c$. For each $\sigma \in G$ there is an $a_\sigma \in A$ such that $ia_\sigma = \sigma b - b$. The function $\sigma \mapsto a_\sigma$ is a 1-cocycle and we define δc to be the cohomology class of this 1-cocycle in $H^1(G,A)$. The definition is easily seen to be independent of the choice of the element b. If $\delta c = 0$, then $a_\sigma = i^{-1}(\sigma b - b) = \sigma a - a$, $a \in A$, so that $b' = b - ia$ is an element of B^G with $jb' = c$. This shows the exactness of the sequence

$$0 \longrightarrow A^G \longrightarrow B^G \longrightarrow C^G \xrightarrow{\ \delta\ } H^1(G,A).$$

We shall meet this again in a larger frame in §3.

The group $H^1(G,A)$ admits a concrete interpretation using the concept of *torsors*. Since this concept may be more fully exploited in the framework of non-abelian groups A, we generalize $H^1(G,A)$ as follows.

---

*) Since G is automatically understood as a topological group, we usually write $\mathrm{Hom}(G,A)$ instead of $\mathrm{Hom}_{cts}(G,A)$.

A ***G*-group** A is a not necessarily abelian group with the discrete topology on which G acts continuously. We denote the action of $\sigma \in G$ on $a \in A$ by ${}^\sigma a$, so that ${}^\sigma(ab) = {}^\sigma a\,{}^\sigma b$. A *cocycle* of G with coefficients in A is a continuous function $\sigma \mapsto a_\sigma$ on G with values in the group A such that

$$a_{\sigma\tau} = a_\sigma\,{}^\sigma a_\tau .$$

The set of cocycles is denoted by $\mathscr{Z}^1(G, A)$. Two cocycles a, a' are said to be cohomologous if there exists a $b \in A$ such that $a'_\sigma = b^{-1} a_\sigma {}^\sigma b$. This is an equivalence relation in $\mathscr{Z}^1(G, A)$ and the quotient set is denoted by $H^1(G, A)$. It has a distinguished element given by the cocycle $a_\sigma = 1$.

A ***G*-set** is a discrete topological space X with a continuous action of G. Let A be a G-group. An ***A*-torsor** is a G-set X with a simply transitive right action $X \times A \longrightarrow X$, $(x, a) \mapsto xa$, of A which is compatible with the G-action on X. This means that for every pair $x, y \in X$ there is a unique $a \in A$ such that $y = xa$, and ${}^\sigma(xa) = {}^\sigma x\,{}^\sigma a$. For example, if

$$1 \longrightarrow A \longrightarrow B \xrightarrow{j} C \longrightarrow 1$$

is an exact sequence of G-groups, then the cosets $j^{-1}(c)$ for $c \in C^G$ are typical A-torsors. It is clear what we mean by an isomorphism of A-torsors. Let now *TORS* (A) denote the set of isomorphism classes of A-torsors. It has a distinguished element given by the A-torsor A, and is thus a pointed set.

(1.2.3) Proposition. *We have a canonical bijection of pointed sets*

$$H^1(G, A) \cong \mathit{TORS}\,(A).$$

Proof: We define a map

$$\lambda : \mathit{TORS}\,(A) \longrightarrow H^1(G, A)$$

as follows. Let X be an A-torsor and let $x \in X$. For every $\sigma \in G$ there is a unique $a_\sigma \in A$ such that ${}^\sigma x = x a_\sigma$. One verifies at once that a_σ is a cocycle. Changing x to xb changes this cocycle to $b^{-1} a_\sigma {}^\sigma b$, which is cohomologous. We define $\lambda(X)$ to be the class of a_σ.

We define an inverse $\mu : H^1(G, A) \longrightarrow \mathit{TORS}\,(A)$ as follows. Let the set X be the group A. We let G act on X in the twisted form

$${}^{\sigma\prime}x = a_\sigma \cdot {}^\sigma x .$$

The action of A on X is given by right multiplication. In this way, X becomes an A-torsor and this defines the map μ. Replacing a_σ by $b^{-1} a_\sigma {}^\sigma b$, we have an isomorphism $x \mapsto b^{-1}x$ of A-torsors. One now checks that $\lambda \circ \mu = 1$ and $\mu \circ \lambda = 1$. $\square$

Remark: If $0 \longrightarrow A \xrightarrow{i} B \xrightarrow{j} C \longrightarrow 0$ is an exact sequence of G-modules and if we identify in the exact sequence

$$0 \longrightarrow A^G \longrightarrow B^G \longrightarrow C^G \xrightarrow{\delta} H^1(G, A)$$

$H^1(G, A)$ with *TORS* (A), then the map δ is given by $\delta c = j^{-1}(c)$.

The group $H^2(G, A)$: We return to the case that A is abelian. The inhomogeneous 2-cocycles are the continuous functions $x : G \times G \longrightarrow A$ such that $\partial x = 0$, i.e.

$$x(\sigma\tau, \rho) + x(\sigma, \tau) = x(\sigma, \tau\rho) + \sigma x(\tau, \rho).$$

Among these we find the inhomogeneous 2-coboundaries as the functions

$$x(\sigma, \tau) = y(\sigma) - y(\sigma\tau) + \sigma y(\tau)$$

with an arbitrary 1-cochain $y : G \longrightarrow A$.

The 2-cocycles had been known before the development of group cohomology as **factor systems** and occurred in connection with group extensions. To explain this, we assume that either A or G is finite, in order to avoid topological problems (but see (2.7.7)).

The question is: how many groups $\hat{G}$ are there, which have the G-module A as a normal subgroup and G as the factor group (we write A multiplicatively). To be more precise, we consider all exact sequences

$$1 \longrightarrow A \longrightarrow \hat{G} \longrightarrow G \longrightarrow 1$$

of topological groups (i.e. of profinite groups if A is finite, and of discrete groups if G is finite), such that the action of G on A is given by

$${}^{\sigma}a = \hat{\sigma} a \hat{\sigma}^{-1},$$

where $\hat{\sigma} \in \hat{G}$ is a pre-image of $\sigma \in G$. If

$$\begin{array}{ccccccccc} 1 & \longrightarrow & A & \longrightarrow & \hat{G}' & \longrightarrow & G & \longrightarrow & 1 \\ & & \| & & \downarrow f & & \| & & \\ 1 & \longrightarrow & A & \longrightarrow & \hat{G} & \longrightarrow & G & \longrightarrow & 1 \end{array}$$

is a commutative diagram of such sequences with a topological isomorphism f, then we call these sequences equivalent, and we denote the set of equivalence classes $[\hat{G}]$ by *EXT*(G, A). This set has a distinguished element given by the semi-direct product $\hat{G} = A \rtimes G$ (see ex.1 below).

(1.2.4) Theorem (SCHREIER). *We have a canonical bijection of pointed sets*

$$H^2(G, A) \cong EXT(G, A).$$

Proof: We define a map

$$\lambda : EXT(G, A) \longrightarrow H^2(G, A)$$

as follows. Let the class $[\hat{G}] \in EXT(G, A)$ be represented by the exact sequence

$$1 \longrightarrow A \longrightarrow \hat{G} \longrightarrow G \longrightarrow 1.$$

We choose a continuous section $s : G \longrightarrow \hat{G}$ of $\hat{G} \longrightarrow G$, and we set $\hat{\sigma} = s(\sigma)$. Such a section exists (see §1, ex.4). Regarding A as a subgroup of $\hat{G}$, every $\hat{\gamma} \in \hat{G}$ has a unique representation

$$\hat{\gamma} = a\hat{\sigma}, \quad a \in A,\ \sigma \in G,$$

and we have

$$\hat{\sigma}a = \hat{\sigma}a\hat{\sigma}^{-1}\hat{\sigma} = {}^{\sigma}a\hat{\sigma}.$$

The elements $\hat{\sigma}\hat{\tau}$ and $\widehat{\sigma\tau}$ are both mapped onto $\sigma\tau$, i.e.

$$\hat{\sigma}\hat{\tau} = x(\sigma, \tau)\widehat{\sigma\tau},$$

with an element $x(\sigma, \tau) \in A$ such that $x(\sigma, 1) = x(1, \sigma) = 1$. Since $\hat{\sigma}$ is a continuous function of σ and A is closed in $\hat{G}$, $x(\sigma, \tau)$ is a continuous map $x : G \times G \longrightarrow A$. The associativity $(\hat{\sigma}\hat{\tau})\hat{\rho} = \hat{\sigma}(\hat{\tau}\hat{\rho})$ yields that $x(\sigma, \tau)$ is a 2-cocycle:

$$(\hat{\sigma}\hat{\tau})\hat{\rho} = x(\sigma, \tau)\widehat{\sigma\tau}\hat{\rho} = x(\sigma, \tau)x(\sigma\tau, \rho)(\sigma\tau\rho)\hat{}\,,$$

$$\hat{\sigma}(\hat{\tau}\hat{\rho}) = \hat{\sigma}x(\tau, \rho)\widehat{\tau\rho} = {}^{\sigma}x(\tau, \rho)\hat{\sigma}\widehat{\tau\rho} = {}^{\sigma}x(\tau, \rho)\, x(\sigma, \tau\rho)(\sigma\tau\rho)\hat{}\,,$$

i.e.

$$x(\sigma, \tau)x(\sigma\tau, \rho) = {}^{\sigma}x(\tau, \rho)\, x(\sigma, \tau\rho).$$

We thus get a cohomology class $c = [x(\sigma, \tau)] \in H^2(G, A)$. This class does not depend on the choice of the continuous section $s : G \longrightarrow \hat{G}$. If $s' : G \longrightarrow \hat{G}$ is another one, and if we set $\tilde{\sigma} = s'(\sigma)$, then $\tilde{\sigma} = y(\sigma)\hat{\sigma}$, $y(\sigma) \in A$, and $\tilde{\sigma}\tilde{\tau} = \tilde{x}(\sigma, \tau)\widetilde{\sigma\tau}$. For the 2-cocycle $\tilde{x}(\sigma, \tau)$ we obtain

$$\begin{aligned} \tilde{\sigma}\tilde{\tau} &= \tilde{x}(\sigma, \tau)y(\sigma\tau)\widehat{\sigma\tau} = \tilde{x}(\sigma, \tau)y(\sigma\tau)x(\sigma, \tau)^{-1}\hat{\sigma}\hat{\tau} \\ &= \tilde{x}(\sigma, \tau)x(\sigma, \tau)^{-1}y(\sigma\tau)y(\sigma)^{-1}\,\tilde{\sigma}y(\tau)^{-1}\tilde{\tau} \\ &= \tilde{x}(\sigma, \tau)x(\sigma, \tau)^{-1}y(\sigma\tau)y(\sigma)^{-1}\,{}^{\sigma}y(\tau)^{-1}\tilde{\sigma}\tilde{\tau}, \end{aligned}$$

i.e. $\tilde{x}(\sigma, \tau) = x(\sigma, \tau)y(\sigma, \tau)$ with the 2-coboundary

$$y(\sigma, \tau) = y(\sigma)y(\sigma\tau)^{-1}\,{}^{\sigma}y(\tau).$$

The cohomology class $c = [x(\sigma, \tau)]$ also does not depend on the choice of the representative $1 \longrightarrow A \longrightarrow \hat{G} \longrightarrow G \longrightarrow 1$ in the class $[\hat{G}]$. Namely, if

$$\begin{array}{ccccccccc} 1 & \longrightarrow & A & \longrightarrow & \hat{G} & \longrightarrow & G & \longrightarrow & 1 \\ & & \| & & \wr\downarrow f & & \| & & \\ 1 & \longrightarrow & A & \longrightarrow & \hat{G}' & \longrightarrow & G & \longrightarrow & 1 \end{array}$$

is a commutative diagram and $\hat{\sigma}' = f(\hat{\sigma})$, then

$$\hat{\sigma}'\hat{\tau}' = f(\hat{\sigma})f(\hat{\tau}) = f(\hat{\sigma}\hat{\tau}) = f(x(\sigma,\tau)\widehat{\sigma\tau}) = x(\sigma,\tau)(\widehat{\sigma\tau})',$$

i.e. the group extensions $\hat{G}'$ and $\hat{G}$ yield the same 2-cocycle $x(\sigma,\tau)$. We thus get a well-defined map

$$\lambda : \mathit{EXT}(G,A) \longrightarrow H^2(G,A).$$

In order to prove the bijectivity, we construct an inverse $\mu : H^2(G,A) \longrightarrow \mathit{EXT}(G,A)$. Every cohomology class $c \in H^2(G,A)$ contains a *normalized* 2-cocycle $x(\sigma,\tau)$, i.e. a cocycle such that

$$x(\sigma,1) = x(1,\sigma) = 1.$$

Namely, if $x(\sigma,\tau)$ is any 2-cocycle in c, then we obtain from the equality $x(\sigma\tau,\rho)x(\sigma,\tau) = x(\sigma,\tau\rho)\,{}^{\sigma}x(\tau,\rho)$ that

$$x(\sigma,1) = {}^{\sigma}x(1,1), \quad x(1,\rho) = x(1,1).$$

Setting $y(\sigma) = x(1,1)$ for all $\sigma \in G$, we obtain a 2-coboundary

$$y(\sigma,\tau) = y(\sigma)y(\sigma\tau)^{-1}\,{}^{\sigma}y(\tau),$$

and the 2-cocycle $x'(\sigma,\tau) = x(\sigma,\tau)y(\sigma,\tau)^{-1}$ has the property that

$$x'(\sigma,1) = x(\sigma,1)({}^{\sigma}x(1,1))^{-1} = 1,\ \ x'(1,\tau) = x(1,\tau)x(1,1)^{-1} = 1.$$

Let now $x(\sigma,\tau)$ be a normalized 2-cocycle in c. On the set $\hat{G} = A \times G$ with the product topology we define the continuous multiplication

$$(a,\sigma)(b,\tau) = (x(\sigma,\tau)a\,{}^{\sigma}b, \sigma\tau).$$

This product is associative because of the cocycle property:

$$((a,\sigma)(b,\tau))(c,\rho) = (x(\sigma,\tau)a\,{}^{\sigma}b, \sigma\tau)(c,\rho)$$
$$= (x(\sigma\tau,\rho)x(\sigma,\tau)a\,{}^{\sigma}b\,{}^{\sigma\tau}c, \sigma\tau\rho) = (x(\sigma,\tau\rho)\,{}^{\sigma}x(\tau,\rho)a\,{}^{\sigma}b\,{}^{\sigma\tau}c, \sigma\tau\rho)$$
$$= (a,\sigma)(x(\tau,\rho)b\,{}^{\tau}c, \tau\rho) = (a,\sigma)((b,\tau),(c,\rho)).$$

(1,1) is an identity element:

$$(a,\sigma)(1,1) = (x(\sigma,1)a,\sigma) = (a,\sigma) = (x(1,\sigma)a,\sigma) = (1,1)(a,\sigma)$$

and $([{}^{\sigma^{-1}}x(\sigma,\sigma^{-1})\,{}^{\sigma^{-1}}a]^{-1},\sigma^{-1})$ is an inverse of (a,σ) since

$$(a,\sigma)([{}^{\sigma^{-1}}x(\sigma,\sigma^{-1})\,{}^{\sigma^{-1}}a]^{-1},\sigma^{-1}) = (a\,{}^{\sigma}({}^{\sigma^{-1}}a)^{-1},\sigma\sigma^{-1}) = (1,1).$$

In this way $\hat{G} = A \times G$ becomes a group with the product topology, and the maps $a \mapsto (a,1)$ and $(a,\sigma) \mapsto \sigma$ yield an exact sequence

$$1 \longrightarrow A \longrightarrow \hat{G} \longrightarrow G \longrightarrow 1.$$

Setting $\hat{\sigma} = (1,\sigma)$, we have $\hat{\sigma}^{-1} = ({}^{\sigma^{-1}}x(\sigma,\sigma^{-1})^{-1},\sigma^{-1})$ and

$$\hat{\sigma}(a,1)\hat{\sigma}^{-1} = (x(\sigma,1)\,{}^{\sigma}a,\sigma)({}^{\sigma^{-1}}x(\sigma,\sigma^{-1})^{-1},\sigma^{-1}) = ({}^{\sigma}a,1).$$

We thus obtain an element $[\hat{G}]$ in $\mathit{EXT}(G,A)$. This element does not depend on the choice of the normalized 2-cocycle $x(\sigma,\tau)$ in c. For, if $x'(\sigma,\tau) = x(\sigma,\tau)y(\sigma,\tau)^{-1}$ is another one, $y(\sigma,\tau) = y(\sigma)y(\sigma\tau)^{-1}\,{}^{\sigma}y(\tau)$ is a 2-coboundary, and if $\hat{G}'$ is the group given by the multiplication on $A \times G$ via $x'(\sigma, \tau)$, then the map $f : (a, \sigma) \mapsto (y(\sigma)a, \sigma)$ is an isomorphism from $\hat{G}$ to $\hat{G}'$ and the diagram

$$\begin{array}{ccccccccc} 1 & \longrightarrow & A & \longrightarrow & \hat{G} & \longrightarrow & G & \longrightarrow & 1 \\ & & \| & & \downarrow f & & \| & & \\ 1 & \longrightarrow & A & \longrightarrow & \hat{G}' & \longrightarrow & G & \longrightarrow & 1 \end{array}$$

is commutative, noting that $y(1) = 1$ because $1 = x'(1, \sigma) = x(1, \sigma)y(1)^{-1} = y(1)^{-1}$. Therefore $[\hat{G}] = [\hat{G}']$, and we get a well-defined map

$$\mu : H^2(G, A) \longrightarrow EXT(G, A).$$

This map is inverse to the map λ constructed before. For, if $x(\sigma, \tau)$ is the 2-cocycle produced by a section $G \longrightarrow \hat{G}, \sigma \mapsto \hat{\sigma}$, of a group extension

$$1 \longrightarrow A \longrightarrow \hat{G} \longrightarrow G \longrightarrow 1,$$

then the map $f : (a, \sigma) \mapsto a\hat{\sigma}$ is an isomorphism of the group $A \times G$, endowed with the multiplication given by $x(\sigma, \tau)$, onto $\hat{G}$. This proves the theorem. □

It is a significant feature of cohomology theory that we don't have concrete interpretations of the groups $H^n(G, A)$ for dimensions $n \geq 3$ in general. This does, however, not at all mean that they are uninteresting. Besides their natural appearance, the importance of the higher dimensional cohomology groups is seen in the fact that the theory endows them with an abundance of homomorphic connections, with which one obtains important isomorphism theorems. These theorems give concrete results for the interesting lower dimensional groups, whose proofs, however, have to take the cohomology groups of all dimensions into account.

Next we show that the cohomology groups $H^n(G, A)$ of a profinite group G with coefficients in a G-module A are built up in a simple way from those of the finite factor groups of G. Let U, V run through the open normal subgroups of G. If $V \subseteq U$, then the projections

$$G^{n+1} \longrightarrow (G/V)^{n+1} \longrightarrow (G/U)^{n+1}$$

induce homomorphisms

$$C^n(G/U, A^U) \longrightarrow C^n(G/V, A^V) \longrightarrow C^n(G, A),$$

which commute with the operators ∂^{n+1}. We therefore obtain homomorphisms

$$H^n(G/U, A^U) \longrightarrow H^n(G/V, A^V) \longrightarrow H^n(G, A).$$

The groups $H^n(G/U, A^U)$ thus form a direct system and we have a canonical homomorphism $\varinjlim_U H^n(G/U, A^U) \longrightarrow H^n(G, A)$.

(1.2.5) Proposition. *The above homomorphism is an isomorphism:*

$$\varinjlim_{U} H^n(G/U, A^U) \xrightarrow{\sim} H^n(G, A).$$

Proof: Already the homomorphism

$$\varinjlim_{U} C^\bullet(G/U, A^U) \longrightarrow C^\bullet(G, A)$$

is an isomorphism of complexes. The injectivity is clear, since the maps

$$C^\bullet(G/U, A^U) \to C^\bullet(G, A)$$

are injective.

Let conversely $x : G^{n+1} \longrightarrow A$ be an n-cochain of G. Since A is discrete, x is locally constant. We conclude that there exists an open normal subgroup U_0 of G such that x is constant on the cosets of U_0^{n+1} in G^{n+1}. It takes values in A^{U_0}, since for all $\sigma \in U_0$ we have

$$x(\sigma_0, \ldots, \sigma_n) = x(\sigma\sigma_0, \ldots, \sigma\sigma_n) = \sigma x(\sigma_0, \ldots, \sigma_n).$$

Hence x is the composite of

$$G^{n+1} \longrightarrow (G/U_0)^{n+1} \xrightarrow{x_{U_0}} A^{U_0}$$

with an n-cochain x_{U_0} of G/U_0, and is therefore the image of the element in $\varinjlim_U C^n(G/U, A^U)$ defined by x_{U_0}. This shows the surjectivity. Since the functor $\varinjlim$ is exact, we obtain the isomorphisms

$$\begin{aligned} \varinjlim_{U} H^n(G/U, A^U) &\cong H^n(\varinjlim_{U} C^\bullet(G/U, A^U)) \\ &\cong H^n(C^\bullet(G, A)) \\ &= H^n(G, A). \end{aligned}$$

□

Finally we introduce **Tate cohomology**. We do this for finite groups here, and will extend the theory to profinite groups in §8. *Let G be a finite group for the remainder of this section.*

We consider the **norm residue group**

$$\hat{H}^0(G, A) = A^G/N_G A,$$

where $N_G A$ is the image of the norm map*)

$$N_G : A \longrightarrow A, \quad N_G\, a = \sum_{\sigma \in G} \sigma a.$$

*)The name "norm" is chosen instead of "trace", because in Galois cohomology this map will often be written multiplicatively, i.e. $N_G\, a = \prod_{\sigma \in G} \sigma a$.

We call the groups

$$\hat{H}^n(G, A) = \begin{cases} A^G/N_G A & \text{for } n = 0, \\ H^n(G, A) & \text{for } n \geq 1 \end{cases}$$

the **modified cohomology groups**. We also obtain these groups from a complex. Namely, we extend the standard complex $(C^n(G, A))_{n \geq 0}$ to

$$\hat{C}^\bullet(G, A): \; C^{-1}(G, A) \xrightarrow{\partial^0} C^0(G, A) \xrightarrow{\partial^1} C^1(G, A) \xrightarrow{\partial^2} \dots,$$

where $C^{-1}(G, A) = C^0(G, A)$ and $\partial^0 x$ is the constant function with value $\sum_{\sigma \in G} x(\sigma)$. We then obtain the modified cohomology groups for all $n \geq 0$ as the cohomology groups of this complex,

$$\hat{H}^n(G, A) = H^n(\hat{C}^\bullet(G, A)).$$

Besides the fixed module A^G, we have also a "cofixed module" $A_G = A/I_G A$, where $I_G A$ is the subgroup of A generated by all elements of the form $\sigma a - a$, $a \in A, \sigma \in G$. A_G is the largest quotient of A on which G acts trivially. We set

$$H_0(G, A) = A_G.$$

If G is a finite group, then $I_G A$ is contained in the group

$$_{N_G}A = \{a \in A \mid N_G a = 0\},$$

and we set

$$\hat{H}_0(G, A) = {}_{N_G}A/I_G A.$$

The norm $N_G : A \longrightarrow A$ induces a map $N_G : H_0(G, A) \longrightarrow H^0(G, A)$, and the proof of the following proposition is obvious.

(1.2.6) Proposition. *We have an exact sequence*

$$0 \longrightarrow \hat{H}_0(G, A) \longrightarrow H_0(G, A) \xrightarrow{N_G} H^0(G, A) \longrightarrow \hat{H}^0(G, A) \longrightarrow 0.$$

The group $\hat{H}_0(G, A)$ is very often denoted by $\hat{H}^{-1}(G, A)$ for the following reason. For a finite group G one can define cohomology groups $\hat{H}^n(G, A)$ for arbitrary integral dimensions $n \in \mathbb{Z}$ as follows:

For $n \geq 0$, let $\mathbb{Z}[G^{n+1}]$ be the abelian group of all formal $\mathbb{Z}$-linear combinations $\sum a_{(\sigma_0, \dots, \sigma_n)}(\sigma_0, \dots, \sigma_n)$, $\sigma_0, \dots, \sigma_n \in G$, with its obvious G-module structure. We consider the (homological) **complete standard resolution of** $\mathbb{Z}$, i.e. the sequence of G-modules $X_\bullet = X_\bullet(G, \mathbb{Z})$

$$\dots \longrightarrow X_2 \xrightarrow{\partial_2} X_1 \xrightarrow{\partial_1} X_0 \xrightarrow{\partial_0} X_{-1} \xrightarrow{\partial_{-1}} X_{-2} \longrightarrow \dots$$

where $X_n = X_{-1-n} = \mathbb{Z}[G^{n+1}]$ for $n \geq 0$, and the differentials are defined for $n > 0$ by

$$\begin{aligned} \partial_n(\sigma_0, \dots, \sigma_n) &= \sum_{i=0}^{n} (-1)^i (\sigma_0, \dots, \sigma_{i-1}, \sigma_{i+1}, \dots, \sigma_n) \\ \partial_{-n}(\sigma_0, \dots, \sigma_{n-1}) &= \sum_{\tau \in G} \sum_{i=0}^{n} (-1)^i (\sigma_0, \dots, \sigma_{i-1}, \tau, \sigma_i, \dots, \sigma_{n-1}), \end{aligned}$$

while $\partial_0 : X_0 \to X_{-1}$ is given by

$$\partial_0(\sigma_0) = \sum_{\tau \in G} \tau .$$

The (cohomological) **complete standard resolution of A** is defined as the sequence of G-modules $X^\bullet = X^\bullet(G, A) = \mathrm{Hom}(X_\bullet, A)$

$$\dots \longrightarrow X^{-2} \xrightarrow{\partial^{-1}} X^{-1} \xrightarrow{\partial^0} X^0 \xrightarrow{\partial^1} X^1 \xrightarrow{\partial^2} X^2 \longrightarrow \dots$$

where $X^{-1-n} = X^n = \mathrm{Hom}(X_n, A) = \mathrm{Map}(G^{n+1}, A)$ for $n \geq 0$ and $\partial^n = \mathrm{Hom}(\partial_n, A)$ for $n \in \mathbb{Z}$. $X^\bullet$ is a complex. Using the maps

$$D^{-n} : X^{-n+1} \longrightarrow X^{-n}$$

given by

$$\begin{aligned} &(D^n x)(\sigma_0, \dots, \sigma_n) = x(1, \sigma_0, \dots, \sigma_n) && \text{for} \quad n \geq 0, \\ &(D^{-1} x)(\sigma_0) = \delta_{\sigma_0,1} x(1) \quad {}^{*)} && \text{for} \quad n = 1, \\ &(D^{-n} x)(\sigma_0, \dots, \sigma_{n-1}) = \delta_{\sigma_0,1} x(\sigma_1, \dots, \sigma_{n-1}) && \text{for} \quad n \geq 2, \end{aligned}$$

we get

$$D^n \circ \partial^{n+1} + \partial^n \circ D^{n-1} = id$$

for all $n \in \mathbb{Z}$. From this we conclude that the above complex $X^\bullet$ is exact.

For every $n \in \mathbb{Z}$, we now define the **n-th Tate cohomology group** $\hat{H}^n(G, A)$ as the cohomology group of the complex

$$\hat{C}^\bullet(G, A) = ((X^n)^G)_{n \in \mathbb{Z}}$$

at the place n:

$$\hat{H}^n(G, A) = H^n(\hat{C}^\bullet(G, A)).$$

Clearly, for $n \geq 0$ we get the previous (modified) cohomology groups, and it is immediate to see that $\hat{H}^{-1}(G, A)$ is our group $\hat{H}_0(G, A) = {}_{N_G}A/I_G A$. More generally, the Tate cohomology in negative dimensions can be identified with *homology* (see §8).

---

*) i.e. $\delta_{\sigma,\tau} = 0$ if $\sigma \neq \tau$ and $\delta_{\sigma,\sigma} = 1$.

Exercise 1. Let G be a profinite group and A a G-group. Assume that either G or A is finite. The *semi-direct product* is a group

$$\hat{G} = A \rtimes G$$

containing A and G such that every element of $\hat{G}$ has a unique presentation $a\sigma$, $a \in A$, $\sigma \in G$, and $(a\sigma)(a'\sigma') = a\, {}^{\sigma}a'\sigma\sigma'$. We then have a group extension

$$1 \to A \to \hat{G} \xrightarrow{\pi} G \to 1$$

and the inclusion $G \hookrightarrow \hat{G}$ is a homomorphic section of π. Two homomorphic sections $s, s' : G \to \hat{G}$ of π are *conjugate* if there is an $a \in A$ such that $s'(\sigma) = as(\sigma)a^{-1}$ for all $\sigma \in G$. Let $SEC\,(\hat{G} \to G)$ be the set of conjugacy classes of homomorphic sections of $\hat{G} \xrightarrow{\pi} G$. Then there is a canonical bijection of pointed sets

$$H^1(G, A) \cong SEC\,(\hat{G} \to G).$$

Exercise 2. There is the following interpretation of $H^3(G, A)$. Consider all possible exact sequences

$$1 \longrightarrow A \xrightarrow{i} N \xrightarrow{\alpha} \hat{G} \xrightarrow{\pi} G \longrightarrow 1,$$

where N is a group with an action $\hat{\sigma} : \nu \mapsto {}^{\hat{\sigma}}\nu$ of $\hat{G}$ satisfying ${}^{\alpha(\nu)}\nu' = \nu\nu'\nu^{-1}$, $\nu, \nu' \in N$, and $\alpha({}^{\hat{\sigma}}\nu) = \hat{\sigma}\alpha(\nu)\hat{\sigma}^{-1}$, $\nu \in N$, $\hat{\sigma} \in \hat{G}$. Impose on the set of all such exact sequences the smallest equivalence relation such that

$$1 \longrightarrow A \longrightarrow N \longrightarrow \hat{G} \longrightarrow G \longrightarrow 1$$

is equivalent to

$$1 \longrightarrow A \longrightarrow N' \longrightarrow \hat{G}' \longrightarrow G \longrightarrow 1,$$

whenever there is a commutative diagram

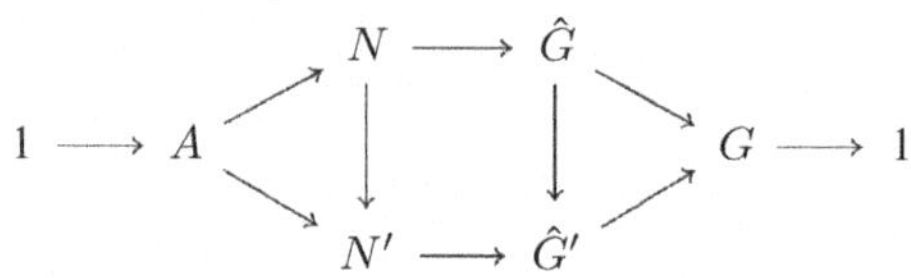

in which the vertical arrows are compatible with the actions of $\hat{G}$ and $\hat{G}'$ on N and N' (but need not be bijective). If $EXT^2(G, A)$ denotes the set of equivalence classes, then we have a canonical bijection

$$EXT^2(G, A) \cong H^3(G, A)$$

(see [18], chap. IV, th. 5.4).

Exercise 3. Let G be finite and let $(A_i)_{i \in I}$ be a family of G-modules. Show that

$$H^r(G, \prod_{i \in I} A_i) = \prod_{i \in I} H^r(G, A_i)$$

for all $r \geq 0$.

Exercise 4. An inhomogeneous cochain $x \in C^n(G, A)$, $n \geq 1$, is called **normalized** if $x(\sigma_1, \ldots, \sigma_n) = 0$ whenever one of the σ_i is equal to 1. Show that every class in $H^n(G, A)$ is represented by a normalized cocycle.

Hint: Construct inductively cochains $x_0, x_1, \ldots, x_n \in C^n(G, A)$ and $y_1, \ldots, y_n \in C^{n-1}(G, A)$ such that

$$x_0 = x,\ x_i = x_{i-1} - \partial y_i, \quad i = 1, \ldots, n,$$

$$y_i(\sigma_1, \ldots, \sigma_{n-1}) = (-1)^{i-1} x_{i-1}(\sigma_1, \ldots, \sigma_{i-1}, 1, \sigma_i, \ldots, \sigma_{n-1}).$$

Then x_n is normalized and $x - x_n$ is a coboundary.

§3. The Exact Cohomology Sequence

Having introduced the cohomology groups $H^n(G, A)$, we now turn to the question of how they behave if we change the G-module A. If

$$f : A \longrightarrow B$$

is a homomorphism of G-modules, i.e. a homomorphism such that $f(\sigma a) = \sigma f(a)$ for $a \in A$, $\sigma \in G$, then we have the induced homomorphism

$$f : C^n(G, A) \to C^n(G, B), \quad x(\sigma_0, \ldots, \sigma_n) \mapsto fx(\sigma_0, \ldots, \sigma_n),$$

and the commutative diagram

$$\begin{array}{ccccccc} \cdots \longrightarrow & C^n(G,A) & \xrightarrow{\partial^{n+1}} & C^{n+1}(G,A) & \longrightarrow \cdots \\ & \downarrow f & & \downarrow f & \\ \cdots \longrightarrow & C^n(G,B) & \xrightarrow{\partial^{n+1}} & C^{n+1}(G,B) & \longrightarrow \cdots \end{array} .$$

In other words, $f : A \longrightarrow B$ induces a homomorphism

$$f : C^\bullet(G, A) \longrightarrow C^\bullet(G, B)$$

of complexes. Taking homology groups of these complexes, we obtain homomorphisms

$$f : H^n(G, A) \longrightarrow H^n(G, B).$$

Besides these homomorphisms there is another homomorphism, the "connecting homomorphism", which is less obvious, but is of central importance in cohomology theory. For its definition we make use of the following general lemma, which should be seen as the crucial point of homological algebra.

(1.3.1) Snake lemma. *Let*

$$\begin{array}{ccccccccc} & & A & \xrightarrow{i} & B & \xrightarrow{j} & C & \longrightarrow & 0 \\ & & \downarrow \alpha & & \downarrow \beta & & \downarrow \gamma & & \\ 0 & \longrightarrow & A' & \xrightarrow{i'} & B' & \xrightarrow{j'} & C' & & \end{array}$$

be a commutative diagram of abelian groups with exact rows. We then have a canonical exact sequence

$$\ker(i) \longrightarrow \ker(\alpha) \xrightarrow{i} \ker(\beta) \xrightarrow{j} \ker(\gamma) \xrightarrow{\delta} \operatorname{coker}(\alpha) \xrightarrow{i'} \operatorname{coker}(\beta) \xrightarrow{j'} \operatorname{coker}(\gamma) \longrightarrow \operatorname{coker}(j').$$

Proof: The existence and exactness of the upper and the lower sequence is evident. The crucial cohomological phenomenon is the natural, but slightly hidden, appearance of the homomorphism

$$\delta : \ker(\gamma) \longrightarrow \operatorname{coker}(\alpha).$$

It is obtained as follows. Let $c \in \ker(\gamma)$. Let $b \in B$ and $a' \in A'$ be elements such that

$$jb = c \quad \text{and} \quad i'a' = \beta b.$$

The element b exists since j is surjective and a' exists (and is uniquely determined by b) since $j'\beta b = \gamma j b = \gamma c = 0$. We define

$$\delta c := a' \mod \alpha(A).$$

This definition does not depend on the choice of b, since if $\tilde{b} \in B$ is another element such that $j\tilde{b} = c$ and $i'\tilde{a}' = \beta\tilde{b}$, $\tilde{a}' \in A'$, then $j(\tilde{b} - b) = 0$, i.e. $\tilde{b} - b = ia$, $a \in A$, so that $i'(\tilde{a}' - a') = \beta(\tilde{b} - b) = \beta ia = i'\alpha a$, and thus $\tilde{a}' - a' = \alpha a$, i.e. $\tilde{a}' \equiv a' \mod \alpha(A)$.

Exactness at $\ker(\gamma)$: $\delta c = 0$ means $a' = \alpha a$, $a \in A$, which implies $\beta(b - ia) = i'a' - i'\alpha a = 0$, i.e. $b - ia \in \ker(\beta)$ and $j(b - ia) = c$.

Exactness at $\operatorname{coker}(\alpha)$: Let $a' \in A'$ such that $i'a' \equiv 0 \mod \beta(B)$, i.e. $i'a' = \beta b$, $b \in B$. Setting $c = jb$, we have by definition $\delta c = a' \mod \alpha(A)$. □

We now show that every exact sequence of G-modules

$$0 \longrightarrow A \xrightarrow{i} B \xrightarrow{j} C \longrightarrow 0$$

gives rise to a canonical homomorphism

$$\delta : H^n(G, C) \longrightarrow H^{n+1}(G, A)$$

for every $n \geq 0$. We consider the commutative diagram

$$\begin{array}{ccccccccc}
0 & \longrightarrow & C^n(G,A) & \longrightarrow & C^n(G,B) & \longrightarrow & C^n(G,C) & \longrightarrow & 0 \\
 & & \downarrow{\scriptstyle \partial_A} & & \downarrow{\scriptstyle \partial_B} & & \downarrow{\scriptstyle \partial_C} & & \\
0 & \longrightarrow & C^{n+1}(G,A) & \longrightarrow & C^{n+1}(G,B) & \longrightarrow & C^{n+1}(G,C) & \longrightarrow & 0.
\end{array}$$

It is exact, which is seen by passing to the inhomogeneous cochains (see also ex.1). By the snake lemma, we obtain a homomorphism

$$\delta : \ker(\partial_C) \longrightarrow \operatorname{coker}(\partial_A).$$

(1.3.2) Theorem. *For every exact sequence* $0 \longrightarrow A \to B \to C \to 0$ *of* G*-modules, the above homomorphism* δ *induces a homomorphism*

$$\delta : H^n(G,C) \longrightarrow H^{n+1}(G,A)$$

and we obtain an exact sequence

$$0 \longrightarrow A^G \longrightarrow B^G \longrightarrow C^G \xrightarrow{\delta} H^1(G,A) \longrightarrow \cdots$$

$$\cdots \longrightarrow H^n(G,A) \longrightarrow H^n(G,B) \longrightarrow H^n(G,C) \xrightarrow{\delta} H^{n+1}(G,A) \longrightarrow \cdots .$$

Proof: Setting $\bar{C}^n(G,A) = C^n(G,A)/B^n(G,A)$ and similarly for B and C in place of A, we obtain from the above diagram the commutative diagram

$$\begin{array}{ccccccc}
 & & \bar{C}^n(G,A) & \longrightarrow & \bar{C}^n(G,B) & \longrightarrow & \bar{C}^n(G,C) & \longrightarrow & 0 \\
 & & \Big\downarrow{\scriptstyle \partial_A} & & \Big\downarrow{\scriptstyle \partial_B} & & \Big\downarrow{\scriptstyle \partial_C} & & \\
0 & \longrightarrow & Z^{n+1}(G,A) & \longrightarrow & Z^{n+1}(G,B) & \longrightarrow & Z^{n+1}(G,C), & &
\end{array}$$

which is obviously exact. Noting that

$$\ker(\partial_A) = H^n(G,A) \text{ and } \operatorname{coker}(\partial_A) = H^{n+1}(G,A),$$

the snake lemma yields an exact sequence

$$H^n(G,A) \longrightarrow H^n(G,B) \longrightarrow H^n(G,C) \xrightarrow{\delta} H^{n+1}(G,A) \longrightarrow H^{n+1}(G,B) \longrightarrow H^{n+1}(G,C).$$

This proves the theorem. □

The homomorphism $\delta : H^n(G,C) \longrightarrow H^{n+1}(G,A)$ is called the **connecting homomorphism**, or simply the **δ-homomorphism**, and the exact sequence in the theorem is called the **long exact cohomology sequence**.

Remark: If the group G is finite, then, using the unrestricted complex

$$\ldots \longrightarrow X^{n-1}(G,A) \longrightarrow X^n(G,A) \longrightarrow X^{n+1}(G,A) \longrightarrow \ldots \qquad (n \in \mathbb{Z})$$

mentioned in §2, we get by the same argument an unrestricted long exact cohomology sequence

$$\cdots \xrightarrow{\delta} \hat{H}^n(G,A) \longrightarrow \hat{H}^n(G,B) \longrightarrow \hat{H}^n(G,C) \xrightarrow{\delta} \hat{H}^{n+1}(G,A) \longrightarrow \cdots .$$

The connecting homomorphism δ has the following compatibility properties.

(1.3.3) Proposition. *If*

$$\begin{array}{ccccccccc} 0 & \longrightarrow & A & \xrightarrow{i} & B & \xrightarrow{j} & C & \longrightarrow & 0 \\ & & \Big\downarrow f & & \Big\downarrow h & & \Big\downarrow g & & \\ 0 & \longrightarrow & A' & \xrightarrow{i'} & B' & \xrightarrow{j'} & C' & \longrightarrow & 0 \end{array}$$

is an exact commutative diagram of G-modules, then the diagrams

$$\begin{array}{ccc} H^n(G,C) & \xrightarrow{\delta} & H^{n+1}(G,A) \\ \Big\downarrow g & & \Big\downarrow f \\ H^n(G,C') & \xrightarrow{\delta} & H^{n+1}(G,A') \end{array}$$

are commutative.

Proof: This follows immediately from the definition of δ. If $\overline{c^n} \in H^n(G,C)$ and if $b^n \in C^n(G,B)$ and $a^{n+1} \in C^{n+1}(G,A)$ are such that $jb^n = c^n$ and $ia^{n+1} = \partial^{n+1}b^n$, then $\delta\overline{c^n} = \overline{a^{n+1}}$, and $f\delta\overline{c^n} = \overline{fa^{n+1}} = \overline{fa^{n+1}}$. On the other hand, setting $c'^n = gc^n$, $b'^n = hb^n$, $a'^{n+1} = fa^{n+1}$, we have $j'b'^n = c'^n$, $i'a'^{n+1} = \partial^{n+1}b'^n$, so that

$$\delta\overline{gc^n} = \delta\overline{c'^n} = \overline{a'^{n+1}} = \overline{fa^{n+1}} = f\delta\overline{c^n}. \qquad \square$$

In order to avoid repeated explanations it is convenient to introduce the notion of δ-functor. Let $\mathcal{A}$ and $\mathcal{B}$ be abelian categories. An **exact $\boldsymbol{\delta}$-functor** from $\mathcal{A}$ to $\mathcal{B}$ is a family $H = \{H^n\}_{n\in\mathbb{Z}}$ of functors $H^n : \mathcal{A} \longrightarrow \mathcal{B}$ together with homomorphisms

$$\delta : H^n(C) \longrightarrow H^{n+1}(A)$$

defined for each short exact sequence $0 \longrightarrow A \longrightarrow B \longrightarrow C \longrightarrow 0$ in $\mathcal{A}$ with the following properties:

(i) δ is *functorial*, i.e. if

$$\begin{array}{ccccccccc} 0 & \longrightarrow & A & \longrightarrow & B & \longrightarrow & C & \longrightarrow & 0 \\ & & \Big\downarrow & & \Big\downarrow & & \Big\downarrow & & \\ 0 & \longrightarrow & A' & \longrightarrow & B' & \longrightarrow & C' & \longrightarrow & 0 \end{array}$$

is a commutative diagram of short exact sequences in $\mathcal{A}$, then

$$\begin{array}{ccc} H^n(C) & \xrightarrow{\delta} & H^{n+1}(A) \\ \Big\downarrow & & \Big\downarrow \\ H^n(C') & \xrightarrow{\delta} & H^{n+1}(A') \end{array}$$

is a commutative diagram in $\mathcal{B}$.

(ii) The sequence

$$\cdots \longrightarrow H^n(A) \longrightarrow H^n(B) \longrightarrow H^n(C) \xrightarrow{\delta} H^{n+1}(A) \longrightarrow \cdots$$

is exact for every exact sequence $0 \to A \to B \to C \to 0$ in $\mathcal{A}$.

If a family of functors H^n is given only for an interval $-\infty \le r \le n \le s \le \infty$, then one completes it tacitly by setting $H^n = 0$ for $n \notin [r, s]$.

In this sense the family of functors $H^n(G, -)$ (completed by $H^n(G, -) = 0$ for $n < 0$) is a δ-functor from the category of G-modules into the category of abelian groups. A curious property of δ-functors is their "anticommutativity".

(1.3.4) Proposition. *Let $\{H^n\}$ be an exact δ-functor from $\mathcal{A}$ to $\mathcal{B}$. If*

$$\begin{array}{ccccccccc} & & 0 & & 0 & & 0 & & \\ & & \downarrow & & \downarrow & & \downarrow & & \\ 0 & \longrightarrow & A' & \longrightarrow & A & \longrightarrow & A'' & \longrightarrow & 0 \\ & & \downarrow & & \downarrow & & \downarrow & & \\ 0 & \longrightarrow & B' & \longrightarrow & B & \longrightarrow & B'' & \longrightarrow & 0 \\ & & \downarrow & & \downarrow & & \downarrow & & \\ 0 & \longrightarrow & C' & \longrightarrow & C & \longrightarrow & C'' & \longrightarrow & 0 \\ & & \downarrow & & \downarrow & & \downarrow & & \\ & & 0 & & 0 & & 0 & & \end{array}$$

is a commutative diagram in $\mathcal{A}$ with exact rows and columns, then

$$\begin{array}{ccc} H^{n-1}(C'') & \xrightarrow{\ \delta\ } & H^n(C') \\ {\scriptstyle\delta}\downarrow & & \downarrow{\scriptstyle -\delta} \\ H^n(A'') & \xrightarrow{\ \delta\ } & H^{n+1}(A') \end{array}$$

is a commutative diagram in $\mathcal{B}$.

Proof: It simplifies the proof if we assume that $\mathcal{A}$ is a category whose objects are abelian groups (together with some extra structure), as then we may prove statements by "diagram chases" with elements. We may do this, since it can be shown that every small abelian category may be fully embedded into a category of modules over an appropriate ring in such a way that exactness relations are preserved, and in any case we shall apply the proposition only to the category of G-modules.

Let D be the kernel of the composite map $B \longrightarrow C''$, so that the sequence

$$0 \longrightarrow D \longrightarrow B \longrightarrow C'' \longrightarrow 0$$

is exact. Let

$$i : A' \longrightarrow A \oplus B'$$

be the direct sum of the maps $A' \longrightarrow A$ and $A' \longrightarrow B'$ and let

$$j : A \oplus B' \longrightarrow B$$

be the difference $d_1 - d_2$ of the maps $d_1 : A \longrightarrow B$ and $d_2 : B' \longrightarrow B$. One checks at once that we get an exact sequence

$$0 \longrightarrow A' \xrightarrow{i} A \oplus B' \xrightarrow{j} D \longrightarrow 0$$

and that the diagram

$$\begin{array}{ccccccccc}
A' & \longrightarrow & A & \longrightarrow & A'' & \longrightarrow & B'' & \longrightarrow & C'' \\
{\scriptstyle id}\uparrow & & \uparrow{\scriptstyle \mathrm{pr}_1} & & \vdots & & \uparrow & & \uparrow{\scriptstyle id} \\
A' & \longrightarrow & A \oplus B' & \longrightarrow & D & \longrightarrow & B & \longrightarrow & C'' \\
{\scriptstyle -id}\downarrow & & \downarrow{\scriptstyle -\mathrm{pr}_2} & & \vdots & & \downarrow & & \downarrow{\scriptstyle id} \\
A' & \longrightarrow & B' & \longrightarrow & C' & \longrightarrow & C & \longrightarrow & C''
\end{array}$$

of solid arrows is commutative. This diagram can be commutatively completed by homomorphisms $D \to A''$ and $D \to C'$, since $\mathrm{im}(D \to B'') \subseteq \mathrm{im}(A'' \to B'')$ and $A'' \to B''$ is injective, and since $\mathrm{im}(D \to C) \subseteq \mathrm{im}(C' \to C)$ and $C' \to C$ is injective. From this we obtain the commutative diagram

$$\begin{array}{ccccc}
H^{n-1}(C'') & \xrightarrow{\delta} & H^n(A'') & \xrightarrow{\delta} & H^{n+1}(A') \\
{\scriptstyle id}\uparrow & & \uparrow & & \uparrow{\scriptstyle id} \\
H^{n-1}(C'') & \xrightarrow{\delta} & H^n(D) & \xrightarrow{\delta} & H^{n+1}(A') \\
{\scriptstyle id}\downarrow & & \downarrow & & \downarrow{\scriptstyle -id} \\
H^{n-1}(C'') & \xrightarrow{\delta} & H^n(C') & \xrightarrow{\delta} & H^{n+1}(A')
\end{array}$$

and the proposition follows. □

From the exact cohomology sequence, we often get important isomorphism theorems. For example, if

$$0 \longrightarrow A \longrightarrow B \longrightarrow C \longrightarrow 0$$

is an exact sequence of G-modules and if $H^n(G,B) = H^{n+1}(G,B) = 0$, then

$$\delta : H^n(G,C) \longrightarrow H^{n+1}(G,A)$$

is an isomorphism. For this reason it is very important to know which G-modules are cohomologically trivial in the following sense.

(1.3.5) Definition. *A G-module A is called* **acyclic** *if $H^n(G,A) = 0$ for all $n > 0$. A is called* **cohomologically trivial** *(***welk** *in German,* **flasque** *in French) if*

$$H^n(H,A) = 0$$

for all closed subgroups H of G and all $n > 0$.

Important examples of cohomologically trivial G-modules are the **induced G-modules** given by

$$\mathrm{Ind}_G(A) = \mathrm{Map}\,(G,A),$$

where A is any G-module. The elements of $\mathrm{Ind}_G(A)$ are the continuous functions $x : G \longrightarrow A$ (with the discrete topology on A) and the action of $\sigma \in G$ on x is given by $(\sigma x)(\tau) = \sigma x(\sigma^{-1}\tau)$.

If G is a finite group, then we have an isomorphism

$$\mathrm{Ind}_G(A) \cong A \otimes \mathbb{Z}[G]$$

given by $x \mapsto \sum\limits_{\sigma\in G} x(\sigma) \otimes \sigma$, where $\mathbb{Z}[G] = \{\sum\limits_{\sigma\in G} n_\sigma \sigma \mid n_\sigma \in \mathbb{Z}\}$ is the **group ring** of G.

(1.3.6) Proposition. (i) *The functor $A \mapsto \mathrm{Ind}_G(A)$ is exact.*

(ii) *An induced G-module A is also an induced H-module for every closed subgroup H of G, and if H is normal, then A^H is an induced G/H-module.*

(iii) *If one of the G-modules A and B is induced, then so is $A \otimes B$. If G is finite, the same holds for $\mathrm{Hom}(A,B)$.*

(iv) *If U runs through the open normal subgroups of G, then*

$$\mathrm{Ind}_G(A) = \varinjlim_U \mathrm{Ind}_{G/U}(A^U).$$

We leave the simple proof to the reader (for (ii) use ex.4 of §1 to find a homeomorphism $G \cong H \times G/H$). As mentioned above, the very importance of the induced G-modules lies in the following fact.

(1.3.7) Proposition. *The induced G-modules $M = \mathrm{Ind}_G(A)$ are cohomologically trivial. If G is finite, we have moreover $\hat{H}^n(G,M) = 0$ for all $n \in \mathbb{Z}$.* *)

Proof: We consider the standard resolutions

$$X^\bullet(G,A) \text{ and } X^\bullet(G,\mathrm{Ind}_G(A))$$

of A and $\mathrm{Ind}_G(A)$. The map

$$X^n(G,\mathrm{Ind}_G(A))^G \longrightarrow X^n(G,A),$$

given by $x(\sigma_0,\dots,\sigma_n) \mapsto y(\sigma_0,\dots,\sigma_n) = x(\sigma_0,\dots,\sigma_n)(1)$ obviously commutes with ∂. Furthermore, it is an isomorphism, since it has the map $y(\sigma_0,\dots,\sigma_n) \mapsto x(\sigma_0,\dots,\sigma_n)(\sigma) = \sigma y(\sigma^{-1}\sigma_0,\dots,\sigma^{-1}\sigma_n)$ as inverse. We thus have an isomorphism

$$C^\bullet(G,\mathrm{Ind}_G(A)) \cong X^\bullet(G,A)$$

of complexes. But $X^\bullet(G,A)$ is exact by (1.2.1), so that

$$H^n(G,\mathrm{Ind}_G(A)) = H^n(C^\bullet(G,\mathrm{Ind}_G(A))) = H^n(X^\bullet(G,A)) = 0$$

for $n \geq 1$. If H is a closed subgroup of G, then by (1.3.6) we may write $\mathrm{Ind}_G(A) = \mathrm{Ind}_H(B)$ and get $H^n(H,\mathrm{Ind}_G(A)) = 0$. If G is finite, then the same argument holds for the extended complex $(X^n)_{n\in\mathbb{Z}}$, hence $\hat{H}^n(G,\mathrm{Ind}_G(A)) = 0$ for all $n \in \mathbb{Z}$. □

The above proposition allows us to adopt a technique, called **dimension shifting**, by which definitions and proofs concerning the cohomology groups for all G-modules A and all n, may be reduced to a single dimension n, e.g. $n = 0$. Given A, define the G-module A_1 by the exact sequence

$$0 \longrightarrow A \xrightarrow{\ i\ } \mathrm{Ind}_G(A) \longrightarrow A_1 \longrightarrow 0,$$

where ia is the constant function $(ia)(\sigma) = a$. This is a sequence of G-modules. If H is a closed subgroup of G, then $H^n(H,\mathrm{Ind}_G(A)) = 0$ for all $n \geq 1$ by (1.3.7), and the exact cohomology sequence shows that the homomorphism

$$\delta : H^n(H,A_1) \longrightarrow H^{n+1}(H,A)$$

is surjective for $n = 0$ and bijective for $n > 0$. If we define $A_0 = A$ and inductively

$$A_p = (A_{p-1})_1 \quad \text{for } p > 0,$$

then (1.3.7) yields inductively the

*) We shall see in §7 that $\hat{H}^n(G,A) = 0$, $n \in \mathbb{Z}$, for any cohomologically trivial G-module A.

(1.3.8) Proposition. *For all $n, p \geq 0$ and all subgroups $H \subseteq G$, we have a canonical homomorphism*

$$\delta^p : H^n(H, A_p) \longrightarrow H^{n+p}(H, A),$$

which is a surjection for $n = 0$ and an isomorphism for $n > 0$.

If G is a finite group, then we may also consider the exact sequence

$$0 \longrightarrow A_{-1} \longrightarrow \mathrm{Ind}_G(A) \xrightarrow{\nu} A \longrightarrow 0,$$

where ν associates the element $\sum_{\sigma \in G} x(\sigma)$ to an element $x \in \mathrm{Ind}_G(A)$. We define

$$A_p = (A_{p+1})_{-1} \quad \text{for } p < 0\,,$$

It is easy to see that

$$A_p \cong A \otimes J_G^{\otimes p} \quad \text{and} \quad A_{-p} \cong A \otimes I_G^{\otimes p}$$

for $p \geq 0$, where the G-modules I_G and J_G are given by the exact sequences

$$0 \longrightarrow I_G \longrightarrow \mathbb{Z}[G] \xrightarrow{\varepsilon} \mathbb{Z} \longrightarrow 0\,,$$

$$0 \longrightarrow \mathbb{Z} \xrightarrow{N_G} \mathbb{Z}[G] \longrightarrow J_G \longrightarrow 0\,.$$

Here ε is the **augmentation map**

$$\varepsilon : \sum_{\sigma \in G} a_\sigma \sigma \mapsto \sum_{\sigma \in G} a_\sigma\,,$$

and $N_G(1) = \sum_{\sigma \in G} \sigma$. The G-module I_G is called the **augmentation ideal** of $\mathbb{Z}[G]$.

Noting that $\hat{H}^n(H, \mathrm{Ind}_G(A)) = 0$, we obtain canonical isomorphisms

$$\hat{H}^n(H, A) \cong \hat{H}^{n-p}(H, A_p)$$

for all $n, p \in \mathbb{Z}$. Furthermore, we observe the following rule for the G-module $\mathrm{Hom}(A, \mathbb{Q}/\mathbb{Z})$: if $p \in \mathbb{Z}$, then

$$\mathrm{Hom}(A, \mathbb{Q}/\mathbb{Z})_p \cong \mathrm{Hom}(A_{-p}, \mathbb{Q}/\mathbb{Z})\,.$$

Let G again be a profinite group. The G-modules $X^n = X^n(G, A)$ in the standard resolution

$$0 \longrightarrow A \longrightarrow X^0 \longrightarrow X^1 \longrightarrow X^2 \longrightarrow \ldots$$

are all induced G-modules, since $X^0 = \mathrm{Ind}_G(A)$ and $X^n = \mathrm{Ind}_G(X^{n-1})$. They are thus cohomologically trivial and, in particular, acyclic. We call a resolution

$$0 \longrightarrow A \longrightarrow Y^0 \longrightarrow Y^1 \longrightarrow Y^2 \longrightarrow \ldots$$

of A **acyclic** (resp. **resolution by cohomologically trivial G-modules**) if the Y^n are acyclic (resp. cohomologically trivial). It is a remarkable fact that the cohomology groups $H^n(G, A)$ can be obtained from any acyclic resolution.

(1.3.9) Proposition. *If*

$$0 \longrightarrow A \longrightarrow Y^0 \xrightarrow{\partial} Y^1 \xrightarrow{\partial} Y^2 \xrightarrow{\partial} \dots$$

is an acyclic resolution of A, then canonically

$$H^n(G,A) \cong H^n(H^0(G,Y^\bullet)).$$

Proof: Setting $K^p = \ker(Y^p \xrightarrow{\partial} Y^{p+1})$, we obtain the short exact sequences

$$\begin{array}{lllll} 0 \longrightarrow A & \longrightarrow Y^0 & \longrightarrow K^1 & \longrightarrow 0, \\ 0 \longrightarrow K^1 & \longrightarrow Y^1 & \longrightarrow K^2 & \longrightarrow 0, \\ & \dots & & \\ 0 \longrightarrow K^{n-2} & \longrightarrow Y^{n-2} & \longrightarrow K^{n-1} & \longrightarrow 0, \\ 0 \longrightarrow K^{n-1} & \longrightarrow Y^{n-1} & \longrightarrow K^n & \longrightarrow 0. \end{array}$$

Since the Y^n are acyclic, the exact cohomology sequence yields for $n \geq 1$ isomorphisms

$$H^1(G,K^{n-1}) \xrightarrow[\sim]{\delta} H^2(G,K^{n-2}) \xrightarrow[\sim]{\delta} \cdots \xrightarrow[\sim]{\delta} H^n(G,A).$$

On the other hand we have the exact sequence

$$H^0(G,Y^{n-1}) \longrightarrow H^0(G,K^n) \longrightarrow H^1(G,K^{n-1}) \longrightarrow H^1(G,Y^{n-1}) = 0$$

and

$$H^0(G,K^n) = \ker(H^0(G,Y^n) \longrightarrow H^0(G,Y^{n+1})),$$

$$\mathrm{im}\,(H^0(G,Y^{n-1}) \longrightarrow H^0(G,K^n)) = \mathrm{im}\,(H^0(G,Y^{n-1}) \longrightarrow H^0(G,Y^n)),$$

which proves that canonically

$$H^n(G,A) \cong H^1(G,K^{n-1}) = H^n(H^0(G,Y^\bullet)). \qquad \square$$

For example, if H is a closed subgroup of G, then the standard resolution $0 \to A \to X^\bullet$ of a G-module A is also an acyclic resolution of the H-module A, hence $H^n(H,A) \cong H^n(H^0(H,X^\bullet))$. This isomorphism is also obtained from the restriction map $X^\bullet(G,A)^H \to X^\bullet(H,A)^H$, as one may see by dimension shifting.

Remark: There exists a variant of (1.3.9) for the modified cohomology. If G is a finite group, then, for all $q \in \mathbb{Z}$ and every G-module A, there are canonical isomorphisms

$$\hat{H}^q(G,A) \cong H^q(H^0(G,Y^\bullet)),$$

where $Y^\bullet$ is a *complete acyclic resolution* of A, i.e. a complex

$$\begin{array}{ccccccccccccc}
\cdots \xrightarrow{\partial^{-2}} & Y^{-2} & \xrightarrow{\partial^{-1}} & Y^{-1} & & \xrightarrow{\partial^{0}} & & Y^{0} & \xrightarrow{\partial^{1}} & Y^{1} & \xrightarrow{\partial^{2}} & Y^{2} & \xrightarrow{\partial^{3}} \cdots \\
& & & & \searrow^{\varepsilon} & & \nearrow_{\mu} & & & & & & \\
& & & & & A & & & & & & & \\
& & & & \nearrow & & \searrow & & & & & & \\
& & & 0 & & & & 0 & & & & &
\end{array}$$

consisting of cohomologically trivial G-modules Y^n, $n \in \mathbb{Z}$, which is exact everywhere and $\partial^0 = \mu \circ \varepsilon$.

Exercise 1. The functor $A \mapsto C^n(G, A)$ is exact.

Hint: $C^n(G, A) = X^n(G, A)^G$, and $X^n(G, A)$ is induced.

Exercise 2. For any pair of maps $A \xrightarrow{f} B \xrightarrow{g} C$ of abelian groups, there is an exact sequence

$$0 \longrightarrow \ker(f) \longrightarrow \ker(g \circ f) \longrightarrow \ker(g) \longrightarrow$$
$$\longrightarrow \operatorname{coker}(f) \longrightarrow \operatorname{coker}(g \circ f) \longrightarrow \operatorname{coker}(g) \longrightarrow 0.$$

Exercise 3. Let G be a finite group and let

$$0 \longrightarrow A \xrightarrow{\alpha} B \xrightarrow{\beta} C \xrightarrow{\gamma} D \longrightarrow 0$$

be an exact sequence of G-modules. Define a homomorphism

$$\delta^2 : \hat{H}^{n-1}(G, D) \longrightarrow \hat{H}^{n+1}(G, A).$$

Show that the following conditions are equivalent:

(i) δ^2 is an isomorphism for all $n \in \mathbb{Z}$,

(ii) $\hat{H}^n(G, B) \to \hat{H}^n(G, C)$ is an isomorphism for all $n \in \mathbb{Z}$.

Hint: Show this first under the assumption $\hat{H}^n(G, B) = 0$ for all $n \in \mathbb{Z}$. Then apply (1.3.4) to the exact commutative diagram

$$\begin{array}{ccccccccc}
 & & 0 & & 0 & & 0 & & \\
 & & \downarrow & & \downarrow & & \downarrow & & \\
0 & \longrightarrow & A & \xrightarrow{\alpha} & B & \xrightarrow{\beta} & \beta B & \longrightarrow & 0 \\
 & & \downarrow{\scriptstyle i\circ\alpha} & & \downarrow{\scriptstyle (i,\beta)} & & \downarrow & & \\
0 & \longrightarrow & \operatorname{Ind}_G B & \xrightarrow{(id,0)} & \operatorname{Ind}_G B \oplus C & \xrightarrow{0+id} & C & \longrightarrow & 0 \\
 & & \downarrow & & \downarrow & & \downarrow{\scriptstyle \gamma} & & \\
0 & \longrightarrow & X & \longrightarrow & Y & \longrightarrow & D & \longrightarrow & 0 \\
 & & \downarrow & & \downarrow & & \downarrow & & \\
 & & 0 & & 0 & & 0 & &
\end{array}$$

where $X = \operatorname{coker}(i \circ \alpha)$ and $Y = \operatorname{coker}(i, \beta)$.

Exercise 4. Let A be a G-module and let A_0 be the trivial G-module with underlying abelian group A. Then we have an isomorphism $\operatorname{Ind}_G(A) \cong \operatorname{Ind}_G(A_0)$ of G-modules.

Exercise 5. Let G be a finite group and let $0 \to A \to I_1 \to I_2 \to \cdots \to I_p \to B \to 0$ be an exact sequence of G-modules, where the $I_1, \ldots, I_p$ are acyclic G-modules. Then $\hat{H}^n(G, A) \cong \hat{H}^{n-p}(G, B)$ for $n \geq p$.

Exercise 6. Let $C^\bullet = (C^n, d_n)_{n\in\mathbb{Z}}$ and $C'^\bullet = (C'^n, d'_n)_{n\in\mathbb{Z}}$ be two complexes in an abelian category and let $f = (f_n)_{n\in\mathbb{Z}}$ and $g = (g_n)_{n\in\mathbb{Z}}$ be two morphisms from $C^\bullet$ to $C'^\bullet$. A **homotopy** from f to g is a family $h = (h_n)_{n\in\mathbb{Z}}$ of morphisms $h_n : C^{n+1} \to C'^n$ such that

$$h_n d_{n+1} + d'_n h_{n-1} = f_n - g_n.$$

We say that f and g are homotopic and write $f \simeq g$ if such a family exists.

Show that in this case f and g induce the same homomorphisms $H^n(C^\bullet) \to H^n(C'^\bullet)$ on the homology.

Exercise 7. If G is finite and $(A_i)_{i\in I}$ is a projective system of *finite* G-modules, then

$$H^n(G, \varprojlim_i A_i) = \varprojlim_i H^n(G, A_i).$$

(Compare exercise 3 in §2.)

Exercise 8. If $1 \to A \to B \to C \to 1$ is an exact sequence of G-groups, then

$$1 \longrightarrow A^G \longrightarrow B^G \longrightarrow C^G \longrightarrow H^1(G, A) \longrightarrow H^1(G, B) \longrightarrow H^1(G, C).$$

is an exact sequence of pointed sets, i.e. the image of a map is equal to the pre-image of the distinguished element. If A is in the center of B, then the sequence extends exactly by an arrow $\xrightarrow{\delta} H^2(G, A)$, given by $c_\sigma \mapsto a_{\sigma,\tau} = c_\sigma {}^\sigma c_\tau c_{\sigma\tau}^{-1}$.

§4. The Cup-Product

If A and B are two G-modules, then $A \otimes_{\mathbb{Z}} B$ is also a G-module (by $\sigma(a \otimes b) = \sigma a \otimes \sigma b$), and we obtain for every pair $p, q \geq 0$ a bilinear map

$$(*) \qquad C^p(G, A) \times C^q(G, B) \xrightarrow{\cup} C^{p+q}(G, A \otimes B)$$

by

$$(a \cup b)(\sigma_0, \ldots, \sigma_{p+q}) = a(\sigma_0, \ldots, \sigma_p) \otimes b(\sigma_p, \ldots, \sigma_{p+q}).$$

For this map, we have the following formula.

(1.4.1) Proposition. $\partial(a \cup b) = (\partial a) \cup b + (-1)^p (a \cup \partial b)$.

Proof: We have

$$\partial(a \cup b)(\sigma_0, \dots, \sigma_{p+q+1}) = \sum_{i=0}^{p+q+1} (-1)^i (a \cup b)(\sigma_0, \dots, \hat{\sigma}_i, \dots, \sigma_{p+q+1})$$

$$= \sum_{i=0}^{p} (-1)^i a(\sigma_0, \dots, \hat{\sigma}_i, \dots, \sigma_{p+1}) \otimes b(\sigma_{p+1}, \dots, \sigma_{p+q+1})$$

$$+ \sum_{i=p+1}^{p+q+1} (-1)^i a(\sigma_0, \dots, \sigma_p) \otimes b(\sigma_p, \dots, \hat{\sigma}_i, \dots, \sigma_{p+q+1}).$$

On the other hand

$$(\partial a \cup b)(\sigma_0, \dots, \sigma_{p+q+1}) = (\partial a)(\sigma_0, \dots, \sigma_{p+1}) \otimes b(\sigma_{p+1}, \dots, \sigma_{p+q+1})$$

$$= \sum_{i=0}^{p+1} (-1)^i a(\sigma_0, \dots, \hat{\sigma}_i, \dots, \sigma_{p+1}) \otimes b(\sigma_{p+1}, \dots, \sigma_{p+q+1})$$

and

$$(a \cup \partial b)(\sigma_0, \dots, \sigma_{p+q+1}) = a(\sigma_0, \dots, \sigma_p) \otimes (\partial b)(\sigma_p, \dots, \sigma_{p+q+1})$$

$$= a(\sigma_0, \dots, \sigma_p) \otimes \sum_{i=0}^{q+1} (-1)^i b(\sigma_p, \dots, \hat{\sigma}_{p+i}, \dots, \sigma_{p+q+1}).$$

Now, in the second of these seven formula lines, let the index i run from 0 to $p+1$ and in the third from p to $p+q+1$. The additional summands appearing cancel to give the result claimed. □

From this proposition, it follows that $a \cup b$ is a cocycle if a and b are cocycles, and a coboundary if one of the cochains a and b is a coboundary and the other a cocycle. Therefore the pairing $(*)$ induces a bilinear map

$$H^p(G, A) \times H^q(G, B) \xrightarrow{\cup} H^{p+q}(G, A \otimes B), \quad (\alpha, \beta) \longmapsto \alpha \cup \beta.$$

This map is called the **cup-product**. For $p = q = 0$, we obtain the map

$$A^G \times B^G \longrightarrow (A \otimes B)^G, \quad (a, b) \longmapsto a \otimes b.$$

We will see below that the cup-product induces a bilinear map

$$\hat{H}^p(G, A) \times \hat{H}^q(G, B) \xrightarrow{\cup} \hat{H}^{p+q}(G, A \otimes B)$$

on the modified cohomology of a finite group G for all $p, q \in \mathbb{Z}$.

Whenever a new cohomological map is introduced, we must check its functoriality properties and also its compatibility with the cohomological maps already defined. Directly from the definition follows the

(1.4.2) Proposition. *For two homomorphisms $A \to A'$, $B \to B'$ of G-modules, we have the commutative diagram*

$$\begin{array}{ccccc} H^p(G,A) & \times & H^q(G,B) & \xrightarrow{\cup} & H^{p+q}(G,A\otimes B) \\ \downarrow & & \downarrow & & \downarrow \\ H^p(G,A') & \times & H^q(G,B') & \xrightarrow{\cup} & H^{p+q}(G,A'\otimes B'). \end{array}$$

The cup-product is very often used in a slightly more general form. Instead of the bilinear map $A \times B \to A \otimes B$, we may consider an arbitrary bilinear pairing of G-modules $A \times B \longrightarrow C$, $(a,b) \longmapsto ab$. It factors through $A \otimes B$, and the composite

$$H^p(G,A) \times H^q(G,B) \xrightarrow{\cup} H^{p+q}(G,A\otimes B) \longrightarrow H^{p+q}(G,C)$$

is also called the *cup-product*. The compatibility with the δ-homomorphism is given in the following proposition.

(1.4.3) Proposition. (i) *Let*

$$0 \to A' \to A \to A'' \to 0 \quad \text{and} \quad 0 \to C' \to C \to C'' \to 0$$

be exact sequences of G-modules. Let B be another G-module and suppose we are given a pairing $A \times B \to C$ which induces pairings $A' \times B \to C'$ and $A'' \times B \to C''$. Then the diagram

$$\begin{array}{ccccc} H^p(G,A'') & \times & H^q(G,B) & \xrightarrow{\cup} & H^{p+q}(G,C'') \\ \downarrow{\scriptstyle\delta} & & \downarrow{\scriptstyle id} & & \downarrow{\scriptstyle\delta} \\ H^{p+1}(G,A') & \times & H^q(G,B) & \xrightarrow{\cup} & H^{p+q+1}(G,C') \end{array}$$

is commutative, i.e.

$$\delta(\alpha'' \cup \beta) = \delta\alpha'' \cup \beta.$$

(ii) *Let*

$$0 \to B' \to B \to B'' \to 0 \quad \text{and} \quad 0 \to C' \to C \to C'' \to 0$$

be exact sequences of G-modules and let $A \times B \to C$ be a pairing which induces pairings $A \times B' \to C'$ and $A \times B'' \to C''$. Then the diagram

$$\begin{array}{ccccc} H^p(G,A) & \times & H^q(G,B'') & \xrightarrow{\cup} & H^{p+q}(G,C'') \\ \downarrow{\scriptstyle id} & & \downarrow{\scriptstyle\delta} & & \downarrow{\scriptstyle(-1)^p\delta} \\ H^p(G,A) & \times & H^{q+1}(G,B') & \xrightarrow{\cup} & H^{p+q+1}(G,C') \end{array}$$

is commutative, i.e.

$$(-1)^p\,\delta(\alpha \cup \beta'') = \alpha \cup \delta\beta''.$$

Proof: We show (ii). Let $\alpha = \bar{a}$, $\beta'' = \bar{b}''$, $a \in Z^p(G, A)$, $b'' \in Z^q(G, B'')$. Let $b \in C^q(G, B)$ be a pre-image of b'' (b exists by §3, ex.1). Identifying B' with its image in B, $\delta\beta''$ is by definition represented by the cocycle $\partial b \in Z^{q+1}(G, B')$ and $\delta(\alpha \cup \beta'')$ by the cocycle $\partial(a \cup b) \in Z^{p+q+1}(G, C)$. Recalling that $\partial a = 0$, we obtain from (1.4.1)

$$\partial(a \cup b) = (\partial a) \cup b + (-1)^p(a \cup \partial b) = (-1)^p(a \cup \partial b).$$

Passing to the cohomology classes, we get $\delta(\alpha \cup \beta'') = (-1)^p(\alpha \cup \delta\beta'')$.
(i) is proven in the same way. □

As before, we make everywhere the identifications

$$(A \otimes B) \otimes C = A \otimes (B \otimes C) \quad \text{and} \quad A \otimes B = B \otimes A.$$

(1.4.4) Proposition. *The cup-product is associative and skew commutative, i.e. for $\alpha \in H^p(G, A)$, $\beta \in H^q(G, B)$, $\gamma \in H^r(G, C)$ we have*

$$(\alpha \cup \beta) \cup \gamma = \alpha \cup (\beta \cup \gamma)$$

and

$$\alpha \cup \beta = (-1)^{pq}(\beta \cup \alpha).$$

Proof: Let a, b, c be cocycles representing α, β, γ. Then

$((a \cup b) \cup c)(\sigma_0, \ldots, \sigma_{p+q+r}) = (a \cup b)(\sigma_0, \ldots, \sigma_{p+q}) \otimes c(\sigma_{p+q}, \ldots, \sigma_{p+q+r})$

$= a(\sigma_0, \ldots, \sigma_p) \otimes b(\sigma_p, \ldots, \sigma_{p+q}) \otimes c(\sigma_{p+q}, \ldots, \sigma_{p+q+r})$

$= a(\sigma_0, \ldots, \sigma_p) \otimes (b \cup c)(\sigma_p, \ldots, \sigma_{p+q+r}) = (a \cup (b \cup c))(\sigma_0, \ldots, \sigma_{p+q+r})$.

Passing to the cohomology classes gives $(\alpha \cup \beta) \cup \gamma = \alpha \cup (\beta \cup \gamma)$.

The formula $\alpha \cup \beta = (-1)^{pq}(\beta \cup \alpha)$ is not so easy to prove at the level of cocycles. We therefore use the method of dimension shifting introduced in §3. By (1.3.8), we have the surjections $\delta^n : H^0(G, A_n) \longrightarrow H^n(G, A)$. Applying (1.4.3) (i) p times (resp. (ii) q times), we obtain a commutative diagram

$$\begin{array}{ccccccc}
H^0(G, A_p) & \times & H^0(G, B_q) & \xrightarrow{\cup} & H^0(G, (A \otimes B_q)_p) & = & H^0(G, A_p \otimes B_q) \\
\downarrow{\scriptstyle \delta^p} & & \downarrow{\scriptstyle id} & & \downarrow{\scriptstyle \delta^p} & & \\
H^p(G, A) & \times & H^0(G, B_q) & \xrightarrow{\cup} & H^p(G, (A \otimes B)_q) & = & H^p(G, A \otimes B_q) \\
\downarrow{\scriptstyle id} & & \downarrow{\scriptstyle \delta^q} & & \downarrow{\scriptstyle (-1)^{pq}\delta^q} & & \\
H^p(G, A) & \times & H^q(G, B) & \xrightarrow{\cup} & H^{p+q}(G, A \otimes B). & &
\end{array}$$

For $p = q = 0$ the rule $\alpha \cup \beta = \beta \cup \alpha$ is clear. Since the vertical arrows are surjective, $\alpha \cup \beta = (-1)^{pq}(\beta \cup \alpha)$ follows for $p, q \geq 0$. □

(1.4.5) Proposition. *Let*

$$0 \longrightarrow A' \longrightarrow A \longrightarrow A'' \longrightarrow 0 \quad \text{and} \quad 0 \longrightarrow B' \longrightarrow B \longrightarrow B'' \longrightarrow 0$$

be exact sequences of G-modules. Suppose we are given a pairing

$$\varphi : A \times B \to C$$

into a G-module C such that $\varphi(A' \times B') = 0$, i.e. φ induces pairings φ' and φ'' such that the diagram

$$\begin{array}{ccccc} A' & \times & B'' & \xrightarrow{\varphi'} & C \\ {\scriptstyle i}\downarrow & & \uparrow{\scriptstyle v} & & \| \\ A & \times & B & \xrightarrow{\varphi} & C \\ {\scriptstyle j}\downarrow & & \uparrow{\scriptstyle u} & & \| \\ A'' & \times & B' & \xrightarrow{\varphi''} & C \end{array}$$

commutes. Then the diagram

$$\begin{array}{ccccc} H^p(G, A'') & \times & H^q(G, B') & \xrightarrow{\cup} & H^{p+q}(G, C) \\ {\scriptstyle \delta}\downarrow & & \uparrow{\scriptstyle \delta} & & \downarrow{\scriptstyle (-1)^{p+1}} \\ H^{p+1}(G, A') & \times & H^{q-1}(G, B'') & \xrightarrow{\cup} & H^{p+q}(G, C) \end{array}$$

is commutative, i.e.

$$(\delta\alpha) \cup \beta + (-1)^p(\alpha \cup \delta\beta) = 0$$

for $\alpha \in H^p(G, A'')$ and $\beta \in H^{q-1}(G, B'')$.

Proof: Let $a'' \in Z^p(G, A'')$ and $b'' \in Z^{q-1}(G, B'')$ be cocycles representing α and β respectively, and let $a \in C^p(G, A)$ and $b \in C^{q-1}(G, B)$ be pre-images. Then there exist $a' \in C^{p+1}(G, A')$ and $b' \in C^q(G, B')$ such that $ia' = \partial a$ and $ub' = \partial b$. Now $\delta\alpha$ is represented by a' and $\delta\beta$ by b'. It follows that

$$(\delta\alpha) \cup \beta + (-1)^p(\alpha \cup \delta\beta) = 0$$

since the left class is represented by

$$\begin{aligned} a' \cup b'' + (-1)^p(a'' \cup b') &= a' \cup vb + (-1)^p(ja \cup b') \\ &= ia' \cup b + (-1)^p(a \cup ub') \\ &= \partial a \cup b + (-1)^p(a \cup \partial b) \\ &= \partial(a \cup b), \end{aligned}$$

which is a coboundary. □

For discrete G-modules A and C, let G act on $\mathrm{Hom}(A,C)$ by

$$(g\phi)(a) = g\phi(g^{-1}a),$$

where $\phi \in \mathrm{Hom}(A,C)$, $g \in G$ and $a \in A$. If G is finite or if A is finitely generated as a $\mathbb{Z}$-module, then $\mathrm{Hom}(A,C)$ is a discrete G-module and the canonical pairing

$$\mathrm{Hom}(A,C) \times A \longrightarrow C$$

induces the cup-product

$$H^p(G, \mathrm{Hom}(A,C)) \times H^q(G,A) \xrightarrow{\cup} H^{p+q}(G,C).$$

We have the following

(1.4.6) Corollary. *Let*

$$0 \longrightarrow A' \longrightarrow A \longrightarrow A'' \longrightarrow 0$$

be an exact sequence of G-modules and suppose that C is another G-module such that the sequence

$$0 \longrightarrow \mathrm{Hom}(A'',C) \longrightarrow \mathrm{Hom}(A,C) \longrightarrow \mathrm{Hom}(A',C) \longrightarrow 0$$

is also exact. Assume further, that G is finite or that A, A' and A'' are finitely generated as $\mathbb{Z}$-modules. Then the diagram

$$\begin{array}{ccccc}
H^p(G,\mathrm{Hom}(A',C)) & \times & H^q(G,A') & \xrightarrow{\cup} & H^{p+q}(G,C) \\
\delta\big\downarrow & & \big\uparrow\delta & & \big\downarrow(-1)^{p+1} \\
H^{p+1}(G,\mathrm{Hom}(A'',C)) & \times & H^{q-1}(G,A'') & \xrightarrow{\cup} & H^{p+q}(G,C)
\end{array}$$

is commutative, i.e.

$$(\delta\hat{\alpha}) \cup \alpha + (-1)^p(\hat{\alpha} \cup \delta\alpha) = 0$$

for $\hat{\alpha} \in H^p(G,\mathrm{Hom}(A',C))$ and $\alpha \in H^{q-1}(G,A'')$.

Proof: This follows from (1.4.5) applied to the natural pairing

$$\mathrm{Hom}(A,C) \times A \longrightarrow C. \qquad \square$$

In the next proposition, whose proof is taken from [7], §7, we define the cup-product in arbitrary integral dimensions if G is a finite group.

(1.4.7) Proposition. *Let G be a finite group. Then there exists a unique family of bilinear maps*

$$\hat{H}^p(G,A) \times \hat{H}^q(G,B) \xrightarrow{\cup} \hat{H}^{p+q}(G,C),$$

defined for all integers $p,q \in \mathbb{Z}$ and all pairings $A \times B \to C$ of G-modules, such that:

(i) *These bilinear maps are functorial with respect to the modules.*

(ii) *For $p = q = 0$ they are induced by the natural map*

$$A^G \times B^G \longrightarrow C^G.$$

(iii) *Let*

$$0 \to A' \to A \to A'' \to 0 \quad \text{and} \quad 0 \to C' \to C \to C'' \to 0$$

be exact sequences of G-modules. Let B be another G-module and suppose we are given a pairing $A \times B \to C$ which induces pairings $A' \times B \to C'$ and $A'' \times B \to C''$. Then the diagram

$$\begin{array}{ccccc} \hat{H}^p(G,A'') & \times & \hat{H}^q(G,B) & \xrightarrow{\cup} & \hat{H}^{p+q}(G,C'') \\ \downarrow{\scriptstyle\delta} & & \downarrow{\scriptstyle id} & & \downarrow{\scriptstyle\delta} \\ \hat{H}^{p+1}(G,A') & \times & \hat{H}^q(G,B) & \xrightarrow{\cup} & \hat{H}^{p+q+1}(G,C') \end{array}$$

is commutative, i.e. $\delta(\alpha'' \cup \beta) = \delta\alpha'' \cup \beta$.

(iv) *Let*

$$0 \to B' \to B \to B'' \to 0 \quad \text{and} \quad 0 \to C' \to C \to C'' \to 0$$

be exact sequences of G-modules and let $A \times B \to C$ be a pairing which induces pairings $A \times B' \to C'$ and $A \times B'' \to C''$. Then the diagram

$$\begin{array}{ccccc} \hat{H}^p(G,A) & \times & \hat{H}^q(G,B'') & \xrightarrow{\cup} & \hat{H}^{p+q}(G,C'') \\ \downarrow{\scriptstyle id} & & \downarrow{\scriptstyle\delta} & & \downarrow{\scriptstyle(-1)^p\delta} \\ \hat{H}^p(G,A) & \times & \hat{H}^{q+1}(G,B') & \xrightarrow{\cup} & \hat{H}^{p+q+1}(G,C') \end{array}$$

is commutative, i.e. $(-1)^p\,\delta(\alpha \cup \beta'') = \alpha \cup \delta\beta''$.

Proof: Without loss of generality, we may assume that $C = A \otimes B$. If $X_\bullet = X_\bullet(G,\mathbb{Z})$ denotes the complete standard resolution of $\mathbb{Z}$, cf. §2, p.22, then we obtain a homomorphism of complexes

$$\mathrm{Hom}_G(X_\bullet, A) \otimes \mathrm{Hom}_G(X_\bullet, B) \longrightarrow \mathrm{Hom}_G(X_\bullet \otimes X_\bullet, A \otimes B).$$

The proof of existence of the cup-product depends on constructing G-module homomorphisms

$$\varphi_{p,q} : X_{p+q} \longrightarrow X_p \otimes X_q ,$$

for all integers p, q, satisfying the following two conditions

(1) $$\varphi_{p,q}\partial = (\partial \otimes 1)\varphi_{p+1,q} + (-1)^p(1 \otimes \partial)\varphi_{p,q+1} ,$$

(2) $$(\varepsilon \otimes \varepsilon)\varphi_{0,0} = \varepsilon ,$$

where $\varepsilon : X_0 \to \mathbb{Z}$ is defined by $\varepsilon(\sigma) = 1$ for all $\sigma \in G$. The induced map of complexes $\varphi : X_\bullet \longrightarrow X_\bullet \otimes X_\bullet$ then defines a homomorphism

$$\begin{aligned} \mathrm{Hom}_G(X_\bullet, A) \otimes \mathrm{Hom}_G(X_\bullet, B) &\longrightarrow \mathrm{Hom}_G(X_\bullet, A \otimes B) \\ (f, g) &\longmapsto f \cdot g = (f \otimes g)\varphi . \end{aligned}$$

It follows from (1) that

$$\partial(f \cdot g) = (\partial f) \cdot g + (-1)^p f \cdot (\partial g) .$$

Hence if f, g are cocycles, so is $f \cdot g$, and the cohomology class of $f \cdot g$ depends only on the classes of f and g. Thus we obtain homomorphisms

$$\hat{H}^p(G, A) \otimes \hat{H}^q(G, B) \xrightarrow{\cup} \hat{H}^{p+q}(G, A \otimes B) ,$$

which obviously satisfy (i), and (ii) is a consequence of (2). The properties (iii) and (iv) are proved as in (1.4.3). This gives us the existence of the cup-product and the uniqueness is proved by starting with (ii) and shifting dimensions by (iii) and (iv), as in §3, p.32. Observe that the exact sequences

$$0 \to A \to \mathrm{Ind}_G(A) \to A_1 \to 0 \quad \text{and} \quad 0 \to A_{-1} \to \mathrm{Ind}_G(A) \to A \to 0$$

split over $\mathbb{Z}$. Thus the result of tensoring these by any G-module B is still exact and $\mathrm{Ind}_G(A) \otimes B = \mathrm{Ind}_G(A \otimes B)$.

It remains to define the maps $\varphi_{p,q}$, which we will do as follows:
If $p \geq 0$ and $q \geq 0$,

$$\varphi_{p,q}(\sigma_0, \ldots, \sigma_{p+q}) = (\sigma_0, \ldots, \sigma_p) \otimes (\sigma_p, \ldots, \sigma_{p+q}) .$$

If $p \geq 1$ and $q \geq 1$,

$$\varphi_{-p,-q}(\sigma_1, \ldots, \sigma_{p+q}) = (\sigma_1, \ldots, \sigma_p) \otimes (\sigma_{p+1}, \ldots, \sigma_{p+q}) .$$

If $p \geq 0$ and $q \geq 1$,

$$\varphi_{p,-p-q}(\sigma_1, \ldots, \sigma_q) = \sum (\sigma_1, \tau_1, \ldots, \tau_p) \otimes (\tau_p, \ldots, \tau_1, \sigma_1, \ldots, \sigma_q) ,$$

$$\varphi_{-p-q,p}(\sigma_1, \ldots, \sigma_q) = \sum (\sigma_1, \ldots, \sigma_q, \tau_1, \ldots, \tau_p) \otimes (\tau_p, \ldots, \tau_1, \sigma_q) ,$$

$$\varphi_{p+q,-q}(\sigma_0, \ldots, \sigma_p) = \sum (\sigma_0, \ldots, \sigma_p, \tau_1, \ldots, \tau_q) \otimes (\tau_q, \ldots, \tau_1) ,$$

$$\varphi_{-q,p+q}(\sigma_0, \ldots, \sigma_p) = \sum (\tau_1, \ldots, \tau_q) \otimes (\tau_q, \ldots, \tau_1, \sigma_0, \ldots, \sigma_p) ,$$

where the τ_i on the right-hand side run independently through G. The verification that the $\varphi_{p,q}$ satisfy the formulae above is straightforward. □

Remark: Under the assumption that G is finite, the propositions (1.4.4) and (1.4.5) hold for all integers $p, q, r \in \mathbb{Z}$.

From the proof of the last proposition, we obtain the following explicit formulae.

(1.4.8) Proposition. *Let G be a finite group. Let $\bar{x} \in \hat{H}^p(G, A)$, $p \geq -1$ and $\bar{y} \in H^q(G, B)$, $q \geq 1$. Let $y \in \mathscr{C}^q(G, B)$ be an inhomogeneous cocycle representing $\bar{y}$. For $p \geq 0$, let $x \in \mathscr{C}^p(G, A)$ be a representing cocycle of $\bar{x}$. For $p = -1$, let $x \in {}_{N_G}A$ represent $\bar{x}$. Then the inhomogeneous cochains*

$$(x \cup y)(\sigma_1, \dots, \sigma_q) = x \otimes y(\sigma_1, \dots, \sigma_q) \qquad \text{for } p = 0,$$

$$(x \cup y)(\sigma_1, \dots, \sigma_{q-1}) = \sum\nolimits_{\sigma \in G} \sigma x \otimes \sigma y(\sigma^{-1}, \sigma_1, \dots, \sigma_{q-1}) \qquad \text{for } p = -1,$$

$$(x \cup y)(\sigma_1, \dots, \sigma_{p+q}) = x(\sigma_1, \dots, \sigma_p) \otimes \sigma_1 \cdots \sigma_p y(\sigma_{p+1}, \dots, \sigma_{p+q}) \qquad \text{for } p > 0,$$

represent the cup-product $\bar{x} \cup \bar{y} \in \hat{H}^{p+q}(G, A \otimes B)$.

Exercise 1. Let R be a G-ring, i.e. a ring with an action of G such that $\sigma(a + b) = \sigma a + \sigma b$ and $\sigma(ab) = \sigma a \sigma b$. Show that

$$H(G, R) := \bigoplus_{n \geq 0} H^n(G, R)$$

is a graded ring with respect to the cup-product which is induced by the multiplication $R \times R \to R$.

Exercise 2. Let A be an R-module with a G-operation compatible with the R-module structure. Show that

$$H(G, A) := \bigoplus_{n \geq 0} H^n(G, A)$$

is in a natural way an $H(G, R)$-module.

Exercise 3. Prove the following generalization of (1.4.3): Let

$$0 \to A' \to A \to A'' \to 0,$$

$$0 \to B' \to B \to B'' \to 0,$$

$$0 \to C' \to C \to C'' \to 0$$

be exact sequences of G-modules. Suppose we are given a pairing $\varphi\colon A \times B \to C$ such that

$$\varphi(A' \times B') = 0, \quad \varphi(A \times B') \subseteq C' \quad \text{and} \quad \varphi(A' \times B) \subseteq C'.$$

Then we get an induced pairing $A'' \times B'' \to C''$, and for $\alpha'' \in H^p(G, A'')$ and $\beta'' \in H^q(G, B'')$ we have

$$\delta(\alpha'' \cup \beta'') = (\delta\alpha'') \cup \beta'' + (-1)^p \alpha'' \cup (\delta\beta'').$$

§5. Change of the Group G

We now turn to the question of what happens to the cohomology groups $H^n(G,A)$ if we change the group G. We put ourselves in the most general situation if we consider two profinite groups G and G', a G-module A, a G'-module A' and two homomorphisms

$$\varphi : G' \longrightarrow G, \quad f : A \longrightarrow A',$$

such that $f(\varphi(\sigma')a) = \sigma' f(a)$. From such a compatible pair of homomorphisms, we obtain a homomorphism

$$C^n(G,A) \longrightarrow C^n(G',A'), \quad a \mapsto f \circ a \circ \varphi.$$

Trivially, this commutes with ∂ and therefore induces a homomorphism

$$H^n(G,A) \longrightarrow H^n(G',A').$$

If we have two compatible pairs of homomorphisms $G'' \to G' \to G$, $A \to A' \to A''$, then the homomorphism

$$H^n(G,A) \longrightarrow H^n(G'',A''),$$

induced by the composites $G'' \to G$ and $A \to A''$, is the composite of the homomorphisms

$$H^n(G,A) \longrightarrow H^n(G',A') \quad \text{and} \quad H^n(G',A') \longrightarrow H^n(G'',A'')$$

given by $G' \to G$, $A \to A'$ and $G'' \to G'$, $A' \to A''$. Thus the cohomology groups $H^n(G,A)$ are functorial in G and A simultaneously.

Let $(G_i)_{i\in I}$ be a projective system of profinite groups and let $(A_i)_{i\in I}$ be a direct system, where each A_i is a G_i-module and the transition maps

$$G_j \to G_i, \ A_i \to A_j$$

form compatible pairs. Then, with the induced homomorphisms

$$H^n(G_i,A_i) \longrightarrow H^n(G_j,A_j),$$

the cohomology groups $H^n(G_i,A_i)$ form a direct system of abelian groups. As a generalization of (1.2.5), we have the

(1.5.1) Proposition. *If* $G = \varprojlim_{i\in I} G_i$ *and* $A = \varinjlim_{i\in I} A_i$, *then*

$$H^n(G,A) \cong \varinjlim_{i\in I} H^n(G_i,A_i).$$

Proof: For every $i \in I$, the pair $G \to G_i$, $A_i \to A$ is compatible. We therefore have a canonical homomorphism $\kappa_i : C^n(G_i,A_i) \to C^n(G,A)$, hence a homomorphism

$$\kappa : \varinjlim_{i\in I} C^n(G_i,A_i) \longrightarrow C^n(G,A),$$

which obviously commutes with the ∂-homomorphism. It therefore suffices to show that κ is an isomorphism.

For the surjectivity, let $y : G^n \to A$ be the inhomogeneous cochain associated to $x \in C^n(G, A)$. Since G^n is compact, A discrete and y continuous, y takes only finitely many values and factors through $\bar{y} : (G/U)^n \to A$ for a suitable open normal subgroup U. The finitely many values are represented by elements of some A_i, i.e. $\bar{y}$ is the composite of a function $\bar{y}_i : (G/U)^n \to A_i$ with $A_i \to A$. On the other hand, there exists a $j \geq i$ such that the projection $G \to G/U$ factors through the canonical map $G_j \to G/U$, i.e. we obtain an inhomogeneous cochain $y_j : G_j^n \to A_j$ as the composite

$$G_j^n \longrightarrow (G/U)^n \xrightarrow{\bar{y}_i} A_i \longrightarrow A_j,$$

such that the composite $G^n \longrightarrow G_j^n \xrightarrow{y_j} A_j \longrightarrow A$ is y. If $x_j \in C^n(G_j, A_j)$ is the homogeneous cochain associated to y_j, then its image in $C^n(G, A)$ is x. This shows the surjectivity of κ.

For the injectivity, let $x_i \in C^n(G_i, A_i)$ be a cochain which becomes zero in $C^n(G, A)$, i.e. the composite

$$G^{n+1} \longrightarrow G_i^{n+1} \xrightarrow{x_i} A_i \longrightarrow A$$

is zero. Since x_i has only finitely many values, there exists a $j \geq i$ such that the composite

$$G_j^{n+1} \longrightarrow G_i^{n+1} \xrightarrow{x_i} A_i \longrightarrow A_j$$

is already zero, i.e. x_i becomes zero in $C^n(G_j, A_j)$ and hence represents the zero class in $\varinjlim_i C^n(G_i, A_i)$. This shows the injectivity of κ. □

We shall have to deal mainly with three special cases of homomorphisms $H^n(G, A) \to H^n(G', A')$ coming from compatible pairs $G' \to G$, $A \to A'$, and an additional case, arising in a different way.

1. Conjugation. Let H be a closed subgroup of G, A a G-module and B an H-submodule of A. For $\sigma, \tau \in G$ we write $\tau^\sigma = \sigma^{-1}\tau\sigma$ and ${}^\sigma H = \sigma H \sigma^{-1}$. The two compatible homomorphisms

$${}^\sigma H \longrightarrow H,\ \tau \longmapsto \tau^\sigma, \qquad B \longrightarrow \sigma B,\ b \longmapsto \sigma b$$

induce isomorphisms

$$\sigma_* : H^n(H, B) \longrightarrow H^n({}^\sigma H, \sigma B),$$

which are called **conjugation**. We have

$$1_* = id \quad \text{and} \quad (\sigma\tau)_* = \sigma_*\tau_*,$$

from what we have said above about composition.

2. Inflation. Let H be a normal closed subgroup of G and A a G-module. Then A^H is a G/H-module. The projection and injection

$$G \longrightarrow G/H, \qquad A^H \hookrightarrow A$$

form a compatible pair of homomorphisms, which induces a homomorphism

$$inf_G^{G/H} : H^n(G/H, A^H) \longrightarrow H^n(G, A),$$

called **inflation**. The inflation is transitive, i.e. for two normal closed subgroups $H \subseteq F$ of G, we have

$$inf_G^{G/H} \circ inf_{G/H}^{G/F} = inf_G^{G/F}.$$

3. Restriction. For an arbitrary closed subgroup H of G and a G-module A, we consider the two homomorphisms

$$H \xhookrightarrow{incl} G, \qquad A \xrightarrow{id} A.$$

On the cochains they induce the restriction maps and we obtain homomorphisms on the cohomology

$$res_H^G : H^n(G, A) \longrightarrow H^n(H, A),$$

called **restriction**. Clearly the restriction is transitive, i.e. for two closed subgroups $F \subseteq H$, we have

$$res_F^H \circ res_H^G = res_F^G.$$

4. Corestriction. If H is an *open* subgroup of G, then besides the restriction, we have another map in the opposite direction, which is a kind of norm map and is called the **corestriction**: it arises from the standard resolution $A \to X^\bullet = X^\bullet(G, A)$ of the G-module A, which is also an acyclic resolution of A as an H-module, i.e.

$$H^n(H, A) = H^n((X^\bullet)^H)$$

(see §3 p.34). For $n \geq 0$, we have for the G-module X^n the *norm* map $N_{G/H} : (X^n)^H \to (X^n)^G$. It obviously commutes with ∂, hence we have a morphism of complexes

$$N_{G/H} : (X^\bullet)^H \longrightarrow (X^\bullet)^G.$$

Taking homology groups of these complexes, we obtain canonical homomorphisms

$$cor_G^H : H^n(H, A) \longrightarrow H^n(G, A).$$

For $n = 0$, this is the usual norm map

$$N_{G/H} : A^H \longrightarrow A^G.$$

For two open subgroups $F \subseteq H$ of G, the equation $N_{G/H} \circ N_{H/F} = N_{G/F}$ implies the transitivity

$$cor_G^H \circ cor_H^F = cor_G^F.$$

On the level of cochains the corestriction is explicitly given as follows. From every right coset $c = H\sigma \in H\backslash G$, we choose a fixed representative $\bar{c} \in c$ and define the homomorphism

$$cor : C^n(H, A) \longrightarrow C^n(G, A),$$

by

$$(cor\, x)(\sigma_0, \ldots, \sigma_n) = \sum_{c \in H\backslash G} \bar{c}^{-1} x(\bar{c}\sigma_0 \overline{c\sigma_0}^{-1}, \ldots, \bar{c}\sigma_n \overline{c\sigma_n}^{-1}).$$

$cor\, x$ is again G-linear. In fact, if $\sigma \in G$, then $\bar{c}\sigma =: \tau_\sigma \overline{c\sigma}$ for some $\tau_\sigma \in H$ and

$$\begin{aligned}
&\sigma^{-1}(cor\, x)(\sigma\sigma_0, \ldots, \sigma\sigma_n)\\
&= \sum_{c \in H\backslash G} \sigma^{-1}\bar{c}^{-1} x(\bar{c}\sigma\sigma_0 \overline{c\sigma\sigma_0}^{-1}, \ldots)\\
&= \sum_{c \in H\backslash G} \overline{c\sigma}^{-1}\tau_\sigma^{-1} x(\tau_\sigma \overline{c\sigma}\sigma_0 \overline{(c\sigma)\sigma_0}^{-1}, \ldots)\\
&= \sum_{c \in H\backslash G} \bar{c}^{-1} x(\bar{c}\sigma_0 \overline{c\sigma_0}^{-1}, \ldots).
\end{aligned}$$

Obviously $cor \circ \partial = \partial \circ cor$, so we get a homomorphism

$$cor : H^n(H, A) \longrightarrow H^n(G, A).$$

It is functorial in A and commutes with the δ-homomorphism, which we will see in a moment. By *dimension shifting* we see that it coincides with the corestriction cor_G^H constructed before: for each $n \geq 0$, we have the commutative diagrams

$$\begin{array}{ccc}
H^0(H, A_n) & \xrightarrow{\delta^n} & H^n(H, A)\\
{\scriptstyle cor_G^H}\Big\downarrow\Big\downarrow{\scriptstyle cor} & & {\scriptstyle cor_G^H}\Big\downarrow\Big\downarrow{\scriptstyle cor}\\
H^0(G, A_n) & \xrightarrow{\delta^n} & H^n(G, A),
\end{array}$$

where the horizontal maps are surjective. The vertical arrows cor_G^H and cor are the norm $N_{G/H}$ for $n = 0$, hence coincide for all $n \geq 0$.

(1.5.2) Proposition. *The maps σ_*, inf, res, cor are functorial in the G-module considered, and they commute with the δ-homomorphism.*

Proof: The functoriality is seen already on the level of cochains. We show the commutativity of the corestriction with δ, leaving the other cases to the reader. Let

$$0 \longrightarrow A'' \xrightarrow{i} A \xrightarrow{j} A' \longrightarrow 0$$

be an exact sequence of G-modules. We then obtain a commutative exact diagram

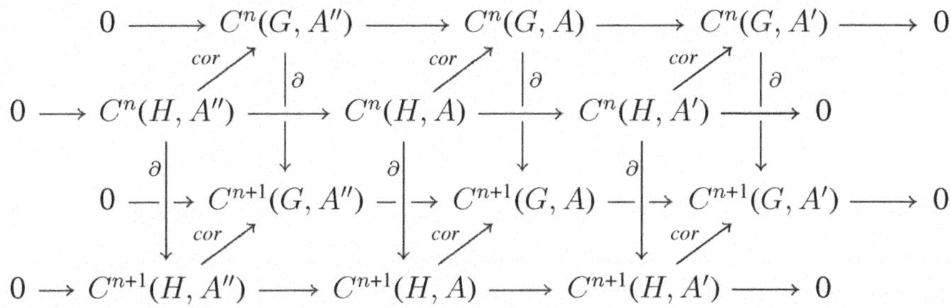

and from this, the commutativity of the diagram

$$\begin{array}{ccc} H^n(H,A') & \xrightarrow{\ \delta\ } & H^{n+1}(H,A'') \\ {\scriptstyle cor}\downarrow & & \downarrow{\scriptstyle cor} \\ H^n(G,A') & \xrightarrow{\ \delta\ } & H^{n+1}(G,A''). \end{array}$$

Namely, let $a' \in Z^n(H,A')$ be a cocycle in the class $\alpha' \in H^n(H,A')$. If $a \in C^n(H,A)$ is a pre-image of a', then ∂a is a cocycle in the class $\delta\alpha' \in H^{n+1}(H,A'')$ and $cor\,\partial a$ is a cocycle in the class $cor\,\delta\alpha'$. On the other hand, $cor\,\partial a = \partial cor\,a$, and $cor\,a$ is a pre-image of the cocycle $cor\,a'$, which represents $cor\,\alpha' \in H^n(G,A')$, so that $\partial\,cor\,a \in Z^{n+1}(G,A'')$ represents the class $\delta\,cor\,\alpha' \in H^{n+1}(G,A'')$. Therefore we have $\delta \circ cor = cor \circ \delta$.

In the same way one proves $\delta\circ\sigma_* = \sigma_*\circ\delta$, $\delta\circ res = res\circ\delta$, $\delta\circ inf = inf\circ\delta$, the latter if the sequence $0 \to A''^H \to A^H \to A'^H \to 0$ is also exact. □

(1.5.3) Proposition. *The maps σ_*, inf, res, cor are compatible with the cup-product as follows.*

(i) $\sigma_*(\alpha\cup\beta) = \sigma_*\alpha\cup\sigma_*\beta$

for $\alpha \in H^p(H,A)$, $\beta \in H^q(H,B)$ and $\sigma \in G$.

(ii) *inf* $(\alpha\cup\beta) = (inf\ \alpha)\cup(inf\ \beta)$

for $\alpha \in H^p(G/H,A^H)$, $\beta \in H^q(G/H,B^H)$, *if H is a normal closed subgroup of G.*

(iii) *res* $(\alpha\cup\beta) = (res\ \alpha)\cup(res\ \beta)$

for $\alpha \in H^p(G,A)$, $\beta \in H^q(G,B)$, *if H is a closed subgroup of G.*

(iv) *cor* $(\alpha\cup res\ \beta) = (cor\ \alpha)\cup\beta$

for $\alpha \in H^p(H,A)$, $\beta \in H^q(G,B)$, *if H is an open subgroup of G.*

Proof. (i), (ii) and (iii) are seen at once on the level of cochains. (iv) is equivalent to the commutativity of the diagram

$$\begin{array}{ccccc} H^p(H,A) & \times & H^q(H,B) & \xrightarrow{\cup} & H^{p+q}(H,A\otimes B) \\ \downarrow{\scriptstyle cor} & & \uparrow{\scriptstyle res} & & \downarrow{\scriptstyle cor} \\ H^p(G,A) & \times & H^q(G,B) & \xrightarrow{\cup} & H^{p+q}(G,A\otimes B). \end{array}$$

We may by (1.2.5) assume that G is finite: apply $\varinjlim_U$ to the diagram with G, H replaced by $G/U,\ H/U$, where U runs through the open normal subgroups U contained in H. By dimension shifting, we may transform the above diagram into the diagram

$$\begin{array}{ccccc} \hat{H}^0(H,A_p) & \times & \hat{H}^0(H,B_q) & \xrightarrow{\cup} & \hat{H}^0(H,A_p\otimes B_q) \\ \downarrow{\scriptstyle cor} & & \uparrow{\scriptstyle res} & & \downarrow{\scriptstyle cor} \\ \hat{H}^0(G,A_p) & \times & \hat{H}^0(G,B_q) & \xrightarrow{\cup} & \hat{H}^0(G,A_p\otimes B_q), \end{array}$$

which comes from the diagram

$$\begin{array}{ccccc} A_p^H & \times & B_q^H & \xrightarrow{\otimes} & (A_p\otimes B_q)^H \\ \downarrow{\scriptstyle N_{G/H}} & & \uparrow & & \downarrow{\scriptstyle N_{G/H}} \\ A_p^G & \times & B_q^G & \xrightarrow{\otimes} & (A_p\otimes B_q)^G. \end{array}$$

But this diagram is commutative:

$$N_{G/H}(a\otimes b) = \sum_{\sigma\in G/H} \sigma a\otimes \sigma b = \sum_{\sigma\in G/H} \sigma a\otimes b = N_{G/H}(a)\otimes b. \qquad \square$$

The compatibilities of the maps σ_*, *inf, res, cor* with each other are described by the following propositions.

(1.5.4) Proposition. *σ_* commutes with inf, res, cor.*

(1.5.5) Proposition. *For two closed subgroups $V \subseteq U \subseteq G$, we have*

(i) $inf_U^{U/V} \circ res_{U/V}^{G/V} = res_U^G \circ inf_G^{G/V}$,

if V is normal in G,

(ii) $inf_G^{G/V} \circ cor_{G/V}^{U/V} = cor_G^U \circ inf_U^{U/V}$,

if V is normal and U is open in G.

All this can be seen directly on the level of cochains. Another useful formula is stated in the following

(1.5.6) Proposition. *If U, V are closed subgroups of G, with V open, and if σ runs through a system of representatives of the (finite) double coset decomposition*

$$G = \dot{\bigcup_{\sigma}} U\sigma V,$$

then we have the **double coset formula**

$$res_U^G \circ cor_G^V = \sum_{\sigma} cor_U^{U\cap\sigma V\sigma^{-1}} \circ \sigma_* \circ res_{V\cap\sigma^{-1}U\sigma}^V .$$

Proof: Because of (1.2.5), (1.5.4) and (1.5.5), we may assume that G is a finite group. By dimension shifting we are then reduced to show the formula only for $n = 0$, i.e. on $H^0(V, A) = A^V$. In this case *res* becomes the inclusion, *cor* the norm and σ_* the map $a \mapsto \sigma a$. Therefore we have to prove the formula

$$N_{G/V}a = \sum_{\sigma} N_{U/U\cap\sigma V\sigma^{-1}}(\sigma a).$$

For every σ, we choose a system τ_σ of left representatives of $U/U\cap\sigma V\sigma^{-1}$, i.e.

$$U = \dot{\bigcup_{\tau_\sigma}} \tau_\sigma(U\cap\sigma V\sigma^{-1}).$$

Then

$$G = \dot{\bigcup_{\sigma,\tau_\sigma}} \tau_\sigma\sigma V,$$

i.e. $\tau_\sigma\sigma$ runs through a system of representatives of G/V, and so

$$N_{G/V}a = \sum_{\sigma}\sum_{\tau_\sigma} \tau_\sigma\sigma a = \sum_{\sigma} N_{U/U\cap\sigma V\sigma^{-1}}(\sigma a). \qquad \square$$

(1.5.7) Corollary. *If U is an open subgroup of G, then*

$$cor_G^U \circ res_U^G = (G:U).$$

If U is normal, then

$$res_U^G \circ cor_G^U = N_{G/U}.$$

The first formula is trivial in dimension zero and follows for arbitrary dimension by dimension shifting. The second formula is the double coset formula for the case $V = U$.

(1.5.8) Corollary. *If $G = UV$, then the diagram*

$$\begin{array}{ccc} H^n(V,A) & \xrightarrow{res} & H^n(U\cap V,A) \\ \downarrow{\scriptstyle cor} & & \downarrow{\scriptstyle cor} \\ H^n(G,A) & \xrightarrow{res} & H^n(U,A) \end{array}$$

is commutative.

For a locally compact abelian group A, let $A^\vee$ denote the Pontryagin dual, see §1, p.9. We have a canonical isomorphism

$$H^2(G,\mathbb{Z})^\vee \cong G^{ab}$$

onto the abelianized group $G^{ab} = G/G'$, the quotient of G by the closure G' of the commutator subgroup. For this, note that $H^n(G,\mathbb{Q}) = 0$ for $n \geq 1$ (see (1.6.2)(c)). Therefore the exact sequence $0 \to \mathbb{Z} \to \mathbb{Q} \to \mathbb{Q}/\mathbb{Z} \to 0$ yields an isomorphism

$$H^2(G,\mathbb{Z}) \cong H^1(G,\mathbb{Q}/\mathbb{Z}) = \mathrm{Hom}_{cts}(G,\mathbb{Q}/\mathbb{Z}) = (G^{ab})^\vee,$$

and, using Pontryagin duality,

$$H^2(G,\mathbb{Z})^\vee \cong (G^{ab})^{\vee\vee} = G^{ab}.$$

For an open subgroup H of G, we may ask what homomorphism

$$G^{ab} \longrightarrow H^{ab}$$

is induced by the dual

$$cor^\vee : H^2(G,\mathbb{Z})^\vee \longrightarrow H^2(H,\mathbb{Z})^\vee$$

of the corestriction. The answer is given by the **transfer map** (Verlagerung) of group theory, which is defined as follows.

For each right coset $c \in H\backslash G$, we choose a fixed representative $\bar{c}$, $\bar{1} = 1$, so that $c = H\bar{c}$. Then the transfer is the continuous homomorphism

$$Ver : G^{ab} \longrightarrow H^{ab}, \quad \sigma G' \longmapsto \prod_{c\in H\backslash G} \bar{c}\sigma\overline{c\sigma}^{-1}H'.$$

(1.5.9) Proposition. *The map $cor^\vee : H^2(G,\mathbb{Z})^\vee \to H^2(H,\mathbb{Z})^\vee$ induces the transfer map*

$$Ver : G^{ab} \longrightarrow H^{ab}.$$

If H is normal in G, the composite $H^{ab} \to G^{ab} \to H^{ab}$ is the norm $N_{G/H}$.

Proof: On the homogeneous 1-cochains, the corestriction

$$cor : C^1(H, \mathbb{Q}/\mathbb{Z}) \to C^1(G, \mathbb{Q}/\mathbb{Z})$$

is given by

$$(cor\, x)(\sigma_0, \sigma_1) = \sum_{c \in H\backslash G} x(\bar{c}\sigma_0\overline{c\sigma_0}^{-1}, \bar{c}\sigma_1\overline{c\sigma_1}^{-1}).$$

On the associated inhomogeneous cochains $y(\sigma) = x(1, \sigma)$, it is thus given by

$$(cor\, y)(\sigma) = (cor\, x)(1, \sigma) = \sum_{c \in H\backslash G} y(\bar{c}\sigma\overline{c\sigma}^{-1}).$$

Hence on the dual G^{ab} of $H^1(G, \mathbb{Q}/\mathbb{Z}) = Z^1(G, \mathbb{Q}/\mathbb{Z}) = \mathrm{Hom}(G^{ab}, \mathbb{Q}/\mathbb{Z})$, it is given by the transfer map.
Let H be normal in G. The composite $H^{ab} \longrightarrow G^{ab} \xrightarrow{Ver} H^{ab}$ is dual to the composite of

$$H^2(H, \mathbb{Z}) \xrightarrow{cor} H^2(G, \mathbb{Z}) \xrightarrow{res} H^2(H, \mathbb{Z}),$$

which by (1.5.7) is $res \circ cor = N_{G/H}$. □

If A is a G-module, then for every pair $V \subseteq U$ of open subgroups, we have two maps

$$A^U \underset{N_{U/V}}{\overset{incl}{\rightleftarrows}} A^V,$$

the inclusion and the norm. For the cohomology groups, we have similarly two maps

$$H^n(U, A) \underset{cor}{\overset{res}{\rightleftarrows}} H^n(V, A),$$

and these satisfy the double coset formula (1.5.6). From this observation we are led to a generalization of G-modules, which gives a conceptual explanation of the double coset formula. We enlarge the totality of open subgroups U of G as follows.

We consider the category $B(G)$ of finite G-sets, i.e. of finite sets X with a continuous action of G. For every open subgroup U of G, the quotient G/U is a finite G-set by left multiplication. It is a *connected*, i.e. transitive, G-set, and every connected finite G-set X is of this form. For, if $x \in X$ and $G_x = \{\sigma \in G \mid \sigma x = x\}$, then $G/G_x \to X$, $\sigma G_x \mapsto \sigma x$, is an isomorphism of G-sets. The category $B(G)$ has the advantage of containing, for every two finite G-sets X, Y, the disjoint union $X \amalg Y$ (as categorical sum), and for every pair of morphisms $f : X \to S$, $g : Y \to S$, the fibre product $X \times_S Y = \{(x, y) \in X \times Y \mid f(x) = g(y)\}$.

(1.5.10) Definition. *A* **G-modulation** *is a pair of functors*

$$A = (A^*, A_*) : B(G) \to \mathcal{A}b$$

into the category $\mathcal{A}b$ of abelian groups, A_* covariant, A^* contravariant, such that

$$A^*(X) = A_*(X) =: A(X)$$

for all $X \in B(G)$, and that the following two conditions are satisfied:

(i) $A(X \amalg Y) = A(X) \oplus A(Y)$.

(ii) *If the left one of the diagrams*

$$\begin{array}{ccccccc} X & \xrightarrow{g'} & X' & & A(X) & \xleftarrow{g'^*} & A(X') \\ {\scriptstyle f}\downarrow & & \downarrow{\scriptstyle f'} & \text{and} & {\scriptstyle f_*}\downarrow & & \downarrow{\scriptstyle f'_*} \\ Y & \xrightarrow{g} & Y' & & A(Y) & \xleftarrow{g^*} & A(Y') \end{array}$$

is cartesian (i.e. $X \cong X' \times_{Y'} Y$), then the right one commutes. Here $\varphi^* = A^*(\varphi)$, $\varphi_* = A_*(\varphi)$ for a morphism φ in $B(G)$. *)

In this form the G-modulations were introduced by A. *Dress* [43] under the name "Mackey functors" (they were defined earlier by A. *Green* under still another name, see [59]).

By condition (i), a G-modulation A is completely determined by its restriction to the full subcategory $B_0(G)$ of the G-sets G/U, where U runs through the open subgroups of G. One then writes $A(U)$ in place of $A(G/U)$. Every morphism in $B_0(G)$ is in a unique way the composite of a projection

$$\pi_U^V : G/V \longrightarrow G/U \qquad (V \subseteq U)$$

and a "conjugation"

$$c(\sigma) : G/U \longrightarrow G/\sigma U \sigma^{-1}, \quad \tau U \longmapsto \tau\sigma^{-1}(\sigma U \sigma^{-1}).$$

We set

$$\begin{aligned} res_V^U &= A^*(\pi_U^V) &&: A(U) \longrightarrow A(V), \\ ind_U^V &= A_*(\pi_U^V) &&: A(V) \longrightarrow A(U), \\ \sigma^* &= A^*(c(\sigma)) &&: A(\sigma U \sigma^{-1}) \longrightarrow A(U), \\ \sigma_* &= A_*(c(\sigma)) &&: A(U) \longrightarrow A(\sigma U \sigma^{-1}). \end{aligned}$$

---

*) Clearly, this notion of G-modulations extends to G-modulations $A : B(G) \to \mathcal{A}$ with values in an arbitrary abelian category $\mathcal{A}$.

(1.5.11) Proposition (Double Coset Formula). *Let U, $V \subseteq W$ be open subgroups of G and let R be a system of representatives of $U\backslash W/V$. Then, for every G-modulation A, we have the formula*

$$res_U^W \circ ind_W^V = \sum_{\sigma\in R} ind_U^{U\cap\sigma V\sigma^{-1}} \circ \sigma_* \circ res_{V\cap\sigma^{-1}U\sigma}^V .$$

Proof: Let $X = G/U$, $Y = G/V$, $S = G/W$. For the fibre product $X \times_S Y$ we have the orbit decomposition

$$X \times_S Y = \dot{\bigcup_{\sigma\in R}} C_\sigma,$$

where C_σ is the G-orbit of the element $(U, \sigma V) \in X \times_S Y$. Let $f = \pi_W^U$, $g = \pi_W^V$ and let $p : X \times_S Y \to X$, $q : X \times_S Y \to Y$ be the projections and $p_\sigma = p\,|_{C_\sigma}$, $q_\sigma = q\,|_{C_\sigma}$. Then by the properties (i) and (ii) in (1.5.10)

$$(*) \qquad f^* \circ g_* = \sum_{\sigma\in R} p_{\sigma *} \circ q_\sigma^* .$$

The isotropy groups of the elements $(U, \sigma V)$ and $(\sigma^{-1}U, V)$ of C_σ are $U \cap \sigma V\sigma^{-1}$ and $\sigma^{-1}U\sigma \cap V$, and we have the commutative diagram

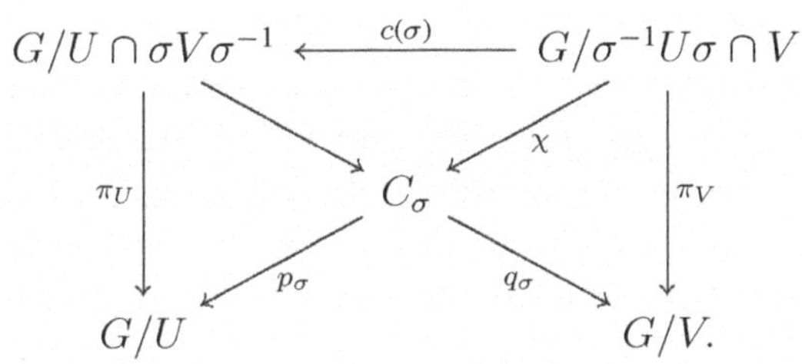

In the formula $(*)$ we have $f^* = res_U^W$, $g_* = ind_W^V$, $p_{\sigma *} = \pi_{U*} \circ c(\sigma)_* \circ \chi_*^{-1} = ind_U^{U\cap\sigma V\sigma^{-1}} \circ \sigma_* \circ \chi^*$ and $q_\sigma^* = (\pi_V \circ \chi^{-1})^* = (\chi^{-1})^* \circ \pi_V^* = \chi^{*-1} \circ res_{V\cap\sigma^{-1}U\sigma}^V$. This gives the desired result. □

Recall that a G-modulation A is completely determined by its restriction to $B_0(G)$, since every finite G-set is the disjoint union of connected G-sets, and every connected G-set is isomorphic to a set G/U. Conversely, we have the

(1.5.12) Proposition. *Let $A = (A^*, A_*) : B_0(G) \to \mathcal{A}b$ be a pair of functors, A_* covariant, A^* contravariant, such that $A^*(U) = A_*(U) := A(U)$. Assume that the double coset formula (1.5.11) holds and that moreover $\sigma^* \circ \sigma_* = id$ for every $\sigma \in G$. Then A extends uniquely to a G-modulation on $B(G)$.*

One obtains the extension of A to $B(G)$ as follows. Let $X \in B(G)$ be an arbitrary finite G-set. We let G act on the group $A_X = \prod_{x \in X} A(G_x)$ by the conjugation $\sigma_x : A(G_x) \to A(G_{\sigma x})$, and we set

$$A(X) = \operatorname{Hom}_G(X, A_X),$$

i.e. $A(X)$ is the group of all G-equivariant maps $X \to A_X$. The proof that this becomes a G-modulation is left to the reader.

In order to give a G-modulation A, it thus suffices to define an abelian group $A(U)$ for every open subgroup U of G and to establish, for every $\sigma \in G$ and every pair $V \subseteq U$ of open subgroups, the maps

$$A(U) \underset{\sigma^*}{\overset{\sigma_*}{\rightleftarrows}} A(\sigma U \sigma^{-1}), \qquad A(V) \underset{res}{\overset{ind}{\rightleftarrows}} A(U),$$

which yield the functors A^* and A_*, and then to verify the double coset formula.

It is clear what is meant by a morphism $A \to B$ of G-modulations. So the G-modulations form a category, which we denote by $\mathcal{M}od(G)$. We mention the following examples of G-modulations.

Example 1: G-modules. Let M be a G-module. For every open subgroup U we set $M(U) = M(G/U) = M^U$. For $\sigma \in G$ we define the maps σ^* and σ_* by $a \mapsto \sigma^{-1}a$ and $a \mapsto \sigma a$, and for every pair $V \subseteq U$ we define res_V^U and ind_U^V as the inclusion $M^U \hookrightarrow M^V$ and the norm $M^V \to M^U$, $a \mapsto N_{U/V}(a) = \prod_{\sigma \in U/V} \sigma a$ (M is written multiplicatively). In this way, every G-module M becomes a G-modulation, denoted again by M, and we obtain an embedding

$$i : Mod(G) \longrightarrow \mathcal{M}od(G)$$

of the category $Mod(G)$ of G-modules. With this embedding, $Mod(G)$ becomes the full subcategory of $\mathcal{M}od(G)$ of the G-modulations A with *Galois descent*, meaning that for every pair $V \trianglelefteq U$ of open subgroups the restriction

$$A(U) \longrightarrow A(V)^{U/V}$$

is an isomorphism.

We note that, in particular, every abelian group A gives rise to a *constant* G-modulation $U \mapsto A(U) = A$ with the maps $\sigma^* = \sigma_* = id$, $res_V^U = id$ and $ind_U^V = (U : V)$.

The embedding $i : Mod(G) \to \mathcal{M}od(G)$ has as left adjoint the functor $\mathcal{M}od(G) \to Mod(G)$, $A \mapsto \bar{A} = \varinjlim_U A(U)$, i.e. for every G-module M we have

$$\operatorname{Hom}_{\mathcal{M}od(G)}(A, iM) = \operatorname{Hom}_{Mod(G)}(\bar{A}, M).$$

Example 2: Cohomology. For every G-module M and every $n \geq 0$, the cohomology H^n yields a G-modulation

$$A : B_0(G) \longrightarrow \mathcal{A}b, \quad G/U \mapsto A(U) = H^n(U, M),$$

if we choose for res_V^U and ind_U^V the cohomological restriction res_V^U and corestriction cor_U^V, and for σ^*, σ_* the cohomological conjugations $(\sigma^{-1})_*$, σ_*.

Example 3: The fundamental G-modulation π^{ab}. We consider the map

$$B_0(G) \longrightarrow \mathcal{A}b, \quad G/U \longmapsto \pi^{ab}(U) := U^{ab},$$

which associates to every open subgroup U of G the abelianized group $U^{ab} = U/U'$, where U' is the closure of the commutator subgroup of U. For $\sigma \in G$ the maps $\sigma_* : U^{ab} \to (\sigma U \sigma^{-1})^{ab}$ and $\sigma^* : U^{ab} \to (\sigma^{-1} U \sigma)^{ab}$ are the conjugations $x \mapsto \sigma x \sigma^{-1}$ and $x \mapsto \sigma^{-1} x \sigma$, and for a pair $V \subseteq U$ the map ind_U^V is induced by the inclusion $V \hookrightarrow U$, whereas res_V^U is given by the *transfer* $Ver : U^{ab} \to V^{ab}$.

Example 4: The representation ring. We consider all finite dimensional complex representations $\mathcal{V}$ of the profinite group G, i.e. all continuous homomorphisms $G \to GL(\mathcal{V})$, where $\mathcal{V}$ is any finite dimensional $\mathbb{C}$-vector space. Such representations have finite images. Let $R(G)$ be the set of isomorphism classes $\{\mathcal{V}\}$ of such representations $\mathcal{V}$. We define an addition in $R(G)$ by

$$\{\mathcal{V}\} + \{\mathcal{V}'\} := \{\mathcal{V} \oplus \mathcal{V}'\}.$$

Then $R(G)$ becomes a commutative monoid. From $R(G)$ we obtain an additive group $Rep(G)$ by setting

$$Rep(G) = (R(G) \times R(G))/\sim,$$

where the equivalence relation $\sim$ is defined by

$$(\{\mathcal{V}\}, \{\mathcal{V}'\}) \sim (\{\mathcal{W}\}, \{\mathcal{W}'\}) \iff \{\mathcal{V}\} + \{\mathcal{W}'\} = \{\mathcal{W}\} + \{\mathcal{V}'\}.$$

$R(G)$ becomes a submonoid of $Rep(G)$ by identifying $\{\mathcal{V}\}$ with $(\{\mathcal{V}\}, O)$. We may even turn $Rep(G)$ into a ring if we define the multiplication by $\{\mathcal{V}\}\{\mathcal{W}\} = \{\mathcal{V} \otimes \mathcal{W}\}$. This ring is called the **representation ring** of G. Forgetting the ring structure, we obtain a G-modulation

$$Rep : B_0(G) \longrightarrow \mathcal{A}b, \quad G/U \longmapsto Rep(U),$$

as follows. Let $\sigma \in G$ and let U be an open subgroup of G. If $U \to GL(\mathcal{V})$ is a representation of U, then the composite with the conjugation $\sigma U \sigma^{-1} \to U$, $\sigma u \sigma^{-1} \mapsto u$, gives a representation $\sigma U \sigma^{-1} \to GL(\mathcal{V})$. This assignment extends to an isomorphism $\sigma_* : Rep(U) \to Rep(\sigma U \sigma^{-1})$, and we set $\sigma^* = \sigma_*^{-1}$. For a pair $V \subseteq U$ of open subgroups, the map

$$res_V^U : Rep(U) \to Rep(V)$$

is obtained by restricting a representation $U \to GL(\mathcal{V})$ to V – this is a homomorphism (even a ring homomorphism). The map

$$ind_U^V : Rep(V) \longrightarrow Rep(U)$$

is obtained by associating to a representation $\mathcal{V}$ of V the induced representation $ind_U^V(\mathcal{V})$ of U. The underlying vector space consists of all continuous maps $f : U \to \mathcal{V}$ such that $f(\tau\sigma) = \tau f(\sigma)$ for $\tau \in V$, $\sigma \in U$. The action of $\sigma \in U$ on $ind_U^V(\mathcal{V})$ is given by $f \mapsto \sigma f$, $(\sigma f)(x) = f(x\sigma)$. In this way we obtain in fact a G-modulation, because the double coset formula is in this case a theorem of *I. Mackey*. This brought about the name *Mackey functor* (see [43]).

For every G-modulation A, the groups $A(U)$ are in a canonical way topological groups: a basis of neighbourhoods of 0 is given by the groups $ind_U^V A(V)$ for $V \subseteq U$. It is easy to see that the maps σ^*, σ_*, res, ind are continuous. A is called *quasi-compact, Hausdorff, compact* etc. if all the groups $A(U)$ have the corresponding property. A is Hausdorff if $\bigcap_{V\subseteq U} ind_U^V A(V) = 0$ for all U. A is compact if it is quasi-compact and Hausdorff, and this is equivalent to being *profinite*, i.e. all $A(U)$ are profinite groups.

To every G-modulation A there is associated a submodulation NA, called the modulation of "universal norms", which is given by

$$NA(U) = \bigcap_{V\subseteq U} ind_U^V A(V).$$

The fact that, for $V \subseteq U$, the homomorphisms $A(V) \underset{res}{\overset{ind}{\rightleftarrows}} A(U)$ induce homomorphisms $NA(V) \underset{res}{\overset{ind}{\rightleftarrows}} NA(U)$ is trivial for *ind* and follows for *res* from the double coset formula. The quotient A/NA is a Hausdorff G-modulation (see exercise 6).

For further results on G-modulations we refer to [147], [161], [235].

Exercise 1. For $n \geq 1$ we have $\varinjlim_U H^n(U, A) = 0$, where U runs over the open subgroups of G and the limit is taken over the restriction maps $res : H^n(U, A) \to H^n(V, A)$, $V \subseteq U$.

Exercise 2. If H is a normal subgroup of G and A a G-module, then res_H^G is a homomorphism

$$res_H^G : H^n(G, A) \longrightarrow H^n(H, A)^{G/H},$$

and if moreover H is open, then cor_G^H yields a homomorphism

$$cor_G^H : H^n(H, A)_{G/H} \longrightarrow H^n(G, A).$$

Exercise 3. Let H be an open subgroup of G, A a G-module and $A \to X^\bullet = X^\bullet(G, A)$ the standard resolution of A. Then the restriction $H^n(G, A) \xrightarrow{res} H^n(H, A)$ is obtained by taking the homology of the restriction map

$$(*) \qquad (X^\bullet)^G \xrightarrow{res} (X^\bullet)^H.$$

Hint: By (1.3.9), we obtain isomorphisms of δ-functors

$$H^n(G, A) = H^n(H^0(G, X^\bullet)), \quad H^n(H, A) \cong H^n(H^0(H, X^\bullet)).$$

$(*)$ induces a functorial map $H^n(G, A) \xrightarrow{\iota} H^n(H, A)$, which commutes with the δ-homomorphism. It coincides with *res* for $n = 0$ and for $n > 0$ by dimension shifting.

Exercise 4. Consider a diagram

$$\begin{array}{ccccccccc} 1 & \longrightarrow & A & \longrightarrow & \hat{G} & \longrightarrow & G & \longrightarrow & 1 \\ & & \downarrow{\scriptstyle f} & & \downarrow{\scriptstyle \hat{\varphi}} & & \downarrow{\scriptstyle \varphi} & & \\ 1 & \longrightarrow & A' & \longrightarrow & \hat{G}' & \longrightarrow & G' & \longrightarrow & 1 \end{array}$$

with exact rows, where G, G' are profinite groups and A (resp. A') is a G-module (resp. a G'-module). Consider moreover the maps

$$H^2(G', A') \xrightarrow{\varphi^*} H^2(G, A') \xleftarrow{f_*} H^2(G, A)$$

given by the pairs $(G \xrightarrow{\varphi} G',\ A' \xrightarrow{id} A')$ and $(G \xrightarrow{id} G,\ A \xrightarrow{f} A')$. Let $u' \in H^2(G', A')$, $u \in H^2(G, A)$ be the classes belonging to the upper and lower group extension respectively.

(i) Show that the diagram of solid arrows can be commutatively completed by an arrow $\hat{\varphi} : \hat{G} \to \hat{G}'$ if and only if f is a G-homomorphism (G acting on A' via φ) and

$$\varphi^*(u') = f_*(u).$$

(ii) Two such arrows $\hat{\varphi}_1, \hat{\varphi}_2$ are called *equivalent* if there exists an $a' \in A'$, such that

$$\hat{\varphi}_2(\hat{\sigma}) = a'\hat{\varphi}_1(\hat{\sigma})a'^{-1} \quad \text{for all } \hat{\sigma} \in \hat{G}.$$

Show that the set of equivalence classes is an $H^1(G, A')$-torsor.

Exercise 5. Let U, V be open subgroups of G, and let A be a U-module. If $V \subseteq U$, then A is also a V-module, which we denote by $\mathrm{Res}_V^U A$. For $\sigma \in G$ we denote by σA the $\sigma U \sigma^{-1}$-module, whose underlying abelian group is A and the action of $\tau \in \sigma U \sigma^{-1}$ is given by $a \mapsto \sigma^{-1}\tau\sigma a$.

Show that for any two open subgroups U, V and any U-module A, we have an isomorphism of V-modules

$$\mathrm{Res}_U^G \mathrm{Ind}_G^V A \cong \bigoplus_{\sigma \in R} \mathrm{Ind}_U^{U \cap \sigma V \sigma^{-1}} \sigma \mathrm{Res}_{V \cap \sigma^{-1} U \sigma}^V A,$$

where R is a set of representatives of $U \backslash G / V$ and where the modules $\mathrm{Ind}\, A$ are defined below on p.61. In particular, if U is a normal open subgroup of G, then

$$\mathrm{Res}_U^G \mathrm{Ind}_G^U A \cong \bigoplus_{\sigma \in G/U} \sigma A.$$

Exercise 6. Let A be a G-modulation and let

$$NA(U) = \bigcap_{V \subseteq U} ind_U^V A(V).$$

Show that $U \mapsto NA(U)$ is a submodulation of A, and that the modulation A/NA is Hausdorff. Show that we get also a G-modulation $\hat{A}$, the "completion of A", by setting

$$\hat{A}(U) = \varprojlim_{V \subseteq U} A(U)/N_{U/V} A^U.$$

Exercise 7. For any two G-modulations A, B we have G-modulations $A \otimes B$ and $\mathcal{H}om(A, B)$. In particular, we have the notion of a "dual" $A^* = \mathcal{H}om(A, \mathbb{Q}/\mathbb{Z})$ of a G-modulation A.

Exercise 8. Let X be a finite G-set. A **complex vector bundle** on X is a continuous representation $G \to GL(\mathcal{V})$ on a finite dimensional $\mathbb{C}$-vector space $\mathcal{V}$ such that the projection $X \times \mathcal{V} \to X$ is a morphism of G-sets. The vector bundle is called a **line bundle** if $\dim \mathcal{V} = 1$. Define an abelian group $\mathrm{Pic}(X)$ of isomorphism classes of line bundles on X. Show that $X \mapsto \mathrm{Pic}(X)$ is actually a G-modulation Pic, and show that $\mathrm{Pic} = \mathcal{H}om(\pi^{ab}, \mathbb{Q}/\mathbb{Z})$.

§6. Basic Properties

We collect in this section some basic properties of cohomology groups, which will be used repeatedly. If we write $\hat{H}^n(G, A)$ in the following for profinite groups G, then for $n \geq 1$ this means, as for finite groups, $\hat{H}^n(G, A) = H^n(G, A)$. From the formula (1.5.7), $cor_G^U \circ res_U^G = (G : U)$, follows the

(1.6.1) Proposition. *Let G be a profinite group and let U be an open subgroup. Assume that G is finite or that $n \geq 1$. Then for every discrete G-module A such that $\hat{H}^n(U, A) = 0$, we have*

$$(G : U)\hat{H}^n(G, A) = 0.$$

In particular, if G is finite, then $\hat{H}^n(G, A)$ is annihilated by the order of G. If, moreover, A is finitely generated as a $\mathbb{Z}$-module, then $\hat{H}^n(G, A)$ is finite.

We conclude that for arbitrary profinite groups G the cohomology groups $H^n(G, A)$, $n \geq 1$, are torsion groups, since by (1.2.5)

$$H^n(G, A) = \varinjlim_U H^n(G/U, A^U),$$

where U runs through the open normal subgroups of G.

(1.6.2) Proposition. *Let G be a finite group and let A be a G-module. Assume that multiplication by p is an automorphism of A for all prime numbers $p \mid \#G$. Then*

$$\hat{H}^i(G, A) = 0 \qquad \textit{for all } i \in \mathbb{Z}.$$

If G is profinite, this remains true for $i \geq 1$. In particular, A is a cohomologically trivial G-module in the following cases:

a) *A is a torsion group whose supernatural order is prime to $\#G$,*

b) *A is an abelian profinite group whose supernatural order is prime to $\#G$,*

c) *A is uniquely divisible.*

Proof: Let us first assume that G is finite. Putting $m = \#G$, the map $A \xrightarrow{m} A$ is an automorphism of A by assumption. It therefore induces an isomorphism

$$m : \hat{H}^i(G, A) \xrightarrow{\sim} \hat{H}^i(G, A)$$

which, by (1.6.1), implies that $\hat{H}^i(G, A) = 0$.

Now let G be profinite. We have $H^i(G, A) = \varinjlim_U H^i(G/U, A^U)$ for $i \geq 1$. Let $U \subseteq G$ be a normal open subgroup and $m = \#(G/U)$. By assumption, the m-multiplication map $m : A \to A$ is an isomorphism. Taking U-invariants, we see that the m-multiplication map $m : A^U \to A^U$ is also an isomorphism. By the first part of the proof, we obtain $H^i(G/U, A^U) = 0$ for all $i \geq 1$, showing that $H^i(G, A) = 0$ for all $i \geq 1$. Finally note that the assumption on the p-multiplication map for all $p \mid \#G$ is satisfied in the cases a)–c), and that everything remains true after replacing G by a closed subgroup. This finishes the proof. □

Now let H be a closed normal subgroup of G. If A is a discrete G-module, then the cohomology group $H^n(H, A)$ is a discrete G-module, too: each $\sigma \in G$ acts on it by conjugation σ_*.

(1.6.3) Proposition. *The closed normal subgroup H acts trivially on the cohomology group $H^n(H, A)$, i.e. $H^n(H, A)$ is a G/H-module. In particular, the conjugation $\sigma_* : H^n(G, A) \to H^n(G, A)$ is the identity for all $\sigma \in G$.*

Proof: The assertion is trivial for $H^0(H, A) = A^H$. For $n > 0$ it follows by dimension shifting from the commutative diagram (see (1.3.8))

$$\begin{array}{ccc} \hat{H}^0(H, A_n) & \xrightarrow{\delta^n} & H^n(H, A) \\ \Big\downarrow{\scriptstyle \sigma_*} & & \Big\downarrow{\scriptstyle \sigma_*} \\ \hat{H}^0(H, A_n) & \xrightarrow{\delta^n} & H^n(H, A). \end{array}$$

□

Now let H be an arbitrary closed subgroup of G. For every H-module A, we consider the G-module

$$M = \mathrm{Ind}_G^H(A)$$

consisting of all continuous maps $x : G \to A$ such that $x(\tau\sigma) = \tau x(\sigma)$ for all $\tau \in H$. The action of $\rho \in G$ on M is given by $x(\sigma) \mapsto (\rho x)(\sigma) = x(\sigma\rho)$. The module M is said to be obtained by *inducing A from H to G.* *)

*) The terminology *induced module* is commonly used, but strictly speaking it is slightly inaccurate. From a categorical point of view the situation is as follows. Given a pair $H \subseteq G$ of abstract groups, the forgetful functor *Res*: *G-Mod*→*H-Mod* admits the left adjoint functor *Ind*: *H-Mod*→*G-Mod*, $A \mapsto \mathbb{Z}[G] \otimes_{\mathbb{Z}[H]} A$, and the right adjoint functor *Coind*: *H-Mod*→*G-Mod*, $A \mapsto Map_H(G, A)$. If H has finite index in G, both functors are isomorphic. However, with respect to the groups, the functor *Ind* is covariant while *Coind* is contravariant. This phenomenon can be viewed as the reason for the existence of *cor*. In the case of profinite groups, we have the functor *Coind* on discrete modules, but we write $\mathrm{Ind}_G^H A$ for *Coind*(A). Furthermore, we have the functor *Ind* ("compact induction") on compact modules .

We have a canonical projection

$$\pi : \mathrm{Ind}_G^H(A) \longrightarrow A, \quad x \mapsto x(1).$$

This is a homomorphism of H-modules, which maps the H-submodule $A' = \{x : G \to A \mid x(\tau) = 0 \text{ for all } \tau \notin H\}$ isomorphically onto A. We identify A' with A. When H is of finite index in G and $\sigma_1, \dots, \sigma_n$ is a system of representatives of G/H, then

$$M = \bigoplus_{i=1}^{n} \sigma_i A\,.$$

If A is a G-module, then $\mathrm{Ind}_G^H(A)$ is canonically isomorphic to the G-module $\mathrm{Map}(G/H, A)$ of all continuous functions $y : G/H \to A$, where the action of $\rho \in G$ on y is given by $(\rho y)(\sigma H) = \rho y(\rho^{-1}\sigma H)$. The isomorphism

$$\mathrm{Ind}_G^H(A) \cong \mathrm{Map}(G/H, A)$$

is given by $x(\sigma) \mapsto y(\sigma H) = \sigma x(\sigma^{-1})$. In particular, when $H = 1$ we have a canonical isomorphism

$$\mathrm{Ind}_G^H(A) \cong \mathrm{Ind}_G(A)$$

with the G-module $\mathrm{Ind}_G(A) = \mathrm{Map}(G, A)$ on which G acts by $(\rho x)(\sigma) = \rho x(\rho^{-1}\sigma)$ (see §3). We have seen in (1.3.7) that

$$H^n(G, \mathrm{Ind}_G(A)) = 0 \quad \text{for all} \quad n > 0.$$

We generalize this fact with the following proposition, commonly cited as **Shapiro's lemma.**

(1.6.4) Proposition. *Let H be a closed subgroup of G and let A be an H-module. Then, for all $n \geq 0$, we have a canonical isomorphism*

$$sh : H^n(G, \mathrm{Ind}_G^H(A)) \xrightarrow{\sim} H^n(H, A).$$

Proof: The groups $H^n(G, \mathrm{Ind}_G^H(A))$ are the homology groups of the complex $X^\bullet(G, \mathrm{Ind}_G^H(A))^G$. The canonical homomorphism

$$\pi : \mathrm{Ind}_G^H(A) \longrightarrow A \quad , \quad x \mapsto x(1),$$

of H-modules induces an isomorphism

$$H^0(G, \mathrm{Ind}_G^H(A)) \xrightarrow{\sim} H^0(H, A)\,,$$

and for each $n \geq 0$, a homomorphism

$$X^n(G, \mathrm{Ind}_G^H(A))^G \longrightarrow X^n(G, A)^H.$$

This is actually an isomorphism, since it has as inverse the map which associates to a function $y(\sigma_0, \dots, \sigma_n)$ in $X^n(G, A)$ the function $x(\sigma_0, \dots, \sigma_n)(\sigma) = y(\sigma\sigma_0, \dots, \sigma\sigma_n)$. This is readily checked. We thus have

$$H^n(G, \mathrm{Ind}_G^H(A)) \cong H^n(X^\bullet(G, A)^H).$$

The complex $0 \to A \to X^0(G,A) \to X^1(G,A) \to \cdots$ is an *acyclic* resolution of the H-module A, since, by the remark following (1.3.8), the $X^n(G,A)$ are induced G-modules. Therefore, by (1.3.9), we obtain a canonical isomorphism

$$H^n(X^\bullet(G,A)^H) \cong H^n(H,A). \qquad \square$$

Remark: If G is finite, the same argument yields isomorphisms

$$\hat{H}^n(G, \mathrm{Ind}_G^H(A)) \cong \hat{H}^n(H,A)$$

for all $n \in \mathbb{Z}$.

If A is a G-module, then we have an injective G-homomorphism

$$i : A \longrightarrow \mathrm{Ind}_G^H(A), \quad (ia)(\sigma) = \sigma a\,.$$

If, moreover, H is an open subgroup of G, then we have a G-homomorphism

$$\nu : \mathrm{Ind}_G^H(A) \longrightarrow A, \quad \nu(x) = \sum_{\sigma \in G/H} \sigma x(\sigma^{-1}),$$

where σ runs through a system of representatives of G/H. *) By this and by the lemma of Shapiro, we get the following interpretation of the restriction and the corestriction.

(1.6.5) Proposition. *We have commutative diagrams*

$$\begin{array}{ccc} H^n(G,\mathrm{Ind}_G^H(A)) & \xrightarrow{sh} & H^n(H,A) \\ \downarrow{\scriptstyle \nu_*} & & \downarrow{\scriptstyle cor} \\ H^n(G,A) & = & H^n(G,A), \end{array} \qquad \begin{array}{ccc} H^n(G,\mathrm{Ind}_G^H(A)) & \xrightarrow{sh} & H^n(H,A) \\ \uparrow{\scriptstyle i_*} & & \uparrow{\scriptstyle res} \\ H^n(G,A) & = & H^n(G,A). \end{array}$$

Proof: For the restriction this is obvious, so we prove it only for the corestriction. By dimension shifting, it suffices to consider the case $n = 1$. For each class $c \in H \setminus G$, let $\bar{c} \in c$ be a fixed representative. Let $x \in Z^1(G, \mathrm{Ind}_G^H(A))$.

*) These homomorphisms are given by the **Frobenius reciprocity**. On the one hand we have the isomorphism

$$\mathrm{Hom}_G(A, \mathrm{Ind}_G^H(B)) \cong \mathrm{Hom}_H(\mathrm{Res}_H^G(A), B)\,,$$

where H is a closed subgroup of G, A is a G-module and B is an H-module. Let $B = \mathrm{Res}_H^G(A)$; then $i : A \to \mathrm{Ind}_G^H \mathrm{Res}_H^G(A)$ is the unit of the adjunction Res $\dashv$ Ind. If, moreover, H is an open subgroup of G, then we have the isomorphism

$$\mathrm{Hom}_G(\mathrm{Ind}_G^H(B), A) \cong \mathrm{Hom}_H(B, \mathrm{Res}_H^G(A))$$

(where $\mathrm{Ind}_G^H(B)$ is identified with $\bigoplus_{\sigma \in G/H} \sigma B$, see the footnote on p.61). Thus we obtain $\nu : \mathrm{Ind}_G^H \mathrm{Res}_H^G(A) \to A$ as the counit of the adjunction Ind $\dashv$ Res.

By definition, sh maps the class of x to the class of the cocycle $y = sh(x) \in Z^1(H, A)$ given by

$$y(\sigma_0, \sigma_1) = x(\sigma_0, \sigma_1)(1).$$

Noting that the action of $\sigma \in G$ on $f \in \mathrm{Ind}_G^H(A)$ is given by $(\sigma f)(\tau) = f(\tau\sigma)$, we obtain

$$\begin{aligned} cor\, sh(x)(\sigma_0, \sigma_1) &= \sum_{\bar{c}} \bar{c}^{-1} x(\bar{c}\sigma_0\overline{c\sigma_0}^{-1}, \bar{c}\sigma_1\overline{c\sigma_1}^{-1})(1) \\ &= \sum_{\bar{c}} \bar{c}^{-1} x(\sigma_0\overline{c\sigma_0}^{-1}, \sigma_1\overline{c\sigma_1}^{-1})(\bar{c}). \end{aligned}$$

On the other hand

$$\nu_*(x)(\sigma_0, \sigma_1) = \sum_{\bar{c}} \bar{c}^{-1} x(\sigma_0, \sigma_1)(\bar{c}).$$

Since x is a cocycle, we have

$$\begin{aligned} x(\sigma_0\overline{c\sigma_0}^{-1}, \sigma_1\overline{c\sigma_1}^{-1}) &= x(\sigma_0\overline{c\sigma_0}^{-1}, \sigma_0) + x(\sigma_0, \sigma_1\overline{c\sigma_1}^{-1}) \\ &= x(\sigma_0\overline{c\sigma_0}^{-1}, \sigma_0) + x(\sigma_0, \sigma_1) + x(\sigma_1, \sigma_1\overline{c\sigma_1}^{-1}) \\ &= x(\sigma_0\overline{c\sigma_0}^{-1}, \sigma_0) - x(\sigma_1\overline{c\sigma_1}^{-1}, \sigma_1) + x(\sigma_0, \sigma_1). \end{aligned}$$

We therefore have to show that the function

$$f(\sigma_0, \sigma_1) = \sum_{\bar{c}} \bar{c}^{-1}[x(\overline{c\sigma_0}^{-1}, 1)(\bar{c}\sigma_0)] - \sum_{\bar{c}} \bar{c}^{-1}[x(\overline{c\sigma_1}^{-1}, 1)(\bar{c}\sigma_1)]$$

is a coboundary. Noting that $x(\overline{c\sigma_i}^{-1}, 1) \in \mathrm{Ind}_G^H(A)$ and $\bar{c}\sigma_i\overline{c\sigma_i}^{-1} \in H$, we have

$$\begin{aligned} &\sum_{\bar{c}} \bar{c}^{-1}\,[x(\overline{c\sigma_i}^{-1}, 1)(\bar{c}\sigma_i)] \\ &\quad = \sum_{\bar{c}} \bar{c}^{-1}\bar{c}\sigma_i\overline{c\sigma_i}^{-1}[x(\overline{c\sigma_i}^{-1}, 1)(\overline{c\sigma_i}\sigma_i^{-1}\bar{c}^{-1}\bar{c}\sigma_i)] \\ &\quad = \sum_{\bar{c}} \sigma_i\overline{c\sigma_i}^{-1}[x(\overline{c\sigma_i}^{-1}, 1)(\overline{c\sigma_i})] = \sigma_i \sum_{\bar{c}} \bar{c}^{-1} x(\bar{c}^{-1}, 1)(\bar{c}) \end{aligned}$$

since $c\sigma_i$ runs through $H\backslash G$ as c does. This shows that $f(\sigma_0, \sigma_1) = \sigma_1 a - \sigma_0 a$ where $a = -\sum \bar{c}^{-1} x(\bar{c}^{-1}, 1)(\bar{c})$, so is a coboundary. □

Proposition (1.6.5) (as well as (1.6.4)) will follow without computation from a general uniqueness theorem for δ-functors, which we will prove in II §6 (see (2.6.3)). Here, it says that ν_* and $cor \circ sh$ are morphisms between the δ-functors $H^n(G, \mathrm{Ind}_G^H(-))$ and $H^n(H, -)$ which coincide for $n = 0$ and hence for all n. In the same way one can show the following fact:

If H is a *normal* closed subgroup of G and A is a G-module, then we have a further action of G on $\mathrm{Ind}_G^H(A)$ given by $x \mapsto \sigma_* x$, $(\sigma_* x)(\rho) = \sigma x(\sigma^{-1}\rho)$. σ_* is even an automorphism of the G-module $\mathrm{Ind}_G^H(A)$ (not only of the abelian group), and is the identity for $\sigma \in H$. It therefore induces an automorphism on the cohomology group $H^n(G, \mathrm{Ind}_G^H(A))$, which in this way becomes a G/H-module. Considering the G/H-action on $H^n(H, A)$ given by conjugation (see §5), the map

$$sh : H^n(G, \mathrm{Ind}_G^H(A)) \longrightarrow H^n(H, A)$$

becomes an isomorphism of G/H-modules.

Besides the maps *inf, res, cor,* we sometimes have to consider still another, more subtle map, whose meaning is best understood in the framework of *spectral sequences* (see II §1), but which can explicitly be defined as follows.

(1.6.6) Proposition. *Let H be a normal subgroup of G and A a G-module. Then there is a canonical homomorphism*

$$tg : H^1(H, A)^{G/H} \longrightarrow H^2(G/H, A^H),$$

called **transgression**, *which is given as follows. If $x : H \to A$ is an inhomogeneous 1-cocycle in a class $[x] \in H^1(H, A)^{G/H}$, there exists a 1-cochain $y : G \to A$ such that $y|_H = x$ and that $(\partial y)(\sigma_1, \sigma_2)$ is contained in A^H and depends only on the cosets $\sigma_1 H, \sigma_2 H$, i.e. may be regarded as a cocycle of G/H. For each such cochain y,*

$$tg[x] = [\partial y].$$

Proof: We construct a cochain $y : G \to A$ with the following properties

(i) $y|_H = x$,

(ii) $y(\sigma\tau) = y(\sigma) + \sigma y(\tau)$ for $\sigma \in G$, $\tau \in H$,

(iii) $y(\tau\sigma) = y(\tau) + \tau y(\sigma)$ for $\sigma \in G$, $\tau \in H$.

Let $s : G/H \to G, \gamma \mapsto s\gamma$, be a continuous section of the projection $G \to G/H$ such that $s1 = 1$. Such a section exists by ex.4 of §1. Since $[x]$ is invariant under every $\gamma \in G/H$, we have

$$(1) \qquad s\gamma x((s\gamma)^{-1}\tau s\gamma) - x(\tau) = \tau y(s\gamma) - y(s\gamma)$$

for an element $y(s\gamma) \in A$. We may assume that $y(1) = 0$ and that $\gamma \mapsto y(s\gamma)$ is continuous. In fact, there exists an open normal subgroup U of G such that $x(\tau)$ depends only on the cosets $\tau(H \cap U)$ and is contained in A^U. Therefore the left side takes the same value for all elements $s\gamma$ in a coset mod U. So we may choose for $y(s\gamma)$ the same value within a coset of G mod U, and this means that $y(s\gamma)$ is continuous as a function of γ. For an arbitrary $\sigma = s\gamma\tau \in G$, we now set

$$y(\sigma) = y(s\gamma) + s\gamma x(\tau).$$

Then $y|_H = x$. Let $\tau, \tau' \in H$, $g = s\gamma$ and $\sigma = g\tau'$. Then

$$\begin{aligned} y(\sigma\tau) &= y(g\tau'\tau) = y(g) + gx(\tau'\tau) \\ &= y(g) + gx(\tau') + g\tau' x(\tau) = y(\sigma) + \sigma x(\tau). \end{aligned}$$

This proves (ii). Using (1), we obtain

$$\begin{aligned} y(\tau g) &= y(g\tau^g) = y(g) + gx(\tau^g) \\ &= y(g) + x(\tau) + \tau y(g) - y(g) = x(\tau) + \tau y(g), \end{aligned}$$

and for arbitrary $\sigma = g\tau'$,

$$\begin{aligned} y(\tau\sigma) &= y(\tau g \tau') = y(\tau g) + \tau g x(\tau') \\ &= x(\tau) + \tau y(g) + \tau g x(\tau') \\ &= y(\tau) + \tau y(\sigma). \end{aligned}$$

This proves (iii).

The function

$$\partial y(\sigma_1, \sigma_2) = \sigma_1 y(\sigma_2) - y(\sigma_1\sigma_2) + y(\sigma_1)$$

depends only on the cosets $\sigma_1 H, \sigma_2 H$, i.e. $\partial y(\sigma_1, \sigma_2\tau)$ and $\partial y(\sigma_1\tau, \sigma_2)$ are independent of $\tau \in H$. In fact,

$$\begin{aligned} \partial y(\sigma_1, \sigma_2\tau) &= \sigma_1 y(\sigma_2\tau) - y(\sigma_1\sigma_2\tau) + y(\sigma_1) \\ &= \sigma_1\sigma_2 y(\tau) + \sigma_1 y(\sigma_2) - y(\sigma_1\sigma_2\tau) + y(\sigma_1) \\ &= \sigma_1 y(\sigma_2) - y(\sigma_1\sigma_2) + y(\sigma_1) = \partial y(\sigma_1, \sigma_2), \\ \partial y(\sigma_1\tau, \sigma_2) &= \sigma_1\tau y(\sigma_2) - y(\sigma_1\tau\sigma_2) + y(\sigma_1\tau) \\ &= \sigma_1 y(\tau\sigma_2) - \sigma_1 y(\tau) - y(\sigma_1\tau\sigma_2) + y(\sigma_1\tau) \\ &= \sigma_1 y(\tau\sigma_2) - y(\sigma_1\tau\sigma_2) + y(\sigma_1) \\ &= \partial y(\sigma_1, \tau\sigma_2) = \partial y(\sigma_1, \sigma_2\tau^{\sigma_2}) = \partial y(\sigma_1, \sigma_2). \end{aligned}$$

From $\partial\partial y = 0$ and $\partial y(\tau, \sigma) = \partial y(1, \sigma) = 0$, we now obtain

$$\tau\partial y(\sigma_1, \sigma_2) = \partial y(\tau\sigma_1, \sigma_2) - \partial y(\tau, \sigma_1\sigma_2) + \partial y(\tau, \sigma_1) = \partial y(\sigma_1, \sigma_2),$$

i.e. $\partial y(\sigma_1, \sigma_2) \in A^H$.

Finally, let $y' : G \to A$ be any cochain such that $y'|_H = x$ and that $\partial y'(\sigma_1, \sigma_2)$ takes values in A^H and depends only on the cosets $\sigma_1 H, \sigma_2 H$. Then, for the function $z = y - y'$, we have $z(\tau) = 0$ for $\tau \in H$ and $\partial z(\tau, \sigma) = \partial z(1, \tau\sigma) = 0$, i.e. $\tau z(\sigma) - z(\sigma) = 0$, so that $z(\sigma) \in A^H$, and $\partial z(1, \tau\sigma) = \partial z(1, \sigma)$, so that $z(\tau\sigma) = z(\sigma)$. Therefore ∂y and $\partial y'$ may be viewed as cocycles in $\mathscr{Z}^2(G/H, A^H)$ and differ by the coboundary $\partial z \in \mathscr{B}^2(G/H, A^H)$, i.e. define the same cohomology class in $H^2(G/H, A^H)$. Thus defining $tg[x] = [\partial y]$ gives a well-defined homomorphism. □

(1.6.7) Proposition (Five Term Exact Sequence). *Let H be a closed normal subgroup of G and let A be a G-module. We then have an exact sequence*

$$0 \longrightarrow H^1(G/H, A^H) \xrightarrow{inf} H^1(G, A) \xrightarrow{res} H^1(H, A)^{G/H}$$
$$\xrightarrow{tg} H^2(G/H, A^H) \xrightarrow{inf} H^2(G, A).$$

Moreover, if $H^i(H, A) = 0$ for $i = 1, \dots, n-1$, we have an exact sequence

$$0 \longrightarrow H^n(G/H, A^H) \xrightarrow{inf} H^n(G, A) \xrightarrow{res} H^n(H, A)^{G/H}$$
$$\xrightarrow{tg} H^{n+1}(G/H, A^H) \xrightarrow{inf} H^{n+1}(G, A).$$

Proof: Consider the first sequence. The image of *res* is contained in $H^1(H, A)^{G/H}$, since the map $H^1(G, A) \xrightarrow{res} H^1(H, A)$ is a homomorphism of G-modules and $H^1(G, A)^G = H^1(G, A)$ by (1.6.3).

Exactness at $H^1(G/H, A^H)$. For the injectivity of the first map *inf*, let $x : G/H \to A^H$ be an inhomogeneous 1-cocycle such that the composite *inf* $x : G \to G/H \to A$ is a coboundary, $(\mathit{inf}\, x)(\sigma) = \sigma a - a$. For all $\tau \in H$, we have $\sigma a - a = \sigma\tau a - a$, hence $a \in A^H$. Therefore $x(\sigma H) = \sigma H a - a$ is a 1-coboundary.

Exactness at $H^1(G, A)$. Let $x : G/H \to A^H$ be an inhomogeneous 1-cocycle. Then for $\tau \in H$,

$$(\mathit{res} \circ \mathit{inf}\, x)(\tau) = (\mathit{inf}\, x)(\tau) = x(\tau H) = x(H) = x(1) = 0,$$

i.e. im(*inf*) $\subseteq$ ker(*res*). Conversely, let $x : G \to A$ be an inhomogeneous 1-cocycle such that *res* x is a coboundary, i.e. $x(\tau) = \tau a - a$ for $\tau \in H$. The 1-cocycle $x'(\sigma) = x(\sigma) - (\sigma a - a)$ of G defines the same cohomology class as x and satisfies $x'(\tau) = 0$ for $\tau \in H$. Hence

$$x'(\sigma\tau) = x'(\sigma) + \sigma x'(\tau) = x'(\sigma),$$

and also

$$x'(\tau\sigma) = x'(\tau) + \tau x'(\sigma) = \tau x'(\sigma).$$

We now define $y : G/H \to A$ by $y(\sigma H) = x'(\sigma)$. Then $y(\sigma H) \in A^H$, because $y(\sigma H) = y(\tau\sigma H) = \tau y(\sigma H)$ for all $\tau \in H$, and we obtain a 1-cocycle with *inf* $y = x'$. This shows ker(*res*) $\subseteq$ im(*inf*).

Exactness at $H^1(H, A)^{G/H}$. If $y \in \mathscr{Z}^1(G, A)$ and $x = \mathit{res}\, y$ represents a class $[x]$ in $H^1(H, A)^{G/H}$, then $tg[x] = [\partial y] = 0$, i.e. im(*res*) $\subseteq$ ker(*tg*). Conversely, let $x \in \mathscr{Z}^1(H, A)$ represent a class $[x] \in H^1(H, A)^{G/H}$ such that $tg[x] = 0$. Let $y \in \mathscr{C}^1(G, A)$ be a cochain as in (1.6.6). Viewing ∂y as a 2-cocycle of G/H, then $[\partial y] = tg[x] = 0$, hence $\partial y = \partial z$, where $z \in \mathscr{C}^1(G/H, A^H)$. Viewing y and z as functions on G, we have $y - z \in \mathscr{Z}^1(G, A)$. Since $\mathit{res}\,(y - z)$ and $\mathit{res}\, y = x$ are 1-cocycles of H, so is $\mathit{res}\, z$, and since z is constant on H, we have $res\, z = 0$. Thus $res\,(y - z) = res\, y = x$, i.e. $[x] = res\,[y - z]$. This proves $\ker(tg) \subseteq \operatorname{im}(res)$.

Exactness at $H^2(G/H, A^H)$. Let $x \in \mathscr{Z}^1(H, A)$ be an inhomogeneous cocycle, representing a class $[x] \in H^1(H, A)^{G/H}$. Then, by (1.6.6), there is a cocycle $z \in \mathscr{Z}^2(G/H, A^H)$ such that $inf\, z = \partial y$ and $tg[x] = [z]$. Thus $inf\, z \in \mathscr{B}^2(G, A)$, hence $inf\, tg[x] = [inf\, z] = 0$, showing $\operatorname{im}(tg) \subseteq \ker(inf)$. Conversely, let $z \in \mathscr{Z}^2(G/H, A^H)$ be a normalized cocycle, i.e. $z(1, \sigma) = z(\sigma, 1) = 0$ (see p.19), such that $inf\,[z] = [inf\, z] = 0$. Then $inf\, z = \partial y$ with $y \in \mathscr{C}^1(G, A)$. Setting $x = res\, y$ we have $\partial x = res\, \partial y = res\, inf\, z = 0$ and, regarding ∂y as the 2-cocycle z of G/H, $tg[x] = [\partial y] = [z]$. This proves $\ker(inf) \subseteq \operatorname{im}(tg)$.

The exact sequence

$$0 \longrightarrow H^n(G/H, A^H) \longrightarrow H^n(G, A) \longrightarrow H^n(H, A)^{G/H}$$
$$\xrightarrow{tg} H^{n+1}(G/H, A^H) \longrightarrow H^{n+1}(G, A)$$

for G-modules A such that $H^i(H, A) = 0$ for $i = 1, \dots, n-1$, is obtained by induction. We have it for $n = 1$, and we assume it for $n \geq 1$. Let A be a G-module, such that $H^i(H, A) = 0$ for $i = 1, \dots, n$. Consider the sequence

$$0 \longrightarrow A^H \longrightarrow \bar{A}^H \longrightarrow A_1^H \longrightarrow 0,$$

where $\bar{A} = \operatorname{Ind}_G(A)$, which is exact since $H^1(H, A) = 0$. We therefore have a commutative diagram

$$\begin{array}{ccccccc}
0 \longrightarrow & H^n(G/H, A_1^H) & \xrightarrow{inf} & H^n(G, A_1) & \xrightarrow{res} & H^n(H, A_1)^{G/H} \\
& \delta\big\downarrow & & \delta\big\downarrow & & \delta\big\downarrow \\
0 \longrightarrow & H^{n+1}(G/H, A^H) & \xrightarrow{inf} & H^{n+1}(G, A) & \xrightarrow{res} & H^{n+1}(H, A)^{G/H} \\
\xrightarrow{tg} & H^{n+1}(G/H, A_1^H) & \xrightarrow{inf} & H^{n+1}(G, A_1) & & \\
& \delta\big\downarrow & & \delta\big\downarrow & & \\
\xrightarrow{tg} & H^{n+2}(G/H, A^H) & \xrightarrow{inf} & H^{n+2}(G, A), & &
\end{array}$$

where the δ's are isomorphisms, since $\bar{A}$ is G-induced and H-induced, and $\bar{A}^H$ is G/H-induced. The lower map tg is defined by the upper one. Furthermore,

$$H^i(H, A_1) \cong H^{i+1}(H, A) = 0 \quad \text{for} \quad i = 1, \dots, n-1.$$

Therefore, by assumption, the upper sequence is exact, hence also the lower one. □

Let p be a prime number. A profinite group G is called a **pro-p-group** if its supernatural order is a p-power, i.e. for every open normal subgroup U of G, the finite group G/U is a p-group. Equivalently, a pro-p-group is the projective limit of finite p-groups.

(1.6.8) Definition. *A* ***p*-Sylow subgroup** *of a profinite group* G *is a closed subgroup* G_p *which is a pro-*p*-group such that the index* $(G : G_p)$ *is prime to* p.

The Sylow theorems for finite groups hold as well for profinite groups.

(1.6.9) Theorem. *Let* G *be a profinite group and* p *a prime number.*

(i) *There exists a* p*-Sylow subgroup* G_p.

(ii) *Every pro-*p*-subgroup is contained in a* p*-Sylow subgroup.*

(iii) *The* p*-Sylow subgroups of* G *are conjugate.*

Proof: Let U run through the open normal subgroups of G and let $\Sigma_p(U)$ denote the finite, nonempty set of all p-Sylow subgroups of G/U. If $V \subseteq U$ are two open normal subgroups, then the projection $G/V \to G/U$ maps p-Sylow subgroups onto p-Sylow subgroups and induces a surjection $\Sigma_p(V) \to \Sigma_p(U)$, so that the $\Sigma_p(U)$ form a projective system of nonempty finite sets. The projective limit $\varprojlim \Sigma_p(U)$ is a compact, nonempty topological space (see [160], chap.IV).

(i) Now let $(S_U) \in \varprojlim \Sigma_p(U)$. For $V \subseteq U$ we have the surjective projection $S_V \to S_U$, i.e. the system (S_U) is a projective system of finite p-groups. The projective limit $G_p = \varprojlim S_U$ is then a p-Sylow subgroup of G by definition.

(ii) Let H be a pro-p-subgroup of G and let H_U be its image under $G \to G/U$. Let $\Sigma_p'(U)$ be the set of p-Sylow subgroups of G/U, containing H_U. Again, $\varprojlim \Sigma_p'(U)$ is nonempty. Let $(S_U) \in \varprojlim \Sigma_p'(U)$. Then the inclusions $H_U \hookrightarrow S_U$ form a morphism of projective systems, and we obtain

$$H = \varprojlim H_U \subseteq \varprojlim S_U = G_p.$$

(iii) Let G_p and G_p' be two p-Sylow subgroups of G and let S_U and S_U' be their images in G/U. Let $C(U)$ be the set of elements $\sigma_U \in G/U$ such that $\sigma_U S_U \sigma_U^{-1} = S_U'$. The $C(U)$ form again a projective system of finite, nonempty sets. The projective limit $\varprojlim C(U)$ is nonempty. If $\sigma = (\sigma_U) \in \varprojlim C(U) \subseteq G$, then clearly $\sigma G_p \sigma^{-1} = G_p'$. □

For an abelian group A, the ***p*-primary part** $A(p)$ is the subgroup consisting of all elements of A of p-power order. If A is a torsion group, then $A = \bigoplus_p A(p)$. For the p-primary part of the torsion group $\hat{H}^n(G, A)$, we have the

(1.6.10) Proposition. *Let A be a discrete G-module and let G_p a p-Sylow subgroup of G. Assume that G is finite or that $n \geq 1$. Then the homomorphism*

$$res : \hat{H}^n(G, A)(p) \longrightarrow \hat{H}^n(G_p, A)$$

is injective, and if G_p is open in G, the homomorphism

$$cor : \hat{H}^n(G_p, A) \longrightarrow \hat{H}^n(G, A)(p)$$

is surjective.

Proof: The proposition holds for any closed subgroup H of G of index $(G : H)$ prime to p. Indeed, if H is open, then by (1.5.7), $cor \circ res = (G : H)$. Since $(G : H)$ is prime to p, the map

$$\hat{H}^n(G, A)(p) \xrightarrow{cor \circ res} \hat{H}^n(G, A)(p)$$

is an automorphism, so that *res* must be injective and *cor* surjective. If H is not open, the injectivity of *res* follows from (1.5.1). □

(1.6.11) Corollary. *Assume that G is finite or that $n \geq 1$. If $\hat{H}^n(G_p, A) = 0$ for all prime numbers p, then*

$$\hat{H}^n(G, A) = 0.$$

Because of the above proposition, we are often reduced to the cohomology of pro-p-groups. Its most frequently used property is the assertion (1.6.13) which we first prove for finite p-groups.

(1.6.12) Proposition. *Let G be a finite p-group and let A be a p-primary G-module. If $H^0(G, A) = 0$ or $H_0(G, A) = 0$, then $A = 0$.*

Proof: For the proof of $H^0(G, A) = 0 \Rightarrow A = 0$, we may assume that A is finite, since every element of the p-primary G-module A generates a finite G-module. $A \smallsetminus A^G$ is a disjoint union of G-orbits Ga not consisting of only one point, hence $\#Ga \equiv 0 \bmod p$ and $\#A \equiv \#A^G \bmod p$. If $A^G = 0$, then $\#A \equiv 1 \bmod p$, and hence $A = 0$. If $H_0(G, A) = A_G = 0$, then, setting $A^* = \mathrm{Hom}(A, \mathbb{Q}_p/\mathbb{Z}_p)$, we get $(A^*)^G \cong (A_G)^* = 0$, whence $A^* = 0$ and $A = 0$. □

Recall that a G-module $A \neq 0$ is said to be *simple* if it does not contain any submodule other than 0 and A itself. Since the set $\{ga \mid g \in G\} \subseteq A$ is finite for every $a \in A$, a simple module is finitely generated as an abelian group.

Therefore there exists a prime number p such that the p-multiplication $A \xrightarrow{p} A$ is not surjective, whence $pA = 0$. We conclude that a simple module is finite and there exists a unique prime number p such that $pA = 0$.

(1.6.13) Corollary. *Let G be a pro-p-group. Then every discrete simple p-primary G-module A is isomorphic to $\mathbb{Z}/p\mathbb{Z}$ (with trivial G-action). In particular, if A is a p-primary G-module, then*

$$A = 0 \quad \textit{if and only if} \quad A^G = 0\,.$$

Proof: Let A be a discrete simple p-primary G-module. Then $pA = 0$. Further, $A^U \neq 0$ for some open normal subgroup $U \subseteq G$, and by (1.6.12) we obtain $0 \neq H^0(G/U, A^U) = A^G \subseteq A$. Hence $A = A^G$ by simplicity. We conclude that A is an $\mathbb{F}_p$-vector space with trivial G-action and therefore of dimension 1. □

(1.6.14) Proposition.

(i) *The maximal closed subgroups of a pro-p-group G are normal of index p.*

(ii) *A homomorphism $G \to G'$ of pro-p-groups is surjective if and only if the induced homomorphism $H^1(G', \mathbb{Z}/p\mathbb{Z}) \to H^1(G, \mathbb{Z}/p\mathbb{Z})$ is injective. In particular,*

$$G = 1 \Leftrightarrow H^1(G, \mathbb{Z}/p\mathbb{Z}) = 0.$$

Proof: (i) Let H be a maximal closed subgroup of G. Then there exists an open normal subgroup U of G such that $HU/U \neq G/U$, since otherwise $H = G$. Clearly, HU/U is a maximal subgroup of the finite group G/U and is therefore normal of index p. This is a well-known result in group theory which follows from the first Sylow theorem for finite groups (see [73], chap.4, (4.2.2)). Since H is maximal, it is the pre-image of HU/U under $G \to G/U$ and is thus also normal of index p.

(ii) If $G \to G'$ is surjective, then obviously $H^1(G', \mathbb{Z}/p\mathbb{Z}) \longrightarrow H^1(G, \mathbb{Z}/p\mathbb{Z})$ is injective. Conversely, assume the latter. If $G \to G'$ were not surjective, the image would be contained in a maximal subgroup H of G' which is normal of index p. The composition $G' \to G'/H \cong \mathbb{Z}/p\mathbb{Z}$ would be an element $\chi \neq 0$ in $H^1(G', \mathbb{Z}/p\mathbb{Z})$ which becomes zero in $H^1(G, \mathbb{Z}/p\mathbb{Z})$, a contradiction. □

(1.6.15) Proposition. *Let*

$$\phi : G' \longrightarrow G$$

be a homomorphism of pro-p-groups. Let A be $\mathbb{Z}/p^r\mathbb{Z}$, $r \in \mathbb{N}$, or $\mathbb{Q}_p/\mathbb{Z}_p$ with trivial action. Then ϕ is an isomorphism if and only if the induced homomorphism

$$H^i(\phi, A) : H^i(G, A) \longrightarrow H^i(G', A)$$

is an isomorphism for $i = 1$ and injective for $i = 2$.

Proof: Let us show the nontrivial direction. $\mathbb{Z}/p\mathbb{Z}$ is a subgroup of A in a canonical way, and we obtain, for any group H, an inclusion

$$H^1(H, \mathbb{Z}/p\mathbb{Z}) \subset H^1(H, A).$$

In particular, $H^1(\phi, \mathbb{Z}/p\mathbb{Z})$ is injective if $H^1(\phi, A)$ is, and therefore (1.6.14) implies that ϕ is surjective. Put $K = \ker(\phi)$ and consider the 5-term exact sequence

$$0 \to H^1(G, A) \xrightarrow{H^1(\phi)} H^1(G', A) \to H^1(K, A)^G \to H^2(G, A) \xrightarrow{H^2(\phi)} H^2(G', A).$$

Our assumptions imply $H^1(K, A)^G = 0$, thus $H^1(K, A) = 0$ by (1.6.13). We conclude that $H^1(K, \mathbb{Z}/p\mathbb{Z}) = 0$, and $K = 0$ by (1.6.14)(ii). □

(1.6.16) Proposition. *Let G and G' be pro-p-groups and assume that G^{ab} and $(G')^{ab}$ are torsion groups. Let p^r be a common exponent of the factor commutator groups and assume that $H^2(G, \mathbb{Q}_p/\mathbb{Z}_p) = 0 = H^2(G', \mathbb{Q}_p/\mathbb{Z}_p)$. Let $\phi : G' \to G$ be a homomorphism. Then the following assertions are equivalent.*

(i) *ϕ is an isomorphism.*

(ii) *$H^1(\phi) : H^1(G, \mathbb{Z}/p^r\mathbb{Z}) \to H^1(G', \mathbb{Z}/p^r\mathbb{Z})$ is an isomorphism.*

(iii) *$H^2(\phi) : H^2(G, \mathbb{Z}/p^r\mathbb{Z}) \to H^2(G', \mathbb{Z}/p^r\mathbb{Z})$ is an isomorphism.*

Proof: Clearly, (i) implies (ii) and (iii) and, by (1.6.15), (ii) and (iii) together imply (i). It remains to show that (ii) and (iii) are equivalent. Since $H^2(G, \mathbb{Q}_p/\mathbb{Z}_p) = 0$, the exact sequence

$$0 \longrightarrow \mathbb{Z}/p^r\mathbb{Z} \longrightarrow \mathbb{Q}_p/\mathbb{Z}_p \xrightarrow{p^r} \mathbb{Q}_p/\mathbb{Z}_p \to 0$$

induces the four term exact sequence

$$0 \to H^1(G, \mathbb{Z}/p^r\mathbb{Z}) \xrightarrow{\alpha} H^1(G, \mathbb{Q}_p/\mathbb{Z}_p) \xrightarrow{\beta}$$

$$H^1(G, \mathbb{Q}_p/\mathbb{Z}_p) \xrightarrow{\gamma} H^2(G, \mathbb{Z}/p^r\mathbb{Z}) \to 0.$$

By the assumption on G^{ab}, the map α is an isomorphism. Hence β is zero and γ is an isomorphism. The same argument also applies to G' and therefore (ii) and (iii) are both equivalent to

(iv) $H^1(\phi, \mathbb{Q}_p/\mathbb{Z}_p) : H^1(G, \mathbb{Q}_p/\mathbb{Z}_p) \to H^1(G', \mathbb{Q}_p/\mathbb{Z}_p)$ is an isomorphism.

This concludes the proof. □

Exercise 1. Let G be a profinite group, H a closed normal subgroup, A a discrete G-module, and let $s : G/H \to G$ be a continuous section of the projection $G \to G/H$. Let $x : H \to A$ be a cocycle representing a class $[x] \in H^1(H, A)^{G/H}$, and set $y(\sigma) = s\gamma x(\tau)$ for $\sigma = s\gamma\tau$, $\gamma \in G/H, \tau \in H$.

Show that if H acts trivially on A, then $y : G \to A$ is a cochain as in (1.6.6), i.e. $y|_H = x$ and $\partial y(\sigma_1, \sigma_2)$ depends only on the cosets $\sigma_1 H, \sigma_2 H$, so that $tg[x] = [\partial y]$.

Exercise 2. Let G be finite, H a subgroup, A a G-module, and let H' be the commutator subgroup of H. The group extension

$$1 \longrightarrow H^{ab} \longrightarrow G/H' \longrightarrow G/H \longrightarrow 1$$

defines a class $u \in H^2(G/H, H^{ab})$.

Assume that H acts trivially on A. Then $H^1(H, A)^{G/H} = H^0(G/H, \mathrm{Hom}\,(H^{ab}, A))$, and the cup-product

$$H^2(G/H, H^{ab}) \times H^0(G/H, \mathrm{Hom}\,(H^{ab}, A)) \xrightarrow{\cup} H^2(G/H, A)$$

yields a homomorphism

$$u\cup : H^1(H, A)^{G/H} \longrightarrow H^2(G/H, A).$$

Show that this homomorphism coincides with $-tg$. Generalize this result to profinite groups.

Hint: Let x be a cocycle representing a class $[x] \in H^1(H, A)^{G/H}$. Then $x^\sigma = x$ for all $\sigma \in G$. Let y be the function $y(\sigma) = s\gamma x(\tau)$ $(\sigma = s\gamma\tau)$ as in ex.1. The class $u \in H^2(G/H, H^{ab})$ is represented by the cocycle $\tau(\gamma_1, \gamma_2) = \bar{s}\gamma_1 \bar{s}\gamma_2 \bar{s}(\gamma_1\gamma_2)^{-1}$, where $\bar{s}$ is the composite of $G/H \xrightarrow{s} G \to G/H'$. Show that

$$\partial y(\gamma_1, \gamma_2) = -x(\tau(\gamma_1, \gamma_2)).$$

Exercise 3. Let G be a finite group, H a subgroup, and let A be a G-module. Consider the exact sequence

$$0 \longrightarrow A \longrightarrow \bar{A} \longrightarrow A_1 \longrightarrow 0$$

with $\bar{A} = \mathrm{Ind}_G(A)$, the induced G-module. The associated long exact cohomology sequence yields the exact sequence

$$0 \longrightarrow A^H \longrightarrow \bar{A}^H \longrightarrow A_1^H \xrightarrow{\delta} H^1(H, A) \longrightarrow 0.$$

We split this up into the two exact sequences

(1) $$0 \longrightarrow A^H \longrightarrow \bar{A}^H \longrightarrow B \longrightarrow 0,$$

(2) $$0 \longrightarrow B \longrightarrow A_1^H \longrightarrow H^1(H, A) \longrightarrow 0,$$

where B denotes the image of $\bar{A}^H$ in A_1^H. Since $\bar{A}^H$ is an induced G/H-module, we get from (1)

$$H^1(G/H, B) \xrightarrow[\sim]{\delta_2} H^2(G/H, A^H).$$

Show that the composite of

$$H^1(H,A)^{G/H} \xrightarrow{\delta_1} H^1(G/H,B) \xrightarrow[\sim]{\delta_2} H^2(G/H,A^H),$$

where δ_1 is obtained from (2), is the *transgression*.

Exercise 4. Define the *cohomology groups of a pair* $H \subseteq G$ and a G-module A by

$$H^0(G,H,A) = 0 \quad\text{and}\quad H^n(G,H,A) = H^{n-1}(G,\Gamma(A)) \quad\text{for } n \geq 1,$$

where $\Gamma(A)$ is defined by the exact sequence

$$0 \longrightarrow A \longrightarrow \mathrm{Ind}_G^H(A) \longrightarrow \Gamma(A) \longrightarrow 0.$$

We then have a "relative exact cohomology sequence" (see [178])

$$\ldots \to H^n(G,A) \to H^n(H,A) \xrightarrow{\delta} H^{n+1}(G,H,A) \to H^{n+1}(G,A) \to H^{n+1}(H,A) \to \ldots.$$

Exercise 5. For two pairs $H \subseteq G$ and $L \subseteq K$ and a homomorphism $\varphi : K \to G$ such that $\varphi L \subseteq H$, we have functorial maps

$$\varphi^n : H^n(G,H,A) \longrightarrow H^n(K,L,A).$$

§7. Cohomology of Cyclic Groups

In this section we will compute the cohomology of cyclic groups and we will introduce the concept of the Herbrand quotient. Let G be a finite cyclic group. Recall that for every G-module A we have

$$\hat{H}^0(G,A) = A^G/N_GA \qquad \text{and} \qquad \hat{H}^{-1}(G,A) = {}_{N_G}A/I_GA.$$

If σ is a generator of the cyclic group G, then $I_GA = (\sigma - 1)A$, since for all $i \geq 1$ we have the equality

$$\sigma^i - 1 = (\sigma - 1)(\sigma^{i-1} + \cdots + \sigma + 1).$$

(1.7.1) Proposition. *Let G be a finite cyclic group. Then the group $\hat{H}^2(G,\mathbb{Z})$ is cyclic of the same order as G. Let $\chi \in \hat{H}^2(G,\mathbb{Z})$ be any generator. Then the cup-product induces isomorphisms*

$$\chi\cup : \hat{H}^n(G,A) \xrightarrow{\sim} \hat{H}^{n+2}(G,A)$$

for all $n \in \mathbb{Z}$ and every G-module A. In particular, we have isomorphisms

$$\hat{H}^{2n}(G,A) \cong A^G/N_GA, \quad \hat{H}^{2n-1}(G,A) \cong {}_{N_G}A/I_GA$$

for all G-modules A and all $n \in \mathbb{Z}$.

Proof: Let $\sigma \in G$ be a generator and let $N = \#G$. Consider the four-term exact sequence

$$(*) \qquad 0 \longrightarrow \mathbb{Z} \xrightarrow{\mu} \mathbb{Z}[G] \xrightarrow{\sigma-1} \mathbb{Z}[G] \xrightarrow{\varepsilon} \mathbb{Z} \longrightarrow 0$$

where ε is the augmentation map $\sum a_i\sigma^i \mapsto \sum a_i$ and $\mu(a) = a(1 + \sigma + \cdots + \sigma^{N-1})$. It induces an isomorphism (see the remark following (1.3.8))

$$\delta^2 : \hat{H}^0(G, \mathbb{Z}) \xrightarrow{\sim} \hat{H}^2(G, \mathbb{Z}).$$

Since

$$\hat{H}^0(G, \mathbb{Z}) = \mathbb{Z}/(1 + \sigma + \cdots + \sigma^{N-1})\mathbb{Z} = \mathbb{Z}/N\mathbb{Z},$$

we see that $\hat{H}^2(G, \mathbb{Z})$ is cyclic of order N and every generator is of the form $\chi = \delta^2(m)$, $m \in (\mathbb{Z}/N\mathbb{Z})^\times$.

Now let A be a G-module. Since all objects in $(*)$ are $\mathbb{Z}$-free, it remains exact when tensored by A. Hence for every $n \in \mathbb{Z}$, we obtain an isomorphism $\delta^2 : \hat{H}^n(G, A) \xrightarrow{\sim} \hat{H}^{n+2}(G, A)$, which fits into the commutative diagram

$$\begin{array}{ccc} \hat{H}^n(G, A) & = & \hat{H}^n(G, A) \\ {\scriptstyle m}\downarrow & & \downarrow{\scriptstyle \chi\cup} \\ \hat{H}^n(G, A) & \xrightarrow{\delta^2} & \hat{H}^{n+2}(G, A). \end{array}$$

It remains to show that multiplication by m induces an automorphism on $\hat{H}^n(G, A)$. But this is clear, because by (1.6.1), $\hat{H}^n(G, A)$ is an abelian group which is annihilated by $N = \#G$, hence a $\mathbb{Z}/N\mathbb{Z}$-module. □

(1.7.2) Proposition. *Let G be a finite cyclic group. If $0 \to A \to B \to C \to 0$ is an exact sequence of G-modules, then we have an exact hexagon*

$$\begin{array}{ccccc} & \hat{H}^0(G, A) & \longrightarrow & \hat{H}^0(G, B) & \\ \nearrow & & & & \searrow \\ \hat{H}^{-1}(G, C) & & & & \hat{H}^0(G, C) \\ \nwarrow & & & & \swarrow \\ & \hat{H}^{-1}(G, B) & \longleftarrow & \hat{H}^{-1}(G, A). & \end{array}$$

Proof: All maps in the hexagon above are the canonical ones, except the homomorphism $\hat{H}^0(G, C) \to \hat{H}^{-1}(G, A)$ which is the composite of the connecting homomorphism $\hat{H}^0(G, C) \to \hat{H}^1(G, A)$ and the inverse of the isomorphism $\hat{H}^{-1}(G, A) \xrightarrow{\sim} \hat{H}^1(G, A)$ obtained in (1.7.1). In order to prove exactness, we only have to show that the diagram

$$\begin{array}{ccc} \hat{H}^{-1}(G,A) & \longrightarrow & \hat{H}^{-1}(G,B) \\ \downarrow \wr & & \downarrow \wr \\ \hat{H}^{1}(G,A) & \longrightarrow & \hat{H}^{1}(G,B) \end{array}$$

commutes, where the vertical maps are the isomorphisms obtained in (1.7.1). But this is obvious. □

A very useful concept for calculating indices and orders of abelian groups is the **Herbrand quotient**, given by

(1.7.3) Definition. *Let A be an abelian group and let f, g be endomorphisms of A such that $fg = gf = 0$, i.e. $\operatorname{im} g \subseteq \ker f$ and $\operatorname{im} f \subseteq \ker g$. Then the Herbrand quotient with respect to f and g is defined to be*

$$q_{f,g}(A) = \frac{(\ker f : \operatorname{im} g)}{(\ker g : \operatorname{im} f)}$$

provided that both indices are finite.

The following special case is of great importance for the cohomology of cyclic groups.

(1.7.4) Definition. *If G is a finite cyclic group and A a G-module, then the* **Herbrand quotient** *of A is defined to be*

$$h(G,A) = \frac{\#\hat{H}^0(G,A)}{\#\hat{H}^{-1}(G,A)},$$

provided that both orders are finite.

If σ is a generator of the cyclic group G of order n, then the endomorphisms

$$D = \sigma - 1 \quad \text{and} \quad N_G = 1 + \sigma + \ldots + \sigma^{n-1}$$

have the property that $D \cdot N_G = N_G \cdot D = 0$. As $\hat{H}^0(G,A) = A^G/N_G A$ and $\hat{H}^{-1}(G,A) = {}_{N_G}A/I_G A$, we see that $h(G,A) = q_{D,N}(A)$.

The salient property of the Herbrand quotient is its multiplicativity.

(1.7.5) Proposition. *Let G be a finite cyclic group. If $0 \to A \to B \to C \to 0$ is an exact sequence of G-modules, then*

$$h(G, B) = h(G, A) \cdot h(G, C)$$

in the sense that, whenever two of these quotients are defined, so is the third and the identity holds.

Proof: Let $f\colon \hat{H}^0(G, A) \to \hat{H}^0(G, B)$. Using (1.7.2), we obtain an exact sequence

$$0 \to \ker f \to \hat{H}^0(G, A) \to \hat{H}^0(G, B) \to \ldots \to \hat{H}^{-1}(G, C) \to \ker f \to 0,$$

and therefore

$$\#\hat{H}^0(G, A) \cdot \#\hat{H}^0(G, C) \cdot \#\hat{H}^{-1}(G, B) = \#\hat{H}^0(G, B) \cdot \#\hat{H}^{-1}(G, A) \cdot \#\hat{H}^{-1}(G, C).$$

At the same time, we see that if any two of the quotients are well-defined, then so is the third. From the last equation we obtain the desired equality. □

Another special case for the Herbrand quotient is the following: let $f = 0$ and let g be the multiplication by a natural number n. Then

$$q_{0,n}(A) = \frac{(A : nA)}{\#_n A}.$$

Considering A as a trivial G-module, where G is a cyclic group of order n, we have $q_{0,n}(A) = h(G, A)$. In particular, for an exact sequence $0 \to A \to B \to C \to 0$ of abelian groups, we obtain

$$q_{0,n}(B) = q_{0,n}(A) \cdot q_{0,n}(C),$$

again in the sense that whenever two of these quotients are defined, so is the third and the identity holds.

(1.7.6) Proposition. *Let A be a finite abelian group and let f, g be endomorphisms of A such that $fg = gf = 0$. Then*

$$q_{f,g}(A) = 1.$$

In particular, if G is a finite cyclic group and A is a finite G-module, then $h(G, A) = 1$.

Proof: Since $\#A = \#\ker f \cdot \#\mathrm{im}\, f = \#\ker g \cdot \#\mathrm{im}\, g$, the result follows. □

From the last result and from (1.7.5) we obtain $h(G, A) = h(G, B)$ for any G-submodule B of A of finite index.

A **procyclic group** is a profinite group G which is topologically generated by a single element σ, i.e. G is the closure of the subgroup $(\sigma) = \{\sigma^n \mid n \in \mathbb{Z}\}$. For example,

$$\mathbb{Z}_p = \varprojlim_{n\in\mathbb{N}} \mathbb{Z}/p^n\mathbb{Z} \qquad \text{and} \qquad \hat{\mathbb{Z}} = \varprojlim_{n\in\mathbb{N}} \mathbb{Z}/n\mathbb{Z}$$

are (additive) procyclic groups, and every procyclic group G is a quotient of $\hat{\mathbb{Z}}$ (see [160], chap.IV, § 2, example 7). As

$$\hat{\mathbb{Z}} = \prod_p \mathbb{Z}_p\,,$$

where p runs through the prime numbers, every procyclic group G is of the form

$$G \cong \prod_{p\in T} \mathbb{Z}/p^{e_p}\mathbb{Z} \times \prod_{p\in S} \mathbb{Z}_p\,,$$

where T and S are disjoint sets of prime numbers and e_p, $p \in T$, are natural numbers.

We assume now that G is torsion-free, i.e. $G = \prod_{p\in S}\mathbb{Z}_p$. Let $\mathbb{N}(S)$ be the set of natural numbers not divisible by prime numbers $p \notin S$. Then the group G^n of n-th powers, $n \in \mathbb{N}(S)$, are the open subgroups of G and thus

$$G = \varprojlim_{n\in\mathbb{N}(S)} G/G^n \cong \varprojlim_{n\in\mathbb{N}(S)} \mathbb{Z}/n\mathbb{Z}.$$

We say that an abelian group X is ***S*-divisible** if $X = nX$ for all $n \in \mathbb{N}(S)$, and is ***S*-torsion** if $X = \bigcup_{n\in\mathbb{N}(S)} {}_nX$, where ${}_nX = \{x \in X \mid nx = 0\}$.

(1.7.7) Proposition. *Let $G = \prod_{p\in S} \mathbb{Z}_p$ be a torsion-free procyclic group and let A be a discrete G-module.*

(i) *If A is S-torsion, then $H^1(G,A) \cong A_G$.*

(ii) *If A is torsion or S-divisible, then $H^n(G,A) = 0$ for $n \geq 2$.*

Proof: (i) Let σ be a topological generator of G. Then $A^G = \ker(A \xrightarrow{\sigma-1} A)$ and $A_G = \operatorname{coker}(A \xrightarrow{\sigma-1} A)$. Let $N_n = 1 + \sigma + \cdots + \sigma^{n-1}$. Then

$$\begin{aligned} H^1(G,A) &= \varinjlim_{n\in\mathbb{N}(S)} H^1(G/G^n, A^{G^n}) \\ &\cong \varinjlim_{n\in\mathbb{N}(S)} {}_{N_n}A^{G^n}/(\sigma-1)A^{G^n} = A'/(\sigma-1)A, \end{aligned}$$

where $A' = \{a \in A \mid N_n a = 0 \text{ for some } n \in \mathbb{N}(S)\}$. The isomorphism

$$H^1(G,A) \cong A'/(\sigma-1)A$$

is given by associating to a 1-cocycle $x : G \to A$ the value $x(\sigma)$. If A is S-torsion, then $A' = A$. In fact, for $a \in A$ there exist $n, m \in \mathbb{N}(S)$, such that $na = 0$ and $\sigma^m a = a$. From this, it follows that

$$(1 + \sigma + \cdots + \sigma^{mn-1})a = n(1 + \sigma + \cdots + \sigma^{m-1})a = 0,$$

i.e. $a \in A'$. This proves (i).

(ii) By (1.7.1), we have

$$H^2(G/G^n, A^{G^n}) \cong A^G/N_n A^{G^n}.$$

A careful analysis of the definition of this isomorphism shows that the inflation map $H^2(G/G^n, A^{G^n}) \to H^2(G/G^{nm}, A^{G^{nm}})$ corresponds to the homomorphism

$$A^G/N_n A^{G^n} \longrightarrow A^G/N_{nm} A^{G^{nm}},$$

given as multiplication by m. If A is finite and m is a multiple of the order of A, then this homomorphism is zero, hence

$$H^2(G, A) = \varinjlim H^2(G/G^n, A^{G^n}) = 0.$$

If A is torsion, then $A = \varinjlim A_\alpha$, where A_α runs through the finite G-submodules of A, hence $H^2(G, A) = \varinjlim H^2(G, A_\alpha) = 0$.

Assume now inductively that $H^n(G, A) = 0$, $n \geq 2$, for all torsion modules A. For any torsion module A, consider the exact sequence

$$0 \longrightarrow A \longrightarrow \mathrm{Ind}_G(A) \longrightarrow A_1 \longrightarrow 0,$$

where the left arrow associates to $a \in A$ the constant function $x(\tau) = a$. Clearly, as A is torsion, so are $\mathrm{Ind}_G(A)$ and A_1 (note that every continuous map $x : G \to A$ has finite image). By (1.3.7), $\mathrm{Ind}_G(A)$ is cohomologically trivial, so that

$$H^{n+1}(G, A) \cong H^n(G, A_1) = 0.$$

Now let A be S-divisible and let $m \in \mathbb{N}(S)$. From the exact sequence

$$0 \longrightarrow {}_mA \longrightarrow A \xrightarrow{m} A \longrightarrow 0,$$

we get the exact cohomology sequence

$$H^n(G, {}_mA) \longrightarrow H^n(G, A) \xrightarrow{m} H^n(G, A).$$

Since ${}_mA$ is torsion, $H^n(G, {}_mA) = 0$ for $n \geq 2$, i.e. multiplication by m is injective on $H^n(G, A)$ for $n \geq 2$. But these groups are S-torsion by (1.6.1), which finishes the proof. □

Exercise 1. Let f and g be two commuting endomorphisms of an abelian group A. Show that

$$q_{0,gf}(A) = q_{0,g}(A) \cdot q_{0,f}(A),$$

provided all quotients are defined.

Exercise 2. Let G be a cyclic group of prime order p, and let A be a G-module such that $q_{0,p}(A)$ is defined. Show that

$$h(G,A)^{p-1} = q_{0,p}(A^G)^p / q_{0,p}(A).$$

Hint: Let σ be a generator of G. Show that in the ring $\mathbb{Z}[G]/\mathbb{Z}N_G \cong \mathbb{Z}[\zeta]$, ζ a primitive p-th root of unity, one has $p = (\sigma - 1)^{p-1}\varepsilon$, where ε is a unit in $\mathbb{Z}[G]/\mathbb{Z}N_G$. Then use the exact sequence $0 \to A^G \to A \to A^{\sigma-1} \to 0$.

§8. Cohomological Triviality

Let G be a profinite group. For every prime number p, let G_p be a p-Sylow subgroup of G. We have called a discrete G-module A *cohomologically trivial* if $H^n(H,A) = 0$ for all $n > 0$ and all closed subgroups H of G. We have seen (cf. (1.3.7)) that induced G-modules are cohomologically trivial. We will give now further criteria for cohomological triviality.

(1.8.1) Proposition. *A discrete G-module A is cohomologically trivial if and only if for every prime number p it is a cohomologically trivial G_p-module.*

Proof: This follows easily from (1.6.11). □

(1.8.2) Proposition. *A discrete G-module A is cohomologically trivial if and only if for every open normal subgroup U of G, the G/U-module A^U is cohomologically trivial.*

Proof: If the A^U are cohomologically trivial G/U-modules, then by (1.5.1),

$$H^n(H,A) = \varinjlim_{U} H^n(HU/U, A^U) = 0$$

for $n > 0$ and every closed subgroup H, i.e. A is a cohomologically trivial G-module. For a closed subgroup H/U of G/U, the sequence

$$0 \longrightarrow H^n(H/U, A^U) \longrightarrow H^n(H,A) \longrightarrow H^n(U,A)$$

is exact if $H^i(U,A) = 0$ for $i = 1, \dots, n-1$ (see (1.6.7)). If A is cohomologically trivial, this is true for all $n > 0$ and we get $H^n(H/U, A^U) = 0$ for $n > 0$, showing that A^U is a cohomologically trivial G/U-module. □

(1.8.3) Proposition. *Let G be a finite p-group and let A be a p-primary G-module. If $pA = 0$ and if $\hat{H}^q(G, A) = 0$ for one q, then A is an induced G-module and, in particular, cohomologically trivial.*

Proof: Let $\Lambda = \mathbb{F}_p[G]$, I a basis of the $\mathbb{F}_p$-vector space A^G and $V = \bigoplus_I \Lambda$. In the exact sequence of G-modules

$$0 \longrightarrow \mathrm{Hom}\,(A/A^G, V) \longrightarrow \mathrm{Hom}\,(A, V) \longrightarrow \mathrm{Hom}(A^G, V) \longrightarrow 0\,,$$

$B = \mathrm{Hom}(A/A^G, V)$ is an induced G-module by (1.3.6)(iii), so that we have $H^1(G, B) = 0$, and the homomorphism

$$\mathrm{Hom}_G(A, V) \longrightarrow \mathrm{Hom}_G(A^G, V) = \mathrm{Hom}(A^G, V^G)$$

is surjective. We have canonically $\Lambda^G \cong \mathbb{F}_p$, hence an isomorphism $A^G \cong V^G$, which by the above argument extends to a G-homomorphism

$$j : A \longrightarrow V.$$

By (1.6.12), the equality $\ker(j|_{A^G}) = \ker(j)^G = 0$ implies that j is injective . If C is the cokernel of j, then we have an exact sequence

$$0 \longrightarrow A^G \xrightarrow{\sim} V^G \longrightarrow C^G \longrightarrow H^1(G, A) \longrightarrow 0.$$

Hence if $H^1(G, A) = 0$, then $C^G = 0$ and thus $C = 0$, i.e. $A \cong V$ is an induced G-module. If $\hat{H}^q(G, A) = 0$ for some q, then $H^1(G, A_{q-1}) \cong \hat{H}^q(G, A) = 0$, i.e., by the argument as above, A_{q-1} is an induced G-module. But this implies $H^1(G, A) = \hat{H}^{2-q}(G, A_{q-1}) = 0$, and hence A is induced. □

(1.8.4) Proposition. *Let G be a finite group and let A be a G-module such that for every prime number p there exists a dimension $n_p \in \mathbb{Z}$ with*

$$\hat{H}^{n_p}(G_p, A) = \hat{H}^{n_p+1}(G_p, A) = 0.$$

Then A is a cohomologically trivial G-module. If A is $\mathbb{Z}$-free, it is a direct summand of a free $\mathbb{Z}[G]$-module. Conversely, for every cohomologically trivial G-module A, we have

$$\hat{H}^n(H, A) = 0$$

for all $n \in \mathbb{Z}$ and all subgroups H of G.

Proof: Let

(1) $$0 \longrightarrow R \longrightarrow F \longrightarrow A \longrightarrow 0$$

be an exact sequence with a free $\mathbb{Z}[G]$-module F. We claim that, for every prime number p, R/pR is an induced G_p-module. In fact, since F is an induced G_p-module and $\hat{H}^{n_p}(G_p, A) = \hat{H}^{n_p+1}(G_p, A) = 0$, we obtain

(2) $$\hat{H}^{n_p+1}(G_p, R) = \hat{H}^{n_p+2}(G_p, R) = 0,$$

and the exact sequence

$$(3) \qquad 0 \longrightarrow R \xrightarrow{p} R \longrightarrow R/pR \longrightarrow 0$$

yields $\hat{H}^{n_p+1}(G_p, R/pR) = 0$, giving the claim by (1.8.3).

Now we assume that A is $\mathbb{Z}$-free. Then from (1) we obtain an exact sequence

$$0 \longrightarrow \mathrm{Hom}(A, R) \longrightarrow \mathrm{Hom}(A, F) \longrightarrow \mathrm{Hom}(A, A) \longrightarrow 0.$$

If we knew that $H^1(G, \mathrm{Hom}(A, R)) = 0$, then the homomorphism

$$\mathrm{Hom}_G(A, F) \longrightarrow \mathrm{Hom}_G(A, A)$$

would be surjective, so that the identity map of A could be extended to a G-embedding $A \hookrightarrow F$, making A a direct summand of F. We thus have to show $H^1(G, M) = 0$ for $M = \mathrm{Hom}(A, R)$. Since A is $\mathbb{Z}$-free, we get from (3) an exact sequence

$$0 \longrightarrow M \xrightarrow{p} M \longrightarrow \mathrm{Hom}(A, R/pR) \longrightarrow 0,$$

i.e. $M/pM \cong \mathrm{Hom}(A, R/pR)$ is an induced G_p-module, since R/pR is an induced G_p-module. From this it follows that

$$H^1(G_p, M) \xrightarrow{p} H^1(G_p, M)$$

is an isomorphism, whence $H^1(G_p, M) = 0$ for all p. By (1.6.11), we get $H^1(G, M) = 0$, as required.

Now let A be an arbitrary G-module. Since R is $\mathbb{Z}$-free, it follows from (2) and from what we have just seen, that R is a direct summand of a free $\mathbb{Z}[G]$-module, i.e. of an induced G-module. Therefore R is cohomologically trivial. As F is cohomologically trivial, so is A.

Conversely, we have seen that for every cohomologically trivial G-module A and every surjection $\varepsilon : F \twoheadrightarrow A$ with F free, the module $R := \ker(\varepsilon)$ is a direct summand in a free module. Hence

$$\hat{H}^n(G, A) \cong \hat{H}^{n+1}(G, R) = 0$$

for all $n \in \mathbb{Z}$. □

(1.8.5) Corollary. *Let G be a finite group and let A, B be G-modules. If either A is cohomologically trivial and B is divisible, or if B is cohomologically trivial and A is $\mathbb{Z}$-free, then* $\mathrm{Hom}(A, B)$ *is cohomologically trivial.*

Proof: Let A be cohomologically trivial and B divisible. We consider an exact sequence

$$0 \longrightarrow R \longrightarrow F \longrightarrow A \longrightarrow 0,$$

where F is a free $\mathbb{Z}[G]$-module. Since F and A are cohomologically trivial, so is R, and since R is $\mathbb{Z}$-free, it is moreover a direct summand of a free $\mathbb{Z}[G]$-module, say F'. Therefore $\mathrm{Hom}(R, B)$ is a direct summand of the induced G-module $\mathrm{Hom}(F', B)$ and is thus cohomologically trivial. Since B is divisible, i.e. an injective $\mathbb{Z}$-module, the sequence

$$0 \longrightarrow \mathrm{Hom}(A, B) \longrightarrow \mathrm{Hom}(F, B) \longrightarrow \mathrm{Hom}(R, B) \longrightarrow 0$$

is exact. The last two modules are cohomologically trivial, hence also the module $\mathrm{Hom}(A, B)$.

If B is cohomologically trivial and A is $\mathbb{Z}$-free, the same argument applies if we exchange the roles of A and B. □

Exercise 1. Let G be a finite group and let A, B be G-modules. Show that $A \otimes B$ is cohomologically trivial if A is cohomologically trivial and either A or B is without p-torsion for all primes $p \mid \#G$.

Exercise 2. Let $f : A \to B$ be a homomorphism of G-modules and let $f_n : H^n(H, A) \to H^n(H, B)$ be the induced homomorphisms on the cohomology ($H \subseteq G$). Assume, for some $q \in \mathbb{Z}$, that f_{q-1} is injective, f_q is bijective, f_{q+1} is surjective, for all subgroups H of G. Then f_n is an isomorphism for all $n \in \mathbb{Z}$ and all subgroups H of G (see [124]).

§9. Tate Cohomology of Profinite Groups

In this section we extend the definition of the modified cohomology groups, defined in §2 for finite groups, to profinite groups. We start with the definition of homology of finite groups.

Let G be a finite group and let A be a G-module. Recall from §2 the cofixed module $A_G = A/I_G A$, which is the largest quotient of A on which G acts trivially. We consider the homological complex $P_\bullet$ given by

$$P_n = \mathbb{Z}[G^{n+1}],$$

and the differential $\partial_n : P_n \to P_{n-1}$ given by

$$\partial_n(\sigma_0, \ldots, \sigma_n) = \sum_{i=0}^{n} (-1)^i (\sigma_0, \ldots, \sigma_{i-1}, \sigma_{i+1}, \ldots, \sigma_n).$$

Moreover, we have the homomorphism $\partial_0 : P_0 \to \mathbb{Z}$, which is defined by $\sum a_\sigma \sigma \mapsto \sum a_\sigma$. Using the contracting homotopy given by $D_n(\sigma_0, \ldots, \sigma_n) = (1, \sigma_0, \ldots, \sigma_n)$, one verifies in a similar way as in the proof of (1.2.1) that the complex

$$\cdots \longrightarrow P_2 \xrightarrow{\partial_2} P_1 \xrightarrow{\partial_1} P_0 \xrightarrow{\partial_0} \mathbb{Z} \longrightarrow 0$$

is exact. Moreover, this complex consists of torsion-free abelian groups and therefore remains exact when tensored by any G-module A. Now we apply the functor "cofixed module". For $n \geq 0$, we set

$$C_n(G, A) = (P_n \otimes A)_G$$

and we call the group

$$H_n(G, A) = H_n(C_\bullet(G, A))$$

the **n-dimensional homology group** of G with coefficients in A. From the exact sequence above it follows that

$$H_0(G, A) = A_G$$

which is consistent with our definition in §2. In contrast to cohomology, which is contravariant, homology is covariantly functorial in the group. As a functor in the module, homology is dual to cohomology in the following sense.

(1.9.1) Proposition. *Let G be a finite group and let A be a G-module. If $A^* = \mathrm{Hom}(A, \mathbb{Q}/\mathbb{Z})$, then we have a natural isomorphism*

$$H^i(G, A^*) \cong H_i(G, A)^*, \quad i \geq 0.$$

Proof: Applying the exact functor $\mathrm{Hom}(-, \mathbb{Q}/\mathbb{Z})$ to the exact sequence

$$\cdots \xrightarrow{\partial_2} P_1 \otimes A \xrightarrow{\partial_1} P_0 \otimes A \xrightarrow{\partial_0} A \longrightarrow 0,$$

we obtain the exact sequence

$$0 \longrightarrow A^* \longrightarrow (P_0 \otimes A)^* \longrightarrow (P_1 \otimes A)^* \longrightarrow \cdots,$$

which, by (1.3.6)(iii), is an induced, hence acyclic, resolution of A^*. By (1.3.9), we obtain isomorphisms

$$H^i(G, A^*) \cong H^i(H^0(G, (P_\bullet \otimes A)^*)) \cong H_i(H_0(G, P_\bullet \otimes A))^* \cong H_i(G, A)^*.$$

□

All cohomological notions have their homological counterpart. For example, for a normal subgroup U of G, we have the **coinflation** maps

$$\mathit{coinf}: H_n(G, A) \longrightarrow H_n(G/U, A_U),$$

which are induced by the projections $\mathbb{Z}[G^{n+1}] \to \mathbb{Z}[(G/U)^{n+1}]$ and $A \to A_U$. Recalling the modified homology group $\hat{H}_0(G, A) = {}_{N_G}A/I_GA$ from §2, we call the groups

$$\hat{H}_n(G, A) = \begin{cases} {}_{N_G}A/I_GA & \text{for } n = 0, \\ H_n(G, A) & \text{for } n \geq 1 \end{cases}$$

the **modified homology groups**. In §2 we identified the group $\hat{H}_0(G, A)$ with $\hat{H}^{-1}(G, A)$. The following proposition extends this to arbitrary dimension.

(1.9.2) Proposition. *Let G be a finite group and let A be a G-module. Then we have natural isomorphisms*

$$\hat{H}^n(G,A) \cong \begin{cases} \hat{H}^n(G,A) & \text{for } n \geq 0, \\ \hat{H}_{-n-1}(G,A) & \text{for } n \leq -1. \end{cases}$$

Proof: We have seen this for $n \geq -1$ in §2. In order to deal with the remaining dimensions, we consider the dual G-modules

$$(X_{-1-n})^+ = \operatorname{Hom}(X_{-1-n}, \mathbb{Z})$$

for $n \geq 0$. Let $((\sigma_0,\dots,\sigma_n)^*)_{\sigma_0,\dots,\sigma_n \in G}$, be the dual basis of $(X_{-1-n-})^+$, i.e. $(\sigma_0,\dots,\sigma_n)^*$ maps $(\sigma_0,\dots,\sigma_n)$ to 1 and all other basis elements of $X_{-1-n} = \mathbb{Z}[G^{n+1}]$ to zero. We consider the homomorphisms

$$\varphi_n : P_n \longrightarrow (X_{-1-n})^+$$

defined by sending $(\sigma_0,\dots,\sigma_n)$ to $(\sigma_0,\dots,\sigma_n)^*$. A straightforward calculation shows that the φ_n are G-homomorphisms and that the diagram

$$\begin{array}{ccc} P_n & \xrightarrow{\varphi_n} & (X_{-1-n})^+ \\ \downarrow{\scriptstyle \partial_n} & & \downarrow{\scriptstyle (\partial_{-n})^+} \\ P_{n-1} & \xrightarrow{\varphi_{n-1}} & (X_{-n})^+ \end{array}$$

commutes for $n \geq 1$. Therefore, for $n \geq 1$, $H_n(G,A)$ is the cohomology of the complex

$$((X_\bullet)^+ \otimes A)_G$$

in dimension $(-1-n)$. Since the abelian groups X_i are free of finite rank, the duality maps $D_i : (X_i)^+ \otimes A \longrightarrow \operatorname{Hom}(X_i, A)$, given by $D_i(f \otimes a)(x) = f(x)a$, are isomorphisms. These isomorphisms are easily seen to be G-invariant, and we obtain an isomorphism of complexes

$$D : ((X_\bullet)^+ \otimes A)_G \xrightarrow{\sim} \operatorname{Hom}(X_\bullet, A)_G.$$

Finally, by (1.3.6)(iii), the modules $\operatorname{Hom}(X_i, A)$ are induced. By (1.2.6), the norm induces an isomorphism of complexes

$$N_G : \operatorname{Hom}(X_\bullet, A)_G \xrightarrow{\sim} \operatorname{Hom}_G(X_\bullet, A) = \hat{C}^\bullet(G,A).$$

This finishes the proof. □

Now let G be a profinite group and let A be a discrete G-module. We want to define Tate cohomology groups $\hat{H}^i(G,A)$, $i \in \mathbb{Z}$. For $i \leq 0$ this requires the **deflation map**

$$def : \hat{H}^i(G/V, A^V) \longrightarrow \hat{H}^i(G/U, A^U), \quad i \leq 0,$$

for open normal subgroups $V \subseteq U$, which is defined as follows:

Consider, for $j \geq 1$, the composition

$$H_j(G/V, A^V) \xrightarrow{coinf} H_j(G/U, (A^V)_U) \xrightarrow{N_*} H_j(G/U, A^U),$$

where N_* is the map which is induced on homology by the norm

$$N_{U/V} : (A^V)_U \longrightarrow A^U .$$

Via the identification $\hat{H}^i = H_{-i-1}$, this defines the deflation in dimension $i \leq -2$. In dimensions $i = 0, -1$, the map *def* is induced via the identifications

$$\hat{H}^0(G/U, A^U) \cong A^G/N_{G/U}A^U \quad \text{and} \quad \hat{H}^{-1}(G/U, A^U) \cong {}_{N_{G/U}}A^U/I_{G/U}A^U$$

by the identity and the norm map, respectively.

(1.9.3) Definition (Tate cohomology). *Let G be a profinite group and let A be a discrete G-module. For $i > 0$, we set $\hat{H}^i(G, A) = H^i(G, A)$. For $i \leq 0$, we set*

$$\hat{H}^i(G, A) = \varprojlim_{U, def} \hat{H}^i(G/U, A^U),$$

where U runs through the open normal subgroups of G.

Our next goal is a cup-product pairing for Tate cohomology. We start by specifying a map on the chain level on the negative part of the complete standard resolution which defines the deflation map on cohomology in dimension ≤ -2. For a finite group G, let $X_\bullet^{(G)}$ denote the (homological) complete standard resolution for G.

(1.9.4) Lemma. *Let G be a finite group and let $U \subseteq G$ be a normal subgroup. In negative dimension, let the map $\alpha_{-i} : X_{-i}^{(G/U)} \longrightarrow X_{-i}^{(G)}$ be given by*

$$(\sigma_0 U, \ldots, \sigma_{i-1} U) \longmapsto \sum_{\tau_0, \ldots, \tau_{i-1} \in U} (\sigma_0 \tau_0, \ldots, \sigma_{i-1} \tau_{i-1}),$$

where the $\tau_0, \ldots, \tau_{i-1}$ on the right-hand side run independently through U. Then in dimension ≤ -2 the deflation map is induced by a cocycle map

$$def : \mathrm{Hom}_G(X_\bullet^{(G)}, A) \longrightarrow \mathrm{Hom}_{G/U}(X_\bullet^{(G/U)}, A^U),$$

which is uniquely defined by the commutative diagram

$$\begin{array}{ccc} \mathrm{Hom}(X_{-i}^{(G)}, A)_G & \xrightarrow[\sim]{N_G} & \mathrm{Hom}_G(X_{-i}^{(G)}, A) \\ \Big\downarrow {\scriptstyle (\alpha_{-i}^*, (N_U)_*)} & & \Big\downarrow {\scriptstyle def} \\ \mathrm{Hom}(X_{-i}^{(G/U)}, A^U)_{G/U} & \xrightarrow[\sim]{N_{G/U}} & \mathrm{Hom}_{G/U}(X_{-i}^{(G/U)}, A^U). \end{array}$$

Proof: First note that the map α_{-i} commutes with the differential on the negative part of the standard complex. Let $((\sigma_0, \ldots, \sigma_{i-1})^*)_{\sigma_0,\ldots,\sigma_{i-1}\in G}$ denote the dual basis of $(X_{-i}^{(G)})^+ = \mathrm{Hom}(X_{-i}^{(G)}, \mathbb{Z})$. The dual basis of $(X_{-i}^{(G/U)})^+$ is denoted by $((\sigma_0 U, \ldots, \sigma_{i-1}U)^*)_{\sigma_0 U,\ldots,\sigma_{i-1}U\in G/U}$. Consider the diagram

$$\begin{array}{ccccc}
((X_{-i}^{(G)})^+ \otimes A)_G & \xrightarrow[\sim]{D_i} & \mathrm{Hom}(X_{-i}^{(G)}, A)_G & \xrightarrow[\sim]{N_G} & \mathrm{Hom}_G(X_{-i}^{(G)}, A) \\
\big\downarrow{\scriptstyle coinf} & & \big\downarrow{\scriptstyle (\alpha_{-i}^*, pr_*)} & & \big\downarrow{\scriptstyle \beta} \\
((X_{-i}^{(G/U)})^+ \otimes A_U)_{G/U} & \xrightarrow[\sim]{D_i} & \mathrm{Hom}(X_{-i}^{(G/U)}, A_U)_{G/U} & \xrightarrow[\sim]{N_{G/U}} & \mathrm{Hom}_{G/U}(X_{-i}^{(G/U)}, A_U) \\
\big\downarrow{\scriptstyle id\otimes N_U} & & \big\downarrow{\scriptstyle (id^*, (N_U)_*)} & & \big\downarrow{\scriptstyle (id^*, (N_U)_*)} \\
((X_{-i}^{(G/U)})^+ \otimes A^U)_{G/U} & \xrightarrow[\sim]{D_i} & \mathrm{Hom}(X_{-i}^{(G/U)}, A^U)_{G/U} & \xrightarrow[\sim]{N_{G/U}} & \mathrm{Hom}_{G/U}(X_{-i}^{(G/U)}, A^U).
\end{array}$$

Here, $pr : A \to A_U$ is the natural projection, *coinf* is the map which is induced by pr and by the map $(X_{-i}^{(G)})^+ \to (X_{-i}^{(G/U)})^+$ which sends $(\sigma_0, \ldots, \sigma_{i-1})^*$ to $(\sigma_0 U, \ldots, \sigma_{i-1}U)^*$. Via the identification

$$H_{i-1}(G, A) = H^{-i}\left(((X_\bullet)^+ \otimes A)_G\right), \quad i \geq 2,$$

(see the proof of (1.9.2)) it induces coinflation on homology. The maps D_{-i} are induced by the canonical duality isomorphisms $(X_{-i})^+ \otimes A \cong \mathrm{Hom}(X_{-i}, A)$. The upper left square commutes and we define β in order to make the upper right square commute. The lower squares obviously commute. By definition, *def* is induced by the composition of the two vertical arrows on the left. This completes the proof of the lemma. □

(1.9.5) Proposition. *Let $i \geq 2$ and $q \geq 1$. Let G be a finite group, U a normal subgroup of G and let A and B be G-modules. Then*

$$\bar{x} \cup \mathit{inf}\, \bar{y} = \mathit{inf}\,(\mathit{def}\, \bar{x} \cup \bar{y}) \in H^q(G, A \otimes B),$$

for $\bar{x} \in \hat{H}^{-i}(G, A)$ and $\bar{y} \in H^{q+i}(G/U, B^U)$.

Proof: Let $x \in \mathrm{Hom}_G(X_{-i}^{(G)}, A)$ and $y \in \mathrm{Hom}_{G/U}(X_{q+i}^{(G/U)}, A)$ be cocycles representing $\bar{x}$ and $\bar{y}$, respectively. We calculate both sides on the level of cochains. Let us start with the left-hand side. For $(\sigma_0, \ldots, \sigma_q) \in X_q^{(G)}$ we obtain

$$(x \cup \mathit{inf}\, y)(\sigma_0, \ldots, \sigma_q) = \sum_{\tau_1,\ldots,\tau_i \in G} x(\tau_1, \ldots, \tau_i) \otimes y(\tau_i U, \ldots, \tau_1 U, \sigma_0 U, \ldots, \sigma_q U),$$

see the definition of the maps $\varphi_{p,q}$ in the proof of (1.4.7). In order to compute the right-hand side, choose a $z \in \mathrm{Hom}(X_{-i}^{(G)}, A)$ with $x = N_G z$. Then, using again the maps $\varphi_{p,q}$ of (1.4.7), by (1.9.4) we have

$$\mathit{inf}\,(\mathit{def}\; x \cup y)(\sigma_0, \ldots, \sigma_q) = \Big((N_{G/U}(\alpha_{-i}^*, (N_U)_*)z) \cup y\Big)\,(\sigma_0 U, \ldots, \sigma_q U).$$

The right-hand side is equal to

$$\sum_{\tau_1 U, \ldots, \tau_i U \in G/U} (N_{G/U}(\alpha_{-i}^*, (N_U)_*)z)(\tau_1 U, \ldots, \tau_i U) \otimes y(\tau_i U, \ldots, \tau_1 U, \sigma_0 U, \ldots, \sigma_q U).$$

Using the definition of α_{-i} and of N_U this transforms to

$$\sum_{\tau U \in G/U} \sum_{u \in U} \sum_{\tau_1 U, \ldots, \tau_i U \in G/U} \sum_{u_1, \ldots, u_i \in U}$$
$$\tau u\, z(\tau^{-1}\tau_1 u_1, \ldots, \tau^{-1}\tau_i u_i) \otimes y(\tau_i U, \ldots, \tau_1 U, \sigma_0 U, \ldots, \sigma_q U),$$

which coincides with

$$\sum_{\tau_1, \ldots, \tau_i \in G} (N_G z)(\tau_1, \ldots, \tau_i) \otimes y(\tau_i U, \ldots, \tau_1 U, \sigma_0 U, \ldots, \sigma_q U).$$

This shows the proposition. □

(1.9.6) Proposition. *Let G be a finite group and let U be a normal subgroup of G. Let A and B be G-modules. Then, for $i \le 0$ and $q \ge 1$, the diagram*

$$(*)\qquad \begin{array}{ccccc} \hat{H}^i(G, A) & \times & H^{q-i}(G, B) & \xrightarrow{\;\cup\;} & H^q(G, A \otimes B) \\ \big\downarrow{\scriptstyle \mathit{def}} & & \big\uparrow{\scriptstyle \mathit{inf}} & & \big\uparrow{\scriptstyle \mathit{inf}} \\ \hat{H}^i(G/U, A^U) & \times & H^{q-i}(G/U, B^U) & \xrightarrow{\;\cup\;} & H^q(G/U, (A \otimes B)^U) \end{array}$$

commutes.

Proof: For $i \le -2$ this follows from (1.9.5). Via the isomorphism

$$\begin{array}{ccl} {}_{N_G}A/I_G A & \xrightarrow{\sim} & \hat{H}^{-1}(G, A) \\ a & \longmapsto & x_a : \mathbb{Z}[G] \to A,\ \sigma \mapsto \sigma(a), \end{array}$$

the cup product is given on the chain level by

$$x_a \cup y(\sigma_0, \ldots, \sigma_q) = \sum_{\sigma \in G} \sigma a \otimes y(\sigma, \sigma_0, \ldots, \sigma_q).$$

The commutativity of $(*)$ for $i = -1$ follows immediately, since *def* is induced by the norm

$$N_U :\ {}_{N_G}A/I_G A \longrightarrow {}_{N_{G/U}}A^U/I_{G/U}A^U.$$

The case $i = 0$ is obvious. □

Now let G be a profinite group and let A be a discrete G-module which is finitely generated as a $\mathbb{Z}$-module. The subgroup of elements of G that act trivially on A is open, since for all $a \in A$ the group $G_a = \{\sigma \in G \mid \sigma a = a\}$ is open. Hence $A^U = A$ for U sufficiently small. In this case

$$\mathrm{Hom}(A, B^U) = \mathrm{Hom}(A, B)^U$$

for any G-module B. By passing to the limit of the diagram $(*)$ of (1.9.6), we obtain the

(1.9.7) Corollary. *Let G be a profinite group and let A and B be discrete G-modules. Then we have a cup-product pairing*

$$\hat{H}^i(G, A) \times H^{q-i}(G, B) \longrightarrow H^q(G, A \otimes B)$$

for $i \le 0$ and $q \ge 1$.

Let G be a profinite group and let A be a discrete G-module which is finitely generated as a $\mathbb{Z}$-module. By (1.6.1), $\hat{H}^i(G/U, A^U)$ is finite for all i and all open normal subgroups U of G. Therefore the groups

$$\hat{H}^i(G, A) = \varprojlim_{U,\, def} \hat{H}^i(G/U, A^U), \quad i \le 0,$$

are naturally equipped with a profinite topology. For $i > 0$ we give $\hat{H}^i(G, A)$ the discrete topology.

(1.9.8) Lemma. *Let G be a profinite group and let A be a discrete G-module which is finitely generated as a $\mathbb{Z}$-module. If p is a prime number with $p^\infty | \#G$, then we have a canonical isomorphism*

$$\hat{H}^0(G, A)(p) \cong A^G \otimes \mathbb{Z}_p.$$

Proof: It suffices to show that for arbitrarily given $n \in \mathbb{N}$ the subgroup $N_{G/U}A^U$ is contained in $p^n A^G$ for sufficiently small U. Let U be an open normal subgroup such that $A^U = A$. Then for every normal open subgroup $V \subseteq U$ with $p^n | (U : V)$ and for every $a \in A^V = A^U = A$, we have

$$N_{G/V}a = N_{G/U}N_{U/V}a = (U : V)N_{G/U}a \in p^n A^G.$$

This shows the lemma. □

(1.9.9) Proposition. *Let*

$$0 \longrightarrow A' \longrightarrow A \longrightarrow A'' \longrightarrow 0$$

be a short exact sequence of discrete G-modules which are finitely generated as $\mathbb{Z}$-modules. Then there is an associated long exact cohomology sequence

$$\cdots \longrightarrow \hat{H}^{-n}(G, A') \longrightarrow \hat{H}^{-n}(G, A) \longrightarrow \hat{H}^{-n}(G, A'') \longrightarrow \cdots$$

ending with $\cdots \to \hat{H}^0(G, A') \to \hat{H}^0(G, A) \to \hat{H}^0(G, A'')$.

For every prime number p with $p^\infty | \#G$, we obtain a long exact sequence

$$\cdots \longrightarrow \hat{H}^i(G, A')(p) \longrightarrow \hat{H}^i(G, A)(p) \longrightarrow \hat{H}^i(G, A'')(p) \longrightarrow \cdots$$

which is unbounded in both directions. The groups are compact for $i \le 0$, discrete for $i > 0$ and all homomorphisms are continuous.

Proof: Since $A^U = A$ for small U, we obtain for small U a long exact sequence

$$\cdots \longrightarrow \hat{H}^0(G/U, A'^U) \longrightarrow \hat{H}^0(G/U, A^U) \longrightarrow \hat{H}^0(G/U, A''^U)$$

of finite abelian groups. Passing to the projective limit, we obtain the negative part of our long exact sequence. Now let p be a prime number with $p^\infty | \#G$ and consider for small U the long exact sequence

$$0 \longrightarrow A'^G \longrightarrow A^G \longrightarrow A''^G \longrightarrow \hat{H}^1(G/U, A'^U) \longrightarrow \cdots .$$

Tensoring by $\mathbb{Z}_p$ and passing to the direct limit over U, we obtain the right-hand part of our long exact sequence. The left-hand part is obtained by taking the p-part of the long exact sequence which we obtained above. Both fit together by (1.9.8). The continuity of the maps is clear from their definitions. $\square$

Let G be a profinite group and let A be a discrete G-module. For each pair of open normal subgroups $V \subseteq U$ of G, the exact sequences

$$(*) \qquad 0 \longrightarrow N_{G/U} A^U \longrightarrow A^G \longrightarrow \hat{H}^0(G/U, A^U) \longrightarrow 0,$$

$$(**) \qquad 0 \longrightarrow I_{G/U} A^U \longrightarrow {}_{N_{G/U}} A^U \longrightarrow \hat{H}^{-1}(G/U, A^U) \longrightarrow 0$$

induce exact sequences

$$0 \longrightarrow N_G A \longrightarrow A^G \longrightarrow \hat{H}^0(G, A),$$

$$0 \longrightarrow I_G A \longrightarrow {}_{N_G} A \longrightarrow \hat{H}^{-1}(G, A),$$

where we have set

$$N_G A = \varprojlim_U N_{G/U} A^U = \bigcap_U N_{G/U} A^U,$$

$${}_{N_G} A = \varprojlim_U {}_{N_{G/U}} A^U \quad , \quad I_G A = \varprojlim_U I_{G/U} A^U .$$

The group $N_G A$ is called the group of **universal norms**. The last exact sequences are in general not exact on the right, unless we make assumptions on the module. This motivates the following

(1.9.10) Definition. *Let G be a profinite group. A* **level-compact** *G-module is a discrete G-module A which is endowed with the following additional topological structure: For each open subgroup $U \subseteq G$ the group A^U carries a compact group topology, such that for all open subgroups U and V and all $\sigma \in G$ with $V \subseteq \sigma U \sigma^{-1}$, the natural map*

$$\sigma : A^U \longrightarrow A^V,\ a \longmapsto \sigma a,$$

is continuous.

In particular, for $V \subseteq U$, the group A^U carries the subgroup topology of A^V with respect to the inclusion $A^U \subseteq A^V$, and the norm map $N_{U/V} : A^V \to A^U$ is continuous.

Assume now that the discrete G-module A is level-compact. In this case, the additional topological structure on A induces a compact group topology on all groups in the above sequences $(*)$ and $(**)$. Furthermore, all maps occurring are continuous. Since the projective limit is exact on compact groups, we obtain the

(1.9.11) Lemma. *Let A be a level-compact G-module. Then*

$$\hat{H}^0(G, A) = A^G/N_G A \quad \text{and} \quad \hat{H}^{-1}(G, A) = {}_{N_G}A/I_G A.$$

If A is level-compact, then the groups

$$\hat{H}^i(G, A) = \varprojlim_{U, def} \hat{H}^i(G/U, A^U), \quad i \le 0,$$

are abelian profinite groups in a natural way: For each open normal subgroup $U \subseteq G$, the group $\hat{H}^i(G/U, A^U)$ inherits a natural compact topology from A^U via the standard complex. Furthermore, this group is annihilated by $(G : U)$, hence $\hat{H}^i(G/U, A^U)$ is profinite. Since *def* is continuous, the group $\hat{H}^i(G, A)$ is profinite as an inverse limit of abelian profinite groups. For $i > 0$ we give $\hat{H}^i(G, A) = H^i(G, A)$ the discrete topology.

We will frequently make use of the following fact: For a level-compact G-module A, the natural map

$$\varprojlim_{U, Norm} A^U \longrightarrow N_G A = \bigcap_U N_{G/U} A^U$$

is surjective. This follows easily from the fact that a filtered inverse limit of nonempty compact spaces is nonempty, cf. [160], chap.IV, (2.3). In particular, the norm map $N_{G/U} : N_U A \to N_G A$ is surjective for all open normal subgroups $U \subseteq G$.

The next lemma shows that we have some flexibility when calculating the Tate cohomology of level-compact modules.

(1.9.12) Lemma. *Let A be a level-compact G-module. Suppose that for every normal open subgroup $U \subseteq G$ a closed (with respect to the additional topology) G-submodule*

$$A(U) \subseteq A^U$$

is given in such a way that the following conditions hold

(i) $N_U A \subseteq A(U)$ *for all* U,

(ii) *for* $V \subseteq U$, $N_{U/V} : A^V \to A^U$ *maps* $A(V)$ *to* $A(U)$.

Then

$$\hat{H}^i(G, A) \cong \varprojlim_U \hat{H}^i(G/U, A(U))$$

for all $i \leq -1$.

Proof: The transition maps in the above inverse system are defined by an obvious modification of *def* (possible by condition (ii)). We have seen in (1.9.4) that the deflation maps in negative dimensions are given by a map on the chain level. As projective limits are exact on compact groups, we see that for $i \leq -2$ the group $\hat{H}^i(G, A)$ can also be calculated as the quotient of the inverse limit of the cocycles modulo the inverse limit of the coboundaries. These, however, take values in the groups of universal norms on the corresponding levels, i.e., by condition (i), we may take the limit over the groups $A(U)$ instead of A^U as well. This shows the lemma for $i \leq -2$.

Now we show the assertion for $i = -1$. We have $\hat{H}^{-1}(G, A) = {}_{N_G}A/I_G A$ with ${}_{N_G}A = \varprojlim_U {}_{N_{G/U}}A^U$ and $I_G A = \varprojlim_U I_{G/U} A^U$. Recalling that the functor $\varprojlim$ is exact on compact groups, we have to show

$$ {}_{N_G}A = \varprojlim_U {}_{N_{G/U}}A(U) \quad \text{and} \quad I_G A = \varprojlim_U I_{G/U} A(U).$$

Obviously, $\varprojlim {}_{N_{G/U}}A(U) \subseteq {}_{N_G}A$. Let $(a_U) \in {}_{N_G}A$. Since $\varprojlim$ is taken over the norm maps $N_{U/V} : A^V \longrightarrow A^U$, we have $a_U \in \bigcap_{V \subseteq U} N_{U/V} A^V = N_U A \subseteq A(U)$. This proves ${}_{N_G}A = \varprojlim {}_{N_{G/U}}A(U)$.

The inclusion $A(U) \subseteq A^U$ yields the injection

$$\varprojlim_U I_{G/U}A(U) \longrightarrow \varprojlim_U I_{G/U}A^U = I_G A.$$

For the surjectivity, let $(a_U) \in \varprojlim_U I_{G/U}A^U$. The projective limit is taken over the maps $N_{U/V} : I_{G/V}A^V \longrightarrow I_{G/U}A^U$ given by $(\sigma - 1)a \longmapsto (\bar{\sigma} - 1)N_{U/V}(a)$. Therefore

$$(*) \qquad a_U \in \bigcap_{V \subseteq U} I_{G/U}N_{U/V}A^V.$$

We show $a_U \in I_{G/U}N_U A$ for any fixed U. Let V run through the open normal subgroups of G contained in U. Let $\tilde{A}^V = \prod_{\sigma \in G/U} A^V$ and let

$$f_V : \tilde{A}^V \longrightarrow I_{G/U}A^U$$

be the composite of the maps

$$\tilde{A}^V \xrightarrow{\tilde{N}_{U/V}} \tilde{A}^U \xrightarrow{f_U} I_{G/U}A^U,$$

where $\tilde{N}_{U/V} = \prod_{\sigma \in G/U} N_{U/V}$ and $f_U : (a_\sigma)_{\sigma \in G/U} \mapsto \sum_{\sigma \in G/U}(\sigma - 1)a_\sigma$. All groups are compact and f_V is continuous. Therefore, using $(*)$, $f_V^{-1}(a_U)$ is a nonempty, closed and thus compact subset of $\tilde{A}^V$. It follows that $\varprojlim_{V \subseteq U} f_V^{-1}(a_U) \neq \varnothing$.
Let $(\tilde{a}_V) \in \varprojlim_{V \subseteq U} f_V^{-1}(a_U)$. Then for $\tilde{a}_U = (a_\sigma)_{\sigma \in G/U}$ we have $a_\sigma \in N_U A$, hence

$$a_U = f_U(\tilde{a}_U) = \sum_{\sigma \in G/U} (\sigma - 1)a_\sigma \in I_{G/U}N_U A.$$

This proves that $(a_U) \in \varprojlim_U I_{G/U}N_U A \subseteq \varprojlim_U I_{G/U}A(U)$, whence $I_G A = \varprojlim_U I_{G/U}A(U)$. □

In everything that follows, we tacitly assume maps between level-compact modules to be continuous with respect to the additional topology.

(1.9.13) Proposition. *Let*

$$0 \longrightarrow A' \xrightarrow{i} A \xrightarrow{j} A''$$

be an exact sequence of level-compact G-modules, such that the induced map

$$N_U A \xrightarrow{N_U j} N_U A''$$

is surjective for a cofinal system of open normal subgroups U of G (and hence for all open normal subgroups). Then there is an associated long exact cohomology sequence

$$\cdots \longrightarrow \hat{H}^{-n}(G, A') \longrightarrow \hat{H}^{-n}(G, A) \longrightarrow \hat{H}^{-n}(G, A'') \longrightarrow \cdots$$

ending with $\cdots \to \hat{H}^0(G, A') \to \hat{H}^0(G, A) \to \hat{H}^0(G, A'')$. Moreover, if j is surjective, we obtain the long exact cohomology sequence unbounded in both directions, i.e. from $-\infty$ to $+\infty$.

Proof: For every open normal subgroup U in G we consider the kernel $A'(U) := \ker(N_U A \xrightarrow{N_U j} N_U A'')$. We have inclusions $N_U A' \subseteq A'(U) \subseteq A'^U$, and obtain the exact and commutative diagram

$$\begin{array}{ccccccccc} 0 & \longrightarrow & A'(U) & \longrightarrow & N_U A & \longrightarrow & N_U A'' & \longrightarrow & 0 \\ & & \downarrow{\scriptstyle N_{G/U}} & & \downarrow{\scriptstyle N_{G/U}} & & \downarrow{\scriptstyle N_{G/U}} & & \\ 0 & \longrightarrow & A'^G & \longrightarrow & A^G & \longrightarrow & A''^G. & & \end{array}$$

Consider the long exact cohomology sequence

$$\cdots \to \hat{H}^i(G/U, A'(U)) \to \hat{H}^i(G/U, N_U A) \to \hat{H}^i(G/U, N_U A'') \to \cdots$$

associated to the upper line. It consists of compact abelian groups, is clearly exact and all homomorphisms including the boundary maps are continuous (for example use the snake lemma in the abelian category of compact abelian groups). Passing to the inverse limit over U, we obtain, using (1.9.12), the asserted long exact sequence up to dimension -1.

By compactness, the image of $N_{G/U} : N_U A \longrightarrow A^G$ is $N_G A$ and hence, by (1.9.11), the cokernel of this map is $\hat{H}^0(G, A)$. We denote its kernel by

$$X(A, U) = {}_{N_{G/U}} N_U A,$$

which contains $Y(A, U) := I_{G/U} N_U A$, and the same holds for A''. The snake lemma implies an exact commutative diagram

$$\begin{array}{ccccccccc} Y(A,U) & \xrightarrow{j} & Y(A'',U) & & & & & & \\ \downarrow & & \downarrow & & & & & & \\ X(A,U) & \to & X(A'',U) & \xrightarrow{\delta} & A'^G/N_{G/U}A'(U) & \to & \hat{H}^0(G,A) & \to & \hat{H}^0(G,A''). \end{array}$$

Observe that

$$\varprojlim_U (A'^G/N_{G/U}A'(U)) = \hat{H}^0(G, A').$$

Furthermore, the upper map $j : I_{G/U} N_U A \to I_{G/U} N_U A''$ is obviously surjective and $N_{G/U} : N_U A \to A^G$ maps $I_{G/U} N_U A$ to zero. Therefore $\delta : X(A'', U) \to A'^G/N_{G/U}A'(U)$ maps $Y(A'', U)$ to zero by the definition of δ. This means that we may replace in the last diagram the group $X(A, U)$ by $\hat{H}^{-1}(G/U, N_U A) = X(A, U)/Y(A, U)$ and $X(A'', U)$ by $\hat{H}^{-1}(G/U, N_U A'') = X(A'', U)/Y(A'', U)$ and obtain an exact sequence of compact groups and continuous homomorphisms. Taking projective limits over U, we obtain the exact sequence

$$\hat{H}^{-1}(G, A) \to \hat{H}^{-1}(G, A'') \xrightarrow{\delta} \hat{H}^0(G, A') \to \hat{H}^0(G, A) \to \hat{H}^0(G, A'').$$

Now suppose that j is surjective. Consider the commutative and exact diagram

$$\begin{array}{ccccccc}
0 & & 0 & & & & \\
\uparrow & & \uparrow & & & & \\
\hat{H}^0(G,A) & \longrightarrow & \hat{H}^0(G,A'') & & & & \\
\uparrow & & \uparrow & \searrow & & & \\
A^G & \longrightarrow & A''^G & \longrightarrow & H^1(G,A') & \longrightarrow & H^1(G,A) \\
\uparrow & & \uparrow & \nearrow_{0} & & & \\
N_G A & \xrightarrow{N_G j} & N_G A'' & & & & \\
\uparrow & & \uparrow & & & & \\
0 & & 0\,. & & & &
\end{array}$$

The vertical sequences are exact by (1.9.11). We deduce the existence of the dotted arrow, which glues the already proven long exact sequence in negative dimension with the long exact sequence

$$H^1(G,A') \longrightarrow H^1(G,A) \longrightarrow H^1(G,A'') \longrightarrow \cdots\,. \qquad \square$$

The homomorphisms in the negative part of the long exact sequence are easily seen to be continuous. The maps in dimensions at least 1 are continuous because the groups are discrete. But the boundary map $\delta^0 : \hat{H}^0(G,A'') \to H^1(G,A')$ may be discontinuous. Therefore the next proposition is nontrivial.

(1.9.14) Proposition. *If p is a prime number, then the long exact sequence of lemma* (1.9.13) *also induces a long exact sequence of the p-parts.*

Proof: All occurring cohomology groups are either abelian profinite groups or abelian discrete torsion groups and therefore they naturally decompose into the direct sum of their p-parts and their prime-to-p-parts. In order to prove the corollary, it suffices to show that also the differentials decompose into a direct sum of homomorphisms. This is trivially true for continuous differentials, hence it remains to consider the homomorphism $\delta^0 : \hat{H}^0(G,A'') \to H^1(G,A')$. Let $x \in \hat{H}^0(G,A'')(p)$. Choose a pre-image $m'' \in A''^G$ of x and let $m \in A$ be a pre-image of m''. Furthermore, let $U \subseteq G$ be an open normal subgroup in G such that $m \in A^U$. The closed subgroup generated by m in A^U maps onto the closed subgroup generated by x which coincides with $\mathbb{Z}_p \cdot x \subseteq \hat{H}^0(G,A'')(p)$. We conclude that $\delta^0(\lambda x) \in H^1(G/U, A'^U) \subseteq H^1(G,A')$ for all $\lambda \in \mathbb{Z}_p$. Writing $\#(G/U) = Np^k$ with $(N,p) = 1$, we obtain that $\delta^0(x) = N\delta^0(N^{-1}x) \in N \cdot H^1(G/U, A'^U) = H^1(G/U, A'^U)(p) \subseteq H^1(G,A')(p)$. A similar argument shows that δ^0 sends the prime-to-p-part of $\hat{H}^0(G,A'')$ to the prime-to-p-part of $H^1(G,A')$. $\square$

Applying the above results to the case of a finite module, we obtain the

(1.9.15) Proposition. *Let A be a finite G-module. Assume that $\ell^\infty \,|\, \#G$ for all prime numbers ℓ dividing the order of A. Then*

$$\hat{H}^0(G, A) = A^G \quad \textit{and} \quad \hat{H}^i(G, A) = 0 \textit{ for all } i \leq -1.$$

Proof: Let $U \subseteq G$ be an open subgroup with $A = A^U$ and let $V \subseteq U$ be an open normal subgroup. Then the norm map $N_{U/V}$ from $A = A^V$ into $A = A^U$ is just the multiplication by $(U : V)$. As $\ell^\infty \,|\, \#G$ for all prime numbers ℓ dividing the order of A, we conclude that $N_U A = 0$ for all open subgroups U of G. Applying (1.9.13) to the exact sequence

$$0 \longrightarrow 0 \longrightarrow 0 \longrightarrow A,$$

we obtain $\hat{H}^i(G, A) = 0$ for all $i \leq -1$. The statement on $\hat{H}^0(G, A)$ follows likewise from (1.9.8) or (1.9.11). □

Remark: Tate cohomology of profinite groups was first introduced by G. Poitou in [168]. The presentation of this section essentially follows [196].

Chapter II
Some Homological Algebra

§1. Spectral Sequences

If G is a profinite group and H a closed normal subgroup, then one may ask whether the cohomology groups $H^n(G,A)$ can be computed from the cohomology groups of the smaller groups H and G/H. We have already seen a relation of this type, namely the isomorphism

$$H^n(G/H, H^0(H,A)) \cong H^n(G,A)$$

if $H^n(H,A) = 0$ for all $n \geq 1$, which follows from (1.6.7). There is a quite general relation, which is denoted by

$$H^p(G/H, H^q(H,A)) \Rightarrow H^n(G,A)$$

and is called a "spectral sequence". The situation is slightly involved. It roughly says that there is a canonical decreasing filtration of $H^n = H^n(G,A)$,

$$H^n = F^0H^n \supseteq F^1H^n \supseteq \cdots \supseteq F^nH^n \supseteq F^{n+1}H^n = 0,$$

such that the quotient $F^pH^n/F^{p+1}H^n$ is isomorphic not directly to the group $H^p(G/H, H^{n-p}(H,A))$, but to a certain subquotient of it. The notion of **spectral sequence** is very general and of utmost importance in cohomology theory. The general set-up, which can be generalized in several directions if the underlying category has inductive limits, is the following.

Let $\mathcal{A}$ be an abelian category. A (decreasing) **filtration** of an object A is a family $(F^pA)_{p\in\mathbb{Z}}$ of subobjects F^pA of A such that $F^pA \supseteq F^{p+1}A$ for all p. Write

$$gr_pA = F^pA/F^{p+1}A.$$

By convention, we put $F^{\infty}A = 0$ and $F^{-\infty}A = A$. We say that the filtration is **finite** if there exist $n, m \in \mathbb{Z}$ with $F^mA = 0$ and $F^nA = A$. Given filtered objects A and B in $\mathcal{A}$, a morphism $f\colon A \to B$ is said to be compatible with the filtration if $f(F^pA) \subseteq F^pB$ for all $p \in \mathbb{Z}$.

Let m be a natural number. An ***E_m*-spectral sequence***) in $\mathcal{A}$ is a system

$$E = (E_r^{pq}, E^n),$$

*) In the applications usually only E_1- and E_2-spectral sequences occur.

J. Neukirch et al., *Cohomology of Number Fields*,
Grundlehren der mathematischen Wissenschaften 323,
DOI 10.1007/978-3-540-37889-1_2, © Springer-Verlag Berlin Heidelberg 2008

consisting of

a) objects $E_r^{pq} \in \mathcal{A}$ for all $(p, q) \in \mathbb{Z} \times \mathbb{Z}$ and any integer $r \geq m$,

b) morphisms $d = d_r^{pq} : E_r^{pq} \longrightarrow E_r^{p+r,q-r+1}$ with $d \circ d = 0$ and such that for each fixed pair $(p, q) \in \mathbb{Z} \times \mathbb{Z}$ the morphisms d_r^{pq} and $d_r^{p-r,q+r-1}$ vanish for sufficiently large r,

c) isomorphisms α_r^{pq}: $\ker(d_r^{pq})/\mathrm{im}(d_r^{p-r,q+r-1}) \xrightarrow{\sim} E_{r+1}^{pq}$,

d) finitely filtered objects $E^n \in \mathcal{A}$ for all $n \in \mathbb{Z}$,

e) isomorphisms β^{pq}: $E_\infty^{pq} \xrightarrow{\sim} gr_p E^{p+q}$.

By b) and c), the objects E_r^{pq} are independent of r for r sufficiently large and are then denoted by E_∞^{pq}. These are the objects occurring in e).

In other words, for each $r \geq m$, the system E_r^{pq} is a system of complexes whose cohomology groups are the objects E_{r+1}^{pq} of the next system. A spectral sequence is like a book with (infinitely many) pages E_m^{pq}, E_{m+1}^{pq}, $E_{m+2}^{pq}, \ldots$ and a limit page E^n at the end.

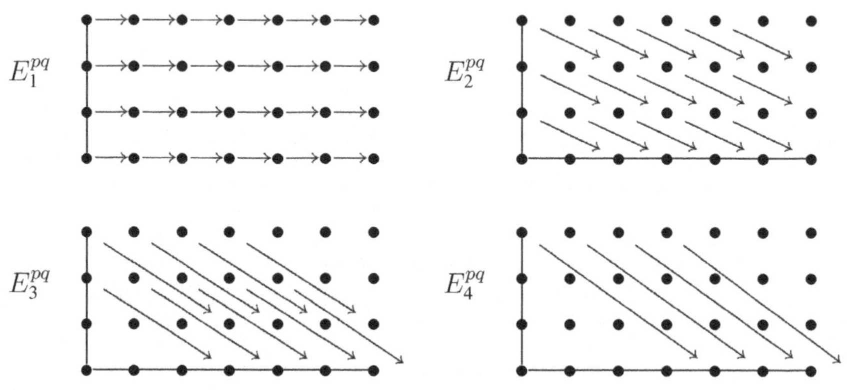

For an E_m-spectral sequence $E = (E_r^{pq}, E^n)$, one usually writes

$$E_m^{pq} \Rightarrow E^{p+q}$$

or $E_m^{pq} \Rightarrow E^n$. The E_m^{pq} are called the **initial terms**, the E^n the **limit terms** and the d_r^{pq} **differentials**. By forgetting the first $m'-m$ pages, an E_m-spectral sequence induces an $E_{m'}$-spectral sequence for all $m' \geq m$ in a natural way.

A **morphism of E_m-spectral sequences**

$$\varphi : E = (E_r^{pq}, E^n) \longrightarrow E' = (E_r'^{pq}, E'^n)$$

in $\mathcal{A}$ is a system of morphisms

$$\varphi_r^{pq} : E_r^{pq} \longrightarrow E_r'^{pq}, \quad \varphi^n : E^n \longrightarrow E'^n,$$

where the φ^n are compatible with the filtrations of E^n and E'^n and the φ_r^{pq}, φ^n commute with d_r^{pq}, α_r^{pq} and β^{pq}.

If $E_r^{pq} = 0$ for $p < 0$ or $q < 0$, one speaks of a **first quadrant** spectral sequence. In this case we have

$$E_r^{pq} = E_\infty^{pq} \quad \text{for } r > \max(p, q+1),\ r \geq m.$$

Once a first quadrant spectral sequence is given, we obtain a realm of homomorphic connections. We restrict to the most important case of an E_2-spectral sequence. Of basic importance are the two homomorphisms

$$E_2^{n,0} \longrightarrow E^n \longrightarrow E_2^{0,n},$$

the so-called **edge morphisms**. The first one is the composite of the morphisms

$$E_2^{n,0} \longrightarrow E_3^{n,0} \longrightarrow \ldots \longrightarrow E_\infty^{n,0} \longrightarrow E^n,$$

which are well-defined because $F^{n+1}E^n = 0$ and $E_r^{n+r,-r+1} = 0$ for $r \geq 2$ (so that $E_{r+1}^{n,0}$ is a quotient of $E_r^{n,0}$). The second one is the composite of the morphisms

$$E^n \longrightarrow E_\infty^{0,n} \longrightarrow \ldots \longrightarrow E_3^{0,n} \longrightarrow E_2^{0,n},$$

which are well-defined because $F^0E^n = E^n$ and $E_r^{-r,n+r-1} = 0$ for $r \geq 2$ (so that $E_{r+1}^{0,n}$ is embedded in $E_r^{0,n}$). A direct consequence of the definition of the edge morphisms is the following

(2.1.1) Proposition. *For any first quadrant E_2-spectral sequence the sequence*

$$0 \longrightarrow E_2^{1,0} \longrightarrow E^1 \longrightarrow E_2^{0,1} \xrightarrow{d} E_2^{2,0} \longrightarrow E^2$$

is exact. It is called the associated **five term exact sequence**.

We get a generalization of this result under the assumption that $E_2^{pq} = 0$ for $0 < q < n$. Namely, in this case we have isomorphisms $E_{n+1}^{0,n} \xrightarrow{\sim} \cdots \xrightarrow{\sim} E_2^{0,n}$ and $E_2^{n+1,0} \xrightarrow{\sim} \cdots \xrightarrow{\sim} E_{n+1}^{n+1,0}$. Therefore the differential $E_{n+1}^{0,n} \to E_{n+1}^{n+1,0}$ induces a homomorphism

$$E_2^{0,n} \xrightarrow{d} E_2^{n+1,0}.$$

We obtain the following

(2.1.2) Lemma. *Assume that, in a first quadrant E_2-spectral sequence, the terms E_2^{pq} vanish for $0 < q < n$ and all p. Then*

$$E_2^{m,0} \cong E^m$$

for $m < n$ and the sequence

$$0 \longrightarrow E_2^{n,0} \longrightarrow E^n \longrightarrow E_2^{0,n} \xrightarrow{d} E_2^{n+1,0} \longrightarrow E^{n+1}$$

is exact.

The proof of this lemma (and also of the results below) is elementary and we refer to [21], chap.XV, §5. The most frequent application of spectral sequences are in the following special cases.

(2.1.3) Lemma. *Assume that a first quadrant E_2-spectral sequence is given.*

(i) *If $E_2^{pq} = 0$ for all $q > 1$ and all p, then we have a long exact sequence*

$$
\begin{array}{l}
0 \longrightarrow E_2^{1,0} \longrightarrow E^1 \longrightarrow E_2^{0,1} \xrightarrow{d} \\
\quad\; E_2^{2,0} \longrightarrow E^2 \longrightarrow E_2^{1,1} \xrightarrow{d} \\
\quad\; E_2^{3,0} \longrightarrow E^3 \longrightarrow E_2^{2,1} \longrightarrow \cdots .
\end{array}
$$

(ii) *If $E_2^{pq} = 0$ for all $p > 1$ and all q, then the sequences*

$$0 \longrightarrow E_2^{1,n-1} \longrightarrow E^n \longrightarrow E_2^{0,n} \longrightarrow 0$$

are exact for all $n \geq 1$.

(2.1.4) Lemma. *Assume that, in a first quadrant E_2-spectral sequence, the term E_2^{pq} vanishes for all (p, q) with $(p - m)\cdot(q - n) < 0$. Then*

$$E_2^{mn} \cong E^{m+n}.$$

In particular, if $E_2^{pq} = 0$ for all $q > 0$, then

$$E_2^{m,0} \cong E^m \qquad \text{for all } m.$$

Proof. If $p > m$ and $q < n$ or if $p < m$ and $q > n$, then $E_r^{pq} = 0$ for all $r \geq 2$, since it is a subquotient of E_2^{pq}, and hence $E_\infty^{pq} = 0$. Therefore on the line $p + q = m + n$, all terms E_∞^{pq} are zero up to E_∞^{mn} and consequently $E_\infty^{mn} \cong E^{m+n}$. The maps

$$E_r^{m-r,n+r-1} \xrightarrow{d_r} E_r^{m,n} \xrightarrow{d_r} E_r^{m+r,n-r+1}$$

are zero for all $r \geq 2$, hence $E_2^{mn} = E_3^{mn} = \cdots = E_r^{mn} = E_\infty^{mn}$. □

In practice, the differentials of a spectral sequence are often difficult to calculate. We are in a rather comfortable situation if they vanish from a certain point on. In this case one says that the spectral sequence *degenerates*. The precise definition is the following.

(2.1.5) Definition. *An E_m-spectral sequence* **degenerates** *at $E_{m'}$ for some $m' \geq m$ if the differentials*

$$d_r^{pq} \colon E_r^{pq} \longrightarrow E_r^{p+r,q-r+1}$$

vanish for all $r \geq m'$ and all $(p, q) \in \mathbb{Z} \times \mathbb{Z}$. Hence, in this case

$$E_{m'}^{pq} = E_{m'+1}^{pq} = E_{m'+2}^{pq} = \cdots = E_\infty^{pq}$$

for all $p, q \in \mathbb{Z} \times \mathbb{Z}$.

§2. Filtered Cochain Complexes

Many spectral sequences arise from filtered cochain complexes. Let $\mathcal{A}$ be an abelian category. By a (cochain) **complex** $A^\bullet$ in $\mathcal{A}$ we understand a sequence $A^\bullet = (A^n, d^n)_{n\in\mathbb{Z}}$ of objects and homomorphisms

$$\cdots \longrightarrow A^{n-1} \xrightarrow{d^{n-1}} A^n \xrightarrow{d^n} A^{n+1} \xrightarrow{d^{n+1}} A^{n+2} \longrightarrow \cdots$$

with $d^{n+1} \circ d^n = 0$ for all $n \in \mathbb{Z}$. The d's are called **differentials**. We say that $A^\bullet$ is **bounded below** (resp. **bounded above**, resp. **bounded**) if $A^n = 0$ for $n \ll 0$ (resp. for $n \gg 0$, resp. for $n \ll 0$ and $n \gg 0$). We set

$$\begin{aligned} Z^n(A^\bullet) &= \ker(A^n \xrightarrow{d^n} A^{n+1}), \\ B^n(A^\bullet) &= \operatorname{im}(A^{n-1} \xrightarrow{d^{n-1}} A^n). \end{aligned}$$

The elements of $Z^n(A^\bullet)$ and $B^n(A^\bullet)$ are called the **n-cocycles** and **n-coboundaries**, respectively. As $d \circ d = 0$, we have $B^n(A^\bullet) \subseteq Z^n(A^\bullet)$. The factor group

$$H^n(A^\bullet) = Z^n(A^\bullet)/B^n(A^\bullet)$$

is called the **n-dimensional cohomology group** of $A^\bullet$.

A homomorphism of complexes $f\colon A^\bullet \to B^\bullet$ is a sequence $f = (f^n)_{n\in\mathbb{Z}}$ of homomorphisms $f^n\colon A^n \to B^n$ with $f^{n+1} \circ d^n = d^n \circ f^n$ for all $n \in \mathbb{Z}$. A homomorphism $f\colon A^\bullet \to B^\bullet$ of complexes induces homomorphisms

$$H^n(f)\colon H^n(A^\bullet) \longrightarrow H^n(B^\bullet)$$

on the cohomology, and we call f a **quasi-isomorphism** if $H^n(f)$ is an isomorphism for all n.

A **filtration by subcomplexes** of $A^\bullet$ is a filtration $F^\bullet A^n$ of A^n for all $n \in \mathbb{Z}$ such that for each n, $F^n A^\bullet$ is a subcomplex of $A^\bullet$. We say that the filtration $F^\bullet A^\bullet$ is **biregular**, if, for each $n \in \mathbb{Z}$, the filtration $F^\bullet A^n$ is finite.

Examples: For any complex $A^\bullet$ we have the following filtrations

1. The **trivial filtration** $tr^\bullet A^\bullet$ defined by

$$tr^p A^\bullet = \begin{cases} A^\bullet, & \text{for } p \le 0, \\ 0, & \text{for } p \ge 1. \end{cases}$$

We have

$$H^q(tr^n A^\bullet) = \begin{cases} H^q(A^\bullet), & \text{for } n \le 0, \\ 0, & \text{for } n \ge 1. \end{cases}$$

2. Consider for $p \in \mathbb{Z}$ the subcomplex $\tau_{\le p}(A^\bullet)$ of $A^\bullet$ given by

$$\tau_{\le p}(A^\bullet)^n = \begin{cases} A^n, & \text{for } n \le p-1, \\ \ker(A^n \xrightarrow{d} A^{n+1}), & \text{for } n = p, \\ 0, & \text{for } n \ge p+1. \end{cases}$$

We have

$$H^q(\tau_{\leq p}(A^\bullet)) = \begin{cases} H^q(A^\bullet), & \text{for } q \leq p, \\ 0, & \text{for } q \geq p+1. \end{cases}$$

The **canonical filtration** $\tau^\bullet A^\bullet$ on $A^\bullet$ is the decreasing filtration defined by $\tau^p A^n = \tau_{\leq -p}(A^\bullet)^n$.

A biregular filtration $F^\bullet A^\bullet$ induces an E_1-spectral sequence in the following way.

(2.2.1) Proposition. *Let $F^\bullet A^\bullet$ be a biregularly filtered cochain complex. For $(p,q) \in \mathbb{Z} \times \mathbb{Z}$ and $r \in \mathbb{Z} \cup \{\infty\}$, we put*

$$\begin{aligned} Z_r^{pq} &= \ker\left(F^p A^{p+q} \to A^{p+q+1}/F^{p+r}A^{p+q+1}\right), \\ B_r^{pq} &= d(F^{p-r}A^{p+q-1}) \cap F^p A^{p+q}, \\ E_r^{pq} &= Z_r^{pq}/(B_{r-1}^{pq} + Z_{r-1}^{p+1,q-1}), \\ F^p H^{p+q}(A^\bullet) &= \operatorname{im}(H^{p+q}(F^p A^\bullet) \to H^{p+q}(A^\bullet)). \end{aligned}$$

Then the differential d of the complex $A^\bullet$ induces homomorphisms

$$d = d_r^{pq}\colon E_r^{pq} \longrightarrow E_r^{p+r,q-r+1}$$

for all $r \in \mathbb{Z}$ in a natural way. There are canonical isomorphisms

$$\alpha_r^{pq}\colon \ker(d_r^{pq})/\operatorname{im}(d_r^{p-r,q+r-1}) \xrightarrow{\sim} E_{r+1}^{pq}$$

for all $r \in \mathbb{Z}$. For fixed $(p,q) \in \mathbb{Z} \times \mathbb{Z}$, the morphisms d_r^{pq} and $d_r^{p-r,q+r-1}$ vanish for sufficiently large r and we have natural isomorphisms

$$E_r^{pq} \xrightarrow{\sim} E_\infty^{pq}, \quad r \gg 0\,.$$

Finally, there exist natural isomorphisms

$$\beta^{pq}\colon E_\infty^{pq} \xrightarrow{\sim} gr_p H^{p+q}(A^\bullet).$$

In particular, these data define a spectral sequence

$$E_1^{pq} \Rightarrow H^{p+q}(A^\bullet).$$

Remark: We have $E_r^{pq} = gr_p A^{p+q}$ for all $r \leq 0$. For $r \leq -1$ the differentials d_r^{pq} are zero and $\alpha_r^{pq} = id_{gr_p A^{p+q}}$. The sequence of isomorphisms α_r^{pq} for $r \geq 0$ starts with

$$\alpha_0^{pq}\colon H^q(E_0^{p\bullet}) = H^q(gr_p A^{p+\bullet}) \xrightarrow{\sim} E_1^{pq}.$$

Proof of (2.2.1): The fact that d induces homomorphisms d_r^{pq} and the existence of natural isomorphism α_r^{pq} can be easily verified using the definition of the objects occurring (cf. [21], I.3.1.5). As the filtration is biregular, for each fixed pair $(p,q) \in \mathbb{Z} \times \mathbb{Z}$ the morphisms d_r^{pq} and $d_r^{p-r,q+r-1}$ vanish for sufficiently large r. The natural isomorphisms $E_r^{pq} = E_\infty^{pq}$ for $r \gg 0$ and $E_\infty^{pq} \xrightarrow{\sim} gr_p H^{p+q}(A^\bullet)$ can be read off directly from the definition, using the biregularity of the filtration. □

(2.2.2) Definition. *We call the spectral sequence of* (2.2.1) *the spectral sequence associated to the biregularly filtered complex* $F^\bullet A^\bullet$.

Examples: 1. For the trivial filtration $tr^\bullet A^\bullet$, the associated spectral sequence has the following shape:

$$E_1^{pq} = \begin{cases} H^q(A^\bullet), & \text{for } p = 0, \\ 0, & \text{for } p \neq 0. \end{cases}$$

The spectral sequence degenerates at E_1, i.e. $E_1^{pq} = E_\infty^{pq}$ for all p, q.

2. For the canonical filtration $\tau^\bullet A^\bullet$, the associated spectral sequence has the following shape:

$$E_1^{pq} = \begin{cases} H^{-p}(A^\bullet), & \text{for } p + q = -p, \\ 0, & \text{for } p + q \neq -p. \end{cases}$$

The spectral sequence degenerates at E_1, i.e. $E_1^{pq} = E_\infty^{pq}$ for all p, q.

In both examples, the filtration on the limit terms has only one nontrivial graded piece, which is not the case in general. Quite often, spectral sequences arise from double complexes.

(2.2.3) Definition. *A* **double complex** $A^{\bullet\bullet}$ *is a collection of objects* $A^{pq} \in \mathcal{A}$, $p, q \in \mathbb{Z}$, *together with differentials* $d'^{pq}: A^{pq} \to A^{p+1,q}$ *and* $d''^{p,q}: A^{pq} \to A^{p,q+1}$ *such that* $d' \circ d' = 0 = d'' \circ d''$ *and* $d' \circ d'' + d'' \circ d' = 0$. *The associated* **total complex**

$$A^\bullet = tot(A^{\bullet\bullet})$$

is the single complex with $A^n = \bigoplus_{p+q=n} A^{pq}$ *whose differential* $d : A^n \to A^{n+1}$ *is given by the sum of the maps*

$$d = d' + d'' : A^{pq} \longrightarrow A^{p+1,q} \oplus A^{p,q+1}, \quad p + q = n.$$

Example: Let $(C^\bullet, d_C)$ and $(D^\bullet, d_D)$ be two complexes of abelian groups. Their **tensor product** is the double complex

$$A^{\bullet\bullet} = C^\bullet \otimes D^\bullet$$

with $A^{pq} = C^p \otimes D^q$, $p, q \in \mathbb{Z}$, and with the following differentials:

$$d'^{pq} = d_C^p \otimes id_{D^q}, \qquad d''^{pq} = (-1)^p \, id_{C^p} \otimes d_D^q.$$

As there is no danger of confusion, the associated total complex $\operatorname{tot}(C^\bullet \otimes D^\bullet)$ is usually also called the tensor product of $C^\bullet$ and $D^\bullet$.

Given a double complex $A^{\bullet\bullet}$, we have the **natural filtration** $F^p A^\bullet$ of the total complex defined by

$$F^p A^n = \bigoplus_{i \geq p} A^{i, n-i}.$$

Example: Let $D^\bullet$ be any complex of abelian groups and let $C^\bullet$ be the complex given by

$$C^i = \begin{cases} \mathbb{Z}, & \text{for } i = 0, \\ 0, & \text{for } i \neq 0. \end{cases}$$

Then $\operatorname{tot}(C^\bullet \otimes D^\bullet) = D^\bullet$ and the induced filtration on $D^\bullet$ is the trivial filtration.

Let $A^{\bullet\bullet}$ be a double complex. We will assume in the following that, for each n, there are only finitely many nonzero A^{pq} on the line $p+q = n$. Then the above filtration on $A^\bullet = \operatorname{tot}(A^{\bullet\bullet})$ is biregular and induces a spectral sequence

$$E_1^{pq} \Rightarrow H^{p+q}(A^\bullet)$$

converging to the cohomology of $A^\bullet$. The initial terms E_1^{pq} are obtained by taking cohomology in direction q:

$$E_1^{pq} = H^q(A^{p\bullet}, d'').$$

The E_1 terms give a complex

$$H^q(A^{\bullet\bullet}) : \cdots \xrightarrow{d'} H^q(A^{p-1,\bullet}) \xrightarrow{d'} H^q(A^{p,\bullet}) \xrightarrow{d'} H^q(A^{p+1,\bullet}) \xrightarrow{d'} \cdots,$$

whose cohomology yields the E_2-terms:

$$E_2^{pq} = H^p(H^q(A^{\bullet\bullet})).$$

For reasons that will become apparent later, one often forgets the E_1-page and calls the spectral sequence

$$E_2^{pq} = H^p(H^q(A^{\bullet\bullet})) \Rightarrow H^{p+q}(\operatorname{tot} A^{\bullet\bullet})$$

the **spectral sequence associated to the double complex $A^{\bullet\bullet}$.**

Remark: For a double complex $A^{\bullet\bullet}$ of abelian groups, the differentials $d_2^{pq} : E_2^{pq} \to E_2^{p+2,q-1}$ may be described as follows. For each class $c \in E_2^{pq}$, there are elements $x \in A^{pq}$ and $y \in A^{p+1,q-1}$ with the following properties:

1) $d''x = 0, d''y = -d'x$,
2) c is represented by x and $d_2^{pq}c$ by $d'y$.

(2.2.4) Lemma. *Let $A^{\bullet\bullet}$ be a first quadrant double complex, i.e. $A^{pq} = 0$ if $p < 0$ or $q < 0$. Assume that for each $q \geq 0$ the horizontal complex $A^{0q} \to A^{1q} \to A^{2q} \to \cdots$ is exact (i.e. trivial cohomology in dimension ≥ 1). Then $E^n = H^n(B^\bullet)$, where $B^\bullet$ is the complex $\ker(A^{0\bullet} \to A^{1\bullet})$.*

Proof: Setting $'A^{pq} = A^{qp}$, we obtain a double complex $'A^{\bullet\bullet}$ with the same total complex $A^\bullet$ as $A^{\bullet\bullet}$, and hence a new spectral sequence $'E_2^{pq} \Rightarrow E^n$ with the same limit terms E^n (with different filtrations, however). Now the vertical sequences $'A^{p0} \to 'A^{p1} \to \cdots$ are exact, so that $'E_2^{pq} = 0$, i.e. $'E_\infty^{pq} = 0$ for $q > 0$ and all $p \geq 0$. This means that $E^n = 'F^n E^n \cong 'E_\infty^{n,0} = 'E_2^{n,0} = H^q(H^p(A^{\bullet\bullet}))$ for $p = 0, q = n$. □

Our first application of spectral sequences is the following

(2.2.5) Proposition. *Let R be a commutative ring with unit. Let $f: D^\bullet \to D'^\bullet$ be a quasi-isomorphism of complexes of R-modules and let $C^\bullet$ be a complex consisting of flat R-modules. Assume that one of the following conditions is fulfilled:*

(i) *$C^\bullet$ is bounded above,*
(ii) *$D^\bullet$ and $D'^\bullet$ are bounded below,*
(iii) *R is a Dedekind domain.*

Then the induced homomorphism

$$id_{C^\bullet} \otimes f : \ tot(C^\bullet \otimes_R D^\bullet) \longrightarrow tot(C^\bullet \otimes_R D'^\bullet)$$

is a quasi-isomorphism.

Proof: We start with the special case $D'^\bullet = 0$, i.e. $D^\bullet$ is exact, and we have to show that $\operatorname{tot}(C^\bullet \otimes_R D^\bullet)$ is exact.

Let us first assume that $D^\bullet$ is bounded above. Then, if (i) or (ii) holds, the natural filtration on $\operatorname{tot}(C^\bullet \otimes_R D^\bullet)$ is biregular, and for the associated spectral sequence we have

$$E_1^{pq} = H^q(C^p \otimes D^\bullet) = 0$$

as C^p is flat for all p. Hence all limit terms $H^{p+q}(\mathrm{tot}\,(C^\bullet \otimes_R D^\bullet))$ vanish.

If $D^\bullet$ is not bounded above, we write $D^\bullet = \varinjlim_n \tau_{\le n}(D^\bullet)$ and obtain

$$H^k(\mathrm{tot}\,(C^\bullet \otimes_R D^\bullet)) = \varinjlim_n H^k(\mathrm{tot}\,(C^\bullet \otimes_R \tau_{\le n}(D^\bullet))) = 0$$

for all k.

Keeping the assumption $D'^\bullet = 0$, we now assume that R is a Dedekind domain. Then each submodule of a flat R-module is again flat. Hence the complexes $\tau_{\le n}(C^\bullet)$ consist of flat R-modules for all n, and we obtain

$$H^k(\mathrm{tot}\,(C^\bullet \otimes_R D^\bullet)) = \varinjlim_n H^k(\mathrm{tot}\,(\tau_{\le n}(C^\bullet) \otimes_R D^\bullet)) = 0$$

for all k. This settles the case $D'^\bullet = 0$.

In the general case we consider the mapping cone $C(f)^\bullet$ of f. It is defined by

$$C(f)^n = D^{n+1} \oplus D'^n$$

with differential $d((a,b)) = (-d(a), d'(b) + f(a))$ for $a \in D^{n+1}$, $b \in D'^n$. The natural long exact sequence

$$\cdots \longrightarrow H^n(D^\bullet) \longrightarrow H^n(D'^\bullet) \longrightarrow H^n(C(f)^\bullet) \longrightarrow \cdots$$

shows that $C(f)^\bullet$ is exact, as f is a quasi-isomorphism. Moreover, the construction of the mapping cone commutes with tensor products, i.e.

$$C(id_{C^\bullet} \otimes f)^\bullet = \mathrm{tot}\,(C^\bullet \otimes C(f)^\bullet).$$

By the first part of the proof, $\mathrm{tot}\,(C^\bullet \otimes C(f)^\bullet)$ is exact if one of the conditions (i)–(iii) is satisfied. Hence $C(id_{C^\bullet} \otimes f)^\bullet$ is exact and therefore $id_{C^\bullet} \otimes f$ is a quasi-isomorphism. □

Exercise 1. Calculate the spectral sequence associated to the *stupid* filtration

$$\sigma_{\ge p}(A^\bullet)^n = \begin{cases} A^n, & \text{for } n \ge p, \\ 0, & \text{for } n \le p-1. \end{cases}$$

Does it degenerate?

Exercise 2. Let p be a prime number and $R = \mathbb{Z}/p^2\mathbb{Z}$. Consider the exact and flat complex of R-modules

$$C^\bullet = \quad \cdots \xrightarrow{p} \mathbb{Z}/p^2\mathbb{Z} \xrightarrow{p} \mathbb{Z}/p^2\mathbb{Z} \xrightarrow{p} \cdots.$$

Show that $H^i(\mathrm{tot}\,(C^\bullet \otimes_R C^\bullet)) \neq 0$ for all i.

§3. Degeneration of Spectral Sequences

In this section we investigate the degeneration of the spectral sequence attached to a biregularly filtered cochain complex. The first and easiest case is degeneration at E_1.

(2.3.1) Theorem. *For the spectral sequence*

$$E_1^{pq} \Rightarrow H^{p+q}(A^\bullet)$$

associated to a biregularly filtered complex $F^\bullet A^\bullet$ the following assertions are equivalent:

(i) *The spectral sequence degenerates at E_1.*
(ii) *For all n, p we have $F^p A^n \cap d(A^{n-1}) = d(F^p A^{n-1})$.*
(iii) *For all n, p the natural map $H^n(F^p A^\bullet) \to H^n(A^\bullet)$ is injective.*

If, moreover, the maps in (iii) *are split-injections, we obtain a (non-canonical) splitting*

$$H^n(A^\bullet) \cong \bigoplus_{p+q=n} E_1^{pq}.$$

Proof: Without loss of generality, we may work in the category of modules over a ring. Assume that (i) holds. For sufficiently large p, we have $F^p A^{n-1} = 0$, hence assertion (ii) holds for $p \gg 0$. We fix $n \in \mathbb{Z}$ and proceed by decreasing induction on p. Let $x \in F^p A^n \cap d(A^{n-1})$. As the filtration is biregular, $x \in d(F^m A^{n-1})$ for some m, which we choose as large as possible ($m = \infty$ allowed). We assume that $m = p - r$ for some $r \geq 1$, and show that this yields a contradiction. Let $y \in F^{p-r} A^{n-1}$ be a pre-image of x. By construction, $y \in Z_r^{p-r,n-1-p+r}$. By assumption, the differential d_r: $E_r^{p-r,n-1-p+r} \to E_r^{p,n-p}$ is the zero map. Hence $x = dy \in B_{r-1}^{p,n-p} + Z_{r-1}^{p+1,n-p-1}$, i.e. we may write x in the form

$$x = dy' + x', \qquad y' \in F^{p-r+1} A^{n-1},\ x' \in F^{p+1} A^n.$$

As x and dy' are coboundaries, the same holds for x'. By our inductive assumption, we have $x' \in d(F^{p+1} A^{n-1})$, which implies $x \in F^p A^n \cap d(F^{p-r+1} A^{n-1})$, contradicting the maximality of $m = p - r$. Hence $m \geq p$, showing that $x \in d(F^p A^{n-1})$. This proves (i)$\Rightarrow$(ii). The implication (ii)$\Rightarrow$(i), as well as the equivalence (ii)$\Leftrightarrow$(iii) are elementary.

Finally, assume that (i)–(iii) hold. By definition, $F^p H^n(A^\bullet)$ is the image of the natural map $H^n(F^p A^\bullet) \to H^n(A^\bullet)$, which is a split-injection. As the filtration on $H^n(A^\bullet)$ is finite, we recursively obtain a splitting

$$H^n(A^\bullet) \cong \bigoplus_p F^p H^n(A^\bullet) / F^{p+1} H^n(A^\bullet)$$

into a finite direct sum. By (i), we have

$$E_1^{p,n-p} = E_\infty^{p,n-p} = \mathrm{gr}_p H^n(A^\bullet) = F^p H^n(A^\bullet)/F^{p+1}H^n(A^\bullet),$$

showing the last assertion of the proposition. □

By a formal reindexing procedure, we can displace a spectral sequence in the following sense: Assume we are given an E_m-spectral sequence $E_m^{pq} \Rightarrow E^n$. Putting

$$\tilde{E}^n = E^n,\ F^p\tilde{E}^n = F^{p+n}E^n \text{ and } \tilde{E}_r^{pq} = E_{r+1}^{2p+q,-p} \text{ for } r \geq m-1,$$

we obtain an E_{m-1}-spectral sequence converging to the same limit terms, but with a shifted filtration, which we call the **displaced spectral sequence**. It is a remarkable fact that, if the spectral sequence E arises from a biregular filtered cochain complex, then the displaced spectral sequence $\tilde{E}$ arises from another filtration on the same complex, the displaced filtration. This will be useful in showing that a spectral sequence degenerates at E_2, just by showing that the displaced spectral sequence $\tilde{E}$ satisfies the conditions of (2.3.1).

(2.3.2) Definition. *Let $F^\bullet A^\bullet$ be a biregular filtered cochain complex. The filtration*

$$\mathrm{Dis}(F)^p A^n = Z_1^{p+n,-p} = \{a \in F^{p+n}A^n \mid da \in F^{p+n+1}A^{n+1}\}$$

is called the **displaced filtration**. *We denote the complex $A^\bullet$, together with the filtration* Dis(F), *by* Dis($A^\bullet$).

One easily verifies that Dis($A^\bullet$) is a filtered complex:

$$d(Z_1^{p+n,-p}) \subseteq F^{p+1+n}A^{n+1} \cap \ker(d) \subseteq Z_\infty^{p+n+1,-p} \subseteq Z_1^{p+n+1,-p},$$

and the filtration Dis(F) is obviously biregular.

Example: Displacing the trivial filtration *tr* on $A^\bullet$, we obtain the canonical filtration τ:

$$\begin{aligned}\mathrm{Dis}(tr)^p A^n &= Z_1^{p+n,-p} = \{a \in tr^{p+n}A^n \mid da \in tr^{p+n+1}A^{n+1}\} \\ &= \begin{cases} A^n, & \text{for } n \leq -p-1, \\ \ker(A^n \to A^{n+1}), & \text{for } n = -p, \\ 0, & \text{for } n \geq -p+1, \end{cases} \\ &= \tau^p A^n.\end{aligned}$$

(2.3.3) Proposition. *For all $r \geq 1$, there are natural isomorphism*

$$E_r^{pq}(\mathrm{Dis}(A^\bullet)) \xrightarrow{\sim} E_{r+1}^{2p+q,-p}(A^\bullet)$$

commuting with the corresponding differentials. The displaced spectral sequence is the spectral sequence associated to the displaced filtration.

Proof: For $r \geq 0$ we have

$$\begin{aligned} {}_{\mathrm{Dis}F}Z_r^{pq} &= \left\{x \in \mathrm{Dis}(F)^p A^{p+q} \mid dx \in \mathrm{Dis}(F)^{p+r} A^{p+q+1}\right\} \\ &= \{x \in F^{2p+q} A^{p+q} \mid dx \in F^{2p+q+r+1} A^{p+q+1}\} \\ &= {}_F Z_{r+1}^{2p+q,-p}. \end{aligned}$$

Analogously,

$$\begin{aligned} {}_{\mathrm{Dis}F}B_r^{pq} &= d(\mathrm{Dis}(F)^{p-r} A^{p+q-1}) \cap \mathrm{Dis}(F)^p A^{p+q} \\ &= d(F^{2p+q-r-1} A^{p+q-1}) \cap F^{2p+q} A^{p+q} \\ &= {}_F B_{r+1}^{2p+q,-p}. \end{aligned}$$

By definition, $E_r^{pq} = Z_r^{pq}/(B_{r-1}^{pq} + Z_{r-1}^{p+1,q-1})$. Hence the natural identifications above induce isomorphisms $E_r^{pq}(\mathrm{Dis}(A^\bullet)) \xrightarrow{\sim} E_{r+1}^{2p+q,-p}(A^\bullet)$ for $r \geq 1$. Therefore, the E_r^{pq}-terms of the displaced spectral sequence and of the spectral sequence associated to the displaced filtration are canonically isomorphic. The same holds for the limit terms, which can easily be seen from their definitions. □

As an application, we obtain the following degeneration result. It should not be mistaken for the well-known Künneth-formula, which arises from another filtration on the tensor product (see the exercise below).

(2.3.4) Theorem. *Let R be a Dedekind domain and let $C^\bullet$ and $D^\bullet$ be complexes of R-modules such that the natural filtration on their tensor product is biregular.*[*] *If $C^\bullet$ consists of flat (i.e. torsion-free) R-modules, then the spectral sequence of the double complex $A^{\bullet\bullet} = C^\bullet \otimes_R D^\bullet$ degenerates at E_2. Furthermore, we have a non-canonical splitting*

$$H^n(\mathrm{tot}(C^\bullet \otimes_R D^\bullet)) \cong \bigoplus_{p+q=n} E_2^{pq}.$$

---

[*] e.g. both complexes are bounded above, or both complexes are bounded below, or one of the complexes is bounded.

Proof: Let $A^\bullet = \operatorname{tot}(C^\bullet \otimes_R D^\bullet)$, and let $F^pA^n = \bigoplus_{i\geq p} C^i \otimes_R D^{n-i}$ be the natural filtration. We want to show that the spectral sequence of the double complex degenerates at E_2. By (2.3.3), it suffices to show that the spectral sequence associated to the displaced filtration degenerates at E_1. Thus, by (2.3.1), we have to show that the natural maps

$$H^n(\operatorname{Dis}(F)^p A^\bullet) \longrightarrow H^n(A^\bullet)$$

are split-injections for all p and n. Using the flatness of $C^\bullet$, we obtain

$$\begin{aligned}
\operatorname{Dis}(F)^p A^n &= \ker\left(F^{p+n}A^n \to F^{p+n}A^{n+1}/F^{p+n+1}A^{n+1}\right),\\
&= \left(\textstyle\bigoplus_{i>p+n} C^i \otimes_R D^{n-i}\right) \oplus C^{p+n} \otimes_R \ker(d_D^{-p})\\
&= \operatorname{tot}\left(C^\bullet \otimes_R \tau_{\leq -p}(D^\bullet)\right)^n.
\end{aligned}$$

As the filtration on $H^n(A^\bullet)$ is finite, it therefore remains to show that for all $n, m \in \mathbb{Z}$ the natural map

$$H^n\left(\operatorname{tot}(C^\bullet \otimes_R \tau_{\leq m}(D^\bullet))\right) \longrightarrow H^n\left(\operatorname{tot}(C^\bullet \otimes_R \tau_{\leq m+1}(D^\bullet))\right)$$

is a split-injection. Let $X^\bullet$ be a complex consisting of projective R-modules together with a quasi-isomorphism $X^\bullet \to \tau_{\leq m+1}(D^\bullet)$. Using the flatness of $C^\bullet$, we obtain a commutative diagram

$$\begin{array}{ccc}
\operatorname{tot}(C^\bullet \otimes_R \tau_{\leq m}(X^\bullet)) & \xrightarrow{\text{quasi-iso}} & \operatorname{tot}(C^\bullet \otimes_R \tau_{\leq m}(D^\bullet))\\
\downarrow & & \downarrow\\
\operatorname{tot}(C^\bullet \otimes_R X^\bullet) & \xrightarrow{\text{quasi-iso}} & \operatorname{tot}(C^\bullet \otimes_R \tau_{\leq m+1}(D^\bullet)).
\end{array}$$

By (2.2.5)(iii) the horizontal maps are quasi-isomorphisms. Since $X^\bullet$ is a complex consisting of projective R-modules, the inclusion of complexes

$$\tau_{\leq m}(X^\bullet) \hookrightarrow X^\bullet$$

has a section for each $m \in \mathbb{Z}$. Indeed, by our assumption on R, $d(X^m) \subseteq X^{m+1}$ is a projective R-module, and the short exact sequence

$$0 \longrightarrow \ker(d^m) \longrightarrow X^m \longrightarrow d(X^m) \longrightarrow 0$$

splits. Using any section $s\colon X^m \to \ker(d^m)$ in dimension m, and the obvious maps in the other dimensions, we obtain a splitting of $\tau_{\leq m}(X^\bullet) \hookrightarrow X^\bullet$.

Therefore the left vertical complex homomorphism in the diagram above has a section, which finishes the proof. □

The technique of displacing, as well as the results (2.3.1) and (2.3.3), are due to *P. Deligne*, cf. [34]. The statement of (2.3.4) is implicitly contained in the article [98] by *U. Jannsen*, and we adopted his idea of proof.

Exercise (Künneth-formula): Let R be a Dedekind domain and let $C^\bullet$ and $D^\bullet$ be complexes of R-modules. Assume that $C^\bullet$ consists of flat R-modules. Consider the biregular filtration on $A^\bullet = \mathrm{tot}(C^\bullet \otimes_R D^\bullet)$ defined by

$$F^p(A^n) = \begin{cases} \bigoplus_{i+j=n} C^i \otimes_R D^j, & \text{for } p \le -1, \\ \bigoplus_{i+j=n} Z^i(C^\bullet) \otimes_R D^j, & \text{for } p = 0, \\ 0, & \text{for } p \ge 1, \end{cases}$$

and show the following assertions regarding the associated spectral sequence:

(i) $E_2^{p,q} = \bigoplus_{i+j=q} \mathrm{Tor}^R_{-p}(H^i(C^\bullet), H^j(D^\bullet))$; in particular $E_2^{pq} = 0$ for $p \neq 0, -1$.

(ii) For each n we obtain a short exact sequence

$$0 \to \bigoplus_{i+j=n} H^i(C^\bullet) \otimes_R H^j(D^\bullet) \to H^n(A^\bullet) \to \bigoplus_{i+j=n+1} \mathrm{Tor}^R_1(H^i(C^\bullet), H^j(D^\bullet)) \to 0.$$

Moreover, these sequences split (non-canonically).

§4. The Hochschild-Serre Spectral Sequence

Relevant for the cohomology of profinite groups is the following

(2.4.1) Theorem. *Let G be a profinite group, H a closed normal subgroup and A a G-module. Then there exists a first quadrant spectral sequence*

$$E_2^{pq} = H^p(G/H, H^q(H, A)) \Rightarrow H^{p+q}(G, A).$$

It is called the **Hochschild-Serre spectral sequence.**

Proof: To the standard resolution $0 \to A \to X^\bullet$ of the G-module A, we apply the functor $H^0(H, -)$, and get the complex

$$H^0(H, X^0) \longrightarrow H^0(H, X^1) \longrightarrow H^0(H, X^2) \longrightarrow \cdots$$

of G/H-modules. For each $H^0(H, X^q)$, we consider the cochain complex

$$H^0(H, X^q)^{G/H} \longrightarrow C^\bullet(G/H, H^0(H, X^q))$$

and we put

$$C^{pq} = C^p(G/H, H^0(H, X^q)) = X^p(G/H, X^q(G, A)^H)^{G/H}, \quad p, q \ge 0.$$

We make $C^{\bullet\bullet}$ into a (anti-commutative) double complex by using the following differentials: We let

$$d'_{pq} : C^{pq} \longrightarrow C^{p+1,q}$$

be the differential of the complex $X^\bullet(G/H, X^q(G,A)^H)^{G/H}$ at the place p. Further, we define

$$d''_{pq} : C^{pq} \longrightarrow C^{p,q+1}$$

as $(-1)^p$ times the differential of the complex $X^p(G/H, X^\bullet(G,A)^H)^{G/H}$ at the place q. Then $(C^{\bullet\bullet}, d', d'')$ is a double complex and we define the Hochschild-Serre spectral sequence as the associated spectral sequence

$$E_2^{pq} \Rightarrow E^n .$$

We compute the terms E_2^{pq} and E^n. By definition, $E_2^{pq} = H^p(H^q(C^{\bullet\bullet}))$. We have $H^q(H^0(H, X^\bullet)) = H^q(H,A)$ (see p.34). The functor $C^p(G/H, -)$ is exact (I §3, ex.1). Therefore

$$\begin{aligned} H^q(C^{p\bullet}) = H^q(C^p(G/H, H^0(H, X^\bullet))) &= C^p(G/H, H^q(H^0(H, X^\bullet))) \\ &= C^p(G/H, H^q(H,A)), \end{aligned}$$

hence

$$E_2^{pq} = H^p(C^\bullet(G/H, H^q(H,A))) = H^p(G/H, H^q(H,A)).$$

As for the limit terms, we note that for every $q \geq 0$ the complexes $C^{\bullet q} = C^\bullet(G/H, H^0(H, X^q))$ are exact. In fact, every X^q is an induced G-module, hence $H^0(H, X^q)$ is an induced, and thus acyclic, G/H-module by (1.3.6) and (1.3.7), i.e. $H^p(C^{\bullet q}) = H^p(G/H, H^0(H, X^q)) = 0$ for $p > 0$. By lemma (2.2.4), we obtain $E^n = H^n(B^\bullet)$, where $B^\bullet$ is the complex

$$B^\bullet = \ker(C^0(G/H, (X^\bullet)^H) \to C^1(G/H, (X^\bullet)^H)) = ((X^\bullet)^H)^{G/H} = (X^\bullet)^G.$$

Therefore

$$E^n = H^n((X^\bullet)^G) = H^n(G,A). \qquad \square$$

From (2.1.4) follows the

(2.4.2) Corollary. *If* $H^q(H,A) = 0$ *for* $q > 0$, *then*

$$H^n(G/H, A^H) \cong H^n(G,A).$$

Another consequence is the five term exact sequence

$$\begin{aligned} 0 \longrightarrow H^1(G/H, A^H) \longrightarrow H^1(G,A) \longrightarrow H^1(H,A)^{G/H} \\ \longrightarrow H^2(G/H, A^H) \longrightarrow H^2(G,A), \end{aligned}$$

which we proved in I §6 in an elementary way, but with some difficulty. It still requires, however, careful checking to show that the maps are the inflation, restriction and transgression respectively.

For *inf* and *res*, this identification is given in the available literature (e.g. [128]). For the transgression, we give the proof here.

(2.4.3) Theorem. *The differential*

$$d_2^{0,1} : H^1(H, A)^{G/H} \longrightarrow H^2(G/H, A^H)$$

is the transgression tg *as defined in* (1.6.6).

Proof (TH. MOSER, J. STIX): In order to calculate $H^1(H, A)$ we use the acyclic resolution $A \to X^\bullet(G, A)$ of the G-module A. Thus an element $z \in H^1(H, A)^{G/H}$ is represented by an H-invariant 1-cocycle $x : G \times G \to A$. The invariance of z under G/H implies that for $\rho, \sigma \in G$ the cocycles ρx and σx differ by a 1-coboundary, i.e. there is a map

$$b : G \times G \longrightarrow X^0(G, A)^H, \quad (\sigma, \rho) \mapsto b_{\sigma,\rho} ,$$

such that

$$d(b_{\sigma,\rho}) = \rho x - \sigma x ,$$

where d denotes the differential of $X^\bullet(G, A)$. We obtain

$$(*) \qquad b_{\sigma,\rho}(\tau_1) - b_{\sigma,\rho}(\tau_0) = (\rho x)_{\tau_0,\tau_1} - (\sigma x)_{\tau_0,\tau_1}$$

for all $\sigma, \rho, \tau_0, \tau_1 \in G$. Therefore we may assume that b is G-invariant and $b_{1,1} = 0$. Furthermore, since $\tau x = x$ for $\tau \in H$, we may also assume that b factors through $G/H \times G/H$. Then, for all $\sigma, \rho, \gamma \in G/H$,

$$b_{\rho,\gamma}(\tau) - b_{\sigma,\gamma}(\tau) + b_{\sigma,\rho}(\tau)$$

is an element of the module A^H independent of $\tau \in G$, and so the 2-cocycle ∂b: $(G/H)^3 \to X^0(G, A)^H$, given by

$$\partial b_{\sigma,\rho,\gamma} = b_{\rho,\gamma} - b_{\sigma,\gamma} + b_{\sigma,\rho} ,$$

is constant with value in A^H. By the remark on page 105 it represents the image of z under $d_2^{0,1}$. The associated inhomogeneous 2-cocycle, which also represents $d_2^{0,1}(z)$, is

$$a : G/H \times G/H \longrightarrow A^H, \quad (\sigma, \rho) \mapsto a_{\sigma,\rho},$$

where

$$(**) \qquad a_{\sigma,\rho} = \partial b_{1,\sigma,\sigma\rho}(\zeta) = b_{\sigma,\sigma\rho}(\zeta) - b_{1,\sigma\rho}(\zeta) + b_{1,\sigma}(\zeta)$$

with $\zeta \in G$ arbitrary.

We now represent $tg(z)$ as follows. We restrict x to $H \times H$, and pass to the associated inhomogeneous 1-cocycle $x_0 : H \to A$, $x_0(\tau) = x_{1,\tau}$, which also represents z. Consider then the function

$$y : G \longrightarrow A, \quad \sigma \mapsto y_\sigma = x_{1,\sigma} + b_{1,\sigma}(\sigma).$$

We will show that y satisfies the properties (i), (ii), (iii) of the proof of proposition (1.6.6),

(i) $y|_H = x_0$,

(ii) $y_{\sigma\tau} = y_\sigma + \sigma y_\tau$ for $\sigma \in G$, $\tau \in H$,

(iii) $y_{\tau\sigma} = y_\tau + \tau y_\sigma$ for $\sigma \in G$, $\tau \in H$,

hence $tg(z) = [\partial y]$. Since $b_{1,\tau} = b_{1,1} = 0$ for $\tau \in H$, property (i) follows. Let $\sigma, \rho \in G$. Then, by $(*)$ and $(**)$ and since x is a cocycle, we obtain

$$\begin{aligned}(\partial y)_{\sigma,\rho} &= (dx)_{1,\sigma,\sigma\rho} + (\sigma x)_{\sigma,\sigma\rho} - x_{\sigma,\sigma\rho} \\ &\quad + b_{1,\sigma}(\sigma) - b_{1,\sigma}(\sigma\rho) + (\partial b)_{1,\sigma,\sigma\rho}(\sigma\rho) \\ &= (\partial b)_{1,\sigma,\sigma\rho}(\sigma\rho) \;=\; a_{\sigma,\rho}.\end{aligned}$$

If $\sigma \in H$ or $\rho \in H$, then the expression above is zero and so (ii) and (iii) follow. Now, for arbitrary $\sigma, \rho \in G$, the equality above shows that $tg(z) = [\partial y] = [a] = d_2^{01}$. This proves the theorem. □

A subtle and useful relation of the Hochschild-Serre spectral sequence to the *cup-product* is obtained as follows. Let G be a profinite group, H an open normal subgroup of G and H' the closure of the commutator subgroup of H. Let A be a G-module on which H acts trivially. We then have, for $p > 0$, two canonical homomorphisms

$$(*) \qquad d_2, u\cup : H^{p-1}(G/H, H^1(H, A)) \longrightarrow H^{p+1}(G/H, A)$$

which are defined as follows. The first map d_2 is the differential $d_2^{p-1,1}$ of the Hochschild-Serre spectral sequence

$$E_2^{pq} = H^p(G/H, H^q(H, A)) \Rightarrow H^{p+q}(G, A).$$

On the other hand, the group extension

$$1 \longrightarrow H^{ab} \longrightarrow G/H' \longrightarrow G/H \longrightarrow 1$$

defines a cohomology class $u \in H^2(G/H, H^{ab})$. Using the equality $H^1(H, A) = \mathrm{Hom}(H^{ab}, A)$, we obtain a canonical pairing

$$H^{ab} \times H^1(H, A) \longrightarrow A\,,$$

which induces a cup-product

$$H^2(G/H, H^{ab}) \times H^{p-1}(G/H, H^1(H, A)) \xrightarrow{\ \cup\ } H^{p+1}(G/H, A).$$

The second map $u\cup$ is given by $x \mapsto u \cup x$.

(2.4.4) Theorem. *Let A be a G-module and H an open normal subgroup of G which acts trivially on A. Then, for $p > 0$, the maps*

$$d_2, u\cup : H^{p-1}(G/H, H^1(H, A)) \longrightarrow H^{p+1}(G/H, A)$$

are the same up to sign, i.e. $d_2(x) = -u \cup x$.

In particular, the transgression $tg : H^1(H, A)^{G/H} \to H^2(G/H, A)$ is given by $tg(x) = -u \cup x$.

Remark: The statement of (2.4.4) remains true for an arbitrary closed normal subgroup H of G, but then one has to use the continuous cohomology class $u \in H^2_{cts}(G/H, H^{ab})$ representing the group extension $1 \to H^{ab} \to G/H' \to G/H \to 1$ (cf. (2.7.7)).

Proof: Suppose first that G is a finite group. Then the projection $H \to H^{ab}$ may be regarded as a G/H-invariant 1-cocycle of H, i.e. as an element $\varepsilon \in H^0(G/H, H^1(H, H^{ab}))$. From the spectral sequence

$$E_2^{pq} = H^p(G/H, H^q(H, H^{ab})) \Rightarrow H^{p+q}(G, H^{ab}),$$

we obtain the differential

$$d_2^{0,1} : H^1(H, H^{ab})^{G/H} \longrightarrow H^2(G/H, H^{ab}),$$

which by (2.4.3) is the transgression tg as defined in (1.6.6). We prove $tg(\varepsilon) = -u$ (in the additive notation of $H^2(G/H, H^{ab})$).

Let $s : G/H \to G$ be a section of the projection $G \to G/H$, $\sigma \mapsto \bar{\sigma}$. Define the 1-cochain $y : G \to H^{ab}$ by

$$y(\sigma) = \sigma(s\bar{\sigma})^{-1} \mod H'.$$

Then $y|_H = \varepsilon$ and

$$\begin{aligned}(\partial y)(\sigma_1, \sigma_2) &\equiv y(\sigma_1\sigma_2)^{-1\sigma_1}y(\sigma_2)y(\sigma_1)\\ &\equiv s(\overline{\sigma_1\sigma_2})\sigma_2^{-1}\sigma_1^{-1}\sigma_1\sigma_2(s\overline{\sigma_2})^{-1}\sigma_1^{-1}\sigma_1(s\overline{\sigma_1})^{-1}\\ &\equiv [(s\overline{\sigma_1})(s\overline{\sigma_2})s(\overline{\sigma_1\sigma_2})^{-1}]^{-1} \mod H',\end{aligned}$$

showing that ∂y depends only on the classes $\overline{\sigma_1}, \overline{\sigma_2} \in G/H$, and hence $tg(\varepsilon) = [\partial y]$ by definition of tg. But the function in square brackets is a 2-cocycle which represents the class u, hence $tg(\varepsilon) = -u$.

The differential

$$d_2^{p,1} : H^p(G/H, H^1(H, A)) \longrightarrow H^{p+2}(G/H, A^H)$$

is obtained from the part

$$\begin{array}{ccc} C^p(G/H, X^1(G,A)^H) & \xrightarrow{\partial'} & C^{p+1}(G/H, X^1(G,A)^H) \\ & & \uparrow \partial'' \\ & & C^{p+1}(G/H, X^0(G,A)^H) \xrightarrow{\partial'} C^{p+2}(G/H, X^0(G,A)^H) \end{array}$$

of the double complex $C^{pq}(A) = C^p(G/H, X^q(G,A)^H)$ as follows. Let $z \in H^p(G/H, H^1(H, A))$. z is given by an element $\alpha \in C^{p,1}(A)$ such that $\partial''\alpha = 0$ and that the induced function $\bar{\alpha} : (G/H)^{p+1} \to H^1(X^\bullet(G,A)^H) = H^1(H, A)$ is a p-cocycle representing z. This means that there exists a $\beta \in C^{p+1,0}$ such that $\partial'\alpha = \partial''\beta$. The image $d_2^{p,1}(z)$ is then represented by the cocycle $\partial'\beta \in C^{p+2,0}$.

This process may also be interpreted as follows. From the complex $0 \to A^H \to X^0(G,A)^H \to X^1(G,A)^H \to X^2(G,A)^H$, we obtain the exact sequence of G/H-modules

$$0 \longrightarrow A^H \longrightarrow X^0(G,A)^H \longrightarrow Z \longrightarrow H^1(H,A) \longrightarrow 0 \tag{1}$$

with $Z = Z^1(X^\bullet(G,A)^H)$. Splitting it up into two short exact sequences

$$0 \longrightarrow I(A) \longrightarrow Z \longrightarrow H^1(H,A) \longrightarrow 0, \tag{2}$$

$$0 \longrightarrow A^H \longrightarrow X^0(G,A)^H \longrightarrow I(A) \longrightarrow 0, \tag{3}$$

we obtain two δ-homomorphisms

$$H^p(G/H, H^1(H,A)) \xrightarrow{\delta} H^{p+1}(G/H, I(A)) \xrightarrow{\delta} H^{p+2}(G/H, A^H),$$

with $d_2^{p,1} = \delta \circ \delta$. In fact, the element α above is a lifting $\alpha : (G/H)^{p+1} \to Z^1(G,A)^H$ of the p-cocycle $\bar{\alpha}$ representing z, hence δz is represented by the $(p+1)$-cocycle $\partial'\alpha$. Since $\partial'\alpha = \partial''\beta$, $\beta : (G/H)^{p+2} \to X^0(G,A)^H$ is a cochain which lifts $\partial'\alpha$, hence $\partial'\beta$ represents $\delta([\partial'\alpha]) = \delta\delta(z)$, showing that $\delta\delta(z) = d_2^{p,1}(z)$.

Let us abbreviate the sequences (2) and (3) by

$$0 \longrightarrow S'(A) \longrightarrow S(A) \longrightarrow S''(A) \longrightarrow 0,$$
$$0 \longrightarrow T'(A) \longrightarrow T(A) \longrightarrow T''(A) \longrightarrow 0,$$

with $S' = T''$. Replacing A by the G-module H^{ab}, we obtain exact sequences of G/H-modules

$$0 \longrightarrow S'(H^{ab}) \longrightarrow S(H^{ab}) \longrightarrow S''(H^{ab}) \longrightarrow 0,$$
$$0 \longrightarrow T'(H^{ab}) \longrightarrow T(H^{ab}) \longrightarrow T''(H^{ab}) \longrightarrow 0.$$

Setting $B = \mathrm{Hom}(H^{ab}, A)$, we have the pairings

$$S(H^{ab}) \times B \longrightarrow S(A), \quad T(H^{ab}) \times B \longrightarrow T(A),$$

which induce pairings $S'(H^{ab}) \times B \to S'(A)$, $S''(H^{ab}) \times B \to S''(A)$, and similarly for T. We now assume that H acts trivially on A, i.e. $\mathrm{Hom}(H^{ab}, A) = H^1(H,A)$. We apply proposition (1.4.3) twice and obtain the commutative diagram

$$\begin{array}{ccccc}
H^0(G/H, S''(H^{ab})) & \times & H^p(G/H, B) & \xrightarrow{\cup} & H^p(G/H, S''(A)) \\
\delta\downarrow & & \| & & \downarrow\delta \\
H^1(G/H, S'(H^{ab})) & \times & H^p(G/H, B) & \xrightarrow{\cup} & H^{p+1}(G/H, S'(A)) \\
\delta\downarrow & & \| & & \downarrow\delta \\
H^2(G/H, T'(H^{ab})) & \times & H^p(G/H, B) & \xrightarrow{\cup} & H^{p+2}(G/H, T'(A)).
\end{array}$$

In the upper pairing, we have $S''(H^{ab}) = H^1(H, H^{ab})$, $S''(A) = H^1(H, A)$ and $\varepsilon \cup z = z$ (recall that $z \in H^p(G/H, \mathrm{Hom}(H^{ab}, A)) = H^p(G/H, H^1(H, A)))$. Since $\delta\delta\varepsilon = d_2^{0,1}(\varepsilon) = tg(\varepsilon) = -u$, it follows that

$$d_2^{p,1}(z) = \delta\delta(z) = \delta\delta(\varepsilon \cup z) = \delta\delta\varepsilon \cup z = -u \cup z.$$

This proves the theorem for a finite group G.

Now let G be an arbitrary profinite group. We denote by G' the commutator subgroup of G. We let W run through the open normal subgroups of G which are contained in H. Setting $\overline{G} = G/W$, $\overline{H} = H/W$, we have a commutative exact diagram

$$\begin{array}{ccccccccc} 1 & \longrightarrow & H^{ab} & \longrightarrow & G/H' & \longrightarrow & G/H & \longrightarrow & 1 \\ & & \downarrow & & \downarrow & & \| & & \\ 1 & \longrightarrow & \overline{H}^{ab} & \longrightarrow & \overline{G}/\overline{H}' & \longrightarrow & G/H & \longrightarrow & 1. \end{array}$$

The lower group extension defines an element $u_W \in H^2(G/H, \overline{H}^{ab})$ which is the image of u under $H^2(G/H, H^{ab}) \xrightarrow{\pi} H^2(G/H, \overline{H}^{ab})$ (see (3.6.1)). On the other hand, we have a commutative diagram

$$\begin{array}{ccc} H^p(G/H, H^1(\overline{H}, A)) & \xrightarrow{d_2} & H^{p+2}(G/H, A) \\ \downarrow{\scriptstyle inf} & & \| \\ H^p(G/H, H^1(H, A)) & \xrightarrow{d_2} & H^{p+2}(G/H, A). \end{array}$$

We have shown that $d_2 x_W = -u_W \cup x_W$ for $x_W \in H^p(G/H, H^1(\bar{H}, A))$. The diagram

$$\begin{array}{ccccc} H^2(G/H, \overline{H}^{ab}) & \times & H^p(G/H, H^1(\overline{H}, A)) & \xrightarrow{\cup} & H^{p+2}(G/H, A) \\ {\scriptstyle \pi}\uparrow & & \downarrow{\scriptstyle inf} & & \| \\ H^2(G/H, H^{ab}) & \times & H^p(G/H, H^1(H, A)) & \xrightarrow{\cup} & H^{p+2}(G/H, A) \end{array}$$

is commutative because of the commutativity of the cup-product with the inflation. So, $d_2 \mathit{inf}\, x_W = d_2 x_W = -u_W \cup x_W = -\pi u \cup x_W = -u \cup \mathit{inf}\, x_W$. Since $H^1(H, A) = \varinjlim_W H^1(H/W, A)$, each $x \in H^p(G/H, H^1(H, A))$ is of the form $x = \mathit{inf}\, x_W$ for some W. This completes the reduction to the case of a finite group G. □

If $1 \to H \to G \to G/H \to 1$ is a split group extension and A is a G-module on which H acts trivially, then (2.4.4) shows that the differentials $d_2^{*,1}$ are trivial. However, the Hochschild-Serre spectral sequence does, in general, *not* degenerate at E_2, even if A is a trivial G-module, see [46], example on page 83. But, we have the following

(2.4.5) Proposition. *Let $1 \to H \to G \to G/H \to 1$ be a split exact sequence of profinite groups, and let A be a discrete G-module on which H acts trivially. Then*

$$\mathit{inf} : H^*(G/H, A) \longrightarrow H^*(G, A)$$

is a monomorphism onto a direct summand, and all differentials into the horizontal edge of the Hochschild-Serre spectral sequence for A vanish, i.e. $d_r^{*,r-1} = 0$ for all $r \geq 2$.

Proof: Let $\iota\colon G/H \hookrightarrow G$ be a splitting of the projection $\pi\colon G \twoheadrightarrow G/H$, i.e. $\pi\iota = id$. Since $A^H = A$, the edge homomorphisms

$$H^*(G/H, A) = E_2^{*,0} \longrightarrow E_\infty^{*,0} \subseteq H^*(G, A)$$

are split injections, so $E_2^{*,0} = E_r^{*,0} = E_\infty^{*,0}$ for $r \geq 2$. Since $E_{r+1}^{*,0} = E_r^{*,0}/\mathrm{im} d_r^{*,r-1}$, we obtain $d_r^{*,r-1} = 0$. □

In case of a *direct* product, the Hochschild-Serre spectral sequence degenerates at E_2. We follow the proof given by *U. Jannsen*, see [98].

(2.4.6) Theorem. *Let G and H be profinite groups and let A be a discrete H-module, regarded as a $(G \times H)$-module via trivial action of the group G. Then the Hochschild-Serre spectral sequence*

$$E_2^{pq} = H^p(G, H^q(H, A)) \Rightarrow H^n(G \times H, A)$$

degenerates at E_2. Furthermore, it splits in the sense that there is a decomposition

$$H^n(G \times H, A) \cong \bigoplus_{p+q=n} H^p(G, H^q(H, A)).$$

Proof: For a trivial G-module B, we have a natural isomorphism of complexes

$$C^\bullet(G, B) \cong C^\bullet(G, \mathbb{Z}) \otimes B.$$

This is easily verified if G is finite, and the result for general G follows by a straightforward limit process. Therefore, under the assumptions of the theorem, the Hochschild-Serre spectral sequence is the spectral sequence associated to the double complex

$$C^\bullet(G, \mathbb{Z}) \otimes X^\bullet(G \times H, A)^H.$$

As the complex $C^\bullet(G, \mathbb{Z})$ consists of flat $\mathbb{Z}$-modules, the result follows from (2.3.4). □

Remark: The decomposition in the theorem is non-canonical: it cannot be made functorial in A. However, for a fixed H-module A, it can be made functorial in G. For a proof of these assertions see [98].

Exercise 1. Show that in the Hochschild-Serre spectral sequence, the edge morphisms

$$H^n(G/H, A^H) \longrightarrow H^n(G,A) \longrightarrow H^n(H,A)^{G/H}$$

are the inflation and the restriction.

Exercise 2. From the exact sequence $0 \to A \to \mathrm{Ind}_G(A) \to A_1 \to 0$, one obtains the four term exact sequence

$$0 \longrightarrow A^H \longrightarrow \mathrm{Ind}_G(A)^H \longrightarrow A_1^H \longrightarrow H^1(H,A) \longrightarrow 0$$

of G/H-modules, and hence a homomorphism

$$\delta^2 : H^p(G/H, H^1(H,A)) \longrightarrow H^{p+2}(G/H, A^H).$$

Show that δ^2 is the differential $d_2^{p,1} : E_2^{p,1} \to E_2^{p+2,0}$.

Exercise 3. Let $E(G,H,A)$ denote the Hochschild-Serre spectral sequence

$$E_2^{pq} = H^p(G/H, H^q(H,A)) \Rightarrow H^n(G,A).$$

If G' is an open subgroup of G and $H' = H \cap G'$, then we have two morphisms of spectral sequences $E(G,H,A) \underset{cor}{\overset{res}{\rightleftarrows}} E(G',H',A)$.

Exercise 4. (i) Assume that $H^q(H,A) = 0$ for $q > 1$. Then we have an exact sequence

$$\begin{array}{lllllll} 0 & \longrightarrow & H^1(G/H, H^0(H,A)) & \longrightarrow & H^1(G,A) & \longrightarrow & H^0(G/H, H^1(H,A)) \\ & \longrightarrow & H^2(G/H, H^0(H,A)) & \longrightarrow & H^2(G,A) & \longrightarrow & H^1(G/H, H^1(H,A)) \\ & \longrightarrow & H^3(G/H, H^0(H,A)) & \longrightarrow & H^3(G,A) & \longrightarrow & H^2(G/H, H^1(H,A)) \longrightarrow \cdots. \end{array}$$

(ii) Assume that $cd\, G/H \le 1$. Then we have exact sequences

$$0 \longrightarrow H^1(G/H, H^{n-1}(H,A)) \longrightarrow H^n(G,A) \longrightarrow H^n(H,A)^{G/H} \longrightarrow 0$$

for all $n \ge 1$ and all discrete G-modules A.

Exercise 5. If $A \times B \to C$ is a pairing of G-modules, then for the terms E_r^{pq} of the Hochschild-Serre spectral sequence, we have a cup-product

$$E_r^{pq}(A) \times E_r^{p'q'}(B) \xrightarrow{\cup} E_r^{p+p',q+q'}(C)$$

such that

$$d_r(\alpha \cup \beta) = (d_r\alpha) \cup \beta + (-1)^{p+q} \alpha \cup d_r(\beta).$$

Exercise 6. (Künneth-formula) Let G and H be profinite groups and let A be a discrete H-module, regarded as a $(G \times H)$-module via trivial action of the group G. Show that there exist natural short exact sequences for all n

$$0 \to \bigoplus_{i+j=n} H^i(G,\mathbb{Z}) \otimes H^j(H,A) \to H^n(G \times H, A) \to \bigoplus_{i+j=n+1} \mathrm{Tor}_1^{\mathbb{Z}}(H^i(G,\mathbb{Z}), H^j(H,A)) \to 0.$$

Moreover, these sequences split.

Hint: Consider the double complex $C^\bullet(G,\mathbb{Z}) \otimes X^\bullet(G \times H, A)^H$, which, by the proof of (2.4.6), calculates the cohomology $H^*(G \times H, A)$. Then use the result of the exercise at the end of §3.

Exercise 7. Let G and H be profinite groups and let F be a field, considered as a trivial module. Show that

$$H^n(G \times H, F) \cong \bigoplus_{i+j=n} H^i(G,F) \otimes_F H^j(H,F).$$

§5. The Tate Spectral Sequence

Another spectral sequence, due to *J. Tate*, which in a sense is dual to the Hochschild-Serre spectral sequence, is obtained as follows. Let G be a profinite group. For an abelian group M we set

$$M^* = \mathrm{Hom}(M, \mathbb{Q}/\mathbb{Z}),$$

which is equipped with the natural G-structure $g(\phi)(m) = g(\phi(g^{-1}(m)))$, if M is a G-module. Let A be a G-module. For two open subgroups $V \subseteq U$ of G, we have the maps

$$cor^* : H^n(U, A)^* \longrightarrow H^n(V, A)^*,$$

dual to the corestriction, by which the family $(H^n(U, A)^*)$ becomes a direct system of abelian groups.

(2.5.1) Definition. *Let G be a profinite group, H a closed subgroup and A a G-module. Then, for every $n \geq 0$, we set*

$$D_n(H, A) = \varinjlim_{U \supseteq H} H^n(U, A)^*,$$

where U runs through the open subgroups of G containing H.

If H is open, then $D_n(H, A) = H^n(H, A)^*$. If H is a normal closed subgroup of G, then it suffices to let U run through the normal open subgroups of G containing H. $D_n(H, A)$ is then a G/H-module. As the functors $X \mapsto X^*$ and $\varinjlim$ are exact, we see that the family $(D_n(H, -))_{n \geq 0}$ is a contravariant *) δ-functor on *Mod*(G). We set

$$D_n(A) = D_n(\{1\}, A).$$

(2.5.2) Definition. *For a G-module A and an integer $n \geq 0$, we write*

$$cd(G, A) \leq n$$

if $H^q(H, A) = 0$ for all $q > n$ and all closed subgroups H of G. (The letters "cd" stand for "cohomological dimension".)

We call the following spectral sequence the **Tate spectral sequence.**

---

*) This means that the arrows in the usual exact cohomology sequence are reversed.

(2.5.3) Theorem. *If $cd(G, A) \le n$, then for every closed normal subgroup H, there exists a first quadrant spectral sequence*

$$E_2^{pq} = H^p(G/H, D_{n-q}(H, A)) \Rightarrow H^{n-(p+q)}(G, A)^*.$$

In particular, for $H = 1$, we have a spectral sequence

$$E_2^{pq} = H^p(G, D_{n-q}(A)) \Rightarrow H^{n-(p+q)}(G, A)^*.$$

Proof: We consider the standard resolution $A \to X^\bullet$ of the G-module A and set $Z^i = \ker(X^i \to X^{i+1})$. Splitting up the complex

$$\tilde{X}^\bullet : \ 0 \to X^0 \to X^1 \to X^2 \to \cdots \to X^{n-1} \to Z^n \to 0$$

into short exact sequences and, recalling that the G-modules X^i are cohomologically trivial, we obtain for $r > 0$

$$H^r(H, Z^n) \cong H^{r+1}(H, Z^{n-1}) \cong \cdots \cong H^{r+n}(H, A).$$

In particular, $H^r(H, Z^n) = 0$ for $r > 0$, since $cd(G, A) \le n$, i.e. Z^n is a cohomologically trivial G-module.

Let U be an open normal subgroup of G. We apply to the complex $\tilde{X}^\bullet$ first the functor $H^0(U, -)$ and then the functor $\mathrm{Hom}(-, \mathbb{Q}/\mathbb{Z})$, and obtain a complex

$$0 \longrightarrow Y^0 \longrightarrow Y^1 \longrightarrow Y^2 \longrightarrow \cdots \longrightarrow Y^n \longrightarrow 0,$$

where $Y^q = H^0(U, \tilde{X}^{n-q})^*$ for $q \ge 0$, and in particular, $Y^0 = H^0(U, Z^n)^*$. This is a complex of G/U-modules, which by (1.8.2) and (1.8.5) are cohomologically trivial, since X^{n-q} and Z^n are cohomologically trivial G-modules. Since the functor $\mathrm{Hom}(-, \mathbb{Q}/\mathbb{Z})$ is exact, we obtain

$$H^q(Y^\bullet) = H^q(H^0(U, \tilde{X}^{n-\bullet})^*) = H^q(H^0(U, \tilde{X}^{n-\bullet}))^* = H^{n-q}(U, A)^*$$

for all q.

For each Y^q, we consider the cochain complex $C^\bullet(G/U, Y^q)$ and obtain a double complex $C^{pq} = C^p(G/U, Y^q)$, $p, q \ge 0$. In order to make it anti-commutative, we multiply the differentials $(p, q) \to (p, q+1)$ by $(-1)^p$, cf. the proof of (2.4.1). As described in §2, this double complex yields a spectral sequence

$$E_2^{pq} \Rightarrow E^{p+q}.$$

We compute the initial terms $E_2^{pq} = H^p(H^q(C^{\bullet\bullet}))$. The functor $C^p(G/U, -)$ is exact (I §3, ex.1), so that

$$\begin{aligned} H^q(C^{\bullet\bullet}) &= H^q(C^\bullet(G/U, Y^\bullet)) = C^\bullet(G/U, H^q(Y^\bullet)) \\ &= C^\bullet(G/U, H^{n-q}(U, A)^*), \end{aligned}$$

hence

$$E_2^{pq} = H^p(C^\bullet(G/U, H^{n-q}(U, A)^*)) = H^p(G/U, H^{n-q}(U, A)^*).$$

As for the limit terms E^{p+q}, we note that for each $q \geq 0$ the complex

$$C^0(G/U, Y^q) \longrightarrow C^1(G/U, Y^q) \longrightarrow C^2(G/U, Y^q) \longrightarrow \cdots$$

is exact; its homology groups are the groups $H^p(G/U, Y^q)$, which are zero for $p > 0$, since Y^q is cohomologically trivial.

Therefore, setting $B^\bullet = \ker(C^{0\bullet} \to C^{1\bullet})$, we have by (2.2.4)

$$E^{p+q} = H^{p+q}(B^\bullet) = H^{p+q}(H^0(G/U, Y^\bullet)).$$

Since Y^q is cohomologically trivial, we have $\hat{H}^i(G/U, Y^q) = 0$ for all i. By (1.2.6), we obtain

$$\begin{aligned} H^0(G/U, Y^q) &= \mathrm{Hom}((\tilde{X}^{n-q})^U, \mathbb{Q}/\mathbb{Z})^{G/U} = \mathrm{Hom}(((\tilde{X}^{n-q})^U)_{G/U}, \mathbb{Q}/\mathbb{Z}) \\ &\xleftarrow[\sim]{N^*_{G/U}} \mathrm{Hom}(((\tilde{X}^{n-q})^U)^{G/U}, \mathbb{Q}/\mathbb{Z}) = H^0(G, \tilde{X}^{n-q})^* \end{aligned}$$

and consequently

$$\begin{aligned} E^{p+q} &= H^{p+q}(H^0(G, \tilde{X}^{n-\bullet})^*) = H^{p+q}(H^0(G, \tilde{X}^{n-\bullet}))^* \\ &= H^{n-(p+q)}(G, A)^*. \end{aligned}$$

We thus obtain a spectral sequence

$$E_2^{pq} = H^p(G/U, H^{n-q}(U, A)^*) \Rightarrow H^{n-(p+q)}(G, A)^*.$$

If we now let U run through the open subgroups of G containing H and take direct limits, we get the spectral sequence

$$E_2^{pq} = H^p(G/H, D_{n-q}(H, A)) \Rightarrow H^{n-(p+q)}(G, A)^*. \qquad \square$$

Remark: Tate gave a proof of this spectral sequence using the cohomology groups in negative dimensions (see [230]), which we have avoided here.

The Tate spectral sequence

$$E(G, H, A): \ E_2^{pq} = H^p(G/H, D_{n-q}(H, A)) \Rightarrow H^{n-(p+q)}(G, A)^*$$

is functorial in G and H in the following sense. Let G' be an open subgroup of G and $H' = H \cap G'$. We then have two morphisms of spectral sequences

$$(*) \qquad E(G, H, A) \underset{cor\ res^*}{\overset{cor^*\ res}{\rightleftarrows}} E(G', H', A),$$

such that the maps on the initial terms and the limit terms are given as follows. The $E_2^{pq} \rightleftarrows E_2'^{pq}$ are given as the composites of

$$H^p(G/H, D_{n-q}(H, A)) \underset{cor}{\overset{res}{\rightleftarrows}} H^p(G'/H', D_{n-q}(H, A))$$

$$\underset{res^*}{\overset{cor^*}{\rightleftarrows}} H^p(G'/H', D_{n-q}(H', A)),$$

where cor^* and res^* are induced by the direct limit of the maps

$$H^{n-q}(U, A)^* \underset{res^*}{\overset{cor^*}{\rightleftarrows}} H^{n-q}(U', A)^*,$$

U running through the open normal subgroups of G containing H and $U' = U \cap G'$. The maps on the limit terms are

$$H^{n-(p+q)}(G,A)^* \underset{res^*}{\overset{cor^*}{\rightleftarrows}} H^{n-(p+q)}(G',A)^*.$$

In particular, for $H = \{1\}$, we obtain for the edge morphisms a commutative diagram

$$\begin{array}{ccccc} H^p(G,D_n(A)) & \longrightarrow & H^{n-p}(G,A)^* & \longrightarrow & H^0(G,D_{n-p}(A)) \\ {\scriptstyle res}\downarrow\uparrow{\scriptstyle cor} & & {\scriptstyle cor^*}\downarrow\uparrow{\scriptstyle res^*} & & {\scriptstyle incl}\downarrow\uparrow{\scriptstyle N_{G/G'}} \\ H^p(G',D_n(A)) & \longrightarrow & H^{n-p}(G',A)^* & \longrightarrow & H^0(G',D_{n-p}(A)). \end{array}$$

All this results from the following consideration. Assume that H is an open subgroup of G. Then the spectral sequence $E(G,H,A)$ is obtained from the double complex

$$C^{pq}(G,H,A) = C^p(G/H, \tilde{X}^{n-q}(G,A)^{H*}),$$

where $\tilde{X}^i = X^i$ for $i = 0, \cdots, n-1$ and $\tilde{X}^n = \ker(X^n \to X^{n+1})$ as in the proof of (2.5.3). We have on the one hand the homomorphisms

$$C^p(G/H, \tilde{X}^{n-q}(G,A)^{H*}) \underset{cor}{\overset{res}{\rightleftarrows}} C^p(G'/H', \tilde{X}^{n-q}(G',A)^{H'*}).$$

On the other hand, the duals of the maps

$$\tilde{X}^{n-q}(G,A) \underset{cor}{\overset{res}{\rightleftarrows}} \tilde{X}^{n-q}(G',A)$$

yield homomorphisms

$$\tilde{X}^{n-q}(G,A)^{H*} \underset{res^*}{\overset{cor^*}{\rightleftarrows}} \tilde{X}^{n-q}(G',A)^{H'*}. \qquad ^{*)}$$

After composing, we obtain two morphisms of double complexes

$$C^{pq}(G,H,A) \underset{cor\ res^*}{\overset{cor^*\ res}{\rightleftarrows}} C^{pq}(G',H',A),$$

and these induce the above morphisms $(*)$ of spectral sequences. The effects on the cohomology groups mentioned are obtained in a straightforward manner by recalling our identifications

$$E_2^{pq} \cong H^p(G/H, H^{n-q}(H,A)^*), \qquad E^{p+q} \cong H^n(G,A)^*.$$

We finish this section on spectral sequences by explicitly determining the edge morphisms $E^q \to E_2^{0,q}$ and $E_2^{p,0} \to E^p$ of the Tate spectral sequence.

---

*) The corestriction cor is defined on all of $\tilde{X}^{n-q}(G',A)$ after choosing a section $s\colon G/G' \to G$ of $G \to G/G'$. It induces a homomorphism $cor\colon \tilde{X}^{n-q}(G',A)^{H'} \to \tilde{X}^{n-q}(G,A)^H$ if this choice is taken in such a way that $s(c)\sigma = \tau_\sigma s(c\sigma)$ for all $c \in G'\backslash G$, $\sigma \in H$ and some $\tau_\sigma \in H'$.

(2.5.4) Theorem. *The edge morphism*

$$H^{n-q}(G,A)^* \longrightarrow \Big(\varinjlim_{U\supseteq H} H^{n-q}(U,A)^*\Big)^{G/H}$$

in the Tate spectral sequence is the direct limit of the maps cor^*, *dual to the corestriction maps*

$$cor : H^{n-q}(U,A) \longrightarrow H^{n-q}(G,A).$$

Proof: We use the notation of the proof of (2.5.3). We may assume that H is open. The spectral sequence is obtained from the double complex

$$C^{pq} = C^p(G/H, Y^q),$$

where $Y^q = H^0(H,\tilde{X}^{n-q})^*$ and $\tilde{X}^i = X^i$ for $i < n$ and $\tilde{X}^n = Z^n$. Let $C^\bullet = \operatorname{tot}(C^{\bullet\bullet})$, $B^\bullet = \ker(C^{0\bullet} \to C^{1\bullet})$ and $K^\bullet = C^{0\bullet}$. Then $B^\bullet$ is a subcomplex of $C^\bullet$ and $H^q(B^\bullet) = H^q(C^\bullet) = E^q$ by (2.2.4). Therefore the edge morphism $E^q \to E_2^{0,q}$ is the map

$$(1) \qquad H^q(B^\bullet) \longrightarrow H^q(K^\bullet)$$

induced by the composite $B^\bullet \to C^\bullet \xrightarrow{\pi} K^\bullet$, which is the inclusion. Identifying $C^{0\bullet} = C^0(G/H, Y^\bullet)$ with $Y^\bullet$, this is the inclusion

$$(2) \qquad (Y^\bullet)^{G/H} \hookrightarrow Y^\bullet.$$

The image of (1) is contained in

$$E_2^{0,q} = \ker(H^q(C^{0\bullet}) \xrightarrow{\partial} H^q(C^{1\bullet})) = H^q(Y^\bullet)^{G/H},$$

and the edge morphism becomes the map

$$edge : H^q((Y^\bullet)^{G/H}) \longrightarrow H^q(Y^\bullet)^{G/H},$$

induced by the inclusion (2). From what we have seen in the proof of (2.5.3), this map is identified with a map

$$H^q(G,A)^* \longrightarrow [H^q(H,A)^*]^{G/H}$$

as follows. We have the canonical isomorphism

$$\operatorname{Hom}((\tilde{X}^{n-\bullet})^G, \mathbb{Q}/\mathbb{Z}) \cong H^0(G/H, Y^\bullet),$$

which is the same as the dual $N_{G/H}^*$ of

$$(\tilde{X}^{n-\bullet})^H \xrightarrow{N_{G/H}} (\tilde{X}^{n-\bullet})^G.$$

We obtain a commutative diagram

$$\begin{array}{ccc}
H^q(H^0(G/H,Y^\bullet)) & \xrightarrow{\ edge\ } & H^q(Y^\bullet)^{G/H} \\
{\scriptstyle N_{G/H}^*}\uparrow & & \| \\
H^q(H^0(G,\tilde{X}^{n-\bullet})^*) & \xrightarrow{N_{G/H}^*} & H^q(H^0(H,\tilde{X}^{n-\bullet})^*)^{G/H} \\
\| & & \| \\
H^{n-q}(G,A)^* & \xrightarrow{\ cor^*\ } & [H^{n-q}(H,A)^*]^{G/H},
\end{array}$$

which identifies the edge homomorphism with the dual of the corestriction.

□

We now consider the edge morphism $E_2^{p,0} \to E^p$ in the Tate spectral sequence for the case $H = \{1\}$. It is a homomorphism

$$H^p(G, D_n(A)) \xrightarrow{edge} H^{n-p}(G, A)^*.$$

In particular, for $p = n$ and $A = \mathbb{Z}$, we have a canonical homomorphism, called the **trace map**,

$$tr : H^n(G, D_n(\mathbb{Z})) \longrightarrow \mathbb{Q}/\mathbb{Z},$$

provided $cd(G, \mathbb{Z}) \le n$. On the other hand, for every pair $V \subseteq U$ of open normal subgroups of G and each $i \ge 0$, we have the canonical pairing

$$H^i(V, A)^* \times A^U \longrightarrow H^i(V, \mathbb{Z})^*, \quad (\chi, a) \longmapsto f(x) = \chi(ax).$$

Taking first the direct limit over V and then over U, we obtain a canonical bilinear map $D_i(A) \times A \longrightarrow D_i(\mathbb{Z})$, which gives us a cup-product

$$H^p(G, D_i(A)) \times H^{n-p}(G, A) \xrightarrow{\cup} H^n(G, D_i(\mathbb{Z})).$$

For $i = n$ this yields, together with the map tr, a homomorphism

$$H^p(G, D_n(A)) \xrightarrow{cup} H^{n-p}(G, A)^* .$$

(2.5.5) Theorem. *Suppose that* $cd(G, \mathbb{Z}) \le n$ *and let* $A \in Mod(G)$ *be finitely generated as a* $\mathbb{Z}$*-module. If* $cd(G, A) \le n$, *then the two maps*

$$H^p(G, D_n(A)) \overset{edge}{\underset{cup}{\rightrightarrows}} H^{n-p}(G, A)^*$$

coincide for all $p \in \mathbb{Z}$.

Proof: The Tate spectral sequence arises from the double complex

$$C^{pq}(A) = C^p(G/U, H^0(U, \tilde{X}^{n-q}(G, A))^*)$$

and the application of $\varinjlim_U$. We have a canonical pairing

$$\varphi : (\tilde{X}^{n-q}(G, A)^U)^* \times A \longrightarrow (\tilde{X}^{n-q}(G, \mathbb{Z})^U)^*, \quad (\chi, a) \longmapsto \varphi(\chi, a) = f,$$

where $f : \tilde{X}^{n-q}(G, \mathbb{Z})^U \longrightarrow \mathbb{Q}/\mathbb{Z}$ is defined by $f(x) = \chi(ax)$.

If $z(\sigma_0, \dots, \sigma_{p-j})$ is a $(p - j)$-cochain with coefficients in $(\tilde{X}^{n-q}(G, A)^U)^*$ and $t(\sigma_0, \dots, \sigma_j)$ is an j-cocycle in $Z^j(G, A)$, then

$$(z \cup t)(\sigma_0, \dots, \sigma_p) = \varphi(z(\sigma_0, \dots, \sigma_{p-j}), t(\sigma_{p-j}, \dots, \sigma_p))$$

is a p-cochain with coefficients in $(\tilde{X}^{n-q}(G, \mathbb{Z})^U)^*$. Thus we get a map

$$C^{p-j,q}(A) \times Z^j(G, A) \longrightarrow C^{p,q}(\mathbb{Z}),$$

i.e. for fixed $t \in Z^j(G, A)$, we have a morphism of double complexes

$$\cup t : C^{\bullet - j, \bullet}(A) \longrightarrow C^{\bullet, \bullet}(\mathbb{Z})$$

of degree $(j, 0)$ and hence a transformation of the associated edge morphisms. Applying $\varinjlim_U$, we obtain a map

$$\cup t : H^{n-j}(G, D_n(A)) \longrightarrow H^n(G, D_n(\mathbb{Z})),$$

which only depends on the cohomology class of the cocycle t. As the map tr is defined via the edge morphism, we obtain a commutative diagram

$$\begin{array}{ccccc} H^{n-j}(G, D_n(A)) & \times & H^j(G, A) & \xrightarrow{\cup} & H^n(G, D_n(\mathbb{Z})) \\ \Big\downarrow{\scriptstyle edge} & & \Big\| & & \Big\downarrow{\scriptstyle tr} \\ H^j(G, A)^* & \times & H^j(G, A) & \longrightarrow & \mathbb{Q}/\mathbb{Z}, \end{array}$$

where the upper arrow is the cup-product with respect to the pairing

$$D_n(A) \times A \longrightarrow D_n(\mathbb{Z}),$$

and the lower arrow is the evaluation map $(\chi, t) \mapsto \chi(t)$. From this diagram we get the commutative diagram

$$\begin{array}{ccc} H^{n-j}(G, D_n(A)) & \xrightarrow{cup} & \mathrm{Hom}(H^j(G, A), H^n(G, D_n(\mathbb{Z}))) \\ \Big\downarrow{\scriptstyle edge} & & \Big\downarrow{\scriptstyle tr} \\ H^j(G, A)^* & = & \mathrm{Hom}(H^j(G, A), \mathbb{Q}/\mathbb{Z}), \end{array}$$

which shows that the edge morphism coincides with the composite $tr \circ cup$. Setting $j = n - p$, we obtain the assertion of the theorem. □

We conclude this section with a vanishing criterion for the terms $D_i(A)$.

(2.5.6) Lemma. *Suppose that $D_i(\mathbb{Z}) = 0$. Assume that $A \in Mod(G)$ is finitely generated as a $\mathbb{Z}$-module and has torsion divisible only by prime numbers ℓ for which $D_{i+1}(\mathbb{Z})$ is ℓ-divisible. Then $D_i(A) = 0$.*

Proof: If A is finitely generated as a $\mathbb{Z}$-module, then $A = A^U$ for some open subgroup U of G. Since in the definition of the D_i the group G may be replaced by U, we may assume that A is a trivial G-module. If A is torsion-free, then $A \cong \mathbb{Z}^n$ as a G-module, hence $D_i(A) \cong D_i(\mathbb{Z})^n = 0$. It remains to consider the case $A = \mathbb{Z}/m\mathbb{Z}$. From the exact sequence $0 \to \mathbb{Z} \to \mathbb{Z} \to \mathbb{Z}/m\mathbb{Z} \to 0$, we obtain the exact sequence

$$D_{i+1}(\mathbb{Z}) \xrightarrow{m} D_{i+1}(\mathbb{Z}) \longrightarrow D_i(\mathbb{Z}/m\mathbb{Z}) \longrightarrow D_i(\mathbb{Z}) = 0,$$

hence our assumptions imply $D_i(\mathbb{Z}/m\mathbb{Z}) = 0$. □

Exercise 1. Compute the Tate spectral sequence for the case $G \cong \hat{\mathbb{Z}}$.

Exercise 2. If $0 < cd\,(G, A) \le n$, then, for all open normal subgroups U of G, we have

$$D_n(A)^U = H^n(U, A)^*.$$

Hint: Use (3.3.11).

Exercise 3. Let G be a profinite group and suppose we are given a direct limit $A = \varinjlim A_\alpha$ of G-modules. Show that there exist natural homomorphisms

$$D_i(G, A) \longrightarrow \varprojlim D_i(A_\alpha)$$

for all i. What are the images of these maps?

§6. Derived Functors

We have constructed the cohomology groups $H^n(G, A)$ in a direct and natural way from the diagram

$$\cdots \;\; \overrightarrow{\overrightarrow{\overrightarrow{\rightarrow}}} \;\; G \times G \times G \;\; \overrightarrow{\overrightarrow{\rightarrow}} \;\; G \times G \;\; \overrightarrow{\rightarrow} \;\; G.$$

The advantage of this definition is that it is concrete, elementary and down to earth. Its disadvantage is that it is difficult to generalize and to get deeper insights. There is another much more general definition of cohomology which we describe now.

We have explained in I §3 the notion of δ-functor $H = (H^n)_{n \ge 0}$ between abelian categories $\mathcal{A}$ and $\mathcal{A}'$. It is a family of additive functors $H^n : \mathcal{A} \to \mathcal{A}'$, which turns a short exact sequence

$$0 \longrightarrow A \longrightarrow B \longrightarrow C \longrightarrow 0$$

in $\mathcal{A}$ functorially into a long exact sequence

$$\cdots \longrightarrow H^n(A) \longrightarrow H^n(B) \longrightarrow H^n(C) \xrightarrow{\;\delta\;} H^{n+1}(A) \longrightarrow \cdots$$

in $\mathcal{A}'$. A morphism between two δ-functors H and H' from $\mathcal{A}$ to $\mathcal{A}'$ is a system $f = (f^n)_{n \ge 0}$ of functorial morphisms

$$f^n : H^n \longrightarrow H'^n,$$

which commute with δ. That is, for any exact sequence

$$0 \to A \to B \to C \to 0$$

in $\mathcal{A}$, the diagram

$$\begin{array}{ccc} H^n(C) & \xrightarrow{\;\delta\;} & H^{n+1}(A) \\ {\scriptstyle f^n(C)}\downarrow & & \downarrow{\scriptstyle f^{n+1}(A)} \\ H'^n(C) & \xrightarrow{\;\delta\;} & H'^{n+1}(A) \end{array}$$

commutes.

(2.6.1) Definition. *A δ-functor $H = (H^n)_{n\ge 0}$ from $\mathcal{A}$ to $\mathcal{A}'$ is called* **universal** *if, for every other δ-functor $H' = (H'^n)_{n\ge 0}$ from $\mathcal{A}$ to $\mathcal{A}'$, each morphism $f^0 : H^0 \to H'^0$ of functors extends uniquely to a morphism $f : H \to H'$ of δ-functors.*

We have the following criterion for the universality of a δ-functor. An additive functor $F : \mathcal{A} \to \mathcal{A}'$ is called **effaceable** if, for each object A in $\mathcal{A}$, there is a monomorphism $u : A \to I$ in $\mathcal{A}$ such that $F(u) = 0$.

(2.6.2) Theorem. *A δ-functor $H = (H^n)_{n\ge 0}$ from $\mathcal{A}$ to $\mathcal{A}'$ is universal if the functors H^n are effaceable for $n > 0$.*

For the proof we refer to [66], chap.I. The idea is the following. Let $H' = (H'^n)_{n\ge 0}$ be an arbitrary δ-functor from $\mathcal{A}$ to $\mathcal{A}'$ and let $f^0 : H^0 \to H'^0$ be a morphism of functors. Assume that we have shown that there exists a uniquely determined morphism of functors $f^i : H^i \to H'^i$, $i = 1, \ldots, n$, which commute with δ. Let $A \in \mathcal{A}$ and let $0 \to A \xrightarrow{u} I \to J \to 0$ be an exact sequence such that $H^{n+1}(u) = 0$. Then we obtain a uniquely determined morphism $f^{n+1} : H^{n+1}(A) \to H'^{n+1}(A)$ using the exact commutative diagram

$$\begin{array}{ccccccc}
H^n(I) & \longrightarrow & H^n(J) & \xrightarrow{\delta} & H^{n+1}(A) & \longrightarrow & 0 \\
\downarrow{\scriptstyle f^n} & & \downarrow{\scriptstyle f^n} & & \downarrow{\scriptstyle f^{n+1}} & & \\
H'^n(I) & \longrightarrow & H'^n(J) & \xrightarrow{\delta} & H'^{n+1}(A). & &
\end{array}$$

It remains to show that f^{n+1} is functorial and commutes with δ.

If G is a profinite group, then the functors $H^n(G, -)$, $n > 0$, are effaceable, since every G-module A embeds into the induced G-module $\mathrm{Ind}_G(A)$, which is acyclic, i.e. has trivial cohomology. We therefore have the

(2.6.3) Theorem. *The δ-functor $(H^n(G, -))_{n\ge 0}$ is universal.*

With this theorem many proofs of isomorphism, uniqueness etc. are obtained automatically.

Example 1. Let H be a closed normal subgroup. We then have the following proof of Shapiro's lemma (1.6.4)

$$sh : H^n(G, \mathrm{Ind}_G^H(A)) \cong H^n(H, A).$$

Noting that Ind_G^H is exact and that $\mathrm{Ind}_G^H(\mathrm{Ind}_H(B)) = \mathrm{Ind}_G(B)$, we see that $(H^n(G, \mathrm{Ind}_G^H(-)))$ and $(H^n(H, -))$ are effaceable δ-functors on *Mod*(H), and are thus universal. They are functorially isomorphic in dimension $n = 0$. Hence they are isomorphic as δ-functors. By the uniqueness assertion, the isomorphism is the composite

$$H^n(G, \mathrm{Ind}_G^H(A)) \xrightarrow{res} H^n(H, \mathrm{Ind}_G^H(A)) \xrightarrow{\pi_*} H^n(H, A),$$

where π_* is induced by $\pi : \mathrm{Ind}_G^H(A) \to A$, $f \mapsto f(1)$. In fact, this composite is a morphism of δ-functors, which in dimension $n = 0$ coincides with the initial isomorphism.

Example 2. For every G-module A, we have the commutative diagram

$$\begin{array}{ccc} & H^n(G, \mathrm{Ind}_G^H(A)) & \\ {\scriptstyle sh}\swarrow & & \searrow{\scriptstyle \nu_*} \\ H^n(H, A) & \xrightarrow{cor} & H^n(G, A) \end{array}$$

as claimed in (1.6.5). In fact, ν_* and $cor \circ sh$ are morphisms of universal δ-functors, which coincide in dimension $n = 0$ by the definition of *sh, cor*, ν_*. Hence they coincide for all $n \geq 0$ by the uniqueness assertion of (2.6.1).

If $F : \mathcal{A} \to \mathcal{A}'$ is an additive functor, there exists up to a canonical isomorphism at most one universal δ-functor H from $\mathcal{A}$ to $\mathcal{A}'$ with $H^0 = F$. This δ-functor, if it exists, is then called the **right derived functor** of F and is denoted by $R^\bullet F = (R^n F)_{n \geq 0}$. Obviously, it is defined up to canonical isomorphism. The question is, when does it exist?

By theorem (2.6.3), the universal δ-functor $H^\bullet(G, -)$ on *Mod*(G) is the right derived functor

$$H^\bullet(G, -) = R^\bullet \Gamma$$

of the functor

$$\Gamma(-) = H^0(G, -) : Mod(G) \longrightarrow Ab, \quad A \longmapsto A^G.$$

Suppose $\mathcal{M}$ is a full abelian subcategory of *Mod*(G) which has the property that for a discrete G-module M, the induced module $\mathrm{Ind}_G(M)$ is also in $\mathcal{M}$. Then the same reasoning as above shows that the restriction of $H^\bullet(G, -)$ to $\mathcal{M}$ is the right derived functor $R^\bullet \Gamma$ of the functor $\Gamma(-) : \mathcal{M} \longrightarrow Ab$, $\quad A \mapsto A^G$. Examples of such subcategories $\mathcal{M}$ are the category $Mod_t(G)$ of discrete G-modules which are torsion groups, or the category $Mod_{(p)}(G)$ of discrete G-modules which are p-torsion groups, where p is a prime number.

Recall that an additive functor $F : \mathcal{A} \to \mathcal{A}'$ is called **left exact** if for each exact sequence $0 \to A \to B \to C$ the sequence

$$0 \longrightarrow F(A) \longrightarrow F(B) \longrightarrow F(C)$$

is also exact.

The left exactness of F is clearly a necessary condition for the existence of the right derived functor $R^\bullet F$, since H^0 is left exact. This condition is already sufficient if $\mathcal{A}$ has *sufficiently many injectives*:

An object A in $\mathcal{A}$ is **injective** if for every monomorphism $B \to C$ in $\mathcal{A}$ the map $\mathrm{Hom}(C, A) \to \mathrm{Hom}(B, A)$ is surjective. $\mathcal{A}$ is said to have **sufficiently many injectives** if, for any object A, there exists a monomorphism $A \to I$ into an injective object.

(2.6.4) Theorem. *Let $\mathcal{A}$ have sufficiently many injectives. Then for each left exact additive functor $F : \mathcal{A} \to \mathcal{A}'$, the right derived functor $R^\bullet F = (R^n F)_{n \geq 0}$ exists.*

For the proof we refer to [21], chap. V, §3, but we explain the idea of it. Since A has sufficiently many injectives, each object $A \in \mathcal{A}$ has an **injective resolution**, i.e. there is an exact complex

$$0 \longrightarrow A \longrightarrow I^0 \longrightarrow I^1 \longrightarrow I^2 \longrightarrow \cdots$$

with injective objects I^n in $\mathcal{A}$. We apply the functor F and get a complex

$$0 \longrightarrow F(A) \longrightarrow F(I^0) \longrightarrow F(I^1) \longrightarrow F(I^2) \longrightarrow \cdots.$$

We define

$$R^n F(A) = H^n(F(I^\bullet)), \qquad n \geq 0;$$

in particular, $R^0 F(A) = \ker(F(I^0) \to F(I^1)) = F(A)$.

The independence of this definition from the injective resolution chosen is seen as follows. If $A \to I^\bullet$ and $A' \to I'^\bullet$ are injective resolutions of A and A', then, because of the injectivity property of the I^n, every morphism $u : A \to A'$ extends to a morphism of complexes

$$\begin{array}{ccc} A & \longrightarrow & I^\bullet \\ {\scriptstyle u}\big\downarrow & & \big\downarrow{\scriptstyle u} \\ A' & \longrightarrow & I'^\bullet, \end{array}$$

and every two such extensions are homotopic (cf. I §3, exercise 6). This means that the induced maps from $F(I^\bullet)$ to $F(I'^\bullet)$ are homotopic, hence induce the same homomorphism $H^n(F(I^\bullet)) \to H^n(F(I'^\bullet))$ on the homology. In particular, if $A = A'$, we find extensions $u : I^\bullet \to I'^\bullet$, $v : I'^\bullet \to I^\bullet$, such that $u \circ v$ and $v \circ u$ are homotopic to the identity, hence induce mutually inverse isomorphisms $H^n(F(I^\bullet)) \rightleftarrows H^n(F(I'^\bullet))$. This shows the independence.

The property of being a δ-functor is seen as follows. Any exact sequence $0 \to A \to B \to C \to 0$ in $\mathcal{A}$ may be extended to an exact sequence

$$\begin{array}{ccccccccc} 0 & \longrightarrow & A & \longrightarrow & B & \longrightarrow & C & \longrightarrow & 0 \\ & & \downarrow & & \downarrow & & \downarrow & & \\ 0 & \longrightarrow & I_A^\bullet & \longrightarrow & I_B^\bullet & \longrightarrow & I_C^\bullet & \longrightarrow & 0 \end{array}$$

of injective resolutions. Since I_A^n is injective, all exact sequences

$$0 \longrightarrow I_A^n \longrightarrow I_B^n \longrightarrow I_C^n \longrightarrow 0$$

split and therefore

$$0 \longrightarrow F(I_A^n) \longrightarrow F(I_B^n) \longrightarrow F(I_C^n) \longrightarrow 0$$

remains exact. The exact sequence

$$0 \longrightarrow F(I_A^\bullet) \longrightarrow F(I_B^\bullet) \longrightarrow F(I_C^\bullet) \longrightarrow 0$$

of complexes yields, in the same way as in I §2, a long exact sequence

$$\cdots \longrightarrow R^n F(A) \longrightarrow R^n F(B) \longrightarrow R^n F(C) \xrightarrow{\delta} R^{n+1} F(A) \longrightarrow \cdots.$$

We have obtained a δ-functor $R^\bullet F = (R^n F)_{n \geq 0}$. For an injective object I, we have $R^n F(I) = 0$ for $n > 0$ since $0 \to I \xrightarrow{id} I \to 0$ is an injective resolution of I. Since $\mathcal{A}$ has sufficiently many injectives, the $R^n F$, $n > 0$, are effaceable, hence $R^\bullet F$ is universal.

(2.6.5) Lemma. *If G is a profinite group, then the category $Mod(G)$ of discrete G-modules has sufficiently many injectives.*

Proof: To every abstract G-module M we can associate the submodule

$$M^\delta := \bigcup_{U \subseteq G} M^U,$$

where U runs through the open subgroups of G. If we endow M with the discrete topology, then M^δ is the maximal submodule on which G acts continuously (compare with the remark after (1.1.8)). One easily verifies that every G-homomorphism from a discrete G-module N to M factors through M^δ. In particular, we see that the discrete module I^δ is an injective object in *Mod*(G) provided the (abstract) G-module I is injective. The category of abstract G-modules has sufficiently many injective objects (see [79], chap. IV: it is canonically equivalent to the category of modules over the group ring $\mathbb{Z}[G]$). Therefore we can embed a given discrete G-module M into an injective abstract module I and then M is automatically contained in the injective discrete module I^δ. □

The Hochschild-Serre spectral sequence (2.4.1) becomes a special case of the following general result.

(2.6.6) Theorem. *Let $\mathcal{A}$ and $\mathcal{A}'$ be abelian categories with sufficiently many injectives and let $\mathcal{A}''$ be another abelian category. Let*

$$\mathcal{A} \xrightarrow{F} \mathcal{A}' \xrightarrow{F'} \mathcal{A}''$$

be left exact additive functors. Assume that F maps injective objects from $\mathcal{A}$ to $\boldsymbol{F'}$**-acyclic** *objects, i.e. those annihilated by R^nF' for $n > 0$. Then there is a cohomological spectral sequence*

$$E_2^{pq} = R^pF'(R^qF(A)) \Rightarrow R^{p+q}(F' \circ F)(A)\,,$$

which is called the **Grothendieck spectral sequence***.*

This spectral sequence is obtained as follows. There exists a homomorphism of the complex $F(I^\bullet)$ into a double complex of $\mathcal{A}'$-injective objects $I^{\bullet\bullet}$ which induces injective resolutions of all groups $F(I^q)$ and also for all cocycle, coboundary and cohomology groups of the complex $F(I^\bullet)$ (a so-called *Cartan-Eilenberg resolution*, cf. [21], chap.XVII). Applying to the double complex $I^{\bullet\bullet} = (I^{pq})_{p,q\geq 0}$ the functor F', we obtain a double complex

$$(A^{pq})_{p,q\geq 0} = (F'(I^{pq}))_{p,q\geq 0}.$$

The spectral sequence $E^{pq} \Rightarrow E^n$ associated with this double complex is the maintained spectral sequence

$$E_2^{pq} = R^pF'(R^qF(A)) \Rightarrow R^n(F' \circ F)(A).$$

For the proof we refer to [66] and [21].

If G is a profinite group and H a closed subgroup, then we have the additive left exact functors

$$\begin{aligned} F = H^0(H,-) &: Mod(G) \to Mod(G/H), && A \mapsto A^H,\\ F' = H^0(G/H,-) &: Mod(G/H) \to \mathcal{A}b, && B \mapsto B^{G/H},\\ F' \circ F = H^0(G,-) &: Mod(G) \to \mathcal{A}b, && A \mapsto A^G. \end{aligned}$$

In this case, the Grothendieck spectral sequence

$$E_2^{pq} = R^pF'(R^qF(A)) \Rightarrow R^{p+q}(F' \circ F)(A)$$

coincides with the Hochschild-Serre spectral sequence. We easily see that the E_2-terms and the limit terms are the same, since

$$R^qF = H^q(H,-),\ R^pF' = H^p(G/H,-),\ R^n(F' \circ F) = H^n(G,-).$$

That both spectral sequences actually coincide follows from the fact that the functor 'homogeneous cochain complex' is a 'resolving functor', see [66].

So far we have dealt with the right derivation of a left exact, covariant functor. In the later applications it will be often useful to work with certain modifications of this concept.

Assume that we are given abelian categories $\mathcal{B}$ and $\mathcal{B}'$. We say that a family $H = (H_n)_{n\in\mathbb{Z}}$ of functors $H_n : \mathcal{B} \to \mathcal{B}'$ is a **homological δ-functor** if the family $K = (K^n)_{n\in\mathbb{Z}}$, defined by $K^n := H_{-n}$, is a (cohomological) δ-functor as defined before. The following notions and statements are dual to those given before for cohomological δ-functors and we leave their verification to the reader. We also note that, up to the obvious modifications, one can also work with contravariant functors.

We say that a homological δ-functor $H = (H_n)_{n\geq 0}$ is **universal** if, for every other homological δ-functor $H' = (H'_n)_{n\geq 0}$, each morphism $f_0 : H'_0 \to H_0$ of functors extends uniquely to a morphism $f : H' \to H$ of homological δ-functors. A functor $G : \mathcal{B} \to \mathcal{B}'$ is called **coeffaceable** if, for every object $B \in \mathcal{B}$, there is an epimorphism $\phi : P \to B$ with $G(\phi) = 0$. A homological δ-functor $H = (H_n)_{n\geq 0} : \mathcal{B} \to \mathcal{B}'$ is universal if the functors H_n are coeffaceable for $n > 0$. If $G : \mathcal{B} \to \mathcal{B}'$ is an additive functor, there exists up to canonical isomorphism at most one universal homological δ-functor H from $\mathcal{B}$ to $\mathcal{B}'$ with $H_0 = G$. This δ-functor, if it exists, is then called the **left derived functor** of G and is denoted by $L_\bullet G = (L_n G)_{n\geq 0}$.

An object P of $\mathcal{B}$ is called **projective** if for every epimorphism $A \to B$ in $\mathcal{B}$ the map $\mathrm{Hom}(P, A) \longrightarrow \mathrm{Hom}(P, B)$ is surjective. We say that $\mathcal{B}$ has **sufficiently many projectives** if for every object B there exists an epimorphism $P \to B$ with projective P.

The left derived functor of $G : \mathcal{B} \to \mathcal{B}'$ exists if G is right exact and $\mathcal{B}$ has sufficiently many projectives.

Now we introduce the **homology of profinite groups**. The homology groups are compact abelian groups and they have compact G-modules as coefficients. In order to prevent confusion, we use the notation $\mathscr{D} = \mathscr{D}(G)$ for the category of discrete G-modules, which so far has been denoted by *Mod*(G). The category of compact G modules will be denoted by $\mathscr{C} = \mathscr{C}(G)$.

(2.6.7) Definition. *Let G be a profinite group and let $A \in \mathscr{C}$ be a compact G-module. The* **cofixed module** *(or* **module of coinvariants***) A_G of A is the largest Hausdorff quotient of A on which G acts trivially, i.e. A_G is the quotient of A by the closed subgroup generated by the elements $(ga - a)$, $g \in G$, $a \in A$.*

We denote the category of compact abelian groups by $\mathcal{A}b^c$ and in order to stress the difference we write $\mathcal{A}b^d$ for the category of (discrete) abelian groups. One easily verifies that $(-)_G$ is a right exact functor from $\mathscr{C}$ to $\mathcal{A}b^c$. Furthermore, the category $\mathscr{C}$ is dual to $\mathscr{D}$ by Pontryagin duality and therefore it has sufficiently many projectives by (2.6.5).

(2.6.8) Definition. *For a profinite group G and a compact module A, the* **homology groups** *are defined as the left derivatives of the cofixed-module functor*

$$H_n(G,A) := L_n(-)_G(A).$$

In particular, we have $H_0(G,A) = A_G$, and if $0 \to A \to B \to C \to 0$ is an exact sequence in $\mathscr{C}$, then we get a long exact homology sequence

$$\cdots \to H_{n+1}(G,C) \to H_n(G,A) \to H_n(G,B) \to H_n(G,C) \to \cdots.$$

The homology theory for profinite groups is dual to the cohomology theory: every cohomological result has its homological analogue. Fortunately we do not have to prove everything twice because of the following

(2.6.9) Theorem. *Let G be a profinite group and A be a compact G-module. Then there are functorial isomorphisms for all $i \geq 0$*

$$H_i(G,A)^\vee \cong H^i(G,A^\vee),$$

where $^\vee$ denotes the Pontryagin dual.

Proof: The theorem is true for $i = 0$ by the definition of the fixed and the cofixed module. Now the following diagram of categories and functors is commutative

$$\begin{array}{ccc} \mathscr{C} & \xrightarrow{\;\vee\;} & \mathscr{D} \\ \big\downarrow{\scriptstyle (-)_G} & & \big\downarrow{\scriptstyle (-)^G} \\ \mathscr{A}b^c & \xrightarrow{\;\vee\;} & \mathscr{A}b^d. \end{array}$$

Furthermore, Pontryagin duality is an exact, contravariant functor that transfers $\mathscr{C}$-projectives to $\mathscr{D}$-injectives. The statement of the theorem now follows from the universal property for the derived functors. □

We see from the last theorem that, in principle, one can avoid the use of homology groups, working only with cohomology. Indeed, the decision whether to work with cohomology or homology, is more or less a question of personal taste.

We finish this section with a spectral sequence for Ext-groups. Let R be a ring (with unit). Then the functor $\mathrm{Hom}_R(-,-)$ is a bifunctor from the category of R-modules to abelian groups, which is contravariant in the first and covariant in the second variable. Its derivations

$$\mathrm{Ext}^i_R(-,-)$$

may likewise be computed using projective resolutions of the first, or using injective resolutions of the second, variable. (See [79] or any textbook about homological algebra for this basic fact.) The Hom-acyclic objects in the first resp. second variable are projective resp. injective R-modules. The following spectral sequence connects the Ext-groups for modules over a group ring $R[G]$ with that over R.

(2.6.10) Theorem. *Let R be a commutative ring with unit, let G be a finite group and let M and N be $R[G]$-modules. Then there exists a natural spectral sequence*

$$E_2^{pq} = H^p(G, \mathrm{Ext}^q_R(M, N)) \Rightarrow \mathrm{Ext}^{p+q}_{R[G]}(M, N)\,.$$

Proof: First we observe that

$$\mathrm{Hom}_{R[G]}(M, N) \cong \mathrm{Hom}_R(M, N)^G,$$

thus the left exact functor $\mathrm{Hom}_{R[G]}(M, -)$ is the composition of the left exact functors $\mathrm{Hom}_R(M, -)$ and $H^0(G, -)$. Now assume that N is injective. Then N is a direct summand of $\mathrm{Ind}_G N$. By (1.3.6)(iii), the G-module $\mathrm{Hom}_R(M, \mathrm{Ind}_G N)$ is induced. Thus $\mathrm{Hom}_R(M, N)$ is cohomologically trivial because it is a direct summand of an induced module. Therefore theorem (2.6.6) gives us the desired spectral sequence. □

(2.6.11) Corollary. *Let G be a finite group whose order is invertible in the commutative ring R. Then an $R[G]$-module M is projective if and only if it is R-projective.*

Proof: A free $R[G]$-module is free as an R-module. If M is a projective $R[G]$-module, then it is a direct summand of a free $R[G]$-module and therefore also projective as an R-module. In order to show the other implication, assume that M is an $R[G]$-module which is R-projective. The cohomology groups $H^i(G, M)$ are R-modules and annihilated by $\#G$ for $i \geq 1$ by (1.6.1). Hence they are trivial for $i \geq 1$ and for an arbitrary $R[G]$-module N, the spectral sequence (2.6.10) degenerates to a sequence of isomorphisms

$$\mathrm{Ext}^i_{R[G]}(M, N) \longrightarrow \mathrm{Ext}^i_R(M, N)^G = 0$$

for $i \geq 1$. Hence M is $R[G]$-projective. □

We obtain the following result, which is known as **Maschke's Theorem**.

(2.6.12) Corollary (MASCHKE). *Let G be a finite group and let K be a field whose characteristic does not divide the order of G. Then the category of $K[G]$-modules is semi-simple.*

Exercise 1. Define for an abstract group G the homology with values in a G-module as the left derivation of the cofixed module functor on the category of abstract G-modules. Assume that G is a finite group and let A be a finite G-module. Then we can view G as an abstract group and A as an abstract module or we can view G as a profinite group and A as a compact module. Show that the corresponding homology groups are the same and that both coincide with the homology groups introduced in I §9.

Exercise 2. Let R be a commutative ring with unit, let G be a finite group and let M and N be $R[G]$-modules such that M is R-projective and N is cohomologically trivial. Show that

$$\mathrm{Ext}^1_{R[G]}(M, N) = 0\,.$$

§7. Continuous Cochain Cohomology

In chapter I, we started considering the simplicial diagram

$$\cdots \;\; \substack{\longrightarrow\\\longrightarrow\\\longrightarrow\\\longrightarrow} \;\; G\times G\times G \;\; \substack{\longrightarrow\\\longrightarrow\\\longrightarrow} \;\; G\times G \;\; \substack{\longrightarrow\\\longrightarrow} \;\; G,$$

which gave rise to a standard complex $C^\bullet(G, A)$, and defined the cohomology of the profinite group G with values in the discrete module A. In the last section we learned that this cohomology can be characterized by a universal property which, in particular, explains its good functorial behaviour. This functorial approach via universal constructions is extremely useful and clarifies the principles behind classical homological notions.

Sometimes, however, the simplicial approach seems to reach further. For instance, the categories of finite, compact, locally compact or of all topological G-modules lack the existence of sufficiently many "good" objects, and the existence of derived functors on these categories is not guaranteed. But we can still define cohomology using the standard complex.

Let G be a profinite group and let A be any *topological* G-module (see I §1). We form the **continuous homogeneous cochain complex of G with coefficients in A**, denoted by $C^\bullet_{cts}(G, A)$, by taking the G-invariants of the continuous standard resolution $X^\bullet_{cts}(G, A)$, which is defined in exactly the same way as in I §2 for discrete modules.

(2.7.1) Definition. *We call the n-th cohomology group of the complex $C^{\bullet}_{cts}(G,A)$ the n-th* **continuous cochain cohomology group** *of G with coefficients in A. We denote this group by $H^n_{cts}(G,A)$.*

If A is discrete, then clearly $H^n_{cts}(G,A) = H^n(G,A)$, but for an arbitrary topological G-module A, the right-hand side of the equation is not defined.

The same arguments as used in I §3 show the following

(2.7.2) Lemma. *Let*

$$0 \longrightarrow A \xrightarrow{\alpha} B \xrightarrow{\beta} C \longrightarrow 0$$

be a short exact sequence of topological G-modules such that the topology of A is induced by that of B and such that β has a continuous section (just a continuous map, not necessarily a homomorphism). Then there exist canonical boundary homomorphisms

$$\delta\colon\ H^n_{cts}(G,C) \longrightarrow H^{n+1}_{cts}(G,A)$$

and we obtain an exact sequence

$$\cdots \to H^n_{cts}(G,A) \to H^n_{cts}(G,B) \to H^n_{cts}(G,C) \xrightarrow{\delta} H^{n+1}_{cts}(G,A) \to \cdots$$

which begins $0 \to A^G \to B^G \to C^G \xrightarrow{\delta} H^1_{cts}(G,A)$.

Note that we can apply this lemma in the particular case that A is an open submodule of B and $C = B/A$ is the quotient module with the quotient topology, which is discrete.

Suppose

$$\mathcal{B}\colon\ A \times B \longrightarrow C$$

is a continuous G-pairing, i.e. a continuous biadditive map such that $\sigma(a \cdot b) = (\sigma a \cdot \sigma b)$ for $a \in A$, $b \in B$, $\sigma \in G$, where $a \cdot b$ denotes $\mathcal{B}(a,b)$. Then $\mathcal{B}$ induces biadditive maps

$$C^p_{cts}(G,A) \times C^q_{cts}(G,B) \longrightarrow C^{p+q}_{cts}(G,C)$$

via the formula

$$(f \cup g)(\sigma_0,\dots,\sigma_{p+q}) = f(\sigma_0,\dots,\sigma_p) \cdot g(\sigma_p,\dots,\sigma_{p+q}).$$

This cup-product of cochains satisfies the identity

$$\partial(f \cup g) = (\partial f) \cup g + (-1)^p f \cup (\partial g)$$

and consequently induces pairings

$$H^p_{cts}(G,A) \times H^q_{cts}(G,B) \longrightarrow H^{p+q}_{cts}(G,C),$$

which enjoy the same properties as the cup-products considered in I §4.

The maps *inf*, *res*, *cor* with respect to a change of groups are defined and have the same properties as the maps introduced in I §5. (Note that we exclusively made calculations there on the chain level!)

In some important special situations, we can still relate the continuous cochain cohomology to derived functors. For this we need some preparations.

Let $\mathcal{A}$ be an abelian category and let $\mathcal{A}^{\mathbb{N}}$ be the category of inverse systems in $\mathcal{A}$ indexed by the set $\mathbb{N}$ of natural numbers with the natural order. Thus objects in $\mathcal{A}^{\mathbb{N}}$ are inverse systems

$$\cdots \longrightarrow A_{n+1} \xrightarrow{d_n} A_n \longrightarrow \cdots \longrightarrow A_1 \xrightarrow{d_0} A_0$$

in $\mathcal{A}$ and morphisms are commutative diagrams

$$\begin{array}{ccccccccc} \cdots \longrightarrow & A_{n+1} & \longrightarrow & A_n & \longrightarrow \cdots \longrightarrow & A_1 & \longrightarrow & A_0 \\ & \downarrow{\scriptstyle f_{n+1}} & & \downarrow{\scriptstyle f_n} & & \downarrow{\scriptstyle f_1} & & \downarrow{\scriptstyle f_0} \\ \cdots \longrightarrow & B_{n+1} & \longrightarrow & B_n & \longrightarrow \cdots \longrightarrow & B_1 & \longrightarrow & B_0. \end{array}$$

The category $\mathcal{A}$ is abelian, with kernels and cokernels taken componentwise. One can show that $\mathcal{A}^{\mathbb{N}}$ has sufficiently many injectives, provided that this is the case for $\mathcal{A}$. Moreover, an object (A_n, d_n) in $\mathcal{A}^{\mathbb{N}}$ is injective if and only if all A_n are injective and all d_n are split surjections (see [96], prop. 1.1). A left exact functor

$$F\colon \mathcal{A} \longrightarrow \mathcal{B}$$

into another abelian category induces a functor $F^{\mathbb{N}}\colon \mathcal{A}^{\mathbb{N}} \longrightarrow \mathcal{B}^{\mathbb{N}}$ in the obvious way. If $\mathcal{A}$ has sufficiently many injectives, then the right derivatives $R^iF^{\mathbb{N}}$ exist and we have (loc.cit. prop. 1.2)

$$R^iF^{\mathbb{N}} = (R^iF)^{\mathbb{N}} \text{ for } i \ge 0, \text{ i.e. } R^iF^{\mathbb{N}}(A_n, d_n) = (R^iFA_n, R^iF(d_n)).$$

If inverse limits over $\mathbb{N}$ exist in $\mathcal{B}$, we can define the functor

$$\varprojlim_n F\colon \mathcal{A}^{\mathbb{N}} \longrightarrow \mathcal{B},$$

which is by definition the composition of the functor $F^{\mathbb{N}}$ with the limit functor $\varprojlim\colon \mathcal{B}^{\mathbb{N}} \longrightarrow \mathcal{B}$.

Assume that $\mathcal{A}$ and $\mathcal{B}$ have sufficiently many injectives, F is left exact and $F^{\mathbb{N}}$ sends injectives to $\varprojlim$-acyclic objects. Then, by (2.6.6), we have a spectral sequence

$$E_2^{pq} = \varprojlim_n{}^p(R^qFA_n) \Rightarrow R^{p+q}(\varprojlim_n F)(A_n, d_n),$$

where $\varprojlim^\bullet$ denotes the right derived functor of $\varprojlim$.

We say that a system (A_n, d_n) satisfies the **Mittag-Leffler property** if for each n the images of the transition maps $A_{n+m} \to A_n$ are the same for all sufficiently large m. Observe that every inverse system of finite abelian groups automatically has the Mittag-Leffler property. We call a system (A_n, d_n) **ML-zero** if for each n there is an $m = m(n)$ such that the transition map $A_{n+m} \to A_n$ is zero.

(2.7.3) Proposition. *Let (A_n, d_n) be a ML-zero system in $\mathcal{A}$. Then*

$$R^p(\underset{\overleftarrow{n}}{\lim}\, F)(A_n, d_n) = 0$$

for all $p \geq 0$.

Proof: Given a system $A = (A_n, d_n)$, we define a new system $A[1]$ by $A[1] = (A_{n+1}, d_{n+1})$. There is a canonical map $A[1] \xrightarrow{can} A$ given by the composition of the transition maps. If $A \hookrightarrow I^\bullet$ is an injective resolution, then $A[1] \hookrightarrow I[1]^\bullet$ is too, and we obtain a commutative diagram

$$\begin{array}{ccc} A[1] & \hookrightarrow & I[1]^\bullet \\ {\scriptstyle can}\downarrow & & \downarrow{\scriptstyle can} \\ A & \hookrightarrow & I^\bullet . \end{array}$$

Since $\underset{\leftarrow}{\lim}\, I[1]^m \xrightarrow{\sim} \underset{\leftarrow}{\lim}\, I^m$ for all $m \geq 0$, we obtain natural isomorphisms for all $p \geq 0$

$$R^p(\underset{\overleftarrow{n}}{\lim}\, F)A[1] \xrightarrow{\sim} R^p(\underset{\overleftarrow{n}}{\lim}\, F)A\ .$$

If all transition maps of A are zero, then this canonical isomorphism is the zero map, and hence $R^p(\underset{\leftarrow}{\lim}\, F)A = 0$ for all $p \geq 0$ in this case.

Let (A_n) be a ML-zero system and let $J = \{n_1, n_2, \ldots\} \subset \mathbb{N}$ be a cofinal subset such that all transition maps in the inverse system $(A_{n_j})_{j\in\mathbb{N}}$ are zero. Now $\underset{\leftarrow}{\lim}\, F$ is also the composition of the exact forgetful functor

$$V_J\colon \mathcal{A}^{\mathbb{N}} \longrightarrow \mathcal{A}^{\mathbb{N}}, \quad (A_n) \mapsto (A_{n_j}),$$

which preserves injectives, with $\underset{\leftarrow}{\lim}\, F$. The spectral sequence implies

$$R^p(\underset{\overleftarrow{n}}{\lim}\, F)(A_n) = R^p(\underset{\overleftarrow{j}}{\lim}\, F)(A_{n_j}) = 0$$

for all $p \geq 0$. □

The following well-known proposition was formulated in [182], prop. 1, with too much generality (see [150] for a counter-example). Therefore we give a complete proof here in the case that $\mathcal{B}$ is the category of modules over a ring.

(2.7.4) Proposition. *Let $\mathcal{B}$ be the category of R-modules for some ring R. Then $\varprojlim^p = 0$ for $p \geq 2$, and there is a canonical exact sequence*

$$0 \longrightarrow \varprojlim_n (B_n) \longrightarrow \prod_n B_n \xrightarrow{id-d_n} \prod_n B_n \longrightarrow \varprojlim_n{}^1 (B_n) \longrightarrow 0.$$

If $B \in \mathrm{Ob}(\mathcal{B}^{\mathbb{N}})$ satisfies the Mittag-Leffler property, then $\varprojlim^1 B = 0$.

Proof: By the definition of $\varprojlim$, we have an exact sequence

$$0 \longrightarrow \varprojlim_n (B_n) \longrightarrow \prod_n B_n \xrightarrow{id-d_n} \prod_n B_n,$$

which is also exact on the right, if all transition maps of (B_n) are surjective (for any element $(b_n) \in \prod_n B_n$ a pre-image under $(id - d_n)$ is easily constructed recursively). Let

$$0 \longrightarrow (A_n) \longrightarrow (B_n) \longrightarrow (C_n) \longrightarrow 0$$

be an exact sequence in $\mathcal{B}^{\mathbb{N}}$ such that all transition maps of the systems (A_n) and (B_n) are surjective. Then the transition maps of (C_n) are also surjective and the snake lemma shows that the sequence of inverse limits

$$0 \longrightarrow \varprojlim_n (A_n) \longrightarrow \varprojlim_n (B_n) \longrightarrow \varprojlim_n (C_n) \longrightarrow 0$$

is exact. Since injective objects in $\mathcal{B}^{\mathbb{N}}$ have surjective transition maps, we obtain $\varprojlim^p (B_n) = 0$ for all $p \geq 1$ and any system (B_n) with surjective transition maps.

If (B_n) is an arbitrary inverse system, we consider the inverse systems (PB_n) and (QP_n) given by $PB_n = \prod_{i=1}^n B_i$ and $QP_n = \prod_{i=1}^{n-1} B_i$, and with the natural projections $p_n \colon \prod_{i=1}^{n+1} B_i \to \prod_{i=1}^n B_i$ as transition maps. We obtain an exact sequence of inverse systems

$$0 \longrightarrow (B_n) \longrightarrow (PB_n) \longrightarrow (QP_n) \longrightarrow 0,$$

given by the commutative exact diagrams

$$\begin{array}{ccccccccc}
0 \to & B_{n+1} & \xrightarrow{(id, d_n, d_n d_{n-1}, \dots)} & \prod_{i=1}^{n+1} B_i & \xrightarrow{p_n - (d_n \mathrm{pr}_{n+1}, d_{n-1} \mathrm{pr}_n, \dots)} & \prod_{i=1}^{n} B_i & \to 0 \\
& \downarrow d_n & & \downarrow p_n & & \downarrow p_{n-1} & \\
0 \longrightarrow & B_n & \xrightarrow{(id, d_{n-1}, d_{n-2} d_{n-1}, \dots)} & \prod_{i=1}^{n} B_i & \xrightarrow{p_{n-1} - (d_{n-1} \mathrm{pr}_n, d_{n-2} \mathrm{pr}_{n-1}, \dots)} & \prod_{i=1}^{n-1} B_i & \to 0,
\end{array}$$

where pr_i denotes the projection onto B_i. Since $\varprojlim^p (PB_n) = 0 = \varprojlim^p (QB_n)$ for $p \geq 1$, we obtain $\varprojlim^p (B_n) = 0$ for $p \geq 2$ and the exact sequence stated in the proposition.

Finally, let (B_n) be a Mittag-Leffler system. We consider the subsystem (B'_n) of (B_n), where, for each n, B'_n is the image of the transition map $B_{n+m} \to B_n$, for sufficiently large m, which is independent of m. Then the transition maps of (B'_n) are surjective and the cokernel (B''_n) of the natural inclusion $(B'_n) \hookrightarrow (B_n)$ is ML-zero. Therefore we obtain isomorphisms

$$0 = \varprojlim_n{}^p(B'_n) \xrightarrow{\sim} \varprojlim_n{}^p(B_n)$$

for all $p \geq 1$. □

Suppose we are given a profinite group G and a compact topological G-module A whose underlying topology is profinite. It is not difficult to see that such a module is the inverse limit of finite G-modules and we call such modules **profinite modules** for short. The continuous cochain cohomology of profinite modules can also be defined by means of derived functors, namely as an Ext-group in the category of profinite G-modules.*) We will not make use of this in the following, but the reader should notice that this is the conceptual reason behind the next theorem.

(2.7.5) Theorem. *Suppose that the compact G-module A has a presentation*

$$A = \varprojlim_{n\in\mathbb{N}} A_n$$

as a countable inverse limit of finite, discrete G-modules. Then there exists a natural exact sequence

$$0 \longrightarrow \varprojlim_n{}^1 H^{i-1}(G, A_n) \longrightarrow H^i_{cts}(G, A) \longrightarrow \varprojlim_n H^i(G, A_n) \longrightarrow 0.$$

Proof: Consider the categories $(G\text{-}\mathcal{Mod})^{\mathbb{N}}$ and $\mathcal{Ab}^{\mathbb{N}}$ of inverse systems of discrete G-modules and of abelian groups, respectively. Both categories are abelian and have sufficiently many injective objects. Furthermore, the functor $H^0(G, -)$ sends injective G-modules to injective (i.e. divisible) abelian groups: indeed, this is obvious for modules of the form $\mathrm{Ind}_G M$ for a divisible module M. If I is injective, then it is divisible and it is a direct summand of $\mathrm{Ind}_G I$. Let us consider the functor $\varprojlim H^0(G, -)$. By (2.7.4), applied to $\mathcal{B} = \mathcal{Ab}$, the spectral sequence for the composition of derived functors degenerates to a series of short exact sequences

$$0 \to \varprojlim_n{}^1 H^{i-1}(G, A_n) \to R^i(\varprojlim_n H^0(G, -))(A_n) \to \varprojlim_n H^i(G, A_n) \to 0.$$

---

*) $H^i_{cts}(G, A) \cong \mathrm{Ext}^i_{\text{profinite } G\text{-modules}}(\hat{\mathbb{Z}}, A)$, where G acts trivially on $\hat{\mathbb{Z}}$. This can easily be deduced from (5.2.14).

It therefore remains to show that there are isomorphisms for all i:

$$R^i(\varprojlim_n H^0(G,-))\,(A_n) \cong H^i_{cts}(G,A).$$

To prove this, let us first replace the system (A_n) by its subsystem (A'_n) with

$$A'_n = \bigcap_m \operatorname{im}(A_{n+m})$$

(since the A_i are finite, all intersections are finite too). The system A'_n has surjective transition maps and $A \cong \varprojlim A'_n$ as a topological G-module. The cokernel of the natural inclusion map $(A'_n) \hookrightarrow (A_n)$ is ML-zero. By (2.7.3), we have isomorphisms

$$R^i(\varprojlim H^0(G,-))\,(A'_n) \xrightarrow{\sim} R^i(\varprojlim H^0(G,-))\,(A_n)$$

for all $i \geq 0$. Therefore, replacing (A_n) by (A'_n), we may assume that all transition maps of the system (A_n) are surjective. By definition, $H^i_{cts}(G,A)$ is the i-th cohomology of the continuous homogeneous cochain complex $C^\bullet_{cts}(G,A)$, which can be identified with the inverse limit over n of $C^\bullet(G,A_n)$. The complexes $C^\bullet(G,A_n)$ are the G-invariants of the complexes $X^\bullet(G,A_n)$, which are acyclic resolutions of A_n for all n, see I §3. Moreover, these complexes are functorial, hence we get a $H^0(G,-)^{\mathbb{N}}$-acyclic resolution

$$(A_n) \hookrightarrow (X^\bullet(G,A_n)),$$

from which $H^i_{cts}(G,A)$ is obtained by applying $\varprojlim H^0(G,-)$ and taking cohomology. Hence it remains to show that the systems $(X^\bullet(G,A_n))$ are $\varprojlim H^0(G,-)$-acyclic, which follows easily from the fact that the systems of abelian groups $(C^i(G,A_n)) = (X^i(G,A_n)^G)$ have surjective transition maps for all i (cf. I §3, ex.1). □

(2.7.6) Corollary. *Let A be a compact G-module having a presentation*

$$A = \varprojlim_{n\in\mathbb{N}} A_n$$

as a countable inverse limit of finite, discrete G-modules. If $H^i(G,A_n)$ is finite for all n, then

$$H^{i+1}_{cts}(G,A) = \varprojlim_n H^{i+1}(G,A_n).$$

If $A = \varprojlim_{n\in\mathbb{N}} A_n$, where A_n is finite, then the corollary applies, for example, to $H^1_{cts}(G,A)$ and, if the group G is finitely generated, also to $H^2_{cts}(G,A)$.

As we did in I §2, p.17, in the case of a finite A, we consider for a profinite G-module A the set $EXT(G,A)$ of equivalence classes of exact sequences of profinite groups

$$1 \longrightarrow A \longrightarrow \hat{G} \longrightarrow G \longrightarrow 1$$

such that the action of G on A is given by

$${}^{\sigma}a = \hat{\sigma} a \hat{\sigma}^{-1},$$

where $\hat{\sigma} \in \hat{G}$ is a pre-image of $\sigma \in G$. The same proof as that of (1.2.4) shows also in this case the

(2.7.7) Theorem. *We have a canonical bijection of pointed sets*

$$H^2_{cts}(G, A) \cong EXT(G, A).$$

Let ℓ be a prime number and T a topological G-module which, as a topological group, is a finitely generated $\mathbb{Z}_\ell$-module with the natural topology, and on which G acts $\mathbb{Z}_\ell$-linearly.

(2.7.8) Proposition. *Let Y be a finitely generated $\mathbb{Z}_\ell$-submodule of $H^n_{cts}(G, T)$. Then the quotient group $H^n_{cts}(G, T)/Y$ contains no nontrivial ℓ-divisible subgroup.*

Proof: (cf. [232], prop. 2.1) Suppose $x_i \in H^n_{cts}(G, T)$, $0 \le i < \infty$, such that $x_i \equiv \ell x_{i+1} \bmod Y$ for all i. We must show $x_0 \in Y$. Let y_j, $1 \le j \le m$, be a finite set of generators for Y. For each i, let f_i be an n-cocycle representing x_i, and for each j, let g_j be an n-cocycle representing y_j. Then there are $(n-1)$-cochains h_i and elements $a_{ij} \in \mathbb{Z}_\ell$ such that

$$f_i = \ell f_{i+1} + \sum_{j=1}^{m} a_{ij} g_j + \partial h_i.$$

Hence

$$f_0 = \sum_{j=1}^{m} a_j g_j + \partial h$$

with $a_j = \sum_{i \ge 0} \ell^i a_{ij}$ and $h = \sum_{i \ge 0} \ell^i h_i$. The use of infinite sums here is formally justified by the fact that T is the inverse limit of its quotients $T/\ell^i T$ and consequently

$$C^\bullet_{cts}(G, T) = \varprojlim_i C^\bullet(G, T/\ell^i T)$$

is a projective limit of modules, each of which is killed by a fixed power of ℓ. This completes the proof. $\square$

(2.7.9) Corollary. *The $\mathbb{Z}_\ell$-module $H^n_{cts}(G,T)$ is finitely generated if and only if $H^n_{cts}(G,T)/\ell H^n_{cts}(G,T)$ is finite.*

Proof: In order to show the nontrivial assertion, assume that $y_1, \dots, y_m$ generate $H^n_{cts}(G,T)$ modulo ℓ. Putting $Y = \langle y_1, \dots, y_m \rangle$, we conclude that the group $H^n_{cts}(G,T)/Y$ is ℓ-divisible, hence trivial by the last proposition. □

(2.7.10) Corollary. *Assume that the cohomology groups of G with coefficients in finite ℓ-primary modules are finite. Then $H^n_{cts}(G,T)$ is a finitely generated $\mathbb{Z}_\ell$-module for all n and the canonical map*

$$H^n_{cts}(G,T) \otimes \mathbb{Q}_\ell/\mathbb{Z}_\ell \longrightarrow H^n(G, T \otimes \mathbb{Q}_\ell/\mathbb{Z}_\ell)$$

is an isogeny, i.e. has finite kernel and finite cokernel.

Proof: Replacing, if necessary, T by an open submodule, we may assume that T is torsion-free. Then the exact sequence $0 \to T \xrightarrow{\ell^m} T \to T/\ell^m \to 0$ implies the exact sequence

$$0 \longrightarrow H^n_{cts}(G,T)/\ell^m \longrightarrow H^n(G, T/\ell^m) \longrightarrow {}_{\ell^m}H^n_{cts}(G,T) \longrightarrow 0\,.$$

Now the statements follow from (2.7.9). □

Suppose now that T is torsion-free. Tensoring it, over $\mathbb{Z}_\ell$, by the exact sequence $0 \to \mathbb{Z}_\ell \to \mathbb{Q}_\ell \to \mathbb{Q}_\ell/\mathbb{Z}_\ell \to 0$ gives an exact sequence

$$(*) \qquad 0 \longrightarrow T \longrightarrow V \longrightarrow W \longrightarrow 0\,,$$

in which V is a finite dimensional $\mathbb{Q}_\ell$-vector space, T is an open compact subgroup and W is a discrete divisible ℓ-primary torsion group.

(2.7.11) Proposition. *There are isomorphisms for all n*

$$H^n_{cts}(G,V) \cong H^n_{cts}(G,T) \otimes_{\mathbb{Z}_\ell} \mathbb{Q}_\ell.$$

In the exact cohomology sequence associated with $(*)$, the kernel of the boundary homomorphism

$$\delta\colon\ H^{n-1}(G,W) \longrightarrow H^n_{cts}(G,T)$$

is the maximal divisible subgroup of $H^{n-1}(G,W)$, and its image is the torsion subgroup of $H^n_{cts}(G,T)$.

Proof: Since V is a vector space over $\mathbb{Q}_\ell$, so is $H^n_{cts}(G,V)$ for all n. Furthermore, $H^n(G,W)$ is an ℓ-torsion group for all n. By (2.7.2), we have a long exact sequence associated to $(*)$. Tensoring over $\mathbb{Z}_\ell$ with $\mathbb{Q}_\ell$ implies the first statement. Clearly,

$$\ker(H^{n-1}(G,W) \to H^n_{cts}(G,T)) = \operatorname{im}(H^{n-1}(G,V) \to H^{n-1}(G,W))$$

is ℓ-divisible. On the other hand, by (2.7.8), each divisible subgroup of $H^{n-1}(G,W)$ must be contained in the kernel. Since W is torsion, the group $\operatorname{im}(H^{n-1}(G,W) \to H^n_{cts}(G,T))$ is a torsion group. On the other hand, it is equal to the kernel of the map $H^n_{cts}(G,T) \to H^n_{cts}(G,V)$ and therefore must contain all torsion elements of $H^n_{cts}(G,T)$. □

(2.7.12) Corollary. *Assume that the cohomology groups of G with coefficients in finite ℓ-primary modules are finite. Then*

$$\begin{aligned} \operatorname{rank}_{\mathbb{Z}_\ell} H^n_{cts}(G,T) &= \dim_{\mathbb{Q}_\ell} H^n_{cts}(G,V) \\ &= \operatorname{corank}_{\mathbb{Z}_\ell} H^n(G, T \otimes \mathbb{Q}_\ell/\mathbb{Z}_\ell) \\ &= \operatorname{rank}_{\mathbb{Z}_\ell} H_n(G, \operatorname{Hom}(T,\mathbb{Z}_\ell))\,. \end{aligned}$$

Proof: The first two equalities follow from (2.7.11) and (2.7.10). Since T is finitely generated, we have

$$\operatorname{Hom}(T \otimes \mathbb{Q}_\ell/\mathbb{Z}_\ell, \mathbb{Q}_\ell/\mathbb{Z}_\ell) = \operatorname{Hom}(T, \operatorname{Hom}(\mathbb{Q}_\ell/\mathbb{Z}_\ell, \mathbb{Q}_\ell/\mathbb{Z}_\ell)) = \operatorname{Hom}(T,\mathbb{Z}_\ell)\,,$$

thus (2.6.9) implies the third equality. □

Continuous cochain cohomology was introduced by *J. Tate* in [232]. In addition, we have taken several arguments from the paper [96] of *U. Jannsen*.

Chapter III
Duality Properties of Profinite Groups

§1. Duality for Class Formations

Let G be a finite group. If A and B are two G-modules, the cup-product associated with the canonical pairing

$$\mathrm{Hom}(A,B)\times A\longrightarrow B,\quad (f,a)\mapsto f(a),$$

yields a pairing

$$\hat{H}^i(G,\mathrm{Hom}(A,B))\times\hat{H}^{n-i}(G,A)\stackrel{\cup}{\longrightarrow}\hat{H}^n(G,B),$$

which we call again the cup-product. When $B=\mathbb{Q}/\mathbb{Z}$, we set

$$A^*=\mathrm{Hom}(A,\mathbb{Q}/\mathbb{Z}).$$

(3.1.1) Proposition. *Let G be a finite group and let A be a G-module. Then for all $i\in\mathbb{Z}$, the pairing*

$$\hat{H}^i(G,A^*)\times\hat{H}^{-i-1}(G,A)\stackrel{\cup}{\longrightarrow}\hat{H}^{-1}(G,\mathbb{Q}/\mathbb{Z})=\tfrac{1}{\#G}\mathbb{Z}/\mathbb{Z}\subseteq\mathbb{Q}/\mathbb{Z}$$

induces an isomorphism

$$\hat{H}^i(G,A^*)\stackrel{\sim}{\longrightarrow}\hat{H}^{-i-1}(G,A)^*.$$

Remark: Via the identification (see (1.9.2)) $H_i=\hat{H}^{-i-1}$, $i>0$, we recover the isomorphism $H^i(G,A^*)\cong H_i(G,A)^*$ from (1.9.1).

Proof: First let $i=0$. A homomorphism $f:A\to\mathbb{Q}/\mathbb{Z}$ is a G-homomorphism, i.e. $f\in H^0(G,A^*)$, if and only if $f(I_GA)=0$. Therefore the map

$$H^0(G,A^*)\longrightarrow H_0(G,A)^*,$$

which associates to a G-homomorphism $f:A\to\mathbb{Q}/\mathbb{Z}$ the induced homomorphism $g:A/I_GA\to\mathbb{Q}/\mathbb{Z}$, is an isomorphism. If $f\in N_GA^*$, i.e. $f=\sum_{\sigma\in G}\sigma h$ for $h\in A^*$, then for $a\in {}_{N_G}A$ we have $f(a)=\sum(\sigma h)(a)=\sum h(\sigma^{-1}a)=h(N_Ga)=0$. This shows that we have a well-defined map

$$A^{*G}/N_GA^*\longrightarrow({}_{N_G}A/I_GA)^*.$$

J. Neukirch et al., *Cohomology of Number Fields*,
Grundlehren der mathematischen Wissenschaften 323,
DOI 10.1007/978-3-540-37889-1_3, © Springer-Verlag Berlin Heidelberg 2008

For the surjectivity, let $g : {}_{N_G}A \to \mathbb{Q}/\mathbb{Z}$ be a homomorphism that vanishes on $I_G A$. Since $\mathbb{Q}/\mathbb{Z}$ is an injective $\mathbb{Z}$-module, it can be extended to a homomorphism $g : A \to \mathbb{Q}/\mathbb{Z}$, which is a G-homomorphism because $g(I_G A) = 0$. This shows the surjectivity.

For the injectivity, let $f \in A^{*G}$ be such that $f({}_{N_G}A) = 0$. As

$$N_G : A/{}_{N_G}A \longrightarrow N_G A$$

is an isomorphism, there exists a $g \in (N_G A)^*$ such that $f(a) = g(N_G a)$ for all $a \in A$. We may extend g to a homomorphism $g : A \to \mathbb{Q}/\mathbb{Z}$ and then $f = N_G g$, since

$$(N_G g)(a) = \sum_{\sigma \in G} g(\sigma^{-1} a) = g(N_G a) = f(a).$$

This proves the injectivity.

Now let $i \in \mathbb{Z}$ be arbitrary. We use dimension shifting (see I §3, p.32) in order to reduce to the case $i = 0$. From (1.4.6) we obtain the commutative diagram

$$\begin{array}{ccccc}
\hat{H}^i(G, \mathrm{Hom}(A, \mathbb{Q}/\mathbb{Z})) & \times & \hat{H}^{-i-1}(G, A) & \stackrel{\cup}{\longrightarrow} & \hat{H}^{-1}(G, \mathbb{Q}/\mathbb{Z}) \\
\delta^i \big\uparrow \wr & & \delta^i \big\downarrow \wr & & (-1)^{i(i+1)/2} \big\downarrow \wr \\
\hat{H}^0(G, \mathrm{Hom}(A, \mathbb{Q}/\mathbb{Z})_i) & \times & \hat{H}^{-1}(G, A_{-i}) & \stackrel{\cup}{\longrightarrow} & \hat{H}^{-1}(G, \mathbb{Q}/\mathbb{Z}).
\end{array}$$

Since $\mathrm{Hom}(A, \mathbb{Q}/\mathbb{Z})_i \cong \mathrm{Hom}(A_{-i}, \mathbb{Q}/\mathbb{Z})$, the desired result follows. □

Next, we consider the case $B = \mathbb{Z}$ and get the following

(3.1.2) Proposition. *If G is a finite group and A is a $\mathbb{Z}$-free G-module, then for all integers $i \in \mathbb{Z}$, the pairing*

$$\hat{H}^i(G, \mathrm{Hom}(A, \mathbb{Z})) \times \hat{H}^{-i}(G, A) \stackrel{\cup}{\longrightarrow} \hat{H}^0(G, \mathbb{Z}) = \mathbb{Z}/\#G\mathbb{Z}$$

yields an isomorphism

$$\hat{H}^i(G, \mathrm{Hom}(A, \mathbb{Z})) \cong \hat{H}^{-i}(G, A)^*.$$

Proof: Since A is $\mathbb{Z}$-free, the sequence

$$0 \longrightarrow \mathrm{Hom}(A, \mathbb{Z}) \longrightarrow \mathrm{Hom}(A, \mathbb{Q}) \longrightarrow \mathrm{Hom}(A, \mathbb{Q}/\mathbb{Z}) \longrightarrow 0$$

is exact. Multiplication by $\#G$ on $\mathrm{Hom}(A, \mathbb{Q})$ is an isomorphism, hence also on the group $\hat{H}^n(G, \mathrm{Hom}(A, \mathbb{Q}))$, which is therefore zero by (1.6.1). For this reason, we get a commutative diagram

$$\begin{array}{ccccc}
\hat{H}^{i-1}(G, \mathrm{Hom}(A, \mathbb{Q}/\mathbb{Z})) & \times & \hat{H}^{-i}(G, A) & \xrightarrow{\cup} & \hat{H}^{-1}(G, \mathbb{Q}/\mathbb{Z}) \\
\downarrow{\scriptstyle\delta} & & \downarrow{\scriptstyle id} & & \downarrow{\scriptstyle\delta} \\
\hat{H}^{i}(G, \mathrm{Hom}(A, \mathbb{Z})) & \times & \hat{H}^{-i}(G, A) & \xrightarrow{\cup} & \hat{H}^{0}(G, \mathbb{Z}),
\end{array}$$

in which the vertical arrows are isomorphisms. Hence the proposition follows from (3.1.1). □

(3.1.3) Definition. *Let G be a finite group. We call a G-module C a* **class module** *if for all subgroups H of G*

(i) *$H^1(H, C) = 0$ and*

(ii) *$H^2(H, C)$ is cyclic of order $\#H$.*

A generator γ of $H^2(G, C)$ is called a **fundamental class**.

Obviously, C is also a class module for every subgroup H of G. Note that if γ is a generator of $H^2(G, C)$, then $\gamma_H = res_H^G \gamma$ is a generator of $H^2(H, C)$: since $cor \circ res\, \gamma = (G : H)\gamma$, the order of γ_H is divisible by $\#H$. If $U \subseteq G$ is normal, then C^U is not necessarily a class module for G/U.

If G is cyclic, then $\mathbb{Z}$ is a class module because of the isomorphisms (cf. (1.7.1)) $\hat{H}^1(H, \mathbb{Z}) \cong \hat{H}^{-1}(H, \mathbb{Z}) = 0$ and $H^2(H, \mathbb{Z}) \cong \hat{H}^0(H, \mathbb{Z}) = \mathbb{Z}/(\#H)\mathbb{Z}$. We have seen in (1.7.1) that the cup-product with a fundamental class is a periodicity operator for the cohomology of G, i.e.

$$\gamma \cup : \hat{H}^n(G, A) \xrightarrow{\sim} \hat{H}^{n+2}(G, A)$$

for every G-module A and for all $n \in \mathbb{Z}$.

We relate the property of being a class module to the property of cohomological triviality. To each G-module C and each class $\gamma \in H^2(G, C)$ we associate a G-module $C(\gamma)$ as follows. Let $B = \bigoplus_{\sigma \neq 1} \mathbb{Z} b_\sigma$ be the free abelian group with basis b_σ, indexed by the elements $\sigma \in G$, $\sigma \neq 1$. We set

$$C(\gamma) = C \oplus B,$$

and we let G act on $C(\gamma)$ by means of an inhomogeneous cocycle $c(\sigma, \tau)$ representing γ as follows: we set $b_1 = c(1, 1)$ and define

$$\sigma b_\tau = b_{\sigma\tau} - b_\sigma + c(\sigma, \tau).$$

This is really a G-action, i.e. $(\rho\sigma)b_\tau = \rho(\sigma b_\tau)$ and $1 b_\tau = b_\tau$, because of the cocycle relation $\rho c(\sigma, \tau) - c(\rho\sigma, \tau) + c(\rho, \sigma\tau) - c(\rho, \sigma) = 0$ *). The G-module $C(\gamma)$ is constructed in such a way that the cocycle $c(\sigma, \tau)$ becomes necessarily

---

*) The reader may check that, up to isomorphism, $C(\gamma)$ depends only on the class γ, not on the cocycle $c(\sigma, \tau)$.

in $C(\gamma)$ a coboundary $c(\sigma,\tau) = \sigma b(\tau) - b(\sigma\tau) + b(\sigma)$, where the 1-cochain $b(\sigma)$ is defined by $b(\sigma) = b_\sigma$. Thus $H^2(G,C) \to H^2(G,C(\gamma))$ maps γ to zero. $C(\gamma)$ is therefore called the **splitting module** of γ. *)

We obtain a four term exact sequence

$$0 \longrightarrow C \xrightarrow{\ i\ } C(\gamma) \xrightarrow{\ \varphi\ } \mathbb{Z}[G] \longrightarrow \mathbb{Z} \longrightarrow 0$$

of G-modules if we define φ by

$$\varphi(c) = 0 \text{ for } c \in C \text{ and } \varphi(b_\sigma) = \sigma - 1 \text{ for } \sigma \neq 1.$$

Splitting up this sequence into two short exact sequences, we obtain, for each $n \in \mathbb{Z}$ and each subgroup $H \subseteq G$, a homomorphism

$$\delta^2 : \hat{H}^n(H,\mathbb{Z}) \longrightarrow \hat{H}^{n+2}(H,C).$$

(3.1.4) Theorem. *Let G be a finite group. For each $n \in \mathbb{Z}$ and each subgroup $H \subseteq G$, the homomorphism*

$$\delta^2 : \hat{H}^n(H,\mathbb{Z}) \to \hat{H}^{n+2}(H,C)$$

is given by the cup-product $\beta \mapsto \gamma_H \cup \beta$, where $\gamma_H = res^G_H(\gamma)$. The following conditions are equivalent.

(i) *$C(\gamma)$ is a cohomologically trivial G-module,*

(ii) *C is a class module with fundamental class γ,*

(iii) *δ^2 is an isomorphism for all $n \in \mathbb{Z}$ and all H.*

Remark: If C is a class module for G, then, by the above theorem, we have isomorphisms

$$(\delta^2)^{-1} : H^2(H,C) \xrightarrow{\ \sim\ } \tfrac{1}{\#H}\mathbb{Z}/\mathbb{Z}, \quad \gamma_H \mapsto \tfrac{1}{\#H} \bmod \mathbb{Z},$$

where $\gamma \in H^2(G,C)$ is a chosen fundamental class. These are called **invariant maps** and denoted by *inv*.

Proof: The map δ^2 arises from the two exact sequences

(1) $$0 \longrightarrow I_G \longrightarrow \mathbb{Z}[G] \longrightarrow \mathbb{Z} \longrightarrow 0,$$

(2) $$0 \longrightarrow C \longrightarrow C(\gamma) \longrightarrow I_G \longrightarrow 0$$

and is the composite of the maps

(3) $$\hat{H}^n(H,\mathbb{Z}) \xrightarrow[\sim]{\ \delta_1\ } \hat{H}^{n+1}(H,I_G) \xrightarrow{\ \delta_2\ } \hat{H}^{n+2}(H,C),$$

where δ_1 is always an isomorphism. For $n = 0$ we have the maps

(4) $$\mathbb{Z}/\#H\mathbb{Z} = \hat{H}^0(H,\mathbb{Z}) \xrightarrow{\ \delta_1\ } H^1(H,I_G) \xrightarrow{\ \delta_2\ } H^2(H,C).$$

*) The module $C(\gamma)$ also arises as the extension corresponding to the class $\gamma \in H^2(G,C)$ via the isomorphisms $H^2(G,C) \cong \mathrm{Ext}^2_{\mathbb{Z}[G]}(\mathbb{Z},C) \cong \mathrm{Ext}^1_{\mathbb{Z}[G]}(I_G,C)$.

For the generator $\bar{1} = 1 \bmod \#H$ of $\hat{H}^0(H,\mathbb{Z})$, we have

$$(5) \qquad \delta_2\delta_1\bar{1} = \gamma_H = res_H^G\gamma.$$

In fact, a pre-image of the 0-cocycle $1 \in \mathscr{Z}^0(H,\mathbb{Z})$ in $\mathscr{C}^0(H,\mathbb{Z}[G])$ is $1 \in \mathbb{Z}[G]$ and $\delta_1\bar{1} \in H^1(H, I_G)$ is represented by the 1-cocycle $(\partial 1)(\sigma) = \sigma - 1$. A lift of $\partial 1$ in $\mathscr{C}^1(H, C(\gamma))$ is given by $x(\sigma) = b_\sigma$, and $\delta_2\delta_1\bar{1}$ is represented by

$$(\partial x)(\sigma,\tau) = \sigma b_\tau - b_{\sigma\tau} + b_\sigma = c(\sigma,\tau)\,.$$

This proves (5). Now let $\beta \in \hat{H}^n(H,\mathbb{Z})$, where n is arbitrary. Applying proposition (1.4.3) to $B = \mathbb{Z}$ and to the two exact sequences (1) and (2), we obtain

$$\delta^2\beta = \delta_2\delta_1(\bar{1}\cup\beta) = \delta_2(\delta_1\bar{1}\cup\beta) = \delta_2\delta_1\bar{1}\cup\beta = \gamma_H\cup\beta.$$

Noting that $\hat{H}^i(H, I_G) = \hat{H}^{i-1}(H,\mathbb{Z}) = 0$ for $i = 0$ and 2, we obtain from (2) the exact sequence

$$0 \longrightarrow H^1(H,C) \longrightarrow H^1(H,C(\gamma)) \longrightarrow H^1(H,I_G)$$
$$\xrightarrow{\delta} H^2(H,C) \longrightarrow H^2(H,C(\gamma)) \longrightarrow 0.$$

If $C(\gamma)$ is cohomologically trivial, then $H^1(H,C) = 0$ and the composite

$$\mathbb{Z}/\#H\mathbb{Z} = \hat{H}^0(H,\mathbb{Z}) \xrightarrow[\sim]{\delta} H^1(H,I_G) \xrightarrow{\delta} H^2(H,C)$$

is an isomorphism which maps $\bar{1}$ onto γ_H. Therefore C is a class module with fundamental class γ. Conversely, if this is true, then $H^1(H,C) = 0$ and $\delta : H^1(H,I_G) \to H^2(H,C)$ is an isomorphism; hence $H^i(H,C(\gamma)) = 0$ for $i = 1,2$, and therefore for all i by (1.8.4).

The equivalence (i) $\Leftrightarrow$ (iii) follows from (3) and from the exact sequence

$$\ldots \longrightarrow \hat{H}^n(H,C) \longrightarrow \hat{H}^n(H,C(\gamma)) \longrightarrow \hat{H}^n(H,I_G)$$
$$\xrightarrow{\delta} \hat{H}^{n+1}(H,C) \longrightarrow \hat{H}^{n+1}(H,C(\gamma)) \longrightarrow \ldots \qquad \square$$

For $\mathbb{Z}$-free G modules A we can now prove the following duality theorem.

(3.1.5) Theorem (Nakayama-Tate). *Let G be a finite group, let C be a class module for G and let $\gamma \in H^2(G,C)$ be a fundamental class. Then, for all integers $i \in \mathbb{Z}$, the cup-product*

$$\hat{H}^i(G,\mathrm{Hom}(A,C)) \times \hat{H}^{2-i}(G,A) \xrightarrow{\cup} H^2(G,C) \cong \tfrac{1}{\#G}\mathbb{Z}/\mathbb{Z},$$

where $H^2(G,C) \cong \frac{1}{\#G}\mathbb{Z}/\mathbb{Z}$ is given by $\gamma \mapsto \frac{1}{\#G} \bmod \mathbb{Z}$, induces an isomorphism

$$\hat{H}^i(G,\mathrm{Hom}(A,C)) \cong \hat{H}^{2-i}(G,A)^*,$$

provided that A is $\mathbb{Z}$-free. If, in addition, A is finitely generated, then this is an isomorphism of finite abelian groups.

Proof: Let $0 \longrightarrow C \longrightarrow C(\gamma) \longrightarrow I_G \longrightarrow 0$ be an exact sequence as in the proof of (3.1.4). As A is $\mathbb{Z}$-free, the sequences

$$0 \longrightarrow \operatorname{Hom}(A,C) \longrightarrow \operatorname{Hom}(A,C(\gamma)) \longrightarrow \operatorname{Hom}(A,I_G) \longrightarrow 0$$

$$0 \longrightarrow \operatorname{Hom}(A,I_G) \longrightarrow \operatorname{Hom}(A,\mathbb{Z}[G]) \longrightarrow \operatorname{Hom}(A,\mathbb{Z}) \longrightarrow 0$$

are exact and the G-modules in the middle are cohomologically trivial by (1.8.5), (3.1.4), and (1.3.7). We now have for $i \in \mathbb{Z}$ a commutative diagram

$$\begin{array}{ccccc}
\hat{H}^{i-2}(G,\operatorname{Hom}(A,\mathbb{Z})) & \times & \hat{H}^{2-i}(G,A) & \xrightarrow{\cup} & \hat{H}^0(G,\mathbb{Z}) \\
\downarrow{\scriptstyle\delta} & & \downarrow{\scriptstyle id} & & \downarrow{\scriptstyle\delta} \\
\hat{H}^{i-1}(G,\operatorname{Hom}(A,I_G)) & \times & \hat{H}^{2-i}(G,A) & \xrightarrow{\cup} & \hat{H}^1(G,I_G) \\
\downarrow{\scriptstyle\delta} & & \downarrow{\scriptstyle id} & & \downarrow{\scriptstyle\delta} \\
\hat{H}^{i}(G,\operatorname{Hom}(A,C)) & \times & \hat{H}^{2-i}(G,A) & \xrightarrow{\cup} & \hat{H}^2(G,C),
\end{array}$$

where the vertical arrows are isomorphisms. Hence the asserted isomorphism follows from (3.1.2). If A is finitely generated, then the cohomology groups $\hat{H}^i(G,A)$ are finite, and so are their duals, by (1.6.1). □

For a subgroup H of G we have commutative diagrams

$$\begin{array}{ccc}
\hat{H}^{i}(H,\operatorname{Hom}(A,C)) & \xrightarrow{\sim} & \hat{H}^{2-i}(H,A)^* \\
{\scriptstyle res}\uparrow\downarrow{\scriptstyle cor} & & {\scriptstyle cor^*}\uparrow\downarrow{\scriptstyle res^*} \\
\hat{H}^{i}(G,\operatorname{Hom}(A,C)) & \xrightarrow{\sim} & \hat{H}^{2-i}(G,A)^*
\end{array}$$

where cor^* and res^* are the maps dual to the corestriction and restriction, and the upper map relies on the choice of the fundamental class $\gamma_H \in \hat{H}^2(H,C)$. The commutativity of the diagram follows at once from $cor(\gamma \cup res\,\beta) = (cor\;\gamma) \cup \beta$ (see (1.5.3)(iv)) and $inv \circ cor = inv$.

Applying the duality theorem to the case $A = \mathbb{Z}$, $i = 0$, and recalling

$$H^2(G,\mathbb{Z})^* \cong H^1(G,\mathbb{Q}/\mathbb{Z})^* = \operatorname{Hom}(G^{ab},\mathbb{Q}/\mathbb{Z})^* = G^{ab},$$

we obtain

(3.1.6) Theorem. *If C is a class module for the finite group G, then we have an isomorphism*

$$\rho = \rho_G : G^{ab} \xrightarrow{\sim} C^G/N_G C,$$

called the **Nakayama map**. *It depends on the choice of a fundamental class $\gamma \in H^2(G,C)$ and satisfies (by definition) the formula*

$$\chi(\sigma) = inv(\rho(\sigma) \cup \delta\chi)$$

for all characters $\chi \in H^1(G,\mathbb{Q}/\mathbb{Z}) \underset{\sim}{\xrightarrow{\delta}} H^2(G,\mathbb{Z})$.

The inverse map

$$\rho^{-1} : C^G/N_G C \xrightarrow{\sim} G^{ab}$$

is often called the reciprocity isomorphism. We call its composite with the natural projection $C^G \twoheadrightarrow C^G/N_G C$, i.e. the map

$$rec = rec_G : C^G \longrightarrow\!\!\!\!\!\!\rightarrow G^{ab},$$

the **reciprocity homomorphism.**

The reciprocity homomorphism is also called the **norm residue symbol** and, if $\alpha \in C^G$, we also write

$$(\alpha, G) := rec(\alpha).$$

The name norm residue symbol reflects the fact that this symbol detects whether $\alpha \in C^G$ is the norm of an element in C. Indeed, $(\alpha, G) = 0$ if and only if $\alpha \in N_G C$.

The isomorphism $G^{ab} \cong C^G/N_G C$ is an abstract version of class field theory. If $L|K$ is a finite Galois extension of local or global fields, then the multiplicative group $L^\times$ is a class module in the local case, and the idèle class group C_L in the global case, as we shall see. Combined with the above theorem this gives the main theorem of local and global class field theory.

The following theorem gives an explicit description of the Nakayama map.

(3.1.7) Theorem. *If C is a class module for the finite group G and $\gamma \in H^2(G, C)$ is a fundamental class, then the Nakayama map*

$$\rho : G^{ab} \xrightarrow{\sim} C^G/N_G C$$

is explicitly given by

$$\sigma \bmod [G, G] \longmapsto \sum_\tau c(\tau, \sigma) \bmod N_G C,$$

where c is a cocycle representing γ.

Proof: We will prove the theorem by means of the diagram

$$(*) \qquad \begin{array}{ccccc} G^{ab} & \times\ H^1(G, \mathbb{Q}/\mathbb{Z}) & \longrightarrow & \frac{1}{\#G}\mathbb{Z}/\mathbb{Z} \\ {\scriptstyle\nu}\downarrow\downarrow{\scriptstyle\rho} & \downarrow{\scriptstyle\delta} & & \uparrow{\scriptstyle inv} \\ \hat{H}^0(G, C) & \times\ H^2(G, \mathbb{Z}) & \xrightarrow{\ \cup\ } & H^2(G, C), \end{array}$$

in which the upper horizontal arrow is the evaluation map $(\sigma,\chi)\mapsto\chi(\sigma)$, $inv(\gamma)=\frac{1}{\#G}\bmod\mathbb{Z}$, where ρ is the Nakayama isomorphism and ν is the map given by

$$\bar\sigma\longmapsto\sum_\tau c(\tau,\sigma)\mod N_GC\,,\qquad {}^{*)}$$

where $\bar\sigma=\sigma\bmod[G,G]$. It suffices to show that the diagram is commutative with both maps ν and ρ, since then theorem (3.1.5) identifies ν and ρ with the dual δ^* of δ. The commutativity of the diagram with the map ρ follows from the definition of ρ and is an equivalent version of the formula

$$\chi(\sigma)=inv(\rho(\sigma)\cup\delta\chi)\,,$$

see (3.1.5). It remains to show the commutativity for ν. For this we consider the larger diagram

$$\begin{array}{ccccc}
G^{ab} & \times\ H^1(G,\mathbb{Q}/\mathbb{Z}) & \longrightarrow & \hat H^{-1}(G,\mathbb{Q}/\mathbb{Z})=\frac{1}{\#G}\mathbb{Z}/\mathbb{Z}\\
{\scriptstyle\delta_1}\downarrow & {\scriptstyle id}\downarrow & & {\scriptstyle\delta_3}\downarrow\\
\hat H^{-1}(G,I_G) & \times\ H^1(G,\mathbb{Q}/\mathbb{Z}) & \stackrel{\cup}{\longrightarrow} & \hat H^0(G,I_G\otimes\mathbb{Q}/\mathbb{Z})\\
{\scriptstyle id}\downarrow & {\scriptstyle\delta}\downarrow & & {\scriptstyle-\delta_4}\downarrow\\
\hat H^{-1}(G,I_G) & \times\ H^2(G,\mathbb{Z}) & \stackrel{\cup}{\longrightarrow} & H^1(G,I_G)\\
{\scriptstyle\delta_2}\downarrow & {\scriptstyle id}\downarrow & & {\scriptstyle\delta_5}\downarrow\\
\hat H^0(G,C) & \times\ H^2(G,\mathbb{Z}) & \stackrel{\cup}{\longrightarrow} & H^2(G,C),
\end{array}$$

in which $\delta_1: G^{ab}\to\hat H^{-1}(G,I_G)=I_G/I_G^2$ is defined by $\bar\sigma\mapsto\sigma-1\bmod I_G^2$ and the other δ_i's arise from the exact sequences

$$(1)\qquad 0\longrightarrow I_G\longrightarrow\mathbb{Z}[G]\longrightarrow\mathbb{Z}\longrightarrow 0,$$

$$(2)\qquad 0\longrightarrow I_G\otimes\mathbb{Q}/\mathbb{Z}\longrightarrow\mathbb{Z}[G]\otimes\mathbb{Q}/\mathbb{Z}\longrightarrow\mathbb{Q}/\mathbb{Z}\longrightarrow 0,$$

$$(3)\qquad 0\longrightarrow I_G\longrightarrow I_G\otimes\mathbb{Q}\longrightarrow I_G\otimes\mathbb{Q}/\mathbb{Z}\longrightarrow 0,$$

$$(4)\qquad 0\longrightarrow C\longrightarrow C(\gamma)\longrightarrow I_G\longrightarrow 0.$$

We make the homomorphisms δ explicit by choosing in (1) - (4) sections of the last arrows, by which the cocycles with coefficients in the last groups are sent to cochains with coefficients in the middle groups, to which we apply the coboundary operator ∂. For example, the map

$$C(\gamma)=C\oplus\bigoplus_{\sigma\neq1}\mathbb{Z}b_\sigma\longrightarrow I_G$$

---

*) Note that the sum is contained in C^G because of the cocycle relation $\rho c(\tau,\sigma)=c(\rho\tau,\sigma)+c(\rho,\tau)-c(\rho,\tau\sigma)$ and the fact that $\rho\tau$ and $\tau\sigma$ run through G if τ does.

is defined by $C \to 0$ and $b_\sigma \mapsto \sigma - 1$. It has the section $\sigma - 1 \mapsto b_\sigma$, and $b : G \to C(\gamma)$, $\sigma \mapsto b_\sigma$, is an inhomogeneous cochain with coboundary $(\partial^2 b)(\sigma, \tau) = c(\sigma, \tau)$, which lifts the 1-cocycle $x : G \to I_G$, $x(\sigma) = \sigma - 1$. We set $g = \#G$, $\bar{a} = a \bmod \mathbb{Z}$ for $a \in \mathbb{Q}$ with $ga \in \mathbb{Z}$. Noting that the elements in $(I_G \otimes \mathbb{Q}/\mathbb{Z})^G$ are of the form $\sum_\tau (\tau - 1) \otimes \bar{a} = N_G \otimes \bar{a}$, the δ's are given as follows

$$\begin{aligned}
\delta_1 &: \bar{\sigma} &&\longmapsto \sigma - 1 \bmod I_G^2,\\
\delta_2 &: \sigma - 1 &&\longmapsto \sum_\tau c(\tau, \sigma) \bmod N_G C,\\
\delta_3 &: \bar{a} &&\longmapsto \sum_\tau (\tau - 1) \otimes \bar{a},\\
-\delta_4 &: N_G \otimes \bar{a} &&\longmapsto gax,\\
\delta_5 &: x &&\longmapsto c.
\end{aligned}$$

The first assertion holds by definition and the others rely on the relations

$$\begin{aligned}
&\partial^0 b_\sigma = \sum_\tau \tau b_\sigma = \sum_\tau b_{\tau\sigma} - \sum_\tau b_\tau + \sum_\tau c(\tau, \sigma) = \sum_\tau c(\tau, \sigma),\\
&\partial^0 (1 \otimes \bar{a}) = N_G (1 \otimes \bar{a}) = \sum_\tau \tau \otimes \bar{a} = \sum_\tau (\tau - 1) \otimes \bar{a},\\
&\partial^1 (\sum_\tau (\tau - 1) \otimes a)(\rho) = (\rho - 1)[(N_G \otimes a) - ga1] = -ga(\rho - 1),\\
&(\partial^2 b)(\sigma, \tau) = \sigma b_\tau - b_{\sigma\tau} + b_\sigma = c(\sigma, \tau).
\end{aligned}$$

This result shows that the composite $-\delta_5\delta_4\delta_3$ is inverse to *inv*, since it maps $\frac{1}{g} \bmod \mathbb{Z}$ to γ.

The upper partial diagram is commutative, because $\delta_1(\sigma) \cup \chi$ is represented by

$$\begin{aligned}
\sum_\tau \tau(\sigma - 1) \otimes \chi(\tau^{-1}) &= \sum_\tau \tau\sigma \otimes \chi(\tau^{-1}) - \sum_\tau \tau \otimes \chi(\tau^{-1})\\
&= \sum_\tau \tau \otimes \chi(\sigma\tau^{-1}) - \sum_\tau \tau \otimes \chi(\tau^{-1})\\
&= \sum_\tau \tau \otimes \chi(\sigma) = \sum_\tau (\tau - 1) \otimes \chi(\sigma)
\end{aligned}$$

(see (1.4.8)), hence by the same element as $\delta_3(\chi(\sigma))$. The middle and the lower partial diagrams are commutative by (1.4.3) and the remark following (1.4.6). This proves that the diagram $(*)$ is commutative. □

From now on let G be a profinite group. We extend the notion of "class module", which we needed for the duality theorem (3.1.5), to profinite groups as follows. We denote open subgroups of G by the letters U, V, W.

(3.1.8) Definition. *Let G be a profinite group. A* **formation module** *for G is a discrete G-module C together with a system of isomorphisms*

$$inv_{U/V} : H^2(U/V, C^V) \xrightarrow{\sim} \tfrac{1}{(U:V)}\mathbb{Z}/\mathbb{Z}$$

for every pair $V \subseteq U$ of open subgroups, V normal in U, such that the following conditions hold.

(i) $H^1(U/V, C^V) = 0$.

(ii) *For open normal subgroups $W \subseteq V$ of an open subgroup U, the diagram*

$$\begin{array}{ccccc} H^2(U/V, C^V) & \xrightarrow{inf} & H^2(U/W, C^W) & \xrightarrow{res} & H^2(V/W, C^W) \\ {\scriptstyle inv}\downarrow\wr & & {\scriptstyle inv}\downarrow\wr & & {\scriptstyle inv}\downarrow\wr \\ \frac{1}{(U:V)}\mathbb{Z}/\mathbb{Z} & \xrightarrow{incl} & \frac{1}{(U:W)}\mathbb{Z}/\mathbb{Z} & \xrightarrow{(U:V)} & \frac{1}{(V:W)}\mathbb{Z}/\mathbb{Z} \end{array}$$

is commutative.

The pair (G, C) is called a **class formation**.

Remarks: 1. For finite G the notion of a formation module is stronger than that of a class module because there is no compatibility condition for the passage to quotients for the latter.

2. From (ii) it follows that the diagram

$$\begin{array}{ccc} H^2(V/W, C^W) & \xrightarrow{inv} & \frac{1}{(V:W)}\mathbb{Z}/\mathbb{Z} \\ {\scriptstyle cor}\downarrow & & \downarrow{\scriptstyle incl} \\ H^2(U/W, C^W) & \xrightarrow{inv} & \frac{1}{(U:W)}\mathbb{Z}/\mathbb{Z} \end{array}$$

is commutative because $cor_U^V \circ res_V^U = (U : V)$.

3. The isomorphisms

$$inv : H^2(G/V, C^V) \xrightarrow{\sim} \tfrac{1}{(G:V)}\mathbb{Z}/\mathbb{Z}$$

form a direct system. Passing to the direct limit, we obtain a homomorphism

$$inv : H^2(G, C) \longrightarrow \mathbb{Q}/\mathbb{Z},$$

which is called the **invariant map**. It is injective and its image is $\frac{1}{\#G}\mathbb{Z}/\mathbb{Z} = \varinjlim_V \frac{1}{(G:V)}\mathbb{Z}/\mathbb{Z}$.

4. In terms of G-modulations (see I §5), a formation module for G is a discrete G-module C together with an isomorphism

$$inv : H^2(C) \to \hat{H}^0(\mathbb{Z})^*$$

of G-modulations, which has the additional property that $H^1(C) = 0$.

5. For every open normal subgroup $U \subseteq G$, the module of invariants C^U is a class module for G/U. The elements $inv_{G/U}^{-1}(\frac{1}{(G:U)} \bmod \mathbb{Z}) \in H^2(G/U, C^U)$ form a compatible system of fundamental classes for varying U. We therefore obtain reciprocity homomorphisms $rec_{G/U} : C^G \twoheadrightarrow (G/U)^{ab}$ which are compatible for open normal subgroups $V \subseteq U$ in the sense that $rec_{G/U}$ is the composition of $rec_{G/V}$ with the natural projection $(G/V)^{ab} \twoheadrightarrow (G/U)^{ab}$. Passing to the projective limit, we therefore obtain the **reciprocity homomorphism**

$$rec = rec_G : C^G \longrightarrow G^{ab},$$

which is also called the **norm residue symbol**. It has dense image and kernel $N_G C$.

Our aim is to prove a version of theorem (3.1.5) for profinite groups. We consider a discrete G-module A which is finitely generated as a $\mathbb{Z}$-module. Then, for any discrete G-module B, $\mathrm{Hom}(A, B)$ is a discrete G-module and the pairing

$$\mathrm{Hom}(A, B) \times A \longrightarrow B, \quad (f, a) \longmapsto f(a),$$

induces a cup-product

$$\hat{H}^i(G, \mathrm{Hom}(A, B)) \times \hat{H}^{2-i}(G, A) \longrightarrow H^2(G, B)$$

for all $i \in \mathbb{Z}$: for $i = 1$ this is the usual cup-product and for $i \neq 1$ it is the cup-product constructed in chap. I, §9, see (1.9.7).

We recall the definition of the topology on the Tate cohomology groups of A from chap. I, §9. By (1.6.1), the groups $\hat{H}^i(G/U, A^U)$ are finite for all i and all open normal subgroups U of G. Therefore the groups

$$\hat{H}^i(G, A) = \varprojlim_{U, def} \hat{H}^i(G/U, A^U), \quad i \leq 0,$$

are naturally equipped with a profinite topology. For $i > 0$ we give $\hat{H}^i(G, A)$ the discrete topology. Now assume that C is a formation module for G and that A is free as a $\mathbb{Z}$-module. For sufficiently small U we have $A^U = A$, and hence $\mathrm{Hom}(A, C^U) = \mathrm{Hom}(A, C)^U$. Therefore the groups

$$\hat{H}^i(G, \mathrm{Hom}(A, C)) = \varprojlim_{U, def} \hat{H}^i(G/U, \mathrm{Hom}(A, C)^U), \quad i \leq 0,$$

are naturally equipped with a profinite topology, as the groups on the right hand side are finite by (3.1.5). For $i \geq 1$, we give $\hat{H}^i(G, \mathrm{Hom}(A, C))$ the discrete topology.

(3.1.9) Theorem. *Let G be a profinite group and let C be a formation module for G. Let A be a discrete G-module which is finitely generated and free as a $\mathbb{Z}$-module. Then the cup-product*

$$\hat{H}^i(G, \mathrm{Hom}(A, C)) \times \hat{H}^{2-i}(G, A) \xrightarrow{\cup} H^2(G, C) \xrightarrow[inv]{\sim} \tfrac{1}{\#G}\mathbb{Z}/\mathbb{Z},$$

induces a topological isomorphism

$$\hat{H}^i(G, \mathrm{Hom}(A, C)) \xrightarrow{\sim} \hat{H}^{2-i}(G, A)^\vee .$$

for all $i \in \mathbb{Z}$.

Proof: For sufficiently small U we have $A^U = A$ and therefore $\mathrm{Hom}(A, C^U) = \mathrm{Hom}(A, C)^U$. Thus, for $i \neq 1$, the theorem follows easily from (3.1.5) and (1.9.6) by passing to the limit over G/U, where U runs through the open normal subgroups in G. For $i = 1$, let $U \subseteq G$ be an open normal subgroup which acts trivially on A. Since A is $\mathbb{Z}$-free, $H^1(U/V, A^V) = 0$ for every open normal subgroup $V \subseteq U$ and therefore

$$H^1(G/U, A^U) \rightarrowtail H^1(G/V, A^V).$$

Using the fact that C is a formation module for G, we obtain

$$H^1(U/V, \mathrm{Hom}(A, C)^V) = H^1(U/V, \mathrm{Hom}(A, C^V)) \cong H^1(U/V, C^V)^{\mathrm{rank}_{\mathbb{Z}} A} = 0.$$

This implies that

$$H^1(G/U, \mathrm{Hom}(A, C)^U) \rightarrowtail H^1(G/V, \mathrm{Hom}(A, C)^V).$$

A stationary limit process shows that (3.1.5) implies the desired result also in dimension 1. □

Recall from (1.9.10) that a level-compact G-module is a discrete G-module C with a natural compact topology on C^U for each open subgroup $U \subseteq G$.

(3.1.10) Corollary. *Let G be a profinite group and let C be a level-compact formation module for G. Then the reciprocity homomorphism*

$$rec : C^G \longrightarrow G^{ab}$$

is surjective.

Proof: We apply (3.1.9) for $i = 0$ and $A = \mathbb{Z}$. The reciprocity map *rec* is the composite

$$C^G \longrightarrow \hat{H}^0(G, C) \xrightarrow{\sim} H^2(G, \mathbb{Z})^\vee = G^{ab}.$$

If C is level-compact, then, by (1.9.11), the first arrow is surjective. □

The central result of this section is the following duality theorem for level-compact class formations, which generalizes (3.1.9) to the case when A is not necessarily free. In what follows the reader should note the following

Topological Remarks: Let A be a discrete G-module which is finitely generated as a $\mathbb{Z}$-module and let C be a level-compact G-module.

For sufficiently small V we have $A^V = A$ and therefore

$$\mathrm{Hom}(A, C)^V = \mathrm{Hom}(A, C^V).$$

Considering A as a group with the discrete topology, $\mathrm{Hom}(A, C^V)$ endowed with the compact open topology is compact. For an arbitrary open subgroup $U \subseteq G$, the group $\mathrm{Hom}(A, C)^U$ is a closed subgroup of the compact group $\mathrm{Hom}(A, C)^V$, where V is chosen sufficiently small. We conclude that $\mathrm{Hom}(A, C)$ is level-compact in a natural way. In particular, the groups $\hat{H}^i(G, \mathrm{Hom}(A, C))$ carry a natural profinite topology for $i \le 0$. If, as happens in the situation of theorem (3.1.9), the groups $\hat{H}^i(G/U, \mathrm{Hom}(A, C)^U)$ are finite for all U, then the knowledge of the topology of the groups $\mathrm{Hom}(A, C)^U$ is redundant for an understanding of the profinite topology of the groups $\hat{H}^i(G, \mathrm{Hom}(A, C))$, $i \le 0$. Finally, recall that $\hat{H}^i(G, \mathrm{Hom}(A, C))$ carries the discrete topology for $i \ge 1$ by convention.

In addition, assume that C is a formation module. Let $U \subseteq G$ be an open normal subgroup such that $A^U = A$. Then, for $i \in \mathbb{Z}$, the cup-product

$$\hat{H}^i(G/U, \mathrm{Hom}(A, C)^U) \times \hat{H}^{2-i}(G/U, A^U) \xrightarrow{\cup} H^2(G/U, C^U) \underset{inv}{\cong} \tfrac{1}{(G:U)}\mathbb{Z}/\mathbb{Z}$$

is continuous with respect to the topologies induced on the cohomology groups by the complete standard complex. As the group $H^2(G/U, C^U)$ is compact and finite, it is discrete. The groups $\hat{H}^{2-i}(G/U, A^U)$ are discrete and finite by (1.6.1). Thus, for each $i \le 0$, we obtain a continuous homomorphism

$$\hat{H}^i(G/U, \mathrm{Hom}(A, C)^U) \longrightarrow \hat{H}^{2-i}(G/U, A^U)^\vee$$

from a compact group to a finite discrete group. Passing to the projective limit over all U, we obtain a continuous homomorphism

$$\hat{H}^i(G, \mathrm{Hom}(A, C)) \longrightarrow H^{2-i}(G, A)^\vee$$

between compact abelian groups. For $i \ge 2$ and U as above, the cup-product defines a homomorphism

$$H^i(G/U, \mathrm{Hom}(A, C)^U) \longrightarrow \hat{H}^{2-i}(G/U, A)^\vee.$$

Passing to the direct limit, we obtain the (continuous) homomorphism

$$H^i(G, \mathrm{Hom}(A, C)) \longrightarrow \hat{H}^{2-i}(G, A)^\vee$$

between discrete groups. Finally, for $i = 1$, the usual cup product induces a (continuous) homomorphism

$$H^1(G, \mathrm{Hom}(A, C)) \longrightarrow H^1(G, A)^\vee$$

from a discrete to a compact group.

(3.1.11) Duality Theorem. *Let G be a profinite group and let C be a level-compact formation module for G such that the group of universal norms $N_U C$ is divisible for all open subgroups U of G. Let A be a discrete G-module which is finitely generated as a $\mathbb{Z}$-module. Consider the cup-product*

$$\hat{H}^i(G, \operatorname{Hom}(A, C)) \times \hat{H}^{2-i}(G, A) \xrightarrow{\cup} H^2(G, C) \underset{inv}{\xrightarrow{\sim}} \tfrac{1}{\#G}\mathbb{Z}/\mathbb{Z}\,.$$

The following assertions are true:

(i) *Let $i \leq 0$. Then the pairing above induces a topological isomorphism*

$$\hat{H}^i(G, \operatorname{Hom}(A, C)) \xrightarrow{\sim} \hat{H}^{2-i}(G, A)^\vee\,.$$

This holds true for all $i \in \mathbb{Z}$ if A is $\mathbb{Z}$-free.

(ii) *Let $i \in \mathbb{Z}$ and let p be a prime number with $p^\infty | \#G$. Assume that C is p-divisible. Then the pairing above induces an isomorphism*

$$\hat{H}^i(G, \operatorname{Hom}(A, C))(p) \xrightarrow{\sim} \hat{H}^{2-i}(G, A)(p)^\vee\,.$$

Remark: Part (ii) is only interesting if A has p-torsion because otherwise the statement follows easily from (i). The isomorphism in (ii) is a topological isomorphism for $i \neq 1$. For $i = 1$ it is an algebraic isomorphism from a discrete group onto a compact group. It is a topological isomorphism if and only if $H^1(G, A)(p)$ is finite. This is true in our main application, when G is the Galois group of the maximal extension of a global field which is unramified outside a given finite set of places.

Proof: If A is $\mathbb{Z}$-free, assertion (i) follows from (3.1.9). For arbitrary A, there exists an exact sequence

$$0 \longrightarrow R \longrightarrow F \longrightarrow A \longrightarrow 0$$

of G-modules, where R and F are finitely generated and $\mathbb{Z}$-free. Applying the functor $\operatorname{Hom}(-, C)$, we obtain an exact sequence of level-compact modules

$$0 \longrightarrow \operatorname{Hom}(A, C) \longrightarrow \operatorname{Hom}(F, C) \longrightarrow \operatorname{Hom}(R, C)\,.$$

Let U run through the open normal subgroups such that $F^U = F$. Then

$$N_U \operatorname{Hom}(F, C) = \operatorname{Hom}(F, N_U C), \qquad N_U \operatorname{Hom}(R, C) = \operatorname{Hom}(R, N_U C)$$

and the map

$$N_U \operatorname{Hom}(F, C) \longrightarrow N_U \operatorname{Hom}(R, C)$$

is surjective since $N_U C$ is divisible. For $i \leq 0$ we obtain a commutative diagram in which we write $\sim$ for $\operatorname{Hom}(-, C)$:

$$\begin{array}{ccccccccc}
\hat{H}^{i-1}(G, \tilde{F}) & \to & \hat{H}^{i-1}(G, \tilde{R}) & \longrightarrow & \hat{H}^i(G, \tilde{A}) & \longrightarrow & \hat{H}^i(G, \tilde{F}) & \longrightarrow & \hat{H}^i(G, \tilde{R}) \\
\downarrow & & \downarrow & & \downarrow & & \downarrow & & \downarrow \\
H^{3-i}(G, F)^\vee & \to & H^{3-i}(G, R)^\vee & \to & H^{2-i}(G, A)^\vee & \to & H^{2-i}(G, F)^\vee & \to & H^{2-i}(G, R)^\vee.
\end{array}$$

The upper row is exact by (1.9.13) and the lower row is exact, too. The vertical arrows, except the middle one, are isomorphisms of compact abelian groups by (3.1.9). Furthermore, using (1.4.6), it is easy to see that all squares commute up to sign. Hence, by the five lemma (in the abelian category of compact abelian groups), the middle one is also a topological isomorphism. This proves assertion (i).

In order to prove (ii), we may, without loss of generality, assume that the torsion part of A consists only of p-torsion. Since C is p-divisible, we now obtain the exact sequence

$$0 \longrightarrow \mathrm{Hom}(A, C) \longrightarrow \mathrm{Hom}(F, C) \longrightarrow \mathrm{Hom}(R, C) \longrightarrow 0,$$

and, using (1.9.14), the long exact cohomology sequence which is unbounded in both directions

$$\cdots \longrightarrow \hat{H}^i(G, \tilde{A})(p) \longrightarrow \hat{H}^i(G, \tilde{F})(p) \longrightarrow \hat{H}^i(G, \tilde{R})(p) \longrightarrow \cdots$$

in which we wrote $\sim$ for $\mathrm{Hom}(-, C)$. Furthermore, (1.9.9) gives us a corresponding long exact cohomology sequence associated to the short exact sequence $0 \to R \to F \to A \to 0$ in which all groups are compact or discrete and all maps are continuous. Hence this sequence remains exact after taking Pontryagin duals. Thus the duality map of (3.1.11) defines a map between two long exact sequences. This map commutes up to signs with the differentials. For the square

$$\begin{array}{ccc} \hat{H}^0(G, \tilde{R})(p) & \longrightarrow & H^1(G, \tilde{A})(p) \\ \downarrow & & \downarrow \\ H^2(G, R)^\vee(p) & \longrightarrow & H^1(G, A)^\vee(p) \end{array}$$

this follows from the surjectivity of the map $H^0(G, \tilde{R}) \to \hat{H}^0(G, \tilde{R})$ by (1.9.11). For the square

$$\begin{array}{ccc} H^1(G, \tilde{R})(p) & \longrightarrow & H^2(G, \tilde{A})(p) \\ \downarrow & & \downarrow \\ H^1(G, R)^\vee(p) & \longrightarrow & \hat{H}^0(G, A)^\vee(p) \end{array}$$

it follows from the isomorphism $\hat{H}^0(G, A)(p) \cong H^0(G, A) \otimes \mathbb{Z}_p$ by (1.9.8). That all other squares commute up to sign is obvious by the definition of the maps occurring and by the compatibilities proved in I §4. Therefore the statement (ii) for A follows from that for F and R and from the five lemma.

□

Remark: Theorem (3.1.11)(i), together with the idea of the proof, was formulated by *G. Poitou* in 1966 (see [168]) in order to prove an important duality theorem over local and global fields that was announced without proof by *J. Tate* in 1962 (see [229]). In 1969, a proof based on the same idea was independently given by *K. Uchida* (see [236]). A similar proof was presented by *K. Haberland* in 1978 (see [72]), but this, however, is not without mistakes. *Uchida*'s proof is correct, but rather terse, and we have given a detailed account of his arguments. These papers, however, all failed to prove the central fact that the cup-product for the Tate cohomology of profinite groups is well-defined (see I §9). This gap was filled in [196], where also the extension of the theorem by assertion (ii) was proven.

We mention a further feature of a class formation (G, C), which plays an important role in "non-abelian class field theory". Assume that C has trivial universal norms, i.e. $N_U C = \bigcap_{V \subseteq U} N_{U/V} C^V = 1$ for every open subgroup U of G, and that C is a *topological* G-module such that C^V/C^U is *compact* for every pair $V \subseteq U$ of open subgroups. For every triple $W \subseteq V \subseteq U$ of open subgroups, W, V normal in U, we may consider the diagram

$$(1) \qquad \begin{array}{ccccccccc} 1 & \longrightarrow & C^W & \longrightarrow & \mathcal{W}(U/W) & \longrightarrow & U/W & \longrightarrow & 1 \\ & & \big\downarrow{\scriptstyle N_{V/W}} & & \big\downarrow{\scriptstyle \varphi_{V/W}} & & \big\downarrow{\scriptstyle \pi} & & \\ 1 & \longrightarrow & C^V & \longrightarrow & \mathcal{W}(U/V) & \longrightarrow & U/V & \longrightarrow & 1, \end{array}$$

where the horizontal sequences are the group extensions defined by the fundamental classes $u_{U/W} \in H^2(U/W, C^W)$ and $u_{U/V} \in H^2(U/V, C^V)$. One can show that the diagram of solid arrows can be commutatively completed by an arrow $\varphi_{V/W}$. Because of the compactness assumption, one can moreover show that there exists a transitive family of arrows $\varphi_{V/W}$, i.e. $\varphi_{W/W'} \circ \varphi_{V/W} = \varphi_{V/W'}$ for $U \supseteq V \supseteq W \supseteq W'$. It is therefore possible to take the (left exact) projective limit, which yields an exact sequence

$$(2) \qquad 1 \longrightarrow C^\times \xrightarrow{\ f\ } \mathcal{W} \xrightarrow{\ g\ } G$$

with $C^\times = \varprojlim_V C^V$; the right arrow has a dense image. The group $\mathcal{W}$ is called the **Weil group** of the class formation (G, C). $\mathcal{W}$ is a topological group and has the following properties.

(i) Let $\mathcal{W}(U) := g^{-1}(U)$ and let $\mathcal{W}(U)^c$ be the closure of the commutator subgroup of $\mathcal{W}(U)$. If V is open and normal in U, then

$$\mathcal{W}(U/V) = \mathcal{W}(U)/\mathcal{W}(V)^c,$$

and we have a commutative exact diagram

$$\begin{array}{ccccccccc} 1 & \longrightarrow & C^\times & \longrightarrow & \mathcal{W}(U) & \longrightarrow & U & & \\ & & \big\downarrow{\scriptstyle N} & & \big\downarrow{\scriptstyle \varphi} & & \big\downarrow{\scriptstyle \pi} & & \\ 1 & \longrightarrow & C^V & \longrightarrow & \mathcal{W}(U/V) & \longrightarrow & U/V & \longrightarrow & 1, \end{array}$$

and, in particular, a canonical isomorphism $\rho_U : C^U \xrightarrow{\sim} \mathcal{W}(U)^{ab}$.

(ii) For every pair of open subgroups $V \subseteq U$ and every $\sigma \in \mathcal{W}$, the diagrams

$$\begin{array}{ccc} C^V & \xrightarrow{\rho_V} & \mathcal{W}(V)^{ab} \\ {\scriptstyle incl}\uparrow & & \uparrow{\scriptstyle Ver} \\ C^U & \xrightarrow{\rho_U} & \mathcal{W}(U)^{ab} \end{array} \quad , \quad \begin{array}{ccc} C^U & \xrightarrow{\rho_U} & \mathcal{W}(U)^{ab} \\ {\scriptstyle \sigma^*}\downarrow & & \downarrow{\scriptstyle c_\sigma} \\ C^{U^\sigma} & \xrightarrow{\rho_{U^\sigma}} & \mathcal{W}(U^\sigma)^{ab} \end{array}$$

are commutative, where $U^\sigma = \sigma U \sigma^{-1}$, σ^* is the action of $g(\sigma) \in G$ on C and $c_\sigma(x) = \sigma x \sigma^{-1}$.

The Weil group $\mathcal{W}$ is determined by these properties up to isomorphism in the following sense. If $\mathcal{W}'$ is another topological group with these properties, then there is an isomorphism $\mathcal{W} \to \mathcal{W}'$, compatible with the above structures, and this isomorphism is uniquely determined up to an inner automorphism of $\mathcal{W}'$ by an element of $C^\times$.

The proofs of these assertions are rather deep. Since we will not make use of the Weil group in this book, we refer for the details to [6], chap.14 and [124], chap.IX, 3.

Exercise 1. If G is a finite group and C a class module, then $H^3(G,C) = 1$ and $H^4(G,C) \cong \mathrm{Hom}\,(G, \mathbb{Q}/\mathbb{Z})$.

Exercise 2. Let G be a finite group, C a class module and A a $\mathbb{Z}$-free G-module. Show that every fundamental class $\gamma \in H^2(G,C)$ defines an isomorphism

$$\gamma_H \cup : \hat{H}^n(H,A) \to \hat{H}^{n+2}(H, A \otimes C)$$

for every $n \in \mathbb{Z}$ and every subgroup H.

Exercise 3. Show that the duality theorem (3.1.11)(i) may be interpreted as an isomorphism of G-modulations

$$\hat{H}^i(\mathrm{Hom}(A,C)) \cong \hat{H}^{2-i}(A)^\vee.$$

Exercise 4. Let G be a finite group, A a G-module and $\alpha \in H^2(G,A)$. Apply to the four term exact sequence

(1) $$0 \longrightarrow A \longrightarrow A(\alpha) \longrightarrow \mathbb{Z}[G] \longrightarrow \mathbb{Z} \longrightarrow 0$$

the exact functor $\mathrm{Hom}\,(\ \ , \mathbb{Q}/\mathbb{Z})$ and obtain the exact sequence

(2) $$0 \longrightarrow \mathbb{Q}/\mathbb{Z} \longrightarrow \mathbb{Z}[G]^* \longrightarrow A(\alpha)^* \longrightarrow A^* \longrightarrow 0.$$

(i) Show that the homomorphism

$$\delta^2 : \hat{H}^n(G, A^*) \longrightarrow \hat{H}^{n+2}(G, \mathbb{Q}/\mathbb{Z}),$$

arising from this sequence, coincides with the cup-product $\beta \mapsto -\alpha \cup \beta$ which is induced by the pairing $A \times A^* \to \mathbb{Q}/\mathbb{Z}$.

(ii) Show that $\delta^2 = -\alpha\cup$ is an isomorphism for all $n \in \mathbb{Z}$ if A is a class module and α is a fundamental class.

Hint: (i) Apply proposition (1.4.6) to the two pairs of exact sequences

$$0 \longrightarrow I_G \longrightarrow \mathbb{Z}[G] \longrightarrow \mathbb{Z} \longrightarrow 0, \quad 0 \longrightarrow A \longrightarrow A(\alpha) \longrightarrow I_G \longrightarrow 0,$$

$$0 \longrightarrow \mathbb{Z}^* \longrightarrow \mathbb{Z}[G]^* \longrightarrow I_G^* \longrightarrow 0, \quad 0 \longrightarrow I_G^* \longrightarrow A(\alpha)^* \longrightarrow A^* \longrightarrow 0.$$

Let $\beta \in \hat{H}^n(G, A^*)$ and consider the element $1 \in H^0(G, \mathbb{Z})$. The homomorphism $\delta^2 : H^0(G, \mathbb{Z}) \to H^2(G, A)$ arising from (1) maps 1 to α (see the proof of (3.1.5)). Therefore

$$\begin{aligned} \delta^2\beta = \delta(\delta\beta) \cup 1 = (-1)^{n+2}(\delta\beta \cup \delta 1) = (-1)^{n+2+n+1}(\beta \cup \delta^2 1) \\ = -(\beta \cup \alpha) = -\alpha \cup \beta. \end{aligned}$$

(ii) If A is a class module, then $A(\alpha)$ is cohomologically trivial by (3.1.4), hence also $A(\alpha)^*$ is by (1.8.5).

§2. An Alternative Description of the Reciprocity Homomorphism

Let G be a profinite group and let C be a G-module. If C is a formation module for G, then we have the reciprocity homomorphism

$$rec_G : C^G \longrightarrow G^{ab},$$

which was obtained from the Nakayama-Tate duality theorem (3.1.5) by passing to the projective limit. If the pair (G, C) satisfies certain conditions described below, we get a simple criterion for C being a formation module and an alternative, non-cohomological, description of the reciprocity homomorphism as presented in [160], chap.IV. This method applies, for instance, to the absolute Galois group of a local or global field and becomes important if one wants to understand the reciprocity homomorphism more explicitly. However, we will not use the results below in the following, so the reader might skip this section on the first reading.

Assume we are given a G-module C and a pair of homomorphisms

$$(G \xrightarrow{d} \hat{\mathbb{Z}},\ C^G \xrightarrow{v} \hat{\mathbb{Z}}),$$

where d is surjective and $Z = v(C^G)$ has the properties

$$\mathbb{Z} \subseteq Z \quad \text{and} \quad Z/nZ \cong \mathbb{Z}/n\mathbb{Z} \ \text{ for all } n \in \mathbb{N},$$

in particular, the cokernel $\hat{\mathbb{Z}}/Z$ of v is uniquely divisible. Furthermore, we assume that, for every open subgroup U of G,

$$v(N_{G/U}C^U) = f_U Z$$

with $f_U = (\hat{\mathbb{Z}} : d(U))$. Then we have the surjective homomorphisms

$$d_U = \tfrac{1}{f_U} d : U \longrightarrow \hat{\mathbb{Z}} \quad , \quad v_U = \tfrac{1}{f_U} v \circ N_{G/U} : C^U \longrightarrow Z.$$

For every pair $V \subseteq U$ of open subgroups, we set $f_{U/V} = f_V/f_U$, $e_{U/V} = (U:V)/f_{U/V}$, and we get the commutative diagrams

$$\begin{array}{ccc} V^{ab} & \xrightarrow{d_V} & \hat{\mathbb{Z}} \\ {\scriptstyle i}\big\downarrow\big\uparrow{\scriptstyle Ver} & & {\scriptstyle f_{U/V}}\big\downarrow\big\uparrow{\scriptstyle e_{U/V}} \\ U^{ab} & \xrightarrow{d_U} & \hat{\mathbb{Z}}, \end{array} \qquad \begin{array}{ccc} C^{V} & \xrightarrow{v_V} & \hat{\mathbb{Z}} \\ {\scriptstyle N_{U/V}}\big\downarrow\big\uparrow{\scriptstyle incl} & & {\scriptstyle f_{U/V}}\big\downarrow\big\uparrow{\scriptstyle e_{U/V}} \\ C^{U} & \xrightarrow{v_U} & \hat{\mathbb{Z}}. \end{array}$$

The situation is most briefly, and best, formulated in the language of G-modulations (see I §5). We have on the one hand the *fundamental G-modulation*

$$\pi^{ab} : U \longmapsto U^{ab},$$

where, for two open subgroups $V \subseteq U$, the maps

$$U^{ab} \underset{ind_U^V}{\overset{res_V^U}{\rightleftarrows}} V^{ab}$$

are the transfer *Ver* : $U^{ab} \to V^{ab}$ and the map induced by the inclusion $V \hookrightarrow U$. On the other hand, we consider C as the G-modulation

$$C : U \longmapsto C^U,$$

where res_V^U and ind_U^V are the inclusion and the norm. C is endowed with the submodulation $NC : U \mapsto N_U C$. The submodulation $N\pi^{ab} : U \mapsto N_U U^{ab}$ of π^{ab} is trivial, because we have a surjection

$$1 = \varprojlim_{V \subseteq U} V \twoheadrightarrow \varprojlim_{V \subseteq U} V[U,U]/[U,U] = N_U U^{ab}.$$

Finally, we consider the G-modulation $U \mapsto \hat{\mathbb{Z}}$, where

$$res_V^U = e_{U/V} \quad \text{and} \quad ind_U^V = f_{U/V}.$$

We denote this G-modulation by $\hat{\mathbb{Z}}$ (observe that it depends on $d : G \to \hat{\mathbb{Z}}$ since the numbers $f_{U/V}$ and $e_{U/V}$ do). Then, in the above situation, we have two morphisms

$$d : \pi^{ab} \longrightarrow \hat{\mathbb{Z}}, \qquad v : C \longrightarrow \hat{\mathbb{Z}}$$

of G-modulations. The main result of abstract class field theory may now be formulated as follows.

(3.2.1) Theorem. *Assume that for every pair $V \subseteq U$ of open subgroups of G, with V normal in U and U/V cyclic, we have*

$$\#\hat{H}^i(U/V, C^V) = \begin{cases} (U:V) & \text{for} \quad i = 0, \\ 1 & \text{for} \quad i = 1, \end{cases}$$

the **class field axiom**. *Then there is a unique morphism*

$$r : C \longrightarrow \pi^{ab}$$

of G-modulations such that $v = d \circ r$.

Proof: The theorem is just a reformulation of the results in [160], chap.IV. For every pair $V \subseteq U$ of open subgroups of G, V normal in U, we have by [160], chap.IV, (6.3) a canonical isomorphism

$$\hat{r}_{U/V} : (U/V)^{ab} \xrightarrow{\sim} C^U/N_{U/V}C^V .$$

We briefly recall the definition of $\hat{r}_{U/V}$ (see [160], chap.IV, (5.6), where, in contrast to our notation, this map was called the reciprocity homomorphism). Let $I = \ker d$ and $I_U = I \cap U$. The semigroup

$$\mathrm{Frob}(U/I_V) = \{\tilde{\sigma} \in U/I_V \mid d_U(\tilde{\sigma}) \in \mathbb{N}\} ,$$

whose elements we call *Frobenius lifts*, maps surjectively onto U/V

$$\mathrm{Frob}(U/I_V) \to U/V, \quad \tilde{\sigma} \mapsto \sigma = \tilde{\sigma} \mod V ,$$

cf. [160], chap.IV, (4.4). If $\sigma \in U/V$ and $\bar{\sigma}$ is its image in $(U/V)^{ab}$, then $\hat{r}_{U/V}(\bar{\sigma})$ is defined by

$$\hat{r}_{U/V}(\bar{\sigma}) = N_{U/S}(\pi_S) \mod N_{U/V}C^V ,$$

where $S = \langle\tilde{\sigma}\rangle I_V \subseteq U$ for some Frobenius lift $\tilde{\sigma}$ of σ and $\pi_S \in C^S$ is any element with $v_S(\pi_S) = 1$. If the class field axiom holds, it was shown in [160], chap.IV that $\hat{r}_{U/V}$ is a well-defined homomorphism which, in particular, does not depend on the various choices and which moreover is an isomorphism.

Taking the projective limit over V of the surjections

$$r_{U/V} : C^U \twoheadrightarrow C^U/N_{U/V}C^V \xrightarrow[\sim]{(\hat{r}_{U/V})^{-1}} (U/V)^{ab}$$

yields a family of homomorphisms

$$r_U : C(U) = C^U \twoheadrightarrow C^U/N_U C \hookrightarrow \varprojlim_V C^U/N_{U/V}C^V \xrightarrow{\sim} U^{ab} = \pi^{ab}(U)$$

with dense image. By [160], chap.IV, (6.4) and (6.5), this family defines a morphism $r : C \longrightarrow \pi^{ab}$ of G-modulations such that the diagram

$$\begin{array}{ccc} C & \xrightarrow{\ \ r\ \ } & \pi^{ab} \\ & {\scriptstyle v}\searrow \quad \swarrow{\scriptstyle d} & \\ & \hat{\mathbb{Z}} & \end{array}$$

is commutative.

In order to prove uniqueness, first observe that any morphism $r : C \to \pi^{ab}$ factors through C/NC, because π^{ab} has trivial universal norms, and the same is true for v. Let U be an open subgroup of G. For an open normal subgroup V of U, we consider the group U/I_V. Obviously, it is enough to show that in the commutative diagram

$$\begin{array}{ccc} C^U & \xrightarrow{\ \ r_{U/I_V}\ \ } & (U/I_V)^{ab} \\ & {\scriptstyle v_U}\searrow \quad \swarrow{\scriptstyle d_U} & \\ & \hat{\mathbb{Z}} & \end{array}$$

the homomorphism $r_{U/I_V} = \varprojlim_{I_V \subseteq W \subseteq U} r_{U/W}$ (W open and normal in U) is uniquely determined by d_U and v_U, since then we can pass to the projective limit over V in order to obtain the desired result. Let $\mathcal{S}$ be the *finite* set of splittings of the surjection $d_U : U/I_V \twoheadrightarrow \hat{\mathbb{Z}}$. We denote the images by W_s/I_V for $s \in \mathcal{S}$, where W_s is an open subgroup of U. Since

$$\langle W_s/I_V, s \in \mathcal{S}\rangle = U/I_V \,,$$

we see that $\langle W_s,\ s \in \mathcal{S}\rangle = U$. By [160], chap.IV, (6.7), we have the equality $N_{U|W}C^W \cdot N_{U|W'}C^{W'} = N_{U|WW'}C^{WW'}$ for two open subgroups W and W' of U. Since $\mathcal{S}$ is finite, we obtain

$$\langle N_{U|W_s}C^{W_s},\ s \in \mathcal{S}\rangle = C^U \,.$$

Now the commutative diagram

$$\begin{array}{ccccc} C^U & & \xrightarrow{\ r_{U/I_V}\ } & & (U/I_V)^{ab} \\ {\scriptstyle N_{U|W_s}}\uparrow & & & & \uparrow \\ C^{W_s} & & \xrightarrow{\ r_{W_s/I_V}\ } & & W_s/I_V \\ & {\scriptstyle v_{W_s}}\searrow & & \swarrow{\scriptstyle d_{W_s}}\ \sim & \\ & & \hat{\mathbb{Z}} & & \end{array}$$

shows that r_{U/I_V} is uniquely determined by d_U and v_U since this is the case for all maps r_{W_s/I_V}. □

The morphism $r : C \to \pi^{ab}$ yields, in particular, a homomorphism

$$r_G : C^G \longrightarrow G^{ab},$$

which we now want to compare with the homomorphism at the beginning of this section.

For every open subgroup U of G, we again set $I_U = \ker(U \xrightarrow{d_U} \hat{\mathbb{Z}})$ and $\Gamma_U = U/I_U$. We have a commutative diagram

$$\begin{array}{ccc} C^{I_U} & \xrightarrow{\ \tilde{v}_U\ } & Z \\ \uparrow & & \| \\ C^U & \xrightarrow{\ v_U\ } & Z \end{array}$$

where $\tilde{v}_U = \varinjlim_{I_U \subseteq U' \subseteq U} v_{U'}$. Recalling that $Z/\mathbb{Z}$ is uniquely divisible, we get the sequence of canonical homomorphisms

$$H^2(\Gamma_U, C^{I_U}) \xrightarrow{(\tilde{v}_U)_*} H^2(\Gamma_U, Z) = H^2(\Gamma_U, \mathbb{Z}) \xrightarrow{\sim} H^1(\Gamma_U, \mathbb{Q}/\mathbb{Z}) \xrightarrow{\sim} \mathbb{Q}/\mathbb{Z}\,,$$

and we denote its composite by

$$inv_U : H^2(\Gamma_U, C^{I_U}) \longrightarrow \mathbb{Q}/\mathbb{Z}.$$

For a pair $V \subseteq U$ of open subgroups, we have the compatible pair of homomorphisms $\Gamma_V \to \Gamma_U$, $C^{I_U} \hookrightarrow C^{I_V}$, which yields a canonical homomorphism

$$res_V^U : H^2(\Gamma_U, C^{I_U}) \longrightarrow H^2(\Gamma_V, C^{I_V}),$$

and we have the commutative diagram

$$\begin{array}{ccccccc} H^2(\Gamma_V, C^{I_V}) & \longrightarrow & H^2(\Gamma_V, \mathbb{Z}) & \xrightarrow{\delta^{-1}} & H^1(\Gamma_V, \mathbb{Q}/\mathbb{Z}) & \longrightarrow & \mathbb{Q}/\mathbb{Z} \\ {\scriptstyle res}\uparrow & & \uparrow{\scriptstyle e_{U/V}\cdot res} & & \uparrow{\scriptstyle e_{U/V}\cdot res} & & \uparrow{\scriptstyle (U:V)} \\ H^2(\Gamma_U, C^{I_U}) & \longrightarrow & H^2(\Gamma_U, \mathbb{Z}) & \xrightarrow{\delta^{-1}} & H^1(\Gamma_U, \mathbb{Q}/\mathbb{Z}) & \longrightarrow & \mathbb{Q}/\mathbb{Z}, \end{array}$$

noting that $e_{U/V} \cdot f_{U/V} = (U : V)$ and that the canonical generator φ_V of Γ_V is mapped onto the $f_{U/V}$-th power of the canonical generator φ_U of Γ_U.

(3.2.2) Proposition. *If C satisfies the class field axiom* (3.2.1), *then, for every open subgroup U of G, the homomorphism*

$$inv_U : H^2(\Gamma_U, C^{I_U}) \longrightarrow \mathbb{Q}/\mathbb{Z}$$

is an isomorphism.

Proof: We have to show that $H^2(\Gamma_U, C^{I_U}) \xrightarrow{(\tilde{v}_U)_*} H^2(\Gamma_U, Z)$ is an isomorphism. Let $U_n := d_U^{-1}(n\hat{\mathbb{Z}})$ and $D_n := \ker(v_{U_n} : C^{U_n} \to Z)$. From [160], chap.IV, (6.2) it follows that the class field axiom (3.2.1) implies $\hat{H}^i(U/U_n, D_n) = 0$ for $i = 0, -1$, hence for all i. Since

$$D := \ker(\tilde{v}_U : C^{I_U} \to Z) = \bigcup_n D_n = \varinjlim_n D_n\,,$$

we get for $i \geq 1$

$$H^i(\Gamma_U, D) = \varinjlim_n H^i(U/U_n, D_n) = 0\,.$$

Taking the cohomology of the exact sequence $0 \longrightarrow D \longrightarrow C^{I_U} \xrightarrow{\tilde{v}_U} Z \longrightarrow 0$ of Γ_U-modules, we get the exact sequence

$$0 = H^2(\Gamma_U, D) \longrightarrow H^2(\Gamma_U, C^{I_U}) \xrightarrow{(\tilde{v}_U)_*} H^2(\Gamma_U, Z) \xrightarrow{\delta} H^3(\Gamma_U, D) = 0.$$

Therefore $(\tilde{v}_U)_*$ is an isomorphism. □

(3.2.3) Proposition. *Under the above assumptions, the following assertions are equivalent.*

(i) *C satisfies the class field axiom* (3.2.1).

(ii) *$H^2(\Gamma_U, C^{I_U}) = H^2(U, C)$ for all open subgroups U of G and C is a formation module with respect to the isomorphisms*

$$inv_U : H^2(U, C) \xrightarrow{\sim} \mathbb{Q}/\mathbb{Z}.$$

Proof: If C is a formation module, then for every open normal subgroup V of U,

$$H^i(U/V, C^V) \cong \begin{cases} \mathbb{Z}/(U:V)\mathbb{Z} & \text{if } \ i = 2\,, \\ 0 & \text{if } \ i = 1\,, \end{cases}$$

and $H^2(U/V, C^V) \cong \hat{H}^0(U/V, C^V)$ if U/V is cyclic, so that the class field axiom holds.

Conversely, assume that C satisfies the class field axiom. We claim that

$$\#H^2(U/V, C^V) \,\big|\, (U:V)$$

for every pair $V \triangleleft U$. In fact, this is true if U/V is cyclic because $H^2 \cong \hat{H}^0$. If U/V is a p-group, it follows inductively from the exact sequence

$$0 \longrightarrow H^2(U/W, C^W) \longrightarrow H^2(U/V, C^V) \longrightarrow H^2(W/V, C^V),$$

where W/V is a normal subgroup of U/V of order p. In the general case, let $(U/V)_p$ be a p-Sylow subgroup of U/V and U_p the pre-image of $(U/V)_p$ in U. Since the restriction map

$$res : H^2(U/V, C^V) \hookrightarrow \bigoplus_p H^2((U/V)_p, C^{U_p})$$

is injective by (1.6.10), we obtain

$$\#H^2(U/V, C^V) \,\Big|\, \prod_p \#H^2((U/V)_p, C^{U_p}) \,\Big|\, \prod_p \#(U/V)_p = \#(U/V)\,.$$

For every open normal subgroup V of U, we have the exact sequence

$$0 \longrightarrow H^2(U/V, C^V) \xrightarrow{inf} H^2(U, C) \xrightarrow{res} H^2(V, C)$$

and we identify $H^2(U/V, C^V)$ with its image in $H^2(U, C)$. Let $n = (U:V)$ and let $U_n = d_U^{-1}(n\hat{\mathbb{Z}})$. Then

$$H^2(U/V, C^V) = H^2(U/U_n, C^{U_n}).$$

In fact, because

$$\#H^2(U/V, C^V) \,\big|\, (U:V) = (U:U_n) = \#H^2(U/U_n, C^{U_n})\,,$$

it suffices to show the inclusion "$\supseteq$". But this follows from the exact commutative diagram

$$\begin{array}{ccccccc}
0 \longrightarrow H^2(U/V, C^V) & \longrightarrow & H^2(U,C) & \xrightarrow{res} & H^2(V,C) \\
& & \uparrow & & \uparrow \\
& & H^2(\Gamma_U, C^{I_U}) & \xrightarrow{res} & H^2(\Gamma_V, C^{I_V}) \\
& & inv_U \downarrow \wr & & inv_V \downarrow \wr \\
& & \mathbb{Q}/\mathbb{Z} & \xrightarrow{(U:V)} & \mathbb{Q}/\mathbb{Z},
\end{array}$$

in which inv_U and inv_V are isomorphisms by (3.2.2). Since $H^2(U/U_n, C^{U_n}) \subseteq H^2(\Gamma_U, C^{I_U})$ has order $n = (U : V)$, it is mapped by the middle arrow *res*, and thus by the upper arrow *res*, to zero, hence

$$H^2(U/U_n, C^{U_n}) \subseteq H^2(U/V, C^V).$$

We therefore obtain

$$H^2(U,C) = \bigcup_V H^2(U/V, C^V) = \bigcup_n H^2(U/U_n, C^{U_n}) = H^2(\Gamma_U, C^{I_U}).$$

For V open and normal in U, we have the commutative diagram

$$\begin{array}{ccc}
H^2(V,C) & \xrightarrow[\sim]{inv_V} & \mathbb{Q}/\mathbb{Z} \\
{\scriptstyle res} \uparrow & & \uparrow {\scriptstyle (U:V)} \\
H^2(U,C) & \xrightarrow[\sim]{inv_U} & \mathbb{Q}/\mathbb{Z},
\end{array}$$

and the induced isomorphisms

$$inv_{U/V} : H^2(U/V, C^V) \longrightarrow \tfrac{1}{(U:V)}\mathbb{Z}/\mathbb{Z}$$

define on C the structure of a formation module in the sense of (3.1.8). □

Let us assume that C satisfies the class field axiom and thus is, in particular, a formation module with respect to the isomorphisms

$$inv_U : H^2(U,C) \xrightarrow{\sim} \mathbb{Q}/\mathbb{Z}.$$

Noting that $H^2(U,\mathbb{Z}) \cong H^1(U,\mathbb{Q}/\mathbb{Z}) = \mathrm{Hom}(U^{ab}, \mathbb{Q}/\mathbb{Z})$, the cup-product

$$\varprojlim_V \hat{H}^0(U/V, C^V) \times H^2(U,\mathbb{Z}) \longrightarrow H^2(U,C) \xrightarrow[\sim]{inv} \mathbb{Q}/\mathbb{Z}$$

yields by (3.1.5) homomorphisms

$$rec_U : C^U \twoheadrightarrow C^U/N_U C \hookrightarrow \varprojlim_V C^U/N_{U/V}C^V \xrightarrow{\sim} U^{ab}.$$

These homomorphisms commute with conjugation, restriction and corestriction by the rules (1.5.3) and (1.5.7), i.e. they form a morphism

$$rec : C \longrightarrow \pi^{ab}$$

of G-modulations.

(3.2.4) Theorem. *The morphisms*

$$r,\ rec : C \longrightarrow \pi^{ab}$$

of G-modulations coincide.

Proof: Because of the uniqueness assertion of (3.2.1), it suffices to show that the diagram

$$\begin{array}{ccc} C & \xrightarrow{\ rec\ } & \pi^{ab} \\ & {\scriptstyle v}\searrow \quad \swarrow{\scriptstyle d} & \\ & \hat{\mathbb{Z}} & \end{array}$$

commutes. Let U be an open subgroup of G. As before, we put $I_U = \ker(U \xrightarrow{d_U} \hat{\mathbb{Z}})$ and $\Gamma_U = U/I_U$. We have the commutative diagram

$$\begin{array}{ccccccc} H^0(U,C) & \times & H^2(U,\mathbb{Z}) & \xrightarrow{\ \cup\ } & H^2(U,C) & \xrightarrow{inv} & \mathbb{Q}/\mathbb{Z} \\ \| & & \uparrow{\scriptstyle inf} & & \wr\uparrow{\scriptstyle inf} & & \| \\ H^0(\Gamma_U, C^{I_U}) & \times & H^2(\Gamma_U,\mathbb{Z}) & \xrightarrow{\ \cup\ } & H^2(\Gamma_U, C^{I_U}) & & \| \\ \downarrow{\scriptstyle v_U} & & \wr\uparrow{\scriptstyle (d_U)^*} & & \wr\downarrow{\scriptstyle (d_U)^*(\tilde{v}_U)_*} & & \| \\ H^0(\hat{\mathbb{Z}}, Z) & \times & H^2(\hat{\mathbb{Z}},\mathbb{Z}) & \xrightarrow{\ \cup\ } & H^2(\hat{\mathbb{Z}}, Z) & \xrightarrow{inv} & \mathbb{Q}/\mathbb{Z}\,, \end{array}$$

which induces the commutative diagram

$$\begin{array}{ccc} C^U & \xrightarrow{rec_U} & U^{ab} \\ \downarrow{\scriptstyle v_U} & & \downarrow{\scriptstyle d_U} \\ Z & \hookrightarrow & \hat{\mathbb{Z}}\,. \end{array}$$

This proves the theorem. □

§3. Cohomological Dimension

Let G be a profinite group, $Mod(G)$ the category of (discrete) G-modules, and $Mod_t(G)$, $Mod_p(G)$, $Mod_f(G)$ the category of G-modules which are torsion, p-torsion, finite respectively as abelian groups. A fundamental numerical invariant of G is the cohomological dimension.

(3.3.1) Definition. *The* **cohomological dimension** *$cd\,G$ (resp.* **strict cohomological dimension** *$scd\,G$) of G is the smallest integer n such that*

$$H^q(G,A) = 0 \quad \text{for all} \quad q > n$$

and all $A \in Mod_t(G)$ (resp. $A \in Mod(G)$), and is ∞ if no such integer exists.

Let p be a prime number. The **cohomological p-dimension** *$cd_p\,G$ (resp.* **strict cohomological p-dimension** *$scd_p\,G$) is the smallest integer n such that the p-primary part*

$$H^q(G,A)(p) = 0 \quad \text{for all} \quad q > n \quad {}^{*)}$$

and all $A \in Mod_t(G)$ (resp. $A \in Mod(G)$), and is ∞ if no such integer exists.

The profinite group G is of **virtual cohomological dimension** *$vcd\,G = n$ resp.* **virtual strict cohomological dimension** *$vscd\,G = n$ if there exists an open subgroup U of G such that $cd\,U = n$ resp. $scd\,U = n$. The p-versions $vcd_p\,G$ and $vscd_p\,G$ are defined analogously.*

Since every abelian torsion group X decomposes into the direct sum $X = \bigoplus_p X(p)$ of its p-primary parts $X(p)$, we have

$$cd\,G = \sup_p cd_p\,G \quad \text{and} \quad scd\,G = \sup_p scd_p\,G.$$

If G has an element of order p – in particular, if G is finite and $p \mid \#G$ – then $cd_p\,G = \infty$. In fact, if H is a subgroup of order p, then by (1.6.4) and (1.7.1),

$$H^{2n}(G, \mathrm{Ind}_G^H(\mathbb{Z}/p\mathbb{Z})) \cong H^{2n}(H, \mathbb{Z}/p\mathbb{Z}) \cong \hat{H}^0(H, \mathbb{Z}/p\mathbb{Z}) = \mathbb{Z}/p\mathbb{Z} \neq 0$$

for all $n \geq 0$.

(3.3.2) Proposition. *The following conditions are equivalent.*

(i) *$cd_p\,G \leq n$,*

(ii) *$H^q(G,A) = 0$ for all $q > n$ and all $A \in Mod_p(G)$,*

(iii) *$H^{n+1}(G,A) = 0$ for all simple G-modules A with $pA = 0$.*

For a pro-p-group G, we have, in particular,

$$cd\,G \leq n \quad \Leftrightarrow \quad H^{n+1}(G, \mathbb{Z}/p\mathbb{Z}) = 0.$$

---

*) The *p-primary part* $X(p)$ of an abelian torsion group X consists of all elements of X which have p-power order.

Proof: (i) $\Leftrightarrow$ (ii) follows from $A = \bigoplus_p A(p)$ if $A \in Mod_t(G)$, and $H^q(G,A)(p) = H^q(G, A(p))$. Assume (iii). If A is finite of p-power order, then $H^{n+1}(G,A) = 0$ follows by induction on $\#A$ by the exact cohomology sequence associated to the exact sequence $0 \to B \to A \to A/B \to 0$, where B is a simple non-zero submodule of A. It then follows for general $A \in Mod_p(G)$ by taking direct limits. From this, we get (ii) by induction on q, considering the exact sequence $0 \to A \to \mathrm{Ind}_G(A) \to A_1 \to 0$, which by (1.3.8) yields the isomorphism $H^{q+1}(G,A) \cong H^q(G,A_1)$. □

(3.3.3) Proposition. $cd_p\, G \le scd_p\, G \le cd_p\, G + 1.$

Proof: $cd_p\, G \le scd_p\, G$ is trivial. Let $A \in Mod(G)$ and consider the exact sequences

$$0 \longrightarrow {}_pA \longrightarrow A \xrightarrow{p} pA \longrightarrow 0, \quad 0 \longrightarrow pA \longrightarrow A \longrightarrow A/pA \longrightarrow 0.$$

Let $q > cd_p\, G + 1$. Then $H^q(G, {}_pA) = H^{q-1}(G, A/pA) = 0$ since ${}_pA$ and $A/pA \in Mod_p(G)$. Therefore

$$H^q(G,A) \longrightarrow H^q(G,pA) \quad \text{and} \quad H^q(G,pA) \longrightarrow H^q(G,A)$$

are injective. The composite is multiplication by p since the composite of $A \xrightarrow{p} pA \to A$ is multiplication by p. It follows that $H^q(G,A)(p) = 0$, showing $scd_p\, G \le cd_p\, G + 1$. □

(3.3.4) Corollary. *Assume that $cd_p\, G = n$ is finite. Then $scd_p\, G = n$ if and only if $H^{n+1}(U, \mathbb{Z})(p) = 0$ for all open subgroups U of G.*

Proof: Assume the latter. If the G-module A is finitely generated as a $\mathbb{Z}$ module, then there is an open subgroup U of G acting trivially on A, and A is a quotient B/C of $B = \mathrm{Ind}_G^U(\mathbb{Z}^m)$ for some m. Since $scd_p\, G \le n+1$, we have $H^{n+1}(G,A)(p) = 0$, and this result extends to an arbitrary G-module A by passing to the direct limit. □

Example: Let $G = \hat{\mathbb{Z}} = \varprojlim_n \mathbb{Z}/n\mathbb{Z}$. Then

$$cd_p\, G = 1 \quad \text{and} \quad scd_p\, G = 2.$$

In fact, if A is a finite G-module of p-power order, then every extension

$$0 \longrightarrow A \longrightarrow \hat{G} \longrightarrow G \longrightarrow 1$$

splits (the closed subgroup $\overline{(\sigma)}$ of $\hat{G}$, topologically generated by a pre-image $\sigma \in \hat{G}$ of $1 \in G$, is mapped isomorphically onto G). By (1.2.4), $H^2(G,A) \cong$

$EXT(G, A)$, hence $H^2(G, A) = 0$ for finite $A \in Mod_p(G)$, and by taking inductive limits, for all $A \in Mod_p(G)$. Noting that $H^1(G, \mathbb{Z}/p\mathbb{Z}) = \mathbb{Z}/p\mathbb{Z} \neq 0$, this shows $cd_p\, G = 1$. On the other hand, $H^2(G, \mathbb{Z}) \cong H^1(G, \mathbb{Q}/\mathbb{Z}) \cong \mathbb{Q}/\mathbb{Z}$, so that $scd_p\, G = 2$.

(3.3.5) Proposition. *If H is a closed subgroup of G, then*

$$cd_p\, H \leq cd_p\, G \quad \text{and} \quad scd_p\, H \leq scd_p\, G.$$

We have equality in each of the following cases:

(i) *$(G : H)$ is prime to p,*

(ii) *H is open and $cd_p\, G < \infty$.*

Proof: The inequalities follow from Shapiro's lemma (1.6.4)

$$H^q(G, \mathrm{Ind}_G^H(A)) \cong H^q(H, A),$$

noting that if A is torsion, then so is $\mathrm{Ind}_G^H(A)$. On the other hand, in case (i),

$$res : H^q(G, A)(p) \longrightarrow H^q(H, A)(p)$$

is injective (see the proof of (1.6.10)). In case (ii), consider for $A \in Mod(G)$ the exact sequence

$$0 \longrightarrow B \longrightarrow \mathrm{Ind}_G^H(A) \xrightarrow{\nu} A \longrightarrow 0$$

of G-modules, where ν is given by $\nu x = \sum_{\sigma \in G/H} \sigma x(\sigma^{-1})$. We obtain a homomorphism

$$H^n(H, A)(p) = H^n(G, \mathrm{Ind}_G^H(A))(p) \longrightarrow H^n(G, A)(p),$$

which is surjective if $H^{n+1}(G, B)(p) = 0$. This is the case if either $n = scd_p\, G$ or $n = cd_p\, G$ and $A \in Mod_t(G)$. Thus in either case, (i) or (ii), we have the implication

$$H^n(H, A)(p) = 0 \;\Rightarrow\; H^n(G, A)(p) = 0$$

and this means $scd_p\, H \geq scd_p\, G$ or $cd_p\, H \geq cd_p\, G$ respectively. □

SERRE has shown that a much weaker condition than $cd_p\, G < \infty$ guarantees the equality $cd_p\, G = cd_p\, H$ for an open subgroup H. One requires only that G contains no element of order p (see [211] and [75]).

(3.3.6) Corollary. *If G_p is a p-Sylow subgroup of G, then*

$$cd_p\, G = cd_p\, G_p = cd\, G_p \quad \text{and} \quad scd_p\, G = scd_p\, G_p = scd\, G_p.$$

(3.3.7) Corollary. *Let G be a profinite group. Then*

$$cd_p\, G = 0 \quad \textit{if and only if} \quad \#G \textit{ is prime to } p.$$

Proof: By (3.3.6), we may assume that G is a pro-p-group, and then the assertion follows from (1.6.14)(ii). □

(3.3.8) Proposition. *If H is a closed normal subgroup of G, then*

$$cd_p\, G \le cd_p\, G/H + cd_p\, H.$$

If $cd_p\, G/H < \infty$ and $cd_p\, H < \infty$, and if $H^n(U, \mathbb{Z}/p\mathbb{Z})$ is finite for $n = cd_p\, H$ and all open subgroups U of H *) *, then equality holds.*

Proof: We may assume that $m = cd_p\, G/H$ and $n = cd_p\, H$ are finite. Consider for $A \in Mod_p(G)$ the Hochschild-Serre spectral sequence

$$E_2^{ij} = H^i(G/H, H^j(H, A)) \Rightarrow H^{i+j}(G, A) = E^{i+j}.$$

Let $q > m + n$. If $i + j = q$, then either $i > m$ or $j > n$, hence $E_2^{ij} = 0$. As $H^q(G, A)$ has a filtration, whose quotients are subquotients E_∞^{ij} of E_2^{ij}, we get $H^q(G, A) = 0$, so that $cd_p\, G \le m + n$.

Now assume that $H^n(U, \mathbb{Z}/p\mathbb{Z})$ is finite for all open subgroups U of H. Since $cd_p\, H = n$, there exists a finite H-module A with $pA = 0$ and $H^n(H, A) \neq 0$. Let G' be an open subgroup of G such that $H' = G' \cap H$ acts trivially on A. The canonical surjection

$$\mathrm{Ind}_H^{H'}(A) \xrightarrow{\nu} A, \quad x \mapsto \sum_{\sigma \in H/H'} \sigma x(\sigma^{-1}),$$

induces a surjection

$$H^n(H', A) = H^n(H, \mathrm{Ind}_H^{H'}(A)) \xrightarrow{\nu_*} H^n(H, A),$$

again because $n = cd_p\, H$. Therefore $H^n(H', A) \neq 0$, and hence the group $H^n(H', \mathbb{Z}/p\mathbb{Z})$ is finite and non-zero.

Now let G''/H' be a p-Sylow subgroup of G'/H'. Then $cd_p\, G''/H' = cd_p\, G'/H' = cd_p\, G/H = m$ and $cd_p\, H' = cd_p\, H = n$ by (3.3.5). Therefore, by (2.1.4),

$$H^{m+n}(G'', \mathbb{Z}/p\mathbb{Z}) = H^m(G''/H', H^n(H', \mathbb{Z}/p\mathbb{Z})).$$

This group is non-zero. Namely, since G''/H' is a pro-p-group, there is a surjective homomorphism $H^n(H', \mathbb{Z}/p\mathbb{Z}) \to \mathbb{Z}/p\mathbb{Z}$ of G''/H'-modules, which induces a surjective homomorphism

$$H^m(G''/H', H^n(H', \mathbb{Z}/p\mathbb{Z})) \longrightarrow H^m(G''/H', \mathbb{Z}/p\mathbb{Z}) \neq 0.$$

We therefore obtain $cd_p\, G \ge cd_p\, G'' \ge m + n$. □

We take the following theorem from [247], see also [45].

---

*) If H is a pro-p-group, this is already true if $H^n(H, \mathbb{Z}/p\mathbb{Z})$ is finite.

(3.3.9) Theorem. *Let N be a closed normal subgroup of a profinite group G and let $cd_p\, G = n$ and $cd_p\, N = k$. Assume that the group $H^k(N, \mathbb{Z}/p\mathbb{Z})$ is finite and nonzero. Then*

$$vcd_p\, G/N = n - k\,.$$

Remark: The condition $H^k(N, \mathbb{Z}/p\mathbb{Z}) \neq 0$ is necessary. If N is a pro-p-group, then $cd_p\, N = k$ implies $H^k(N, \mathbb{Z}/p\mathbb{Z}) \neq 0$, see (3.3.2). This is not true for a general profinite group N.

Proof: In the following we write $H^*(-)$ for $H^*(-, \mathbb{Z}/p\mathbb{Z})$. Replacing G by the pre-image of a p-Sylow subgroup of G/N under the canonical projection and using (3.3.6), we may assume that G/N is a pro-p-group. Since $H^k(N) = \varinjlim_{N'} H^k(N/N')$ is finite, there exists an open subgroup M of N, which is normal in G, such that the inflation map

$$\mathit{inf}\, : H^k(N/M) \longrightarrow H^k(N)$$

is surjective.

Now there exists an open subgroup U_0 of G containing M such that

$$(*) \qquad U_0 \cap N = M \quad \text{and} \quad U_0N/M \cong N/M \times U_0/M\,.$$

This is a consequence of the following group theoretical

Claim: Let N be a finite normal subgroup of the profinite group G. Then there exists an open subgroup U of G, such that U is contained in the centralizer of N in G and $U \cap N = 1$. In particular, UN is open in G and $UN \cong U \times N$.

Proof of the claim: Let K be the kernel of the canonical homomorphism $G \to \mathrm{Aut}(N)$ which is induced by the conjugation. Then K is open and normal in G, and $K \cap N$ is the center $Z(N)$ of N. Choosing for every element $z \in Z(N)\backslash\{1\}$ an open normal subgroup U_z of K such that $z \notin U_z$, then the group $U = \bigcap_{z \in Z(N)\backslash\{1\}} U_z$ has the desired properties.

Using the claim for the finite subgroup N/M of G/M, we obtain U_0 as in $(*)$. Since $H^k(N/M)$ is finite, replacing U_0 by a suitable open subgroup, we may further assume that U_0 acts trivially on $H^k(N/M)$. Then $H^k(N)$ is a finite trivial U-module where $U = U_0N$.

Now we consider the Hochschild-Serre spectral sequences

$$E_2^{p,q} = H^p(U/N, H^q(N)) \Rightarrow H^{p+q}(U)$$

and

$$\bar{E}_2^{p,q} = H^p(U/N, H^q(N/M)) \Rightarrow H^{p+q}(U/M)\,,$$

which are associated to the group extensions

$$1 \to N \to U \to U/N \to 1 \ \text{ and } \ 1 \to N/M \to U/M \to U/N \to 1\,.$$

By (2.4.6), the second spectral sequence degenerates at $\bar{E}_2$.

Let $X^\bullet(U,\mathbb{Z}/p\mathbb{Z})$, $X^\bullet(U/M,\mathbb{Z}/p\mathbb{Z})$ and $X^\bullet(U/N,\mathbb{Z}/p\mathbb{Z})$ be the standard resolutions of $\mathbb{Z}/p\mathbb{Z}$ as a trivial U-module, U/M-module and U/N-module, respectively. Then the canonical map $X^\bullet(U/M,\mathbb{Z}/p\mathbb{Z})\to X^\bullet(U,\mathbb{Z}/p\mathbb{Z})$ induces a homomorphism of double complexes

$$X^\bullet(U/N, X^\bullet(U/M,\mathbb{Z}/p\mathbb{Z})^{N/M})^{U/N} \to X^\bullet(U/N, X^\bullet(U,\mathbb{Z}/p\mathbb{Z})^{N})^{U/N},$$

and so we obtain a morphism of the associated E_2-spectral sequences

$$(\bar{E}_r^{pq},\bar{d}_r)\longrightarrow(E_r^{pq},d_r).$$

Since $cd_p\, N = k$, and as the second spectral sequence degenerates, we have a commutative diagram

$$\begin{array}{ccc} E_\infty^{n+1-k,k} & \hookrightarrow & E_2^{n+1-k,k} \\ \uparrow & & \uparrow \\ \bar{E}_\infty^{n+1-k,k} & \xrightarrow{\sim} & \bar{E}_2^{n+1-k,k}. \end{array}$$

Next we prove that $E_2^{n+1-k,k}$ vanishes. By (3.3.5)(ii), we have $cd_p\, U = n$, which implies $E_\infty^{n+1-k,k} = 0$. Therefore it suffices to show that the right-hand vertical arrow in the commutative diagram above is surjective. This can be seen as follows: By construction of M, the map

$$inf : H^k(N/M)\longrightarrow H^k(N)$$

is an epimorphism of trivial U/N-modules, which splits, as both groups are $\mathbb{F}_p$-vector spaces. Hence the map

$$\bar{E}_2^{n+1-k,k} = H^{n+1-k}(U/N, H^k(N/M))\longrightarrow E_2^{n+1-k,k} = H^{n+1-k}(U/N, H^k(N))$$

is surjective.

We obtain $E_2^{n+1-k,k} = H^{n+1-k}(U/N, H^k(N)) = 0$. By assumption, $H^k(N)\cong(\mathbb{Z}/p\mathbb{Z})^m$ for some $m\geq 1$, and by construction of U, $H^k(N)$ is a trivial U/N-module. Hence

$$0 = H^{n+1-k}(U/N, H^k(N)) \cong H^{n+1-k}(U/N)^m.$$

This implies $H^{n+1-k}(U/N) = 0$, and so $cd_p\,(U/N)\leq n-k$, as U/N is a pro-p-group, see (3.3.2). On the other hand, (3.3.8) implies $cd_p\,(U/N)\geq n-k$, which finishes the proof. □

(3.3.10) Corollary. *Let G be a pro-p-group, $cd_p\, G = n < \infty$, and let N be a closed normal subgroup of G such that $cd_p\, N = cd_p\, G$ and $H^n(N,\mathbb{Z}/p\mathbb{Z})$ is finite. Then N is open in G.*

Proof: By (3.3.9) and the subsequent remark, we obtain $vcd_p\, G/N = 0$. Hence G/N is finite by (3.3.7). □

The cohomological dimension has the following effect on the corestriction.

(3.3.11) Proposition. *Let U be an open subgroup of G. If $scd\, G = n$ (resp. $cd_p\, G = n$), then for every G-module A (resp. for every p-torsion G-module), the corestriction*

$$cor : H^n(U, A) \longrightarrow H^n(G, A)$$

is surjective. If U is normal in G, then

$$cor : H^n(U, A)_{G/U} \longrightarrow H^n(G, A)$$

is an isomorphism.

Proof: The first assertion follows from the second. Indeed, if U_0 is an open normal subgroup of G contained in U, then the composition

$$cor\,{}^{U_0}_{G} : H^n(U_0, A) \xrightarrow{cor\,{}^{U_0}_{U}} H^n(U, A) \xrightarrow{cor\,{}^{U}_{G}} H^n(G, A)$$

shows that the surjectivity of $cor\,{}^{U_0}_{G}$ implies the surjectivity of $cor\,{}^{U}_{G}$. Therefore we may assume U to be normal in G.

Let $A \to X^\bullet$ be the standard resolution of the G-module A. Then the map $cor : H^i(U, A) \to H^i(G, A)$ is given by taking cohomology of the map of complexes $N = N_{G/U} : X^{\bullet U} \to X^{\bullet G}$. The G-module $Y^n = \ker(X^n \to X^{n+1})$ is cohomologically trivial by our assumption on the cohomological dimension. Therefore

$$0 \longrightarrow A \longrightarrow X^0 \longrightarrow \cdots \longrightarrow X^{n-1} \longrightarrow Y^n \longrightarrow 0$$

is a finite resolution of A by cohomologically trivial G-modules. By (1.3.9), we obtain the exact commutative diagram

$$\begin{array}{ccccccc}
(X^{n-1})^U & \longrightarrow & (Y^n)^U & \longrightarrow & H^n(U, A) & \longrightarrow & 0 \\
\downarrow{\scriptstyle N} & & \downarrow{\scriptstyle N} & & \downarrow{\scriptstyle cor} & & \\
(X^{n-1})^G & \longrightarrow & (Y^n)^G & \longrightarrow & H^n(G, A) & \longrightarrow & 0.
\end{array}$$

Taking G/U-coinvariants of the upper row, we obtain the exact commutative diagram

$$\begin{array}{ccccccc}
((X^{n-1})^U)_{G/U} & \longrightarrow & ((Y^n)^U)_{G/U} & \longrightarrow & H^n(U, A)_{G/U} & \longrightarrow & 0 \\
\downarrow{\scriptstyle N} & & \downarrow{\scriptstyle N} & & \downarrow{\scriptstyle cor} & & \\
((X^{n-1})^U)^{G/U} & \longrightarrow & ((Y^n)^U)^{G/U} & \longrightarrow & H^n(G, A) & \longrightarrow & 0.
\end{array}$$

By (1.8.2), the G/U-modules $(X^{n-1})^U$ and Y^U are cohomologically trivial. Therefore, by (1.2.6), the maps N are bijective, and hence also *cor*, by the five lemma. □

We will now consider the *Euler-Poincaré characteristic* of a pro-p-group.

(3.3.12) Definition. *Let G be a pro-p-group of finite cohomological dimension. Assume that the groups $H^i(G, \mathbb{F}_p)$ are finite for all $i \geq 0$. Then the* **Euler-Poincaré characteristic** *of G is the alternating sum*

$$\chi(G) = \sum_i (-1)^i \dim_{\mathbb{F}_p} H^i(G, \mathbb{F}_p)\,.$$

If A is a finite p-primary G-module with $pA = 0$, then we put

$$\chi(G, A) = \sum_i (-1)^i \dim_{\mathbb{F}_p} H^i(G, A)\,.$$

Let G be as above and let $0 \longrightarrow A_1 \longrightarrow A_2 \longrightarrow A_3 \longrightarrow 0$ be an exact sequence of finite $\mathbb{F}_p[G]$-modules. Then obviously

$$\chi(G, A_2) = \chi(G, A_1) + \chi(G, A_3)\,;$$

in particular, from (1.6.13), it follows for an $\mathbb{F}_p[G]$-module A of order p^r that

$$\chi(G, A) = r \cdot \chi(G)\,.$$

(3.3.13) Proposition. *Let G be a pro-p-group of finite cohomological dimension and assume that the groups $H^i(G, \mathbb{F}_p)$ are finite for all $i \geq 0$. If U is an open subgroup of G, then*

$$\chi(U) = (G : U)\chi(G)\,.$$

Proof: Using (1.6.4), we obtain $\chi(U) = \chi(G, \mathrm{Ind}_G^U \mathbb{F}_p) = (G : U)\chi(G)$. □

If we drop the assumption that the pro-p-group G is of finite cohomological dimension (but keep the assumption on the finiteness of $H^i(G, \mathbb{F}_p)$ for $i \leq n$ for some n), then we define the *partial Euler-Poincaré characteristic* of G.

(3.3.14) Definition. *Let $n \geq 0$ and let G be a pro-p-group such that the groups $H^i(G, \mathbb{F}_p)$ are finite for all i with $0 \leq i \leq n$. Then the* **n-th partial Euler-Poincaré characteristic** *of G is the alternating sum*

$$\chi_n(G) = \sum_{i=0}^{n}(-1)^i \dim_{\mathbb{F}_p} H^i(G, \mathbb{F}_p)\,.$$

For a finite $\mathbb{F}_p[G]$-module A, we define

$$\chi_n(G, A) = \sum_{i=0}^{n}(-1)^i \dim_{\mathbb{F}_p} H^i(G, A)\,.$$

Induction on the length of a composition series of the $\mathbb{F}_p[G]$-module A yields the formula:

(3.3.15) Lemma. *If G and A are as above, then*

$$(-1)^n \chi_n(G, A) \leq (-1)^n \dim_{\mathbb{F}_p} A \cdot \chi_n(G)\,.$$

The following theorem, due to *H. Koch* [110], can be considered as a converse of (3.3.13).

(3.3.16) Theorem. *Let G be a pro-p-group such that the groups $H^i(G, \mathbb{F}_p)$ are finite for $0 \leq i \leq n$. Let $\mathfrak{U}$ be a cofinal set of open subgroups of G. Then the following assertions are equivalent:*

(i) $\chi_n(U) = (G : U)\chi_n(G)$ *for all* $U \in \mathfrak{U}$.

(ii) $cd_p\, G \leq n$.

Proof: The implication (ii) $\Rightarrow$ (i) is just (3.3.13), so let us assume that (i) holds. By (3.3.2)(iii), it suffices to show that $H^{n+1}(G, \mathbb{F}_p) = 0$. Let $\bar{a} \in H^{n+1}(G, \mathbb{F}_p)$ with $a \in C^{n+1}(G, \mathbb{F}_p)$. Then there exists a $U \in \mathfrak{U}$ such that a depends only on the cosets of G/U. Define the finite $\mathbb{F}_p[G]$-module A by the exact sequence

$$0 \longrightarrow \mathbb{F}_p \xrightarrow{\ \varphi\ } \mathrm{Ind}_G^U \mathbb{F}_p \longrightarrow A \longrightarrow 0\,.$$

We obtain a commutative diagram

$$\begin{array}{ccc} H^{n+1}(G, \mathbb{F}_p) & \xrightarrow{\qquad \varphi_* \qquad} & H^{n+1}(G, \mathrm{Ind}_G^U \mathbb{F}_p) \\ & {\scriptstyle res}\searrow \qquad \swarrow{\scriptstyle sh}\,\wr & \\ & H^{n+1}(U, \mathbb{F}_p) & \end{array}$$

which, by the choice of U, shows that $res\,\bar{a} = 0$, and so $\varphi_*\bar{a} = 0$. Furthermore, we get from the short exact sequence above the exact cohomology sequence

$$0 \longrightarrow H^0(G, \mathbb{F}_p) \longrightarrow \cdots \longrightarrow H^n(G, A) \longrightarrow \ker(\varphi_*) \longrightarrow 0\,,$$

which implies, using (3.3.15), (1.6.4) and the assumption (i),

$$\begin{aligned} \dim_{\mathbb{F}_p} \ker(\varphi_*) &= (-1)^n(\chi_n(G) + \chi_n(G, A) - \chi_n(G, \mathrm{Ind}_G^U \mathbb{F}_p)) \\ &\le (-1)^n(\chi_n(G) + \dim_{\mathbb{F}_p} A \cdot \chi_n(G) - \chi_n(U)) \\ &= (-1)^n((G:U)\chi_n(G) - \chi_n(U)) = 0\,. \end{aligned}$$

Thus $\ker(\varphi_*) = 0$ and therefore $\bar{a} = 0$, which proves the theorem. □

Remark: One can prove the following generalization of (3.3.16), cf. [194]: If there exists a number N such that $(-1)^n\chi_n(U) + N \ge (-1)^n(G:U)\chi_n(G)$ for all $U \in \mathfrak{U}$, then G is finite or $cd\, G \le n$.

Exercise 1. If $cd_p\, G = 1$, then $scd_p\, G = 2$.

Exercise 2. Let $p \ne 2$ and let G be the group of affine transformations $x \mapsto ax + b$ with $b \in \mathbb{Z}_p$ and $a \in \mathbb{Z}_p^\times$. Then $cd_p\, G = scd_p\, G = 2$.

Exercise 3. If $cd_p\, G \ne 0, \infty$, then the exponent of p in the order of G is infinite.

Exercise 4. $cd_p\, G = cd_p\, G/H + cd_p\, H$ if H is contained in the center of G and $cd_p\, G/H < \infty$.

Exercise 5. Let H be a closed normal subgroup of G. If $cd_p\, G/H \ne 0$, then $scd_p\, G \le cd_p\, G/H + scd_p\, H$.

Exercise 6. If $H^{n+1}(U, \mathbb{Z}) = H^{n+2}(U, \mathbb{Z}) = 0$ for all open subgroups U of G, then $scd_p\, G \le n$.

Hint: Using the exact sequence $0 \to \mathbb{Z} \xrightarrow{p} \mathbb{Z} \to \mathbb{Z}/p\mathbb{Z} \to 0$, one sees that $H^{n+1}(G_p, \mathbb{Z}/p\mathbb{Z}) = 0$ for a p-Sylow subgroup G_p, hence $cd_p\, G = cd\, G_p \le n$, and $scd_p\, G \le n$ by (3.3.4).

§4. Dualizing Modules

Let G be a profinite group and let $Mod(G)$ be the category of discrete G-modules A. In II §5, we introduced the G-modules

$$D_i(A) = \varinjlim_U H^i(U, A)^*,$$

where U runs through the open normal subgroups of G and the limit is taken over the maps cor^* dual to the corestriction. It comes equipped with the canonical homomorphism

$$H^i(G, A)^* \longrightarrow D_i(A)$$

and the pairing of G-modules

$$D_i(A) \times A \longrightarrow D_i(\mathbb{Z})$$

(cf. p.125). We obtain a canonical homomorphism

$$\varphi_A : H^i(G, A)^* \longrightarrow \mathrm{Hom}_G(A, D_i(\mathbb{Z})).$$

(3.4.1) Theorem. *If $n = scd\, G < \infty$, then the map*

$$\varphi_A : H^n(G, A)^* \longrightarrow \mathrm{Hom}_G(A, D_n(\mathbb{Z}))$$

is an isomorphism for all $A \in Mod(G)$. We call the G-module $D = D_n(\mathbb{Z})$ the **dualizing module** *of G.*

Remark: By definition, the dualizing module of an open subgroup U of G is the dualizing module of G regarded as a U-module.

Proof: By a straightforward limit argument using (1.5.1), we are reduced to the case where A is a finitely generated $\mathbb{Z}$-module. Let U be an open normal subgroup of G acting trivially on A, i.e. A is a $\mathbb{Z}[G/U]$-module.

Suppose the theorem is proven for free $\mathbb{Z}[G/U]$-modules. Then we may choose an exact sequence

$$0 \leftarrow A \leftarrow F_0 \leftarrow F_1$$

with free $\mathbb{Z}[G/U]$-modules F_0, F_1 and obtain the bijectivity of φ_A from the exact commutative diagram

$$\begin{array}{ccccccc}
0 \longrightarrow & H^n(G, A)^* & \longrightarrow & H^n(G, F_0)^* & \longrightarrow & H^n(G, F_1)^* \\
& \downarrow{\scriptstyle \varphi_A} & & \downarrow{\scriptstyle \varphi_{F_0}} & & \downarrow{\scriptstyle \varphi_{F_1}} \\
0 \longrightarrow & \mathrm{Hom}_G(A, D) & \longrightarrow & \mathrm{Hom}_G(F_0, D) & \longrightarrow & \mathrm{Hom}_G(F_1, D).
\end{array}$$

So let $A = \mathbb{Z}[G/U]$ for an open normal subgroup U of G. Using Shapiro's lemma (1.6.4), we obtain the commutative diagram

$$\begin{array}{ccc}
H^n(G, \mathbb{Z}[G/U])^* & \xrightarrow{\varphi_{G,\mathbb{Z}[G/U]}} & \mathrm{Hom}_G(\mathbb{Z}[G/U], D) \\
\uparrow{\scriptstyle \wr\; sh^*} & & \| \\
H^n(U, \mathbb{Z})^* & \xrightarrow{\varphi_{U,\mathbb{Z}}} & \mathrm{Hom}_U(\mathbb{Z}, D).
\end{array}$$

Finally, the map $\varphi_{U,\mathbb{Z}}$ is the direct limit over V of the maps

$$cor^* : H^n(U, \mathbb{Z})^* \longrightarrow \mathrm{Hom}_U(\mathbb{Z}, H^n(V, \mathbb{Z})^*) = (H^n(V, \mathbb{Z})_{U/V})^*,$$

which are isomorphisms by (3.3.11). □

If U is an open subgroup of G, then, by construction, the diagram

$$\begin{array}{ccccc} H^i(U,A)^* & \xrightarrow{\varphi_A} & \mathrm{Hom}_U(A, D_i(\mathbb{Z})) & \hookrightarrow & \mathrm{Hom}(A, D_i(\mathbb{Z})) \\ \uparrow{\scriptstyle cor^*} & & \uparrow & & \| \\ H^i(G,A)^* & \xrightarrow{\varphi_A} & \mathrm{Hom}_G(A, D_i(\mathbb{Z})) & \hookrightarrow & \mathrm{Hom}(A, D_i(\mathbb{Z})) \end{array}$$

is commutative. Passing to the direct limit, we obtain the

(3.4.2) Corollary. *For $A \in Mod(G)$, we have a functorial homomorphism*

$$D_i(A) \longrightarrow \mathrm{Hom}(A, D_i(\mathbb{Z})),$$

which is an isomorphism for $i = scd\, G$ if A is a finitely generated $\mathbb{Z}$-module.

We defined the **trace map**

$$tr : H^n(G, D) \longrightarrow \mathbb{Q}/\mathbb{Z}$$

as the first edge morphism of the Tate spectral sequence $H^i(G, D_{n-j}(\mathbb{Z})) \Rightarrow H^{n-(i+j)}(G,\mathbb{Z})^*$ (see also ex.2). The pairing

$$\mathrm{Hom}(A, D) \times A \longrightarrow D$$

gives us a cup-product

$$H^i(G, \mathrm{Hom}(A, D)) \times H^{n-i}(G, A) \xrightarrow{\cup} H^n(G, D)\,,$$

which, together with the map *tr*, yields a homomorphism

$$H^i(G, \mathrm{Hom}(A, D)) \longrightarrow H^{n-i}(G, A)^*.$$

We can now prove the central result of this section.

(3.4.3) Duality Theorem. *Let G be a profinite group of strict cohomological dimension $scd\, G = n < \infty$ such that $D_k(\mathbb{Z}) = 0$ for $k < n$. Let $A \in Mod(G)$ be finitely generated over $\mathbb{Z}$ and assume that the ℓ-primary part of A is nontrivial only for prime numbers ℓ such that $D = D_n(\mathbb{Z})$ is ℓ-divisible.*

*Then for all $i \in \mathbb{Z}$ *) the cup-product and the trace map*

$$H^i(G, \mathrm{Hom}(A, D)) \times H^{n-i}(G, A) \xrightarrow{\cup} H^n(G, D) \xrightarrow{tr} \mathbb{Q}/\mathbb{Z}$$

yield an isomorphism

$$H^i(G, \mathrm{Hom}(A, D)) \cong H^{n-i}(G, A)^*.$$

*) $H^i = 0$ for $i < 0$ by definition.

Proof: Because of (3.4.2), we may replace $\mathrm{Hom}(A, D)$ by $D_n(A)$ and consider the map

$$(*) \qquad H^i(G, D_n(A)) \longrightarrow H^{n-i}(G, A)^*$$

induced by the cup-product with respect to the canonical pairing $D_n(A) \times A \to D_n(\mathbb{Z})$. But by (2.5.5) this map is the edge morphism $E_2^{i,0} \to E^i$ of the Tate spectral sequence

$$E_2^{ij} = H^i(G, D_{n-j}(A)) \Rightarrow H^{n-(i+j)}(G, A)^* = E^{i+j}.$$

From the assumption in the theorem, it follows that $D_k(A) = 0$ for $k < n$ by lemma (2.5.6), hence the spectral sequence degenerates, showing that $(*)$ is an isomorphism. □

Remarks: 1. Setting $i = n$ and $A = \mathbb{Z}$, we see that the trace map $tr : H^n(G, D) \to \mathbb{Q}/\mathbb{Z}$ is an isomorphism. It is clear that the duality theorem remains valid if we replace it by any other isomorphism $H^n(G, D) \cong \mathbb{Q}/\mathbb{Z}$. This can be useful in the applications, where we may have a canonical such isomorphism without it being clear that it is the edge morphism.

2. The condition $D_0 = \varinjlim_U \mathbb{Q}/\mathbb{Z} = 0$ is equivalent to the assertion that every prime number p divides the order of G infinitely often, in the sense that all Sylow subgroups G_p are infinite.

We obtain an important variant of the dualizing module and the duality theorem as follows. Let P be a nonempty set of prime numbers. Let $\mathbb{N}(P)$ denote the set of all natural numbers which are divisible only by prime numbers in P. Let $Mod_P(G)$ denote the category of all P-torsion G-modules, i.e. modules A consisting of elements a such that $na = 0$ for some $n \in \mathbb{N}(P)$.

We define the *cohomological P-dimension* $cd_P\, G$ as the smallest number $n \geq 0$ such that $H^i(G, A) = 0$ for all $i > n$ and all $A \in Mod_P(G)$. If no such number exists, we set $cd_P\, G = \infty$. In other words, $cd_P\, G = \sup_{p \in P}\{cd_p\, G\}$.

We now set

$$D_i(\mathbb{Z}_P) = \varinjlim_{m \in \mathbb{N}(P)} D_i(\mathbb{Z}/m\mathbb{Z}) = \varinjlim_{m \in \mathbb{N}(P)} \varinjlim_U H^i(U, \mathbb{Z}/m\mathbb{Z})^*,$$

and, in particular,

$$D_i(\hat{\mathbb{Z}}) = \varinjlim_{m \in \mathbb{N}} D_i(\mathbb{Z}/m\mathbb{Z}).$$

Again we have for $A \in Mod_P(G)$ a functorial homomorphism

$$\varphi_A : H^i(G, A)^* \longrightarrow \mathrm{Hom}_G(A, D_i(\mathbb{Z}_P))$$

which is obtained as above from the canonical pairing

$$H^i(G, A)^* \times {}_mA^U \longrightarrow H^i(V, \mathbb{Z}/m\mathbb{Z})^*, \ (\chi, a) \mapsto \chi_a(x) = \chi(cor(ax)),$$

and by taking direct limits over V and $m \in \mathbb{N}(P)$, and then over U.

(3.4.4) Theorem. *If $cd_P\, G = n < \infty$, then for all $A \in Mod_P(G)$, the map*

$$\varphi_A : H^n(G, A)^* \longrightarrow \mathrm{Hom}_G(A, D_n(\mathbb{Z}_P))$$

is an isomorphism, and, if A is finite, we obtain an isomorphism of G-modules

$$D_n(A) \cong \mathrm{Hom}(A, D_n(\mathbb{Z}_P)).$$

The proof is the same as that of (3.4.1) (resp. (3.4.2)); we have just to replace $\mathbb{Z}$ by $\mathbb{Z}/m\mathbb{Z}$ ($m \in \mathbb{N}(P)$) and $\varinjlim_U$ by $\varinjlim_{m\in\mathbb{N}(P)} \varinjlim_U$.

The G-module $D_n(\mathbb{Z}_P)$ represents the functor

$$T : Mod_P(G) \longrightarrow \mathcal{A}b, \quad A \longmapsto T(A) = H^n(G, A)^*,$$

and is called the **dualizing module** of G at P (or of $Mod_P(G)$). We denote it briefly by D_P. If P consists of a single prime number p, then we write $D_{(p)}$.

(3.4.5) Corollary (Serre Criterion). *Let p be a prime number and let $1 \leq n = cd_p\, G < \infty$. We have $scd_p\, G = n + 1$ if and only if there exists an open subgroup U of G such that $D_{(p)}^U$ contains a subgroup isomorphic to $\mathbb{Q}_p/\mathbb{Z}_p$.*

Proof: The existence of a subgroup of $D_{(p)}^U$ isomorphic to $\mathbb{Q}_p/\mathbb{Z}_p$ is equivalent to $\mathrm{Hom}_U(\mathbb{Q}_p/\mathbb{Z}_p, D_{(p)}) \neq 0$. Since $D_{(p)}$ is also the dualizing module of U at p, this means by (3.4.4) that $H^n(U, \mathbb{Q}_p/\mathbb{Z}_p) \neq 0$. But this last group is the p-primary component of $H^n(U, \mathbb{Q}/\mathbb{Z})$. The exact sequence $0 \to \mathbb{Z} \to \mathbb{Q} \to \mathbb{Q}/\mathbb{Z} \to 0$ yields $H^n(U, \mathbb{Q}/\mathbb{Z})(p) \cong H^{n+1}(U, \mathbb{Z})(p)$, and the corollary follows now from (3.3.4) and (3.3.5) (ii). □

The Tate spectral sequence

$$H^i(G, D_{n-j}(\mathbb{Z}/m\mathbb{Z})) \Rightarrow H^{n-(i+j)}(G, \mathbb{Z}/m\mathbb{Z})^*,$$

$m \in \mathbb{N}(P)$, gives an edge morphism

$$H^n(G, D(\mathbb{Z}/m\mathbb{Z})) \longrightarrow H^0(G, \mathbb{Z}/m\mathbb{Z})^* = \tfrac{1}{m}\mathbb{Z}/\mathbb{Z}.$$

Taking limits over $m \in \mathbb{N}(P)$, we obtain a homomorphism

$$tr : H^n(G, D_P) \longrightarrow \mathbb{Q}_P/\mathbb{Z}_P := \bigoplus_{p\in P} \mathbb{Q}_p/\mathbb{Z}_p\,,$$

again called the **trace map**. We now obtain the following variant of the duality theorem (3.4.3).

(3.4.6) Duality Theorem. *For a profinite group with $cd_P\, G = n < \infty$, the following assertions are equivalent.*

(i) $D_i(\mathbb{Z}/p\mathbb{Z}) = 0$ *for all* $p \in P$ *and all* $i < n$.

(ii) *For all* $i \in \mathbb{Z}$ *and all finite G-modules* $A \in Mod_P(G)$, *the cup-product and the trace map*

$$H^i(G, \mathrm{Hom}(A, D_P)) \times H^{n-i}(G, A) \xrightarrow{\cup} H^n(G, D_P) \xrightarrow{tr} \mathbb{Q}_P/\mathbb{Z}_P$$

yield an isomorphism

$$H^i(G, \mathrm{Hom}(A, D_P)) \cong H^{n-i}(G, A)^*.$$

In this case G, is called a **duality group at P of dimension n.**

Proof: The implication (i) ⇒ (ii) follows by the same argument as in the proof of (3.4.3). So we have only to show the implication (ii) ⇒ (i). For every pair $V \subseteq U$ of open normal subgroups of G, we have a commutative diagram of G-modules

$$\begin{array}{ccc} \mathrm{Ind}_G^V(A) & \xrightarrow{\nu_U^V} & \mathrm{Ind}_G^U(A) \\ & {\scriptstyle \nu_G^V}\searrow \qquad \swarrow{\scriptstyle \nu_G^U} & \\ & A & \end{array}$$

where $\nu_U^V(x)(\sigma) = \sum_{\tau \in U/V} \tau x(\tau^{-1}\sigma)$. By the lemma of Shapiro (1.6.4) and by (1.6.5) and the subsequent remark, we obtain from this a commutative diagram of G-modules

$$\begin{array}{ccccc} H^i(G, \mathrm{Ind}_G^V(A)) & & \xrightarrow{\quad sh \quad} & & H^i(V, A) \\ & {\scriptstyle \nu_*}\searrow & & \swarrow{\scriptstyle cor} & \\ {\scriptstyle \nu_*}\downarrow & & H^i(G, A) & & \downarrow{\scriptstyle cor} \\ & {\scriptstyle \nu_*}\nearrow & & \nwarrow{\scriptstyle cor} & \\ H^i(G, \mathrm{Ind}_G^U(A)) & & \xrightarrow{\quad sh \quad} & & H^i(U, A). \end{array}$$

In this diagram the maps sh are isomorphisms, and hence define an isomorphism of projective systems $(H^n(G, \mathrm{Ind}_G^U(A))) \cong (H^n(U, A))$ of G-modules. We therefore have a canonical isomorphism

$$D_i(A) \cong \varinjlim_U H^i(G, \mathrm{Ind}_G^U(A))^*.$$

Applying (ii) and (1.5.1), we obtain for all $i < n$ and $p \in P$,

$$\begin{aligned} D_i(\mathbb{Z}/p\mathbb{Z}) &\cong \varinjlim_{U} H^i(G, \mathrm{Ind}_G^U(\mathbb{Z}/p\mathbb{Z}))^* \\ &\cong \varinjlim_{U} H^{n-i}(G, \mathrm{Hom}(\mathrm{Ind}_G^U(\mathbb{Z}/p\mathbb{Z}), D_P)) \\ &\cong \varinjlim_{U} H^{n-i}(G, \mathrm{Ind}_G^U(\mathrm{Hom}(\mathbb{Z}/p\mathbb{Z}, D_P)) \\ &\cong \varinjlim_{U,res} H^{n-i}(U, {}_pD_P) = 0. \end{aligned}$$

□

(3.4.7) Corollary. *If G is a duality group at P of dimension n, then for every $p \in P$ there exists an exact sequence*

$$0 \longrightarrow D_n(\mathbb{Z}/p\mathbb{Z}) \longrightarrow D_P \xrightarrow{p} D_P \longrightarrow 0.$$

In particular, D_P is p-divisible for all $p \in P$.

Proof: Let $p \in P$, $m > 1$, and consider the exact diagram

$$\begin{array}{ccccccccc} 0 & \longrightarrow & \mathbb{Z}/p^m\mathbb{Z} & \longrightarrow & \mathbb{Z}/p^{m+1}\mathbb{Z} & \longrightarrow & \mathbb{Z}/p\mathbb{Z} & \longrightarrow & 0 \\ & & \downarrow & & \downarrow & & \| & & \\ 0 & \longrightarrow & \mathbb{Z}/p^{m-1}\mathbb{Z} & \longrightarrow & \mathbb{Z}/p^m\mathbb{Z} & \longrightarrow & \mathbb{Z}/p\mathbb{Z} & \longrightarrow & 0. \end{array}$$

Applying the functor $\varinjlim_U H^n(U, -)^*$ and passing to the limit over m, we obtain the exact sequence

$$0 \longrightarrow D_n(\mathbb{Z}/p\mathbb{Z}) \longrightarrow D_{(p)} \xrightarrow{p} D_{(p)} \longrightarrow 0.$$

This proves the corollary since $D_{(p)}$ is the p-torsion subgroup of the torsion group D_P. □

(3.4.8) Corollary. *Assume that $scd\, G = n$ and $D_k(\mathbb{Z}) = 0$ for $k = 0, \dots, n-1$. Let p be a prime number. Then $D_n(\mathbb{Z})$ is p-divisible if and only if G is a duality group at p of dimension n. In this case*

$$D_{(p)} = D_n(\mathbb{Z})(p)\,.$$

Proof: Consider the exact sequence

$$0 \longrightarrow \mathbb{Z} \xrightarrow{p} \mathbb{Z} \longrightarrow \mathbb{Z}/p\mathbb{Z} \longrightarrow 0$$

and apply the functor $\varinjlim H^k(U, -)^*$. We obtain $D_k(\mathbb{Z}/p\mathbb{Z}) = 0$ for $k \le n-2$ and an exact sequence

$$0 \to D_n(\mathbb{Z}/p\mathbb{Z}) \to D_n(\mathbb{Z}) \xrightarrow{p} D_n(\mathbb{Z}) \to D_{n-1}(\mathbb{Z}/p\mathbb{Z}) \to 0\,.$$

Hence $D_{n-1}(\mathbb{Z}/p\mathbb{Z}) = 0$, i.e. G is a duality group at p of dimension n by (3.4.6) if and only if $D_n(\mathbb{Z})$ is p-divisible. Further, in this case,

$$\begin{aligned} D_{(p)} &= \varinjlim_{U,m} H^n(U, \mathbb{Z}/p^m\mathbb{Z})^* \\ &= \varinjlim_{U,m} H^0(U, \mathrm{Hom}(\mathbb{Z}/p^m\mathbb{Z}, D_n(\mathbb{Z}))) \\ &= D_n(\mathbb{Z})(p)\,. \end{aligned}$$

□

Remark: The duality theorems (3.4.3) and (3.4.6) are due to *J. Tate* (see [230]). Tate gave the duality isomorphism as the edge morphism of the Tate spectral sequence, which by (2.5.5) is induced by the cup-product. Another slightly different method, due to *J. L. Verdier*, is to prove (3.4.3) and (3.4.6) by analyzing the group of "local homomorphisms" (see [240]). The concept of duality groups was also inspired by the work of *R. Bieri* and *B. Eckmann*. in the case of discrete groups (see [11]).

Exercise 1. Let $n = scd\, G$. If the homomorphism

$$H^p(G, D_n(A)) \longrightarrow H^{n-p}(G, A)^*$$

is bijective for all $p \in \mathbb{Z}$ and all $A \in Mod(G)$ finitely generated over $\mathbb{Z}$, then $D_i = 0$ for $i < n$ and D_n is divisible.

Exercise 2. Applying (3.4.1) to $A = D$, we obtain a canonical isomorphism

$$H^n(G, D)^* \cong \mathrm{Hom}_G(D, D).$$

The trace map $tr: H^n(G, D) \longrightarrow \mathbb{Q}/\mathbb{Z}$, defined as the first edge morphism of the Tate spectral sequence, is an element of the left-hand group, which corresponds to *id* in the right-hand group, and $\mathrm{Hom}_G(D, D) \cong \mathbb{Z}$.

Exercise 3. Let $G = \hat{\mathbb{Z}}$. Show that $scd\, G = 2$, $cd\, G = 1$. Calculate D and D_P for the set P of all prime numbers. Show that there are isomorphisms

$$H^i(G, \mathrm{Hom}(A, \mathbb{Q}/\mathbb{Z})) \cong H^{1-i}(G, A)^*$$

for all i and all finite G-modules A.

Exercise 4. Let $F_n = \langle x_1, \dots, x_n \rangle$ be a free pro-p-group on n generators. Show that the dualizing module $D_{(p)}$ of F_n is given by the exact sequence

$$0 \longrightarrow \mathbb{Q}_p/\mathbb{Z}_p \xrightarrow{diag} \bigoplus_{i=0}^n \mathrm{Ind}_{F_n}^{\langle x_i \rangle} \mathbb{Q}_p/\mathbb{Z}_p \longrightarrow D_{(p)} \longrightarrow 0,$$

where $x_0 := x_n^{-1} x_{n-1}^{-1} \cdots x_1^{-1}$ and $\langle x_i \rangle$ is the closed procyclic group generated by x_i in F_n ($i = 0, \dots, n$).
(The characterization of $D_{(p)}$ given in ex.4 on p.V-24 (Cinquième éd. p.76) in [240] is incorrect!)

§5. Projective pro-$\mathfrak{c}$-groups

In this section we introduce the concept of embedding problems and consider profinite groups G of cohomological dimension $cd\, G \le 1$. We start by considering the abstract embedding problem for a profinite group G.

(3.5.1) Definition. (i) *An* **embedding problem** $\mathscr{E}(G) = \mathscr{E}(G, \varphi, \alpha)$ *for a profinite group* G *is a diagram*

$$\begin{array}{ccccccccc} & & & & & & G & & \\ & & & & & & \downarrow{\scriptstyle\varphi} & & \\ 1 & \longrightarrow & N & \longrightarrow & E & \xrightarrow{\alpha} & \bar{G} & \longrightarrow & 1 \end{array}$$

with an exact sequence of profinite groups and a surjective homomorphism φ.

(ii) *A* **solution** *of the embedding problem* $\mathscr{E}(G)$ *is a homomorphism* $\psi\colon G \to E$ *such that* $\alpha \circ \psi = \varphi$. *A solution is called* **proper** *if* ψ *is surjective.*

(iii) *Two solutions* ψ *and* ψ' *are called* **equivalent** *if*

$$\psi'(\sigma) = a^{-1}\psi(\sigma)a$$

for all $\sigma \in G$ *with a fixed element* $a \in N$. *The set of all solutions of* $\mathscr{E}(G)$ *modulo equivalence is denoted by* $\mathscr{S}_{\mathscr{E}(G)}$ *and is considered as a discrete topological space.*

An embedding problem for a Galois group G corresponds to a field theoretical problem. Let $G = G(\tilde{k}|k)$ be the Galois group of a Galois extension $\tilde{k}$ of a field k and let $K|k$ be a Galois subextension. If $E \twoheadrightarrow G(K|k)$ is a group extension, then a proper solution of the embedding problem $\mathscr{E}(G, G \xrightarrow{can} G(K|k), \alpha)$ defines a Galois extension $L \supseteq K \supseteq k$ with Galois group $G(L|k)$ isomorphic to E such that α is the canonical projection $G(L|k) \twoheadrightarrow G(K|k)$. If the solution is not proper, then one obtains only a Galois algebra with group E instead of a field L.

(3.5.2) Definition. (i) *A class* $\mathfrak{c}$ *of finite groups is called a* **full class** *if it is closed under taking subgroups, homomorphic images and group extensions. (i.e. if* $1 \to G' \to G \to G'' \to 1$ *is an exact sequence of finite groups, then* G *is in* $\mathfrak{c}$ *if* G' *and* G'' *are).*

(ii) *If* $\mathfrak{c}$ *is a full class of finite groups, a* **pro-$\mathfrak{c}$-group** *is a projective limit of groups in* $\mathfrak{c}$.

(iii) *For a full class* $\mathfrak{c}$ *of finite groups,* $S(\mathfrak{c})$ *denotes the set of prime numbers* p *with* $\mathbb{Z}/p\mathbb{Z} \in \mathfrak{c}$. *For a set* S *of prime numbers,* $\mathfrak{c}(S)$ *denotes the full class of finite groups whose orders are divisible only by prime numbers in* S.

Remark: We have $S(\mathfrak{c}(S)) = S$ and an inclusion $\mathfrak{c} \subseteq \mathfrak{c}(S(\mathfrak{c}))$ which is, in general, strict.

If $\mathfrak{c}$ is the class of all finite groups, finite p-groups, or finite solvable groups, we get the profinite groups, the pro-p-groups, or the prosolvable groups, respectively. The class of pro-p-groups does not contain a nontrivial, proper full subclass, because every finite p-group G has a series of subgroups $1 = G_0 \subseteq G_1 \subseteq \cdots \subseteq G_n = G$ such that G_{i-1} is normal in G_i and $G_i/G_{i-1} \cong \mathbb{Z}/p\mathbb{Z}$ for $i = 1, \ldots, n$. In particular, the class of pro-$\mathfrak{c}$-groups contains the class of pro-p-groups if and only if $\mathbb{Z}/p\mathbb{Z} \in \mathfrak{c}$.

(3.5.3) Definition. *Let $\mathfrak{c}$ be a full class of finite groups.*

(i) *For a profinite group G, the group*

$$G(\mathfrak{c}) = \varprojlim_{G/U \in \mathfrak{c}} G/U,$$

where U runs through the open normal subgroups of G such that $G/U \in \mathfrak{c}$, is called the **maximal pro-$\mathfrak{c}$-factor group** *of G.*

(ii) *For an abelian torsion group A, the subgroup*

$$A(\mathfrak{c}) = \bigcup_{B \in \mathfrak{c}} B,$$

where the union is taken over all finite subgroups $B \subseteq A$ with $B \in \mathfrak{c}$, is called the **$\mathfrak{c}$-torsion subgroup** *of A.*

If G is a finite abelian torsion group, then both possible interpretations of $G(\mathfrak{c})$ coincide in the sense that the $\mathfrak{c}$-torsion subgroup maps isomorphically onto the maximal pro-$\mathfrak{c}$-factor group.

(3.5.4) Proposition. *Let G be a profinite group and let $\mathfrak{c}$ be a full class of finite groups.*

(i) *For the kernel $H = \ker(G \twoheadrightarrow G(\mathfrak{c}))$ we have $H(\mathfrak{c}) = 1$.*

(ii) *Let $G' \to G \to G'' \to 1$ be an exact sequence of profinite groups. Then the sequence*

$$G'(\mathfrak{c}) \to G(\mathfrak{c}) \to G''(\mathfrak{c}) \to 1$$

is exact.

(iii) *Let $1 \to G' \to G \to G'' \to 1$ be an exact sequence of profinite groups. If G'' is a pro-$\mathfrak{c}$-group, then the sequence*

$$1 \to G'(\mathfrak{c}) \to G(\mathfrak{c}) \to G''(\mathfrak{c}) \to 1$$

is exact.

Proof: We start by showing (i). Suppose H has a proper open normal subgroup N with $H/N \in \mathfrak{c}$. We let G act on H by conjugation. The subgroup G_0 of all $\sigma \in G$ such that $\sigma N \sigma^{-1} = N$ is open in G, since it consists of all elements $\sigma \in G$ which maps a compact set (namely N) into an open set (namely N). Therefore G/G_0 is finite and

$$\tilde{N} := \bigcap_{\sigma \in G/G_0} \sigma N \sigma^{-1}$$

is open in H and normal in G. The injective homomorphism

$$H/\tilde{N} \longrightarrow \prod_{\sigma \in G/G_0} H/\sigma N \sigma^{-1}$$

shows that $H/\tilde{N} \in \mathfrak{c}$. The group extension $1 \to H/\tilde{N} \to G/\tilde{N} \to G/H \to 1$ shows that $G/\tilde{N}$ is a pro-$\mathfrak{c}$-group, contradicting the definition of H.

Now let $G' \to G \to G'' \to 1$ be exact. The surjectivity of $G(\mathfrak{c}) \to G''(\mathfrak{c})$ is immediate. Therefore, in order to show (ii), we may replace G' by its image in G and suppose that $1 \to G' \to G \to G'' \to 1$ is exact. Let H be the kernel of $G \twoheadrightarrow G(\mathfrak{c})$. We have an exact sequence of pro-$\mathfrak{c}$-groups

$$1 \to G'/(G' \cap H) \to G/H \to G/G'H \to 1$$

and a surjection $G'(\mathfrak{c}) \twoheadrightarrow G'/(G' \cap H)$. Consider the exact sequence

$$1 \to G'H/G' \to G/G' \to G/G'H \to 1.$$

The group $G'H/G'$ is a quotient of H and has no nontrivial homomorphisms to $\mathfrak{c}$-groups by (i). We therefore obtain an isomorphism $G''(\mathfrak{c}) = (G/G')(\mathfrak{c}) \xrightarrow{\sim} G/G'H$, showing (ii).

If G'' is a pro-$\mathfrak{c}$-group, then $H \subseteq G'$ and, again using (i), the surjection $G' \to G'/H$ induces an isomorphism $G'(\mathfrak{c}) \xrightarrow{\sim} G'/H$. This shows (iii). □

(3.5.5) Definition. *Let $\mathfrak{c}$ be a full class of finite groups.*

(i) A profinite group G is called **$\mathfrak{c}$-projective** *if every embedding problem for G where the kernel is a pro-$\mathfrak{c}$-group has a solution.*

(ii) A profinite group G is **projective,** *if it is $\mathfrak{f}$-projective, where $\mathfrak{f}$ is the class of all finite groups.*

(iii) If S is a set of prime numbers, then a $\mathfrak{c}(S)$-projective group is called **S-projective**. *If S consists only of one prime number p, i.e. $\mathfrak{c}(S)$ is the class of finite p-groups, we call such a group* **p-projective**.

Remark: If G is a pro-$\mathfrak{c}$-group which is $\mathfrak{c}$-projective in the above sense, then G is a projective object in the category of pro-$\mathfrak{c}$-groups.

Our aim is to prove the following result due to *K. W. Gruenberg* [71].

(3.5.6) Theorem. *Let $\mathfrak{c}$ be a full class of finite groups. For a profinite group G the following assertions are equivalent:*

(i) *G is $\mathfrak{c}$-projective.*

(ii) *Every group extension $1 \to N \to E \to G \to 1$ where N is a pro-$\mathfrak{c}$-group splits.*

(iii) *Every group extension $1 \to A \to E \to G \to 1$ where A is a finite abelian p-group with $p \in S(\mathfrak{c})$ splits.*

(iv) *$cd_p\, G \le 1$ for all prime numbers $p \in S(\mathfrak{c})$.*

(v) *G is $S(\mathfrak{c})$-projective.*

The crucial step in the proof of (3.5.6) is the following result.

(3.5.7) Proposition. *A profinite group G is $\mathfrak{c}$-projective if every embedding problem*

$$(*) \qquad \begin{array}{ccccccccc} & & & & & & G & & \\ & & & & & & \downarrow & & \\ 1 & \longrightarrow & N & \longrightarrow & E & \longrightarrow & \bar{G} & \longrightarrow & 1 \end{array}$$

for G is solvable, where E is finite and $N \in \mathfrak{c}$ is a minimal abelian normal subgroup of E.

Proof: We have to show that every embedding problem $(*)$ with N a pro-$\mathfrak{c}$-group is solvable. Firstly, we reduce to the case when N is finite. Assume that all embedding problems $(*)$ with N a finite $\mathfrak{c}$-group are solvable and consider an embedding problem $(*)$ with N an arbitrary pro-$\mathfrak{c}$-group. Let X be the set of all pairs (N', ψ') consisting of a closed subgroup N' of N which is normal in E and a solution $\psi'\colon G \to E/N'$ of the induced embedding problem

$$\begin{array}{ccccccccc} & & & & & & G & & \\ & & & & & \swarrow^{\psi'} & \downarrow{\scriptstyle\varphi} & & \\ 1 & \longrightarrow & N/N' & \longrightarrow & E/N' & \xrightarrow{\bar{\alpha}} & \bar{G} & \longrightarrow & 1. \end{array}$$

We write $(N'', \psi'') \ge (N', \psi')$ if $N'' \subseteq N'$ and if ψ' coincides with the composite $G \xrightarrow{\psi''} E/N'' \twoheadrightarrow E/N'$. Then X is inductively ordered and nonempty. By Zorn's lemma, there exists a maximal element (N', ψ'). We have to show that $N' = 1$. Assume the contrary. Then there exists a proper open subgroup U of N'. As the topology of N' is induced by that of E, there exists a normal open subgroup $\tilde{E} \subseteq E$ such that $N'' = \tilde{E} \cap N'$ is contained in U. The group N'' is a proper open subgroup in N' and normal in E. By assumption, the embedding problem with finite kernel

$$\begin{array}{ccccccccc}
 & & & & & & G & & \\
 & & & & & \swarrow & \downarrow{\scriptstyle \psi'} & & \\
1 & \longrightarrow & N'/N'' & \longrightarrow & E' & \longrightarrow & \mathrm{im}(\psi') & \longrightarrow & 1 \\
 & & \| & & \downarrow & & \downarrow & & \\
1 & \longrightarrow & N'/N'' & \longrightarrow & E/N'' & \longrightarrow & E/N' & \longrightarrow & 1,
\end{array}$$

where E' is the pre-image of $\mathrm{im}(\psi')$ in E/N'', has a solution. Therefore we obtain a homomorphism ψ'': $G \to E/N''$ such that the pair (N'', ψ'') is strictly larger than the pair (N', ψ'). This contradicts the maximality of (N', ψ').

Next we show that we can also assume E to be finite. Assume all embedding problems $(*)$ with E finite and $N \in \mathfrak{c}$ are solvable, and consider an arbitrary embedding problem with finite kernel $N \in \mathfrak{c}$. Since N is finite, there exists an open normal subgroup U of E with $U \cap N = 1$. We obtain the commutative and exact diagram

$$\begin{array}{ccccccccc}
 & & & & & & G & & \\
 & & & & & & \downarrow{\scriptstyle \varphi} & & \\
1 & \longrightarrow & N & \longrightarrow & E & \longrightarrow & \overline{G} & \longrightarrow & 1 \\
 & & \| & & \downarrow & \swarrow{\scriptstyle \psi} & \downarrow & & \\
1 & \longrightarrow & N & \longrightarrow & E/U & \longrightarrow & \overline{G}/\mathrm{im}(U) & \longrightarrow & 1\,,
\end{array}$$

where the broken arrow is obtained by the assumption of this reduction. By (3.5.8) below, the right hand square is cartesian, i.e. the natural map from E to the fibre product $E/U \times_{\overline{G}/\mathrm{im}(U)} \overline{G}$ is an isomorphism. Therefore the pair (ψ, φ) induces a homomorphism $G \longrightarrow E$, completing the diagram commutatively.

Now we prove by induction on the (finite) order of $N \in \mathfrak{c}$ that every embedding problem $(*)$ with E finite has a solution. We distinguish between the following three cases:

(1) N is not a minimal normal subgroup of E. Then we can choose a normal subgroup M of E such that $1 \subsetneqq M \subsetneqq N$. By induction,

$$\begin{array}{ccccccccc}
 & & & & & & G & & \\
 & & & & & {\scriptstyle \psi}\swarrow & \downarrow & & \\
1 & \longrightarrow & N/M & \longrightarrow & E/M & \longrightarrow & \overline{G} & \longrightarrow & 1
\end{array}$$

has a solution ψ: $G \to E/M$. Again by induction, the embedding problem

$$\begin{array}{ccccccccc}
 & & & & & & G & & \\
 & & & & & & \downarrow{\scriptstyle \psi} & & \\
1 & \longrightarrow & M & \longrightarrow & E & \longrightarrow & E/M & \longrightarrow & 1
\end{array}$$

has a solution.

(2) Let N be a minimal normal subgroup of E and assume that N is not contained in the Frattini subgroup $\Phi(E)$ of E (which is the intersection of all maximal subgroups of E). Then there exists a maximal subgroup E_1 of E such that $N \nsubseteq E_1$, hence $NE_1 = E$, i.e. E_1 projects onto $\bar{G}$. Then $N_1 = E_1 \cap N$ is a proper subgroup of N and thus, by induction, the embedding problem

$$\begin{array}{ccccccccc} & & & & & & G & & \\ & & & & & \swarrow & \downarrow & & \\ 1 & \longrightarrow & N_1 & \longrightarrow & E_1 & \longrightarrow & \bar{G} & \longrightarrow & 1 \\ & & \downarrow & & \downarrow & & \| & & \\ 1 & \longrightarrow & N & \longrightarrow & E & \longrightarrow & \bar{G} & \longrightarrow & 1 \end{array}$$

has a solution.

(3) Let N be a minimal normal subgroup of E and assume that $N \subseteq \Phi(E)$. Since $\Phi(E)$ is nilpotent (see [81], III Satz 3.6), the group N is abelian. By assumption, we get a solution of the embedding problem in this case. □

(3.5.8) Lemma. *Let E be a profinite group and let U and V be closed normal subgroups of G. Then the diagram of profinite groups*

$$\begin{array}{ccc} E/U\cap V & \xrightarrow{p_U} & E/U \\ \downarrow{\scriptstyle p_V} & & \downarrow \\ E/V & \longrightarrow & E/UV \end{array}$$

is cartesian, i.e. the projections define an isomorphism

$$E/U\cap V \xrightarrow[\sim]{(p_U,p_V)} E/U \times_{E/UV} E/V,$$

where the group on the right hand side is the fibre product, i.e. the subgroup of elements (a,b) in the product $E/U \times E/V$ such that a and b project to the same element in E/UV.

Proof: Injectivity is trivial. Let $a, b \in E$ be elements with the same image in E/UV and consider the element $(p_U(a), p_V(b)) \in E/U \times_{E/UV} E/V$. By assumption, $ab^{-1} = uv$ for elements $u \in U$, $v \in V$. Then the element $g = u^{-1}a = vb \in E$ maps to $((p_U(a), p_V(b))$, which shows surjectivity. □

In order to detect $\mathfrak{c}$-projective groups, we are, by (3.5.7), reduced to a special case of embedding problems with finite abelian kernel. Next we investigate arbitrary embedding problems with finite abelian kernel $A = \ker(\alpha)$. The group $\overline{G}$ acts on A by conjugation and the group extension $1 \to A \to E \overset{\alpha}{\to} \overline{G} \to 1$ is determined by its class $\varepsilon \in H^2(\overline{G}, A)$, see (1.2.4).

Let $\mathscr{E}(G, \varphi, \alpha)$ be an embedding problem with finite abelian kernel A for the profinite group G and let $H = \ker\, \varphi$. Consider the diagram

$$\begin{array}{ccccccccc} & & & & & & H & & \\ & & & \psi_0 \swarrow & & & \downarrow & & \\ & & & & & & G & & \\ & & & & \psi \swarrow & & \downarrow \varphi & & \\ 1 & \longrightarrow & A & \longrightarrow & E & \underset{\alpha}{\longrightarrow} & \overline{G} & \longrightarrow & 1\,. \end{array}$$

We consider A as a G-module via φ. Then H acts trivially on A. A solution ψ induces a $\bar{G}$-homomorphism $\psi_0\colon H \to A$. From the Hochschild-Serre spectral sequence we get the exact sequence (cf. (1.6.7))

$$0 \longrightarrow H^1(\overline{G}, A) \longrightarrow H^1(G, A) \xrightarrow{res} H^1(H, A)^{\bar{G}} \xrightarrow{tg} H^2(\overline{G}, A) \xrightarrow{inf} H^2(G, A).$$

(3.5.9) Proposition (Hoechsmann). *Let $\varepsilon \in H^2(\overline{G}, A)$ be the cohomology class corresponding to the group extension of the embedding problem $\mathscr{E}(G) = \mathscr{E}(G, \varphi, \alpha)$, where $A = \ker(\alpha)$ is a finite $\bar{G}$-module.*

Then $\mathscr{E}(G)$ has a solution if and only if $inf(\varepsilon) = 0$, i.e. if there exists a $\bar{G}$-homomorphism $\psi_0\colon H \to A$ with $tg(\psi_0) = \varepsilon$.

Proof: (see [80], 1.1). Consider the commutative exact diagram

$$\begin{array}{ccccccccc} 1 & \longrightarrow & A & \longrightarrow & E \times_{\overline{G}} G & \xrightarrow{pr_2} & G & \longrightarrow & 1 \\ & & \| & & \downarrow pr_1 & & \downarrow \varphi & & \\ 1 & \longrightarrow & A & \longrightarrow & E & \underset{\alpha}{\longrightarrow} & \overline{G} & \longrightarrow & 1 \end{array}$$

where $E \times_{\overline{G}} G = \{(e, \sigma) \in E \times G \mid \alpha(e) = \varphi(\sigma)\}$ is the fibre product of α and φ. The group extension above corresponds to $inf(\varepsilon) \in H^2(G, A)$. If $inf(\varepsilon) = 0$, then this extension splits, i.e. there exists a homomorphism

$$s\colon G \to E \times_{\overline{G}} G$$

such that $pr_2 \circ s = id$. Obviously, $\psi = pr_1 \circ s$ is then a solution of $\mathscr{E}(G)$. Conversely, a solution ψ defines a section s of pr_2 by $s(\sigma) = (\psi(\sigma), \sigma)$, and therefore $inf(\varepsilon) = 0$. □

(3.5.10) Corollary. *An embedding problem $\mathscr{E}(G,\varphi,\alpha)$,*

$$\begin{array}{ccccccccc} & & & & & & G & & \\ & & & & & & \downarrow{\scriptstyle\varphi} & & \\ 1 & \longrightarrow & A & \longrightarrow & E & \xrightarrow{\alpha} & \overline{G} & \longrightarrow & 1, \end{array}$$

where A is a finite abelian p-group, is solvable if and only if the corresponding ***p*-Sylow embedding problem**

$$\begin{array}{ccccccccc} & & & & & & G_p & & \\ & & & & & & \downarrow{\scriptstyle\varphi_p} & & \\ 1 & \longrightarrow & A & \longrightarrow & E_p & \xrightarrow{\alpha_p} & \overline{G}_p & \longrightarrow & 1 \end{array}$$

is solvable. Here the index p indicates the corresponding p-Sylow subgroups, which are chosen in a compatible way.

Proof: The 2-cocycle ε_p of the Sylow problem is just the restriction of the 2-cocycle ε of the initial problem. Therefore the result follows from (3.5.9) and the commutative diagram

$$\begin{array}{ccccc} [\varepsilon_p] \in & H^2(\overline{G}_p, A) & \longrightarrow & H^2(G_p, A) \\ & \uparrow & & \uparrow \\ [\varepsilon] \in & H^2(\overline{G}, A) & \longrightarrow & H^2(G, A), \end{array}$$

where the restriction maps are injective by (1.6.10). □

(3.5.11) Proposition. *Let $\mathscr{E}(G,\varphi,\alpha)$ be an embedding problem with finite abelian kernel A which has a solution. Then $\mathscr{S}_{\mathscr{E}(G)}$ is a principal homogeneous space over $H^1(G,A)$.**)

The homogeneous subspaces over $H^1(\overline{G},A) \subseteq H^1(G,A)$ consist of all solutions ψ modulo equivalence whose restrictions to $H = \ker(\varphi)$ induce a fixed $\overline{G}$-homomorphism $\psi_0\colon H^{ab} \to A$.

Proof: Let ψ be a solution of $\mathscr{E}(G)$ and let $[x] \in H^1(G,A)$, where $x\colon G \to A$ is a 1-cocycle. Then

$${}^x\psi\colon G \longrightarrow E\ ,\ \sigma \longmapsto x(\sigma)\cdot\psi(\sigma)$$

---

*)If X is a topological space and Γ a topological group acting continuously and simply transitively on X, then X is called a **principal homogeneous space over Γ**.

is a solution of $\mathscr{E}(G)$. Indeed, ${}^x\psi$ is a homomorphism since

$$x(\sigma\tau)\cdot\psi(\sigma\tau) = x(\sigma)x(\tau)^{\psi(\sigma)}\psi(\sigma)\psi(\tau) = x(\sigma)\psi(\sigma)x(\tau)\psi(\tau)$$

and $\alpha \circ {}^x\psi = \varphi$. Another 1-cocycle $x' \in [x]$ induces an equivalent solution of ${}^x\psi$. Thus $H^1(G,A)$ acts on $\mathscr{S}_{\mathscr{E}(G)}$. This action is transitive since for any two solutions ψ and ψ' we get a 1-cocycle $x(\sigma) = \psi'(\sigma)\psi(\sigma)^{-1}$, hence a class in $H^1(G,A)$. Furthermore, the action is simply transitive since two solutions ψ and ψ' are equivalent if and only if x is a 1-coboundary. Finally, from the exact 5-term sequence, it follows that $\psi|_H = \psi'|_H$ if and only if $[x] \in \mathit{inf}\,(H^1(\bar{G},A))$. □

Now we are able to prove the theorem.

Proof of (3.5.6): (i) ⇒ (ii) is trivial and (ii) ⇒ (iii) is also trivial, noting that the classes $\mathfrak{c}$ and $\mathfrak{c}(S(\mathfrak{c}))$ contain the same abelian groups.

(iii) ⇒ (iv). Let $p \in S(\mathfrak{c})$ and let A be a finite G-module in $\mathit{Mod}_p(G)$. By assumption, every group extension $1 \to A \to E \to G \to 1$ splits and therefore, by (1.2.4),

$$H^2(G,A) \cong \mathit{EXT}\,(A,G) = 0.$$

An arbitrary G-module $A \in \mathit{Mod}_p(G)$ is the union of its finite submodules. Taking the inductive limit, we see that $H^2(G,A) = 0$, and hence $cd_p\,G \le 1$.

(iv) ⇒ (v). Let us assume that $cd_p\,G \le 1$ for all $p \in S(\mathfrak{c})$. Then $H^2(G,A) = 0$ for all finite p-primary G-modules A, $p \in S(\mathfrak{c})$. By (3.5.9), every embedding problem (∗) with a minimal abelian normal subgroup $A \in \mathfrak{c}(S(\mathfrak{c}))$ of E is solvable, since such an A is necessarily of the form $(\mathbb{Z}/p\mathbb{Z})^m$ for some prime number $p \in S(\mathfrak{c})$ and some integer $m \ge 1$. By (3.5.7), G is $S(\mathfrak{c})$-projective. Finally, (v) ⇒ (i) is trivial, as $\mathfrak{c} \subseteq \mathfrak{c}(S(\mathfrak{c}))$. □

(3.5.12) Corollary. *If $\mathfrak{c} \subseteq \mathfrak{d}$ are two full classes of finite groups and if G is a pro-$\mathfrak{c}$-group which is $\mathfrak{c}$-projective, then it is $\mathfrak{d}$-projective.*

Proof: We use the equivalence (3.5.6) (i)⇔(iii), thus we have to show that every group extension

$$1 \to A \to E \to G \to 1$$

splits, where A is a finite abelian p-group, $p \in S(\mathfrak{d})$. By assumption, this is true for $p \in S(\mathfrak{c})$. If $p \in S(\mathfrak{d}) \smallsetminus S(\mathfrak{c})$, then the orders of the groups G and A are relatively prime, hence $\mathit{EXT}\,(A,G) \cong H^2(G,A) = 0$ by (1.6.2). □

(3.5.13) Corollary. *Denoting by $(solv)$ the full class of finite solvable groups, the following assertions are equivalent for a profinite group G.*

(i) *G is projective.*

(ii) *G is $(solv)$-projective.*

(iii) *$cd\, G \le 1$.*

Proof: This follows easily from (3.5.6) (i)⇔(iv). □

Let p be a prime number. From (3.3.7) we know that the profinite groups G with $cd_p\, G = 0$ are exactly the groups of order prime to p. By (3.5.6), the profinite groups with $cd_p\, G \le 1$ are exactly the p-projective groups. Prototypes of projective groups are the free pro-$\mathfrak{c}$-groups, which are defined as follows.

(3.5.14) Definition. *A free pro-$\mathfrak{c}$-group over a set X is a pro-$\mathfrak{c}$-group F together with a map $i: X \to F$ satisfying the following properties.*

(1) *Every open subgroup contains almost all elements of $i(X)$ (i.e. all up to a finite number).*

(2) *If $j : X \to G$ is any other map with the property (1) into a pro-$\mathfrak{c}$-group G, then there exists a unique homomorphism $f: F \to G$ with $j = f \circ i$.*

The free pro-$\mathfrak{c}$-group F over a set X always exists and is unique up to unique isomorphism. One obtains it by starting with the ordinary free group F_0 over X (see [73]) with the inclusion $X \hookrightarrow F_0$. Let U run through all normal subgroups containing almost all elements $x \in X$ and such that $F_0/U \in \mathfrak{c}$. Then $F = \varprojlim_U F_0/U$, together with the induced map $i : X \to F$, is a free pro-$\mathfrak{c}$-group over X. In fact, if $j : X \to G$ is as in (2), then the universal property of the free group F_0 gives a homomorphism $f_0 : F_0 \to G$ such that $j = f_0 \circ incl_X$. If U runs through all open normal subgroups of G, then we obtain $f : F \to G$ as the composite of the homomorphisms

$$F \longrightarrow \varprojlim_U F_0/f_0^{-1}(U) \longrightarrow \varprojlim_U G/U = G.$$

f is unique as it is continuous and F is topologically generated by $i(X)$.

Observe that the map $i: X \to F$ in (3.5.14) is necessarily injective. The elements of $i(X) \subseteq F$ are called the **free generators** of F, the set $i(X)$ is called a **basis** of F, and $\mathrm{rk}(F) = \#X$ is the **rank** of the free pro-$\mathfrak{c}$-group. We also write $F_X(\mathfrak{c})$ for F and $F_n(\mathfrak{c})$, resp. $F_\omega(\mathfrak{c})$, if $\mathrm{rk}(F) = n \in \mathbb{N}$, resp. if X has countable cardinality.

The free pro-$\mathfrak{c}$-group of rank 0 is the trivial group. The free pro-$\mathfrak{c}$-group of rank 1 is the product $\hat{\mathbb{Z}}(\mathfrak{c}) = \prod_{p \in S(\mathfrak{c})} \mathbb{Z}_p$.

(3.5.15) Proposition. *Let $\mathfrak{c}$ be a full class of finite groups and let F be the free pro-$\mathfrak{c}$-group over a set X. Then F is projective.*

Proof: By (3.5.12), it suffices to show that F is $\mathfrak{c}$-projective. Let

$$1 \to A \to E \xrightarrow{\pi} F \to 1$$

be an exact sequence with A a finite abelian p-group for some prime number $p \in S(\mathfrak{c})$. Then $A \in \mathfrak{c}$ and E is a pro-$\mathfrak{c}$-group. We choose a continuous section $F \to E$ of π and obtain a lift $s\colon X \to E$ of $i\colon X \to F$. It satisfies condition (1) of (3.5.14), since A is finite. By the universal property for F, s extends to a homomorphism $s\colon F \to E$ such that $\pi \circ s$ is the identity on X, hence on F. This shows that the extension $1 \to A \to E \to F \to 1$ splits. By (3.5.6), F is a projective pro-$\mathfrak{c}$-group. □

(3.5.16) Corollary. *Let F be the free pro-$\mathfrak{c}$-group over a set X. If the class $\mathfrak{c}$ contains the p-groups, then $cd_p F = 1$, otherwise $cd_p F = 0$.*

Proof: By (3.3.7), $cd_p F = 0$ if and only if the class $\mathfrak{c}$ does not contain the p-groups. As F is projective, (3.5.6) implies $cd_p F \leq 1$ for all p. □

For pro-p-groups we have a converse:

(3.5.17) Proposition. *A pro-p-group G is free if and only if $cd\, G \leq 1$.*

We shall prove this in (3.9.5). See also (4.1.5) for a more general result.

(3.5.18) Definition. *The* **rank** rk(G) *of a pro-$\mathfrak{c}$-group G is the smallest cardinal number α such that there exists a set X of cardinality α and a surjection $F_X(\mathfrak{c}) \twoheadrightarrow G$ from a free pro-$\mathfrak{c}$-group over X onto G.*

The rank of a pro-p-group G is often denoted by $d(G)$. In (3.9.1) we will see that $d(G) = \dim_{\mathbb{F}_p} H^1(G, \mathbb{Z}/p\mathbb{Z})$.

In the following, we will prove that a pro-$\mathfrak{c}$-group G of at most countable infinite rank, for which every embedding problem with E a finite $\mathfrak{c}$-group has a proper solution, is isomorphic to the free pro-$\mathfrak{c}$-group $F_\omega(\mathfrak{c})$ of countable infinite rank. We start by showing that $F_\omega(\mathfrak{c})$ itself has this property.

(3.5.19) Lemma. *Every embedding problem*

$$\begin{array}{ccccccccc} & & & & & & F_\omega(\mathfrak{c}) & & \\ & & & & & & \Big\downarrow{\scriptstyle\varphi} & & \\ 1 & \longrightarrow & N & \longrightarrow & E & \xrightarrow{\alpha} & \overline{G} & \longrightarrow & 1 \end{array}$$

with E a finite $\mathfrak{c}$-group has a proper solution.

Proof: Let X be a basis of $F_\omega(\mathfrak{c})$. Then $Y = X \smallsetminus \ker(\varphi)$ is finite. For each $y \in Y$ we choose an element $e_y \in E$ such that $\alpha(e_y) = \varphi(y)$. Furthermore, we choose a surjective map ψ_0': $X \cap \ker(\varphi) \to N$ (observe that $X \cap \ker(\varphi)$ is infinite) and define ψ_0: $X \to E$ by $\psi_0(y) = e_y$ for $y \in Y$ and by ψ_0' on $X \cap \ker(\varphi)$. Now ψ_0 extends to a surjective homomorphism ψ: $F_\omega(\mathfrak{c}) \twoheadrightarrow E$ such that $\alpha \circ \psi = \varphi$. □

The following result is due to *K. Iwasawa* (see [86], th. 4).

(3.5.20) Proposition. *Let G be a pro-$\mathfrak{c}$-group of at most countable infinite rank such that every embedding problem*

$$\begin{array}{ccccccccc} & & & & & & G & & \\ & & & & & & \Big\downarrow{\scriptstyle\varphi} & & \\ 1 & \longrightarrow & N & \longrightarrow & E & \xrightarrow{\alpha} & \overline{G} & \longrightarrow & 1 \end{array}$$

with E a finite $\mathfrak{c}$-group has a proper solution. Then G is isomorphic to $F_\omega(\mathfrak{c})$.

Proof: Let $F = F_\omega(\mathfrak{c})$. Since G and F have at most countable rank, there exist sequences

$$F = F_0 \supseteq F_1 \supseteq F_2 \supseteq \dots$$

$$G = G_0 \supseteq G_1 \supseteq G_2 \supseteq \dots$$

of open normal subgroups with trivial intersection. We inductively define two additional sequences

$$F = F'_0 \supseteq F'_1 \supseteq F'_2 \supseteq \dots$$
$$G = G'_0 \supseteq G'_1 \supseteq G'_2 \supseteq \dots$$

of open normal subgroups and a sequence of isomorphisms $\varphi_n\colon G/G'_n \xrightarrow{\sim} F/F'_n$ such that the diagram

$$\begin{array}{ccc}
G/G'_{n+1} & \xrightarrow[\sim]{\varphi_{n+1}} & F/F'_{n+1} \\
\downarrow{\scriptstyle can} & & \downarrow{\scriptstyle can} \\
G/G'_n & \xrightarrow[\sim]{\varphi_n} & F/F'_n
\end{array}$$

commutes for every n. Let n be given and suppose all objects with indices up to n have already been defined. If n is even, then $F'_{n+1} := F_n \cap F'_n$ is open in F and $F/F'_{n+1} \in \mathfrak{c}$. The embedding problem

$$\begin{array}{ccccc}
 & & & & G \\
 & & \swarrow{\scriptstyle \varphi'_{n+1}} & & \downarrow{\scriptstyle can} \\
F/F'_{n+1} & \xrightarrow{can} & F/F'_n & \xrightarrow[(\varphi_n)^{-1}]{\sim} & G/G'_n
\end{array}$$

has a proper solution φ'_{n+1} by assumption. If $G'_{n+1} = \ker(\varphi'_{n+1})$, then φ'_{n+1} induces an isomorphism $\varphi_{n+1}\colon G/G'_{n+1} \xrightarrow{\sim} F/F'_{n+1}$ which commutes with φ_n. If n is odd, exchange the roles of G and F and use (3.5.19) in order to find an isomorphism $\psi_{n+1}\colon F/F'_{n+1} \xrightarrow{\sim} G/G'_{n+1}$. Let $\varphi_{n+1} = \psi_{n+1}^{-1}$. Since $F'_{n+1} \subseteq F_n$ and $G'_{n+1} \subseteq G_n$, the intersection of all groups F'_n (resp. G'_n) is equal to 1. Thus the isomorphisms φ_n define an isomorphism $\varphi\colon G \xrightarrow{\sim} F$. □

Exercise: Consider the diagram

$$\begin{array}{ccccccccc}
 & & & & & & G & & \\
 & & & & & & \downarrow{\scriptstyle f} & & \\
1 & \longrightarrow & P & \longrightarrow & E & \xrightarrow{\pi} & \Gamma & \longrightarrow & 1
\end{array}$$

with P an abelian group. Two liftings $f', f'' : G \to E$ of $f : G \to \Gamma$ are *conjugate* if there exists a $\sigma \in P$ such that $f'' = \sigma \circ f' \circ \sigma^{-1}$. Show that the group $H^1(G,P)$ acts simply transitively on the set of conjugacy classes of liftings, provided some liftings exist. Formulate and prove this also for non-abelian P.

§6. Profinite Groups of $scd\, G = 2$

The profinite groups G of strict cohomological dimension $scd\, G = 2$ are especially important for the theory of number fields, because of their close connection with class field theory. Trivial examples are all groups of cohomological dimension 1 (see §3, ex.1).

Let G be any profinite group. We denote open subgroups of G by U, V, W and we write $V \lhd U$ if V is a normal subgroup of U. In this case V^{ab} is a U/V-module by conjugation, and we have a group extension

$$1 \longrightarrow V^{ab} \longrightarrow U/V' \longrightarrow U/V \longrightarrow 1,$$

where V' denotes the closure of the commutator subgroup $[V, V]$ of V. By (1.2.4), the group extension defines a canonical cohomology class

$$u_{U/V} \in H^2(U/V, V^{ab}),$$

which plays an important role in the sequel.

(3.6.1) Lemma. *Let $W \subseteq V \subseteq U$ be open subgroups of G.*

(i) *If $W \lhd U$, then $u_{V/W}$ is the image of $u_{U/W}$ under*

$$res : H^2(U/W, W^{ab}) \longrightarrow H^2(V/W, W^{ab}).$$

(ii) *If $W \lhd U$ and $V \lhd U$, then $(V : W)u_{U/W}$ is the image of $u_{U/V}$ under the map*

$$i : H^2(U/V, V^{ab}) \longrightarrow H^2(U/W, W^{ab}),$$

induced by the pair $U/W \to U/V$, $Ver : V^{ab} \to W^{ab}$.

Proof: Assertion (i) follows from the commutative diagram

$$\begin{array}{ccccccccc} 1 & \longrightarrow & W^{ab} & \longrightarrow & V/W' & \longrightarrow & V/W & \longrightarrow & 1 \\ & & \| & & \downarrow & & \downarrow & & \\ 1 & \longrightarrow & W^{ab} & \longrightarrow & U/W' & \longrightarrow & U/W & \longrightarrow & 1. \end{array}$$

For (ii) we consider the commutative diagram

$$\begin{array}{ccccccccc} 1 & \longrightarrow & W^{ab} & \longrightarrow & U/W' & \longrightarrow & U/W & \longrightarrow & 1 \\ & & \downarrow & & \downarrow & & \downarrow & & \\ 1 & \longrightarrow & V^{ab} & \longrightarrow & U/V' & \longrightarrow & U/V & \longrightarrow & 1. \end{array}$$

From this, we conclude that $u_{U/W}$ and $u_{U/V}$ have the same image u' in $H^2(U/W, V^{ab})$. The composite of $W^{ab} \longrightarrow V^{ab} \xrightarrow{Ver} W^{ab}$ is the norm $N_{V/W}$ by (1.5.9). Therefore we have a commutative diagram

$$\begin{array}{ccccc} H^2(U/V, V^{ab}) & \xrightarrow{inf} & H^2(U/W, V^{ab}) & \longleftarrow & H^2(U/W, W^{ab}) \\ & {\searrow}^{i} & \downarrow{\scriptstyle Ver} & {\swarrow}^{\nu} & \\ & & H^2(U/W, W^{ab}), & & \end{array}$$

where ν is induced by $N_{V/W} : W^{ab} \to W^{ab}$. This yields $i(u_{U/V}) = Ver(u') = \nu(u_{U/W})$. But if g is a finite group, h a normal subgroup, A a g-module, then the map $\hat{H}^n(g, A) \to \hat{H}^n(g, A)$ induced by $N_{g/h} : A \to A$ is multiplication by $(g : h)$. This is trivial for $n = 0$ and follows for $n \geq 0$ by dimension shifting. Therefore $i(u_{U/V}) = (V : W)u_{U/W}$. □

(3.6.2) Lemma. *For every pair $V \lhd U$ of open subgroups of G, we have an exact commutative diagram*

$$\begin{array}{ccccccccc} H_2(U/V, \hat{\mathbb{Z}}) & \longrightarrow & H_0(U/V, V^{ab}) & \xrightarrow{i} & U^{ab} & \longrightarrow & (U/V)^{ab} & \longrightarrow & 1 \\ & & \downarrow{\scriptstyle N_{U/V}} & & \downarrow{\scriptstyle Ver} & & \downarrow{\scriptstyle \rho} & & \\ 1 & \longrightarrow & N_{U/V}V^{ab} & \longrightarrow & (V^{ab})^{U/V} & \longrightarrow & \hat{H}^0(U/V, V^{ab}) & \longrightarrow & 1, \end{array}$$

where the homomorphism ρ is given by

$$\sigma \longmapsto \prod_{\tau \in U/V} u(\tau, \sigma),$$

$u(\tau, \sigma)$ being a 2-cocycle representing the canonical class $u_{U/V}$. If U/V is cyclic, the homomorphism i is injective.

Proof: The Hochschild-Serre spectral sequence

$$H^i(U/V, H^j(V, \mathbb{Q}/\mathbb{Z})) \Rightarrow H^{i+j}(U, \mathbb{Q}/\mathbb{Z})$$

induces the exact sequence

$$0 \to H^1(U/V, \mathbb{Q}/\mathbb{Z}) \to H^1(U, \mathbb{Q}/\mathbb{Z}) \to H^1(V, \mathbb{Q}/\mathbb{Z})^{U/V} \to H^2(U/V, \mathbb{Q}/\mathbb{Z}).$$

Dualizing this sequence, we obtain the upper exact sequence of the diagram. The lower sequence is trivially exact.

The composite of $V^{ab} \longrightarrow U^{ab} \xrightarrow{Ver} V^{ab}$ is the norm map $N_{U/V}$ by (1.5.9), which shows that the left partial diagram of (3.6.2) is commutative. Therefore *Ver* induces a homomorphism ρ: $(U/V)^{ab} \to \hat{H}^0(U/V, V^{ab})$. In order to describe it explicitly, we choose a representative $\hat{\sigma} \in U$ for each coset $\sigma \in U/V$, i.e. $\sigma = \hat{\sigma}V = V\hat{\sigma}$. Then the function

$$u(\tau, \sigma) = \hat{\tau}\hat{\sigma}\widehat{\tau\sigma}^{-1} \mod V'$$

is a cocycle representing the class $u_{U/V}$. On the other hand, by definition of the transfer *Ver*,

$$Ver(\hat{\sigma} \mod U') = \prod_{\tau \in U/V} \hat{\tau}\hat{\sigma}\widehat{\tau\sigma}^{-1} \mod V' = \prod_{\tau \in U/V} u(\tau, \sigma).$$

Passing from U to U/V and then to $(U/V)^{ab}$, this map induces the map $\rho : (U/V)^{ab} \to \hat{H}^0(U/V, V^{ab})$, given by

$$\sigma \longmapsto \prod_{\tau \in U/V} u(\tau, \sigma)$$

as claimed.

If U/V is cyclic, then

$$H_2(U/V, \hat{\mathbb{Z}}) \cong H^2(U/V, \mathbb{Q}/\mathbb{Z})^* \cong \hat{H}^0(U/V, \mathbb{Q}/\mathbb{Z})^* = 0,$$

and therefore the homomorphism i in (3.6.2) is injective. □

(3.6.3) Lemma. *Consider the conditions*

(1) $H^1(U/V, V^{ab}) = 1$,

(2) $H^2(U/V, V^{ab})$ *is of order* $\#(U/V)$ *and is generated by* $u_{U/V}$.

If they hold for all pairs $V \lhd U$ *such that* U/V *is cyclic of a prime order* p, *then they hold for all pairs* $V \lhd U$.

Proof: Assume the conditions hold whenever $U/V \cong \mathbb{Z}/p\mathbb{Z}$. We then prove them for the case $\#U/V = p^n$ by induction on n. Since the p-group U/V has a nontrivial center, it sits in an exact sequence

$$1 \longrightarrow W/V \longrightarrow U/V \longrightarrow U/W \longrightarrow 1$$

with $W/V \cong \mathbb{Z}/p\mathbb{Z}$ and $\#U/W = p^{n-1}$. We may therefore assume the conditions (1) and (2) for W/V and U/W. From (3.6.2), with U replaced by W, it follows that

$$Ver : W^{ab} \longrightarrow (V^{ab})^{W/V}$$

is an isomorphism, since $N_{W/V}$ is an isomorphism as the kernel of $N_{W/V}$ is equal to $\hat{H}^{-1}(W/V, V^{ab}) \cong H^1(W/V, V^{ab}) = 1$ and ρ is the Nakayama map (cf. III §1). Therefore, by (1.6.7), we obtain exact sequences

$$1 \longrightarrow H^q(U/W, W^{ab}) \xrightarrow{i} H^q(U/V, V^{ab}) \xrightarrow{res} H^q(W/V, V^{ab})$$

for $q = 1$ and for $q = 2$ since $H^1(W/V, V^{ab}) = 1$.

For $q = 1$ we obtain $H^1(U/V, V^{ab}) = \{1\}$, and for $q = 2$

$$\#H^2(U/V, V^{ab}) \le p^n = (U : V).$$

Furthermore, $u_{U/V}$ must be of order p^n, since otherwise $1 = p^{n-1}u_{U/V} = i(p^{n-2}u_{U/V})$ by (3.6.1). As i is injective, this implies $p^{n-2}u_{U/V} = 1$, contradicting the induction hypothesis. This proves condition (2).

We now prove the conditions (1) and (2) for a general pair $V \lhd U$. For every prime number p, let U_p/V be a p-Sylow subgroup of U/V. By (1.6.10), the restriction map

$$res : H^q(U/V, V^{ab}) \longrightarrow \bigoplus_p H^q(U_p/V, V^{ab})$$

for $q = 1, 2$, is injective. The case $q = 1$ yields $H^1(U/V, V^{ab}) = \{1\}$. When $q = 2$, $u_{U/V}$ is mapped to the element $\bigoplus_p u_{U_p/V}$ by (3.6.1), which generates the direct sum. Therefore *res* is an isomorphism, $u_{U/V}$ generates $H^2(U/V, V^{ab})$ and

$$\#H^2(U/V, V^{ab}) = \prod_p \#H^2(U_p/V, V^{ab}) = \prod_p \#U_p/V = \#U/V. \qquad \square$$

We are now able to prove the following theorem, which gives a four-fold characterization of the profinite groups of strict cohomological dimension 2.

(3.6.4) Theorem. *For a profinite group $G \neq 1$, the following conditions are equivalent:*

(i) *$scd\, G = 2$.*

(ii) *For every pair $V \lhd U$, the transfer*

$$Ver : U^{ab} \longrightarrow (V^{ab})^{U/V}$$

*is an isomorphism.**)

(iii) *For every pair $V \lhd U$, the U/V-module V^{ab} is a class module with fundamental class $u_{U/V}$, i.e. the conditions (1) and (2) of (3.6.3) hold.*

(iv) *The G-module $D_2 = \varinjlim U^{ab}$ is a formation module with $D_2^U = U^{ab}$ and with $(u_{U/V})$ as a system of fundamental classes.*

(v) *There exists a level-compact formation module C with trivial universal norm groups $N_U C$ for all open subgroups $U \subseteq G$.*

Proof: (i) $\Rightarrow$ (ii): This follows from (3.3.11) and (1.5.9).

(ii) $\Rightarrow$ (iii): Assume (ii) and let $V \lhd U$ be a pair such that U/V is cyclic of prime order p. Consider the exact commutative diagram (3.6.2), in which the homomorphism

*) In other words, the fundamental G-modulation $\pi^{ab} : U \mapsto U^{ab}$ is a G-module (see I §5).

$$i : H_0(U/V, V^{ab}) \to U^{ab}$$

is injective and $\ker(N_{U/V}) = \hat{H}^{-1}(U/V, V^{ab})$. By the snake lemma, the bijectivity of *Ver* implies

$$H^1(U/V, V^{ab}) \cong \hat{H}^{-1}(U/V, V^{ab}) = 1$$
$$H^2(U/V, V^{ab}) \cong \hat{H}^0(U/V, V^{ab}) \cong U/V \cong \mathbb{Z}/p\mathbb{Z}.$$

In order to show that $u_{U/V}$ generates $H^2(U/V, V^{ab})$, i.e. $u_{U/V} \neq 0$, consider the exact commutative diagram

$$\begin{array}{ccccccccc} 1 & \longrightarrow & V^{ab} & \longrightarrow & U/V' & \longrightarrow & U/V & \longrightarrow & 1 \\ & & \downarrow & & \downarrow & & \downarrow & & \\ 1 & \longrightarrow & H_0(U/V, V^{ab}) & \longrightarrow & U^{ab} & \longrightarrow & U/V & \longrightarrow & 1. \end{array}$$

If $u_{U/V} = 0$, then the upper group extension splits, i.e. U/V' is a semi-direct product $V^{ab} \rtimes \Gamma$ with a group Γ which is mapped isomorphically onto U/V. Its image in U^{ab} is nontrivial. But in the case of a semi-direct product $V^{ab} \rtimes \Gamma$, the transfer $Ver : V^{ab} \rtimes \Gamma \to V^{ab}$ maps Γ to 1, since for $\gamma \in \Gamma$, $Ver(\gamma) = \prod_{\beta \in \Gamma} \beta\gamma(\gamma\beta)^{-1} = 1$. Therefore $Ver : U^{ab} \to (V^{ab})^{U/V}$ is not injective, a contradiction. This proves (iii) in the case $U/V \cong \mathbb{Z}/p\mathbb{Z}$, and the general case follows because of lemma (3.6.3).

(iii) $\Rightarrow$ (i): If (iii) holds, then by §1, ex.4, the cup-product

$$u_{U/V}\cup : H^p(U/V, \mathrm{Hom}(V^{ab}, \mathbb{Q}/\mathbb{Z})) \longrightarrow H^{p+2}(U/V, \mathbb{Q}/\mathbb{Z}),$$

induced by the pairing $V^{ab} \times \mathrm{Hom}(V^{ab}, \mathbb{Q}/\mathbb{Z}) \to \mathbb{Q}/\mathbb{Z}$ is an isomorphism for $p > 0$ and a surjection for $p = 0$. By (2.4.4), this map coincides up to sign with the differential

$$d_2^{p,1} : E_2^{p,1} \longrightarrow E_2^{p+2,0}$$

of the Hochschild-Serre spectral sequence

$$E_2^{p,q} = H^p(U/V, H^q(V, \mathbb{Q}/\mathbb{Z})) \Rightarrow H^{p+q}(U, \mathbb{Q}/\mathbb{Z}).$$

The surjectivity implies $E_3^{p+2,0} = 0$ for $p \geq 0$ and thus $E_\infty^{p,0} = 0$ for $p \geq 2$. Therefore the edge morphism

$$E_2^{p,0} = H^p(U/V, \mathbb{Q}/\mathbb{Z}) \longrightarrow H^p(U, \mathbb{Q}/\mathbb{Z})$$

is the zero map since it factors through $E_\infty^{p,0}$. But this edge morphism is the inflation map, and we obtain

$$H^{p+1}(U, \mathbb{Z}) \cong H^p(U, \mathbb{Q}/\mathbb{Z}) = \varinjlim_{V} H^p(U/V, \mathbb{Q}/\mathbb{Z}) = 0$$

for $p \geq 2$. From this it follows that $scd\, G = 2$ (see §3, ex.6), noting that $scd\, G = 0$ or 1 is impossible.

(ii) ⇒ (iv): The inclusion functor

$$i : Mod\,(G) \longrightarrow \mathscr{M}od\,(G)$$

from the category of G-modules into the category of G-modulations identifies *Mod* (G) with the full subcategory of G-modulations M with "Galois descent", i.e. for which $res : M(U) \to M(V)^{U/V}$ is an isomorphism. The functor i has as left adjoint the functor

$$j : M \longmapsto \varinjlim_{U} M^U .$$

We have, in particular, $D_2 = j(\pi^{ab})$, and the condition (ii) means that the fundamental G-modulation π^{ab} has Galois descent, i.e. $\pi^{ab} = i(D_2)$. Therefore

$$D_2^U = (iD_2)(U) = \pi^{ab}(U) = U^{ab}.$$

Since (ii) is equivalent to (iii), the G-module D_2 is a formation module with respect to the isomorphisms

$$inv_{U/V} : H^2(U/V, V^{ab}) \longrightarrow \tfrac{1}{(U:V)}\mathbb{Z}/\mathbb{Z}$$

given by $u_{U/V} \mapsto \frac{1}{(U:V)} \bmod \mathbb{Z}$, i.e. with $(u_{U/V})$ as a system of fundamental classes.

(iv) ⇒ (v): Assuming (iv), D_2 is a level-compact formation module, since $D_2^U = U^{ab}$. Therefore it suffices to show that the universal norm groups $N_U D_2$ are trivial. For this we consider, for every pair $V \triangleleft U$, the diagram

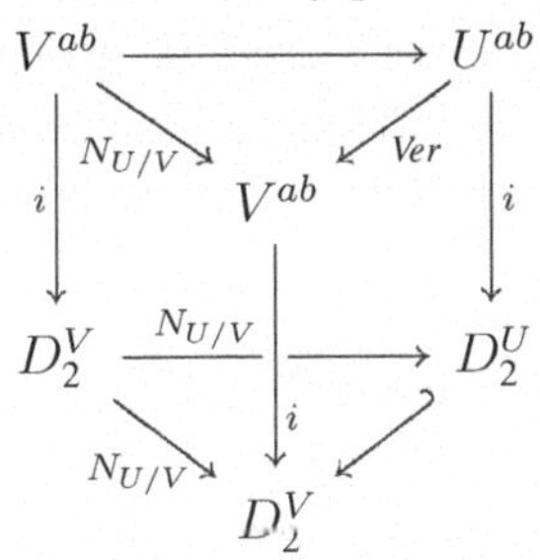

The top of this diagram is commutative by (1.5.9), the bottom and the left-hand side diagram are trivially commutative and the right-hand side diagram by the definition of D_2. Therefore the back diagram is commutative, i.e. we have a commutative diagram

$$\begin{array}{ccc} V & \hookrightarrow & U \\ \downarrow & & \downarrow \\ D_2^V & \xrightarrow{N_{U/V}} & D_2^U \end{array}$$

with surjective vertical arrows. Since $\bigcap_{V \subseteq U} V = \{1\}$, we obtain

$$N_U D_2 = \bigcap_{V \subseteq U} N_{U/V} D_2^V = \{1\},$$

what we wanted to prove.

(v) $\Rightarrow$ (ii): Let C be a level-compact formation module with trivial universal norm groups. By class field theory, we have, for every open subgroup U in G, a canonical continuous homomorphism $C^U \to U^{ab}$ with dense image. It is surjective, since C^U is compact, and injective, since the kernel $N_U C$ is 1. If V is an open normal subgroup of U, then we have the commutative diagram

$$\begin{array}{ccc} C^U & = & (C^V)^{U/V} \\ \downarrow & & \downarrow \\ U^{ab} & \xrightarrow{Ver} & (V^{ab})^{U/V}, \end{array}$$

which shows that *Ver* is an isomorphism. □

Remark: The equivalence (i) $\Leftrightarrow$ (iii) in the above theorem is due to *J. Tate*, as *Y. Kawada* remarks. Kawada gives Tate's proof in [101] (see ex.4). The equivalence was independently proven by *A. Brumer* (see [19]). Brumer's proof, however, is rather involved. Another proof is given in [72]. The equivalence (ii) $\Leftrightarrow$ (iii) was first proven by *J.-P. Serre* (see [208], chap. VI), and another proof of it is found in [72], 2.3. These proofs rely in an essential way on the use of negative dimensional cohomology.

In the proof presented here, we have also shown that the (level-compact) dualizing module $D_2 = \varinjlim U^{ab}$ of a profinite group G of $scd\, G = 2$ has trivial universal norm groups $N_U D_2$. Therefore we may apply *Poitou*'s duality theorem (3.1.11)(i) to the G-module D_2 and obtain the

(3.6.5) Corollary. *Let G be a profinite group of $scd\, G = 2$. We then have a canonical isomorphism*

$$inv : H^2(G, D_2) \xrightarrow{\sim} \tfrac{1}{\#G}\mathbb{Z}/\mathbb{Z}$$

and the cup-product

$$\hat{H}^i(G, \mathrm{Hom}(A, D_2)) \times \hat{H}^{2-i}(G, A) \xrightarrow{\cup} H^2(G, D_2) \cong \tfrac{1}{\#G}\mathbb{Z}/\mathbb{Z}$$

induces an isomorphism

$$\hat{H}^i(G, \mathrm{Hom}(A, D_2)) \xrightarrow{\sim} \hat{H}^{2-i}(G, A)^\vee$$

for all $i \in \mathbb{Z}$ and every G-module A which is a finitely generated free $\mathbb{Z}$-module. For $i \le 0$ this is true also for every G-module A, finitely generated over $\mathbb{Z}$.

By the equivalence (i) $\Leftrightarrow$ (v) in theorem (3.6.4), class field theory seems to be very strongly restricted to cohomological dimension 2. This, however, is not really true. The restriction is due to the fact that we insist on the presence of a G-module rather than a *G-modulation* with trivial universal norms and the properties of a class formation. For example, the fundamental group $G = \pi_1(X, x)$ of a smooth proper curve X over a finite field $\mathbb{F}$ has $scd\,G = 3$. Thus, by the equivalence (i) $\Leftrightarrow$ (ii) of (3.6.4), its fundamental G-modulation π^{ab} is not a G-module. We have, on the other hand, the G-modulation $Pic : U \mapsto \mathrm{Pic}\,(X(U))$, which associates to an open subgroup the Picard group $\mathrm{Pic}\,(X(U))$ of the unramified covering $X(U) \to X$ determined by U. Passing to the profinite completion $\widehat{Pic}$, we obtain a canonical isomorphism

$$\pi^{ab} \cong \widehat{Pic}$$

of G-modulations (see [208]).

Exercise 1. A profinite group G of $scd\,G \le 2$ contains no closed abelian subgroups other than procyclic ones.

Exercise 2. The group $G = \pi_1(X, x)$, where X is a smooth proper curve over a finite field, has the property of the groups of ex. 1 but has $scd\,G = 3$.

Exercise 3. Not every group of $scd\,G = 3$ has the property of the groups of ex. 1.

Exercise 4. Prove the implication (v) $\Rightarrow$ (i) in (3.6.4) directly by means of Poitou's duality theorem (3.1.11)(i) for $i = -1$.

Hint: If C is level-compact with trivial universal norms, then by (3.1.11)(i) and (1.9.12), we have for $A = \mathbb{Z}$ and $A = \mathbb{Z}/p\mathbb{Z}$

$$\begin{aligned} H^3(U, A)^* &\cong \hat{H}^{-1}(U, \mathrm{Hom}(A, C)) \\ &= \hat{H}^{-1}(U, N_U \mathrm{Hom}(A, C)) \\ &= \hat{H}^{-1}(U, \mathrm{Hom}(A, N_U C)) = 0. \end{aligned}$$

Let G_p be a p-Sylow subgroup of G. Then

$$H^3(G_p, \mathbb{Z}/p\mathbb{Z}) = \varinjlim_{U \supseteq G_p} H^3(U, \mathbb{Z}/p\mathbb{Z}) = 0,$$

since

$$G_p = \bigcap_{U \supseteq G_p} U = \varprojlim_{U \supseteq G_p} U,$$

hence $cd_p\,G = cd\,G_p \le 2$ by (3.3.2). Since $H^3(U, \mathbb{Z}) = 0$ for all open subgroups U, we obtain $scd\,G \le 2$ by (3.3.4), and $scd\,G = 2$, since

$$H^2(G, \mathbb{Z}) = H^1(G, \mathbb{Q}/\mathbb{Z}) \neq 0.$$

§7. Poincaré Groups

Poincaré groups are groups for which a cohomological duality theorem of Poincaré type $H^i \times H^{n-i} \to H^n \hookrightarrow \mathbb{Q}/\mathbb{Z}$ holds with a special dualizing module. These groups play an important role in topology as well as in number theory. For example, the pro-p-completion of the fundamental group of a compact Riemann surface, and the Galois group of the maximal p-extension of a p-adic local field are typical examples of Poincaré groups. For these and other reasons they are of special interest. We fix a prime number p for our investigations.

In II §5 we have introduced, for every G-module A, the G-modules

$$D_i(A) = \varinjlim_{U} H^i(U, A)^*,$$

and we have called the G-module

$$D_{(p)} = \varinjlim_{\nu} D_n(\mathbb{Z}/p^\nu\mathbb{Z})$$

the **dualizing module** of G at p if $n = cd_p\, G$. Its importance lies in the functorial isomorphism

$$H^n(G, A)^* \cong \mathrm{Hom}_G(A, D_{(p)})$$

for all $A \in Mod_p(G)$. For this section we set $I = D_{(p)}$.

We have called G a **duality group at p of dimension n** if $D_i(\mathbb{Z}/p\mathbb{Z}) = 0$ for $i < n$ (see (3.4.6)). In this case, the edge morphism of the Tate spectral sequence

$$H^i(G, D_{n-j}(A)) \Rightarrow H^{n-(i+j)}(G, A)^*$$

for $A \in Mod_p(G)$ is a functorial isomorphism

$$H^i(G, \mathrm{Hom}(A, I)) \cong H^{n-i}(G, A)^*,$$

which is also obtained from the cup-product

$$H^i(G, \mathrm{Hom}(A, I)) \times H^{n-i}(G, A) \xrightarrow{\cup} H^n(G, I) \xrightarrow{tr} \mathbb{Q}_p/\mathbb{Z}_p.$$

(3.7.1) Definition. *A duality group G at p of dimension n is called a* **Poincaré group at p** *if $I \cong \mathbb{Q}_p/\mathbb{Z}_p$ as an abelian group.*

Since

$$\mathrm{Aut}(I) = \mathrm{Aut}(\mathbb{Q}_p/\mathbb{Z}_p) = \mathbb{Z}_p^\times,$$

the group G acts on I via a continuous character

$$\chi : G \longrightarrow \mathbb{Z}_p^\times,$$

and I is determined by χ (up to isomorphism). The Serre criterion (3.4.5) implies that $scd_p\, G = n + 1$ if $\chi(G)$ is finite and $scd_p\, G = n$ if $\chi(G)$ is infinite.

(3.7.2) Theorem. *For a finitely generated profinite group G the following assertions are equivalent.*

(i) *G is a Poincaré group at p of dimension 2,*

(ii) *$cd_p\,G = 2$ and $I \cong \mathbb{Q}_p/\mathbb{Z}_p$ (as an abelian group),*

(iii) *$cd_p\,G = 2$ and ${}_pI \cong \mathbb{Z}/p\mathbb{Z}$ (as an abelian group).*

Proof: The implications (i) ⇒ (ii) ⇒ (iii) are trivial. Now we show that (iii) implies $D_i(\mathbb{Z}/p\mathbb{Z}) = 0$ for $i = 0, 1$. The p-Sylow subgroups G_p of G are infinite, since $0 < cd_p\,G = cd_p\,G_p < \infty$. This means that for every open subgroup U, there is an open subgroup $V \subseteq U$ such that $p \mid (U : V)$. From this, it follows that

$$D_0(\mathbb{Z}/p\mathbb{Z}) = \varinjlim_U H^0(U, \mathbb{Z}/p\mathbb{Z})^* = \varinjlim_U \mathbb{Z}/p\mathbb{Z} = 0,$$

since the transition maps in the last direct limit are multiplication by $(U : V)$.

Let $A \in Mod_p(G)$ be a finite G-module with $pA = 0$ and let us set $A' = \mathrm{Hom}\,(A, I)$. Every open subgroup U of G has the same cohomological dimension 2 as G (see (3.3.5)) and the same dualizing module I. Noting that $A'' = A$, we obtain a canonical isomorphism

$$(*) \qquad \varphi^*_{A'} : H^0(U, A)^* \xrightarrow{\sim} H^2(U, A').$$

Let $U^* = U^p[U, U]$, where $[U, U]$ is the closure of the commutator subgroup of U. The group U/U^* is the largest profinite quotient of U which is abelian and of exponent p. The restriction

$$res : H^1(U, \mathbb{Z}/p\mathbb{Z}) \longrightarrow H^1(U^*, \mathbb{Z}/p\mathbb{Z})$$

is obviously the zero map. Since G is finitely generated, so is U, and thus U/U^* is finite, i.e. U^* is open. Now let A be the finite U-module defined by the exact sequence

$$0 \longrightarrow \mathbb{Z}/p\mathbb{Z} \longrightarrow \mathrm{Ind}_U^{U^*}(\mathbb{Z}/p\mathbb{Z}) \longrightarrow A \longrightarrow 0.$$

Applying the exact functor $\mathrm{Hom}\,(-, I)$, we obtain another exact sequence

$$0 \longrightarrow A' \longrightarrow \mathrm{Ind}_U^{U^*}(\mathbb{Z}/p\mathbb{Z})' \longrightarrow {}_pI \longrightarrow 0.$$

By (1.6.5), the composite of the maps

$$H^1(U, \mathbb{Z}/p\mathbb{Z}) \longrightarrow H^1(U, \mathrm{Ind}_U^{U^*}(\mathbb{Z}/p\mathbb{Z})) \xrightarrow[\sim]{sh} H^1(U^*, \mathbb{Z}/p\mathbb{Z})$$

is the restriction, hence the zero map. Therefore we obtain from the two exact sequences the exact commutative diagram

$$\begin{array}{ccccccc}
 & & H^1(U, {}_pI) & \xrightarrow{\alpha_U} & H^2(U, A') & \longrightarrow & H^2(U, \mathrm{Ind}_U^{U^*}(\mathbb{Z}/p\mathbb{Z})') \\
 & & & & \uparrow\wr & & \uparrow\wr \\
0 & \longrightarrow & H^1(U, \mathbb{Z}/p\mathbb{Z})^* & \longrightarrow & H^0(U, A)^* & \longrightarrow & H^0(U, \mathrm{Ind}_U^{U^*}(\mathbb{Z}/p\mathbb{Z}))^*
\end{array}$$

where the vertical arrows are the isomorphisms $(*)$. From this diagram, we get on the one hand an injection

$$D_1(\mathbb{Z}/p\mathbb{Z}) = \varinjlim_{U, cor^*} H^1(U, \mathbb{Z}/p\mathbb{Z})^* \hookrightarrow \varinjlim_{U} \operatorname{im}(\alpha_U),$$

and on the other hand a surjection

$$\varinjlim_{U, res} H^1(U, {}_pI) \longrightarrow \varinjlim_{U} \operatorname{im}(\alpha_U).$$

The left-hand group is zero by (1.5.1), hence $D_1(\mathbb{Z}/p\mathbb{Z}) = 0$. Hence G is a duality group at p of dimension 2. Further, $I \cong \mathbb{Q}_p/\mathbb{Z}_p$ or $\mathbb{Z}/p^k\mathbb{Z}$, $k \geq 1$ (as an abelian group) since ${}_pI \cong \mathbb{Z}/p\mathbb{Z}$. But I is p-divisible by (3.4.7) and therefore G is a Poincaré group. □

(3.7.3) Corollary. *Let G be a finitely generated pro-p-group of cohomological dimension equal to 2. Then G is a Poincaré group if and only if*

$$\dim_{\mathbb{F}_p} H^2(N, \mathbb{Z}/p\mathbb{Z}) = 1$$

for every open normal subgroup N of G.

Proof: Since $cd_p\, G = 2$, we have for open normal subgroups $N' \subseteq N$ of the group G

$$H^2(N, \mathbb{Z}/p^m\mathbb{Z})/p \xrightarrow{\sim} H^2(N, \mathbb{Z}/p\mathbb{Z})$$

and by (3.3.11) the surjectivity of the corestriction map

$$H^2(N', \mathbb{Z}/p\mathbb{Z}) \longrightarrow H^2(N, \mathbb{Z}/p\mathbb{Z}).$$

Thus we obtain for the dualizing module I of G

$${}_pI = {}_p(\varinjlim_{m, cor^*} H^2(N, \mathbb{Z}/p^m\mathbb{Z})^*) \cong \varinjlim_{cor^*} H^2(N, \mathbb{Z}/p\mathbb{Z})^*$$

and ${}_pI$ is isomorphic to $\mathbb{Z}/p\mathbb{Z}$ if and only if $\dim_{\mathbb{F}_p} H^2(N, \mathbb{Z}/p\mathbb{Z}) = 1$ for every open normal subgroup N of G. The result follows from (3.7.2). □

Later, in §9, theorem (3.9.15), we will see that it is enough to consider only open subgroups N of G with $(G : N) \leq p$.

The following theorem shows that the class of Poincaré groups is closed under group extensions. It was first proved by *A. Pletch* (see [167]) and later (independently) by *K. Wingberg* (see [253]).

(3.7.4) Theorem. *Let*

$$1 \longrightarrow H \longrightarrow G \longrightarrow G/H \longrightarrow 1$$

be an exact sequence of profinite groups such that

a) $H^i(U, \mathbb{Z}/p\mathbb{Z})$ *is finite for all open subgroups* U *of* H *and all* $i \ge 0$,

b) $cd_p\, G/H < \infty$.

Then if two of the three groups are Poincaré groups at p, *so is the third. Moreover, in this case we have:*

(i) $cd_p\, G = cd_p\, H + cd_p\, G/H$.

(ii) *There is a canonical* G*-isomorphism* $I(G)^* \cong I(H)^* \otimes_{\mathbb{Z}_p} I(G/H)^*$.

Remark: Theorem (3.7.4) can be generalized in several directions, cf. [167], [253]. The most general version known to us is the following: we call a profinite group **virtual duality group at p** if some open subgroup is a duality group at p. Then, assuming a) above, G is a virtual duality group if and only if H and G/H are. The (virtual) cohomological dimension and the dualizing module of G can be computed from that of H and of G/H as in (i) and (ii) (see [SW]).

Proof of (3.7.4) (see [253]): Let $d = cd_p\, G$, $m = cd_p\, H$, $n = cd_p\, G/H$. By our assumption that two of the groups are Poincaré groups, d, m and n are finite, recalling that $n < \infty$, $m \le d$ and $d \le m + n$. Because of a) we obtain from (3.3.8)

$$d = m + n.$$

Let g run through the open normal subgroups of G and let $h = g \cap H$. Then $g/h = gH/H$ runs through the open normal subgroups of G/H. For a G-module $A \in Mod_p(G)$, we consider the Hochschild Serre spectral sequence

$$E(g,h,A):\; E_2^{ij}(g,h,A) = H^i(g/h, H^j(h,A)) \Rightarrow H^{i+j}(g,A).$$

If $g' \subseteq g$ is another open normal subgroup of G, then the corestriction yields a morphism

$$cor\, :\, E(g',h',A) \longrightarrow E(g,h,A)$$

of spectral sequences, where $h' = h \cap g'$ (see II §4, ex.3). The map

$$E_2^{ij}(g',h',A) \to E_2^{ij}(g,h,A)$$

is the composite of the maps

$$H^i(g'/h', H^j(h',A)) \xrightarrow{cor_h^{h'}} H^i(g'/h', H^j(h,A)) \xrightarrow{cor_{g/h}^{g'/h'}} H^i(g/h, H^j(h,A)),$$

and the map between the limit terms is the corestriction

$$cor_g^{g'}\, :\, H^{i+j}(g',A) \longrightarrow H^{i+j}(g,A).$$

For $2 \le r \le \infty$ we set

$$D^r_{ij}(G,A) = \varinjlim_{g} E^{ij}_r(g,h,A)^*.$$

If h' runs through the open subgroups of H which are normal in G, then the $H^j(h',A)$ and thus also $\varinjlim_{h'} H^j(h',A)$ are G/H-modules. As in the proof of (1.5.1), we see that

$$(*) \qquad D^2_{ij}(G,A) = \varinjlim_{h'} \varinjlim_{g/h} H^i(g/h, H^j(h',A))^*,$$

where for both limits the transition maps are (induced by) cor^*.

By (3.4.6), G is a duality group at p of dimension $d = n+m$ if and only if

$$D_{i+j}(G,\mathbb{Z}/p\mathbb{Z}) = \varinjlim_{g} H^{i+j}(g,\mathbb{Z}/p\mathbb{Z})^* = 0 \quad \text{for } i+j \ne d \quad {}^{*)},$$

and this is equivalent to

$$(**) \qquad D^\infty_{ij}(G,\mathbb{Z}/p\mathbb{Z}) = 0 \quad \text{for} \quad i+j \ne d.$$

Assume now that G/H is a Poincaré group at p. Then by (3.4.6)(i) the open subgroups g/h are also Poincaré groups at p of the same dimension n and the same dualizing module $I(G/H)$. Furthermore, with condition a) it follows by a standard argument that the groups $H^j(h',A)$ are finite for every finite $A \in Mod_p(G)$. Therefore we obtain by (3.4.6) (ii) (using (1.5.3)(iv) and (1.5.1))

$$\begin{aligned} D^2_{ij}(G,A) &= \varinjlim_{g/h,res}\ \varinjlim_{h',cor^*} H^{n-i}(g/h, \mathrm{Hom}(H^j(h',A), I(G/H))) \\ (***) \qquad &= \begin{cases} \mathrm{Hom}(\varprojlim_{h'} H^j(h',A), I(G/H)) & \text{for } i=n, \\ 0 & \text{otherwise,} \end{cases} \\ &= \begin{cases} \mathrm{Hom}(D_j(H,A)^*, I(G/H)) & \text{for } i=n, \\ 0 & \text{otherwise.} \end{cases} \end{aligned}$$

Here $D_j(H,A)^*$ should be seen in the topological sense, i.e. as a compact abelian group, and $\mathrm{Hom}(D_j(H,A)^*, I(G/H))$ are the continuous homomorphisms, i.e. the homomorphisms with finite image. From this, we deduce that the following assertions are equivalent.

(1) H is a duality group at p,

(2) $D_j(H,\mathbb{Z}/p\mathbb{Z}) = 0$ for $j \ne m$,

(3) $D^2_{ij}(G,\mathbb{Z}/p\mathbb{Z}) = 0$ for $(i,j) \ne (n,m)$,

(4) $D^\infty_{ij}(G,\mathbb{Z}/p\mathbb{Z}) = 0$ for $(i,j) \ne (n,m)$,

(5) G is a duality group at p.

---

*) $D_{i+j}(G,\mathbb{Z}/p\mathbb{Z})$ has here a different meaning from II §1.

Thus we have proved the following

(3.7.5) Lemma. *Suppose that assumptions* a) *and* b) *of* (3.7.4) *are fulfilled and assume that G/H is a Poincaré group at p of dimension n.*

Then H is a duality group at p of dimension m if and only if G is a duality group at p of dimension $m+n$. In this case the dualizing module of H is isomorphic to the dualizing module of G regarded as an H-module.

We proceed with the proof of (3.7.4) and assume now that G and H are duality groups at p. Since the groups $H^{m-j}(h,\mathbb{Z}/p\mathbb{Z})$ are finite, and since $\varinjlim_{h'} H^{m-j}(h',\mathbb{Z}/p\mathbb{Z}) = 0$ for $j \neq m$ if h' runs through the open subgroups h' of h, we find for every h an $h' \subseteq h$ such that

$$\mathit{res} : H^{m-j}(h,\mathbb{Z}/p\mathbb{Z}) \longrightarrow H^{m-j}(h',\mathbb{Z}/p\mathbb{Z})$$

is the zero map for all $j \neq m$. Since h and h' are duality groups at p of dimension m (as H is), we have a commutative diagram of non-degenerate pairings

$$\begin{array}{ccccc}
H^j(h, {}_pI(H)) & \times & H^{m-j}(h,\mathbb{Z}/p\mathbb{Z}) & \xrightarrow{\ \cup\ } & \mathbb{Z}/p\mathbb{Z} \\
\uparrow{\scriptstyle cor} & & \downarrow{\scriptstyle res} & & \Big\| \\
H^j(h', {}_pI(H)) & \times & H^{m-j}(h',\mathbb{Z}/p\mathbb{Z}) & \xrightarrow{\ \cup\ } & \mathbb{Z}/p\mathbb{Z},
\end{array}$$

which shows that the left corestriction maps are zero. Therefore we see from $(*)$ that

$$D^2_{ij}(G,A) = 0, \qquad \text{for all } i \text{ and all } j \neq m,$$

and consequently

$$D^{\infty}_{im}(G, {}_pI(H)) = D^2_{im}(G, {}_pI(H)), \qquad \text{for all } i.$$

If we assume, in addition, that H is a Poincaré group, it follows that

$${}_pI(H) \cong \mathbb{Z}/p\mathbb{Z}$$

(as an abelian group), and there exists an open subgroup g of G which acts trivially on ${}_pI(H)$. Therefore $(**)$ implies

$$D^2_{im}(G, {}_pI(H)) = D^2_{im}(G,\mathbb{Z}/p\mathbb{Z}) = 0, \qquad \text{for all } i \neq n.$$

Together with $(*)$, this implies

$$\begin{aligned} D_i(G/H, \mathbb{Z}/p\mathbb{Z}) &= \varinjlim_{g/h} H^i(g/h, \mathbb{Z}/p\mathbb{Z})^* \\ &= \varinjlim_{g/h, cor^*} (\varprojlim_{h', res^*} H^i(g/h, H^0(h', \mathbb{Z}/p\mathbb{Z})^*))^* \\ &= \varinjlim_{g/h, cor^*} (\varprojlim_{h', cor} H^i(g/h, H^m(h', {}_pI(H))))^* \\ &= \varinjlim_{g/h, cor^*} \varinjlim_{h', cor^*} H^i(g/h, H^m(h', {}_pI(H)))^* \\ &= D^2_{im}(G, {}_pI(H)) = 0. \end{aligned}$$

We have thus shown that G/H is a duality group at p of dimension n.

The proof of (ii) goes as follows. As before let h' run through all open subgroups of H which are normal in G. Since $m = cd_p\, H$, we have by (2.1.4),

$$H^{m+n}(g, A) = H^n(g/h, H^m(h, A)),$$

and we obtain similarly as in $(***)$

$$\begin{aligned} I(G) &= \varinjlim_{\nu} \varinjlim_{g} H^{m+n}(g, \mathbb{Z}/p^\nu\mathbb{Z})^* \\ &= \varinjlim_{\nu} \varinjlim_{h'} \varinjlim_{g/h} H^n(g/h, H^m(h', \mathbb{Z}/p^\nu\mathbb{Z}))^* \\ &= \varinjlim_{\nu} \varinjlim_{h'} \varinjlim_{g/h, res} H^0(g/h, \operatorname{Hom}(H^m(h', \mathbb{Z}/p^\nu\mathbb{Z}), I(G/H))) \\ &= \varinjlim_{\nu} \operatorname{Hom}(\varprojlim_{h'} H^m(h', \mathbb{Z}/p^\nu\mathbb{Z}), I(G/H)) \\ &= \varinjlim_{\nu} \operatorname{Hom}(D_m(H, \mathbb{Z}/p^\nu\mathbb{Z})^*, I(G/H)) \\ &= \operatorname{Hom}(I(H)^*, I(G/H)) = (I(H)^* \otimes_{\mathbb{Z}_p} I(G/H)^*)^*. \end{aligned}$$

In particular, it follows that

$$\operatorname{rank}_{\mathbb{Z}_p} I(G)^* = \operatorname{rank}_{\mathbb{Z}_p} I(H)^* \cdot \operatorname{rank}_{\mathbb{Z}_p} I(G/H)^*.$$

This completes the proof of the theorem. □

Let us finally consider the case of a pro-p-group G. We set

$$H^i(G) = H^i(G, \mathbb{Z}/p\mathbb{Z})$$

and consider these groups as $\mathbb{F}_p$-vector spaces.

(3.7.6) Proposition (SERRE). *For an infinite pro-p-group G the following statements are equivalent:*

(i) *G is a Poincaré group of dimension n.*

(ii) $\dim_{\mathbb{F}_p} H^i(G) < \infty$ *for all $i \le n$, $H^n(G) \cong \mathbb{F}_p$, and the cup-product yields a non-degenerate pairing* *)

$$H^i(G) \times H^{n-i}(G) \longrightarrow H^n(G) \cong \mathbb{F}_p \qquad \text{for all } 0 \le i \le n.$$

Proof: (i) $\Rightarrow$ (ii). Let G be a Poincaré group of dimension n. Since $I \cong \mathbb{Q}_p/\mathbb{Z}_p$, we have ${}_pI \cong \mathbb{Z}/p\mathbb{Z}$ as abelian groups, and consequently also as G-modules, since G is a pro-p-group and $\#\mathrm{Aut}({}_pI) = p-1$ is prime to p. The G-module $\mathrm{Hom}(\mathbb{Z}/p\mathbb{Z}, {}_pI)$ is also isomorphic to $\mathbb{Z}/p\mathbb{Z}$, hence the duality isomorphism (3.4.6) (ii) yields an isomorphism

$$H^i(G) \cong \mathrm{Hom}\,(H^{n-i}(G), \mathbb{F}_p) = H^{n-i}(G)^*,$$

given by the cup-product

$$H^i(G) \times H^{n-i}(G) \xrightarrow{\cup} H^n(G) \xrightarrow{tr} \mathbb{F}_p.$$

For $i = n$, we obtain $H^n(G) \cong \mathbb{F}_p$ and for $i \ge 0$,

$$H^i(G)^{**} \cong H^{n-i}(G)^* \cong H^i(G),$$

from which follows that $H^i(G)$ is finite. (The canonical homomorphism from a vector space to its bidual is an isomorphism if and only if the dimension is finite!)

(ii) $\Rightarrow$ (i). Let A be any finite G-module such that $pA = 0$ and let $0 \le i \le n$. We claim that the pairing

$$H^i(G, A) \times H^{n-i}(G, A^*) \xrightarrow{\cup} H^n(G) \cong \mathbb{F}_p$$

induces an isomorphism $\alpha_i^A : H^i(G, A^*) \xrightarrow{\sim} H^{n-i}(G, A)^*$. This is true for $A = \mathbb{Z}/p\mathbb{Z}$ by the assumption (ii). We first show that for an arbitrary A (with $pA = 0$) α_i^A is surjective for $i = 0$, bijective for $i = 1, \dots n-1$ and injective for $i = n$. We proceed by induction on $\dim_{\mathbb{F}_p} A$.

Since $\mathbb{Z}/p\mathbb{Z}$ is the only simple p-primary G-module, there is an exact sequence of G-modules

$$0 \longrightarrow A_0 \longrightarrow A \longrightarrow \mathbb{Z}/p\mathbb{Z} \longrightarrow 0.$$

This, together with the dual sequence

$$0 \longrightarrow (\mathbb{Z}/p\mathbb{Z})^* \longrightarrow A^* \longrightarrow A_0^* \longrightarrow 0\,,$$

yields an exact diagram for $0 \le i \le n$

*) A pairing $A \times B \to \mathbb{Q}/\mathbb{Z}$ is **non-degenerate** if it induces injections $A \hookrightarrow B^*$ and $B \hookrightarrow A^*$. Clearly, these are isomorphisms if A and B are finite.

$$\begin{array}{ccccccccc}
H^{i-1}(A_0^*) & \to & H^i((\mathbb{Z}/p\mathbb{Z})^*) & \to & H^i(A^*) & \to & H^i(A_0^*) & \to & H^{i+1}((\mathbb{Z}/p\mathbb{Z})^*) \\
\downarrow{\scriptstyle \alpha_{i-1}^{A_0}} & & \downarrow{\scriptstyle \alpha_i^{\mathbb{Z}/p\mathbb{Z}}} & & \downarrow{\scriptstyle \alpha_i^{A}} & & \downarrow{\scriptstyle \alpha_i^{A_0}} & & \downarrow{\scriptstyle \alpha_{i+1}^{\mathbb{Z}/p\mathbb{Z}}} \\
H^{j+1}(A_0)^* & \to & H^j(\mathbb{Z}/p\mathbb{Z})^* & \to & H^j(A)^* & \to & H^j(A_0)^* & \to & H^{j-1}(\mathbb{Z}/p\mathbb{Z})^*,
\end{array}$$

where $i + j = n$ and $H^k(A)$ means $H^k(G, A)$, $H^k = 0$ for negative k and the map α_k is the zero map for $k < 0$ and $k > n$. Now the induction step follows by diagram chasing.

It remains to show duality in dimension $i = 0$ and $i = n$ and by symmetry it suffices to deal with the case $i = 0$. Let A be a finite G-module with $pA = 0$ and let U be an open subgroup with $A^U = A$. Choose an open subgroup $V \subsetneqq U$ (this is where we use that G has infinite order) strictly contained in U. Consider the exact sequence

$$0 \longrightarrow A_{-1} \longrightarrow \bar{A} \xrightarrow{\phi} A \longrightarrow 0,$$

where $\bar{A} := \mathrm{Map}(G/H, A)$, ϕ is the map $f \mapsto \sum_{g \in G/H} f(g)$ and $A_{-1} := \ker \phi$ (compare with the remark following (1.3.8)). We claim that the map

$$H^0(\phi) : H^0(G, \bar{A}) \to H^0(G, A)$$

is zero. Indeed, via the identification $\bar{A} \cong \mathrm{Ind}_G^H(A)$ (see I §6), it corresponds to the corestriction map $cor : H^0(V, A) \to H^0(G, A)$, which factors through $cor : H^0(V, A) \to H^0(U, A)$. Since U acts trivially on A, the last map is multiplication by $(U : V)$, hence trivial as $pA = 0$.

We know that the functor $H^0(G, -)$ is coeffaceable (see II §2) on G-modules annihilated by p. Now choose an injection $A \hookrightarrow B$ into a G-module B with $pB = 0$ such that the map $H^0(G, B^*) \to H^0(G, A^*)$ is the zero map. Then we get a commutative exact diagram

$$\begin{array}{ccccccc}
H^0(B^*) & \xrightarrow{0} & H^0(A^*) & \longrightarrow & H^1((B/A)^*) & \longrightarrow & H^1(B^*) \\
\downarrow{\scriptstyle \alpha_0^{B}} & & \downarrow{\scriptstyle \alpha_0^{A}} & & \downarrow{\scriptstyle \alpha_1^{B/A}} & & \downarrow{\scriptstyle \alpha_1^{B}} \\
H^n(B)^* & \longrightarrow & H^n(A)^* & \longrightarrow & H^{n-1}(B/A)^* & \longrightarrow & H^{n-1}(B)^*.
\end{array}$$

The α_1's are bijective, hence α_0^A is injective. So we have proved duality for modules annihilated by p and for $i = 0, \dots, n$.

We apply this result to $A = \mathbb{Z}/p\mathbb{Z}[G/U]$, where U runs through the open normal subgroups of G. By (3.4.1), Shapiro's lemma and (1.6.5), we obtain for $0 \le i < n$

$$
\begin{aligned}
D_i(\mathbb{Z}/p\mathbb{Z}) &= \varinjlim_{U} H^i(G, \mathbb{Z}/p\mathbb{Z}[G/U])^* \\
&= \varinjlim_{U} H^{n-i}(G, \operatorname{Hom}(\mathbb{Z}/p\mathbb{Z}[G/U], \mathbb{Z}/p\mathbb{Z})) \\
&= \varinjlim_{U} H^{n-i}(G, \operatorname{Ind}_G^U(\mathbb{Z}/p\mathbb{Z})) \\
&= \varinjlim_{U,res} H^{n-i}(U, \mathbb{Z}/p\mathbb{Z}) = 0,
\end{aligned}
$$

by (1.5.1).

Next we show $cd_p\, G \le n$. Let $x \in H^{n+1}(G, A)$, $A \in Mod_p(G)$, $pA = 0$. By (1.5.1), we have $\varinjlim_U H^{n+1}(U, A) = 0$, i.e. there exists an open subgroup U of G such that x becomes zero in $H^{n+1}(U, A) = H^{n+1}(G, \operatorname{Ind}_G^U(A))$. From the exact sequence

$$0 \longrightarrow A \longrightarrow \operatorname{Ind}_G^U(A) \longrightarrow B \longrightarrow 1,$$

it follows that there is an exact sequence

$$H^n(G, \operatorname{Ind}_G^U(A)) \longrightarrow H^n(G, B) \longrightarrow H^{n+1}(G, A) \longrightarrow H^{n+1}(G, \operatorname{Ind}_G^U(A)).$$

The functor $H^n(G, -)$ is right exact on G-modules annihilated by p since it is dual to the functor $H^0(G, -)$ which is left exact. Therefore the last arrow is injective, and $x = 0$. This proves $cd_p\, G \le n$.

Thus we have shown that G is a duality group at p of dimension n. It remains to determine the dualizing module I. By (3.4.7), I is a divisible p-torsion group and therefore it suffices to show ${}_pI \cong \mathbb{Z}/p\mathbb{Z}$. By (3.4.7), ${}_pI \cong D_n(\mathbb{Z}/p\mathbb{Z})$ and the same calculation as above for D_i, $i < n$, shows

$$D_n(\mathbb{Z}/p\mathbb{Z}) = \varinjlim_{U,res} H^0(U, \mathbb{Z}/p\mathbb{Z}) = \mathbb{Z}/p\mathbb{Z}.$$

This proves the proposition. □

The pro-p-groups G which are Poincaré groups of dimension 2 are called Demuškin groups. We give an explicit description and classification of them in §9.

Exercise 1. Let U be an open subgroup of the profinite group G. If $cd_p\, G < \infty$ then G is a duality group at p of dimension n if and only if U is.

Exercise 2. If a p-Sylow subgroup G_p of a profinite group G is a duality group at p of dimension n, then so is G. Is the converse true?

Exercise 3. Let G be a Poincaré group of dimension $n > 0$.

(i) If H is a proper closed subgroup of G, then the restriction map $H^n(G, \mathbb{Z}/p\mathbb{Z}) \longrightarrow H^n(H, \mathbb{Z}/p\mathbb{Z})$ is 0.

(ii) If H is a closed, but not open subgroup of G, then $cd\, H \le n - 1$.

§8. Filtrations

In this section G is always a pro-p-group. In the following we introduce the notion of the q-central series of G.

(3.8.1) Definition. *Let G be a pro-p-group and let q be a power of p. Then the* **descending q-central series** *of G is the filtration $\{G^i\}_{i\geq 1}$ recursively defined by*

$$G^1 = G, \quad G^{i+1} = (G^i)^q[G^i, G],$$

where $[G^i, G]$ and $(G^i)^q$ are the closed subgroups topologically generated by the commutators

$$(x, y) = x^{-1}y^{-1}xy,$$

$x \in G^i$, $y \in G$, and by the q-th powers of elements of G^i, respectively. *)

If $q = 0$, then we denote this series by $\{G_i\}_{i\geq 1}$, i.e.

$$G_1 = G, \quad G_{i+1} = [G_i, G].$$

It is called **descending central series** *of G.*

(3.8.2) Proposition. *Let G be a pro-p-group. Then the subgroups G^i of the q-central series are normal in G and G^i/G^{i+1} is contained in the center of G/G^{i+1}. Furthermore,*

$$\bigcap_i G^i = 1.$$

If G is finitely generated and $q \neq 0$, then the subgroups G^i form a fundamental system of open neighbourhoods of 1.

Proof: The first two statements are obvious. For the third, let U be an open normal subgroup of G. The projection $G \to G/U$ maps the q-central series (G^i) of G into a subseries of the q-central series of G/U, which terminates with $\{1\}$. Therefore $G^i \subseteq U$ for i sufficiently large.

Now assume that G is finitely generated and $q \neq 0$. We will show that the subgroups G^i are of finite index, hence open, and proceed by induction on i. For $i = 1$, this is trivial. Assume that $(G : G^i)$ is finite. Then G^i is a finitely generated pro-p-group and $G^i/(G^i)^q[G^i, G^i]$ is a finitely generated abelian group of exponent q, thus finite, and has $G^i/(G^i)^q[G^i, G] = G^i/G^{i+1}$ as quotient. This shows that $(G : G^{i+1})$ is finite, so G^{i+1} is open in G. □

---

*) The reader should take care about the different meaning of the superscripts i and q.

We collect some formulae for commutators and p-powers of the pro-p-group G.

(3.8.3) Proposition. *If $x \in G^i$, $y \in G^j$, $a \in q^r\mathbb{Z}_p$, then*

(i) $(xy)^a \equiv x^a y^a (y,x)^{\binom{a}{2}} \bmod G^{i+j+\max(1,r)}$,

(ii) $(x^a, y) \equiv (x,y)^a ((x,y),x)^{\binom{a}{2}} \bmod G^{i+j+1+\max(1,r)}$,

(iii) $(x, y^a) \equiv (x,y)^a ((x,y),y)^{\binom{a}{2}} \bmod G^{i+j+1+\max(1,r)}$.

Proof: For $a \in \mathbb{N}$, this is proved by induction using the identities

$$(uv, w) = (u,w)((u,w),v)(v,w),$$
$$(u, vw) = (u,w)(u,v)((u,v),w).$$

The general result is obtained by passing to the limit. □

(3.8.4) Corollary. *For $i, j \geq 1$ and $r, s \geq 0$, where $r + s \neq 0$, we have*

$$[(G_i)^{q^r}, (G_j)^{q^s}] \subseteq (G_{i+j})^{q^{r+s}} \cdot (G_{i+j+1})^{q^{r+s-1}} \cdot G^{i+j+r+s+1}.$$

The quotient G^i/G^{i+1} is an abelian group, which we now write additively. We denote this additive group by $\mathrm{gr}_i(G)$. The direct sum

$$\mathrm{gr}(G) = \bigoplus_{i=1}^{\infty} \mathrm{gr}_i(G)$$

has the structure of a *Lie algebra* over $k = \mathbb{Z}_p/q\mathbb{Z}_p$. The Lie bracket [,] is induced by the commutator, that is, if $\xi = \bar{x} \in \mathrm{gr}_i(G)$ and $\eta = \bar{y} \in \mathrm{gr}_j(G)$, then $[\xi, \eta]$ is the image of $(x,y) = x^{-1}y^{-1}xy$ in $\mathrm{gr}_{i+j}(G)$. It is convenient to carry out commutator calculations additively with respect to the filtration $\{G^i\}$ in the Lie algebra $\mathrm{gr}(G)$.

Proposition (3.8.3) shows that the map $x \mapsto x^q$ of G^i into G^{i+1} induces a mapping $\pi_i : \mathrm{gr}_i(G) \to \mathrm{gr}_{i+1}(G)$. The family (π_i) then induces a map $\pi_* : \mathrm{gr}(G) \to \mathrm{gr}(G)$. Let π be an indeterminate over the ring $k = \mathbb{Z}/q\mathbb{Z}$ if $q \neq 0$ and the zero element of $k = \mathbb{Z}_p$ if $q = 0$. Then there exists a unique mapping

$$\Phi : k[\pi] \times \mathrm{gr}(G) \longrightarrow \mathrm{gr}(G),$$

which is k-linear in the first variable, such that $\Phi(\pi^i, \xi) = \pi_*^i(\xi)$. If we set $\alpha \cdot \xi = \Phi(\alpha, \xi)$, we have $\pi^i \cdot (\pi^j \cdot \xi) = \pi^{i+j} \cdot \xi$. Proposition (3.8.3) with $a = q$ now yields the

(3.8.5) Proposition. *Let* $\xi \in \mathrm{gr}_i(G)$ *and* $\eta \in \mathrm{gr}_j(G)$. *Then*

(i) $\pi \cdot (\xi + \eta) = \pi \cdot \xi + \pi \cdot \eta$ *if* $i = j > 1$,

(ii) $\pi \cdot (\xi + \eta) = \pi \cdot \xi + \pi \cdot \eta + \binom{q}{2} [\xi, \eta]$ *if* $i = j = 1$,

(iii) $\pi \cdot [\xi, \eta] = [\pi \cdot \xi, \eta]$ *if* $i \neq 1$,

$\pi \cdot [\xi, \eta] = [\xi, \pi \cdot \eta]$ *if* $j \neq 1$,

(iv) $[\pi \cdot \xi, \eta] = \pi \cdot [\xi, \eta] + \binom{q}{2} [[\xi, \eta], \xi]$ *if* $i = j = 1$,

(v) $[\xi, \pi \cdot \eta] = \pi \cdot [\xi, \eta] + \binom{q}{2} [[\xi, \eta], \eta]$ *if* $i = j = 1$.

Remark: If $G = F$ is a free pro-p-group and if q is not a power of 2, then $\binom{q}{2} \equiv 0 \bmod q$ and $\mathrm{gr}(F)$ is a free Lie algebra over $k[\pi]$ (see [125]).

For a free pro-p-group F, we obtain the following formula for the intersection of the p-central series and the central series of F:

(3.8.6) Proposition. *Let F be a free pro-p-group and let $\{F^i\}$ be the p-central series and $\{F_i\}$ the central series of F. Then for all $i \geq j \geq 1$, there is the equality*

$$F^i \cap F_j = (F_j)^{p^{i-j}} \cdot (F_{j+1})^{p^{i-j-1}} \cdot \ldots \cdot F_i .$$

Proof: We proceed by induction on j. For $j = 1$ we have to prove that

$$F^i = F^{p^{i-1}} \cdot \ldots \cdot F_i .$$

First, we show that

$$F^i = F^{p^{i-1}} \cdot \ldots \cdot F_i \cdot F^{i+1} .$$

This is obviously true for $i = 1$. So let $i > 1$ and assume that the assertion is true for $i - 1$. Then, using (3.8.4),

$$\begin{aligned} F^i &= (F^{i-1})^p [F^{i-1}, F] \\ &= (F^{p^{i-2}} \cdot \ldots \cdot F_{i-1} \cdot F^i)^p [F^{p^{i-2}} \cdot \ldots \cdot F_{i-1} \cdot F^i, F] \\ &= F^{p^{i-1}} \cdot \ldots \cdot F_{i-1}^p \cdot F_i \cdot F^{i+1} . \end{aligned}$$

Assume we have proved that $F^i = F^{p^{i-1}} \cdot \ldots \cdot F_i \cdot F^k$ for $k \geq i + 1$. Then

$$\begin{aligned} F^i &= F^{p^{i-1}} \cdot \ldots \cdot F_i \cdot (F^{k-1})^p [F^{k-1}, F] \\ &= F^{p^{i-1}} \cdot \ldots \cdot F_i \cdot (F^{p^{k-2}} \cdot \ldots \cdot F_{k-1} \cdot F^k)^p [F^{p^{k-2}} \cdot \ldots \cdot F_{k-1} \cdot F^k, F] \\ &= F^{p^{i-1}} \cdot \ldots \cdot F_i \cdot F^{k+1} . \end{aligned}$$

Since $\bigcap_k F^k = 1$, we have proved the case $j = 1$.

Now let $i \geq j > 1$. Then

$$F^i \cap F_j = F^i \cap F_{j-1} \cap F_j = (F_{j-1})^{p^{i-j+1}} \cdot \ldots \cdot F_i \cap F_j$$

by the induction assumption. Let

$$x = z_{j-1}^{p^{i-j+1}} y \in F_j \quad \text{with} \quad z_{j-1} \in F_{j-1}, y \in (F_j)^{p^{i-j}} \cdot \ldots \cdot F_i .$$

Claim: F_{j-1}/F_j is $\mathbb{Z}_p$-torsion-free.

Proof: Using a limit argument we may assume that F is finitely generated. Since the free pro-p-group F is the completion of a free discrete group, F_j/F_{j+1} is the completion of a free (finitely generated) abelian group by [260], Satz 4. Hence it is a free $\mathbb{Z}_p$-module.

Using this claim, we see $z_{j-1} \in F_j$, so that $x \in (F_j)^{p^{i-j}} \cdot \ldots \cdot F_i$. This proves the proposition. □

We now define a refinement $\{G^{(i,j)}\}$ of the descending p-central series of G.

(3.8.7) Definition. *Let $\{G^i\}$ and $\{G_j\}$ be the p-central series and the central series of the pro-p-group G, respectively. We set, for $i, j \geq 1$,*

$$G^{(i,j)} := (G^i \cap G_j)\, G^{i+1}.$$

Obviously we have

$$G^{(i,1)} = G^i \text{ and } G^{(i,j)} = G^{i+1} \text{ for } j > i \geq 1.$$

We introduce the following *notational convention*:

The letter ν always stands for a pair (i, j), $i \geq j \geq 1$, and we order these pairs lexicographically. We say that

$$\begin{aligned} \nu + 1 &= (i, j + 1) && \textit{if } i > j, \\ \nu + 1 &= (i + 1, 1) && \textit{if } \nu = (i, i). \end{aligned}$$

The descending chain $\{G^{(\nu)}\}$ of normal characteristic subgroups is a refinement of the descending p-central series. In particular, $G^{(\nu)}/G^{(\nu+1)}$ is an $\mathbb{F}_p$-vector space for all ν.

(3.8.8) Proposition. *For every $\nu = (i, j)$, the $\mathbb{F}_p$-vector space homomorphism*

$$\begin{aligned} \psi_\nu : \quad (G/G^2)^{\otimes j} &\longrightarrow G^{(\nu)}/G^{(\nu+1)} \\ \bar{x}_1 \otimes \cdots \otimes \bar{x}_j &\longmapsto ([x_1, [x_2, [\cdots, x_j]] \cdots])^{p^{i-j}} \bmod G^{(\nu+1)} \end{aligned}$$

is well-defined and surjective.

Proof: Let $S = \{x_\alpha\}_{\alpha \in I}$ be a minimal system of generators of G. Then their image $\bar{S} := S \bmod G^p[G,G]$ in $G/G^2 = G/G^p[G,G]$ is an $\mathbb{F}_p$-basis of G/G^2. (In fact, if R is any proper subset of S, then R generates a closed subgroup $\neq G$, which sits in a maximal, i.e. normal, subgroup M of index p, so that $\bar{R} \subseteq M/G^2 \neq G/G^2$ cannot generate G/G^2.) Thus the tensors $\bar{x}_{\alpha_1} \otimes \cdots \otimes \bar{x}_{\alpha_j}$, $\alpha_1, \ldots, \alpha_j \in I$ define a basis of $(G/G^2)^{\otimes j}$. Recalling the definition of the (ν)-filtration, it follows from (3.8.6) (for free groups and then for all pro-p-groups) that the elements

$$([x_{\alpha_1}, [x_{\alpha_2}, [\cdots, x_{\alpha_j}]] \cdots])^{p^{i-j}}, \qquad \alpha_1, \ldots, \alpha_j \in I$$

generate $G^{(\nu)}$ modulo $G^{(\nu+1)}$. Finally, an inductive application of (3.8.3) shows that ψ_ν is well-defined. □

Remark: The same proof shows that we can define ψ_ν as a homomorphism

$$(G/G^2)^{\otimes j} \longrightarrow G^{(\nu)}/G^{i+1} \subseteq G^i/G^{i+1}$$

if either $j > 1$ or if p is odd (hence $p \mid \binom{p}{2}$).

§9. Generators and Relations

For this entire section G is a pro-p-group. We set

$$H^n(G) = H^n(G, \mathbb{Z}/p\mathbb{Z})$$

and regard these groups as $\mathbb{F}_p$-vector spaces. By the dimension, $\dim H^n(G)$, we mean the cardinality of a basis. $H^1(G)$ is the Pontryagin dual of the group G/G^*, where $G^* = G^p[G,G]$, i.e. the closure of the subgroup generated by commutators and p-th powers.*) Indeed, the group G/G^* is the largest profinite abelian quotient of exponent p of G, so that

$$H^1(G) = \mathrm{Hom}(G, \mathbb{Z}/p\mathbb{Z}) = \mathrm{Hom}(G/G^*, \mathbb{Q}/\mathbb{Z}).$$

A **generator system** of G is a convergent family $\mathscr{S} = (g_i)_{i \in I}$, $g_i \in G$, which generates G as a topological group. By **convergent** we mean convergent to 1, i.e. every open subgroup of G contains g_i for almost all i. The **rank** of a pro-p-group G is the infimum over the cardinalities $\mathrm{card}(\mathscr{S}) = \mathrm{card}(I)$ of generator systems $\mathscr{S}$ of G, and is denoted by $d(G)$. A generator system is **minimal** if no proper subfamily is a generator system.

*) G^* is the Frattini subgroup of G, i.e. the intersection of all maximal closed subgroups.

(3.9.1) Proposition. *A convergent family $\mathscr{S}$ is a generator system of G if and only if the family $\overline{\mathscr{S}} = (\overline{g}_i)_{i\in I}$ of residue classes $\overline{g}_i = g_i \bmod G^*$ generates G/G^*. $\mathscr{S}$ is minimal if and only if $\overline{\mathscr{S}}$ is. Every minimal system of generators has cardinality $d(G)$ and we have the equality*

$$d(G) = \dim H^1(G).$$

In particular, G is finitely generated if and only if $H^1(G)$ is finite.

The first assertion of the proposition is often called **Frattini argument**.

Proof: Let H be the closed subgroup of G generated by $\mathscr{S}$. Then we have the equivalences

$$\begin{array}{lll} & \text{the inclusion } H \to G \text{ is surjective} & \\ \Leftrightarrow & H^1(G) \to H^1(H) \text{ is injective} & \text{(by (1.6.14)(ii))} \\ \Leftrightarrow & H/H^* \to G/G^* \text{ is surjective} & \text{(by Pontryagin duality).} \end{array}$$

This shows the statement. □

Let N be a normal subgroup of G. A **generator system of N as a normal subgroup** is a convergent family $\mathscr{S}$ such that N is the smallest closed normal subgroup of G containing all elements of $\mathscr{S}$. In the case $N = G$, (3.9.1) shows that $\mathscr{S}$ generates G as a normal subgroup (of itself) if and only if $\mathscr{S}$ generates G as a pro-p-group.

(3.9.2) Proposition. *Let N be a normal closed subgroup of G, $\mathscr{S}$ a convergent family of elements of N and H the closed subgroup generated by $\mathscr{S}$. Then the restriction map*

$$res : H^1(N)^G \longrightarrow H^1(H)$$

is injective if and only if $\mathscr{S}$ is a generator system of N as a normal subgroup.

Proof: Let N' be the normal closed subgroup of G generated by $\mathscr{S} = (n_i)_{i\in I}$ (as a normal subgroup), and consider the commutative diagram

$$\begin{array}{ccc} H^1(N)^G & \xrightarrow{\ res_1\ } & H^1(H) \\ & {\scriptstyle res_2}\searrow \quad \nearrow{\scriptstyle res_3} & \\ & H^1(N')^G . & \end{array}$$

As a pro-p-group, N' is generated by the family $\tilde{\mathscr{S}} = (\sigma n_i \sigma^{-1})_{i\in I, \sigma\in G}$. For $\chi \in H^1(N')^G$, we have

$$\begin{aligned} res_3(\chi) = 0 &\Rightarrow \chi(\sigma n_i \sigma^{-1}) = \chi^\sigma(n_i) = \chi(n_i) = 0 \text{ for all } \sigma \text{ and all } i \\ &\Rightarrow \chi(\tilde{\mathscr{S}}) = 0 \Rightarrow \chi = 0, \end{aligned}$$

i.e. res_3 is injective. Therefore res_1 is injective if and only if res_2 is injective. But this is equivalent with the injectivity of the homomorphism $res^N_{N'} : H^1(N) \to H^1(N')$ since

$$\ker(res_2) = \ker(res^N_{N'})^G = 0 \iff \ker(res^N_{N'}) = 0$$

by (1.6.12). Finally, the injectivity of $res^N_{N'}$ is equivalent with $N' = N$ by (1.6.14). □

(3.9.3) Corollary. *Let N be a normal closed subgroup of G. Then a convergent family $\mathscr{S}$ of elements of N generates N as a normal subgroup if and only if the family $\overline{\mathscr{S}}$ of residue classes modulo $N^p[G,N]$ generates $N/N^p[G,N]$. In particular, $\mathscr{S}$ is minimal if and only if $\overline{\mathscr{S}}$ is, and in this case*

$$\mathrm{card}(\mathscr{S}) = \dim H^1(N)^G.$$

Proof: Let H be the closed subgroup in N generated by $\mathscr{S}$. Taking into account that $N/N^p[G,N] = (H^1(N)^G)^\vee$, we the have following equivalences

$$\begin{array}{lll} & \mathscr{S} \text{ generates } N \text{ as a normal subgroup} & \\ \Leftrightarrow & H^1(N)^G \to H^1(H) \text{ is injective} & \text{(by (3.9.2))} \\ \Leftrightarrow & H/H^* \to N/N^p[G,N] \text{ is surjective} & \text{(by Pontryagin duality).} \end{array}$$

This shows the statement. □

Let $\mathscr{S} = (g_i)_{i\in I}$ be a system of generators of the pro-p-group G and let F be the free pro-p-group with basis $(x_i)_{i\in I}$, see (3.5.14). Sending x_i to g_i, we obtain an exact sequence

$$1 \longrightarrow R \longrightarrow F \longrightarrow G \longrightarrow 1.$$

A **relation system**, also called a **system of defining relations** with respect to $\mathscr{S}$, is a generator system $\mathscr{R}$ of R as a normal subgroup of F.

(3.9.4) Proposition. *Suppose that the pro-p-group G is finitely generated and let $\mathscr{S}$ be a finite system of generators of G. Then a finite system $\mathscr{R}$ of defining relations with respect to $\mathscr{S}$ exists if and only if $H^2(G)$ is finite. If $\mathscr{R}$ is finite and minimal, we have the equality*

$$\mathrm{card}(\mathscr{S}) - \mathrm{card}(\mathscr{R}) = \dim H^1(G) - \dim H^2(G).$$

Proof: Applying (3.9.3) to the closed subgroup R of F, we obtain

$$\mathrm{card}(\mathscr{R}) = \dim H^1(R)^F = \dim H^1(R)^G.$$

Consider the five term exact sequence

$$0 \longrightarrow H^1(G) \longrightarrow H^1(F) \longrightarrow H^1(R)^G \longrightarrow H^2(G) \longrightarrow H^2(F).$$

Since F is free, $H^2(F) = 0$, and we obtain the result by counting $\mathbb{F}_p$-dimensions. □

In particular, given G, the cardinality of a minimal system of defining relations only depends on the cardinality of the system $\mathscr{S}$ of generators. If $\mathscr{S}$ is a minimal system of generators, we call the cardinality of a minimal relation system the **relation rank** of G. We denote it by $r(G)$.

(3.9.5) Proposition. *Let G be a pro-p-group. Then the relation rank of G satisfies the formula*

$$r(G) = \dim H^2(G).$$

In particular, G is a free pro-p-group if and only if $cd\, G \le 1$.

Proof: Let $\mathscr{S}$ be a minimal system of generators of G. Then the exact sequence

$$0 \longrightarrow H^1(G) \xrightarrow{\sim} H^1(F) \longrightarrow H^1(R)^G \longrightarrow H^2(G) \longrightarrow 0$$

yields the isomorphism $H^1(R)^G \xrightarrow{\sim} H^2(G)$, and so $r(G) = \dim H^1(R)^G = \dim H^2(G)$. □

From (3.3.16) we obtain the

(3.9.6) Corollary. *Let G be a finitely generated pro-p-group of rank $d(G)$ and let $\mathfrak{U}$ be a cofinal set of open subgroups of G. Then G is a free pro-p-group if and only if*

$$d(U) - 1 = (G : U)(d(G) - 1) \quad \textit{for all } U \in \mathfrak{U}.$$

Remark: If G is a finitely generated pro-p-group, i.e. $d(G) < \infty$, then the relation rank $r(G)$ need not be finite. We have the following example occurring in number theory. Let $G = G(k(p)|k)$ be the Galois group of the maximal p-extension of a p-adic local field k (cf. VII §5) and let G^t, $t \in \mathbb{R}, t \ge 0$, be the higher ramification groups of G (see [130] for the definition). Since G is finitely generated (cf. (7.5.11)), the quotients G/G^t are also finitely generated. But *N. L. Gordeev* proved in [58] that $r(G/G^t)$ is infinite if $t > 1$. This extended the result of *E. Maus* who showed this for $t \notin \mathbb{Z}[\frac{1}{p}], t > 1$, see [131].

The rank and the relation rank of a pro-p-group G played an important role in the solution of the famous and long standing **class field tower problem** in number theory. The question is whether the maximal unramified p-extension $K(p)|K$ of a number field K need always be of finite degree. It was a great surprise when, in 1964, the Russian mathematicians *E. S. Golod* and *I. R. Šafarevič* proved that this is not always the case. The Galois group G

of $K(p)|K$ is a pro-p-group, and the crucial point in the proof (presented in X §8) was the discovery that for a *finite* p-group G the relation rank has to be very large in comparison with the rank. More precisely, we have the

(3.9.7) Theorem. *If G is a finite p-group, then*

$$r(G) > \tfrac{1}{4}\, d(G)^2 .$$

For the proof we need the following

(3.9.8) Lemma. *Let G be a finite p-group and $\Lambda = \mathbb{F}_p[G]$. For every finite G-module A such that $pA = 0$, there is a resolution*

$$0 \longrightarrow A \longrightarrow \Lambda^{b_0} \xrightarrow{\partial} \Lambda^{b_1} \xrightarrow{\partial} \Lambda^{b_2} \xrightarrow{\partial} \cdots,$$

where $b_n = \dim H^n(G,A)$, and $\partial((\Lambda^{b_n})^G) = 0$.

Proof: We have canonically $\Lambda^G \cong \mathbb{F}_p$. Let $a_1, \dots, a_{b_0}$ be a basis of the $\mathbb{F}_p$-vector space A^G. Then the isomorphism $A^G \to (\Lambda^{b_0})^G = \mathbb{F}_p^{b_0}$, $a_i \mapsto e_i$, extends to an injective G-homomorphism

$$j : A \longrightarrow \Lambda^{b_0}.$$

Indeed, the map $\mathrm{Hom}_G(A, \Lambda^{b_0}) \to \mathrm{Hom}_G(A^G, \Lambda^{b_0})$ is surjective, since in the exact sequence

$$0 \longrightarrow \mathrm{Hom}(A/A^G, \Lambda^{b_0}) \longrightarrow \mathrm{Hom}(A, \Lambda^{b_0}) \longrightarrow \mathrm{Hom}(A^G, \Lambda^{b_0}) \longrightarrow 0$$

of induced G-modules, H^1 of the first term is zero. The extension j is automatically injective, since from $\ker(j|_G) = \ker(j)^G = 0$, it follows that $\ker(j) = 0$ by (1.6.12). Since Λ^{b_0} is an induced G-module, from the exact sequence

$$0 \longrightarrow A \longrightarrow \Lambda^{b_0} \longrightarrow B \longrightarrow 0 ,$$

we obtain isomorphisms

$$H^i(G,B) \cong H^{i+1}(G,A)$$

for $i \geq 1$. The same holds for $i = 0$, since in the exact sequence

$$A^G \longrightarrow (\Lambda^{b_0})^G \longrightarrow B^G \longrightarrow H^1(G,A) \longrightarrow 0$$

the first arrow is bijective, i.e. $(\Lambda^{b_0})^G$ is mapped to zero.

Proceeding in the same way with the G-module B in place of A and noting that $\dim H^0(G,B) = \dim H^1(G,A) = b_1$, we obtain an exact sequence

$$0 \longrightarrow B \longrightarrow \Lambda^{b_1} \longrightarrow C \longrightarrow 0$$

such that $B^G \to (\Lambda^{b_1})^G$ is an isomorphism, i.e. $(\Lambda^{b_1})^G$ is mapped to zero. If we define $\partial : \Lambda^{b_0} \to \Lambda^{b_1}$ to be the composite map $\Lambda^{b_0} \to B \to \Lambda^{b_1}$, then $\partial((\Lambda^{b_0})^G) = 0$. Continuing this process, the lemma follows by induction. □

Proof of theorem (3.9.7): For every finite G-module A such that $pA = 0$, we define the "ascending central series"

$$0 = A_0 \subseteq A_1 \subseteq A_2 \subseteq \cdots \subseteq A_m = A$$

by $A_0 = 0$, $A_1 = A^G$ and $A_{n+1}/A_n = (A/A_n)^G$ for $n \geq 1$. By (1.6.12), $A_n \neq A_{n+1}$, unless $A_n = A$. If $h : A \to B$ is an injective G-homomorphism, then we see inductively

$$A_n = h^{-1}(B_n).$$

We set $c_n(A) = \dim(A_{n+1}/A_n)$ and form the *Poincaré polynomial*

$$P_A(t) = \sum_{n \geq 0} c_n(A) t^n.$$

If $0 < t < 1$ is a real variable, then

$$P_A(t)\frac{1}{1-t} = \sum_{n \geq 0} s_n(A) t^n,$$

where $s_n(A) = \sum_{i=0}^{n} c_i(A) = \dim(A_{n+1})$.

We apply the lemma to the G-module $A = \mathbb{F}_p$. Recalling that $\mathbb{F}_p = \Lambda^G = \Lambda_1$, we obtain an exact sequence

$$(*) \qquad 0 \longrightarrow E \xrightarrow{\partial} D \xrightarrow{\partial} R,$$

where $E = \Lambda/\Lambda_1$, $D = \Lambda^d$, $R = \Lambda^r$,

$$d = \dim H^1(G) = d(G), \quad r = \dim H^2(G) = r(G),$$

such that $\partial(D_1) = R_0 = 0$. From this it follows inductively that $\partial(D_n) \subseteq R_{n-1}$, and we obtain the sequences

$$(**) \qquad 0 \longrightarrow E_n \longrightarrow D_n \longrightarrow R_{n-1}, \quad n \geq 1.$$

These sequences are again exact, since $(*)$ is exact and $E_n = \partial^{-1}(D_n)$. Now consider the Poincaré polynomials associated to E, D, R. Setting $P(t) = P_\Lambda(t)$, we obtain

$$P_E(t) = \frac{P(t)-1}{t}, \quad P_D(t) = dP(t), \quad P_R(t) = rP(t).$$

From $(**)$ we get the inequalities

$$s_n(D) \leq s_n(E) + s_{n-1}(R),$$

(where $s_{-1}(R) = 0$), so that

$$P_D(t)\frac{1}{1-t} \leq P_E(t)\frac{1}{1-t} + P_R(t)\frac{t}{1-t}.$$

It follows that

$$dP(t) \leq \frac{P(t)-1}{t} + rtP(t) \quad \text{for} \quad 0 < t < 1,$$

and so

$$1 \leq P(t)(rt^2 - dt + 1) \quad \text{if} \quad 0 < t < 1.$$

Since $P(t)$ has positive coefficients, we obtain

$$0 < rt^2 - dt + 1 \quad \text{if} \quad 0 < t < 1.$$

Substituting $t \mapsto \frac{d}{2r}$, we get

$$r > \tfrac{1}{4}d^2,$$

as asserted. This substitution is valid, since the exact sequence

$$0 \to H^1(G) \to H^2(G,\mathbb{Z}) \xrightarrow{p} H^2(G,\mathbb{Z}) \to H^2(G)$$

shows that $d \le r < 2r$, i.e. $0 < \frac{d}{2r} < 1$. □

Remarks: 1. *Golod* and *Šafarevič* had originally proved only that $r(G) > \frac{1}{4}(d(G)-1)^2$ (see [57]). The sharper inequality $r(G) > \frac{1}{4}d(G)^2$ was obtained independently by *W. Gaschütz* and *E. B. Vinberg* [241]. The proof given here is dual to the proof presented by *P. Roquette* in [183], who uses homology instead of cohomology. There is another more general proof (based, however, on the same idea as in [183]) with more far reaching results given by *H. Koch* in [110] and [72], and yet another proof was given by *J.-P. Serre* in [210].

2. We want to present the result of *Koch* without proof. For that we need the notion of the Zassenhaus filtration of a pro-p-group. Let G be a finitely generated pro-p-group and for $n \ge 1$ let the ideal $I^n(G)$ of $\mathbb{F}_p[[G]]$ be the n-th power of the augmentation ideal $I(G)$. The filtration

$$G_{(n)} = \{g \mid g - 1 \in I^n(G)\}, \qquad n \ge 1,$$

is called the **Zassenhaus filtration** of G. The normal subgroups $G_{(n)}$ form a full system of neighbourhoods of the identity of G. For the basic properties of $\{G_{(n)}\}_{n\ge1}$, see [110], §7.4. We mention only that

$$(G_{(n)})^p \subseteq G_{(np)} \quad \text{and} \quad [G_{(n)}, G] \subseteq G_{(n+1)}.$$

Now let

$$1 \longrightarrow R \longrightarrow F \longrightarrow G \longrightarrow 1$$

be a minimal presentation of the *finite* p-group G by a free pro-p-group F and assume that $R \subset F_{(m)}$ for some m, where $\{F_{(n)}\}_{n\ge1}$ is the Zassenhaus filtration of F. Then

$$r(G) > \frac{d(G)^m}{m^m}(m-1)^{m-1}.$$

This formula shows that for a finite p-group we get a better bound for $r(G)$ if we have information on the complexity of the relations.

We are now aiming at the classification of the Demuškin groups, which are defined as follows.

(3.9.9) Definition. *A pro-p-group G is called a* **Demuškin group** *if its cohomology $H^i(G)$ has the following properties:*

(i) $\dim_{\mathbb{F}_p} H^1(G) < \infty$,

(ii) $\dim_{\mathbb{F}_p} H^2(G) = 1$,

(iii) *the cup-product $H^1(G) \times H^1(G) \to H^2(G)$ is non-degenerate.*

By (3.7.6), an infinite Demuškin group has necessarily cohomological dimension 2 and is precisely a Poincaré group of dimension 2. The finite Demuškin groups are classified by the following

(3.9.10) Proposition. *The group $G = \mathbb{Z}/2\mathbb{Z}$ is the only finite Demuškin group.*

Proof: The relation rank $r = r(G)$ of a Demuškin group G is equal to 1 by definition. Therefore if G has generator rank $d = d(G) \geq 2$, then its abelianization G^{ab} has a $\mathbb{Z}_p$-rank of at least $d - 1 \geq 1$. We conclude that $d = 1$ for a finite Demuškin group, i.e. G is cyclic. Hence $H^1(G)$ is a one dimensional $\mathbb{F}_p$-vector space and for every $x \in H^1(G)$ we have $2 \cdot (x \cup x) = x \cup x + x \cup x = 0$ by the anti-symmetry of the cup-product. This gives a contradiction for odd p [*)] and therefore $G \cong \mathbb{Z}/2^k\mathbb{Z}$ for some $k \geq 1$. A straightforward computation, which we leave to the reader, shows that the cup-product $H^1(G) \times H^1(G) \xrightarrow{\cup} H^2(G) \cong \mathbb{F}_2$ is trivial for $k > 1$, and nontrivial, hence non-degenerate, for $k = 1$. □

The appearance of $G = \mathbb{Z}/2\mathbb{Z}$ as the only finite Demuškin group is interesting from the arithmetic point of view, since G is the absolute Galois group of the field $\mathbb{R}$.

Let us consider more generally finitely generated pro-p-groups G with only one defining relation, i.e.

$$n = \dim_{\mathbb{F}_p} H^1(G) < \infty \qquad \text{and} \qquad \dim_{\mathbb{F}_p} H^2(G) = 1.$$

Such a group is called a **one-relator pro-p-group**. We have an exact sequence involving G,

$$1 \longrightarrow R \longrightarrow F \longrightarrow G \longrightarrow 1,$$

[*)] In other words: the cup-product is always anti-symmetric, which implies in odd characteristics that it is alternating. A vector space with a non-degenerate alternating bilinear form is necessarily of even dimension. In characteristic 2 the cup-product may induce a symmetric, non alternating form.

where F is the free pro-p-group of rank n and R is generated by one element ρ as normal subgroup of F. We also write $R = (\rho)$. Passing to the abelianized groups, we obtain G^{ab} as a quotient of $F^{ab} \cong \mathbb{Z}_p^n$ by a subgroup which is either isomorphic to $\mathbb{Z}_p$ or zero. Noting that $H^1(G) = H^1(G^{ab}) = (\mathbb{Z}/p\mathbb{Z})^n$, we conclude that

$$G^{ab} \cong \mathbb{Z}_p^n \qquad \text{or} \qquad G^{ab} \cong \mathbb{Z}/p^f\mathbb{Z} \times \mathbb{Z}_p^{n-1} \qquad (f \geq 1).$$

We set $q = p^f$ in the second case and $q = 0$ in the first case. The numbers n and q are invariants of the group G. Since $F/F^q[F,F] \to G/G^q[G,G]$ is an isomorphism by definition of q, we have

$$R \subseteq F^q[F,F] = F^2\,,$$

where F^i is the q-central series of F.

The central result which we want to prove is the following

(3.9.11) Theorem (Demuškin). *Let G be a one-relator pro-p-group. Suppose that the invariant q of G is $\neq 2$. Then G is a Demuškin group if and only if it is isomorphic to the pro-p-group defined by n generators $x_1, \dots, x_n$ subject to the one relation*

$$x_1^q(x_1,x_2)(x_3,x_4)\cdots(x_{n-1},x_n) = 1,$$

where (x,y) is the commutator $x^{-1}y^{-1}xy$. In particular, G is then determined by the two invariants n and q.

The proof of the theorem is not easy. We follow in essence the presentation in [117] by *J. Labute*, where the case $q = 2$ is also treated.

Before considering one-relator pro-p-groups and proving the result stated above, we consider the case where G is an arbitrary finitely generated pro-p-group of rank n. If G^{ab} has a nontrivial torsion subgroup, then we set $k = \mathbb{Z}_p/q\mathbb{Z}_p$, where q is the smallest elementary divisor of G^{ab}, i.e. $q = p^f$ is the maximal p-power such that G^{ab}/q is a free $\mathbb{Z}_p/q\mathbb{Z}_p$-module. When $G^{ab} \cong \mathbb{Z}_p^n$ we set $k = \mathbb{Z}_p$ and change our usual notation and denote by $H^i(G,\mathbb{Z}_p)$ the continuous cochain cohomology (see II §7). Since G is finitely generated, (2.7.6) implies $H^i(G,\mathbb{Z}_p) = \varprojlim_\nu H^i(G,\mathbb{Z}/p^\nu\mathbb{Z})$ for $i = 0,1,2$. Let

$$1 \longrightarrow R \longrightarrow F \longrightarrow G \longrightarrow 1$$

be a minimal presentation of the pro-p-group G. The inflation

$$\mathit{inf} : H^1(G,k) \longrightarrow H^1(F,k)$$

is an isomorphism with which we identify the two groups.

Since F is free, we have $H^2(F,k) = 0$, and the five term exact sequence (1.6.7) shows that also the transgression $tg : H^1(R,k)^G \to H^2(G,k)$ is an isomorphism. More generally, we obtain isomorphisms

$$tg : \mathrm{Hom}_{cts}(R_G^{ab}, \mathbb{Z}/p^n\mathbb{Z}) = H^1(R, \mathbb{Z}/p^n\mathbb{Z})^G \xrightarrow{\sim} H^2(G, \mathbb{Z}/p^n\mathbb{Z})$$

for all $n \le f$. Therefore every element $\rho \in R$ gives rise to trace maps

$$tr = tr_\rho : H^2(G, \mathbb{Z}/p^n\mathbb{Z}) \longrightarrow \mathbb{Z}/p^n\mathbb{Z}$$

which are defined by $\varphi \mapsto (tg^{-1}\varphi)(\rho)$. If $q = 0$, then these maps are defined for all $n \ge 1$ and also with $\mathbb{Z}_p$-coefficients.

(3.9.12) Proposition. *Let*

$$1 \longrightarrow R \longrightarrow F \longrightarrow G \longrightarrow 1$$

be a minimal presentation of the finitely generated pro-p-group G.

(i) *Every basis $x_1, \dots, x_n$ of F defines a k-basis $\chi_1, \dots, \chi_n$ of the k-module $H^1(F,k) = H^1(G,k)$ such that $\chi_i(x_j) = \delta_{ij}$.*

(ii) *Given an element $\rho \in R$, the bilinear form induced by the cup-product*

$$H^1(G,k) \times H^1(G,k) \xrightarrow{\cup} H^2(G,k) \xrightarrow{tr_\rho} k$$

is non-degenerate if and only if it is non-degenerate with k replaced by $\mathbb{Z}/p\mathbb{Z}$, and in this case tr_ρ is surjective.

(iii) *If G is a one-relator group and if ρ generates R as a normal subgroup, then tr_ρ is injective.*

Proof: (i) By definition of q, a minimal generator system $x_1, \dots, x_n$ of F defines an isomorphism $F/F^q[F,F] \cong k^n$, where $[F,F]$ is the closure of the commutator subgroup of F. (i) follows from this.

(ii) Consider the commutative diagram

$$\begin{array}{ccccccc}
H^1(G,k) & \times & H^1(G,k) & \xrightarrow{\cup} & H^2(G,k) & \xrightarrow{tr_\rho} & k \\
\downarrow & & \downarrow & & \downarrow & & \downarrow \\
H^1(G,\mathbb{Z}/p\mathbb{Z}) & \times & H^1(G,\mathbb{Z}/p\mathbb{Z}) & \xrightarrow{\cup} & H^2(G,\mathbb{Z}/p\mathbb{Z}) & \xrightarrow{tr_\rho} & \mathbb{Z}/p\mathbb{Z},
\end{array}$$

where the vertical arrows are induced by the reduction map $k \to \mathbb{Z}/p\mathbb{Z}$. If B is a matrix for the upper bilinear form, then $\overline{B} = B \bmod p$ is a matrix for the lower one. The upper one is non-degenerate if and only if B is invertible over k, i.e. $\det(B) \in k^\times$, and this is the case if and only if $\det(\overline{B}) \neq 0$, i.e. if the lower one is non-degenerate. Clearly tr_ρ must be surjective in this case.

(iii) If ρ generates R as a normal subgroup in G, then it generates R_G^{ab}. Therefore if $tg^{-1}(\varphi)(\rho) = 0$, then $\varphi = 0$, so that tr_ρ is injective. □

Let $\mathscr{R} = \{\rho_i \mid i \in I\}$ be a minimal system of defining relations of the finitely generated pro-p-group $G = F/R$.

(3.9.13) Proposition. *Let*

$$1 \longrightarrow R \longrightarrow F \longrightarrow G \longrightarrow 1$$

be a minimal presentation of the finitely generated pro-p-group G. Let $x_1, \dots, x_n$ be a basis of F and let $\chi_1, \dots, \chi_n$ be the corresponding k-basis of $H^1(F,k) = \mathrm{Hom}(F,k) = \mathrm{Hom}(G,k)$, i.e. $\chi_i(x_j) = \delta_{ij}$.

(i) *Every element $\rho \in F^2$ has a representation*

$$\rho = \prod_{j=1}^{n} x_j^{qa_j} \cdot \prod_{1 \le k < l \le n} (x_k, x_l)^{a_{kl}} \cdot \rho', \qquad \rho' \in F^3,\ a_j, a_{kl} \in k.$$

The a_{kl} are uniquely determined by ρ, and if $q \neq 0$ so are the a_j.

(ii) *If $\mathscr{R} = \{\rho_i \mid i \in I\}$ is a minimal system of defining relations of G and*

$$\rho_i = \prod_{j=1}^{n} x_j^{qa_j^i} \cdot \prod_{1 \le k < l \le n} (x_k, x_l)^{a_{kl}^i} \cdot \rho_i', \qquad \rho_i' \in F^3,\ a_j^i, a_{kl}^i \in k, i \in I\,,$$

then the bilinear form

$$H^1(G,k) \times H^1(G,k) \xrightarrow{\cup} H^2(G,k) \xrightarrow{tr_{\rho_i}} k$$

is given by the matrix $B_i = (b_{kl}^i)$ with respect to the basis $\chi_1, \dots, \chi_n$, where

$$b_{kl}^i = tr_{\rho_i}(\chi_k \cup \chi_l) = \begin{cases} -a_{kl}^i & \text{if} \quad k < l, \\ a_{lk}^i & \text{if} \quad k > l, \\ -\binom{q}{2} a_k^i & \text{if} \quad k = l. \end{cases}$$

Proof: The existence of a representation (i) for a finite word in the letters $x_1, \dots, x_n$ is obtained by a simple collecting process (see M. HALL [73], chap. 11.1). For an arbitrary $\rho \in F^2$ it follows by a limit process, noting that the discrete free group generated by the $x_1, \dots, x_n$ is a dense subgroup of F. If $q \neq 0$, then $k = \mathbb{Z}/q\mathbb{Z}$ and

$$F/F^{q^2}[F,F] = (\mathbb{Z}/q^2\mathbb{Z})^n.$$

The image of ρ in this last group is $(qa_1, \dots, qa_n)$, so that the a_j are uniquely determined in k by ρ. The uniqueness of the a_{kl} follows from the proof of

(ii) The cohomology class $\chi_k \cup \chi_l \in H^2(G,k)$ is represented by the inhomogeneous 2-cocycle

$$c_0(\sigma, \tau) = \chi_k(\sigma)\chi_l(\tau).$$

Let c be the inflation of c_0 to F. Since $H^2(F, k) = 0$, there exists an inhomogeneous cochain

$$u = u_{kl} : F \longrightarrow k$$

such that $c = \partial u$ and, moreover, by subtracting the homomorphism $h : F \to k$, $h(x_j) = u(x_j)$, we can suppose $u(x_j) = 0$, $j = 1, \dots, n$. Then

(1) $$u(xy) = u(x) + u(y) - \chi_k(x)\chi_l(y), \quad x, y \in F.$$

In particular, $u(xy) = u(x) + u(y)$ whenever x or y is contained in $F^2 = F^q[F, F]$. Therefore u is a homomorphism on F^2 which vanishes on F^3, since F^3 is topologically generated by the products $x^q(x, y)$, $x \in F^2$, $y \in F$, on which u behaves multiplicatively.

The restriction v of u to R is an element of $H^1(R, k)^G$ since $(y, x) \in F^3$ for $y \in R$ and $x \in F$, and so

$$v(x^{-1}yx) = u(y(y, x)) = u(y) + u((y, x)) = v(y).$$

By the definition (1.6.6) of the transgression, we now obtain

$$tg(v) = [\partial u] = \chi_k \cup \chi_l,$$

and by definition of the matrix $B_i = (b^i_{kl})$ and of tr_{ρ_i}, we get

$$b^i_{kl} = tr_{\rho_i}(\chi_k \cup \chi_l) = v(\rho_i) = u_{kl}(\rho_i).$$

Since B_i is anti-symmetric, it therefore remains to show that

(2) $$u(\rho_i) = \begin{cases} -a^i_{kl} & \text{if} \quad k < l, \\ -\binom{q}{2} a^i_k & \text{if} \quad k = l. \end{cases}$$

We compute the values $u(x_j^m)$ and $u((x_\nu, x_\mu))$ by means of (1).

If $k \neq l$, we have $u(x_j^{m+1}) = u(x_j^m)$ and $u(x_j^{-1}) = 0$, which implies $u(x_j^m) = 0$ for any $m \in \mathbb{Z}$. If $k = l$, we have

$$u(x_j^{m+1}) = u(x_j^m) - \chi_k(x_j^m)\chi_l(x_j) = u(x_j^m) - m\delta_{kj}.$$

This implies

$$u(x_j^m) = -\binom{m}{2}\delta_{kj} \quad \text{for} \quad m = 1, 2, 3, \dots$$

Furthermore, noting that $u(x^{-1}) + u(x) + \chi_k(x)\chi_l(x) = 0$, we have for $\nu < \mu$

$$\begin{aligned} u((x_\nu, x_\mu)) &= u(x_\nu^{-1}) + u(x_\mu^{-1}x_\nu x_\mu) + \chi_k(x_\nu)\chi_l(x_\nu) \\ &= -\delta_{k\nu} + u(x_\mu^{-1}x_\nu x_\mu) + \delta_{k\nu}\delta_{l\nu} \\ &= u(x_\mu^{-1}) + u(x_\nu x_\mu) + \chi_k(x_\mu)\chi_l(x_\nu) + \chi_k(x_\mu)\chi_l(x_\mu) \\ &= -\delta_{k\nu}\delta_{l\mu} + u(x_\nu) + u(x_\mu) - \chi_k(x_\nu)\chi_l(x_\mu) + \delta_{k\mu}\delta_{l\nu} \\ &= \delta_{k\mu}\delta_{l\nu} - \delta_{k\nu}\delta_{l\mu}, \end{aligned}$$

i.e. for $\nu < \mu$ and $k < l$

$$u((x_\nu, x_\mu)) = \begin{cases} -1 & \text{if } k = \nu,\ l = \mu, \\ 0 & \text{otherwise.} \end{cases}$$

Now applying the homomorphism $u : F^2/F^3 \to k$ to

$$\rho_i \equiv \prod_{j=1}^{n} x_j^{qa_j^i} \prod_{1\le\nu<\mu\le n} (x_\nu, x_\mu)^{a^i_{\nu\mu}} \mod F^3$$

yields the desired result (2) and also the uniqueness of the a^i_{kl}. □

If $q \neq 0$ and $k = \mathbb{Z}/q\mathbb{Z}$, we define the **Bockstein homomorphism**

$$B : H^1(G,k) \longrightarrow H^2(G,k)$$

as the connecting homomorphism in the long exact cohomology sequence associated to

$$0 \longrightarrow \mathbb{Z}/q\mathbb{Z} \xrightarrow{q} \mathbb{Z}/q^2\mathbb{Z} \longrightarrow \mathbb{Z}/q\mathbb{Z} \longrightarrow 0\,.$$

We get a second interpretation of the exponents a^i_j in the expression for the relations of the group G in (3.9.13) (ii):

(3.9.14) Proposition. *With the notation of* (3.9.13) *and assuming that* $q \neq 0$, *we have*

$$tr_{\rho_i}(B(\chi_j)) = -a^i_j\,.$$

We leave the proof to the reader, since it follows the same lines as the proof of (3.9.13).

Before we determine explicitly the defining relation of a Demuškin group, we first sharpen the result (3.7.3) concerning the characterization of a Demuškin group by properties of its subgroups.

(3.9.15) Theorem. *Let G be a finitely generated one-relator pro-p-group of rank $d(G) > 1$. Then the following assertions are equivalent:*

(i) *G is a Demuškin group.*

(ii) *$cd_p G = 2$ and the dualizing module of G is isomorphic to $\mathbb{Q}_p/\mathbb{Z}_p$ as an abelian group.*

(iii) *$cd_p G = 2$ and $\dim_{\mathbb{F}_p} H^2(N, \mathbb{Z}/p\mathbb{Z}) = 1$ for every open subgroup N of G.*

(iv) *$cd_p G = 2$ and $d(N) - 2 = (G : N)(d(G) - 2)$ for every open subgroup N of G.*

(v) *$cd_p G = 2$ and $\dim_{\mathbb{F}_p} H^2(N, \mathbb{Z}/p\mathbb{Z}) = 1$ for every open subgroup N of G with $(G : N) = p$.*

(vi) *$d(N) - 2 = (G : N)(d(G) - 2)$ for every open subgroup N of G with $(G : N) = p$.*

Remark: Observe that in assertion (vi) we do not assume that $cd_p\,G = 2$. The equivalences between (i), (v) and (vi) were first proved by *I. V. Andožskii* and later independently by *J. Dummit* and *J. Labute* (without the assertion $cd_p\,G = 2$ in (v)) - see [3] and [44].

Proof: The equivalences between (i), (ii) and (iii) follow from (3.7.2) and (3.7.3). Using (3.3.13), we get the implications (iii) $\Rightarrow$ (iv) and (v) $\Rightarrow$ (vi). The assertions (v) and (vi) are trivial consequences of (iii) and (iv), respectively. So let us assume that (vi) holds, and try to prove (i).

Let

$$G^{ab} \cong \mathbb{Z}_p/q\mathbb{Z}_p \times \mathbb{Z}_p^{d(G)-1},$$

where q is equal 0 if G^{ab} is torsion-free. Furthermore, let

$$1 \longrightarrow R \longrightarrow F \longrightarrow G \longrightarrow 1$$

be a minimal presentation of G by a free pro-p-group F of rank $d = d(G)$. Suppose that the cup-product pairing

$$H^1(G,\mathbb{Z}/p\mathbb{Z}) \times H^1(G,\mathbb{Z}/p\mathbb{Z}) \to H^2(G,\mathbb{Z}/p\mathbb{Z}) \cong \mathbb{Z}/p\mathbb{Z}$$

is degenerate, and let χ_1 be an element of the radical of this pairing. Extend this element to a basis $\chi_1,\dots,\chi_d$ of $H^1(G,\mathbb{Z}/p\mathbb{Z}) \cong H^1(F,\mathbb{Z}/p\mathbb{Z})$ and let $\bar{x}_1,\dots,\bar{x}_d$ be the corresponding dual basis of F/F^2 (F^i denotes the p-central series of F). We define

$$E = (x_1^p, x_2, \dots, x_d)_F \lhd F$$

to be the subgroup of F generated by all elements between the brackets and their conjugates, where x_i is an arbitrary lifting of $\bar{x}_i$ to F. Then $R \subseteq E$ and if we put $N = E/R \subseteq G$ and $\overline{G} = G/N = F/E = \langle \bar{x}_1 \rangle \cong \mathbb{Z}/p\mathbb{Z}$, we get the commutative exact diagram

$$\begin{array}{ccccccccc}
 & & & & \overline{G} & = & \overline{G} & & \\
 & & & & \uparrow & & \uparrow & & \\
1 & \longrightarrow & R & \longrightarrow & F & \longrightarrow & G & \longrightarrow & 1 \\
 & & \| & & \uparrow & & \uparrow & & \\
1 & \longrightarrow & R & \longrightarrow & E & \longrightarrow & N & \longrightarrow & 1 .
\end{array}$$

We consider the $\overline{G}$-module $\bar{R} := RE^2/E^2$ which fits into the exact sequence

$$(*) \qquad 0 \longrightarrow \bar{R} \longrightarrow E/E^2 \longrightarrow N/N^2 \longrightarrow 0 .$$

Since $R/R^p[R,F] \cong \mathbb{Z}/p\mathbb{Z}$ surjects onto $\bar{R}_{\overline{G}}$, the $\overline{G}$-module $\bar{R}$ is generated by one element. Observe that

$$F^2 \subseteq E \quad \text{and} \quad F^3 \subseteq [[E,F],F] \cdot E^2 .$$

It follows from (3.9.13)(ii) and the fact that $\chi_1 \cup \chi_i = 0$ for all $i \geq 1$, that

$$\bar{R} \subseteq \begin{cases} \langle x_1^p \rangle [[E, \langle x_1 \rangle], \langle x_1 \rangle] E^2/E^2 & \text{if } q = p \neq 2\,, \\ [[E, \langle x_1 \rangle], \langle x_1 \rangle] E^2/E^2 & \text{otherwise}\,, \end{cases}$$

and so

$$\bar{R} \subseteq \begin{cases} \langle x_1^p \rangle E^2/E^2 \cdot (E/E^2)^{(\bar{x}_1 - 1)^2} & \text{if } q = p \neq 2\,, \\ (E/E^2)^{(\bar{x}_1 - 1)^2} & \text{otherwise}\,. \end{cases}$$

Thus in both cases

$$\bar{R}^{(\bar{x}_1 - 1)^{p-2}} = 0$$

(since $(E/E^2)^{(\bar{x}_1 - 1)^p} = 0$) and therefore

$$\dim_{\mathbb{F}_p} \bar{R} \leq p - 2\,.$$

From the exact sequence $(*)$, we obtain using (3.9.6)

$$\begin{aligned} \dim_{\mathbb{F}_p} N/N^2 &= \dim_{\mathbb{F}_p} E/E^2 - \dim_{\mathbb{F}_p} \bar{R} \\ &\geq p(d-1) + 1 - (p-2) \\ &> p(d-2) + 2\,. \end{aligned}$$

This contradicts the assumption (vi) and therefore the cup-product pairing has to be non-degenerate, i.e. G is a Demuškin group. □

Now we are going to determine a defining relation of a Demuškin group explicitly. Let $G = F/R$ be a one-relator pro-p-group of rank n where $R = (\rho)$. First we want to put the relation ρ in the shape

$$\rho \equiv x_1^q (x_1, x_2)(x_3, x_4) \cdots (x_{n-1}, x_n) \mod F^3$$

and for this, we have to find an appropriate basis of $H^1(F, k)$.

If $q \neq 0$, then there is a uniquely determined class $\sigma \bmod [F, F] \in F^{ab} = F/[F, F]$ such that

$$\rho \equiv \sigma^q \mod [F, F].$$

We deduce this from the fact that $F^{ab} \cong \mathbb{Z}_p^n$ and that the image $\bar{\rho}$ of ρ in F^{ab} topologically generates a closed subgroup $(\bar{\rho})$ such that $F^{ab}/(\bar{\rho}) \cong \mathbb{Z}/q\mathbb{Z} \times \mathbb{Z}_p^{n-1}$.

We need a result concerning symplectic bilinear forms in which we write $\chi \cup \chi'$ for $tr(\chi \cup \chi')$.

(3.9.16) Proposition. *Assume that $q \neq 2$ and that the bilinear pairing induced by the cup-product $H^1(G, k) \times H^1(G, k) \to H^2(G, k) \xrightarrow{tr} k$ is non-degenerate.*

Then n is even and there exists a k-basis $\chi_1, \dots, \chi_n$ of $H^1(G, k)$ such that

$$\chi_1 \cup \chi_2 = \chi_3 \cup \chi_4 = \ldots = \chi_{n-1} \cup \chi_n = 1$$

and $\chi_i \cup \chi_j = 0$ for all other $i < j$, and $\chi_i(\sigma) = \delta_{1i}$ when $q \neq 0$.

Proof: By (3.9.9), the cup-product $H^1(G) \times H^1(G) \xrightarrow{\cup} H^2(G) \cong \mathbb{F}_p$ is a non-degenerate bilinear form over the field $\mathbb{F}_p$. If $p \neq 2$, it is alternating, hence $n = \dim_{\mathbb{F}_p} H^1(G)$ is even. When $q = 2^f$, $f > 1$, n is also even, since the cup-product on $H^1(G, \mathbb{Z}/2^f\mathbb{Z})$ is still alternating and non-degenerate, which implies that $H^1(G, \mathbb{Z}/2^f\mathbb{Z})$ decomposes into a direct sum of hyperbolic planes. For this, we refer to [106], chap.I, §4.

We start with any k-basis $\chi_1, \dots, \chi_n$ of $H^1(G, k)$ such that $\chi_i(\sigma) = \delta_{1i}$ when $q \neq 0$. To find such a basis when $q \neq 0$, one only has to extend the image of σ in $F/F^q[F, F]$ to a basis of $F/F^q[F, F]$ and then take the dual basis. The nondegeneracy of the cup-product means that the matrix $B = (\chi_k \cup \chi_l)$ is invertible over k. Therefore one of the elements $\chi_1 \cup \chi_i$ with $i > 1$ must be a unit of k. After a permutation, we may assume that $\chi_1 \cup \chi_2$ is a unit, and after multiplying χ_2 by a unit, we may even assume $\chi_1 \cup \chi_2 = 1$. If $\chi_1 \cup \chi_i = a_i \neq 0$ for some $i > 2$, replace χ_i by $\chi_i - a_i\chi_2$. Since the condition $\chi_i(\sigma) = \delta_{1i}$ when $q \neq 0$ is not altered, we may assume $\chi_1 \cup \chi_i = 0$ for $i > 2$.

Now if V is the subspace spanned by $\chi_3, \dots, \chi_n$, our cup-product restricted to $V \times V$ is non-degenerate and alternating. Hence we may inductively choose $\chi_3, \dots, \chi_n$ such that

$$\chi_3 \cup \chi_4 = \chi_5 \cup \chi_6 = \dots = \chi_{n-1} \cup \chi_n = 1$$

and $\chi_i \cup \chi_j = 0$ for all other $2 < i < j$. When $q \neq 0$ the condition $\chi_i(\sigma) = \delta_{1i}$ is still satisfied, $\chi_1 \cup \chi_2 = 1$ and $\chi_1 \cup \chi_i = 0$ for $i > 2$. If we replace χ_2 by

$$\chi_2 + a_3\chi_3 + \dots + a_n\chi_n$$

with $a_{2i} = \chi_2 \cup \chi_{2i-1}$ and $a_{2i-1} = -\chi_2 \cup \chi_{2i}$, we have, in addition, $\chi_2 \cup \chi_i = 0$ for $i > 2$. This proves the proposition. □

The following corollary is a first approximation to our main theorem (3.9.11) on Demuškin groups.

(3.9.17) Corollary. *Let $G = F/(\rho)$ be a finitely generated one-relator pro-p-group with the invariants (n, q). Let $q \neq 2$.*

Then G is a Demuškin group if and only if there exists a basis $x_1, \dots, x_n$ of F such that

$$\rho \equiv x_1^q(x_1, x_2)(x_3, x_4)\cdots(x_{n-1}, x_n) \mod F^3.$$

Proof: For any minimal generator system $x_1, \dots, x_n$ of F with corresponding k-basis $\chi_1, \dots, \chi_n$ of $H^1(G, k)$, we have by (3.9.13)(i)

$$\rho \equiv \prod_{i=1}^{n} x_i^{qa_i} \cdot \prod_{1 \le k < l \le n} (x_k, x_l)^{a_{kl}} \mod F^3$$

with $a_{kl} \in k$ uniquely determined (and $a_i \in k$ also, when $q \neq 0$). Furthermore, by (3.9.13) (ii), the matrix $B = (b_{kl})$ of the cup-product pairing $\chi \cup \chi'$ is given by

$$b_{kl} = \chi_k \cup \chi_l = \begin{cases} -a_{kl} & \text{if} \quad k < l, \\ a_{lk} & \text{if} \quad k > l, \\ -\binom{q}{2} a_k & \text{if} \quad k = l. \end{cases}$$

Assume that

$$\rho \equiv x_1^q (x_1, x_2)(x_3, x_4) \cdots (x_{n-1}, x_n) \mod F^3.$$

Then $b_{12} = b_{34} = \cdots = b_{n-1,n} = -1$ and $b_{kl} = 0$ for all other $k < l$, and $b_{kk} = -\binom{q}{2}$, i.e. B is the matrix

$$B = \begin{pmatrix} -\binom{q}{2} & 1 & & & \\ -1 & -\binom{q}{2} & & 0 & \\ & & \ddots & & \\ & 0 & & -\binom{q}{2} & 1 \\ & & & -1 & -\binom{q}{2} \end{pmatrix}.$$

It has determinant $\det(B) = (1 + \binom{q}{2}^2)^{n/2} \mod q\mathbb{Z}_p$, which is 1 if $p \neq 2$, or if $q = 2^f \neq 2$. Hence the cup-product is non-degenerate and G is thus a Demuškin group.

Conversely, if G is a Demuškin group, then the cup-product $\chi \cup \chi'$ is non-degenerate and we may choose a k-basis $\chi_1, \ldots, \chi_n$ of $H^1(G, k)$ as in proposition (3.9.16). Let $\xi_1, \ldots, \xi_n$ be the dual basis of $F/F^q[F, F]$ and $x_1, \ldots, x_n$ a lift to F. Then $x_1, \ldots, x_n$ is a minimal generator system of F with corresponding basis $\chi_1, \ldots, \chi_n$ of $H^1(G, k)$. When $q \neq 0$, let

$$\sigma \equiv \prod_{i=1}^{n} x_i^{\hat{a}_i} \mod [F, F], \quad \hat{a}_i \in \mathbb{Z}_p.$$

In the representation

$$\rho \equiv \prod_{i=1}^{n} x_i^{q a_i} \cdot \prod_{1 \le k < l \le n} (x_k, x_l)^{a_{kl}} \mod F^3, \; a_i, a_{kl} \in k,$$

we have

$$a_i \equiv \hat{a}_i \mod q\mathbb{Z}_p = \chi_i(\sigma) = \delta_{1i}$$

when $q \neq 0$, because of the uniqueness of the a_i, and

$$a_{12} = a_{34} = \cdots = a_{n-1,n} = 1$$

and $a_{kl} = 0$ for all other $k < l$. We therefore get

$$\rho \equiv x_1^q (x_1, x_2)(x_3, x_4) \cdots (x_{n-1}, x_n) \mod F^3. \qquad \square$$

Our last task is to make an equation from the congruence

$$\rho \equiv x_1^q(x_1, x_2)(x_3, x_4) \cdots (x_{n-1}, x_n) \mod F^3 .$$

This is achieved by a successive approximation process recalling that by (3.8.2), the q-central series (F^i) of the finitely generated free pro-p-group F, i.e. $F^1 = F$, $F^{i+1} = (F^i)^q[F^i, F]$, where q is any nontrivial p-power, form a fundamental system of open neighbourhoods of 1.

From now on we assume that the rank of F, i.e. the dimension $n = \dim_{\mathbb{F}_p} H^1(F)$, is *even*. For every n-tuple $y = (y_1, \dots, y_n) \in F^n$, we set

$$r(y) = y_1^q(y_1, y_2)(y_3, y_4) \cdots (y_{n-1}, y_n).$$

The following lemma is the most subtle part of the proof of theorem (3.9.11).

(3.9.18) Lemma. *Let $q \neq 2$ and let $\rho \in F^2$. Then for every $j \geq 3$ there exists a basis $x = (x_1, \dots, x_n)$ of F such that*

$$\rho \equiv r(x) \mod F^j,$$

provided this is true for $j = 3$.

Proof: Let $x = (x_1, \dots, x_n)$ be any minimal generator system of F. Let $t_1, \dots, t_n \in F^{j-1}$ and set $y_i = x_i t_i^{-1}$. Then $y = (y_1, \dots, y_n)$ is again a minimal generator system of F since $y_i \equiv x_i \mod F^q[F, F]$. We may write

$$r(x) = r(y) d_{j-1}(t_1, \dots, t_n),$$

where $d_{j-1}(t_1, \dots, t_n)$ is a uniquely determined element of F^j. A simple calculation using proposition (3.8.5) shows that if τ_i is the image of t_i in $\mathrm{gr}_{j-1}(F)$, then the image of $d_{j-1}(t_1, \dots, t_n)$ in $\mathrm{gr}_j(F)$ is

$$(*) \qquad \pi \cdot \tau_1 + \binom{q}{2}[\tau_1, \xi_1] + [\tau_1, \xi_2] + [\xi_1, \tau_2] + \cdots + [\tau_{n-1}, \xi_n] + [\xi_{n-1}, \tau_n],$$

where ξ_i is the image of x_i in $\mathrm{gr}_1(F)$. Hence d_{j-1} induces a k-linear homomorphism

$$\delta_{j-1} : \mathrm{gr}_{j-1}(F)^n \longrightarrow \mathrm{gr}_j(F), \quad j \geq 3.$$

Claim: The map δ_{j-1} is surjective.

Let $H_j = \mathrm{im}(\delta_{j-1})$. In order to prove that $H_j = \mathrm{gr}_j(F)$, it suffices to show $\pi \cdot \tau \in H_j$ for every $\tau \in \mathrm{gr}_{j-1}(F)$. In fact, $\mathrm{gr}_j(F)$ is generated by the elements $\pi \cdot \tau$ and $[\tau, \xi_i]$ with $\tau \in \mathrm{gr}_{j-1}(F)$; the explicit expression $(*)$ for $\delta_{j-1}(\tau_1, \dots, \tau_n)$ tells us that we have $[\tau, \xi_i] \in \mathrm{im}(\delta_{j-1})$ for $i \geq 3$ and $\pi \cdot \tau + [\tau, \xi_2], [\tau, \xi_1] \in \mathrm{im}(\delta_{j-1})$.

We now proceed by induction. Assume that we have shown that $H_j = \mathrm{im}(\delta_{j-1})$ for some $j \geq 3$. If $\tau \in \mathrm{gr}_{j-1}(F)$, then

$$\tau = \delta_{j-1}(\tau_1, \ldots, \tau_n)$$

with $\tau_1, \ldots, \tau_n \in \mathrm{gr}_{j-1}(F)$. But then, using (3.8.5),

$$\pi \cdot \tau = \delta_{j-1}(\pi \cdot \tau_1, \ldots, \pi \cdot \tau_n),$$

which implies $\pi \cdot \tau \in H_{j+1}$ for $\tau \in H_j$.

Thus we are reduced to proving the lemma for $j = 3$, that is, $\pi \cdot \tau \in H_3$ for $\tau \in H_2$. Moreover, it suffices to take τ in the form $\pi \cdot \xi_i, [\xi_i, \xi_j]$, since these elements generate $\mathrm{gr}_2(F)$ by (3.9.13) (i). The ring k is a local ring with maximal ideal $\mathfrak{m} = pk$. Hence by Nakayama's lemma, it suffices to prove $\pi \cdot \mathrm{gr}_2(F) \subseteq H_3 + \mathfrak{m}\, \mathrm{gr}_3(F)$, since then we would have $\mathrm{gr}_3(F) = H_3 + \mathfrak{m}\, \mathrm{gr}_3(F)$. Set $M = \mathfrak{m}\, \mathrm{gr}_3(F)$. Then by proposition (3.8.5), we have

$$\pi \cdot [\xi_i, \xi_j] = [\pi \cdot \xi_i, \xi_j] + m = [\xi_i, \pi \cdot \xi_j] + m',$$

where $m, m' \in M$. Therefore, since $[\tau, \xi_i] \in \mathrm{im}(\delta_2)$ if $i \neq 2$, we have $\pi \cdot [\xi_i, \xi_j] \in H_3 + M$ for any i, j. Moreover, as $\pi \cdot \tau + [\tau, \xi_2] \in H_3$ for any $\tau \in \mathrm{gr}_2(F)$, we have $\pi^2 \cdot \xi_i \in H_3 + M$ for any i. This proves the surjectivity of δ_{j-1}.

Now let $j \geq 3$ and assume that there exists a minimal generator system $x = (x_1, \ldots, x_n)$ of F such that $\rho \equiv r(x) \bmod F^j$, i.e. $\rho = r(x)e_j$, $e_j \in F^j$. Let ε_j be the image of e_j in $\mathrm{gr}_j(F)$. Since δ_{j-1} is surjective, there exists $t_1, \ldots, t_n \in F^{j-1}$ such that $\delta_{j-1}(\tau_1, \ldots, \tau_n) = -\varepsilon_j$, i.e. $d_{j-1}(t_1, \ldots, t_j) = e_j^{-1} \bmod F^{j+1}$. Put $y_i = x_i t_i^{-1}$. Then $y = (y_1, \ldots, y_n)$ is a minimal generator system of F such that

$$\rho = r(x)e_j = r(y)d_{j-1}(t_1, \ldots, t_n)e_j \equiv r(y) \mod F^{j+1}. \qquad \square$$

Proof of theorem (3.9.11): The proof is an immediate consequence of the corollary (3.9.17) and the above lemma. Let $G = F/(\rho)$ be a one-relator pro-p-group with invariants (n, q). If there exists a basis $x = (x_1, \ldots, x_n)$ of F such that $\rho = r(x)$, then a fortiori $\rho \equiv r(x) \bmod F^3$, and from (3.9.17) it follows that G is a Demuškin group.

Conversely, let G be a Demuškin group. Then by (3.9.17) and (3.9.18), there exists for every $j \geq 3$ a minimal generator system $x^{(j)} = (x_1^{(j)}, \ldots, x_n^{(j)})$ such that

$$\rho \equiv r(x^{(j)}) \mod F^j.$$

When $q = 0$ this remains valid if we replace the central series F^j by the p-central series, i.e. it is always true that

$$(*) \qquad \rho \equiv r(x^{(j)}) \mod U(j),$$

where $(U(j))$ is a series of open normal subgroups which form a basis of neighbourhoods of 1 by (3.8.2). If $X^{(j)}$ denotes the finite nonempty set of minimal generator systems $x^{(j)} = (x_1^{(j)}, \dots, x_n^{(j)})$ of $F/U(j)$ satisfying $(*)$, then $\varprojlim X^{(j)}$ is nonempty and any $x = (x_1, \dots, x_n)$ in this projective limit is a basis of F such that $\rho = r(x)$. This concludes the proof of theorem (3.9.11). $\square$

Remark: We have proved a sharper result than the assertion of theorem (3.9.11). Namely, we have shown that for *any* defining relation $\rho \in F$ of G, i.e. $G = F/(\rho)$, there exists a basis $x_1, \dots, x_n$ of F, such that

$$\rho = x_1^q (x_1, x_2)(x_3, x_4) \cdots (x_{n-1}, x_n).$$

Along the same lines, but with more complicated modifications in the individual steps, one can also settle the case $q = 2$. This was originally proved by *J.-P. Serre* in the case when n is odd, and later *J. Labute* obtained the general case. For this we refer the reader to *Labute*'s article [117]. The result is the following

(3.9.19) Theorem. *Let $G = F/(\rho)$ be a Demuškin group of rank n such that $q = 2$.*

(i) *If n is odd, there exists a basis $x_1, \dots, x_n$ of F such that*

$$\rho = x_1^2 x_2^{2^f} (x_2, x_3)(x_4, x_5) \cdots (x_{n-1}, x_n)$$

for some $f = 2, 3, \dots, \infty$ ($f = \infty$ means $2^f = 0$).

(ii) *If n is even, there exists a basis $x_1, \dots, x_n$ of F such that*

$$\rho = x_1^{2+\alpha} (x_1, x_2) x_3^{2^f} (x_3, x_4)(x_5, x_6) \cdots (x_{n-1}, x_n)$$

for some $f - 2, 3, \dots, \infty$ and $\alpha \subset 4\mathbb{Z}_2$.

For further results on one-relator pro-p-groups, see [110], [118], [255].

In arithmetic applications one sometimes meets the situation that one is given a canonical system of generators of a pro-p-group G. Even if the group G is free, we are interested in a presentation of G in terms of generators and relations, which is not necessarily minimal, but uses the given system of generators (or at least generators which are closely related to the given ones). The following theorem provides a typical example of such a situation. Its proof, for which we refer the reader to [251], Lemma 2.4, uses an approximation process similar to that in the proof of theorem (3.9.11).

(3.9.20) Theorem. *Let*

$$1 \longrightarrow H \longrightarrow G \longrightarrow D \longrightarrow 1$$

be an exact sequence of pro-p-groups. Assume that G is free and that D is a Demuškin group of rank n with torsion-free abelianization. Let D be generated by $\bar{x}_1, \ldots, \bar{x}_n$ with the one defining relation

$$(\bar{x}_1, \bar{x}_2) \cdots (\bar{x}_{n-1}, \bar{x}_n) = 1.$$

Suppose that the free $\mathbb{Z}_p$-module $H/[H,G]$ has the basis

$$\{\tilde{y}_j \bmod [H,G] \,|\, j = 1, \ldots, s\}$$

and that

$$\prod_{j=1}^{s} \tilde{y}_j \in [G,G].$$

Then there exist generators $x_1, \ldots, x_n, y_1, \ldots, y_s$ of G with

(i) $x_i \bmod H = \bar{x}_i$, $i = 1, \ldots, n$,

(ii) $y_j = (\tilde{y}_j)^{\alpha\tau_j}$ *with* $\alpha \in \mathbb{Z}_p^\times$, $\tau_j \in G$, $j = 1, \ldots, s$,

subject to the one relation

$$(x_1, x_2) \cdots (x_{n-1}, x_n) \prod_{j=1}^{s} y_j = 1.$$

Exercise: Let G be a pro-p-group. Consider the filtration $G = G_1 \supseteq G_2 \supseteq \ldots$ defined by $G_n = (G_{n-1})^*$. Show that $\bigcap_{n \geq 1} G_n = \{1\}$ and that the G_n are open if and only if $d(G) < \infty$.

Chapter IV
Free Products of Profinite Groups

§1. Free Products

As in III §5, we let $\mathfrak{c}$ be a full class of finite groups and we consider the category of pro-$\mathfrak{c}$-groups. A family

$$\kappa_i : G_i \longrightarrow G, \; i \in I,$$

of homomorphisms of pro-$\mathfrak{c}$-groups is called **convergent** (to 1) if every open subgroup U of G contains the images $\kappa_i(G_i)$ for almost all i, i.e. all but a finite number. The free products of pro-$\mathfrak{c}$-groups are defined by the following universal property.

(4.1.1) Definition. *The free pro-$\mathfrak{c}$-product of a family G_i, $i \in I$, of pro-$\mathfrak{c}$-groups is a pro-$\mathfrak{c}$-group G together with a convergent family of homomorphisms*

$$\kappa_i : G_i \longrightarrow G, \quad i \in I,$$

such that for every other convergent family of homomorphisms $\kappa_i' : G_i \to G'$ there is a unique homomorphism of pro-$\mathfrak{c}$-groups $\kappa' : G \to G'$ such that the diagrams

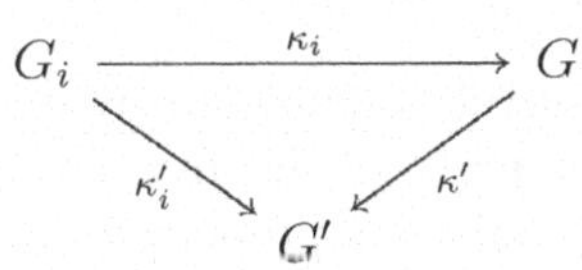

are commutative. The group G is then denoted by

$$G = \mathop{\ast}_{i \in I} G_i.$$

The free pro-$\mathfrak{c}$-product is clearly unique up to isomorphism, if it exists. Its existence is based on the usual free product $\mathfrak{G} = \overset{discr}{\underset{i \in I}{\ast}} G_i$ of the G_i in the category of groups (see [73]), which contains the G_i as subgroups. Let $\mathfrak{B}_{\mathfrak{G}}$ denote the family of all normal subgroups $\mathfrak{N}$ of $\mathfrak{G}$ such that

(1) $\mathfrak{G}/\mathfrak{N} \in \mathfrak{c}$,

(2) $\mathfrak{N} \supseteq G_i$ for almost all $i \in I$,

(3) $\mathfrak{N} \cap G_i$ is an open subgroup of the pro-$\mathfrak{c}$-group G_i.

J. Neukirch et al., *Cohomology of Number Fields*,
Grundlehren der mathematischen Wissenschaften 323,
DOI 10.1007/978-3-540-37889-1_4, © Springer-Verlag Berlin Heidelberg 2008

Taking $\mathfrak{B}_{\mathfrak{G}}$ as a basis of open neighbourhoods of the identity of $\mathfrak{G}$ defines a group topology $T_{\mathfrak{G}}$ on $\mathfrak{G}$ *), which induces the profinite topology on G_i. Indeed, if $\mathfrak{N} \in \mathfrak{B}_{\mathfrak{G}}$, then $\mathfrak{N} \cap G_i$ is open in G_i, and, conversely, if U is any open normal subgroup of G_i, we obtain a homomorphism $\mathfrak{G} \to G_i/U$ whose kernel $\mathfrak{N}$ is in $\mathfrak{B}_{\mathfrak{G}}$ and has the property that $\mathfrak{N} \cap G_i = U$. This homomorphism is obtained when we map G_j onto $1 \in G_i/U$ for $j \neq i$ and G_i onto G_i/U by the canonical projection. Since G_i is compact, it is a closed subgroup of $\mathfrak{G}$ in the topology $T_{\mathfrak{G}}$. We now take the completion

$$G = \varprojlim_{\mathfrak{N} \in \mathfrak{B}_{\mathfrak{G}}} \mathfrak{G}/\mathfrak{N},$$

which comes equipped with a canonical continuous homomorphism $\lambda \colon \mathfrak{G} \to G$, and with the family of composite maps

$$\kappa_i : G_i \hookrightarrow \mathfrak{G} \xrightarrow{\lambda} G.$$

These satisfy the universal condition of a free pro-$\mathfrak{c}$-product. In fact, if we have any convergent family of homomorphisms $\kappa_i' : G_i \to G'$ into a pro-$\mathfrak{c}$-group G', then the diagram

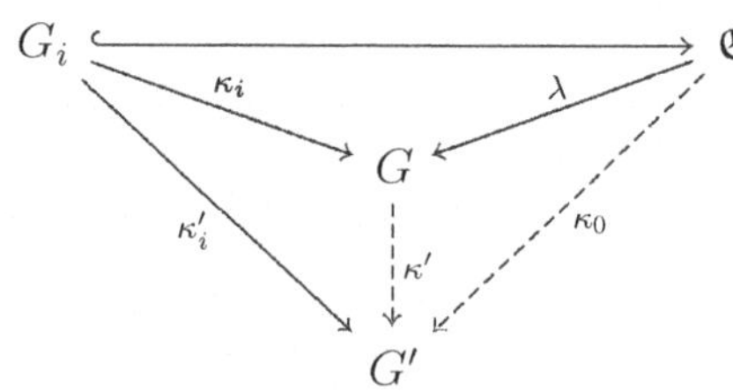

of solid arrows is commutatively completed by the dotted arrows; κ_0 exists because of the universal property of the free product $\mathfrak{G}$ in the category of groups, and κ' exists by the universal property of the completion: because of the conditions (1), (2), (3) for the normal subgroups $\mathfrak{N} \in \mathfrak{B}_{\mathfrak{G}}$, κ_0 is continuous and thus defines a homomorphism $\kappa' : G \to G'$ of pro-$\mathfrak{c}$-groups.

Example: Let $\hat{\mathbb{Z}}(\mathfrak{c})$ be the free pro-$\mathfrak{c}$-group of rank 1 (see III §5). The free pro-$\mathfrak{c}$-group F over a set X is the free pro-$\mathfrak{c}$-product

$$F = \mathop{\ast}_{x \in X} \hat{\mathbb{Z}}(\mathfrak{c}).$$

This follows at once from the universal properties, as do the following remarks and propositions.

---

*) The topology $T_{\mathfrak{G}}$ is Hausdorff, by remark 1 at the end of §3. But we don't need this here.

If J is a subset of I, then we have an exact sequence which splits canonically

$$(*) \qquad 1 \longrightarrow N \longrightarrow \mathop{\ast}_{i\in I} G_i \underset{s}{\overset{\pi}{\rightleftarrows}} \mathop{\ast}_{i\in J} G_i \longrightarrow 1.$$

The homomorphism π is obtained by choosing the canonical maps $\kappa_i : G_i \to \mathop{\ast}_{i\in J} G_i$ for $i \in J$ and the trivial homomorphism $G_i \to \{1\} \subseteq \mathop{\ast}_{i\in J} G_i$ for $i \in I \smallsetminus J$. The homomorphism s is obtained by the canonical maps $\kappa_j : G_j \to \mathop{\ast}_{i\in I} G_i$ for $j \in J$. We have $\pi \circ s = id$ and consider $\mathop{\ast}_{i\in J} G_i$ as embedded in $\mathop{\ast}_{i\in I} G_i$,

$$\mathop{\ast}_{i\in J} G_i \subseteq \mathop{\ast}_{i\in I} G_i.$$

In particular, we view the G_i as subgroups of $G = \mathop{\ast}_{i\in I} G_i$. They generate G topologically, and the kernel N in $(*)$ is the smallest closed normal subgroup of G containing the G_i, $i \in I \smallsetminus J$.

For an arbitrary index set I, we consider the family of finite subsets of I partially ordered by inclusion.

(4.1.2) Proposition. *The free pro-$\mathfrak{c}$-product $G = \mathop{\ast}_{i\in I} G_i$ is the projective limit*

$$G = \varprojlim_{S} \mathop{\ast}_{i\in S} G_i\,,$$

where S runs through the finite subsets of I.

Proof: It is sufficient to show that the projective limit on the right-hand side satisfies the universal property of the free product with respect to homomorphisms to finite groups in $\mathfrak{c}$. But for a finite group H we have

$$\mathrm{Hom}(\varprojlim_{S\subseteq I} \mathop{\ast}_{i\in S} G_i, H) = \varinjlim_{S\subseteq I} \mathrm{Hom}(\mathop{\ast}_{i\in S} G_i, H) = \varinjlim_{S\subseteq I} \bigoplus_{i\in S} \mathrm{Hom}(G_i, H).$$

This concludes the proof, because for every convergent family of homomorphisms $(\kappa_i : G_i \to H)_{i\in I}$, there exists a finite subset $S \subseteq I$ such that $\kappa_i = 0$ for all $i \notin S$. □

If $\mathfrak{c}' \subseteq \mathfrak{c}$ is a full subclass of $\mathfrak{c}$, then to each pro-$\mathfrak{c}$-group G we may associate the maximal pro-$\mathfrak{c}'$-factor group $G(\mathfrak{c}')$. This is the projective limit

$$G(\mathfrak{c}') = \varprojlim_{U} G/U$$

over all open normal subgroups U of G such that $G/U \in \mathfrak{c}'$. It is characterized by the universal property that every homomorphism of G into a pro-$\mathfrak{c}'$-group factors through $G(\mathfrak{c}')$. From this, we obtain the

(4.1.3) Proposition. *Denoting by $\ast'$ the free pro-$\mathfrak{c}'$-product, we have*

$$(\underset{i\in I}{\ast} G_i)(\mathfrak{c}') = \underset{i\in I}{\ast}{}' G_i(\mathfrak{c}').$$

We now turn to the cohomology of free pro-$\mathfrak{c}$-products. It behaves in a particularly nice way. Let $G = \underset{i\in I}{\ast} G_i$ and let A be a G-module. Then the restriction maps $res_i : H^n(G,A) \to H^n(G_i,A)$ define a homomorphism into the direct sum

$$H^n(G,A) \longrightarrow \bigoplus_{i\in I} H^n(G_i,A).$$

In fact, if S runs through the finite subsets of I, then by (4.1.2) and (1.5.1) we obtain

$$\begin{aligned} H^n(G,A) &= H^n(\varprojlim_S \underset{i\in S}{\ast} G_i, A) = \varinjlim_S H^n(\underset{i\in S}{\ast} G_i, A) \\ &\to \varinjlim_S \bigoplus_{i\in S} H^n(G_i,A) = \bigoplus_{i\in I} H^n(G_i,A). \end{aligned}$$

We will now prove the following remarkable isomorphism theorem.

(4.1.4) Theorem. *Let $G = \underset{i\in I}{\ast} G_i$. Then for every torsion G-module A we have a canonical exact sequence*

$$0 \to A^G(\mathfrak{c}) \to A(\mathfrak{c}) \to \bigoplus_{i\in I} (A/A^{G_i})(\mathfrak{c}) \xrightarrow{\delta} H^1(G,A) \xrightarrow{res} \bigoplus_{i\in I} H^1(G_i,A) \to 0,$$

and for all $n \geq 2$

$$H^n(G,A) \cong \bigoplus_{i\in I} H^n(G_i,A).$$

Proof: By a direct limit argument we may assume that A is finite. Moreover, we may assume $A \in \mathfrak{c}$. Namely, if we decompose A into its p-primary components $A(p)$, then the cohomology groups decompose into direct summands. These are zero if p does not divide the order of G (i.e. the order of a finite quotient G/U). Otherwise one sees at once that $A \in \mathfrak{c}$. Moreover we have $A/A^{G_i} = 0$ for almost all $i \in I$, since there is an open subgroup U of G which acts trivially on A.

The exact sequence of the theorem will follow by the snake lemma from the exact commutative diagram

$$\begin{array}{ccccccccc} 0 & \longrightarrow & A/A^G & \xrightarrow{\partial} & \mathscr{Z}^1(G,A) & \longrightarrow & H^1(G,A) & \longrightarrow & 0 \\ & & \downarrow & & \downarrow{\scriptstyle res} & & \downarrow{\scriptstyle res} & & \\ 0 & \longrightarrow & \bigoplus\limits_{i\in I} A/A^{G_i} & \xrightarrow{\partial} & \bigoplus\limits_{i\in I} \mathscr{Z}^1(G_i,A) & \longrightarrow & \bigoplus\limits_{i\in I} H^1(G_i,A) & \longrightarrow & 0, \end{array}$$

once we have shown that the restriction map

$$(*) \qquad res : \mathscr{Z}^1(G, A) \longrightarrow \bigoplus_{i \in I} \mathscr{Z}^1(G_i, A)$$

on inhomogeneous 1-cocycles is bijective (its image is contained in the direct sum by the same argument as for $H^n(G, A)$). To this end, consider the exact commutative diagram

$$(**) \qquad \begin{array}{ccccccccc} 1 & \longrightarrow & A & \longrightarrow & \hat{G}_i & \overset{\pi_i}{\underset{s_i}{\rightleftarrows}} & G_i & \longrightarrow & 1 \\ & & \| & & \downarrow & & \downarrow & & \\ 1 & \longrightarrow & A & \longrightarrow & \hat{G} & \overset{\pi}{\underset{s}{\rightleftarrows}} & G & \longrightarrow & 1, \end{array}$$

where $\hat{G}$ is the semi-direct product of A by G with the canonical homomorphic section s, $\hat{G}_i = \pi^{-1}(G_i)$ and $s_i = s|_{G_i}$. The map $\chi \mapsto s' = \chi s$ gives a 1-1 correspondence between the 1-cocycles $\chi \in \mathscr{Z}^1(G, A)$ and the homomorphic sections $s' : G \to \hat{G}$ of π. If $res_i \chi = 1$ for all $i \in I$, then $s'|_{G_i}$ coincides with s_i for all $i \in I$, hence $s' = s$ by the universal property of the pro-$\mathfrak{c}$-product, and so $\chi = 1$. This proves the injectivity of $(*)$.

Conversely let $(\chi_i)_{i \in I}$ be a family of cocycles $\chi_i \in \mathscr{Z}^1(G_i, A)$ such that $\chi_i = 1$ for almost all $i \in I$. Then we obtain new sections $s'_i = \chi_i s_i : G_i \to \hat{G}_i$, hence a family of homomorphisms $s'_i : G_i \to \hat{G}$ which is convergent since $s'_i = s_i$ for almost all $i \in I$. This family defines a homomorphism $s' : G \to \hat{G}$ such that $s'|_{G_i} = s'_i$ by the universal property of the free pro-$\mathfrak{c}$-product. From $\pi_i \circ s'_i = id$ it follows that $\pi \circ s' = id$, i.e. s' is a new section of π and defines a 1-cocycle $\chi(\sigma) = s'(\sigma)s(\sigma)^{-1}$ such that $\chi|_{G_i} = \chi_i$ for all $i \in I$. This proves the surjectivity of $(*)$.

It remains to prove the second assertion of the theorem. By dimension shifting, we obtain the surjectivity of the map

$$res : H^n(G, A) \longrightarrow \bigoplus_{i \in I} H^n(G_i, A)$$

for $n \geq 1$. It suffices to prove the injectivity for $n = 2$, again by dimension shifting. And again we may assume that A is finite and in $\mathfrak{c}$, and moreover that I is finite, by (4.1.2). Let $x \in H^2(G, A)$ be such that $x_i = res_i x = 0$ for all $i \in I$. By (1.2.4), x and the x_i define group extensions, and we have a commutative diagram $(**)$ with homomorphic sections $s_i : G_i \to \hat{G}_i \subseteq \hat{G}$, since the upper group extensions split. By the universal property of the free pro-$\mathfrak{c}$-product $G = \ast_{i \in I} G_i$, the s_i define a homomorphic section $s : G \to \hat{G}$ such that $s|_{G_i} = s_i$. Hence the lower group extension splits, i.e. $x = 0$. This shows the injectivity. □

For pro-p-groups G we have a converse result to the above theorem. We set $H^n(G) = H^n(G, \mathbb{Z}/p\mathbb{Z})$.

(4.1.5) Theorem. *Let G be a pro-p-group and let G_i, $i \in I$, be a convergent subgroup family of G. Then the following conditions are equivalent.*

(i) $G = \mathop{\ast}\limits_{i\in I} G_i$ *(free pro-p-product).*

(ii) *The homomorphism*

$$res : H^n(G) \longrightarrow \bigoplus_{i\in I} H^n(G_i)$$

is bijective for $n = 1$ and injective for $n = 2$.

Proof: The implication (i) $\Rightarrow$ (ii) follows from (4.1.4) since G acts trivially on $\mathbb{Z}/p\mathbb{Z}$.

Conversely assume that (ii) holds. The inclusions $G_i \to G$ are by assumption a convergent family of homomorphisms, and thus define a homomorphism of the free pro-p-product

$$\hat{G} = \mathop{\ast}\limits_{i\in I} G_i \xrightarrow{\pi} G.$$

By (4.1.4) we have

$$H^j(\hat{G}) = \bigoplus_{i\in I} H^j(G_i)$$

for $j = 1, 2$, and so, by assumption,

$$H^j(G) \longrightarrow H^j(\hat{G})$$

is an isomorphism for $j = 1$ and injective for $j = 2$. Now the result follows from (1.6.15). □

The concept of free pro-$\mathfrak{c}$-product has generalizations in many directions (see [154]). One can define *restricted* free pro-$\mathfrak{c}$-products $\mathop{\ast}\limits_{i\in I}(G_i, H_i)$ with respect to subgroups $H_i \subseteq G_i$ and one obtains an *idèle* version. Further, one can consider free pro-$\mathfrak{c}$-groups *with amalgamated subgroup* $G_1 \mathop{\ast}\limits_{H} G_2$. We mention some facts about the last concept in the exercises.

However, one important generalization will be introduced in §3. While the groups occurring as factors in the free product have been independent of each other so far, we will introduce there the free pro-$\mathfrak{c}$-product over a family of pro-$\mathfrak{c}$-groups which vary continuously over a topological base space.

Exercise 1. Let $i_1 : H \hookrightarrow G_1, i_2 : H \hookrightarrow G_2$ be two injective homomorphisms of pro-$\mathfrak{c}$-groups. The **push-out** of i_1 and i_2 is a commutative diagram of pro-$\mathfrak{c}$-groups

$$\begin{array}{ccc} H & \xrightarrow{i_2} & G_2 \\ {\scriptstyle i_1}\downarrow & & \downarrow{\scriptstyle \kappa_2} \\ G_1 & \xrightarrow[\kappa_1]{} & G \end{array}$$

with the universal property that any two homomorphisms $\kappa_1' : G_1 \to G'$, $\kappa_2' : G_2 \to G'$ into a pro-$\mathfrak{c}$-group G' such that $\kappa_1' \circ i_1 = \kappa_2' \circ i_2$ determine a unique homomorphism $f : G \to G'$ such that $\kappa_1 \circ f = \kappa_1'$ and $\kappa_2 \circ f = \kappa_2'$. One speaks of an **amalgamated free pro-$\mathfrak{c}$-product** along H if κ_1 and κ_2 are injections. One writes in this case

$$G = G_1 \underset{H}{*} G_2.$$

Show that the push-out always exists. The amalgamated product does not always exist (see [179] for an example).

Exercise 2. If $\mathfrak{c}$ is the class of all finite groups, the amalgamated free profinite product exists in the following cases:

1) H is central in either G_1 or G_2;
2) H is normal in both G_1 and G_2 and is topologically finitely generated;
3) H is finite.

The amalgamated free pro-$\mathfrak{c}$-product $G_1 \underset{H}{*} G_2$ of two pro-p-groups G_1, G_2 along a common procyclic subgroup H exists (see [179]).

For the next exercises (see [178]), assume that the amalgamated free pro-$\mathfrak{c}$-product $G = G_1 \underset{H}{*} G_2$ exists and consider the cohomology groups *of pairs* as defined in I §6, ex.4 for a torsion G-module A.

Exercise 3. (Excision) The canonical homomorphism

$$H^n(G, G_1, A) \longrightarrow H^n(G_2, H, A)$$

induced by the inclusion $(G_2, H) \hookrightarrow (G, G_1)$ is an isomorphism for all $n \geq 0$.

Exercise 4. The inclusions $(G_1, H) \hookrightarrow (G, H)$ and $(G_2, H) \hookrightarrow (G, H)$ induce an isomorphism

$$H^n(G, H, A) \cong H^n(G_1, H, A) \oplus H^n(G_2, H, A) \quad (n \geq 0).$$

Exercise 5. (Mayer-Vietoris sequence) We have an exact sequence

$$\begin{aligned} \cdots \longrightarrow H^n(H, A) \xrightarrow{\Delta} \; & H^{n+1}(G, A) \longrightarrow H^{n+1}(G_1, H, A) \oplus H^{n+1}(G_2, H, A) \\ \xrightarrow{\Psi} \; & H^{n+1}(H, A) \longrightarrow H^{n+2}(G, A) \longrightarrow \cdots, \end{aligned}$$

where Δ is the composite of the canonical maps

$$H^n(H, A) \xrightarrow{\delta} H^{n+1}(G_2, H, A) \xrightarrow{\sim} H^{n+1}(G, G_1, A) \xrightarrow{\iota} H^{n+1}(G, A),$$

and $\Psi(x_1 \oplus x_2) = h_1(x_1) - h_2(x_2)$, where h_1 and h_2 are induced by the inclusions $H \hookrightarrow G_1$ and $H \hookrightarrow G_2$.

§2. Subgroups of Free Products

A very useful result is the following analogue of the well-known Kurosh subgroup theorem in (discrete) group theory. In the form presented here it is due to *E. Binz*, *J. Neukirch* and *G. Wenzel* (see [12]). There are different proofs and generalizations by several other authors (see [56], [74], [135]).

(4.2.1) Theorem. *Let $G = \mathop{\ast}\limits_{i\in I} G_i$ be the free pro-$\mathfrak{c}$-product of the G_i and let H be an open subgroup of G. Then there exist systems S_i of representatives s_i of the double coset decomposition $G = \dot\bigcup_{s_i\in S_i} Hs_iG_i$ for all i and a free pro-$\mathfrak{c}$-group $F \subseteq G$ of the finite rank*

$$\mathrm{rk}(F) = \sum_{i\in I}[(G:H) - \#S_i] - (G:H) + 1,$$

such that the natural inclusions induce a free product decomposition

$$H = \mathop{\ast}\limits_{i,s_i}(G_i^{s_i} \cap H) \ast F,$$

where $G_i^{s_i}$ $(= s_iG_is_i^{-1})$ denotes the conjugate subgroup.

Remarks: 1. If $N \subseteq H$ is an open subgroup which is normal in G, then $G_i \subseteq N$ for almost all $i \in I$, and for such i we have $\#S_i = (G:H)$. Hence $\mathrm{rk}(F)$ is finite.
2. It is not true in general that we may choose *any* systems S_i of representatives of the double coset decompositions. Only if $\mathfrak{c}$ is the class of p-groups the theorem is true for any choice of the S_i by (4.1.5).

Proof of theorem (4.2.1): Let $\mathfrak{G} = \mathop{\ast}\limits_{i\in I}^{discr} G_i$ be the (discrete) free product of the G_i in the category of groups. We give $\mathfrak{G}$ the topology $T_{\mathfrak{G}}$ which is defined by the family $\mathfrak{B}_{\mathfrak{G}}$ of open normal subgroups $\mathfrak{N} \subseteq \mathfrak{G}$ satisfying the three conditions

(1) $\mathfrak{G}/\mathfrak{N} \in \mathfrak{c}$,
(2) $\mathfrak{N} \supseteq G_i$ for almost all i and
(3) $\mathfrak{N} \cap G_i$ is an open subgroup of the pro-$\mathfrak{c}$-group G_i for all i.

Then G is the completion of $\mathfrak{G}$ with respect to $T_{\mathfrak{G}}$, and the restriction of $T_{\mathfrak{G}}$ to G_i is the given profinite topology on G_i (compare the existence proof for free products in §1). Let $\kappa : \mathfrak{G} \to G$ be the canonical completion homomorphism. We denote the isomorphic images of the groups G_i in G and $\mathfrak{G}$ also by G_i.

Now let H be an open subgroup in G. We denote the open subgroup $\kappa^{-1}(H) \subseteq \mathfrak{G}$ by $\mathfrak{H}$. In particular, $\mathfrak{H}$ is of finite index in $\mathfrak{G}$ and by the Kurosh subgroup theorem for discrete groups (see [215], chap.I §5.5 th. 14), there exist systems $\mathfrak{S}_i$ of representatives σ_i of the double coset decomposition $\mathfrak{G} = \dot{\bigcup} \mathfrak{H}\sigma_i G_i$ and a (discrete) free group $\mathfrak{F} \subseteq \mathfrak{G}$ of rank

$$\mathrm{rk}(\mathfrak{F}) = \sum_{i\in I} [(\mathfrak{G} : \mathfrak{H}) - \#\mathfrak{S}_i] - (\mathfrak{G} : \mathfrak{H}) + 1,$$

such that the natural inclusions induce a free product decomposition

$$\mathfrak{H} = \mathop{\ast}_{i,\sigma_i}^{discr} (G_i^{\sigma_i} \cap \mathfrak{H}) \overset{discr}{\ast} \mathfrak{F}.$$

By the argument of remark 1 above, we conclude that the sum on the right side of the rank equation is finite.

Since $T_{\mathfrak{G}}$ induces on G_i, and therefore also on $G_i^{\sigma_i}$, the pro-$\mathfrak{c}$-topology, and since $\mathfrak{H}$ is open, $G_i^{\sigma_i} \cap \mathfrak{H}$ is an open subgroup of the pro-$\mathfrak{c}$-group $G_i^{\sigma_i}$. Now let $T_{\mathfrak{H}}$ be the topology on $\mathfrak{H}$ which is induced by the family $\mathfrak{B}_{\mathfrak{H}}$ of open normal subgroups $\mathfrak{I} \subseteq \mathfrak{H}$ satisfying the three conditions

(1) $\mathfrak{H}/\mathfrak{I} \in \mathfrak{c}$,
(2) $\mathfrak{I} \supseteq G_i^{\sigma_i} \cap \mathfrak{H}$ for almost all i, σ_i and
(3) $\mathfrak{I} \cap G_i^{\sigma_i} \cap \mathfrak{H}$ is an open subgroup of the pro-$\mathfrak{c}$-group $G_i^{\sigma_i} \cap \mathfrak{H}$ for all i, σ_i.

We first note that $T_{\mathfrak{H}}$ induces on $G_i^{\sigma_i} \cap \mathfrak{H}$ the pro-$\mathfrak{c}$-topology of $G_i^{\sigma_i} \cap \mathfrak{H}$, and induces on $\mathfrak{F}$ the topology given by all normal subgroups $\mathfrak{I}_{\mathfrak{F}} \subseteq \mathfrak{F}$ with $\mathfrak{F}/\mathfrak{I}_{\mathfrak{F}} \in \mathfrak{c}$. Therefore the completion of $\mathfrak{H}$ with respect to the topology $T_{\mathfrak{H}}$ has a free product decomposition as in the statement of the theorem. Thus we have to show:

Claim: The restriction of the topology $T_{\mathfrak{G}}$ to $\mathfrak{H}$ is equal to $T_{\mathfrak{H}}$.

For an $\mathfrak{N} \in \mathfrak{B}_{\mathfrak{G}}$, one easily observes $\mathfrak{N} \cap \mathfrak{H} \in \mathfrak{B}_{\mathfrak{H}}$ and therefore the topology $T_{\mathfrak{H}}$ is finer than $T_{\mathfrak{G}} \cap \mathfrak{H}$. To show that also $T_{\mathfrak{G}} \cap \mathfrak{H}$ is finer than $T_{\mathfrak{H}}$, let $\mathfrak{I} \in \mathfrak{B}_{\mathfrak{H}}$. We have to find a group $\mathfrak{N} \in \mathfrak{B}_{\mathfrak{G}}$ such that $\mathfrak{N} \cap \mathfrak{H} \subseteq \mathfrak{I}$. We claim that we can take for $\mathfrak{N}$ the intersection $\tilde{\mathfrak{I}}$ of all conjugates of $\mathfrak{I}$ in $\mathfrak{G}$. Since $\mathfrak{I}$ is of finite index in $\mathfrak{H}$ and $\mathfrak{H}$ is of finite index in $\mathfrak{G}$, $\mathfrak{I}$ is of finite index in $\mathfrak{G}$. Therefore $\mathfrak{I}$ has only finitely many conjugates $\mathfrak{I}^{\tau_j}$ in $\mathfrak{G}$, i.e. $\tilde{\mathfrak{I}}$ is open and of finite index in $\mathfrak{G}$. Denoting by $\tilde{\mathfrak{H}} \in \mathfrak{B}_G$ the intersection of the finitely many conjugates of $\mathfrak{H}$ in $\mathfrak{G}$, we obtain an exact sequence

$$1 \to \tilde{\mathfrak{H}}/\tilde{\mathfrak{I}} \to \mathfrak{G}/\tilde{\mathfrak{I}} \to \mathfrak{G}/\tilde{\mathfrak{H}} \to 1.$$

The group $\tilde{\mathfrak{H}}/\tilde{\mathfrak{I}}$ is canonically embedded into the group $\prod_{\tau_j} \mathfrak{H}^{\tau_j}/\mathfrak{I}^{\tau_j} \in \mathfrak{c}$, and the class $\mathfrak{c}$ is full. Therefore $\mathfrak{G}/\tilde{\mathfrak{I}} \in \mathfrak{c}$. Next we have to show that $G_i \cap \tilde{\mathfrak{I}}$ is open in the pro-$\mathfrak{c}$-topology of G_i. It is sufficient to show that $G_i \cap \mathfrak{I}^{\tau_j}$ is open in G_i or, equivalently, that $G_i^{\tau_j^{-1}} \cap \mathfrak{I}$ is open in $G_i^{\tau_j^{-1}}$. If we write $\tau_j^{-1} = h \cdot \sigma_i \cdot g_i$,

$h \in \mathfrak{H}$, $\sigma_i \in \mathfrak{S}_i$ and $g_i \in G_i$, we get $G_i^{\tau_j^{-1}} \cap \mathfrak{I} = (G_i^{\sigma_i} \cap \mathfrak{I})^h$ and this group is open in the pro-$\mathfrak{c}$-group $(G_i^{\sigma_i})^h = G_i^{h\sigma_i g_i} = G_i^{\tau_j^{-1}}$ *). At the same time, we see that $G_i^{\tau_j^{-1}} \cap \mathfrak{I} = (G_i^{\sigma_i} \cap \mathfrak{I})^h = (G_i^{\sigma_i})^h = G_i^{h\sigma_i g_i} = G_i^{\tau_j^{-1}}$ for almost all $i \in I$. This proves the claim. The proof of the theorem is then completed by the observation that the images $s_i \in H$ of the σ_i under κ form systems S_i of representatives of the double coset decomposition $\bigcup Hs_iG_i$ in G. □

The next corollary is the profinite analogue of the Nielson-Schreier formula for discrete groups. If G is a pro-p-group, we have already proved it in III §5 using partial Euler-Poincaré characteristics (see (3.9.6)).

(4.2.2) Corollary. *An open subgroup H of a free pro-$\mathfrak{c}$-group G is again a free pro-$\mathfrak{c}$-group. The ranks satisfy the equation*

$$\mathrm{rk}(H) - 1 = (\mathrm{rk}(G) - 1)\,(G : H).$$

Proof: A free pro-$\mathfrak{c}$-group G is the free pro-$\mathfrak{c}$-product $G = \ast_{i\in I} G_i$, where $G_i \cong \hat{\mathbb{Z}}(\mathfrak{c})$ denotes the free pro-$\mathfrak{c}$-groups of rank 1 (see III §5). Since (loc.cit.) $\hat{\mathbb{Z}}(\mathfrak{c}) = \prod_{p\in S} \mathbb{Z}_p$, where S is the set of prime numbers which divide the order of a group in $\mathfrak{c}$, we see that every open subgroup of the free pro-$\mathfrak{c}$-group of rank 1 is isomorphic to the free pro-$\mathfrak{c}$-group of rank 1 again. By (4.2.1), there exists a system S_i of representatives s_i of the double coset decomposition $G = \bigcup_{s_i\in S_i} Hs_iG_i$ and a free pro-$\mathfrak{c}$ group $F \subseteq G$ of the finite rank $\mathrm{rk}(F) = \sum_{i\in I}[(G:H) - \#S_i] - (G:H) + 1$, such that the natural inclusions induce a free product decomposition $H = \ast_{i,s_i}(G_i^{s_i} \cap H) \ast F$. Therefore H is a free pro-$\mathfrak{c}$-group (as a free pro-$\mathfrak{c}$-product of free pro-$\mathfrak{c}$-groups) and

$$\begin{aligned}
\mathrm{rk}(H) &= \sum_{i,s_i} \mathrm{rk}(G_i^{s_i} \cap H) + \mathrm{rk}(F) \\
&= \sum_{i\in I} \#S_i + \mathrm{rk}(F) \\
&= \sum_{i\in I} (G:H) - (G:H) + 1 \\
&= (G:H)\,(\mathrm{rk}(G) - 1) + 1.
\end{aligned}$$

□

Having proved several results which are quite similar to those in discrete group theory, the reader should be warned of some curious pathologies in the profinite case, which are (more or less) due to the existence of infinite words in the profinite products.

---

*) Observe the rule $(G^a)^b = G^{ba}$.

For example, if a group P is the free product of its subgroups G and H, then one should expect that for every pair $\sigma, \tau \in P$, the conjugates G^σ and H^τ also form a free product. However, look at the following example due to *D. Haran* (see [74]).

(4.2.3) Proposition. *Assume that $\mathbb{Z}/2\mathbb{Z}, \mathbb{Z}/3\mathbb{Z} \in \mathfrak{c}$ and let G (resp. H) be a finitely generated pro-$\mathfrak{c}$-group having a normal subgroup of index 2 (resp. 3). Let*

$$P = G * H.$$

Then there exist elements $\sigma, \tau \in P$ such that the subgroups G^σ and H^τ do not form a free pro-$\mathfrak{c}$-product in P.

Proof: Let $\mathcal{Q}$ be the set of closed subgroups $Q \subseteq P$ such that there exist $\sigma, \tau \in P$ with $G^\sigma, H^\tau \subseteq Q$. For $Q \in \mathcal{Q}$ the sets $\{\sigma \in P \mid G^\sigma \subseteq Q\}$ and $\{\tau \in P \mid H^\tau \subseteq Q\}$ are closed in P. We deduce that $\mathcal{Q}$ is closed under descending chains. By Zorn's lemma, it has a minimal element, say Q. Assume that $Q = G^\sigma * H^\tau$ for suitable $\sigma, \tau \in P$. Then, by our assumptions on G and H and by the universal property of the free product, there would exist a homomorphism $\phi : Q \to \mathfrak{S}_4$ (the symmetric group on four elements), such that $\phi(G^\sigma) = \langle(12)\rangle$ and $\phi(H^\tau) = \langle(134)\rangle$. Since $\mathfrak{S}_4$ is generated by the cycles (12) and (134), ϕ is surjective. Choose $\rho \in Q$ with $\phi(\rho) = (1234)$. Then

$$\phi(\langle G^\sigma, H^{\rho\tau}\rangle) = \langle(12), (134)^{(1234)}\rangle = \langle(12), (124)\rangle \subsetneqq \mathfrak{S}_4.$$

This shows that $\langle G^\sigma, H^{\rho\tau}\rangle \subsetneqq Q$ is contained in $\mathcal{Q}$. This contradicts the minimality of Q. □

Remarks: 1. It follows from the construction of the free pro-$\mathfrak{c}$-product that $P = G^\sigma * H^\tau$ whenever σ and τ are *finite* words in elements of G and H (i.e. σ and τ are the products of finitely many elements which are in G or H).
2. This kind of pathology does not occur if $\mathfrak{c}$ is the class of pro-p-groups. Then $P = G * H$ is the free pro-p-product of G^σ and H^τ for every pair $\sigma, \tau \in P$. This follows from (4.1.5). However, by the result of exercise 3 below, this is the only good case.

Exercise 1. Assume that we are given pro-$\mathfrak{c}$-groups G, H and an open normal subgroup $N \subseteq G$. Show that the kernel of the homomorphism

$$G * H \longrightarrow G/N\,,$$

which, by the universal property of the free pro-$\mathfrak{c}$-product, is given by the projection $G \to G/N$ and the trivial homomorphism $H \to \{1\} \subseteq G/N$, is of the form

$$N * (\mathop{*}_{\sigma \in \mathcal{R}} H^\sigma),$$

where $\mathcal{R}$ is a system of representatives of the cosets of N in G.

Exercise 2. Let p be a prime number and assume $\mathbb{Z}/p\mathbb{Z} \in \mathfrak{c}$. Determine the kernels of the homomorphisms

$$\mathbb{Z}/p\mathbb{Z} * \mathbb{Z}/p\mathbb{Z} \to \mathbb{Z}/p\mathbb{Z},$$

which, by the universal property of the free pro-$\mathfrak{c}$-product, are associated to the pairs (id, id) and $(0, id)$.

Exercise 3. Assume that the full class $\mathfrak{c}$ is strictly larger than the class (p-groups) and assume that G, H are pro-p-groups. Let $P = G * H$ be the free pro-$\mathfrak{c}$-product of G and H.

(i) Show that P is not a pro-p-group if G and H are nontrivial.

(ii) Show that there are $\sigma, \tau \in P$ such that the closed subgroup $U = \langle G^\sigma, H^\tau \rangle$, generated by G^σ and H^τ in P, is a pro-p-group.

(iii) Show that the closed subgroup U constructed in (ii) is isomorphic to the free pro-p-product of G and H.

Hint: Use (4.2.1) for (i), the Sylow theorems (1.6.9) for (ii) and (4.1.5) for (iii).

Exercise 4. Assume that the finite group G is the semi-direct product of its subgroup H and its normal subgroup N. Consider the canonical surjection

$$\varepsilon : N * H \longrightarrow G,$$

which is induced by the inclusions of N and H into G. Show that the kernel of ε is a free profinite group of rank $(\#N - 1)(\#H - 1)$ with basis

$$\{hnh^{-1}n_h^{-1} \in N * H \mid h \in H \backslash \{1\}, n \in N \backslash \{1\}\},$$

where $n_h := (hnh^{-1}) \in N$, i.e. $\ker(\varepsilon)$ is the free profinite group on the symbols which represent the difference between formal and real conjugation of nontrivial elements of N by nontrivial elements of H.

§3. Generalized Free Products

In this section we define a vast generalization of the free products of §1. Instead of (discrete) families of profinite groups we now consider families of profinite groups which vary continuously over a topological base space. Such generalizations were introduced by *D. Gildenhuys* and *L. Ribes* ([56]) for locally constant families and by *D. Haran* ([74]) and (independently) by *O. Melnikov* ([135]) in its most general form.

As before, let $\mathfrak{c}$ be a full class of finite groups. Given continuous maps $f : X \to Z$, $g : Y \to Z$ between profinite spaces, we will write

$$X \times_Z Y = \{(x, y) \in X \times Y | f(x) = g(y)\}$$

for the fibre product of X and Y over Z, which is a profinite space.

(4.3.1) Definition. *Let T be a profinite space. A* **bundle of pro-$\mathfrak{c}$-groups** *over T is a group object in the category of profinite spaces over T, such that the fibre over every point of T is a pro-$\mathfrak{c}$-group. In other words, a bundle of pro-$\mathfrak{c}$-groups over T is a tuple $(\mathcal{G}, p, T, m, e, \iota)$, where $\mathcal{G}$ is a profinite space together with continuous maps*

$p\ : \mathcal{G} \to T$ *(the structure map),*
$m : \mathcal{G} \times_T \mathcal{G} \to \mathcal{G}$ *(the multiplication),*
$e\ : T \to \mathcal{G}$ *a section to p (the unit),*
$\iota\ : \mathcal{G} \to \mathcal{G}$ *(the inversion),*

such that the fibre $G_t := p^{-1}(t)$, together with the induced maps $m_t : G_t \times G_t \to G_t$, $\iota_t : G_t \to G_t$ and the unit element $e_t := e(t)$, is a pro-$\mathfrak{c}$-group for every point $t \in T$.

Because of the existence of the unit e, the structure map p is necessarily surjective. Usually we will denote such a bundle by $(\mathcal{G}, p, T)$, or by $\mathcal{G}$ if no confusion is possible.

Example 1. If G is a pro-$\mathfrak{c}$-group and T is a profinite space, then we always have the **constant bundle** $(G \times T, \mathrm{pr}_T, T)$, where pr_T is the projection $G \times T \to T$ and the maps m, ι, e are those induced by the group operations in G.

Example 2. Assume that $(G_i)_{i \in I}$ is a (discrete) family of pro-$\mathfrak{c}$-groups. We define the associated bundle of pro-$\mathfrak{c}$-groups as the bundle over the one point compactification $\bar{I} := I \,\dot\cup\, \{*\}$ of I which is given as a set by

$$\mathcal{G} := \dot\bigcup_{i \in I} G_i \,\dot\cup\, \{*\}\,,$$

and which has the following topology: $G_i \subseteq \mathcal{G}$ (together with its profinite topology) is open in $\mathcal{G}$ for all i, and for every open neighbourhood $U \subseteq \bar{I}$ of $* \in \bar{I}$, let

$$\dot\bigcup_{i \in U} G_i \,\dot\cup\, \{*\}$$

be an open neighbourhood of $* \in \mathcal{G}$. The space $\mathcal{G}$ is profinite and one checks that the map

$$p : \mathcal{G} \to \bar{I};\ \ G_i \ni g_i \mapsto i,\ \ * \mapsto *$$

is continuous. Viewing $\{*\}$ as the group with one element, we see that the group operations on the G_i's induce the structure of a bundle of pro-$\mathfrak{c}$-groups on the triple $(\mathcal{G}, p, \bar{I})$.

Assume that we are given a pro-$\mathfrak{c}$-group G and a profinite space T.

(4.3.2) Definition. *A family $(G_t)_{t\in T}$ of subgroups of the profinite group G, indexed over the profinite space T, is a* **continuous family of subgroups** *if for every open subgroup $U \subseteq G$ the set $T(U) := \{t \in T \mid G_t \subseteq U\}$ is open in T.*

Given such a continuous family of subgroups, we define a bundle of pro-$\mathfrak{c}$-groups $(\mathcal{G}, p, T)$ over T by putting

$$\mathcal{G} = \{(g,t) \in G \times T \,\big|\, g \in G_t\}$$

and defining p to be the restriction to $\mathcal{G}$ of the projection of $G \times T$ onto T. We define the maps m, ι, e by restricting the corresponding maps from the constant bundle.

(4.3.3) Lemma. *$(\mathcal{G}, p, T)$ is a bundle of pro-$\mathfrak{c}$-groups.*

Proof: Once we have proved that $\mathcal{G}$ is a profinite space, all properties follow from the corresponding properties of the constant bundle $G \times T \to T$. Hence it remains to show that $\mathcal{G}$ is closed in $G \times T$. Let $(h,s) \notin \mathcal{G}$, i.e. $h \notin G_s$. Then there exist open subgroups $V, W \subseteq G$ with $G_s \subseteq W$ and $hV \cap W = \varnothing$. Since the family (G_t) is continuous, $T(W)$ is open in T. But then $(h,s) \in hV \times T(W) \subseteq G \times T \setminus \mathcal{G}$, which finishes the proof. $\square$

(4.3.4) Definition. *A morphism of bundles*

$$\phi : (\mathcal{G}, p_{\mathcal{G}}, T) \to (\mathcal{H}, p_{\mathcal{H}}, S)$$

is a pair $\phi_{\mathcal{G}} : \mathcal{G} \to \mathcal{H}$, $\phi_T : T \to S$ of continuous maps such that

(i) *the diagram*

$$\begin{array}{ccc} \mathcal{G} & \xrightarrow{\phi_{\mathcal{G}}} & \mathcal{H} \\ \downarrow{\scriptstyle p_{\mathcal{G}}} & & \downarrow{\scriptstyle p_{\mathcal{H}}} \\ T & \xrightarrow{\phi_T} & S \end{array}$$

commutes and

(ii) *for every $t \in T$ the associated map $\phi_t : G_t \to H_{\phi_T(t)}$ is a group homomorphism.*

We say that ϕ is surjective if $\phi_{\mathcal{G}}$ (and hence also ϕ_T) is surjective.

We will not distinguish between the pro-$\mathfrak{c}$-group G and the bundle $(G, p, \{*\})$ over the one point space $\{*\}$. In particular, a morphism from a bundle $(\mathcal{G}, p, T)$ to a group G is a continuous map $\phi : \mathcal{G} \to G$ such that the induced maps $\phi_t : G_t \to G$ are group homomorphisms for every $t \in T$. One easily verifies that the family ($\phi(G_t)$)$_{t\in T}$ is a continuous family of subgroups of G indexed by T.

(4.3.5) Definition. *The free pro-$\mathfrak{c}$-product of a bundle $(\mathcal{G}, p, T)$ of pro-$\mathfrak{c}$-groups is a pro-$\mathfrak{c}$-group*

$$G = \mathop{\ast}_{T} \mathcal{G}$$

together with a morphism $\omega : \mathcal{G} \to G$, which has the following universal property: for every morphism $f : \mathcal{G} \to H$ from $\mathcal{G}$ to a pro-$\mathfrak{c}$-group H there exists a unique homomorphism of pro-$\mathfrak{c}$-groups $\phi : G \to H$ with $f = \phi \circ \omega$.

In categorical language this means that the functor "free product over T" is a left adjoint to the functor "constant bundle" from (pro-$\mathfrak{c}$-groups) to (bundles of pro-$\mathfrak{c}$-groups over T).

The map ω induces a canonical map $\omega_t : G_t \to G$ for every $t \in T$. We will see in (4.3.12) that ω_t is an injective group homomorphism for every $t \in T$ and that the images of the G_t in G are in some sense independent from each other.

(4.3.6) Proposition. *The free pro-$\mathfrak{c}$-product $\mathop{\ast}_{T} \mathcal{G}$ exists and is unique up to unique isomorphism.*

Proof: The uniqueness assertion is clear by the universal property. The existence follows from the following construction:

Let L be the abstract free product of the family of groups $(G_t)_{t\in T}$ and let $\lambda : \mathcal{G} \to L$ be the map which is given on every G_t as the natural inclusion of G_t into L. Then define the free product $G = \ast_T \mathcal{G}$ as the completion of L with respect to the topology which is given by the family of normal subgroups $N \subseteq L$ of finite index for which (a) $L/N \in \mathfrak{c}$ and (b) the composition $p_N \circ \lambda$ of λ with the natural projection $p_N : L \to L/N$ is continuous. It is easily verified that G has the required universal property. (Compare with the construction of the free product in §1.) □

(4.3.7) Definition. *Assume that $(G_t)_{t\in T}$ is a continuous family of subgroups of the pro-$\mathfrak{c}$-group G indexed over the profinite space T and let $\mathcal{G}$ be the pro-$\mathfrak{c}$-group bundle which is associated to the family $(G_t)_{t\in T}$ by* (4.3.3). *We say that G is the* **free pro-$\mathfrak{c}$-product** *over the family $(G_t)_{t\in T}$ if the canonical homomorphism $\mathop{\ast}_{T} \mathcal{G} \to G$ is an isomorphism.*

Example 3. Assume that G is a pro-$\mathfrak{c}$-group and that T is a profinite space. Then we denote the free pro-$\mathfrak{c}$-product over the constant bundle $G \times T \to T$ by $\ast_T G$.

Assume, in addition, that a profinite group Γ acts continuously on T. Then, by the universal property of the free product, Γ also acts continuously on $\ast_T G$, i.e. $\ast_T G$ becomes a pro-$\mathfrak{c}$-Γ operator group, which is defined as follows.*)

(4.3.8) Definition. *Let Γ be a profinite group. A* **pro-$\mathfrak{c}$-Γ operator group** *is a pro-$\mathfrak{c}$-group G together with a homomorphism*

$$\phi : \Gamma \longrightarrow \mathrm{Aut}(G),$$

such that the action: $\Gamma \times G \to G$, $(\gamma, g) \mapsto \phi(\gamma)(g)$, is continuous.

If $\Gamma = T$ acts on itself by left multiplication and if $G = F_r$ is the free pro-$\mathfrak{c}$-group of rank r, then $\underset{\Gamma}{\ast} F_r$ is the **free pro-$\mathfrak{c}$-Γ operator group** *of rank r.*

Free pro-$\mathfrak{c}$-Γ operator groups satisfy the obvious universal property in the category of pro-$\mathfrak{c}$-Γ operator groups.

Example 4. Assume that we are given a (discrete) family $(G_i)_{i \in I}$ of pro-$\mathfrak{c}$-groups and let $\mathcal{G}$ be the associated bundle over the one point compactification $\bar{I}$ of I (see example 2). Then to give a morphism $\phi : \mathcal{G} \to G$ is the same as to give a convergent (see §1) family of group homomorphisms $\phi_i : G_i \to G$. Therefore we obtain a canonical isomorphism

$$\underset{\bar{I}}{\ast} \mathcal{G} \cong \underset{i \in I}{\ast} G_i.$$

If I is finite, it is compact and we can omit the point $* \in \bar{I}$.

More generally, from the universal property follows the

(4.3.9) Proposition. *Let $(\mathcal{G}, p, T)$ be a bundle of pro-$\mathfrak{c}$-groups and assume that $T = T_1 \dot\cup \cdots \dot\cup T_k$ is a finite disjoint decomposition of T into closed and open subsets. Then there is a canonical isomorphism*

$$\underset{T}{\ast} \mathcal{G} = \underset{T_1}{\ast} \mathcal{G}_1 * \cdots * \underset{T_k}{\ast} \mathcal{G}_k,$$

where $\mathcal{G}_i$ denotes the restricted bundle $p^{-1}(T_i)$ for $i = 1, \ldots, k$.

*) The notion of a profinite operator group was introduced by *H. Koch*

One of the basic facts of the theory of profinite groups is the assertion that a topological group which is profinite as a topological space is already the inverse limit of finite groups (see (1.1.3)). The following theorem extends this fact to continuous families of profinite groups.

We call a bundle of pro-$\mathfrak{c}$-groups *finite* if it is finite as a space, i.e. it is a finite disjoint union of groups in $\mathfrak{c}$.

(4.3.10) Theorem. *Every pro-$\mathfrak{c}$-group bundle is the inverse limit of a system of finite bundles with surjective transition morphisms.*

Proof: Denote the given bundle by $(\mathcal{G}, p, T)$. As a topological spaces, $\mathcal{G}$ and T are inverse limits of finite discrete spaces. We will make use of the following claim:

For every pair of disjoint decompositions $\mathcal{G} = U_1 \dot\cup \cdots \dot\cup U_n$ of $\mathcal{G}$ and $T = V_1 \dot\cup \cdots \dot\cup V_m$ of T into open and closed subsets there exists a morphism

$$\phi : (\mathcal{G}, p, T) \to (\mathcal{H}, p', S)$$

to a finite bundle $(\mathcal{H}, p', S)$, such that the decompositions of $\mathcal{G}$ (resp. T) into the pre-images of the (finitely many) points of $\mathcal{H}$ (resp. S) are finer than the given decomposition.

The proof of this claim will be given below.

Observe that the set of finite quotients of $(\mathcal{G}, p, T)$ is filtered. Indeed, assume that we are given two finite quotients $(\mathcal{H}_i, p_i, S_i)$ $(i = 1, 2)$ of $(\mathcal{G}, p, T)$. Then we first take their pullbacks to a suitable chosen common finite basis S and then we take the fibre product over S. Now it follows from the claim that the canonical morphism from $(\mathcal{G}, p, T)$ to the inverse limit of its finite quotients is an isomorphism.

It remains to show the claim. We construct ϕ in several steps.

Step 1. Fix a point $t \in T$. Then, by (1.1.3), there exists a finite group H_t and a homomorphism $\phi_t : G_t \to H_t$, such that the decomposition of G_t into the pre-images of the elements of H_t is finer than the decomposition $G_t = \bigcup_{i=1}^{n} (U_i \cap G_t)$. (In order to be accurate, we should remove the indices i with $U_i \cap G_t = \varnothing$.)

Step 2. We can extend $\phi_t : G_t \to H_t$ to a bundle morphism $\phi_t : \mathcal{G}_{W_t} \to H_t$, where W_t is a sufficiently small open and closed neighbourhood of $t \in T$ and $\mathcal{G}_{W_t}$ denotes the restriction of $\mathcal{G}$ to W_t (see exercise 1 below).

Step 3. Making W_t smaller, if necessary, we may assume that the decomposition of $\mathcal{G}_{W_t}$ into the pre-images of the elements of H_t is finer than the decomposition $\mathcal{G}_{W_t} = \bigcup_{i=1}^n (\mathcal{G}_{W_t} \cap U_i)$.

Step 4. The open and closed sets W_t ($t \in T$) cover T. Replacing W_t by $W_t \cap V_j$ ($j = 1, \dots, m$), we may assume that every W_t is contained in one of the closed and open subsets $V_1, \dots, V_m$. Since T is compact, there exists a finite subcovering $T = W_{t_1} \cup \dots \cup W_{t_k}$, which we may assume to be disjoint, by replacing W_{t_j} by $W_{t_j} \setminus \bigcup_{\nu=1}^{j-1} W_{t_\nu}$.

Step 5. The bundle $H_{t_1} \dot\cup \cdots \dot\cup H_{t_k}$ (together with the obvious structure map to the set of k elements) is a finite bundle to which we have a bundle morphism from $(\mathcal{G}, p, T)$ of the required type. □

The next proposition is the natural generalization of (4.1.2).

(4.3.11) Proposition. *Let*

$$(\mathcal{G}, p, T) = \varprojlim_{i \in I} (\mathcal{G}_i, p_i, T_i)$$

be the inverse limit of the pro-$\mathfrak{c}$-group bundles $\mathcal{G}_i$. Then

$$\mathop{\ast}_{T} \mathcal{G} = \varprojlim_{i \in I} \mathop{\ast}_{T_i} \mathcal{G}_i .$$

Proof: It suffices to show that the inverse limit on the right-hand side satisfies the universal property of the free product with respect to homomorphisms into finite groups in $\mathfrak{c}$. To begin with, we show the statement in the special case that all bundles $\mathcal{G}_i$ are finite and all transition maps are surjective. In this case, we have equalities for every finite group $H \in \mathfrak{c}$:

$$\begin{aligned}
\operatorname{Hom}(\mathop{\ast}_{T} \mathcal{G}, H) &= \operatorname{Mor}((\mathcal{G}, p, T), H) = \operatorname{Mor}(\varprojlim_{i} (\mathcal{G}_i, p_i, T_i), H) \\
&= \varinjlim_{i} \operatorname{Mor}((\mathcal{G}_i, p_i, T_i), H) = \varinjlim_{i} \operatorname{Hom}(\mathop{\ast}_{T_i} \mathcal{G}_i, H) \\
&= \operatorname{Hom}(\varprojlim_{i} \mathop{\ast}_{T_i} \mathcal{G}_i, H).
\end{aligned}$$

Returning to the general case, represent every $\mathcal{G}_i$ as the inverse limit of its finite quotients: $\mathcal{G}_i = \varprojlim_{j} \mathcal{G}_{ij}$. Every morphism

$$\phi_{i_1, i_2} : \mathcal{G}_{i_1} \to \mathcal{G}_{i_2}$$

is realized as a limit of morphisms on finite levels. Therefore we can replace $\mathcal{G}_i$ by the system $\mathcal{G}_{ij}$ without changing either side in the stated equality. In other words (changing the index set), we may assume that all $\mathcal{G}_i$ are finite. Again, without changing the projective limit, we can add the images of all transition maps to the inverse system and since all $\mathcal{G}_i$ are finite, the intersections

$$\bigcap_{j\geq i} \mathrm{im}(\mathcal{G}_j) \subseteq \mathcal{G}_i$$

stabilize and set up a cofinal subsystem. Finally, we arrive at the point that all bundles in the system are finite and all transition maps are surjective, and this special case was already solved at the beginning of the proof. □

Using (4.3.10) and (4.3.11), many results for finite free products carry over to the case of generalized free products. For example, we have the

(4.3.12) Proposition. *Let $\omega : \mathcal{G} \to G$, $G = \mathop{*}\limits_{T} \mathcal{G}$ be a free pro-$\mathfrak{c}$-product. Then*

(i) $\omega|_{G_t} : G_t \to G$ *is an injective group homomorphism for every* $t \in T$,

(ii) *if* $s, t \in T$ *and* $g \in G$ *satisfy* $\omega(G_s) \cap \omega(G_t)^g \neq 1$, *then* $s = t$.

Proof: Since all statements are compatible with inverse limits, we may assume by (4.3.10), (4.3.11) that $\mathcal{G}$ is finite. Then (i) is trivial by the results of §1. In order to prove (ii), assume $s \neq t$. It suffices to construct a group H and a homomorphism $f : G \to H$ whose restriction to $\omega(G_s)$ is injective and such that the images of $\omega(G_s)$ and $g\omega(G_t)g^{-1}$ have trivial intersection in H. For example, we can choose $H := G_s \times G_t$. Let f be the homomorphism which corresponds via the universal property to the bundle homomorphism which is given by $G_s \to H$, $g_s \mapsto (g_s, 1)$, $G_t \to H$, $g_t \mapsto (1, g_t)$, and the trivial homomorphism on G_u for every other u. Then we have $f(\omega(G_s)) = G_s \times 1$ and $f(g\omega(G_t)g^{-1}) = f(g)f(\omega(G_t))f(g)^{-1} = 1 \times G_t$. □

Remarks: 1. In fact a much stronger statement than (4.3.12) is true: the canonical homomorphism from the discrete free product of the G_t to $G = \mathop{*}\limits_{T} \mathcal{G}$ is injective. Using (4.3.10) and (4.3.11), this can be deduced from the results in [56].

2. There are also variants of the Kurosh subgroup theorem for the generalized free products (see [56], [74], [135], [261]).

Now we calculate the cohomology of free products. We start with the definition of restricted direct sums.

(4.3.13) Definition. *Let $(\mathcal{G}, T)$ be a bundle of pro-$\mathfrak{c}$-groups, $G = *_T\mathcal{G}$ and A a torsion G-module. For each $n \geq 0$ we denote by*

$$\bigoplus_{t\in T}{}' H^n(G_t, A)$$

the subgroup of elements in $\prod_{t\in T} H^n(G_t, A)$ which are in the image of some map

$$\bigoplus_{t'\in T'} H^n(G'_{t'}, A') \longrightarrow \prod_{t\in T} H^n(G_t, A),$$

where $(\mathcal{G}', T')$ is a finite quotient bundle of $(\mathcal{G}, T)$ and A' is a finite submodule of A which is a $*_{T'}\mathcal{G}'$-module.

Remark: If $(G_i)_{i\in I}$ is a discrete family of pro-$\mathfrak{c}$-groups and $\mathcal{G}$ is the associated bundle over the one point compactification $\bar{I} = I \cup \{*\}$ of the discrete set I, see example 2 page 257, then

$$\bigoplus_{i\in\bar{I}}{}' H^n(G_i, A) = \bigoplus_{i\in I} H^n(G_i, A).$$

In many arithmetical applications the considered bundle will be the projective limit of bundles which are associated to discrete families $(G_{\lambda,i})_{i\in I_\lambda}$, $\lambda \in \Lambda$, of pro-$\mathfrak{c}$-groups, i.e. $(\mathcal{G}, T)$ is the projective limit of the associated bundles $(\mathcal{G}_\lambda, \bar{I}_\lambda)$ over the one point compactifications $\bar{I}_\lambda = I_\lambda \cup \{*\}$ of the discrete sets I_λ, $\lambda \in \Lambda$.

Assume in this situation that the torsion $*_T\mathcal{G}$-module A is a $*_{\bar{I}_\lambda}\mathcal{G}_\lambda$-module for all $\lambda \in \Lambda$. Since $\bigoplus'_{t\in T} H^n(G_t, A) = \varinjlim_{\lambda\in\Lambda} \bigoplus'_{i\in\bar{I}_\lambda} H^n(G_{\lambda,i}, A)$, we have in this case

$$\bigoplus_{t\in T}{}' H^n(G_t, A) = \varinjlim_{\lambda\in\Lambda} \bigoplus_{i\in I_\lambda} H^n(G_{\lambda,i}, A).$$

(4.3.14) Theorem. *Let $(\mathcal{G}, T)$ be a bundle of pro-$\mathfrak{c}$-groups, $G = *_T\mathcal{G}$ and A a torsion G-module. Then we have a canonical exact sequence*

$$0 \to A^G(\mathfrak{c}) \to A(\mathfrak{c}) \to \bigoplus_{t\in T}{}'(A/A^{G_t})(\mathfrak{c}) \to H^1(G, A) \to \bigoplus_{t\in T}{}' H^1(G_t, A) \to 0,$$

and for all $n \geq 2$

$$H^n(G, A) \cong \bigoplus_{t\in T}{}' H^n(G_t, A).$$

Proof: By a direct limit argument, we may assume that A is finite. The composite map $\mathcal{G} \to G \to \mathrm{Aut}(A)$ is continuous, hence locally constant. Using the claim in the proof of (4.3.10), we find a presentation

$$(\mathcal{G}, p, T) = \varprojlim_{i\in I} (\mathcal{G}_i, p_i, T_i)$$

of the bundle $\mathcal{G}$ as an inverse limit of finite bundles $\mathcal{G}_i = \dot{\bigcup}_{s\in T_i} G_{i,s}$, $i \in I$, such that A is a $*_{T_i}\mathcal{G}_i$-module for all i. We set $G_i = *_{T_i}\mathcal{G}_i$. By (4.3.11), $G = \varprojlim_{i\in I} G_i$, and by (4.1.4), we obtain an exact sequence

$$0 \to A/A^G \to \varinjlim_{i\in I} \bigoplus_{s\in T_i} A/A^{G_{i,s}} \to H^1(G,A) \to \varinjlim_{i\in I} \bigoplus_{s\in T_i} H^1(G_{i,s},A) \to 0$$

and isomorphisms for all $n \geq 2$

$$H^n(G,A) \xrightarrow{\sim} \varinjlim_{i\in I} \bigoplus_{s\in T_i} H^n(G_{i,s},A).$$

For $n \geq 1$ we consider the natural map

$$\phi\colon \varinjlim_{i\in I} \bigoplus_{s\in T_i} H^n(G_{i,s},A) \longrightarrow {\bigoplus_{t\in T}}' H^n(G_t,A).$$

Each finite quotient bundle $(\mathcal{G}', p', T')$ of $(\mathcal{G}, p, T)$ is a quotient of $(\mathcal{G}_i, p_i, T_i)$ for sufficiently large $i \in I$. Therefore ϕ is surjective. Let $x \in \bigoplus_{s\in T_{i_0}} H^n(G_{i_0,s},A)$ represent a nonzero element in $\varinjlim_{i\in I} \bigoplus_{s\in T_i} H^n(G_{i,s},A)$. Then, for each $i \geq i_0$, the set $S_i \subseteq T_i$ of elements $s \in T_i$ such that the s-component of the image of x in $\bigoplus_{s\in T_i} H^n(G_{i,s},A)$ is nonzero is nonempty. As the inverse limit of finite nonempty sets is nonempty, ϕ is injective, hence an isomorphism. Similarly, the natural map

$$\varinjlim_{i\in I} \bigoplus_{s\in T_i} A/A^{G_{i,s}} \longrightarrow {\bigoplus_{t\in T}}' A/A^{G_t}$$

is an isomorphism. This completes the proof of the theorem. □

(4.3.15) Corollary. *Let $(\mathcal{G}, p, T)$ be a bundle of pro-$\mathfrak{c}$-groups and let p be a prime number. Then*

$$cd_p \mathop{*}_{T} \mathcal{G} = \sup_{t\in T} \{cd_p\, G_t\} \quad \text{and} \quad scd_p \mathop{*}_{T} \mathcal{G} = \sup_{t\in T} \{scd_p\, G_t\}.$$

Proof: By (4.3.14), we obtain the inequality $cd_p\, G \leq \sup_{t\in T} \{cd_p\, G_t\}$. By (4.3.12), each G_t is a closed subgroup of G, hence, by (3.3.5), $cd_p\, G \geq \sup_{t\in T} \{cd_p\, G_t\}$. It remains to show the statement on the strict cohomological dimension. By the same argument as for cd_p, we obtain the inequality $scd_p\, G \geq \sup_{t\in T} \{scd_p\, G_t\} =: n$. Assume that n is finite. If all G_t are trivial, so is the free product. As profinite groups of strict cohomological p-dimension 1 do not exist, we may assume $n \geq 2$. We claim that $H^m(G,A)(p) = 0$ for all discrete G-modules A and all $m \geq n+1$. By the first part of the proof,

$$cd_p\, G = \sup_{t\in T} \{cd_p\, G_t\} \leq \sup_{t\in T} \{scd_p\, G_t\} = n.$$

Therefore we may assume that A is torsion-free, hence $\mathbb{Z}$-flat. Tensoring the exact sequence $0 \to \mathbb{Z}_p \to \mathbb{Q}_p \to \mathbb{Q}_p/\mathbb{Z}_p \to 0$ by A and using the fact that uniquely divisible modules are cohomologically trivial, we obtain

$$\begin{aligned} H^m(G,A)(p) &\cong H^m(G,A)\otimes \mathbb{Z}_p \cong H^m(G,A\otimes \mathbb{Z}_p) \\ &\cong H^{m-1}(G,A\otimes \mathbb{Q}_p/\mathbb{Z}_p) \\ &\cong \bigoplus_{t\in T}{}' H^{m-1}(G_t,A\otimes \mathbb{Q}_p/\mathbb{Z}_p). \end{aligned}$$

For each $t \in T$, we have $H^{m-1}(G_t, A \otimes \mathbb{Q}_p/\mathbb{Z}_p) \cong H^m(G_t, A)(p) = 0$. Hence $H^m(G, A)(p) = 0$, and $scd_p\, G = n$. □

Exercise 1. Let $(\mathcal{G}, p, T)$ be a bundle of pro-$\mathfrak{c}$-groups and assume that for some $t \in T$ we are given a homomorphism $f : G_t \to H$ into a finite group $H \in \mathfrak{c}$. Show that there exists a closed and open neighbourhood U of t in T and a morphism $F : \mathcal{G}_U \to H$ extending f, i.e. $F|_{G_t} = f$.
Hint: By I §1, ex.2 there exists a continuous map $F : \mathcal{G} \to H$ extending f. The two maps from $\mathcal{G} \times_T \mathcal{G} \to H$ which are given by $(g_1, t) \times (g_2, t) \mapsto F(g_1, t) \cdot F(g_2, t)$ and by $(g_1, t) \times (g_2, t) \mapsto F((g_1, t) \cdot (g_2, t))$ coincide on an open neighbourhood of $G_t \subseteq \mathcal{G} \times_T \mathcal{G}$.

Exercise 2. Assume that the profinite space T is the topological quotient of the profinite space T' by the action of the profinite group P. Let $\mathcal{G}$ be a bundle of profinite groups over T and let

$$\mathcal{G}' := \mathcal{G} \times_T T'$$

be the pull-back of $\mathcal{G}$ to T'. We set $G := \mathop{*}\limits_{T} \mathcal{G}$ and $G' := \mathop{*}\limits_{T'} \mathcal{G}'$. Then for every abelian group A, the cohomology groups $H^i(G', A)$ are discrete P-modules in a natural way. Show that there exist isomorphisms for all i

$$H^i(G', A) = \mathrm{Ind}_P H^i(G, A).$$

The next exercises deal with profinite operator groups. For more information about operator groups, we refer the reader to [256] and [136].

Exercise 3. Let Γ be a profinite group, G a pro-$\mathfrak{c}$-group and let

$$\phi : \Gamma \to \mathrm{Aut}(G)$$

be a homomorphism. Show that the action: $\Gamma \times G \to G$, $(\gamma, g) \mapsto \phi(\gamma)(g)$, is continuous if and only if G possesses a system of neighbourhoods of the identity consisting of open Γ-invariant normal subgroups.

Exercise 4. Prove the assertion that the free pro-$\mathfrak{c}$-Γ operator group $*_\Gamma F_r$ is equal to the maximal pro-$\mathfrak{c}$-factor group of the kernel of the surjection $F_r * \Gamma \twoheadrightarrow \Gamma$, where $*$ denotes the free product in the category of profinite groups.
Hint: See [94] Satz 3.4.

Exercise 5. Let p be a prime number and let $\Gamma = \mathbb{Z}_p$ be the additive group of p-adic integers. Let G be a pro-p-Γ operator group. Show that the following conditions are equivalent:

(i) G is a free pro-p-Γ operator group of rank r,
(ii) G is a free pro-p-group, the fixed module $(G^{ab})^\Gamma$ is trivial and the cofixed module G^{ab}_Γ is a free $\mathbb{Z}_p$-module of rank r.

Hint: In order to show the difficult implication (ii) ⇒ (i), choose a homomorphism

$$\phi : *_\Gamma F_r \to G$$

which induces an isomorphism $\mathbb{Z}_p^r \xrightarrow{\sim} G^{ab}_\Gamma$. Use (1.6.14) and the topological Nakayama lemma (see 5.2.18) in order to show that ϕ is surjective. Then apply Γ-homology to the exact sequence $0 \to (\ker\phi)^{ab}_G \to \mathbb{Z}_p[\![\Gamma]\!]^r \to G^{ab} \to 0$ and observe that $H_1(\Gamma, G^{ab}) = (G^{ab})^\Gamma = 0$.

Chapter V

Iwasawa Modules

The Iwasawa algebra, usually denoted by the Greek letter Λ, is the complete group algebra $\mathbb{Z}_p[\![\Gamma]\!]$ of a group Γ, which is *noncanonically* isomorphic to $\mathbb{Z}_p$. This means that we will not specify a particular isomorphism $\phi : \Gamma \xrightarrow{\sim} \mathbb{Z}_p$ or, equivalently, we will not fix a topological generator γ of the procyclic group Γ.

We can view Λ from many different algebraic aspects. By definition, it is a complete group algebra and we will see that, choosing a generator $\gamma \in \Gamma$, it can be identified with a power series ring over $\mathbb{Z}_p$. However, Λ is also a commutative, two-dimensional regular local ring, it is a complete local ring and it is a compact $\mathbb{Z}_p$-algebra. Using all these properties of Λ, we will develop the structure theory for Λ-modules, so-called Iwasawa modules. With an eye on applications to generalized Iwasawa algebras, we will treat each aspect of Λ separately and in greatest possible generality as long as this does not require additional work. We have, however, omitted the interpretation of the elements of Λ as $\mathbb{Z}_p$-valued measures on Γ. For this point of view, which is of particular importance for the analytic part of Iwasawa theory, we refer the reader to [122] or [246].

In the first and third sections (after preparing some basic properties of complete group rings in section 2) we will treat the classification of modules *up to pseudo-isomorphism* which was developed by *K. Iwasawa* and *J.-P. Serre*. To obtain a finer structure theory, we will apply the homotopy theory of modules, developed in section 4, to the case of Λ-modules. This will be done in section 5 closely following work of *U. Jannsen*. Finally, in section 6 we will investigate some further properties of complete group algebras which will be needed in the arithmetic applications.

Throughout this chapter, we will assume basic knowledge of commutative algebra, as can be found in [16] or in many other text books about this subject.

J. Neukirch et al., *Cohomology of Number Fields*,
Grundlehren der mathematischen Wissenschaften 323,
DOI 10.1007/978-3-540-37889-1_5,

§1. Modules up to Pseudo-Isomorphism

I. In this subsection, let A be a commutative, noetherian and integrally closed domain with quotient field K. For every prime ideal $\mathfrak{p}$ in A, one has a canonical embedding $A_\mathfrak{p} \hookrightarrow K$ and $A_\mathfrak{p}$ is integrally closed.

Let $\boldsymbol{P(A)}$ be the set of prime ideals of height $\mathrm{ht}(\mathfrak{p}) = \dim A_\mathfrak{p} = 1$. Since A is integrally closed, the localization $A_\mathfrak{p}$ with respect to $\mathfrak{p} \in P(A)$ is a discrete valuation ring and

$$A = \bigcap_{\mathfrak{p}\in P(A)} A_\mathfrak{p}$$

(see [16], chap. VII, §1, no.6, th. 4).

(5.1.1) Definition. *An A-module M is called* **reflexive** *if the canonical map*

$$\varphi_M : M \longrightarrow M^{++} = \mathrm{Hom}_A(\mathrm{Hom}_A(M, A), A)\,,$$

$$m \longmapsto \varphi_M(m) : \alpha \longmapsto \alpha(m)\,,$$

of M to its bidual is an isomorphism.

Remark: A reflexive module is torsion-free because the dual $M^+ = \mathrm{Hom}_A(M, A)$ of an A-module M is always torsion-free.

If M is a finitely generated and torsion-free A-module, then the localization $M_\mathfrak{p}$ of M with respect to a prime ideal $\mathfrak{p}$ of A is a torsion-free $A_\mathfrak{p}$-module and we have injections

$$M \hookrightarrow M_\mathfrak{p} \hookrightarrow M_\mathfrak{p} \otimes_{A_\mathfrak{p}} K = M \otimes_A K =: V$$

$$M^+ \hookrightarrow (M^+)_\mathfrak{p} \hookrightarrow (M^+)_\mathfrak{p} \otimes_{A_\mathfrak{p}} K = M^+ \otimes_A K = \mathrm{Hom}_K(V, K) =: V^\wedge.$$

We see that

$$M^+ \cong \{\lambda \in V^\wedge \mid \lambda(m) \in A \ \text{ for all } \ m \in M\}$$

$$(M^+)_\mathfrak{p} \cong \{\lambda \in V^\wedge \mid \lambda(m) \in A_\mathfrak{p} \text{ for all } \ m \in M_\mathfrak{p}\} \cong (M_\mathfrak{p})^+.$$

Identifying $(M^+)_\mathfrak{p}$ and $(M_\mathfrak{p})^+$, we write $M_\mathfrak{p}^+$ for this module and consider it as a submodule of $V^\wedge$.

(5.1.2) Lemma. *Let M be a finitely generated torsion-free A-module. Then*

(i) $M^+ = \bigcap_{\mathfrak{p}\in P(A)} M_\mathfrak{p}^+$.

(ii) $M^{++} = \bigcap_{\mathfrak{p}\in P(A)} M_\mathfrak{p}$,

(iii) $M = \bigcap_{\mathfrak{p}\in P(A)} M_\mathfrak{p}$ *if and only if M is reflexive.*

Proof: Let $\lambda \in \bigcap M_{\mathfrak{p}}^+$. Then for every $m \in M$ we get $\lambda(m) \in A_{\mathfrak{p}}$ for all $\mathfrak{p} \in P(A)$, so that $\lambda(m) \in A$ and therefore $\lambda \in M^+$. This proves (i) because the other inclusion is obvious.

Since $M_{\mathfrak{p}}$ is a finitely generated torsion-free module over the discrete valuation ring $A_{\mathfrak{p}}$, $\mathfrak{p} \in P(A)$, $M_{\mathfrak{p}}$ is free and $M_{\mathfrak{p}} \to M_{\mathfrak{p}}^{++}$ is an isomorphism. Via the identification of V with $V^{\wedge\wedge}$, we prove (ii), from which (iii) follows immediately. □

(5.1.3) Corollary. *If M is finitely generated, then M^+ is reflexive.*

(5.1.4) Definition. *A finitely generated A-module M is called* **pseudo-null** *if the following equivalent conditions are fulfilled:*

(i) *$M_{\mathfrak{p}} = 0$ for all prime ideals $\mathfrak{p}$ in A of height* $\operatorname{ht}(\mathfrak{p}) \leq 1$.

(ii) *If $\mathfrak{p}$ is a prime ideal with $\mathfrak{a} = \operatorname{ann}_A(M) \subseteq \mathfrak{p}$, then* $\operatorname{ht}(\mathfrak{p}) \geq 2$.

Remarks: 1. We have $M_{\mathfrak{p}} = 0$ if and only if there is an $s \in A \setminus \mathfrak{p}$ such that $sM = 0$, hence $\operatorname{ann}_A(M) \not\subseteq \mathfrak{p}$. This shows the above equivalence.

2. A pseudo-null module is torsion because $M_{(0)} = M \otimes_A K = 0$.

3. If A is a Dedekind domain, then M is pseudo-null if and only if $M = 0$.

4. If A is a 2-dimensional, noetherian, integrally closed local domain with finite residue field, then M is pseudo-null if and only if M is finite. Indeed, if M is finite, then there exists an $r \in \mathbb{N}$ such that $\mathfrak{m}^r M = 0$, hence $\operatorname{supp}(M) \subseteq \{\mathfrak{m}\}$,where $\mathfrak{m}$ denotes the maximal ideal of A. Conversely, if $\operatorname{supp}(M) = \{\mathfrak{p} \in \operatorname{Spec} A \mid \mathfrak{a} \subseteq \mathfrak{p}\}$ is contained in $\{\mathfrak{m}\}$, then $\mathfrak{m}^r \subseteq \mathfrak{a}$ for some $r \in \mathbb{N}$ and M is a finitely generated $A/\mathfrak{m}^r$-module. But $A/\mathfrak{m}^r$ is finite, thus M is finite.

(5.1.5) Definition. *A homomorphism $f : M \to N$ of finitely generated A-modules is called a* **pseudo-isomorphism** *if* $\ker(f)$ *and* $\operatorname{coker}(f)$ *are pseudo-null or, equivalently, if*

$$f_{\mathfrak{p}} : M_{\mathfrak{p}} \xrightarrow{\sim} N_{\mathfrak{p}}$$

is an isomorphism for all $\mathfrak{p}$ of height ≤ 1. We write

$$f : M \xrightarrow{\approx} N.$$

(5.1.6) Lemma. *Let M be a finitely generated A-torsion module and let $\alpha \in A$ be a non-zero element such that $\operatorname{supp}(A/\alpha A)$ is disjoint to $\operatorname{supp}(M) \cap P(A)$. Then the multiplication on M by α is a pseudo-isomorphism.*

Proof: This is clear, since α is a unit in $A_{\mathfrak{p}}$ for every $\mathfrak{p} \in \mathrm{supp}(M) \cap P(A)$ and $M_{\mathfrak{p}} = 0$ for $\mathfrak{p} = (0)$ or $\mathfrak{p} \in P(A) \smallsetminus \mathrm{supp}(M)$. □

For a finitely generated A-module M let

$T_A(M)$ be the torsion submodule and

$F_A(M) = M/T_A(M)$ be the maximal torsion-free quotient of M.

(5.1.7) Proposition. *If the A-module M is finitely generated, then*

(i) *there exists a pseudo-isomorphism*

$$f : M \xrightarrow{\approx} T_A(M) \oplus F_A(M),$$

(ii) *there exists a finite family $(\mathfrak{p}_i)_{i\in I}$ of prime ideals of height 1 in A, a finite family $(n_i)_{i\in I}$ of natural numbers and a pseudo-isomorphism*

$$g : T_A(M) \xrightarrow{\approx} \bigoplus_{i\in I} A/\mathfrak{p}_i^{n_i}.$$

The families $(\mathfrak{p}_i)$ and (n_i) are uniquely determined by $T_A(M)$ up to renumbering.

Proof: Let $\{\mathfrak{p}_1, \dots, \mathfrak{p}_h\} = \mathrm{supp}(T_A(M)) \cap P(A)$. If $h = 0$, then $T_A(M)$ is pseudo-null and the maps $f : M \xrightarrow{can} F_A(M)$ and $g : T_A(M) \longrightarrow 0$ are of the required form. Now let $h > 0$ and $S = \bigcap_{i=1}^h A \smallsetminus \mathfrak{p}_i = A \smallsetminus \bigcup_{i=1}^h \mathfrak{p}_i$. Then $S^{-1}A$ is a semi-local Dedekind domain and therefore a principal ideal domain.

The $S^{-1}A$-module $S^{-1}T_A(M)$ is the torsion module of $S^{-1}M$. Using the structure theorem for modules over principal ideal domains, we see that $S^{-1}T_A(M)$ is a direct summand of $S^{-1}M$. Since M is finitely generated, we have

$$\mathrm{Hom}_{S^{-1}A}(S^{-1}M, S^{-1}T_A(M)) = S^{-1}\mathrm{Hom}_A(M, T_A(M)).$$

Hence there exists a morphism $f_0 : M \to T_A(M)$ and $s_0 \in S$ such that

$$\frac{f_0}{s_0} : S^{-1}M \longrightarrow S^{-1}T_A(M)$$

is the projector of $S^{-1}M$ onto its direct summand $S^{-1}T_A(M)$. Therefore $\frac{f_0}{s_0}\big|_{S^{-1}T_A(M)} = id_{S^{-1}T_A(M)}$ and thus there exists an $s_1 \in S$ such that for $f_1 = s_1 f_0$

$$f_1\big|_{T_A(M)} = s_1 s_0\, id_{T_A(M)}\,.$$

Now let

$$f = (f_1, can) : M \longrightarrow T_A(M) \oplus F_A(M)\,.$$

The commutative exact diagram

$$\begin{array}{ccccccccc}
0 & \longrightarrow & T_A(M) & \longrightarrow & M & \longrightarrow & F_A(M) & \longrightarrow & 0 \\
 & & \downarrow{\scriptstyle f_1|T_A(M)} & & \downarrow{\scriptstyle f} & & \| & & \\
0 & \longrightarrow & T_A(M) & \longrightarrow & T_A(M) \oplus F_A(M) & \longrightarrow & F_A(M) & \longrightarrow & 0
\end{array}$$

shows $\ker(f) = \ker(f_1|_{T_A(M)})$ and $\operatorname{coker}(f) = \operatorname{coker}(f_1|_{T_A(M)})$. But $f_1|_{T_A(M)}$ is a pseudo-isomorphism by (5.1.6). This proves (i).
In order to prove (ii), let

$$E := \bigoplus_{i=1}^{h} \bigoplus_{j=1}^{r_i} A/\mathfrak{p}_i^{n_{ij}}$$

for natural numbers n_{ij} such that there exists an isomorphism

$$g_0 : S^{-1}T_A(M) \xrightarrow{\sim} S^{-1}E.$$

We use again the structure theorem for modules over principal ideal domains, and the fact that $S^{-1}A$ is a semi-local ring with maximal ideals $S^{-1}\mathfrak{p}_i$, $i = 1, \ldots, n$. Using

$$\operatorname{Hom}_{S^{-1}A}(S^{-1}T_A(M), S^{-1}E) = S^{-1}\operatorname{Hom}_A(T_A(M), E),$$

we obtain a morphism $g : T_A(M) \to E$ and an $s \in S$ such that $g = sg_0$. Again by (5.1.6) we see that g is a pseudo-isomorphism. □

Remarks: 1. The same argument as in the proof of (ii) shows that for a pseudo-isomorphism

$$f : M \xrightarrow{\approx} N$$

of finitely generated torsion modules there exists a pseudo-isomorphism

$$g : N \xrightarrow{\approx} M.$$

Therefore we will use in this case the notion $M \approx N$.

For general finitely generated A-modules, the existence of a pseudo-isomorphism $M \xrightarrow{\approx} N$ does not imply the existence of a pseudo-isomorphism in the other direction; see §3, ex.1 for an example.

2. Let

$$0 \longrightarrow M' \longrightarrow M \longrightarrow M'' \longrightarrow 0$$

be an exact sequence of finitely generated A-torsion modules such that the associated sets of prime ideals of height 1 of M' and M'' are disjoint. Then there exists a pseudo-isomorphism

$$M \xrightarrow{\approx} M' \oplus M''.$$

The proof is similar to the proof of (5.1.7) (ii).

(5.1.8) Proposition. *Let M be a finitely generated torsion-free A-module. Then there exists an injective pseudo-isomorphism of M into a reflexive A-module M'.*

Proof: The canonical morphism

$$\varphi_M : M \longrightarrow M^{++}$$

is a pseudo-isomorphism because $M_\mathfrak{p}$ is a free finitely generated $A_\mathfrak{p}$-module, hence $M_\mathfrak{p} \xrightarrow{\sim} M_\mathfrak{p}^{++}$ for all $\mathfrak{p}$ of height ≤ 1. Furthermore, by (5.1.3), M^{++} is reflexive and $\ker(\varphi_M) \otimes_A K = 0$, thus $\ker(\varphi_M)$ is torsion and therefore zero. $\square$

II. Now let A be a 2-dimensional regular local ring *). The following proposition is essential for the structure theory of Iwasawa modules.

(5.1.9) Proposition. *Let A be an n-dimensional regular local ring, $2 \leq n < \infty$, let $(p_1, \ldots, p_n)$ be a regular system of parameters generating the maximal ideal of A and let $p_0 := 0$. For a finitely generated A-module M, the following assertions are equivalent.*

(i) *For every $i = 0, \ldots, n-2$, the $A/(p_0, \ldots, p_i)$-module $M/(p_0, \ldots, p_i)M$ is reflexive.*

(ii) *M is a free A-module.*

In particular, a reflexive A-module M over a 2-dimensional regular local ring A is free.

Proof (*Diekert* [40]): In order to prove the nontrivial implication we assume (i). In particular, M is reflexive, hence torsion-free. Therefore multiplication by p_1 is injective on M. If $\varphi : A^r \twoheadrightarrow M$ is a minimal free presentation of M, then we obtain the following commutative exact diagram

$$\begin{array}{ccccccccc} 0 & \longrightarrow & A^r & \xrightarrow{p_1} & A^r & \longrightarrow & (A/p_1)^r & \longrightarrow & 0 \\ & & \downarrow{\scriptstyle\varphi} & & \downarrow{\scriptstyle\varphi} & & \downarrow{\scriptstyle\bar\varphi} & & \\ 0 & \longrightarrow & M & \xrightarrow{p_1} & M & \longrightarrow & M/p_1 & \longrightarrow & 0\,. \end{array}$$

*) For the definition and properties of regular local rings and regular systems of parameters see [129].

Assume that M/p_1 is a free A/p_1-module. Then Nakayama's lemma implies that $\bar{\varphi}$ is an isomorphism, hence multiplication by p_1 on $\ker(\varphi)$ is an isomorphism. Again by Nakayama's lemma, we obtain $\ker(\varphi) = 0$.
It remains to show that M/p_1 is a free A/p_1-module. We are reduced to the case $n = 2$ since for $n > 2$ we argue by induction applied to the $(n-1)$-dimensional regular local ring A/p_1, the regular system of parameters $(\bar{p}_2, \dots, \bar{p}_n)$, where $\bar{p}_i = p_i + p_1 A$, and the module M/p_1.*)
So let $n = 2$. Thus A/p_1 is regular of dimension 1, i.e. a discrete valuation ring, in particular, an integral domain. Therefore the A/p_1-module $\mathrm{Hom}_A(M^+, A/p_1)$ is torsion-free. Since A is an integral domain, the map

$$M^{++}/p_1 = \mathrm{Hom}_A(M^+, A) \otimes A/p_1 \hookrightarrow \mathrm{Hom}_A(M^+, A/p_1)$$

is injective and, because M is reflexive, we see that $M/p_1 = M^{++}/p_1$ is a torsion-free module over the discrete valuation ring A/p_1, so that M/p_1 is free. This finishes the proof of the proposition. □

From (5.1.7), (5.1.8) and (5.1.9) we obtain the

(5.1.10) Structure Theorem. *Let A be a 2-dimensional regular local ring and let M be a finitely generated A-module. Then there exist finitely many prime ideals $\mathfrak{p}_i$, $i \in I$, of height 1, a nonnegative integer r, natural numbers $n_i \in \mathbb{N}$ and a pseudo-isomorphism*

$$f : M \xrightarrow{\approx} A^r \oplus \bigoplus_{i \in I} A/\mathfrak{p}_i^{n_i}.$$

The prime ideals $\mathfrak{p}_i$ and the numbers r, n_i are uniquely determined by M:

$$r = \dim_K M \otimes_A K, \quad \{\mathfrak{p}_i \mid i \in I\} = \mathrm{supp}(M) \cap P(A).$$

§2. Complete Group Rings

In this section, let $\mathcal{O}$ be a commutative local ring which is complete in its $\mathfrak{m}$-adic topology, where $\mathfrak{m}$ is the maximal ideal. We further assume that $\mathcal{O}/\mathfrak{m}^n$ is finite for all n; in particular, $\mathcal{O}$ is compact. Let p be the characteristic of the finite residue field $k = \mathcal{O}/\mathfrak{m}$.**). Furthermore, let G be a profinite group.

*) see [129], th. 14.2.

**) In the applications $\mathcal{O}$ will always be the ring of integers in a finite extension field $K|\mathbb{Q}_p$.

(5.2.1) Definition. *The* **complete group algebra of G over $\mathcal{O}$** *is the topological inverse limit*

$$\mathcal{O}[\![G]\!] := \varprojlim_{U} \mathcal{O}[G/U],$$

where U runs through the open normal subgroups of G.

Since $\mathcal{O}[\![G]\!]$ is a compact $\mathcal{O}$-algebra, the assignment $G \to \mathcal{O}[\![G]\!]$ defines a covariant functor from the category of profinite groups to that of compact $\mathcal{O}$-algebras. In particular, if $N \subseteq G$ is a closed normal subgroup, then we have an epimorphism $\mathcal{O}[\![G]\!] \twoheadrightarrow \mathcal{O}[\![G/N]\!]$, whose kernel $I(N)$ is a two-sided closed ideal in $\mathcal{O}[\![G]\!]$. It is the closed left (right) ideal generated by the elements $x-1$, $x \in N$. For the particular case $N = G$, we set

(5.2.2) Definition. *The kernel $I_G := I(G)$ of the canonical epimorphism, the* **augmentation map**

$$\mathcal{O}[\![G]\!] \longrightarrow\!\!\!\!\!\!\rightarrow \mathcal{O},$$

is called the **augmentation ideal** *of G.*[*)]

By a (left) $\mathcal{O}[\![G]\!]$-module M, we always understand a separated topological module, i.e. M carries the structure of a Hausdorff abelian topological group and the structure of an $\mathcal{O}[\![G]\!]$-module such that the action $\mathcal{O}[\![G]\!] \times M \to M$ is continuous. In other words, M is a Hausdorff topological $\mathcal{O}$-module with a continuous G-action. By

$$M_G = M/I_G M,$$

we denote the maximal quotient module of M on which G acts trivially. We call M_G the **module of coinvariants** of M (cf. II §6). If M is an $\mathcal{O}[\![G]\!]$-module, then a finitely generated submodule $N \subseteq M$ is the continuous homomorphic image of the compact module $\mathcal{O}[\![G]\!]^n$ for some n, and hence closed. Thus the notions 'topologically finitely generated' and 'finitely generated' coincide for $\mathcal{O}[\![G]\!]$-modules, and finitely generated modules are compact.

If we are given a left $\mathcal{O}[\![G]\!]$-module M, we can define a right $\mathcal{O}[\![G]\!]$-module M^0 keeping the $\mathcal{O}$-module structure and letting $g \in G$ act as g^{-1}. This establishes an equivalence between the categories of left and right $\mathcal{O}[\![G]\!]$-modules, and we will sometimes ignore the difference between left and right modules using this natural equivalence.

The category $\mathscr{C}$ of compact $\mathcal{O}[\![G]\!]$-modules and the category $\mathscr{D}$ of discrete $\mathcal{O}[\![G]\!]$-modules will be of particular importance. Both are abelian categories,

---

[*)] The reader should not confuse $I_N \subseteq \mathcal{O}[\![N]\!]$ with $I(N) \subseteq \mathcal{O}[\![G]\!]$ for a closed normal subgroup $N \subseteq G$.

and Pontryagin duality defines a contravariant equivalence of categories between $\mathscr{C}$ and $\mathscr{D}$.

(5.2.3) Definition. *For a set X we define the* **free compact $\mathcal{O}[\![G]\!]$-module** *with basis X as*

$$F(X) = \mathcal{O}[\![G]\!]^X = \prod_{x \in X} \mathcal{O}[\![G]\!]$$

with the product topology. If $F(X) \twoheadrightarrow M$ is a continuous surjection onto a compact $\mathcal{O}[\![G]\!]$-module M, then we call the image of X in M a **set of topological generators**.

Remark: The compact $\mathcal{O}[\![G]\!]$-module $F(X)$ has the following universal property: for every compact module $M \in \mathscr{C}$ and every convergent family $(m_x)_{x \in X}$ of elements of M (i.e. for every open neighbourhood U of $0 \in M$, one has $m_x \in U$ for all but finitely many $x \in X$) there exists a unique continuous $\mathcal{O}[\![G]\!]$-module homomorphism $f : F(X) \to M$ such that $f(1_x) = m_x$ (compare with the non-abelian situation in IV §1).

(5.2.4) Proposition. (i) *Every compact $\mathcal{O}[\![G]\!]$-module is the projective limit of finite modules, in particular, it is an abelian pro-p-group. The category $\mathscr{C}$ has sufficiently many projectives and exact inverse limits.*

(ii) *Every discrete $\mathcal{O}[\![G]\!]$-module is the direct limit of finite modules, in particular, it is an abelian p-torsion group. The category $\mathscr{D}$ has sufficiently many injectives and exact direct limits.*

Proof: Assume that $N \in \mathscr{D}$ and $n \in N$. Then $\mathrm{ann}_{\mathcal{O}[\![G]\!]}(n)$ is an open ideal in $\mathcal{O}[\![G]\!]$. Therefore $\mathcal{O}[\![G]\!] \cdot n \subseteq N$ is an $\mathcal{O}/\mathfrak{m}^k[G/U]$-module for some k and some open normal subgroup $U \subseteq G$. This shows the first statement of (ii) and therefore also the first statement of (i) by duality. Hence every $M \in \mathscr{C}$ has a convergent set of topological generators. This implies that free compact modules are projective and that every compact module is the quotient of a free module. By duality we find that $\mathscr{D}$ has sufficiently many injectives. The statement about the exactness of limits only depends on the underlying abelian topological groups and is well-known for discrete modules. It follows for compact modules by duality. □

(5.2.5) Corollary. *A compact $\mathcal{O}[\![G]\!]$-module has a fundamental system of neighbourhoods of zero consisting of open submodules.*

A tensor product for compact $\mathcal{O}[\![G]\!]$-modules is defined by its universal property. Explicitly, let M be a compact right and N be a compact left $\mathcal{O}[\![G]\!]$-module. Then the **complete tensor product** is a compact $\mathcal{O}$-module $M\hat{\otimes}_{\mathcal{O}[\![G]\!]}N$ coming along with an $\mathcal{O}[\![G]\!]$-bihomomorphism *)

$$\alpha : M \times N \longrightarrow M\hat{\otimes}_{\mathcal{O}[\![G]\!]}N$$

with the following property: given any $\mathcal{O}[\![G]\!]$-bihomomorphism f of $M \times N$ into a compact $\mathcal{O}$-module R, there is a unique $\mathcal{O}$-module homomorphism $g : M\hat{\otimes}N \to R$ such that $f = g \circ \alpha$.
The complete tensor product is constructed as follows:

$$M\hat{\otimes}_{\mathcal{O}[\![G]\!]}N = \varprojlim_{U,V} \; M/U \otimes_{\mathcal{O}[\![G]\!]} N/V,$$

where U (resp. V) run through the open $\mathcal{O}[\![G]\!]$-submodules of M (resp. N). Observe that M/U and N/V are finite, so that $M\hat{\otimes}N$ is a compact $\mathcal{O}$-module. The natural bihomomorphisms $M \times N \to M/U \otimes N/V$ induce the desired bihomomorphism $\alpha : M \times N \to M\hat{\otimes}N$ by passing to the limit. The exact sequence

$$0 \longrightarrow \mathrm{im}(M \otimes V + U \otimes N) \longrightarrow M \otimes N \longrightarrow M/U \otimes N/V \longrightarrow 0$$

shows that $M\hat{\otimes}N$ is the completion of $M \otimes N$ in the topology induced by taking $\mathrm{im}(M \otimes V + U \otimes N)$ as a fundamental system of open neighbourhoods of 0.

We denote the left derived functors of the (right exact) functor $-\hat{\otimes}_{\mathcal{O}[\![G]\!]}-$ by $\mathrm{Tor}_{\bullet}^{\mathcal{O}[\![G]\!]}(-,-)$. The canonical isomorphism $M_G \cong \mathcal{O}\hat{\otimes}_{\mathcal{O}[\![G]\!]}M$ induces the

(5.2.6) Proposition. *There are canonical isomorphisms for all $i \geq 0$ and all $M \in \mathscr{C}$:*

$$H_i(G, M) \cong \mathrm{Tor}_i^{\mathcal{O}[\![G]\!]}(\mathcal{O}, M).$$

Proof: Since both functors agree for $i = 0$, it suffices to show that a free $\mathcal{O}[\![G]\!]$-module F has trivial G-homology and one easily reduces to the case $F = \mathcal{O}[\![G]\!]$. We have

$$H_i(G, \mathcal{O}[\![G]\!]) = \varprojlim_{U \subseteq G} H_i(G/U, \mathcal{O}[G/U]),$$

where U runs through the open normal subgroups in G. But the G/U-module $\mathcal{O}[G/U] \cong \mathcal{O} \otimes_{\mathbb{Z}} \mathbb{Z}[G/U]$ is induced, hence homologically trivial. □

---

*) i.e. α is a continuous $\mathcal{O}$-homomorphism such that $\alpha(m\lambda, n) = \alpha(m, \lambda n)$ for $m \in M$, $n \in N$ and $\lambda \in \mathcal{O}[\![G]\!]$.

Since $\mathscr{C}$ (resp. $\mathscr{D}$) is an abelian category with sufficiently many projectives (resp. injectives), we have Ext-functors

$$\mathrm{Ext}^{\bullet}_{\mathcal{O}[\![G]\!]}(-,-) : \mathscr{C} \times \mathscr{C} \longrightarrow \mathcal{A}b$$
$$\mathrm{Ext}^{\bullet}_{\mathcal{O}[\![G]\!]}(-,-) : \mathscr{D} \times \mathscr{D} \longrightarrow \mathcal{A}b$$

in the usual way. In the following, we will make use of intermediate homomorphism groups. Let $M \in \mathscr{C}$, $N \in \mathscr{D}$ and let $f : M \to N$ be an $\mathcal{O}[\![G]\!]$-homomorphism. Then f has finite image which is therefore invariant under an open ideal in $\mathcal{O}[\![G]\!]$. Thus we have a functor

$$\mathscr{H}om_{\mathcal{O}[\![G]\!]}(-,-) : \mathscr{C} \times \mathscr{D} \longrightarrow (\text{discrete } \mathcal{O}\text{-modules})$$

and we can use either projective resolutions in $\mathscr{C}$ or injective resolutions in $\mathscr{D}$ to define the functors

$$\mathscr{E}xt^{\bullet}_{\mathcal{O}[\![G]\!]}(-,-) : \mathscr{C} \times \mathscr{D} \longrightarrow (\text{discrete } \mathcal{O}\text{-modules}).$$

Noting the canonical isomorphism

$$N^G \cong \mathscr{H}om_{\mathcal{O}[\![G]\!]}(\mathcal{O}, N)$$

for $N \in \mathscr{D}$, we have the

(5.2.7) Proposition. *There are canonical isomorphisms*

$$H^i(G, N) \cong \mathscr{E}xt^i_{\mathcal{O}[\![G]\!]}(\mathcal{O}, N)$$

for all $i \geq 0$ and all $N \in \mathscr{D}$.

Proof: For $N \in \mathscr{D}$ the induced module $\mathrm{Ind}_G(N)$ is also in $\mathscr{D}$, i.e. carries the structure of an $\mathcal{O}[\![G]\!]$-module in a natural way. By the arguments of II §6, we therefore see that the functor $H^\bullet(G,-)$ is universal as a δ-functor on $\mathscr{D}$. The same is true for $\mathscr{E}xt^i_{\mathcal{O}[\![G]\!]}(\mathcal{O},-)$ and both functors agree in degree 0. □

(5.2.8) Proposition. *If $M = \varprojlim_{i\in I} M_i \in \mathscr{C}$ and $N = \varinjlim_{j \in J} N_j \in \mathscr{D}$, then*

$$\mathscr{E}xt^n_{\mathcal{O}[\![G]\!]}(M, N) = \varinjlim_{i,j} \mathscr{E}xt^n_{\mathcal{O}[\![G]\!]}(M_i, N_j)$$

for every $n \geq 0$.

Proof: It suffices to show that the functor commutes with limits in the first and in the second variable separately. Let $N \in \mathscr{D}$ be fixed and represent every M_i as an inverse limit over its finite quotients: $M_i = \varprojlim_k M_{i,k}$. Every homomorphism from M_i into N has finite image and therefore factors through some $M_{i,k}$, i.e.

$$\mathscr{H}\!om_{\mathcal{O}[\![G]\!]}(M_i, N) = \varinjlim_k \mathscr{H}\!om_{\mathcal{O}[\![G]\!]}(M_{i,k}, N).$$

Thus the statement of the proposition is true for $n = 0$ in the first variable if we have a surjective system of finite modules. If $i_1 \geq i_2$, then for every finite quotient M_{i_2,k_2} of M_{i_2}, the map $M_{i_1} \longrightarrow M_{i_2} \longrightarrow\!\!\!\!\!\rightarrow M_{i_2,k_2}$ factors through some finite quotient M_{i_1,k_1} of M_{i_1}. Therefore we can write M in the form $M = \varprojlim_{i,k} M_{i,k}$ and we have

$$\varinjlim_i \mathscr{H}\!om_{\mathcal{O}[\![G]\!]}(M_i, N) = \varinjlim_{i,k} \mathscr{H}\!om_{\mathcal{O}[\![G]\!]}(M_{i,k}, N)$$

from what we have already shown. By this consideration, it suffices to consider the case that all M_i are finite. But then we may change the projective system M_i to $\bigcap_{i' \geq i} \mathrm{im}(M_{i'} \to M_i)$ again without changing the limits. Thus we have reduced to the case of a system of finite modules with surjective transition maps. As shown above, in this case $\mathscr{H}\!om(-,-)$ commutes with inverse limits in the first argument and using a fixed injective resolution of the second argument shows the same for $\mathscr{E}\!xt(-,-)$. The proof for limits in the second variable is formally dual to that for the first variable, and will be omitted. □

(5.2.9) Corollary. *For $M \in \mathscr{C}$ and $N \in \mathscr{D}$, there are canonical isomorphisms for all $i \geq 0$*

$$\mathrm{Tor}_i^{\mathcal{O}[\![G]\!]}(M, N^\vee) \cong \mathscr{E}\!xt^i_{\mathcal{O}[\![G]\!]}(M, N)^\vee,$$

where $^\vee$ denotes the Pontryagin dual.

Proof: Let U (resp. V) run through the open submodules of M (resp. $N^\vee$). Then we obtain by (5.2.8)

$$\begin{aligned}\left(M \hat{\otimes}_{\mathcal{O}[\![G]\!]} N^\vee\right)^\vee &= \Big(\varprojlim_{U,V} M/U \otimes_{\mathcal{O}[\![G]\!]} N^\vee/V\Big)^\vee \\ &= \varinjlim_{U,V} (M/U \otimes_{\mathcal{O}[\![G]\!]} N^\vee/V)^\vee \\ &= \varinjlim_{U,V} \mathrm{Hom}_{\mathcal{O}[\![G]\!]}(M/U, (N^\vee/V)^\vee) \\ &= \mathscr{H}\!om_{\mathcal{O}[\![G]\!]}(M, N).\end{aligned}$$

This proves the corollary using the methods of homological algebra. □

From now on we will use the following *notational convention*:

If a given ring (with unit) $\mathcal{A}$ has a natural topology, we tacitly assume that all $\mathcal{A}$-modules are Hausdorff topological modules. If $\mathcal{A}$ is compact, the unspecified term *module* will always mean compact $\mathcal{A}$-module.

(5.2.10) Definition. *Let $\mathcal{A}$ be a ring and let M be an $\mathcal{A}$-module. The* **projective dimension** $pd_{\mathcal{A}}M$ *of M is the minimal number n such that there exists a projective resolution*

$$0 \longrightarrow P_n \longrightarrow P_{n-1} \longrightarrow \ldots \longrightarrow P_0 \longrightarrow M \longrightarrow 0$$

for M of length n. We set $pd_{\mathcal{A}}M = \infty$ if no such resolution exists. For the trivial $\mathcal{A}$-module 0, we set $pd_{\mathcal{A}}\, 0 = -1$.

The **projective dimension of the ring** $\mathcal{A}$, *denoted by $pd(\mathcal{A})$, is defined as* $\sup\{pd_{\mathcal{A}}M \mid M$ *an $\mathcal{A}$-module*$\}$.

Observe that simple discrete or compact $\mathcal{O}[\![G]\!]$-modules are finite, i.e. compact *and* discrete.

(5.2.11) Proposition. *For $M \in \mathscr{C}$ the following assertions are equivalent:*

(i) $pd_{\mathcal{O}[\![G]\!]}M \le n$,

(ii) $\mathscr{E}xt^{n+1}_{\mathcal{O}[\![G]\!]}(M,N) = 0$ *for all simple N,*

(iii) $\mathrm{Ext}^{n+1}_{\mathcal{O}[\![G]\!]}(M,N) = 0$ *for all simple N,*

(iv) $\mathrm{Tor}^{\mathcal{O}[\![G]\!]}_{n+1}(M,N) = 0$ *for all simple N.*

Proof: Using a projective resolution of M, we have for N simple

$$\mathscr{E}xt^{n+1}_{\mathcal{O}[\![G]\!]}(M,N) = \mathrm{Ext}^{n+1}_{\mathcal{O}[\![G]\!]}(M,N).$$

Thus (ii) $\Leftrightarrow$ (iii). The equivalence (ii) $\Leftrightarrow$ (iv) follows from (5.2.9). Using dimension shifting, the remaining equivalence (i) $\Leftrightarrow$ (iii) reduces to the statement

$$P \in \mathscr{C} \text{ is projective } \Longleftrightarrow \mathrm{Ext}^1_{\mathcal{O}[\![G]\!]}(P,N) = 0 \text{ for all simple } N.$$

Therefore we have to show that $P \in \mathscr{C}$ is projective if every exact sequence

$$0 \longrightarrow N \longrightarrow N' \longrightarrow P \longrightarrow 0$$

in $\mathscr{C}$ with N simple splits. Assume that $0 \to A \to B \to P \to 0$ is any exact sequence in $\mathscr{C}$ and consider the collection S of pairs (C,s) consisting of a closed submodule $C \subseteq A$ and a splitting morphism $s : P \to B/C$ such that $\pi \circ s = id_P$, where $\pi : B/C \twoheadrightarrow P$ is the canonical projection. Obviously, S

is not empty and, since $B/\bigcap C_i = \varprojlim B/C_i$, it is an inductively ordered set. By Zorn's lemma, there exists a minimal element (C, s) in S. If $C \neq 0$, we can find an open submodule $C' \subseteq C$ such that C/C' is simple. By assumption there exists a morphism $t : P \to B/C'$ which makes the diagram

$$\begin{array}{ccccccccc} & & & & & & P & & \\ & & & & & \swarrow^{t} & \downarrow{\scriptstyle s} & & \\ 0 & \longrightarrow & C/C' & \longrightarrow & B/C' & \longrightarrow & B/C & \longrightarrow & 0 \end{array}$$

commutative. The element $(C', t) \in S$ is strictly smaller than (C, s). This contradiction proves that $C = 0$, thus P is projective. □

Using (5.2.11) and (5.2.8), we obtain the

(5.2.12) Corollary. *For $M = \varprojlim_i M_i \in \mathscr{C}$, one has*

$$pd_{\mathcal{O}[\![G]\!]} M \le \sup_i \{pd_{\mathcal{O}[\![G]\!]} M_i\}.$$

In particular, the inverse limit of projective modules is projective.

(5.2.13) Corollary. $pd_{\mathcal{O}[\![G]\!]} \mathcal{O} = cd_p\, G.$

Proof: From (5.2.7) and (3.3.2) we obtain the inequality $pd_{\mathcal{O}[\![G]\!]} \mathcal{O} \le cd_p\, G$. The other inequality does not follow directly. The difficulty (which does not occur for $\mathcal{O} \cong \mathbb{Z}_p$) lies in the fact that a simple G-module need not be an $\mathcal{O}[\![G]\!]$-module. We overcome the problem as follows. Let N be a simple G-module with $pN = 0$. Then N is a finite dimensional $\mathbb{F}_p$-vector space. By our assumptions, $k = \mathcal{O}/\mathfrak{m}$ is a finite extension of $\mathbb{F}_p$. Then $N \otimes_{\mathbb{F}_p} k$ is an $\mathcal{O}[\![G]\!]$-module and we obtain for all i

$$H^i(G, N) \otimes_{\mathbb{F}_p} k \cong H^i(G, N \otimes_{\mathbb{F}_p} k) \cong \mathscr{E}xt^i_{\mathcal{O}[\![G]\!]}(\mathcal{O}, N \otimes_{\mathbb{F}_p} k).$$

Hence $H^i(G, N) = 0$ for $i > pd_{\mathcal{O}[\![G]\!]} \mathcal{O}$. This finishes the proof. □

Now we are in the position to identify continuous cochain cohomology of a compact $\mathcal{O}[\![G]\!]$-module with Ext-groups.

(5.2.14) Proposition. *For $M \in \mathscr{C}$, we have isomorphisms*

$$H^i_{cts}(G, M) \cong \mathrm{Ext}^i_{\mathcal{O}[\![G]\!]}(\mathcal{O}, M).$$

for every $i \ge 0$.

Proof: Consider, for every open normal subgroup $U \subseteq G$, the complex $P_\bullet^U$ of free $\mathcal{O}[G/U]$-modules given by

$$P_n^U = \mathcal{O}[(G/U)^{n+1}]$$

with differentials $d_n : P_n^U \to P_{n-1}^U$ defined for $n > 0$ by

$$d_n(g_0, \ldots, g_n) = \sum_{i=0}^{n} (-1)^i (g_0, \ldots, g_{i-1}, g_{i+1}, \ldots, g_n).$$

The complex $P_\bullet^U$ is a free resolution of the trivial $\mathcal{O}[G/U]$-module $\mathcal{O}$. For a finite $\mathcal{O}[G/U]$-module A, we have a natural isomorphism of complexes

$$\mathrm{Hom}^\bullet_{\mathcal{O}[G/U]}(P_\bullet^U, A) \cong C^\bullet(G/U, A),$$

where $C^\bullet(G/U, A)$ is the homogeneous cochain complex of G/U with coefficients in A as defined in I §2. Passing to the inverse limit over U, we obtain from the complexes $P_\bullet^U$ a complex $P_\bullet$ which is a resolution of the trivial $\mathcal{O}[\![G]\!]$-module $\mathcal{O}$ by compact $\mathcal{O}[\![G]\!]$-modules. Since every compact $\mathcal{O}[\![G]\!]$-module is the inverse limit of finite ones, we obtain natural isomorphisms of complexes for every $M \in \mathscr{C}$

$$\mathrm{Hom}^\bullet_{\mathcal{O}[\![G]\!]}(P_\bullet, M) \cong C^\bullet_{cts}(G, M),$$

where $C^\bullet_{cts}(G, M)$ is the continuous homogeneous cochain complex of G with coefficients in M as defined in II §7. By definition, the cohomology of the complex on the right is continuous cochain cohomology. Thus it remains to show that P_n is a projective $\mathcal{O}[\![G]\!]$-module for every $n \geq 0$, because then the cohomology of the left complex is the required Ext-group.

In order to prove that P_n is projective, it suffices to show that $\mathrm{Ext}^1_{\mathcal{O}[\![G]\!]}(P_n, N)$ vanishes for every finite simple N (see (5.2.11)). Let $U \subseteq G$ be open and normal. A class $x \in \mathrm{Ext}^1_{\mathcal{O}[\![G]\!]}(P_n^U, N)$ corresponds to an extension of $\mathcal{O}[\![G]\!]$-modules

$$0 \longrightarrow N \longrightarrow N' \longrightarrow P_n^U \longrightarrow 0.$$

Since N is finite, N' is an $\mathcal{O}[G/V]$-module for some open subgroup $V \subseteq U$ which is normal in G. Consider the pull-back of the above sequence via the natural surjection

$$P_n^V \to P_n^U.$$

We obtain an exact sequence of $\mathcal{O}[G/V]$-modules which splits because P_n^V is a free $\mathcal{O}[G/V]$-module. In other words, the image of x in $\mathrm{Ext}^1_{\mathcal{O}[\![G]\!]}(P_n^V, N)$ vanishes. Using a projective resolution of the first argument, we see that for finite N and compact A the groups $\mathrm{Ext}_{\mathcal{O}[\![G]\!]}(A, N)$ and $\mathscr{E}xt_{\mathcal{O}[\![G]\!]}(A, N)$ coincide. By (5.2.8), we therefore obtain

$$\mathrm{Ext}^1_{\mathcal{O}[\![G]\!]}(P_n, N) \cong \varinjlim_U \mathrm{Ext}^1_{\mathcal{O}[\![G]\!]}(P_n^U, N) = 0. \qquad \square$$

The next result is the natural analogue of the **universal coefficient theorem** for abstract groups (compare, for example, [79], VI, th.15.1).

(5.2.15) Theorem (Universal Coefficient Theorem). *Denoting the homology group $H_n(G, \mathcal{O})$ by $H_n(G)$, the following holds:*

(i) *Let N be a discrete $\mathcal{O}$-module considered as an element in $\mathcal{D}$ with G acting trivially on N. Then we have a natural cohomological spectral sequence*

$$E_2^{pq} = \mathcal{E}xt^p_{\mathcal{O}}(H_q(G), N) \Rightarrow H^{p+q}(G, N).$$

In particular, if $\mathcal{O}$ is a discrete valuation ring, then we obtain a natural exact sequence for all n:

$$0 \longrightarrow \mathcal{E}xt^1_{\mathcal{O}}(H_{n-1}(G), N) \longrightarrow H^n(G, N) \longrightarrow \mathcal{H}om_{\mathcal{O}}(H_n(G), N) \longrightarrow 0.$$

(ii) *Let M be a compact $\mathcal{O}$-module considered as an element in $\mathcal{C}$ with G acting trivially on M. Then we have a natural homological spectral sequence*

$$E^2_{pq} = \mathrm{Tor}^{\mathcal{O}}_p(H_q(G), M) \Rightarrow H_{p+q}(G, M).$$

In particular, if $\mathcal{O}$ is a discrete valuation ring, then we obtain a natural exact sequence for all n:

$$0 \longrightarrow H_n(G) \hat{\otimes}_{\mathcal{O}} M \longrightarrow H_n(G, M) \longrightarrow \mathrm{Tor}^{\mathcal{O}}_1(H_{n-1}(G), M) \longrightarrow 0.$$

Proof: Let $P^\bullet \to \mathcal{O}$ be a compact $\mathcal{O}[[G]]$-projective resolution of $\mathcal{O}$ (concentrated in negative degrees) and let $N \to I^\bullet$ be a discrete $\mathcal{O}$-injective resolution of N. Consider the spectral sequence associated to the double complex

$$A^{pq} = \mathcal{H}om_{\mathcal{O}[[G]]}(P^{-q}, I^p) = \mathcal{H}om_{\mathcal{O}}(P_G^{-q}, I^p).$$

By (2.2.4), its limit term E^n is $\mathcal{E}xt^n_{\mathcal{O}[[G]]}(\mathcal{O}, N) \cong H^n(G, N)$ (see (5.2.7)). Its initial terms E_2^{pq} are easily computed as $\mathcal{E}xt^p_{\mathcal{O}}(H_q(G), N)$, which shows the spectral sequence in (i). If $\mathcal{O}$ is a discrete valuation ring, then $pd\,\mathcal{O} = 1$. Therefore $E_2^{pq} = 0$ for $p \geq 2$, and the spectral sequence induces the asserted short exact sequences. This proves (i).

The proof of (ii) is similar: one chooses a compact $\mathcal{O}$-projective resolution $Q^\bullet \to M$ and considers the double complex

$$A_{pq} = P^{-q} \hat{\otimes}_{\mathcal{O}[[G]]} Q^{-p} = P_G^{-q} \hat{\otimes}_{\mathcal{O}} Q^{-p}. \qquad \square$$

Remark: As in the case of abstract groups, one can show that the exact sequences in (5.2.15) split by an unnatural splitting.

Recall that $\mathcal{O}/\mathfrak{m}^n$ is finite for all n. Therefore $\mathcal{O}[\![G]\!]$ is an inverse limit of finite discrete (hence artinian) rings and the ideals

$$\mathfrak{m}^n\mathcal{O}[\![G]\!] + I(U), \quad n \in \mathbb{N},\ U \subseteq G \text{ open normal},$$

are a fundamental system of neighbourhoods of $0 \in \mathcal{O}[\![G]\!]$. We denote by $\mathrm{Rad}_G \subseteq \mathcal{O}[\![G]\!]$ the **radical** of $\mathcal{O}[\![G]\!]$, i.e. the inverse limit of the radicals of $\mathcal{O}/\mathfrak{m}^n[G/U]$. *) Then Rad_G is a closed two-sided ideal which is the intersection of all open left (right) maximal ideals. The powers $(\mathrm{Rad}_G)^n$, $n \geq 1$, define a topology on $\mathcal{O}[\![G]\!]$, which we will call the ***R*-topology**.

(5.2.16) Proposition. (i) *The R-topology is finer than the canonical topology on $\mathcal{O}[\![G]\!]$; in particular, it is Hausdorff.*

(ii) *The following assertions are equivalent:*

a) $\mathrm{Rad}_G \subseteq \mathcal{O}[\![G]\!]$ *is open.*
b) $\mathcal{O}[\![G]\!]$ *is a semi-local ring.*
c) $(G : G_p) < \infty$, *where G_p is a p-Sylow subgroup in G.*

If assertions (a)–(c) *hold, then the finitely many left (right) maximal ideals of $\mathcal{O}[\![G]\!]$ are open.*

(iii) *$\mathcal{O}[\![G]\!]$ is a local ring if and only if G is a pro-p-group. In this case, the maximal ideal of $\mathcal{O}[\![G]\!]$ is equal to $\mathfrak{m}\mathcal{O}[\![G]\!] + I_G$.*

Proof: For arbitrary n and U, the radical of $\mathcal{O}/\mathfrak{m}^n[G/U]$ is nilpotent, since this ring is artinian ([17], chap.8, §6, no. 4, th. 3). This shows (i).

In order to prove a) $\Rightarrow$ b), assume that a) holds. Then $\mathcal{O}[\![G]\!]/\mathrm{Rad}_G$ is finite, hence there are only finitely many *open* left maximal ideals in $\mathcal{O}[\![G]\!]$, $\mathfrak{M}_1, \ldots, \mathfrak{M}_r$ say. Let $\mathfrak{M}$ be any left maximal ideal. If $\mathfrak{M}$ were not open, we could find elements $x_1, \ldots, x_r$ with $x_i \notin \mathfrak{M}_i$, $i = 1, \ldots, r$. The elements $x_1, \ldots, x_r$ generate a left ideal, I say, which is necessarily closed being a homomorphic image of a free (hence compact) module of finite rank. Applying (5.2.5) to the module $\mathcal{O}[\![G]\!]/I$, we see that I must be contained in an open left maximal ideal. But this is obviously not possible, hence $\mathfrak{M}$ must be open. This shows a) $\Rightarrow$ b) and the final assertion of (ii), which obviously also holds for the right maximal ideals. The implication b) $\Rightarrow$ a) is trivial.

To show a) $\Rightarrow$ c) assume that Rad_G is open and choose n and U such that $\mathfrak{m}^n\mathcal{O}[\![G]\!] + I(U) \subseteq \mathrm{Rad}_G$. We conclude that for every open $V \subseteq U$ which is normal in G and for every $u \in U$ the image of $u-1$ in $\mathcal{O}/\mathfrak{m}[G/V]$ is contained in the radical, hence is nilpotent. This implies $(u-1)^{p^r} \equiv u^{p^r} - 1 \equiv 0$, i.e. $u^{p^r} \in V$, for some r. Hence U is a pro-p-group, showing a) $\Rightarrow$ c). If $\mathcal{O}[\![G]\!]$ is local, then Rad_G is open and maximal, hence $\mathrm{Rad}_G \supseteq \mathfrak{m}\,\mathcal{O}[\![G]\!] + I_G$. Now the

*) The radical of an abstract ring is the intersection of all left (right) maximal ideals.

argument above shows that G is a pro-p-group. Thus we proved the 'only if' part of (iii).

In order to show c) $\Rightarrow$ a), let $(G : G_p) < \infty$. Then the intersection U of all (finitely many) conjugates of G_p is an open normal subgroup of G. We will prove that $\mathfrak{m}\,\mathcal{O}[\![G]\!] + I(U)$ is contained in Rad_G. Let $\mathfrak{M} \subseteq \mathcal{O}[\![G]\!]$ be a left open maximal ideal. Then $N = \mathcal{O}[\![G]\!]/\mathfrak{M}$ is a simple $\mathcal{O}[\![G]\!]$-module; in particular, N is finite and p-torsion by (5.2.4). We conclude that $\mathfrak{m}^n N = 0$ for some n and thus $\mathfrak{m}N = 0$ because N is simple. Hence $\mathfrak{m} \subseteq \mathfrak{M}$, and so $\mathfrak{m}\mathcal{O}[\![G]\!] \subseteq \mathrm{Rad}_G$.

Let $V \subseteq U$ be any open subgroup which is normal in G. We follow the argument given in the proof of [32], (5.26): the augmentation ideal of the artinian ring $\mathcal{O}/\mathfrak{m}[U/V]$ is the only left maximal ideal, hence equal to the radical and therefore nilpotent. Using the identity $g(u-1)g'(u'-1) = gg'(u^{g'}-1)(u'-1)$, we see that the image of $I(U)$ in $\mathcal{O}/\mathfrak{m}[G/V]$ is a nilpotent ideal, hence contained in $\mathrm{Rad}(\mathcal{O}/\mathfrak{m}[G/V])$. Varying V and passing to the limit, we obtain $I(U) \subseteq \mathrm{Rad}_G$, showing the implication c) $\Rightarrow$ a).

If G is itself a pro-p-group, the arguments above show $\mathfrak{m}\mathcal{O}[\![G]\!] + I_G \subseteq \mathrm{Rad}_G$. But $\mathfrak{m}\mathcal{O}[\![G]\!] + I_G$ is an open maximal ideal, so that $\mathcal{O}[\![G]\!]$ is a local ring, showing the remaining assertion of (iii). □

Now let M be a compact $\mathcal{O}[\![G]\!]$-module. Then, in addition to the given topology, there are two other topologies on M:

1. The topology given by the sequence of submodules $\{\mathfrak{m}^n M + I(U)M\}_{n,U}$, where $n \in \mathbb{N}$, and U runs through the open normal subgroups of G. We call this topology the **($\mathfrak{m}$, I)-topology**.
2. The R-topology, which is given by the sequence of submodules $\{(\mathrm{Rad}_G)^n M\}_{n\in\mathbb{N}}$.

By (5.2.16), the R-topology is finer than the $(\mathfrak{m}, I)$-topology.

(5.2.17) Proposition. (i) *The $(\mathfrak{m}, I)$-topology is finer than the original topology on M. In particular, the $(\mathfrak{m}, I)$- and the R-topology are Hausdorff.*

(ii) *If M is finitely generated, then the $(\mathfrak{m}, I)$-topology coincides with the original topology on M.*

Proof: Assume that $N \subseteq M$ is an open submodule. Then by continuity, for every $x \in M$ there exists a neighbourhood $V_x \subseteq M$ of x, an $n_x \in \mathbb{N}$ and an open $U_x \subseteq G$, such that $(\mathfrak{m}^{n_x}\mathcal{O}[\![G]\!] + I(U_x))V_x \subseteq N$. Since M is compact, it is covered by finitely many $V_{x_1}, \ldots, V_{x_r}$ and therefore

$$\left(\mathfrak{m}^n\mathcal{O}[\![G]\!] + I(U)\right)M \subseteq N$$

for $n = \max(n_{x_1}, \dots, n_{x_r})$ and $U = \bigcap_{i=1}^r U_{x_i}$. This shows (i).

If M is finitely generated, then there exists a surjection

$$\mathcal{O}[\![G]\!]^r \longrightarrow\!\!\!\!\rightarrow \big(M \text{ with } (\mathfrak{m}, I)\text{-topology}\big)$$

for some $r \in \mathbb{N}$, which is automatically continuous, as $\mathcal{O}[\![G]\!]$ carries the $(\mathfrak{m}, I)$-topology. This shows that M with the $(\mathfrak{m}, I)$-topology is quasi-compact and Hausdorff by (i). Hence the identity map $(M$ with $(\mathfrak{m}, I)$-topology$) \to M$ is a continuous bijection between compact spaces and therefore a homeomorphism. □

The following lemma is called the **Topological Nakayama Lemma** for complete group rings.

(5.2.18) Lemma. (i) *If $M \in \mathscr{C}$ and $\mathrm{Rad}_G M = M$, then $M = 0$.*

(ii) *$M \in \mathscr{C}$ is generated by $x_1, \dots, x_r$ if and only if $x_i + \mathrm{Rad}_G M$, $i = 1, \dots, r$, generate $M/\mathrm{Rad}_G M$ as an $\mathcal{O}[\![G]\!]/\mathrm{Rad}_G$-module.*

Proof: (i) By (5.2.17), the R-topology is Hausdorff. Hence

$$0 = \bigcap_{n=1}^{\infty} (\mathrm{Rad}_G)^n M = M.$$

In order to show the nontrivial implication in (ii), assume that we have $x_1, \dots, x_r \in M$ such that $x_i + \mathrm{Rad}_G M$, $i = 1, \dots, r$, generate $M/\mathrm{Rad}_G M$. The map

$$\varphi : \bigoplus_{i=1}^{r} \mathcal{O}[\![G]\!]\, e_i \longrightarrow (M \text{ with } (\mathfrak{m}, I)\text{-topology}) \xrightarrow{id} M,$$
$$e_i \longmapsto x_i,$$

is continuous by (5.2.17) (i). Since $\mathcal{O}[\![G]\!]$ is compact, the image N of φ is closed and therefore $M/N \in \mathscr{C}$. But by construction

$$\mathrm{Rad}_G(M/N) = (\mathrm{Rad}_G M + N)/N = M/N,$$

hence (i) yields $M = N$, i.e. φ is surjective. □

(5.2.19) Corollary. *Assume that the profinite group G is topologically finitely generated and that $(G : G_p) < \infty$, where G_p is a p-Sylow subgroup in G. Then the R-topology coincides with the canonical topology on the semi-local ring $\mathcal{O}[\![G]\!]$. For any compact $\mathcal{O}[\![G]\!]$-module M the R- and the $(\mathfrak{m}, I)$-topology coincide. If M is finitely generated, both coincide with the original topology on M.*

Proof: By (5.2.16), the R-topology is finer than the canonical topology, and $\mathcal{O}[[G]]$ is semi-local. Furthermore, Rad_G is open. In order to show that both topologies coincide, we have to show that all powers of Rad_G are open. Choose $n \in \mathbb{N}$ and an open normal subgroup $U \subseteq G$ with

$$I := \mathfrak{m}^n \mathcal{O}[[G]] + I(U) \subseteq \mathrm{Rad}_G.$$

It suffices to show that the powers of I are open. As $\mathfrak{m}/\mathfrak{m}^2 \subseteq \mathcal{O}/\mathfrak{m}^2$ is finite, the topological Nakayama lemma (5.2.18) (for $G = 1$) implies that $\mathfrak{m}$ and hence also $\mathfrak{m}^n$ is finitely generated. If G is topologically finitely generated, then so is U and therefore $I(U)$ is a finitely generated ideal in $\mathcal{O}[[G]]$. We conclude that I and all its powers are finitely generated; in particular, they are closed ideals in $\mathcal{O}[[G]]$. It therefore remains to show that I^m has finite index in $\mathcal{O}[[G]]$ for all m. To begin with, the ring $\mathcal{O}[[G]]/I$ is finite. Furthermore, for all m, the ideal I^m is finitely generated. Hence I^m/I^{m+1} is finitely generated over $\mathcal{O}[[G]]/I$, and thus finite. Now the assertion follows by induction. By what we have just seen, the R- and the $(\mathfrak{m}, I)$-topology coincide on any $\mathcal{O}[[G]]$-module. If, finally, M is finitely generated, then, by (5.2.17), the original topology on M coincides with the $(\mathfrak{m}, I)$-topology. □

(5.2.20) Corollary. *Let G be a pro-p-group and let $P \in \mathscr{C}$ be finitely generated. Then P is a free $\mathcal{O}[[G]]$-module if and only if P is projective.*

Proof: For the nontrivial implication let P be projective and let $P/\mathfrak{M}P \cong K^r$, where $\mathfrak{M}$ is the maximal ideal of the local ring $\mathcal{O}[[G]]$ and $K = \mathcal{O}[[G]]/\mathfrak{M}$. By (5.2.18) (ii), we get a surjection

$$\varphi : \mathcal{O}[[G]]^r \twoheadrightarrow P,$$

so that $\mathcal{O}[[G]]^r \cong P \oplus \ker(\varphi)$ and $\ker(\varphi)/\mathfrak{M}\ker(\varphi) = 0$. Now (5.2.18) (i) yields $\ker(\varphi) = 0$. □

(5.2.21) Corollary. *Let G be a finite p-group and let P be a finitely generated $\mathcal{O}[G]$-module which is free as an $\mathcal{O}$-module and cohomologically trivial as a G-module. Then P is a free $\mathcal{O}[G]$-module.*

Proof: From (1.8.4) it follows that P is $\mathcal{O}[G]$-projective (actually in that proposition we considered $\mathbb{Z}[G]$-modules, but the proof is the same for $\mathcal{O}[G]$-modules). Now the result follows from the previous corollary. □

We will finish this section with some considerations about abstract $\mathcal{O}[\![G]\!]$-modules. We saw in (5.2.17) that a given finitely generated compact $\mathcal{O}[\![G]\!]$-module always carries the $(\mathfrak{m}, I)$-topology. However, if we consider an abstract finitely generated $\mathcal{O}[\![G]\!]$-module M, it is not clear whether it can be endowed with a topology such that it becomes a compact $\mathcal{O}[\![G]\!]$-module. By (5.2.17), this is possible if and only if the $(\mathfrak{m}, I)$-topology on M is Hausdorff. This is always true if M is projective, because then it is a direct summand in a free module of finite rank and the $(\mathfrak{m}, I)$-topology on $\mathcal{O}[\![G]\!]$ is Hausdorff by definition. We make the following

(5.2.22) Definition. *Let $\mathcal{A}$ be a ring (with unit) and let M be an $\mathcal{A}$-module (abstract, resp. compact if $\mathcal{A}$ is a topological ring). Then M is called* **finitely presented** *if there exists an exact sequence (a* **presentation** *of M)*

$$P_1 \longrightarrow P_0 \longrightarrow M \longrightarrow 0$$

with finitely generated (abstract, resp. compact) projective $\mathcal{A}$-modules P_1 and P_0.

If M is a finitely presented abstract $\mathcal{O}[\![G]\!]$-module and $P_1 \to P_0 \to M \to 0$ is a presentation, then P_1 and P_0 are naturally endowed with the $(\mathfrak{m}, I)$-topology and they are compact. Furthermore, every $\mathcal{O}[\![G]\!]$-homomorphism is automatically continuous for the $(\mathfrak{m}, I)$-topology. Hence the image of P_1 in P_0 is closed and we can give M the quotient topology. Then, a posteriori by (5.2.17), this topology is the $(\mathfrak{m}, I)$-topology which is therefore Hausdorff. Thus we have proved the

(5.2.23) Proposition. *The forgetful functor from compact $\mathcal{O}[\![G]\!]$-modules to abstract $\mathcal{O}[\![G]\!]$-modules defines an equivalence of categories:*

$$\left\{\begin{array}{c}\text{finitely presented}\\ \text{compact } \mathcal{O}[\![G]\!]\text{-modules}\\ \text{with continuous}\\ \mathcal{O}[\![G]\!]\text{-homomorphisms}\end{array}\right\} \overset{1-1}{\longleftrightarrow} \left\{\begin{array}{c}\text{finitely presented}\\ \text{abstract } \mathcal{O}[\![G]\!]\text{-modules}\\ \text{with abstract}\\ \mathcal{O}[\![G]\!]\text{-homomorphisms}\end{array}\right\}.$$

If we assume that $\mathcal{O}[\![G]\!]$ is noetherian (e.g. $\mathcal{O} = \mathbb{Z}_p$ and G is a compact Lie group over $\mathbb{Q}_p$, cf. [126], V 2.2.4), then every finitely generated module is finitely presented and the category of finitely generated modules has sufficiently many projectives. Therefore we can calculate Tor and $\mathcal{E}xt$ in either of the categories (compare exercise 1), and also the topological projective dimension coincides with the usual one.

Remark: Most of the material of this section is contained in the article [19] by A. BRUMER, where the slightly more general notion of pseudocompact algebras is considered.

Exercise 1. If one of the compact $\mathcal{O}[\![G]\!]$-modules M and N is finitely presented, then

$$M\hat{\otimes}_{\mathcal{O}[\![G]\!]}N \cong M \otimes_{\mathcal{O}[\![G]\!]} N.$$

Exercise 2. If $M = \varprojlim_{i\in I} M_i$ and $N = \varprojlim_{j\in J} N_j$ are in $\mathscr{C}$, then

$$\mathrm{Tor}_n^{\mathcal{O}[\![G]\!]}(M,N) = \varprojlim_{i,j} \mathrm{Tor}_n^{\mathcal{O}[\![G]\!]}(M_i,N_j)$$

for all $n \in \mathbb{N}$.

Exercise 3. For a compact right $\mathcal{O}[\![G]\!]$-module M and a compact left $\mathcal{O}[\![G]\!]$-module N, define a continuous G-action on $M\hat{\otimes}_{\mathcal{O}}N$ by $g(m\otimes n) = mg^{-1}\otimes gn$. Show that there is a natural spectral sequence of homological type

$$E^2_{i,j} = H_i\big(G, \mathrm{Tor}_j^{\mathcal{O}}(M,N)\big) \Rightarrow \mathrm{Tor}_{i+j}^{\mathcal{O}[\![G]\!]}(M,N).$$

Exercise 4. Let H be a closed normal subgroup of G. Show the existence of a spectral sequence for $M \in \mathscr{C}$ and $N \in \mathscr{D}$ of cohomological type

$$E_2^{i,j} = H^i\big(G/H, \mathscr{E}xt^j_{\mathcal{O}[\![H]\!]}(M,N)\big) \Rightarrow \mathscr{E}xt^{i+j}_{\mathcal{O}[\![G]\!]}(M,N).$$

Exercise 5. Show that

$$pd\,\mathcal{O}[\![G]\!] = pd\,\mathcal{O} + cd_p\,G,$$

where $p = \mathrm{char}(\mathcal{O}/\mathfrak{m})$. In particular,

$$pd\,\mathcal{O}[\![G]\!] = pd\,\mathcal{O}[\![G_p]\!]$$

if G_p is a p-Sylow subgroup in G.

Exercise 6. (Topological Nakayama Lemma) Let $\mathcal{A}$ be a local ring with maximal ideal $\mathfrak{m}$ which is compact in its $\mathfrak{m}$-adic topology. Let M be a compact $\mathcal{A}$-module. Then the following is true.

(i) The $\mathfrak{m}$-adic topology of M is finer than the given one. In particular, M is Hausdorff with respect to the $\mathfrak{m}$-adic topology.

(ii) If M is finitely generated, then both topologies coincide.

(iii) If $\mathfrak{m}M = M$, then $M = 0$.

(iv) M is finitely generated if and only if $M/\mathfrak{m}M$ is a finite dimensional $\mathcal{A}/\mathfrak{m}$-vector space.

Hint: Imitate the proofs of (5.2.17) and (5.2.18).

Exercise 7. (Generalization of Maschke's theorem) Let H be a closed, normal subgroup in G of (not necessarily finite) index prime to $p = \mathrm{char}(\mathcal{O}/\mathfrak{m})$. Then a finitely generated $\mathcal{O}[\![G]\!]$-module is projective if and only if it is $\mathcal{O}[\![H]\!]$-projective.

Hint: See (2.6.11) and use the spectral sequence in exercise 4.

§3. Iwasawa Modules

As in the last section, we assume that $\mathcal{O}$ is a commutative noetherian local ring with maximal ideal $\mathfrak{m}$, finite residue field $k = \mathcal{O}/\mathfrak{m}$ and complete in its $\mathfrak{m}$-adic topology. Let $\mathcal{O}[[T]]$ be the power series ring in one variable over $\mathcal{O}$. Then $\mathcal{O}[[T]]$ is a local ring with maximal ideal $(\mathfrak{m}, T)$, residue field $\mathcal{O}[[T]]/(\mathfrak{m}, T) = k$, noetherian (see [16], chap.III, §2, no.10) and complete with respect to its $(\mathfrak{m}, T)$-adic topology (loc. cit. no.6).

A well-known and useful technical result is the so-called division lemma. For the proof, we refer the reader to [16], chap.VII, §3, no.8.

(5.3.1) Division Lemma. *Let* $f = \sum_{n=0}^{\infty} a_n T^n \in \mathcal{O}[[T]]$ *and let*

$$s := \inf\{n \mid a_n \notin \mathfrak{m}\}$$

be finite. The number s is called the **reduced degree** *of f. Then every $g \in \mathcal{O}[[T]]$ can be written uniquely as*

$$g = fq + r$$

with $q \in \mathcal{O}[[T]]$ and a polynomial $r \in \mathcal{O}[T]$ of degree $\leq s-1$. In particular, $\mathcal{O}[[T]]/(f)$ is a free $\mathcal{O}$-module of rank s with basis $\{T^i \bmod f \mid i = 0, \dots, s-1\}$.

(5.3.2) Definition. *A polynomial $F \in \mathcal{O}[T]$ is called a* **Weierstraß polynomial** *if it is of the form*

$$F = T^s + a_{s-1}T^{s-1} + \cdots + a_1 T + a_0$$

with coefficients $a_0, \dots, a_{s-1}$ contained in $\mathfrak{m}$.

(5.3.3) Corollary. *Let F be a Weierstraß polynomial. Then the injection $\mathcal{O}[T] \hookrightarrow \mathcal{O}[[T]]$ induces an isomorphism*

$$\mathcal{O}[T]/F\mathcal{O}[T] \longrightarrow \mathcal{O}[[T]]/F\mathcal{O}[[T]].$$

Proof: Let $s = \deg(F)$. Then s is the reduced degree of F. Using (5.3.1), the commutative diagram

$$\mathcal{O}[T]/(F) \longrightarrow \mathcal{O}[[T]]/(F)$$

$$\mathcal{O}[T]/(F) \xrightarrow{\sim} \sum_{i=0}^{s-1} T^i \mathcal{O} \xleftarrow{\sim} \mathcal{O}[[T]]/(F)$$

gives the result. □

(5.3.4) Weierstraß Preparation Theorem. *Let $f \in \mathcal{O}[[T]]$ with finite reduced degree s. Then there exists a unique decomposition*

$$f = F \cdot u$$

into a Weierstraß polynomial F of degree s and a unit $u \in \mathcal{O}[[T]]$. Furthermore, F is the characteristic polynomial of the endomorphism on the free $\mathcal{O}$-module $\mathcal{O}[[T]]/(f)$ given by the multiplication by T.

Proof: We apply the division lemma to f and T^s: There exists a unique $v \in \mathcal{O}[[T]]$ and a unique polynomial $G = \sum_{i=0}^{s-1} a_i T^i$ such that

$$T^s = f \cdot v - G.$$

Since f has reduced degree s and

$$T^s + \bar{a}_{s-1} T^{s-1} + \cdots + \bar{a}_0 = \bar{f} \cdot \bar{v}$$

(here $\bar{\ }$ denotes the reduction $\bmod \mathfrak{m}$), it follows that $\bar{a}_i = 0$ for all $i = 0, \ldots, s-1$, and $\deg(\bar{v}) = 0$. Therefore $v \in \mathcal{O}[[T]]^\times$ and $T^s + G$ is a Weierstraß polynomial. Using corollary (5.3.3), we obtain

$$\mathcal{O}[[T]]/(f) = \mathcal{O}[[T]]/(F) \cong \mathcal{O}[T]/(F)$$

thus proving the last assertion. □

Now let $p = \mathrm{char}(\mathcal{O}/\mathfrak{m})$ and assume that Γ is a free pro-p-group of rank 1, i.e. Γ is (noncanonically) isomorphic to the additive group $\mathbb{Z}_p$.

(5.3.5) Proposition. *Assume that γ is a topological generator of $\Gamma \cong \mathbb{Z}_p$. Then the map*

$$\begin{aligned} \mathcal{O}[[T]] &\xrightarrow{\sim} \mathcal{O}[\![\Gamma]\!]\,, \\ T &\longmapsto \gamma - 1\,, \end{aligned}$$

is an isomorphism of topological $\mathcal{O}$-algebras.

Proof: Consider the Weierstraß polynomials

$$\omega_n = (T+1)^{p^n} - 1 = T^{p^n} + \sum_{i=1}^{p^n - 1} \binom{p^n}{i} T^{p^n - i}, \quad n \geq 0\,,$$

and let Γ_n be the unique subgroup of Γ of index p^n. Using corollary (5.3.3), we see that the map

$$\begin{aligned} \mathcal{O}[[T]]/(\omega_n) \xrightarrow{\sim} \mathcal{O}[T]/(\omega_n) &\longrightarrow \mathcal{O}[\Gamma/\Gamma_n]\,, \\ T \bmod \omega_n &\longmapsto \gamma - 1 \bmod \Gamma_n\,, \end{aligned}$$

is an isomorphism of $\mathcal{O}$-algebras with inverse map $\gamma \bmod \Gamma_n \mapsto T{+}1 \bmod \omega_n$. Since

$$\omega_{n+1} = \omega_n((T+1)^{p^n(p-1)} + \cdots + (T+1)^{p^n} + 1),$$

we obtain a commutative diagram

$$\begin{array}{ccc} \mathcal{O}[[T]]/(\omega_{n+1}) & \xrightarrow{\sim} & \mathcal{O}[\Gamma/\Gamma_{n+1}] \\ \downarrow & & \downarrow \\ \mathcal{O}[[T]]/(\omega_n) & \xrightarrow{\sim} & \mathcal{O}[\Gamma/\Gamma_n] \end{array}$$

and hence an isomorphism

$$\varprojlim_n \mathcal{O}[[T]]/(\omega_n) \xrightarrow{\sim} \varprojlim_n \mathcal{O}[\Gamma/\Gamma_n] = \mathcal{O}[\![\Gamma]\!].$$

Finally, the natural homomorphism

$$\mathcal{O}[[T]] \longrightarrow \varprojlim_n \mathcal{O}[[T]]/(\omega_n)$$

is an isomorphism: the compactness of $\mathcal{O}[[T]]$ implies its surjectivity and the inclusions

$$\omega_n \mathcal{O}[[T]] \subseteq (\mathfrak{m}, T)^{n+1}$$

show that its kernel $\bigcap_n \omega_n \mathcal{O}[[T]]$ is zero. Thus we obtain the desired result. □

Now we specialize to the case $\mathcal{O} = \mathbb{Z}_p$.

(5.3.6) Definition. *We call the complete group ring $\Lambda = \mathbb{Z}_p[\![\Gamma]\!]$ the* **Iwasawa algebra** *and a compact Λ-module an* **Iwasawa module**.

By (5.3.5), the Iwasawa algebra Λ is isomorphic to the power series ring $\mathbb{Z}_p[[T]]$. The isomorphism depends on the choice of a topological generator γ of Γ. In the following we will identify $\mathbb{Z}_p[\![\Gamma]\!]$ with $\mathbb{Z}_p[[T]]$ using any fixed generator γ.

(5.3.7) Lemma. *The prime ideals of height 1 in Λ are*

$$\mathfrak{p} = (p) \quad \text{and} \quad \mathfrak{p} = (F),$$

where F is an irreducible Weierstraß polynomial over $\mathbb{Z}_p$.

Proof: Since $\mathbb{Z}_p[\![\Gamma]\!]$ is factorial, the prime ideals of height 1 are of the form $\mathfrak{p} = (f)$ where f is an irreducible element in Λ. Let $(f) \neq (p)$; then the reduced element $\bar{f} \in \mathbb{Z}/p\mathbb{Z}[\![\Gamma]\!]$ is not trivial, and using the Weierstraß preparation theorem (5.3.4) we get

$$(f) = (F),\ F \text{ an irreducible Weierstraß polynomial over } \mathbb{Z}_p.$$

But a Weierstraß polynomial is irreducible in $\mathbb{Z}_p[[T]]$ if and only if it is irreducible in $\mathbb{Z}_p[T]$, see [16], chap.VII, §3.8, cor. of prop. 7. □

Applying the structure theorem (5.1.10) to Λ and using (5.3.5) and remark 4 after (5.1.4), we obtain the

(5.3.8) Structure Theorem for Iwasawa Modules. *Let M be a finitely generated Iwasawa module. Then there exist irreducible Weierstraß polynomials F_j, numbers r, m_i, n_j, and a homomorphism*

$$M \xrightarrow{\approx} \Lambda^r \oplus \bigoplus_{i=1}^{s} \Lambda/p^{m_i} \oplus \bigoplus_{j=1}^{t} \Lambda/F_j^{n_j}$$

with finite kernel and cokernel. The numbers r, m_i, n_j and the prime ideals $F_j\Lambda$ are uniquely determined by M.

(5.3.9) Definition. *With the notation of (5.3.8) we call*

$r(M) = \operatorname{rank}_\Lambda(M) = r$ *the* **Λ-rank** *of M,*

$\mu(M) = \sum_{i=1}^{s} m_i$ *the* **Iwasawa μ-invariant** *of M,*

$\lambda(M) = \sum_{j=1}^{t} n_j \deg(F_j)$ *the* **Iwasawa λ-invariant** *of M,*

$F_{M,\gamma} = \prod_{j=1}^{t} F_j^{n_j}$ *the* **characteristic polynomial** *of M.*

Furthermore, we call a finitely generated Λ-module of the form

$$E = \Lambda^r \oplus \bigoplus_{i=1}^{s} \Lambda/p^{m_i} \oplus \bigoplus_{j=1}^{t} \Lambda/F_j^{n_j}$$

an **elementary** *Λ-module.*

Remarks: 1. The invariants defined above depend on M only up to pseudo-isomorphism and $r(M)$, $\mu(M)$ and $\lambda(M)$ are independent of the chosen generator γ, in contrast to the characteristic polynomial. Furthermore, observe that $F_{M,\gamma} = F_{T_\Lambda(M),\gamma}$, where $T_\Lambda(M)$ is the Λ-torsion submodule of M.

2. The invariants $\mu(M)$ and $\lambda(M)$ are additive and $F_{M,\gamma}$ is multiplicative in short exact sequences of finitely generated Λ-torsion modules. Furthermore, a finitely generated Λ-torsion module M is finite if and only if $\lambda(M) = 0 = \mu(M)$.

3. Let M be a finitely generated Λ-torsion module, then

$$\lambda(M) = \dim_{\mathbb{Q}_p}(M \otimes_{\mathbb{Z}_p} \mathbb{Q}_p)$$

and F_M is the characteristic polynomial of the endomorphism on the $\mathbb{Q}_p$-vector space $M \otimes_{\mathbb{Z}_p} \mathbb{Q}_p$ given by multiplication by T.

As before, M_Γ denotes the module of Γ-coinvariants of M:

$$M_\Gamma = M/I_\Gamma M \cong M/TM,$$

where I_Γ is the augmentation ideal of Λ. The topological Nakayama lemma (5.2.18) implies the

(5.3.10) Proposition. *Let M be an Iwasawa module. Then the following assertions are equivalent.*

(i) *M is a finitely generated Λ-module.*

(ii) *M_Γ is a finitely generated $\mathbb{Z}_p$-module.*

(iii) *$M/\mathfrak{m}M$ is a finite-dimensional $\mathbb{F}_p$-vector space.*

Very useful is the following

(5.3.11) Lemma. *Let*

$$0 \longrightarrow M_1 \longrightarrow M_2 \longrightarrow M_3 \longrightarrow 0$$

be an exact sequence of Λ-modules. Then there is an exact sequence of $\mathbb{Z}_p$-modules

$$0 \longrightarrow M_1^\Gamma \longrightarrow M_2^\Gamma \longrightarrow M_3^\Gamma \longrightarrow (M_1)_\Gamma \longrightarrow (M_2)_\Gamma \longrightarrow (M_3)_\Gamma \longrightarrow 0 .$$

Proof: Since the sequences

$$0 \longrightarrow M_i^\Gamma \longrightarrow M_i \xrightarrow{\gamma-1} M_i \longrightarrow (M_i)_\Gamma \longrightarrow 0$$

are exact for $i = 1, 2, 3$, the result follows from the snake lemma. □

We denote the unique subgroup of Γ of index p^n by Γ_n.

(5.3.12) Definition. *Let M be an Iwasawa module. Then M^δ denotes the maximal Λ-submodule of M on which Γ acts discretely:*

$$M^\delta = \bigcup_n M^{\Gamma_n}.$$

Let $M_0 = \operatorname{tor}_{\mathbb{Z}_p} M^\delta$. If $M^\delta/M_0 \neq 0$, then let $d = d(M)$ be the minimal number such that Γ_d acts trivially on M^δ/M_0. If this module is zero, we put $d = -1$.

(5.3.13) Definition. *We denote the p^n-th cyclotomic polynomial by*

$$\xi_n = \frac{\omega_n}{\omega_{n-1}}, \qquad n \geq 0,$$

where we put $\omega_{-1} = 1$ and

$$\omega_n = (T+1)^{p^n} - 1 = T^{p^n} + \sum_{i=1}^{p^n-1} \binom{p^n}{i} T^{p^n-i}, \qquad n \geq 0.$$

Hence

$$\omega_n = \xi_0 \cdot \xi_1 \cdots \xi_n \text{ and } \xi_0 = \omega_0 = T,\ \xi_k = \sum_{i=0}^{p-1} (1+T)^{ip^{k-1}} \quad \text{for } k \geq 1.$$

(5.3.14) Lemma. *Let M be a finitely generated Λ-module. Then*

(i) *M^δ is a Λ-torsion module and finitely generated as a $\mathbb{Z}_p$-module, thus $d(M) < \infty$.*

(ii) *M_0 is the maximal finite Λ-submodule of M.*

(iii) *$\operatorname{supp}(M^\delta) \cap P(\Lambda) \subseteq \{(\xi_n) \mid n \geq 0\}$, i.e. the prime ideals of height 1 in the support of M^δ are principal ideals generated by cyclotomic polynomials. Moreover, there is a pseudo-isomorphism*

$$M^\delta \approx \bigoplus_i \Lambda/\xi_{n_i}$$

with $n_i \leq d(M)$.

(iv) *$(M/M^\delta)^\delta$ is $\mathbb{Z}_p$-torsion-free.*

(v) *Let, in addition, M be Λ-torsion. Then M_{Γ_n} is finite for all n if and only if $d(M) = -1$.*

Proof: Since M is a finitely generated Λ-module and Λ is noetherian, M^δ is also finitely generated. Therefore there exists an $n \in \mathbb{N}$ such that Γ_n acts trivially on M^δ, showing (i) and (ii). Since $\omega_{d(M)}(M^\delta/M_0) = 0$, we get (iii).

From the exact sequence

$$0 \longrightarrow M^\delta \longrightarrow M \longrightarrow M/M^\delta \longrightarrow 0$$

we obtain (using (5.3.11) and passing to the limit) the exact sequence

$$0 \longrightarrow M^\delta \xrightarrow{id} M^\delta \longrightarrow (M/M^\delta)^\delta \longrightarrow \varinjlim_n (M^\delta)_{\Gamma_n}.$$

The module $\varinjlim_n (M^\delta)_{\Gamma_n}$ is $\mathbb{Z}_p$-torsion-free since the transition maps

$$(M^\delta)_{\Gamma_n} \xrightarrow{\frac{\omega_m}{\omega_n}} (M^\delta)_{\Gamma_m}, \quad m \ge n,$$

coincide with multiplication by p^{m-n} if n is large enough. Hence $(M/M^\delta)^\delta$ is $\mathbb{Z}_p$-torsion-free. This proves (iv).

Finally, (v) follows from the fact that for a finitely generated Λ-torsion module M, the group M_{Γ_n} is finite if and only if M^{Γ_n} is finite. □

For a finitely generated Λ-module M, we define the Λ-submodule M^{cycl} of M in the following way. Let M_0 be as above, $M_1 = M^\delta$ and we define inductively

$$M_{i+1} = \ker\left(M \longrightarrow (M/M_i)/(M/M_i)^\delta\right) \text{ for } i \ge 1\,.$$

From the commutative exact diagram

$$\begin{array}{ccccccccc} 0 & \longrightarrow & M_{i+1} & \longrightarrow & M & \longrightarrow & (M/M_i)/(M/M_i)^\delta & \longrightarrow & 0 \\ & & & & \downarrow & & \| & & \\ 0 & \longrightarrow & (M/M_i)^\delta & \longrightarrow & M/M_i & \longrightarrow & (M/M_i)/(M/M_i)^\delta & \longrightarrow & 0, \end{array}$$

we obtain an exact sequence

$$0 \longrightarrow M_i \longrightarrow M_{i+1} \longrightarrow (M/M_i)^\delta \longrightarrow 0 \quad \text{for } i \ge 0\,.$$

The submodules M_i are Λ-torsion modules whose support $\mathrm{supp}(M_i) \cap P(\Lambda)$ is contained in $\{(\xi_n) \mid n \ge 0\}$, and $(M/M_{i+1})^\delta = ((M/M_i)/(M/M_i)^\delta)^\delta$ is $\mathbb{Z}_p$-torsion-free by (5.3.14)(iv). Therefore, since M is finitely generated, the sequence of submodules

$$M_0 \subseteq M_1 \subseteq \cdots \subseteq M_i \subseteq M_{i+1} \subseteq \cdots$$

stabilizes.

(5.3.15) Definition. *Let M be a finitely generated Λ-module. Then*

$$M^{cycl} := \bigcup_i M_i\,.$$

(5.3.16) Lemma. *Let M be a finitely generated Λ-module. Then*

(i) *M^{cycl} is a Λ-torsion module and finitely generated as a $\mathbb{Z}_p$-module with $M_0 \subseteq M^\delta \subseteq M^{cycl}$,*

(ii) *$\mathrm{supp}(M^{cycl}) \cap P(\Lambda) = \mathrm{supp}(M^\delta) \cap P(\Lambda) \subseteq \{(\xi_n) \mid 0 \le n \le d(M)\}$, $\mathrm{supp}(M/M^{cycl}) \cap P(\Lambda)$ is disjoint to $\{(\xi_n) \mid n \ge 0\}$,*

(iii) *$(M/M^{cycl})^\delta = (M/M^{cycl})^{cycl} = 0$.*

Proof: The first assertion is obvious. If

$$M \approx \Lambda^r \oplus \bigoplus_i \Lambda/p^{m_i} \oplus \bigoplus_j \Lambda/F_j^{n_j} \oplus \bigoplus_k \Lambda/(\xi_{n_k})^{t_k},$$

where F_j are irreducible Weierstraß polynomials different to ξ_n for all $n \geq 0$, then

$$M_i \approx \bigoplus_k \Lambda/(\xi_{n_k})^{\min(t_k,i)} \qquad \text{for } i \geq 0,$$

and

$$M^{cycl} \approx \bigoplus_k \Lambda/(\xi_{n_k})^{t_k}$$

with $n_k \leq d(M)$. This proves (ii).

As we have seen above, the Λ-module $(M/M^{cycl})^{cycl}$ is $\mathbb{Z}_p$-torsion-free. Furthermore, supp$(M/M^{cycl}) \cap P(\Lambda)$ is disjoint to the set of prime ideals $\{(\xi_n) \mid n \geq 0\}$ by (ii), and so $(M/M^{cycl})^{\delta} \subseteq (M/M^{cycl})^{cycl} = 0$. □

One classical feature of Iwasawa theory is the description of the asymptotic behaviour of $\#M_{\Gamma_n}$ when $n \to \infty$, provided these orders are finite. This finiteness is equivalent to $d(M) = -1$ as we saw above. The following proposition covers the case of nonnegative d.

(5.3.17) Proposition. *Let M be a finitely generated Λ-torsion module and let $n_0 \geq d(M)$ be a fixed number. Then*

$$\#(M/\tfrac{\omega_n}{\omega_{n_0}}M) = p^{\mu p^n + \lambda n + \nu}$$

for all n large enough, where $\mu = \mu(M)$, $\lambda = \lambda(M)$ and ν is a constant not depending on n.

Proof: Let $n \geq n_0$ and put

$$\begin{aligned} \nu_n &:= \tfrac{\omega_n}{\omega_{n_0}} = \xi_{n_0+1} \cdots \xi_n \\ &= 1 + (1+T)^{p^{n_0}} + \cdots + (1+T)^{p^{n_0}(p^{n-n_0}-1)}. \end{aligned}$$

First, we observe that $M/\nu_n M$ is finite for all $n \geq n_0$. Indeed, since ν_n is disjoint to supp$(M) \cap P(\Lambda)$ for $n \geq n_0$, the homomorphism $M \xrightarrow{\nu_n} M$ is a pseudo-isomorphism by (5.1.6). Furthermore, we obtain the

Claim: Multiplication by ν_n is injective on M/M_0.

Applying the snake lemma to the exact sequence

$$0 \longrightarrow M_0 \longrightarrow M \longrightarrow M/M_0 \longrightarrow 0,$$

we obtain the exactness of

$$0 \longrightarrow M_0/\nu_n \longrightarrow M/\nu_n \longrightarrow (M/M_0)/\nu_n \longrightarrow 0,$$

since $\ker(M/M_0 \xrightarrow{\nu_n} M/M_0) = 0$ by the claim, and so

$$\#M/\nu_n = \#(M/M_0)/\nu_n \cdot \#M_0/\nu_n.$$

In order to calculate $\#(M/M_0)/\nu_n$, we put $N = M/M_0$. Using the structure theorem, we obtain an exact sequence

$$0 \longrightarrow N \longrightarrow E \longrightarrow C \longrightarrow 0,$$

where C is finite and E is an elementary Λ-module, i.e.

$$E = \bigoplus_{i=1}^{s} \Lambda/p^{m_i} \oplus \bigoplus_{j=1}^{t} \Lambda/F_j^{n_j}.$$

Since multiplication by ν_n is injective on E, we have an exact sequence

$$0 \longrightarrow {}_{\nu_n}C \longrightarrow N/\nu_n \longrightarrow E/\nu_n \longrightarrow C/\nu_n \longrightarrow 0,$$

where ${}_{\nu_n}C$ is defined by the exact sequence $0 \to {}_{\nu_n}C \to C \xrightarrow{\nu_n} C \to C/\nu_n \to 0$. It follows that

$$\#N/\nu_n = \#E/\nu_n.$$

Let $E_i = \Lambda/p^{m_i}$. Obviously, $\#E_i/\omega_n = p^{m_i p^n}$, and the exact sequence

$$0 \longrightarrow E_i/\omega_{n_0} \xrightarrow{\nu_n} E_i/\omega_n \longrightarrow E_i/\nu_n \longrightarrow 0$$

shows that

$$\#E_i/\nu_n = p^{m_i(p^n - p^{n_0})}.$$

For the other summands of E, we need the

(5.3.18) Lemma. *Let M be a finitely generated Λ-torsion module which is free of rank λ as a $\mathbb{Z}_p$-module. Then*

$$\frac{\omega_{n+1}}{\omega_n} M = pM \qquad \text{for } n > \frac{\lambda(\lambda-1)}{2}.$$

We proceed with the proof of (5.3.17). If $E_j = \Lambda/F_j(T)^{n_j}$, then the sequence

$$0 \longrightarrow E_j/\nu_n \xrightarrow{\frac{\omega_{n+1}}{\omega_n}} E_j/\nu_{n+1} \longrightarrow E_j/p \longrightarrow 0$$

is exact for $n \gg 0$ by (5.3.18) and the fact that multiplication by $\xi_{n+1} = \frac{\omega_{n+1}}{\omega_n}$ is injective on E_j. Therefore

$$\#E_j/\nu_{n+1} = p^{n_j \deg(F_j)} \#E_j/\nu_n,$$

showing that

$$\#E_j/\nu_n = p^{n_j \deg(F_j)(n-n_1)} \#E_j/\nu_{n_1},$$

where $n_1 > \max(n_0, \lambda(\lambda-1)/2)$. Putting everything together, we obtain

$$\#E/\nu_n = p^{\lambda(N)n + \mu(N)p^n} \cdot \text{const},$$

where the constant does not depend on n if n is large enough. Because $\lambda(M) = \lambda(N)$ and $\mu(M) = \mu(N)$, we have finished the proof of the asymptotic formula. □

Proof of (5.3.18): The endomorphism on $M \otimes_{\mathbb{Z}_p} \mathbb{F}_p$ given by multiplication by T has the characteristic polynomial X^λ. Hence T is nilpotent, or equivalently γ acts unipotently. Choosing a suitable basis of $M \otimes_{\mathbb{Z}_p} \mathbb{F}_p$, the matrix representing the action of γ is of the form

$$\begin{pmatrix} 1 & & * \\ & \ddots & \\ & & 1 \end{pmatrix} \in GL(\lambda, \mathbb{F}_p),$$

hence contained in a p-Sylow subgroup of $GL(\lambda, \mathbb{F}_p)$. Therefore $\gamma^{p^m} = 1$ on $M \otimes_{\mathbb{Z}_p} \mathbb{F}_p$ for $m \geq \frac{\lambda(\lambda-1)}{2}$. Now let $n > \frac{\lambda(\lambda-1)}{2}$ and let $A \in M(\lambda \times \lambda, \mathbb{Z}_p)$ be the matrix corresponding to the action of γ on M with respect to some basis. Then

$$A^{p^{n-1}} \equiv I \mod p$$

and so

$$A^{p^n} \equiv I \mod p^2,$$

where I denotes the unit matrix. It follows that

$$A^{p^n(p-1)} + \cdots + A^{p^n} + I \equiv pI \mod p^2,$$

so that

$$A^{p^n(p-1)} + \cdots + A^{p^n} + I = pU \quad \text{with } U \in GL(\lambda, \mathbb{Z}_p)\,.$$

Therefore

$$(\gamma^{p^n(p-1)} + \cdots + \gamma^{p^n} + 1)M = pM\,. \qquad \square$$

The following proposition gives a criterion for freeness of a Λ-module.

(5.3.19) Proposition. *Let M be a finitely generated Λ-module.*

(i) *The following assertions are equivalent:*
   a) $pd_\Lambda M \leq 1$.
   b) *M^{Γ_n} is $\mathbb{Z}_p$-free for some n (resp. every n).*
   c) *M has no finite nontrivial Λ-submodule.*

(ii) *M is a free Λ-module if and only if $M^\Gamma = 0$ and M_Γ is $\mathbb{Z}_p$-free.*

Proof: First we prove (ii). In order to show the nontrivial implication, let

$$0 \longrightarrow K \longrightarrow \Lambda^d \longrightarrow M \longrightarrow 0$$

be a minimal presentation of M by a free module. Using (5.3.11), we obtain the exact sequence

$$0 = M^{\Gamma} \longrightarrow K_{\Gamma} \longrightarrow \mathbb{Z}_p^d \longrightarrow M_{\Gamma} \longrightarrow 0.$$

Since d was chosen minimal and M_{Γ} is $\mathbb{Z}_p$-free, the map on the right is an isomorphism. Using Nakayama's lemma, $K_{\Gamma} = 0$ implies $K = 0$.

Now we prove the equivalence of the assertions in (i). Assuming a), there exists an exact sequence

$$0 \longrightarrow \Lambda^{d_1} \longrightarrow \Lambda^{d_0} \longrightarrow M \longrightarrow 0.$$

Using (5.3.11), it follows that $M^{\Gamma_n} \subseteq (\Lambda^{d_1})_{\Gamma_n}$ showing b). Let M_0 be the maximal finite Λ-module of M. Since $M_0 = 0$ if and only if $M_0^{\Gamma_n} = 0$ for some n, the inclusion $M_0^{\Gamma_n} \subseteq M^{\Gamma_n}$ shows the equivalence between b) and c). Suppose b) is true and let

$$0 \longrightarrow N \longrightarrow \Lambda^{d_0} \longrightarrow M \longrightarrow 0$$

be a presentation of M with kernel N. Since $(\Lambda^{d_0})^{\Gamma} = 0$, we obtain by (5.3.11) the exact sequence

$$0 \longrightarrow M^{\Gamma} \longrightarrow N_{\Gamma} \longrightarrow (\Lambda^{d_0})_{\Gamma}.$$

From our assumption it follows that M^{Γ} is $\mathbb{Z}_p$-free, thus N_{Γ} is also $\mathbb{Z}_p$-free. Hence the Λ-module N is free by (ii) and therefore $pd_{\Lambda} M \leq 1$. □

For a finitely generated Λ-module M, we define

$$\begin{aligned} d_0(M) &= \dim_{\mathbb{F}_p} H_0(\Gamma, M)/p, \\ d_1(M) &= \dim_{\mathbb{F}_p} {}_pH_0(\Gamma, M) + \dim_{\mathbb{F}_p} H_1(\Gamma, M)/p, \\ d_2(M) &= \dim_{\mathbb{F}_p} {}_pH_1(\Gamma, M). \end{aligned}$$

(5.3.20) Proposition. *Let M be a finitely generated Λ-module. Then there exists an exact sequence*

$$0 \longrightarrow \Lambda^{d_2(M)} \longrightarrow \Lambda^{d_1(M)} \longrightarrow \Lambda^{d_0(M)} \longrightarrow M \longrightarrow 0.$$

In particular,

$$\operatorname{rank}_{\Lambda}(M) = d_0(M) - d_1(M) + d_2(M) = \operatorname{rank}_{\mathbb{Z}_p} M_{\Gamma} - \operatorname{rank}_{\mathbb{Z}_p} M^{\Gamma}.$$

Proof: By Nakayama's lemma (5.2.18), we have a minimal presentation $\varphi : \Lambda^{d_0(M)} \twoheadrightarrow M$. Let $N = \ker \varphi$, i.e. we have an exact sequence

$$0 \longrightarrow N \longrightarrow \Lambda^{d_0(M)} \longrightarrow M \longrightarrow 0.$$

Using (5.3.11), we obtain an exact sequence

$$0 \longrightarrow M^{\Gamma} \longrightarrow N_{\Gamma} \longrightarrow \mathbb{Z}_p^{d_0(M)} \longrightarrow M_{\Gamma} \longrightarrow 0$$

which induces the exact sequence

$$0 \longrightarrow M^{\Gamma}/p \longrightarrow N_{\Gamma}/p \longrightarrow {}_p(M_{\Gamma}) \longrightarrow 0$$

(note that $\mathbb{Z}_p^{d_0(M)}/p \xrightarrow{\sim} M_{\Gamma}/p$) and an isomorphism

$${}_p(M^{\Gamma}) \xrightarrow{\sim} {}_p(N_{\Gamma}).$$

It follows that $\dim_{\mathbb{F}_p} N_{\Gamma}/p = d_1(M)$. Hence, again by Nakayama's lemma, we get a minimal presentation $\psi : \Lambda^{d_1(M)} \twoheadrightarrow N$. Let N' be the kernel of ψ. Since $N^{\Gamma} = 0$, we find that N'_{Γ} is $\mathbb{Z}_p$-free and therefore N' is Λ-free by (5.3.19)(ii). Furthermore,

$$\operatorname{rank}_{\Lambda}(N') = \dim_{\mathbb{F}_p} N'_{\Gamma}/p = \dim_{\mathbb{F}_p} {}_p(N_{\Gamma}) = \dim_{\mathbb{F}_p} {}_p(M^{\Gamma}) = d_2(M).$$

□

Exercise 1. Let M be the kernel of the natural projection $\Lambda \longrightarrow \Lambda_{(p,\Gamma)} \cong \mathbb{Z}/p\mathbb{Z}$. Show that M is a torsion-free Λ-module of rank 1 which is not free, and that no pseudo-isomorphism $\Lambda \xrightarrow{\approx} M$ exists.

Exercise 2. Let M be a finitely generated Λ-module. Show the following statements.

a) For every $n = 0, 1, \dots$, the $\mathbb{Z}_p$-module $M/\omega_n M$, $\omega_n = (1+T)^{p^n} - 1$, is finitely generated and
$$\operatorname{rank}_{\mathbb{Z}_p} M/\omega_{n+1} M \geq \operatorname{rank}_{\mathbb{Z}_p} M/\omega_n M .$$

b) M is a Λ-torsion module if and only if $\operatorname{rank}_{\mathbb{Z}_p} M/\omega_n M$ is bounded for $n \to \infty$.

c) Let M be Λ-torsion, then the following assertions are equivalent:

(i) $\mu(M) = 0$.

(ii) M is a finitely generated $\mathbb{Z}_p$-module.

(iii) $\dim_{\mathbb{F}_p} M \otimes_{\mathbb{Z}_p} \mathbb{F}_p < \infty$.

(iv) $\dim_{\mathbb{F}_p}(M/\omega_n M \otimes_{\mathbb{Z}_p} \mathbb{F}_p)$ is bounded for $n \to \infty$.

Exercise 3. Let M be a finitely generated Λ-torsion module. Then the following assertions are equivalent:

(i) $\# M^{\Gamma_n} < \infty$,

(ii) $\# M_{\Gamma_n} < \infty$,

(iii) $F_{M,\gamma}(\zeta - 1) \neq 0$ for all ζ with $\zeta^{p^n} = 1$.

Note that the statement (iii) is independent of the choice of the generator γ of Γ. If the conditions above are fulfilled, then

$$\frac{\# M^{\Gamma_n}}{\# M_{\Gamma_n}} = \frac{1}{p^{\mu(M)p^n}} \prod_{\zeta^{p^n}=1} \left| F_{M,\gamma}(\zeta - 1) \right|_p ,$$

where $|\cdot|_p$ denotes the p-adic valuation on $\mathbb{C}_p$ normalized by $|\, p \,|_p = \frac{1}{p}$.

Exercise 4. Let M be a finitely generated Λ-module. Show that the following assertions are equivalent:

(i) M is a free Λ-module.

(ii) There exist two elements $a,\, b \in \Lambda$ which generate the maximal ideal of Λ, such that

$$_aM = 0 \quad \text{and} \quad {}_b(M/aM) = 0\,,$$

where for a Λ-module N and an element $c \in \Lambda$ the module $_cN$ is defined by $\{x \in N \mid cx = 0\}$. Show that M is Λ-free if and only if M is $\mathbb{Z}_p$-torsion-free and M/pM has no Γ-invariants.

Hint: Use a minimal presentation of M by a free Λ-module and imitate the proof of (5.3.19).

§4. Homotopy of Modules

In this and the following two sections we will follow closely the work of *U. Jannsen* [97]. The reader is strongly advised to consult the original paper for many more results than are presented here; in particular, classification theorems of $\mathbb{Z}_p[[T]]$-modules up to isomorphism.

Let Λ be a ring with unit, not necessarily commutative. Recall that $pd_\Lambda M$ denotes the projective dimension of a Λ-module M.

(5.4.1) Definition. *We denote the full subcategory of $Mod(\Lambda)$ of modules M with $pd_\Lambda M \le 1$ by $Mod_1(\Lambda)$.*

(5.4.2) Definition. (i) *A homomorphism $f : M \to N$ of Λ-modules is* **homotopic to zero** *($f \simeq 0$) if it factors through a projective module P*

$$f : M \longrightarrow P \longrightarrow N.$$

Two homomorphisms $f, g : M \to N$ are **homotopic** *($f \simeq g$) if $f - g$ is homotopic to zero. We denote the* **homotopy category of Λ-modules** *by $Ho(\Lambda)$, i.e. the category whose objects are Λ-modules and in which the homomorphism groups are given by $\mathrm{Hom}_\Lambda(M, N)/\{f \simeq 0\}$.**) *We denote the full subcategory of $Ho(\Lambda)$ whose objects are in $Mod_1(\Lambda)$ by $Ho_1(\Lambda)$.*

(ii) *A homomorphism $f : M \to N$ of Λ-modules is a* **homotopy equivalence** *if there exists a homomorphism $g : N \to M$ such that $fg \simeq id_N$ and $gf \simeq id_M$, i.e. an isomorphism in $Ho(\Lambda)$. In this case, we say that M and N are* **homotopy equivalent** *($M \simeq N$).*

---

*) Observe that the notion of homotopy is compatible with composition of homomorphisms.

(5.4.3) Proposition. *Let $f, g : M \to N$ be homomorphisms of Λ-modules.*

(i) *The following assertions are equivalent.*

a) $f \simeq g$,

b) $f^*, g^* : \mathrm{Ext}^i_\Lambda(N, R) \longrightarrow \mathrm{Ext}^i_\Lambda(M, R)$ *are equal for all Λ-modules R and all $i \geq 1$,*

c) $f^*, g^* : \mathrm{Ext}^1_\Lambda(N, R) \longrightarrow \mathrm{Ext}^1_\Lambda(M, R)$ *are equal for all Λ-modules R.*

(ii) *The following assertions are equivalent.*

a) *f is a homotopy equivalence,*

b) $f^* : \mathrm{Ext}^i_\Lambda(N, R) \longrightarrow \mathrm{Ext}^i_\Lambda(M, R)$ *is an isomorphism for all R and all $i \geq 1$,*

c) $f^* : \mathrm{Ext}^1_\Lambda(N, R) \longrightarrow \mathrm{Ext}^1_\Lambda(M, R)$ *is an isomorphism for all R,*

d) *There are projective Λ-modules P and Q and an isomorphism σ such that f factors as*

$$f : M \xhookrightarrow{\;can\;} M \oplus P \xrightarrow[\sim]{\;\sigma\;} N \oplus Q \xrightarrow{\;can\;} N .$$

Furthermore, if $f : M \to N$ is a homotopy equivalence between finitely generated Λ-modules M and N, then the projective Λ-modules P and Q in d) *may also be chosen finitely generated.*

Proof: (i) Obviously, we may assume that $g = 0$. Since $\mathrm{Ext}^i_\Lambda(P, R) = 0$, $i \geq 1$, for a projective module P, $f \simeq 0$ implies $f^* = 0$. This proves a) $\Rightarrow$ b). The implication b) $\Rightarrow$ c) is trivial. In order to show c) $\Rightarrow$ a), let $\pi : P \twoheadrightarrow N$ be a surjection with P a projective module and let $K = \ker \pi$. Taking the pull-back via f, we obtain the commutative exact diagram

$$\begin{array}{ccccccccc} 0 & \longrightarrow & K & \longrightarrow & X & \longrightarrow & M & \longrightarrow & 0 \\ & & \| & & \downarrow & & \downarrow{\scriptstyle f} & & \\ 0 & \longrightarrow & K & \longrightarrow & P & \xrightarrow{\pi} & N & \longrightarrow & 0 . \end{array}$$

Since $f^* : \mathrm{Ext}^1_\Lambda(N, K) \to \mathrm{Ext}^1_\Lambda(M, K)$ is zero by assumption c), the upper sequence in the diagram splits, thus f factors through the projective Λ-module P.

(ii) Assuming a), there exists $g : N \to M$ such that $fg \simeq id$ and $gf \simeq id$. By (i), it follows that $g^* f^* = id$ and $f^* g^* = id$, and so we obtain b). The implication b) $\Rightarrow$ c) is trivial. In order to prove c) $\Rightarrow$ d), let $\pi : P \twoheadrightarrow N$ be a surjective Λ-homomorphism with P a projective module and let the module K be defined by the exactness of the sequence

$$(*) \qquad 0 \longrightarrow K \longrightarrow M \oplus P \xrightarrow{f+\pi} N \longrightarrow 0 .$$

Applying $\mathrm{Ext}_\Lambda(-,K)$, we obtain the exact sequence

$$\mathrm{Hom}_\Lambda(M\oplus P,K)\longrightarrow \mathrm{Hom}_\Lambda(K,K)\longrightarrow \mathrm{Ext}^1_\Lambda(N,K)\xrightarrow{\sim}\mathrm{Ext}^1_\Lambda(M\oplus P,K)$$

where the right-hand map is an isomorphism by assumption c). A pre-image of $id\in \mathrm{Hom}_\Lambda(K,K)$ gives us a splitting of the exact sequence $(*)$ and we obtain an isomorphism

$$\sigma: M\oplus P\xrightarrow{\sim} N\oplus K$$

which induces f, using the canonical injection and projection respectively. Applying $\mathrm{Ext}_\Lambda(-,R)$ for an arbitrary Λ-module R and using c) we see that $Q:=K$ is projective and we obtain d). Obviously, d) implies a), since the injection $M\hookrightarrow M\oplus P$ and the projection $N\oplus Q\twoheadrightarrow N$ are homotopy equivalences (having the canonical projection and injection as inverses respectively).

Finally, the proof of the implication c) $\Rightarrow$ d) shows that we can choose P finitely generated if N is. The isomorphism σ then implies that Q is also finitely generated, if M is. □

(5.4.4) Definition. *Let M be a (left) Λ-module. Then*

$$E^i(M):=\mathrm{Ext}^i_\Lambda(M,\Lambda),\quad i\ge 0,$$

are (right) Λ-modules by functoriality and the right Λ-module structure of the bimodule Λ. By convention, we set $E^i(M)=0$ for $i<0$. The Λ-dual $E^0(M)$ will also be denoted by M^+.

Remarks: 1. By proposition (5.4.3), the functor E^i factors through a functor

$$Ho(\Lambda)\xrightarrow{E^i} Mod\,(\Lambda)\quad \text{for } i\ge 1.$$

2. Clearly, E^i can also be viewed as functor from right modules to left modules. Also several functors defined below will interchange left and right action. In the case of a group ring, there is a natural equivalence between right and left modules induced by the involution of the group ring given by passing to the inverses of the group elements. In general this is not possible, but for the theory it is not necessary, and in the following we will not specify if we are talking about left and right Λ-modules or if a functor interchanges left and right Λ-actions. This would only cause notational complications and it will always be clear where one has to insert “left” or “right” .

(5.4.5) Definition. *We denote the full subcategory of $Mod\,(\Lambda)$ of modules M with $M^+=0$ by $Mod_+(\Lambda)$. We denote the full subcategory of $Ho(\Lambda)$ whose objects are in $Mod_+(\Lambda)$ by $Ho_+(\Lambda)$.*

(5.4.6) Lemma. *The canonical localization functor $ho : Mod(\Lambda) \to Ho(\Lambda)$ induces an equivalence of categories*

$$ho : Mod_+(\Lambda) \xrightarrow{\sim} Ho_+(\Lambda).$$

Proof: Recall that for a projective module P, the canonical homomorphism $\varphi_P : P \to P^{++}$ is injective (see [17], chap.II, §2, no.8). If $M \in Mod_+(\Lambda)$ and $f : M \to P$ is any homomorphism from M to a projective module P, then the commutative diagram

$$\begin{array}{ccc} M & \xrightarrow{f} & P \\ \downarrow & & \downarrow \\ 0 = M^{++} & \longrightarrow & P^{++} \end{array}$$

shows that $f = 0$. Therefore a homomorphism in $Mod_+(\Lambda)$ which is homotopic to zero *is* zero. This proves the lemma. □

Recall that a Λ-module is called finitely presented if there exists an exact sequence

$$P_1 \longrightarrow P_0 \longrightarrow M \longrightarrow 0$$

with finitely generated projective modules P_1 and P_0.

(5.4.7) Definition. *We denote the full subcategory of $Mod(\Lambda)$ whose objects are finitely presented Λ-modules by $Mod^{fp}(\Lambda)$. The notation $Mod_1^{fp}(\Lambda)$, $Ho^{fp}(\Lambda)$, $Ho_1^{fp}(\Lambda), \dots$ have their obvious meaning.*

Now we will construct a contravariant duality functor

$$D : Ho^{fp}(\Lambda) \longrightarrow Ho^{fp}(\Lambda)$$

as follows:

For every finitely presented module M, we choose a presentation $P_1 \to P_0 \to M \to 0$ of M by finitely generated projectives. Then we define $DM \in \mathcal{O}b(Mod^{fp}(\Lambda)) = \mathcal{O}b(Ho^{fp}(\Lambda))$ by the exact sequence

$$0 \longrightarrow M^+ \longrightarrow P_0^+ \longrightarrow P_1^+ \longrightarrow DM \longrightarrow 0.$$

If $f : M \to N$ is a homomorphism and $Q_1 \to Q_0 \to N \to 0$ is the chosen presentation of N, then we choose $\alpha : P_0 \to Q_0$ and $\beta : P_1 \to Q_1$ to make the diagram

$$\begin{array}{ccccccc} P_1 & \longrightarrow & P_0 & \longrightarrow & M & \longrightarrow & 0 \\ \downarrow{\scriptstyle\beta} & & \downarrow{\scriptstyle\alpha} & & \downarrow{\scriptstyle f} & & \\ Q_1 & \longrightarrow & Q_0 & \longrightarrow & N & \longrightarrow & 0 \end{array}$$

commutative (this is possible since P_0 and P_1 are projective). We define $Df : DN \to DM$ by the commutative diagram

$$\begin{array}{ccccccccc} 0 & \longrightarrow & N^+ & \longrightarrow & Q_0^+ & \longrightarrow & Q_1^+ & \longrightarrow & DN & \longrightarrow & 0 \\ & & \downarrow{\scriptstyle f^+} & & \downarrow{\scriptstyle \alpha^+} & & \downarrow{\scriptstyle \beta^+} & & \downarrow{\scriptstyle Df} & & \\ 0 & \longrightarrow & M^+ & \longrightarrow & P_0^+ & \longrightarrow & P_1^+ & \longrightarrow & DM & \longrightarrow & 0. \end{array}$$

(5.4.8) Definition. *We call the functor*

$$D : Ho^{fp}(\Lambda) \longrightarrow Ho^{fp}(\Lambda)$$

the **transpose**.

(5.4.9) Proposition.

(i) *The functor D is well-defined and (up to canonical functor isomorphism) independent of the chosen presentations.*

(ii) *D is a (contravariant) autoduality of $Ho^{fp}(\Lambda)$, i.e. $D \circ D \cong id$.*

(iii) *For $M \in Mod^{fp}(\Lambda)$, there exists a canonical exact sequence*

$$0 \longrightarrow E^1(DM) \longrightarrow M \xrightarrow{\varphi_M} M^{++} \longrightarrow E^2(DM) \longrightarrow 0,$$

where φ_M is the canonical homomorphism of M to its bidual.

(5.4.10) Definition. *For a finitely presented Λ-module M, we set*

$$T_i(M) := E^i(DM), \quad i = 1, 2.$$

Note that, by (5.4.3), the groups $E^i(DM)$ are well-defined in $Mod(\Lambda)$, while DM is defined only up to homotopy equivalence. Thus $T_i(M)$ only depends on the homotopy equivalence class of M.

Proof of (5.4.9): Let R be a Λ-module. Consider the commutative exact diagram

$$\begin{array}{ccccccc} & & P_1 \otimes_\Lambda R & \longrightarrow & P_0 \otimes_\Lambda R & \longrightarrow & M \otimes_\Lambda R & \longrightarrow & 0 \\ & & \vdots & & \wr\downarrow{\scriptstyle \varphi_{P_0,R}} & & \downarrow{\scriptstyle \varphi_{M,R}} & & \\ 0 & \longrightarrow & \mathrm{Hom}(K, R) & \longrightarrow & \mathrm{Hom}(P_0^+, R) & \longrightarrow & \mathrm{Hom}(M^+, R) & & \end{array}$$

where $K = \ker(P_1^+ \twoheadrightarrow DM)$ and $\varphi_{M,R}$ is the canonical homomorphism

$$M \otimes_\Lambda R \xrightarrow{\varphi_M \otimes R} M^{++} \otimes_\Lambda R \longrightarrow \mathrm{Hom}(M^+, R).$$

Observe that $\varphi_{P_0,R}$ is an isomorphism, since P_0 is finitely generated and projective, and that the dotted arrow factors as

$$P_1 \otimes R \xrightarrow[\sim]{\varphi_{P_1,R}} \operatorname{Hom}(P_1^+, R) \longrightarrow \operatorname{Hom}(K, R).$$

The snake lemma (1.3.1) implies that

$$\begin{aligned} \ker \varphi_{M,R} &\cong \operatorname{coker}(\operatorname{Hom}(P_1^+, R) \longrightarrow \operatorname{Hom}(K, R)) \cong \operatorname{Ext}^1_\Lambda(DM, R), \\ \operatorname{coker} \varphi_{M,R} &\cong \operatorname{coker}(\operatorname{Hom}(P_0^+, R) \longrightarrow \operatorname{Hom}(M^+, R)) \cong \operatorname{Ext}^1_\Lambda(K, R) \\ &\cong \operatorname{Ext}^2_\Lambda(DM, R). \end{aligned}$$

Therefore we obtain an exact sequence

$$(*)\quad 0 \to \operatorname{Ext}^1_\Lambda(DM, R) \xrightarrow{u} M \otimes_\Lambda R \xrightarrow{\varphi_{M,R}} \operatorname{Hom}(M^+, R) \xrightarrow{v} \operatorname{Ext}^2_\Lambda(DM, R) \to 0,$$

where the homomorphisms u and v a priori depend on the chosen presentation $P_1 \to P_0 \to M \to 0$. Now assume that we are given a homomorphism $f : M \to N$ in $\mathit{Mod}^{fp}(\Lambda)$. Let the diagram

$$\begin{array}{ccccccc} P_1 & \longrightarrow & P_0 & \longrightarrow & M & \longrightarrow & 0 \\ \downarrow{\scriptstyle \beta} & & \downarrow{\scriptstyle \alpha} & & \downarrow{\scriptstyle f} & & \\ Q_1 & \longrightarrow & Q_0 & \longrightarrow & N & \longrightarrow & 0 \end{array}$$

be as in the definition of Df, which we denote for the moment by $Df_{(\alpha,\beta)}$. Then for every Λ-module R, we obtain an exact commutative diagram

$$\begin{array}{ccccccc} 0 & \longrightarrow & \operatorname{Ext}^1_\Lambda(DM, R) & \xrightarrow{u_{P_1,P_0}} & M \otimes_\Lambda R & \xrightarrow{\varphi_{M,R}} & \operatorname{Hom}(M^+, R) \\ & & \downarrow{\scriptstyle Df^*_{(\alpha,\beta)}} & & \downarrow{\scriptstyle f\otimes id} & & \downarrow{\scriptstyle (f^+)^*} \\ 0 & \longrightarrow & \operatorname{Ext}^1_\Lambda(DN, R) & \xrightarrow{u_{Q_1,Q_0}} & N \otimes_\Lambda R & \xrightarrow{\varphi_{N,R}} & \operatorname{Hom}(N^+, R). \end{array}$$

Therefore Df is well-defined, i.e. independent of α, β (use (5.4.3)(i)). Furthermore, setting $M = N$ and $f = id$, we see that DM does not depend (up to canonical homotopy equivalence) on the chosen projective presentation. Thus D is a functor from $\mathit{Mod}^{fp}(\Lambda)$ to $\mathit{Ho}^{fp}(\Lambda)$.

If $f : M \to N$ is homotopic to zero, then by definition it factors through a projective module, P say. But then choose (compare with the construction above) $\alpha : P_0 \to Q_0$ also factorizing through P, and $\beta : P_1 \to Q_1$ to be zero:

$$\begin{array}{ccccccc} P_1 & \longrightarrow & P_0 & \longrightarrow & M & \longrightarrow & 0 \\ \downarrow & & \downarrow & & \downarrow & & \\ 0 & \longrightarrow & P & = & P & \longrightarrow & 0 \\ \downarrow & & \downarrow & & \downarrow & & \\ Q_1 & \longrightarrow & Q_0 & \longrightarrow & N & \longrightarrow & 0. \end{array}$$

This shows $Df \simeq 0$. Hence D is well-defined as a functor

$$D : Ho^{fp}(\Lambda) \longrightarrow Ho^{fp}(\Lambda).$$

The assertion (ii) is now trivial since we can choose the sequence $P_0^+ \to P_1^+ \to DM \to 0$ as a projective presentation of DM, showing that $D(DM) \simeq M$. The exact sequence in (iii) follows from $(*)$ on setting $R = \Lambda$ and from the remark that a posteriori the maps u and v do not depend on the chosen presentation. □

(5.4.11) Corollary. *The transpose D defines an equivalence of categories*

$$Ho_1^{fp}(\Lambda) \underset{D}{\overset{D}{\rightleftarrows}} Ho_+^{fp}(\Lambda),$$

and D restricted to $Ho_1^{fp}(\Lambda)$ coincides with E^1.

Proof: If $M \in Ho_+^{fp}(\Lambda)$, then the exact sequence $0 \to P_0^+ \to P_1^+ \to DM \to 0$ shows that $DM \in Ho_1^{fp}(\Lambda)$. Conversely, if $M \in Ho_1^{fp}(\Lambda)$ and we choose the presentation in the form $0 \to P_1 \to P_0 \to M \to 0$, then

$$DM \cong \operatorname{coker}(P_0^+ \to P_1^+) \cong \operatorname{Ext}^1_\Lambda(M, \Lambda) \quad \text{and}$$

$$(DM)^+ \cong \ker(P_1^{++} \to P_0^{++}) \cong \ker(P_1 \to P_0) = 0.$$ □

Together with (5.4.6), the last result implies the

(5.4.12) Corollary. *The functor E^1 defines an equivalence of categories*

$$Ho_1^{fp}(\Lambda) \xrightarrow[\sim]{E^1} Mod_+^{fp}(\Lambda).$$

Now we assume that

$$\Lambda = \mathbb{Z}_p[\![G]\!],$$

where G is a profinite group. Recall from the discussion in §2 that every finitely presented Λ-module carries a natural compact topology and that Λ-homomorphisms of such modules are continuous. If P is finitely generated and projective, then so is P^+. Suppose that the Λ-module M has a resolution

$$\cdots \longrightarrow P_2 \longrightarrow P_1 \longrightarrow P_0 \longrightarrow M \longrightarrow 0$$

by finitely generated projective modules P_i, $i = 0, 1, \ldots$; in particular, M is finitely presented. Then the complex of finitely generated projectives

$$0 \longrightarrow P_0^+ \longrightarrow P_1^+ \longrightarrow P_2^+ \longrightarrow \cdots$$

computes $E^i(M)$. Hence the groups $E^i(M)$ are canonically endowed with a compact topology. If Λ is noetherian, this applies to every finitely generated module M. The following theorem relates the Λ-modules $E^r(M)$ to the discrete G-modules

$$D_r(M^\vee) = \varinjlim_{U \subseteq G} H^r(U, M^\vee)^* , \quad r \geq 0,$$

where

$$M^\vee = \mathrm{Hom}_{cts}(M, \mathbb{Q}_p/\mathbb{Z}_p) = \varinjlim_{U} \mathrm{Hom}\,(M_U, \mathbb{Q}_p/\mathbb{Z}_p) = \varinjlim_{U} (M_U)^*$$

is a discrete G-module (see (2.5.1) for the definition of $D_r(A)$ with a discrete G-module A). We set $D_r(M^\vee) = 0$ if $r < 0$.

(5.4.13) Theorem. *Assume that the Λ-module M has a resolution by finitely generated projective modules.*

(i) *There exists a functorial exact sequence*

$$0 \longrightarrow D_r(M^\vee) \otimes_{\mathbb{Z}_p} \mathbb{Q}_p/\mathbb{Z}_p \longrightarrow E^r(M)^\vee \longrightarrow \mathrm{tor}_{\mathbb{Z}_p} D_{r-1}(M^\vee) \longrightarrow 0$$

for all r.

If, in addition, $\mathrm{tor}_{\mathbb{Z}_p} M$ and $M/\mathrm{tor}_{\mathbb{Z}_p} M$ also have a resolution by finitely generated projectives, then the following hold for all r:

(ii) $$E^r(M/\mathrm{tor}_{\mathbb{Z}_p} M)^\vee \cong \varinjlim_{m} D_r({}_{p^m}(M^\vee)),$$

(iii) $$E^r(\mathrm{tor}_{\mathbb{Z}_p} M)^\vee \cong \varinjlim_{m} D_{r-1}(M^\vee/p^m).$$

(iv) *There is a long exact sequence*

$$\to E^r(M)^\vee \to \varinjlim_{m} D_r({}_{p^m}(M^\vee)) \to \varinjlim_{m} D_{r-2}(M^\vee/p^m) \to E^{r-1}(M)^\vee \to,$$

functorial in M and in G.

Proof: Assume for the moment that Λ is noetherian. The functor $M \rightsquigarrow (M^+)^\vee$ from the category of finitely generated Λ-modules to abelian groups is the composition of the right exact functors $M \rightsquigarrow D_0(M^\vee)$ (this functor takes values in $\mathbb{Z}_p$-modules) and $N \rightsquigarrow N \otimes_{\mathbb{Z}_p} \mathbb{Q}_p/\mathbb{Z}_p$ because

$$\begin{aligned} M^+ &= \varprojlim_{U \subseteq G} \mathrm{Hom}_\Lambda\,(M, \mathbb{Z}_p[G/U]) = \varprojlim_{U \subseteq G} \mathrm{Hom}_{\mathbb{Z}_p[G/U]}\,(M_U, \mathbb{Z}_p[G/U]) \\ &= \varprojlim_{U \subseteq G} \mathrm{Hom}_{\mathbb{Z}_p}\,(M_U, \mathbb{Z}_p), \end{aligned}$$

where the limit is taken over all normal subgroups $U \subseteq G$ with respect to the duals of the norm maps, so that

$$(M^+)^\vee = \varinjlim_{U\subseteq G} \operatorname{Hom}_{\mathbb{Z}_p}(M_U, \mathbb{Z}_p)^\vee = \varinjlim_{U\subseteq G} M_U \otimes_{\mathbb{Z}_p} \mathbb{Q}_p/\mathbb{Z}_p.$$

The r-th derived functor of $M \rightsquigarrow D_0(M^\vee)$ is $M \rightsquigarrow D_r(M^\vee)$ and the first functor in the composition sends projectives to acyclics for the second functor. We get a Grothendieck spectral sequence of homological type

$$E^2_{ij} = \operatorname{Tor}_i^{\mathbb{Z}_p}(D_j(M^\vee), \mathbb{Q}_p/\mathbb{Z}_p) \Rightarrow E_{i+j} = E^{i+j}(M)^\vee .$$

Since

$$\operatorname{Tor}_i^{\mathbb{Z}_p}(N, \mathbb{Q}_p/\mathbb{Z}_p) = \begin{cases} N \otimes \mathbb{Q}_p/\mathbb{Z}_p, & i = 0, \\ \operatorname{tor}_{\mathbb{Z}_p} N, & i = 1, \\ 0, & i \geq 2, \end{cases}$$

we obtain (i) under the assumption that Λ is noetherian.

If Λ is not noetherian, then the category of modules having a resolution by finitely generated projectives has no good properties. Therefore we cannot use the general machinery and have to construct the sequence by hand.

Let $\cdots \to P_1 \to P_0 \to M \to 0$ be a resolution of M by finitely generated projectives. Then the complex

$$(*) \qquad \cdots \longrightarrow D_0(P_2^\vee) \longrightarrow D_0(P_1^\vee) \longrightarrow D_0(P_0^\vee) \longrightarrow 0$$

calculates $D_\bullet(M^\vee)$ and consists of torsion-free, hence flat, $\mathbb{Z}_p$-modules. Choosing a Cartan-Eilenberg resolution of the complex $(*)$ by a $\mathbb{Z}_p$-projective double complex, we obtain (i) from the spectral sequence associated to the double complex tensored by $\mathbb{Q}_p/\mathbb{Z}_p$.

The exact sequence

$$(**) \qquad 0 \longrightarrow (M/\operatorname{tor}_{\mathbb{Z}_p} M)^\vee \longrightarrow M^\vee \longrightarrow (\operatorname{tor}_{\mathbb{Z}_p} M)^\vee \longrightarrow 0$$

and the fact that $(M/\operatorname{tor}_{\mathbb{Z}_p} M)^\vee/p^m = 0$ imply that

$$\varinjlim_m D_{r-1}((\operatorname{tor}_{\mathbb{Z}_p} M)^\vee/p^m) \xrightarrow{\sim} \varinjlim_m D_{r-1}(M^\vee/p^m).$$

This shows that it suffices to show (iii) for the case that M is $\mathbb{Z}_p$-torsion. Next we show that we may assume M to be $\mathbb{Z}_p$-torsion-free in (ii).

Consider, for every m, the p^m-torsion sequence associated to $(**)$

$$0 \longrightarrow {}_{p^m}(M/\operatorname{tor}_{\mathbb{Z}_p} M)^\vee \longrightarrow {}_{p^m}M^\vee \longrightarrow {}_{p^m}(\operatorname{tor}_{\mathbb{Z}_p} M)^\vee \longrightarrow 0.$$

Since $\operatorname{tor}_{\mathbb{Z}_p} M$ is finitely generated, there exists an n such that the p^n-multiplication map

$${}_{p^{m+n}}(\operatorname{tor}_{\mathbb{Z}_p} M)^\vee \xrightarrow{p^n} {}_{p^m}(\operatorname{tor}_{\mathbb{Z}_p} M)^\vee$$

is the zero map for all m. Hence $\varinjlim_m D_r({}_{p^m}((\operatorname{tor}_{\mathbb{Z}_p} M)^\vee)) = 0$ and thus

$$\varinjlim_m D_r({}_{p^m}(M^\vee)) \xrightarrow{\sim} \varinjlim_m D_r({}_{p^m}((M/\operatorname{tor}_{\mathbb{Z}_p} M)^\vee)).$$

In order to prove (ii), let

$$P_\bullet : \qquad \cdots \longrightarrow P_1 \longrightarrow P_0 \longrightarrow M \longrightarrow 0$$

be a resolution of the $\mathbb{Z}_p$-torsion-free module M by finitely generated projectives. Then

$$0 \longrightarrow (M/p^m)^\vee \longrightarrow (P_0/p^m)^\vee \longrightarrow (P_1/p^m)^\vee \longrightarrow \cdots$$

is a resolution of the discrete G-module $(M/p^m)^\vee$ by cohomologically trivial G-modules. Therefore

$$\begin{aligned} E^r(M)^\vee &= H^r((P_\bullet^+)^\vee) \\ &= H^r(\varinjlim_m \varinjlim_{U\subseteq G} (((P_\bullet/p^m)^\vee)^U)^*) \\ &= \varinjlim_m \varinjlim_{U\subseteq G} H^r(((P_\bullet/p^m)^\vee)^U)^* \\ &= \varinjlim_m \varinjlim_{U\subseteq G} H^r(U,(M/p^m)^\vee)^* \\ &= \varinjlim_m D_r({}_{p^m}(M^\vee)). \end{aligned}$$

This shows (ii).

In order to prove (iii), we may assume that $M = \mathrm{tor}_{\mathbb{Z}_p} M$. Then M is already annihilated by some power of p and hence the same is true for $D_{r-1}(M^\vee)$. Therefore $D_{r-1}(M^\vee) \otimes \mathbb{Q}_p/\mathbb{Z}_p = 0$ and (i) implies that

$$\begin{aligned} E^r(M)^\vee &\cong \mathrm{tor}_{\mathbb{Z}_p} D_{r-1}(M^\vee) \\ &= D_{r-1}(M^\vee) \\ &= \varinjlim_m D_{r-1}(M^\vee/p^m). \end{aligned}$$

(The last limit becomes stationary.) This shows (iii).

Finally, (iv) follows from (ii),(iii) and the long exact Ext-sequence

$$\cdots \to E^r(M)^\vee \to E^r(M/\mathrm{tor}_{\mathbb{Z}_p} M)^\vee \to E^{r-1}(\mathrm{tor}_{\mathbb{Z}_p} M)^\vee \to \cdots . \qquad \square$$

(5.4.14) Corollary. *If $cd_p\, G = n$ is finite and if M has a resolution by finitely generated projectives, then $E^r(M) = 0$ for $r > n + 1$.*

Proof: Obviously, we have $D_r(A) = 0$ for $r > n$ and for all p-primary discrete G-modules A. Using (5.4.13)(i), we obtain the result. $\square$

(5.4.15) Corollary. *Assume that G is a duality group* [*)] *at p of dimension n with dualizing module $D_n^{(p)} = \varinjlim_i D_n(\mathbb{Z}/p^i\mathbb{Z})$. Then the following hold for every Λ-module M which has a resolution by finitely generated projectives:*

(i) *If M is free of finite rank as a $\mathbb{Z}_p$-module, then*

$$E^r(M)^\vee \cong \begin{cases} \varinjlim_m D_n((M/p^m)^\vee) \cong M \otimes_{\mathbb{Z}_p} D_n^{(p)} & \text{if } r = n, \\ 0 & \text{otherwise.} \end{cases}$$

(ii) *If M is a finite p-primary G-module, then*

$$E^r(M)^\vee \cong \begin{cases} \mathrm{Hom}_{\mathbb{Z}_p}(M^\vee, D_n^{(p)}) & \text{if } r = n+1, \\ 0 & \text{otherwise.} \end{cases}$$

Proof: By (5.4.13)(ii), we have in case (i)

$$E^r(M)^\vee = \varinjlim_m D_r((M/p^m)^\vee),$$

which is zero for $r \neq n$ by (3.4.6), and

$$\begin{aligned} E^n(M)^\vee &\cong \varinjlim_m \varinjlim_{U \subseteq G} H^0(U, \mathrm{Hom}_{\mathbb{Z}_p}((M/p^m)^\vee, D_n^{(p)})) \\ &\cong \varinjlim_m \mathrm{Hom}_{\mathbb{Z}_p}((M/p^m)^\vee, D_n^{(p)}) \cong M \otimes_{\mathbb{Z}_p} D_n^{(p)}. \end{aligned}$$

In order to prove (ii), we use (5.4.13)(iii):

$$E^r(M)^\vee \cong D_{r-1}(M^\vee),$$

and hence $E^{n+1}(M)^\vee \cong D_n(M^\vee) \cong \mathrm{Hom}_{\mathbb{Z}_p}(M^\vee, D_n^{(p)})$ and $E^r(M) = 0$ for $r \neq n+1$. □

We finish this section with some remarks concerning the change of the group. Let H be an open subgroup in G. We consider the forgetful functor from abstract (resp. compact) $\mathbb{Z}_p[\![G]\!]$-modules to abstract (resp. compact) $\mathbb{Z}_p[\![H]\!]$-modules.

(5.4.16) Lemma. *The forgetful functor sends projectives to projectives, i.e. an abstract (resp. compact) $\mathbb{Z}_p[\![G]\!]$-module which is projective as an abstract (resp. compact) $\mathbb{Z}_p[\![G]\!]$-module is also projective as an abstract (resp. compact) $\mathbb{Z}_p[\![H]\!]$-module. If H is normal and of prime-to-p index in G, then also the converse statement is true.*

[*)] see III §4.

Proof: A projective module is a direct summand in a free module. It thus suffices to show that $\mathbb{Z}_p[[G]]$ is a free $\mathbb{Z}_p[[H]]$-module. But this is clear because

$$\mathbb{Z}_p[[G]] \cong \mathbb{Z}_p[[G]] \otimes_{\mathbb{Z}_p[[H]]} \mathbb{Z}_p[[H]] \cong \mathrm{Ind}_G^H(\mathbb{Z}_p[[H]]).$$

Now suppose that H is normal in G and that $p \nmid (G:H)$. Since $\#(G/H)^{-1} \in \mathbb{Z}_p$, the functor $M \rightsquigarrow M^{G/H}$ is exact on $\mathbb{Z}_p[G/H]$-modules. Therefore the equality

$$\mathrm{Hom}_{\mathbb{Z}_p[[G]]}(M, N) = \mathrm{Hom}_{\mathbb{Z}_p[[H]]}(M, N)^{G/H}$$

implies isomorphisms

$$\mathrm{Ext}^i_{\mathbb{Z}_p[[G]]}(M, N) \cong \mathrm{Ext}^i_{\mathbb{Z}_p[[H]]}(M, N)^{G/H}$$

for all abstract (compact) $\mathbb{Z}_p[[G]]$-modules M, N and all $i \geq 0$. This implies the remaining statement. □

The following proposition shows that the functors E^i commute with the forgetful functor:

(5.4.17) Proposition. *Suppose M is a $\mathbb{Z}_p[[G]]$-module which has a resolution by finitely generated projectives. Let H be an open subgroup in G. Then we have natural isomorphisms of (right) $\mathbb{Z}_p[[H]]$-modules*

$$\mathrm{Ext}^i_{\mathbb{Z}_p[[G]]}(M, \mathbb{Z}_p[[G]]) \cong \mathrm{Ext}^i_{\mathbb{Z}_p[[H]]}(M, \mathbb{Z}_p[[H]])$$

for all $i \geq 0$.

Proof: The isomorphism $\mathbb{Z}_p[[G]] \cong \mathrm{Ind}_G^H(\mathbb{Z}_p[[H]])$ (see the proof of (5.4.16)), together with Frobenius reciprocity, shows that

$$\mathrm{Hom}_{\mathbb{Z}_p[[G]]}(M, \mathbb{Z}_p[[G]]) \cong \mathrm{Hom}_{\mathbb{Z}_p[[H]]}(M, \mathbb{Z}_p[[H]]).$$

By (5.4.16) and since the index of H in G is finite, the forgetful functor sends finitely generated projectives to finitely generated projectives. Therefore the above isomorphism extends to Ext^i for all $i \geq 0$. □

§5. Homotopy Invariants of Iwasawa Modules

In this section, we specialize again to the case $G = \Gamma \cong \mathbb{Z}_p$, so that $\Lambda = \mathbb{Z}_p[[\Gamma]]$ is the Iwasawa algebra. Then Λ is a commutative 2-dimensional regular local ring and therefore its projective dimension $pd(\Lambda)$ is equal to 2 (see [129], th. 19.2; this also follows from §2, ex.5). Thus for every Λ-module M the projective dimension $pd_\Lambda M$ is equal or less than 2, i.e. there exists a projective resolution

$$0 \longrightarrow P_2 \longrightarrow P_1 \longrightarrow P_0 \longrightarrow M \longrightarrow 0$$

of M of length 2 (or smaller). This implies that $E^i(M) = 0$ for $i \geq 3$. By (5.4.9)(iii), the following definition agrees with the one given in (5.4.10).

(5.5.1) Definition. *For a finitely generated Λ-module M we set*

$$T_0(M) := \text{the maximal finite submodule of } M\,,$$

$$T_1(M) := \ker(\varphi_M : M \to M^{++})\,,$$

$$T_2(M) := \operatorname{coker}(\varphi_M : M \to M^{++})\,.$$

(5.5.2) Lemma. *For a Λ-module M the following assertions are equivalent.*

(i) *M is projective,*

(ii) *$M \simeq 0$.*

If M is finitely generated, then M is free if these hold.

Proof: Obviously, (i) $\Rightarrow$ (ii). If $M \simeq 0$, then the identity map $id : M \to M$ factors through a projective module, hence M is itself projective, being a direct summand of a projective. The last assertion follows from (5.2.20). □

(5.5.3) Proposition. *Let M be a finitely generated Λ-module. Then*

(i) *$T_1(M)$ is the Λ-torsion submodule of M.*

(ii) *$E^1(M)$ is a Λ-torsion module. If M is Λ-torsion, then $E^1(M)$ has no nontrivial finite submodules, i.e. $T_0E^1(M) = 0$. If M is finite, then $E^1(M) = 0$.*

(iii) *$T_2(M)$ is finite, and $T_2(M) = 0$ if and only if $M/T_1(M)$ is free (hence $M \cong \Lambda^r \oplus T_1(M)$ for some r in this case).*
In particular, $T_2(M) = 0$ for a Λ-torsion module M.

(iv) *$E^2(M)$ is finite. One has*

$$E^2(M) \cong E^2(T_0(M)) \cong T_0(M)^\vee$$

and the following assertions are equivalent:

a) *$E^2(M) = 0$,*

b) *$T_0(M) = 0$,*

c) *$pd_\Lambda M \leq 1$.*

Proof: Let K be the field of fractions of Λ. By definition, see (5.5.1), we have the exact sequence

$$0 \longrightarrow T_1(M) \otimes_\Lambda K \longrightarrow M \otimes_\Lambda K \xrightarrow{\sim} M^{++} \otimes_\Lambda K \longrightarrow T_2(M) \otimes_\Lambda K \longrightarrow 0.$$

Thus $T_1(M)$ and $T_2(M)$ are Λ-torsion modules. Since M^{++} is torsion-free, we have proved (i) and the first statement of (ii) because $E^1(M) \cong E^1(DDM) \cong T_1(DM)$. Now let M be a Λ-torsion module and let

$$P_1 \longrightarrow P_0 \longrightarrow M \longrightarrow 0$$

be a presentation of M by free Λ-modules. We obtain the exact sequence

$$0 = M^+ \longrightarrow P_0^+ \longrightarrow P_1^+ \longrightarrow DM \longrightarrow 0$$

showing that $pd_\Lambda DM \le 1$, and so DM has no finite nontrivial Λ-submodules by (5.3.19)(i) .*) Thus $E^1(M) = E^1(DDM) = T_1(DM)$ has this property. Finally, if M is finite, then p^n and ω_n annihilate M for some n large enough and the same is true for $E^1(M)$. Thus $E^1(M)$ is finite and therefore zero.
In order to prove (iii), we observe that $(M/T_1(M))_\mathfrak{p}$ is a free $\Lambda_\mathfrak{p}$-module of finite rank by (i). Hence

$$(M/T_1(M))_\mathfrak{p} \xrightarrow{\sim} (M/T_1(M))_\mathfrak{p}^{++} = M_\mathfrak{p}^{++}$$

for all prime ideals $\mathfrak{p}$ of Λ of height ≤ 1. Thus $T_2(M)$ is pseudo-null, i.e. finite. Furthermore, M^{++} is free and the exact sequence

$$0 \longrightarrow M/T_1(M) \longrightarrow M^{++} \longrightarrow T_2(M) \longrightarrow 0$$

shows that $(M/T_1(M))^\Gamma \subseteq (M^{++})^\Gamma = 0$ and proves the exactness of

$$0 \longrightarrow T_2(M)^\Gamma \longrightarrow (M/T_1(M))_\Gamma \longrightarrow (M^{++})_\Gamma.$$

Therefore $(M/T_1(M))_\Gamma$ is $\mathbb{Z}_p$-free if and only if $T_2(M)^\Gamma = 0$ or equivalently $T_2(M) = 0$. Now (5.3.19) (ii) implies assertion (iii).

Since $E^2(M) = T_2(DM)$, this module is finite by (iii). From the structure theorem for Λ-modules, we obtain an exact sequence

$$0 \longrightarrow T_0(M) \longrightarrow M \xrightarrow{f} E \longrightarrow C \longrightarrow 0,$$

where C is finite and

$$E \cong \Lambda^r \oplus \bigoplus_{i=1}^{s} \Lambda/p^{m_i} \oplus \bigoplus_{j=1}^{t} \Lambda/F_j^{n_j}.$$

The long exact Ext-sequence implies the exactness of

$$E^2(\mathrm{im} f) \longrightarrow E^2(M) \longrightarrow E^2(T_0(M)) \longrightarrow 0$$

and that

$$E^2(E) \longrightarrow\!\!\!\!\rightarrow E^2(\mathrm{im} f)$$

*) In fact, DM is only defined up to homotopy equivalence, but we see that DM has no nontrivial finite submodules for every choice of DM.

is surjective. Since $pd_\Lambda E \le 1$, we get $E^2(E) = 0$, and so $E^2(M) \cong E^2(T_0(M))$. Now we recall that Γ is a Poincaré group of dimension 1 with dualizing module $\mathbb{Q}_p/\mathbb{Z}_p$ (trivial action). Therefore (5.4.15)(ii) implies $E^2(T_0(M)) = T_0(M)^\vee$. The equivalence between a) and b) is now trivial and the equivalence between b) and c) is contained in (5.3.19)(i). □

(5.5.4) Corollary. *Let M be a finitely generated Λ-module. Then*

(i) $E^1(M/T_0(M)) \xrightarrow{\sim} E^1(M)$.

(ii) $E^1(M) = 0$ *if and only if* $M/T_0(M)$ *is free*
(hence $M \cong \Lambda^r \oplus T_0(M)$ for some $r \ge 0$ in this case).

Proof: The exact sequence

$$0 = E^0(T_0(M)) \longrightarrow E^1(M/T_0(M)) \longrightarrow E^1(M) \longrightarrow E^1(T_0(M)),$$

together with (5.5.3)(ii), implies assertion (i). By (5.5.3)(iv), we have $pd_\Lambda(M/T_0(M)) \le 1$. Therefore the following assertions are equivalent:

$$\begin{aligned} M/T_0(M) \text{ is free} &\iff M/T_0(M) \simeq 0 && \text{(by lemma (5.5.2))}\\ &\iff E^1(M/T_0(M)) = 0 && \text{(by corollary (5.4.12))}\\ &\iff E^1(M) = 0 && \text{(by (i)).} \end{aligned}$$

□

(5.5.5) Definition. *Let M be a finitely generated Λ-torsion module and let $\{\pi_n\}$ be a sequence of non-zero elements of Λ such that*

$$\pi_0 \in \mathfrak{m}\,, \qquad \pi_{n+1} \in \pi_n \mathfrak{m}\,,$$

and such that the set of prime ideals dividing the principal ideals $\pi_n\Lambda$, $n \ge 0$, is disjoint to the set of prime ideals of height 1 in $\operatorname{supp}(M)$. Let

$$\alpha(M) := \varprojlim_n \operatorname{Hom}(M/\pi_n M, \mathbb{Q}_p/\mathbb{Z}_p)$$

with respect to the inductive system

$$\begin{aligned} M/\pi_n &\longrightarrow M/\pi_m && \text{for } m \ge n \ge 0\,,\\ x \bmod \pi_n &\longmapsto \tfrac{\pi_m}{\pi_n} x \bmod \pi_m. \end{aligned}$$

The Λ-module $\alpha(M)$ is called the **Iwasawa-adjoint** *of M .*

(5.5.6) Proposition. *For a finitely generated Λ-torsion module M one has a canonical isomorphism*

$$\alpha(M) \cong E^1(M).$$

In particular, $\alpha(M)$ is independent of the choice of the sequence $\{\pi_n\}$.

Proof: From the exact sequence

$$T_0(M)/\pi_n \longrightarrow M/\pi_n \longrightarrow (M/T_0(M))/\pi_n \longrightarrow 0$$

and $\varinjlim_n T_0(M)/\pi_n = 0$, we obtain $\alpha(M/T_0(M)) \cong \alpha(M)$. Thus using (5.5.4)(i), we may assume $T_0(M) = 0$. Then the sequence

$$0 \longrightarrow M \xrightarrow{\pi_n} M \longrightarrow M/\pi_n \longrightarrow 0$$

is exact and M/π_n is finite for all $n \geq 0$, because the multiplication on M by π_n is a pseudo-isomorphism by (5.1.6). Using (5.5.3)(iv), we get isomorphisms

$$E^1(M)/\pi_n \xrightarrow{\sim} E^2(M/\pi_n) \cong \mathrm{Hom}_{\mathbb{Z}_p}(M/\pi_n, \mathbb{Q}_p/\mathbb{Z}_p).$$

Since $\pi_n \to 0$ in Λ when $n \to \infty$ and $E^1(M)$ is finitely generated, it follows that

$$E^1(M) = \varprojlim_n E^1(M)/\pi_n \cong \varprojlim_n \mathrm{Hom}_{\mathbb{Z}_p}(M/\pi_n, \mathbb{Q}_p/\mathbb{Z}_p) = \alpha(M). \qquad \square$$

(5.5.7) Corollary. *Let M be a finitely generated Λ-torsion module. Assume that $\mu(M) = 0$. Then*

$$E^1(M) \cong \mathrm{Hom}_{\mathbb{Z}_p}(M, \mathbb{Z}_p).$$

Proof: Since $\mu(M) = 0$, we can take $\pi_n = p^{n+1}$. From (5.5.6) we obtain

$$\begin{aligned} E^1(M) &\cong \varprojlim_n \mathrm{Hom}_{\mathbb{Z}_p}(M/p^n, \mathbb{Q}_p/\mathbb{Z}_p) \\ &= \varprojlim_n \mathrm{Hom}_{\mathbb{Z}_p}(M, \mathbb{Z}/p^n) = \mathrm{Hom}_{\mathbb{Z}_p}(M, \mathbb{Z}_p). \end{aligned} \qquad \square$$

(5.5.8) Proposition. *Let M be a finitely generated Λ-module.*

(i) *There exists an exact sequence*

$$0 \longrightarrow E^2(T_2(M)) \longrightarrow E^1(M) \longrightarrow E^1(T_1(M)) \longrightarrow 0$$

inducing isomorphisms

(ii) $E^2(T_2(M)) \cong E^1(M/T_1(M)) \cong T_0(E^1(M))$,

(iii) $E^1(T_1(M)) \cong E^1(M)/T_0(E^1(M))$.

Furthermore, there are canonical isomorphisms

(iv) $E^1(E^1(M)) \cong T_1(M)/T_0(M)$,

(v) $E^2(E^1(M)) \cong T_2(M)$,

(vi) $E^2(E^2(M)) \cong T_0(M)$.

Proof: From the exact sequence

$$0 \longrightarrow T_1(M) \longrightarrow M \xrightarrow{\varphi_M} M^{++} \longrightarrow T_2(M) \longrightarrow 0,$$

we obtain exact sequences

$$0 = E^1(M^{++}) \to E^1(\operatorname{im}\varphi_M) \xrightarrow{\sim} E^2(T_2(M)) \to E^2(M^{++}) = 0,$$

$$0 = E^0(T_1(M)) \to E^1(\operatorname{im}\varphi_M) \to E^1(M) \to E^1(T_1(M)) \to E^2(\operatorname{im}\varphi_M) = 0.$$

This implies (i) and the first isomorphism of (ii). Using $T_0(E^1(T_1(M))) = 0$ (by (5.5.3)(ii)) and the finiteness of $E^2(T_2(M))$, we obtain the second isomorphism in (ii) and assertion (iii).

In order to prove (iv), we first observe that by (5.5.4)(i) and (iii) above,

$$\begin{aligned} E^1(E^1(M)) &\cong E^1(E^1(M)/T_0(E^1(M))) \\ &\cong E^1(E^1(T_1(M))) \cong E^1(E^1(T_1(M)/T_0(M))). \end{aligned}$$

In the proof of (5.4.11) we saw that E^1 coincides with D on $Ho_1^{f_p}(\Lambda)$. Since $pd_\Lambda(T_1(M)/T_0(M)) \leq 1$ by (5.5.3)(iv), we therefore obtain

$$\begin{aligned} E^1(E^1(T_1(M)/T_0(M))) &= E^1(D(T_1(M)/T_0(M))) \\ &= T_1(T_1(M)/T_0(M)) = T_1(M)/T_0(M). \end{aligned}$$

Finally, from (i) we get the isomorphism

$$0 = E^2(E^1(T_1(M))) \longrightarrow E^2(E^1(M)) \xrightarrow{\sim} E^2(E^2(T_2(M))) \longrightarrow 0$$

and using (5.5.3)(iv),

$$E^2(E^2(T_2(M))) \cong T_2(M)^{**} = T_2(M),$$

thus proving (v). Again by (5.5.3)(iv) we obtain (vi). □

(5.5.9) Corollary. *We have the following equivalences:*

$$E^1(M) \text{ is finite} \iff T_1(M) \text{ is finite} \iff E^1(E^1(M)) = 0.$$

(5.5.10) Proposition. *Let M be a finitely generated Λ-module.*

(i) $E^0(M) \cong \varprojlim_{n,m} {}_{p^m}(M^\vee)^{\Gamma_n}$ *is free of the same rank as* M,

(ii) $E^1(\mathrm{tor}_{\mathbb{Z}_p} M) \cong \varprojlim_{n,m} (M^\vee/p^m)^{\Gamma_n}$,

(iii) $E^1(M/\mathrm{tor}_{\mathbb{Z}_p} M) \cong \varprojlim_{n,m} ({}_{p^m}(M^\vee))_{\Gamma_n}$
($\cong \mathrm{Hom}_{\mathbb{Z}_p}(M, \mathbb{Z}_p)$ *if* M *is a* Λ*-torsion module*),

(iv) $E^1(M^\delta) \cong \varprojlim_{n,m} {}_{p^m}((M^\vee)_{\Gamma_n}) \cong \mathrm{Hom}_{\mathbb{Z}_p}(M^\delta, \mathbb{Z}_p)$,

(v) $E^1(M/M^\delta) \cong \varprojlim_{n,m} ((M^\vee)^{\Gamma_n})/p^m$,

(vi) $E^2(M) \cong \varprojlim_{n,m} (M^\vee)/(p^m, \Gamma_n)$,

where the transition maps are the obvious ones. Recall that $M^\delta = \bigcup_n M^{\Gamma_n}$ *is the maximal* Λ*-submodule of* M *on which* Γ *acts discretely.*

Proof: Since $H^0(\Gamma_n, A) = A^{\Gamma_n}$ and $H^1(\Gamma_n, A) = A_{\Gamma_n}$ for a discrete Γ-module A, the assertions (i), (ii), (iii) and (vi) follow immediately from (5.4.13)(ii)-(iv). If M is Λ-torsion, then $M/\mathrm{tor}_{\mathbb{Z}_p} M$ is a torsion module with trivial μ-invariant and we obtain from (5.5.7) that

$$E^1(M/\mathrm{tor}_{\mathbb{Z}_p} M) = \mathrm{Hom}_{\mathbb{Z}_p}(M/\mathrm{tor}_{\mathbb{Z}_p} M, \mathbb{Z}_p) = \mathrm{Hom}_{\mathbb{Z}_p}(M, \mathbb{Z}_p)\,,$$

showing the additional statement in (iii). Using (5.4.13)(i), we obtain an exact sequence

$$0 \longrightarrow \varprojlim_{n,m} (M^\vee)^{\Gamma_n}/p^m \longrightarrow E^1(M) \longrightarrow \varprojlim_{n,m} {}_{p^m}((M^\vee)_{\Gamma_n}) \longrightarrow 0\,,$$

where the cokernel is isomorphic to $\mathrm{Hom}_{\mathbb{Z}_p}(M^\delta, \mathbb{Z}_p)$ while the kernel vanishes for $M = M^\delta$. By (5.5.3)(iv) we have $E^2(M/M^\delta) = 0$, thus we obtain an exact sequence

$$0 \longrightarrow E^1(M/M^\delta) \longrightarrow E^1(M) \longrightarrow E^1(M^\delta) \longrightarrow 0.$$

Since the first exact sequence is functorial in M, it must be isomorphic to the second one, and we get (iv) and (v). □

For the rest of this section we consider Λ-modules up to pseudo-isomorphism (denoted by $\approx$).

(5.5.11) Proposition. *Let M and M' be finitely generated Λ-torsion modules. Then*

$$M \approx M' \quad \text{if and only if} \quad E^1(M') \approx E^1(M).$$

Proof: We may assume $T_0(M) = 0 = T_0(M')$. From the exact sequence

$$0 \longrightarrow M \longrightarrow M' \longrightarrow C \longrightarrow 0$$

with a finite Λ-module C, we obtain

$$0 = E^1(C) \longrightarrow E^1(M') \longrightarrow E^1(M) \longrightarrow E^2(C),$$

and hence $E^1(M') \approx E^1(M)$. Conversely, since $E^1(M') \approx E^1(M)$, (5.5.8)(iv) implies the pseudo-isomorphisms $M \approx E^1(E^1(M)) \approx E^1(E^1(M')) \approx M'$.

□

(5.5.12) Definition. *Let M be a Λ-module. We define the Λ-module M° to be the Λ-module M with the inverse Γ-action:*

$$\text{if } x \in M^\circ \text{ then } \gamma \circ x := \gamma^{-1}x \text{ for } \gamma \in \Gamma.$$

Remark: If $F \in \Lambda$ is a Weierstraß polynomial, then

$$(\Lambda/F)^\circ \cong \mathrm{Hom}\,(\Lambda/F, \mathbb{Z}_p).$$

We have $(\Lambda/F)^\circ \cong \Lambda/F'$ for the Weierstraß polynomial F' which is obtained from F by substituting $(1+T)^{-1} - 1$ for T and multiplying by $(1+T)^{\deg F}$.

(5.5.13) Proposition. *Let M be a finitely generated Λ-module. Then there exists a pseudo-isomorphism*

$$E^1(M) \approx T_1(M)^\circ.$$

Proof: From the exact sequence

$$0 \longrightarrow \mathrm{tor}_{\mathbb{Z}_p} M \longrightarrow M \longrightarrow M/\mathrm{tor}_{\mathbb{Z}_p} M \longrightarrow 0,$$

we obtain the exact sequence

$$0 \to E^1(M/\mathrm{tor}_{\mathbb{Z}_p} M) \to E^1(M) \to E^1(\mathrm{tor}_{\mathbb{Z}_p} M) \to E^2(M/\mathrm{tor}_{\mathbb{Z}_p} M).$$

Using (5.5.8)(i) and (5.5.7), we see that

$$\begin{aligned} E^1(M/\mathrm{tor}_{\mathbb{Z}_p}M) &\approx E^1(T_1(M)/\mathrm{tor}_{\mathbb{Z}_p}M) \\ &\cong \mathrm{Hom}_{\mathbb{Z}_p}(T_1(M)/\mathrm{tor}_{\mathbb{Z}_p}M, \mathbb{Z}_p) \\ &\cong (T_1(M)/\mathrm{tor}_{\mathbb{Z}_p}M)^\circ. \end{aligned}$$

Furthermore,

$$\begin{aligned} E^1(\mathrm{tor}_{\mathbb{Z}_p}M) &\approx E^1(\bigoplus_{i=1}^s \Lambda/p^{m_i}) \qquad \text{for some } m_i \geq 0 \\ &\cong \bigoplus_{i=1}^s E^1(\Lambda/p^{m_i}) \\ &\cong \bigoplus_{i=1}^s (\Lambda/p^{m_i})^\circ \approx (\mathrm{tor}_{\mathbb{Z}_p}M)^\circ, \end{aligned}$$

so that

$$\begin{aligned} E^1(M) \approx{}& E^1(M/\mathrm{tor}_{\mathbb{Z}_p}M) \oplus E^1(\mathrm{tor}_{\mathbb{Z}_p}M) \\ &\approx (T_1(M)/\mathrm{tor}_{\mathbb{Z}_p}M)^\circ \oplus (\mathrm{tor}_{\mathbb{Z}_p}M)^\circ \approx T_1(M)^\circ. \end{aligned}$$ □

Exercise: Let M be a finitely generated Λ-torsion module and let Y be defined by the exact sequence

$$0 \longrightarrow T_0(M) \longrightarrow M \longrightarrow \prod_{\mathrm{ht}(\mathfrak{p})=1} M_{\mathfrak{p}} \longrightarrow Y \longrightarrow 0.$$

Then there is a canonical isomorphism

$$\alpha(M) \cong \mathrm{Hom}_{cts}(Y, \mathbb{Q}_p/\mathbb{Z}_p).$$

This was the original definition of the adjoint due to Iwasawa.

Hint: Let $\{\pi_n\}$ be a sequence of elements of Λ as considered in (5.5.5) and let $\mathfrak{a}$ be the annihilator ideal of M. Let S be the multiplicatively closed subset in $\bar{\Lambda} := \Lambda/\mathfrak{a}$ which is generated by the images of $\pi_0, \pi_1, \ldots$. Prove the isomorphism

$$A := S^{-1}\bar{\Lambda} \cong \varinjlim_i \bar{\Lambda}_{(i)},$$

where the inductive system is given by $\bar{\Lambda}_{(i)} = \bar{\Lambda}$, $\bar{\Lambda}_{(i)} \to \bar{\Lambda}_{(j)}$, $\lambda \mapsto \frac{\pi_j}{\pi_i}\lambda$. By the assumptions on $\pi_0, \pi_1, \ldots$, the ring A is artinian, and thus

$$A \cong \prod_{\substack{\mathfrak{p} \in \mathrm{supp}(M) \\ \mathrm{ht}(\mathfrak{p})=1}} \Lambda_{\mathfrak{p}}/\mathfrak{a}\Lambda_{\mathfrak{p}}.$$

Conclude that

$$\varinjlim_i M \otimes_\Lambda \bar{\Lambda}_{(i)} \cong M \otimes_\Lambda A \cong \prod_{\mathrm{ht}(\mathfrak{p})=1} M_{\mathfrak{p}}.$$

Finally, consider the exact commutative diagram

$$\begin{array}{ccccccccc} 0 & \longrightarrow & \ker(\pi_i) & \longrightarrow & M & \xrightarrow{\pi_i} & M & \longrightarrow & M/\pi_i M & \longrightarrow & 0 \\ & & \downarrow & & \| & & \downarrow \frac{\pi_j}{\pi_i} & & \downarrow \frac{\pi_j}{\pi_i} & & \\ 0 & \longrightarrow & \ker(\pi_j) & \longrightarrow & M & \xrightarrow{\pi_j} & M & \longrightarrow & M/\pi_j M & \longrightarrow & 0\,. \end{array}$$

§6. Differential Modules and Presentations

R. Fox, in 1953, established a theory of derivations in the group ring of a free discrete group, the "free differential calculus". In the following we consider a topological version of this theory for the completed group ring of a profinite group.

Throughout this section, let p be a prime number, $\mathfrak{c}$ a class of finite groups containing $\mathbb{Z}/p\mathbb{Z}$ and closed under taking subgroups, homomorphic images and group extensions. Let G be a pro-$\mathfrak{c}$-group and let $\alpha : \mathbb{Z}_p[\![G]\!] \to \mathbb{Z}_p$ denote the augmentation map (cf. §2).

(5.6.1) Definition. *A* **derivation** *of $\mathbb{Z}_p[\![G]\!]$ into a compact $\mathbb{Z}_p[\![G]\!]$-module A is a continuous $\mathbb{Z}_p$-linear map*

$$D : \mathbb{Z}_p[\![G]\!] \longrightarrow A$$

satisfying

$$D(fg) = D(f)\cdot\alpha(g) + f\cdot D(g)$$

for all $f, g \in \mathbb{Z}_p[\![G]\!]$.

Remark: A derivation is clearly determined by its restriction to $G \subseteq \mathbb{Z}_p[\![G]\!]$ and this restriction is a continuous inhomogeneous 1-cocycle of G with values in the compact G-module A (cf. II §7). On the other hand, one easily verifies that every continuous inhomogeneous 1-cocycle of G with values in A extends to a derivation of $\mathbb{Z}_p[\![G]\!]$ into A. Note that

(i) $D(a) = 0$ for $a \in \mathbb{Z}_p$,
(ii) $D(\sigma^{-1}) = -\sigma^{-1}D(\sigma)$ for $\sigma \in G$.

Now we construct a $\mathbb{Z}_p[\![G]\!]$-module $\Omega_{\mathbb{Z}_p[\![G]\!]}$ as follows: take the free compact $\mathbb{Z}_p[\![G]\!]$-module which is (topologically) generated by the symbols df, $f \in \mathbb{Z}_p[\![G]\!]$, and then take the quotient by the closed submodule generated by the relations

$$d(af + bg) - adf - bdg$$
$$d(fg) - df\cdot\alpha(g) - f\cdot dg$$

where $f, g \in \mathbb{Z}_p[\![G]\!]$, $a, b \in \mathbb{Z}_p$. (In particular, these relations imply that $d1 = 0$ for the unit element $1 \in G$.) The map

$$d : \mathbb{Z}_p[\![G]\!] \longrightarrow \Omega_{\mathbb{Z}_p[\![G]\!]}\,, \quad f \longmapsto df\,,$$

is a derivation and the pair $(\Omega_{\mathbb{Z}_p[\![G]\!]}, d)$ satisfies the universal property:

For every derivation $D : \mathbb{Z}_p[\![G]\!] \to A$, there exists a unique continuous $\mathbb{Z}_p[\![G]\!]$-homomorphism $\varphi : \Omega_{\mathbb{Z}_p[\![G]\!]} \to A$ with $D = \varphi \circ d$.

(5.6.2) Definition. *We call* $\Omega_{\mathbb{Z}_p[\![G]\!]}$ *the module of (noncommutative)* **differential forms** *of* $\mathbb{Z}_p[\![G]\!]$.

Recall that the kernel of the augmentation map: $\alpha : \mathbb{Z}_p[\![G]\!] \longrightarrow \mathbb{Z}_p$ is called the augmentation ideal and denoted by $I_G \subseteq \mathbb{Z}_p[\![G]\!]$.

(5.6.3) Proposition. *There is a canonical isomorphism*

$$\varphi : \Omega_{\mathbb{Z}_p[\![G]\!]} \xrightarrow{\sim} I_G ,$$

satisfying $\varphi(d\sigma) = \sigma - 1$ *for every* $\sigma \in G \subseteq \mathbb{Z}_p[\![G]\!]$.

Proof: For $\sigma, \tau \in G$, we have the identity

$$\sigma\tau - 1 = (\sigma - 1) + \sigma(\tau - 1)$$

in I_G, so φ defines a homomorphism. Furthermore, the inverse map φ^{-1} : $I_G \longrightarrow \Omega_{\mathbb{Z}_p[\![G]\!]}$, $\sigma - 1 \longmapsto d\sigma$, is well-defined. □

(5.6.4) Proposition. *If* F *is a free pro-*$\mathfrak{c}$*-group of rank* r *on the generators* $x_1, \ldots, x_r$, *then* $\Omega_{\mathbb{Z}_p[\![F]\!]}$ *is a free* $\mathbb{Z}_p[\![F]\!]$*-module of rank* r *on the generators* $dx_1, \ldots, dx_r$.

Proof: Let G be a pro-$\mathfrak{c}$-group and let A be a compact $\mathbb{Z}_p[\![G]\!]$-module. We have a split exact sequence

$$1 \longrightarrow A \longrightarrow E \xrightarrow{\pi} G \longrightarrow 1$$

where E is the semi-direct product of G by A and the action of G on A is the natural one.[*] Since A is a pro-p-group, E is a pro-$\mathfrak{c}$-group. Therefore we have a 1–1-correspondence between the continuous homomorphic sections to π and the derivations from $\mathbb{Z}_p[\![G]\!]$ to A.[**]

Now let $G = F$ be a free pro-$\mathfrak{c}$-group. Then, by the observation above, a derivation from $\mathbb{Z}_p[\![F]\!]$ to A is uniquely determined by the arbitrarily chosen images of a set of free generators of F. This proves the proposition. □

Returning to the general case, assume that we are given an exact sequence

$$1 \longrightarrow \mathscr{H} \longrightarrow \mathscr{G} \longrightarrow G \longrightarrow 1$$

of pro-$\mathfrak{c}$-groups with the corresponding exact sequence

$$0 \longrightarrow I \longrightarrow \mathbb{Z}_p[\![\mathscr{G}]\!] \longrightarrow \mathbb{Z}_p[\![G]\!] \longrightarrow 0,$$

[*] The sequence corresponds to the zero element in $H^2(G, A)$, cf. I §2.

[**] cf. ex.1 in I §2.

where

$$I = I_{\mathcal{H}}\mathbb{Z}_p[\![\mathcal{G}]\!].$$

The following proposition is the profinite analogue of what is called *the second fundamental sequence* in commutative algebra.

(5.6.5) Proposition. *There exists an exact sequence of $\mathbb{Z}_p[\![G]\!]$-modules*

$$I/I^2 \xrightarrow{\theta} \mathbb{Z}_p[\![G]\!] \otimes_{\mathbb{Z}_p[\![\mathcal{G}]\!]} \Omega_{\mathbb{Z}_p[\![\mathcal{G}]\!]} \longrightarrow \Omega_{\mathbb{Z}_p[\![G]\!]} \longrightarrow 0$$

and the image of θ is canonically isomorphic to $\mathcal{H}^{ab}(p) = H_1(\mathcal{H}, \mathbb{Z}_p)$.

Proof: Recall that for a compact $\mathbb{Z}_p[\![\mathcal{H}]\!]$-module M,

$$M_{\mathcal{H}} := M/I_{\mathcal{H}}M = \mathbb{Z}_p \otimes_{\mathbb{Z}_p[\![\mathcal{H}]\!]} M.$$

Furthermore, $\mathbb{Z}_p[\![\mathcal{G}]\!]$ is a (compactly) induced $\mathcal{H}$-module and therefore homologically trivial.[*] Now we apply $H_\bullet(\mathcal{H}, -) = \mathrm{Tor}_\bullet^{\mathbb{Z}_p[\![\mathcal{G}]\!]}(\mathbb{Z}_p[\![G]\!], -)$ to the exact sequence

$$0 \longrightarrow I_{\mathcal{G}} \longrightarrow \mathbb{Z}_p[\![\mathcal{G}]\!] \xrightarrow{aug} \mathbb{Z}_p \longrightarrow 0.$$

Via the identification $\mathbb{Z}_p[\![G]\!] = \mathbb{Z}_p[\![\mathcal{G}]\!]/I$ and by (5.6.3), we obtain the commutative diagram with exact rows:

$$\begin{array}{ccccccc} 0 \to I_{\mathcal{H}}\mathbb{Z}_p[\![\mathcal{G}]\!]/I_{\mathcal{H}}I_{\mathcal{G}} & \longrightarrow & I_{\mathcal{G}}/I_{\mathcal{H}}I_{\mathcal{G}} & \longrightarrow \\ \| & & \| & \\ 0 \longrightarrow H_1(\mathcal{H}, \mathbb{Z}_p) & \longrightarrow & \mathbb{Z}_p[\![G]\!] \otimes_{\mathbb{Z}_p[\![\mathcal{G}]\!]} \Omega_{\mathbb{Z}_p[\![\mathcal{G}]\!]} & \longrightarrow \end{array}$$

$$\begin{array}{ccccc} \mathbb{Z}_p[\![\mathcal{G}]\!]/I_{\mathcal{H}}\mathbb{Z}_p[\![\mathcal{G}]\!] & \longrightarrow & \mathbb{Z}_p & \to & 0 \\ \| & & \| & & \\ \mathbb{Z}_p[\![G]\!] & \longrightarrow & \mathbb{Z}_p & \to & 0. \end{array}$$

The inclusion $(I_{\mathcal{H}}\mathbb{Z}_p[\![\mathcal{G}]\!])^2 \subseteq I_{\mathcal{H}}I_{\mathcal{G}}$ then finally implies the existence of

$$\theta : I/I^2 \longrightarrow I_{\mathcal{H}}\mathbb{Z}_p[\![\mathcal{G}]\!]/I_{\mathcal{H}}I_{\mathcal{G}}.$$ □

Now assume that G is a finitely generated pro-$\mathfrak{c}$-group. Then we have an exact sequence

$$1 \longrightarrow R \longrightarrow F_d \longrightarrow G \longrightarrow 1$$

for some $d \in \mathbb{N}$, where F_d is a free pro-$\mathfrak{c}$-group of rank d and R is the normal subgroup of F_d generated by the *relations* of G (with respect to the chosen

[*] In other words, the Pontryagin dual of $\mathbb{Z}_p[\![\mathcal{G}]\!]$ is an induced discrete $\mathcal{G}$-module, hence a cohomologically trivial $\mathcal{G}$-module.

generators of G, i.e. the images of a basis of F_d in G). The abelian pro-p-group $R^{ab}(p)$ is a $\mathbb{Z}_p[\![G]\!]$-module in a natural way and we call it the ***p*-relation module** of G with respect to the given presentation of G. The following theorem is a profinite analogue of a theorem of Lyndon for discrete groups.

(5.6.6) Theorem. *There is a canonical exact sequence*

$$0 \longrightarrow R^{ab}(p) \longrightarrow \mathbb{Z}_p[\![G]\!]^d \longrightarrow \mathbb{Z}_p[\![G]\!] \longrightarrow \mathbb{Z}_p \longrightarrow 0.$$

In particular, if $cd_p\, G \le 2$, then $R^{ab}(p)$ is a projective $\mathbb{Z}_p[\![G]\!]$-module.

Proof: We apply (5.6.5) to the exact sequence $1 \to R \to F_d \to G \to 1$ and observe that $\Omega_{\mathbb{Z}_p[\![G]\!]} = I_G$ by (5.6.3) and that $\Omega_{\mathbb{Z}_p[\![F_d]\!]}$ is a free $\mathbb{Z}_p[\![F_d]\!]$-module of rank d by (5.6.4). The last assertion follows from (5.2.13). □

We now consider the following general problem: Given an exact sequence of pro-$\mathfrak{c}$-groups

$$1 \longrightarrow \mathscr{H} \longrightarrow \mathscr{G} \longrightarrow G \longrightarrow 1,$$

what can we say about the structure of $\mathscr{H}^{ab}(p)$ as a $\mathbb{Z}_p[\![G]\!]$-module?

Theorem (5.6.6) gives us information in the case that $\mathscr{G}$ is a free pro-$\mathfrak{c}$-group. In the general case, assume that $\mathscr{G}$ is finitely generated and choose a presentation $F \twoheadrightarrow \mathscr{G}$ of $\mathscr{G}$ by a free pro-$\mathfrak{c}$-group F of rank d. Then we obtain a commutative diagram

$$\begin{array}{ccccccccc}
 & & 1 & & 1 & & & & \\
 & & \downarrow & & \downarrow & & & & \\
 & & N & = & N & & & & \\
 & & \downarrow & & \downarrow & & & & \\
1 & \longrightarrow & R & \longrightarrow & F & \longrightarrow & G & \longrightarrow & 1 \\
 & & \downarrow & & \downarrow & & \| & & \\
1 & \longrightarrow & \mathscr{H} & \longrightarrow & \mathscr{G} & \longrightarrow & G & \longrightarrow & 1 \\
 & & \downarrow & & \downarrow & & & & \\
 & & 1 & & 1 & & & &
\end{array}$$

where R and N are defined by the exactness of the corresponding sequences. In addition, we set

$$X := \mathscr{H}^{ab}(p),$$

$$Y := \mathbb{Z}_p[\![G]\!] \otimes_{\mathbb{Z}_p[\![\mathscr{G}]\!]} \Omega_{\mathbb{Z}_p[\![\mathscr{G}]\!]} = I_{\mathscr{G}} / I_{\mathscr{H}} I_{\mathscr{G}}\,.$$

(5.6.7) Proposition. *With the notation above, we have a commutative exact diagram*

$$\begin{array}{ccccccccc}
 & & & & & & 0 & & 0 & & \\
 & & & & & & \downarrow & & \downarrow & & \\
0 & \longrightarrow & H_2(\mathscr{H},\mathbb{Z}_p) & \longrightarrow & N^{ab}_{\mathscr{H}}(p) & \longrightarrow & R^{ab}(p) & \longrightarrow & X & \longrightarrow & 0\\
 & & \| & & \| & & \downarrow & & \downarrow & & \\
0 & \longrightarrow & H_2(\mathscr{H},\mathbb{Z}_p) & \longrightarrow & N^{ab}_{\mathscr{H}}(p) & \longrightarrow & \mathbb{Z}_p[\![G]\!]^d & \longrightarrow & Y & \longrightarrow & 0\\
 & & & & & & \downarrow & & \downarrow & & \\
 & & & & & & I_G & = & I_G & & \\
 & & & & & & \downarrow & & \downarrow & & \\
 & & & & & & 0 & & 0. & &
\end{array}$$

In particular, there is an exact sequence

$$0 \longrightarrow R^{ab}(p) \longrightarrow X \oplus \mathbb{Z}_p[\![G]\!]^d \longrightarrow Y \longrightarrow 0.$$

Furthermore, if $cd_p\mathscr{G} \le 2$, then $N^{ab}(p)$ is a projective $\mathbb{Z}_p[\![\mathscr{G}]\!]$-module and $N^{ab}_{\mathscr{H}}(p)$ is a projective $\mathbb{Z}_p[\![G]\!]$-module. If $cd_p\mathscr{G} \le 2$ and $cd_p G \le 1$, then $H_2(\mathscr{H},\mathbb{Z}_p)$ is a projective $\mathbb{Z}_p[\![G]\!]$-module.

Proof: The upper horizontal sequence is the homological form of the five term exact sequence (1.6.7) for the group extension $1 \to N \to R \to \mathscr{H} \to 1$ and the module $\mathbb{Z}_p$. The zero on the left follows because $cd_pR \le cd_pF = 1$. We obtain the lower horizontal sequence by taking $\mathscr{H}$-homology of the exact sequence

$$(*) \qquad 0 \longrightarrow N^{ab}(p) \longrightarrow \mathbb{Z}_p[\![\mathscr{G}]\!]^d \longrightarrow I_{\mathscr{G}} \longrightarrow 0$$

(apply (5.6.6) to $1 \to N \to F \to \mathscr{G} \to 1$), using the homological $\mathscr{H}$-triviality of $\mathbb{Z}_p[\![\mathscr{G}]\!]$ and the resulting isomorphism $H_1(\mathscr{H}, I_{\mathscr{G}}) \cong H_2(\mathscr{H},\mathbb{Z}_p)$.

The right-hand vertical sequence is the $\mathscr{H}$-homology sequence associated to $0 \to I_{\mathscr{G}} \to \mathbb{Z}_p[\![\mathscr{G}]\!] \to \mathbb{Z}_p \to 0$ (using the same argument as above). Finally, the left-hand vertical sequence is the sequence from theorem (5.6.6). The diagram commutes since all arrows are the natural ones and we use the compatible identifications

$$\begin{aligned}\mathbb{Z}_p[\![G]\!]^d &= \mathbb{Z}_p[\![G]\!] \otimes_{\mathbb{Z}_p[\![F]\!]} \Omega_{\mathbb{Z}_p[\![F]\!]}\\ &= \mathbb{Z}_p[\![G]\!] \otimes_{\mathbb{Z}_p[\![\mathscr{G}]\!]} \mathbb{Z}_p[\![\mathscr{G}]\!] \otimes_{\mathbb{Z}_p[\![F]\!]} \Omega_{\mathbb{Z}_p[\![F]\!]} = H_0(\mathscr{H}, \mathbb{Z}_p[\![\mathscr{G}]\!]^d).\end{aligned}$$

Considering the pull back extension X' of

$$\begin{array}{ccccccccc} 0 & \longrightarrow & R^{ab}(p) & \longrightarrow & \mathbb{Z}_p[\![G]\!]^d & \longrightarrow & I_G & \longrightarrow & 0 \\ & & & & \uparrow & & \uparrow & & \\ & & & & X' & \longrightarrow\!\!\!\!\rightarrow & Y, & & \end{array}$$

we see that $X' \cong \mathbb{Z}_p[\![G]\!]^d \oplus X$, which gives us the exact sequence asserted in the proposition.

Finally, we assume $cd_p\, \mathcal{G} \le 2$. Then by (5.2.13) and the exact sequence $(*)$, we conclude that $N^{ab}(p)$ is a projective $\mathbb{Z}_p[\![\mathcal{G}]\!]$-module, so that $N^{ab}_{\mathcal{H}}(p)$ is a projective $\mathbb{Z}_p[\![G]\!]$ -module. If, in addition, $cd_p\, G \le 1$, then $pd\, \mathbb{Z}_p[\![G]\!] \le 2$ (see §2 ex.5 and (5.2.13)), and so $pd_{\mathbb{Z}_p[\![G]\!]} H_2(\mathcal{H}, \mathbb{Z}_p) = pd_{\mathbb{Z}_p[\![G]\!]} X - 2 = 0$. □

(5.6.8) Corollary. *With the notation above, we assume in addition that $\mathcal{G}$ is a finitely generated pro-p-group with finitely many defining relations. Let*

$$1 \longrightarrow N \longrightarrow F \longrightarrow \mathcal{G} \longrightarrow 1$$

be a minimal presentation of $\mathcal{G}$ by a free pro-p-group F. Then $N^{ab}_{\mathcal{H}}$ is a finitely generated free $\mathbb{Z}_p[\![G]\!]$-module and

$$\operatorname{rank}_{\mathbb{Z}_p[\![G]\!]} N^{ab}_{\mathcal{H}} = \dim_{\mathbb{F}_p} H^2(\mathcal{G}, \mathbb{F}_p).$$

Proof: From the exact sequence

$$0 \longrightarrow H^1(\mathcal{G}, \mathbb{F}_p) \xrightarrow{\sim} H^1(F, \mathbb{F}_p) \longrightarrow H^1(N, \mathbb{F}_p)^{\mathcal{G}} \longrightarrow H^2(\mathcal{G}, \mathbb{F}_p) \longrightarrow 0$$

it follows that

$$\dim_{\mathbb{F}_p} (N^{ab}_{\mathcal{H}})/G = \dim_{\mathbb{F}_p} H^1(N, \mathbb{F}_p)^{\mathcal{G}} = \dim_{\mathbb{F}_p} H^2(\mathcal{G}, \mathbb{F}_p),$$

and so $N^{ab}_{\mathcal{H}}$ is finitely generated as a $\mathbb{Z}_p[\![G]\!]$-module. By (5.6.7) the $\mathbb{Z}_p[\![G]\!]$-module $N^{ab}_{\mathcal{H}}(p)$ is projective, and so it follows from (5.2.20) that it is free.

□

Now we give a description of the $\mathbb{Z}_p[\![G]\!]$-module $Y = I_{\mathcal{G}}/I_{\mathcal{H}} I_{\mathcal{G}}$ up to homotopy equivalence in terms of the p-dualizing module $D_2^{(p)} = \varinjlim_m D_2(\mathbb{Z}/p^m\mathbb{Z})$ of $\mathcal{G}$.

(5.6.9) Proposition. *Assume that $cd_p\, \mathcal{G} = 2$ and that $N^{ab}(p)$ is a finitely generated $\mathbb{Z}_p[\![\mathcal{G}]\!]$-module. Let $Z = (D_2^{(p)\mathcal{H}})^\vee$. Then*

$$Y \simeq DZ,$$

so Y is determined by Z up to projective summands. If, in addition, the group $H_2(\mathcal{H}, \mathbb{Z}_p)$ vanishes, then

$$E^1(Y) \cong Z\,.$$

Proof: Applying (5.6.6) to the exact sequence $1 \to N \to F \to \mathcal{G} \to 1$, we see that the finitely generated $\mathbb{Z}_p[\![\mathcal{G}]\!]$-module $N^{ab}(p)$ is projective. Thus $\mathbb{Z}_p$ possesses a resolution by finitely generated projective $\mathbb{Z}_p[\![\mathcal{G}]\!]$-modules and we get an exact sequence

$$(\mathbb{Z}_p[\![\mathcal{G}]\!]^d)^+ \longrightarrow N^{ab}(p)^+ \longrightarrow E^1(I_{\mathcal{G}}) \longrightarrow 0.$$

Applying (5.4.13)(ii) to $\mathbb{Z}_p$, we obtain

$$E^1(I_{\mathcal{G}}) \cong E^2(\mathbb{Z}_p) = (\varinjlim_m D_2(\mathbb{Z}/p^m\mathbb{Z}))^\vee = (D_2^{(p)})^\vee.$$

Taking $\mathcal{H}$-coinvariants of the sequence above, yields the exact sequence

$$(\mathbb{Z}_p[\![G]\!]^d)^+ \longrightarrow N^{ab}_{\mathcal{H}}(p)^+ \longrightarrow Z \longrightarrow 0$$

(here we use $\mathrm{Hom}_{\mathbb{Z}_p[\![\mathcal{G}]\!]}(M, \mathbb{Z}_p[\![\mathcal{G}]\!])_{\mathcal{H}} = \mathrm{Hom}_{\mathbb{Z}_p[\![G]\!]}(M_{\mathcal{H}}, \mathbb{Z}_p[\![G]\!])$ for a finitely generated projective $\mathbb{Z}_p[\![\mathcal{G}]\!]$-module M). From the exact sequence

$$N^{ab}_{\mathcal{H}}(p) \longrightarrow \mathbb{Z}_p[\![G]\!]^d \longrightarrow Y \longrightarrow 0$$

and the fact that $N^{ab}_{\mathcal{H}}(p)$ is projective, we now obtain $DY \simeq Z$ and $Y \simeq DZ$. If $H_2(\mathcal{H}, \mathbb{Z}_p) = 0$, then we have the exact sequence

$$0 \longrightarrow N^{ab}_{\mathcal{H}}(p) \longrightarrow \mathbb{Z}_p[\![G]\!]^d \longrightarrow Y \longrightarrow 0$$

inducing the exact sequence

$$(\mathbb{Z}_p[\![G]\!]^d)^+ \longrightarrow N^{ab}_{\mathcal{H}}(p)^+ \longrightarrow E^1(Y) \longrightarrow 0\,.$$

Thus $E^1(Y) \cong Z$. □

In the following, we will make use of some well-known facts about group algebras of finite groups, which we briefly recall.

(5.6.10) Theorem. *Let R be a complete discrete valuation ring with quotient field K of characteristic zero and residue field k of positive characteristic, and let G be a finite group. Furthermore, let L, M, N be finitely generated $R[G]$-modules. Then the following hold:*

(i) *If $M \oplus L \cong N \oplus L$, then $M \cong N$.*

(ii) *If M and N are projective and $M \otimes K \cong N \otimes K$ as $K[G]$-modules, then $M \cong N$.*

(iii) *If M and N are projective and $M \otimes k \cong N \otimes k$ as $k[G]$-modules, then $M \cong N$.*

Assertion (i) is a consequence of the Krull-Schmidt-Azuyama theorem, see [32], §6, cor.6.15. For (ii) we refer the reader to [214], chap.16.1, cor. 2 of thm. 34. Finally, (iii) follows in a straightforward manner from Nakayama's lemma.

Returning to the profinite situation, one often wants to determine the $\mathbb{Z}_p[[G]]$-module structure of $Y = I_{\mathcal{G}}/I_{\mathcal{H}}I_{\mathcal{G}}$ not only up to homotopy equivalence but up to isomorphism. The following proposition is useful in the applications.

(5.6.11) Proposition. *Let G be a profinite group. Let M and N be finitely generated $\mathbb{Z}_p[[G]]$-modules such that*

(i) $M \simeq N$,

(ii) $M \otimes \mathbb{Q}_p \cong (N \oplus \mathbb{Z}_p[[G]]^m) \otimes \mathbb{Q}_p$ *for some* $m \in \mathbb{N}$.

Then

$$M \cong N \oplus \mathbb{Z}_p[[G]]^m .$$

In particular, a finitely generated projective $\mathbb{Z}_p[[G]]$-module P is free if and only if $P \otimes \mathbb{Q}_p$ is $(\mathbb{Z}_p[[G]] \otimes \mathbb{Q}_p)$-free. Furthermore, instead of (ii), *it suffices to assume the weaker condition*

(ii)′ $M_U \otimes \mathbb{Q}_p \cong N_U \otimes \mathbb{Q}_p \oplus \mathbb{Q}_p[G/U]^m$

for all open normal subgroups U of G.

Proof: Since $M \simeq N \simeq N \oplus \mathbb{Z}_p[[G]]^m$, it follows from (5.4.3)(ii) that there are finitely generated projective $\mathbb{Z}_p[[G]]$-modules P_1 and P_2 such that

$$M \oplus P_1 \cong N \oplus \mathbb{Z}_p[[G]]^m \oplus P_2.$$

From the second assumption (ii) (or from (ii)′) and by (5.6.10)(i), we get

$$(P_1)_U \otimes \mathbb{Q}_p \cong (P_2)_U \otimes \mathbb{Q}_p$$

for all open normal subgroups U of G. Hence the projective $\mathbb{Z}_p[G/U]$-modules $(P_1)_U$ and $(P_2)_U$ are isomorphic by (5.6.10)(ii). Again by (5.6.10)(i), we obtain

$$M_U \cong N_U \oplus \mathbb{Z}_p[G/U]^m .$$

In particular, we have, for every $n \in \mathbb{N}$ and every open normal subgroup $U \subseteq G$, an isomorphism of finite $\mathbb{Z}/p^n\mathbb{Z}[G/U]$-modules

$$(M/p^n)_U \cong (N/p^n)_U \oplus \mathbb{Z}/p^n\mathbb{Z}[G/U]^m .$$

Passing to the projective limit, the result follows by the usual compactness argument. □

(5.6.12) Proposition. *Let*

$$1 \longrightarrow H \longrightarrow G \longrightarrow \mathbb{Z}_p \longrightarrow 1$$

be an exact sequence of a profinite groups such that H is of order prime to p. Then the augmentation ideal I_G of $\mathbb{Z}_p[\![G]\!]$ is a free $\mathbb{Z}_p[\![G]\!]$-module of rank 1, i.e.

$$I_G \cong \mathbb{Z}_p[\![G]\!]\,.$$

Proof: Since $cd_p\, G = 1$, the $\mathbb{Z}_p[\![G]\!]$-module I_G is projective by (5.2.13). Let $U \subseteq G$ be an open normal subgroup. From the exact sequence

$$0 \longrightarrow H_1(U, \mathbb{Z}_p) \longrightarrow (I_G)_U \longrightarrow \mathbb{Z}_p[G/U] \longrightarrow \mathbb{Z}_p \longrightarrow 0\,,$$

using Maschke's theorem (2.6.12) and observing that $H_1(U, \mathbb{Z}_p) = U^{ab}(p)$ is isomorphic to $\mathbb{Z}_p$, we obtain

$$\mathbb{Q}_p \oplus (I_G)_U \otimes \mathbb{Q}_p \cong \mathbb{Q}_p[G/U] \oplus \mathbb{Q}_p\,,$$

and by (5.6.10)(i), we get $(I_G)_U \otimes \mathbb{Q}_p \cong \mathbb{Q}_p[G/U]$. The result now follows from (5.6.11). □

(5.6.13) Proposition. *Let*

$$1 \longrightarrow \mathscr{H} \longrightarrow \mathscr{G} \longrightarrow G \longrightarrow 1$$

be an exact sequence of profinite groups, where $\mathscr{G}$ is finitely generated of $cd_p\, \mathscr{G} \le 2$ and G is a pro-p-group of $cd_p\, G \le 2$ with finite relation rank $r(G) = \dim_{\mathbb{F}_p} H^2(G, \mathbb{Z}/p\mathbb{Z}) < \infty$. Furthermore, assume that

$$H_2(\mathscr{G}, \mathbb{Z}_p) = 0 = H_2(\mathscr{H}, \mathbb{Z}_p).$$

In particular, this is fulfilled if $scd_p\, \mathscr{G} \le 2$.

Then $X = \mathscr{H}^{ab}(p)$ is a finitely generated $\mathbb{Z}_p[\![G]\!]$-module with $pd_{\mathbb{Z}_p[\![G]\!]} X \le 1$ and the following assertions are equivalent:

(i) *X is free as a $\mathbb{Z}_p[\![G]\!]$-module.*

(ii) *X_G is free as a $\mathbb{Z}_p$-module.*

(iii) *The map ${}_p\mathscr{G}^{ab} \longrightarrow {}_pG^{ab}$ is injective.*

Proof: Since $cd_p\, G \le 2$, we have an exact sequence

$$0 \longrightarrow H_2(G, \mathbb{Z}_p) \xrightarrow{\ p\ } H_2(G, \mathbb{Z}_p) \longrightarrow H_2(G, \mathbb{Z}/p\mathbb{Z})\,,$$

and from our assumption, we know that $\dim_{\mathbb{F}_p} H_2(G, \mathbb{Z}/p\mathbb{Z}) = r(G)$ is finite. Thus $H_2(G, \mathbb{Z}_p)$ is free of finite rank as a $\mathbb{Z}_p$-module. The Hochschild-Serre sequence

$$0 \longrightarrow H_2(G, \mathbb{Z}_p) \longrightarrow X_G \longrightarrow \mathscr{G}^{ab}(p) \longrightarrow G^{ab} \longrightarrow 0$$

(using $H_2(\mathcal{G}, \mathbb{Z}_p) = 0$) shows that X_G is finitely generated as a $\mathbb{Z}_p$-module and that X_G is $\mathbb{Z}_p$-free if and only if $\ker(\mathcal{G}^{ab}(p) \to G^{ab})$ is $\mathbb{Z}_p$-torsion-free. This proves the equivalence (ii) $\Longleftrightarrow$ (iii).

Since $H_2(\mathcal{H}, \mathbb{Z}_p) = 0$ and noting that $cd_p\,\mathcal{G}, cd_p\,G \le 2$, proposition (5.6.7) shows that

$$0 \longrightarrow N^{ab}_{\mathcal{H}}(p) \longrightarrow R^{ab}(p) \longrightarrow X \longrightarrow 0$$

is a projective resolution of X of length 1, i.e. $pd_{\mathbb{Z}_p[\![G]\!]}X \le 1$.

Recall that $\mathbb{Z}_p[\![G]\!]$ is a local ring since G is a pro-p-group by (5.2.16)(iii), and therefore finitely generated projective $\mathbb{Z}_p[\![G]\!]$-modules are free by (5.2.20). Since X_G is finitely generated as a $\mathbb{Z}_p$-module, Nakayama's lemma (5.2.18) implies that X is a finitely generated $\mathbb{Z}_p[\![G]\!]$-module.

Now assume that X_G is $\mathbb{Z}_p$-free and let

$$0 \longrightarrow P \longrightarrow \mathbb{Z}_p[\![G]\!]^r \longrightarrow X \longrightarrow 0$$

be a minimal resolution of X, i.e. $(\mathbb{Z}_p[\![G]\!]^r)_G = \mathbb{Z}_p^r \xrightarrow{\sim} X_G$. We obtain an isomorphism $H_1(G, X) \cong P_G$. Since $cd_p\,G \le 2$ and $H_2(\mathcal{G}, \mathbb{Z}_p) = 0$, the Hochschild-Serre spectral sequence shows that

$$H_1(G, H_1(\mathcal{H}, \mathbb{Z}_p)) = E^2_{1,1} = E^\infty_{1,1} = 0.$$

Thus $P_G = 0$ and therefore the compact module P is trivial by Nakayama's lemma . This proves the nontrivial implication (ii) $\Rightarrow$ (i). □

(5.6.14) Corollary. *Assertions* (i) – (iii) *of* (5.6.13) *are true if* $cd_p\,\mathcal{G} \le 1$, *and in this case*

$$\operatorname{rank}_{\mathbb{Z}_p[\![G]\!]} X = \dim_{\mathbb{F}_p} H^1(\mathcal{G}, \mathbb{F}_p) - d(G) + r(G)\,,$$

where $d(G) = \dim_{\mathbb{F}_p} H^1(G, \mathbb{F}_p)$ *and* $r(G) = \dim_{\mathbb{F}_p} H^2(G, \mathbb{F}_p)$.

Proof: Since $cd_p\,\mathcal{G} \le 1$, we have $scd_p\,\mathcal{H}, scd_p\,\mathcal{G} \le 2$. Furthermore,

$$_p\mathcal{G}^{ab} = (H^2(\mathcal{G}, \mathbb{F}_p))^\vee = 0$$

showing that (iii) is true. The Hochschild-Serre sequence

$$0 \longrightarrow H^1(G, \mathbb{F}_p) \longrightarrow H^1(\mathcal{G}, \mathbb{F}_p) \longrightarrow H^1(\mathcal{H}, \mathbb{F}_p)^G \longrightarrow H^2(G, \mathbb{F}_p) \longrightarrow 0$$

implies that

$$\begin{aligned} \dim_{\mathbb{F}_p} X_G/p &= \dim_{\mathbb{F}_p} H^1(\mathcal{H}, \mathbb{F}_p)^G \\ &= \dim_{\mathbb{F}_p} H^1(\mathcal{G}, \mathbb{F}_p) - d(G) + r(G). \end{aligned}$$

Since X is $\mathbb{Z}_p[\![G]\!]$-free, we have $\dim_{\mathbb{F}_p} X_G/p = \operatorname{rank}_{\mathbb{Z}_p[\![G]\!]} X$. This finishes the proof. □

We now consider the following situation, which is related to Iwasawa theory. Let

$$1 \longrightarrow \mathcal{H} \longrightarrow \mathcal{G} \longrightarrow \Gamma \longrightarrow 1$$

be an exact sequence of profinite groups, where $\mathcal{G}$ is finitely generated and $\Gamma \cong \mathbb{Z}_p$. Let $\Lambda = \mathbb{Z}_p[\![\Gamma]\!]$. Then $X = \mathcal{H}^{ab}(p) = H_1(\mathcal{H}, \mathbb{Z}_p)$ is a finitely generated Iwasawa module, since

$$\dim_{\mathbb{F}_p}(X/p)_\Gamma = \dim_{\mathbb{F}_p} H^1(\mathcal{G}, \mathbb{Z}/p\mathbb{Z}) - 1$$

is finite by (5.3.10). Furthermore, we assume that the dimensions

$$h_i = \dim_{\mathbb{F}_p} H^i(\mathcal{G}, \mathbb{Z}/p\mathbb{Z}), i \le 2,$$

are finite, and we set

$$\chi_2(\mathcal{G}) = \sum_{i=0}^{2} (-1)^i h_i.$$

(5.6.15) Lemma.

(i) *$pd_\Lambda X \le 1$ if and only if $H^1(\Gamma, H^1(\mathcal{H}, \mathbb{Q}_p/\mathbb{Z}_p))$ is p-divisible.*

(ii) *If $cd_p \mathcal{G} \le 2$, then $H_2(\mathcal{H}, \mathbb{Z}_p) = H^2(\mathcal{H}, \mathbb{Q}_p/\mathbb{Z}_p)^\vee$ is a free Λ-module of finite rank.*

Proof: From (5.3.19)(i), we know that $pd_\Lambda X \le 1$ is equivalent to the statement that X^Γ is $\mathbb{Z}_p$-free, or dually that $H^1(\Gamma, H^1(\mathcal{H}, \mathbb{Q}_p/\mathbb{Z}_p))$ is p-divisible. This proves (i).

The surjection $H^2(\mathcal{G}, \mathbb{Q}_p/\mathbb{Z}_p) \twoheadrightarrow H^2(\mathcal{H}, \mathbb{Q}_p/\mathbb{Z}_p)^\Gamma$ (which is obtained from the Hochschild-Serre spectral sequence noting that $cd_p\Gamma = 1$) shows that the $\mathbb{Z}_p$-module $H_2(\mathcal{H}, \mathbb{Z}_p)_\Gamma$ is finitely generated, and so $H_2(\mathcal{H}, \mathbb{Z}_p)$ is a finitely generated Λ-module by (5.3.10). It follows from (5.6.7) that $H_2(\mathcal{H}, \mathbb{Z}_p)$ is projective, hence Λ-free by (5.2.20).

□

(5.6.16) Proposition. *Suppose $pd_\Lambda X \le 1$. Then there exists an exact sequence*

$$0 \to \Lambda^{h_2 - t} \to \Lambda^{h_1 - 1} \to X \to 0,$$

where $t = \dim_{\mathbb{F}_p}(H_2(\mathcal{H}, \mathbb{Z}_p)/p)_\Gamma$. In particular,

$$\mathrm{rank}_\Lambda(X) = -\chi_2(\mathcal{G}) + t.$$

If $cd_p \mathcal{G} \le 2$, then $t = \mathrm{rank}_\Lambda H_2(\mathcal{H}, \mathbb{Z}_p)$.

Proof: This follows from (5.3.20). Indeed, we have

$$\begin{aligned} d_0(X) &= \dim_{\mathbb{F}_p}(X/p)_\Gamma = h_1 - 1, \\ d_1(X) &= \dim_{\mathbb{F}_p}{}_p(X_\Gamma) + \dim_{\mathbb{F}_p} X^\Gamma/p \\ &= \dim_{\mathbb{F}_p} H^1(\mathscr{H}, \mathbb{Q}_p/\mathbb{Z}_p)^\Gamma/p + \dim_{\mathbb{F}_p}{}_p H^1(\Gamma, H^1(\mathscr{H}, \mathbb{Q}_p/\mathbb{Z}_p)) \\ &= \dim_{\mathbb{F}_p} H^1(\mathscr{G}, \mathbb{Q}_p/\mathbb{Z}_p)/p + \dim_{\mathbb{F}_p}{}_p H^1(\Gamma, H^1(\mathscr{H}, \mathbb{Q}_p/\mathbb{Z}_p)), \\ d_2(X) &= \dim_{\mathbb{F}_p}{}_p X^\Gamma = 0, \end{aligned}$$

where the last equality follows from $pd_\Lambda X \leq 1$ and (5.3.19)(i). The exact sequence

$$0 \to H^1(\Gamma, H^1(\mathscr{H}, \mathbb{Q}_p/\mathbb{Z}_p)) \to H^2(\mathscr{G}, \mathbb{Q}_p/\mathbb{Z}_p) \to H^2(\mathscr{H}, \mathbb{Q}_p/\mathbb{Z}_p)^\Gamma \to 0,$$

which is induced by the Hochschild-Serre spectral sequence noting that $cd_p\Gamma = 1$, and the fact that $H^1(\Gamma, H^1(\mathscr{H}, \mathbb{Q}_p/\mathbb{Z}_p))/p = 0$ by (5.6.15), shows that

$$\begin{aligned} d_1(X) &= \dim_{\mathbb{F}_p} H^1(\mathscr{G}, \mathbb{Q}_p/\mathbb{Z}_p)/p + \dim_{\mathbb{F}_p}{}_p H^2(\mathscr{G}, \mathbb{Q}_p/\mathbb{Z}_p) \\ &\quad -\dim_{\mathbb{F}_p}{}_p H^2(\mathscr{H}, \mathbb{Q}_p/\mathbb{Z}_p)^\Gamma. \end{aligned}$$

The exact cohomology sequence

$$0 \to H^1(\mathscr{G}, \mathbb{Q}_p/\mathbb{Z}_p)/p \to H^2(\mathscr{G}, \mathbb{Z}/p\mathbb{Z}) \to {}_p H^2(\mathscr{G}, \mathbb{Q}_p/\mathbb{Z}_p) \to 0$$

obtained from $0 \to \mathbb{Z}/p\mathbb{Z} \to \mathbb{Q}_p/\mathbb{Z}_p \xrightarrow{p} \mathbb{Q}_p/\mathbb{Z}_p \to 0$ now implies that

$$\begin{aligned} d_1(X) &= \dim_{\mathbb{F}_p} H^2(\mathscr{G}, \mathbb{Z}/p\mathbb{Z}) - \dim_{\mathbb{F}_p}{}_p H^2(\mathscr{H}, \mathbb{Q}_p/\mathbb{Z}_p))^\Gamma \\ &= h_2 - t. \end{aligned}$$

Finally, t is equal to $\mathrm{rank}_\Lambda H_2(\mathscr{H}, \mathbb{Z}_p)$ if $cd_p\mathscr{G} \leq 2$ by (5.6.15). □

(5.6.17) Theorem. *With the notation above, the following assertions are equivalent:*

(i) *X contains no finite nontrivial Λ-submodule and* $\mathrm{rank}_\Lambda X = -\chi_2(\mathscr{G})$.

(ii) *$H^2(\mathscr{H}, \mathbb{Q}_p/\mathbb{Z}_p) = 0$ and $H^2(\mathscr{G}, \mathbb{Q}_p/\mathbb{Z}_p)$ is p-divisible.*

Proof: By (5.3.19)(i) and (5.6.16), assertion (i) is equivalent to $pd_\Lambda X \leq 1$ and $t = 0$, hence to the fact that $H^1(\Gamma, H^1(\mathscr{H}, \mathbb{Q}_p/\mathbb{Z}_p))$ is p-divisible and that $H^2(\mathscr{H}, \mathbb{Q}_p/\mathbb{Z}_p) = 0$. But this is precisely statement (ii), which can be seen using the exact sequence

$$0 \to H^1(\Gamma, H^1(\mathscr{H}, \mathbb{Q}_p/\mathbb{Z}_p)) \to H^2(\mathscr{G}, \mathbb{Q}_p/\mathbb{Z}_p) \to H^2(\mathscr{H}, \mathbb{Q}_p/\mathbb{Z}_p)^\Gamma \to 0.$$

□

(5.6.18) Theorem. *Assume that $\mathscr{G}$ (and hence also $\mathscr{H}$) is a pro-p-group. Then the following assertions are equivalent:*

(i) *$\mathscr{H}$ is a free pro-p-group.*

(ii) *$H^2(\mathscr{G}, \mathbb{Q}_p/\mathbb{Z}_p)$ is p-divisible, $H^2(\mathscr{H}, \mathbb{Q}_p/\mathbb{Z}_p) = 0$ and the μ-invariant of X is zero.*

Proof: A pro-p-group G is free if and only if $H^2(G, \mathbb{Z}/p\mathbb{Z}) = 0$ (see (3.5.17) and (3.3.2)(ii)), thus if and only if G^{ab} is $\mathbb{Z}_p$-torsion-free and $H^2(G, \mathbb{Q}_p/\mathbb{Z}_p) = 0$. Hence (i) is equivalent to the statement that X is $\mathbb{Z}_p$-torsion-free and $H^2(\mathscr{H}, \mathbb{Q}_p/\mathbb{Z}_p) = 0$. The latter occurs precisely when $\mu(X) = 0$ and $X^\Gamma = H^1(\Gamma, H^1(\mathscr{H}, \mathbb{Q}_p/\mathbb{Z}_p))^\vee$ is $\mathbb{Z}_p$-torsion-free. Again using the exact sequence in the proof of (5.6.17), we obtain the result. □

(5.6.19) Corollary. *In the situation of theorem (5.6.18), assume, in addition, that $cd_p \mathscr{G} \le 2$. Let $\mathscr{U}$ be an open subgroup of $\mathscr{G}$, $\mathscr{V} = \mathscr{H} \cap \mathscr{U}$ and $\Gamma' = \mathscr{U}/\mathscr{V} \subseteq \Gamma$. Then $Y = \mathscr{V}^{ab}$ is a finitely generated $\Lambda' = \mathbb{Z}_p[[\Gamma']]$-module and the following assertions are equivalent:*

(i) $\mu(X) = 0$ *and* $H^2(\mathscr{H}, \mathbb{Q}_p/\mathbb{Z}_p) = 0$.

(ii) $\mu(Y) = 0$ *and* $H^2(\mathscr{V}, \mathbb{Q}_p/\mathbb{Z}_p) = 0$.

Proof: Since $cd_p \mathscr{U} = cd_p \mathscr{G} \le 2$, the cohomology groups $H^2(\mathscr{G}, \mathbb{Q}_p/\mathbb{Z}_p)$ and $H^2(\mathscr{U}, \mathbb{Q}_p/\mathbb{Z}_p)$ are p-divisible. Thus (i) (resp. (ii)) is equivalent to the freeness of $\mathscr{H}$ (resp. $\mathscr{V}$) by (5.6.18). Since $\mathscr{V}$ is open in $\mathscr{H}$ and $cd_p \mathscr{H} < \infty$, we have $cd_p \mathscr{V} = cd_p \mathscr{H}$ by (3.3.5)(ii). Now (3.5.17) implies the result. □

Arithmetic Theory

Chapter VI

Galois Cohomology

§1. Cohomology of the Additive Group

The Galois groups $G = G(L|K)$ of Galois extensions $L|K$ are profinite groups and we may such use cohomology theory which we have developed in the preceding chapters. We are particularly interested in the meaning of the cohomology groups $H^n(G, A)$ for extensions of local and global fields, but we first study their properties for general Galois extensions $L|K$.

Of particular importance is the *absolute Galois group* $G_K = G(\bar{K}|K)$ of a field K. It depends on the choice of a separable closure $\overline{K}|K$, and is therefore unique only up to inner automorphisms (we will return to this point in chapter XII). By (1.6.3), however, its cohomology is independent of the choice and we write

$$H^n(K, A) := H^n(G_K, A).$$

In the following we will assume that all extension fields are contained in a fixed separable closure.

The first $G(L|K)$-module which comes to mind is the additive group of a Galois extension L of K. It is cohomologically trivial. In order to show this, we may assume that $G = G(L|K)$ is finite, see (1.2.5). Then the assertion is trivial if $\operatorname{char}(K) = 0$, because in this case L is uniquely divisible and the result follows from (1.6.2)(c). In general, one argues as follows: the G-module L is induced, because of the existence of a normal basis, i.e. of an element $\theta \in L$ such that

$$L = \bigoplus_{\sigma \in G} K\sigma\theta.$$

Because of (1.3.7) and (1.2.5), we obtain the

(6.1.1) Proposition. *If $L|K$ is an arbitrary Galois extension with Galois group G, then*

$$H^q(G, L) = 0 \quad \text{for all } q > 0.$$

J. Neukirch et al., *Cohomology of Number Fields*,
Grundlehren der mathematischen Wissenschaften 323,
DOI 10.1007/978-3-540-37889-1_6, © Springer-Verlag Berlin Heidelberg 2008

We call a field L **p-closed** if it has no Galois extensions of degree p. For example, the separable closure $\bar{K}|K$ is p-closed and also the maximal p-extension $K(p)|K$, i.e. the composite of all Galois extensions of p-power order.

Let us assume now that the characteristic $p = \text{char}(K)$ is positive. If the field L is p-closed then the homomorphism

$$\wp : L \longrightarrow L, \quad \wp(x) = x^p - x,$$

is surjective, which can be seen as follows. Consider, for $a \in L$, the separable polynomial $f(x) = x^p - x - a$. If α is a root of f, then $\alpha + 1, \dots, \alpha + p - 1$ are the other roots. Therefore, if a were not in the image of $\wp$, then the splitting field of this polynomial would be a cyclic extension of L of degree p. But we assumed L to be p-closed. We thus obtain the exact sequence

$$0 \longrightarrow \mathbb{Z}/p\mathbb{Z} \longrightarrow L \xrightarrow{\wp} L \longrightarrow 0,$$

and the associated exact cohomology sequence yields, together with (6.1.1), the

(6.1.2) Corollary. *Let $p = \text{char}(K) > 0$. If $L|K$ is a p-closed extension, then*

$$H^n(G(L|K), \mathbb{Z}/p\mathbb{Z}) = \begin{cases} K/\wp K & \text{for} \quad n = 1, \\ 0 & \text{for} \quad n \geq 2. \end{cases}$$

Applying the last corollary to the fixed field of L with respect to a p-Sylow group of $G(L|K)$, (3.3.6) implies the

(6.1.3) Corollary. *Let $p = \text{char}(K) > 0$. If $L|K$ is a p-closed extension, then*

$$cd_p\, G(L|K) \leq 1\,.$$

From the above corollaries and (3.9.1), (3.9.5), we obtain the following theorem, which in a sense gives a complete survey of the Galois extensions of K of p-power degree.

(6.1.4) Theorem. *If $p = \text{char}(K) > 0$, then the Galois group G of the maximal p-extension $K(p)|K$ is a free pro-p-group of rank*

$$\text{rk}(G) = \dim_{\mathbb{F}_p} K/\wp K.$$

(6.1.5) Corollary. *If K is a countable field of characteristic $p > 0$ then $G(K(p)|K)$ is a free pro-p-group of finite or countable infinite rank.*

Next we determine the Galois groups of the maximal p-extensions of local and global fields of characteristic p.

(6.1.6) Lemma. *Let $K = k((t))$ be the field of Laurent series over a perfect field k with $p = \mathrm{char}(k) > 0$. Then $tk[[t]] \subseteq \wp K$, in particular, $\wp K$ is open in the topology of K as a discrete valuation field. If $\mathcal{R}$ is a system of representatives of k modulo $\wp k$, then the set $\mathcal{L}$ of Laurent polynomials,*

$$\mathcal{L} = \{ \sum_{\substack{(i,p)=1\\ i<0}} a_i t^i + a_0,\ a_i \in k \text{ for } i < 0 \text{ and } a_0 \in \mathcal{R}\}$$

is a system of representatives of K modulo $\wp K$.

Proof: An easy computation shows that all power series of the form $\sum_{i=1}^{\infty} a_i t^i$ are in the image of $\wp$, and for $a = b^p$ and $i < 0$ we have

$$at^{pi} - bt^i = \wp(bt^i).$$

With these observations in mind, one immediately verifies that the set $\mathcal{L}$ is a system of representatives of K modulo $\wp K$. □

We obtain the following

(6.1.7) Proposition. *Let $K = k((t))$ be the field of Laurent series over a field k with $p = \mathrm{char}(k) > 0$. If k is finite or countable, then $G(K(p)|K)$ is a free pro-p-group of countable infinite rank. If k is uncountable, then $G(K(p)|K)$ is free of uncountable rank.*

Proof: In order to prove the statement, we may assume that k is perfect: otherwise we replace k by its perfect hull without changing the absolute Galois group and the cardinalities in question. Now the result follows from (6.1.6) and (6.1.4). □

In the global case we obtain the

(6.1.8) Proposition. *Let K be a global field of positive characteristic $p > 0$. Then $G(K(p)|K)$ is a free pro-p-group of countable infinite rank.*

Proof: By (6.1.5), the group in question is a free pro-p-group of at most countable rank. In order to prove that the rank is infinite, we may restrict to the case $K = \mathbb{F}(t)$, where $\mathbb{F}$ is a finite field. All the infinitely many representatives of $\mathbb{F}((t))$ modulo $\wp\mathbb{F}((t))$ given in (6.1.6) are in K and obviously also independent modulo $\wp K$. Therefore, by (6.1.4), the rank of $G(K(p)|K)$ is infinite. □

Let $K_p|K$ denote the maximal abelian extension of exponent $p = \mathrm{char}(K)$, i.e. the composite of all cyclic extensions of degree p. Its Galois group is

$$G(K_p|K) = G/G^p[G,G],$$

where $[G,G]$ is the closure of the commutator subgroup of $G = G(K(p)|K)$. For the Pontryagin dual of this Galois group we obtain the isomorphism

$$G(K_p|K)^\vee \cong H^1(G, \mathbb{Z}/p\mathbb{Z}) \cong K/\wp K$$

The isomorphism associates to $a \in K$ the character $\chi_a : G(K_p|K) \to \mathbb{F}_p$, given by $\chi_a(\sigma) = \sigma\alpha - \alpha$, where α is a root of the equation $x^p - x = a$. Dually, we obtain the isomorphism of *Artin-Schreier theory* (cf. [160], chap.IV, §3)

$$G(K_p|K) \cong \mathrm{Hom}(K/\wp K, \mathbb{Q}/\mathbb{Z}).$$

This explicit description of $K_p|K$ has the following generalization to the maximal abelian extension $K_{p^n}|K$ of exponent p^n, whose Galois group is

$$G(K_{p^n}|K) = G/G^{p^n}[G,G].$$

For every $n \geq 1$, there exists a unique functor W_n from the category of rings to itself, such that for every commutative ring R with unit the underlying set of $W_n(R)$ is R^n and the map

$$gh : W_n(R) \longrightarrow R^n,$$

$$gh(a_0, \dots, a_{n-1}) =$$
$$(a_0, a_0^p + pa_1, \dots, a_0^{p^{n-1}} + pa_1^{p^{n-2}} + p^2 a_2^{p^{n-3}} + \cdots + p^{n-1}a_{n-1}),$$

is a homomorphism of rings (see [212], chap.II, §6 or [160], chap.II, §4, ex. 2-4). The elements $(a_0, \dots, a_{n-1}) \in W_n(R)$ are called the **Witt vectors of length n** and the elements $a_0,\ a_0^p + pa_1, \dots,\ a_0^{p^{n-1}} + \cdots + p^{n-1}a_{n-1}$ the **ghost components** of $(a_0, \dots, a_{n-1})$. For $R = \mathbb{F}_p$, we have a canonical isomorphism

$$W_n(\mathbb{F}_p) \xrightarrow{\sim} \mathbb{Z}/p^n\mathbb{Z},$$

$$(a_0, \dots, a_{n-1}) \longmapsto \tilde{a}_0 + \tilde{a}_1 p + \cdots + \tilde{a}_{n-1}p^{n-1} \bmod p^n,$$

where $\tilde{a}_i \in \mathbb{Z}$ is the unique representative of a_i with $0 \leq \tilde{a}_i < p$. The Witt vectors with respect to a general ring R should be seen as an abstract version of this p-adic expansions.

For an arbitrary ring R we have $W_1(R) = R$, and for each $n \geq 1$ an exact sequence of abelian groups

$$0 \longrightarrow W_n(R) \xrightarrow{V} W_{n+1}(R) \longrightarrow R \longrightarrow 0,$$

where V is given by $(a_0,\ldots,a_{n-1}) \mapsto (0,a_0,\ldots,a_{n-1})$ and is called the **Verschiebung**. If R is an integral domain of characteristic p, i.e. $pR = 0$, then we have an exact sequence

$$0 \longrightarrow W_n(\mathbb{F}_p) \longrightarrow W_n(R) \xrightarrow{\wp} W_n(R),$$

where $\wp = F - id$, $F((a_0\ldots,a_{n-1})) = (a_0^p,\ldots,a_{n-1}^p)$, so that $\ker(\wp) = W_n(\mathbb{F}_p) \cong \mathbb{Z}/p^n\mathbb{Z}$. We obtain the following generalization of Artin-Schreier theory.

(6.1.9) Theorem (Artin-Schreier-Witt). *Let K be a field of characteristic $p > 0$. Then for the Galois group of the maximal abelian extension $K_{p^n}|K$ of exponent p^n, we have a canonical isomorphism*

$$G(K_{p^n}|K) \cong \mathrm{Hom}(W_n(K)/\wp W_n(K), \mathbb{Q}/\mathbb{Z}).$$

Proof: Consider the separable closure $\bar{K}|K$ and the exact sequence of G_K-modules

$$(*) \qquad 0 \longrightarrow \mathbb{Z}/p^n\mathbb{Z} \longrightarrow W_n(\bar{K}) \xrightarrow{\wp} W_n(\bar{K}) \longrightarrow 0;$$

here the surjectivity of $\wp$ is obvious for $n = 1$, and for $n > 1$ it follows recursively from the commutative and exact diagrams

$$\begin{array}{ccccccccc} 0 & \longrightarrow & W_n(\bar{K}) & \xrightarrow{V} & W_{n+1}(\bar{K}) & \longrightarrow & W_1(\bar{K}) & \longrightarrow & 0 \\ & & \downarrow{\scriptstyle\wp} & & \downarrow{\scriptstyle\wp} & & \downarrow{\scriptstyle\wp} & & \\ 0 & \longrightarrow & W_n(\bar{K}) & \xrightarrow{V} & W_{n+1}(\bar{K}) & \longrightarrow & W_1(\bar{K}) & \longrightarrow & 0. \end{array}$$

The exact sequence

$$0 \longrightarrow W_n(\bar{K}) \xrightarrow{V} W_{n+1}(\bar{K}) \longrightarrow \bar{K} \longrightarrow 0$$

shows by induction on n that the G_K-module $W_n(\bar{K})$ is cohomologically trivial, since $\bar{K}$ is cohomologically trivial by (6.1.1). The cohomology sequence associated to $(*)$ is therefore an exact sequence

$$0 \longrightarrow \mathbb{Z}/p^n\mathbb{Z} \longrightarrow W_n(K) \xrightarrow{\wp} W_n(K) \longrightarrow H^1(G_K, \mathbb{Z}/p^n\mathbb{Z}) \longrightarrow 0$$

which yields the isomorphism

$$H^1(G_K, \mathbb{Z}/p^n\mathbb{Z}) = \mathrm{Hom}(G(K_{p^n}|K), \mathbb{Q}/\mathbb{Z}) \cong W_n(K)/\wp W_n(K),$$

and, dually, the isomorphism

$$G(K_{p^n}|K) \cong \mathrm{Hom}(W_n(K)/\wp W_n(K), \mathbb{Q}/\mathbb{Z}). \qquad \square$$

If $K|k$ is a Galois extension of local or global fields, it is natural to consider also the ring of integers $\mathcal{O}_K$ as a Galois module. We formulate the next results in such generality that they apply to the local and the global case as well.

Recall that the normalization B of a Dedekind domain A in a finite Galois extension K of its quotient field k is called *tamely ramified* if for each prime ideal $\mathfrak{p}$ of A the unique prime factorization of $\mathfrak{p}B$ in B has the form

$$\mathfrak{p}B = (\mathfrak{P}_1 \cdots \mathfrak{P}_g)^e,$$

with e prime to the characteristic of $A/\mathfrak{p}$, and such that the residue field extensions $(B/\mathfrak{P}_i) \mid (A/\mathfrak{p})$, $i = 1, \dots, g$, are separable. If $K|k$ is an infinite Galois extension, we say that $B|A$ is tamely ramified if the normalization of A in every finite normal subextension of $K|k$ is tamely ramified.

(6.1.10) Theorem. *Let A be a Dedekind domain with quotient field k, $K|k$ a Galois extension and B the integral closure of A in K. Then the following conditions are equivalent*

(i) *B is a cohomologically trivial $G(K|k)$-module.*

(ii) *$B|A$ is tamely ramified.*

Proof: By (1.8.2), we may assume that $K|k$ is finite. We consider the trace map

$$Tr_{B|A}\colon B \longrightarrow A, \quad x \mapsto \sum_{\sigma \in G(K|k)} \sigma x .$$

Claim: $Tr_{B|A}$ is surjective if and only if $B|A$ is tamely ramified.

Proof of the claim: We start with the case that A, and hence also B, is a complete discrete valuation ring and denote the residue fields of A and B by κ and κ', respectively. By Nakayama's Lemma, the homomorphism of finitely generated A-modules $Tr_{B|A}$ is surjective if and only the induced map $Tr\colon \kappa' \to \kappa$ is surjective. As the inertia group acts trivially on κ', we have $Tr = e \cdot Tr_{\kappa'|\kappa}$, where e is the ramification index. As $Tr_{\kappa'|\kappa}$ is surjective if $\kappa'|\kappa$ is a separable field extension and zero otherwise, we see that $Tr_{B|A}$ is surjective if and only if $\kappa'|\kappa$ is separable and $p \nmid e$, where $p = \mathrm{char}(\kappa)$.

In the general situation, $Tr_{B|A}$ is surjective if and only if for every prime ideal $\mathfrak{p}$ of A the induced map

$$Tr_{B|A} \otimes \hat{A}_{\mathfrak{p}}\colon B \otimes_A \hat{A}_{\mathfrak{p}} \longrightarrow \hat{A}_{\mathfrak{p}}$$

is surjective. As

$$B \otimes_A \hat{A}_{\mathfrak{p}} = \prod_{\mathfrak{P}|\mathfrak{p}} \hat{B}_{\mathfrak{P}},$$

the general case reduces to the case of complete discrete valuation rings. This proves the claim.

We put $G = G(K|k)$ and start by showing the implication (i)⇒(ii). If B is cohomologically trivial, then $\hat{H}^0(G, B) = 0$, hence $Tr_{B|A}$ is surjective. By the claim, $B|A$ is tamely ramified. It remains to show (ii)⇒(i). As the trace is surjective, we have $\hat{H}^0(G,B) = 0$. Next we prove that $H^1(G,B) = 0$. Let $a(\sigma) \in B$ be a 1-cocycle and let $x \in B$ be such that $Tr_{B|A}(x) = 1$. Setting

$$b := \sum_{\sigma\in G} a(\sigma)\sigma x\,,$$

we obtain for $\tau \in G$,

$$\tau b = \sum_{\sigma\in G} \tau a(\sigma)(\tau\sigma x) = \sum_{\sigma\in G} (a(\tau\sigma) - a(\tau))(\tau\sigma x) = b - a(\tau)Tr_{B|A}(x)\,.$$

Therefore $a(\tau) = (1-\tau)b$, hence $a(\tau)$ is a 1-coboundary.

Since the arguments above hold in the same way for all subgroups of $G(K|k)$, the result follows from (1.8.4). □

In the case of a finite, cyclic Galois group, it is easy to calculate the Herbrand index.

(6.1.11) Proposition. *Let A be a Dedekind domain with quotient field k, $K|k$ a finite, cyclic Galois extension and B the integral closure of A in K. If $A/\mathfrak{m}$ is finite for all maximal ideals $\mathfrak{m} \subseteq A$, then the Herbrand index $h(G(K|k), B)$ is trivial, i.e.*

$$\#\hat{H}^{-1}(G(K|k), B) = \#\hat{H}^0(G(K|k), B),$$

and both groups are finite.

Proof: Let $G(K|k) = \{\sigma_1, \ldots, \sigma_n\}$. We consider a normal basis of $K|k$, i.e. we choose an $\alpha \in K$ such that $\sigma_1\alpha, \ldots, \sigma_n\alpha$ is a k-basis of K. For suitably chosen $a \in k^\times$, we have an inclusion

$$B \subseteq Aa\sigma_1\alpha + \ldots + Aa\sigma_n\alpha$$

of B into a free $A[G(K|k)]$-module. The quotient has finite A-length and is therefore finite. By (1.7.5) and (1.7.6), we obtain $h(G(K|k), B) = 1$. □

§2. Hilbert's Satz 90

Again let $L|K$ be a Galois extension and G its Galois group. The multiplicative group $L^\times$ is also a G-module. It is only cohomologically trivial in exceptional cases, but we always have

(6.2.1) Theorem (Hilbert's Satz 90). $H^1(G, L^\times) = 1$.

Proof: By (1.2.5), we may assume that $L|K$ is finite. Let $a : G \to L^\times$ be an inhomogeneous 1-cocycle. For $c \in L^\times$ we put

$$b = \sum_{\sigma \in G} a(\sigma)\sigma c.$$

Since the automorphisms $\sigma \in G$ are linearly independent (see [17], chap.5, §7, No.5), we may choose $c \in L^\times$ in such a way that b is non-zero. For τ in G we obtain

$$\tau(b) = \sum_{\sigma \in G} \tau(a(\sigma))\tau\sigma c = \sum a(\tau)^{-1}a(\tau\sigma)\tau\sigma c = a(\tau)^{-1}b.$$

Thus $a(\tau) = b\tau(b)^{-1}$ for all $\tau \in G$, i.e. a is a coboundary. □

In the preceding section we obtained from the equality $H^1(G, L) = 0$ the theorem of Artin-Schreier-Witt for abelian extensions of exponent p^n, where $p = \mathrm{char}(K)$. From the vanishing of $H^1(G, L^\times)$ we obtain a similar result, called *Kummer theory*.

Let n be a natural number, not divisible by the characteristic of K. We denote the group of n-th (resp. of all) roots of unity in the separable closure $\bar{K}$ of K by μ_n (resp. μ). As $H^1(G_K, \bar{K}^\times) = 1$, we obtain from the exact sequence

$$1 \longrightarrow \mu_n \longrightarrow \bar{K}^\times \xrightarrow{\ n\ } \bar{K}^\times \longrightarrow 1$$

the exact sequence

$$K^\times \xrightarrow{\ n\ } K^\times \xrightarrow{\ \delta\ } H^1(G_K, \mu_n) \longrightarrow 0,$$

and hence an isomorphism

$$H^1(G_K, \mu_n) \cong K^\times / K^{\times n}.$$

Let $K_n|K$ be the maximal abelian extension of exponent n, i.e. the composite of all finite abelian extensions $L|K$ in $\bar{K}$ of degree dividing n. If $\mu_n \subseteq K$, we obtain isomorphisms

$$\mathrm{Hom}_{cts}(G(K_n|K), \mu_n) \cong H^1(G_K, \mu_n) \cong K^\times / K^{\times n}.$$

Dually, we obtain the

(6.2.2) Theorem. *If $n \geq 1$ is prime to the characteristic of K and $\mu_n \subseteq K$, then*

$$G(K_n|K) \cong \mathrm{Hom}(K^\times / K^{\times n}, \mu_n).$$

Hilbert's Satz 90 may be extended from the multiplicative group $L^\times$ of a Galois extension $L|K$ to the G-group $GL_n(L)$ of all invertible $n \times n$-matrices over L in terms of non-abelian cohomology (see I §2).

(6.2.3) Theorem. *For every Galois extension $L|K$ with Galois group $G = G(L|K)$ we have*

$$H^1(G, GL_n(L)) = 1.$$

Proof: By proposition (1.2.5) (which also holds by the same arguments for non-abelian cohomology), we may assume that $L|K$ is finite. Let a be an 1-cocycle of G with values in $GL_n(L)$ and consider for a vector $x \in L^n$ the vector

$$b(x) = \sum_{\sigma \in G} a(\sigma)\sigma x.$$

The set $\{b(x),\ x \in L^n\}$ generates the L-vector space L^n. In fact, if f is a linear form on L^n which vanishes on the $b(x)$, then for all $\lambda \in L$,

$$0 = f(b(\lambda x)) = \sum_{\sigma \in G} f(a(\sigma)\sigma\lambda\sigma x) = \sum_{\sigma \in G} f(a(\sigma)\sigma x)\sigma\lambda,$$

i.e. we have a linear relation between the $\sigma\lambda$. But the automorphisms σ are linearly independent, so that $f(a(\sigma)\sigma x) = 0$, and since the matrices $a(\sigma)$ are invertible, we find $f = 0$.

From this observation, we can find vectors $x_1, \dots, x_n$ such that the $y_i = b(x_i)$ are linearly independent. If c is the matrix with columns $x_1, \dots, x_n$, then $b := b(c) = (b(x_1), \dots, b(x_n)) = (y_1, \dots, y_n)$ is an invertible matrix and

$$b = \sum_{\sigma \in G} a(\sigma)\sigma c.$$

As in the case $n = 1$, we conclude that $a(\sigma) = b(\sigma b)^{-1}$, i.e. a is a coboundary. □

We denote the G-group of all invertible $n \times n$-matrices over L with determinant equal to 1 by $SL_n(L)$.

(6.2.4) Corollary. $H^1(G, SL_n(L)) = 1.$

Proof: From the exact sequence

$$1 \longrightarrow SL_n(L) \longrightarrow GL_n(L) \xrightarrow{\det} L^\times \longrightarrow 1,$$

we obtain the exact sequence of pointed sets

$$GL_n(K) \xrightarrow{det} K^\times \longrightarrow H^1(G, SL_n(L)) \longrightarrow H^1(G, GL_n(L)) = 1$$

(see I §3 ex.8), in which *det* is surjective. □

The non-abelian cohomology groups H^1 have various interesting meanings, all of which arise from the bijection

$$H^1(G, A) \cong \mathit{TORS}(A)$$

(see (1.2.3)). We mention the following application to the theory of *quadratic forms*.

Let V be a vector space over K equipped with a quadratic form φ. Denote by φ_L the canonical extension of φ to the L-vector space $V_L = V \otimes_K L$. The **orthogonal group** $O(\varphi_L)$ of φ_L is the group of automorphisms of the pair (V_L, φ_L), i.e. the L-automorphisms $f : V_L \to V_L$ with $\varphi_L(f(v)) = \varphi_L(v)$ for all $v \in V_L$. The Galois group $G = G(L|K)$ acts on V_L by $\sigma(v \otimes \lambda) = v \otimes \sigma\lambda$ and on $O(\varphi_L)$ by $\sigma(f) = \sigma \circ f \circ \sigma^{-1}$ for $\sigma \in G$ and $f : V_L \to V_L$ in $O(\varphi_L)$. Thus $O(\varphi_L)$ is a G-group.

We say that (V, φ) and (V', φ') are *isomorphic over* L if the pairs (V_L, φ_L) and (V'_L, φ'_L) are isomorphic. We denote by $E_\varphi(L|K)$ the set of isomorphism classes of pairs (V', φ') which become isomorphic to (V, φ) over L.

(6.2.5) Proposition. *We have a canonical bijection of pointed sets*

$$E_\varphi(L|K) \cong H^1(G, O(\varphi_L)).$$

Proof: Let (V', φ') be a representative of a class in $E_\varphi(L|K)$ and let $X(V', \varphi')$ be the nonempty set of isomorphisms $f : (V_L, \varphi_L) \to (V'_L, \varphi'_L)$. This is a G-set by $\sigma(f) = \sigma \circ f \circ \sigma^{-1}$ and is equipped with a simply transitive right action of the G-group $O(\varphi_L)$ compatible with the G-action. In other words, $X(V', \varphi')$ is an $O(\varphi_L)$-*torsor* in the sense of I §2 and we obtain a map

$$(*) \qquad E_\varphi(L|K) \longrightarrow \mathit{TORS}\,(O(\varphi_L)).$$

We prove the bijectivity of this map by constructing an inverse as follows. Let X be an $O(\varphi_L)$-torsor and let $x \in X$. Then for every $\sigma \in G$ we have a unique $A_\sigma \in O(\varphi_L)$ such that $\sigma x = x A_\sigma$. The function $\sigma \mapsto A_\sigma$ is a 1-cocycle. Since $O(\varphi_L) \subseteq GL(V_L)$ and $H^1(G, GL(V_L)) = \{1\}$ by (6.2.3), there exists an L-automorphism f of V_L such that

$$A_\sigma = f^{-1} \circ \sigma(f) \qquad \text{for all } \sigma \in G.$$

For the quadratic form $\varphi' = f(\varphi)$ we have

$$\sigma(\varphi') = \sigma(f)(\sigma(\varphi)) = \sigma(f)(\varphi) = (f \circ A_\sigma)(\varphi) = f(\varphi) = \varphi'.$$

Hence φ' is rational over K, i.e. a quadratic form on V. Associating to X the isomorphism class of the pair (V, φ'), we get a map $\mathit{TORS}\,(O(\varphi_L)) \to E_\varphi(L|K)$. A straightforward check, which we leave as an exercise for the reader, shows that this map is inverse to $(*)$. The bijection

$$E_\varphi(L|K) \cong H^1(G, O(\varphi_L))$$

is now the composite of $(*)$ with the bijection $\mathit{TORS}\,(O(\varphi_L)) \cong H^1(G, O(\varphi_L))$ of (1.2.4). □

Remark: The proof shows that the bijection $E_\varphi(L|K) \cong H^1(G, O(\varphi_L))$ is explicitly given by associating to a pair (V', φ') the class of the 1-cocycle A_σ given by $\sigma(f) = fA_\sigma$, where f is an isomorphism $(V_L, \varphi_L) \to (V'_L, \varphi'_L)$.

Another application of non-abelian cohomology is to the G-group $PGL_n(L)$ which may be defined by the exact sequence

$$1 \longrightarrow L^\times \longrightarrow GL_n(L) \longrightarrow PGL_n(L) \longrightarrow 1.$$

Let $\mathbb{P}_K^{n-1}$ be the $(n-1)$-dimensional projective space over K, considered as a K-variety. The group $PGL_n(L)$ is the automorphism group

$$PGL_n(L) = \mathrm{Aut}_L(\mathbb{P}_L^{n-1}),$$

where $\mathbb{P}_L^{n-1} = \mathbb{P}_K^{n-1} \otimes_K L$. The Galois group $G = G(L|K)$ acts on $\mathbb{P}_L^{n-1}$ via the action on L and the action on $PGL_n(L)$ is induced by the action on $\mathbb{P}_L^{n-1}$: for $\sigma \in G$ and $A \in PGL_n(L)$, we have ${}^\sigma A = \sigma \circ A \circ \sigma^{-1}$.

A **Brauer-Severi variety** over K of dimension $n-1$ is a K-variety X which becomes isomorphic to $\mathbb{P}^{n-1}$ over a Galois extension $L|K$. This means that $X \otimes_K L$ and $\mathbb{P}_L^{n-1}$ are isomorphic as L-varieties. We then say that X *splits* over L. Let us denote by $BS_n(L|K)$ the pointed set of isomorphism classes $[X]$ of Brauer-Severi varieties over K of dimension $n-1$ which split over L. We define a canonical map

$$BS_n(L|K) \longrightarrow H^1(G, PGL_n(L))$$

as follows. Let X be an $(n-1)$-dimensional Brauer-Severi variety over K which splits over L. Then the set $T(X)$ of all isomorphisms of L-schemes

$$f : \mathbb{P}_L^{n-1} \longrightarrow X \otimes_K L$$

is a $PGL_n(L)$-*torsor*. In fact, on the one hand G acts on $X \otimes_K L$ via the second factor, and hence on $T(X)$ by $\sigma(f) = \sigma \circ f \circ \sigma^{-1}$, and on the other hand $T(X)$ is equipped with a simply transitive right action of $PGL_n(L)$, compatible with the G-action:

$$\sigma(f \circ A) = \sigma \circ f \circ A \circ \sigma^{-1} = \sigma f \sigma^{-1} \sigma A \sigma^{-1} = \sigma(f)\, {}^\sigma A.$$

Therefore we obtain a canonical map

$$(1) \qquad BS_n(L|K) \longrightarrow \mathit{TORS}(PGL_n(L)), \quad X \longmapsto T(X).$$

The composite with the bijection (1.2.4)

$$\mathit{TORS}(PGL_n(L)) \cong H^1(G, PGL_n(L))$$

gives a map

$$(2) \qquad BS_n(L|K) \longrightarrow H^1(G, PGL_n(L)).$$

It is explicitly given by associating to a Brauer-Severi variety X the class of the cocycle $A_\sigma = f^{-1} \circ \sigma(f)$, where f is an element of $T(X)$.

(6.2.6) Theorem. $H^1(G, PGL_n(L)) \cong BS_n(L|K)$.

Proof: We construct an inverse of the map (1) as follows. Let T be any $PGL_n(L)$-torsor and let $f \in T$. For every $\sigma \in G$ there is a unique $A_\sigma \in PGL_n(L)$ such that $\sigma(f) = fA_\sigma$. A_σ is a 1-cocycle of G with values in $PGL_n(L)$. We change the action of $\sigma \in G$ on $\mathbb{P}_L^{n-1}$ by $\sigma^* = A_\sigma \circ \sigma$ (note that $(\sigma\tau)^* = A_{\sigma\tau}\sigma\tau = A_\sigma {}^\sigma A_\tau \sigma\tau = A_\sigma \sigma A_\tau \sigma^{-1}\sigma\tau = A_\sigma\sigma A_\tau \tau = \sigma^*\tau^*$). In order to indicate the changed G-action on $\mathbb{P}_L^{n-1}$, we write $Y(T)$ instead. The action of G on $Y(T)$ is a so-called "descent datum" and by *descent theory* of algebraic geometry, there exists a K-variety $X(T) = Y(T)/G$, unique up to isomorphism, such that we have an isomorphism of L-varieties $X(T) \otimes_K L \cong Y(T)$ compatible with the G-action (see [67], chap.VIII, 7.8). This isomorphism induces an isomorphism of $PGL_n(L)$-torsors

$$T(X(T)) = \mathrm{Isom}_L(\mathbb{P}_L^{n-1}, Y(T)) \cong T,$$

which is given by $A \mapsto fA$. If we start with another $f \in T$, then we obtain a Brauer-Severi variety over K isomorphic to $X(T)$. If, conversely, we start with a Brauer-Severi variety X which splits over L, then $X(T(X)) \cong X$, since we have an isomorphism

$$X(T(X)) \otimes_K L \cong X \otimes_K L$$

of L-varieties compatible with the G-action, and so $X(T(X))$ and X are both solutions of the same descent problem. □

Exercise 1. Let $M|K$ be a Galois extension and $L|K$ a finite subextension. Denote by $c \mapsto \bar{c}$ a section $G(M|L)\backslash G(M|K) \to G(M|K)$ of $G(M|K) \to G(M|L)\backslash G(M|K)$, i.e. a system of right representatives. Associate to every inhomogeneous 1-cocycle $x : G(M|L) \to PGL_n(L)$ the function $cor\, x : G(M|K) \to PGL_n(M)$ given by

$$(cor\, x)(\sigma) = \prod_c \bar{c}^{-1} x(\bar{c}\sigma\overline{c\sigma}^{-1}).$$

Show that $cor\, x$ is a 1-cocycle and that we obtain a canonical map

$$cor_K^L : H^1(G(M|L), PGL_n(M)) \longrightarrow H^1(G(M|K), PGL_n(M)),\ [x] \longmapsto [cor\, x],$$

which does not depend on the choice of the section $c \mapsto \bar{c}$. PGL_n may be replaced by any other algebraic group over K.

Exercise 2. Let $M \supseteq L \supseteq K$ be as above and let $1 \to A \to B \to C \to 1$ be an exact sequence of algebraic groups over K with A in the center of B. Show that we have a commutative diagram

$$\begin{array}{ccc} H^1(G(M|L), C(M)) & \xrightarrow{\delta} & H^2(G(M|L), A(M)) \\ \downarrow{\scriptstyle cor} & & \downarrow{\scriptstyle cor} \\ H^1(G(M|K), C(M)) & \xrightarrow{\delta} & H^2(G(M|K), A(M)). \end{array}$$

§3. The Brauer Group

Again let $L|K$ be a Galois extension with group $G = G(L|K)$. We now study the group $H^2(G, L^\times)$. It is linked with a classical theory which began with the discovery of the quaternion algebra

$$H = \mathbb{R} + \mathbb{R}i + \mathbb{R}j + \mathbb{R}k$$

by the Irish mathematician *W. R. Hamilton* (1805–1865). The multiplication in H is given by $i^2 = j^2 = k^2 = -1$, $ij = -ji = k$ (implying $jk = -kj = i$ and $ki = -ik = j$). H is also given as the matrix algebra $H = \left\{\left(\begin{smallmatrix} z & u \\ -\bar{u} & \bar{z}\end{smallmatrix}\right) \mid z, u \in \mathbb{C}\right\}$. H has center $\mathbb{R}$ and is a skew field, i.e. has no nontrivial two-sided ideal, and it is in essence the only $\mathbb{R}$-algebra with these properties. This example raised the question of **central simple algebras** over arbitrary fields K, i.e. finite dimensional K-algebras with center K and without nontrivial two-sided ideals. For these algebras we have the following equivalent characterizations.

(6.3.1) Proposition. *For a finite dimensional K-algebra A the following conditions are equivalent:*

(i) *A is a central simple algebra.*

(ii) *If $\overline{K}|K$ is the separable closure of K, then the $\overline{K}$-algebra $A_{\overline{K}} = A \otimes_K \overline{K}$ is isomorphic to a full matrix algebra $M_n(\overline{K})$.*

(iii) *There exists a finite Galois extension $L|K$ such that $A_L = A \otimes_K L$ is a full matrix algebra $M_n(L)$.*

(iv) *$A \cong M_m(D)$, where D is a skew field over K of finite degree.*

For the proof of this proposition and for the basic properties of central simple algebras, we refer to [102], [42] and [14], chap.VIII, §5, §10. We say that the central simple K-algebra A **splits** over the extension $L|K$, or that L is a **splitting field** for A, if $A \otimes_K L \cong M_n(L)$ for some n. A is called a **cyclic** algebra if it has a cyclic splitting field $L|K$, of degree $\sqrt{\dim_K A}$.

Two central simple K-algebras A and B are called **similar** if

$$A \otimes_K M_r(K) \cong B \otimes_K M_s(K)$$

for some r, s and we write $A \sim B$. This is the same to say that the skew fields associated with A and B are isomorphic. The tensor product $A \otimes_K B$ of two central simple K-algebras is again central simple. This leads to the

(6.3.2) Definition. *The* **Brauer group** $Br(K)$ *of a field K is the set of all similarity classes $[A]$ of central simple K-algebras A, endowed with the multiplication*

$$[A][B] = [A \otimes_K B].$$

The product is well-defined since

$$[A \otimes_K M_r(K) \otimes_K B \otimes_K M_s(K)] = [A \otimes_K B \otimes_K M_{rs}(K)] = [A \otimes_K B].$$

The identity of $Br(K)$ is $1 = [K]$ and the inverse of $[A]$ is $[A]^{-1} = [A^{op}]$, where A^{op} is A as a K-vector space and the multiplication of A^{op} is given by $A^{op} \times A^{op} \to A^{op}$, $(a, b) \mapsto ba$.

If $L|K$ is an extension of fields, then we have the homomorphism

$$res_L^K : Br(K) \longrightarrow Br(L), \ A \longmapsto [A \otimes_K L],$$

and we denote the kernel by $Br(L|K)$. This is the group of central simple K-algebras which split over L. If $L|K$ runs through the finite Galois subextensions of $\bar{K}|K$, then by (6.3.1)(iii)

$$Br(K) = \bigcup_L Br(L|K).$$

In the finite Galois case the classes of $Br(L|K)$ are explicitly represented by the following K-algebras. Let $G = G(L|K)$ and $n = [L : K]$. Let $x : G \times G \to L^\times$ be a normalized (i.e. $x(\sigma, 1) = x(1, \sigma) = 1$) inhomogeneous 2-cocycle. On the n^2-dimensional K-vector space

$$A = \bigoplus_{\sigma \in G} Le_\sigma \qquad (e_\sigma \text{ formal symbols})$$

we define a multiplication by

$$(\sum_\sigma x_\sigma e_\sigma)(\sum_\tau y_\tau e_\tau) = \sum_{\sigma,\tau} x_\sigma \sigma y_\tau x(\sigma, \tau) e_{\sigma\tau}.$$

This multiplication is associative because of the cocycle relation

$$x(\sigma, \tau) x(\sigma\tau, \rho) = \sigma x(\tau, \rho) x(\sigma, \tau\rho),$$

and makes A a K-algebra with the identity $1 = e_1$. This K-algebra is called the **crossed product** of L and G by x and is denoted by

$$A = C(L, G, x).$$

The crossed products have the following properties

(6.3.3) Proposition.

(i) *$C(L, G, x)$ is a central simple K-algebra which splits over L.*

(ii) *The normalized cocycles x and y are cohomologous if and only if $C(L, G, x) \cong C(L, G, y)$.*

(iii) *$C(L, G, xy) \sim C(L, G, x) \otimes_K C(L, G, y)$.*

(iv) *Every simple central K-algebra which splits over L is similar to a crossed product.*

For the proof of this proposition we refer to [102]. Associating now to a cohomology class $[x] \in H^2(G, L^\times)$ the class $[C(L, G, x)]$ (x normalized), we obtain a map

$$H^2(G, L^\times) \longrightarrow Br(L|K),$$

which is well-defined and injective by (ii), multiplicative by (iii) and surjective by (iv), hence an isomorphism of groups. If $M \supseteq L \supseteq K$ are two finite Galois extensions, then it is evident that the diagram

$$\begin{array}{ccc} H^2(G(M|K), M^\times) & \longrightarrow & Br(M|K) \\ {\scriptstyle inf}\uparrow & & \uparrow{\scriptstyle incl} \\ H^2(G(L|K), L^\times) & \longrightarrow & Br(L|K) \end{array}$$

is commutative. Taking direct limits, we obtain the

(6.3.4) Theorem. *For every Galois extension $L|K$ we have a canonical isomorphism*

$$H^2(G(L|K), L^\times) \cong Br(L|K).$$

In particular,

$$H^2(K, \bar{K}^\times) \cong Br(K),$$

showing that $Br(K)$ is a torsion group.

In general, Brauer groups are difficult to calculate. We will determine them for local and global fields in later chapters. For finite fields, we have the following result.

(6.3.5) Proposition. *Let K be a finite field. Then $Br(K) = 0$.*

Proof: Let $L|K$ be a finite extension. Then $G(L|K)$ is a cyclic group, and $h(G(L|K), L^\times) = 1$ by (1.7.6). Therefore

$$\#Br(L|K) = \#H^2(G(L|K), L^\times) = \#H^1(G(L|K), L^\times) = 1$$

by Hilbert's Satz 90. Passing to the limit over all L, we obtain $Br(K) = 0$.

□

For a natural number n prime to the characteristic of K, we consider the exact sequence

$$1 \longrightarrow \mu_n \longrightarrow \bar{K}^\times \xrightarrow{\ n\ } \bar{K}^\times \longrightarrow 1$$

of G_K-modules. The associated exact cohomology sequence, Hilbert's Satz 90 and the above theorem yield an isomorphism

$$H^2(K, \mu_n) \cong {}_nBr(K).$$

Taking direct limits over n and denoting by $Br(K)^{(p')}$ the subgroup of elements of order prime to p, we obtain the

(6.3.6) Corollary. *We have*

$$H^2(K,\mu) \cong Br(K) \quad \textit{or} \quad H^2(K,\mu) \cong Br(K)^{(p')}$$

according to whether char$(K) = 0$ *or* char$(K) = p > 0$.

The Brauer group $Br(K)$ is functorial in K in a twofold sense. If $\rho : K \to K'$ is a homomorphism of fields, then K' may also be viewed as a K-algebra and we obtain for every K-algebra A, a K-algebra

$$A_\rho = A \otimes_{K,\rho} K'$$

together with a homomorphism $\rho : A \to A_\rho$, $a \mapsto a \otimes 1'$. A_ρ is also a K'-algebra and we obtain a homomorphism

$$\rho_* : Br(K) \longrightarrow Br(K'), \quad [A] \longmapsto [A_\rho].$$

If ρ is an inclusion $K \subseteq L$, then ρ_* is the restriction map

$$res_L^K : Br(K) \longrightarrow Br(L).$$

If $L|K$ is a finite separable extension, we have on the other hand also a corestriction homomorphism

$$cor_K^L : Br(L) \longrightarrow Br(K),$$

which is obtained as follows. Let $\rho : L \to \bar{K}$ run through the K-embeddings of L into the separable closure $\bar{K}$ of K. If A is any L-algebra, then we form the tensor product of the L-algebras A_ρ,

$$T(A) = \bigotimes_\rho A_\rho.$$

For every $\sigma \in G_K$ we have the K-isomorphism $\sigma : A_{\sigma^{-1}\rho} \to A_\rho$ and we get a G_K-action on $T(A)$, viewed as a K-algebra, given by

$$\sigma(\otimes_\rho a_\rho) = \otimes_\rho b_\rho \quad \text{with } b_\rho = \sigma a_{\sigma^{-1}\rho}.$$

We now take the fixed ring and obtain a K-algebra

$$cor_K^L(A) = T(A)^{G_K}.$$

If A is a central simple L-algebra, then $cor_K^L(A)$ is a central simple K-algebra (see [102]). It is obvious that $cor_K^L(A \otimes_L B) \cong cor_K^L(A) \otimes_K cor_K^L(B)$ and $cor_K^L(M_n(L)) = M_{n^d}(K)$ with $d = [L : K]$. Therefore we obtain a canonical homomorphism

$$cor_K^L : Br(L) \longrightarrow Br(K), \quad [A] \longmapsto [cor_K^L(A)],$$

the **corestriction** for Brauer groups.

(6.3.7) Proposition. *Let $L|K$ be a finite subextension of the separable closure $\bar{K}|K$ and let $\sigma \in G_K = G(\bar{K}|K)$. We then have the commutative diagrams*

$$\begin{array}{ccc} H^2(L,\bar{K}^\times) & \xrightarrow{\sim} & Br(L) \\ {\scriptstyle res}\uparrow\downarrow{\scriptstyle cor} & & {\scriptstyle res}\uparrow\downarrow{\scriptstyle cor} \\ H^2(K,\bar{K}^\times) & \xrightarrow{\sim} & Br(K), \end{array} \qquad \begin{array}{ccc} H^2(L,\bar{K}^\times) & \xrightarrow{\sim} & Br(L) \\ {\scriptstyle \sigma_*}\downarrow & & {\scriptstyle \sigma_*}\downarrow \\ H^2(\sigma L,\bar{K}^\times) & \xrightarrow{\sim} & Br(\sigma L). \end{array}$$

One has to show that, for a finite Galois extension $M|K$ containing L, the diagrams

$$\begin{array}{ccc} H^2(G(M|L),M^\times) & \longrightarrow & Br(M|L) \\ {\scriptstyle res}\uparrow\downarrow{\scriptstyle cor} & & {\scriptstyle res}\uparrow\downarrow{\scriptstyle cor} \\ H^2(G(M|K),M^\times) & \longrightarrow & Br(M|K), \end{array} \qquad \begin{array}{ccc} H^2(G(M|L),M^\times) & \longrightarrow & Br(M|L) \\ {\scriptstyle \sigma_*}\downarrow & & {\scriptstyle \sigma_*}\downarrow \\ H^2(G(M|\sigma L),M^\times) & \longrightarrow & Br(M|\sigma L) \end{array}$$

are commutative, where the horizontal maps are given by forming crossed products. Let x be a normalized 2-cocycle of $G(M|K)$, resp. of $G(M|L)$, with values in $M^\times$. Then by a straightforward argument

$$C(M,G(M|K),x)\otimes_K L = C(M,G(M|L),\mathit{res}\,x),$$

resp.

$$C(M,G(M|L),x)\otimes_{K,\sigma}\sigma L = C(M,G(M|\sigma L),\sigma_* x),$$

giving the commutativity for the maps *res* and σ_*. For *cor* the proof is more involved and needs a more comprehensive study of non-abelian cohomology. We refer to [181], th. 11 or [91], th. 3.13.20.

In I §5 we have seen that the cohomology functor

$$H^2(\bar{K}^\times): L \longmapsto H^2(L,\bar{K}^\times)$$

may be interpreted as a G-*modulation* for the absolute Galois group $G = G_K$ with respect to the maps *res* , *cor* , σ_* in the sense of (1.5.12). The isomorphism (6.3.6) and the above proposition (6.3.7) show that also the Brauer group may be interpreted as a G-modulation and that we have the

(6.3.8) Theorem. *The functor*

$$Br: L \longmapsto Br(L)$$

is a G-modulation and the family of isomorphisms $H^2(L,\bar{K}^\times)\cong Br(L)$ is an isomorphism

$$H^2(\bar{K}^\times) \xrightarrow{\sim} Br$$

of G-modulations.

The main assertion of this theorem is the *double coset formula* (1.5.11) for the classes [A] of central simple algebras, i.e. the similarity

$$res_M^K \circ cor_K^L(A) \sim \prod_{\sigma\in R} cor_M^{M\sigma L} \circ \sigma_* \circ res_{L\sigma^{-1}M}^L(A),$$

for two finite separable extensions $M, L \supseteq K$ and a central simple L-algebra A. Here R is a system of representatives of $G_M\backslash G_K/G_L$. When $L = M$ one may replace $\prod$ by $\bigotimes$ and the similarity $\sim$ by an isomorphism

$$cor_K^L(A) \otimes_K L \cong \bigotimes_\sigma \sigma_*(A) = \bigotimes_\sigma (A \otimes_{K,\sigma} \sigma L).$$

For a K-algebra A, we have

$$cor_K^L(A \otimes_K L) \cong A^{\otimes d}, \quad d = [L : K].$$

We finish this section by giving the following geometric interpretation for the Brauer group. In the preceding section we defined the *Brauer-Severi varieties* over K as the K-varieties X which become isomorphic to $\mathbb{P}^{n-1}$ for some n over some Galois extension L, i.e.

$$X \otimes_K L \cong \mathbb{P}_L^{n-1}$$

as L-varieties. We denote by $BS(K)$ the set of isomorphism classes of all Brauer-Severi varieties over K, by $BS(L|K)$ the subset of classes which split over L and by $BS_n(L|K)$ the subset of $BS(L|K)$, consisting of the isomorphism classes of K-varieties which become isomorphic to $\mathbb{P}^{n-1}$ over L. Then

$$BS(L|K) = \bigcup_{n\in\mathbb{N}} BS_n(L|K).$$

(6.3.9) Theorem. *For every Galois extension $L|K$ and every $n \geq 1$ there exists a natural injective map*

$$BS_n(L|K) \hookrightarrow Br(L|K)$$

whose image is the set of classes in $Br(L|K)$ which are represented by a central simple algebra of dimension n^2 over K. If $[L : K]$ is finite and divides n, then the above map is bijective.

Proof: The exact sequence of $G(L|K)$-groups

$$1 \longrightarrow L^\times \longrightarrow GL_n(L) \longrightarrow PGL_n(L) \longrightarrow 1$$

induces a map (see I §3 ex.8)

$$\delta : BS_n(L|K) \cong H^1(G(L|K), PGL_n(L)) \longrightarrow H^2(G(L|K), L^\times).$$

A K-algebra A is central simple of dimension n^2 and splits over L if and only if $A \otimes_K L \cong M_n(L)$. Since $\mathrm{Aut}_L(M_n(L)) = PGL_n(L)$, the same arguments as in the case of Brauer-Severi varieties show that the set $CSA_n(L|K)$ of K-isomorphism classes of such algebras is naturally isomorphic to $H^1(G(L|K), PGL_n(L))$. Thus we obtain a diagram

$$\begin{array}{ccc} H^1(G(L|K), PGL_n(L)) & \xrightarrow{\delta} & H^2(G(L|K), L^\times) \\ \downarrow \wr & & \downarrow \wr \\ CSA_n(L|K) & \xrightarrow{\quad can \quad} & Br(L|K) \end{array}$$

where the map can sends a central simple algebra to its similarity class in $Br(L|K)$. Two similar central simple algebras of the same dimension are isomorphic and therefore can is injective. The statement of the theorem now follows from the fact (see [102] V §30) that the above diagram commutes. Finally, if $n = m \cdot [L : K]$, then the class in $H^2(G(L|K), L^\times)$ of a 2-cocycle α is represented by the algebra $C(L, G(L|K), \alpha) \otimes_K M_m(K)$. □

Exercise 1. If $M \supseteq L \supseteq K$ are two finite separable extensions, then the corestriction of Brauer groups obeys the rule $cor_K^L \circ cor_L^M = cor_K^M$.

Exercise 2. Let $L|K$ be a cyclic extension of degree n, σ a generator of the Galois group G and $a \in K^\times$. Show that the ring

$$(a, L|K, \sigma) := \bigoplus_{i=0}^{n-1} L\, e^i$$

with multiplication $e^n = a$, $e\lambda = (\sigma\lambda)e$ $(\lambda \in L)$, is a cyclic central simple algebra.

Exercise 3. Under the assumptions of ex.2 show that $(a, L|K, \sigma) \cong (b, L|K, \sigma)$ provided $a/b \in N_{L|K}(L^\times)$, and

$$(a, L|K, \sigma) \otimes_K (b, L|K, \sigma) \cong M_n((ab, L|K, \sigma)).$$

Exercise 4. Keeping the assumptions of ex.2, show that the assignment $a \mapsto (a, L|K, \sigma)$ induces an isomorphism

$$\theta_\sigma : K^\times / N_{L|K} L^\times \xrightarrow{\sim} Br(L|K).$$

Exercise 5. Deduce from ex.4 that the quaternion algebra H is the only central skew field over $\mathbb{R}$ different from $\mathbb{R}$ (Theorem of Frobenius).

Exercise 6. (Theorem of Albert-Hochschild). If $K|k$ is a totally inseparable extension, then the restriction map $Br(k) \to Br(K)$ is surjective.

Exercise 7. If K is a field of characteristic $p > 0$, then the Brauer group $Br(K)$ is p-divisible.

§4. The Milnor K-Groups

The **Hilbert symbol** is the main ingredient in the formulation of the general reciprocity law of n-th power residues (see [160], chap.V, §3). It is defined for a p-adic local field $K_\mathfrak{p}$, and is a map

$$\left(\frac{\,,\,}{\mathfrak{p}}\right) : K_\mathfrak{p}^\times \times K_\mathfrak{p}^\times \longrightarrow \mu_n$$

which is multiplicative in both arguments a and b and satisfies the relation $(\frac{a,1-a}{\mathfrak{p}}) = 1$ for all $a \neq 0, 1$.*) This example leads naturally to a general notion of a **symbol** for any field F as a multi-multiplicative map

$$\underbrace{F^\times \times \cdots \times F^\times}_{n \text{ times}} \longrightarrow A, \quad (a_1, \ldots, a_n) \longmapsto [a_1, \ldots, a_n],$$

into a (multiplicatively written) abelian group A such that

$$[a_1, \ldots, a_n] = 1 \text{ whenever } a_i + a_j = 1 \text{ for some } i \neq j.$$

Every such symbol factors through a group $K_n^M(F)$, which is the universal target of symbols and is defined as follows.

(6.4.1) Definition. *The n-th* **Milnor K-group** *of a field F is the quotient*

$$K_n^M(F) = (F^\times \otimes_\mathbb{Z} \cdots \otimes_\mathbb{Z} F^\times)/I_n,$$

where I_n is the subgroup of $F^\times \otimes_\mathbb{Z} \cdots \otimes_\mathbb{Z} F^\times$ generated by the elements $a_1 \otimes \cdots \otimes a_n$ such that $a_i + a_j = 1$ for some $i \neq j$.

We have the canonical symbol

$$F^\times \times \cdots \times F^\times \longrightarrow K_n^M(F), \quad (a_1, \ldots, a_n) \longmapsto \{a_1, \ldots, a_n\},$$

where $\{a_1, \ldots, a_n\} = a_1 \otimes \cdots \otimes a_n \bmod I_n$. Every other symbol of n arguments is obtained from this by composition with a homomorphism of the group $K_n^M(F)$.

It is convenient to put $K_0^M(F) = \mathbb{Z}$. The multiplication in $K_n^M(F)$ will be written additively, i.e.

$$\{\ldots, a_i b_i, \ldots\} = \{\ldots, a_i, \ldots\} + \{\ldots, b_i, \ldots\},$$

although for $K_1^M(F) = F^\times$ the multiplicative notation will also be used. For $n, m \in \mathbb{N}$ the images of $I_n \otimes \underbrace{F^\times \otimes \cdots \otimes F^\times}_{m \text{ times}}$ and $\underbrace{F^\times \otimes \cdots \otimes F^\times}_{n \text{ times}} \otimes I_m$ in

---

*) The reciprocity law is the product formula $\prod_\mathfrak{p} (\frac{a,b}{\mathfrak{p}}) = 1$ for two numbers $a, b \in K^\times$ in a number field K with the completions $K_\mathfrak{p}$ (infinite primes included).

$\underbrace{F^\times \otimes \cdots \otimes F^\times}_{n+m \text{ times}}$ belong to I_{n+m}. Therefore we have a homomorphism

$$K_n^M(F) \times K_m^M(F) \longrightarrow K_{n+m}^M(F),$$
$$(\{a_1, \ldots, a_n\}, \{b_1, \ldots, b_m\}) \longmapsto \{a_1, \ldots, a_n, \; b_1, \ldots, b_m\},$$

and we obtain a graded ring

$$K^M(F) = \bigoplus_{n=0}^{\infty} K_n^M(F).$$

One knows that $K_n^M(\mathbb{F}) = 0$ for a finite field $\mathbb{F}$ and $n \geq 2$, and that

$$K_2^M(\mathbb{Q}) = \mu_2 \oplus \mathbb{F}_3^\times \oplus \mathbb{F}_5^\times \oplus \mathbb{F}_7^\times \oplus \mathbb{F}_{11}^\times \oplus \ldots$$

and $K_n^M(\mathbb{Q}) \cong \mu_2$ for $n \geq 3$. If F is algebraically closed, then it is not difficult to show that the groups $K_n^M(F)$ are uniquely divisible for $n \geq 2$. In general it is a difficult and usually most cumbersome problem to compute the K-groups or investigate their properties. On the other hand K-theory has nowadays attained a most prominent position in algebra, number theory and arithmetic geometry. But it is beyond the scope of this book to go into the K-groups more closely (we refer to [142] and [232] for further literature). We mention them here because of their close relation to Galois cohomology. This connection arises as follows.

Let N be a natural number prime to the characteristic of F. From the exact sequence

$$1 \longrightarrow \mu_N \longrightarrow \bar{F}^\times \xrightarrow{N} \bar{F}^\times \longrightarrow 1,$$

we obtain a surjective homomorphism

$$\delta_F : F^\times \longrightarrow H^1(F, \mu_N)$$

with kernel $F^{\times N}$. On the other hand, we have for every $n \geq 1$ the cup-product

$$H^1(F, \mu_N) \times \cdots \times H^1(F, \mu_N) \xrightarrow{\cup} H^n(F, \mu_N^{\otimes n}), \text{ *)}$$

hence a map

$$(*) \qquad \underbrace{F^\times \times \cdots \times F^\times}_{n \text{ times}} \longrightarrow H^n(F, \mu_N^{\otimes n}).$$

Let us denote the image of $(a_1, \ldots, a_n) \in F^\times \times \cdots \times F^\times$ by

$$(a_1, \ldots, a_n)_F := \delta_F a_1 \cup \cdots \cup \delta_F a_n.$$

*) Note that $\mu_N^{\otimes n}$ is isomorphic to μ_N as an abelian group, but as a G_F-module it is different: if $\chi : G_F \to (\mathbb{Z}/N\mathbb{Z})^\times$ is the character that gives the action of G_F on μ_N, $\sigma : \zeta \mapsto \zeta^{\chi(\sigma)}$, than χ^n is the character for $\mu_N^{\otimes n}$.

(6.4.2) Theorem (TATE). *The map* $(*)$ *induces a homomorphism*

$$h_F : K_n^M(F) \longrightarrow H^n(F, \mu_N^{\otimes n}),$$

called the **Galois symbol**.

Proof: The multiplicativity in each argument is clear from the definition and we only have to show that $(a_1, \ldots, a_n)_F = 1$ if $a_i + a_j = 1$ for $i \neq j$. It suffices to consider the case $n = 2$: if $n > 2$ and, say $i = 1$, $j = 2$, then $(a_1, \ldots, a_n)_F = (a_1, a_2)_F \cup (a_3, \ldots, a_n)_F$.

So let $n = 2$ and $a \in F^\times$, $a \neq 1$. Let $X^N - a = \prod_i f_i(X)$ with $f_i(X)$ monic and irreducible in $F[X]$. For each i let a_i be a root of $f_i(X)$ and let $F_i = F(a_i)$. Then

$$1 - a = \prod_i f_i(1) = \prod_i N_{F_i|F}(1 - a_i).$$

Hence

$$(1 - a, a)_F = (\prod_i N_{F_i|F}(1 - a_i), a)_F = \prod_i (N_{F_i|F}(1 - a_i), a)_F.$$

Because of the formula $cor(\alpha \cup res\, \beta) = (cor\, \alpha) \cup \beta$ (see (1.5.3) (iv)), and because cor is the norm on H^0 and commutes with δ, we have

$$\begin{array}{rcccl} (N_{F_i|F}(1 - a_i), a)_F & = & cor_F^{F_i}(1 - a_i, a)_{F_i} & = & cor_F^{F_i}(1 - a_i, a_i^N)_{F_i} \\ & = & cor_F^{F_i}(1 - a_i, a_i)_{F_i}^N & = & 1, \end{array}$$

hence $(1 - a, a)_F = 1$. □

For the Galois symbol, we have the fundamental

(6.4.3) Theorem (VOEVODSKY-ROST). *For every field* F *and every* $N \in \mathbb{N}$ *prime to the characteristic of* F*, the Galois symbol yields an isomorphism*

$$h_F : K_n^M(F)/NK_n^M(F) \xrightarrow{\sim} H^n(F, \mu_N^{\otimes n}).$$

For $N = 2$ this was first conjectured by *J. MILNOR* in [142] (in the language of quadratic forms) and the general form was stated by *S. BLOCH* and *K. KATO* in [13]. For many years it was known as the *Bloch-Kato conjecture*.

Theorem (6.4.3) is trivial for $n = 0$ and arbitrary $N \in \mathbb{N}$, as well as for $N = 1$ and arbitrary n. For $n = 1$ it is just a reformulation of Kummer theory.

For a proof of (6.4.3), which lies far beyond the scope of this book, we refer to *V. Voevodsky*'s paper [Vo]. The special case $n = 2$ had been settled already in 1982 by *A. S. Merkur'ev* and *A. A. Suslin*, see [138], [239] or [137]. The next steps towards the general conjecture was the case $n = 3$ and N a power of 2 which was proved by Merkur'ev and Suslin (see [MS]) and, independently, by *M. Rost* (see [Ro]) in 1986. In 2001, Voevodsky proved the case with arbitrary n and N a power of 2, see [242], [243]. Finally, using results of Rost, Voevodsky proved the general statement in 2010. There also exists a version of (6.4.3) for N a power of the characteristic of F, which had been proven already by Bloch and Kato themselves, see [13].

For local and global fields and $n = 2$ the Bloch-Kato Conjecture was proven earlier by *J. Tate* (see [231] or [232]), who also constructed an ℓ-adic variant of the Galois symbol. Let $\mathbb{Z}_\ell(n) = \varprojlim_m \mu_{\ell^m}^{\otimes n}$ as a compact G_F-module. Then there exists an ℓ-adic Galois symbol

$$h_F : K_n^M(F) \longrightarrow H^n_{cts}(F, \mathbb{Z}_\ell(n)).$$

Tate showed that the following theorem follows from (6.4.3) (for $n = 2$).

(6.4.4) Theorem. *For every prime number ℓ, the ℓ-primary part of $K_2^M(F)$ is the direct sum of its maximal divisible subgroup, which is killed by h_F, and a subgroup which is mapped isomorphically by h_F onto the torsion subgroup of $H^2_{cts}(F, \mathbb{Z}_\ell(2))$.*

If F is a global field, then the group $K_2^M(F)$ is a torsion group with no non-zero divisible subgroup; so, in this case, its ℓ-primary part is mapped isomorphically onto the torsion subgroup of $H^2_{cts}(F, \mathbb{Z}_\ell(2))$.

Exercise 1. Let K be a field with a normalized discrete valuation v. Let $\mathcal{O}$ be its valuation ring, $\mathfrak{p}$ the maximal ideal, π a prime element and $\kappa(v) = \mathcal{O}/\mathfrak{p}$ the residue field. Define a map

$$\partial_v : \underbrace{K^\times \times \cdots \times K^\times}_{n \text{ times}} \longrightarrow K_{n-1}^M(\kappa(v))$$

as follows. Let $\alpha_1, \dots, \alpha_n \in K^\times$, $\alpha_i = \pi^{a_i}\varepsilon_i$, $a_i = v(\alpha_i)$, $\varepsilon_i \in \mathcal{O}^\times$, and put $\bar{\varepsilon}_i = \varepsilon_i \mod \mathfrak{p} \in \kappa(v)$.

For $n = 1$, put $\partial(\alpha_1) = a_1$. For $n > 1$ and $k_1, \dots, k_m$ with $1 \le k_1 < \dots < k_m \le n$, $m \le n$, put

$$\partial^{k_1,\dots,k_m}(\alpha_1, \dots, \alpha_n) = a_{k_1} \dots a_{k_m} xy,$$

where x is the symbol $\{\bar{\varepsilon}_1, \dots, \bar{\varepsilon}_n\} \in K_{n-m}^M(\kappa(v))$ with omitted elements at the $k_1, \dots, k_m$-th places if $m < n$, and equal to 1 if $m = n$; y is equal to $\{-1, \dots, -1\} \in K_{m-1}^M(\kappa(v))$

if $m > 1$, and equal to 1 otherwise. Now put

$$\partial_v(\alpha_1, \ldots, \alpha_n) = \sum_{\substack{k_1,\ldots,k_m \\ 1\le m\le n}} (-1)^{n-k_1-\cdots-k_m} \partial^{k_1,\ldots,k_m}(\alpha_1, \ldots, \alpha_n),$$

and show that this function induces a homomorphism

$$\partial_v : K_n^M(K) \longrightarrow K_{n-1}^M(\kappa(v)),$$

which is called the **tame symbol**.

Hint: [50], chap. IV, 2.

Exercise 2. Keeping the assumptions of ex.1, let $\{\alpha, \beta\} \in K_2^M(K)$. Then

$$\partial_v(\alpha, \beta) = (-1)^{v(\alpha)v(\beta)} \alpha^{v(\beta)} \beta^{-v(\alpha)} \mod \mathfrak{p}.$$

Exercise 3. Let $K = F(X)$ be a rational function field in one variable and let v run through the normalized discrete valuations of K which are trivial on F and which are different from the valuation v_∞ given by $v_\infty(f(X)/g(X)) = \deg(g(X)) - \deg(f(X))$ for $f(X), g(X) \in F[X]$. The sequence

$$0 \longrightarrow K_n^M(F) \longrightarrow K_n^M(K) \xrightarrow{\oplus\partial_v} \bigoplus_{v\neq v_\infty} K_{n-1}^M(\kappa(v)) \longrightarrow 0$$

is exact and splits (Theorem of Tate-Milnor, cf. [142]).

Hint: Study the proof in [142] or in [50] chap. IX, 2.4.

§5. Dimension of Fields

An important invariant of a field k is the cohomological dimension $cd\, G_k$ of its absolute Galois group. In order to study its properties we have to make use of a remarkable principle, which roughly says that a homogeneous polynomial equation $f(x_1, \ldots, x_n) = 0$ over a field k has automatically a non-zero solution in k if the number of variables is large compared to its degree. This principle gives rise to another notion of dimension, which is directly attached to a field and which we are now going to introduce.

By an ***n*-form** in k we mean a homogeneous polynomial $f \in k[x_1, \ldots, x_n]$. A *nontrivial zero* in k is an n-tuple $(\alpha_1, \ldots, \alpha_n) \in k^n$, different from $(0, \ldots, 0)$, such that $f(\alpha_1, \ldots, \alpha_n) = 0$.

We obtain special n-forms as follows. Let $K|k$ be a finite extension of degree n and let $x_1, \ldots, x_n$ be indeterminates. Then the extension $K(x_1, \ldots, x_n)$ of $k(x_1, \ldots, x_n)$ is also finite and we may consider the norm N of this extension;

$N(\xi)$ is the determinant of the transformation $a \mapsto \xi a$ of the $k(x)$-vector space $K(x)$. Choosing a basis $\omega_1, \ldots, \omega_n$ of $K|k$,

$$N(x_1, \ldots, x_n) := N\Big(\sum_{i=1}^{n} x_i \omega_i\Big)$$

defines an n-form in k of degree n which induces the norm map from K to k:

$$N_{K|k} : K \longrightarrow k\,, \quad (a_1, \ldots, a_n) \mapsto N(a_1, \ldots, a_n)\,.$$

These forms are called **norm forms**. They have the special property of having only the trivial zero in k (since $N_{K|k}(\alpha) \neq 0$ for $\alpha \neq 0$). More generally, we call an n-form f of degree d a **normic form of order i** if $n = d^i$ and if f has only the trivial zero in k.

(6.5.1) Lemma. *If k is not algebraically closed, then k admits normic forms of arbitrarily large degree.*

Proof: Since k is not algebraically closed, there exists a finite extension $K|k$ of degree $n > 1$ for some $n \in \mathbb{N}$, thus a normic n-form of order 1, as we saw above.

Let f, g be forms in n_1, n_2 variables of degrees d_1, d_2 respectively. We denote by $f(g \mid \ldots \mid g)$ the form which is obtained by inserting g for each variable and using new variables after each occurrence of $|$. We obtain in this way an n_1n_2-form of degree d_1d_2, which is normic of order i if both f and g are normic of order i. From this observation we obtain for each normic n-form f of degree d and order i and each $\nu \in \mathbb{N}$ a normic n^ν-form $f^{(\nu)}$ of degree d^ν and order i by setting

$$f^{(1)} = f \quad \text{and} \quad f^{(\nu+1)} = f^{(\nu)}(f \mid \ldots \mid f).$$

□

(6.5.2) Definition. *The* **diophantine dimension** *$dd(k)$ of a field k is the smallest integer $r \geq 0$ such that any n-form f of degree $d \geq 1$ has a nontrivial zero in k, whenever $n > d^r$. We set $dd(k) = \infty$ if no such number r exists.*

Clearly, the fields with $dd(k) = 0$ are the algebraically closed fields. Fields with $dd(k) \leq 1$ are called **quasi-algebraically closed** and fields with $dd(k) \leq i$ are called **C_i-fields**. Thus a field is a C_i-field if every n-form of degree d with $n > d^i$ has a nontrivial zero.

(6.5.3) Theorem (Artin, Lang, Nagata). *Let $dd(k) = r$ and let $f_1, \dots, f_s$ be n-forms of degree d. If $n > sd^r$, then these forms have a common nontrivial zero in k.*

Proof:*) If k is algebraically closed, i.e. $r = 0$, then the equations $f_1 = 0, \dots, f_s = 0$ define a projective variety in the projective space $\mathbb{P}^{n-1}(k)$ of dimension greater than or equal to $(n-1) - s \geq 0$. The result follows from this.

Assume then that k is not algebraically closed. Let Φ be a normic form of order 1 and of degree $e \geq s$, which exists by (6.5.1) . Consider the sequence of forms

$$\Phi^{(1)} = \Phi^{(1)}(f) = \Phi(f_1, \dots, f_s \mid f_1, \dots, f_s \mid \dots \mid f_1, \dots, f_s \mid 0, \dots, 0)$$
$$\Phi^{(2)} = \Phi^{(2)}(f) = \Phi^{(1)}(f_1, \dots, f_s \mid f_1, \dots, f_s \mid \dots \mid f_1, \dots, f_s \mid 0, \dots, 0)$$

etc., where after each vertical line we use new variables, and we insert as many complete sets of f's as possible. We put $\Phi^{(0)} = \Phi$.

Thus $\Phi^{(1)}$ has $n[\frac{e}{s}]$ variables and has degree $de \leq ds([\frac{e}{s}] + 1)$. (If x is a real number, the symbol $[x]$ will always mean the largest integer less than or equal to x.) If $r = 1$, we want

$$n[\tfrac{e}{s}] > ds([\tfrac{e}{s}] + 1) \quad \text{or} \quad (n - ds)[\tfrac{e}{s}] > ds.$$

Since $n - ds > 0$, this can be arranged by taking e large. Then $\Phi^{(1)}$ has a nontrivial zero, which, since Φ is normic, is a common zero of all the f_i.

If $r > 1$, we have to use the higher $\Phi^{(m)}$. Now $\Phi^{(m)}$ has degree $D_m = d^m e$, and if N_m is the number of variables in $\Phi^{(m)}$, then $N_0 = e$ and

$$N_{m+1} = n[\tfrac{N_m}{s}]\,.$$

We want to choose m so large that $N_m > (D_m)^r$. Now $[\frac{N_m}{s}] = \frac{N_m}{s} - \frac{t_m}{s}$, where $0 \leq t_m < s$. Hence

$$\begin{aligned}\frac{N_{m+1}}{(D_{m+1})^r} &= \frac{n[\frac{N_m}{s}]}{d^r(D_m)^r} = \frac{n}{sd^r}\frac{N_m}{(D_m)^r} - \frac{n}{sd^r}\frac{t_m}{e^r(d^r)^m} \\ &\geq \frac{n}{sd^r}\frac{N_m}{(D_m)^r} - \frac{n}{sd^r}\frac{s}{e^r(d^r)^m}.\end{aligned}$$

Using the same inequality for $m, m-1, \dots, 2, 1$, we get

*)The proof is taken from [60].

$$\begin{aligned}
\frac{N_{m+1}}{(D_{m+1})^r} &\geq \left(\frac{n}{sd^r}\right)^2 \left(\frac{N_{m-1}}{(D_{m-1})^r} - \frac{s}{e^r(d^r)^{m-1}}\right) - \left(\frac{n}{sd^r}\right)\left(\frac{s}{e^r(d^r)^m}\right) \\
&\vdots \qquad\qquad \vdots \qquad\qquad \vdots \\
&\geq \left(\frac{n}{sd^r}\right)^{m+1} \frac{N_0}{(D_0)^r} - \frac{s}{e^r}\frac{n}{s}\frac{1}{(d^r)^{m+1}}\left(\left(\frac{n}{s}\right)^m + \left(\frac{n}{s}\right)^{m-1} + \cdots + 1\right) \\
&= \left(\frac{n}{sd^r}\right)^{m+1} \frac{e}{e^r} - \frac{s}{e^r}\frac{n}{s}\frac{1}{(d^r)^{m+1}} \frac{\left(\frac{n}{s}\right)^m - 1}{\left(\frac{n}{s}\right) - 1}. \\
&\geq \left(\frac{n}{sd^r}\right)^{m+1} \frac{1}{e^r}\left(e - \frac{s^2}{n-s}\right).
\end{aligned}$$

Since e can be chosen large enough that $(n-s)e - s^2 > 0$, and $\left(\frac{n}{sd^r}\right) > 1$, we see that $\frac{N_m}{(D_m)^r} \to \infty$ as $m \to \infty$ and we are done. □

Note: Lang [121] generalized this theorem to the case where the f_i have different degrees $d_1, \dots, d_s$ and $n > d_1^i + \cdots + d_s^i$, but only under the extra hypothesis that k has a normic form of order i of every degree. It would be interesting to remove this hypothesis if possible.

We are now able to prove the following theorem, which in this generality is due to *S. Lang* and in the essential case to *C. Tsen*.

(6.5.4) Theorem. *If $K|k$ is an extension of finite transcendence degree n, then*

$$dd(K) \leq dd(k) + n.$$

Proof: Suppose first that $K|k$ is algebraic, i.e. $n = 0$. Let $dd(k) = r < \infty$ and let $f(x_1, \dots, x_m)$ be an m-form in K of degree d with $m > d^r$. We have to show that f has a nontrivial zero in K. Since the coefficients of f lie in a finite extension of k, we may assume that $K|k$ is finite. Let $\omega_1, \dots, \omega_s$ be a basis of $K|k$. Introduce new variables y_{ij} with

$$x_i = \omega_1 y_{i1} + \cdots + \omega_s y_{is},$$

$i = 1, \dots, m$. Then $f(x) = f_1(y)\omega_1 + \cdots + f_s(y)\omega_s$, where $f_1, \dots, f_s$ are sm-forms of degree d in k. Finding a nontrivial zero of f is equivalent to finding a common nontrivial zero of $f_1, \dots, f_s$ in k. But this can be done by the previous theorem, since $sm > sd^r$.

Now let $n > 0$. Then K is an algebraic extension of a purely transcendental extension $k(t_1, \dots, t_n)$. By the above proof and by induction, we are reduced to the case $K = k(t)$ with an indeterminate t. By homogeneity, it suffices to consider forms with coefficients in the polynomial ring $k[t]$.

Suppose $f(x_1, \dots, x_m)$ is an m-form of degree d with $m > d^{r+1}$ and with coefficients in $k[t]$. We have to show that it has a nontrivial zero in K. Introduce new variables y_{ij} with

$$x_i = y_{i0} + y_{i1}t + y_{i2}t^2 + \cdots + y_{is}t^s,$$

$i = 1, \dots, m$, where s will be specified later. If ℓ is the highest degree of the coefficients of f, we get

$$f(x) = f_0(y) + f_1(y)t + \cdots + f_{ds+\ell}(y)t^{ds+\ell},$$

where $f_0, \dots, f_{ds+\ell}$ are $m(s+1)$-forms of degree d in k. We can apply theorem (6.5.3) to these forms provided that

$$m(s+1) > (ds + \ell + 1)d^r$$

or

$$(m - d^{r+1})s > (\ell + 1)d^r - m.$$

This can be satisfied by taking s large. The common nontrivial zero of the f_μ's in k gives a nontrivial zero of f in $K = k(t)$. □

(6.5.5) Corollary (Tsen). *If K is a function field in one variable over an algebraically closed field k, then $dd(K) = 1$.*

In fact, $K|k$ is of transcendence degree 1, hence $dd(K) \le dd(k) + 1 = 1$, and thus $dd(K) = 1$, since K is not algebraically closed.

We mention the following two further results without giving the proofs. Both may be found in [60], 6.25 and 2.3, and the second one can also be found in [213], chap.1, §2, th.3.

(6.5.6) Theorem (Lang). *A field K which is complete with respect to a discrete valuation with algebraically closed residue field has $dd(K) = 1$.*

(6.5.7) Theorem (Chevalley - Warning). *A finite field $\mathbb{F}$ is of diophantine dimension one, i.e. $dd(\mathbb{F}) = 1$.*

For the cohomological applications the fields K with $dd(K) = 1$ play a particularly important role.

(6.5.8) Proposition. *Let K be a field with $dd(K) = 1$. Then the Brauer group $Br(K)$ is zero, and for every finite Galois extension $L|K$ the norm map*

$$N_{L|K} : L^\times \longrightarrow K^\times$$

is surjective.

Proof: Let $L|K$ be a Galois extension of degree n. Let $\omega_1, \ldots, \omega_n$ be a basis and $\alpha \in K^\times$. The norm form $N(x_1, \ldots, x_n) = N\left(\sum_{i=1}^n x_i\omega_i\right)$ is an n-form of degree n, and

$$f(x_1, \ldots, x_n, x) = N(x_1, \ldots, x_n) - \alpha x^n$$

is an $(n+1)$-form of degree n. It has therefore a nontrivial zero $(a_1, \ldots, a_n, a) \in K^n$. We have $a \neq 0$, since otherwise $(a_1, \ldots, a_n)$ would be a nontrivial zero of the norm form. Setting $b_i = a_i/a$ we obtain

$$N(b_1\omega_1 + \cdots + b_n\omega_n) = N(b_1, \ldots, b_n) = \alpha.$$

This shows that the norm is surjective.

We now have $\hat{H}^n(G(L|K), L^\times) = 1$ for $n = 0$, and for $n = 1$ this is true by Hilbert's Satz 90. Since every intermediate field K' of $L|K$ has $dd(K') \leq 1$, this holds also for the extension $L|K'$. From this it follows by (1.8.4) that $L^\times$ is a cohomologically trivial G-module, i.e. $H^2(G(L|K), L^\times) = 0$. Therefore

$$Br(K) \cong H^2(K, \bar{K}^\times) = \varinjlim_{L} H^2(G(L|K), L^\times) = 0. \qquad \square$$

We now turn to the *cohomological dimension* of fields.

(6.5.9) Definition. *The* **cohomological dimensions** *$cd_p(k)$, $cd(k)$, $scd_p(k)$ and $scd(k)$ of a field k are defined as the cohomological dimensions $cd_p\,G$, $cd\,G$, $scd_p\,G$, $scd\,G$ of its absolute Galois group $G = G_k$ (see (3.3.1)).*

The cohomological dimension $cd_p(k)$ for a field of $\mathrm{char}(k) = p > 0$ plays an exceptional role:

(6.5.10) Proposition. *If k is a field of characteristic $p > 0$, then $cd_p(k) \leq 1$.*

Proof: This is a special case of (6.1.3). $\square$

We now consider a prime number p different from the characteristic of the field in question.

(6.5.11) Proposition. *For a field k such that* $\mathrm{char}(k) \neq p$, *and for a natural number $n \in \mathbb{N}$, the following conditions are equivalent.*

(i) $cd_p(k) \leq n$,

(ii) $H^{n+1}(K, \overline{K}^\times)(p) = 0$ *and* $H^n(K, \overline{K}^\times)$ *is p-divisible for every algebraic extension $K|k$,*

(iii) $H^{n+1}(K, \overline{K}^\times)(p) = 0$ *and* $H^n(K, \overline{K}^\times)$ *is p-divisible for every finite separable extension $K|k$.*

Proof: Let $K|k$ be algebraic. The exact cohomology sequence associated to the exact Kummer sequence

$$1 \longrightarrow \mu_p \longrightarrow \overline{K}^\times \xrightarrow{p} \overline{K}^\times \longrightarrow 1 \quad {}^{*)}$$

says that the condition

$$H^{n+1}(K, \overline{K}^\times)(p) = 0 \text{ and } H^n(K, \overline{K}^\times) \text{ is } p\text{-divisible}$$

is equivalent to $H^{n+1}(K, \mu_p) = 0$. Assume $cd_p(k) \leq n$. Since G_K is isomorphic to a closed subgroup of G_k, we have by (3.3.5) $cd_p\, G_K \leq n$, hence $H^{n+1}(K, \mu_p) = 0$. This yields (i) $\Rightarrow$ (ii). The implication (ii) $\Rightarrow$ (iii) is trivial.

Assume that (iii) holds. Let K be the fixed field of a p-Sylow subgroup G_p. The degree $[K(\mu_p) : K]$ divides $p-1$ as well as p. This means that $\mu_p \subseteq K$, i.e. $\mu_p \cong \mathbb{Z}/p\mathbb{Z}$ as a G_K-module. Writing K as a union of finite separable extension of k which contain μ_p, we obtain $H^{n+1}(K, \mathbb{Z}/p\mathbb{Z}) = 0$, hence $cd_p k = cd_p K \leq n$ by (3.3.6) and (3.3.2)(iii). □

(6.5.12) Corollary. *If $dd(K) \leq 2$ and $cd(K) = 2$, then $dd(K) = 2$.*

Proof: If $dd(K) \leq 1$, then $dd(K') \leq 1$ for every finite extension by (6.5.4), hence $Br(K') = 0$ by (6.5.8) and thus $cd(K) \leq 1$ by (6.5.11). □

If $K|k$ is an algebraic field extension, then by (3.3.5)

$$cd_p(K) \leq cd_p(k),$$

since G_K is isomorphic to a closed subgroup of G_k. If $K|k$ is a finite extension, then the inequality $cd_p(K) \neq cd_p(k)$ is an exceptional case, as in the example

$$0 = cd_2(\mathbb{C}) < cd_2(\mathbb{R}) = \infty.$$

Namely, from (3.3.5) we obtain the

---

*) As before, $\overline{K}$ denotes always the separable closure of K.

(6.5.13) Proposition. *If $K|k$ is a finite extension such that $cd_p(k) < \infty$ or $p \nmid [K:k]$, then $cd_p(K) = cd_p(k)$ and also $scd_p(K) = scd_p(k)$.*

The main result on cohomological dimension of general fields is the following

(6.5.14) Theorem. *Let k be a field such that $cd_p(k) < \infty$ for a prime number $p \neq \mathrm{char}(k)$. If $K|k$ is a finitely generated extension of transcendence degree n, then*

$$cd_p(K) = cd_p(k) + n.$$

Proof:*) The field K is a finite extension of a purely transcendental extension $k(t_1, \ldots, t_n)$ of k. By (6.5.13) and induction on n, we are reduced to the case $K = k(t)$. Consider the diagram of fields

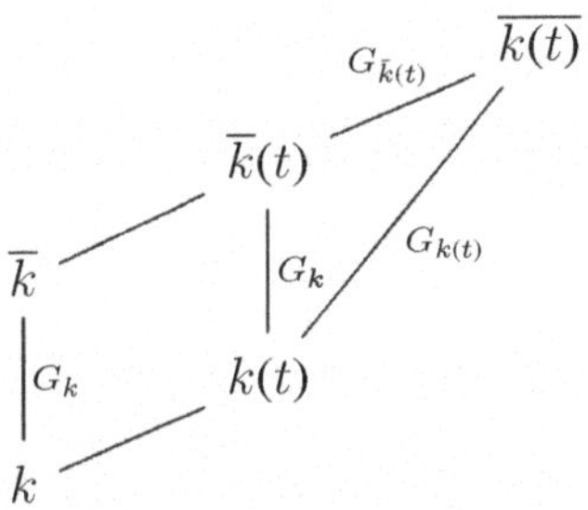

where $\bar{k}$ (resp. $\overline{k(t)}$) denotes the separable closure of k (resp. $k(t)$). We have $G(\bar{k}(t)|k(t)) = G(\bar{k}|k) = G_k$ and we obtain a group extension

$$(*) \qquad 1 \longrightarrow G_{\bar{k}(t)} \longrightarrow G_{k(t)} \longrightarrow G_k \longrightarrow 1.$$

From (3.3.8) it follows that

$$cd_p(k(t)) \le cd_p(k) + cd_p(\bar{k}(t)).$$

Every finite extension K of $\bar{k}(t)$ has $dd(K) = 1$ by (6.5.5) and thus has a trivial Brauer group $Br(K) \cong H^2(K, \bar{K}^\times) = 0$. Since $H^1(K, \bar{K}^\times) = 0$, we obtain from (6.5.11) $cd_p(\bar{k}(t)) = 1$, hence

$$cd_p(k(t)) \le cd_p(k) + 1.$$

For the proof of the equality we replace k by the fixed field of a p-Sylow subgroup of G_k. The dimensions $cd_p(k)$ and $cd_p(k(t))$ do not change by (3.3.6). Hence we may assume that G_k is a pro-p-group. It follows that the G_k-module μ_p of p-th roots of unity is contained in k, and is hence isomorphic to $\mathbb{Z}/p\mathbb{Z}$.

*) The proof is taken from [210], chap.II, § 4.

Let $d = cd_p(k)$. We have to show $H^{d+1}(k(t), \mu_p) \neq 0$. To this end we consider the Hochschild-Serre spectral sequence

$$E_2^{ij} = H^i(G_k, H^j(H, \mu_p)) \Rightarrow H^{i+j}(G_{k'}, \mu_p),$$

associated to the exact sequence $(*)$, where $k' = k(t)$ and $H = G(\overline{k(t)} \mid \bar{k}(t))$. We have proved already that $cd_p\, H = 1$, i.e. $H^j(H, \mu_p) = 0$ for $j > 1$. Therefore, by (2.1.4),

$$H^d(G_k, H^1(H, \mu_p)) \cong H^{d+1}(G_{k'}, \mu_p).$$

From the exact sequence $1 \to \mu_p \to \overline{k(t)}^{\times} \xrightarrow{p} \overline{k(t)}^{\times} \to 1$ we obtain an isomorphism of G_k-modules $H^1(H, \mu_p) = \bar{k}(t)^\times/\bar{k}(t)^{\times p}$, and hence an isomorphism

$$H^{d+1}(G_{k'}, \mu_p) \cong H^d(G_k, \bar{k}(t)^\times/\bar{k}(t)^{\times p}).$$

Now let $v_{\mathfrak{p}} : \bar{k}(t)^\times \to \mathbb{Z}$ be the $\mathfrak{p}$-adic valuation associated to the prime ideal $\mathfrak{p} = (t)$ of $\bar{k}[t]$. The valuation $v_{\mathfrak{p}}$ induces a surjective homomorphism $\bar{k}(t)^\times/\bar{k}(t)^{\times p} \to \mathbb{Z}/p\mathbb{Z}$ of G_k-modules, hence a homomorphism

$$H^d(G_k, \bar{k}(t)^\times/\bar{k}(t)^{\times p}) \longrightarrow H^d(G_k, \mathbb{Z}/p\mathbb{Z}),$$

which is again surjective because $cd_p\, G_k = d$. Since G_k is a nontrivial pro-p-group, $H^d(G_k, \mathbb{Z}/p\mathbb{Z}) \neq 0$ by (3.3.2)(iii). This implies $H^d(G_k, \bar{k}(t)^\times/\bar{k}(t)^{\times p})$ $\neq 0$, and so $H^{d+1}(G_k, \mu_p) \neq 0$. □

Remark. The formula $cd_p(k(t_1, \dots, t_n)) = cd_p(k) + n$ holds true if $cd_p(k) = \infty$ (see [8]).

Using Lang's theorem (6.5.6), we obtain in a similar way the

(6.5.15) Theorem. *Let K be a field, complete with respect to a discrete valuation with perfect residue field k. If $p \neq \mathrm{char}(K)$ is a prime number and $cd_p(k) < \infty$, then*

$$cd_p(K) = cd_p(k) + 1.$$

The proof is completely analogous to the above proof. One uses the exact sequence $1 \to G_{\tilde{K}} \to G_K \to G_k \to 1$, where $\tilde{K}|K$ is the maximal unramified extension with Galois group $G(\tilde{K}|K) \cong G_k$.

From the last two theorems, we obtain $\mathbb{C}(X, Y)$, $\mathbb{F}_q(X)$, $\mathbb{Q}_p$ as examples of fields K with $cd(K) = 2$. The first two fields have also diophantine dimension 2, which follows from (6.5.8) and (6.5.12). *E. Artin* conjectured that $dd(\mathbb{Q}_p) = 2$ also. But this is not true as was shown by *G. Terjanian* [234]. However, $\mathbb{Q}_p$ comes close to the C_2-property in the following sense. A field k is said to have property $C_i(d)$ if every form of degree d in more than

d^i variables has a nontrivial zero in k. *J. Ax* and *S. Kochen* [9] proved the following result:

> *For every positive integer* d *there exists a finite set of primes,* $A(d)$*, such that* $\mathbb{Q}_p$ *has property* $C_2(d)$ *for all* $p \notin A(d)$.

One knows that $\mathbb{Q}_p$ has the properties $C_2(2)$ and $C_2(3)$ for all p. One also knows that for given p and d there exists an integer $i \geq 2$ such that $\mathbb{Q}_p$ has property $C_i(d)$. But $\mathbb{Q}_p$ is not C_i for any i (see [AK]). In contrast, notice that the fields $\mathbb{F}_p((X))$ are C_2, see [60], cor. 4.9. For general fields k it is conjectured by *J.-P. Serre*, cf. [210], 5th edition, chap. II §4.5, that the inequality

$$cd(k) \leq dd(k)$$

should hold. In this direction Serre showed (in an exercise) that $cd_2(k) \leq dd(k)$ if the Bloch-Kato conjecture holds for $N = 2$. As mentioned before, this conjecture has been proven in the meantime (cf. (6.4.3)). Furthermore, we have the following result.

(6.5.16) Theorem. *If* k *is a* C_2*-field, then* $cd(k) \leq dd(k)$.

If $dd(k) \leq 1$, then $dd(K) \leq 1$ for every finite extension $K|k$ by (6.5.4), hence $Br(K) = 0$ by (6.5.8), and therefore $cd(k) \leq 1$ by (6.5.11). The implication $dd(k) \leq 2 \Rightarrow cd(k) \leq 2$ is a result of *A. S. Merkur'ev* and *A. A. Suslin*, see [223], cor. 24.9.

Exercise: Assume k is a C_2-field, i.e. $dd(k) \leq 2$.

(i) Every quadratic form of 5 variables has a nontrivial zero.

(ii) If D is a skew field with center k and finite over k, then the *reduced norm* $Nrd : D^\times \to k^\times$ is surjective.
(The reduced norm Nrd is the composite of $D \longrightarrow D \otimes_k K \cong M_n(K) \xrightarrow{\det} K^\times$, where $K|k$ is a splitting field of D and $n^2 = \dim_k D$.)

Chapter VII
Cohomology of Local Fields

§1. Cohomology of the Multiplicative Group

We now begin the development of cohomology in number theory. As a ground field we take a nonarchimedean **local field** k, i.e. a field which is complete with respect to a discrete valuation and has a finite residue field. This covers two cases, namely ***p*-adic local fields**, i.e. finite extensions of $\mathbb{Q}_p$ for some prime number p, and **fields of formal Laurent series** $\mathbb{F}((t))$ in one variable over a finite field. For the basic properties of local fields we refer to [160], chapters II and V. As always, $\bar{k}|k$ denotes a separable closure of k and $K|k$ the subextensions of $\bar{k}|k$. v_k denotes the valuation of k, normalized by $v_k(k^\times) = \mathbb{Z}$, and κ the residue field. For every Galois extension $K|k$ we set

$$H^i(K|k) := H^i(G(K|k), K^\times), \quad i \geq 0.$$

If $K|k$ is finite, we also set $\hat{H}^i(K|k) = \hat{H}^i(G(K|k), K^\times)$ for $i \in \mathbb{Z}$. The basis for the results in this chapter is the following theorem.

(7.1.1) Theorem. (Class Field Axiom). *For a finite cyclic extension $K|k$ we have*

$$\#\hat{H}^i(K|k) = \begin{cases} [K:k] & \textit{for} \quad i = 0, \\ 1 & \textit{for} \quad i - 1. \end{cases}$$

Proof: We first show that for any finite Galois extension $K|k$ there exists a submodule $V_1 \subseteq U_K^1$ which has finite index and is cohomologically trivial. Let $G = G(K|k) = \{\sigma_1, \ldots, \sigma_n\}$. We consider a normal basis of $K|k$, i.e. we choose an $\alpha \in K$ such that $\sigma_1\alpha, \ldots, \sigma_n\alpha$ is a k-basis of K. For suitably chosen $a \in k^\times$, we find an inclusion of G-modules

$$M := \mathcal{O}_k a\sigma_1\alpha + \ldots + \mathcal{O}_k a\sigma_n\alpha \subseteq \mathcal{O}_K.$$

We have an isomorphism $M \cong \mathcal{O}_k[G]$ of G-modules. Furthermore, M has finite index in $\mathcal{O}_K$, hence $\pi^m\mathcal{O}_K \subseteq M$ for some $m \geq 1$, where $\pi \in \mathcal{O}_k$ is a uniformizer. This implies $(\pi^{m+i}M) \cdot (\pi^{m+i}M) \subseteq \pi^{2m+2i}\mathcal{O}_K \subseteq \pi^{m+2i}M$. Therefore the subsets

J. Neukirch et al., *Cohomology of Number Fields*,
Grundlehren der mathematischen Wissenschaften 323,
DOI 10.1007/978-3-540-37889-1_7, © Springer-Verlag Berlin Heidelberg 2008

$$V_i := 1 + \pi^{m+i} M \subseteq U_K^1$$

are G-submodules of finite index and we have isomorphisms of G-modules

$$V_i/V_{i+1} \xrightarrow{\sim} M/\pi M, \quad v_i \longmapsto \pi^{-m-i}(v_i - 1) \bmod \pi M$$

for all $i \geq 1$. Since $V_i/V_{i+1} \cong (\mathcal{O}_k/\pi)[G]$ is cohomologically trivial, the finite G-modules V_1/V_i are cohomologically trivial for all $i \geq 1$, and so is their inverse limit $V_1 \cong \varprojlim_i V_1/V_i$.

Now assume that G is cyclic. For the Herbrand index of U_K, we obtain

$$h(G, U_K) = h(G, V_1) \cdot h(G, U_K/V_1) = 1$$

by (1.7.5) and (1.7.6). The exact sequence $0 \to U_K \to K^\times \to \mathbb{Z} \to 0$ implies $h(G, K^\times) = h(G, \mathbb{Z}) = \#G$. Finally, $H^1(G, K^\times) = 0$ by Hilbert's Satz 90, and so $\#\hat{H}^0(G, K^\times) = \#G = [K : k]$. □

Next we show the vanishing of the cohomology of the unit group in unramified extensions.

(7.1.2) Proposition. *Let $K|k$ be a Galois extension. Then the following holds.*

(i) *If $K|k$ is unramified, then the group of units U_K and the group of principal units U_K^1 are cohomologically trivial $G(K|k)$-modules.*

(ii) *If $K|k$ is tamely ramified, then the group of principal units U_K^1 is a cohomologically trivial $G(K|k)$-module.*

Proof: By a direct limit argument, we may assume that $K|k$ is finite. Let $K|k$ be unramified and let $\lambda|\kappa$ be the associated residue extension. Then $G(K|k) \cong G(\lambda|\kappa)$, and

$$U_K^i/U_K^{i+1} \cong \lambda$$

is a cohomologically trivial $G(K|k)$-module for all $i \geq 1$ by (6.1.1). It follows by induction that the finite G-module U_K^1/U_K^i is cohomologically trivial for all $i \geq 1$. Since the cohomology of a finite group with values in an inverse limit of finite modules is the inverse limit of the cohomology groups, the isomorphism $U_K^1 \cong \varprojlim_i U_K^1/U_K^i$ implies that

$$H^n(G(K|k), U_K^1) = \varprojlim_i H^n(G(K|k), U_K^1/U_K^i) = 0 \quad \text{for all } n \geq 1.$$

Hence U_K^1 is cohomologically trivial. Finally, by Hilbert's Satz 90 and the vanishing of the Brauer group of a finite field (6.3.5), $\lambda^\times$ is cohomologically trivial, and the exact sequence

$$0 \longrightarrow U_K^1 \longrightarrow U_K \longrightarrow \lambda^\times \longrightarrow 0$$

shows that U_K is also cohomologically trivial.

If $K|k$ is tamely ramified, let K' be the maximal unramified extension of k inside K. Since U_K^1 is a $\mathbb{Z}_p$-module and the order of $G(K|K')$ is prime to p, we have $H^j(K|K', U_K^1) = 0$ for $j > 0$, and the Hochschild-Serre spectral sequence

$$E_2^{ij} = H^i(K'|k, H^j(K|K', U_K^1)) \Rightarrow H^{i+j}(K|k, U_K^1)$$

induces natural isomorphisms

$$H^i(K'|k, U_{K'}^1) \xrightarrow{\sim} H^i(K|k, U_K^1)$$

for all $i \geq 0$. Therefore (i) implies (ii). $\square$

We now consider the *maximal unramified extension* $\tilde{k}|k$. Its Galois group $\Gamma_k = G(\tilde{k}|k)$ is topologically generated by the Frobenius automorphism φ_k and is canonically isomorphic to $\hat{\mathbb{Z}}$; φ_k corresponds to $1 \in \hat{\mathbb{Z}}$. From the two exact sequences

$$\begin{array}{l} 0 \longrightarrow U_{\tilde{k}} \longrightarrow \tilde{k}^\times \xrightarrow{v_{\tilde{k}}} \mathbb{Z} \longrightarrow 0, \\ 0 \longrightarrow \mathbb{Z} \longrightarrow \mathbb{Q} \longrightarrow \mathbb{Q}/\mathbb{Z} \longrightarrow 0, \end{array}$$

in which $U_{\tilde{k}}$ and $\mathbb{Q}$ are cohomologically trivial Γ_k-modules, we obtain the isomorphisms

$$H^2(\tilde{k}|k) \xrightarrow[\sim]{v_{\tilde{k}}} H^2(\Gamma_k, \mathbb{Z}) \xrightarrow[\sim]{\delta^{-1}} H^1(\Gamma_k, \mathbb{Q}/\mathbb{Z}) \xrightarrow[\sim]{\varphi} \mathbb{Q}/\mathbb{Z},$$

where φ is given by $\varphi(\chi) = \chi(\varphi_k)$. We denote the composition by

$$inv_{\tilde{k}|k} : H^2(\tilde{k}|k) \xrightarrow{\sim} \mathbb{Q}/\mathbb{Z}.$$

For a finite separable extension $K|k$ we have a commutative diagram

$$\begin{array}{ccc} H^2(\tilde{K}|K) & \xrightarrow{inv} & \mathbb{Q}/\mathbb{Z} \\ {\scriptstyle res}\uparrow & & \uparrow{\scriptstyle [K:k]} \\ H^2(\tilde{k}|k) & \xrightarrow{inv} & \mathbb{Q}/\mathbb{Z}, \end{array}$$

where the map *res* is induced by the compatible pair $\Gamma_K \to \Gamma_k$, $\tilde{k}^\times \hookrightarrow \tilde{K}^\times$. Indeed, this follows directly from the definition of the map *inv* and the diagram

$$\begin{array}{ccccccc} H^2(\tilde{K}|K) & \longrightarrow & H^2(\Gamma_K, \mathbb{Z}) & \xrightarrow{\delta^{-1}} & H^1(\Gamma_K, \mathbb{Q}/\mathbb{Z}) & \longrightarrow & \mathbb{Q}/\mathbb{Z} \\ {\scriptstyle res}\uparrow & & \uparrow{\scriptstyle e_{K|k}\cdot res} & & \uparrow{\scriptstyle e_{K|k}\cdot res} & & \uparrow{\scriptstyle e_{K|k}\cdot f_{K|k}} \\ H^2(\tilde{k}|k) & \longrightarrow & H^2(\Gamma_k, \mathbb{Z}) & \xrightarrow{\delta^{-1}} & H^1(\Gamma_k, \mathbb{Q}/\mathbb{Z}) & \longrightarrow & \mathbb{Q}/\mathbb{Z}, \end{array}$$

which is commutative since the Frobenius automorphism $\varphi_K \in \Gamma_K$ is mapped onto the $f_{K|k}$-th power of the Frobenius automorphism $\varphi_k \in \Gamma_k$. Further observe that $e_{K|k} \cdot f_{K|k} = [K : k]$. This, and what follows, was already seen in III §2 in a general setting.

If $K|k$ is a Galois extension, then, because $H^1(K|k) = 0$, the sequence

$$(*) \qquad 0 \longrightarrow H^2(K|k) \xrightarrow{inf} H^2(\bar{k}|k) \xrightarrow{res} H^2(\bar{k}|K)$$

is exact.*) We identify $H^2(K|k)$ with its image in $H^2(\bar{k}|k)$. Of crucial importance is the following

(7.1.3) Theorem. $H^2(\bar{k}|k) = H^2(\tilde{k}|k)$.

Proof: First, we claim that

$$\#H^2(K|k) \,\Big|\, [K : k]$$

for a finite Galois extension K of k. In fact, this is true if $K|k$ is cyclic because of $H^2 \cong \hat{H}^0$. If $G(K|k)$ is a p-group, it follows inductively from the exact sequence

$$0 \longrightarrow H^2(L|k) \longrightarrow H^2(K|k) \longrightarrow H^2(K|L),$$

where $G(K|L)$ is a normal subgroup of $G(K|k)$ of order p. In the general case, let Σ_p be a p-Sylow subfield of $K|k$. Since the restriction map

$$res : H^2(K|k) \hookrightarrow \bigoplus_p H^2(K|\Sigma_p)$$

is injective by (1.6.10), we obtain

$$\#H^2(K|k) \,\Big|\, \prod_p \#H^2(K|\Sigma_p) \,\Big|\, \prod_p [K : \Sigma_p] = [K : k]\,.$$

Let $n = [K : k]$ and let k_n be the unramified extension of k of degree n. Then

$$H^2(K|k) = H^2(k_n|k)\,,$$

where we identify $H^2(K|k)$ and $H^2(k_n|k)$ with their images in $H^2(\bar{k}|k)$. In fact, because

$$\#H^2(K|k) \,\Big|\, [K : k] = [k_n : k] = \#H^2(k_n|k),$$

it suffices to show the inclusion "$\supseteq$". But this follows from the exact commutative diagram

$$\begin{array}{ccccccc}
0 & \longrightarrow & H^2(K|k) & \longrightarrow & H^2(\bar{k}|k) & \xrightarrow{res} & H^2(\bar{k}|K) \\
 & & & & \big\uparrow & & \big\uparrow \\
 & & & & H^2(\tilde{k}|k) & \xrightarrow{res} & H^2(\tilde{K}|K) \\
 & & & & {\scriptstyle inv_{\tilde{k}|k}}\big\downarrow\wr & & {\scriptstyle inv_{\tilde{K}|K}}\big\downarrow\wr \\
 & & & & \mathbb{Q}/\mathbb{Z} & \xrightarrow{[K:k]} & \mathbb{Q}/\mathbb{Z},
\end{array}$$

*)The group $H^2(\bar{k}|k)$ is often called the "Brauer group", because it is isomorphic to the group $Br(k)$ of central simple algebras by (6.3.4).

in which $inv_{\bar{k}|k}$ and $inv_{\bar{K}|K}$ are isomorphisms as shown above. Since $H^2(k_n|k) \subseteq H^2(\tilde{k}|k)$ has order $n = [K:k]$, it is mapped by the middle arrow *res*, and thus by the upper arrow *res*, to zero, hence

$$H^2(k_n|k) \subseteq H^2(K|k).$$

We therefore obtain

$$H^2(\bar{k}|k) = \bigcup_K H^2(K|k) = \bigcup_n H^2(k_n|k) = H^2(\tilde{k}|k).$$

□

(7.1.4) Corollary. *We have a canonical isomorphism*

$$inv_k : H^2(\bar{k}|k) \xrightarrow{\sim} \mathbb{Q}/\mathbb{Z},$$

called the **invariant map**. *For every finite separable extension $K|k$ we have the commutative diagrams*

$$\begin{array}{ccc} H^2(\bar{k}|K) & \xrightarrow{inv} & \mathbb{Q}/\mathbb{Z} \\ {\scriptstyle res}\uparrow\downarrow{\scriptstyle cor} & & {\scriptstyle [K:k]}\uparrow\downarrow{\scriptstyle id} \\ H^2(\bar{k}|k) & \xrightarrow{inv} & \mathbb{Q}/\mathbb{Z}. \end{array}$$

The commutativity for *cor* follows from $cor \circ res = [K:k]$. If $K|k$ is Galois, then the exact sequence $(*)$ shows that inv_k induces an isomorphism

$$inv_{K|k} : H^2(K|k) \xrightarrow{\sim} \tfrac{1}{[K:k]}\mathbb{Z}/\mathbb{Z}.$$

In other words (see (3.1.8)):

(7.1.5) Corollary. *The pair $(G_k, \bar{k}^\times)$ is a class formation.*

As a first application, we obtain (by (6.3.4)) the

(7.1.6) Corollary. *For the Brauer group $Br(k)$ of central simple k-algebras we have a canonical isomorphism*

$$inv_k : Br(k) \xrightarrow{\sim} \mathbb{Q}/\mathbb{Z},$$

and $Br(L)(p) = 0$ for any extension $L|k$ of degree divisible by p^∞.

The last assertion follows from (7.1.4): if $L_\alpha|k$ runs through the finite subextensions of $L|k$, then

$$H^2(\bar{k}|L)(p) = \varinjlim_{res} H^2(\bar{k}|L_\alpha)(p) \cong \varinjlim_{[L_\alpha:k]} (\mathbb{Q}/\mathbb{Z})(p) = 0.$$

From (3.1.4) and III §1 ex.4, we obtain the

(7.1.7) Corollary. *Let $K|k$ be a finite Galois extension with Galois group G. The cup-product with the fundamental class $u_{K|k} \in H^2(K|k)$ yields isomorphisms for $i \geq 0$:*

$$u_{K|k}\cup : \hat{H}^i(G,\mathbb{Z}) \xrightarrow{\sim} \hat{H}^{i+2}(G,K^\times),$$

$$u_{K|k}\cup : \hat{H}^i(G,\mathrm{Hom}(K^\times,\mathbb{Q}/\mathbb{Z})) \xrightarrow{\sim} \hat{H}^{i+2}(G,\mathbb{Q}/\mathbb{Z}).$$

(7.1.8) Theorem. *The following assertions hold:*

(i) *The cohomological dimension of G_k is*

$$cd_p(k) = \begin{cases} 2, & \text{if } p \neq \mathrm{char}(k), \\ 1, & \text{if } p = \mathrm{char}(k), \end{cases}$$

and $cd_p(L) \leq 1$ for every extension $L|k$ of degree divisible by p^∞.

(ii) *Suppose the characteristic of k does not divide $n \in \mathbb{N}$. Then we have*

$$H^i(k,\mu_n) = \begin{cases} k^\times/k^{\times n} & \text{for } i = 1, \\ \frac{1}{n}\mathbb{Z}/\mathbb{Z} & \text{for } i = 2, \\ 0 & \text{for } i \geq 3. \end{cases}$$

(iii) *If A is a finite G_k-module of order prime to char(k), then the groups $H^i(k,A)$ are finite for all $i \geq 0$.*

(iv) *Let A be a G_k-module which is finitely generated as a $\mathbb{Z}$-module and assume that the order of tor(A) is prime to char(k). Then $H^1(k,A)$ is finite.*

Proof: (i) Since $H^1(L,\bar{k}^\times) = 0$ and $H^2(L,\bar{k}^\times)(p) = 0$ as we have just seen, we obtain from (6.5.11) that $cd_p(L) \leq 1$. We apply this to the maximal unramified extension $\tilde{k}|k$. Let $\Gamma = G(\tilde{k}|k)$ and let A be a p-torsion G_k-module. We have $cd\,(\tilde{k}) \leq 1$, hence $H^j(\tilde{k},A) = 0$ for $j > 1$. Therefore the Hochschild-Serre spectral sequence

$$E_2^{ij} = H^i(\Gamma, H^j(\tilde{k}, A)) \Rightarrow H^{i+j}(k, A)$$

together with $cd_p\, \Gamma = 1$ yields by (2.1.4)

$$H^{i+1}(k, A) \cong H^i(\Gamma, H^1(\tilde{k}, A)) = 0 \quad \text{for} \quad i \geq 2,$$

hence $cd_p(k) \leq 2$. If $p \neq \text{char}(k)$, the equality $cd_p(k) = 2$ follows from $H^2(k, \mu_p) \cong \mathbb{Z}/p\mathbb{Z} \neq 0$, see below. If $p = \text{char}(k)$, then $cd_p(k) = 1$ by (6.1.3).

(ii) Recalling that $H^0(k, \bar{k}^\times) = k^\times$, $H^1(k, \bar{k}^\times) = 1$, $H^2(k, \bar{k}^\times) \cong \mathbb{Q}/\mathbb{Z}$, the Kummer sequence (for n prime to char(k))

$$1 \longrightarrow \mu_n \longrightarrow \bar{k}^\times \xrightarrow{\;n\;} \bar{k}^\times \longrightarrow 1$$

yields $H^1(k, \mu_n) \cong k^\times/k^{\times n}$ and $H^2(k, \mu_n) \cong \frac{1}{n}\mathbb{Z}/\mathbb{Z}$. $H^i(k, \mu_n) = 0$ for $i > 2$ follows from (i).

(iii) Let A be a finite G_k-module of order m prime to char(k). Then $H^i(k, A) = 0$ for $i > 2$ by (i). Choose a finite Galois extension $K|k$ in $\bar{k}$ over which A and μ_m become trivial Galois modules. As a G_K-module, A is isomorphic to a finite direct sum of G_K-modules of type μ_n, $n|m$. Since $H^j(K, \mu_n)$ is finite for all $j \geq 0$, so is $H^j(K, A)$, and the spectral sequence

$$E_2^{ij} = H^i(G(K|k), H^j(K, A)) \Rightarrow H^n(k, A)$$

shows that also $H^n(k, A)$ is finite as a group with a finite filtration, whose quotients are subquotients of the finite groups on the left-hand side.

(iv) Let tor(A) denote the torsion subgroup of A. The cohomology sequence associated to $0 \to \text{tor}(A) \to A \to A/\text{tor}(A) \to 0$ and (iii) show that we may assume A to be $\mathbb{Z}$-free. Let $K|k$ be a finite Galois extension over which A is trivial, i.e. $A \cong \mathbb{Z}^n$ as a G_K-module for some n. Then $H^1(K, A) = H^1(K, \mathbb{Z})^n = 0$ and $H^1(k, A) = H^1(G(K|k), A)$ is finite. □

Using (6.3.6), we obtain the following corollary.

(7.1.9) Corollary. *If k is a p-adic local field, then canonically*

$$H^2(k, \mu) \cong \mathbb{Q}/\mathbb{Z},$$

and, if $p = \text{char}(k) > 0$, then

$$H^2(k, \mu) = \bigcup_{p \nmid n} \tfrac{1}{n}\mathbb{Z}/\mathbb{Z} = \bigoplus_{\ell \neq p} \mathbb{Q}_\ell/\mathbb{Z}_\ell\,.$$

§2. The Local Duality Theorem

Local fields are topological fields and topological questions enter into the game. These questions become even more important in the global theory because of the appearance of the idèle and idèle class groups. For this reason we premise the discussion of local and global theory with the following preparatory

Topological Remarks: Assume we are given an abelian topological group M on which a profinite group G acts in such a way that the action is continuous with respect to the given topology as well as with respect to the discrete topology. Then let

$$A = H^i(G, M)$$

be the i-th cohomology group of M as a discrete G-module. If $i \geq 1$, we consider these groups always as *discrete* topological groups. Hence the dual $H^i(G, M)^\vee$ for $i \geq 1$ is always a profinite group. But for $i = 0$ the group $H^0(G, M)$ inherits the initial topology of M. Examples of this situation are $M = k^\times$, where k is a local field, and the level-compact modules considered in I §9. Suppose that

$$A \times B \xrightarrow{(\ ,\ )} C$$

is a continuous bilinear pairing of topological groups. We then obtain continuous homomorphisms

$$A \xrightarrow{\alpha} \operatorname{Hom}(B, C), \quad a \longmapsto f_a : b \longmapsto (a, b),$$

$$B \xrightarrow{\beta} \operatorname{Hom}(A, C), \quad b \longmapsto g_b : a \longmapsto (a, b),$$

inducing bijective continuous homomorphisms

$$A/\ker(\alpha) \longrightarrow \operatorname{im}(\alpha) \ \text{ and } \ B/\ker(\beta) \longrightarrow \operatorname{im}(\beta),$$

where the left groups have the quotient topology and the right ones the induced topology. We say that the pairing $A \times B \to C$ is **non-degenerate** if both maps α and β are injective. Furthermore, we want to recall the notation $A^* = \operatorname{Hom}(A, \mathbb{Q}/\mathbb{Z})$ for an abelian group A. If A is a discrete abelian torsion group, then A^* coincides with the Pontryagin dual $A^\vee$, but we consider A^* as a discrete group, while $A^\vee$ is equipped with a natural compact topology. We always have $A^{\vee\vee} = A$ but the equality $A^{**} = A$ holds if and only if A is finite.

Corollary (7.1.5) says that for every finite Galois extension $K|k$ of a local field k the multiplicative group $K^\times$ is a *class module* for the Galois group $G = G(K|k)$ in the sense of (3.1.3). By (7.1.4), we have even a canonical fundamental class $\gamma \in H^2(G, K^\times)$. Therefore, by the theorem of Nakayama-Tate (3.1.5), we obtain the

(7.2.1) Theorem. *Let $K|k$ be a finite extension of local fields with Galois group G. Let A be a finitely generated $\mathbb{Z}$-free G-module and $A' = \mathrm{Hom}(A, K^\times)$. Then for all $i \in \mathbb{Z}$ the cup-product*

$$\hat{H}^i(G, A') \times \hat{H}^{2-i}(G, A) \xrightarrow{\cup} H^2(G, K^\times) = \tfrac{1}{\#G}\mathbb{Z}/\mathbb{Z}$$

induces an isomorphism of finite abelian groups

$$\hat{H}^i(G, A') \cong \hat{H}^{2-i}(G, A)^*.$$

The case $A = \mathbb{Z}$ and $i = 3$ yields the

(7.2.2) Corollary. $H^3(G, K^\times) = 0$.

In the case $i = 0$ and $A = \mathbb{Z}$, we have $H^2(G, \mathbb{Z})^* \cong H^1(G, \mathbb{Q}/\mathbb{Z})^* = (G^{ab})^{**} = G^{ab}$, and we obtain the main theorem of local class field theory, the "local reciprocity law" (for a proof of the second statement see [212], XIV, §6):

(7.2.3) Theorem. *Let k be a local field and let $K|k$ be a finite Galois extension. Then there is a canonical isomorphism*

$$k^\times/N_{K|k}K^\times \cong G(K|k)^{ab}.$$

The norm groups $N_{K|k}K^\times$ for finite Galois extensions $K|k$ are precisely the open subgroups of finite index in $k^\times$.

It was this isomorphism which initiated the development of cohomology theory in number theory. Before this development the isomorphism was obtained only in a complicated and obscure way via a detour to the theory of global fields. It was *J. Tate* who put the reciprocity law on a conceptual basis in the above cohomological setting. The law may also be proved in a direct group theoretical way (see [160], chap. IV and V).

However, theorem (7.2.1) is still too rigid for our intended applications as it holds only for $\mathbb{Z}$-free Galois modules and applies only to finite Galois groups. Our aim is to prove a duality theorem of the above type for the absolute Galois group G_k of a local field and for G_k-modules which are finitely generated as $\mathbb{Z}$-modules. The essential step on this path is the explicit determination of the *dualizing module* $D' = D_2(\hat{\mathbb{Z}})$ of the category $Mod_t(G_k)$ of discrete G_k-modules which are torsion as abelian groups (see III §4). It is defined by

$$D' = \varinjlim_{K,n} H^2(K, \mathbb{Z}/n\mathbb{Z})^*,$$

where $K|k$ runs through the finite subextensions of $\bar{k}|k$, and it is characterized by a functorial isomorphism $H^2(K, A)^* \cong \mathrm{Hom}_{G_K}(A, D')$ for $A \in Mod_t(G_k)$.

(7.2.4) Theorem. *The G_k-module D' is canonically isomorphic to the G_k-module μ of all roots of unity in $\bar{k}$.*

Proof: Let $n \in \mathbb{N}$, and suppose $(n,p) = 1$ if $p = \text{char}(k) > 0$. Let ${}_nI = \ker(D' \xrightarrow{n} D')$. Since D' is also the dualizing module of every open subgroup V of G, we obtain canonically for the G_k-module μ_n

$$\text{Hom}_V(\mu_n, {}_nI) = \text{Hom}_V(\mu_n, D') \cong H^2(V, \mu_n)^* \cong \mathbb{Z}/n\mathbb{Z}$$

by (7.1.8). This shows that $\text{Hom}_V(\mu_n, {}_nI)$ is independent of V, and we obtain a canonical isomorphism of G_k-modules

$$\text{Hom}(\mu_n, {}_nI) \cong \mathbb{Z}/n\mathbb{Z}.$$

Let $f_n : \mu_n \to {}_nI$ be the element corresponding to 1 mod $n\mathbb{Z}$. The other elements of $\text{Hom}(\mu_n, {}_nI)$ are $f_n^i(\zeta) = f_n(\zeta^i)$, $i = 0, \ldots, n-1$. f_n is a G_k-homomorphism. It is injective, since it has order precisely n. It is also surjective, since otherwise we would have an $x \in {}_nI \smallsetminus \text{im}(f_n)$, and hence a homomorphism $\mu_n \to (x) \subseteq {}_nI$, different from the f_n^i. If k has characteristic zero, we obtain a G_k-isomorphism

$$f : \mu = \bigcup_{n\in\mathbb{N}} \mu_n \xrightarrow{\sim} \bigcup_{n\in\mathbb{N}} {}_nI = D'.$$

If $p = \text{char}(k) > 0$, this remains true, since the p-primary components of μ and D' are trivial, the latter because $cd_p\, G_k = 1$. $\square$

(7.2.5) Corollary. *For every prime number p we have $scd_p(k) = 2$.*

In fact, if $p \neq \text{char}(k)$, this is true by Serre's criterion (3.4.5), and if $p = \text{char}(k) > 0$, then $cd_p(k) = 1$ by (6.5.10), and hence $scd_p(k) = 2$ (cf. III §3 ex.1).

As the main result of the cohomology theory of local fields we obtain the

(7.2.6) Theorem (Tate Duality). *Let k be a p-adic local field. Let A be a finite G_k-module and set $A' = \text{Hom}(A, \mu)$. Then the cup-product*

$$H^i(k, A') \times H^{2-i}(k, A) \xrightarrow{\cup} H^2(k, \mu) \cong \mathbb{Q}/\mathbb{Z}$$

induces for $0 \le i \le 2$ an isomorphism of finite abelian groups

$$H^i(k, A') \xrightarrow{\sim} H^{2-i}(k, A)^*.$$

If k is a local field of characteristic $p > 0$, the same holds true for the finite G_k-modules A of order prime to p, except that $H^2(k, \mu) \cong \bigcup_{p\nmid n} \frac{1}{n}\mathbb{Z}/\mathbb{Z}$.

Proof: This theorem is a special case of the abstract duality theorem (3.4.6). We have only to show that for $p \neq \mathrm{char}(k)$

$$D_i(\mathbb{Z}/p\mathbb{Z}) = \varinjlim_{K|k} H^i(K, \mathbb{Z}/p\mathbb{Z})^* = 0$$

for $i = 0, 1$. Since every subextension $K|k$ of $\bar{k}|k$ admits a subextension $K'|K$ of a degree divisible by p, we have $D_0(\mathbb{Z}/p\mathbb{Z}) = \varinjlim_K \mathbb{Z}/p\mathbb{Z} = 0$. Let $K|k$ run through the subextensions of $\bar{k}|k$. Then from the main theorem of local class field theory (7.2.3) we obtain a surjection (which is in fact an isomorphism)

$$K^\times/K^{\times p} \twoheadrightarrow G_K^{ab}/p.$$

Passing to the direct limit, we obtain a surjection from $\varinjlim_K K^\times/K^{\times p} = \bar{k}^\times/\bar{k}^{\times p} = 1$ onto $\varinjlim_K G_K^{ab}/p = \varinjlim_K H^1(K, \mathbb{Z}/p\mathbb{Z})^* = D_1(\mathbb{Z}/p\mathbb{Z})$, and so $D_1(\mathbb{Z}/p\mathbb{Z}) = 0$. □

The duality theorem (7.2.6) deals with finite G_k-modules. It is desirable to extend it to G_k-modules A which are finitely generated as $\mathbb{Z}$-modules. The main reason is that it may then be applied to *algebraic tori*: a G_k-module A which is finitely generated and free as a $\mathbb{Z}$-module is the character group of the algebraic torus $T = \mathrm{Hom}(A, \mathbb{G}_m)$ over k, and $A' = \mathrm{Hom}(A, \bar{k}^\times) = T(\bar{k})$ is the G_k-module of $\bar{k}$-rational points of T. This generalization is accomplished by the following computation of the dualizing module

$$D_2(\mathbb{Z}) = \varinjlim_U H^2(U, \mathbb{Z})^*$$

of G_k, i.e. the dualizing object for the category $Mod(G_k)$ of *all* G_k-modules.

Let $K|k$ be a finite separable extension and let U_K denote its group of units. U_K is compact and we consider the profinite completion of $K^\times$,

$$\hat{K}^\times = \varprojlim_N K^\times/N,$$

where N runs through the open subgroups of finite index. We then have an exact commutative diagram

$$\begin{array}{ccccccccc}
0 & \longrightarrow & U_K & \longrightarrow & K^\times & \xrightarrow{v_K} & \mathbb{Z} & \longrightarrow & 0 \\
 & & \| & & \downarrow & & \downarrow & & \\
0 & \longrightarrow & U_K & \longrightarrow & \hat{K}^\times & \xrightarrow{\hat{v}_K} & \hat{\mathbb{Z}} & \longrightarrow & 0,
\end{array}$$

hence $\hat{K}^\times/K^\times \cong \hat{\mathbb{Z}}/\mathbb{Z}$, which is a uniquely divisible group. The inclusion $k^\times \hookrightarrow K^\times$ is continuous and induces an injection $\hat{k}^\times \hookrightarrow \hat{K}^\times$. We will consider the injective limit $\varinjlim \hat{K}^\times$, where K runs through the finite subextensions of $\bar{k}|k$. If e is the ramification index of $K|k$, then $v_{K|k} = ev_k$ and we have a commutative diagram of topological isomorphisms

$$\begin{array}{ccc} \hat{K}^\times/K^\times & \xrightarrow[\sim]{\hat{v}_K} & \hat{\mathbb{Z}}/\mathbb{Z} \\ \uparrow & & \uparrow e \\ \hat{k}^\times/k^\times & \xrightarrow[\sim]{\hat{v}_k} & \hat{\mathbb{Z}}/\mathbb{Z}, \end{array}$$

noting that $\hat{\mathbb{Z}}/\mathbb{Z}$ is uniquely divisible. Applying $\varinjlim_K$ to the exact sequence

$$1 \longrightarrow K^\times \longrightarrow \hat{K}^\times \longrightarrow \hat{\mathbb{Z}}/\mathbb{Z} \longrightarrow 1,$$

we obtain an exact sequence of G_k-modules

$$1 \longrightarrow \bar{k}^\times \longrightarrow \varinjlim_K \hat{K}^\times \longrightarrow \hat{\mathbb{Z}}/\mathbb{Z} \longrightarrow 1,$$

since $\varinjlim_e \hat{\mathbb{Z}}/\mathbb{Z} \cong \hat{\mathbb{Z}}/\mathbb{Z}$. Since $v_K(x) = v_K(\sigma x)$ for all $\sigma \in G(K|k)$, the latter group is a trivial G_k-module and, since it is uniquely divisible, it is a cohomologically trivial G_k-module. The exact cohomology sequence therefore yields the

(7.2.7) Proposition.

$$\begin{aligned} H^0(k, \varinjlim_K \hat{K}^\times) &= \hat{k}^\times, \\ H^i(k, \varinjlim_K \hat{K}^\times) &= H^i(k, \bar{k}^\times) \ \ \textit{for } i > 0. \end{aligned}$$

(7.2.8) Theorem. *The dualizing module $D_2 = D_2(\mathbb{Z})$ of G_k is canonically isomorphic to the G_k-module $\varinjlim_K \hat{K}^\times$.*

Proof: For every finite Galois extension $K|k$ we have the norm residue symbol

$$(\ ,K|k) : k^\times/N_{K|k}K^\times \longrightarrow G(K|k)^{ab},$$

which is an isomorphism, as above. By (7.2.3), the norm groups $N_{K|k}K^\times$ are precisely the open subgroups of finite index in $k^\times$. Hence, passing to the projective limit, we obtain a canonical isomorphism

$$(\ ,k) : \hat{k}^\times \longrightarrow G_k^{ab} \cong H^2(k, \mathbb{Z})^\vee.$$

For a finite Galois extension $K|k$, we obtain the commutative diagram

$$\begin{array}{ccccc}
\hat{K}^\times & \longrightarrow & G_K^{ab} & \xrightarrow{\sim} & H^2(K,\mathbb{Z})^\vee \\
\uparrow & & \uparrow{\scriptstyle Ver} & & \uparrow{\scriptstyle cor^\vee} \\
\hat{k}^\times & \longrightarrow & G_k^{ab} & \xrightarrow{\sim} & H^2(k,\mathbb{Z})^\vee
\end{array}$$

(cf. [160], chap.IV, (6.4) and (1.5.9)). Taking direct limits, we get an isomorphism

$$(\ ,\bar{k}) : \varinjlim_K \hat{K}^\times \xrightarrow{\sim} D_2(\mathbb{Z}). \qquad \square$$

Let A be a G_k-module which is finitely generated as a $\mathbb{Z}$-module. By $\mathrm{tor}(A)$ we denote the torsion subgroup of A, which is a G_k-module. Furthermore, we will use the following notation for the rest of this chapter:

$$\begin{aligned}
A' &= \mathrm{Hom}(A, \bar{k}^\times), \\
A^D &= \mathrm{Hom}(A, \varinjlim_K \hat{K}^\times) \;=\; \mathrm{Hom}(A, D_2).
\end{aligned}$$

As a generalization of Tate's local duality theorem (7.2.6) we have the

(7.2.9) Theorem. *Let A be a G_k-module which is finitely generated as a $\mathbb{Z}$-module and assume that $\#\mathrm{tor}(A)$ is prime to $\mathrm{char}(k)$. Then, for $0 \le i \le 2$, the cup-product and the map inv*

$$H^i(k, A^D) \times H^{2-i}(k, A) \xrightarrow{\cup} H^2(k, \bar{k}^\times) \xrightarrow{inv} \mathbb{Q}/\mathbb{Z}$$

induce an isomorphism

$$\Delta^i : H^i(k, A^D) \xrightarrow{\sim} H^{2-i}(k, A)^*.$$

For $i = 1$ these groups are finite.

Proof: We know by (7.2.5) that $scd(k) = 2$. We have $D_0(\mathbb{Z}) = 0$ by the remark following (3.4.3), and, trivially, $D_1(\mathbb{Z}) = \varinjlim H^1(K,\mathbb{Z})^* = 0$. If $\mathrm{char}(k) = 0$, then $\bar{k}^\times$ is divisible and so is $D_2(\mathbb{Z})$. If $p = \mathrm{char}(k) > 0$, then $D_2(\mathbb{Z})$ is ℓ-divisible for all prime numbers $\ell \neq \mathrm{char}(k)$. Therefore we obtain the isomorphisms by Tate's duality theorem (3.4.3) and the subsequent remark. The finiteness of $H^1(k, A)$ is part (iv) of (7.1.8). $\square$

The G_k-module $D_2 = \varinjlim_K \hat{K}^\times$ looks rather awkward and one may ask whether it can be replaced by $\bar{k}^\times$ in the theorem. This is clear for a finite module A, since $\mathrm{Hom}(A, D_2) = \mathrm{Hom}(A, \bar{k}^\times) = \mathrm{Hom}(A, \mu)$. Furthermore, the groups $A^D = \mathrm{Hom}(A, D_2)$ and $A' = \mathrm{Hom}(A, \bar{k}^\times)$ inherit the natural topologies of $D_2 \cong \varinjlim \hat{K}^\times$ *) and of $\bar{k}^\times$. Therefore the groups $H^0(k, A')$ and $H^0(k, A^D)$ are also topological groups in a natural way.

Since $H^2(k, A)$ is a discrete torsion group, we may replace $H^2(k, A)^*$ by the compact topological group $H^2(k, A)^\vee$ (which has the same underlying abstract group) and obtain an abstract isomorphism $H^0(k, A^D) \xrightarrow{\sim} H^2(k, A)^\vee$ between abelian topological groups.

(7.2.10) Proposition. *The duality isomorphism*

$$H^0(k, A^D) \xrightarrow{\sim} H^2(k, A)^\vee$$

is a homeomorphism. The natural injection

$$H^0(k, A') \hookrightarrow H^0(k, A^D)$$

is continuous and $H^0(k, A^D)$ is the profinite completion of $H^0(k, A')$ with respect to the open subgroups of finite index. We have isomorphisms

$$H^i(k, A') \xrightarrow{\sim} H^i(k, A^D) \ \text{ for } i > 0$$

and the groups are finite for $i = 1$.

Proof: The exact cohomology sequence associated to the sequence

$$0 \longrightarrow \mathrm{tor}(A) \longrightarrow A \longrightarrow A/\mathrm{tor} \longrightarrow 0,$$

the finiteness of the cohomology of tor(A), and the identity $\mathrm{tor}(A)' = \mathrm{tor}(A)^D$, allow us to assume A to be $\mathbb{Z}$-free. Let $K|k$ be a finite Galois extension over which A becomes a trivial Galois module, and let $G = G(K|k)$ be its Galois group. We have an isomorphism of G_K-modules $A \cong \mathbb{Z}^N$ for some $N \in \mathbb{N}$. Now consider the commutative diagram

$$\begin{array}{ccccc} H^0(K, A')^{G(K|k)} & \hookrightarrow & H^0(K, A^D)^{G(K|k)} & \xrightarrow{\sim} & (H^2(K, A)^\vee)^{G(K|k)} \\ \| & & \| & & \uparrow\wr \\ H^0(k, A') & \hookrightarrow & H^0(k, A^D) & \xrightarrow{\sim} & H^2(k, A)^\vee. \end{array}$$

The vertical arrow on the right is an isomorphism by (3.3.11) and the right upper horizontal isomorphism is a homeomorphism by the definition of the topology on D_2 and because A is a trivial G_K-module. The right commutative square now gives the first statement.

By definition of $\hat{K}^\times$, the group $H^0(K, A^D) = \mathrm{Hom}(A, \hat{K}^\times) \cong (\hat{K}^\times)^N$ is the completion of $H^0(K, A') = \mathrm{Hom}(A, K^\times) \cong (K^\times)^N$ with respect to the open

*) The topology depends on the choice of this isomorphism. We use the natural isomorphism constructed in (7.2.8).

subgroups of finite index, and this remains valid for the fixed modules under $G(K|k)$. This shows the second statement.

We now prove the maintained isomorphisms for $i > 0$. From the exact sequence

$$0 \longrightarrow \bar{k}^\times \longrightarrow D_2 \longrightarrow \hat{\mathbb{Z}}/\mathbb{Z} \longrightarrow 0,$$

we obtain the exact sequence

$$0 \longrightarrow A' \longrightarrow A^D \longrightarrow \mathrm{Hom}(A, \hat{\mathbb{Z}}/\mathbb{Z}) \longrightarrow 0.$$

Since $\hat{\mathbb{Z}}/\mathbb{Z}$ is uniquely divisible, so is the G_k-module $\mathrm{Hom}(A, \hat{\mathbb{Z}}/\mathbb{Z})$, and hence it is cohomologically trivial. From the exact cohomology sequence it follows that

$$(*) \qquad H^i(k, A') \cong H^i(k, A^D)$$

for $i > 1$. For $i = 1$ we need a little additional argument: for any open subgroup $H \subseteq G_k$ which acts trivially on A, the sequence

$$0 \longrightarrow (A')^H \longrightarrow (A^D)^H \longrightarrow \mathrm{Hom}(A, \hat{\mathbb{Z}}/\mathbb{Z}) \longrightarrow 0$$

is exact, since $H^1(H, A') \cong H^1(H, \mathrm{Hom}(\mathbb{Z}^N, \bar{k}^\times)) = H^1(H, \bar{k}^\times)^N = 0$. Because $\hat{H}^i(G_k/H, \mathrm{Hom}(A, \hat{\mathbb{Z}}/\mathbb{Z})) = 0$ for $i \geq 0$, we obtain

$$H^1(G_k/H, (A')^H) = H^1(G_k/H, (A^D)^H),$$

and from this follows $(*)$, by taking direct limits. Finally, the finiteness of $H^1(k, A') \cong H^1(k, A^D)$ follows from (7.2.9). □

The above duality theorem contains local class field theory as a special case. Namely, taking $i = 0$ and $A = \mathbb{Z}$, we get an isomorphism

$$\hat{k}^\times = H^0(k, D_2) \xrightarrow{\sim} H^2(G_k, \mathbb{Z})^\vee \cong H^1(G_k, \mathbb{Q}/\mathbb{Z})^\vee = G_k^{ab},$$

and from this we get the following theorem.

(7.2.11) Theorem. *Let k be a local field. Then there is an exact sequence*

$$0 \longrightarrow k^\times \xrightarrow{(\ ,k)} G_k^{ab} \longrightarrow \hat{\mathbb{Z}}/\mathbb{Z} \longrightarrow 0,$$

where $(\ ,k)$ is the **norm residue symbol** *of local class field theory.*

The homomorphism $(\ ,k)$ is by definition (cf. (3.1.6)) characterized by the

(7.2.12) Proposition. *For every $\chi \in H^1(G_k, \mathbb{Q}/\mathbb{Z})$ the norm residue symbol satisfies the formula*

$$\chi((a,k)) = inv(a \cup \delta\chi),$$

where $\delta : H^1(G_k, \mathbb{Q}/\mathbb{Z}) \xrightarrow{\sim} H^2(G_k, \mathbb{Z})$.

We have seen that there are two ways to construct abelian extensions of exponent m of the ground field k, where m is prime to char(k). First, using Kummer theory (if μ_m is contained in k), we obtain an isomorphism

$$k^\times/k^{\times m} \xrightarrow[\sim]{\delta} H^1(k,\mu_m).$$

Secondly, class field theory yields

$$k^\times/k^{\times m} \xrightarrow[\sim]{rec} G_k^{ab}/m.$$

We want to compare these maps. Using (1.4.6) for the exact sequence

$$0 \longrightarrow \mathbb{Z} \xrightarrow{m} \mathbb{Z} \longrightarrow \mathbb{Z}/m\mathbb{Z} \longrightarrow 0$$

and the G_k-module $\bar{k}^\times$ (observe that the sequence $0 \to \mu_m \to \bar{k}^\times \xrightarrow{m} \bar{k}^\times \to 0$ is also exact), we get the commutative diagram

$$\begin{array}{ccccc} H^0(G,\bar{k}^\times) & \times & H^2(G,\mathbb{Z}) & \xrightarrow{\cup} & H^2(G,\bar{k}^\times) \\ \delta\downarrow & & \uparrow\delta & & \downarrow -id \\ H^1(G,\mu_m) & \times & H^1(G,\mathbb{Z}/m\mathbb{Z}) & \xrightarrow{\cup} & H^2(G,\bar{k}^\times). \end{array}$$

From (7.2.12) it follows that

$$\chi((a,k)) = inv(a \cup \delta\chi) = inv(-\delta(a) \cup \chi)$$

for $a \in H^0(G,\bar{k}^\times)$ and $\chi \in H^1(G,\mathbb{Z}/m\mathbb{Z}) \subseteq H^1(G,\mathbb{Q}/\mathbb{Z})$. Thus we obtain

(7.2.13) Corollary. *Let $m \in \mathbb{N}$ be not divisible by the characteristic of k. Then the diagram*

$$\begin{array}{ccccc} G_k^{ab}/m & \times & \mathrm{Hom}(G_k,\mathbb{Z}/m\mathbb{Z}) & \xrightarrow{\cup} & \mathbb{Z}/m\mathbb{Z} \\ \wr\uparrow rec & & \| & & \\ k^\times/k^{\times m} & & \| & & \wr\uparrow inv \\ \wr\downarrow -\delta & & \| & & \\ H^1(k,\mu_m) & \times & H^1(k,\mathbb{Z}/m\mathbb{Z}) & \xrightarrow{\cup} & H^2(k,\mu_m) \end{array}$$

is commutative. The cup-product on top is given by applying a character $\chi \in \mathrm{Hom}(G_k,\mathbb{Z}/m\mathbb{Z})$ to an element of G_k^{ab}/m.

An important addendum to the local duality theorem arises from the presence of the *maximal unramified extension* $\tilde{k}|k$. Let $\Gamma = G(\tilde{k}|k)$ be its Galois group and $T = G(\bar{k}|\tilde{k})$ the inertia group. Let $\tilde{\mathcal{O}}$ be the valuation ring of $\tilde{k}$ and observe that $\tilde{\mathcal{O}}^\times$ is a G_k-submodule of D_2. A G_k-module M is called **unramified** if $M^T = M$.

Let A be a G_k-module which is finitely generated as a $\mathbb{Z}$-module. We assume that the order of the $\mathbb{Z}$-torsion subgroup $\mathrm{tor}(A)$ of A is prime to the characteristic of the residue field of k. If A is unramified, we write

$$A^d = \mathrm{Hom}(A, \tilde{\mathcal{O}}^\times),$$

which is a submodule of $A^D = \mathrm{Hom}(A, D_2)$. Obviously, A^d is also unramified. If A is finite and unramified, then

$$A^D = A' = A^d = \mathrm{Hom}(A, \mu)\,.$$

(7.2.14) Definition. *For a G_k-module A, we set*

$$H^i_{nr}(k, A) = \mathrm{im}(H^i(\Gamma, A^T) \longrightarrow H^i(G_k, A))$$

and these groups are called the **unramified cohomology groups** . *)

If A is unramified, then we have

$$H^0_{nr}(k, A) = H^0(k, A), \quad H^0_{nr}(k, A^d) \subseteq H^0(k, A')\,,$$

and if in addition A is finite, then

$$H^2_{nr}(k, A) = H^2_{nr}(k, A^d) = 0$$

as $cd\,\Gamma = 1$.

(7.2.15) Theorem. *Let A be an unramified G_k-module which is finitely generated as a $\mathbb{Z}$-module and assume that $\#\mathrm{tor}(A)$ is prime to the characteristic of the residue field of k. Then the groups $H^i_{nr}(k, A^d)$ and $H^{2-i}_{nr}(k, A)$ annihilate each other in the pairing*

$$H^i(k, A^D) \times H^{2-i}(k, A) \xrightarrow{\cup} H^2(k, \bar{k}^\times) \hookrightarrow \mathbb{Q}/\mathbb{Z}\,.$$

They are mutually orthogonal complements for $i = 1$, and for $0 \le i \le 2$ if A is finite.

Proof: The Γ-module $\tilde{\mathcal{O}}^\times$ is cohomologically trivial by (7.1.2)(i) and the commutative diagram

$$\begin{array}{ccccc}
H^i(\Gamma, A^d) & \times & H^{2-i}(\Gamma, A) & \xrightarrow{\cup} & H^2(\Gamma, \tilde{\mathcal{O}}^\times) = 0 \\
\downarrow & & \downarrow & & \downarrow \\
H^i(k, A^D) & \times & H^{2-i}(k, A) & \xrightarrow{\cup} & H^2(\bar{k}, D_2) \hookrightarrow \mathbb{Q}/\mathbb{Z}
\end{array}$$

*) The letters nr stand for “non ramifié”.

shows that the images $H^i_{nr}(k, A^d)$ and $H^{2-i}_{nr}(k, A)$ of the upper groups are orthogonal. If A is finite, then

$$H^0_{nr}(k, A^d) = H^0(k, A^D) \quad \text{and} \quad H^2_{nr}(k, A) = 0$$

recalling that $cd\,\Gamma = 1$,

$$H^2_{nr}(k, A^d) = 0 \quad \text{and} \quad H^0_{nr}(k, A) = H^0(k, A).$$

We have therefore to investigate only the case $i = 1$ for a G_k-module A which is finitely generated as a $\mathbb{Z}$-module. Using the exact sequence

$$0 \longrightarrow \mathrm{tor}(A) \longrightarrow A \longrightarrow A/\mathrm{tor} \longrightarrow 0,$$

we obtain a commutative diagram

$$\begin{array}{ccc} H^1(k, \mathrm{tor}(A)) & \twoheadrightarrow & H^1(T, \mathrm{tor}(A))^\Gamma \\ \downarrow & & \downarrow \wr \\ H^1(k, A) & \longrightarrow & H^1(T, A)^\Gamma, \end{array}$$

where the upper map is surjective, since $cd\,\Gamma = 1$, and the map on the right-hand side is an isomorphism, since A is a trivial T-module and therefore $H^1(T, A/\mathrm{tor}) = \mathrm{Hom}(T, A/\mathrm{tor}) = 0$. Thus in the commutative diagram

$$\begin{array}{ccccccccc} 0 & \longrightarrow & H^1(\Gamma, A) & \longrightarrow & H^1(k, A) & \longrightarrow & H^1(T, A)^\Gamma & \longrightarrow & 0 \\ & & & & \wr\downarrow & & \downarrow \varphi_A & & \\ & & & & H^1(k, A^D)^* & \longrightarrow & H^1(\Gamma, A^d)^* & \longrightarrow & 0 \end{array}$$

the upper sequence is exact. The map on the bottom is surjective, since

$$H^1(\Gamma, A^d) \hookrightarrow H^1(\Gamma, (A^D)^T) \xrightarrow{inf} H^1(k, A^D)$$

is injective. Since we already saw that $H^1(\Gamma, A)$ and $H^1(\Gamma, A^d)$ annihilate each other, we obtain the dotted map φ_A, which is necessarily surjective. We have to show that φ_A is bijective.

In the commutative diagram

$$\begin{array}{ccc} H^1(T, \mathrm{tor}(A))^\Gamma & \xrightarrow{\sim} & H^1(T, A)^\Gamma \\ \downarrow \varphi_{\mathrm{tor}(A)} & & \downarrow \varphi_A \\ H^1(\Gamma, (\mathrm{tor}(A))^d)^* & \hookrightarrow & H^1(\Gamma, A^d)^*, \end{array}$$

the lower map is injective, which one sees as follows: let Γ' be an open subgroup of Γ acting trivially on A/tor. Because $scd\,\Gamma = 2$, we have a surjection

$$H^2(\Gamma', \mathrm{Hom}(A/\mathrm{tor}, \tilde{\mathcal{O}}^\times)) \xrightarrow{cor} H^2(\Gamma, \mathrm{Hom}(A/\mathrm{tor}, \tilde{\mathcal{O}}^\times)) \longrightarrow 0$$

and the group on the left is zero, since $\tilde{\mathcal{O}}^\times$ is cohomologically trivial. Thus $H^2(\Gamma,(A/\text{tor})^d)=0$ and we are reduced to the case of a finite module: if $\varphi_{\text{tor}(A)}$ is injective, then φ_A is also injective.

Since the order of $\text{tor}(A)$ is prime to the residue characteristic p, we can replace T by its tame part T_0. Using Kummer theory, we get a natural isomorphism

$$T_0 \cong \varprojlim_{(n,p)=1} \mu_n$$

(cf. the remarks at the beginning of §5). As an abelian group, T_0 is isomorphic to $\prod_{\ell\neq p}\mathbb{Z}_\ell$ and is therefore a duality group of dimension 1 for torsion modules of order prime to p with dualizing module $I\cong\prod_{\ell\neq p}\mathbb{Q}_\ell/\mathbb{Z}_\ell$. As a G_k-module, I is naturally isomorphic to the prime-to-p part of the module μ of all roots of unity. Therefore we obtain

$$H^1(T_0,\text{tor}(A))\cong H^0(T_0,\text{Hom}(\text{tor}(A),\mu))^* = \text{Hom}(\text{tor}(A),\mu)^*,$$

so that, using $H^1(T,\text{tor}(A))=H^1(T_0,\text{tor}(A))$, we get

$$H^1(T,\text{tor}(A))^\Gamma \cong (\text{Hom}(\text{tor}(A),\mu)_\Gamma)^* = H^1(\Gamma,(\text{tor}(A))^d)^*\,.$$

This finishes the proof of the theorem. □

If A is an unramified G_k-module which is finitely generated and free as a $\mathbb{Z}$-module, then we can consider A^d as a torus over $\mathcal{O}_k$ with character group A. Using the technique of smooth base change, the next corollary can also be derived (at least for the part prime to the residue characteristic) from a theorem of *S.* LANG which asserts that a connected algebraic group over a finite field is cohomologically trivial.

(7.2.16) Corollary. *Let A be an unramified G_k-module which is finitely generated and free as a $\mathbb{Z}$-module. Then*

$$H^i(\Gamma,A^d)=0 \quad \textit{for all} \quad i\geq 1\,.$$

Proof: This is clear for $i\geq 3$ and we saw the assertion for $i=2$ in the proof of (7.2.15). Finally,

$$H^1(\Gamma,A^d)\cong(H^1(T,A)^\Gamma)^* = \text{Hom}_\Gamma(T,A)^*=0. \qquad \square$$

For the cohomology theory of global fields, we will need also the following duality theorem for the extension $\mathbb{C}|\mathbb{R}$.

(7.2.17) Theorem. *Let $G = G(\mathbb{C}|\mathbb{R})$. Let A be a finitely generated G-module and set $A' = \mathrm{Hom}(A, \mathbb{C}^\times)$. Then $H^2(G, \mathbb{C}^\times) \cong \frac{1}{2}\mathbb{Z}/\mathbb{Z}$, and the cup-product*

$$\hat{H}^i(G, A') \times \hat{H}^{2-i}(G, A) \xrightarrow{\cup} H^2(G, \mathbb{C}^\times)$$

induces for all $i \in \mathbb{Z}$ an isomorphism

$$\hat{H}^i(G, A') \xrightarrow{\sim} \hat{H}^{2-i}(G, A)^*.$$

Proof: The exact sequence $1 \to \mu_2 \to \mathbb{C}^\times \xrightarrow{2} \mathbb{C}^\times \to 1$ yields $H^2(G, \mathbb{C}^\times) \cong H^3(G, \mu_2) \cong H^1(G, \mu_2) \cong \frac{1}{2}\mathbb{Z}/\mathbb{Z}$. The theorem is true for the G-modules

$$(*) \qquad A = \mathbb{Z}/2\mathbb{Z},\ \mathbb{Z},\ \mathbb{Z}(1),$$

where $\mathbb{Z}(1) = \mathbb{Z}$ as abelian group, and the generator σ of G acts as multiplication by -1 (A' is then $\mathbb{C}^\times$ with the action $\sigma z = \bar{z}^{-1}$).

In fact, for $A = \mathbb{Z}$ the cohomology groups $\hat{H}^i(G, A)$ and $\hat{H}^i(G, A')$ are trivial for odd i and the same holds for $A = \mathbb{Z}(1)$ with even i. The cohomology groups that occur in all other cases have order 2. If x and y are the nontrivial elements of $\hat{H}^i(G, A')$ and $\hat{H}^{2-i}(G, A)$, then a direct elementary computation (and then using the periodicity for the cohomology of cyclic groups) shows that $x \cup y$ is the nontrivial element of $H^2(G, \mathbb{C}^\times)$. This proves the theorem for the modules $(*)$.

Suppose we are given an exact sequence $0 \to B \to A \to C \to 0$ of finitely generated G-modules. Setting $\hat{H}^i(M) = \hat{H}^i(G, M)$, we obtain a diagram

$$\begin{array}{ccccccccc}
\hat{H}^{i-1}(B') & \longrightarrow & \hat{H}^i(C') & \longrightarrow & \hat{H}^i(A') & \longrightarrow & \hat{H}^i(B') & \longrightarrow & \hat{H}^{i+1}(C') \\
\downarrow{\scriptstyle\alpha} & & \downarrow{\scriptstyle\beta} & & \downarrow{\scriptstyle\gamma} & & \downarrow{\scriptstyle\delta} & & \downarrow{\scriptstyle\varepsilon} \\
\hat{H}^{2-i+1}(B)^* & \rightarrow & \hat{H}^{2-i}(C)^* & \rightarrow & \hat{H}^{2-i}(A)^* & \rightarrow & \hat{H}^{2-i}(B)^* & \rightarrow & \hat{H}^{2-i-1}(C)^*
\end{array}$$

with exact rows. The partial diagrams are commutative or anti-commutative. The five-lemma now implies that the assertion is true for A if it holds for B and C.

If we want to prove the theorem for an arbitrary finitely generated G-module A, we may assume that its torsion submodule is 2-primary. Since the theorem is true for $\mathbb{Z}/2\mathbb{Z}$, which is the only finite simple 2-primary G-module, we therefore obtain the result for finite G-modules.

Now consider the general case. First of all, we may assume the module A to be torsion-free. Let $0 \neq a \in A$ be arbitrary. If $a + \sigma a \neq 0$, it generates a submodule isomorphic to $\mathbb{Z}$ and otherwise $a = -\sigma a$ generates a submodule isomorphic to $\mathbb{Z}(1)$. Therefore the general case follows by induction on the $\mathbb{Z}$-rank of the module A. $\square$

We finish this section by fixing conventions in the archimedean case. Firstly, we consider the extension $\mathbb{C}|\mathbb{R}$ as ramified, hence the inertia group is the full group:

$$T(\mathbb{C}|\mathbb{R}) = G(\mathbb{C}|\mathbb{R}).$$

We therefore have $H^i_{nr}(K, A) = 0$ for all $i \geq 0$, any archimedean local field K and any $G(\bar{K}|K)$-module A. Furthermore, we put $U_K = K^\times$ and

$$U_K^1 = \begin{cases} \mathbb{R}_+^\times & \text{if } K = \mathbb{R}, \\ \mathbb{C}^\times & \text{if } K = \mathbb{C}. \end{cases}$$

Exercise 1. Let k be a local field and let $\tilde{k}|k$ be the maximal unramified extension. Let $d : G_k \to \hat{\mathbb{Z}}$ be the surjective homomorphism coming from the canonical isomorphism $G(\tilde{k}|k) \cong \hat{\mathbb{Z}}$.

Show that the Weil group W_k of the class formation $(G_k, \bar{k}^\times)$ (see III §1) is the pre-image $W_k = d^{-1}(\mathbb{Z})$.

Hint: For every finite Galois extension $K|k$ we have a commutative exact diagram

$$\begin{array}{ccccccccc} 1 & \longrightarrow & K^\times & \longrightarrow & W(K|k) & \longrightarrow & G(K|k) & \longrightarrow & 1 \\ & & \downarrow{\scriptstyle\rho} & & \downarrow{\scriptstyle g} & & \| & & \\ 1 & \longrightarrow & G(\tilde{K}|K) & \longrightarrow & G(\tilde{K}|k) & \longrightarrow & G(K|k) & \longrightarrow & 1, \end{array}$$

where $\rho(a) = \varphi_K^{v_K(a)}$. Show that the image of g is the pre-image $d^{-1}(\mathbb{Z})$ under $d : G(\tilde{K}|k) \to \hat{\mathbb{Z}}$, and then pass to the limit.

Exercise 2. Let k be a local field, let $\tilde{k}$ be the maximal unramified extension of k and let $T = G(\bar{k}|\tilde{k})$ be the inertia subgroup of G_k. Let A be a finite G_k-module of order prime to the residue characteristic of k (not necessarily unramified). Show that

$$H^1_{nr}(k, A)^\perp = H^1_{nr}(k, A')$$

with respect to the Tate-pairing.

§3. The Local Euler-Poincaré Characteristic

Let k be a local field and let p be its residue characteristic. If A is a finite G_k-module of order prime to char(k) (in the case char(k) > 0), then the cohomology groups $H^i(k, A)$ are finite groups by (7.1.8)(iii). We set $h^i(k, A) = \#H^i(k, A)$ and define the **Euler-Poincaré characteristic** of A by

$$\chi(k, A) = \prod_i h^i(k, A)^{(-1)^i} = \frac{h^0(k, A)h^2(k, A)}{h^1(k, A)}.$$

If $0 \to A \to B \to C \to 0$ is an exact sequence of finite G_k-modules of order prime to char(k), then the alternating product of the orders of the groups in the exact cohomology sequence

$$\ldots \longrightarrow H^i(k,A) \longrightarrow H^i(k,B) \longrightarrow H^i(k,C) \longrightarrow H^{i+1}(k,A) \longrightarrow \ldots$$

is 1, and from this it follows that

$$\chi(k,B) = \chi(k,A)\chi(k,C).$$

Our aim is to prove the

(7.3.1) Theorem (TATE). *For every finite G_k-module A of order a prime to* char(k) *we have*

$$\chi(k,A) = \|a\|_k\,,$$

*where $\| \;\|_k$ is the normalized absolute value of k.**)

The formula is simple, but the proof is surprisingly difficult. It mirrors an explicit description of the multiplicative group $K^\times$ of a finite Galois extension $K|k$ as a $G(K|k)$-module.

Let $G = G(K|k)$ and let ℓ be a prime number $\neq$ char(k). We consider the *Grothendieck group* $K_0'(\mathbb{F}_\ell[G])$ of finite $\mathbb{F}_\ell[G]$-modules (cf. [32], chap.II, §16b):

Let $F(G)$ be the free abelian group generated by the set of isomorphism classes of finite $\mathbb{F}_\ell[G]$-modules. Denoting the isomorphism class of A by $\{A\}$, let $R(G)$ be the subgroup in $F(G)$ generated by all elements

$$\{B\} - \{A\} - \{C\}$$

arising from short exact sequences $0 \to A \to B \to C \to 0$. Then $K_0'(\mathbb{F}_\ell[G])$ is defined as the quotient $F(G)/R(G)$. We denote by $[A]$ the class of $\{A\}$ in $K_0'(\mathbb{F}_\ell[G])$. $K_0'(\mathbb{F}_\ell[G])$ becomes a ring by linear extension of the product $[A][B] = [A \otimes_{\mathbb{F}_\ell} B]$.

If A is a finite $\mathbb{F}_\ell[G]$-module, then we can view it as a G_k-module which becomes a trivial Galois module over K, and the cohomology groups $H^i(K,A)$ are also finite $\mathbb{F}_\ell[G]$-modules. We define

$$h(K,A) = \sum_{i=0}^{2} (-1)^i [H^i(K,A)],$$

which is an element in $K_0'(\mathbb{F}_\ell[G])$. Theorem (7.3.1) will be a consequence of the following theorem, which is a sharpening of it.

*) That is, $\|x\|_k = q^{-v(x)}$, where q is the cardinality of the residue field κ and v is the additive valuation of k with value group $\mathbb{Z}$.

(7.3.2) Theorem (SERRE). *Let A be a finite $\mathbb{F}_\ell[G]$-module. If ℓ is not equal to the residue characteristic p of k, then $h(K,A)=0$. If $\ell=p$ and* char$(k)=0$, *then*

$$h(K,A) = -\dim_{\mathbb{F}_p}(A)[k:\mathbb{Q}_p]\big[\mathbb{F}_p[G]\big].$$

For the proof of Serre's theorem we need the following

(7.3.3) Lemma.

(i) *Let A be a finite $\mathbb{Z}_\ell[G]$-module. Then*

$$[{}_\ell A] = [A_\ell]\,.\ ^{*)}$$

(ii) *Let V be a $\mathbb{Z}_\ell[G]$-module such that V_ℓ and ${}_\ell V$ are finite. If $W\subseteq V$ is a submodule of finite index, then*

$$[V_\ell]-[{}_\ell V] = [W_\ell]-[{}_\ell W]\,.$$

Proof: We prove (ii) first. Using a filtration of V/W by $\mathbb{Z}_\ell[G]$-modules such that the subquotients are $\mathbb{F}_\ell[G]$-modules, we may assume that V/W is an $\mathbb{F}_\ell[G]$-module. Consider the diagram

$$\begin{array}{ccccccccc}
0 & \longrightarrow & W & \longrightarrow & V & \longrightarrow & V/W & \longrightarrow & 0\\
 & & \Big\downarrow{\scriptstyle\ell} & & \Big\downarrow{\scriptstyle\ell} & & \Big\downarrow{\scriptstyle\ell} & & \\
0 & \longrightarrow & W & \longrightarrow & V & \longrightarrow & V/W & \longrightarrow & 0\,.
\end{array}$$

The snake lemma gives the exact sequence

$$0\longrightarrow {}_\ell W\longrightarrow {}_\ell V\longrightarrow V/W\longrightarrow W_\ell\longrightarrow V_\ell\longrightarrow V/W\longrightarrow 0,$$

and hence the result. Assertion (i) follows by applying (ii) to $V=A$, $W=0$.

□

Proof of (7.3.2): First suppose that $A=\mu_\ell$. Using the Kummer sequence, we get

$$\begin{aligned}
H^0(K,\mu_\ell) &= \mu_\ell(K),\\
H^1(K,\mu_\ell) &= K^\times/K^{\times\ell},\\
H^2(K,\mu_\ell) &= H^0(K,\mathbb{Z}/\ell\mathbb{Z})^* = \mathbb{Z}/\ell\mathbb{Z}.
\end{aligned}$$

Therefore

$$h(K,\mu_\ell) = [\mu_\ell(K)] - [K^\times/K^{\times\ell}] + [\mathbb{Z}/\ell\mathbb{Z}].$$

---

*) For a $\mathbb{Z}_\ell$-module V, the modules V_ℓ and ${}_\ell V$ are defined by the exact sequence $0\to {}_\ell V\to V\xrightarrow{\ell} V\to V_\ell\to 0$.

Let U denote the group of units of the valuation ring $\mathcal{O}$ of K. The exact sequence

$$0 \longrightarrow U/U^{\ell} \longrightarrow K^{\times}/K^{\times \ell} \longrightarrow \mathbb{Z}/\ell\mathbb{Z} \longrightarrow 0$$

gives us $[K^{\times}/K^{\times\ell}] = [\mathbb{Z}/\ell\mathbb{Z}] + [U/U^{\ell}]$, hence

(1) $$h(K, \mu_\ell) = [\mu_\ell(K)] - [U/U^{\ell}] = [{}_\ell U] - [U_\ell].$$

Let $\mathfrak{p}$ be the maximal ideal of $\mathcal{O}$ and $V = 1 + \mathfrak{p}$ the group of principal units. V is a finitely generated $\mathbb{Z}_p[G]$-module, and we have the exact sequence

$$1 \longrightarrow V \longrightarrow U \longrightarrow \kappa^{\times} \longrightarrow 1,$$

with $\kappa^{\times}$ the multiplicative group of the residue field of K. Since $\kappa^{\times}$ is finite we obtain from (1) and (7.3.3)

(2) $$h(K, \mu_\ell) = [{}_\ell U] - [U_\ell] = [{}_\ell V] - [V_\ell].$$

Now if $\ell \neq p$, then ${}_\ell V = V_\ell = \{1\}$ since V is a pro-p-group, hence $h(K, \mu_\ell) = 0$.

So let $\ell = p$ and $\mathrm{char}(k) = 0$. Let $W = 1 + \mathfrak{p}^n$ be the group of n-th principal units. It is a $\mathbb{Z}_p[G]$-submodule of finite index in V, hence by lemma (7.3.3) again, we obtain from (2)

$$h(K, \mu_\ell) = [{}_p W] - [W_p].$$

If n is sufficiently large, then the logarithm $\log : W \to \mathfrak{p}^n$ is an isomorphism (see [160], chap.II, (5.5)) and, since $\mathfrak{p}^n$ is a subgroup of finite index in $\mathcal{O}$, we get

(3) $$h(K, \mu_\ell) = [{}_p\mathcal{O}] - [\mathcal{O}/p\mathcal{O}].$$

The extension $K|k$ has a normal basis, $K = \bigoplus_{\sigma \in G} k\sigma\theta$, and we may choose θ in $\mathcal{O}$. If R denotes the valuation ring of k, then $M = \bigoplus_{\sigma \in G} R\sigma\theta$ is a $\mathbb{Z}_p[G]$-submodule of $\mathcal{O}$ of finite index and ${}_pM = 0$ and $M/pM \cong R/pR[G] \cong \mathbb{F}_p[G]^{[k:\mathbb{Q}_p]}$. We now obtain from (3) via (7.3.3)

$$h(K, \mu_\ell) = -[M/pM] = -[k : \mathbb{Q}_p][\mathbb{F}_p[G]].$$

Now let A be an arbitrary finite $\mathbb{F}_\ell[G]$-module. The cup-product on the cochain groups $C^i(K, \mathbb{Z}/\ell\mathbb{Z}) \otimes A \to C^i(K, A)$ is obviously an isomorphism, hence we have an isomorphism of G-modules $H^i(K, \mathbb{Z}/\ell\mathbb{Z}) \otimes A \cong H^i(K, A)$, and so

$$h(K, A) = h(K, \mathbb{Z}/\ell\mathbb{Z}) \cdot [A].$$

The functor $M \mapsto M^* = \mathrm{Hom}(M, \mathbb{F}_\ell)$ is exact on finite $\mathbb{F}_\ell[G]$-modules and thus defines an endomorphism $\xi \mapsto \xi^*$ of the group $K_0'(\mathbb{F}_\ell[G])$. By the duality theorem (7.2.6), we have

$$h(K, \mathbb{Z}/\ell\mathbb{Z})^* = h(K, \mu_\ell),$$

and therefore, since $[\mathbb{F}_\ell[G]] = [\mathbb{F}_\ell[G]^*]$,

$$h(K,A) = \begin{cases} -[k:\mathbb{Q}_p][\mathbb{F}_p[G]]\cdot[A], & \text{if } p=\ell\,,\\ 0\,, & \text{if } p\neq\ell\,. \end{cases}$$

Furthermore, if A_0 denotes the trivial G-module with underlying group A, then $\mathbb{F}_\ell[G]\otimes A_0 \to \mathbb{F}_\ell[G]\otimes A$, $\sigma\otimes a \mapsto \sigma\otimes\sigma a$, is an isomorphism of $\mathbb{F}_\ell[G]$-modules (this is just the dual statement to $\mathrm{Ind}_G^{\{1\}}A \cong \mathrm{Ind}_G A$, see I §6, p.62), so that

$$[\mathbb{F}_\ell[G]]\cdot[A] = [\mathbb{F}_\ell[G]]\cdot[A_0] = \dim_{\mathbb{F}_\ell}(A)[\mathbb{F}_\ell[G]].$$

This finishes the proof of theorem (7.3.2). □

For the deduction of Tate's formula

$$(*) \qquad \chi(A) = \|a\|_k\,,$$

we need from the representation theory of finite groups another

(7.3.4) Lemma. *The group $K_0'(\mathbb{F}_\ell[G])\otimes\mathbb{Q}$ is generated by the images of $K_0'(\mathbb{F}_\ell[H])\otimes\mathbb{Q}$ under Ind_G^H, where H runs through all cyclic subgroups of G of order prime to ℓ.*

Proof: A theorem of E. Artin (see [214], 12.5 th. 26) asserts that the map

$$\mathrm{Ind}\otimes\mathbb{Q} : \bigoplus_{H\in\mathcal{T}} K_0'(\mathbb{Q}_\ell[H])\otimes\mathbb{Q} \longrightarrow K_0'(\mathbb{Q}_\ell[G])\otimes\mathbb{Q}$$

is surjective where $\mathcal{T}$ is the set of all cyclic subgroups of G. (This holds with any field of characteristic zero replacing $\mathbb{Q}_\ell$.) By [214], 16.1, th. 33, there exists a surjective homomorphism $K_0'(\mathbb{Q}_\ell[G])\otimes\mathbb{Q} \twoheadrightarrow K_0'(\mathbb{F}_\ell[G])\otimes\mathbb{Q}$ which is natural with respect to the group G. Therefore the above map $\mathrm{Ind}\otimes\mathbb{Q}$ remains surjective if we replace $\mathbb{Q}_\ell$ by $\mathbb{F}_\ell$. Let $H = H^{(\ell')}\times H_\ell$ be a cyclic subgroup of G where H_ℓ is the ℓ-Sylow subgroup of H. Let M be a simple $\mathbb{F}_\ell[H]$-module. By (1.6.13), $M^{H_\ell}\neq 0$. Since M^{H_ℓ} is also an H-module, we obtain $M = M^{H_\ell}$, hence an isomorphism of $\mathbb{F}_\ell[H]$-modules $\mathrm{Ind}_H^{H^{(\ell')}}\mathrm{Res}_{H^{(\ell')}}^H M \cong M\otimes_{\mathbb{Z}}\mathbb{F}_\ell[H_\ell]$. Since $\mathbb{Z}/\ell\mathbb{Z}$ is the only simple $\mathbb{F}_\ell[H_\ell]$-module, the class of $\mathrm{Ind}_H^{H^{(\ell')}}\mathrm{Res}_{H^{(\ell')}}^H M$ in $K_0'(\mathbb{F}_\ell[H])$ is equal to $n[M]$, where $\#H_\ell = \ell^n$. Therefore the images of $K_0'(\mathbb{F}_\ell[H])\otimes\mathbb{Q}$ and $K_0'(\mathbb{F}_\ell[H^{(\ell')}])\otimes\mathbb{Q}$ under $\mathrm{Ind}\otimes\mathbb{Q}$ in $K_0'(\mathbb{F}_\ell[G])\otimes\mathbb{Q}$ are the same. □

Proof of theorem (7.3.1): Let A be a finite G_k-module of order a. We may assume $\ell A = 0$ for some prime number $\ell\neq\mathrm{char}(k)$. The general case follows from this via the exact sequence $0\to {}_\ell A\to A\to A/_\ell A\to 0$ by induction on the order of A, since χ and $\|\ \|_k$ are multiplicative with respect to short exact sequences. Every finite G_k-module becomes a trivial Galois module over some finite Galois extension $K|k$. It thus suffices to prove the formula $(*)$ for finite $\mathbb{F}_\ell[G]$-modules with $G = G(K|k)$. The functions $\chi(k, A)$ and $\varphi(A) = \|a\|_k$ are *additive*, i.e. they define homomorphisms

$$\chi, \varphi : K_0'(\mathbb{F}_\ell[G]) \longrightarrow \mathbb{Q}_+^\times ,$$

and we have to show $\chi = \varphi$. Using lemma (7.3.4) and observing that $\mathbb{Q}_+^\times$ is torsion-free, it suffices to check this equality on elements of the form $B = \mathrm{Ind}_G^H(A)$, where A is a finite $\mathbb{F}_\ell[H]$-module and H is a cyclic subgroup of G of order prime to ℓ. Let k' be the fixed field of H. Shapiro's lemma yields

$$\chi(k, B) = \chi(k', A)$$

and for $b = \#B$ we have

$$\|a\|_{k'} = \|a\|_k^{[k':k]} = \|b\|_k .$$

This reduces our problem to the case where G is a cyclic group of order prime to ℓ. Then the Hochschild-Serre spectral sequence for the group extension

$$1 \to G_K \to G_k \to G \to 1$$

and the G-module A degenerates. Thus $H^i(k, A) = H^0(G, H^i(K, A))$. Therefore if

$$d : K_0'(\mathbb{F}_\ell[G]) \to \mathbb{Z}$$

is the homomorphism given by

$$d([M]) = \dim_{\mathbb{F}_\ell}(H^0(G, M)),$$

then

$$\chi(k, A) = \ell^{d(h(K,A))} .$$

If $\ell \neq p$, then by (7.3.2) $h(K, A) = 0$, so that $\chi(k, A) = 1$, and if $\ell = p$, then $h(K, A) = -\dim(A)[k : \mathbb{Q}_p][\mathbb{F}_p[G]]$, and since $d([\mathbb{F}_p[G]]) = \dim(\mathbb{F}_p[G]^G) = \dim(\mathbb{F}_p) = 1$, we obtain

$$\chi(k, A) = p^{-[k:\mathbb{Q}_p]\dim A} = \|a\|_k . \qquad \square$$

For the fields $\mathbb{R}$ and $\mathbb{C}$ we have a similar statement.

(7.3.5) Theorem. *If $k = \mathbb{R}$ or $\mathbb{C}$, then for any finite G_k-module A of order a we have*

$$\frac{h^0(k, A) h^0(k, A')}{h^1(k, A)} = \|a\|_k ,$$

where $\|x\|_{\mathbb{R}} = |x|$ and $\|x\|_{\mathbb{C}} = |x|^2$.

Proof: If $k = \mathbb{C}$, then $H^0(k,A) = A$ and $H^0(k,A') = A'$ both have order a, $H^1(k,A) = 0$ and $\|a\|_{\mathbb{C}} = a^2$. Let $k = \mathbb{R}$ and $G = G(\mathbb{C}|\mathbb{R})$, and let σ be the generator of G. For $x \in A$ and $f \in A'$, we have

$$((1-\sigma)f)(x) = f(x)/\sigma(f(\sigma x)) = f(x)(f(\sigma x)) = f((1+\sigma)x),$$

noting that $\sigma\zeta = \zeta^{-1}$ for a root of unity in $\mathbb{C}$. Therefore $1-\sigma : A' \to A'$ is adjoint to $1+\sigma : A \to A$, and so, in the pairing $A' \times A \to \mathbb{C}^\times$, $(A')^G$ and $N_G A$ are exact annihilators. Therefore

$$\|a\|_{\mathbb{R}} = \#A = \#(A')^G \# N_G A = \#H^0(G,A')\#(H^0(G,A)/\hat{H}^0(G,A)).$$

Since A is finite, we have by [160], chap.IV, (7.3) and (1.7.1)

$$\#\hat{H}^0(G,A) = \#\hat{H}^{-1}(G,A) = \#\hat{H}^1(G,A),$$

and the theorem follows. □

We saw in VII §2 that the Galois module μ plays an important role in the case of local fields. Now we introduce the following general terminology. Assume that k is any field. The subgroup μ of roots of unity contained in the separable closure $\bar{k}$ of k is a G_k-module in a natural way. There exists a canonical isomorphism

$$h : \mathrm{Aut}(\mu) \xrightarrow{\sim} \prod_{\ell \neq \mathrm{char}(k)} \mathbb{Z}_\ell^\times,$$

given by $\varphi(\zeta) = \zeta^{h(\varphi)}$. The right side of the equality is defined as follows: if $\zeta^n = 1$, $n \in \mathbb{N}$, then $\zeta^\alpha := \zeta^a$ for any $a \in \mathbb{Z}$ with $a \equiv \alpha \bmod n$. The action of G_k on μ is given by a character

$$\chi_{\mathrm{cycl}} : G_k \to \prod_{\ell \neq \mathrm{char}(k)} \mathbb{Z}_\ell^\times .$$

(7.3.6) Definition. *The character χ_{cycl} is called the* **cyclotomic character**. *Let A be a finite G_k-module whose order is prime to* char(k). *For $i \in \mathbb{Z}$ we denote by $A(i)$ the G_k-module which is equal to A as an abelian group and which is endowed with the (twisted) action*

$$\sigma(a) := \chi_{\mathrm{cycl}}(\sigma)^i \cdot \sigma a,$$

where the action on the right-hand side is the original action of G_k on A. We call $A(i)$ the ***i*-th Tate twist** *of A. We apply the same definition to a discrete resp. compact G-module which is a direct resp. projective limit of finite modules of order prime to* char(k).

In particular, $A = A(0)$ and for $i, j \in \mathbb{Z}$ we have the rule

$$A(i + j) = A(i)(j) = A(j)(i).$$

Note that the above definition can only be applied to modules A such that the multiplication by $\chi_{\mathrm{cycl}}(\sigma)$ is a well-defined automorphism. In particular, the module $\mathbb{Z}(i)$ does not exist (unless we consider archimedean local fields).

For $(n, \mathrm{char}(k)) = 1$ and $i \geq 1$ we have

$$\mathbb{Z}/n\mathbb{Z}\,(i) \cong \mu_n^{\otimes i} \quad \text{and} \quad \mathbb{Z}/n\mathbb{Z}\,(-i) \cong \mathrm{Hom}(\mu_n^{\otimes i}, \mathbb{Z}/n\mathbb{Z}).$$

For a prime number ℓ and $i \in \mathbb{Z}$ we have

$$\mathbb{Z}_\ell(i) = \varprojlim_m \mathbb{Z}/\ell^m\mathbb{Z}(i) \quad \text{and} \quad \mathbb{Q}_\ell/\mathbb{Z}_\ell(i) = \varinjlim_m \mathbb{Z}/\ell^m\mathbb{Z}(i)\,.$$

If A is finite with $nA = 0$, then $A(i) = A \otimes_{\mathbb{Z}} \mathbb{Z}/n\mathbb{Z}\,(i)$ for all $i \in \mathbb{Z}$.

Returning to the case of a local field, we have the

(7.3.7) Theorem. *Let k be a local field with residue characteristic p and let $\ell \neq \mathrm{char}(k)$ be a prime number. Then for all $j \in \mathbb{Z}$*

$$\sum_{i=0}^{2}(-1)^i \dim_{\mathbb{F}_\ell} H^i(G_k, \mathbb{Z}/\ell\mathbb{Z}\,(j)) = \begin{cases} -[k:\mathbb{Q}_p] & \text{if} \quad \ell = p, \\ 0 & \text{otherwise.} \end{cases}$$

Proof: This follows directly from (7.3.1), since

$$\|\ell\|_k = \begin{cases} p^{-[k:\mathbb{Q}_p]} & \text{if} \quad \ell = p, \\ 1 & \text{otherwise.} \end{cases}$$

□

(7.3.8) Corollary. *With the notation as in (7.3.7), the following equalities hold for all $j \in \mathbb{Z}$:*

$$\sum_{i=0}^{2}(-1)^i \mathrm{rank}_{\mathbb{Z}_\ell} H_i(G_k, \mathbb{Z}_\ell(j)) = \begin{cases} -[k:\mathbb{Q}_p] & \text{if} \quad \ell = p, \\ 0 & \text{otherwise,} \end{cases}$$

and

$$\sum_{i=0}^{2}(-1)^i \mathrm{rank}_{\mathbb{Z}_\ell} H^i_{cts}(G_k, \mathbb{Z}_\ell(j)) = \begin{cases} -[k:\mathbb{Q}_p] & \text{if} \quad \ell = p, \\ 0 & \text{otherwise.} \end{cases}$$

Proof: The second equality follows from the first and (2.7.12). In order to prove the first statement, we observe that

$$\mathrm{rank}_{\mathbb{Z}_\ell} H_i(G_k, \mathbb{Z}_\ell(j)) = \dim_{\mathbb{F}_\ell} H_i(G_k, \mathbb{Z}_\ell(j))/\ell - \dim_{\mathbb{F}_\ell} {}_\ell H_i(G_k, \mathbb{Z}_\ell(j))$$

and

$$\dim_{\mathbb{F}_\ell} H_i(G_k, \mathbb{Z}/\ell\mathbb{Z}(j)) = \dim_{\mathbb{F}_\ell} H^i(G_k, \mathbb{Z}/\ell\mathbb{Z}(-j)) .$$

by (2.6.9). The exact sequence

$$0 \longrightarrow \mathbb{Z}_\ell(j) \xrightarrow{\ \ell\ } \mathbb{Z}_\ell(j) \longrightarrow \mathbb{Z}/\ell\mathbb{Z}(j) \longrightarrow 0$$

shows that

$$\begin{aligned} \dim_{\mathbb{F}_\ell} H_0(G_k, \mathbb{Z}/\ell\mathbb{Z}(j)) &= \dim_{\mathbb{F}_\ell} H_0(G_k, \mathbb{Z}_\ell(j))/\ell , \\ \dim_{\mathbb{F}_\ell} H_i(G_k, \mathbb{Z}/\ell\mathbb{Z}(j)) &= \dim_{\mathbb{F}_\ell} H_i(G_k, \mathbb{Z}_\ell(j))/\ell \\ &\quad + \dim_{\mathbb{F}_\ell} {}_\ell H_{i-1}(G_k, \mathbb{Z}_\ell(j)) \end{aligned}$$

for $i \geq 1$. Since $cd_\ell\, G_k = 2$, we obtain

$$\begin{aligned} \sum_{i=0}^{2}(-1)^i \operatorname{rank}_{\mathbb{Z}_\ell} H_i(G_k, \mathbb{Z}_\ell(j)) &= \sum_{i=0}^{\infty}(-1)^i \operatorname{rank}_{\mathbb{Z}_\ell} H_i(G_k, \mathbb{Z}_\ell(j)) \\ &= \sum_{i=0}^{\infty}(-1)^i \dim_{\mathbb{F}_\ell} H^i(G_k, \mathbb{Z}/\ell\mathbb{Z}(-j)) \\ &= \sum_{i=0}^{2}(-1)^i \dim_{\mathbb{F}_\ell} H^i(G_k, \mathbb{Z}/\ell\mathbb{Z}(-j)). \end{aligned}$$

Now the corollary follows from (7.3.7). □

(7.3.9) Corollary. *With the notation as in* (7.3.7), *we have*

$$\dim_{\mathbb{F}_\ell} H^1(G_k, \mathbb{Z}/\ell\mathbb{Z}) = \begin{cases} 1 + \delta + [k : \mathbb{Q}_p] & \text{if} \quad \ell = p , \\ 1 + \delta & \text{otherwise,} \end{cases}$$

where $\delta = 1$ *or* 0 *according to whether the* ℓ*-th roots of unity are contained in* k *or not.*

Proof: By duality, we have $\dim_{\mathbb{F}_\ell} H^2(G_k, \mathbb{Z}/\ell\mathbb{Z}) = \dim_{\mathbb{F}_\ell} H^0(G_k, \mu_\ell) = \delta$. Since $\dim_{\mathbb{F}_\ell} H^0(G_k, \mathbb{Z}/\ell\mathbb{Z}) = 1$, the assertion follows from (7.3.7) with $j = 0$. □

For a prime number $\ell \neq \operatorname{char}(k)$ and $i \in \mathbb{Z}$, $i \neq 0$, we introduce the numbers

$$w_\ell^i := \max\left\{\ell^n \;\middle|\; \exp\left(G(k(\mu_{\ell^n})|k)\right) \text{ divides } i\right\}.$$

Here $\exp(G)$ denotes the exponent of a finite group G. In particular, we have $w_\ell^1 = \#\mu_{\ell^\infty}(k)$ and $w_\ell^i = w_\ell^{-i}$.

Remark: If $k(\mu_{\ell^\infty})|k$ is pro-cyclic (this can only fail if k is a dyadic number field, i.e. a finite extension of $\mathbb{Q}_2$, and $\ell = 2$), then obviously

$$w_\ell^i = \max\left\{\ell^n \;\middle|\; [k(\mu_{\ell^n}) : k] \mid i\right\}.$$

(7.3.10) Proposition. *Assume that k is a finite extension of degree d of $\mathbb{Q}_p$ and let ℓ be a prime number. Then*

$$\text{(i)}\quad H^0(G_k,\mathbb{Q}_\ell/\mathbb{Z}_\ell(i)) \cong \begin{cases} \mathbb{Q}_\ell/\mathbb{Z}_\ell & \text{for } i=0,\\ \mathbb{Z}/w_\ell^i\mathbb{Z} & \text{for } i\neq 0.\end{cases}$$

$$\text{(ii)}\quad H^1(G_k,\mathbb{Q}_\ell/\mathbb{Z}_\ell(i)) \cong \begin{cases} \mathbb{Z}/w_\ell^1\mathbb{Z}\oplus(\mathbb{Q}_\ell/\mathbb{Z}_\ell)^{d+1} & \text{for } i=0, & \ell=p,\\ (\mathbb{Q}_\ell/\mathbb{Z}_\ell)^{d+1} & \text{for } i=1, & \ell=p,\\ \mathbb{Z}/w_\ell^{1-i}\mathbb{Z}\oplus(\mathbb{Q}_\ell/\mathbb{Z}_\ell)^{d} & \text{for } i\neq 0,1, & \ell=p,\\ \mathbb{Z}/w_\ell^1\mathbb{Z}\oplus(\mathbb{Q}_\ell/\mathbb{Z}_\ell) & \text{for } i=0, & \ell\neq p,\\ \mathbb{Q}_\ell/\mathbb{Z}_\ell & \text{for } i=1, & \ell\neq p,\\ \mathbb{Z}/w_\ell^{1-i}\mathbb{Z} & \text{for } i\neq 0,1, & \ell\neq p.\end{cases}$$

$$\text{(iii)}\quad H^2(G_k,\mathbb{Q}_\ell/\mathbb{Z}_\ell(i)) \cong \begin{cases} H^2(G_k,\bar{k}^\times)(\ell)\cong\mathbb{Q}_\ell/\mathbb{Z}_\ell & \text{for } i=1,\\ 0 & \text{for } i\neq 1.\end{cases}$$

Remark: From the proposition above, the continuous cochain cohomology groups with values in $\mathbb{Z}_p(i)$ can be easily calculated by the rule

$$H^j_{cts}(G_k,\mathbb{Z}_\ell(i)) \cong H^{2-j}(G_k,\mathbb{Q}_\ell/\mathbb{Z}_\ell(1-i))^\vee \cong H_{2-j}(G_k,\mathbb{Z}_\ell(i-1)).$$

Proof of (7.3.10): If $i\neq 0$, then

$$\begin{aligned} H^0(G_k,\mathbb{Q}_\ell/\mathbb{Z}_\ell(i)) &= \{\zeta\in\mu_{\ell^\infty}\mid \zeta^{\chi_{\text{cycl}}(\sigma)^i}=\zeta \text{ for all } \sigma\in G_k\}\\ &= \{\zeta\in\mu_{\ell^\infty}\mid \sigma^i\zeta=\zeta \text{ for all } \sigma\in G(k(\mu_{\ell^\infty})|k)\},\end{aligned}$$

and therefore

$$\#H^0(G_k,\mathbb{Q}_\ell/\mathbb{Z}_\ell(i)) = \max\left\{\ell^n \;\middle|\; \sigma^i=1 \text{ for all } \sigma\in G(k(\mu_{\ell^n})|k)\right\} = w_\ell^i.$$

This proves (i) and the assertion for H^2 follows from the result for H^0 and from the local duality theorem (7.2.6):

$$H^2(G_k,\mathbb{Q}_\ell/\mathbb{Z}_\ell(i)) \cong \varinjlim_m H^0(G_k,\mathbb{Z}/\ell^m\mathbb{Z}(1-i))^* \cong \begin{cases}\mathbb{Q}_\ell/\mathbb{Z}_\ell & \text{if } i=1,\\ 0 & \text{otherwise.}\end{cases}$$

Furthermore, from (7.3.8) and the statements for H^0 and H^2, we obtain

$$\operatorname{corank}_{\mathbb{Z}_\ell} H^1(G_k,\mathbb{Q}_\ell/\mathbb{Z}_\ell(i)) = \begin{cases} d+1 & \text{for } i=0,1 & \ell=p,\\ d & \text{for } i\neq 0,1, & \ell=p,\\ 1 & \text{for } i=0,1 & \ell\neq p,\\ 0 & \text{for } i\neq 0,1, & \ell\neq p.\end{cases}$$

It remains to calculate the cotorsion of the group $H^1(G_k,\mathbb{Q}_\ell/\mathbb{Z}_\ell(i))$. Let $m\geq 1$. From the exact sequence $0\to\mathbb{Z}/\ell^m\mathbb{Z}(i)\to\mathbb{Q}_\ell/\mathbb{Z}_\ell(i)\to\mathbb{Q}_\ell/\mathbb{Z}_\ell(i)\to 0$ follows the exact sequence

$$H^1(G_k,\mathbb{Q}_\ell/\mathbb{Z}_\ell(i))/\ell^m \hookrightarrow H^2(G_k,\mathbb{Z}/\ell^m\mathbb{Z}(i)) \twoheadrightarrow {}_{\ell^m}H^2(G_k,\mathbb{Q}_\ell/\mathbb{Z}_\ell(i)).$$

Applying the projective limit, this yields for $i \neq 1$

$$\begin{aligned}\varprojlim_m H^1(G_k, \mathbb{Q}_\ell/\mathbb{Z}_\ell(i))/\ell^m &\cong \varprojlim_m H^2(G_k, \mathbb{Z}/\ell^m\mathbb{Z}(i)) \\ &\cong (\varinjlim_m H^0(G_k, \mathbb{Z}/\ell^m\mathbb{Z}(1-i)))^* \\ &\cong \mathbb{Z}/w_\ell^{1-i}\mathbb{Z},\end{aligned}$$

since $H^2(G_k, \mathbb{Q}_\ell/\mathbb{Z}_\ell(i)) = 0$. If $i = 1$, then the group $H^1(G_k, \mathbb{Q}_\ell/\mathbb{Z}_\ell(1)) \cong k^\times \otimes \mathbb{Q}_\ell/\mathbb{Z}_\ell$ is ℓ-divisible. Therefore

$$\operatorname{tor}\Big(H^1(G_k, \mathbb{Q}_\ell/\mathbb{Z}_\ell(i))^*\Big) \cong \begin{cases} \mathbb{Z}/w_\ell^{1-i}\mathbb{Z} & \text{for } i \neq 1, \\ 0 & \text{for } i = 1. \end{cases}$$

This finishes the proof of (7.3.10). □

§4. Galois Module Structure of the Multiplicative Group

In this section we combine some of the results of chapter V §6 with those of the last sections in order to determine the structure of the p-adic completion of the multiplicative group of a local field and of relation modules of certain extensions of local fields.

(7.4.1) Theorem. *Let k be a p-adic local field and let $n = [k : \mathbb{Q}_p]$. Then the absolute Galois group $\mathscr{G} = G(\bar{k}|k)$ of k is generated by $n + 2$ elements. If*

$$1 \longrightarrow N \longrightarrow F_{n+2} \longrightarrow \mathscr{G} \longrightarrow 1$$

is a presentation of $\mathscr{G} = G(\bar{k}|k)$ by a free profinite group F_{n+2} of rank $n + 2$, then

$$N^{ab}(p) \cong \mathbb{Z}_p[[\mathscr{G}]]$$

as $\mathbb{Z}_p[[\mathscr{G}]]$-modules.

If $K|k$ is a Galois extension of p-adic local fields with Galois group $G = G(K|k)$, then let $\mathscr{H} = G(\bar{k}|K)$. We want to determine the structure of the $\mathbb{Z}_p[[G]]$-module

$$X = \mathscr{H}^{ab}(p) \cong \varprojlim_L A(L) = \varprojlim_{L,m} L^\times/L^{\times p^m},$$

where L runs through all finite subextensions of $K|k$ and the projective limit is taken with respect to the norm maps. $A(L) = \varprojlim_m L^\times/L^{\times p^m}$ is the p-completion of the multiplicative group of the local field L, which is (via the reciprocity map) isomorphic to $G(\bar{k}|L)^{ab}(p)$. We denote the group of roots of unity of p-power order in L by $\mu_{p^\infty}(L)$.

(7.4.2) Theorem. *Let k be a p-adic local field and let $n = [k : \mathbb{Q}_p]$.*

(i) *Let $K|k$ be a Galois extension with $G = G(K|k)$ and let*

$$1 \longrightarrow R_{n+2} \longrightarrow F_{n+2} \longrightarrow G \longrightarrow 1$$

be a presentation of G. If X denotes the $\mathbb{Z}_p[\![G]\!]$-module $G(\bar{k}|K)^{ab}(p)$, then there exists an exact sequence

$$0 \longrightarrow \mathbb{Z}_p[\![G]\!] \longrightarrow R^{ab}_{n+2}(p) \longrightarrow X \longrightarrow 0\,.$$

If $\mu_p \notin K$, then G is generated by $n+1$ elements and there is an isomorphism

$$R^{ab}_{n+1}(p) \cong X\,.$$

(ii) *Let $\sigma_1, \dots, \sigma_{n+2}$ be topological generators of $\mathcal{G} = G(\bar{k}|k)$. Let $a_i \in \mathbb{Z}_p$ with $\sigma_i(\zeta) = \zeta^{a_i}$ for all $\zeta \in \mu_{p^\infty}(\bar{k})$ and let $\bar{\sigma}_i$ be the image of σ_i in $G = G(K|k)$, $i = 1, \dots, n+2$. Then there exists an exact sequence*

$$\begin{array}{ccccccc} 0 \longrightarrow & \mathbb{Z}_p[\![G]\!] & \longrightarrow & \mathbb{Z}_p[\![G]\!]^{n+2} & \longrightarrow Y \longrightarrow 0, \\ & 1 & \longmapsto & (\bar{\sigma}_i - a_i)_i, & \end{array}$$

where $Y = I_{\mathcal{G}}/I_{\mathcal{H}} I_{\mathcal{G}}$, as in V §6. If $\mu_{p^\infty}(K) = 1$, then Y is a free $\mathbb{Z}_p[\![G]\!]$-module of rank $n+1$.

Before we prove the two theorems, which are taken from [97], we provide two lemmas.

(7.4.3) Lemma. *Let k be a local field of characteristic p and let $K|k$ be a finite tamely ramified Galois extension. Then $U^1_K/(U^1_K)^p$ is a cohomologically trivial $G(K|k)$-module, and the norm map induces an isomorphism*

$$\left(U^1_K/(U^1_K)^p\right)_{G(K|k)} \xrightarrow{\sim} U^1_k/(U^1_k)^p\,.$$

Proof: Since the group of principal units U^1_K is $\mathbb{Z}_p$-torsion-free, we have the exact sequence

$$0 \longrightarrow U^1_K \xrightarrow{\ p\ } U^1_K \longrightarrow U^1_K/(U^1_K)^p \longrightarrow 0.$$

By (7.1.2)(ii), U_K^1 is a cohomologically trivial $G(K|k)$-module. It follows that $U_K^1/(U_K^1)^p$ is also cohomologically trivial, and the commutative diagram

$$\begin{array}{ccc} \left(U_K^1/(U_K^1)^p\right)_{G(K|k)} & \xrightarrow[\sim]{N_{G(K|k)}} & \left(U_K^1/(U_K^1)^p\right)^{G(K|k)} \\ & {\scriptstyle N_{K|k}} \searrow \quad \nearrow {\scriptstyle \sim} & \\ & U_k^1/(U_k^1)^p & \end{array}$$

gives the desired result. □

(7.4.4) Lemma. *Let k be a local field of residue characteristic p and let $K|k$ be a finite Galois extension with Galois group $G = G(K|k)$.*

(i) *If k is a p-adic field of degree n over $\mathbb{Q}_p$, then there are isomorphisms of $\mathbb{Q}_p[G]$-modules*

$$\begin{aligned} A(K) \otimes \mathbb{Q} &\cong \mathbb{Q}_p[G]^n \oplus \mathbb{Q}_p\,, \\ U_K^1 \otimes \mathbb{Q} &\cong \mathbb{Q}_p[G]^n\,. \end{aligned}$$

(ii) *If k is the field of Laurent series over a finite field and $K|k$ is tamely ramified, then there are isomorphisms of $\mathbb{Z}_p[G]$-modules*

$$\begin{aligned} A(K) &\cong \mathbb{Z}_p[G]^{\mathbb{N}} \oplus \mathbb{Z}_p\,, \\ U_K^1 &\cong \mathbb{Z}_p[G]^{\mathbb{N}}\,. \end{aligned}$$

Proof: (i) Let U_K^m be the group of principal units of level m of K. For m large enough, the p-adic logarithm induces a G-invariant isomorphism

$$\log : U_K^m \xrightarrow{\sim} \mathfrak{p}^m \subseteq \mathcal{O}_K\,,$$

see [160], chap.II, (5.5). Tensoring by $\mathbb{Q}$ and noting that U_K^m has finite index in U_K^1, the existence of a normal basis for $K|k$ gives us a G-isomorphism

$$U_K^1 \otimes \mathbb{Q} \cong \mathbb{Q}_p[G]^n\,.$$

Applying p-completion to the exact sequence

$$0 \longrightarrow U_K \longrightarrow K^\times \longrightarrow \mathbb{Z} \longrightarrow 0,$$

we obtain the exact sequence

$$0 \longrightarrow U_K^1 \longrightarrow A(K) \longrightarrow \mathbb{Z}_p \longrightarrow 0.$$

Using the semi-simplicity of the category of finitely generated $\mathbb{Q}_p[G]$-modules (2.6.12), we obtain

$$A(K) \otimes \mathbb{Q} \cong U_K^1 \otimes \mathbb{Q} \oplus \mathbb{Q}_p,$$

hence the result.

(ii) Let $k = \mathbb{F}_q((t))$, where $q = p^{f_0}$, and let $K|k$ be a finite tamely ramified Galois extension. By the properties of tamely ramified extensions, one knows that there exists an unramified extension K' of K which is of the form $K' = k(\zeta, \sqrt[e]{t})$, where ζ is a primitive e-th root of unity and $e = q^f - 1$, $f \geq 1$. Using (7.4.3), we may assume that $K = K'$. Then the Galois group $G = G(K|k)$ has the following structure:

$$G = \langle \sigma, \tau \mid \sigma\tau\sigma^{-1} = \tau^q,\ \sigma^f = 1,\ \tau^e = 1 \rangle,$$

where

$$\begin{aligned} \sigma(\zeta) &= \zeta^q \quad \text{and} \quad \sigma(\sqrt[e]{t}) = \sqrt[e]{t}\,, \\ \tau(\zeta) &= \zeta \quad\ \ \text{and} \quad \tau(\sqrt[e]{t}) = \zeta\sqrt[e]{t}\,. \end{aligned}$$

If $H = \langle \tau \rangle$, then $G/H = G(k(\zeta)|k) \cong G(\lambda|\kappa)$, where $\lambda \cong \mathbb{F}_{q^f}$ is the residue field of K and $\kappa \cong \mathbb{F}_q$ is the residue field of k. The existence of a normal basis

$$\bar{\sigma}^i(\alpha), \quad i = 0, \ldots, f-1,$$

of λ gives us an isomorphism $\lambda \cong \kappa[G/H]$ of $\mathbb{F}_q[G/H]$-modules.

For $0 \leq r \leq e-1$, let $\lambda(r) = \varepsilon_r\lambda \subseteq \lambda[H]$ be the 1-dimensional eigenspace corresponding to the idempotent

$$\varepsilon_r = \frac{1}{e}\sum_{j=0}^{e-1} \zeta^{-rj}\tau^j,$$

i.e. $\tau(\beta) = \zeta^r\beta$ for $\beta \in \lambda(r)$. Obviously, $\lambda(r)$ is an $\mathbb{F}_q[H]$-module. However, it is even an $\mathbb{F}_q[G]$-module, as $\sigma\varepsilon_r = \varepsilon_r\sigma$ in $\mathbb{F}_q[G]$, i.e. $\sigma(\beta) = \beta^q$ for $\beta \in \lambda(r)$. Therefore the map

$$\bigoplus_{r=0}^{e-1} \lambda(r) \longrightarrow \kappa[G],$$

defined by $\sigma^i(\alpha) \mapsto \sigma^i\varepsilon_r$ for $\sigma^i(\alpha) \in \lambda(r)$, $i = 0, \ldots, f-1$, $r = 0, \ldots, e-1$, is an isomorphism of $\mathbb{F}_q[G]$-modules. In particular, the modules $\lambda(r)$ are $\mathbb{F}_q[G]$-projective.

Let $i_1 < i_2 < \cdots < i_{e(p-1)}$ be natural numbers such that $1 \leq i_\nu < ep$ and $p \nmid i_\nu$, and let $i_{e(p-1)+1} = ep$. For each $0 \leq r \leq e-1$, there are exactly $p-1$ numbers i_ν such that $i_\nu \equiv r \bmod e$. Therefore we obtain an isomorphism of $\mathbb{F}_q[G]$-modules

$$\bigoplus_{\nu=1}^{e(p-1)} \lambda(i_\nu) \xrightarrow{\ \sim\ } \kappa[G]^{p-1}\,.$$

For each $j \geq 0$, we have an $\mathbb{F}_q[G]$-isomorphism

$$U_K^{epj+i_\nu}(U_K^1)^p/U_K^{epj+i_{\nu+1}}(U_K^1)^p \xrightarrow{\ \sim\ } \lambda(i_\nu), \quad [1 + \alpha\,(\sqrt[e]{t})^{epj+i_\nu}] \mapsto \alpha\,.$$

Since the modules $\lambda(i_\nu)$ are $\mathbb{F}_q[G]$-projective, it follows that

$$U_K^{epj}(U_K^1)^p/U_K^{ep(j+1)}(U_K^1)^p \cong \kappa[G]^{(p-1)},$$

and so we get $\mathbb{F}_p[G]$-isomorphisms

$$(*) \qquad U_K^1/U_K^{epj}(U_K^1)^p \cong \mathbb{F}_p[G]^{f_0(p-1)j}$$

for all $j \geq 1$. Passing to the projective limit, we get $U_K^1/(U_K^1)^p \cong \mathbb{F}_p[G]^{\mathbb{N}}$.

Since $\mathbb{Z}_p[G]^{\mathbb{N}}$ is the free compact $\mathbb{Z}_p[G]$-module (on countably many generators), the diagram

$$\begin{array}{ccc} \mathbb{Z}_p[G]^{\mathbb{N}} & \overset{\phi}{\dashrightarrow} & U_K^1 \\ \downarrow & & \downarrow \\ \mathbb{F}_p[G]^{\mathbb{N}} & \xrightarrow{\sim} & U_K^1/(U_K^1)^p \end{array}$$

can be commutatively completed by a continuous homomorphism ϕ, which is surjective by the topological Nakayama lemma (5.2.18)(i) (for $\mathbb{Z}_p$). Using the fact that U_K^1 is $\mathbb{Z}_p$-torsion-free, a second application of the topological Nakayama lemma implies that $\ker(\phi) = 0$, hence ϕ is an isomorphism.

In order to show the second isomorphism, we consider the exact sequence $0 \to U_K^1 \to A(K) \to \mathbb{Z}_p \longrightarrow 0$. Using (5.2.14) and (2.7.6), we have

$$\mathrm{Ext}^1_{\mathbb{Z}_p[G]}(\mathbb{Z}_p, U_K^1) \cong H^1_{cts}(G, \mathbb{Z}_p[G]^{\mathbb{N}}) = 0,$$

and so an isomorphism $A(K) \cong U_K^1 \oplus \mathbb{Z}_p$, which is the desired result. □

Proof of (7.4.1): First let $K|k$ be a finite tamely ramified Galois extension. Then $G = G(K|k)$ is generated by two elements, σ and τ say, acting on $\mu_{p^\infty}(K)$ by $\zeta^\sigma = \zeta^a$ and $\zeta^\tau = \zeta^b$, $a, b \in \mathbb{Z}_p$. We obtain an exact sequence

$$\mathbb{Z}_p[G]^2 \longrightarrow \mathbb{Z}_p[G] \xrightarrow{\varphi} \mu_{p^\infty}(K)^\vee \longrightarrow 0,$$

where $\ker(\varphi) = (\sigma - a, \tau - b)$, which induces the exact sequence

$$(*) \qquad 0 \longrightarrow \mathbb{Z}_p[G] \longrightarrow \mathbb{Z}_p[G]^2 \longrightarrow M_0 \longrightarrow 0,$$

with $M_0 \simeq D(\mu_{p^\infty}(K)^\vee)$ since $(\mu_{p^\infty}(K)^\vee)^+ = 0$. Setting $Y = I_{\mathcal{G}}/I_{\mathcal{H}} I_{\mathcal{G}}$, where $\mathcal{H} = G(\bar{k}|K)$, we get from (5.6.9) and (7.2.4) the homotopy equivalence

$$M_0 \simeq Y,$$

and for $X = G(\bar{k}|K)^{ab}(p)$ we have the exact sequence

$$0 \longrightarrow X \longrightarrow Y \longrightarrow I_G \longrightarrow 0$$

by (5.6.5). By Maschke's theorem, finitely generated $\mathbb{Q}_p[G]$-modules are projective, so that

$$Y \otimes \mathbb{Q} \cong X \otimes \mathbb{Q} \oplus I_G \otimes \mathbb{Q},$$

and given (7.4.4)(i) and the exact sequence $(*)$, it follows that

$$Y \otimes \mathbb{Q} \cong \mathbb{Q}_p[G]^n \oplus \mathbb{Q}_p \oplus I_G \otimes \mathbb{Q} \cong \mathbb{Q}_p[G]^{n+1} \cong M_0 \otimes \mathbb{Q} \oplus \mathbb{Q}_p[G]^n.$$

From (5.6.11), we get the isomorphism

$$(**) \qquad Y \cong M_0 \oplus \mathbb{Z}_p[G]^n.$$

If

$$1 \longrightarrow R_{n+2} \longrightarrow F_{n+2} \longrightarrow G \longrightarrow 1$$

is a presentation of the finite group G by a free profinite group F_{n+2}, then we obtain from (5.6.6) an exact sequence $0 \to R^{ab}_{n+2}(p) \to \mathbb{Z}_p[G]^{n+2} \to I_G \to 0$. Since $\mathbb{Z}_p[G]^{n+2}$ is projective, we get, using the isomorphism $(**)$ and the exact sequence $(*)$, an commutative exact diagram

$$\begin{array}{ccccccccc} 0 & \longrightarrow & R^{ab}_{n+2}(p) & \longrightarrow & \mathbb{Z}_p[G]^{n+2} & \longrightarrow & I_G & \longrightarrow & 0 \\ & & \downarrow & & \downarrow{\scriptstyle\alpha} & & \| & & \\ 0 & \longrightarrow & X & \longrightarrow & Y & \longrightarrow & I_G & \longrightarrow & 0, \end{array}$$

where the kernel of α is isomorphic to $\mathbb{Z}_p[G]$. Thus we obtain a G-invariant surjection $\beta : R^{ab}_{n+2}(p) \twoheadrightarrow X$, whose kernel is isomorphic to $\mathbb{Z}_p[G]$, and an isomorphism

$$(\mathbb{Z}/\#G\,\mathbb{Z})(p) \cong H^2(G, R^{ab}_{n+2}(p)) \xrightarrow[\sim]{\beta_*} H^2(G, X).$$

Now the exact diagram

$$\begin{array}{ccccccccc} 1 & \longrightarrow & R^{ab}_{n+2}(p) & \longrightarrow & F_{n+2}/[R_{n+2}, R_{n+2}]R^{(p)} & \longrightarrow & G & \longrightarrow & 1 \\ & & \downarrow{\scriptstyle\beta} & & \downarrow & & \| & & \\ 1 & \longrightarrow & X & \longrightarrow & \mathcal{G}/[\mathcal{H}, \mathcal{H}]\mathcal{H}^{(p)} & \longrightarrow & G & \longrightarrow & 1, \end{array}$$

where $R^{(p)} = \ker(R_{n+2} \twoheadrightarrow R_{n+2}(p))$ and $\mathcal{H}^{(p)} = \ker(\mathcal{H} \twoheadrightarrow \mathcal{H}(p))$, can be completed to a commutative diagram, since the corresponding 2-cocycles of the group extensions are mapped by β_* onto each other (after possibly multiplying with a unit in $\mathbb{Z}_p$); use $scd\, F_{n+2} = scd\, \mathcal{G} = 2$ and (3.6.4)(iii) and I §5 ex.4.

Let k_{tr} be the maximal tamely ramified extension of k. Passing to the projective limit over all finite Galois extensions $K|k$ inside k_{tr}, we obtain by the usual compactness argument a surjection

$$F_{n+2} \twoheadrightarrow \mathcal{G}/[G(\bar{k}|k_{tr}), G(\bar{k}|k_{tr})].$$

Thus the profinite group $\mathcal{G}/[G(\bar{k}|k_{tr}), G(\bar{k}|k_{tr})]$ is generated by $n+2$ elements. Since $G(\bar{k}|k_{tr})$ is a pro-p-group, the Frattini argument (3.9.1) implies that $\mathcal{G}$ itself is generated by $n+2$ elements.

In order to prove the second assertion, we first observe that $N^{ab}(p)$ is $\mathbb{Z}_p[[\mathcal{G}]]$-projective by (5.6.7), so that $N^{ab}_{\mathcal{H}}(p)$ is $\mathbb{Z}_p[G]$-projective for all open normal subgroups $\mathcal{H}$ of $\mathcal{G}$, where $G = \mathcal{G}/\mathcal{H}$. From the exact sequence

$$(*) \qquad 0 \longrightarrow N^{ab}_{\mathcal{H}}(p) \longrightarrow R^{ab}_{n+2}(p) \longrightarrow X \longrightarrow 0$$

for $X = \mathcal{H}^{ab}(p)$ (recalling that $scd\, \mathcal{G} = 2$), we obtain

$$R^{ab}_{n+2}(p) \otimes \mathbb{Q} \cong X \otimes \mathbb{Q} \oplus N^{ab}_{\mathcal{H}}(p) \otimes \mathbb{Q},$$

and hence, using again (7.4.4)(i),

$$\mathbb{Q}_p \oplus \mathbb{Q}_p[G]^{n+1} \cong \mathbb{Q}_p \oplus \mathbb{Q}_p[G]^n \oplus N^{ab}_{\mathscr{H}}(p) \otimes \mathbb{Q}$$

so that

$$N^{ab}_{\mathscr{H}}(p) \otimes \mathbb{Q} \cong \mathbb{Q}_p[G] .$$

It follows that $N^{ab}_{\mathscr{H}}(p) \cong \mathbb{Z}_p[G]$ using (5.6.15). We complete the proof of (7.4.1) by passing to the limit over all finite quotients G of $\mathscr{G}$, □

Proof of (7.4.2): Let $G = \mathscr{G}/\mathscr{H}$ for some closed normal subgroup $\mathscr{H}$ of $\mathscr{G}$ (not necessarily of finite index). Again we obtain the exact sequence $(*)$ used in the proof of (7.4.1) and, since $N^{ab}_{\mathscr{H}}(p) \cong \mathbb{Z}_p[\![G]\!]$, we obtain the first statement of (i).

If $\mu_p \not\subseteq K$, then the $\mathbb{Z}_p[\![G]\!]$-module Y is free of rank $n+1$, which can be shown in the same manner as in the proof of (7.4.1). Thus we get an isomorphism $R^{ab}_{n+1}(p) \xrightarrow{\sim} X$. As above, this implies that G can be generated by $n+1$ elements.

Now we prove assertion (ii). Since $a_i \in \mathbb{Z}_p^{\times}$, we get the exact sequence

$$\begin{array}{ccccccc} \mathbb{Z}_p[\![\mathscr{G}]\!]^{n+2} & \longrightarrow & \mathbb{Z}_p[\![\mathscr{G}]\!] & \longrightarrow & \mathbb{Z}_p & \longrightarrow & 0, \\ e_i & \longmapsto & a_i(\bar{\sigma}_i^{-1} - 1) & & & & \end{array}$$

where $\{e_i \mid i = 1, \dots, n+2\}$ is a basis of $\mathbb{Z}_p[\![\mathscr{G}]\!]^{n+2}$, see (5.6.6). Tensoring by $\mathbb{Z}_p(-1)$, using the isomorphism

$$\begin{array}{ccc} \mathbb{Z}_p[\![\mathscr{G}]\!] \otimes_{\mathbb{Z}_p} \mathbb{Z}_p(-1) & \xrightarrow{\sim} & \mathbb{Z}_p[\![\mathscr{G}]\!] \\ g \otimes 1 & \longmapsto & \chi_{\mathrm{cycl}}(g) \cdot g \, , \end{array}$$

and taking $G(\bar{k}|K)$-coinvariants, we obtain the exact sequence

$$\begin{array}{ccccccc} \mathbb{Z}_p[\![G]\!]^{n+2} & \longrightarrow & \mathbb{Z}_p[\![G]\!] & \longrightarrow & \mu_{p^\infty}(K)^\vee & \longrightarrow & 0 . \\ e_i & \longmapsto & (\sigma_i^{1} - a_i) & & & & \end{array}$$

Since $(\mu_{p^\infty}(K)^\vee)_U$ is finite for every open normal subgroup U of G, we have $(\mu_{p^\infty}(K)^\vee)^+ = 0$. Thus the last sequence implies an exact sequence

$$\begin{array}{ccccccccc} 0 & \longrightarrow & \mathbb{Z}_p[\![G]\!] & \longrightarrow & \mathbb{Z}_p[\![G]\!]^{n+2} & \longrightarrow & M_0 & \longrightarrow & 0 \\ & & 1 & \longmapsto & (\bar{\sigma}_i - a_i)_i \, , & & & & \end{array}$$

with $M_0 \simeq D(\mu_{p^\infty}(K)^\vee)$. As in the proof of (7.4.1), we obtain isomorphisms

$$Y_U \cong (M_0)_U \oplus \mathbb{Z}_p[G/U]^n$$

for any open normal subgroup U of G. By (5.6.11), this implies the existence of an isomorphism $Y \cong M_0 \oplus \mathbb{Z}_p[\![G]\!]^n$, showing the assertion. □

For a local field k the absolute Galois group, the inertia group and the ramification group is denoted by G_k, T_k and V_k, respectively. Furthermore, $\mathcal{G}_k = T_k/V_k$ is the Galois group of the maximal tamely ramified extension k_{tr} of k. The next theorem is taken from [94].

(7.4.5) Theorem. *Let k be a local field with residue characteristic p.*

(i) *If k is a p-adic field with $n = [k : \mathbb{Q}_p]$, then there is an exact sequence of $\mathbb{Z}_p[[\mathcal{G}_k]]$-modules*

$$0 \longrightarrow \mathbb{Z}_p[[\mathcal{G}_k]] \longrightarrow \mathbb{Z}_p[[\mathcal{G}_k]]^{n+1} \longrightarrow V_k^{ab} \longrightarrow 0\,.$$

(ii) *If k is a local field of characteristic p, then there is an isomorphism of $\mathbb{Z}_p[[\mathcal{G}_k]]$-modules*

$$V_k^{ab} \cong \mathbb{Z}_p[[\mathcal{G}_k]]^{\mathbb{N}}\,,$$

i.e. V_k^{ab} is a free compact $\mathbb{Z}_p[[\mathcal{G}_k]]$-module of countably infinite rank.

Proof: (i) By (7.4.2), we have an exact sequence of $\mathbb{Z}_p[[\mathcal{G}_k]]$-modules

$$0 \longrightarrow \mathbb{Z}_p[[\mathcal{G}_k]] \longrightarrow \mathbb{Z}_p[[\mathcal{G}_k]]^{n+2} \longrightarrow Y \longrightarrow 0\,,$$

where $Y = I_{G_k}/I_{G_{k_{tr}}} I_{G_k}$, and with $X = G(\bar{k}|k_{tr})^{ab} = V_k^{ab}$ we have the exact sequence

$$0 \longrightarrow X \longrightarrow Y \longrightarrow I_{\mathcal{G}_k} \longrightarrow 0\,.$$

By (5.6.12), the augmentation ideal $I_{\mathcal{G}_k}$ is a free $\mathbb{Z}_p[[\mathcal{G}_k]]$-module of rank 1, hence

$$Y \cong X \oplus \mathbb{Z}_p[[\mathcal{G}_k]]\,.$$

This gives the desired result.

(ii) Let $K|k$ be a finite tamely ramified Galois extension. Then, by (7.4.3), the norm map

$$N_{G(K|k)} : U_K^1/(U_K^1)^p \twoheadrightarrow U_k^1/(U_k^1)^p$$

is surjective, and furthermore we have $N_{G(K|k)} U_K^{e_K pj} = U_k^{e_k pj}$; see [212] chap.V. Therefore

$$N_{G(K|k)} : U_K^1/U_K^{e_K pj}(U_K^1)^p \twoheadrightarrow U_k^1/U_k^{e_k pj}(U_k^1)^p$$

is well-defined and surjective. In the proof of (7.4.4)(ii) we have shown the assertion $(*)$: For $j \geq 1$ there is an isomorphism

$$U_K^1/U_K^{e_K pj}(U_K^1)^p \xrightarrow{\;\sim\;} \mathbb{F}_p[G(K|k)]^{f_0(p-1)j}$$

of $\mathbb{F}_p[G(K|k)]$-modules. Passing to the projective limit over all finite tamely ramified Galois extension $K|k$, we obtain

$$\begin{aligned} V_k^{ab}/p &\cong \varprojlim_{K|k} U_K^1/(U_K^1)^p \\ &\cong \varprojlim_{K|k} \varprojlim_j U_K^1/U_K^{e_K pj}(U_K^1)^p \\ &\cong \varprojlim_{K|k} \varprojlim_j \mathbb{F}_p[G(K|k)]^{f_0(p-1)j} \\ &\cong \varprojlim_j \mathbb{F}_p[\![\mathcal{G}_k]\!]^{f_0(p-1)j} \cong \mathbb{F}_p[\![\mathcal{G}_k]\!]^{\mathbb{N}} \end{aligned}$$

Since V_k^{ab} is $\mathbb{Z}_p$-torsion-free, we may apply the topological Nakayama lemma in a similar way as in the proof of (7.4.4)(ii), to obtain the desired isomorphism $V_k^{ab} \cong \mathbb{Z}_p[\![\mathcal{G}_k]\!]^{\mathbb{N}}$. □

§5. Explicit Determination of Local Galois Groups

The absolute Galois group G_k of a local field is prosolvable. It is the projective limit of the groups $G(K|k)$ of the finite Galois extensions $K|k$ which contain the inertia group $T(K|k)$ and the ramification group $V(K|k)$ as normal subgroups. $V(K|k)$ is a p-group, and $T(K|k)/V(K|k)$ and $G(K|k)/T(K|k)$ are cyclic, hence $G(K|k)$ is a solvable group. Furthermore, the inertia group $T_k = \varprojlim_K T(K|k)$ and the ramification group $V_k = \varprojlim_K V(K|k)$ are normal subgroups of G_k. From (7.1.8)(i) and (7.4.5) we obtain:

(7.5.1) Proposition. *The ramification group V_k of the absolute Galois group G_k of a local field k with residue characteristic p is a free pro-p-group of countably infinite rank.*

The group G_k has several interesting quotients. The simplest is the quotient by the inertia group T_k. This is the Galois group $\Gamma = G(\tilde{k}|k)$ of the *maximal unramified* extension $\tilde{k}|k$. It is canonically isomorphic to the absolute Galois group G_κ of the finite residue field κ, hence to $\hat{\mathbb{Z}}$, and has the Frobenius automorphism σ_k as a canonical topological generator.

We next consider the quotient $\mathcal{G}_k$ of G_k by the ramification group V_k. This is the Galois group

$$\mathcal{G}_k = G(k_{tr}|k)$$

of the *maximal tamely ramified extension* $k_{tr}|k$. $\mathcal{G}_k$ is a group extension of $\mathcal{G}_{\tilde{k}} = G(k_{tr}|\tilde{k})$ by Γ, i.e. we have an exact sequence

$$1 \longrightarrow \mathcal{G}_{\tilde{k}} \longrightarrow \mathcal{G}_k \longrightarrow \Gamma \longrightarrow 1,$$

which splits after choosing of a pre-image σ of $\sigma_k \in \Gamma$. In other words, $\mathcal{G}_k$ is the semi-direct product of $\mathcal{G}_{\tilde{k}}$, which is an abelian group, and Γ. We obtain a complete description of $\mathcal{G}_k$ by determining explicitly $\mathcal{G}_{\tilde{k}}$ as a Γ-module. This is easily achieved:

Let p be the residue characteristic of k and let $\mu^{(p')}$ be the group of roots of unity in $\bar{k}$ of order prime to p. $\mu^{(p')}$ is a Γ-module, which is canonically isomorphic to $\bar{\kappa}^\times$, via the reduction map. The value group of the normalized valuation of k is $\mathbb{Z}$, and that of its extension to k_{tr} is $\Delta = \bigcup\limits_{p\nmid n} \frac{1}{n}\mathbb{Z}$, hence

$$\Delta/\mathbb{Z} = \bigoplus_{\ell \neq p} \mathbb{Q}_\ell/\mathbb{Z}_\ell =: (\mathbb{Q}/\mathbb{Z})^{(p')}.$$

By [160], chap.II,(9.15), we have canonically

$$\mathcal{G}_{\tilde{k}} \cong \operatorname{Hom}(\Delta/\mathbb{Z}, \bar{\kappa}^\times). \;^{*)}$$

The Frobenius automorphism $\sigma_k \in \Gamma$ acts on both groups by $x \mapsto x^q$ ($q = \#\kappa$), since this is its action on $\bar{\kappa}^\times$. Thus $\mathcal{G}_{\tilde{k}}$ is canonically isomorphic to the additive Γ-module $\hat{\mathbb{Z}}^{(p')}(1)$, which is isomorphic to $\hat{\mathbb{Z}}^{(p')} = \prod\limits_{\ell \neq p} \mathbb{Z}_\ell$ as an abelian group, and on which σ_k acts as multiplication by q. We have thus obtained the

(7.5.2) Proposition. *If $\mathcal{G}_k$ denotes the Galois group of the maximal tamely ramified extension k_{tr} of a local field k, then we have a split group extension*

$$1 \longrightarrow \hat{\mathbb{Z}}^{(p')}(1) \longrightarrow \mathcal{G}_k \longrightarrow \Gamma \longrightarrow 1.$$

We may reformulate this result as follows (cf. [87]).

(7.5.3) Theorem (Iwasawa). *The Galois group $\mathcal{G}_k$ of the maximal tamely ramified extension of a local field k is isomorphic to the profinite group $\mathcal{G}$ generated by two elements σ, τ with the only relation*

$$\sigma\tau\sigma^{-1} = \tau^q.$$

---

*) This isomorphism comes from Kummer theory via the following observation: let $K|\tilde{k}$ be a finite extension of degree e with $(e,p) = 1$ and let Π, π be uniformizers of K, $\tilde{k}$. Then $\Pi^e = \pi \cdot u$, where $u \in K$ is a unit. Note that $u^{1/e} \in K$ and $(\Pi/u^{1/e})^e = \pi$. Hence $K = \tilde{k}(\pi^{1/e})$.

Proof: Let F be the free profinite group generated by two elements σ and τ. Let N be the normal closed subgroup of F generated by $\sigma\tau\sigma^{-1}\tau^{-q}$, and set $\mathcal{G} = F/N$. $\mathcal{G}$ is the profinite group generated by the images $\bar\sigma, \bar\tau$ of σ, τ with the defining relation $\bar\sigma\bar\tau\bar\sigma^{-1} = \bar\tau^q$. The homomorphism $F \to \Gamma$, given by $\sigma \mapsto \sigma_k$, $\tau \mapsto 1$, induces a surjection $\mathcal{G} \to \Gamma$. The kernel is the closed normal subgroup Z topologically generated by $\bar\tau$. In fact, Z is normal in $\mathcal{G}$ because $\bar\sigma\bar\tau\bar\sigma^{-1} = \bar\tau^q$, and $\mathcal{G}/Z$ is generated by the image of $\bar\sigma$, i.e. is procyclic with a surjection $\mathcal{G}/Z \to \Gamma$ which must be an isomorphism.

Writing Z additively, the action of σ_k on Z becomes multiplication by q. Since it is an automorphism of Z, the p-Sylow subgroup of Z must be trivial. In other words, Z is a quotient of $\hat{\mathbb{Z}}^{(p')} = \prod\limits_{\ell \neq p} \mathbb{Z}_\ell$.

Now consider the Galois group $\mathcal{G}_k = G(k_{tr}|k)$. Let τ' be a topological generator of $\mathcal{G}_{\tilde k} = G(k_{tr}|\tilde k) \cong \hat{\mathbb{Z}}^{(p')}(1)$ and σ' a pre-image of $\sigma_k \in \Gamma$. Then $\sigma'\tau'\sigma'^{-1} = \tau'^q$, and the surjective homomorphism $F \to \mathcal{G}_k$, given by $\sigma \mapsto \sigma'$, $\tau \mapsto \tau'$, factors through $\mathcal{G}$. We obtain an exact commutative diagram

$$\begin{array}{ccccccccc} 1 & \longrightarrow & Z & \longrightarrow & \mathcal{G} & \longrightarrow & \Gamma & \longrightarrow & 1 \\ & & \downarrow{\scriptstyle\alpha} & & \downarrow{\scriptstyle\beta} & & \| & & \\ 1 & \longrightarrow & \hat{\mathbb{Z}}^{(p')}(1) & \longrightarrow & \mathcal{G}_k & \longrightarrow & \Gamma & \longrightarrow & 1, \end{array}$$

where α is surjective. But α is necessarily an isomorphism, since $\hat{\mathbb{Z}}^{(p')}(1)$ is the procyclic group $\hat{\mathbb{Z}}^{(p')} = \prod_{\ell \neq p} \mathbb{Z}_\ell$, and Z is a quotient of $\hat{\mathbb{Z}}^{(p')}$. Therefore β is an isomorphism, and the theorem is proved. □

We want to take a closer look at the group $\mathcal{G}_k$. For this we introduce the following notation.

If G is a profinite group and g an element of G, then we define the α-power g^α of g for $\alpha \in \hat{\mathbb{Z}}$ as follows: Consider the homomorphism

$$\varphi : \hat{\mathbb{Z}} \twoheadrightarrow \langle g \rangle \subseteq G,$$

which is given by $1 \mapsto g$. Then we define $g^\alpha = \varphi(\alpha)$. Observe that for $\alpha \in \mathbb{Z}$ we obtain the usual powers of g.

For a prime number ℓ consider the projectors $\pi_\ell, \Delta_\ell \in \hat{\mathbb{Z}} \cong \prod_r \mathbb{Z}_r$ given by

$$(\pi_\ell)_r = \begin{cases} 1 \text{ for } r = \ell \\ 0 \text{ for } r \neq \ell\,, \end{cases} \qquad (\Delta_\ell)_r = \begin{cases} 0 \text{ for } r = \ell \\ 1 \text{ for } r \neq \ell\,. \end{cases}$$

Then

$$\pi_\ell \hat{\mathbb{Z}} = \mathbb{Z}_\ell \quad \text{and} \quad \Delta_\ell \hat{\mathbb{Z}} = \hat{\mathbb{Z}}^{(\ell')},$$

where $\mathbb{Z}_\ell$ and $\hat{\mathbb{Z}}^{(\ell')}$ are embedded as direct factors into $\hat{\mathbb{Z}}$. Further, $\Delta_\ell + \pi_\ell = 1$, $\Delta_\ell \pi_\ell = 0$ and for every $\alpha \in \hat{\mathbb{Z}}$:

$$\pi_\ell^\alpha = \pi_\ell \quad \text{and } \Delta_\ell^\alpha = \Delta_\ell.$$

If G is a profinite abelian group, which (written additively) is a $\mathbb{Z}_r$-module, then raising an element to the π_ℓ-power is the zero map if $r \neq \ell$ and is the identity if $r = \ell$.

For a prime number $\ell \neq p$ and a prime number $r|(\ell - 1)$ we define an $(\ell - 1)$-th root of unity of r-power order

$$e(\ell, r) \in \mu_{\ell-1}(r) \subseteq \mathbb{Z}_\ell$$

by

$$p \equiv \prod_{r|\ell-1} e(\ell, r) \mod \ell$$

($e(\ell, r)$ depends on p although we do not indicate this in the notation). We make the convention that $e(\ell, r) = 1$ if $r \nmid (\ell - 1)$ and we put

$$e(\ell) = \prod_{r|\ell-1} e(\ell, r) \in \mathbb{Z}_\ell\,,$$

i.e. $\pi_\ell p = e(\ell) \cdot u$, where $u \in \mathbb{Z}_\ell$ is a principal unit.

(7.5.4) Lemma. (i) *Let r be a prime number. Then for every $n \in \mathbb{N}$, there exists a prime number $\ell \neq p$ such that the r-power root of unity $e(\ell, r)$ is at least of order r^n.*

(ii) *For $q = p^f$, $1 \leq f \in \mathbb{N}$, the homomorphism*

$$\psi : \hat{\mathbb{Z}} \longrightarrow \operatorname{Aut}(\mathbb{Z}^{(p')}) \quad \alpha \longmapsto (x \mapsto q^\alpha x)$$

is injective.

Proof: (i) Consider the fields $k_0 = \mathbb{Q}(\mu_{r^n})$ and $k = k_0(\sqrt[r]{p})$. The set of prime numbers ℓ which are completely decomposed in k is

$$\{\ell \,|\, \ell \equiv 1 \mod r^n\,,\; p^{\frac{\ell-1}{r}} \equiv 1 \mod \ell\}\,,$$

since a prime number ℓ splits completely in k if and only if $\ell \nmid pr$, $\ell \equiv 1$ mod r^n and $\sqrt[r]{p} \in \mathbb{Z}_\ell$. But

$$\begin{aligned} \sqrt[r]{p} \in \mathbb{Z}_\ell \quad &\Leftrightarrow \quad \sqrt[r]{e(\ell)} \in \mathbb{Z}_\ell \\ &\Leftrightarrow \quad e(\ell)^{\frac{\ell-1}{r}} = 1 \\ &\Leftrightarrow \quad p^{\frac{\ell-1}{r}} \equiv 1 \mod \ell\,. \end{aligned}$$

Therefore the density of the set

$$S = \{\ell \,|\, \ell \equiv 1 \mod r^n\,,\; p^{\frac{\ell-1}{r}} \not\equiv 1 \mod \ell\}$$

is $[k_0 : \mathbb{Q}]^{-1} - [k : \mathbb{Q}]^{-1} = 1/r^n$ and, in particular, S is not empty. Thus for $\ell \in S$, the root of unity $e(\ell, r)$ has order r^m with $m \geq n$. This proves (i).

In order to prove (ii), assume that α is an element in the kernel of ψ, i.e. for all $\ell \neq p$ the equality $q^{\alpha}\,\pi_\ell = \pi_\ell$ holds. Then, in particular,

$$\prod_{r|\ell-1} e(\ell, r)^{f\alpha} = (e(\ell)^f)^{\alpha} = 1 \in \mathbb{Z}_\ell$$

and hence also $e(\ell, r)^{f\alpha} = 1$ for all $\ell \neq p$, $r|(\ell - 1)$.
For an arbitrary prime number r and an arbitrary $n \in \mathbb{N}$, we can use (i) in order to find a prime number $\ell \neq p$ such that

$$e(\ell, r)^{r^{n-1}} \neq 1.$$

Hence $f\alpha \in r^n\hat{\mathbb{Z}}$ and since r and n were arbitrary, we conclude that $f\alpha = 0$. Finally, since $1 \leq f \in \mathbb{N}$ and since $\hat{\mathbb{Z}}$ is torsion-free (as an abelian group), we obtain $\alpha = 0$.

An alternative possibility to see assertion (ii) is the following: the homomorphism ψ (after replacing q by q^{-1}) describes the action of $G(\bar{\kappa}|\kappa)$ on the dual $\mathrm{Hom}(\bar{\kappa}^\times, \mathbb{Q}/\mathbb{Z})$ of the multiplicative group $\bar{\kappa}^\times$. Obviously, this action is faithful. □

For a prime number r (possibly equal to p), we set

$$\sigma_r = \sigma^{\pi_r}, \quad \tau_r = \tau^{\pi_r}$$

where σ and τ are generators of the group $\mathcal{G}_k = \langle \sigma, \tau \mid \tau^\sigma = \tau^q \rangle$, $q = p^f$.

(7.5.5) Proposition. *With the notation as above, we have for arbitrary prime numbers r, and $\ell \neq p$*

$$[\sigma_r, \tau_\ell] = \begin{cases} \tau_\ell^{qe(\ell)^{-f}-1} & \text{if } r = \ell\,, \\ \tau_\ell^{e(\ell,r)^f-1} & \text{if } r|\ell-1\,, \\ 1 & \text{if } r \neq \ell\,,\ r \nmid \ell - 1\,. \end{cases}$$

Proof: For $r \neq \ell$ we have

$$\pi_\ell\, q^{\pi_r} = \pi_\ell\, p^{\pi_r f} = (\pi_\ell\, p)^{\pi_r f} = e(\ell, r)^f\,,$$

since all other components in the decomposition of $\pi_\ell\, p$ are annihilated (i.e. sent to 1) when raised to the π_r-th power. Similarly,

$$\pi_\ell\, q^{\pi_\ell} = \pi_\ell\, p^{\pi_\ell f} = (\pi_\ell\, p)^{\pi_\ell f} = \pi_\ell\, q e(\ell)^{-f}.$$

It follows that

$$[\sigma_r, \tau_\ell] = \tau_\ell^{q^{\pi_r}-1} = \tau_\ell^{\pi_\ell(q^{\pi_r}-1)} = \begin{cases} \tau_\ell^{qe(\ell)^{-f}-1} & \text{if } r = \ell\,, \\ \tau_\ell^{e(\ell,r)^f-1} & \text{if } r|\ell - 1\,, \\ 1 & \text{if } r \neq \ell\,,\ r \nmid \ell - 1\,. \end{cases}$$

This proves the proposition. □

(7.5.6) Corollary. *For every prime number r there exists a prime number $\ell \neq p$ such that σ_r and τ_ℓ do not commute. In particular, σ_p does not commute with τ.*

Proof: By (7.5.5), we have the equality

$$[\sigma_r, \tau_\ell] = \tau_\ell^{e(\ell,r)^f-1}$$

for every prime number r with $r|\ell - 1$, and from (7.5.4)(i) it follows that there are prime numbers ℓ such that the order of $e(\ell, r)$ is bigger than the r-part of the fixed number f. □

(7.5.7) Corollary.

(i) *The ramification group V_k of the absolute Galois group G_k of a local field k is the maximal normal pro-p subgroup of G_k.*

(ii) *The subgroup $T_k/V_k = \langle\tau\rangle$ of $\mathcal{G}_k = G_k/V_k$ is the unique maximal abelian normal subgroup of $\mathcal{G}_k$ (which exists in $\mathcal{G}_k$!).*

Proof: (i) Since V_k is a normal pro-p subgroup of G_k and $\langle V_k, \sigma_p\rangle$ is a pro-p Sylow subgroup of G_k, the result follows from the previous corollary.

(ii) Every abelian normal subgroup of $\mathcal{G}_k$ is contained in $\langle\tau\rangle$ by (7.5.5) and (7.5.4)(i). Thus the result follows. □

We next study the maximal pro-p-factor group $G_k(p)$ of G_k, where p is the residue characteristic of the local field k. This is the Galois group $G(k(p)|k)$ of the *maximal p-extension* $k(p)|k$, i.e. of the composite of all finite Galois extensions of p-power degree.

First, we have to compare the cohomology for a $G_k(p)$-module A with respect to $G_k(p)$ and G_k. This will be done in the next proposition in a slightly more general form. Recall the notion $G(\mathfrak{c})$ for the maximal pro-$\mathfrak{c}$-quotient of a profinite group G with respect to a full class $\mathfrak{c}$ of finite groups.

(7.5.8) Proposition. *Let $\mathfrak{c}$ be a full class of finite groups, ℓ a prime number such that $\mathbb{Z}/\ell\mathbb{Z} \in \mathfrak{c}$ and A a $G_k(\mathfrak{c})$-module. Then for every $i \geq 0$ the inflation map*

$$H^i(G_k(\mathfrak{c}), A)(\ell) \longrightarrow H^i(G_k, A)(\ell)$$

is an isomorphism.

Proof: The degree of $k(\mathfrak{c})|k$ is infinitely divisible by ℓ, since the maximal unramified ℓ-extension of k is contained in $k(\mathfrak{c})$ and has Galois group $\mathbb{Z}_\ell$. Using (7.1.8)(i) and (6.1.3), we obtain $cd_\ell(k(\mathfrak{c})) \leq 1$. Let $H = G(\bar{k}|k(\mathfrak{c}))$. Then $H^i(H, A)(\ell) = 0$ for $i \geq 1$. Indeed, since cohomology commutes with direct limits and since A is a trivial H-module, one reduces to the cases $A = \mathbb{Z}$, $\mathbb{Z}/\ell\mathbb{Z}$ and, using the exact sequence $0 \to \mathbb{Z} \to \mathbb{Q} \to \mathbb{Q}/\mathbb{Z} \to 0$, we finally reduce to the case $A = \mathbb{Z}/\ell\mathbb{Z}$. By $cd_\ell H \leq 1$, the assertion is obvious for $i \geq 2$. Since $\mathbb{Z}/\ell\mathbb{Z} \in \mathfrak{c}$, the group H has no nontrivial homomorphism to an ℓ-group, showing the case $i = 1$.

Now the spectral sequence

$$H^i(G_k(\mathfrak{c}), H^j(H, A))(\ell) \Rightarrow H^{i+j}(G_k, A)(\ell)$$

gives isomorphisms

$$H^i(G_k(\mathfrak{c}), A)(\ell) \cong H^i(G_k, A)(\ell) \text{ for all } i \geq 0. \qquad \square$$

We can now explicitly determine the structure of the pro-ℓ-group $G_k(\ell)$. A consequence of (7.5.3) is the

(7.5.9) Proposition. *Let k be a local field with residue characteristic p and let ℓ be a prime number different to p.*

If $\mu_\ell \not\subseteq k$, then $G_k(\ell) \cong \mathbb{Z}_\ell$; if $\mu_\ell \subseteq k$, then $G_k(\ell)$ is a Poincaré group of dimension 2 (i.e. a Demuškin group) of rank 2 with dualizing module $\mu(\ell)$, the group of all ℓ-power roots of unity in $k(\ell)$. Furthermore,

$$G_k(\ell)^{ab} \cong \mathbb{Z}/\ell^s\mathbb{Z} \oplus \mathbb{Z}_\ell\,,$$

where $\ell^s = \#\mu(k)(\ell)$.

From (6.1.7), we obtain the

(7.5.10) Theorem. *Let k be a local field with* char$(k) = p$. *Then $G_k(p)$ is a free pro-p-group of countably infinite rank.*

(7.5.11) Theorem. *Let k be a p-adic local field.*

(i) *If μ_p is not contained in k, then $G_k(p)$ is a free pro-p-group of rank $N+1$ with $N = [k : \mathbb{Q}_p]$.*

(ii) *If $\mu_p \subseteq k$, then $G_k(p)$ is a Poincaré group of dimension* 2 (i.e. a Demuškin group) of rank $N+2$. The dualizing module of $G_k(p)$ is the group $\mu(p)$ of all p-power roots of unity in $k(p)$.

Proof: By the (7.5.8) and by the duality theorem (7.2.6), we obtain

(1) $H^1(G_k(p), \mathbb{Z}/p\mathbb{Z}) = H^1(G_k, \mathbb{Z}/p\mathbb{Z}) \cong H^1(G_k, \mu_p)^* \cong (k^\times/k^{\times p})^*$,

(2) $H^2(G_k(p), \mathbb{Z}/p\mathbb{Z}) = H^2(G_k, \mathbb{Z}/p\mathbb{Z}) \cong H^0(G_k, \mu_p)^*$.

From [160], chap.II, (5.7)(i), we get $k^\times/k^{\times p} \cong (\mathbb{Z}/p\mathbb{Z})^r$ with $r = N+1$ if $\mu_p \nsubseteq k$, or $N+2$ if $\mu_p \subseteq k$. Therefore, by (3.9.1), the pro-p-group $G_k(p)$ has rank

$$\mathrm{rk}(G_k(p)) = N+1 \text{ or } N+2,$$

according to whether $\mu_p \nsubseteq k$ or $\mu_p \subseteq k$.

If $\mu_p \nsubseteq k$, then from (2) it follows that $H^2(G_k(p), \mathbb{Z}/p\mathbb{Z}) = 0$, and $G_k(p)$ is a free pro-p-group by (3.9.5). This proves (i).

Assume $\mu_p \subseteq k$. Then (2) implies that $H^2(G_k(p), \mathbb{Z}/p\mathbb{Z}) \cong \mathbb{Z}/p\mathbb{Z}$ and $H^i(G_k(p), \mathbb{Z}/p\mathbb{Z}) = 0$ for $i > 2$ by (7.5.8) and (7.1.8), so that $cd_p\, G_k(p) = 2$. By the duality theorem (7.2.6), the cup-product yields a non-degenerate pairing

$$H^1(G_k(p), \mathbb{Z}/p\mathbb{Z}) \times H^1(G_k(p), \mathbb{Z}/p\mathbb{Z}) \to H^2(G_k(p), \mathbb{Z}/p\mathbb{Z}) \cong \mathbb{Z}/p\mathbb{Z}.$$

Since $\mu_p \subseteq k$, the p-part $\mu(p)$ of the dualizing module of μ of G_k (see (7.2.4)) is a $G_k(p)$-module. Therefore (7.5.8) shows that $\mu(p)$ is the dualizing module of $G_k(p)$ for the category of p-torsion G_k-modules. From (3.7.2) it follows now that $G_k(p)$ is a Poincaré group. □

When $\mu_p \subseteq k$ we get an explicit description of $G_k(p)$ by applying the results of (3.9.11) and (3.9.19).

(7.5.12) Theorem (Demuškin). *Let k be a p-adic local field of degree $N = [k : \mathbb{Q}_p]$ and let p^s be the largest p-power such that $\mu_{p^s} \subseteq k$. If $p^s > 2$, then $G_k(p)$ is the pro-p-group defined by $N+2$ generators $x_1, \ldots, x_{N+2}$ subject to the one relation*

$$x_1^{p^s}(x_1, x_2)(x_3, x_4) \cdots (x_{N+1}, x_{N+2}) = 1.$$

Proof: By class field theory the abelianized group $G_k(p)^{ab}$ is isomorphic to the pro-p-completion $\hat{k}^\times$ of the multiplicative group $k^\times$ which, by [160], chap.II, (5.7), is isomorphic to $\mathbb{Z} \oplus \mathbb{Z}/(p^f - 1)\mathbb{Z} \oplus \mathbb{Z}/p^s\mathbb{Z} \oplus \mathbb{Z}_p^N$, i.e. $\hat{k}^\times \cong \mathbb{Z}/p^s\mathbb{Z} \oplus \mathbb{Z}_p^{N+1}$. Therefore p^s is the number which we have denoted by q in (3.9.11). Moreover, we have

$$H^1(G_k(p), \mathbb{Z}/p\mathbb{Z}) \cong \operatorname{Hom}(G_k(p), \mathbb{Z}/p\mathbb{Z}) \cong (\mathbb{Z}/p\mathbb{Z})^{N+2},$$

so that $\operatorname{rk}(G_k(p)) = N + 2$. The theorem follows now from (3.9.11). □

The structure of $G_k(2)$ when $p^s = 2$ has been determined by *J.-P. Serre* if N is odd and in general by *J. Labute* (cf. (3.9.19) and [209], [117]). $G_k(2)$ is again generated by $N+2$ generators $x_1, \ldots, x_{N+2}$ with one defining relation ρ, but the shape of this relation depends on further conditions:

If N is odd, then

$$\rho = x_1^2 x_2^4 (x_2, x_3)(x_4, x_5) \cdots (x_{N+1}, x_{N+2}).$$

In particular, if $k = \mathbb{Q}_2$, then $G_k(2)$ is generated by three elements with the defining relation $x^2 y^4 (y, z) = 1$.

If N is even, then we have to take the 2-part of the cyclotomic character $\chi : G_k \to \mathbb{Z}_2^\times$ into account, which is obtained from the action of G_k on the group μ_{2^∞} of all roots of unity of 2-power order. We have $\operatorname{End}(\mu_{2^\infty}) = \mathbb{Z}_2$ and the action of G_k gives the homomorphism $\chi : G_k \to \operatorname{Aut}(\mu_{2^\infty}) = \mathbb{Z}_2^\times$. The structure of $G_k(2)$ depends now on the image of χ:

If $\operatorname{im}(\chi)$ is the closed subgroup of $\mathbb{Z}_2^\times$ generated by $-1 + 2^f$ ($f \geq 2$), then

$$\rho = x_1^{2+2^f} (x_1, x_2)(x_3, x_4) \cdots (x_{N+1}, x_{N+2}).$$

If $\operatorname{im}(\chi)$ is generated by -1 and $1 + 2^f (f \geq 2)$, then

$$\rho = x_2^2 (x_1, x_2) x_3^{2^f} (x_3, x_4) \cdots (x_{N+1}, x_{N+2}).$$

The cohomological methods leading to these results were extended to give an explicit description of the *entire* absolute Galois group G_k of a local field. In the case that k is the field of Laurent series over a finite field, this was done by *H. Koch* [108]. If k is a p-adic local field, the structure of G_k was determined by *U. Jannsen* and *K. Wingberg* [99] for $p \neq 2$; and by *V. Diekert* [39] for $p = 2$ under the condition that $k(\mu_4)|k$ is unramified. The structure of $G_{\mathbb{Q}_2}$ is not known.

The result for local fields of positive characteristic is the following.

(7.5.13) Theorem (*Koch*). *Let k be a local field of characteristic p with residue field κ of cardinality q. Then G_k is isomorphic to the semi-direct product*

$$G_k \cong \mathop{\ast}_{\mathcal{G}_k} F_\omega \rtimes \mathcal{G}_k ,$$

where $\mathcal{G}_k = G(k_{tr}|k)$ is the Galois group of the maximal tamely ramified extension of k, and $\ast_{\mathcal{G}_k} F_\omega$ is the free pro-p-$\mathcal{G}_k$ operator group of countably infinite rank (cf. (4.3.8)).

In other words, the group G_k has generators σ, τ and x_i, $i \in \mathbb{N}$, subject to the following defining conditions resp. relations:

A) *The closed normal subgroup topologically generated by x_i, $i \in \mathbb{N}$, is a pro-p-group.*[*)]

B) *The elements σ, τ satisfy the relation $\sigma\tau\sigma^{-1} = \tau^q$.*

Proof: By (7.5.1) and (7.4.5)(ii), the ramification group V_k is a free pro-p-group, and we have an isomorphism of compact $\mathbb{Z}_p[\![\mathcal{G}_k]\!]$-modules

$$V_k^{ab} \cong \mathbb{Z}_p[\![\mathcal{G}_k]\!]^{\mathbb{N}},$$

i.e. V_k^{ab} is a free $\mathbb{Z}_p[\![\mathcal{G}_k]\!]$-module of countably infinite rank. Since the order of $G(k_{tr}|\tilde{k})$ is prime to p, we have $cd_p\, \mathcal{G}_k = cd_p\, G(\tilde{k}|k) = 1$. Therefore the group $\mathcal{G}_k$ is p-projective by (3.5.6), and the exact sequence

$$1 \longrightarrow V_k \longrightarrow G_k \longrightarrow \mathcal{G}_k \longrightarrow 1$$

splits. Fixing a splitting $G_k \cong V_k \rtimes \mathcal{G}_k$, the ramification group V_k becomes a pro-p-$\mathcal{G}_k$ operator group. Therefore we find a surjective homomorphism of pro-p-$\mathcal{G}_k$ operator groups

$$\psi : \mathop{\ast}_{\mathcal{G}_k} F_\omega \twoheadrightarrow V_k ,$$

where $\ast_{\mathcal{G}_k} F_\omega$ is the free pro-p-$\mathcal{G}_k$ operator group of countably infinite rank and ψ is defined by mapping the generators of F_ω to lifts of the $\mathbb{Z}_p[\![\mathcal{G}_k]\!]$-generators of V_k^{ab}. Since V_k is a free pro-p-group and the map

$$\bar{\psi} : (\mathop{\ast}_{\mathcal{G}_k} F_\omega)^{ab} = \mathbb{Z}_p[\![\mathcal{G}_k]\!]^{\mathbb{N}} \xrightarrow{\sim} V_k^{ab}$$

induced by ψ is an isomorphism, ψ is bijective by (1.6.15). Therefore we obtain an isomorphism

$$(\psi, id) : \mathop{\ast}_{\mathcal{G}_k} F_\omega \rtimes \mathcal{G}_k \xrightarrow{\sim} V_k \rtimes \mathcal{G}_k \cong G_k . \qquad \square$$

[*)] This topological condition could be replaced by an infinite set of algebraic relations.

Now let k be a p-adic field. We describe the structure of G_k in the case $p \neq 2$. It depends on the degree $N = [k : \mathbb{Q}_p]$, the cardinality $q = \#\kappa$ of the residue field κ, the order p^s of the group μ_{p^s} of all p-power roots of unity in the maximal tamely ramified extension $k_{tr}|k$ and on two further numbers $g, h \in \mathbb{Z}_p$ which are defined as follows. By (7.5.3), the Galois group $\mathcal{G}_k = G(k_{tr}|k)$ is generated by two elements σ, τ with defining relation $\sigma\tau\sigma^{-1} = \tau^q$. The actions of σ and τ on μ_{p^s} are given by two numbers $g, h \in \mathbb{Z}_p$ such that

$$\zeta^\sigma = \zeta^g\ ,\ \zeta^\tau = \zeta^h \ \text{ for } \ \zeta \in \mu_{tr}.$$

In the following we denote the commutator $xyx^{-1}y^{-1}$ by $[x, y]$ in contrast to the commutator $(x, y) = x^{-1}y^{-1}xy$ used in III §8. If one had taken there the commutator $[x, y]$ instead of (x, y), the shape of the Demuškin relation would not change because it is the same modulo F^3 (the third filtration step of the descending p-central series), and then the iteration process works as well. Which commutator to use is just a matter of personal taste.

(7.5.14) Theorem (JANNSEN-WINGBERG). *The group G_k is isomorphic to the profinite group generated by $N + 3$ generators $\sigma, \tau, x_0, \dots, x_N$, subject to the following defining conditions resp. relations.*

A) *The closed normal subgroup, topologically generated by $x_0, \dots, x_N$ is a pro-p-group.*

B) *The elements σ, τ satisfy the "tame" relation*

$$\sigma\tau\sigma^{-1} = \tau^q.$$

C) *In addition, the generators satisfy one further relation:*

(i) *for even N*

$$x_0^\sigma = \langle x_0, \tau\rangle^g x_1^{p^s}[x_1, x_2][x_3, x_4] \cdots [x_{N-1}, x_N],$$

(ii) *for odd N*

$$x_0^\sigma = \langle x_0, \tau\rangle^g x_1^{p^s}[x_1, y_1][x_2, x_3][x_4, x_5] \cdots [x_{N-1}, x_N],$$

where

$$\langle x_0, \tau\rangle = (x_0^{h^{p-1}}\tau x_0^{h^{p-2}}\tau \cdots x_0^h\tau)^{\frac{\pi}{p-1}}$$

($\pi = \pi_p$ *being the element of $\hat{\mathbb{Z}}$ with $\pi\hat{\mathbb{Z}} = \mathbb{Z}_p$), and where y_1 is a certain element in the subgroup generated by x_1, σ, τ, described below.*

The definition of the element y_1 is a little bit subtle. Let $\alpha : \mathcal{G}_k \to (\mathbb{Z}/p^s\mathbb{Z})^\times$ be the character describing the action of $\mathcal{G}_k = G(k_{tr}|k)$ on μ_{p^s}, and let $\beta : \mathcal{G}_k \to \mathbb{Z}_p^\times$ be a lift of α (not necessarily homomorphic). For $\rho \in \langle\sigma, \tau\rangle \subseteq G_k$ and $x \in G_k$ set

$$\{x, \rho\} := (x^{\beta(1)}\rho^2 x^{\beta(\rho)}\rho^2 \cdots x^{\beta(\rho^{p-2})}\rho^2)^{\frac{\pi}{p-1}}.$$

Writing $\tau_2 = \tau^{\pi_2}$ and $\sigma_2 = \sigma^{\pi_2}$, where π_2 is the element of $\hat{\mathbb{Z}}$ with $\pi_2\hat{\mathbb{Z}} = \mathbb{Z}_2$, y_1 is given by

$$y_1 = x_1^{\tau_2^{\frac{p+1}{2}}} \{x_1, \tau_2^{p+1}\}^{\sigma_2\tau_2^a} \{\{x_1, \tau_2^{p+1}\}, \sigma_2\tau_2^a\}^{\sigma_2\tau_2^b + \tau_2^{\frac{p+1}{2}}}.$$

Here $a, b \in \mathbb{Z}$ are chosen in such a way that

$$-\alpha(\sigma\tau^a) \bmod p \in (\mathbb{F}_p^\times)^2 \text{ and } -\alpha(\sigma\tau^b) \bmod p \notin (\mathbb{F}_p^\times)^2.$$

For the proof we refer to [99] and [39]. It is based on a theory of *H. Koch* [111] which axiomizes the fact that for every finite, tamely ramified extension $K|k$ the group $G_K(p)$ is a Demuškin group. We would like to mention the following special cases.

For $p > 2$, the group $G_{\mathbb{Q}_p}$ has four generators σ, τ, x_0, x_1 satisfying the relations

$$\tau^\sigma = \tau^p,$$

$$x_0^\sigma = \langle x_0, \tau\rangle\, x_1^p\, [x_1, x_1^{\tau_2^{\frac{p+1}{2}}} \{x_1, \tau_2^{p+1}\}^{\sigma_2\tau_2^{\frac{p-1}{2}}} \{\{x_1, \tau_2^{p+1}\}, \sigma_2\tau_2^{\frac{p-1}{2}}\}^{\sigma_2\tau_2^{\frac{p+1}{2}} + \tau_2^{\frac{p+1}{2}}}],$$

and for the group $G_{\mathbb{Q}_p(\zeta_p)}$ there are generators σ, τ, $x_0, \ldots, x_{p-1}$ satisfying

$$\tau^\sigma = \tau^p,$$

$$x_0^\sigma = (x_0\tau)^\pi\, x_1^p\, [x_1, x_2] \cdots [x_{p-2}, x_{p-1}].$$

Remarks: 1. Although we know by (7.4.1) that G_k can be generated by $N+2$ elements, it is more convenient to use $N+3$ generators in order to obtain "nice" relations.

In the case that $\mu_p \subseteq k$, there is also a satisfactory description with $N+2$ generators, cf. [99], §1.4(d).

2. Generators and relations for G_k were also assigned by *A. V. Jakovlev* in [92]. However, some mistakes required a comprehensive correction. This has been sketched only for the case of even N, and produced three relations, one of them being a somewhat complicated limit (cf. [93]).

We finish this section by showing that the absolute Galois group G_k of a p-adic local field k, $p \neq 2$, possesses nontrivial outer automorphisms. This is easy to see if $N = [k : \mathbb{Q}_p] > 1$. Let $\sigma, \tau, x_0, x_1, \ldots, x_N$ be the generators of G_k described in (7.5.14), satisfying the tame relation $\sigma\tau\sigma^{-1} = \tau^q$ and the wild relation

$$x_0^\sigma = \langle x_0, \tau\rangle^g x_1^{p^s}[x_1, -]\cdots[x_{N-1}, x_N].$$

If we define $\psi : G_k \longrightarrow G_k$ by

$$\psi(y) = y \text{ for } y = \sigma, \tau, x_0, \ldots, x_{N-1} \text{ and } \psi(x_N) = x_N \cdot x_{N-1},$$

then ψ is an automorphism of G_k. Indeed, since

$$[x_{N-1}, x_N] = [x_{N-1}, x_N x_{N-1}],$$

the generators $\sigma, \tau, x_0, \ldots, x_{N-1}, x_N \cdot x_{N-1}$ satisfy both relations if $N > 1$. Now suppose that ψ is an inner automorphism, i.e. there is an element $\rho \in G_k$ such that $\psi(z) = z^\rho$ for all $z \in G_k$. Recall that V_k^i denotes the i-th term of the p-central series of the ramification group V_k with $V_k^1 = V_k$. Since $x_{N-1}^\rho = \psi(x_{N-1}) = x_{N-1}$ and since $x_{N-1}V_k^2$ generates a free $\mathbb{F}_p[\![\mathcal{G}_k]\!]$-module in V_k/V_k^2, by [99], §2, we obtain that $\rho \in V_k$. It follows that $x_N V_k^2 = x_N \cdot x_{N-1} V_k^2$ which is a contradiction. Thus, if $N = [k : \mathbb{Q}_p] > 1$, we have constructed a nontrivial outer automorphism of G_k.

The case $k = \mathbb{Q}_p$ is more difficult. We use the following result from [250].

(7.5.15) Theorem. *Let*

$$\psi_3 : G_k/V_k^3 \xrightarrow{\sim} G_k/V_k^3$$

be an automorphism of G_k/V_k^3 which induces the identity on the factor group $\mathcal{G}_k = G_k/V_k$. Then there exists an automorphism ψ of G_k which coincides with ψ_3 modulo V_k^2.

Remark: This result was proven in [250], Satz 2, but was stated there in an incorrect manner. The result above is exactly what was needed for all results in the two papers [99] and [250], except for the statement in [99], §5.1. There an automorphism of $G_{\mathbb{Q}_p}$ was defined and claimed to be a nontrivial outer automorphism. But the argument used the incorrect formulation of [250], Satz 2 and therefore the constructed automorphism might be inner.

In order to proceed, we need some more notation. The homomorphism $\alpha : \mathcal{G}_k \to (\mathbb{Z}/p^s\mathbb{Z})^\times$ induces an involution * on $\mathbb{F}_p[\![\mathcal{G}_k]\!]$ which is defined by $\rho^* = \alpha(\rho)\rho^{-1}$ for $\rho \in \mathcal{G}_k$. We define the element E of $\mathbb{F}_p[\![\mathcal{G}_k]\!]$ by

$$E = \lim_{K} \frac{1}{e} \sum_{i=0}^{e-1} \tau^{2i} \alpha(\tau)^{-i},$$

where K runs through all finite tamely ramified Galois extensions of k and e denotes the ramification index of $K|k$. Then $E^* = E$ and E is an idempotent which is central in $\mathbb{F}_p[[\mathcal{G}_k]]$, because

$$\sigma E = \lim_K \frac{1}{e}\sum_{i=0}^{e-1} \tau^{2iq}\alpha(\tau)^{-i}\,\sigma = E^q\sigma = E\sigma.$$

From now on, let $k = \mathbb{Q}_p$. Then $q = p$, $\alpha(\sigma) = 1$ and $\alpha(\tau)$ is a primitive $(p-1)$-th root of unity. For

$$\varepsilon = 1 - 2E$$

we have $\varepsilon^* = \varepsilon$ and $\varepsilon^2 = 1$. In particular, ε is a central unit in $\mathbb{F}_p[[\mathcal{G}_{\mathbb{Q}_p}]]$. We write G, $\mathcal{G}$ and V for $G_{\mathbb{Q}_p}$, $\mathcal{G}_{\mathbb{Q}_p}$ and $V_{\mathbb{Q}_p}$, respectively.

Since $cd_p\,\mathcal{G} = 1$, the group $\mathcal{G}$ is p-projective by (3.5.6), and the exact sequence

$$1 \longrightarrow V \longrightarrow G \longrightarrow \mathcal{G} \longrightarrow 1$$

splits. Fixing a splitting $G \cong V \rtimes \mathcal{G}$, the ramification group V becomes a pro-p-$\mathcal{G}$ operator group. Therefore we find a surjective homomorphism of pro-p-$\mathcal{G}$ operator groups

$$f : \mathop{\ast}_{\mathcal{G}} F_2 \twoheadrightarrow V,$$

where $\ast_{\mathcal{G}} F_2$ is the free pro-p-$\mathcal{G}$ operator group of rank 2 with basis $\{x_0, x_1\}$ and f is defined by mapping the generators x_0 and x_1 to lifts of the $\mathbb{Z}_p[[\mathcal{G}]]$-generators of V^{ab}. We obtain a surjection

$$\mathcal{F} := \mathop{\ast}_{\mathcal{G}} F_2 \rtimes \mathcal{G} \twoheadrightarrow V \rtimes \mathcal{G} \cong G.$$

Let $\psi : \mathcal{F} \to \mathcal{F}$ be defined by

$$\psi(y) = y \ \text{ for } y = \sigma, \tau, x_0 \quad \text{and } \psi(x_1) = x_1^{\varepsilon}$$

(where the "sum" ε is chosen in some ordering). Then ψ is an automorphism of $\mathcal{F}$ since it is an automorphism modulo $\mathcal{F}^2$ (ε is a unit in $\mathbb{F}_p[[\mathcal{G}]]$).

We will show that for every finite tamely ramified Galois extension $K|k$ the automorphism ψ induces an automorphism

$$\psi_K : G_K(p)/G_K(p)^3 \xrightarrow{\sim} G_K(p)/G_K(p)^3\,.$$

In order to prove this, let κ_K and λ_K be elements of $\mathbb{F}_p[[\mathcal{G}]]$ defined as

$$\begin{aligned}\kappa_K &= \sum_{i=0}^{f-1} \sigma^i\alpha(\sigma)^{-i}\,,\\ \lambda_K &= \frac{1}{e}\sum_{i=0}^{e-1} \tau^i\alpha(\tau)^{-i}\,,\end{aligned}$$

where e and f denote the ramification index and the residue degree of K respectively. Let $\mathcal{F}_K$ be the pre-image of G_K in $\mathcal{F}$. By [99], §2 (observe that the index of the p-central series differs there from our notation by -1), the relation r_K in $G_K(p)$ satisfies

$$r_K \equiv (x_0^{-\sigma}\langle x_0, \tau\rangle x_1^p[x_1, y_1])^{\kappa_K \lambda_K} \mod \mathcal{F}_K(p)^3 .$$

Let $\delta \in \mathbb{F}_p[\![\mathcal{G}]\!]$ be such that $y_1 = x_1^\delta \mod \mathcal{F}^2$. Then we get

$$\begin{aligned}
\psi(r_K) &\equiv (x_0^{-\sigma}\langle x_0, \tau\rangle)^{\kappa_K\lambda_K} x_1^{p\varepsilon\kappa_K\lambda_K}[x_1^\varepsilon, x_1^{\varepsilon\delta}]^{\kappa_K\lambda_K}\\
&\equiv (x_0^{-\sigma}\langle x_0, \tau\rangle)^{\kappa_K\lambda_K} x_1^{p\kappa_K\lambda_K}[x_1, x_1^{\varepsilon\delta\varepsilon^*}]^{\kappa_K\lambda_K}\\
&\equiv (x_0^{-\sigma}\langle x_0, \tau\rangle)^{\kappa_K\lambda_K} x_1^{p\kappa_K\lambda_K}[x_1, x_1^{\delta}]^{\kappa_K\lambda_K}\\
&\equiv r_K \mod \mathcal{F}_K(p)^3 ,
\end{aligned}$$

since $\varepsilon\lambda_K = \lambda_K$ and ε is central in $\mathbb{F}_p[\![\mathcal{G}]\!]$, so that $\varepsilon\delta\varepsilon^* = \delta\varepsilon\varepsilon^* = \delta\varepsilon^2 = \delta$. This gives us the automorphism ψ_K.

Now, in the limit over all K, ψ induces an automorphism of V/V^3 which is $\mathcal{G}$-invariant because $\psi|_{\mathcal{G}} = id$. Thus we obtain an automorphism of $G/V^3 \cong V/V^3 \rtimes \mathcal{G}$, which is also denoted by ψ. By (7.5.15), it extends to an automorphism φ of G which coincides with ψ modulo V^2.

Suppose that φ is an inner automorphism, i.e. there exists an element $\rho \in G$ such that $\varphi(z) = z^\rho$ for all $z \in G$. Then $x_1^\rho \equiv \varphi(x_1) \equiv \psi(x_1) \equiv x_1^\varepsilon \mod V^2$. Since x_1 generates a free $\mathbb{F}_p[\![\mathcal{G}]\!]$-module in V/V^2, by [99], §2, we obtain $\varepsilon = \rho \mod V \in \mathcal{G}$, which is a contradiction. Thus we have shown that $G_{\mathbb{Q}_p}$ possesses nontrivial outer automorphisms.

Chapter VIII
Cohomology of Global Fields

§1. Cohomology of the Idèle Class Group

Having established the cohomology theory for local fields, we now begin its development for global fields, i.e. algebraic number fields and function fields in one variable over a finite field. The cohomology theory treats both types of fields equally.

The role that the multiplicative group of fields played in the local theory is now taken over for a global field k by the *idèle class group*. Let k be a global field, $\bar{k}$ a separable closure of k and $G_k = G(\bar{k}|k)$ its absolute Galois group. The **idèle group** I_k of k is defined as the restricted product

$$I_k = \prod_{\mathfrak{p}} k_{\mathfrak{p}}^{\times},$$

where $\mathfrak{p}$ runs through all primes of k including the archimedean ones if k is a number field, $k_{\mathfrak{p}}$ is the completion of k at $\mathfrak{p}$ and the restricted product is taken with respect to the unit groups $U_{\mathfrak{p}}$ in $k_{\mathfrak{p}}^{\times}$. The multiplicative group $k^{\times}$ of k injects diagonally into I_k and we define the **idèle class group** of k as the quotient

$$C_k := I_k/k^{\times}.$$

We refer the reader to [160], chap.VI, §1 for basic properties of the idèle and the idèle class group.

If K is a finite separable extension of k, then I_k naturally injects into I_K and we call the direct limit

$$I = \varinjlim_{K|k} I_K$$

the **idèle group of $\bar{k}$**. I is a discrete G_k-module and one easily observes that

$$I_K = I^{G_K}$$

for every finite separable extension $K|k$. The quotient

$$C := I/\bar{k}^{\times}$$

is the **idèle class group of $\bar{k}$**. We have

$$C = \varinjlim_{K|k} C_K$$

J. Neukirch et al., *Cohomology of Number Fields*,
Grundlehren der mathematischen Wissenschaften 323,
DOI 10.1007/978-3-540-37889-1_8,

and a straightforward application of Hilbert's Satz 90 implies that for every finite separable extension $K|k$,

$$C_K = C^{G_K}.$$

For every (possibly infinite) separable extension K of k we put $I_K = I^{G_K}$, $C_K = C^{G_K}$, and if $K|k$ is Galois, we set

$$H^i(K|k) = H^i(G(K|k), C_K),$$

and we also write $\hat{H}^i(K|k)$ for $\hat{H}^i(G(K|k), C_K)$.

The basis of our results in this chapter is the following theorem, called the class field axiom. It is an immediate consequence of the so-called first and second fundamental inequalities for global fields. For a proof of these results, we refer the reader to [6], chap.5,6, or, for the number field case, to [160], chap.VI, (4.4).

(8.1.1) Theorem (Class Field Axiom). *For a finite cyclic extension $K|k$ we have*

$$\#\hat{H}^i(K|k) = \begin{cases} [K:k] & \text{for} \quad i = 0, \\ 1 & \text{for} \quad i = 1. \end{cases}$$

Since $H^1(K|k) \cong \hat{H}^{-1}(K|k)$ for $K|k$ cyclic, the class field axiom is a statement about the kernel and cokernel of the norm map

$$C_K/I_{G(K|k)}C_K \xrightarrow{N_{K|k}} C_k$$

(here $I_{G(K|k)}$ is the augmentation ideal in $\mathbb{Z}[G(K|k)]$), and can therefore be considered as a noncohomological assertion. Starting with this input, we will calculate the cohomology groups of C_K for arbitrary Galois extensions.

Let $K|k$ be a finite separable extension. Setting

$$I_K(\mathfrak{p}) = \prod_{\mathfrak{P}|\mathfrak{p}} K_{\mathfrak{P}}^{\times},$$

we obtain a decomposition

$$I_K = \prod_{\mathfrak{p}} I_K(\mathfrak{p}),$$

where $\mathfrak{p}$ runs through all primes of k and the restricted product is taken with respect to the subgroups

$$U_K(\mathfrak{p}) := \prod_{\mathfrak{P}|\mathfrak{p}} U_{K,\mathfrak{P}},$$

where $U_{K,\mathfrak{P}}$ denotes the group of units in $K_{\mathfrak{P}}^{\times}$.

Passing to various extension fields, we frequently have to choose compatible prolongations of primes. Therefore we make the following fixed choice. For every prime $\mathfrak{p}$ of k, we choose a k-embedding $i_{\mathfrak{p}} : \bar{k} \hookrightarrow \bar{k}_{\mathfrak{p}}$, in particular a prime $\bar{\mathfrak{p}}$ of $\bar{k}$ above $\mathfrak{p}$, see [160], chap.II, (8.1). We denote the subextensions of $\bar{k}|k$ by $K|k$ and for each such extension the prime of K lying under $\bar{\mathfrak{p}}$ by the pointed letter $\mathfrak{P}^{\boldsymbol{\cdot}}$. Thus every separable extension $K|k$ comes equipped with a distinguished prime $\mathfrak{P}^{\boldsymbol{\cdot}}$ above $\mathfrak{p}$. By abuse of notation we write

$$K_{\mathfrak{p}} := i_{\mathfrak{p}}(K)k_{\mathfrak{p}}$$

for the extension $K_{\mathfrak{P}^{\boldsymbol{\cdot}}}$ of the local field $k_{\mathfrak{p}}$ which corresponds to the prime $\mathfrak{P}^{\boldsymbol{\cdot}}$ of K (and which is the completion of K at $\mathfrak{P}^{\boldsymbol{\cdot}}$ if $K|k$ is finite). We adopt the same convention for the unit groups and write $U_{K,\mathfrak{p}}$ for group of units in $K^{\times}_{\mathfrak{P}^{\boldsymbol{\cdot}}}$. Furthermore, we write $G_{\mathfrak{p}}(K|k)$ (or simply $G_{\mathfrak{p}}$ if K is clear from the context) for the decomposition group of $\mathfrak{P}^{\boldsymbol{\cdot}}$ in $G(K|k)$, i.e. $G_{\mathfrak{p}}(K|k) = G(K_{\mathfrak{p}}|k_{\mathfrak{p}})$.

Now let $K|k$ be a finite Galois extension and let $G = G(K|k)$ be its Galois group. The idèle group I_K is then a G-module and

$$I_K = \prod_{\mathfrak{p}} I_K(\mathfrak{p})$$

is a decomposition into G-modules. Then

$$I_K(\mathfrak{p}) = \prod_{\sigma \in G/G_{\mathfrak{p}}} K^{\times}_{\sigma\mathfrak{P}^{\boldsymbol{\cdot}}} = \prod_{\sigma \in G/G_{\mathfrak{p}}} \sigma K^{\times}_{\mathfrak{P}^{\boldsymbol{\cdot}}} = \mathrm{Ind}_G^{G_{\mathfrak{p}}}(K^{\times}_{\mathfrak{p}}),$$

and similarly $U_K(\mathfrak{p}) = \mathrm{Ind}_G^{G_{\mathfrak{p}}}(U_{\mathfrak{P}^{\boldsymbol{\cdot}}})$. By Shapiro's lemma, we obtain isomorphisms

$$\hat{H}^i(G, I_K(\mathfrak{p})) \cong \hat{H}^i(G_{\mathfrak{p}}, K^{\times}_{\mathfrak{p}})$$

for all $i \in \mathbb{Z}$. Furthermore, if $\mathfrak{p}$ is unramified in K, we have for all $i \in \mathbb{Z}$

$$\hat{H}^i(G, U_K(\mathfrak{p})) \cong \hat{H}^i(G_{\mathfrak{p}}, U_{\mathfrak{P}^{\boldsymbol{\cdot}}}) = 0$$

by (7.1.2)(i). From this follows the

(8.1.2) Proposition. *For a finite Galois extension $K|k$, we have*

$$\hat{H}^i(G, I_K) \cong \bigoplus_{\mathfrak{p}} \hat{H}^i(G_{\mathfrak{p}}, K^{\times}_{\mathfrak{p}})$$

for all $i \in \mathbb{Z}$, where $\mathfrak{p}$ runs through all primes of k.

Proof: Let S run through the finite sets of primes of k containing the primes $\mathfrak{p}$ ramified in K and the infinite primes if k is a number field. Setting

$$I_K^S = \prod_{\mathfrak{p} \in S} I_K(\mathfrak{p}) \times \prod_{\mathfrak{p} \notin S} U_K(\mathfrak{p}),$$

we have $I_K = \varinjlim_S I_K^S$ and

$$\begin{aligned}\hat{H}^i(G,I_K) &= \varinjlim_S \hat{H}^i(G,I_K^S)\\ &= \varinjlim_S \Big(\hat{H}^i(G,\prod_{\mathfrak{p}\in S} I_K(\mathfrak{p}))\times \hat{H}^i(G,\prod_{\mathfrak{p}\notin S} U_K(\mathfrak{p}))\Big)\\ &= \varinjlim_S \Big(\prod_{\mathfrak{p}\in S}\hat{H}^i(G,I_K(\mathfrak{p}))\times \prod_{\mathfrak{p}\notin S}\hat{H}^i(G,U_K(\mathfrak{p}))\Big)\\ &\cong \varinjlim_S \Big(\prod_{\mathfrak{p}\in S}\hat{H}^i(G_\mathfrak{p},K_\mathfrak{p}^\times)\Big) = \bigoplus_\mathfrak{p} \hat{H}^i(G_\mathfrak{p},K_\mathfrak{p}^\times).\end{aligned}$$ □

By Hilbert's Satz 90 and by (7.2.2), we obtain the

(8.1.3) Corollary. $H^1(G,I_K) = H^3(G,I_K) = 0.$

By proposition (8.1.2), we may associate to every cohomology class $c \in \hat{H}^i(G,I_K)$ its *local components* $c_\mathfrak{p} \in \hat{H}^i(G_\mathfrak{p},K_\mathfrak{p}^\times)$: $c_\mathfrak{p}$ is the image of c under the composite of the maps

$$\hat{H}^i(G,I_K) \xrightarrow{res_\mathfrak{p}} \hat{H}^i(G_\mathfrak{p},I_K) \xrightarrow{\pi_\mathfrak{p}} \hat{H}^i(G_\mathfrak{p},K_\mathfrak{p}^\times),$$

where $\pi_\mathfrak{p}$ is induced by the projection $I_K \to K_{\mathfrak{P}\cdot}^\times = K_\mathfrak{p}^\times$. Since these maps commute with *inf*, *res*, *cor*, we obtain in a straightforward way the

(8.1.4) Proposition. *Let $L \supseteq K \supseteq k$ be finite Galois subextensions of $\bar{k}|k$. Then for $i \geq 1$*

(i) $inf_{L|k}^{K|k}(c)_\mathfrak{p} = inf_{L_\mathfrak{p}|k_\mathfrak{p}}^{K_\mathfrak{p}|k_\mathfrak{p}}(c_\mathfrak{p})$ *for* $c \in H^i(G(K|k),I_K)$,

(ii) $res_K^k(c)_\mathfrak{P} = res_{K_\mathfrak{P}}^{k_\mathfrak{p}}(c_\mathfrak{p})$ *for* $c \in H^i(G(L|k),I_L)$,

(iii) $cor_k^K(c)_\mathfrak{p} = \sum_{\sigma\in G/G_\mathfrak{p}} \sigma_*^{-1} cor_{k_\mathfrak{p}}^{K_{\sigma\mathfrak{P}\cdot}}(c_{\sigma\mathfrak{P}\cdot})$ *for* $c \in H^i(G(L|K),I_L)$,

where σ_* is the map $H^i(G(K_{\mathfrak{P}\cdot}|k_\mathfrak{p}),K_{\mathfrak{P}\cdot}^\times) \longrightarrow H^i(G(K_{\sigma\mathfrak{P}\cdot}|k_\mathfrak{p}),K_{\sigma\mathfrak{P}\cdot}^\times)$ induced by $\sigma : K_{\mathfrak{P}\cdot} \to \sigma K_{\mathfrak{P}\cdot} = K_{\sigma\mathfrak{P}\cdot}$. For the last two formulae it suffices to require only that $L|k$ is Galois.

Although the idèle group $I = \varinjlim_{K|k} I_K$ of $\bar{k}$ is not the restricted product $\prod_\mathfrak{p} \bar{k}_\mathfrak{p}^\times$, we obtain the following direct decomposition for its cohomology by applying $\varinjlim_{K|k}$ with respect to the inflation maps to the formula (8.1.2):

$$H^i(k, I) \cong \bigoplus_{\mathfrak{p}} H^i(G_{\mathfrak{p}}(\bar{k}|k), (\bar{k})_{\mathfrak{p}}^{\times}) \qquad \text{for all } i \geq 1.$$

Now the question naturally occurs of whether the canonical surjection $G(\bar{k}_{\mathfrak{p}}|k_{\mathfrak{p}}) \twoheadrightarrow G_{\mathfrak{p}}(\bar{k}|k)$ is an isomorphism, or equivalently, whether $(\bar{k})_{\mathfrak{p}} = (\overline{k_{\mathfrak{p}}})$.

(8.1.5) Proposition. *Let $\mathfrak{p}$ be a prime of the global field k. Then*

$$(\bar{k})_{\mathfrak{p}} = (\overline{k_{\mathfrak{p}}}).$$

For the proof we need

(8.1.6) Krasner's Lemma. *Let κ be a complete field with respect to a nonarchimedean valuation and let Ω be an algebraic closure of κ. Let $\alpha \in \Omega$ be separable over κ and let $\alpha = \alpha_1, \ldots, \alpha_n$ be the conjugates of α over κ. Suppose that for $\beta \in \Omega$ we have*

$$|\alpha - \beta| < |\alpha - \alpha_i| \qquad \text{for } i = 2, \ldots, n,$$

where $|\ |$ denotes the unique extension of the valuation to Ω. Then $\kappa(\alpha) \subseteq \kappa(\beta)$.

Proof: Consider the extension $\kappa(\alpha, \beta)|\kappa(\beta)$ and let $K|\kappa(\beta)$ be its Galois closure. Let $\sigma \in G(K|\kappa(\beta))$. Then $\sigma(\beta - \alpha) = \beta - \sigma(\alpha)$. Since $|\sigma(x)| = |x|$ for all x (by the uniqueness of the extension of the absolute value), we have

$$|\beta - \sigma(\alpha)| = |\beta - \alpha| < |\alpha_i - \alpha|$$

for $i = 2, \ldots, n$. Therefore

$$|\alpha - \sigma(\alpha)| < \max\{|\alpha - \beta|, |\beta - \sigma(\alpha)|\} < |\alpha - \alpha_i|$$

for $i = 2, \ldots, n$. It follows that $\sigma(\alpha) = \alpha$, and so $\alpha \in \kappa(\beta)$. □

Proof of (8.1.5): We have the natural inclusion $(\bar{k})_{\mathfrak{p}} \subseteq (\overline{k_{\mathfrak{p}}})$. The proposition is trivial if $\mathfrak{p}$ is archimedean, so assume that $\mathfrak{p}$ is finite. Let α be in $(\overline{k_{\mathfrak{p}}})$ and let $f \in k_{\mathfrak{p}}[X]$ be its minimal polynomial. Since k is dense in $k_{\mathfrak{p}}$, we can choose a polynomial $g \in k[X]$ near to f. Then $|g(\alpha)| = |g(\alpha) - f(\alpha)|$ is small. Writing $g(X) = \prod(X - \beta_j)$ with $\beta_j \in \bar{k} \subseteq (\bar{k})_{\mathfrak{p}}$, we see that $|\alpha - \beta|$ is small for some root β of $g(X)$. In particular, we can choose $g(X)$ and then β such that $|\beta - \alpha| < |\alpha_i - \alpha|$ for all conjugates $\alpha_i \in (\overline{k_{\mathfrak{p}}})$ of α, $\alpha_i \neq \alpha$. By Krasner's lemma, we obtain $\alpha \in k_{\mathfrak{p}}(\beta) = (k(\beta))_{\mathfrak{p}} \subseteq (\bar{k})_{\mathfrak{p}}$. □

Passing to the limit over all finite extensions K of k inside $\bar{k}$, (8.1.2) and (8.1.5) immediately imply

(8.1.7) Proposition. $H^i(k, I) \cong \bigoplus_{\mathfrak{p}} H^i(k_{\mathfrak{p}}, \bar{k}_{\mathfrak{p}}^{\times})$ *for all* $i \geq 1$.

Since $H^1(k, I) = 0$, the inflation maps

$$H^2(G(K|k), I_K) \to H^2(k, I)$$

are injective. We identify the first group with its image. Then the equality $H^2(k, I) = \varinjlim_{K|k} H^2(G(K|k), I_K)$ becomes

$$H^2(k, I) = \bigcup_{K|k} H^2(G(K|k), I_K).$$

In the local case we have seen this with the multiplicative groups in place of I. Like the unramified extensions in the local case, a decisive role is played here by the *cyclic* extensions $K|k$ because of their periodic cohomological behaviour. For this reason the following analogue (8.1.9) of the local situation is of crucial importance. But first we observe the

(8.1.8) Proposition. *Let $L|k$ be a Galois extension and let p be a prime number. Suppose in the number field case that L is totally imaginary if $p = 2$. If p^∞ divides the local degrees $[L_{\mathfrak{p}} : k_{\mathfrak{p}}]$ for all finite primes $\mathfrak{p}$ of k, then*

$$H^2(L, I)(p) = 0$$

and

$$H^2(G(L|k), I_L)(p) \cong H^2(k, I)(p).$$

Proof: Let $K|k$ run through the finite subextensions of $L|k$. Then $H^1(L, I) = \varinjlim_{K,res} H^1(K, I) = 0$, hence the sequence

$$0 \longrightarrow H^2(G(L|k), I_L) \longrightarrow H^2(k, I) \longrightarrow H^2(L, I)$$

is exact. Therefore it suffices to prove $H^2(L, I)(p) = 0$. Using (8.1.7) and passing to the direct limit, we obtain

$$H^2(L, I)(p) = \bigoplus_{\mathfrak{p}} \mathrm{Ind}_{G(L|k)}^{G_{\mathfrak{p}}(L|k)} H^2(L_{\mathfrak{p}}, \bar{L}_{\mathfrak{p}}^{\times})(p).$$

For finite primes we have $H^2(L_{\mathfrak{p}}, \bar{L}_{\mathfrak{p}}^{\times})(p) = 0$ by (7.1.6), and for the archimedean primes the group $H^2(L_{\mathfrak{p}}, \bar{L}_{\mathfrak{p}}^{\times})(p)$ is trivially zero for $p \neq 2$ and by assumption also for $p = 2$. □

Let $k(\mu)$ be the extension of k obtained by adjoining all roots of unity in $\bar{k}$ to k. This extension has an abelian Galois group. In the function field case, $k(\mu)$ is the extension obtained by passing to the algebraic closure of the constant field and $G(k(\mu)|k) \cong \hat{\mathbb{Z}}$. We use the notation $\tilde{k} = k(\mu)$ in this case. If k is a number field, let T be the torsion subgroup of $G(k(\mu)|k)$. Then the field $\tilde{k} := k(\mu)^T$ is an extension of k with Galois group $\hat{\mathbb{Z}}$ which we call the **cyclotomic $\hat{\mathbb{Z}}$-extension** of k *). The decomposition group of a finite prime $G_{\mathfrak{p}}(\tilde{k}|k) \subseteq G(\tilde{k}|k)$ is open and, in particular, isomorphic to $\hat{\mathbb{Z}}$. Since $\hat{\mathbb{Z}}$ is torsion-free, archimedean primes are unramified in $\tilde{k}|k$.

(8.1.9) Proposition. $$H^2(k, I) = \bigcup_{\substack{K|k \\ \text{cyclic}}} H^2(G(K|k), I_K)\,.$$

Moreover, it suffices to take the union over all cyclic extensions of k inside $\tilde{k}$ if k is a function field or a totally imaginary number field. If k is a number field having a real place, it suffices to take the union over all cyclic extensions of k inside $\tilde{k}_1$, where $k_1|k$ is an arbitrarily chosen totally imaginary quadratic extension of k.

Proof: Let $x \in H^2(k, I)$. Using (8.1.7), we decompose x into an archimedean and a nonarchimedean part: $x = x_a + x_n$. By (8.1.8), we find a finite extension K of k inside $\tilde{k}$ such that $res^k_K\, x_n \in H^2(K, I)$ vanishes. This finishes the proof if k is a function field or a totally imaginary number field. If k has a real place, let k_1 be an arbitrarily chosen totally imaginary quadratic extension of k. Let $K_1|k$ be the uniquely defined cyclic extension of k inside $\tilde{k}$ of degree $2 \cdot [K : k]$; in particular, K_1 is a quadratic extension of K. The extensions $k_1|k$ and $K_1|k$ are linearly disjoint because the first is totally ramified and the second is unramified at all real places. The extension $k_1K_1|K$ has Galois group $\mathbb{Z}/2\mathbb{Z} \oplus \mathbb{Z}/2\mathbb{Z}$ and thus contains a unique subextension $L|K$ of degree 2 distinct from K_1 and k_1K. We have the following diagram of fields.

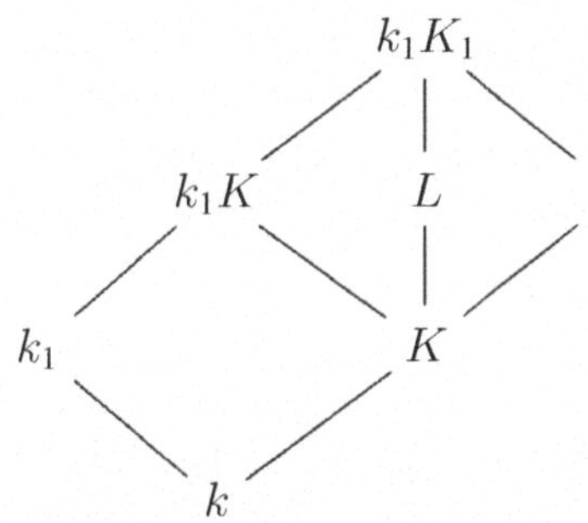

*) The cyclotomic $\hat{\mathbb{Z}}$-extension is the composite of the cyclotomic $\mathbb{Z}_p$-extensions for all prime numbers p. We will recall the definition of the cyclotomic $\mathbb{Z}_p$-extension in XI §1.

Since L is totally imaginary and cyclic over k, we conclude that $res^k_L x \in H^2(L, I)$ is zero. □

Again let $K|k$ be a finite Galois extension with Galois group G and decomposition groups $G_\mathfrak{p} = G(K_\mathfrak{p}|k_\mathfrak{p})$. For every prime $\mathfrak{p}$, we have by VII §1 the invariant map

$$inv_{K_\mathfrak{p}|k_\mathfrak{p}} : H^2(G_\mathfrak{p}, K_\mathfrak{p}^\times) \xrightarrow{\sim} \tfrac{1}{[K_\mathfrak{p}:k_\mathfrak{p}]}\mathbb{Z}/\mathbb{Z}.$$

From the decomposition

$$H^2(G, I_K) \cong \bigoplus_\mathfrak{p} H^2(G_\mathfrak{p}, K_\mathfrak{p}^\times),$$

we obtain a canonical homomorphism

$$inv_{K|k} : H^2(G, I_K) \longrightarrow \tfrac{1}{[K:k]}\mathbb{Z}/\mathbb{Z},$$

given by

$$inv_{K|k}(c) = \sum_\mathfrak{p} inv_{K_\mathfrak{p}|k_\mathfrak{p}}(c_\mathfrak{p}).$$

This invariant map is compatible with *inf*, *res*, *cor* as the following proposition shows.

(8.1.10) Proposition. *If $L \supseteq K \supseteq k$ are finite separable extensions with $L|k$ Galois, then we have the commutative diagrams*

$$\begin{array}{ccc} H^2(G(L|K), I_L) & \xrightarrow{inv_{L|K}} & \frac{1}{[L:K]}\mathbb{Z}/\mathbb{Z} \\ {\scriptstyle res}\big\downarrow\big\uparrow{\scriptstyle cor} & & {\scriptstyle [K:k]}\big\downarrow\big\uparrow{\scriptstyle incl} \\ H^2(G(L|k), I_L) & \xrightarrow{inv_{L|k}} & \frac{1}{[L:k]}\mathbb{Z}/\mathbb{Z}\ . \end{array}$$

Moreover, if $K|k$ is Galois, then $inv_{L|k}$ is an extension of $inv_{K|k}$.

Proof: The proposition is an immediate consequence of (8.1.3) and of the analogous properties of the local *inv*'s as invariant maps of a class formation (see (3.1.8)). If

$$c \in H^2(G(K|k), I_K) \subseteq H^2(G(L|k), I_L),$$

then

$$inv_{L|k}(c) = \sum_\mathfrak{p} inv_{L_\mathfrak{p}|k_\mathfrak{p}}(c_\mathfrak{p}) = \sum_\mathfrak{p} inv_{K_\mathfrak{p}|k_\mathfrak{p}}(c_\mathfrak{p}) = inv_{K|k}(c).$$

Letting $\mathfrak{P}$ run through the primes of K and choosing always a prime $\mathfrak{P}'|\mathfrak{P}$ of L, we obtain for $c \in H^2(G(L|k), I_L)$

$$\begin{aligned}
inv_{L|K}(res\, c) &= \sum_{\mathfrak{P}} inv_{L_{\mathfrak{P}'}|K_{\mathfrak{P}}}(res_K^k(c)_{\mathfrak{P}}) = \sum_{\mathfrak{P}} inv_{L_{\mathfrak{P}'}|K_{\mathfrak{P}}}(res_{K_{\mathfrak{P}}}^{k_{\mathfrak{p}}}(c_{\mathfrak{p}}))\\
&= \sum_{\mathfrak{P}} [K_{\mathfrak{P}} : k_{\mathfrak{p}}]\, inv_{L_{\mathfrak{P}'}|k_{\mathfrak{p}}}(c_{\mathfrak{p}})\\
&= \sum_{\mathfrak{p}} \sum_{\mathfrak{P}|\mathfrak{p}} [K_{\mathfrak{P}} : k_{\mathfrak{p}}]\, inv_{L_{\mathfrak{P}'}|k_{\mathfrak{p}}}(c_{\mathfrak{p}})\\
&= \sum_{\mathfrak{p}} [K : k]\, inv_{L_{\mathfrak{P}'}|k_{\mathfrak{p}}}(c_{\mathfrak{p}})\\
&= [K : k]\, inv_{L|k}(c).
\end{aligned}$$

Finally, for $c \in H^2(G(L|K), I_L)$ we obtain

$$\begin{aligned}
inv_{L|k}(cor\ c) &= \sum_{\mathfrak{p}} inv_{L_{\mathfrak{P}'}|k_{\mathfrak{p}}}(cor_k^K(c)_{\mathfrak{p}}) = \sum_{\mathfrak{p}} \sum_{\mathfrak{P}|\mathfrak{p}} inv_{L_{\mathfrak{P}'}|k_{\mathfrak{p}}}(cor_{k_{\mathfrak{p}}}^{K_{\mathfrak{P}}}(c_{\mathfrak{P}}))\\
&= \sum_{\mathfrak{p}} \sum_{\mathfrak{P}|\mathfrak{p}} inv_{L_{\mathfrak{P}'}|K_{\mathfrak{P}}}(c_{\mathfrak{P}}) \quad = inv_{L|K}(c).
\end{aligned}$$ □

By the compatibility of *inv* with the inflation, we also obtain invariant maps for infinite Galois subextensions $K|k$ of $\bar{k}|k$,

$$inv_{K|k} : H^2(G(K|k), I_K) \longrightarrow \tfrac{1}{[K:k]}\mathbb{Z}/\mathbb{Z},$$

by passing to the direct limit over the finite Galois subextensions $K_\alpha|k$ of $K|k$ and setting

$$\tfrac{1}{[K:k]}\mathbb{Z}/\mathbb{Z} = \bigcup_{\alpha} \tfrac{1}{[K_\alpha:k]}\mathbb{Z}/\mathbb{Z}.$$

In particular, we have an invariant map

$$inv_k : H^2(k, I) \longrightarrow \mathbb{Q}/\mathbb{Z}$$

for the idèle group I of the separable closure $\bar{k}$.

For every finite Galois extension $K|k$ with Galois group G, local class field theory provides us with the norm residue symbol

$$(\quad, K|k) : I_k \longrightarrow G^{ab}$$

given by

$$(\alpha, K|k) = \prod_{\mathfrak{p}} (\alpha_{\mathfrak{p}}, K_{\mathfrak{p}}|k_{\mathfrak{p}}).$$

The product on the right is finite, i.e. well-defined, because $(\alpha_{\mathfrak{p}}, K_{\mathfrak{p}}|k_{\mathfrak{p}}) = 1$ for all primes $\mathfrak{p}$ such that $\alpha_{\mathfrak{p}} \in U_{\mathfrak{p}}$ and $\mathfrak{p}$ is unramified in $K|k$. This norm residue symbol is linked with the invariant map $inv_{K|k}$ as follows. For every character $\chi \in H^1(G, \mathbb{Q}/\mathbb{Z})$, the exact sequence

$$0 \longrightarrow \mathbb{Z} \longrightarrow \mathbb{Q} \longrightarrow \mathbb{Q}/\mathbb{Z} \longrightarrow 0$$

yields an element $\delta\chi \in H^2(G,\mathbb{Z})$. Moreover, we have for every idèle $\alpha \in I_k = H^0(G, I_K)$, the cup-product $\alpha \cup \delta\chi \in H^2(G(K|k), I_K)$.

(8.1.11) Proposition. *For every $\chi \in H^1(G,\mathbb{Q}/\mathbb{Z})$ and $\alpha \in I_k$, we have*

$$\chi((\alpha, K|k)) = inv_{K|k}(\alpha \cup \delta\chi).$$

Proof: The result follows from its local analogue (7.2.12). If $\chi_\mathfrak{p}$ is the restriction of χ to $G_\mathfrak{p} = G(K_\mathfrak{p}|k_\mathfrak{p})$, then $\alpha_\mathfrak{p} \cup \delta\chi_\mathfrak{p} \in H^2(G_\mathfrak{p}, K_\mathfrak{p}^\times)$ is obviously the local component of $\alpha \cup \delta\chi$, so that

$$\begin{aligned} inv_{K|k}(\alpha \cup \delta\chi) &= \sum_\mathfrak{p} inv_{K_\mathfrak{p}|k_\mathfrak{p}}(\alpha_\mathfrak{p} \cup \delta\chi_\mathfrak{p}) \\ &= \sum_\mathfrak{p} \chi_\mathfrak{p}((\alpha_\mathfrak{p}, K_\mathfrak{p}|k_\mathfrak{p})) = \chi((\alpha, K|k)). \end{aligned}$$ □

The next proposition is a first step towards theorem (8.1.22) below, which claims that the pair (G_k, C) is a class formation.

(8.1.12) Proposition. *For every finite Galois extension $K|k$*

$$H^1(G(K|k), C_K) = 0.$$

Proof: If $K|k$ is cyclic, this is part of the class field axiom (8.1.1). If $[K : k] = p^n$ is a prime power, then there exists a cyclic subextension $L|k$ of $K|k$ of degree p, and we can proceed by induction on $[K : k]$ using the exact sequence

$$0 \longrightarrow H^1(G(L|k), C_L) \longrightarrow H^1(G(K|k), C_K) \longrightarrow H^1(G(K|L), C_K).$$

If $K|k$ is arbitrary, we consider the fixed fields Σ_p of the p-Sylow subgroups of $G(K|k)$. Then, by (1.6.10), the restriction map

$$res : H^1(G(K|k), C_K) \hookrightarrow \prod_p H^1(G(K|\Sigma_p), C_K)$$

is injective and we are done. □

(8.1.13) Corollary. $H^1(k, C) = 0.$

By (6.3.4), the group $H^2(k, \bar{k}^\times)$ is canonically isomorphic to the Brauer group $Br(k)$ of central simple algebras over the global field k.

(8.1.14) Proposition. *Let k be a global field. Then*

(i)
$$Br(k) = \bigcup_{\substack{K|k \\ \text{cyclic}}} Br(K|k).$$

(ii) *Let $K|k$ be an infinite Galois extension such that p^∞ divides the local degrees $[K_\mathfrak{p} : k_\mathfrak{p}]$ for all finite primes $\mathfrak{p}$ of K and assume in the number field case that K is totally imaginary if $p = 2$. Then*
$$Br(K)(p) = 0.$$

Remark: We could have sharpened assertion (i) in a similar manner to (8.1.9).

Proof: For every finite Galois extension $K|k$, (8.1.13) and $H^1(K, I) = 0 = H^1(K, \bar{k}^\times)$ give the exact commutative diagram

$$\begin{array}{ccccc}
0 & \longrightarrow & H^2(K, \bar{k}^\times) & \longrightarrow & H^2(K, I) \\
 & & \uparrow & & \uparrow \\
0 & \longrightarrow & H^2(k, \bar{k}^\times) & \longrightarrow & H^2(k, I) \\
 & & \uparrow & & \uparrow \\
0 & \longrightarrow & H^2(G(K|k), K^\times) & \longrightarrow & H^2(G(K|k), I_K) \\
 & & \uparrow & & \uparrow \\
 & & 0 & & 0.
\end{array}$$

Considering this diagram for cyclic extensions $K|k$, we obtain assertion (i) from (8.1.9). Since $Br(K)(p)$ injects into $H^2(K, I)(p)$, the second statement follows from (8.1.8). □

(8.1.15) Proposition. *Let $K|k$ be a cyclic extension with Galois group G. Then the sequence*
$$0 \longrightarrow H^2(G, K^\times) \longrightarrow H^2(G, I_K) \xrightarrow{inv_{K|k}} \tfrac{1}{[K:k]}\mathbb{Z}/\mathbb{Z} \longrightarrow 0$$
is exact.

Proof: We start by showing that $inv_{K|k} : H^2(G, I_K) \to \frac{1}{[K:k]}\mathbb{Z}/\mathbb{Z}$ is surjective. We first assume that $[K : k]$ is a prime power, say p^n. Let K' be the unique extension of degree p inside $K|k$. If all primes of k would split in K', then the norm map $N_{K'|k} : I_{K'} \to I_k$ would be surjective, and so would be the norm map $N_{K'|k} : C_{K'} \to C_k$. Since this contradicts $\#\hat{H}^0(G(K'|k), C_{K'}) = p$, we find a prime $\mathfrak{p}$ of k which is inert in K', and hence also in K. Therefore $G = G_\mathfrak{p} = G(K_\mathfrak{p}|k_\mathfrak{p})$, and we know that

$$inv_{K_\mathfrak{p}|k_\mathfrak{p}} : H^2(G_\mathfrak{p}, K_\mathfrak{p}^\times) \longrightarrow \tfrac{1}{[K_\mathfrak{p}:k_\mathfrak{p}]}\mathbb{Z}/\mathbb{Z}$$

is bijective. We conclude that

$$inv_{K|k} : H^2(G, I_K) \to \tfrac{1}{[K:k]}\mathbb{Z}/\mathbb{Z}$$

is surjective if $[K : k]$ is a prime power. The general case now easily follows from (8.1.10).

Since G is cyclic, $H^3(G, K^\times) = H^1(G, K^\times) = 0$. Moreover, $H^1(G, C_K) = 0$, and so the exact sequence

$$0 \to K^\times \to I_K \to C_K \to 0$$

yields the exact sequence

$$0 \longrightarrow H^2(G, K^\times) \longrightarrow H^2(G, I_K) \longrightarrow H^2(G, C_K) \longrightarrow 0;$$

in particular, the map $H^2(G, K^\times) \longrightarrow H^2(G, I_K)$ is injective and, by (8.1.1), its cokernel has order

$$\#H^2(G, C_K) = \#\hat{H}^0(G, C_K) = [K : k].$$

It remains to show that *inv* is trivial on the image of $H^2(G, K^\times)$. We prove this without the assumption of $K|k$ being cyclic. So let $\alpha \in Br(k)$ be arbitrary. If k is a function field, then, by (8.1.14) and the following remark, we may assume that $\alpha \in Br(K|k)$, where $K = k(\zeta_n)$ for some n prime to $char(k)$. If k is number field, let K be a finite extension of k which is Galois over $\mathbb{Q}$ and such that $\alpha \in Br(K|k)$. Then

$$inv_k(\alpha) = inv_{K|k}(\alpha) = inv_{K|\mathbb{Q}}(cor_\mathbb{Q}^k(\alpha)).$$

Hence we may assume that $k = \mathbb{Q}$ and, by the same argument as in the function field case, $\alpha \in Br(K|\mathbb{Q})$ for a cyclic subextension K of $\mathbb{Q}(\zeta_n)|\mathbb{Q}$ for some n. In both cases, let $G = G(K|k)$ and let χ be a generator of $H^1(G, \mathbb{Q}/\mathbb{Z})$. Then $\delta\chi$ is a generator of $H^2(G, \mathbb{Z})$, and the cup-product

$$\delta\chi\cup : \hat{H}^0(G, K^\times) \longrightarrow H^2(G, K^\times)$$

is the periodicity isomorphism (see (1.7.1)). Therefore every element of $H^2(G, K^\times)$ is of the form $\bar{a} \cup \delta\chi$ with $a \in k^\times$. By (8.1.11), we have

$$inv_{K|k}(\bar{a} \cup \delta\chi) = \chi((a, K|k)).$$

It therefore remains to show that $(a, K|k) = 1$ for $a \in k^\times$. Hence it suffices to show that $(a, k(\zeta_n)|k)\zeta_n = \zeta_n$, where k is a function field and $(n, \mathrm{char}(k)) = 1$ or $k = \mathbb{Q}$ and n arbitrary. Let k be a function field. Then, for any place $\mathfrak{p}$ of k, we have

$$(a, k_\mathfrak{p})\zeta_n = \zeta_n^{\#k(\mathfrak{p})^{v_\mathfrak{p}(a)}}.$$

The 'product formula' yields

$$\prod_\mathfrak{p} \#k(\mathfrak{p})^{v_\mathfrak{p}(a)} = 1,$$

hence

$$(a, k)\zeta_n = \zeta_n^{\prod_\mathfrak{p} \#k(\mathfrak{p})^{v_\mathfrak{p}(a)}} = \zeta_n.$$

The argument in the case $k = \mathbb{Q}$ is similar, but the computation of the local norm residue symbols at the primes $p \mid n$ is much more involved, see, e.g., [160], chap.VI, (5.3) and chap.V, (2.4). This proves the proposition. □

Using (8.1.14)(i) and (8.1.9), we obtain from (8.1.15) the following

(8.1.16) Corollary. *We have an exact sequence*

$$0 \longrightarrow H^2(k, \bar{k}^\times) \longrightarrow H^2(k, I) \xrightarrow{inv_k} \mathbb{Q}/\mathbb{Z} \longrightarrow 0.$$

By (8.1.7), we have

$$H^2(k, I) \cong \bigoplus_\mathfrak{p} H^2(k_\mathfrak{p}, \bar{k}_\mathfrak{p}^\times),$$

and the summands are the Brauer groups $Br(k_\mathfrak{p})$ of the local fields $k_\mathfrak{p}$. Hence corollary (8.1.16) gives us the famous **Hasse principle** for central simple algebras:

(8.1.17) Theorem. *We have an exact sequence*

$$0 \longrightarrow Br(k) \longrightarrow \bigoplus_\mathfrak{p} Br(k_\mathfrak{p}) \xrightarrow{inv_k} \mathbb{Q}/\mathbb{Z} \longrightarrow 0,$$

where inv_k is the sum of the local invariant maps $inv_{k_\mathfrak{p}} : Br(k_\mathfrak{p}) \to \mathbb{Q}/\mathbb{Z}$.

(8.1.18) Corollary. *Let p be a prime number and let $K|k$ be an infinite separable extension of the global field k which we assume to be totally imaginary if k is a number field and $p=2$. If p^∞ divides the local degrees $[K_\mathfrak{p}:k_\mathfrak{p}]$ for all nonarchimedean primes $\mathfrak{p}$ of k, then*

$$cd_p\, G_K \le 1.$$

Proof: If $p = \mathrm{char}(K)$, this is always true by (6.5.10). Otherwise we have to show that $Br(L)(p) = 0$ for every finite separable extension $L|K$ (see (6.5.11)). Passing to inductive limits, (8.1.17) implies that $Br(L)(p)$ injects into $\prod_\mathfrak{p} Br(L_\mathfrak{p})(p)$. Furthermore, $Br(L_\mathfrak{p})(p) \cong H^2(L_\mathfrak{p}, \mu_{p^\infty}) = 0$ by (7.1.8)(i) or by assumption if $p=2$ and $\mathfrak{p}|\infty$. □

We now carry over the results on the cohomology of the idèles to the cohomology of the idèle class group. Recall the notation

$$H^i(K|k) = H^i(G(K|k), C_K).$$

(8.1.19) Lemma. *For every finite Galois extension $K|k$, we have*

$$\#H^2(K|k)\,|\,[K:k].$$

Proof: If $K|k$ is cyclic, the assertion follows from the class field axiom (8.1.1). If $K|k$ is a p-extension and $L|k$ a cyclic subextension of degree p, then the exact sequence (observe that $H^1(K|L)=0$)

$$0 \longrightarrow H^2(L|k) \longrightarrow H^2(K|k) \longrightarrow H^2(K|L)$$

shows, using induction on $[K:k]$,

$$\#H^2(K|k)\,|\,\#H^2(K|L)\cdot \#H^2(L|k)\,|\,[K:L]\cdot[L:k] = [K:k].$$

In the general case, let Σ_p be a p-Sylow field of $K|k$. Since the restriction map

$$res: H^2(K|k) \hookrightarrow \bigoplus_p H^2(K|\Sigma_p)$$

is injective, we obtain

$$\#H^2(K|k)\,\Big|\,\prod_p \#H^2(K|\Sigma_p)\,\Big|\,\prod_p\,[K:\Sigma_p] = [K:k].$$ □

If $N \supseteq K \supseteq k$ are two finite Galois extensions in $\bar{k}|k$, then the sequence

$$0 \longrightarrow H^2(K|k) \xrightarrow{inf} H^2(N|k) \xrightarrow{res} H^2(N|K)$$

is exact by (1.6.7), since $H^1(N|K)=0$. As for the idèles, we identify $H^2(K|k)$ with its image in $H^2(k, C)$ so that

$$H^2(k, C) = \bigcup_{K|k} H^2(K|k),$$

where $K|k$ varies over the finite Galois subextensions of $\bar{k}|k$.

We would like to deduce from the idèlic invariant maps

$$inv_{K|k} : H^2(G(K|k), I_K) \to \tfrac{1}{[K:k]}\mathbb{Z}/\mathbb{Z}$$

invariant maps for the groups $H^2(K|k)$. This is not possible in a direct way, since the map $H^2(G(K|k), I_K) \to H^2(G(K|k), C_K)$ is in general not surjective. But it becomes possible if we first pass to the direct limit.

As before let $\tilde{k}|k$ be the cyclotomic $\hat{\mathbb{Z}}$-extension in the number field case and the $\hat{\mathbb{Z}}$-extension obtained by passing to the algebraic closure of the constant field in the function field case. Since $scd\,\hat{\mathbb{Z}} = 2$, we have $H^3(\tilde{k}|k, \tilde{k}^\times) = 0$, and therefore the exact sequence

$$0 \longrightarrow \tilde{k}^\times \longrightarrow I_{\tilde{k}} \longrightarrow C_{\tilde{k}} \longrightarrow 0$$

induces the exact cohomology sequence

$$0 \longrightarrow H^2(\tilde{k}|k, \tilde{k}^\times) \longrightarrow H^2(\tilde{k}|k, I_{\tilde{k}}) \longrightarrow H^2(\tilde{k}|k, C_{\tilde{k}}) \longrightarrow 0\,.$$

Passing to the limit over all finite subextensions of $\tilde{k}|k$, (8.1.15) induces the exact sequence

$$0 \longrightarrow H^2(\tilde{k}|k, \tilde{k}^\times) \longrightarrow H^2(\tilde{k}|k, I_{\tilde{k}}) \longrightarrow \mathbb{Q}/\mathbb{Z} \longrightarrow 0\,.$$

Therefore we obtain an isomorphism

$$inv_{\tilde{k}|k} : H^2(\tilde{k}|k, C_{\tilde{k}}) \xrightarrow{\sim} \mathbb{Q}/\mathbb{Z}\,.$$

For a finite separable extension $K|k$ we have a commutative diagram by (8.1.10)

$$\begin{array}{ccc}
H^2(\tilde{K}|K, C_{\tilde{K}}) & \xrightarrow[\sim]{inv_{\tilde{K}|K}} & \mathbb{Q}/\mathbb{Z} \\
{\scriptstyle res}\uparrow & & \Big\uparrow \\
H^2(\tilde{K}|k, C_{\tilde{K}}) & & \Big| {\scriptstyle [K:k]} \\
{\scriptstyle inf}\uparrow & & \Big| \\
H^2(\tilde{k}|k, C_{\tilde{k}}) & \xrightarrow[\sim]{inv_{\tilde{k}|k}} & \mathbb{Q}/\mathbb{Z}\,.
\end{array}$$

(8.1.20) Proposition. *The sequence*

$$0 \longrightarrow H^2(k, \bar{k}^\times) \longrightarrow H^2(k, I) \longrightarrow H^2(k, C) \longrightarrow 0$$

is exact.

Proof: Let $K|k$ be an arbitrary finite Galois extension of degree $n = [K : k]$ and let k_n be the unique extension of k in $\tilde{k}$ of degree n. We claim that

$$H^2(K|k) = H^2(k_n|k)\,,$$

where we identify $H^2(K|k)$ and $H^2(k_n|k)$ with their images in $H^2(k, C)$ under the inflation maps. In fact, (8.1.19) and the class field axiom (8.1.1) imply that

$$\#H^2(K|k)\,\Big|\,[K : k] = [k_n : k] = \#H^2(k_n|k),$$

and it therefore suffices to show the inclusion "$\supseteq$". But this follows from the exact commutative diagram

$$\begin{array}{ccccccc}
0 & \longrightarrow & H^2(K|k) & \longrightarrow & H^2(k,C) & \xrightarrow{\ res\ } & H^2(K,C) \\
 & & & & \uparrow & & \uparrow \\
 & & & & H^2(\tilde{k}|k, C_{\tilde{k}}) & \xrightarrow{\ res\ } & H^2(\tilde{K}|K, C_{\tilde{K}}) \\
 & & & & {\scriptstyle inv_{\tilde{k}|k}}\downarrow\wr & & {\scriptstyle inv_{\tilde{K}|K}}\downarrow\wr \\
 & & & & \mathbb{Q}/\mathbb{Z} & \xrightarrow{\ [K:k]\ } & \mathbb{Q}/\mathbb{Z},
\end{array}$$

in which $inv_{\tilde{k}|k}$ and $inv_{\tilde{K}|K}$ are isomorphisms as shown above. Since $H^2(k_n|k) \subseteq H^2(\tilde{k}|k)$ has order $n = [K : k]$, it is mapped by the middle arrow *res*, and thus by the upper arrow *res* , to zero, so that

$$H^2(k_n|k) \subseteq H^2(K|k),$$

which shows the claim. Hence we obtain

$$H^2(k,C) = \bigcup_K H^2(K|k) = \bigcup_n H^2(k_n|k) = H^2(\tilde{k}|k, C_{\tilde{k}}).$$

In particular, this implies

$$H^2(k,C) = \bigcup_{\substack{K|k \\ \text{cyclic}}} H^2(K|k)\,.$$

The same assertion holds if we replace C by I or $\bar{k}^\times$, from (8.1.9) or (8.1.14)(i), respectively. Now for $K|k$ cyclic, we have the exact sequence

$$0 \longrightarrow H^2(K|k, K^\times) \longrightarrow H^2(K|k, I_K) \longrightarrow H^2(K|k, C_K) \longrightarrow 0,$$

because $H^1(K|k, C_K) = 0$ and $H^3(K|k, K^\times) \cong H^1(K|k, K^\times) = 0$. From this and from the above observation now follows the statement of the proposition. □

(8.1.21) Corollary. *We have a canonical isomorphism*

$$inv_k : H^2(k, C) \xrightarrow{\sim} \mathbb{Q}/\mathbb{Z}.$$

From this corollary we obtain for every finite Galois extension $K|k$ a canonical invariant map

$$inv_{K|k} : H^2(K|k) \xrightarrow{\sim} \tfrac{1}{[K:k]}\mathbb{Z}/\mathbb{Z}$$

using the commutative diagram

$$\begin{array}{ccccccccc} 0 & \longrightarrow & H^2(K|k) & \longrightarrow & H^2(k,C) & \xrightarrow{res} & H^2(K,C) & \longrightarrow & 0 \\ & & \downarrow{\scriptstyle inv_{K|k}} & & \downarrow{\scriptstyle inv_k} & & \downarrow{\scriptstyle inv_K} & & \\ 0 & \longrightarrow & \tfrac{1}{[K:k]}\mathbb{Z}/\mathbb{Z} & \longrightarrow & \mathbb{Q}/\mathbb{Z} & \xrightarrow{[K:k]} & \mathbb{Q}/\mathbb{Z} & \longrightarrow & 0. \end{array}$$

The results above and the compatibility of *inv* with *inf* and *res*, which follows from (8.1.10), therefore imply the

(8.1.22) Theorem. *The G-module C is a formation module with respect to the invariant maps $inv_{K|k}$* (see (3.1.8)).

We now obtain the main theorem of global class field theory from (3.1.6): the "global reciprocity law".

(8.1.23) Theorem. *Let $K|k$ be a finite Galois extension of global fields with Galois group $G(K|k)$. Then there is a canonical isomorphism*

$$C_k/N_{K|k}C_K \cong G(K|k)^{ab}.$$

By the results of III §1 (see the remark 5 on p.157), we obtain a **reciprocity homomorphism**

$$rec : C_k \longrightarrow G_k^{ab},$$

which has a dense image, and whose kernel is the group of **universal norms**, i.e. the intersection

$$\bigcap_K N_{K|k}C_K = N_{G_k}C\,.$$

As before we also call *rec* the **norm residue symbol** and we write it in the form $rec(\alpha) = (\alpha, k)$.

Essential is the following theorem, which is called the existence theorem for global class field theory. For a proof, see [6], chap. 8, th. 1. For number fields, see also [123], chap.XI §2 th. 1 or [160], chap.VI th. 6.1.

(8.1.24) Theorem. *The norm groups $N_{K|k}C_K$, where $K|k$ is a finite separable extension, are exactly the open subgroups of finite index in C_k.*

The idèle group $I_k = \prod_{\mathfrak{p}} k_{\mathfrak{p}}^\times$ is a locally compact topological group, $k^\times$ is a discrete, closed subgroup (see [160], chap.VI, (1.5)), and so C_k is also a locally compact topological group. We have a canonical homomorphism, called the "absolute value",

$$| \ | : I_k \longrightarrow \mathbb{R}_+^\times , \quad |\alpha| = \prod_{\mathfrak{p}} |\alpha_{\mathfrak{p}}|_{\mathfrak{p}} ,$$

which is trivial on $k^\times$, and hence induces a homomorphism

$$| \ | : C_k \longrightarrow \mathbb{R}_+^\times .$$

The kernel $C_k^0 = \{x \in C_k \mid |x| = 1\}$ is a compact group (cf. [22], chap.II, §16, Theorem, or [160], chap.VI, (1.6) for the number field case).

If k is a number field, then the homomorphism $| \ |$ is surjective: Since $\mathbb{R}_+^\times$ has no nontrivial finite quotient, C_k and C_k^0 have the same image under the reciprocity map, which is all of G_k^{ab} since C_k^0 is compact and (C_k, k) is dense in G_k^{ab}. This yields the

(8.1.25) Proposition. *In the number field case we have an exact sequence of topological groups*

$$0 \longrightarrow N_{G_k}C \longrightarrow C_k \xrightarrow{(\ ,k)} G_k^{ab} \longrightarrow 0.$$

If k is a function field, with constant field κ of cardinality q, we have the following situation. The group I_k, and hence C_k, is totally disconnected by (1.1.9)(iii); in particular, C_k^0 is an abelian profinite group. For $\alpha \in C_k$ we have $|\alpha| = q^{-\deg(\alpha)}$, where

$$\deg : C_k \to \mathbb{Z}, \qquad \deg(\alpha \bmod k^\times) = \sum_{\mathfrak{p}} v_{\mathfrak{p}}(\alpha_{\mathfrak{p}}) \cdot [\kappa(\mathfrak{p}) : \kappa] ,$$

is the degree map. Let $\pi : G_k^{ab} \twoheadrightarrow G(k\bar{\kappa}|k) \cong G_\kappa \cong \hat{\mathbb{Z}}$ be the canonical projection. Then, for all $\alpha \in C_k$, we have $\pi((\alpha, k)) = \mathrm{Frob}^{\deg(\alpha)}$, where Frob denotes the Frobenius automorphism $x \mapsto x^q$ on $\bar{\kappa}$. Therefore deg is surjective*) and we obtain an exact sequence of topological groups

$$0 \longrightarrow C_k^0 \longrightarrow C_k \xrightarrow{\deg} \mathbb{Z} \longrightarrow 0.$$

Since C_k^0 is profinite the intersection of all open subgroups of finite index in C_k is zero. The degree map extends to a map from $\overline{C}_k := \varprojlim_{K|k} C_k/N_{K|k}C_K$

---

*) This can also be seen directly: one can construct an injection coker(deg) $\hookrightarrow Br(\kappa) = 0$.

onto $\hat{\mathbb{Z}}$. We therefore obtain a commutative exact diagram

$$\begin{array}{ccccccccc} 0 & \longrightarrow & C_k^0 & \longrightarrow & C_k & \xrightarrow{\deg} & \mathbb{Z} & \longrightarrow & 0 \\ & & \| & & \downarrow & & \downarrow & & \\ 0 & \longrightarrow & C_k^0 & \longrightarrow & \overline{C}_k & \xrightarrow{\deg} & \hat{\mathbb{Z}} & \longrightarrow & 0\,, \end{array}$$

which yields, using (8.1.23), the

(8.1.26) Proposition. *In the function field case we have an exact sequence of topological groups*

$$0 \longrightarrow C_k \xrightarrow{(\ ,k)} G_k^{ab} \longrightarrow \hat{\mathbb{Z}}/\mathbb{Z} \longrightarrow 0.$$

§2. The Connected Component of C_k

We now study the topological properties of the idèle class group C_k and, in particular, the connected component D_k of 1 in C_k. The results are due to J. TATE (see [6], chap.9) and are quoted in [160], chap.VI, §1, ex.1–10.

We denote the connected component (of 1) of the idèle class group C_k by D_k. If k is a function field, then I_k, and hence C_k, is totally disconnected, so that $D_k = 1$.

Therefore we assume for the rest of this section that k is a number field.

We introduce the following notation:

$$\begin{array}{lcll} S_\infty & = & S_\infty(k) & \text{the set of archimedean primes of } k\,, \\ S_{\mathbb{R}} & = & S_{\mathbb{R}}(k) & \text{the set of real primes of } k\,, \\ S_{\mathbb{C}} & = & S_{\mathbb{C}}(k) & \text{the set of complex primes of } k\,, \\ r_1 & = & r_1(k) & \text{the number of real primes of } k\,, \\ r_2 & = & r_2(k) & \text{the number of complex primes of } k\,, \\ r & = & r(k) & = r_1(k) + r_2(k)\,. \end{array}$$

The continuous surjection

$$|\ | : C_k \longrightarrow \mathbb{R}_+^\times$$

has a continuous section $s : \mathbb{R}_+^\times \to C_k$: for any infinite prime $\mathfrak{p}$, the restriction of $|\ |$ to the subgroup $\mathbb{R}_+^\times$ of $k_\mathfrak{p}^\times \subseteq C_k$ is an isomorphism, and its inverse gives a section of the homomorphism $|\ |$. Since $\mathbb{R}_+^\times$ is connected, the image of s is contained in D_k and we obtain compatible isomorphisms of topological groups

$$C_k \cong C_k^0 \times \mathbb{R}_+^\times, \quad D_k \cong D_k^0 \times \mathbb{R}_+^\times,$$

where $D_k^0 = C_k^0 \cap D_k$ is the connected component of C_k^0. The group D_k has the following characterizations.

(8.2.1) Theorem.

(i) *D_k is the closure of the image of $\tilde{U} = \prod\limits_{\mathfrak{p}|\infty} U_\mathfrak{p}^1$ under the projection $I_k \to C_k$, where $U_\mathfrak{p}^1 = \mathbb{R}_+^\times$ if $\mathfrak{p}$ is real and $U_\mathfrak{p}^1 = \mathbb{C}^\times$ if $\mathfrak{p}$ is complex.*

(ii) *D_k is the intersection of all open subgroups of C_k of finite index.*

(iii) *D_k is the group of universal norms $N_{\bar{k}|k}C = \bigcap\limits_{K|k} N_{K|k}C_K$ and D_k^0 is the group of the universal norms $N_{\bar{k}|k}C^0 = \bigcap\limits_{K|k} N_{K|k}C_K^0$.*

(iv) *For any finite extension $K|k$ the norm maps $N_{K|k} : D_K \to D_k$ and $N_{K|k} : D_K^0 \to D_k^0$ are surjective.*

(v) *D_k is the group of all divisible elements of C_k, i.e. elements x which are n-th powers, $x = y^n$, for every $n \in \mathbb{N}$.*

(vi) *D_k is divisible, i.e. it is the maximal divisible subgroup in C_k.**)

Remark: All open subgroups in C_k have finite index. We will prove this fact in the context of restricted ramification in (8.3.14).

Proof: (i) Since $\tilde{U}$ is the connected component of 1 in I, the assertion follows from (1.1.9)(iii).

(ii) Since C_k/D_k is totally disconnected, the intersection of all its open subgroups $\bar{U}_i$ is trivial by (1.1.9)(i). But $C_k/D_k = C_k^0/D_k^0$ is compact, i.e. the $\bar{U}_i$ are of finite index. The pre-images U_i in C_k are open subgroups of finite index and D_k is their intersection. On the other hand, any open subgroup contains D_k. This proves (ii).

(iii) By the existence theorem (8.1.24), the open subgroups of finite index in C_k are just the norm groups $N_{K|k}C_K$ of the finite Galois extensions $K|k$, so that (ii) implies that $D_k = N_{\bar{k}|k}C$. As $\mathbb{R}_+^\times$ is cohomologically trivial, we have an isomorphism $C_k^0/N_{K|k}C_K^0 \cong C_k/N_{K|k}C_K$ for every finite Galois extension $K|k$. This implies

$$N_{\bar{k}|k}C^0 = C_k^0 \cap N_{\bar{k}|k}C = C_k^0 \cap D_k = D_k^0.$$

*) If A is an abelian group, then the subgroup of divisible elements contains the maximal divisible subgroup of A, but in general is *not* divisible itself.

(iv) By assertion (iii), D_k^0 is the group of universal norms of the level-compact module C^0, hence $N_{K|k}D_K^0 = D_k^0$ for every finite Galois extension $K|k$. The analogous assertion for D follows from the commutative diagram

$$\begin{array}{ccccccccc} 0 & \longrightarrow & D_K^0 & \longrightarrow & D_K & \longrightarrow & \mathbb{R}_+^\times & \longrightarrow & 0 \\ & & \downarrow{\scriptstyle N_{K|k}} & & \downarrow{\scriptstyle N_{K|k}} & & \downarrow{\scriptstyle N_{K|k}} & & \\ 0 & \longrightarrow & D_k^0 & \longrightarrow & D_k & \longrightarrow & \mathbb{R}_+^\times & \longrightarrow & 0 \end{array}$$

noting that the norm map $N_{K|k} : \mathbb{R}_+^\times \to \mathbb{R}_+^\times$ is just the map $x \mapsto x^{[K:k]}$ and is therefore surjective.

(v) $C_k/D_k = C_k^0/D_k^0$ is a compact, totally disconnected Hausdorff group, i.e. a profinite group. If $x \in C_k$ is divisible, then its image in C_k/D_k is contained in every open subgroup, i.e. the image is 1 and hence $x \in D_k$. Hence (v) follows from (vi).

(vi) Since $\mathbb{R}_+^\times$ is divisible, it suffices to show that D_k^0 is divisible. Let ℓ be a prime number. If the finite extension K of k contains the ℓ-th roots of unity, then for sufficiently large S containing all primes above ℓ and all archimedean primes, the group $(C_K)^\ell \cdot \bar{U}_K^S$ is the norm group of the Kummer extension $K(\sqrt[\ell]{\mathcal{O}_S^\times})|K$. Here $\bar{U}_K^S = U_K^S K^\times/K^\times$ and $U_K^S = \prod_{\mathfrak{p}\notin S} U_\mathfrak{p} \times \prod_{\mathfrak{p}\in S} 1$ (see [123], chap.XI §2 th. 1 or the proof of [160], chap.VI th. 6.1). Hence, for any such S,

$$D_K \subseteq (C_K)^\ell \bar{U}^S.$$

As the absolute value $|\ |$ is trivial on $\bar{U}^S$, we obtain

$$D_K^0 \subseteq \bigcap_S (C_K^0)^\ell \bar{U}^S = (C_K^0)^\ell.$$

Using (iv) this implies

$$(*) \qquad D_k^0 = N_{K|k} D_K^0 \subseteq (N_{K|k} C_K^0)^\ell.$$

For each $a \in D_k^0$, let $a^{1/\ell}$ denote the set of all elements of C_k^0 whose ℓ-th power is a. From $(*)$ we see that the sets $X_K = (N_{K|k}C_K^0) \cap (a^{1/\ell})$ are nonempty. The kernel of the map $C_k^0 \xrightarrow{\ell} C_k^0$, $x \mapsto x^\ell$, is compact and $N_{K|k}C_K^0$ is closed, and so X_K is compact. Therefore the intersection $\bigcap_{K|k} X_K$ is not empty. By (iii), an element of this intersection is an element of D_k^0 whose ℓ-th power is a. This proves (vi). □

By (8.2.1), we obtain from (8.1.25) the

(8.2.2) Corollary. *We have an exact sequence of topological groups*

$$0 \longrightarrow D_k \longrightarrow C_k \xrightarrow{(\ ,k)} G_k^{ab} \longrightarrow 0.$$

In order to give an explicit description of the connected component $D_k \cong D_k^0 \times \mathbb{R}_+^\times$, we have first to consider the group $\bar{\mathbb{Z}}/\mathbb{Z}$, where

$$\bar{\mathbb{Z}} = \mathbb{R} \times \prod_p \mathbb{Z}_p = \mathbb{R} \times \hat{\mathbb{Z}}$$

is the group of integral elements in the adèle ring $A_{\mathbb{Q}}$ of $\mathbb{Q}$.

(8.2.3) Proposition. *The topological group $\bar{\mathbb{Z}}/\mathbb{Z}$ is compact, connected and uniquely divisible. It is called the* **solenoid**.*)

Proof: The canonical projection $\bar{\mathbb{Z}} = \hat{\mathbb{Z}} \times \mathbb{R} \twoheadrightarrow \hat{\mathbb{Z}}$ induces an exact sequence

$$0 \longrightarrow \mathbb{R} \longrightarrow \bar{\mathbb{Z}}/\mathbb{Z} \longrightarrow \hat{\mathbb{Z}}/\mathbb{Z} \longrightarrow 0\,.$$

This shows that $\bar{\mathbb{Z}}/\mathbb{Z}$ is uniquely divisible, since $\mathbb{R}$ and $\hat{\mathbb{Z}}/\mathbb{Z}$ have this property. Furthermore, $\mathbb{R}$ is dense in $\bar{\mathbb{Z}}/\mathbb{Z}$: let $\lambda = (z, x) \bmod \mathbb{Z}$ be an arbitrary element in $(\hat{\mathbb{Z}} \times \mathbb{R})/\mathbb{Z}$, let $m \in \mathbb{Z}$ be given (describing a neighbourhood of $\hat{\mathbb{Z}}$) and let $a \in \mathbb{Z}$ such that $a \equiv z \bmod m$. Then $\lambda = (z-a, x-a) \bmod \mathbb{Z}$ and $(z-a, x-a)$ is contained in the set $(m\hat{\mathbb{Z}}, 0) + (0, x-a) \subseteq m\hat{\mathbb{Z}} \times \mathbb{R}$, which is mapped into a neighbourhood of $x - a \in \mathbb{R} \subseteq \bar{\mathbb{Z}}/\mathbb{Z}$. Now, since $\mathbb{R}$ is connected and the closure of a connected set is connected, we conclude that the solenoid is connected. (Another possible method to show the connectedness of $\bar{\mathbb{Z}}/\mathbb{Z}$ is to use the exact sequence at the beginning of the proof and to show the connectedness of $\hat{\mathbb{Z}}/\mathbb{Z}$.)

Since $\mathbb{Z}$ is closed in $\bar{\mathbb{Z}}$, the quotient $\bar{\mathbb{Z}}/\mathbb{Z}$ is a Hausdorff topological group. Finally, the compact set $\hat{\mathbb{Z}} \times [0,1] \subseteq \bar{\mathbb{Z}}$ is mapped under the canonical projection $\bar{\mathbb{Z}} \twoheadrightarrow \bar{\mathbb{Z}}/\mathbb{Z}$ onto $\bar{\mathbb{Z}}/\mathbb{Z}$. Hence the solenoid is compact. □

We consider the group

$$U_k = \bar{U} \times \tilde{U}\,,$$

where

$$\bar{U} = \prod_{\mathfrak{p}\nmid\infty} U_{\mathfrak{p}} \quad \text{and} \quad \tilde{U} = \prod_{\mathfrak{p}\mid\infty} U_{\mathfrak{p}}^1,$$

and we write every idèle $\alpha \in U_k$ as a product $\alpha = \bar{\alpha}\tilde{\alpha}$ with $\bar{\alpha} \in \bar{U}$, $\tilde{\alpha} \in \tilde{U}$.

The abelian profinite group $\bar{U}$ is a (multiplicative) $\hat{\mathbb{Z}}$-module in a natural way: The exponentiation

$$\mathbb{N} \times \bar{U} \to \bar{U}\;,\; (n, \alpha) \mapsto \alpha^n,$$

extends continuously to an exponentiation

$$\hat{\mathbb{Z}} \times \bar{U} \longrightarrow \bar{U}\;,\; \lambda \longmapsto \alpha^\lambda.$$

---

*) For another description of the solenoid, see ex.1.

(8.2.4) Lemma. *If $\varepsilon_1, \dots, \varepsilon_{r-1}$ are $\mathbb{Z}$-independent units of k, $r = r_1 + r_2$, then the corresponding idèles $\bar\varepsilon_1, \dots, \bar\varepsilon_{r-1} \in \bar U = \prod_{\mathfrak{p}\nmid\infty} U_\mathfrak{p}$ are $\hat{\mathbb{Z}}$-independent.*

Proof: By Dirichlet's unit theorem, the group generated by $\varepsilon_1, \dots, \varepsilon_{r-1}$ (with ordinary integers as exponents) has finite index d in $\mathcal{O}_k^\times$. Consider a relation

$$(*) \qquad \bar\varepsilon_1^{z_1} \cdots \bar\varepsilon_{r-1}^{z_{r-1}} = 1\ , \quad z_i \in \hat{\mathbb{Z}}.$$

We have to show $z_1 = \cdots = z_{r-1} = 0$. Let $n \in \mathbb{N}$. For each z_i we find an integer $\nu_i \in \mathbb{Z}$ such that $\nu_i \equiv z_i \bmod 2dn$. Then $\varepsilon = \varepsilon_1^{\nu_1} \cdots \varepsilon_{r-1}^{\nu_{r-1}}$ is an element of k and we may write

$$\varepsilon = \bar\varepsilon_1^{\nu_1} \cdots \bar\varepsilon_{r-1}^{\nu_{r-1}} \tilde\varepsilon_1^{\nu_1} \cdots \tilde\varepsilon_{r-1}^{\nu_{r-1}}.$$

Dividing the right-hand side by the left-hand side of $(*)$, we obtain

$$\varepsilon = \bar\varepsilon_1^{\nu_1 - z_1} \cdots \bar\varepsilon_{r-1}^{\nu_{r-1}-z_{r-1}} \tilde\varepsilon_1^{\nu_1} \cdots \tilde\varepsilon_{r-1}^{\nu_{r-1}}.$$

At each finite prime each idèle $\tilde\varepsilon_i^{\nu_i}$ has component 1; the remaining factors have $\hat{\mathbb{Z}}$-exponents divisible by $2dn$. This means that ε is a $2dn$-th power in $k_\mathfrak{p}^\times$ for each finite prime $\mathfrak{p}$. But from this it follows that ε is a dn-th power in $k^\times$, i.e. $\varepsilon = \eta^{dn}$, $\eta \in k^\times$, see (9.1.11)(ii). This η must be a unit and consequently η^d is contained in the subgroup generated by the $\varepsilon_1, \dots, \varepsilon_{r-1}$, i.e. $\eta^d = \varepsilon_1^{\mu_1} \cdots \varepsilon_{r-1}^{\mu_{r-1}}$, $\mu_i \in \mathbb{Z}$. We now get

$$\varepsilon = \varepsilon_1^{\nu_1} \cdots \varepsilon_{r-1}^{\nu_{r-1}} = \varepsilon_1^{\mu_1 n} \cdots \varepsilon_{r-1}^{\mu_{r-1} n}.$$

Since the ε_i are $\mathbb{Z}$-independent, we must have $\nu_i = \mu_i n$. Because $\nu_i \equiv z_i \bmod 2dn$, this shows that each z_i is divisible by n, i.e. $z_i \in n\hat{\mathbb{Z}}$. But n was arbitrary, hence $z_i = 0$. □

We shall now determine the structure of D_k^0. It contains a torus T_k which is defined as follows: for each complex prime $\mathfrak{p}$ we have the embedding $k_\mathfrak{p}^\times \hookrightarrow C_k$ and we obtain an embedding

$$\prod_{\mathfrak{p} \in S_\mathbb{C}} k_\mathfrak{p}^\times \hookrightarrow C_k.$$

Each factor $k_\mathfrak{p}^\times = \mathbb{C}^\times$ contains the unit circle $S_\mathfrak{p}^1 = \{z \in k_\mathfrak{p}^\times \mid |z| = 1\}$ and we put

$$T_k = \prod_{\mathfrak{p} \in S_\mathbb{C}} S_\mathfrak{p}^1.$$

It is a compact connected subgroup of D_k^0, also described by the topological isomorphism

$$\exp : \prod_{\mathfrak{p} \in S_\mathbb{C}} \mathbb{R}/\mathbb{Z} \longrightarrow T_k\ , \qquad \theta = (\theta_\mathfrak{p}) \longmapsto (e^{2\pi i \theta_\mathfrak{p}})_{\mathfrak{p} \in S_\mathbb{C}}.$$

We may regard T_k as a subgroup of the idèle group $\tilde U \subseteq U_k$.

For $\mathfrak{p} \in S_\infty$, we have an exponentiation $\mathbb{R} \times U_\mathfrak{p} \to U_\mathfrak{p}$, $(x, \alpha_\mathfrak{p}) \mapsto \alpha_\mathfrak{p}^x = e^{x \log \alpha_\mathfrak{p}}$, where log is the natural logarithm of $\mathbb{R}_+^\times$ if $\mathfrak{p}$ is real, and is any fixed branch of the logarithm of $\mathbb{C}^\times$ if $\mathfrak{p}$ is complex. For fixed $\alpha \in U_k$, the map $\mathbb{R} \to \tilde{U}$, $x \mapsto \tilde{\alpha}^x$, is continuous, and so is the map

$$\bar{\mathbb{Z}} = \hat{\mathbb{Z}} \times \mathbb{R} \longrightarrow U_k, \qquad \lambda = (z, x) \longmapsto \bar{\alpha}^z \tilde{\alpha}^x.$$

Let $\mathcal{O}^+(k)$ be the subgroup of $\mathcal{O}_k^\times$ of totally positive units of k, i.e. $\varepsilon \in \mathcal{O}^+(k)$ if and only if $\varepsilon_\mathfrak{p} > 0$ for every real prime $\mathfrak{p}$. Since $\mathcal{O}^+(k)$ has finite index in $\mathcal{O}_k^\times$, Dirichlet's unit theorem implies that it is a finitely generated group of rank

$$\mathrm{rk}(\mathcal{O}^+(k)) = \#S_\infty(k) - 1 = r - 1,$$

and we choose a fixed $r - 1$-tuple

$$\varepsilon = (\varepsilon_1, \ldots, \varepsilon_{r-1})$$

of $\mathbb{Z}$-independent units. In addition to T_k, we consider the subgroup

$$I(\varepsilon) = \{\varepsilon_1^{\lambda_1} \cdots \varepsilon_{r-1}^{\lambda_{r-1}} \mid \lambda_1, \ldots, \lambda_{r-1} \in \bar{\mathbb{Z}}\}$$

of U_k together with the continuous surjective homomorphism

$$\exp_\varepsilon : \bar{\mathbb{Z}}^{r-1} \longrightarrow I(\varepsilon), \quad (\lambda_1, \ldots, \lambda_{r-1}) \longmapsto \varepsilon_1^{\lambda_1} \cdots \varepsilon_{r-1}^{\lambda_{r-1}}.$$

Note that T_k is a subgroup of $U_k^0 = \{x \in U_k \mid |x| = 1\}$ and the same holds for $I(\varepsilon)$, since $\varepsilon_i^{z_i} \in U_\mathfrak{p}$ for every finite prime $\mathfrak{p}$, so that $|\varepsilon_i^{\lambda_i}| = |\tilde{\varepsilon}_i^{x_i}| = |\tilde{\varepsilon}_i|^{x_i} = |\varepsilon_i|^{x_i} = 1$ (where $\lambda_i = (z_i, x_i) \in \hat{\mathbb{Z}} \times \mathbb{R} = \bar{\mathbb{Z}}$). We get a continuous homomorphism

$$(*) \qquad \exp : (\mathbb{R}/\mathbb{Z})^{r_2} \times \bar{\mathbb{Z}}^{r-1} \longrightarrow U_k^0$$

with image $T_k I(\varepsilon)$.

(8.2.5) Theorem (Tate). *The homomorphism* $(*)$ *induces a topological isomorphism*

$$\exp : (\mathbb{R}/\mathbb{Z})^{r_2} \times (\bar{\mathbb{Z}}/\mathbb{Z})^{r-1} \xrightarrow{\sim} D_k^0.$$

For the connected component D_k of C_k, this yields a topological isomorphism

$$D_k \cong (\mathbb{R}/\mathbb{Z})^{r_2} \times (\bar{\mathbb{Z}}/\mathbb{Z})^{r-1} \times \mathbb{R}.$$

Proof: The second statement follows from the first since $D_k \cong D_k^0 \times \mathbb{R}_+^\times$. In order to prove the first, observe that $\mathcal{O}^+(k)$ is a subgroup of U_k^0, and we obtain a homomorphism

$$\bar{\mathbb{Z}}^{r-1} \longrightarrow U_k^0/\mathcal{O}^+(k).$$

Since $\varepsilon_i^{\lambda_i} \in \mathcal{O}^+(k)$ when $\lambda_i = (n, n) \in \mathbb{Z} \subseteq \bar{\mathbb{Z}}$, this homomorphism factors through $(\bar{\mathbb{Z}}/\mathbb{Z})^{r-1}$. We obtain a continuous homomorphism

$$\exp_\varepsilon : (\bar{\mathbb{Z}}/\mathbb{Z})^{r-1} \longrightarrow C_k^0,$$

whose image is contained in the connected component D_k^0 since $\bar{\mathbb{Z}}/\mathbb{Z}$ is connected by (8.2.3).

I. *Injectivity* of exp: Let $\alpha \in T_k$ and $\beta = \exp_\varepsilon(\lambda) = \varepsilon_1^{\lambda_1} \cdots \varepsilon_{r-1}^{\lambda_{r-1}}$, where $\lambda = (\lambda_1, \dots, \lambda_{r-1}) \in \bar{\mathbb{Z}}^{r-1}$, $\lambda_i = (z_i, x_i)$. For the injectivity of exp we have to show: suppose that $\alpha\beta \in U_k^0$ is a principal idèle,

$$\alpha\beta = a \in k^\times,$$

then $\lambda \in \mathbb{Z}^{r-1}$. Looking only at the components at the finite primes, we have

$$(1) \qquad \bar{\varepsilon}_1^{z_1} \cdots \bar{\varepsilon}_{r-1}^{z_{r-1}} = a.$$

The group generated by $\varepsilon_1, \dots, \varepsilon_{r-1}$ has finite index in $\mathcal{O}^+(k)$ so that $a^d = \varepsilon_1^{\mu_1} \cdots \varepsilon_{r-1}^{\mu_{r-1}}$ with $\mu_i \in \mathbb{Z}$. Raising (1) into the d-th power, we obtain

$$\bar{\varepsilon}_1^{dz_1 - \mu_1} \cdots \bar{\varepsilon}_{r-1}^{dz_{r-1} - \mu_{r-1}} = 1$$

and by lemma (8.2.4), $dz_i - \mu_i = 0$ in $\hat{\mathbb{Z}}$. From this it follows that the z_i are contained in $\mathbb{Z}$. Now

$$\varepsilon = \varepsilon_1^{z_1} \cdots \varepsilon_{r-1}^{z_{r-1}}$$

is an element in k which coincides with a in $k_\mathfrak{p}$ for $\mathfrak{p} \nmid \infty$, hence $\varepsilon = a$. Looking at the components at the infinite primes, we obtain from $\alpha\beta = \varepsilon$

$$\tilde{\alpha}\tilde{\varepsilon}_1^{x_1} \cdots \tilde{\varepsilon}_{r-1}^{x_{r-1}} = \varepsilon_1^{z_1} \cdots \varepsilon_{r-1}^{z_{r-1}}.$$

Taking absolute values $|\ |_\mathfrak{p}$ for every $\mathfrak{p} \in S_\infty$, we get

$$|\varepsilon_1|_\mathfrak{p}^{x_1 - z_1} \cdots |\varepsilon_{r-1}|_\mathfrak{p}^{x_{r-1} - z_{r-1}} = 1.$$

Taking the logarithms of these equations, we conclude that $x_i - z_i = 0$, since the vectors $(\log|\varepsilon_i|_\mathfrak{p})_{\mathfrak{p} \in S_\infty}$, $i = 1, \dots, r-1$, are linearly independent by Dirichlet's unit theorem. This proves $\lambda_i = (z_i, x_i) \in \mathbb{Z}$, i.e. $\lambda \in \mathbb{Z}^{r-1}$, and thus the injectivity, as desired.

II. *Surjectivity* of exp: Let $D_0 \subseteq D_k^0$ be the image of exp and let $a \in D_k^0$. For the proof that $a \in D_0$ we use the fact that a is divisible by (8.2.1)(v). So we may write $a = b^{2hm}$ where h is the class number of k and m a highly divisible integer. The ideal class group Cl_k is the quotient $I_k/V_k k^\times$ with

$$V_k = \prod_{\mathfrak{p} \nmid \infty} U_\mathfrak{p} \times \prod_{\mathfrak{p} | \infty} k_\mathfrak{p}^\times,$$

so the class b^h can be represented by an idèle in V_k, hence b^{2h} by an idèle $\beta \in U_k$. Since $a \in D_k^0$, we have $\beta \in U_k^0$. The element a is therefore represented by the idèle $\beta^m = \bar{\beta}^m \tilde{\beta}^m$. Choosing a suitable highly divisible m, the idèle $\bar{\beta}^m$ is contained in a neighbourhood of 1 which is as small as we like. We shall prove that the idèle class c of $\tilde{\beta}^m$ belongs to D_0. From this it follows that a is in the closure of D_0. But D_0 is compact, hence closed and consequently $a \in D_0$.

Set $\tilde{\alpha} = \tilde{\beta}^m \in U_k^0$. Let $\mathfrak{p}_\nu$, $\nu = 1, \dots, r$, range over the archimedean primes. From the independence of the units $\varepsilon_1, \dots, \varepsilon_{r-1}$ it follows that the vectors $\mathfrak{a}_i = (\log|\varepsilon_i|_{\mathfrak{p}_1}, \ldots, \log|\varepsilon_i|_{\mathfrak{p}_{r-1}})$, $i = 1, \ldots, r-1$, in $\mathbb{R}^{r-1}$ are linearly independent. Setting $\mathfrak{a} = (\log|\tilde{\alpha}|_{\mathfrak{p}_1}, \ldots, \log|\tilde{\alpha}|_{\mathfrak{p}_{r-1}})$, we find $x_1, \ldots, x_{r-1} \in \mathbb{R}$ such that

$$x_1\mathfrak{a}_1 + \cdots + x_{r-1}\mathfrak{a}_{r-1} = \mathfrak{a},$$

i.e.

$$|\tilde{\alpha}|_{\mathfrak{p}_\nu} = |\tilde{\varepsilon}_1|_{\mathfrak{p}_\nu}^{x_1} \cdots |\tilde{\varepsilon}_{r-1}|_{\mathfrak{p}_\nu}^{x_{r-1}} \qquad \text{for } \nu = 1, \ldots, r-1.$$

Since $|\tilde{\alpha}| = 1$, we also have

$$|\tilde{\alpha}|_{\mathfrak{p}_r} = |\tilde{\varepsilon}_1|_{\mathfrak{p}_r}^{x_1} \cdots |\tilde{\varepsilon}_{r-1}|_{\mathfrak{p}_r}^{x_{r-1}}.$$

Let $\lambda_i = (0, x_i) \in \bar{\mathbb{Z}}$. Since $\tilde{\alpha}$ is totally positive, we have

$$\tilde{\alpha}_{\mathfrak{p}} = (\tilde{\varepsilon}_1^{x_1} \cdots \tilde{\varepsilon}_{r-1}^{x_{r-1}})_{\mathfrak{p}}$$

for every real prime $\mathfrak{p}$. For complex $\mathfrak{p}$, both sides differ by an element of value 1. Hence we can write

$$\tilde{\alpha} = \tilde{\varepsilon}_1^{x_1} \cdots \tilde{\varepsilon}_{r-1}^{x_{r-1}} \tilde{\gamma} = \varepsilon_1^{\lambda_1} \cdots \varepsilon_{r-1}^{\lambda_{r-1}} \tilde{\gamma}$$

with $\tilde{\gamma} \in T_k$. This proves that the class c of $\tilde{\alpha}$ is contained in the image D_0 of exp as contended. □

It is not unimportant to know also the cohomology of the connected component D_K of a finite Galois extension $K|k$. It is given by the

(8.2.6) Corollary. *Let $K|k$ be a finite Galois extension with Galois group $G = G(K|k)$ and let m be the number of real primes of k that become complex in K. Then*

$$\hat{H}^i(G, D_K) \cong \begin{cases} (\mathbb{Z}/2\mathbb{Z})^m & \text{if } i \text{ is even,} \\ 0 & \text{if } i \text{ is odd,} \end{cases}$$

and $H^0(G, D_K) = D_k$.

Proof: We may consider the product $\prod_{\mathfrak{P} \in S_{\mathbb{C}}} K_{\mathfrak{P}}^\times$ as a G-submodule of the idèle group I_K as well as of $D_K \subseteq C_K$. In this product we have the canonical G-submodule

$$T_K = \prod_{\mathfrak{P} \in S_{\mathbb{C}}} S_{\mathfrak{P}}^1,$$

where $S_{\mathfrak{P}}^1$ is the unit circle in $K_{\mathfrak{P}}^\times = \mathbb{C}^\times$. By the above theorem and by (8.2.3), the quotient

$$D_K^0/T_K \cong (\bar{\mathbb{Z}}/\mathbb{Z})^{r(K)-1}$$

is uniquely divisible, hence cohomologically trivial. Therefore the exact cohomology sequence yields

$$\hat{H}^i(G, D_K) = \hat{H}^i(G, D_K^0) = \hat{H}^i(G, T_K).$$

Let $\mathfrak{M}$ be the set of primes $\mathfrak{p}$ of k lying under the complex primes of K. For every $\mathfrak{p} \in \mathfrak{M}$ we choose a fixed prime $\mathfrak{P}_0|\mathfrak{p}$ of K and consider the decomposition group $G_{\mathfrak{P}_0}$. We regard the group $\mathbb{R}/\mathbb{Z}$ as a $G_{\mathfrak{P}_0}$-module by letting $\sigma \in G_{\mathfrak{P}_0}$ act by $\sigma\theta = \theta$ if $\sigma = 1$, and $\sigma\theta = -\theta$ if $\sigma \neq 1$. Then

$$\mathbb{R}/\mathbb{Z} \longrightarrow K_{\mathfrak{P}_0}^{\times}\,, \ \theta \longmapsto e^{2\pi i\theta},$$

is an injective $G_{\mathfrak{P}_0}$-homomorphism with image $S^1_{\mathfrak{P}_0}$, and we obtain an injective G-homomorphism

$$\operatorname{Ind}_G^{G_{\mathfrak{P}_0}}(\mathbb{R}/\mathbb{Z}) \longrightarrow \operatorname{Ind}_G^{G_{\mathfrak{P}_0}}(K_{\mathfrak{P}_0}^{\times}) = \prod_{\mathfrak{P}|\mathfrak{p}} K_{\mathfrak{P}}^{\times} = I_K(\mathfrak{p})$$

and hence an injective G-homomorphism

$$\bigoplus_{\mathfrak{p}\in\mathfrak{M}} \operatorname{Ind}_G^{G_{\mathfrak{P}_0}}(\mathbb{R}/\mathbb{Z}) \longrightarrow \prod_{\mathfrak{p}\in\mathfrak{M}} I_K(\mathfrak{p})$$

with image T_K. Shapiro's lemma now yields

$$\begin{aligned}\hat{H}^i(G, T_K) &= \bigoplus_{\mathfrak{p}\in\mathfrak{M}} \hat{H}^i(G, \operatorname{Ind}_G^{G_{\mathfrak{P}_0}}(\mathbb{R}/\mathbb{Z})) \cong \bigoplus_{\mathfrak{p}\in\mathfrak{M}} \hat{H}^i(G_{\mathfrak{P}_0}, \mathbb{R}/\mathbb{Z}) \\ &= \bigoplus_{\mathfrak{p}\in\mathfrak{N}} \hat{H}^i(G_{\mathfrak{P}_0}, \mathbb{R}/\mathbb{Z}),\end{aligned}$$

where $\mathfrak{N}$ is the set of real primes of k becoming complex in K. Let $\mathfrak{p} \in \mathfrak{N}$; then $G_{\mathfrak{P}_0}$ is generated by an element σ of order 2. If i is even, then

$$\hat{H}^i(G_{\mathfrak{P}_0}, \mathbb{R}/\mathbb{Z}) \cong \hat{H}^0(G_{\mathfrak{P}_0}, \mathbb{R}/\mathbb{Z}) = (\mathbb{R}/\mathbb{Z})^{G_{\mathfrak{P}_0}}/N_{G_{\mathfrak{P}_0}}(\mathbb{R}/\mathbb{Z}).$$

The fixed module consists of the elements $\theta \in \mathbb{R}/\mathbb{Z}$ with $\sigma\theta = -\theta = \theta$, i.e. is $\frac{1}{2}\mathbb{Z}/\mathbb{Z}$, and the norm group consists of the elements $N_{G_{\mathfrak{P}_0}}\theta = \theta + \sigma\theta = \theta - \theta = 0$. This proves $\hat{H}^i(G_{\mathfrak{P}_0}, \mathbb{R}/\mathbb{Z}) = \mathbb{Z}/2\mathbb{Z}$.

If i is odd, then

$$\hat{H}^i(G_{\mathfrak{P}_0}, \mathbb{R}/\mathbb{Z}) \cong \hat{H}^{-1}(G_{\mathfrak{P}_0}, \mathbb{R}/\mathbb{Z}) = {}_{N_{G_{\mathfrak{P}_0}}}(\mathbb{R}/\mathbb{Z})/(\sigma - 1)(\mathbb{R}/\mathbb{Z}).$$

For $\theta \in \mathbb{R}/\mathbb{Z}$, we have $(\sigma - 1)\theta = \sigma\theta - \theta = -2\theta$, so that $(\sigma - 1)(\mathbb{R}/\mathbb{Z}) = 2(\mathbb{R}/\mathbb{Z}) = \mathbb{R}/\mathbb{Z}$, and thus $\hat{H}^i(G_{\mathfrak{P}_0}, \mathbb{R}/\mathbb{Z}) = 0$.

Finally, since $T_K^G = T_k$ and $N_{K|k}D_K \subseteq D_k$, we obtain from $\hat{H}^0(G, D_K) = \hat{H}^0(G, D_K^0) = \hat{H}^0(G, T_K)$ the assertion $D_K^G = T_K N_{K|k} D_K \subseteq D_k$, which gives $H^0(G, D_K) = D_k$. □

Exercise 1. Show that the solenoid $\hat{\mathbb{Z}}/\mathbb{Z}$ is the Pontryagin dual $\operatorname{Hom}_{cts}(\mathbb{Q}, \mathbb{R}/\mathbb{Z})$ of $\mathbb{Q}$. It may also be identified with the projective limit $\varprojlim_n S^1$ over the maps $S^1 \xrightarrow{n} S^1$, $z \mapsto z^n$ $(n \in \mathbb{N})$, of the unit circle $S^1 = \{z \in \mathbb{C} \mid |z| = 1\}$.

Exercise 2. The Pontryagin dual $\operatorname{Hom}(D_k, \mathbb{R}/\mathbb{Z})$ of the connected component D_k of C_k is topologically isomorphic to $\mathbb{Z}^{r_2} \times \mathbb{Q}^{r-1} \times \mathbb{R}$, where r_2 is the number of complex primes and r is the number of all infinite primes.

§3. Restricted Ramification

In the foreground of the Galois cohomological considerations in the previous chapters has been the *absolute* Galois group G_k of the fields. In the case of a global field k, cohomology theory gives much more subtle and deeper lying arithmetic laws if we study not just the absolute Galois groups, but also Galois groups "with restricted ramification". This theory is of great importance, so that we pay particular attention to its development.

Let k be a global field, which we regard as a fixed ground field. Let S be a *nonempty* set of primes of k containing the set S_∞ of infinite (i.e. archimedean) primes if k is a number field. In some of the following formulas for a global field k the expressions S_∞, $S_\mathbb{R}$ and $S_\mathbb{C}$ will occur. If k is a function field, then these terms should be redundant.

In place of the separable closure $\bar{k}|k$, we now consider the maximal subextension $k_S|k$ which is unramified outside S. We denote its Galois group by

$$G_S = G(k_S|k).$$

If $K|k$ is any finite subextension of $k_S|k$, we set

$$G_S(K) = G(k_S|K).$$

We also let the symbol S stand for the set of primes of K which lie above the primes in S. Thus $\mathfrak{P} \in S$ means that $\mathfrak{P}$ is a prime of K lying above a prime $\mathfrak{p} \in S$ of k. The **ring of S-integers** of K is defined by

$$\mathcal{O}_{K,S} = \{a \in K \;\big|\; v_\mathfrak{P}(a) \geq 0 \quad \text{for all } \mathfrak{P} \notin S\}.$$

Its group of units $\mathcal{O}_{K,S}^\times$ and its ideal class group $Cl_S(K)$ play a particularly important role. The last group is the quotient of the usual ideal class group Cl_K of K by the subgroup generated by the classes of all prime ideals in S. It is finite (in the function field case note that S is nonempty) and is called the ***S*-ideal class group**. The ring of integers of k_S is denoted by

$$\mathcal{O}_S = \bigcup_{K|k} \mathcal{O}_{K,S}\,,$$

where the union is taken over all finite extensions K of k inside k_S. Finally, we set

$$\mathbb{N}(S) = \{n \in \mathbb{N} \mid n \in \mathcal{O}_{k,S}^\times\}.$$

If k is a number field, then these are the natural numbers such that $v_\mathfrak{p}(n) = 0$ for all $\mathfrak{p} \notin S$. If k is a function field, then they are the numbers prime to char(k).

The following proposition is a special case of (6.1.10).

(8.3.1) Proposition. *Let $K|k$ be a Galois extension with $K \subseteq k_S$. Then*

$$H^i(K|k, \mathcal{O}_{K,S}) = 0 \quad \textit{for all} \quad i > 0\,.$$

(8.3.2) Corollary. *If $p = \mathrm{char}(k) > 0$ and $L|k$ is a p-closed extension inside k_S, then*

$$H^i(L|k, \mathbb{Z}/p\mathbb{Z}) = \begin{cases} \mathcal{O}_{k,S}/\wp\mathcal{O}_{k,S} & \textit{for } i = 1, \\ 0 & \textit{for } i > 1, \end{cases}$$

where $\wp(x) = x^p - x$.

Proof: Let $K|k$ be a finite subextension of $L|k$ and let $a \in \mathcal{O}_{K,S}$. If a is not of the form $\alpha^p - \alpha$ for some $\alpha \in \mathcal{O}_{K,S}$, consider the cyclic p-extension $K(\alpha)|K$ given by $f(\alpha) = 0$, $f = X^p - X - a$. We have

$$d(\alpha) = \mathrm{disc}(f) = (-1)^{p(p-1)/2}\mathrm{Res}(f, f') = \begin{cases} -1 & \text{if } p \equiv 1 \bmod 4, \\ +1 & \text{else.} \end{cases}$$

Obviously, α lies in $\mathcal{O}_{K(\alpha),S}$, i.e. α is an integer in $K(\alpha)_{\mathfrak{P}}$ for $\mathfrak{P}|\mathfrak{p}$ and $\mathfrak{p} \notin S$. For these primes the discriminant of the field extension $K(\alpha)_{\mathfrak{P}}|K_{\mathfrak{p}}$ divides $d(\alpha) = \pm 1$, whence the extension is unramified. Thus $K(\alpha)|K$ is unramified outside S. Since $L|k$ is a p-closed extension inside k_S, we conclude that $\wp : \mathcal{O}_{L,S} \to \mathcal{O}_{L,S}$ is surjective and obtain the exact sequence

$$0 \longrightarrow \mathbb{Z}/p\mathbb{Z} \longrightarrow \mathcal{O}_{L,S} \xrightarrow{\ \wp\ } \mathcal{O}_{L,S} \longrightarrow 0\,.$$

Taking cohomology, the result follows from (8.3.1). □

Since constant field extensions are unramified, p^∞ divides the order of $G(L|k)$. Applying the last corollary to the fixed field of L with respect to a p-Sylow group of $G(L|k)$, (3.3.6) implies the

(8.3.3) Corollary. *If $p = \mathrm{char}(k) > 0$ and if $L|k$ is a p-closed extension inside k_S, then*

$$cd_p\, G(L|k) = 1.$$

For the group of units of $\mathcal{O}_S$, we have the exact **Kummer sequence**:

(8.3.4) Proposition. *If $n \in \mathbb{N}(S)$, then $\mu_n \subseteq k_S$ and the group of units $\mathcal{O}_S^\times$ of $\mathcal{O}_S$ is n-divisible, i.e. the map $\mathcal{O}_S^\times \xrightarrow{\ n\ } \mathcal{O}_S^\times$, $\varepsilon \longmapsto \varepsilon^n$, is surjective. In other words, we have an exact sequence of G_S-modules*

$$0 \longrightarrow \mu_n \longrightarrow \mathcal{O}_S^\times \xrightarrow{\ n\ } \mathcal{O}_S^\times \longrightarrow 0.$$

Proof: If $\mathfrak{p} \notin S$, then $v_\mathfrak{p}(n) = 0$, hence $k_\mathfrak{p}(\mu_n)|k_\mathfrak{p}$ is unramified. Therefore $k(\mu_n)|k$ is unramified outside S, i.e. $\mu_n \subseteq k_S$. Let $\varepsilon \in \mathcal{O}_S^\times$ and let $K = k(\mu_n, \varepsilon)$. For every finite prime $\mathfrak{P} \notin S$, ε and n are units in $K_\mathfrak{P}$, hence $K_\mathfrak{P}(\sqrt[n]{\varepsilon}) \mid K_\mathfrak{P}$ is unramified. Therefore the extension $K(\sqrt[n]{\varepsilon})|K$ is unramified outside S, i.e. $\sqrt[n]{\varepsilon} \in \mathcal{O}_S^\times$. □

In the sequel we denote the finite subextensions of $k_S|k$ by $K|k$. As in §1 we select a fixed embedding $i_\mathfrak{p}\colon k_S \hookrightarrow \bar{k}_\mathfrak{p}$ for every prime $\mathfrak{p}$ of k, in particular a prime $\bar{\mathfrak{p}}$ of k_S, and denote by $\mathfrak{P}^\cdot$ the prime of K lying under $\bar{\mathfrak{p}}$. We set $k_{S,\mathfrak{p}} = i_\mathfrak{p}(k_S)k_\mathfrak{p}$ and

$$K_\mathfrak{p} = K_{\mathfrak{P}^\cdot} = i_\mathfrak{p}(K)k_\mathfrak{p}.$$

We consider the ***S*-idèle group**

$$I_{K,S} := \prod_{\mathfrak{P} \in S} K_\mathfrak{P}^\times .$$

It contains the group of S-units $\mathcal{O}_{K,S}^\times$ as a discrete subgroup and we set

$$C_{K,S} = I_{K,S}/\mathcal{O}_{K,S}^\times .$$

In spite of the analogy with the formation of the idèle class group $C_K = I_K/K^\times$, it is not this group which takes the role of C_K in the "S-theory". The reason is the failure of *Galois descent*, i.e. if $K|k$ is Galois, then $C_{k,S}$ is not always the fixed module $C_{K,S}^{G(K|k)}$. It will become clear in a moment that we have to consider the group

$$C_S(K) = I_K/K^\times U_{K,S}$$

instead of $C_{K,S}$, where $U_{K,S}$ is the compact subgroup

$$U_{K,S} = \prod_{\mathfrak{P} \in S} \{1\} \times \prod_{\mathfrak{P} \notin S} U_\mathfrak{P}$$

of the full idèle group I_K. Since $K^\times \cap U_{K,S} = 1$, we may regard $U_{K,S}$ as a subgroup of $C_K = I_K/K^\times$ and may also write

$$C_S(K) = C_K/U_{K,S}.$$

This group is called the ***S*-idèle class group**. Note that if $K|k$ is Galois, then $U_{K,S}$ is a cohomologically trivial $G(K|k)$-module. Namely, by the same argument as for (8.1.2), we have

$$\hat{H}^i(G(K|k), U_{K,S}) = \bigoplus_{\mathfrak{p} \notin S} \hat{H}^i(G(K_{\mathfrak{P}^\cdot}|k_\mathfrak{p}), U_{\mathfrak{P}^\cdot}),$$

and since $K_{\mathfrak{P}^\cdot}|k_\mathfrak{p}$ is unramified for $\mathfrak{p} \notin S$, $\hat{H}^i(G(K_{\mathfrak{P}^\cdot}|k_\mathfrak{p}), U_{\mathfrak{P}^\cdot}) = 1$ by (7.1.2)(i). The difference of the groups $C_{K,S}$ and $C_S(K)$ is given by the

(8.3.5) Proposition. *$C_{K,S}$ is an open subgroup of $C_S(K)$ and there is an exact sequence*

$$0 \longrightarrow C_{K,S} \longrightarrow C_S(K) \xrightarrow{\pi} Cl_S(K) \longrightarrow 0.$$

In particular, $C_{K,S} = C_S(K)$ if S omits only finitely many primes.

Proof: Consider the canonical injection $I_{K,S} \to I_K$, $\alpha \mapsto \tilde{\alpha}$, and the induced homomorphism

$$j : I_{K,S} \longrightarrow I_K/K^\times U_{K,S} = C_S(K).$$

The image of j is closed. If $\alpha \in I_{K,S}$ is an idèle such that $j(\alpha) = 1$, then $\tilde{\alpha} = a\tilde{u}$ with $a \in K^\times$, $\tilde{u} \in U_{K,S}$. This means that

$$\alpha_{\mathfrak{P}} = \tilde{\alpha}_{\mathfrak{P}} = a \qquad \text{for } \mathfrak{P} \in S \quad \text{and} \quad 1 = \tilde{\alpha}_{\mathfrak{P}} = a u_{\mathfrak{P}} \qquad \text{for } \mathfrak{P} \notin S,$$

hence $\alpha = a$ and $a \in \mathcal{O}_{K,S}^\times$. Therefore $\ker(j) = \mathcal{O}_{K,S}^\times$ and j induces an injection $C_{K,S} \hookrightarrow C_S(K)$ with closed image. Its cokernel is

$$I_K/I_{K,S}U_{K,S}K^\times = \Big(\bigoplus_{\mathfrak{P}\notin S} (K_{\mathfrak{P}}^\times/U_{\mathfrak{P}})\Big)/\mathrm{im}(K^\times) = \Big(\bigoplus_{\mathfrak{P}\notin S} \mathbb{Z}\Big)/\mathrm{im}(K^\times),$$

which can be identified with the finite ideal class group $Cl_S(K)$. In particular, $C_{K,S}$ has finite index in $C_S(K)$ and is thus open.

If S omits only finitely many prime ideals, then $\mathcal{O}_{K,S}$ is a Dedekind ring with only finitely many prime ideals and is hence a principal ideal domain by [16], chap.VII, §2, prop.1. Therefore in this case $Cl_S(K) = 0$ and $C_{K,S} = C_S(K)$. □

The proposition shows that the groups $C_{K,S}$ and $C_S(K)$ are in general different. But they become equal in the limit over all $K \subseteq k_S$. We set

$$I_S = \varinjlim_{K|k} I_{K,S}\,, \qquad U_S = \varinjlim_{K|k} U_{K,S}\,,$$
$$C_S = \varinjlim_{K|k} C_{K,S}\,, \qquad C(k_S) = \varinjlim_{K|k} C_K\,,$$

where $K|k$ runs through all finite subextensions of $k_S|k$. These are G_S-modules and U_S is a cohomologically trivial G_S-module, since $U_{K,S}$ is a cohomologically trivial $G(K|k)$-module if $K|k$ is Galois. Taking the fixed module of C_S under $G_S(K) = G(k_S|K)$, we do not recover $C_{K,S}$ but $C_S(K)$.

(8.3.6) Proposition. *The G_S-module C_S has the following properties:*

(i) $C_S = \varinjlim_{K|k} C_S(K) = C(k_S)/U_S$,

(ii) $C_S^{G_S(K)} = C_S(K)$.

Proof: (i) We have $\varinjlim_{K|k} Cl_S(K) = 0$, since every ideal in $\mathcal{O}_{K,S}$ becomes a principal ideal in a suitable finite unramified Galois extension $L|K$; if K is a number field, we take the maximal subfield of the Hilbert class field of K which is completely split at the primes in S (the principal ideal theorem, see [160], chap. VI, (7.5)). The same argument holds in the function field case since $Cl_S(K)$ is finite for $S \neq \varnothing$. Therefore (i) follows from (8.3.5).

(ii) The exact sequence $0 \to U_S \to C(k_S) \to C_S \to 0$ yields the exact cohomology sequence

$$0 \longrightarrow U_{K,S} \longrightarrow C_K \longrightarrow C_S^{G_S(K)} \longrightarrow H^1(G_S(K), U_S) = 0,$$

hence $C_S^{G_S(K)} = C_K/U_{K,S} = C_S(K)$. □

If $K|k$ is Galois, then we have the exact sequence of $G(K|k)$-modules

$$0 \longrightarrow U_{K,S} \longrightarrow C_K \longrightarrow C_S(K) \longrightarrow 0.$$

Since $U_{K,S}$ is cohomologically trivial, we obtain the

(8.3.7) Proposition. *For every finite Galois subextension $K|k$ of $k_S|k$ and every $i \in \mathbb{Z}$,*

$$\hat{H}^i(G(K|k), C_S(K)) \cong \hat{H}^i(G(K|k), C_K).$$

We now compute the cohomology of the three G_S-modules in the exact sequence

$$0 \longrightarrow \mathcal{O}_S^\times \longrightarrow I_S \longrightarrow C_S \longrightarrow 0.$$

(8.3.8) Proposition (Cohomology of I_S).

(i) $H^0(G_S, I_S) = I_{k,S}$.

(ii) $H^1(G_S, I_S) = 0$.

(iii) $H^2(G_S, I_S)(p) = \bigoplus_{\mathfrak{p} \in S} Br(k_\mathfrak{p})(p)$ *for every prime number $p \in \mathbb{N}(S)$.*

(iv) $H^3(G_S, I_S) = 0$.

Proof: (i) is a consequence of (8.1.2). For $i \geq 1$ and $K|k$ Galois, we obtain just as in §1

$$H^i(G(K|k), I_{K,S}) = \bigoplus_{\mathfrak{p} \in S} H^i(G(K_\mathfrak{P}|k_\mathfrak{p}), K_\mathfrak{P}^\times)$$

and furthermore

$$H^i(G_S, I_S) = \varinjlim_{K|k} H^i(G(K|k), I_{K,S}).$$

From this (ii) follows because of Hilbert's Satz 90. The assertion (iv) is a consequence of $H^3(G(K_{\mathfrak{P}}|k_{\mathfrak{p}}), K_{\mathfrak{P}}^\times) \cong H^1(G(K_{\mathfrak{P}}|k_{\mathfrak{p}}), \mathbb{Z}) = 0$ (see (7.1.7)). For (iii) we have the invariant map of local class field theory

$$(*) \qquad inv_{\mathfrak{p}} : H^2(G(K_{\mathfrak{P}}|k_{\mathfrak{p}}), K_{\mathfrak{P}}^\times) \xrightarrow{\sim} \tfrac{1}{[K_{\mathfrak{P}}:k_{\mathfrak{p}}]}\mathbb{Z}/\mathbb{Z}$$

for nonarchimedean primes $\mathfrak{p}$. If k is a function field, then we have constant extensions of every degree, i.e. $\varinjlim_{K|k}$ of $(*)$ gives $\mathbb{Q}/\mathbb{Z}$. If k is a number field and $p \in \mathbb{N}(S)$, then $k(\mu_{p^\infty})|k$ is an extension inside $k_S|k$ of degree divisible by p^∞, hence

$$\varinjlim_{K|k} H^2(G(K_{\mathfrak{P}}|k_{\mathfrak{p}}), K_{\mathfrak{P}}^\times)(p) = \mathbb{Q}_p/\mathbb{Z}_p$$

for nonarchimedean primes $\mathfrak{p}$. For $\mathfrak{p} \in S_\infty$ the group $G(K_{\mathfrak{P}}|k_{\mathfrak{p}})$ is of order 1 or 2 depending on whether $K_{\mathfrak{P}} = k_{\mathfrak{p}}$ or $K_{\mathfrak{P}} \neq k_{\mathfrak{p}}$. If $2 \in \mathbb{N}(S)$, then $K = k(\sqrt{-1})$ is contained in k_S, and so $H^2(G(K_{\mathfrak{P}}|k_{\mathfrak{p}}), K_{\mathfrak{P}}^\times) = \frac{1}{2}\mathbb{Z}/\mathbb{Z}$ if $\mathfrak{p}$ is real. This proves the proposition. □

(8.3.9) Proposition (Cohomology of C_S). *The pair (G_S, C_S) is a class formation, and we have*

(i) $H^0(G_S, C_S) = C_S(k)$,

(ii) $H^1(G_S, C_S) = 0$,

(iii) $H^2(G_S, C_S) \cong \frac{1}{\#G_S}\mathbb{Z}/\mathbb{Z}$,

(iv) $H^3(G_S, C_S) = 0$.

Proof: (i) is the assertion (ii) of (8.3.6). The other equalities follow from

$$H^i(G(K|k), C_S(K)) = H^i(G(K|k), C_K)\ , \ i \geq 1,$$

(see (8.3.7)) and from the fact that the G_k-module $C = \varinjlim_{K|k} C_K$, where $K|k$ runs through *all* finite Galois subextensions of the separable closure $\bar{k}|k$, is a formation module by (8.1.22). □

As a corollary of the proposition above and (3.1.6), we obtain the global reciprocity law for restricted ramification.

(8.3.10) Theorem *Let $K|k$ be a finite Galois extension inside k_S with Galois group $G(K|k)$. Then there is a canonical isomorphism*

$$C_S(k)/N_{K|k}C_S(K) \cong G(K|k)^{ab}.$$

(8.3.11) Proposition (Cohomology of $\mathcal{O}_S^\times$).

(i) $H^0(G_S, \mathcal{O}_S^\times) = \mathcal{O}_{k,S}^\times$.

(ii) $H^1(G_S, \mathcal{O}_S^\times) \cong Cl_S(k)$.

(iii) *For every prime number $p \in \mathbb{N}(S)$*

$$\begin{aligned} H^2(G_S, \mathcal{O}_S^\times)(p) &= \ker\Big(\bigoplus_{\mathfrak{p}\in S} H^2(k_\mathfrak{p}, \mu_{p^\infty}) \longrightarrow H^2(G_S, C_S)(p)\Big) \\ &\cong \ker\Big(\bigoplus_{\mathfrak{p}\in S\setminus S_\infty} \mathbb{Q}_p/\mathbb{Z}_p \xrightarrow{\Sigma} \mathbb{Q}_p/\mathbb{Z}_p\Big) \quad \text{or} \\ &\cong \ker\Big(\bigoplus_{\mathfrak{p}\in S\setminus S_\infty} \mathbb{Q}_2/\mathbb{Z}_2 \oplus \bigoplus_{\mathfrak{p}\in S_\mathbb{R}} \tfrac{1}{2}\mathbb{Z}/\mathbb{Z} \xrightarrow{\Sigma} \mathbb{Q}_2/\mathbb{Z}_2\Big) \end{aligned}$$

according to whether $p \neq 2$ or $p = 2$ (if k is a function field, then the expressions S_∞ and $S_\mathbb{R}$ are redundant).

(iv) $H^3(G_S, \mathcal{O}_S^\times)(p) = 0$ *for every prime number $p \in \mathbb{N}(S)$.*

Proof: (i) is trivial and the other statements follow from the exact cohomology sequence

$$\longrightarrow H^{i-1}(G_S, C_S) \xrightarrow{\delta} H^i(G_S, \mathcal{O}_S^\times) \longrightarrow H^i(G_S, I_S) \longrightarrow H^i(G_S, C_S) \longrightarrow .$$

For $i = 1$ we obtain by (8.3.8) and (8.3.9) the exact sequence

$$I_{k,S} \longrightarrow C_S(k) \xrightarrow{\delta} H^1(G_S, \mathcal{O}_S^\times) \longrightarrow 0.$$

The cokernel of the left arrow is $Cl_S(k)$ by (8.3.5), proving (ii).

If k is a function field or $p \neq 2$, then (iii) follows from the commutative diagram

$$\begin{array}{ccccccc} 0 \longrightarrow & H^2(G_S, \mathcal{O}_S^\times)(p) & \longrightarrow & H^2(G_S, I_S)(p) & \longrightarrow & H^2(G_S, C_S)(p) \\ & & & \Big\downarrow{\scriptstyle \oplus\, inv_\mathfrak{p}} \wr & & \Big\downarrow \wr \\ & & & \bigoplus_{\mathfrak{p}\in S\setminus S_\infty} \mathbb{Q}_p/\mathbb{Z}_p & \xrightarrow{\ \Sigma\ } & \mathbb{Q}_p/\mathbb{Z}_p, \end{array}$$

where the lower horizontal map is the summation over the components and the zero on the left in the upper row is a consequence of $H^1(G_S, C_S) = 0$. If $p = 2$ and k is a number field, then the same argument gives the result using $H^2(G_S, I_S)(2) \cong \bigoplus_{\mathfrak{p}\in S\setminus S_\infty} \mathbb{Q}_2/\mathbb{Z}_2 \oplus \bigoplus_{\mathfrak{p}\in S_\mathbb{R}} \frac{1}{2}\mathbb{Z}/\mathbb{Z}$.

Finally, (iv) follows from the surjectivity of the upper right-hand arrow and from the equality $H^3(G_S, I_S) = 0$. □

(8.3.12) Corollary. *Let k be a global field and let $p \in S$ be a prime number.*

(i) *If k is a function field or $p \neq 2$, then $H^2(G_S, \mathcal{O}_S^\times)$ is p-divisible.*

(ii) *If k is a number field and $p = 2$, then the canonical map*

$$H^2(G_S, \mathcal{O}_S^\times)/2 \xrightarrow{\sim} \bigoplus_{\mathfrak{p} \in S_{\mathbb{R}}} H^2(k_{\mathfrak{p}}, \bar{k}_{\mathfrak{p}}^\times)$$

is an isomorphism. In particular, $H^2(G_S, \mathcal{O}_S^\times)$ is 2-divisible if and only if k is totally imaginary.

(iii) *If k is a number field and $p = 2$, then the restriction map*

$$H^3(G_S, \mathbb{Z}/2\mathbb{Z}) \xrightarrow{\sim} \bigoplus_{\mathfrak{p} \in S_{\mathbb{R}}} H^3(k_{\mathfrak{p}}, \mathbb{Z}/2\mathbb{Z})$$

is an isomorphism.

Proof: The first two assertions follow immediately from the fact that the group $H^2(G_S, \mathcal{O}_S^\times)(p)$ is isomorphic to the kernel of the summation over the local components.

Using (8.3.11)(iv), it follows from the exact sequence

$$0 \longrightarrow \mu_2 \longrightarrow \mathcal{O}_S^\times \xrightarrow{2} \mathcal{O}_S^\times \longrightarrow 0$$

that $H^2(G_S, \mathcal{O}_S^\times)/2 \cong H^3(G_S, \mathbb{Z}/2\mathbb{Z})$. Since $H^2(k_{\mathfrak{p}}, \bar{k}_{\mathfrak{p}}^\times) \cong H^3(k_{\mathfrak{p}}, \mathbb{Z}/2\mathbb{Z})$ for $\mathfrak{p} \in S_{\mathbb{R}}$, we obtain the last assertion from (ii). □

The groups $C_S(k)$ are locally compact topological groups exactly like C_k. Concerning its connected component we have the

(8.3.13) Theorem. *Assume that k is a number field. The connected component $D_S(k)$ of $C_S(k) = C_k/U_{k,S}$ is given by*

$$D_S(k) = D_k U_{k,S}/U_{k,S}.$$

It is divisible, and there is an exact sequence of topological groups

$$0 \longrightarrow D_S(k) \longrightarrow C_S(k) \xrightarrow{(\ ,k_S|k)} G_S^{ab} \longrightarrow 0.$$

In particular, $D_S(k)$ is the group of universal norms $N_{G_S} C_S$.

Proof: Consider the continuous projection

$$(*) \qquad C_k \longrightarrow C_S(k) = C_k/U_{k,S}.$$

For general topological reasons, the connected component $D_S(k)$ of $C_S(k)$ is the closure of the image $D_k U_{k,S}/U_{k,S}$ of the connected component D_k of C_k.

But this image is already closed, since $U_{k,S}$ is compact and hence $(*)$ is a proper map. This proves the first assertion. The divisibility follows from that of D_k.

For the second statement we use the exact sequence (8.2.2):

$$0 \longrightarrow D_k \longrightarrow C_k \xrightarrow{(\ ,k)} G^{ab} \longrightarrow 0\,.$$

For each nonarchimedean prime $\mathfrak{p}$ the norm residue symbol $(\ ,k)$ maps the subgroup $k_\mathfrak{p}^\times \subseteq C_k$ onto the decomposition group of G_k^{ab} with respect to $\mathfrak{p}$ and the group of units $U_\mathfrak{p} \subseteq k_\mathfrak{p}^\times$ onto the inertia group (see [160], chap.VI, (5.6) and chap.V, (6.2)). Therefore the compact group $U_{k,S}$ is mapped onto the subgroup H of $G_k^{ab} = G(k^{ab}|k)$ which is generated by the inertia groups for the primes not in S. The fixed field of H is therefore the maximal subextension $k_S^{ab}|k$ of $k^{ab}|k$ which is unramified outside S. This gives us an exact commutative diagram

$$\begin{array}{ccccccccc} & & U_{k,S} & \longrightarrow & G(k^{ab}|k_S^{ab}) & \longrightarrow & 0 \\ & & \downarrow & & \downarrow & & \\ 0 \longrightarrow & D_k \longrightarrow & C_k & \longrightarrow & G(k^{ab}|k) & \longrightarrow & 0 \end{array}$$

of topological groups. The snake lemma yields an exact sequence of topological groups

$$0 \longrightarrow D_k U_{k,S}/U_{k,S} \longrightarrow C_k/U_{k,S} \longrightarrow G(k_S^{ab}|k) \longrightarrow 0,$$

and since $G(k_S^{ab}|k) = G(k_S|k)^{ab} = G_S^{ab}$, the theorem is proved. □

(8.3.14) Corollary. *Let k be a number field. For a subgroup N of $C_S(k)$ the following conditions are equivalent:*

(i) *N is the norm group $N_{K|k}C_S(K)$ of a finite Galois extension $K|k$ inside $k_S|k$.*

(ii) *N is an open subgroup of $C_S(k)$.*

Each open subgroup N of $C_S(k)$ contains $D_S(k)$ and has finite index.

Proof: By (8.3.9), the pair (G_S, C_S) is a class formation. The isomorphism

(1) $$C_S(k)/D_S(k) \xrightarrow{\sim} G_S^{ab}$$

is the projective limit of the isomorphisms of finite groups

(2) $$C_S(k)/N_{K|k}C_S(K) \cong G(K|k)^{ab},$$

where $K|k$ runs through the finite Galois subextensions of $k_S|k$. Therefore the norm groups are precisely the open subgroups of finite index in $C_S(k)$ which contain $D_S(k)$. It remains to show that each open subgroup in $C_S(K)$ has finite index and contains $D_S(k)$. As $D_S(k)$ is divisible, it suffices to show that each open subgroup in $C_S(k)$ has finite index.

As $C_S(k)$ is a topological quotient of C_k, we may assume that S is the set of all primes. Let $W \subseteq C_k$ be an open subgroup. The pre-image of W in I_k is open and therefore contains a subgroup of the form

$$V = \prod_{\mathfrak{p}} V_{\mathfrak{p}},$$

where $V_{\mathfrak{p}} \subseteq U_{\mathfrak{p}}$ is an open subgroup for all $\mathfrak{p}$, $V_{\mathfrak{p}} = U_{\mathfrak{p}}$ for almost all $\mathfrak{p}$ and $V_{\mathfrak{p}} \subseteq U_{\mathfrak{p}}^1$ for archimedean $\mathfrak{p}$. For nonarchimedean $\mathfrak{p}$, an open subgroup in $U_{\mathfrak{p}}$ has finite index and for archimedean $\mathfrak{p}$ the group $U_{\mathfrak{p}}^1$ has no proper open subgroup. Hence V has finite index in the group

$$U_k = \prod_{\mathfrak{p}\nmid\infty} U_{\mathfrak{p}} \times \prod_{\mathfrak{p}|\infty} U_{\mathfrak{p}}^1 \subseteq I_k,$$

and it suffices to show that the image of U_k in C_k has finite index. We have

$$I_k/U_k \cong \bigoplus_{\mathfrak{p}\notin S_\infty} \mathbb{Z} \oplus \bigoplus_{\mathfrak{p}\in S_{\mathbb{R}}} \mathbb{Z}/2\mathbb{Z},$$

which gives an isomorphism

$$I_k/k^\times U_k \cong Cl_0(k)$$

where $Cl_0(k)$ is the ideal class group of k in the narrow sense, which is known to be finite. This concludes the proof. □

For a function field k we have a continuous injection

$$C_k \xrightarrow{(\ ,k)} G_k^{ab}.$$

It is not surjective, since in C_k we have the group $C_k^0 = \{x \in C_k \mid |x| = 1\}$, and the quotient $C_k/C_k^0 \cong \mathbb{Z}$ is not profinite. With the same arguments as above in this situation, we obtain the

(8.3.15) Theorem. *If k is a function field, then we have an exact sequence of topological groups*

$$0 \longrightarrow C_S(k) \xrightarrow{(\ ,k_S|k)} G_S^{ab} \longrightarrow \hat{\mathbb{Z}}/\mathbb{Z} \longrightarrow 0.$$

(8.3.16) Corollary. *Let k be a function field. For a subgroup N of $C_S(k)$ the following conditions are equivalent:*

(i) *N is the norm group $N_{K|k}C_S(K)$ of a finite Galois subextension $K|k$ of $k_S|k$.*

(ii) *N is an open subgroup of finite index.*

(iii) *N is an open subgroup which is not contained in the group $C_S^0(k) = \{x \in C_S(k) \mid |x| = 1\}$.*

As before let k be a global field, S a nonempty set of primes of k containing the set S_∞ of infinite primes if k is a number field, and let $k_S|k$ be the maximal extension unramified outside S. Now we study the cohomological dimension of G_S. In the function field case we have the following result (for the case $S = \varnothing$, which is not considered here, we refer to X §1).

(8.3.17) Theorem. *If k is a function field, then for all prime numbers p*

$$scd_p\, G_S = 2.$$

Proof: Since the constant field extensions are contained in k_S, every prime number p divides the order of G_S, hence $cd_p\, G_S > 0$ and therefore $scd_p\, G_S \geq 2$, because the strict cohomological dimension is never 1. It therefore suffices to prove $scd\, G_S \leq 2$. By (3.6.4), this is equivalent to the existence of a level-compact formation module for G_S with trivial universal norms. We construct such a formation module from the G_S-module $C_S = C(k_S)/U_S$, which is a formation module by (8.3.9).

Let $L|K$ be a finite normal intermediate extension in $k_S|k$. The norm group $N = N_{L|K}C_L$ is an open subgroup of finite index in C_K. The property of being unramified at a prime $\mathfrak{p}$ is equivalent to $U_\mathfrak{p} \subseteq N_{L_\mathfrak{P}|K_\mathfrak{p}} L_\mathfrak{P}^\times = N \cap K_\mathfrak{p}^\times$ by class field theory (see [160], chap.VI, (5.8)). This means that the norm groups of finite Galois subextensions $L|K$ of $k_S|K$ are open subgroups N of finite index in C_K containing $U_{K,S} = U_{K,S}K^\times/K^\times$. By the existence theorem of class field theory, they are precisely all such groups (8.3.16), and hence $U_{K,S}K^\times/K^\times$ is their intersection. Therefore $C_S(K) = C_K/U_{K,S}$ becomes a dense subgroup of the *compact* group

$$\overline{C}_S(K) = \varprojlim_N\, C_K/N.$$

The canonical surjective map

$$\deg : C_S(K) \to \mathbb{Z}, \quad \deg(\alpha) = \sum_\mathfrak{p} v_\mathfrak{p}(\alpha_\mathfrak{p}) \deg(\mathfrak{p}),$$

extends to $\overline{C}_S(K)$ and yields an exact commutative diagram

$$\begin{array}{ccccccccc} 0 & \longrightarrow & C_S^0(K) & \longrightarrow & C_S(K) & \xrightarrow{\deg} & \mathbb{Z} & \longrightarrow & 0 \\ & & \| & & \downarrow & & \downarrow & & \\ 0 & \longrightarrow & C_S^0(K) & \longrightarrow & \overline{C}_S(K) & \xrightarrow{\deg} & \hat{\mathbb{Z}} & \longrightarrow & 0, \end{array}$$

which shows that the quotient $\overline{C}_S(K)/C_S(K)$ is isomorphic to $\hat{\mathbb{Z}}/\mathbb{Z}$ and is consequently uniquely divisible. Because of this fact, for a normal intermediate extension $L|K$ with group G, the G-module $\overline{C}_S(L)/C_S(L)$ has trivial cohomology, so that

$$H^i(G, \bar{C}_S(L)) = H^i(G, C_S(L)) \quad \text{for } i \geq 1,$$

and from $C_S(L)^G = C_S(K)$ it follows that $\bar{C}_S(L)^G = \bar{C}_S(K)$. This shows that the G-module $\bar{C}_S = \varinjlim_{K|k} \bar{C}_S(K)$ is a formation module since C_S is.*) It is level-compact because $\bar{C}_S^{G_S(K)} = \bar{C}_S(K)$, and it has trivial universal norms because

$$\bigcap_{K \subseteq L \subseteq k_S} N_{L|K} C_L = U_{K,S} K^\times / K^\times, \qquad \text{i.e.} \quad \bigcap_{K \subseteq L \subseteq k_S} N_{L|K} \bar{C}_S(L) = 0.$$

$\bar{C}_S$ is the desired formation module and the theorem is proved. □

Now we assume that k is a number field. In this case the existence of the infinite primes causes severe difficulties for the determination of the *strict cohomological dimension* $scd_p\, G_S$. However, the cohomological dimension $cd_p\, G_S$ is more easy to handle.

For a prime number p we put

$$S_p = S_p(k) = \{\mathfrak{p} \text{ a prime of } k \text{ dividing } p\}\,.$$

(8.3.18) Proposition. *Let $S_p \cup S_\infty \subseteq S$ and assume that the number field k is totally imaginary if $p = 2$. Then*

$$cd_p\, G_S \leq 2.$$

Proof: k_S contains the group μ_p of p-th roots of unity since $\mathbb{Q}(\mu_p)|\mathbb{Q}$, and hence also $k(\mu_p)|k$, is unramified outside p. The group $\mathcal{O}_S^\times$ of S-units of k_S contains μ_p and is p-divisible by (8.3.4), i.e. the sequence

$$(*) \qquad 0 \longrightarrow \mu_p \longrightarrow \mathcal{O}_S^\times \xrightarrow{p} \mathcal{O}_S^\times \longrightarrow 0$$

is exact. Let $K|k$ be any finite subextension of $k_S|k$. Proposition (8.3.11) shows that $H^3(G_S(K), \mathcal{O}_S^\times)(p) = 0$, and by (8.3.12) the group $H^2(G_S(K), \mathcal{O}_S^\times)$ is p-divisible. Therefore the exact cohomology sequence associated to $(*)$ yields $H^3(G_S(K), \mu_p) = 0$. Now let G_p be a p-Sylow subgroup of G_S. Let Σ_p be its fixed field and let $K|k$ run through the finite subextensions of $\Sigma_p|k$. Noting that $\mu_p \subseteq \Sigma_p$, we get

$$H^3(G_p, \mathbb{Z}/p\mathbb{Z}) \cong H^3(G_p, \mu_p)(-1) = \varinjlim_{K} H^3(G_S(K), \mu_p)(-1) = 0,$$

hence $cd_p\, G_S = cd_p\, G_p \leq 2$. □

*) If $S = \varnothing$, then C_S is no longer a formation module.

(8.3.19) Proposition. *Let $S_2 \cup S_\infty \subseteq S$ and let K be an (possibly infinite) extension of the number field k inside k_S. Then $cd_2\, G_S(K) = \infty$ if and only if K has a real place, and $cd_2\, G_S(K) \le 2$ otherwise.*

Proof: As $S_2 \cup S_\infty \subseteq S$ the field k_S contains $K(\sqrt{-1})$. Therefore, if K has a real place, then $G_S(K)$ contains its decomposition group which is isomorphic to $\mathbb{Z}/2\mathbb{Z}$. Hence $cd_2\, G_S(K) = \infty$ by (3.3.5).

If $cd_2\, G_S(K) = \infty$, then $H^n((G_S(K))_2, \mathbb{Z}/2\mathbb{Z}) \neq 0$ for all n and every 2-Sylow subgroup $(G_S(K))_2$ of $G_S(K)$ by (3.3.6) and (3.3.2)(iii). We therefore find a finite extension $K'|K$ inside k_S such that $H^3(G_S(K'), \mathbb{Z}/2\mathbb{Z}) \neq 0$. From (8.3.12)(iii) we obtain

$$0 \neq H^3(G_S(K'), \mathbb{Z}/2\mathbb{Z}) \cong \varinjlim_{L} \bigoplus_{\mathfrak{p} \in S_{\mathbb{R}}(L)} H^3(L_\mathfrak{p}, \mathbb{Z}/2\mathbb{Z}),$$

where L runs through the finite extensions of k in K'. We conclude that K', and hence also K, has at least one real place.

Finally, assume that $cd_2\, G_S(K)$ is finite. Then $cd_2\, G_S(K)$ is equal to $cd_2\, G_S(K(\sqrt{-1}))$ by (3.3.5)(ii). But $K(\sqrt{-1})$ is the union of totally imaginary number fields, hence $cd_2\, G_S(K(\sqrt{-1})) \le 2$ by (8.3.18). $\square$

Remark: The assumptions of (8.3.18) may be weakened to $S_p \subseteq S$ and, if $p = 2$, S contains no real primes. The statement of (8.3.19) may be sharpened in the following way: Assume that $S_2 \subseteq S$; then $cd_2(G_S(K)) = \infty$ if and only if S contains a real prime, and $cd_2(G_S(K)) \le 2$ otherwise, see (10.6.1).

Together with (8.3.11), we obtain from (8.3.18) the following finiteness theorem.

(8.3.20) Theorem. *Let A be a G_S-module which is finitely generated as a $\mathbb{Z}$-module and whose torsion submodule* tor(A) *is of an order which is a unit in $\mathcal{O}_{k,S}$. Then the following hold:*

(i) *If S and A are finite, then the cohomology groups $H^n(G_S, A)$ are finite for all $n \ge 0$.*

(ii) *If S is finite or A is torsion-free, then $H^1(G_S, A)$ is finite.*

Proof: (i) Let $K|k$ be a finite Galois subextension of $k_S|k$ over which A becomes a trivial Galois module and which contains the n-th roots of unity, where $n = \#A$, and which is totally imaginary in the number field case. Then A is isomorphic over K to a direct sum of modules μ_{p^ν}, $p|n$. If the assertion

holds for $A = \mu_{p^\nu}$ and base field K, the general case follows using the spectral sequence
$$H^i(G(K|k), H^j(G_S(K), A)) \Rightarrow H^{i+j}(G_S, A).$$
Therefore let $A = \mu_{p^\nu}$. By induction, using the exact sequence $0 \to \mu_p \to \mu_{p^\nu} \to \mu_{p^{\nu-1}} \to 0$ and its associated exact cohomology sequence, we may assume $A = \mu_p$ with $p \in \mathbb{N}(S)$. Therefore, as in the proof of (8.3.18), we have the exact sequence
$$0 \longrightarrow \mu_p \longrightarrow \mathcal{O}_S^\times \xrightarrow{p} \mathcal{O}_S^\times \longrightarrow 0,$$
and the associated exact cohomology sequence
$$H^{i-1}(G_S, \mathcal{O}_S^\times) \xrightarrow{p} H^{i-1}(G_S, \mathcal{O}_S^\times) \longrightarrow H^i(G_S, \mu_p) \longrightarrow \dots$$
$$\dots \longrightarrow H^i(G_S, \mathcal{O}_S^\times) \xrightarrow{p} H^i(G_S, \mathcal{O}_S^\times).$$
For $i = 1$ we obtain the finiteness of $H^1(G_S, \mu_p)$ since $\mathcal{O}_{k,S}^\times/(\mathcal{O}_{k,S}^\times)^p$ and $H^1(G_S, \mathcal{O}_S^\times) \cong Cl_S(k)$ are finite. For $i = 2$ we use the finiteness of the set S and see by (8.3.11)(iii) that the kernel of $H^2(G_S, \mathcal{O}_S^\times) \xrightarrow{p} H^2(G_S, \mathcal{O}_S^\times)$ is finite, and thus also $H^2(G_S, \mu_p)$, again using $H^1(G_S, \mathcal{O}_S^\times) \cong Cl_S(k)$. Since $H^n(G_S, \mu_p) = 0$ by (8.3.18) for $n \geq 3$, part (i) of the theorem is proved.

(ii) Let A be torsion-free and let $K|k$ be a finite Galois subextension of $k_S|k$ over which A becomes a trivial Galois module, i.e. $A = \mathbb{Z}^r$ for some $r \geq 0$. Then the exact sequence
$$0 \longrightarrow H^1(G(K|k), A) \longrightarrow H^1(G_S, A) \longrightarrow H^1(G_S(K), \mathbb{Z}^r) = 0$$
gives the desired result. If S is finite and is A arbitrary, then the exact sequence
$$H^1(G_S, \mathrm{tor}(A)) \longrightarrow H^1(G_S, A) \longrightarrow H^1(G_S, A/\mathrm{tor}(A))$$
gives the result using (i) and the result for torsion-free modules. □

Let N run through the open subgroups of finite index in $I_{k,S}$ and put
$$\bar{I}_{k,S} = \varprojlim_N I_{k,S}/N,$$
i.e. $\bar{I}_{k,S}$ is the profinite completion of $I_{k,S}$. We can collect many of the results obtained in the form of two commutative exact diagrams.

(8.3.21) Corollary.

(i) *Let k be a function field. Then we have the commutative exact diagram*
$$\begin{array}{ccccccccc} 0 & \longrightarrow & \mathcal{O}_{k,S}^\times & \longrightarrow & I_{k,S} & \longrightarrow & C_S(k) & \longrightarrow & Cl_S(k) & \longrightarrow & 0 \\ & & \downarrow{\scriptstyle dense} & & \downarrow{\scriptstyle dense} & & \downarrow{\scriptstyle dense} & & \| & & \\ 0 & \longrightarrow & \overline{\mathcal{O}_{k,S}^\times} & \longrightarrow & \bar{I}_{k,S} & \longrightarrow & G_S(k)^{ab} & \longrightarrow & Cl_S(k) & \longrightarrow & 0. \end{array}$$

(ii) *Let* k *be a number field. Then we have the commutative exact diagram*

$$
\begin{array}{ccccccccc}
 & & \bigoplus\limits_{\mathfrak{p}\in S_\infty} N_{G_\mathfrak{p}}\bar{k}_\mathfrak{p}^\times & \hookrightarrow & D_S(k) & & & & \\
 & & \downarrow & & \downarrow & & & & \\
0 \longrightarrow & \mathcal{O}_{k,S}^\times & \longrightarrow & I_{k,S} & \longrightarrow & C_S(k) & \longrightarrow & Cl_S(k) & \longrightarrow 0 \\
 & \downarrow{\scriptstyle dense} & & \downarrow{\scriptstyle dense} & & \downarrow & & \| & \\
0 \longrightarrow & \overline{\mathcal{O}_{k,S}^\times} & \longrightarrow & \bar{I}_{k,S} & \longrightarrow & G_S(k)^{ab} & \longrightarrow & Cl_S(k) & \longrightarrow 0.
\end{array}
$$

Here $\overline{\mathcal{O}_{k,S}^\times}$ *denotes the closure of the image of* $\mathcal{O}_{k,S}^\times$ *in* $\bar{I}_{k,S}$. *The left-hand vertical arrow in* (ii) *is injective if* S *contains at least one nonarchimedean prime.*

Remark: If S is finite, local class field theory induces an isomorphism $\bar{I}_{k,S} \xrightarrow{\sim} \prod_{\mathfrak{p}\in S} G_{k_\mathfrak{p}}^{ab}$. For infinite S, we obtain an isomorphism $\bar{I}_{k,S} \xrightarrow{\sim} \hat{\prod}_{\mathfrak{p}\in S} G_{k_\mathfrak{p}}^{ab}$, where the latter group is the *free corestricted proabelian product* of the groups $G_{k_\mathfrak{p}}^{ab}$ with respect to their inertia subgroups; see [154].

Exercise 1. Show the isomorphism $H^2(G_S, \mathcal{O}_S^\times) \cong Br(k_S|k)$.

Exercise 2. Let k be a number field. Let $K|k$ be a finite Galois subextension of $k_S|k$ with Galois group $\overline{G}$, and let $D^S(K) = D_K \cap U_{K,S}$. If k is totally imaginary, then

(i) $H^0(\overline{G}, D^S(K)) = D^S(k)$,

(ii) $H^0(\overline{G}, D_S(K)) = D_S(k) \iff H^1(\overline{G}, D^S(K)) = 0$,

(iii) $\hat{H}^i(\overline{G}, D_S(K)) \cong \hat{H}^i(\overline{G}, D^S(K))$ for all $i \in \mathbb{Z}$.

Hint: D_K is a cohomologically trivial $\overline{G}$-module.

Exercise 3. Show that the group $C_{k,S} = I_{k,S}/\mathcal{O}_{k,S}^\times$ contains the connected component D_k of C_k.

§4. The Global Duality Theorem

In this section we consider the profinite group

$$G_S = G(k_S|k)$$

of the maximal extension $k_S|k$ which is unramified outside the given nonempty set of primes S (where S_∞ is contained in S if k is a number field). The abstract duality theorem (3.1.9) applies at once to any Galois extension $K|k$ inside k_S in view of (8.3.9).

(8.4.1) Theorem. *Let $K|k$ be a Galois extension which is unramified outside S, let G be its Galois group and let A be a finitely generated and $\mathbb{Z}$-free G-module. Then for all $i \in \mathbb{Z}$ the cup-product*

$$\hat{H}^i(G, \mathrm{Hom}(A, C_S(K))) \times \hat{H}^{2-i}(G, A) \xrightarrow{\cup} H^2(G, C_S(K)) \xrightarrow{inv} \mathbb{Q}/\mathbb{Z}$$

yields a topological isomorphism

$$\hat{H}^i(G, \mathrm{Hom}(A, C_S(K))) \cong \hat{H}^{2-i}(G, A)^\vee .$$

If $K|k$ is finite, these are isomorphisms of finite abelian groups.

Remark: Recall that if $K|k$ is infinite, the topologies on the occurring cohomology groups are defined as the direct resp. projective limit topologies of the groups on the finite levels, which are finite and endowed with the discrete topology. The topology of the S-idèle class group does not play any role here.

The theorem (8.4.1), concerning only $\mathbb{Z}$-free G-modules, is too rigid for the applications that we have in mind. In the local case we have proved a cohomological duality for finite Galois modules A and $A' = \mathrm{Hom}(A, \bar{k}^\times)$. We would like to prove an analogous theorem for the global Galois group G_S, where the idèle class group C_S offers itself as an analogue of the multiplicative group $\bar{k}^\times$. However, for several reasons C_S cannot take over the role of $\bar{k}^\times$: we cannot apply the duality theorem (3.4.3) of Tate, since on the one hand it is not known whether *scd* $G_S \le 2$ if S is finite and k is a number field, and on the other hand for large sets S (e.g. S has density equal to 1) the module C_S is not divisible. The duality theorem (3.4.6) seems to be more convenient, since one only needs the cohomological dimension of G_S, but again there is a problem with the divisibility of the dualizing module of G_S. We will see in X §9 that in fact C_S is p-divisible if $p \in S$ and S is finite. But this will be a consequence of the results of this and the following sections.

We have still the abstract duality theorem (3.1.11)(i), which does not require the divisibility of C_S. Instead it requires a *level-compact* formation module and *divisible universal norm groups.* C_S satisfies the last condition, not the first one. But we may pass to a modified formation module satisfying both conditions without changing the cohomology. This was the idea of G. Poitou and K. Uchida. In what follows the reader should recall the topological remarks before (3.1.11).

As before, $K|k$ denotes the finite subextensions of $k_S|k$. We consider the G_S-module $C_S = C(k_S)/U_S$ with the fixed modules

$$C_S(K) = C_S^{G_S(K)} = C_K/U_{K,S}.$$

For the following discussion we have to distinguish between the case of number fields and of function fields. First let k be a number field. $U_{K,S}$ is contained in the compact group

$$C_K^0 = \ker(C_K \xrightarrow{|\ |} \mathbb{R}_+^\times),$$

and we replace $C_S(K)$ by

$$C_S^0(K) = C_K^0 / U_{K,S}\,.$$

The quotient $C_S(K)/C_S^0(K) \cong \mathbb{R}_+^\times$ is uniquely divisible. We set

$$C_S^0 = \varinjlim_{K|k} C_S^0(K) = C^0(k_S)/U_S$$

where $K|k$ runs through the finite subextensions of $k_S|k$.

(8.4.2) Proposition. *Let k be a number field. Then the following is true.*

(i) *The G_S-module C_S^0 is a level-compact formation module with divisible universal norm group $N_{k_S|k}C_S^0$.*

(ii) *If A is a G_S-module which is finitely generated as a $\mathbb{Z}$-module, then for all $i \in \mathbb{Z}$ the natural map $C_S^0 \to C_S$ induces an isomorphism*

$$\hat{H}^i(G_S, \operatorname{Hom}(A, C_S^0)) \xrightarrow{\sim} \hat{H}^i(G_S, \operatorname{Hom}(A, C_S)).$$

Proof: The G_S-module U_S is cohomologically trivial as mentioned on several previous occasions. Therefore, applying $H^0(G(k_S|K), -)$ to the exact sequence

$$0 \longrightarrow U_S \longrightarrow C^0(k_S) \longrightarrow C_S^0 \longrightarrow 0,$$

we obtain the exact sequence

$$0 \longrightarrow U_{K,S} \longrightarrow C_K^0 \longrightarrow (C_S^0)^{G(k_S|K)} \longrightarrow 0,$$

and so

$$(C_S^0)^{G(k_S|K)} = C_S^0(K).$$

Thus C_S^0 is a level-compact G_S-module. By (8.3.13), $N_{k_S|K}C_S = D_S(K)$ is divisible, hence also $D_S^0(K) = \ker(D_S(K) \xrightarrow{|\ |} \mathbb{R}_+^\times)$. Since $|N_{K'|K}(x)| = 1$ if and only if $|x| = 1$ for $x \in C_S(K')$, where $K'|K$ is a finite subextension of $k_S|K$, we get $N_{k_S|K}C_S^0 = D_S^0(K)$, and hence $N_{k_S|K}C_S^0$ is divisible.

Now let A be a G_S-module which is finitely generated as a $\mathbb{Z}$-module and let $K|k$ be a finite Galois extension inside k_S over which A becomes trivial. Consider the exact sequence of $G(K|k)$-modules

$$0 \longrightarrow C_S^0(K) \longrightarrow C_S(K) \longrightarrow Q(K) \longrightarrow 0$$

in which $Q(K) = C_S(K)/C_S^0(K) \cong \mathbb{R}$ is uniquely divisible. It follows that $\mathrm{Hom}(A, Q(K))$ is also uniquely divisible and the sequence

$$0 \longrightarrow \mathrm{Hom}(A, C_S^0(K)) \longrightarrow \mathrm{Hom}(A, C_S(K)) \longrightarrow \mathrm{Hom}(A, Q(K)) \longrightarrow 0$$

is exact. We therefore obtain natural isomorphisms

$$\hat{H}^i(G(K|k), \mathrm{Hom}(A, C_S^0(K))) \xrightarrow{\sim} \hat{H}^i(G(K|k), \mathrm{Hom}(A, C_S(K)))$$

for all $i \in \mathbb{Z}$. □

For *function fields* there is the following similar result. For each finite subextension $K|k$ of $k_S|k$ consider the compact group

$$\overline{C}_S(K) = \varprojlim_N \, C_K/N,$$

where N runs through the open subgroups of finite index in C_K which contain $U_{K,S}$. As before let

$$\overline{C}_S = \varinjlim_{K|k} \, \overline{C}_S(K).$$

(8.4.3) Proposition. *Let k be a function field. Then the following is true.*

(i) The G_S-module $\overline{C}_S$ is a level-compact formation module with trivial universal norm group $N_{k_S|k}\overline{C}_S = 0$.

(ii) If A is a G_S-module which is finitely generated as a $\mathbb{Z}$-module, then for all $i \in \mathbb{Z}$ the natural homomorphism $C_S \to \overline{C}_S$ induces an isomorphism

$$\hat{H}^i(G_S, \mathrm{Hom}(A, C_S)) \xrightarrow{\sim} \hat{H}^i(G_S, \mathrm{Hom}(A, \overline{C}_S)).$$

Proof: The first assertion was already shown in the proof of (8.3.17) and the second follows in the same way as in (8.4.2) using the exact sequence

$$0 \longrightarrow C_S(K) \longrightarrow \overline{C}_S(K) \longrightarrow \hat{\mathbb{Z}}/\mathbb{Z} \longrightarrow 0,$$

where $K|k$ is a finite Galois extension inside k_S trivializing A. Since $\hat{\mathbb{Z}}/\mathbb{Z}$ is uniquely divisible, we obtain isomorphisms

$$\hat{H}^i(G(K|k), \mathrm{Hom}(A, C_S(K))) \xrightarrow{\sim} \hat{H}^i(G(K|k), \mathrm{Hom}(A, \overline{C}_S(K)))$$

for all $i \in \mathbb{Z}$. □

Applying the duality theorem (3.1.11)(i) to an arbitrary global field, we obtain the

(8.4.4) Theorem. *Let A be a discrete G_S-module which is finitely generated as a $\mathbb{Z}$-module. If k is a number field, then the pairing*

$$\hat{H}^i(G_S, \mathrm{Hom}(A, C_S^0)) \times \hat{H}^{2-i}(G_S, A) \xrightarrow{\cup} H^2(G_S, C_S^0) \cong \tfrac{1}{\#G_S}\mathbb{Z}/\mathbb{Z}$$

induces topological isomorphisms

$$\hat{H}^i(G_S, \mathrm{Hom}(A, C_S^0)) \cong \hat{H}^{2-i}(G_S, A)^\vee$$

for all $i \le 0$. This assertion holds for all $i \in \mathbb{Z}$ if A is assumed to be $\mathbb{Z}$-free.

If k is a function field, then the statements above hold with C_S^0 replaced by $\overline{C}_S$.

Remark: In order to apply also part (ii) of (3.1.11) to (G_S, C_S^0) (resp. $(G_S, \overline{C}_S)$) one has to investigate the question of finding those primes p for which the group C_S is p-divisible. We will do this in chapter X, see (10.11.8).

We formulated the theorem with C_S^0 (resp. $\overline{C}_S$) instead of C_S to obtain the additional information that the duality isomorphisms are topological. If one wants to define a topology on the cohomology groups of $C_S(A)$ directly, one runs into the problem that the category of locally compact abelian groups is not abelian. While it is possible to overcome this technical difficulty, we restrict our considerations here to the case of cohomology in dimension zero:

As before, let A be a discrete G_S-module which is finitely generated as a $\mathbb{Z}$-module. For $U \subseteq G_S$ open, we endow the groups $C_S(A)^U$ with a natural (locally compact) topology: For sufficiently small V we have $A^V = A$ and therefore

$$C_S(A)^V = \mathrm{Hom}(A, C_S^V)\,.$$

Considering A as a group with the discrete topology, $\mathrm{Hom}(A, C_S^V)$ endowed with the compact open topology is locally compact. For an arbitrary open subgroup $U \subseteq G_S$, the group $\mathrm{Hom}(A, C_S)^U$ is a closed subgroup of the locally compact group $\mathrm{Hom}(A, C_S)^V$, where V is chosen sufficiently small. For $U \subseteq G_S$ the norm map $N_{G_S/U} : C_S(A)^U \to C_S(A)^{G_S}$ is continuous. If U is normal in G_S, we endow $\hat{H}^0(G_S/U, C_S(A)^U)$ with the quotient topology from $H^0(G_S/U, C_S(A)^U) = C_S(A)^{G_S}$. Finally, we endow

$$\hat{H}^0(G_S, C_S(A)) = \varprojlim_U \hat{H}^0(G_S/U, C_S(A)^U)$$

with the projective limit topology. Note that the natural homomorphism

$$H^0(G_S, C_S(A)) \longrightarrow \hat{H}^0(G_S, C_S(A))$$

is continuous and has dense image.

(8.4.5) Proposition. *The algebraic isomorphisms*

$$\hat{H}^0(G_S, C_S^0(A)) \xrightarrow{\sim} \hat{H}^0(G_S, C_S(A)),$$

given by (8.4.2) *if k is a number field, and*

$$\hat{H}^0(G_S, C_S(A)) \xrightarrow{\sim} \hat{H}^0(G_S, \overline{C}_S(A)),$$

given by (8.4.3) *if k is a function field, are homeomorphisms. The composite*

$$C_S(A)^{G_S} \longrightarrow \hat{H}^0(G_S, C_S(A)) \xrightarrow{\sim} H^2(G_S, A)^\vee$$

is continuous and has dense image. It is surjective if k is a number field or if A is finite.

Proof: Let $K = k_S^U$ be a finite Galois extension of k inside k_S such that $A^U = A$. We first show that the norm map $N_{G_S/U} : C_S(A)^U \to C_S(A)^{G_S}$ has closed image. From this follows that $\hat{H}^0(G_S, C_S(A))$ is Hausdorff.

We start with the number field case. Let v be an archimedean prime of k and let Γ_k be the image of the subgroup of positive real numbers under the injection $k_v^\times \hookrightarrow C_S(k)$. Let Γ_K be the image of Γ_k in $C_S(K)$. Then we have an algebraic decomposition of $C_S(A)^U$ into a product of closed $G(K|k)$-submodules

$$C_S(A)^U \cong C_S^0(A)^U \times \mathrm{Hom}(A, \Gamma_K)\,.$$

As $C_S^0(A)^U$ is compact and the norm is continuous, $N_{G_S/U} C_S^0(A)^U$ is a compact subgroup of $C_S(A)^{G_S}$. The norm map $N_{G_S/U} : \mathrm{Hom}(A, \Gamma_K) \to \mathrm{Hom}_{G_S}(A, \Gamma_K)$ is a homomorphism of finite dimensional $\mathbb{R}$-vector spaces and has closed image. Therefore

$$N_{G_S/U} C_S(A)^U = N_{G_S/U} C_S^0(A)^U \times N_{G_S/U} \mathrm{Hom}(A, \Gamma_K)$$

is a closed subgroup of $C_S(A)^{G_S}$.

In the function field case, we proceed in a similar way. Let $C_S^0 \subseteq C_S$ be the subgroup of idèle classes of degree zero. C_S^0 is a level-compact G_S-module. Let $x \in C_S(k)$ be a pre-image of 1 under the degree map $\deg : C_S(k) \twoheadrightarrow \mathbb{Z}$, and let $\Gamma_k \cong \mathbb{Z}$ be the discrete subgroup in $C_S(k)$ generated by x. Denote by Γ_K the image of Γ_k in $C_S(K)$. Then

$$C_S^0(A)^U \times \mathrm{Hom}(A, \Gamma_K)$$

is an open subgroup of finite index in $C_S(A)^U$. The norm group $N_{G_S/U} C_S^0(A)^U$ is compact and $N_{G_S/U}\mathrm{Hom}(A, \Gamma_K)$ is discrete as a subgroup of the discrete group $\mathrm{Hom}_{G_S}(A, \Gamma_k)$. We conclude that

$$N_{G_S/U} C_S^0(A)^U \times N_{G_S/U} \mathrm{Hom}(A, \Gamma_K) \subseteq \mathrm{Hom}(A, C_S)^{G_S}$$

is closed, as it is the product of a compact subgroup and a discrete subgroup. Therefore $N_{G_S/U} C_S(A)^U$, containing a closed subgroup of finite index, is closed.

In the number field case, we conclude that the isomorphism

$$(*) \qquad \hat{H}^0(G(K|k), \mathrm{Hom}(A, C_S^0(K))) \xrightarrow{\sim} \hat{H}^0(G(K|k), \mathrm{Hom}(A, C_S(K)))$$

obtained in the proof of (8.4.2) is a continuous bijection from a compact group onto a Hausdorff group, and therefore a homeomorphism.

Now let k be a function field. In the proof of (8.4.3) we obtained the continuous bijection

$$(**) \qquad \hat{H}^0(G(K|k), \mathrm{Hom}(A, C_S(K))) \xrightarrow{\sim} \hat{H}^0(G(K|k), \mathrm{Hom}(A, \overline{C}_S(K))).$$

The group $C_S(K)/C_S^0(K)$ is a trivial $G(K|k)$-module isomorphic to $\mathbb{Z}$. Therefore the image of the natural homomorphism

$$\hat{H}^0(G(K|k), \mathrm{Hom}(A, C_S^0(K))) \longrightarrow \hat{H}^0(G(K|k), \mathrm{Hom}(A, C_S(K)))$$

is a compact subgroup of finite index. Hence $\hat{H}^0(G(K|k), \mathrm{Hom}(A, C_S(K)))$ is compact and $(**)$ is a homeomorphism.

Passing to the projective limit over all K in $(*)$ or $(**)$, respectively, we obtain the first result, and the composite $C_S(A)^{G_S} \to \hat{H}^0(G_S, C_S(A)) \xrightarrow{\sim} H^2(G_S, A)^\vee$ is continuous and has dense image.

If k is number field, then the module $\mathrm{Hom}(A, C_S^0)$ is level-compact, and so $C_S^0(A)^{G_S} \to \hat{H}^0(G_S, \mathrm{Hom}(A, C_S^0))$ is surjective by (1.9.11). It follows that $C_S(A)^{G_S} \to \hat{H}^0(G_S, C_S(A))$ is surjective. If k is a function field and A is finite, then the module $\mathrm{Hom}(A, C_S) = \mathrm{Hom}(A, \overline{C}_S)$ is level-compact, and $C_S(A)^{G_S} \to \hat{H}^0(G_S, C_S(A))$ is also surjective in this case. □

§5. Local Cohomology of Global Galois Modules

As a bridge from local to global cohomology, we shall now introduce an idèlic cohomology for G_S-modules, G_S again being the Galois group $G(k_S|k)$. We start with some general observations concerning the passage from global to local Galois modules.

Let k be a global field, $\mathfrak{p}$ a prime of k, $K|k$ a (possibly infinite) Galois extension and $\mathfrak{P}$ a prime of K above $\mathfrak{p}$. The field

$$K_{\mathfrak{P}} = \bigcup_{k'} k'_{\mathfrak{p}'} ,$$

where k' runs through the finite extensions of k inside K and $\mathfrak{p}' = \mathfrak{P} \cap k'$, is a (possibly infinite) Galois extension of the local field $k_{\mathfrak{p}}$. The Galois group $G(K_{\mathfrak{P}}|k_{\mathfrak{p}})$ is naturally isomorphic to the decomposition group $G_{\mathfrak{P}}(K|k)$ of $\mathfrak{P}$ in $G(K|k)$. We fix a separable closure $\overline{K}_{\mathfrak{P}}$ of $K_{\mathfrak{P}}$ (which is also a separable closure of $k_{\mathfrak{p}}$).

Now let A be a discrete $G(K|k)$-module. We consider A as a $G(\bar{K}_{\mathfrak{P}}|k_{\mathfrak{p}})$-module via the natural projection

$$G(\bar{K}_{\mathfrak{P}}|k_{\mathfrak{p}}) \twoheadrightarrow G_{\mathfrak{P}}(K|k).$$

As any two extensions $\mathfrak{P}_1$ and $\mathfrak{P}_2$ of $\mathfrak{p}$ to K are conjugate, and as inner automorphisms of groups act trivially on group cohomology, the groups $H^i(G(\bar{K}_{\mathfrak{P}_1}|k_{\mathfrak{p}}), A)$ and $H^i(G(\bar{K}_{\mathfrak{P}_2}|k_{\mathfrak{p}}), A)$ are *canonically* isomorphic for all i. We denote them by

$$H^i(k_{\mathfrak{p}}, A),$$

hiding all choices from the notation. We also write the inclusion of K to $\bar{K}_{\mathfrak{P}}$ in the form $K \hookrightarrow \bar{k}_{\mathfrak{p}}$. There are canonical homomorphisms

$$H^i(G(K|k), A) \longrightarrow H^i(k_{\mathfrak{p}}, A)$$

for all $i \geq 0$. If A is trivialized by a finite extension k' of k, which happens for example if A is finitely generated as a $\mathbb{Z}$-module, then A is an unramified $G_{k_{\mathfrak{p}}}$-module for all primes $\mathfrak{p}$ which are unramified in $k'|k$. In particular, A is an unramified $G_{\mathfrak{p}}$-module for almost all primes $\mathfrak{p}$ of k.

We will apply the above considerations to the case $K = k_S$. Recall the notation

$$\mathbb{N}(S) = \{n \in \mathbb{N} \mid n \in \mathcal{O}_{k,S}^\times\}.$$

(8.5.1) Definition. *We denote by $Mod_S(G_S)$ the category of discrete G_S-modules A which are finitely generated as $\mathbb{Z}$-modules and whose torsion has order $\#\mathrm{tor}(A) \in \mathbb{N}(S)$.*

For a *finite* $A \in Mod_S(G_S)$, we define the **dual G_S-module** of A by

$$A' = \mathrm{Hom}(A, \mathcal{O}_S^\times) = \mathrm{Hom}(A, \mu),$$

where $\mathcal{O}_S^\times$ is the group of S-units of k_S and μ is the group of roots of unity in k_S. As $n = \#A$ is in $\mathbb{N}(S)$ by assumption, k_S contains all n-th roots of unity. Therefore A' is finite of the same order as A and contained in $Mod_S(G_S)$. Furthermore, we have a canonical identification $A'' = A$.

We set

$$P^i(G_S, A) = \prod_{\mathfrak{p} \in S}{}' H^i(k_{\mathfrak{p}}, A), \quad i \geq 0,$$

where the restricted product is taken with respect to the subgroups $H^i_{nr}(k_{\mathfrak{p}}, A) = \mathrm{im}(H^i(\tilde{k}_{\mathfrak{p}}|k_{\mathfrak{p}}, A) \to H^i(k_{\mathfrak{p}}, A))$ of $H^i(k_{\mathfrak{p}}, A)$, where $\tilde{k}_{\mathfrak{p}}$ is the maximal unramified extension of $k_{\mathfrak{p}}$, see (7.2.14). In view of the duality theorem (7.2.17) for the groups $\hat{H}^i(k_{\mathfrak{p}}, A)$ at the archimedean primes $\mathfrak{p}$, we make the following

Convention: In the case $i = 0$ we agree that at archimedean primes the groups $H^0(k_\mathfrak{p}, A)$ denote the modified cohomology groups $\hat{H}^0(k_\mathfrak{p}, A)$.

Since the groups $H^i(k_\mathfrak{p}, A)$ are all finite, we see that

$$\begin{array}{lcll}
P^0(G_S, A) & = & \prod\limits_{\mathfrak{p}\in S\setminus S_\infty} H^0(k_\mathfrak{p}, A) \times \prod\limits_{\mathfrak{p}\in S_\infty} \hat{H}^0(k_\mathfrak{p}, A) & \text{is compact,} \\
P^1(G_S, A) & = & \prod\limits_{\mathfrak{p}\in S}' H^1(k_\mathfrak{p}, A) & \text{is locally compact,} \\
P^2(G_S, A) & = & \bigoplus\limits_{\mathfrak{p}\in S} H^2(k_\mathfrak{p}, A) & \text{is discrete,} \\
P^i(G_S, A) & = & \bigoplus\limits_{\mathfrak{p}\in S_\infty} H^i(k_\mathfrak{p}, A) & \text{is finite for } i \geq 3.
\end{array}$$

The direct sums in the last two lines occur because $cd\, G(\tilde{k}_\mathfrak{p}|k_\mathfrak{p}) = cd\, \hat{\mathbb{Z}} = 1$ and $cd\, G_{k_\mathfrak{p}} = 2$. If S is finite, all groups above are finite and the restricted product is just the product.

The local duality theorem (7.2.9) has the following effect on the groups P^i. Let $A \in Mod_S(G_S)$ be finite. Then for $\mathfrak{p} \nmid \infty$, the pairing

$$H^i(k_\mathfrak{p}, A) \times H^{2-i}(k_\mathfrak{p}, A') \xrightarrow{\ \cup\ } H^2(k_\mathfrak{p}, \bar{k}_\mathfrak{p}^\times) \xrightarrow{inv_\mathfrak{p}} \mathbb{Q}/\mathbb{Z},$$

$0 \leq i \leq 2$, yields an isomorphism

$$(*) \qquad \Xi_\mathfrak{p}^i : H^i(k_\mathfrak{p}, A) \longrightarrow H^{2-i}(k_\mathfrak{p}, A')^\vee.$$

For $\mathfrak{p} \mid \infty$ we have to replace H^0 by $\hat{H}^0$. If A is unramified at $\mathfrak{p} \nmid \infty$ and $\mathfrak{p} \nmid \#A$, then, by (7.2.15), the groups $H^{2-i}_{nr}(k_\mathfrak{p}, A')$ and $H^i_{nr}(k_\mathfrak{p}, A)$ are exact orthogonal complements under the pairing, i.e. $H^i_{nr}(k_\mathfrak{p}, A)$ is mapped by $\Xi_\mathfrak{p}^i$ isomorphically onto the subgroup

$$(H^{2-i}(k_\mathfrak{p}, A')/H^{2-i}_{nr}(k_\mathfrak{p}, A'))^\vee \subseteq H^{2-i}(k_\mathfrak{p}, A')^\vee.$$

For the restricted product of the right-hand groups with respect to the left-hand subgroups we have the equality (cf. (1.1.13))

$$\prod_{\mathfrak{p}\in S}' H^{2-i}(k_\mathfrak{p}, A')^\vee = (\prod_{\mathfrak{p}\in S}' H^{2-i}(k_\mathfrak{p}, A'))^\vee.$$

Since A is unramified for almost all $\mathfrak{p}$, we obtain the

(8.5.2) Proposition. *If $A \in Mod_S(G_S)$ is finite, the maps $\Xi_\mathfrak{p}^i$ define homeomorphic isomorphisms*

$$\Xi^i : P^i(G_S, A) \longrightarrow P^{2-i}(G_S, A')^\vee$$

for $0 \leq i \leq 2$.

Now let $A \in Mod_S(G_S)$ be arbitrary. We again define the **dual G_S-module** of A by

$$A' = \mathrm{Hom}(A, \mathcal{O}_S^\times).$$

However, this is not completely satisfactory if A is not finite. Firstly, A' is not necessarily in $Mod_S(G_S)$, and in particular, the situation is asymmetric; also an equality like $A'' = A$ is no longer true. Looking at this situation from the point of view of sheaves, we are interested in the *dual sheaf*, which is no longer locally constant, i.e. is not represented by a Galois module. That is why we make the following notational convention:

For every prime $\mathfrak{p}$ we write

$$H^i(k_\mathfrak{p}, A') \stackrel{\text{def}}{=} H^i(k_\mathfrak{p}, \mathrm{Hom}(A, \bar{k}_\mathfrak{p}^\times)),$$

i.e. we consider A as a local module first, and then take the dual in the local sense. Note that we have an injection $A' \hookrightarrow \mathrm{Hom}(A, \bar{k}_\mathfrak{p}^\times)$ of $G(\bar{k}_\mathfrak{p}|k_\mathfrak{p})$-modules, which induces maps for all i

$$H^i(G_S, A') \longrightarrow H^i(k_\mathfrak{p}, A').$$

If A (and hence also A') is finite, both possible interpretations of $H^i(k_\mathfrak{p}, A')$ coincide. Now assume that A is unramified at the finite prime $\mathfrak{p}$. Then we define

$$H^i_{nr}(k_\mathfrak{p}, A') \stackrel{\text{def}}{=} H^i_{nr}(k_\mathfrak{p}, A^d),$$

where $A^d = \mathrm{Hom}(A, \tilde{\mathcal{O}}_\mathfrak{p}^\times)$ (see VII §2). Again, no confusion occurs if A is finite. If M is A or A', we set, as in the case of finite modules,

$$P^i(G_S, M) = \prod_{\mathfrak{p} \in S}' H^i(k_\mathfrak{p}, M)\,, \quad i \geq 0,$$

where the restricted product is taken with respect to the subgroups $H^i_{nr}(k_\mathfrak{p}, M)$ of $H^i(k_\mathfrak{p}, M)$, and we keep our convention concerning $\hat{H}^0$ at archimedean primes.

However, we have to be careful about the topologies: the cohomology groups of A are considered as topological groups with the discrete topology and the same holds for $H^i(k_\mathfrak{p}, A')$, $i > 0$. But for $i = 0$ the group $H^0(k_\mathfrak{p}, A')$ inherits the topology of $\mathrm{Hom}(A, \bar{k}_\mathfrak{p}^\times)$ and is not discrete. Now for the topology of $P^i(G_S, M)$, where $M = A$ or A', a basis of neighbourhoods of the identity is given by the subgroups

$$\prod_{\mathfrak{p} \in T} X_\mathfrak{p} \times \prod_{\mathfrak{p} \in S \setminus T} H^i_{nr}(k_\mathfrak{p}, M),$$

where T varies over the finite subsets of S and $X_\mathfrak{p}$ varies over a basis of neighbourhoods of the identity in $H^i(k_\mathfrak{p}, M)$; so if $i > 0$ or $M = A$, we may take $X_\mathfrak{p} = 1$.

Therefore if S and A are infinite, among the groups $P^i(G_S, A)$ only $P^1(G_S, A)$ is locally compact because $H^1(k_\mathfrak{p}, A)$ is finite by (7.2.9). Considering the cohomology of A', we see that $H^0_{nr}(k_\mathfrak{p}, A') = H^0(k_\mathfrak{p}, A^d)$ is an open compact subgroup of the locally compact group $H^0(k_\mathfrak{p}, A')$. By (1.1.13), we conclude that $P^0(G_S, A')$ is locally compact. Since $H^1(k_\mathfrak{p}, A')$ is finite, $P^1(G_S, A')$ is also locally compact.

Let us finally mention the special case when A is $\mathbb{Z}$-free. Then, for all $\mathfrak{p}$ such that A is an unramified $G_{k_\mathfrak{p}}$-module, we have

$$H^1_{nr}(k_\mathfrak{p}, A) = H^1(k_\mathfrak{p}, A) \quad \text{and} \quad H^i_{nr}(k_\mathfrak{p}, A') = 0,\ i \geq 1,$$

see (7.2.16). Therefore, if A is $\mathbb{Z}$-free, $P^1(G_S, A)$ is a product of finite groups, hence compact, and $P^i(G_S, A')$, $i \geq 1$, is discrete, being a direct sum of discrete groups.

Again, let $A \in Mod_S(G_S)$ be arbitrary. For each $\mathfrak{p}$, the local duality theorem (7.2.9) together with (7.2.10) induces a continuous composite homomorphism for $i = 1, 2$

$$H^{2-i}(k_\mathfrak{p}, A') \longrightarrow H^{2-i}(k_\mathfrak{p}, A^D) \longrightarrow H^i(k_\mathfrak{p}, A)^\vee.$$

This is an isomorphism of finite groups for $i = 1$, and for $i = 2$ it is an injection with dense image of a locally compact group into a compact group. Dualizing, we obtain continuous homomorphisms for $i = 1, 2$

$$H^i(k_\mathfrak{p}, A) \longrightarrow H^{2-i}(k_\mathfrak{p}, A')^\vee.$$

This is an isomorphism of finite groups for $i = 1$. For $i = 2$ it is an injection with dense image of a discrete group into a locally compact group. If A is unramified at $\mathfrak{p}$, then, by (7.2.15), this map induces a homomorphism of subgroups

$$H^i_{nr}(k_\mathfrak{p}, A) \longrightarrow (H^{2-i}(k_\mathfrak{p}, A')/H^{2-i}_{nr}(k_\mathfrak{p}, A'))^\vee,$$

which is an isomorphism for $i = 1$. By the compatibility of Pontryagin duality with restricted products, cf. (1.1.13), we therefore obtain continuous homomorphisms for $i = 1, 2$.

$$\Xi^i : P^i(G_S, A) \longrightarrow P^{2-i}(G_S, A')^\vee.$$

From the above considerations, we obtain the following

(8.5.3) Proposition. *Let $A \in Mod_S(G_S)$. Then for $i = 1, 2$ the homomorphism Ξ^i is continuous and*

(i) *Ξ^1 is an isomorphism of locally compact groups,*

(ii) *Ξ^2 is a continuous injection with dense image.*

We will now compare the groups $P^i(G_S, A')$ for $A \in Mod_S(G_S)$ with the cohomology groups $H^i(G_S, I_S(A))$ of the G_S-module

$$I_S(A) = \mathrm{Hom}(A, I_S),$$

where I_S is the S-idèle group of k_S.

Let $G_\mathfrak{p} = G(k_{S,\mathfrak{p}}|k)$ be the decomposition group of $\mathfrak{p}$ in $G(k_S|k)$ with respect to some prolongation $\bar{\mathfrak{p}}$ of $\mathfrak{p}$ to k_S, see the discussion at the beginning of this section. If $K|k$ is a finite Galois subextension of k_S with Galois group $\overline{G} = G(K|k)$, then let $\overline{G}_\mathfrak{P} = G(K_\mathfrak{P}|k_\mathfrak{p})$ where $\mathfrak{P} = \bar{\mathfrak{p}}|_K$. Finally, let

$$I_\mathfrak{P}(A) = \mathrm{Hom}(A, K_\mathfrak{P}^\times) \quad \text{and} \quad U_\mathfrak{P}(A) = \mathrm{Hom}(A, U_{K,\mathfrak{P}}),$$

where $U_{K,\mathfrak{P}}$ is the group of units in $K_\mathfrak{P}$. In the limit we use the notation $I_\mathfrak{p}(A) = \mathrm{Hom}(A, k_{S,\mathfrak{p}}^\times)$ and $U_\mathfrak{p}(A) = \mathrm{Hom}(A, U_{k_S,\mathfrak{p}})$.

Now let $K|k$ be large enough so that $G(k_S|K)$ acts trivially on A. Then

$$I_S(A)^{G(k_S|K)} = \mathrm{Hom}(A, I_S)^{G(k_S|K)} = \mathrm{Hom}(A, I_{K,S})$$

and

$$I_{K,S} = \prod_{\mathfrak{p}\in S} \mathrm{Ind}_{\overline{G}}^{\overline{G}_\mathfrak{P}}(K_\mathfrak{P}^\times).$$

The functor $\mathrm{Hom}(A, -)$ commutes with $\prod$, since it commutes with $\prod$ and with $\varinjlim$, recalling that A is finitely generated, and hence with restricted products. Thus we obtain a bijective homomorphism

$$\Psi_K\colon\ H^i(\overline{G}, \mathrm{Hom}(A, I_{K,S})) \xrightarrow{\sim} \prod_{\mathfrak{p}\in S} H^i(\overline{G}_\mathfrak{P}, \mathrm{Hom}(A, K_\mathfrak{P}^\times)),$$

where the restricted product is taken with respect to the subgroups

$$\mathrm{im}\Big(H^i(\overline{G}_\mathfrak{P}, U_\mathfrak{P}(A)) \longrightarrow H^i(\overline{G}_\mathfrak{P}, I_\mathfrak{P}(A))\Big).$$

Note that Ψ_K is continuous. For $i = 0$ it is homeomorphic, but not for $i > 0$, as $H^i(\overline{G}, \mathrm{Hom}(A, I_{K,S}))$ carries the discrete topology by definition. Taking the direct limit over K, we obtain a commutative diagram

$$\begin{array}{ccc}
\varinjlim_K H^i(\overline{G}, \mathrm{Hom}(A, I_{K,S})) & \xrightarrow{\sim} & \varinjlim_K \prod_{\mathfrak{p}\in S} H^i(\overline{G}_\mathfrak{P}, \mathrm{Hom}(A, K_\mathfrak{P}^\times)) \\
\wr\downarrow\ \mathit{inf} & & \downarrow (\mathit{inf}_\mathfrak{p})_{\mathfrak{p}\in S} \\
H^i(G_S, I_S(A)) & \dashrightarrow & \prod_{\mathfrak{p}\in S} H^i(k_\mathfrak{p}, A') = P^i(G_S, A')\ .
\end{array}$$

As any finite extension $K|k$ is unramified at almost all primes, the restricted product $\prod_{\mathfrak{p}\in S} H^i(\overline{G}_\mathfrak{P}, \mathrm{Hom}(A, K_\mathfrak{P}^\times))$ remains unchanged if we take it with respect to the subgroups

$$\mathrm{im}\Big(H^i(\overline{G}_\mathfrak{P}/\overline{H}_\mathfrak{P}, U_\mathfrak{P}(A)^{\overline{H}_\mathfrak{P}}) \longrightarrow H^i(\overline{G}_\mathfrak{P}, I_\mathfrak{P}(A))\Big),$$

where $\overline{H}_\mathfrak{P}$ is the inertia subgroup of $\overline{G}_\mathfrak{P}$. Therefore the right vertical arrow exists and is continuous. As the left vertical map is a topological isomorphism, we obtain the dotted continuous map from $H^i(G_S, I_S(A))$ into $P^i(G_S, A')$.

(8.5.4) Definition. *The continuous homomorphisms*

$$sh^i : H^i(G_S, I_S(A)) \longrightarrow P^i(G_S, A'), \quad i \geq 0,$$

are called **Shapiro maps**.

(8.5.5) Proposition. *Let $A \in Mod_S(G_S)$. Then the following hold:*

(i) sh^0 *is a bijection for function fields. If k is a number field, then sh^0 is surjective with kernel*

$$\ker sh^0 = \prod_{\mathfrak{p} \in S_\infty(k)} N_{\bar{k}_\mathfrak{p}|k_\mathfrak{p}} \operatorname{Hom}(A, \mathbb{C}^\times).$$

(ii) sh^1 *is injective, and bijective if A is $\mathbb{Z}$-free.*

(iii) *If A is $\mathbb{Z}$-free, then sh^2 is injective and for each $\ell \in \mathbb{N}(S)$ the induced map on ℓ-torsion subgroups*

$$H^2(G_S, I_S(A))(\ell) \longrightarrow P^2(G_S, A')(\ell)$$

is an isomorphism.

Remarks: 1. For $\mathbb{Z}$-free A, the bijection sh^1 of (ii) is a homeomorphism, as both groups are discrete. The same is true for the isomorphism stated in (iii).
2. In (10.11.9) we will show the following additional facts for an arbitrary $A \in Mod_S(G_S)$:

- sh^2 is injective and induces an isomorphism on the ℓ-torsion part for all $\ell \in \mathbb{N}(S)$,
- if S is finite, then sh^1 is bijective.

This will follow with the same proof as below using the additional fact that the local fields $(k_S)_\mathfrak{p}$ are ℓ-closed for $\ell \in \mathbb{N}(S)$ and $\mathfrak{p} \in S(k_S)$; see (10.11.6).

Proof: We start with the case $i = 0$. Let $K|k$ be a finite Galois subextension of k_S such that $G(k_S|K)$ acts trivially on A. Putting $\overline{G} = G(K|k)$, we obtain

$$\begin{aligned} H^0(G_S, I_S(A)) &= \operatorname{Hom}(A, I_{K,S})^{\overline{G}} \\ &= \Big(\prod_{\mathfrak{p} \in S} \operatorname{Ind}_{\overline{G}}^{\overline{G}_\mathfrak{P}} \operatorname{Hom}(A, K_\mathfrak{P}^\times)\Big)^{\overline{G}} \\ &= \prod_{\mathfrak{p} \in S} \operatorname{Hom}(A, K_\mathfrak{P}^\times)^{\overline{G}_\mathfrak{P}} \\ &= \prod_{\mathfrak{p} \in S} I_\mathfrak{p}(A)^{G_\mathfrak{P}} = \prod_{\mathfrak{p} \in S} H^0(G_\mathfrak{p}, I_\mathfrak{p}(A)). \end{aligned}$$

Let $N_\mathfrak{p} = G(\bar{k}_\mathfrak{p}|k_{S,\mathfrak{p}})$, i.e. $G_\mathfrak{p} = G_{k_\mathfrak{p}}/N_\mathfrak{p}$. Then

$$H^0(k_\mathfrak{p}, A') = (\operatorname{Hom}(A, \bar{k}_\mathfrak{p}^\times)^{N_\mathfrak{p}})^{G_\mathfrak{p}} = \operatorname{Hom}(A, k_{S,\mathfrak{p}}^\times)^{G_\mathfrak{p}} = H^0(G_\mathfrak{p}, I_\mathfrak{p}(A)).$$

If A is unramified at $\mathfrak{p} \nmid \infty$, i.e. the inertia group with respect to $\mathfrak{p}$ acts trivially on A, then we get $H^0_{nr}(k_\mathfrak{p}, A') = H^0_{nr}(G_\mathfrak{p}, I_\mathfrak{p}(A))$ in the same way. Therefore sh^0 is surjective, and recalling the difference between H^0 and $\hat{H}^0$ for $\mathfrak{p} \in S_\infty$, we obtain the description of ker sh^0.

Since

$$\varinjlim_{K} H^i(G(K|k), \mathrm{Hom}(A, I_{K,S})) = H^i(G_S, I_S(A))$$

and $(inf_\mathfrak{p})_{\mathfrak{p}\in S}$ is injective for $i = 1$, we immediately obtain that sh^1 is injective, i.e. the first assertion of (ii).

Now let A be $\mathbb{Z}$-free. Since $A \in Mod_S(G_S)$, we have an isomorphism $A \cong \mathbb{Z}^r$ for some r, of trivial $G(\bar{k}_\mathfrak{p}|(k_S)_\mathfrak{p})$-modules. Using Hilbert's Satz 90, we obtain

$$H^1(G(\bar{k}_\mathfrak{p}|(k_S)_\mathfrak{p}), \mathrm{Hom}(A, \bar{k}_\mathfrak{p}^\times)) \cong H^1(G(\bar{k}_\mathfrak{p}|(k_S)_\mathfrak{p}), \bar{k}_\mathfrak{p}^\times)^r = 0.$$

The Hochschild-Serre spectral sequence induces an isomorphism

$$H^1(G((k_S)_\mathfrak{p}|k_\mathfrak{p}), \mathrm{Hom}(A, (k_S)_\mathfrak{p}^\times)) \xrightarrow{\sim} H^1(G(\bar{k}_\mathfrak{p}|k_\mathfrak{p}), \mathrm{Hom}(A, \bar{k}_\mathfrak{p}^\times))$$

and an injection

$$H^2(G((k_S)_\mathfrak{p}|k_\mathfrak{p}), \mathrm{Hom}(A, \bar{k}_\mathfrak{p}^\times)) \hookrightarrow H^2(G(\bar{k}_\mathfrak{p}|k_\mathfrak{p}), \mathrm{Hom}(A, \bar{k}_\mathfrak{p}^\times)).$$

Using the commutative diagram

$$\begin{array}{ccc} \bigoplus_{\mathfrak{p}\in S} H^2(G((k_S)_\mathfrak{p}|k_\mathfrak{p}), \mathrm{Hom}(A, (k_S)_\mathfrak{p}^\times)) & \hookrightarrow & \bigoplus_{\mathfrak{p}\in S} H^2(G(\bar{k}_\mathfrak{p}|k_\mathfrak{p}), \mathrm{Hom}(A, \bar{k}_\mathfrak{p}^\times)) \\ \uparrow \wr & & \uparrow \wr \\ H^2(G_S, I_S(A)) & \xrightarrow{sh^2} & P^2(G_S, A'), \end{array}$$

we obtain the first assertion of (iii). Now let $\ell \in \mathbb{N}(S)$. Then $\ell^\infty \mid G((k_S)_\mathfrak{p}|k_\mathfrak{p})$ for all nonarchimedean primes $\mathfrak{p} \in S$. As k_S is totally imaginary, we obtain $H^i(\bar{k}_\mathfrak{p}|(k_S)_\mathfrak{p}, A')(\ell) = 0$ for all $\mathfrak{p} \in S$ and $i \geq 1$. Hence the horizontal injection in the last diagram is an isomorphism on the ℓ-torsion subgroups. This finishes the proof of (iii).

Finally, we show (ii). For an arbitrary $A \in Mod_S(G_S)$ we consider the commutative diagram

$$\begin{array}{ccc} \prod_{\mathfrak{p}\in S} H^1(G((k_S)_\mathfrak{p}|k_\mathfrak{p}), \mathrm{Hom}(A, (k_S)_\mathfrak{p}^\times)) & \xrightarrow{\sim} & \prod_{\mathfrak{p}\in S} H^1(G(\bar{k}_\mathfrak{p}|k_\mathfrak{p}), \mathrm{Hom}(A, \bar{k}_\mathfrak{p}^\times)) \\ \uparrow & & \uparrow \\ H^1(G_S, I_S(A)) & \xrightarrow{sh^1} & P^1(G_S, A'), \end{array}$$

which shows that sh^1 is injective. If A is $\mathbb{Z}$-free, then $P^1(G_S, A')$ maps isomorphically to the direct sum $\bigoplus_{\mathfrak{p}\in S} H^1(G(\bar{k}_\mathfrak{p}|k_\mathfrak{p}), \mathrm{Hom}(A, \bar{k}_\mathfrak{p}^\times))$ and the same is true for $H^1(G_S, I_S(A))$. Hence sh^1 is an isomorphism if A is $\mathbb{Z}$-free. This finishes the proof of (ii). □

§6. Poitou-Tate Duality

We proved in (7.2.9) a cohomological duality theorem for local fields and in (8.4.4) an analogous duality theorem for global fields. The latter, however, does not play the same important role for global fields as its local prototype for local fields. But both together lead to the main result of arithmetic Galois cohomology: a duality principle for a local-global statement. The present section is devoted to the formulation and the proof of this main theorem.

Recall that $G_S = G(k_S|k)$ is the Galois group of the maximal extension $k_S|k$ which is unramified outside the given nonempty set of primes S, where S_∞ is contained in S if k is a number field. $Mod_S(G_S)$ is the category of discrete G_S-modules which are finitely generated as $\mathbb{Z}$-modules and whose torsion has order $\#\mathrm{tor}(A) \in \mathbb{N}(S)$, and

$$A' = \mathrm{Hom}(A, \mathcal{O}_S^\times)$$

is the dual G_S-module of $A \in Mod_S(G_S)$, where $\mathcal{O}_S^\times$ is the group of S-units of k_S. Further recall the convention that the cohomology groups $H^*(k_\mathfrak{p}, A')$ are defined by first considering A as a local Galois module and then taking the dual in the local sense, i.e. applying $\mathrm{Hom}(-, \bar{k}_\mathfrak{p}^\times)$.

(8.6.1) Proposition. *Let $A \in Mod_S(G_S)$ and let M be equal to A or A'. Then the homomorphism*

$$res^i\colon H^i(G_S, M) \to \prod_{\mathfrak{p}\in S} H^i(k_\mathfrak{p}, M)$$

maps $H^i(G_S, M)$ into $P^i(G_S, M) = \prod'_{\mathfrak{p}\in S} H^i(k_\mathfrak{p}, M)$.

Proof: First let $M = A$. Since A is finitely generated, it becomes a trivial Galois module over a finite Galois subextension $L^A|k$ of $k_S|k$. If $x \in H^i(G_S, A)$, then there exists a finite Galois subextension $K|k$ of $k_S|k$ such that x is the image of an element $y \in H^i(K|k, A^{G_K})$ under the inflation map. It follows that for all nonarchimedean primes $\mathfrak{p}$ of S which are unramified in $KL^A|k$, the class $x_\mathfrak{p} = res_\mathfrak{p} x$ is contained in the subgroup $H^i_{nr}(k_\mathfrak{p}, A)$ of $H^i(k_\mathfrak{p}, A)$.

Now let $M = A'$ and $x \in H^i(G_S, A')$. As above we have $x = \mathit{inf}\, y$ with $y \in H^i(K|k, (A')^{G_K})$ for a finite Galois subextension $K|k$ of $k_S|k$. For all nonarchimedean primes $\mathfrak{p}$ of S which are unramified in $KL^A|k$, the extension $K_\mathfrak{p}|k_\mathfrak{p}$ is a subextension of the maximal unramified extension $\tilde{k}_\mathfrak{p}$ of $k_\mathfrak{p}$.

Therefore the class $x_\mathfrak{p} = res_\mathfrak{p} x$ is contained in the image of

$$H^i(\tilde{k}_\mathfrak{p}|k_\mathfrak{p}, \mathrm{Hom}(A, \mathcal{O}_S^\times)^{G_{\tilde{k}_\mathfrak{p}}}) = H^i(\tilde{k}_\mathfrak{p}|k_\mathfrak{p}, \mathrm{Hom}(A, (\mathcal{O}_S^\times)^{G_{\tilde{k}_\mathfrak{p}}}))$$

in $H^i(k_\mathfrak{p}, A')$, i.e. in $H^i_{nr}(k_\mathfrak{p}, A') = H^i_{nr}(k_\mathfrak{p}, A^d) = H^i_{nr}(k_\mathfrak{p}, \mathrm{Hom}(A, \tilde{\mathcal{O}}_\mathfrak{p}^\times))$. This proves the proposition. □

(8.6.2) Definition. *The homomorphisms*

$$\lambda^i : H^i(G_S, M) \longrightarrow P^i(G_S, M)$$

for $M = A$ *and* $M = A'$, *which exist by* (8.6.1), *are called the* **localization maps**. *We set*

$$Ш^i(G_S, M) = \ker(\lambda^i)\ . \quad ^{*)}$$

It is evident that for $M = A'$, the map λ^i is the composition

$$H^i(G_S, A') \longrightarrow H^i(G_S, I_S(A)) \xrightarrow{sh^i} P^i(G_S, A'),$$

where the first arrow is induced by $A' = \mathrm{Hom}(A, \mathcal{O}_S^\times) \to \mathrm{Hom}(A, I_S) = I_S(A)$, and the second is the Shapiro map.

The groups $Ш^i(G_S, M)$ are of great interest. Their vanishing would imply a strict "local-global principle". As a rule, the groups $Ш^i(G_S, M)$ are non-zero, but we shall see that they are always finite.

(8.6.3) Lemma. *For* $m \in \mathbb{N}(S)$, *we have natural isomorphisms*

$$\begin{aligned} Ш^1(G_S, \mathbb{Z}/m\mathbb{Z}) &\cong \mathrm{Hom}(Cl_S(k), \mathbb{Z}/m\mathbb{Z})\,, \\ Ш^1(G_S, \mathbb{Z}') &\cong Cl_S(k)\,. \end{aligned}$$

Proof: By class field theory, $Cl_S(k)$ is naturally isomorphic to the Galois group of the maximal abelian extension of k inside k_S in which all primes of S split completely. Hence

$$\begin{aligned} Ш^1(G_S, \mathbb{Z}/m\mathbb{Z}) &= \ker\Big(H^1(G_S, \mathbb{Z}/m\mathbb{Z}) \longrightarrow \prod_{\mathfrak{p}\in S} H^1(k_\mathfrak{p}, \mathbb{Z}/m\mathbb{Z})\Big) \\ &\cong \mathrm{Hom}(Cl_S(k), \mathbb{Z}/m\mathbb{Z})\,. \end{aligned}$$

Furthermore, $Ш^1(G_S, \mathbb{Z}') = H^1(G_S, \mathcal{O}_S^\times) \cong Cl_S(k)$ by Hilbert's Satz 90 and (8.3.11)(ii). □

*) If A is the G_S-module $\mathcal{A}(k_S)$ of k_S-rational points of an abelian variety $\mathcal{A}$ over k, then this kernel is classically called the **Šafarevič-Tate group**. This explains the Russian letter Ш (sha).

From the exact sequence

$$0 \longrightarrow \mathcal{O}_S^\times \longrightarrow I_S \longrightarrow C_S \longrightarrow 0,$$

we obtain a sequence of G_S-modules

$$0 \longrightarrow A' \longrightarrow I_S(A) \longrightarrow C_S(A) \longrightarrow 0\,,$$

which is exact as $\mathrm{Ext}^1_{\mathbb{Z}}(A, \mathcal{O}_S^\times) = 0$ because $\mathcal{O}_S^\times$ is $\#\mathrm{tor}(A)$-divisible by (8.3.4).

(8.6.4) Theorem. *Let $A \in Mod_S(G_S)$. Then the boundary map*

$$C_S(A)^{G_S} \xrightarrow{\ \delta\ } H^1(G_S, A')$$

is continuous and has finite image. It maps a norm group $N_{G/U}C_S(A)^U$ to zero for some open normal subgroup $U \subseteq G_S$. The group $Ш^1(G_S, A')$ is finite.

Proof: First note that $\mathrm{im}(\delta) = Ш^1(G_S, A')$, since

$$sh^1 : H^1(G_S, I_S(A)) \longrightarrow P^1(G_S, A')$$

is injective by (8.5.5)(ii). Let $K = k_S^H$ be a finite Galois extension of k inside k_S which trivializes A and μ_n, where $n = \#\mathrm{tor}(A)$. Furthermore, let $K_1 = k_S^{H_1}$ be a finite Galois extension of k inside k_S which contains the maximal unramified abelian extension of K in which all primes $\mathfrak{p} \in S(K)$ are completely decomposed. As an H-module, A is isomorphic to a direct sum of modules of the form $\mathbb{Z}$ or μ_m with $m|n$. We have $Ш^1(H, \mathbb{Z}') = Cl_S(K)$ and

$$Ш^1(H, (\mu_m)') = \mathrm{Hom}(Cl_S(K), \mathbb{Z}/m\mathbb{Z})$$

by (8.6.3), and we also have the corresponding statement for the group H_1. By class field theory and the principal ideal theorem, we conclude that the maps

$$res^H_{H_1} : Ш^1(H, A') \to Ш^1(H_1, A'),\ cor^{H_1}_H : Ш^1(H_1, A') \to Ш^1(H, A')$$

are zero. The commutative diagram

$$\begin{array}{ccccccc}
I_S(A)^{H_1} & \longrightarrow & C_S(A)^{H_1} & \xrightarrow{\delta_{H_1}} & Ш^1(H_1, A') & \longrightarrow & 0 \\
\downarrow{\scriptstyle N_{H/H_1}} & & \downarrow{\scriptstyle N_{H/H_1}} & & \downarrow{\scriptstyle 0} & & \\
I_S(A)^{H} & \longrightarrow & C_S(A)^{H} & \xrightarrow{\delta_{H}} & Ш^1(H, A') & \longrightarrow & 0 \\
\downarrow{\scriptstyle N_{G_S/H}} & & \downarrow{\scriptstyle N_{G_S/H}} & & \downarrow{\scriptstyle cor^H_{G_S}} & & \\
I_S(A)^{G_S} & \longrightarrow & C_S(A)^{G_S} & \xrightarrow{\delta} & Ш^1(G_S, A') & \longrightarrow & 0
\end{array}$$

shows that δ annihilates the subgroup $N_{G_S/H_1}C_S(A)^{H_1} \subseteq C_S(A)^{G_S}$. Furthermore, the diagram

$$\begin{array}{ccccccc}
I_S(A)^{H_1} & \longrightarrow & C_S(A)^{H_1} & \xrightarrow{\delta_{H_1}} & Ш^1(H_1, A') & \longrightarrow & 0 \\
{\scriptstyle res^H_{H_1}}\uparrow & & {\scriptstyle res^H_{H_1}}\uparrow & & {\scriptstyle 0}\uparrow & & \\
I_S(A)^{H} & \longrightarrow & C_S(A)^{H} & \xrightarrow{\delta_{H}} & Ш^1(H, A') & \longrightarrow & 0 \\
{\scriptstyle res^{G_S}_{H}}\uparrow & & {\scriptstyle res^{G_S}_{H}}\uparrow & & {\scriptstyle res^{G_S}_{H}}\uparrow & & \\
I_S(A)^{G_S} & \longrightarrow & C_S(A)^{G_S} & \xrightarrow{\delta} & Ш^1(G_S, A') & \longrightarrow & 0
\end{array}$$

shows that $C_S(A)^{G_S}$ is a subgroup of the group

$$B := \mathrm{im}(I_S(A)^{H_1} \longrightarrow C_S(A)^{H_1}).$$

By (8.3.5), the image $C_{K_1,S}$ of the homomorphism $I_{K_1,S} \to C_S(K_1)$ is open of finite index. Its kernel $\mathcal{O}^\times_{K_1,S}$ is a discrete subgroup of $I_{K_1,S}$. For $m \in \mathbb{N}(S)$, $\mathrm{Hom}(\mu_m, I_{K_1,S})$ is compact. As $Ш^1(H_1, \mathbb{Z}/m\mathbb{Z}) = \mathrm{Hom}(Cl_S(K_1), \mathbb{Z}/m\mathbb{Z})$ is finite, the image of the homomorphism

$$\mathrm{Hom}(\mu_m, I_{K_1,S}) \longrightarrow \mathrm{Hom}(\mu_m, C_S(K_1))$$

is a closed subgroup of finite index, hence open. The kernel of this homomorphism is the finite group $\mathrm{Hom}(\mu_m, \mathcal{O}^\times_{K_1,S}) = \mathbb{Z}/m\mathbb{Z}$. Therefore $B \subseteq C_S(A)^{H_1}$ is open of finite index and

$$0 \longrightarrow A'^{H_1} \longrightarrow I_S(A)^{H_1} \longrightarrow B \longrightarrow 0$$

is an exact sequence of locally compact groups with continuous G_S/H_1-action. Furthermore, there exists a G_S/H_1-invariant open neighbourhood U of 1 in $I_S(A)^{H_1}$ which maps homeomorphically onto its image U' in B. Let $V \subseteq U$ be a G/H_1-invariant open neighbourhood of 1 with $V \cdot V^{-1} \subseteq U$ and let V' be the homeomorphic image of V in B. The subset $V' \cap C_S(A)^{G_S}$ is an open neighbourhood of 1 in $C_S(A)^{G_S}$.

Let $x \in V' \cap C_S(A)^{G_S}$. There exists a unique pre-image $y \in V$ of x. The inner derivation $G_S/H_1 \to I_S(A)^{H_1}$, $g \mapsto y^{-1}gy$, takes values in A'^{H_1} and induces a derivation $c_y : G_S/H_1 \to A'^{H_1}$. The class $\bar{c}_y \in H^1(G_S/H_1, A'^{H_1})$ of c_y maps to $\delta(x) \in H^1(G_S, A')$ under inflation. For $g \in G_S/H_1$, we have $y^{-1}gy \in U \cap A'^{H_1} = \{1\}$. Hence $\delta(x) = 0$. This shows that the subgroup $\ker(\delta)$ contains an open neighbourhood of 1 in $C_S(A)^{G_S}$ and is therefore open. Consequently, δ is locally constant. Furthermore, δ factors through the compact quotient $C_S(A)^{G_S}/N_{G_S/H_1}C_S(A)^{H_1}$. Therefore $\mathrm{im}(\delta) = Ш^1(G_S, A')$ is finite. □

In the following we want to construct a canonical non-degenerate pairing between the groups $Ш^1(G_S, A')$ and $Ш^2(G_S, A)$ for each finitely generated G_S-module $A \in Mod_S(G_S)$. We need two lemmas. As before, write

$$A' = \mathrm{Hom}(A, \mathcal{O}_S^\times)\ ,\ I_S(A) = \mathrm{Hom}(A, I_S)\ ,\ C_S(A) = \mathrm{Hom}(A, C_S).$$

(8.6.5) Lemma. *Let $A \in Mod_S(G_S)$ and $i = 1$ or $i = 2$. Then the homomorphism*

$$\xi^i : H^i(G_S, A) \xrightarrow{\lambda^i} P^i(G_S, A) \xrightarrow{\Xi^i} P^{2-i}(G_S, A')^\vee ,$$

where Ξ^i was defined in (8.5.2), *has kernel* $\ker \xi^i = Ш^i(G_S, A)$.

Proof: By (8.5.3), the homomorphisms Ξ^i, $i = 1, 2$, are injective. □

Recall that $P^{2-i}(G_S, A')^\vee$ is a locally compact group for $i = 1, 2$. Therefore we can dualize the composite map $\xi^i = \Xi^i \circ \lambda^i$ for $i = 1, 2$ (but not Ξ^2, since "$P^2(G_S, A)^\vee$" does not exist at least if S and A are infinite).

(8.6.6) Lemma. *For $i = 0$ and $i = 1$, we have a commutative diagram of topological groups*

$$\begin{array}{ccc} H^i(G_S, I_S(A)) & \xrightarrow{can} & H^i(G_S, C_S(A)) \\ {\scriptstyle sh^i}\downarrow & & \downarrow{\scriptstyle \Delta^i} \\ P^i(G_S, A') & \xrightarrow{(\xi^{2-i})^\vee} & H^{2-i}(G_S, A)^\vee, \end{array}$$

where Δ^i is induced by the cup-product pairing

$$H^i(G_S, C_S(A)) \times H^{2-i}(G_S, A) \xrightarrow{\cup} H^2(G_S, C_S) \xrightarrow{inv} \mathbb{Q}/\mathbb{Z}.$$

The map $(\xi^{2-i})^\vee$ on the bottom is the dual to the map ξ^{2-i} obtained in (8.6.5).

Proof: Let $\mathfrak{p} \in S$ be a prime. We make the convention that $H^0(k_\mathfrak{p}, -)$ denotes $\hat{H}^0(k_\mathfrak{p}, -)$ if $\mathfrak{p}$ is archimedean. Using the notation of the preceding section, we have the commutative diagram

$$\begin{array}{ccccccc} H^i(k_\mathfrak{p}, \mathrm{Hom}(A, \bar{k}_\mathfrak{p}^\times)) & \times & H^{2-i}(k_\mathfrak{p}, A) & \xrightarrow{\cup} & H^2(k_\mathfrak{p}, \bar{k}_\mathfrak{p}^\times) & \xrightarrow{inv_\mathfrak{p}} & \mathbb{Q}/\mathbb{Z} \\ \| & & \uparrow{\scriptstyle inf} & & \uparrow{\scriptstyle inf} & & \| \\ H^i(G_\mathfrak{p}, \mathrm{Hom}(A, k_{S,\mathfrak{p}}^\times)) & \times & H^{2-i}(G_\mathfrak{p}, A) & \xrightarrow{\cup} & H^2(G_\mathfrak{p}, k_{S,\mathfrak{p}}^\times) & & \| \\ {\scriptstyle res_\mathfrak{p}}\uparrow & & \uparrow{\scriptstyle res_\mathfrak{p}} & & \uparrow{\scriptstyle res_\mathfrak{p}} & & \| \\ H^i(G_S, \mathrm{Hom}(A, I_S)) & \times & H^{2-i}(G_S, A) & \xrightarrow{\cup} & H^2(G_S, I_S) & \xrightarrow{inv} & \mathbb{Q}/\mathbb{Z}. \end{array}$$

For $x \in H^i(G_S, I_S(A))$, we therefore obtain a commutative diagram

$$\begin{array}{ccccc}
H^{2-i}(G_S,A) & \longrightarrow & \prod\limits_{\mathfrak{p}\in S} H^{2-i}(k_\mathfrak{p},A) & \longrightarrow & \prod\limits_{\mathfrak{p}\in S} H^i(k_\mathfrak{p},A')^\vee \\
\downarrow{\scriptstyle \cup x} & & & & \downarrow{\scriptstyle \cup sh^i(x)} \\
H^2(G_S,I_S) & & \xrightarrow{\qquad \oplus inv_\mathfrak{p} \qquad} & & \bigoplus\limits_{\mathfrak{p}\in S}\mathbb{Q}/\mathbb{Z} \\
\downarrow & & & & \downarrow{\scriptstyle \Sigma} \\
H^2(G_S,C_S) & & \xrightarrow{\qquad inv \qquad} & & \mathbb{Q}/\mathbb{Z}\,.
\end{array}$$

The map from $H^{2-i}(G_S,A)$ to $\mathbb{Q}/\mathbb{Z}$ via the upper right corner is exactly $(\xi^{2-i})^\vee \circ sh^i(x) \in H^{2-i}(G_S,A)^\vee$, while the map via the lower left corner is $\Delta^i \circ can(x)$. This shows that the diagram of the lemma commutes. □

Now we are ready to prove one of the main results of this section.

(8.6.7) Theorem (Poitou-Tate Duality). *Let A be a finitely generated G_S-module with $\#\mathrm{tor}(A) \in \mathbb{N}(S)$. Then there is a perfect pairing* *)

$$Ш^1(G_S,A') \times Ш^2(G_S,A) \longrightarrow \mathbb{Q}/\mathbb{Z}$$

of finite groups, which is induced by the cup-product, i.e. the diagram

$$\begin{array}{ccccccc}
H^0(G_S,C_S(A)) & \times & H^2(G_S,A) & \xrightarrow{\cup} & H^2(G_S,C_S) & \longrightarrow & \mathbb{Q}/\mathbb{Z} \\
\downarrow & & \uparrow & & & \nearrow & \\
Ш^1(G_S,A') & \times & Ш^2(G_S,A) & & & &
\end{array}$$

commutes.

Proof: Consider the diagram with exact upper row

$$(1)\qquad \begin{array}{ccccccc}
I_S(A)^{G_S} & \xrightarrow{\gamma} & C_S(A)^{G_S} & \xrightarrow{\delta} & H^1(G_S,A') & \xrightarrow{i} & H^1(G_S,I_S(A)) \\
\downarrow{\scriptstyle sh^0} & & \downarrow{\scriptstyle \Delta^0} & & \| & & \downarrow{\scriptstyle sh^1} \\
P^0(G_S,A') & \xrightarrow{(\xi^2)^\vee} & H^2(G_S,A)^\vee & \overset{\varepsilon}{\dashrightarrow} & H^1(G_S,A') & \xrightarrow{\lambda^1} & P^1(G_S,A')\;.
\end{array}$$

---

*) A pairing $A \times B \to \mathbb{Q}/\mathbb{Z}$ is **perfect** if it induces isomorphisms $A \xrightarrow{\sim} B^\vee$ and $B \xrightarrow{\sim} A^\vee$.

The map Δ^0 is the composition of the global duality map (8.4.4), (8.4.5),

$$\hat{\Delta}^0 : \hat{H}^0(G_S, C_S(A)) \xrightarrow{\sim} H^2(G_S, A)^\vee ,$$

with the canonical homomorphism $H^0(G_S, C_S(A)) \to \hat{H}^0(G_S, C_S(A))$ (which has dense image). The left-hand square is commutative by (8.6.6) and the right-hand one by the remark following (8.6.2). By (8.6.4), the homomorphism δ maps a norm group $N_{G_S/U}C_S(A)^U$ with U sufficiently small to zero, and we obtain a commutative diagram

$$\begin{array}{ccccc}
C_S(A)^{G_S} & \longrightarrow & \hat{H}^0(G_S, C_S(A)) & \xrightarrow[\sim]{\hat{\Delta}^0} & H^2(G_S, A)^\vee \\
& \searrow & \Big\downarrow \pi & & \Big\downarrow \varepsilon \\
& & C_S(A)^{G_S}/N_{G_S/U}C_S(A)^U & \xrightarrow{\delta} & H^1(G_S, A')
\end{array}$$

where $\varepsilon = \delta \circ \pi \circ (\hat{\Delta}^0)^{-1}$. This map ε completes diagram (1). We have $\mathrm{im}(\varepsilon) = \mathrm{im}(\delta) = \ker(i) = \ker(\lambda^1) = Ш^1(G_S, A')$ (observe that sh^1 is injective by (8.5.5)(ii)). By (8.5.5)(i), the map sh^0 is surjective. Since

$$\varepsilon(\xi^2)^\vee sh^0 = \delta\pi(\hat{\Delta}^0)^{-1}(\xi^2)^\vee sh^0 = \delta\pi(\hat{\Delta}^0)^{-1}\Delta^0\gamma = \delta\gamma = 0 ,$$

we obtain $\varepsilon(\xi^2)^\vee = 0$, i.e. we have an inclusion $\mathrm{im}((\xi^2)^\vee) \subseteq \ker(\varepsilon)$. Since ε is continuous and has finite image by (8.6.4) and (8.4.4), $\ker(\varepsilon)$ is an open subgroup in $H^2(G_S, A)^\vee$. In particular, $\ker(\varepsilon)$ contains the closure of $\mathrm{im}((\xi^2)^\vee)$ (note that $\mathrm{im}((\xi^2)^\vee)$ is closed if A is finite or if k is a number field).

Let $x \in \ker(\varepsilon)$ and let $V \subseteq \ker(\varepsilon)$ be an arbitrarily chosen open subgroup. As Δ^0 has dense image by (8.4.5), there exists $y \in C_S(A)^{G_S}$ with $\Delta^0(y) \in x + V$. Hence $\delta(y) = \varepsilon\Delta^0(y) = 0$. Thus there exists an element $z \in I_S(A)^{G_S}$ such that $\gamma(z) = y$. It follows that $\Delta^0(y) = (\xi^2)^\vee(sh^0(z))$, and so $x + V$ contains an element of $\mathrm{im}((\xi^2)^\vee)$. As $H^2(G_S, A)^\vee$ is profinite, it has a basis of neighbourhoods of zero consisting of subgroups. Therefore $\mathrm{im}((\xi^2)^\vee)$ is dense in $\ker(\varepsilon)$ and the induced map $\ker(\varepsilon)^\vee \to P^0(G_S, A')^\vee$ is injective. We conclude that ξ^2 factors as

$$\xi^2 : \; H^2(G_S, A) \twoheadrightarrow \ker(\varepsilon)^\vee \hookrightarrow P^0(G_S, A')^\vee .$$

This implies that there is an isomorphism

$$Ш^1(G_S, A')^\vee = \mathrm{im}(\varepsilon)^\vee \cong \ker(\xi^2) = Ш^2(G_S, A),$$

which obviously is induced from the diagram

$$\begin{array}{ccccc}
H^0(G_S, C_S(A)) & \times & H^2(G_S, A) & \xrightarrow{\cup} & \mathbb{Q}/\mathbb{Z} \\
\Big\downarrow \delta & & \Big\uparrow & \nearrow & \\
Ш^1(G_S, A') & \times & Ш^2(G_S, A) . & &
\end{array}$$

□

For finite modules we can describe the pairing obtained in (8.6.7) explicitly. Let $A \in Mod_S(G_S)$ be finite and let us consider the "new" pairing

$$Ш^1(G_S, A') \times Ш^2(G_S, A) \xrightarrow{\sqcup} \mathbb{Q}/\mathbb{Z}$$

which is defined as follows: let x and x' be cocycles representing classes $[x] \in Ш^2(G_S, A)$ and $[x'] \in Ш^1(G_S, A')$. As $H^3(G_S, \mathcal{O}_S^\times)(p) = 0$ for every prime number $p \in \mathbb{N}(S)$ (see (8.3.11)), there is a cochain $z \in C^2(G_S, \mathcal{O}_S^\times)$ such that

$$x' \cup x = \partial z.$$

Moreover, for every $\mathfrak{p}$ there are cochains $y_\mathfrak{p} \in C^1(k_\mathfrak{p}, A)$ and $y'_\mathfrak{p} \in C^0(k_\mathfrak{p}, A')$ such that for the components $x_\mathfrak{p} = \lambda_\mathfrak{p}(x)$, $x'_\mathfrak{p} = \lambda_\mathfrak{p}(x')$ we have

$$x_\mathfrak{p} = \partial y_\mathfrak{p} \quad \text{and} \quad x'_\mathfrak{p} = \partial y'_\mathfrak{p}.$$

Then $y'_\mathfrak{p} \cup x_\mathfrak{p} - z_\mathfrak{p}$ and $x'_\mathfrak{p} \cup y_\mathfrak{p} - z_\mathfrak{p}$ are 2-cocycles with values in $\bar{k}_\mathfrak{p}^\times$ which differ by a coboundary. In fact,

$$\partial(y'_\mathfrak{p} \cup x_\mathfrak{p}) = \partial y'_\mathfrak{p} \cup x_\mathfrak{p} = x'_\mathfrak{p} \cup x_\mathfrak{p} = \partial z_\mathfrak{p}$$

and analogously $\partial(x'_\mathfrak{p} \cup y_\mathfrak{p}) = \partial z_\mathfrak{p}$, and moreover

$$\partial(y'_\mathfrak{p} \cup y_\mathfrak{p}) = x'_\mathfrak{p} \cup y_\mathfrak{p} - y'_\mathfrak{p} \cup x_\mathfrak{p}.$$

The "new" pairing is now defined by

$$[x'] \sqcup [x] = \sum_{\mathfrak{p} \in S} inv_\mathfrak{p}[y'_\mathfrak{p} \cup x_\mathfrak{p} - z_\mathfrak{p}] = \sum_{\mathfrak{p} \in S} inv_\mathfrak{p}[x'_\mathfrak{p} \cup y_\mathfrak{p} - z_\mathfrak{p}].$$

One checks in a straightforward manner that this definition does not depend on the choice of the representing cocycles x and x' and the choice of the cochains z, $y_\mathfrak{p}$ and $y'_\mathfrak{p}$.

(8.6.8) Proposition. *The pairing defined above coincides with the pairing obtained in* (8.6.7).

Proof: Let $[x'] \in Ш^1(G_S, A')$ and $[x] \in Ш^2(G_S, A)$. The set of 0-cochains $y'_\mathfrak{p} \in H^0(k_\mathfrak{p}, A')$ with $\partial y'_\mathfrak{p} = x'_\mathfrak{p}$, $\mathfrak{p} \in S$, can be interpreted as a 0-cochain in $H^0(G_S, I_S(A))$. Under the projection $I_S \to C_S$ the element $(y'_\mathfrak{p}) \in H^0(G_S, I_S(A))$ becomes a 0-cochain $y' \in H^0(G_S, C_S(A))$ such that $\delta y' = [x']$ and the value of the pairing in (8.6.7) is $inv[y' \cup x]$. Consider the commutative diagram (8.6.6)

$$\begin{array}{ccc} H^0(G_S, I_S(A)) & \longrightarrow & H^0(G_S, C_S(A)) \\ \downarrow & & \downarrow \\ P^0(G_S, A') & \longrightarrow & H^2(G_S, A)^\vee. \end{array}$$

With the notation of the "new" pairing, we obtain

$$[x'] \sqcup [x] = \sum_{\mathfrak{p} \in S} inv_{\mathfrak{p}}[y'_{\mathfrak{p}} \cup x_{\mathfrak{p}} - z_{\mathfrak{p}}] = inv[y' \cup x]\,,$$

because the image of the 2-cochain $(z_{\mathfrak{p}}) \in C^2(G_S, \mathcal{O}_S^\times)$ in $H^2(G_S, C_S)$ is zero. □

For $A \in Mod_S(G_S)$, we have constructed a natural perfect pairing between $Ш^1(G_S, A')$ and $Ш^2(G_S, A)$. If A is finite, we can interchange the roles of A and A' to obtain a perfect pairing between $Ш^1(G_S, A)$ and $Ш^2(G_S, A')$ as well. If A is infinite, the situation is no longer symmetric. If A is $\mathbb{Z}$-free, i.e. A' is a Spec$(\mathcal{O}_{k,S})$-torus, we have the following

(8.6.9) Theorem. *Let $A \in Mod_S(G_S)$ be torsion-free. Then there is a perfect pairing of finite groups*

$$Ш^1(G_S, A) \times Ш^2(G_S, A') \longrightarrow \mathbb{Q}/\mathbb{Z}\,,$$

which is induced by the cup-product, i.e. the diagram

$$\begin{array}{ccccccc}
H^1(G_S, A) & \times & H^1(G_S, C_S(A)) & \xrightarrow{\cup} & H^2(G_S, C_S) & \longrightarrow & \mathbb{Q}/\mathbb{Z} \\
\uparrow & & \downarrow & & & \nearrow & \\
Ш^1(G_S, A) & \times & Ш^2(G_S, A') & & & &
\end{array}$$

commutes.

Proof: Since $H^1(G_S, A)$ is a finite group by (8.3.20)(ii), its subgroup $Ш^1(G_S, A)$ is also finite. By (8.6.6), the diagram

$$\begin{array}{ccccccc}
H^1(G_S, I_S(A)) & \longrightarrow & H^1(G_S, C_S(A)) & \xrightarrow{\delta} & H^2(G_S, A') & \longrightarrow & H^2(G_S, I_S(A)) \\
{\scriptstyle sh^1}\downarrow & & {\scriptstyle \Delta^1}\downarrow\wr & & \| & & \downarrow{\scriptstyle sh^2} \\
P^1(G_S, A') & \xrightarrow{(\xi^1)^\vee} & H^1(G_S, A)^\vee & \dashrightarrow{\varphi} & H^2(G_S, A') & \xrightarrow{\lambda^2} & P^2(G_S, A')
\end{array}$$

commutes. The map Δ^1 is an isomorphism by (8.4.4), and φ is defined so that the square in the middle commutes. Obviously, the upper sequence is exact. By (8.5.5)(ii), (iii) the map sh^1 is bijective and sh^2 is injective. Therefore the lower sequence in the commutative diagram above is also exact. Since coker$\,(\xi^1)^\vee = Ш^1(G_S, A)^\vee$ by (8.6.5), we obtain $Ш^1(G_S, A)^\vee \cong Ш^2(G_S, A')$. Finally, the associated pairing to $\mathbb{Q}/\mathbb{Z}$ is induced by the cup-product, as stated. This finishes the proof of the theorem. □

Remarks: 1. If S is a finite set containing the set S_p for some prime number p and $A \in Mod_S(G_S)$ is arbitrary, then one can prove the duality between $Ш^1(G_S, A)$ and $Ш^2(G_S, A')$ at least for the p-components of these groups; see (10.11.10).

2. The duality between $Ш^i(G_S, A)$ and $Ш^{3-i}(G_S, A')$, $i = 1, 2$, proven above for torsion-free A (resp. for a torus), has a vast generalization if S is the set of all primes: it holds for 1-motives; see [76].

3. If S is the set of all places, then (8.6.9) holds for an arbitrary finitely generated G_k-module A. This can be seen as follows (statement and proof are due to *J.-L. Colliot-Thélène*): There exists an exact sequence $0 \to C \to B \to A \to 0$ with B torsionfree and $C \cong \bigoplus_{i=1}^n \mathrm{Ind}_{G_k}^{U_i} \mathbb{Z}$, where $U_i \subseteq G_k$ are open subgroups; see [CS], Lemma 0.6. As the groups $H^1(k, C)$, $H^1(k, C')$, $Ш^2(k, C)$ and $Ш^2(k, C')$ vanish, we obtain isomorphisms $Ш^1(k, B) \cong Ш^1(k, A)$ and $Ш^2(k, B') \cong Ш^2(k, A')$, and so the statement for general A follows from that in the torsionfree case. This argument extends to the case when S has Dirichlet density 1.

We proved a duality theorem for global fields in (8.4.4) which, together with its analogue (7.2.6) for local fields, leads to the main result of arithmetic Galois cohomology: a 9-term exact sequence connecting the local and global cohomology groups.

(8.6.10) Long Exact Sequence of Poitou-Tate. *Let S be a nonempty set of primes of a global field k and assume that $S \supseteq S_\infty$ if k is a number field. Let A be a finite G_S-module of order $\#A \in \mathbb{N}(S)$.*

(i) *There is a canonical exact sequence of topological groups*

$$\begin{array}{ccccccccc}
0 \longrightarrow & H^0(G_S, A) & \longrightarrow & P^0(G_S, A) & \to & H^2(G_S, A')^\vee & \to & H^1(G_S, A) & \\
& \text{(finite)} & & \text{(compact)} & & \text{(compact)} & & \text{(discrete)} & \searrow \\
& & & & & & & \text{(locally compact)} & P^1(G_S, A) \\
0 \leftarrow & H^0(G_S, A')^\vee & \leftarrow & P^2(G_S, A) & \longleftarrow & H^2(G_S, A) & \leftarrow & H^1(G_S, A')^\vee & \swarrow \\
& \text{(finite)} & & \text{(discrete)} & & \text{(discrete)} & & \text{(compact)} &
\end{array}$$

(ii) *For $i \geq 3$, the restriction map*

$$H^i(G_S, A) \xrightarrow{\sim} \bigoplus_{\mathfrak{p} \in S_\mathbb{R}} H^i(k_\mathfrak{p}, A)$$

is an isomorphism.

Remark: Some people consider it more suggestive to arrange the terms of the exact 9-term sequence in the following form.

$$\begin{aligned} 0 \longrightarrow H^0(G_S, A) &\longrightarrow P^0(G_S, A) \longrightarrow H^2(G_S, A')^\vee \longrightarrow \\ \longrightarrow H^1(G_S, A) &\longrightarrow P^1(G_S, A) \longrightarrow H^1(G_S, A')^\vee \longrightarrow \\ \longrightarrow H^2(G_S, A) &\longrightarrow P^2(G_S, A) \longrightarrow H^0(G_S, A')^\vee \longrightarrow 0. \end{aligned}$$

Proof of (8.6.10)(ii): For function fields the assertion is obviously true by (8.3.17), so we may assume that k is a number field. If $A = 0$ the statement is trivial. Therefore we may assume that $S \supseteq S_p$ for at least one prime number p. In particular, k_S is totally imaginary.

As $\ell^\infty \mid \#G_S$ for all $\ell \in \mathbb{N}(S)$, (1.9.15) implies $\hat{H}^{-1}(G_S, A') = 0$. From the exact sequence

$$0 \longrightarrow A' \longrightarrow I_S(A) \longrightarrow C_S(A) \longrightarrow 0$$

of level-compact G_S-modules (observe that $C_S(A) = \mathrm{Hom}(A, C_S^0)$) we obtain the exact sequence

$$0 \to \hat{H}^{-1}(G_S, I_S(A)) \to \hat{H}^{-1}(G_S, C_S(A)) \to \hat{H}^0(G_S, A') \to \hat{H}^0(G_S, I_S(A))$$

by (1.9.13). Again by (1.9.15), we have $\hat{H}^0(G_S, A') = A'^{G_S}$. For $\mathfrak{p} \in S$ put $\overline{G}_\mathfrak{p} = G(k_{S,\mathfrak{p}}|k_\mathfrak{p})$. For $\mathfrak{p} \in S \smallsetminus S_\infty$ and $\ell \in \mathbb{N}(S)$ we have $\ell^\infty \mid \#\overline{G}_\mathfrak{p}$ and (1.9.15) implies $\hat{H}^0(\overline{G}_\mathfrak{p}, A') = A'^{\overline{G}_\mathfrak{p}}$. We obtain

$$\hat{H}^0(G_S, I_S(A)) \cong \prod_{\mathfrak{p} \in S \setminus S_\infty} A'^{\overline{G}_\mathfrak{p}} \times \prod_{\mathfrak{p} \in S_\infty} \hat{H}^0(G_\mathfrak{p}, A').$$

Therefore the map $\hat{H}^0(G_S, A') \to \hat{H}^0(G_S, I_S(A))$ is injective, yielding the isomorphism

$$\hat{H}^{-1}(G_S, I_S(A)) \xrightarrow{\sim} \hat{H}^{-1}(G_S, C_S(A)).$$

By (1.9.15), we have $\hat{H}^{-1}(\overline{G}_\mathfrak{p}, A') = 0$ for all $\mathfrak{p} \in S \smallsetminus S_\infty$ and therefore a canonical isomorphism

$$\hat{H}^{-1}(G_S, I_S(A)) \cong \prod_{\mathfrak{p} \in S} \hat{H}^{-1}(\overline{G}_\mathfrak{p}, A') \cong \bigoplus_{\mathfrak{p} \in S_\mathbb{R}} \hat{H}^{-1}(k_\mathfrak{p}, A').$$

Using the duality theorem for archimedean primes (7.2.17)

$$\hat{H}^{-1}(k_\mathfrak{p}, A') \cong H^3(k_\mathfrak{p}, A)^\vee,$$

we obtain the commutative diagram

$$\begin{array}{ccc} \hat{H}^{-1}(G_S, I_S(A)) & \xrightarrow{\ \sim\ } & \hat{H}^{-1}(G_S, C_S(A)) \\ \downarrow \wr & & \downarrow \wr \\ \bigoplus_{\mathfrak{p} \in S_\mathbb{R}} H^3(k_\mathfrak{p}, A)^\vee & \xrightarrow{(\lambda^3)^\vee} & H^3(G_S, A)^\vee \end{array}$$

where the vertical isomorphism on the right-hand side is given by (8.4.4). Hence

$$\lambda^3 : H^3(G_S, A) \xrightarrow{\sim} \bigoplus_{\mathfrak{p} \in S_{\mathbb{R}}} H^3(k_{\mathfrak{p}}, A).$$

is an isomorphism.

In order to prove the assertion for all $i \geq 3$ we use the exact sequence

$$0 \longrightarrow A \longrightarrow \mathrm{Ind}_{\overline{G}} A \longrightarrow B \longrightarrow 0,$$

where $\overline{G} = G(K|k)$, for K a finite totally imaginary extension of k inside k_S such that A is a trivial $G(k_S|K)$-module, and the finite module B is defined as the quotient of $\mathrm{Ind}_{\overline{G}} A$ by A. For $i \geq 3$ we have $H^i(G_S, \mathrm{Ind}_{\overline{G}} A) = H^i(G(k_S|K), A) = 0$ and

$$H^i(k_{\mathfrak{p}}, \mathrm{Ind}_{\overline{G}} A) = \bigoplus_{\mathfrak{P}|\mathfrak{p}} H^i(K_{\mathfrak{P}}, A) = 0, \quad \mathfrak{p} \in S_{\mathbb{R}}.$$

Therefore the horizontal maps in the commutative diagram

$$\begin{array}{ccc} H^i(G_S, B) & \xrightarrow{\sim} & H^{i+1}(G_S, A) \\ \downarrow{\scriptstyle \lambda^i} & & \downarrow{\scriptstyle \lambda^{i+1}} \\ \bigoplus_{\mathfrak{p} \in S_{\mathbb{R}}} H^i(k_{\mathfrak{p}}, B) & \xrightarrow{\sim} & \bigoplus_{\mathfrak{p} \in S_{\mathbb{R}}} H^{i+1}(k_{\mathfrak{p}}, A) \end{array}$$

are isomorphisms for $i \geq 3$. Now the result follows by induction. □

In order to prove part (i) of theorem (8.6.10), we start with the following observations.

(8.6.11) Proposition. *For $A \in Mod_S(G_S)$, the localization map*

$$\lambda^1 : H^1(G_S, A) \longrightarrow P^1(G_S, A)$$

*is proper.**)

Proof: Since A is finitely generated, A is a G_T-module for all sufficiently large finite subsets $T \subseteq S$. The groups

$$P_T = P^1(G_T, A) \times \prod_{\mathfrak{p} \in S \setminus T} H^1_{nr}(k_{\mathfrak{p}}, A)$$

are compact, and each compact neighbourhood of 1 in $P^1(G_S, A)$ is contained in some P_T for $T \subseteq S$ finite. Therefore we have to show that $(\lambda^1)^{-1}(P_T) \subseteq H^1(G_S, A)$ is compact and, since $H^1(G_S, A)$ is discrete, this comes down to showing that $(\lambda^1)^{-1}(P_T)$ is finite. Consider the canonical injections

*) A map of topological spaces is called proper if the pre-images of compact subsets are compact.

$$\bigoplus_{\mathfrak{p} \in S \setminus T(k)} H^1(k_\mathfrak{p}, A)/H^1_{nr}(k_\mathfrak{p}, A) \hookrightarrow \bigoplus_{\mathfrak{p} \in S \setminus T(k)} H^1(T_\mathfrak{p}, A) \overset{diag}{\hookrightarrow} \prod_{\mathfrak{P} \in S \setminus T(k_T)} H^1(T_\mathfrak{P}, A),$$

where $T_\mathfrak{p}$ denotes the inertia group of the local group $G(\bar{k}_\mathfrak{p}|k_\mathfrak{p})$ for a prime $\mathfrak{p}$. We obtain the commutative and exact diagram

$$\begin{array}{ccc}
H^1(G(k_S|k_T), A) & \longrightarrow & \prod\limits_{\mathfrak{P} \in S \setminus T(k_T)} H^1(T_\mathfrak{P}, A) \\
\uparrow & & \uparrow \\
H^1(G_S, A) & \longrightarrow & P^1(G_S, A) \\
\uparrow & & \uparrow \\
H^1(G_T, A) & \longrightarrow & P^1(G_T, A) \times \prod\limits_{\mathfrak{p} \in S \setminus T(k)} H^1_{nr}(k_\mathfrak{p}, A) = P_T\,.
\end{array}$$

Since A is a trivial $G(k_S|k_T)$-module and since the images of the inertia groups $T_\mathfrak{P}$, $\mathfrak{P} \in S \smallsetminus T(k_T)$, generate $G(k_S|k_T)$ as a normal subgroup, the upper horizontal map is injective. Thus

$$(\lambda^1)^{-1}(P_T) \subseteq H^1(G_T, A)\,.$$

We will show that $H^1(G_T, A)$ is a finite group. Recalling (8.3.20), we may assume that A is torsion-free. Then the exact sequence

$$0 \longrightarrow H^1(G_T/H, A) \longrightarrow H^1(G_T, A) \longrightarrow H^1(H, A) = 0,$$

where H is an open normal subgroup of G_T acting trivially on A, shows that $H^1(G_T, A)$ is finite. □

(8.6.12) Lemma. *Let $A \in Mod_S(G_S)$ be finite. Then there is a commutative exact diagram*

$$\begin{array}{ccccccc}
\prod\limits_{\mathfrak{p} \in S_\infty} N_{\bar{k}_\mathfrak{p}|k_\mathfrak{p}} \mathrm{Hom}(A, \mathbb{C}^\times) & \xrightarrow[\sim]{\varphi} & N_{G_S} C_S(A) & & & & \\
\downarrow & & \downarrow & & & & \\
H^0(G_S, I_S(A)) & \longrightarrow & H^0(G_S, C_S(A)) & \xrightarrow{\delta} & H^1(G_S, A') & \rightarrow & H^1(G_S, I_S(A)) \\
\downarrow & & \downarrow & & \| & & \downarrow \\
P^0(G_S, A') & \longrightarrow & H^2(G_S, A)^\vee & \xrightarrow{\varepsilon} & H^1(G_S, A') & \longrightarrow & P^1(G_S, A')
\end{array}$$

where the map φ is an isomorphism. We have a natural exact sequence

$$0 \rightarrow H^0(G_S, A') \rightarrow P^0(G_S, A') \rightarrow H^2(G_S, A)^\vee \rightarrow Ш^1(G_S, A') \rightarrow 0\,.$$

Proof: As we can represent k_S as the union of a countable tower of finite subextensions, we can apply the results of II §7 to projective systems indexed by the set of finite extensions K of k inside k_S. Consider the projective system which attaches to each K the group $H^0(G(k_S|K), A')$ with the norm maps as transition maps. If K trivializes A' and $L|K$ is a finite extension, then the norm

$$N_{L|K} : A' = H^0(G(k_S|L), A') \longrightarrow A' = H^0(G(k_S|K), A')$$

is just multiplication by $[L : K]$. For all prime numbers $\ell \mid \#A' \in \mathbb{N}(S)$, ℓ^∞ divides the order of G_S (observe that $k(\mu_{\ell^\infty}) \subseteq k_S$). Therefore the above projective system is a ML-zero system, and (2.7.3) implies

$$R^i(\varprojlim)H^0(G(k_S|K), A') = 0, \quad i \geq 0 .$$

We let K run through the extensions of k inside k_S which trivialize A' and consider the exact sequences

$$0 \longrightarrow A' \longrightarrow \mathrm{Hom}(A, I_{K,S}) \longrightarrow \mathrm{Hom}(A, I_{K,S})/A' \longrightarrow 0.$$

Passing to the projective limit over K and using the ML-zero property, we obtain an isomorphism

$$\varprojlim_K \mathrm{Hom}(A, I_{K,S}) \xrightarrow{\sim} \varprojlim_K \mathrm{Hom}(A, I_{K,S})/A' .$$

By (8.5.5) the map sh^1 is injective, hence

$$Ш^1(G(k_S|K), A') = \ker\left(H^1(G(k_S|K), A') \longrightarrow H^1(G(k_S|K), I_S(A))\right).$$

Therefore the exact sequence

$$0 \longrightarrow A' \longrightarrow I_S(A) \longrightarrow C_S(A) \longrightarrow 0$$

implies the exact sequence

$$0 \longrightarrow \mathrm{Hom}(A, I_{K,S})/A' \longrightarrow \mathrm{Hom}(A, C_S(K)) \longrightarrow Ш^1(G(k_S|K), A') \longrightarrow 0 .$$

Since

$$\varprojlim_K Ш^1(G(k_S|K), A') = \left(\varinjlim_K Ш^2(G(k_S|K), A)\right)^\vee = 0 ,$$

we obtain the isomorphism $\varprojlim_K \mathrm{Hom}(A, I_{K,S}) \xrightarrow{\sim} \varprojlim_K \mathrm{Hom}(A, C_{K,S})$. Consider the commutative diagram

$$\begin{array}{ccc} \varprojlim_K \mathrm{Hom}(A, I_{K,S}) & \xrightarrow{\sim} & \varprojlim_K \mathrm{Hom}(A, C_{K,S}) \\ \big\downarrow & & \big\downarrow \\ N_{G_S} I_S(A) & \longrightarrow & N_{G_S} C_S(A). \end{array}$$

As the G_S-module $C_S(A) = C_S^0(A)$ is level-compact, the right vertical map is surjective, and hence the natural map

$$N_{G_S} I_S(A) \longrightarrow N_{G_S} C_S(A)$$

is also surjective. Let K be an extension of k inside k_S trivializing A' and let $\mathfrak{p} \in S \smallsetminus S_\infty(K)$. For a sufficiently large finite extension L of K inside k_S (adjoin roots of unity) the norm map

$$N_{L|K}\colon \prod_{\mathfrak{P}|\mathfrak{p}} \mathrm{Hom}(A, L_{\mathfrak{P}}^\times) \longrightarrow \mathrm{Hom}(A, K_{\mathfrak{p}}^\times)$$

is the zero map. Therefore the universal norms of $I_S(A)$ have trivial components at the nonarchimedean primes and we obtain an isomorphism

$$\prod_{\mathfrak{p}\in S_\infty} N_{\bar{k}_\mathfrak{p}|k_\mathfrak{p}} \mathrm{Hom}(A, \mathbb{C}^\times) \xrightarrow{\sim} N_{G_S} I_S(A)\,.$$

We define φ as the composite of this isomorphism with the natural map $N_{G_S} I_S(A) \longrightarrow N_{G_S} C_S(A)$, which is surjective, as shown above.

Let us explain the other objects and maps in the diagram of the lemma. The second row of the diagram is part of the long exact cohomology sequence associated to the short exact sequence $A' \hookrightarrow I_S(A) \twoheadrightarrow C_S(A)$. The first arrow in the third row is the dual of the localization map $\lambda^2(A)$, while the third arrow is the localization map $\lambda^1(A')$. For the commutativity of the left- and right-hand lower squares, see (8.6.6) and the definition of the localization map (8.6.2). The map ε is given by the perfect pairing between $Ш^1(G_S, A')$ and $Ш^2(G_S, A)$; see (8.6.7). In particular, the lower row is exact. The square in the middle commutes by the construction of ε. The exactness of the two columns follows from (8.5.5)(i) and (8.4.4).

Consider the commutative diagram

$$\begin{array}{ccccccc}
 & & & & \prod_{\mathfrak{p}\in S_\infty} N_{\bar{k}_\mathfrak{p}|k_\mathfrak{p}} \mathrm{Hom}(A, \mathbb{C}^\times) & \xrightarrow{\varphi} & N_{G_S} C_S(A) \\
 & & & & \downarrow & & \downarrow \\
0 & \longrightarrow & H^0(G_S, A') & \longrightarrow & H^0(G_S, I_S(A)) & \longrightarrow & H^0(G_S, C_S(A)) \\
 & & \| & & \downarrow & & \downarrow \\
0 & \longrightarrow & H^0(G_S, A') & \longrightarrow & P^0(G_S, A') & \longrightarrow & H^2(G_S, A)^\vee
\end{array}$$

The third row is a complex; the middle row and the columns are exact. A diagram chase shows that φ is an isomorphism and that the lower row is exact. This finishes the proof. □

In order to prove the exactness of the 9-term sequence, we apply (8.6.12) to A and A' and get the two horizontal exact sequences in the diagram (where we write G for G_S)

$$(*)\qquad \begin{array}{ccccccccccc}
0 & \longrightarrow & H^0(G,A) & \longrightarrow & P^0(G,A) & \longrightarrow & H^2(G,A')^\vee & \longrightarrow & H^1(G,A) & \longrightarrow & P^1(G,A) \\
 & & & & & & & & & \overset{\psi}{\swarrow} & \Big\downarrow \wr\, \Xi^1 \\
0 & \longleftarrow & H^0(G,A')^\vee & \longleftarrow & P^2(G,A) & \longleftarrow & H^2(G,A) & \longleftarrow & H^1(G,A')^\vee & \xleftarrow{\lambda^\vee} & P^1(G,A')^\vee .
\end{array}$$

We define the map ψ as the composition $\lambda^\vee \circ \Xi^1$. The pairings

$$\begin{array}{ccccccccc}
H^1(G_S,A) & \times & H^1(G_S,A') & \xrightarrow{\cup} & H^2(G_S,\mathcal{O}_S^\times) & & \\
\Big\downarrow \lambda & & \Big\downarrow \lambda' & & \Big\downarrow & & \\
P^1(G_S,A) & \times & P^1(G_S,A') & \xrightarrow{\cup} & H^2(G_S,I_S) & \xrightarrow{\oplus\, inv_\mathfrak{p}} & \bigoplus_{\mathfrak{p}\in S} \mathbb{Q}/\mathbb{Z} \\
 & & & & \Big\downarrow \iota & & \Big\downarrow \Sigma \\
 & & & & H^2(G_S,C_S) & \xrightarrow{inv} & \mathbb{Q}/\mathbb{Z}
\end{array}$$

show that im(λ) and im(λ') annihilate each other. Therefore the map ψ induces a 9-term complex which we denote by

$$PT(k,S,A),$$

and we know that it is exact at all places but the middle one, which remains to be shown. If the set S is finite, then all groups occurring are finite and the complex is exact if and only if the alternating product of the orders of the groups in the complex is equal to 1. For a finite G_S-module A, we define the second partial Euler-Poincaré characteristic by

$$\chi_2(G_S,A) = \frac{\#H^0(G_S,A)\cdot \#H^2(G_S,A)}{\#H^1(G_S,A)},$$

and put

$$\varphi(G_S,A) = \chi_2(G_S,A) / \prod_{\mathfrak{p}\in S_\infty} \frac{\#H^0(k_\mathfrak{p},A)}{\|\#A\|_\mathfrak{p}},$$

where for function fields the set S_∞ is empty by definition.

(8.6.13) Lemma. *Let S be a finite nonempty set of primes of the global field k containing S_∞ if k is a number field. Let A be a finite G_S-module of order $\#A \in \mathbb{N}(S)$. Then*

(i) $\varphi(G_S,A)\cdot\varphi(G_S,A') = 1$ *if and only if the complex $PT(k,S,A)$ is exact.*

(ii) $\varphi(G_S,-)$ *is multiplicative on short exact sequences of finite modules in $Mod_S(G_S)$.*

Proof: Using the product formula (7.3.5) and (7.3.1), we obtain

$$\begin{aligned}\prod_{\mathfrak{p}\in S_\infty}\frac{\#H^0(k_\mathfrak{p},A)\cdot\#H^0(k_\mathfrak{p},A')}{(\|\#A\|_\mathfrak{p})^2} &= \prod_{\mathfrak{p}\in S_\infty}\frac{\#\hat{H}^0(k_\mathfrak{p},A)}{\|\#A\|_\mathfrak{p}}\\ &= \prod_{\mathfrak{p}\in S\setminus S_\infty}\|\#A\|_\mathfrak{p}\prod_{\mathfrak{p}\in S_\infty}\#\hat{H}^0(k_\mathfrak{p},A)\\ &= \prod_{\mathfrak{p}\in S\setminus S_\infty}\|\#A\|_\mathfrak{p}\prod_{\mathfrak{p}\in S_\infty}\frac{\#\hat{H}^0(k_\mathfrak{p},A)\cdot\#\hat{H}^2(k_\mathfrak{p},A)}{\#\hat{H}^1(k_\mathfrak{p},A)}\\ &= \frac{\#P^0(G_S,A)\cdot\#P^2(G_S,A)}{\#P^1(G_S,A)}\end{aligned}$$

It follows that

$$\varphi(G_S,A)\cdot\varphi(G_S,A') = \chi_2(G_S,A)\cdot\chi_2(G_S,A')/\frac{\#P^0(G_S,A)\cdot\#P^2(G_S,A)}{\#P^1(G_S,A)},$$

thus showing (i). In order to prove (ii), let

$$0\longrightarrow A_1\longrightarrow A_2\longrightarrow A_3\longrightarrow 0$$

be an exact sequence of finite modules in $Mod_S(G_S)$. From the exact cohomology sequence and (8.6.10)(ii) we obtain

$$\begin{aligned}\frac{\chi_2(G_S,A_1)\chi_2(G_S,A_3)}{\chi_2(G_S,A_2)} &= \frac{\#P^3(G_S,A_2)}{\#P^3(G_S,A_1)\cdot\#P^3(G_S,A_3)}\\ &\quad\cdot\frac{\#P^4(G_S,A_1)\cdot\#P^4(G_S,A_3)}{\#P^4(G_S,A_2)}\cdot\#\delta P^4(G_S,A_3)\\ &= \#\delta P^4(G_S,A_3)\end{aligned}$$

because the Herbrand index of finite modules is equal to 1 by (1.7.6), and so $\#P^3(G_S,A)=\#P^4(G_S,A)$. For $\mathfrak{p}\in S_\infty$ the sequence

$$0\longrightarrow H^0(k_\mathfrak{p},A_1)\longrightarrow H^0(k_\mathfrak{p},A_2)\longrightarrow H^0(k_\mathfrak{p},A_3)\longrightarrow \delta H^0(k_\mathfrak{p},A_3)\longrightarrow 0$$

is exact and

$$\#\delta H^0(k_\mathfrak{p},A_3)=\#\delta\hat{H}^0(k_\mathfrak{p},A_3)=\#\delta H^4(k_\mathfrak{p},A_3).$$

Therefore

$$\frac{\chi_2(G_S,A_1)\cdot\chi_2(G_S,A_3)}{\chi_2(G_S,A_2)}=\prod_{\mathfrak{p}\in S_\infty(k)}\frac{\#H^0(k_\mathfrak{p},A_1)\cdot\#H^0(k_\mathfrak{p},A_3)}{\#H^0(k_\mathfrak{p},A_2)},$$

and using $\|\#A_2\|_\mathfrak{p}=\|\#A_1\|_\mathfrak{p}\cdot\|\#A_3\|_\mathfrak{p}$, this proves assertion (ii). □

Proof of (8.6.10)(i): We proceed with several reduction steps.

Step 1. $PT(k,S,A)$ is exact for $A=\mathbb{Z}/p\mathbb{Z}$ and $A=\mu_p$, $p\in\mathbb{N}(S)$.

Proof of step 1: As the Poitou-Tate sequence is self-dual, it suffices to deal with the case $A=\mu_p$. Using (8.3.4), we find the isomorphism

$$I_S/p \xrightarrow{\sim} C_S/p$$

and the exact sequence

$$0\longrightarrow \mu_p \longrightarrow {}_pI_S \longrightarrow {}_pC_S \longrightarrow 0\,.$$

Taking cohomology, we obtain the exact sequence

$$(1)\qquad H^1(G_S,\mu_p)\longrightarrow H^1(G_S,{}_pI_S)\longrightarrow H^1(G_S,{}_pC_S).$$

Since $H^1(G_S,C_S)=0$, we get from the exact sequences

$$0\to {}_pC_S\to C_S\to (C_S)^p\to 0 \quad\text{and}\quad 0\to (C_S)^p\to C_S\to C_S/p\to 0$$

the commutative and exact diagram

$$\begin{array}{ccccccc}
H^0(G_S,C_S) & \xrightarrow{p} & H^0(G_S,(C_S)^p) & \xrightarrow{\delta} & H^1(G_S,{}_pC_S) & \longrightarrow & 0\\
\| & & \downarrow & & \downarrow{\scriptstyle\vartheta} & & \\
H^0(G_S,C_S) & \xrightarrow{p} & H^0(G_S,C_S) & \longrightarrow & C_S(k)/C_S(k)^p & \longrightarrow & 0\\
 & & \downarrow & & & & \\
 & & H^0(G_S,C_S/p) & & & &
\end{array}$$

and therefore an exact sequence

$$(2)\qquad 0\longrightarrow H^1(G_S,{}_pC_S)\xrightarrow{\vartheta} C_S(k)/p\longrightarrow H^0(G_S,C_S/p).$$

Replacing C_S by I_S, we obtain an analogous sequence

$$(3)\qquad 0\longrightarrow H^1(G_S,{}_pI_S)\xrightarrow{\vartheta} I_S(k)/p\longrightarrow H^0(G_S,I_S/p),$$

and all three sequences (1)–(3) fit into the commutative and exact diagram

$$\begin{array}{ccccc}
H^1(G_S,\mu_p) & \longrightarrow & H^1(G_S,{}_pI_S) & \longrightarrow & H^1(G_S,{}_pC_S)\\
 & \searrow{\scriptstyle\tilde\vartheta} & \downarrow{\scriptstyle\vartheta} & & \downarrow{\scriptstyle\vartheta}\\
 & & I_S(k)/p & \longrightarrow & C_S(k)/p\\
 & & \downarrow & & \downarrow\\
 & & H^0(G_S,I_S/p) & \xrightarrow{\sim} & H^0(G_S,C_S/p).
\end{array}$$

We get an exact sequence

$$H^1(G_S,\mu_p)\xrightarrow{\tilde\vartheta} I_S(k)/p\longrightarrow C_S(k)/p\,.$$

By (7.2.13), the diagram

$$\begin{array}{ccc} H^1(G_S,\mu_p) & \xrightarrow{\tilde{\vartheta}} & I_S(k)/p \\ \Big\downarrow{\scriptstyle \lambda^1} & & \wr\Big\downarrow{\scriptstyle rec} \\ P^1(G_S,\mu_p) & \xrightarrow[\sim]{\Xi^1} & P^1(G_S,\mathbb{Z}/p\mathbb{Z})^\vee = \coprod\limits_{\mathfrak{p}\in S} G_{k_\mathfrak{p}}^{ab}/p\,. \end{array}$$

is anti-commutative. As $D_S(k)$ is divisible by (8.3.13), the reciprocity map induces an isomorphism

$$rec : C_S(k)/p \xrightarrow{\sim} G_S^{ab}/p\,.$$

We obtain the commutative diagram

$$\begin{array}{ccccc} H^1(G_S,\mu_p) & \xrightarrow{\tilde{\vartheta}} & I_S(k)/p & \longrightarrow & C_S(k)/p \\ & \searrow{\scriptstyle -\Xi^1\cdot\lambda^1} & {\scriptstyle rec}\Big\downarrow\wr & & {\scriptstyle rec}\Big\downarrow\wr \\ & & \coprod\limits_{\mathfrak{p}\in S} G_{k_\mathfrak{p}}^{ab}/p & \longrightarrow & G_S^{ab}/p \\ & & & \searrow{\scriptstyle (\lambda^1)^\vee} & \Big\| \\ & & & & H^1(G_S,\mathbb{Z}/p\mathbb{Z})^\vee. \end{array}$$

The sequence with the diagonal arrows is up to sign the central part of $PT(k,S,\mu_p)$, which is therefore exact.

Step 2. $PT(k,S,A)$ is exact if $PT(k,T,A)$ is exact for all finite subsets $T\subseteq S$.

Proof of step 2: We have seen that $\operatorname{im}\lambda'_S \subseteq (\operatorname{im}\lambda_S)^\perp$, where

$$\lambda_S\colon H^1(G_S,A)\longrightarrow P^1(G_S,A) \quad\text{and}\quad \lambda'_S\colon H^1(G_S,A')\longrightarrow P^1(G_S,A'),$$

and the orthogonal complement $^\perp$ is taken with respect to the pairing

$$P^1(G_S,A)\times P^1(G_S,A') \xrightarrow{\cup} H^2(G_S,I_S) \longrightarrow \bigoplus_{\mathfrak{p}\in S}\mathbb{Q}/\mathbb{Z} \xrightarrow{\Sigma} \mathbb{Q}/\mathbb{Z}.$$

Assume that we have proved

$$\operatorname{im}\lambda'_T = (\operatorname{im}\lambda_T)^\perp$$

for all finite subsets T of S with $\#A\in\mathbb{N}(T)$ (and $S_\infty\subseteq T$ in the number field case) for the corresponding localization maps λ'_T and λ_T. Let $\pi_T : P^1(G_S,A') \twoheadrightarrow P^1(G_T,A')$ be the canonical projection. Then the commutative diagram

$$\begin{array}{ccc} H^1(G_T,A') & \overset{inf}{\hookrightarrow} & H^1(G_S,A') \\ \Big\downarrow{\scriptstyle \lambda'_T} & & \Big\downarrow{\scriptstyle \lambda'_S} \\ P^1(G_T,A') & \overset{\pi_T}{\twoheadleftarrow} & P^1(G_S,A') \end{array}$$

shows that

$$\operatorname{im}\lambda'_T = \pi_T(\operatorname{im}\lambda'_S).$$

In order to prove the inclusion $(\operatorname{im}\lambda_S)^\perp \subseteq \operatorname{im}\lambda'_S$, let $x \in (\operatorname{im}\lambda_S)^\perp$. Then for all sufficiently large finite subsets $T \subseteq S$, we have $\pi_T(x) \in (\operatorname{im}\lambda_T)^\perp = \operatorname{im}\lambda'_T$, so that

$$x \in \bigcap_T \pi_T^{-1}(\operatorname{im}\lambda'_T) = \bigcap_T \pi_T^{-1}\pi_T(\operatorname{im}\lambda'_S) = \overline{\operatorname{im}\lambda'_S}.$$

Since λ_S is proper by (8.6.11), $\operatorname{im}\lambda_S$ is closed in $P^1(G_S, A)$ and analogously $\operatorname{im}\lambda'_S$ is closed in $P^1(G_S, A')$, and so $x \in \overline{\operatorname{im}\lambda'_S} = \operatorname{im}\lambda'_S$. We obtain

$$\operatorname{im}\lambda'_S = (\operatorname{im}\lambda_S)^\perp.$$

Hence, in order to prove the exactness of $PT(k, S, A)$, we may assume that S is finite.

Step 3. Let $0 \to A \to B \to C \to 0$ be an exact sequence of finite modules in $Mod_S(G_S)$. If $PT(k, S, A)$ and $PT(k, S, C)$ are exact, then $PT(k, S, B)$ is also exact.

Proof of step 3: Since we may assume that S is finite, this follows directly from (8.6.13).

Step 4: $PT(k, S, A)$ is exact if $pA = 0$ and A is trivialized by a finite p-extension of k inside k_S, where p is any prime number.

Proof of step 4: By (1.6.13), A has a composition series whose graded pieces are isomorphic to $\mathbb{Z}/p\mathbb{Z}$. Therefore the exactness of $PT(k, S, A)$ follows from steps 1 and 3.

Step 5: $PT(k, S, A)$ is exact for any finite $A \in Mod_S(G_S)$.

Proof of step 5: By step 3, we may assume that A is a simple module; in particular, $pA = 0$ for some prime number p. Let $K|k$ be a finite Galois subextension of $k_S|k$ which trivializes A. Let G_p be a p-Sylow subgroup of $G = G(K|k)$ and $K_p \subseteq K$ the subfield associated to G_p. By step 4, $PT(K_p, S, A)$ is exact. The composition of *cor* and *res* between the complexes $PT(k, S, A)$ and $PT(K_p, S, A)$ is multiplication by $[K_p : k]$, which is prime to p. Hence $PT(k, S, A)$ can be identified with a direct summand of the exact complex $PT(K_p, S, A)$ and is therefore exact. This finishes the proof. □

There is an alternative way to prove the exactness of the Poitou-Tate sequence which avoids the reduction step to finite S and counting the orders of all cohomology groups occurring. The following proof of step 3 uses some calculations with double complexes and is taken from [169] (dévissage). We start with the following

Claim: Let $(A^{\bullet\bullet}, d', d'')$ be a (commutative or anti-commutative) double complex of abelian groups. For $(i,j) \in \mathbb{Z} \times \mathbb{Z}$ we put

$$Z^{ij} = \ker(d' \colon A^{ij} \to A^{i+1,j}) \cap \ker(d'' \colon A^{ij} \to A^{i,j+1}),$$

$$B^{ij} = \operatorname{im}(d'' \circ d' \colon A^{i-1,j-1} \to A^{ij}) \quad \text{and} \quad H^{ij} = Z^{ij}/B^{ij}.$$

If for $(i,j) \in \mathbb{Z} \times \mathbb{Z}$ the groups $H^i(A^{\bullet,j})$ and $H^{i+1}(A^{\bullet,j})$ vanish, then we have an exact sequence

$$H^{ij} \longrightarrow H^j(A^{i,\bullet}) \longrightarrow H^{i+1,j} \xrightarrow{\ h\ } H^{i,j+1} \longrightarrow H^{j+1}(A^{i,\bullet}).$$

Proof: The map $H^j(A^{i,\bullet}) \to H^{i+1,j}$ is induced by d'. The map h is defined as follows. Let $x \in H^{i+1,j}$ and let $a \in A^{i+1,j}$ be a representing element. We choose $b \in A^{ij}$ with $d'(b) = a$ and define $h(x) \in H^{i,j+1}$ as the class of $d''(b)$. Our assumptions imply that h is well-defined. The verification of the exactness of the sequence is a straightforward diagram chase.

Now we build a double complex $A^{\bullet\bullet}$ in the following way. In column $i = -1 + 3k$, $k \in \mathbb{Z}$, we put $PT(k, S, A)$ shifted by $3k$ positions downwards. In column $i = 3k$, $k \in \mathbb{Z}$, we put $PT(k, S, B)$ shifted by $3k$ positions downwards and in column $i = 1 + 3k$, $k \in \mathbb{Z}$, we put $PT(k, S, C)$ shifted by $3k$ positions downwards. Let us look at a part of this unbounded double complex.

$$\begin{array}{ccccccccc}
 & \uparrow & & \uparrow & & \uparrow & & \uparrow & \\
\longrightarrow & H^1(A')^\vee & \longrightarrow & H^1(B')^\vee & \longrightarrow & H^1(C')^\vee & \longrightarrow & H^0(A')^\vee & \longrightarrow \\
 & \uparrow & & \uparrow & & \uparrow & & \uparrow & \\
\longrightarrow & P^1(A) & \longrightarrow & P^1(B) & \longrightarrow & P^1(C) & \longrightarrow & P^2(A) & \longrightarrow \\
 & \uparrow & & \uparrow & & \uparrow & & \uparrow & \\
\longrightarrow & H^1(A) & \longrightarrow & H^1(B) & \longrightarrow & H^1(C) & \longrightarrow & H^2(A) & \longrightarrow \\
 & \uparrow & & \uparrow & & \uparrow & & \uparrow & \\
\longrightarrow & H^2(A')^\vee & \longrightarrow & H^2(B')^\vee & \longrightarrow & H^2(C')^\vee & \longrightarrow & H^1(A')^\vee & \longrightarrow \\
 & \uparrow & & \uparrow & & \uparrow & & \uparrow &
\end{array}$$

The object $P^1(B)$ is situated at position $(0,0)$ and the double complex is periodic with period $(3,-3)$. The rows are the long exact sequences (truncated if k is not totally imaginary) attached to the short exact sequence $0 \to A \to B \to C \to 0$. The columns are exact, except the $i = 0$ column at $j = 0$, where we want to show the exactness.

Applying the claim, we obtain an exact sequence

$$H^{1,-1} \xrightarrow{h} H^{0,0} \longrightarrow H^0(A^{0,\bullet}) \longrightarrow H^{1,0} \xrightarrow{h'} H^{0,1}$$

and homomorphisms

$$H^{2,-2} \twoheadrightarrow H^{1,-1} \overset{h}{\hookrightarrow} H^{0,0} \twoheadrightarrow H^{-1,1}, \qquad H^{2,-1} \twoheadrightarrow H^{1,0} \overset{h'}{\twoheadrightarrow} H^{0,1} \twoheadrightarrow H^{-1,2}.$$

Since $H^{2,-1} = H^{-1,2}$ is a subquotient of $Ш^2(G_S, A)$, it is finite. Hence h' is an isomorphism. As $H^{1,-1}$ is a subquotient of $Ш^1(G_S, C)$, the group $H^{2,-2} = H^{-1,1}$ is finite, and so h is an isomorphism. The exact sequence above shows that $H^0(A^{0,\bullet}) = 0$, hence $PT(k, S, B)$ is exact.

For finitely generated modules we have the following statement which is weaker than (8.6.10)(i).

(8.6.14) Theorem. *Let S be a nonempty set of primes of a global field k and assume that $S \supseteq S_\infty$ if k is a number field. Let $A \in Mod_S(G_S)$. Then we have an exact sequence*

$$H^1(G_S, A) \to P^1(G_S, A) \to H^1(G_S, A')^\vee \to H^2(G_S, A) \to P^2(G_S, A)\,.$$

Proof: The existence and exactness of the partial sequence

$$P^1(G_S, A) \to H^1(G_S, A')^\vee \to H^2(G_S, A) \to P^2(G_S, A)$$

follows from (8.6.7). The existence of the complex

$$H^1(G_S, A) \to P^1(G_S, A) \to H^1(G_S, A')^\vee$$

follows in exactly the same way as for finite A, cf. the proof of (8.6.10)(i). We denote by A_I the cokernel of the map $sh^1 : H^1(G_S, I_S(A)) \to P^1(G_S, A')$. Recall that sh^1 is injective by (8.5.5). Denote by A_C the cokernel of the map $\alpha : H^1(G_S, C_S(A)) \to H^1(G_S, A)^\vee$ induced by the cup-product and the invariant map $H^2(G_S, C_S) \to \mathbb{Q}/\mathbb{Z}$. By (8.4.4), α is an isomorphism if A is torsion-free. The commutative diagram

$$\begin{array}{ccc} H^1(G_S, I_S(A)) & \longrightarrow & P^1(G_S, A') \\ \big\downarrow & & \big\downarrow \\ H^1(G_S, C_S(A)) & \longrightarrow & H^1(G_S, A)^\vee \end{array}$$

induces a homomorphism $A_I \to A_C$, and our next aim is to show that it is injective.

Let $T \subseteq A$ be the torsion submodule of A and let $F = A/T$. The statement of the theorem is true for $A = T$ by (8.6.10). Therefore we have the commutative exact diagram

$$
\begin{array}{ccccccc}
H^1(G_S,T') & \longrightarrow & H^1(G_S,I_S(T)) & \longrightarrow & H^1(G_S,C_S(T)) & \longrightarrow & H^2(G_S,T') \\
\| & & \Big\downarrow{\scriptstyle sh^1} & & \Big\downarrow{\scriptstyle \alpha_T} & & \| \\
H^1(G_S,T') & \longrightarrow & P^1(G_S,T') & \longrightarrow & H^1(G_S,T)^\vee & \longrightarrow & H^2(G_S,T') \\
 & & \Big\downarrow & & \Big\downarrow & & \\
 & & T_I & \longrightarrow & T_C & &
\end{array}
$$

We conclude that the maps $\alpha_T : H^1(G_S, C_S(T)) \to H^1(G_S, T)^\vee$ and $T_I \to T_C$ are injective. We proceed by considering the commutative and exact diagram

$$
\begin{array}{ccccccc}
H^1(G_S,I_S(F)) & \to & H^1(G_S,I_S(A)) & \to & H^1(G_S,I_S(T)) & \to & H^2(G_S,I_S(F)) \\
\Big\downarrow{\scriptstyle sh^1}\wr & & \Big\downarrow{\scriptstyle sh^1} & & \Big\downarrow{\scriptstyle sh^1} & & \Big\downarrow{\scriptstyle sh^2} \\
P^1(G_S,F') & \longrightarrow & P^1(G_S,A') & \longrightarrow & P^1(G_S,T') & \longrightarrow & P^2(G_S,F') \\
 & & \Big\downarrow & & \Big\downarrow & & \\
 & & A_I & \longrightarrow & T_I & &
\end{array}
$$

By (8.5.5) the map sh^2 on the right-hand side induces an isomorphism on the ℓ-torsion part for all prime numbers $\ell \mid \#T$. Therefore the map $A_I \to T_I$ is injective and the commutative diagram

$$
\begin{array}{ccc}
A_I & \hookrightarrow & T_I \\
\Big\downarrow & & \Big\downarrow \\
A_C & \longrightarrow & T_C
\end{array}
$$

shows that the map $A_I \to A_C$ is also injective. Next we consider the commutative and exact diagram

$$
\begin{array}{ccccccc}
H^0(G_S,C_S(T)) & \to & H^1(G_S,C_S(F)) & \to & H^1(G_S,C_S(A)) & \to & H^1(G_S,C_S(T)) \\
\Big\downarrow & & \Big\downarrow\wr & & \Big\downarrow{\scriptstyle \alpha_A} & & \Big\downarrow{\scriptstyle \alpha_T} \\
H^2(G_S,T)^\vee & \longrightarrow & H^1(G_S,F)^\vee & \longrightarrow & H^1(G_S,A)^\vee & \longrightarrow & H^1(G_S,T)^\vee.
\end{array}
$$

The left-hand vertical arrow is surjective by (8.4.5). We conclude that the map α_A is injective. Finally, consider the diagram

$$\begin{array}{ccccc}
H^1(G_S, A') & \longrightarrow & H^1(G_S, I_S(A)) & \longrightarrow & H^1(G_S, C_S(A)) \\
\| & & \downarrow & & \downarrow \\
H^1(G_S, A') & \longrightarrow & P^1(G_S, A') & \longrightarrow & H^1(G_S, A)^\vee \\
 & & \downarrow & & \downarrow \\
 & & A_I & \hookrightarrow & A_C .
\end{array}$$

All sequences but the middle row are exact, hence the middle row is also exact. As Ξ^1 is an isomorphism by (8.5.3), we obtain the statement of the theorem by dualizing the middle row. □

Theorem (8.6.10) was announced by *J. Tate* in [229], but he never published a proof. *G. Poitou* gave a proof in the case that S is the set of all places in [169], and we have used several of his arguments here. The extension to finitely generated modules in (8.6.14) is due to *T. Takahashi* [226].

§7. The Global Euler-Poincaré Characteristic

In this section we calculate the Euler-Poincaré characteristic of a finite module $A \in Mod_S(G_S)$, where S is finite. We start by proving an assertion concerning the Galois module structure of $\mathcal{O}_{K,S}^\times$ for a finite Galois extension $K|k$ of number fields.

(8.7.1) Lemma. *Let G be a finite group and $E|F$ an extension of fields. Let M_1 and M_2 be finite dimensional $F[G]$-modules such that $M_1 \otimes E$ and $M_2 \otimes E$ are isomorphic as $E[G]$-modules. Then M_1 and M_2 are $F[G]$-isomorphic.*

Proof: *) For finite dimensional $F[G]$-modules M_1 and M_2 consider the isomorphism

$$\mathrm{Hom}_{F[G]}(M_1, M_2) \otimes E \cong \mathrm{Hom}_{E[G]}(M_1 \otimes E, M_2 \otimes E)\,,$$

*) The essential part of the proof is taken from [7], chap.IV, §8, lemma.

which is induced from the canonical isomorphism $\mathrm{Hom}_F(M_1, M_2) \otimes E \cong \mathrm{Hom}_E(M_1 \otimes E, M_2 \otimes E)$, cf. [17], chap.II, §5, prop.7(ii), by taking G-invariants. If $M_1 \otimes E \cong M_2 \otimes E$, then, in particular, $\dim_F M_1 = \dim_F M_2$ and we can speak of the determinant of a homomorphism between M_1 and M_2 (by choosing bases of M_1 and M_2). Let η_i, $i = 1, \dots, m$, be an F-basis of $\mathrm{Hom}_{F[G]}(M_1, M_2)$. Then $\{\eta_i\}$ is also an E-basis of the vector space $\mathrm{Hom}_{E[G]}(M_1 \otimes E, M_2 \otimes E)$. Since $M_1 \otimes E$ and $M_2 \otimes E$ are isomorphic by assumption, there exist $a_i \in E$ such that $\det(\sum a_i\eta_i) \neq 0$. Let

$$f(t) = \det(\sum t_i\eta_i) \in F[t_1, \dots, t_m].$$

$f(t)$ is not the zero polynomial because $f(a_1, \dots, a_m) \neq 0$. Then there exists a finite extension F' of F and an element $b = (b_1, \dots, b_m) \in F'^m$ such that $f(b) \neq 0$ (if F is infinite, then we can take $F' = F$). It follows that $\sum b_i\eta_i$ is an $F'[G]$-isomorphism of $M_1 \otimes F'$ onto $M_2 \otimes F'$. Considered as an $F[G]$-homomorphism, this is an isomorphism $M_1^n \xrightarrow{\sim} M_2^n$, where $n = [F' : F]$. By the Krull-Schmidt theorem, see [32], §6, thm. 6.12, we obtain $M_1 \cong M_2$ as $F[G]$-modules. □

(8.7.2) Proposition. *Let $K|k$ be a finite Galois extension of number fields with Galois group $G = G(K|k)$. Let r_1 and r_2 be the number of real and complex primes of k, respectively, and let $r_1' \le r_1$ be the cardinality of the set S'_∞ of real primes of k which become complex in K. Finally, let $S \supseteq S_\infty$ be a finite set of primes of k and let $S^f = S \setminus S_\infty$. Then there are isomorphisms of $\mathbb{Q}[G]$-modules*

$$\begin{aligned} \mathbb{Q} \oplus \mathcal{O}_{K,S}^\times \otimes \mathbb{Q} &\cong \bigoplus_{\mathfrak{P} \in S(K)} \mathbb{Q} \\ &\cong \bigoplus_{\mathfrak{p} \in S(k)} \mathrm{Ind}_G^{G_\mathfrak{P}} \mathbb{Q} \\ &\cong \mathbb{Q}[G]^{r_2 + r_1 - r_1'} \oplus \bigoplus_{\mathfrak{p} \in (S'_\infty \cup S^f)(k)} \mathrm{Ind}_G^{G_\mathfrak{P}} \mathbb{Q}. \end{aligned}$$

Proof: Consider the G-invariant map

$$Lg = (\Delta, \log) : \mathbb{Z} \oplus \mathcal{O}_{K,S}^\times \longrightarrow \bigoplus_{\mathfrak{P} \in S(K)} \mathbb{R}$$

where Δ is the diagonal embedding of the trivial G-module $\mathbb{Z}$ and $\log(a) = (\log |a|_\mathfrak{P})_{\mathfrak{P} \in S(K)}$. Then by Dirichlet's unit theorem $\ker Lg = \mu(K)$ and the image of Lg is a G-lattice Γ of rank $s = \#S(K)$ in $\bigoplus_{\mathfrak{P} \in S(K)} \mathbb{R}$, cf. [160], chap.VI, (1.1). It follows that

$$\mathbb{R} \oplus \mathcal{O}_{K,S}^\times \otimes \mathbb{R} \cong \bigoplus_{\mathfrak{P} \in S(K)} \mathbb{R}$$

as $\mathbb{R}[G]$-modules, hence by (8.7.1) for $F = \mathbb{Q}$ and $E = \mathbb{R}$, we obtain the desired result. □

(8.7.3) Corollary. *Let $K|k$ be a finite extension of number fields with Galois group $G = G(K|k)$ and let p be a prime number not dividing the order of G. Then there exists a $\mathbb{Z}_p[G]$-isomorphism*

$$\mathbb{Z}_p \oplus \mathcal{O}_{K,S}^{\times} \otimes \mathbb{Z}_p \cong \mu_{p^\infty}(K) \oplus \bigoplus_{\mathfrak{P} \in S(K)} \mathbb{Z}_p$$

where $S \supseteq S_\infty$ is a finite set of primes and G acts on the sum on the right via its action on $S(K)$.

Proof: Consider the $\mathbb{Z}_p[G]$-lattices

$$\Gamma_1 = \mathbb{Z}_p \oplus (\mathcal{O}_{K,S}^{\times} \otimes \mathbb{Z}_p)/\mu_{p^\infty}(K) \quad \text{and} \quad \Gamma_2 = \bigoplus_{\mathfrak{P} \in S(K)} \mathbb{Z}_p \,.$$

Since the order of G is prime to p, every finitely generated, $\mathbb{Z}_p$-free $\mathbb{Z}_p[G]$-module is projective by (2.6.11). From the $\mathbb{Q}_p[G]$-isomorphism $\Gamma_1 \otimes \mathbb{Q}_p \cong \Gamma_2 \otimes \mathbb{Q}_p$ obtained in (8.7.2), it follows from representation theory (5.6.10)(ii) that $\Gamma_1 \cong \Gamma_2$. Therefore we obtain an exact G-invariant sequence

$$0 \longrightarrow \mu_{p^\infty}(K) \longrightarrow \mathbb{Z}_p \oplus \mathcal{O}_{K,S}^{\times} \otimes \mathbb{Z}_p \longrightarrow \bigoplus_{\mathfrak{P} \in S(K)} \mathbb{Z}_p \longrightarrow 0$$

which splits, since Γ_2 is $\mathbb{Z}_p[G]$-projective. □

Recall that the second partial Euler-Poincaré characteristic of a finite G_S-module A is defined by

$$\chi_2(G_S, A) = \frac{\#H^0(G_S, A) \cdot \#H^2(G_S, A)}{\#H^1(G_S, A)} \,.$$

(8.7.4) Global Euler-Poincaré Characteristic Formula. *Let S be a finite nonempty set of primes of the global field k containing S_∞ if k is a number field. Let A be a finite G_S-module of order $\#A \in \mathbb{N}(S)$. Then*

$$\chi_2(G_S, A) = \prod_{\mathfrak{p} \in S_\infty} \frac{\#H^0(k_\mathfrak{p}, A)}{\|\#A\|_\mathfrak{p}} = \prod_{\mathfrak{p} \in S_\infty} \frac{\#\hat{H}^0(k_\mathfrak{p}, A')}{\#H^0(k_\mathfrak{p}, A')}.$$

If k is a function field, this simply says $\chi_2(G_S, A) = 1$.

Remarks: 1. Recall that $\|\#A\|_\mathfrak{p}$ is $\#A$ if $\mathfrak{p}$ is real and $(\#A)^2$ if $\mathfrak{p}$ is complex.
2. The second equality in the statement (8.7.4) follows from (7.3.5).
3. If k is a function field or a totally imaginary number field and A a finite module in $Mod_S(G_S)$, then $\chi(G_S, A) = \chi_2(G_S, A)$ by (8.3.17) and (8.3.18).

Proof: First we prove the formula for the Euler-Poincaré characteristic for the G_S-module $A = \mu_p$, where $p \in \mathbb{N}(S)$. Let $K = k(\mu_p)$ and let $\overline{G} = G(K|k)$. We consider the Grothendieck group $K_0'(\mathbb{F}_p[\overline{G}])$ of finitely generated $\mathbb{F}_p[\overline{G}]$-modules M and we denote the corresponding class of M by $[M]$. We obtain

$$\begin{aligned}
&\text{(i)} & [H^0(k_S|K,\mu_p)] &= [\mu_p],\\
&\text{(ii)} & [H^1(k_S|K,\mu_p)] &= [\mathcal{O}_{K,S}^\times/p] + [{}_pCl_S(K)],\\
&\text{(iii)} & [H^2(k_S|K,\mu_p)] &= [Cl_S(K)/p] - [\mathbb{F}_p]\\
&&&\quad + [\bigoplus_{\mathfrak{p}\in S\setminus S_\infty(K)} \mathbb{F}_p] + [\bigoplus_{\mathfrak{p}\in S_\infty(K)} \hat{H}^0(G_\mathfrak{p}, \mathbb{Z}/p\mathbb{Z})],
\end{aligned}$$

where S_∞ is redundant in the function field case. (i) is obvious and (ii) follows from the exact Kummer sequence (8.3.4), which induces the exact sequence

$$0 \longrightarrow \mathcal{O}_{K,S}^\times/p \longrightarrow H^1(k_S|K,\mu_p) \longrightarrow {}_pCl_S(K) \longrightarrow 0$$

where we used (8.3.11)(ii). From (8.6.12) we obtain the exact sequence

$$\begin{aligned}
0 \to H^0(k_S|K,\mathbb{Z}/p\mathbb{Z}) \to P^0(k_S|K,\mathbb{Z}/p\mathbb{Z}) \to H^2(k_S|K,\mu_p)^\vee\\
\to Ш^1(G(k_S|K),\mathbb{Z}/p\mathbb{Z}) \to 0\,.
\end{aligned}$$

We have

$$Ш^1(G(k_S|K),\mathbb{Z}/p\mathbb{Z})^\vee = Cl_S(K)/p$$

by (8.6.3), and since

$$P^0(k_S|K,\mathbb{Z}/p\mathbb{Z}) = \bigoplus_{\mathfrak{p}\in S\setminus S_\infty(K)} \mathbb{F}_p \oplus \bigoplus_{\mathfrak{p}\in S_\infty(K)} \hat{H}^0(G_\mathfrak{p},\mathbb{Z}/p\mathbb{Z})\,,$$

we obtain (iii).

If k is a function field, then

$$[\mathcal{O}_{K,S}^\times/p \oplus \mathbb{F}_p] = [\bigoplus_{\mathfrak{p}\in S(K)} \mathbb{F}_p] + [\mu_p]$$

and we obtain

$$\sum_{i=0}^{2}(-1)^i[H^i(G_S,\mu_p)] = 0$$

since $[{}_p\,Cl_S(K)] = [Cl_S(K)/p]$. This proves the formula in the function field case for $A = \mu_p$.

We proceed with the proof for number fields in the case that $A = \mu_p$. From (8.7.3) we obtain

$$[\mathcal{O}_{K,S}^\times/p] = [\bigoplus_{\mathfrak{p}\in S(K)} \mathbb{F}_p] + [\mu_p] - [\mathbb{F}_p].$$

Combining this with (i)–(iii), we obtain

$$\sum_{i=0}^{2}(-1)^i[H^i(k_S|K,\mu_p)] = [\bigoplus_{\mathfrak{p}\in S_\infty(K)} \hat{H}^0(G_\mathfrak{p},\mathbb{Z}/p\mathbb{Z})] - [\bigoplus_{\mathfrak{p}\in S_\infty(K)} \mathbb{F}_p].$$

Now we consider the homomorphism

$$\Theta : K_0'(\mathbb{F}_p[\overline{G}]) \longrightarrow \mathbb{Z}\,,\ [M] \longmapsto \dim_{\mathbb{F}_p} M^{\overline{G}}$$

(observe that $\#\overline{G}$ is equal to 1 for $p = 2$ and is prime to p for $p \neq 2$, so that $M \mapsto M^{\overline{G}}$ is an exact functor). Obviously,

$$\chi_2(G_S, M) = p^{\Theta\left(\sum_{i=0}^{2}(-1)^i[H^i(G(k_S|K),M)]\right)}.$$

Therefore

$$\chi_2(G_S, \mu_p) = 1$$

for function fields and

$$\chi_2(G_S, \mu_p) = \prod_{\mathfrak{p}\in S_\infty(k)} \frac{\#\hat{H}^0(k_\mathfrak{p}, \mathbb{F}_p)}{\#H^0(k_\mathfrak{p}, \mathbb{F}_p)}$$

for number fields. This finishes the proof in the case that $A = \mu_p$.

Now we will prove the formula for the Euler-Poincaré characteristic for an arbitrary finite G_S-module whose order is in $\mathbb{N}(S)$. Recall that the function $\varphi(G_S, -)$ defined by

$$\begin{aligned}\varphi(G_S, A) &= \chi_2(G_S, A)/\textstyle\prod_{\mathfrak{p}\in S_\infty} \frac{\#H^0(k_\mathfrak{p},A)}{\|\#A\|_\mathfrak{p}}\\ &= \chi_2(G_S, A)/\textstyle\prod_{\mathfrak{p}\in S_\infty} \frac{\#\hat{H}^0(k_\mathfrak{p},A')}{\#H^0(k_\mathfrak{p},A')}\,.\end{aligned}$$

is multiplicative on short exact sequences, see (8.6.13)(ii). A second property of φ is

$$\varphi(G_S, \mathrm{Ind}_{G_S}^U A) = \varphi(U, A) \qquad \text{for} \quad U \subseteq G_S \text{ open}\,.$$

For the χ_2-part of φ, this follows from Shapiro's lemma. Furthermore,

$$\|\#\mathrm{Ind}_{G_S}^U A\|_\mathfrak{p} = \prod_{\mathfrak{P}|\mathfrak{p}} \|\#A\|_\mathfrak{P}\,,$$

because $\#\mathrm{Ind}_{G_S}^U A = (\#A)^{(G_S:U)}$, and

$$\#H^0(k_\mathfrak{p}, \mathrm{Ind}_{G_S}^U A) = \prod_{\mathfrak{P}|\mathfrak{p}} \#H^0(K_\mathfrak{P}, A)$$

for $\mathfrak{p} \in S_\infty(k)$, where K denotes the fixed field of U. The last equality is trivial for complex primes, and if $\mathfrak{p}$ is real, we have by I §5 ex. 5

$$H^0(G_\mathfrak{p}, \mathrm{Ind}_{G_S}^U A) = \prod_{\sigma\in G_\mathfrak{p}\backslash G_S/U} H^0(G_{\sigma\mathfrak{P}_0}, A^\sigma).$$

Thus the desired property of φ for induced modules is proved.

The reduction from the general statement (8.7.4) to the case $A = \mu_p$ now goes as follows. Let $K|k$ be a finite subextension of $k_S|k$ trivializing A and A' with Galois group $\overline{G} = G(K|k)$. Since φ is multiplicative, we may assume that $pA = 0$ for some prime number $p \in \mathbb{N}(S)$, and we consider φ as a homomorphism from the Grothendieck group $K_0'(\mathbb{F}_p[\bar{G}])$ of finitely generated $\mathbb{F}_p[\bar{G}]$-modules into $\mathbb{Q}_+^\times$:

$$\varphi : K_0'(\mathbb{F}_p[\bar{G}]) \longrightarrow \mathbb{Q}_+^\times .$$

We want to prove that φ is identical to 1. The group $\mathbb{Q}_+^\times$ is torsion-free. Thus, using lemma (7.3.4), we have to show that $\varphi(G_S, \mathrm{Ind}_{\bar{G}}^{\bar{C}} N) = \varphi(C, N) = 1$, where $\bar{C}$ is a cyclic subgroup of $\bar{G}$ of order prime to p, C is the open subgroup of G_S given by $\bar{C} = C/G(k_S|K)$ and N is a finitely generated $\mathbb{F}_p[\bar{C}]$-module. Hence we may assume that $K|k$ is a cyclic extension of degree prime to p trivializing A and A'. We have

$$H^i(k_S|k, A) \cong H^i(k_S|K, A)^{\bar{G}} \qquad \text{for all } i \geq 0.$$

Let

$$\chi' : K_0'(\mathbb{F}_p[\bar{G}]) \longrightarrow K_0'(\mathbb{F}_p[\bar{G}]),\ [M] \longmapsto \sum_{i=0}^{2} (-1)^i [H^i(k_S|K, M)],$$

and

$$\Theta : K_0'(\mathbb{F}_p[\bar{G}]) \longrightarrow \mathbb{Z},\ [M] \longmapsto \dim_{\mathbb{F}_p} M^{\bar{G}};$$

then

$$\chi_2(G_S, M) = p^{\Theta\chi'([M])}$$

(observe that $M \mapsto M^{\bar{G}}$ is an exact functor). *)

Claim: For a finite $\mathbb{F}_p[\bar{G}]$-module M, we have

$$\chi'([M]) = [M'^{\vee}]\, \chi'([\mu_p]).$$

Proof: The pairing

$$\mu_p \times \mathrm{Hom}(M', \mathbb{F}_p) \longrightarrow \mathrm{Hom}(M', \mu_p) = M, \quad (\zeta, f) \mapsto (x \mapsto \zeta^{f(x)}),$$

defines $\bar{G}$-isomorphisms via the cup-product

$$H^i(k_S|K, \mu_p) \otimes \mathrm{Hom}(M', \mathbb{F}_p) \xrightarrow{\sim} H^i(k_S|K, M)$$

(recall that μ_p and M are trivial $G(k_S|K)$-modules). This proves the claim.

Since $\chi'([\mu_p]) = 0$ for function fields, we obtain $\chi_2(G_S, A) = 1$ in this case.

For the rest of the proof, let k be a number field. For $\mathfrak{p} \in S_\infty(K)$ we define

$$\psi_{\mathfrak{p}}' : K_0'(\mathbb{F}_p[\bar{G}]) \longrightarrow K_0'(\mathbb{F}_p[\bar{G}]),\ [M] \longmapsto \sum_{\mathfrak{P}|\mathfrak{p}} [\hat{H}^0(K_{\mathfrak{P}}, M)] - [H^0(K_{\mathfrak{P}}, M')],$$

and

$$\psi' : K_0'(\mathbb{F}_p[\bar{G}]) \longrightarrow K_0'(\mathbb{F}_p[\bar{G}]),\ [M] \longmapsto \sum_{\mathfrak{p} \in S_\infty(k)} \psi_{\mathfrak{p}}'([M]).$$

*) Θ is not a *ring*-homomorphism with respect to the ring structure of $K_0'(\mathbb{F}_p[\bar{G}])$ given by the tensor product.

Obviously,

$$\psi'_{\mathfrak{p}}([M]) = [M']\,\psi'_{\mathfrak{p}}([\mu_p]).$$

Using the claim and the assertion $\chi'([\mu_p]) = \psi'([\mu_p])$ proven at the beginning of this part, we find for the finite G_S-module A that

$$\begin{aligned}\varphi(G_S, A) &= p^{\Theta\chi'([A]) - \Theta\psi'([A])} \\ &= \prod_{\mathfrak{p}\in S_\infty(k)} p^{\Theta([A'^{\vee}]\psi'_{\mathfrak{p}}([\mu_p]) - [A']\psi'_{\mathfrak{p}}([\mu_p]))}.\end{aligned}$$

It remains to prove that for $\mathfrak{p} \in S_\infty(k)$ and for an $\mathbb{F}_p[\overline{G}]$-module A,

$$\Theta([A^{\vee}]\psi'_{\mathfrak{p}}([\mu_p])) = \Theta([A]\psi'_{\mathfrak{p}}([\mu_p]))\,.$$

It is easy to see that

$$\psi'_{\mathfrak{p}}([\mu_p]) = \begin{cases} 0, & p = 2,\ \mathfrak{p} \text{ is real and remains real in } K, \\ -[\operatorname{Ind}_{\overline{G}}^{\overline{G}_{\mathfrak{P}}}\mathbb{F}_p], & \text{otherwise.}\end{cases}$$

This proves the equality above if $p = 2$ and $\mathfrak{p}$ is real and remains real in K. Therefore we assume that we are in one of the other cases. Then it follows from $\operatorname{Ind}_{\overline{G}}^{\overline{G}_{\mathfrak{P}}}\mathbb{F}_p \otimes A \cong \operatorname{Ind}_{\overline{G}}^{\overline{G}_{\mathfrak{P}}} A$ and

$$H^0(\overline{G}, \operatorname{Ind}_{\overline{G}}^{\overline{G}_{\mathfrak{P}}} A) = H^0(\overline{G}_{\mathfrak{P}}, A)$$

that

$$\begin{aligned}\Theta([A]\psi'_{\mathfrak{p}}([\mu_p])) &= \Theta(-[\operatorname{Ind}_{\overline{G}}^{\overline{G}_{\mathfrak{p}}} A]) \\ &= -\dim_{\mathbb{F}_p}(\operatorname{Ind}_{\overline{G}}^{\overline{G}_{\mathfrak{p}}} A)^{\overline{G}} \\ &= -\dim_{\mathbb{F}_p} A^{\overline{G}_{\mathfrak{P}}}.\end{aligned}$$

Since

$$\dim_{\mathbb{F}_p} A^{\overline{G}_{\mathfrak{P}}} = \dim_{\mathbb{F}_p} A_{\overline{G}_{\mathfrak{P}}} = \dim_{\mathbb{F}_p}(A^{\vee})^{\overline{G}_{\mathfrak{P}}}\,,$$

the equality holds in all cases. This finishes the proof. □

(8.7.5) Corollary. *Let p be a prime number and let S be a finite set of primes of the number field k containing S_∞ and all primes above p. If $p \neq 2$, then for all $j \in \mathbb{Z}$,*

$$\sum_{i=0}^{2}(-1)^i \dim_{\mathbb{F}_p} H^i(G_S, \mathbb{Z}/p\mathbb{Z}(j)) = \begin{cases} -r_1(k) - r_2(k) & \text{for } j \text{ odd}, \\ -r_2(k) & \text{for } j \text{ even},\end{cases}$$

and if $p = 2$, then

$$\sum_{i=0}^{2}(-1)^i \dim_{\mathbb{F}_2} H^i(G_S, \mathbb{Z}/2\mathbb{Z}) = -r_2(k)\,.$$

Here $r_1(k)$ and $r_2(k)$ denote the number of real and complex places of the number field k. In the function field case, this alternating sum of dimensions is zero.

Proof: If p is odd, then for a number field k and an archimedean prime $\mathfrak{p}$, the valuation $\|p\|_\mathfrak{p}$ is equal to p resp. p^2 if $\mathfrak{p}$ is real resp. complex, and the order of $H^0(k_\mathfrak{p}, \mathbb{Z}/p\mathbb{Z}(j))$ is 1 for $\mathfrak{p}$ real, j odd, and p in all other cases. In the case $p = 2$, observe that all twists are equal, and we obtain the result as above. □

(8.7.6) Corollary. *With the notation of* (8.7.5), *the following equalities hold for* $p \geq 2$ *and all* $j \in \mathbb{Z}$:

$$\sum_{i=0}^{2}(-1)^i \operatorname{rank}_{\mathbb{Z}_p} H_i(G_S, \mathbb{Z}_p(j)) = \begin{cases} -r_1(k) - r_2(k) & \text{for } j \text{ odd},\\ -r_2(k) & \text{for } j \text{ even}, \end{cases}$$

and

$$\sum_{i=0}^{2}(-1)^i \operatorname{rank}_{\mathbb{Z}_p} H^i_{cts}(G_S, \mathbb{Z}_p(j)) = \begin{cases} -r_1(k) - r_2(k) & \text{for } j \text{ odd},\\ -r_2(k) & \text{for } j \text{ even}. \end{cases}$$

Proof: The second equality follows from the first using (2.7.12), and with exactly the same arguments as in the proof of (7.3.8) (replacing G_k by G_S), we obtain

$$\begin{aligned}\sum_{i=0}^{2}(-1)^i \operatorname{rank}_{\mathbb{Z}_p} H_i(G_S, \mathbb{Z}_p(j)) &= \sum_{i=0}^{2}(-1)^i \dim_{\mathbb{F}_p} H^i(G_S, \mathbb{Z}/p\mathbb{Z}(-j))\\ &\quad -\dim_{\mathbb{F}_p} {}_pH_2(G_S, \mathbb{Z}_p(j)).\end{aligned}$$

Now the corollary follows from (8.7.5) and the following lemma. □

(8.7.7) Lemma. *Let k be a global field, let $p \neq \operatorname{char}(k)$ be a prime number and let S be a finite nonempty set of primes of k containing S_∞ and all primes above p if k is a number field.*

If p is odd, then for all $j \in \mathbb{Z}$, the homology group $H_2(G_S, \mathbb{Z}_p(j))$ is $\mathbb{Z}_p$-torsion-free, and if $p = 2$, then

$$\operatorname{tor}(H_2(G_S, \mathbb{Z}_2(j))) = \begin{cases} (\mathbb{Z}/2\mathbb{Z})^{r_1(k)} & \text{for } j \text{ odd},\\ 0 & \text{for } j \text{ even}. \end{cases}$$

Proof: We prove the dual statement, i.e. $H^2(G_S, \mathbb{Q}_p/\mathbb{Z}_p(-j))$ is p-divisible if p is odd or j is even, and its co-torsion is equal to $(\mathbb{Z}/2\mathbb{Z})^{r_1(k)}$ otherwise. This is clear for function fields by (8.3.17) and in the number field case for $p \neq 2$ since $cd_p\, G_S \leq 2$ in these cases. We have the commutative diagram

$$\begin{array}{ccc} H^3(G_S, \mathbb{Z}/p^m\mathbb{Z}(-j)) & \twoheadrightarrow & {}_{p^m}H^3(G_S, \mathbb{Q}_p/\mathbb{Z}_p(-j)) \\ \downarrow \wr & & \downarrow \wr \\ \prod\limits_{\mathfrak{p}\in S_\infty(k)} H^3(k_\mathfrak{p}, \mathbb{Z}/p^m\mathbb{Z}(-j)) & \twoheadrightarrow & \prod\limits_{\mathfrak{p}\in S_\infty(k)} {}_{p^m}H^3(k_\mathfrak{p}, \mathbb{Q}_p/\mathbb{Z}_p(-j)) \end{array}$$

where the vertical maps are isomorphisms by (8.6.10)(ii). Thus

$$H^2(G_S, \mathbb{Q}_p/\mathbb{Z}_p(-j))/p^m \xrightarrow{\sim} \prod_{\mathfrak{p}\in S_\infty(k)} H^2(k_\mathfrak{p}, \mathbb{Q}_p/\mathbb{Z}_p(-j))/p^m .$$

Since we have

$$H^2(k_\mathfrak{p}, \mathbb{Q}_2/\mathbb{Z}_2(-j)) \cong \hat{H}^0(k_\mathfrak{p}, \mathbb{Q}_2/\mathbb{Z}_2(-j)) \cong \begin{cases} \mathbb{Z}/2\mathbb{Z} & \text{for } j \text{ odd}, \\ 0 & \text{for } j \text{ even}, \end{cases}$$

for a real prime $\mathfrak{p} \in S_\infty(k)$, this yields the lemma. □

A very useful application of the Poitou-Tate duality theorem and the global Euler characteristic formula is a duality statement for certain subgroups of $H^1(G_k, A)$ which are defined by local conditions. In particular, this is important for the theory of elliptic curves where one considers the so-called *Selmer group* of an elliptic curve.

(8.7.8) Definition. *Let k be a global field and A a finite G_k-module of order prime to the characteristic of k. A* **collection $\mathcal{L}$ of local conditions** *for A is a family $\mathcal{L}_\mathfrak{p}$ of subgroups of $H^1(G_\mathfrak{p}, A)$ such that*

$$\mathcal{L}_\mathfrak{p} = H^1(G_\mathfrak{p}/T_\mathfrak{p}, A^{T_\mathfrak{p}}) = \ker(H^1(G_\mathfrak{p}, A) \to H^1(T_\mathfrak{p}, A))$$

for almost all $\mathfrak{p}$, where $T_\mathfrak{p}$ denotes the inertia subgroup of $G_\mathfrak{p}$.

If $\mathcal{L}$ is a collection of local conditions for A, then $\mathcal{L}^D$ is a collection of local conditions for the G_k-module $A' = \mathrm{Hom}(A, \mu)$ defined by

$\mathcal{L}_\mathfrak{p}^D :=$ *the orthogonal complement of $\mathcal{L}_\mathfrak{p}$ with respect to the pairing* $H^1(G_\mathfrak{p}, A) \times H^1(G_\mathfrak{p}, A') \xrightarrow{\cup} \mathbb{Q}/\mathbb{Z}$.

The corresponding global groups are defined by

$$H^1_\mathcal{L}(k, A) := \ker\Big(H^1(G_k, A) \longrightarrow \prod_\mathfrak{p} H^1(G_\mathfrak{p}, A)/\mathcal{L}_\mathfrak{p}\Big),$$

$$H^1_{\mathcal{L}^D}(k, A') := \ker\Big(H^1(G_k, A') \longrightarrow \prod_\mathfrak{p} H^1(G_\mathfrak{p}, A')/\mathcal{L}_\mathfrak{p}^D\Big).$$

If $\mathcal{L}$ is a collection of local conditions for the finite G_k-module A, then we choose a finite nonempty set S of primes of k containing all primes $\mathfrak{p}$ such that $A^{T_\mathfrak{p}} \neq A$ or $\mathcal{L}_\mathfrak{p} \neq H^1(G_\mathfrak{p}/T_\mathfrak{p}, A^{T_\mathfrak{p}})$, and S_∞ and all divisors of $\#A$ if k is number field. Hence A and A' are G_S-modules and we have

$$H^1_{\mathcal{L}}(k,A) = \ker\Big(H^1(G_S,A) \longrightarrow \prod_{\mathfrak{p}\in S} H^1(G_\mathfrak{p},A)/\mathcal{L}_\mathfrak{p}\Big),$$

$$H^1_{\mathcal{L}^D}(k,A') = \ker\Big(H^1(G_S,A') \longrightarrow \prod_{\mathfrak{p}\in S} H^1(G_\mathfrak{p},A')/\mathcal{L}^D_\mathfrak{p}\Big).$$

The following theorem is due to A. WILES.

(8.7.9) Theorem. *Let k be a global field and A a finite G_k-module of order prime to* char(k), *and let $\mathcal{L}$ be a collection of local conditions for A. Then*

$$\frac{\#H^1_{\mathcal{L}}(k,A)}{\#H^1_{\mathcal{L}^D}(k,A')} = \frac{\#H^0(k,A)}{\#H^0(k,A')} \cdot \prod_{\mathfrak{p}} \frac{\#\mathcal{L}_\mathfrak{p}}{\#H^0(k_\mathfrak{p},A)} .$$

Proof: Let S be a finite set of primes of k chosen as above. Recall, by convention we have $S_\infty = \varnothing$ in the function field case. From the commutative exact diagram

$$\begin{array}{ccccccc}
 & & \prod_{\mathfrak{p}\in S} H^1(k_\mathfrak{p},A)/\mathcal{L}_\mathfrak{p} & = & \prod_{\mathfrak{p}\in S} H^1(k_\mathfrak{p},A)/\mathcal{L}_\mathfrak{p} & & H^1_{\mathcal{L}^D}(k,A')^\vee \\
 & & \uparrow & & \twoheaduparrow & & \twoheaduparrow \\
Ш^1(G_S,A) & \hookrightarrow & H^1(G_S,A) & \longrightarrow & \prod_{\mathfrak{p}\in S} H^1(k_\mathfrak{p},A) & \longrightarrow & H^1(G_S,A')^\vee \\
 & & \uparrow & & \uparrow & & \uparrow \\
 & & H^1_{\mathcal{L}}(k,A) & & \prod_{\mathfrak{p}\in S} \mathcal{L}_\mathfrak{p} & \xrightarrow{\ \sim\ } & \prod_{\mathfrak{p}\in S} (H^1(k_\mathfrak{p},A')/\mathcal{L}^D_\mathfrak{p})^\vee,
\end{array}$$

we obtain the exact sequence

$$0 \to Ш^1(G_S,A) \to H^1_{\mathcal{L}}(k,A) \to \prod_{\mathfrak{p}\in S} \mathcal{L}_\mathfrak{p} \to H^1(G_S,A')^\vee \to H^1_{\mathcal{L}^D}(k,A')^\vee \to 0,$$

and so the exact sequence

$$\begin{aligned} 0 \to H^0(G_S,A) \to P^0(G_S,A) \to H^2(G_S,A')^\vee \\ \to H^1_{\mathcal{L}}(k,A) \to \prod_{\mathfrak{p}\in S} \mathcal{L}_\mathfrak{p} \to H^1(G_S,A')^\vee \to H^1_{\mathcal{L}^D}(k,A')^\vee \to 0. \end{aligned}$$

Recall that in the number field case, for $\mathfrak{p} \in S_\infty$, the group $H^0(G_\mathfrak{p}, A)$ is replaced by $\hat{H}^0(G_\mathfrak{p}, A)$ in the term $P^0(G_S, A)$. Using (8.7.4), we get

$$\frac{\#H^1_{\mathcal{L}}(k,A)}{\#H^1_{\mathcal{L}^D}(k,A')} = \frac{\#H^0(G_S,A)}{\#H^0(G_S,A')} \cdot \chi_2(G_S,A') \cdot \prod_{\mathfrak{p}\in S_\infty} \frac{\#\mathcal{L}_\mathfrak{p}}{\#\hat{H}^0(k_\mathfrak{p},A)} \cdot \prod_{\mathfrak{p}\in S\backslash S_\infty} \frac{\#\mathcal{L}_\mathfrak{p}}{\#H^0(k_\mathfrak{p},A)}$$

$$= \frac{\#H^0(k,A)}{\#H^0(k,A')} \cdot \prod_{\mathfrak{p}\in S_\infty} \frac{\#\mathcal{L}_\mathfrak{p}}{\#H^0(k_\mathfrak{p},A)} \cdot \prod_{\mathfrak{p}\in S\backslash S_\infty} \frac{\#\mathcal{L}_\mathfrak{p}}{\#H^0(k_\mathfrak{p},A)}\ .$$

But this is the desired result, since for $\mathfrak{p} \notin S$ the group $\mathcal{L}_\mathfrak{p}$ is equal to $H^1(G_\mathfrak{p}/T_\mathfrak{p}, A^{T_\mathfrak{p}})$ and for all $\mathfrak{p}$ we have $\#H^1(G_\mathfrak{p}/T_\mathfrak{p}, A^{T_\mathfrak{p}}) = \#H^0(G_\mathfrak{p}, A)$ because of the exact sequence

$$0 \longrightarrow H^0(G_\mathfrak{p},A) \longrightarrow A^{T_\mathfrak{p}} \xrightarrow{1-\mathrm{Frob}_\mathfrak{p}} A^{T_\mathfrak{p}} \longrightarrow H^1(G_\mathfrak{p}/T_\mathfrak{p}, A^{T_\mathfrak{p}}) \longrightarrow 0\,. \qquad \square$$

§8. Duality for Unramified and Tamely Ramified Extensions

In this section we prove duality theorems for unramified and tamely ramified extensions of a global field.

Let us fix some notation. Let k be a global field and let S and T be sets of primes of k. Further, let $\mathfrak{c}$ be a full class of finite groups. We set

k_S^T	the maximal extension of k unramified outside S in which all primes of T are completely decomposed,
$k_S^T(\mathfrak{c})$	the maximal $\mathfrak{c}$-extension of k inside k_S^T.

From now on we assume that T is nonempty and contains S_∞ if k is a number field. Furthermore, we assume that $S \cap T = \varnothing$; in particular, S contains only nonarchimedean primes. We set

$k_S^{T,tr}$	the maximal extension of k unramified outside S in which all primes of T are completely decomposed and all primes of S are at most tamely ramified,
$k_S^{T,tr}(\mathfrak{c})$	the maximal $\mathfrak{c}$-extension of k inside $k_S^{T,tr}$.

Remarks: 1. We have $k_\varnothing^T = k_\varnothing^{T,tr}$ and the field $k_\varnothing^\varnothing$ is the maximal unramified extension of k. If k is a number field, we have $k_\varnothing^\varnothing = k_\varnothing^{S_\infty}$.

2. Let k be a number field and let S be a set of finite primes of k such that $S \cap S_p = \varnothing$ for all prime numbers p with $\mathbb{Z}/p\mathbb{Z} \in \mathfrak{c}$. Then a $\mathfrak{c}$-extension is automatically tamely ramified at the primes in S, and we have $k_S^{S_\infty,tr}(\mathfrak{c}) = k_S(\mathfrak{c})$.

For any extension K of k inside $k_S^{T,tr}$ we put

$$E_{K,T}^S = \ker\Big(\mathcal{O}_{K,T}^\times \longrightarrow \prod_{\mathfrak{p}\in S(K)} U_{K,\mathfrak{p}}/U_{K,\mathfrak{p}}^1\Big),$$

i.e. $E_{K,T}^S$ is the group of T-units which are principal units at all places of S. If $T = S_\infty$, we write $E_K^S = E_{K,S_\infty}^S$.

The main result of this section is the following theorem (8.8.1). It generalizes a duality theorem for unramified extensions contained in [257] and a duality theorem for tamely ramified extensions obtained by *G. Wiese* in [248].

(8.8.1) Theorem. *Let k be a global field and let T be a nonempty set of primes of k which contains S_∞ if k is a number field. Let S be a finite set of primes disjoint from T and let $\mathfrak{c}$ be any full class of finite groups. Let A be a discrete $G(k_S^{T,tr}(\mathfrak{c})|k)$-module which is finitely generated and free as a $\mathbb{Z}$-module. Then there are canonical topological isomorphisms*

$$\hat{H}^i(G(k_S^{T,tr}(\mathfrak{c})|k), \mathrm{Hom}(A, E^S_{k_S^{T,tr}(\mathfrak{c}),T})) \cong \hat{H}^{3-i}(G(k_S^{T,tr}(\mathfrak{c})|k), A)^\vee$$

for all $i \in \mathbb{Z}$.

Before proving the theorem, we deduce some corollaries.

(8.8.2) Corollary. *Let k be a number field and let T be a set of primes of k containing S_∞. For the maximal unramified extension $k_\varnothing^T$ in which all primes of T are completely decomposed there are natural isomorphisms*

$$\hat{H}^i(G(k_\varnothing^T|k), \mathcal{O}^\times_{k_\varnothing^T,T}) \cong \hat{H}^{3-i}(G(k_\varnothing^T|k), \mathbb{Z})^\vee$$

for all $i \in \mathbb{Z}$. In particular, $H^2(G(k_\varnothing^T|k), \mathcal{O}^\times_{k_\varnothing^T,T}) = 0$.

Proof: This follows from (8.8.1) by setting $\mathfrak{c}$ to be the class of all finite groups, $S = \varnothing$ and $A = \mathbb{Z}$. □

(8.8.3) Corollary. *Let k be a number field and let S be a finite set of nonarchimedean primes of k. Let p be a prime number such that $S \cap S_p = \varnothing$. Then there are natural isomorphisms for all $i \in \mathbb{Z}$*

$$\hat{H}^i(G(k_S(p)|k), E^S_{k_S(p)}) \cong \hat{H}^{3-i}(G(k_S(p)|k), \mathbb{Z})^\vee.$$

In particular, $H^2(G(k_S(p)|k), E^S_{k_S(p)}) = 0$.

Proof: By remark 2 above, we have $k_S^{S_\infty,tr}(p) = k_S(p)$. Therefore, applying (8.8.1) to the class of finite p-groups with $T = S_\infty$ and $A = \mathbb{Z}$, we obtain the result. □

In order to prove (8.8.1), we introduce some further objects. Let K be a finite extension of k inside $k_S^{T,tr}$. We consider the group

$$U^S_{K,T} = \prod_{\mathfrak{p} \notin (S \cup T)} U_{K,\mathfrak{p}} \times \prod_{\mathfrak{p} \in T} K_\mathfrak{p}^\times \times \prod_{\mathfrak{p} \in S} U^1_{K,\mathfrak{p}},$$

which is a subgroup of the idèle group I_K. Let $Cl^S_T(K)$ be the cokernel of the composite map $U^S_{K,T} \longrightarrow I_K \longrightarrow C_K$. The notation is motivated by the identification $Cl^\varnothing_T(K) = Cl_T(K)$. The reciprocity homomorphism of global class field theory induces an isomorphism of finite abelian groups

$$Cl^S_T(K) \xrightarrow{\sim} G(k_S^{T,tr}|K)^{ab}.$$

For an infinite extension K of k in $k_S^{T,tr}$ we define the above groups as the direct limit over the finite layers. For any K, we obtain an exact sequence

$$0 \longrightarrow E^S_{K,T} \longrightarrow U^S_{K,T} \longrightarrow C_K \longrightarrow Cl^S_T(K) \longrightarrow 0.$$

We have $Cl^S_T(k_S^{T,tr}(\mathfrak{c}))(\mathfrak{c}) = 0$ by the principal ideal theorem.

In the following we use the notation $\hat{H}^i(K|k, -)$ for $\hat{H}^i(G(K|k), -)$. We start with a lemma.

(8.8.4) Lemma. *We have a natural isomorphism*

$$H^1(k_S^{T,tr}(\mathfrak{c})|k, E^S_{k_S^{T,tr}(\mathfrak{c}),T}) \cong Cl^S_T(k)(\mathfrak{c}).$$

Proof: Let $\mathbb{Z}_{(\mathfrak{c})}$ be the subring of $\mathbb{Q}$ obtained from $\mathbb{Z}$ by inverting all primes p with $\mathbb{Z}/p\mathbb{Z} \notin \mathfrak{c}$. For an abelian torsion group A, we have a natural isomorphism $A(\mathfrak{c}) \xrightarrow{\sim} A \otimes \mathbb{Z}_{(\mathfrak{c})}$. Since the extension $k_S^{T,tr}(\mathfrak{c})|k$ is unramified outside S, tamely ramified at S and completely decomposed at T, the $G(k_S^{T,tr}(\mathfrak{c})|k)$-module

$U^S_{k_S^{T,tr}(\mathfrak{c}),T}$ is cohomologically trivial; see (7.1.2). Since $Cl^S_T(k_S^{T,tr}(\mathfrak{c}))(\mathfrak{c}) = 0$, we have an exact sequence

$$0 \to E^S_{k_S^{T,tr}(\mathfrak{c}),T} \otimes \mathbb{Z}_{(\mathfrak{c})} \to U^S_{k_S^{T,tr}(\mathfrak{c}),T} \otimes \mathbb{Z}_{(\mathfrak{c})} \to C_{k_S^{T,tr}(\mathfrak{c})} \otimes \mathbb{Z}_{(\mathfrak{c})} \to 0\ ,$$

which induces a natural isomorphism

$$H^1(k_S^{T,tr}(\mathfrak{c})|k, E^S_{k_S^{T,tr}(\mathfrak{c}),T}) \xrightarrow{\sim} \operatorname{coker}\left(U^S_{k,T} \to C_k\right) \otimes \mathbb{Z}_{(\mathfrak{c})} \cong Cl^S_T(k)(\mathfrak{c}).$$

This shows the statement. □

Proof of (8.8.1): Applying (8.4.1) to $K = k_S^{T,tr}(\mathfrak{c})$, we obtain natural isomorphisms

$$\hat{H}^i(k_S^{T,tr}(\mathfrak{c})|k, \operatorname{Hom}(A, C_{k_S^{T,tr}(\mathfrak{c})})) \cong \hat{H}^{2-i}(k_S^{T,tr}(\mathfrak{c})|k, A)^\vee, \quad i \in \mathbb{Z}.$$

Let $K|k$ be a finite Galois extension inside $k_S^{T,tr}(\mathfrak{c})$ which trivializes A. We consider the exact sequence

$$(*) \qquad 0 \longrightarrow E^S_{K,T} \longrightarrow U^S_{K,T} \longrightarrow C_K \longrightarrow Cl^S_T(K) \longrightarrow 0.$$

The association $K \mapsto Cl^S_T(K)$ is merely a $G(k_S^{T,tr}(\mathfrak{c})|k)$-modulation but not a module, as we are lacking Galois descent; see (1.5.10) and page 56. However, for each fixed K, the group $G(K|k)$ acts on $Cl^S_T(K)$, and we will consider $Cl^S_T(K)$ as a $G(K|k)$-module. Thus $(*)$ is an exact sequence of discrete $G(K|k)$-modules. The group $U^S_{K,T}$ is a cohomologically trivial $G(K|k)$-module. Applying $\operatorname{Hom}(A, -)$ and taking cohomology, we therefore obtain isomorphisms

$$\hat{H}^{i+1}(K|k, \operatorname{Hom}(A, E^S_{K,T})) \cong \hat{H}^i(K|k, \operatorname{Hom}(A, X(K))),$$

where $X(K)$ denotes the kernel of the surjection $C_K \twoheadrightarrow Cl^S_T(K)$, and a long exact sequence

$$\cdots \longrightarrow \hat{H}^i(K|k, \operatorname{Hom}(A, X(K))) \longrightarrow \hat{H}^i(K|k, \operatorname{Hom}(A, C_K))$$
$$\longrightarrow \hat{H}^i(K|k, \operatorname{Hom}(A, Cl^S_T(K))) \longrightarrow \cdots.$$

As the cohomology groups of $\operatorname{Hom}(A, C_K)$ and of $\operatorname{Hom}(A, Cl^S_T(K))$ are finite, the same is true for the cohomology groups of $\operatorname{Hom}(A, X(K))$.

Now let K' be another finite Galois extension of k inside $k_S^{T,tr}(\mathfrak{c})$ which contains K. We consider $Cl^S_T(K)$ as a $G(K'|k)$-module via the projection $G(K'|k) \twoheadrightarrow G(K|k)$. Then the natural maps $Cl^S_T(K) \xrightarrow{can} Cl^S_T(K')$ and $Cl^S_T(K') \xrightarrow{norm} Cl^S_T(K)$ are homomorphisms of $G(K'|k)$-modules and we obtain associated homomorphisms

$$inf: \hat{H}^i(K|k, \operatorname{Hom}(A, Cl^S_T(K))) \to \hat{H}^i(K'|k, \operatorname{Hom}(A, Cl^S_T(K'))),\ i \ge 1,$$

$$def: \hat{H}^i(K'|k, \operatorname{Hom}(A, Cl^S_T(K'))) \to \hat{H}^i(K|k, \operatorname{Hom}(A, Cl^S_T(K))),\ i \le 0.$$

Similarly, we obtain associated maps between the cohomology groups of $\mathrm{Hom}(A, X(K'))$ and $\mathrm{Hom}(A, X(K))$.

Let K' be the maximal abelian extension of K in $k_S^{T,tr}(\mathfrak{c})$. By class field theory, K' is a finite extension of K and we have a natural isomorphism $rec : Cl_T^S(K)(\mathfrak{c}) \xrightarrow{\sim} G(K'|K)$. The commutative diagram

$$\begin{array}{ccc} Cl_T^S(K')(\mathfrak{c}) & \xrightarrow{norm} & Cl_T^S(K)(\mathfrak{c}) \\ {\scriptstyle rec}\downarrow\wr & & {\scriptstyle rec}\downarrow\wr \\ G(k_S^{T,tr}(\mathfrak{c})|K')^{ab} & \xrightarrow{can} & G(k_S^{T,tr}(\mathfrak{c})|K)^{ab} \end{array}$$

shows, since can is the zero map, that

$$Cl_T^S(K')(\mathfrak{c}) \xrightarrow{norm} Cl_T^S(K)(\mathfrak{c})$$

is the zero map. Noting that

$$\hat{H}^i(K|k, \mathrm{Hom}(A, Cl_T^S(K))) = \hat{H}^i(K|k, \mathrm{Hom}(A, Cl_T^S(K)(\mathfrak{c}))),$$

we obtain

$$\varprojlim_K \hat{H}^i(K|k, \mathrm{Hom}(A, Cl_T^S(K))) = 0 \quad \text{for } i \le 0.$$

Furthermore,

$$\varinjlim_K \hat{H}^i(K|k, \mathrm{Hom}(A, Cl_T^S(K))) = 0 \quad \text{for } i \ge 1,$$

since

$$Cl_T^S(K)(\mathfrak{c}) \xrightarrow{can} Cl_T^S(K')(\mathfrak{c})$$

is the zero map by the principal ideal theorem. Passing to the inverse, resp. direct, limit, we obtain isomorphisms

$$\varprojlim_{K, def} \hat{H}^i(K|k, \mathrm{Hom}(A, X(K))) \cong \hat{H}^i(k_S^{T,tr}(\mathfrak{c})|k, \mathrm{Hom}(A, C_{k_S^{T,tr}(\mathfrak{c})})),\ i \le 0,$$

$$\varinjlim_{K, inf} \hat{H}^i(K|k, \mathrm{Hom}(A, X(K))) \cong \hat{H}^i(k_S^{T,tr}(\mathfrak{c})|k, \mathrm{Hom}(A, C_{k_S^{T,tr}(\mathfrak{c})})),\ i \ge 1.$$

Therefore we obtain canonical isomorphisms for $i \ge 2$:

$$\begin{aligned} \hat{H}^i(k_S^{T,tr}(\mathfrak{c})|k, \mathrm{Hom}(A, E_T^S)) &\cong \varinjlim_K H^i(K|k, \mathrm{Hom}(A, E_{K,T}^S)) \\ &\cong \varinjlim_K \hat{H}^{i-1}(K|k, \mathrm{Hom}(A, X(K))) \\ &\cong \hat{H}^{i-1}(k_S^{T,tr}(\mathfrak{c})|k, \mathrm{Hom}(A, C_{k_S^{T,tr}(\mathfrak{c})})) \\ &\cong \hat{H}^{3-i}(k_S^{T,tr}(\mathfrak{c})|k, A)^\vee, \end{aligned}$$

and for $i \le 0$:

$$\begin{aligned} \hat{H}^i(k_S^{T,tr}(\mathfrak{c})|k, \mathrm{Hom}(A, E_T^S)) &\cong \varprojlim_K H^i(K|k, \mathrm{Hom}(A, E_{K,T}^S)) \\ &\cong \varprojlim_K \hat{H}^{i-1}(K|k, \mathrm{Hom}(A, X(K))) \\ &\cong \hat{H}^{i-1}(k_S^{T,tr}(\mathfrak{c})|k, \mathrm{Hom}(A, C_{k_S^{T,tr}(\mathfrak{c})})) \\ &\cong \hat{H}^{3-i}(k_S^{T,tr}(\mathfrak{c})|k, A)^\vee, \end{aligned}$$

where $E_T^S = E^S_{k_S^{T,tr}(\mathfrak{c}),T}$.

For $i = 1$, we need an additional argument. Let $K_0|k$ be a finite Galois extension inside $k_S^{T,tr}(\mathfrak{c})$ which trivializes A and let K be a finite Galois extension of k inside $k_S^{T,tr}(\mathfrak{c})$ containing the maximal abelian extension K_0^{ab} of K_0 in $k_S^{T,tr}(\mathfrak{c})$. Since $H^1(k_S^{T,tr}(\mathfrak{c})|K_0, E_T^S) \cong Cl_T^S(K_0)(\mathfrak{c})$ by (8.8.4), we obtain from the Hochschild-Serre spectral sequence for the group extension

$$1 \longrightarrow G(k_S^{T,tr}(\mathfrak{c})|K) \longrightarrow G(k_S^{T,tr}(\mathfrak{c})|K_0) \longrightarrow G(K|K_0) \longrightarrow 1$$

the exact sequence

$$0 \longrightarrow H^1(K|K_0, E_{K,T}^S) \longrightarrow Cl_T^S(K_0)(\mathfrak{c}) \longrightarrow Cl_T^S(K)(\mathfrak{c}).$$

The map on the right is zero as K contains the field K_0^{ab}, and we get an isomorphism

$$H^1(K|K_0, E_{K,T}^S) \xrightarrow[\sim]{inf} H^1(k_S^{T,tr}(\mathfrak{c})|K_0, E_T^S) \cong Cl_T^S(K_0)(\mathfrak{c}).$$

Therefore we obtain an exact sequence

$$0 \longrightarrow H^1(K_0|k, \mathrm{Hom}(A, E_{K_0,T}^S)) \longrightarrow H^1(K|k, \mathrm{Hom}(A, E_{K,T}^S)) \longrightarrow$$

$$H^0(K_0|k, \mathrm{Hom}(A, Cl_T^S(K_0)(\mathfrak{c}))) \longrightarrow H^2(K_0|k, \mathrm{Hom}(A, E_{K_0,T}^S)),$$

which shows, by comparing it with the same sequences for all extensions of K inside $k_S^{T,tr}(\mathfrak{c})$, that

$$H^1(K|k, \mathrm{Hom}(A, E_{K,T}^S)) \xrightarrow[\sim]{inf} H^1(k_S^{T,tr}(\mathfrak{c})|k, \mathrm{Hom}(A, E_T^S))$$

is an isomorphism. In exactly the same way one proves that the injective map $H^2(K|k, A) \xrightarrow{inf} H^2(k_S^{T,tr}(\mathfrak{c})|k, A)$ induces an isomorphism

$$H^2(K|k, A) \xrightarrow[\sim]{inf} H^2(k_S^{T,tr}(\mathfrak{c})|k, A).$$

Furthermore, since the groups $\mathrm{Hom}(A, U_{K,T}^S)^{G(K|k)}$, $\mathrm{Hom}(A, E_{K,T}^S)^{G(K|k)}$ and $H^1(K|k, \mathrm{Hom}(A, E_{K,T}^S))$ do not depend on $K \supseteq K_0^{ab}$, the exact sequence

$$0 \longrightarrow \mathrm{Hom}(A, E_{K,T}^S)^{G(K|k)} \longrightarrow \mathrm{Hom}(A, U_{K,T}^S)^{G(K|k)} \longrightarrow$$

$$\mathrm{Hom}(A, X(K))^{G(K|k)} \longrightarrow H^1(K|k, \mathrm{Hom}(A, E_{K,T}^S)) \longrightarrow 0$$

shows that the natural map

$$\mathrm{Hom}(A, X(K))^{G(K|k)} \xrightarrow{\sim} \mathrm{Hom}(A, X(K'))^{G(K'|k)}$$

is an isomorphism for any finite Galois extension K' inside $k_S^{T,tr}(\mathfrak{c})$ containing K. If K' is the maximal abelian extension of K in $k_S^{T,tr}(\mathfrak{c})$, then in the commutative and exact diagram

$$\begin{array}{ccccc} \mathrm{Hom}(A, X(K'))^{G(K'|k)} & \hookrightarrow & \mathrm{Hom}(A, C_{K'})^{G(K'|k)} & \rightarrow & \mathrm{Hom}(A, Cl_T^S(K'))^{G(K'|k)} \\ \uparrow\wr & & \| & & \uparrow\varphi \\ \mathrm{Hom}(A, X(K))^{G(K|k)} & \hookrightarrow & \mathrm{Hom}(A, C_K)^{G(K|k)} & \longrightarrow & \mathrm{Hom}(A, Cl_T^S(K))^{G(K|k)} \end{array}$$

the canonical map φ is zero on the $\mathfrak{c}$-part. This implies an isomorphism

$$\mathrm{Hom}(A, X(K))^{G(K|k)} \otimes \mathbb{Z}_{(\mathfrak{c})} \xrightarrow{\sim} \mathrm{Hom}(A, C_K)^{G(K|k)} \otimes \mathbb{Z}_{(\mathfrak{c})},$$

where $\mathbb{Z}_{(\mathfrak{c})}$ is the subring of $\mathbb{Q}$ obtained from $\mathbb{Z}$ by inverting all primes p with $\mathbb{Z}/p\mathbb{Z} \notin \mathfrak{c}$. Now we consider the commutative and exact diagram

$$\begin{array}{ccc}
N_{K|k}\mathrm{Hom}(A, X(K)) \otimes \mathbb{Z}_{(\mathfrak{c})} & \hookrightarrow & N_{K|k}\mathrm{Hom}(A, C_K) \otimes \mathbb{Z}_{(\mathfrak{c})} \\
\downarrow & & \downarrow \\
\mathrm{Hom}(A, X(K))^{G(K|k)} \otimes \mathbb{Z}_{(\mathfrak{c})} & \xrightarrow{\sim} & \mathrm{Hom}(A, C_K)^{G(K|k)} \otimes \mathbb{Z}_{(\mathfrak{c})} \\
\downarrow & & \downarrow \\
\hat{H}^0(K|k, \mathrm{Hom}(A, X(K))) & \twoheadrightarrow & \hat{H}^0(K|k, \mathrm{Hom}(A, C_K)).
\end{array}$$

Replacing k by K_0 (i.e. $A \cong \mathbb{Z}^{rank A}$), the lower map is an isomorphism since

$$\begin{aligned}
\hat{H}^0(K|K_0, X(K)) &\cong \hat{H}^1(K|K_0, E_{K,T}^S) \\
&\cong H^1(k_S^{T,tr}(\mathfrak{c})|K_0, E_{K,T}^S) \\
&\cong \hat{H}^2(k_S^{T,tr}(\mathfrak{c})|K_0, \mathbb{Z})^\vee \\
&\cong \hat{H}^2(K|K_0, \mathbb{Z})^\vee \\
&\cong G(K|K_0)^{ab} \\
&\cong \hat{H}^0(K|K_0, C_K).
\end{aligned}$$

Therefore

$$N_{K|K_0}\mathrm{Hom}(A, X(K)) \otimes \mathbb{Z}_{(\mathfrak{c})} \xrightarrow{\sim} N_{K|K_0}\mathrm{Hom}(A, C_K) \otimes \mathbb{Z}_{(\mathfrak{c})}$$

is an isomorphism. As $N_{K|k} = N_{K_0|k} \circ N_{K|K_0}$, we obtain an isomorphism

$$N_{K|k}\mathrm{Hom}(A, X(K)) \otimes \mathbb{Z}_{(\mathfrak{c})} \xrightarrow{\sim} N_{K|k}\mathrm{Hom}(A, C_K) \otimes \mathbb{Z}_{(\mathfrak{c})}.$$

This shows that

$$\hat{H}^0(K|k, \mathrm{Hom}(A, X(K))) \xrightarrow{\sim} \hat{H}^0(K|k, \mathrm{Hom}(A, C_K))$$

is an isomorphism and therefore

$$\begin{aligned}
H^1(K|k, \mathrm{Hom}(A, E_{K,T}^S)) &\cong \hat{H}^0(K|k, \mathrm{Hom}(A, X(K))) \\
&\cong \hat{H}^0(K|k, \mathrm{Hom}(A, C_K)) \\
&\cong \hat{H}^2(K|k, A)^\vee.
\end{aligned}$$

As the inflation maps

$$H^1(K|k, \mathrm{Hom}(A, E_{K,T}^S)) \xrightarrow{inf} H^1(k_S^{T,tr}(\mathfrak{c})|k, \mathrm{Hom}(A, E_T^S))$$

and $H^2(K|k, A) \xrightarrow{inf} H^2(k_S^{T,tr}(\mathfrak{c})|k, A)$ are isomorphisms, this shows the existence of the required isomorphism for $i = 1$.

It remains to show that this isomorphism is natural, i.e. independent of the choice of K. For typesetting reasons, we prove this below in the case $\mathfrak{c}$ = (all finite groups), leaving the necessary modification in the general case to the reader. We consider the perfect pairing of finite groups

$$(*)\qquad H^1(K|k,\mathrm{Hom}(A,E^S_{K,T}))\times H^2(K|k,A)\to\mathbb{Q}/\mathbb{Z}$$

associated to the isomorphism $H^1(K|k,\mathrm{Hom}(A,E^S_{K,T}))\xrightarrow{\sim} H^2(K|k,A)^\vee$ obtained above. By construction, it is induced via the vertical isomorphisms in the commutative diagram

$$\begin{array}{ccccccc}
\hat H^0(K|k,\mathrm{Hom}(A,X(K))) & \times & H^2(K|k,A) & \xrightarrow{\cup} & H^2(K|k,X(K)) & \xrightarrow{inv'} & \mathbb{Q}/\mathbb{Z}\\
\wr\downarrow\delta & & \| & & \downarrow\delta & & \\
H^1(K|k,\mathrm{Hom}(A,E^S_{K,T})) & \times & H^2(K|k,A) & \xrightarrow{\cup} & H^3(K|k,E^S_{K,T}), & &
\end{array}$$

followed by the pairing in the upper line, where *inv'* denotes the composite map

$$H^2(K|k,X(K))\longrightarrow H^2(K|k,C_K)\xrightarrow{inv}\mathbb{Q}/\mathbb{Z}.$$

Consider the commutative diagram

$$\begin{array}{ccccccc}
H^3(K|k,E^S_{K,T}) & \xleftarrow{\ \delta\ } & H^2(K|k,X(K)) & \longrightarrow & H^2(K|k,C_K) & \xrightarrow{inv} & \mathbb{Q}/\mathbb{Z}\\
\downarrow inf & & \downarrow inf & & \downarrow inf & & \|\\
H^3(k_S^{T,tr}|k,E^S_T) & \xleftarrow[\sim]{\delta} & H^2(k_S^{T,tr}|k,X(k_S^{T,tr})) & = & H^2(k_S^{T,tr}|k,C_{k_S^{T,tr}}) & \xrightarrow{inv} & \mathbb{Q}/\mathbb{Z}.
\end{array}$$

Denoting the map $H^3(k_S^{T,tr}|k,E^S_T)\to\mathbb{Q}/\mathbb{Z}$ induced by the lower line by *inv''*, we conclude that the perfect pairing $(*)$ is the cup-product pairing

$$H^1(K|k,\mathrm{Hom}(A,E^S_{K,T}))\times H^2(K|k,A)\xrightarrow{\cup}H^3(K|k,E^S_{K,T})$$

composed with the composite $H^3(K|k,E^S_{K,T})\xrightarrow{inf}H^3(k_S^{T,tr}|k,E^S_T)\xrightarrow{inv''}\mathbb{Q}/\mathbb{Z}$. It is therefore compatible via the inflation maps with the analogous diagram for an extension K' of K in $k_S^{T,tr}$. This concludes the proof. □

Chapter IX

The Absolute Galois Group of a Global Field

In this chapter we will deal with two fundamental questions in algebraic number theory. The first of these questions is the determination of all extensions of a fixed base field k (where the most important case is $k = \mathbb{Q}$), which means exploring how these extensions are built up over each other, how they are related and how they can be classified. In other words, we want to study the structure of the absolute Galois group G_k of k as a profinite group. But in contrast to the local Galois groups we are far from a complete understanding of the global situation and there are many conjectures but only a few conceptual results. For example, there is a famous conjecture due to *I. R. Šafarevič* which asserts that the subgroup $G_{k(\mu)}$ of G_k is a free profinite group, where $k(\mu)$ is the field obtained from k by adjoining all roots of unity. This was proved by *F. Pop* [171] for function fields, but the conjecture is open in the number field case.

However, this first question reflects only one side of the situation which we consider in algebraic number theory. Namely, it is of purely algebraic nature and can be asked for any field. But number fields are arithmetic fields, endowed with primes, completions etc. Therefore, the second fundamental problem is the question of the decomposition of prime ideals in an extension, and it can be interpreted as the problem of determining the local groups $G_{k_\mathfrak{p}}$ with respect to the global group G_k, i.e. how they lie inside G_k and how they interact.

As we are far from being able to answer these two questions concerning the "algebraic structure" and the "arithmetic structure" of G_k in complete generality, we will only give partial answers to these questions. In the first section we deal with the so-called **Hasse principle**, also called the **local-global principle**. This is the question for the kernel of the natural maps

$$res^i : H^i(G_k, A) \longrightarrow \prod_\mathfrak{p} H^i(G_{k_\mathfrak{p}}, A)\,, \quad i = 1, 2,$$

where A is a G_k-module and res^i is given by the restriction maps $res^i_\mathfrak{p}$. The kernel is just the obstruction to a local-global principle (i.e. that the global properties of $H^i(G_k, A)$ are determined by the corresponding local ones). One

J. Neukirch et al., *Cohomology of Number Fields*,
Grundlehren der mathematischen Wissenschaften 323,
DOI 10.1007/978-3-540-37889-1_9, © Springer-Verlag Berlin Heidelberg 2008

also wants to know this in the case of restricted ramification, so that one wants to determine the kernel

$$Ш^i(k_S, T, A) = \ker\Big(H^i(G_S(k), A) \longrightarrow \prod_{\mathfrak{p}\in T} H^i(G_{k_\mathfrak{p}}, A)\Big)$$

where $T \subseteq S$ are sets of primes of k.

In the second section we deal with the cokernels of these maps. For a finite set T and a "large" set S, the surjectivity is (under certain assumptions) the famous theorem of *Grunwald-Wang*. As a corollary, one obtains that given finitely many abelian local extensions $K_\mathfrak{p}|k_\mathfrak{p}$, there exists an abelian global extension $K|k$ realizing the given local ones (except for one very special case).

In §3 we extend the surjectivity results of §2 by showing the existence of global cohomology classes which realize finitely many given local classes and are of a very special kind. This will be used in an essential way in the proofs of the theorems of Neukirch and Šafarevič later in this chapter.

In §4 we deal with the question of whether a finite number of local groups $G_{k_\mathfrak{p}}$ can form a free product inside G_k (actually we only consider the corresponding pro-p-factor groups for a prime number p). Here the theorem of Grunwald-Wang will play an important role.

A partial answer to the conjecture of Šafarevič is given by *K. Iwasawa*: the maximal prosolvable quotient of $G_{k(\mu)}$ is a free prosolvable group. The same is true for the Galois group of the maximal prosolvable extension $\tilde{k}$ of the maximal abelian extension k^{ab} of k. This will be proved in §5 by considering embedding problems. In this section we also prove the theorem of Neukirch which gives a sufficient condition under which finitely many given local extensions can be simultaneously realized by a solvable global extension.

We conclude the chapter with a famous theorem of Šafarevič which asserts that, for a given global field k, every finite solvable group G occurs as a Galois group of a Galois extension K of k, i.e. $G(K|k) = G$.

§1. The Hasse Principle

The name Hasse principle has its origin in the classical theorem of Hasse-Minkowski, which asserts that a quadric has a rational point over a number field k if and only if it has a rational point over all completions of k. Another example of a local-global principle is the theorem of Hasse on the Brauer group of a global field (8.1.17).

Let k be a global field and S a set of primes of k. We do not assume that S is nonempty nor that S contains S_∞ if k is a number field. We recall the convention that an archimedean prime is unramified if and only if it is completely decomposed, i.e. we consider the extension $\mathbb{C}|\mathbb{R}$ of archimedean local fields as ramified. Given any separable closure $\bar{k}$ of k, we denote the maximal extension of k inside $\bar{k}$ which is unramified outside S by k_S and we set $G_S(k) = G(k_S|k)$. The ring of S-integers of k is defined by

$$\mathcal{O}_{k,S} = \{a \in k \mid v_\mathfrak{p}(a) \ge 0 \quad \text{for all } \mathfrak{p} \notin S \cup S_\infty\}.$$

If k is a function field and $S = \varnothing$, then $\mathcal{O}_{k,S}$ is the algebraic closure of the prime field in k. Recall the notation

$$\mathbb{N}(S) = \{n \in \mathbb{N} \mid n \in \mathcal{O}_{k,S}^\times\}.$$

For a finite $G_S(k)$-module A with $\#A \in \mathbb{N}(S)$ we set $A' = \mathrm{Hom}(A, \mathcal{O}_S^\times)$, where $\mathcal{O}_S^\times = \mathcal{O}_{k_S,S}^\times$.

We are interested in the kernel $Ш^1(k_S, T, A)$ given by the exact sequence

$$0 \longrightarrow Ш^1(k_S, T, A) \longrightarrow H^1(k_S|k, A) \longrightarrow \prod_{\mathfrak{p}\in T} H^1(k_\mathfrak{p}, A),$$

where $T \subseteq S$ and A is a finite $G_S(k)$-module. If $T = S$, we also use the notation $Ш^1(k_S, A) := Ш^1(k_S, S, A) = Ш^1(G_S, A)$.

(9.1.1) Definition. *We say the* **Hasse principle** *holds (in dimension* 1*) for the* G_S*-module* A *and the set of primes* $T \subseteq S$ *if*

$$Ш^1(k_S, T, A) = 0.$$

In this generality the Hasse principle is obviously not true even if $T = S$. To give an example, let $A = \mathbb{Z}/p\mathbb{Z}$ and let $S \supseteq S_p \cup S_\infty$ be a finite set of primes of the number field k. Then $Ш^1(k_S, S, A)$ is equal to the dual of the p-primary part of the S-ideal class group $Cl_S(k)$ of k which is nontrivial in general. One might think that the Hasse principle would hold for a G_S-module A if $S = T$ is the set of all primes of k. But for a reducible G_k-module A there are counterexamples; see [72], 7.3, and [95]. In this section we will deduce several criteria on S, T and A that imply the vanishing of $Ш^1(k_S, T, A)$.

If $k(A)$ denotes the **minimal trivializing extension** of k for the module A, i.e. if $G_{k(A)}$ is the kernel of the homomorphism $G_k \to \mathrm{Aut}(A)$ given by the action of G_k on A, then we define, for a Galois extension $K|k$ inside k_S and containing $k(A)$, the group $Ш^1(K|k, T, A)$ by the exactness of

$$0 \longrightarrow Ш^1(K|k, T, A) \longrightarrow H^1(K|k, A) \xrightarrow{res} \prod_{\mathfrak{p}\in T} H^1(K_\mathfrak{p}|k_\mathfrak{p}, A).$$

As above, we put $Ш^1(K|k, A) := Ш^1(K|k, S, A)$ if $T = S$. We need some further notation.

(9.1.2) Definition. *Let k be a global field and let S be a set of primes of k. The* **Dirichlet-density** *$\delta(S)$ of S is the limit*

$$\delta(S) = \lim_{s \to 1^+} \frac{\sum_{\mathfrak{p} \in S} N\mathfrak{p}^{-s}}{\sum_{\text{all } \mathfrak{p}} N\mathfrak{p}^{-s}}$$

if it exists. We have $0 \le \delta(S) \le 1$. For sets S_1 and S_2 of primes, we use the notation

$$S_1 \precsim S_2 :\Longleftrightarrow \delta(S_1 \smallsetminus S_2) = 0,$$

i.e. S_1 is contained in S_2 up to a set of primes of density zero. If $\Omega|k$ is a finite separable extension of k, then we denote by

$$\delta_\Omega(S) = \delta_\Omega(S(\Omega))$$

the Dirichlet density of the set $S(\Omega)$ of primes of Ω given by all extensions of $S = S(k)$. Furthermore, we set

$$\begin{aligned} cs(\Omega|k) &:= \{\mathfrak{p} \text{ a prime of } k \mid \mathfrak{p} \text{ splits completely in } \Omega|k\},\\ \mathrm{Ram}(\Omega|k) &:= \{\mathfrak{p} \text{ a prime of } k \mid \mathfrak{p} \text{ ramifies in } \Omega|k\} \end{aligned}$$

for a finite Galois extension $\Omega|k$.

Let $\Omega|k$ be a finite Galois extension and let $\mathfrak{p}$ vary over the primes of k which are unramified in $\Omega|k$. Then the set $\mathrm{Frob}_\mathfrak{p}$ of the Frobenius automorphisms with respect to the prolongations of $\mathfrak{p}$ to Ω is a conjugacy class in $G(\Omega|k)$. For a conjugacy class $\mathcal{C}$ in $G(\Omega|k)$ we put

$$P_{\Omega|k}(\mathcal{C}) = \{\mathfrak{p} \text{ a prime of } k \mid \mathrm{Frob}_\mathfrak{p} = \mathcal{C}\}.$$

For a proof of the following theorem refer to [160], chap. VII, (13.6), and [53], chap. 5, §4 in the function field case.

(9.1.3) Čebotarev's Density Theorem. *Let $\Omega|k$ be a finite Galois extension with group G and let $\mathcal{C}$ be a conjugacy class in G. Then*

$$\delta(P_{\Omega|k}(\mathcal{C})) = \frac{\#\mathcal{C}}{\#G}.$$

For a set S of primes of k and a finite Galois extension $\Omega|k$ we mention the relation

$$\delta_\Omega(S) = \delta_k\Big(S \cap cs(\Omega|k)\Big) \cdot [\Omega : k].$$

We start with some calculations.

(9.1.4) Lemma. *Let p be a prime number, $m \geq 1$ a natural number and let $G \subseteq (\mathbb{Z}/p^m\mathbb{Z})^\times = \mathrm{Aut}(\mathbb{Z}/p^m\mathbb{Z})$ be a subgroup. Let A be the G-module which is isomorphic to $\mathbb{Z}/p^m\mathbb{Z}$ as an abelian group and on which G acts in the canonical way. Then*

$$\hat{H}^i(G, A) = 0 \quad \textit{for all } i \in \mathbb{Z},$$

unless $p = 2$, $m \geq 2$ and $-1 \in G$, in which case

$$\hat{H}^i(G, A) \cong \mathbb{Z}/2\mathbb{Z} \quad \textit{for all } i \in \mathbb{Z}.$$

Proof: Let $\alpha = 1 + p^s u \in U^1$, $s \geq 1$, $u \in \mathbb{Z}_p^\times$, be a principal unit of $\mathbb{Z}_p$. If p is odd or if $s \geq 2$, then $v_p(\alpha^r - 1) = s + v_p(r)$ for all $r \in \mathbb{N}$.

Let us start with the case p odd. Then

$$G_1 = \ker(G \to (\mathbb{Z}/p\mathbb{Z})^\times) = G \cap \mathrm{im}(U^1 \to (\mathbb{Z}/p^m\mathbb{Z})^\times)$$

is a p-Sylow subgroup of G and in order to show the first statement, we may replace G by the cyclic group G_1. Let p^{m-s}, $s \geq 1$ be the order of G. We use the same letter for an element $\alpha \in \mathbb{Z}_p^\times$ and its image in $(\mathbb{Z}/p^m\mathbb{Z})^\times$. Fix a generator $\alpha \in U^s \smallsetminus U^{s+1}$ of G. Then $A^G = \ker(\alpha - 1\colon A \to A) = p^{m-s}A$. As

$$\sum_{i=0}^{p^{m-s}-1} \alpha^i = \frac{\alpha^{p^{m-s}} - 1}{\alpha - 1} = p^{m-s}v,$$

with $v \in \mathbb{Z}_p^\times$, we have $N_G(A) = p^{m-s}A = A^G$. Since finite cyclic groups have trivial Herbrand index by (1.7.6), we obtain $\hat{H}^i(G, A) = 0$ for all $i \in \mathbb{Z}$.

Now we assume $p = 2$ and we leave the trivial case $m = 1$ aside. Then there are three possibilities for G:

a) $G \cong \{\pm 1\} \times \langle \alpha \rangle$ with $\alpha \in U^2$,
b) $G \cong \langle \alpha \rangle$ with $\alpha \in U^2$.
c) $G \cong \langle \alpha \rangle$ with $-\alpha \in U^s \smallsetminus U^{s+1}$, $2 \leq s \leq m - 1$.

Note that $-1 \in G$ only in case a). In case b) the same argument as for odd p shows $\hat{H}^i(G, A) = 0$ for all $i \in \mathbb{Z}$. In case c) we have the equality $A^G = \ker(\alpha - 1\colon A \to A) = 2^{m-1}A$. The order of G is equal to $2^{m-s} \geq 2$. As

$$\sum_{i=0}^{2^{m-s}-1} \alpha^i = \frac{\alpha^{2^{m-s}} - 1}{\alpha - 1} = \frac{(-\alpha)^{2^{m-s}} - 1}{\alpha - 1} = 2^{m-1}v,$$

with $v \in \mathbb{Z}_2^\times$, we have $N_G(A) = 2^{m-1}A$. In the same way as in case b) we conclude $\hat{H}^i(G, A) = 0$ for all $i \in \mathbb{Z}$. Let us finally consider case a) and let 2^{m-s} be the order of the subgroup $\langle \alpha \rangle \subseteq G$. Since

$$\sum_{i=0}^{2^{m-s}-1} \alpha^i + \sum_{i=0}^{2^{m-s}-1} -\alpha^i = 0,$$

the norm map $N_G\colon A \to A$ is the zero map. We have $A^G = 2^{m-1}A$ and $I_G A = 2A$, hence $\hat{H}^i(G, A) \cong \mathbb{Z}/2\mathbb{Z}$ for $i = -1, 0$. By case b), we have

$\hat{H}^i(\langle\alpha\rangle, A) = 0$ for all i. For $i \geq 1$ we obtain

$$\#H^i(G, A) = \#H^i(\{\pm 1\}, A^{\langle\alpha\rangle}) = \#\hat{H}^0(\{\pm 1\}, 2^{m-s}A) = 2.$$

In the same way, using homology, we obtain $\#\hat{H}^i(G, A) = 2$ for $i \leq -2$. □

We will use the lemma above in the following situation: Let G be a subgroup of the Galois group $G(k(\mu_{p^m})|k)$, where k is any prime field with $\mathrm{char}(k) \neq p$. Then G acts on μ_{p^m} via the cyclotomic character $\chi_{cycl}\colon G \hookrightarrow (\mathbb{Z}/p^m\mathbb{Z})^\times$. In the case when $k = \mathbb{Q}$ and $p = 2$ the group G corresponds to a field in the diagram

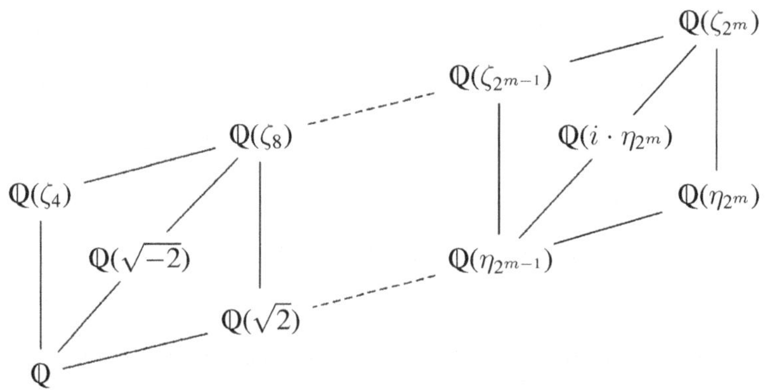

where $\eta_{2^m} = \zeta_{2^m} + \zeta_{2^m}^{-1}$ and ζ_{2^m} is a primitive 2^m-th root of unity.

(9.1.5) Definition. *Let k be a field and $m = 2^r m'$, m' odd, a natural number prime to* $\mathrm{char}(k)$. *We say that we are in the* **special case (k, m)** *if $r \geq 2$ and -1 is in the image of the cyclotomic character $\chi_{cycl} : G(k(\mu_{2^r})|k) \to (\mathbb{Z}/2^r\mathbb{Z})^\times$.*

(9.1.6) Proposition. *Let p be a prime number, $r \in \mathbb{N}$, and let k be any field with* $\mathrm{char}(k) \neq p$. *Then*

$$\hat{H}^i(G(k(\mu_{p^r})|k), \mu_{p^r}) = 0 \qquad \textit{for all } i \in \mathbb{Z},$$

except when $p = 2$, $r \geq 2$ and we are in the special case $(k, 2^r)$. In this case

$$\hat{H}^i(G(k(\mu_{2^r})|k), \mu_{2^r}) \cong \mathbb{Z}/2\mathbb{Z} \qquad \textit{for all } i \in \mathbb{Z}.$$

Let $p = 2$ and $r \geq 2$. Then the special case $(k, 2^r)$ occurs if and only if

$$\mathrm{char}(k) = 0 \quad \textit{and} \quad \mathbb{Q}(\mu_{2^r}) \cap k \textit{ is real},$$
$$\textit{or} \quad \mathrm{char}(k) = \ell \equiv -1 \bmod 2^r \quad \textit{and} \quad \mathbb{F}_\ell(\mu_{2^r}) \cap k = \mathbb{F}_\ell.$$

Remark: The special case in positive characteristic occurs if and only if $k(\mu_{2^{r+1}}) = k(\mu_{2^r})$ and this field is a quadratic extension of k (cf. ex. 3 below).

Proof: As the group $G = G(k(\mu_{p^r})|k)$ acts on μ_{p^r} via the cyclotomic character $\chi_{cycl}\colon G \hookrightarrow (\mathbb{Z}/p^r\mathbb{Z})^\times$, we may apply (9.1.4). It therefore suffices to identify the case $-1 \in \operatorname{im}(\chi_{cycl})$ if $p = 2$, $r \geq 2$. If char$(k) = 0$, then the cyclotomic character factors as

$$\chi_{cycl}\colon\ G \xrightarrow{\sim} G(\mathbb{Q}(\mu_{2^r})|k \cap \mathbb{Q}(\mu_{2^r})) \hookrightarrow (\mathbb{Z}/2^r\mathbb{Z})^\times$$

and, since $-1 \in (\mathbb{Z}/2^r\mathbb{Z})^\times \cong G(\mathbb{Q}(\mu_{2^r})|\mathbb{Q})$ is the complex conjugation, we have $-1 \in \operatorname{im}(\chi_{cycl})$ if and only if $k \cap \mathbb{Q}(\mu_{2^r})$ is real.

Now let char$(k) = \ell \notin \{0, 2\}$. Then the cyclotomic character factors as

$$\chi_{cycl}\colon\ G \xrightarrow{\sim} G(\mathbb{F}_\ell(\mu_{2^r})|k \cap \mathbb{F}_\ell(\mu_{2^r})) \hookrightarrow (\mathbb{Z}/2^r\mathbb{Z})^\times.$$

The group $G(\mathbb{F}_\ell(\mu_{2^r})|k \cap \mathbb{F}_\ell(\mu_{2^r}))$ is cyclic, generated by the relative Frobenius automorphism σ and

$$\chi_{cycl}(\sigma) = \ell^{[\mathbb{F}_\ell(\mu_{2^r})\cap k:\mathbb{F}_\ell]}.$$

Hence $-1 \in \operatorname{im}(\chi_{cycl})$ if and only if $\ell^{[\mathbb{F}_\ell(\mu_{2^r})\cap k:\mathbb{F}_\ell]\cdot n} \equiv -1 \bmod 2^r$ for some $n \in \mathbb{N}$. The last condition is equivalent to $\ell \equiv -1 \bmod 2^r$ and $\mathbb{F}_\ell(\mu_{2^r}) \cap k = \mathbb{F}_\ell$. □

(9.1.7) Definition. *Let k be a global field, $m = 2^r m'$, m' odd, a natural number prime to char(k) and T a set of places of k. We say that we are in the* **special case (k, m, T)** *if we are in the special case (k, m) and all primes $\mathfrak{p} \in T$ decompose in $k(\mu_{2^r})|k$.*

(9.1.8) Lemma. *Let k be a global field. Assume that* char$(k) \neq 2$, *$k(\mu_{2^r})|k$ is cyclic and $\delta(T) > 1/2$. Then we are not in the special case $(k, 2^r, T)$.*

Proof: If we are in the special case $(k, 2^r)$ and $k(\mu_{2^r})|k$ is cyclic, then we have $[k(\mu_{2^r}) : k] = 2$. By Čebotarev's density theorem, the set of primes which do not decompose in $k(\mu_{2^r})|k$ has density $1/2$ and therefore has a nontrivial intersection with T. □

Remarks: 1. The extension $k(\mu_{2^r})|k$ can be non-cyclic only if k is a number field, $r \geq 3$ and $k \cap \mathbb{Q}(\mu_{2^r})$ is real.
2. If $k(\mu_{2^r})|k$ is not cyclic, then all primes $\mathfrak{p} \nmid 2$ are decomposed in $k(\mu_{2^r})|k$. Indeed, these primes are unramified and therefore have a cyclic decomposition group.

3. If $\delta(T) > 1/2$, then, by the first two remarks and (9.1.8), the special case (k, m, T) is equivalent to the statement

k is a number field, $m = 2^r m'$, m' odd, $r \geq 3$,
$k(\mu_{2^r})|k$ is not cyclic, and
all primes $\mathfrak{p}$ dividing 2 in T decompose in $k(\mu_{2^r})|k$,

Now we can formulate some first cases where the Hasse principle holds.

(9.1.9) Theorem. *Let k be a global field and let S be a set of primes of k. The Hasse principle holds for a finite $G_S(k)$-module A and the sets of primes $T \subseteq S$ of k, i.e.*

$$Ш^1(k_S, T, A) = 0,$$

in the following cases:

(i) *A is a trivial $G_S(k)$-module and $\delta(T) > 1/p$, where p is the smallest prime divisor of $\#A$.*

(ii) *$A = \mu_m$ with $m = p_1^{r_1} \cdots p_n^{r_n}$, where p_i are pairwise different prime numbers in $\mathbb{N}(S)$, and*

$$\delta(cs(k(\mu_{p_i^{r_i}})|k) \cap T) > \frac{1}{p_i \cdot [k(\mu_{p_i^{r_i}}) : k]}$$

for all $i = 1, \ldots, n$, except we are in the special case (k, m, T), where $Ш^1(k_S, T, \mu_m) \cong \mathbb{Z}/2\mathbb{Z}$.

(iii) *$cs(k(A)|k) \subsetneq T$ and $\#G(k(A)|k) = lcm\{\#G(k(A)_\mathfrak{p}|k_\mathfrak{p}) \mid \mathfrak{p} \in T\}$.*

Proof: (i) We may assume that $A = \mathbb{Z}/p^r\mathbb{Z}$ and want to show the injectivity of the homomorphism

$$H^1(k_S|k, \mathbb{Z}/p^r\mathbb{Z}) \longrightarrow \prod_{\mathfrak{p} \in T} H^1(k_\mathfrak{p}, \mathbb{Z}/p^r\mathbb{Z}).$$

If $\varphi : G_S(k) \to \mathbb{Z}/p^r\mathbb{Z}$ is in the kernel, then ker(φ) corresponds to a cyclic Galois extension $L|k$ of degree p^s, $0 \leq s \leq r$, which is unramified outside S and completely decomposed at T. Since $\delta(T) > 1/p$, Čebotarev's density theorem (9.1.3) implies $s = 0$, hence $\varphi = 0$.

(ii) We may assume that $m = p^r$ and therefore $K := k(A) = k(\mu_{p^r})$. The commutative and exact diagram

$$
(*)\qquad
\begin{array}{ccccccc}
 & & & & H^1(k_S|K,A) & \longrightarrow & \prod\limits_{T(K)} H^1(K_\mathfrak{p},A) \\
 & & & & \uparrow & & \uparrow \\
0 & \longrightarrow & \text{Ш}^1(k_S,T,A) & \longrightarrow & H^1(k_S|k,A) & \longrightarrow & \prod\limits_{T(k)} H^1(k_\mathfrak{p},A) \\
 & & \uparrow & & \uparrow & & \uparrow \\
0 & \longrightarrow & \text{Ш}^1(K|k,T,A) & \longrightarrow & H^1(K|k,A) & \longrightarrow & \prod\limits_{T(k)} H^1(K_\mathfrak{p}|k_\mathfrak{p},A)
\end{array}
$$

shows, using

$$\delta_K(T) = \delta_k\Big(T \cap cs(K|k)\Big)\cdot [K:k] > \frac{1}{p}$$

and (i), that

$$\text{Ш}^1(k_S,T,A) = \text{Ш}^1(K|k,T,A).$$

By (9.1.6) we have $H^1(K|k,\mu_{p^r}) = 0$ and therefore the group $\text{Ш}^1(k_S,T,A)$ is trivial, unless we are in the special case $(k,2^r)$, where

$$H^1(k(\mu_{2^r})|k,\mu_{2^r}) \cong \mathbb{Z}/2\mathbb{Z}.$$

If there exists a prime $\mathfrak{p} \in T$ such that $\mathfrak{p}$ does not decompose in $k(\mu_{2^r})$, then $G(k(\mu_{2^r})|k) = G(k_\mathfrak{p}(\mu_{2^r})|k_\mathfrak{p})$ and therefore $\text{Ш}^1(k(\mu_{2^r})|k,T,\mu_{2^r}) = 0$.
On the other hand, the homomorphism

$$\mathbb{Z}/2\mathbb{Z} \cong H^1(k(\mu_{2^r})|k,\mu_{2^r}) \xrightarrow{\;res_\mathfrak{p}\;} H^1(k_\mathfrak{p}(\mu_{2^r})|k_\mathfrak{p},\mu_{2^r})$$

is the zero map if the prime $\mathfrak{p}$ decomposes in $k(\mu_{2^r})$. Indeed, in this case $G(k_\mathfrak{p}(\mu_{2^r})|k_\mathfrak{p})$ is a proper subgroup of $G(k(\mu_{2^r})|k)$. Let k' be its fixed field. Then

$$\ker res_\mathfrak{p} = H^1(k'|k,\mu_{2^r}^{G(k(\mu_{2^r})|k')}),$$

which is isomorphic to $\mathbb{Z}/2\mathbb{Z}$ since $\mu_{2^r}^{G(k(\mu_{2^r})|k')} = \mu_2$ or $k' = k(\mu_{2^t})$ for some $t \geq 2$, in which case we apply (9.1.6) again. This shows assertion (ii).

(iii) Let, for a prime number p, p^s be the maximal p-power dividing the order of $G = G(K|k)$, where $K = k(A)$. By assumption there exists at least one $\mathfrak{p} \in T$ such that $p^s \mid \#G_\mathfrak{p}$, hence $G_\mathfrak{p} = G(K_\mathfrak{p}|k_\mathfrak{p})$ contains a Sylow group G_p of G. Since

$$H^1(G,A) \longrightarrow \prod_p H^1(G_p,A)$$

is injective by (1.6.10), the kernel $\text{Ш}^1(K|k,T,A)$ is zero. Using the diagram $(*)$ of the proof of part (ii), we obtain the result. □

Using Poitou-Tate duality, we obtain injectivity results for the second cohomology group.

(9.1.10) Corollary. *Let k be a global field and let S be a nonempty set of primes such that $S_\infty \subseteq S$ if k is a number field. Let A be a finite $G_S(k)$-module with $\#A \in \mathbb{N}(S)$. Then the canonical homomorphism*

$$H^2(k_S|k, A) \longrightarrow \bigoplus_S H^2(k_\mathfrak{p}, A)$$

is injective in the following cases:

(i) *$A = \mathbb{Z}/m\mathbb{Z}$ with $m = p_1^{r_1} \cdots p_n^{r_n}$, where p_i are pairwise different prime numbers in $\mathbb{N}(S)$, and*

$$\delta(cs(k(\mu_{p_i^{r_i}})|k) \cap S) > \frac{1}{p_i \cdot [k(\mu_{p_i^{r_i}}) : k]}$$

for all $i = 1, \ldots, n$, except we are in the special case (k, m, S), where $\text{Ш}^2(k_S, \mathbb{Z}/m\mathbb{Z}) \cong \mathbb{Z}/2\mathbb{Z}$.

(ii) *A' is a trivial $G_S(k)$-module and $\delta(S) > 1/p$, where p is the smallest prime divisor of $\#A$.*

(iii) *$cs(k(A')|k) \subsetneq S$ and $\#G(k(A')|k) = lcm\{\#G(k(A')_\mathfrak{p}|k_\mathfrak{p}) \mid \mathfrak{p} \in S\}$.*

Proof: From the Poitou-Tate duality theorem (8.6.7), we know that

$$\text{Ш}^2(k_S, A) \cong \text{Ш}^1(k_S, A')^\vee.$$

Now everything follows from (9.1.9) with $T = S$ considering A' instead of A. □

A first application is a local-global principle for m-th powers.

(9.1.11) Theorem. *Let k be a global field, m a natural number prime to char(k) and T a set of primes of k of density $\delta(T) = 1$. Then the following holds.*

(i) *The localization homomorphism*

$$k^\times/k^{\times m} \longrightarrow \prod_{\mathfrak{p}\in T} k_\mathfrak{p}^\times/k_\mathfrak{p}^{\times m}$$

is injective, except in the following case:

k is a number field, $m = 2^r m'$, m' odd, $r \geq 3$,
$k(\mu_{2^r})|k$ is not cyclic, and
all primes $\mathfrak{p}$ dividing 2 in T decompose in $k(\mu_{2^r})|k$,

where the kernel is cyclic of order 2.

(ii) *If $\alpha \in k^\times$ is a $2m$-th power in $k_\mathfrak{p}^\times$ for all $\mathfrak{p} \in T$, then α is an m-th power in $k^\times$.*

Proof: In view of $k^\times/k^{\times m} \cong H^1(k,\mu_m)$ assertion (i) follows from (9.1.9)(ii), where S is the set of all primes of k together with (9.1.8) and the subsequent Remark 3.

In order to prove (ii), we may reduce to the case that $m = 2^r$, $r \geq 3$ and we are in the special case (k,m,T). In particular, $\sqrt{-1} \notin k$. Let $K = k(\mu_{2m})$. Then, by (9.1.9)(i), the localization homomorphism

$$K^\times/K^{\times 2m} \longrightarrow \prod_{\mathfrak{p}\in T(K)} K_\mathfrak{p}^\times/K_\mathfrak{p}^{\times 2m}$$

is injective. Let $\alpha \in k^\times$ be a $2m$-th power in $k_\mathfrak{p}^\times$ for all $\mathfrak{p} \in T$. Then $\alpha \in K^{\times 2m}$. The commutative and exact diagram

$$\begin{array}{ccc}
K^\times/K^{\times 2m} & \xrightarrow{\sim} & H^1(G_K,\mu_{2m}) \\
\uparrow & & \uparrow \\
k^\times/k^{\times 2m} & \xrightarrow{\sim} & H^1(G_k,\mu_{2m}) \\
 & & \uparrow \\
 & & H^1(K|k,\mu_{2m})
\end{array}$$

together with (9.1.6) shows that $\alpha^2 = \beta^{2m}$ for some $\beta \in k^\times$. Hence $\alpha = \pm\beta^m$. Suppose $\alpha = -\beta^m$. As α in an m-th power in $k_\mathfrak{p}^\times$ for all $\mathfrak{p} \in T$, the same holds for $-1 \in k^\times$. Hence the quadratic extension $k(\sqrt{-1})|k$ is decomposed at a set of primes of density 1 which contradicts the Čebotarev density theorem. We conclude $\alpha = \beta^m$, showing (ii). □

Now let S be an arbitrary set of primes of k. For a natural number m prime to char(k), we consider the Kummer group*)

$$V_S(k,m) = \ker\Big(k^\times/k^{\times m} \longrightarrow \prod_{\mathfrak{p}\in S} k_\mathfrak{p}^\times/k_\mathfrak{p}^{\times m} \times \prod_{\mathfrak{p}\notin S} k_\mathfrak{p}^\times/U_\mathfrak{p} k_\mathfrak{p}^{\times m}\Big).$$

(Recall the convention $U_\mathfrak{p} = k_\mathfrak{p}^\times$ if $\mathfrak{p}$ is archimedean.) We denote the dual of this group by the letter Б (the Russian "B"), i.e.

$$Б_S(k,m) := V_S(k,m)^\vee .$$

*) The group $V_S(k,m)$ is a localization kernel for flat cohomology. We have

$$V_S(k,m) = \ker\big(H^1_{\mathrm{fl}}(\mathrm{Spec}(\mathcal{O}_{k,S}),\mu_m) \to \textstyle\prod_{\mathfrak{p}\in S} H^1(k_\mathfrak{p},\mu_m)\big).$$

(9.1.12) Proposition. *Let k be a global field, m a natural number prime to* char(k) *and S a set of primes. Then the following holds:*

(i) *The group $V_S(k,m)$ is finite.*

(ii) *If S has density $\delta(S)=1$, then $V_S(k,m)=0$ if we are not in the special case (k,S,m).*

(iii) *Assume that $m \in \mathbb{N}(S)$ and $S \supseteq S_\infty$ if k is a number field. Then*

$$V_S(k,m) \cong Ш^1(G_S,\mu_m).$$

Proof: The isomorphism $V_S(k,m) \cong Ш^1(G_S,\mu_m)$ stated in (iii) follows immediately from Kummer theory. Hence, by (8.6.4), $V_S(k,m)$ is finite if S is nonempty, $m \in \mathbb{N}(S)$ and $S \supseteq S_\infty$ if k is a number field. In the general case, let T be a nonempty finite set containing all primes dividing $m\infty$. Then we have an exact sequence

$$0 \longrightarrow V_{S\cup T}(k,m) \longrightarrow V_S(k,m) \longrightarrow \bigoplus_{\mathfrak{p}\in T\setminus S} U_\mathfrak{p}/U_\mathfrak{p}^m,$$

showing (i). Finally, (ii) follows from (9.1.11)(i) and the definition of $V_S(k,m)$. □

Next we are going to discuss the Hasse principle for the cohomology of simple modules.

(9.1.13) Lemma. *Let G be a profinite group and let A be a finite simple G-module. Assume there exists a normal subgroup N of G of order prime to $\#A$ which acts nontrivially on A. Then*

$$H^i(G,A)=0 \quad \text{for all} \quad i \ge 0.$$

Proof: Since A is simple and N is normal in G acting nontrivially on A, we have $A^N=0$. As N is of order prime to $\#A$, we have $H^i(N,A)=0$ for $i\ge 1$. Now the result follows from the Hochschild-Serre spectral sequence $E_2^{st}=H^s(G/N,H^t(N,A)) \Rightarrow H^{s+t}(G,A)$. □

The next proposition is due to *W. Gaschütz*, see [220], lemma 1.

(9.1.14) Proposition. *Let A be a simple G_k-module, let $K=k(A)$ and suppose that $G=G(K|k)\neq 1$ is solvable. Then*

$$H^i(G,A)=0 \quad \text{for all} \quad i \ge 0.$$

Proof: Since A is simple, it is an $\mathbb{F}_p$-vector space for some prime number p. The solvable group G has a nontrivial abelian normal subgroup N. We claim that its p-Sylow subgroup N_p is trivial. Indeed, N_p is a normal subgroup in G and if $N_p \neq 1$, we obtain $A^{N_p} = 0$ or $A^{N_p} = A$ since A is simple. In the first case it would follow that $A = 0$ by (1.6.12) and in the second that $k(A)$ was not the minimal A-trivializing extension of k.
Thus G has a nontrivial normal subgroup N of order prime to p which acts nontrivially on A (as $k(A)$ is the minimal A-trivializing extension of k). The result follows from (9.1.13). □

Now we can prove that the Hasse principle holds for simple modules under certain conditions.

(9.1.15) Theorem. *Let k be a global field, S a set of primes of k and let A be a finite simple $G_S(k)$-module. Let p be the prime number with $pA = 0$ and let $T \subseteq S$ be a subset. Then*

$$Ш^1(k_S, T, A) = 0,$$

in the following cases:

(i) *$G(k(A)|k)$ is solvable and $cs(k(A)|k) \precsim T$.*

(ii) *$p \in \mathbb{N}(S)$, $G(k(A')|k)$ is solvable and $cs(k(A')|k) \precsim T$.*

(iii) *$p \in \mathbb{N}(S)$ and there are finite Galois extensions $K \subseteq \Omega$ of k inside k_S such that $p \nmid [\Omega : K]$, and*

$$\mu_p \nsubseteq K \text{ and } k(A') \subseteq K,$$
$$\mu_p \subseteq \Omega \text{ and } cs(\Omega|k) \precsim T.$$

Proof: (i) Put $K = k(A)$. If $K \neq k$, then $H^1(K|k, A) = 0$, by (9.1.14). If $K = k$ this is trivially true. In the commutative diagram

$$\begin{array}{ccc} H^1(k_S|K, A) & \hookrightarrow & \prod\limits_{T(K)} H^1(K_\mathfrak{p}, A) \\ \uparrow & & \uparrow \\ H^1(k_S|k, A) & \longrightarrow & \prod\limits_{T} H^1(k_\mathfrak{p}, A) \end{array}$$

the upper map is injective by (9.1.9)(i), since $cs(K|k) \precsim T$ and hence $\delta_K(T) = 1$. This shows assertion (i).

(ii) Note that $A' = \mathrm{Hom}(A, \mu_p)$ is simple, since A is simple. Let $K = k(A)$ and $K' = k(A')$. Then $K' = k(A') \subseteq k_S$ and

$$K'(\mu_p) = K(\mu_p)\,.$$

Therefore the group $H^1(K'(\mu_p)|K, A) = H^1(K(\mu_p)|K, A)$ is zero. By (9.1.14), we have $H^1(K|k, A) = 0$, since A is simple and $G(K|k)$ is solvable as $G(K(\mu_p)|k) = G(K'(\mu_p)|k)$ is. As $cs(K'|k) \underset{\sim}{\subset} T$ we have $\delta_{K'}(T) = 1$ and $\delta_{K(\mu_p)}(T) = 1$. This explains the injectivity of the two left-hand side vertical and the upper horizontal arrow in the commutative diagram

$$\begin{array}{ccc}
H^1(k_S|K(\mu_p), A) & \hookrightarrow & \prod\limits_{T(K(\mu_p))} H^1(K(\mu_p)_{\mathfrak p}, A) \\
\big\uparrow & & \big\uparrow \\
H^1(k_S|K, A) & \longrightarrow & \prod\limits_{T(K)} H^1(K_{\mathfrak p}, A) \\
\big\uparrow & & \big\uparrow \\
H^1(k_S|k, A) & \longrightarrow & \prod\limits_{T(k)} H^1(k_{\mathfrak p}, A)\,,
\end{array}$$

which therefore implies the desired result.

(iii) Since $k(A') \subseteq K \subseteq \Omega$ and $\mu_p \subseteq \Omega$, we have $k(A) \subseteq \Omega$. In addition, $cs(\Omega|k) \underset{\sim}{\subset} T$ implies $\delta_\Omega(T) = 1$. Therefore in the commutative and exact diagram

$$\begin{array}{ccc}
H^1(k_S|\Omega, A) & \hookrightarrow & \prod\limits_{T(\Omega)} H^1(\Omega_{\mathfrak p}, A) \\
\big\uparrow & & \big\uparrow \\
H^1(k_S|k, A) & \longrightarrow & \prod\limits_{T} H^1(k_{\mathfrak p}, A) \\
\big\uparrow & & \\
H^1(\Omega|k, A) & &
\end{array}$$

the upper horizontal map is injective by (9.1.9)(i). Consider the normal subgroup $N = G(\Omega|K) \subseteq G(\Omega|k)$. By assumption $\#N$ is prime to p. As $k(A') \subseteq K$ and $\mu_p \nsubseteq K$, N acts nontrivially on the simple module A. This implies $H^1(\Omega|k, A) = 0$ by (9.1.13) and shows (iii). □

Using Poitou-Tate duality, we obtain the

(9.1.16) Corollary. *Let k be a global field and let S be a nonempty set of primes such that $S_\infty \subseteq S$ if k is a number field. Let A be a finite simple $G_S(k)$-module with $\#A \in \mathbb{N}(S)$ and let p be the prime number with $pA = 0$. Then the canonical homomorphism*

$$H^2(k_S|k, A) \longrightarrow \bigoplus_S H^2(k_\mathfrak{p}, A)$$

is injective in the following cases:

(i) *$G(k(A)|k)$ is solvable and $cs(k(A)|k) \subsetneqq S$.*

(ii) *$G(k(A')|k)$ is solvable and $cs(k(A')|k) \subsetneqq S$.*

(iii) *There are finite Galois extensions $K \subseteq \Omega$ of k inside k_S such that $p \nmid [\Omega : K]$,*

$$\mu_p \nsubseteq K \text{ and } k(A) \subseteq K,$$
$$\mu_p \subseteq \Omega \text{ and } cs(\Omega|k) \subsetneqq S.$$

Proof: Using the Poitou-Tate duality theorem (8.6.7),

$$Ш^2(k_S, A) \cong Ш^1(k_S, A')^\vee,$$

the assertion follows from (9.1.15) with $T = S$ by considering A' instead of A. □

Exercise 1. Let k be a number field, $m = 2^r m'$, m' odd, and let T be a set of primes of k of density $\delta(T) > 1/2$. Let ζ_s be a primitive 2^s-th root of unity and let $\eta_s = \zeta_s + \zeta_s^{-1}$. Assume $\eta_s \in k$ but $\eta_{s+1} \notin k$. Show that the special case (k, m, T) is equivalent to the following properties:

1) $-1, 2 + \eta_s$ and $-(2 + \eta_s)$ are not squares in k,
2) $r > s$,
3) $\{\mathfrak{p} \mid \mathfrak{p}$ divides 2 and $-1, 2 + \eta_s, -(2 + \eta_s)$ are not squares in $k_\mathfrak{p}\} \cap T = \varnothing$.

Hint: [6], chap.10.

Exercise 2. Show that $(\mathbb{Q}(\sqrt{7}), 2^3, \mathcal{P})$, where $\mathcal{P}$ is the set of all places, is a special case, i.e. for $k = \mathbb{Q}(\sqrt{7})$ the map

$$k^\times/k^{\times 8} \longrightarrow \prod_\mathfrak{p} k_\mathfrak{p}^\times/k_\mathfrak{p}^{\times 8}$$

is not injective.

Exercise 3. Let $m \geq 2$ and let k be a field of positive characteristic $\ell \neq 2$. Show that the following conditions are equivalent.

(i) $\ell \equiv -1 \bmod 2^m$ and $\mathbb{F}_\ell(\mu_{2^m}) \cap k = \mathbb{F}_\ell$.
(ii) $k(\mu_{2^{m+1}}) = k(\mu_{2^m})$ is a quadratic extension of k.

§2. The Theorem of Grunwald-Wang

Let, as in the last section, k be a global field and S a set of primes of k, $T \subseteq S$ a subset and A a finite $G_S(k)$-module.

In this section we are interested in the cokernel of the restriction map $res^i = res^i(S,T) = (res^i_{\mathfrak{p}})_{\mathfrak{p}\in T}$

$$H^i(k_S|k, A) \xrightarrow{res^i} \prod_T H^i(k_{\mathfrak{p}}, A) \twoheadrightarrow \operatorname{coker}^i(k_S, T, A)$$

for $i = 1, 2$. If $i = 2$ then the restricted product is just a direct sum and the group $\operatorname{coker}^2(k_S, T, A)$ is often trivial by the following

(9.2.1) Proposition. *Let k be a global field and let S be a nonempty set of primes of k such that $S_\infty \subseteq S$ if k is a number field. Let A be a finite $G_S(k)$-module with $\#A \in \mathbb{N}(S)$ and let $T \subseteq S$ be a subset. If $S \smallsetminus T$ contains a nonarchimedean prime, then the map*

$$res^2(S,T,A):\ H^2(k_S|k, A) \longrightarrow \bigoplus_T H^2(k_{\mathfrak{p}}, A)$$

is surjective.

Proof: The long exact sequence of Poitou-Tate induces the commutative exact diagram

$$\begin{array}{ccccccc}
H^2(k_S|k,A) & \longrightarrow & \bigoplus_S H^2(k_{\mathfrak{p}},A) & \longrightarrow & H^0(k_S|k,A')^\vee & \longrightarrow & 0\\
\| & & \downarrow & & \downarrow & & \\
H^2(k_S|k,A) & \longrightarrow & \bigoplus_T H^2(k_{\mathfrak{p}},A) & \longrightarrow & \operatorname{coker}^2(k_S,T,A) & \longrightarrow & 0.
\end{array}$$

We obtain the exact sequence

$$0 \longrightarrow \operatorname{coker}^2(k_S,T,A)^\vee \longrightarrow H^0(k_S|k,A') \longrightarrow \bigoplus_{S\smallsetminus T} H^2(k_{\mathfrak{p}},A)^\vee .$$

If $\mathfrak{p} \in S \smallsetminus T$ is nonarchimedean, then $H^0(k_S|k, A')$ injects into $H^0(k_{\mathfrak{p}}, A') \cong H^2(k_{\mathfrak{p}}, A)^\vee$. This shows the statement. □

Now we investigate res^1 assuming that T is finite. If S is nonempty, contains S_∞ if k is a number field and $\#A \in \mathbb{N}(S)$, then the local and global duality theorems induce the following commutative exact diagram:

$$\begin{array}{ccccccc}
 & & & & 0 & & \\
 & & & & \uparrow & & \\
Ш^1(k_S, S \smallsetminus T, A') & \hookrightarrow & H^1(k_S|k, A') & \to & \prod\limits_{S \smallsetminus T} H^1(k_\mathfrak{p}, A') & & \\
\uparrow & & \| & & \uparrow & & \\
Ш^1(k_S, A') & \hookrightarrow & H^1(k_S|k, A') & \to & \prod\limits_{S} H^1(k_\mathfrak{p}, A') & \longrightarrow & H^1(k_S|k, A)^\vee \\
 & & & & \uparrow & & \uparrow \\
 & & & & \prod\limits_{T} H^1(k_\mathfrak{p}, A') & \xrightarrow{\sim} & \prod\limits_{T} H^1(k_\mathfrak{p}, A)^\vee \\
 & & & & \uparrow & & \uparrow \\
 & & & & 0 & & \mathrm{coker}^1(k_S, T, A)^\vee \\
 & & & & & & \uparrow \\
 & & & & & & 0\,.
\end{array}$$

This diagram implies the

(9.2.2) Lemma. *Assume that S is nonempty and contains S_∞ if k is a number field. If T is finite and $\#A \in \mathbb{N}(S)$, then there is a canonical exact sequence*

$$0 \longrightarrow Ш^1(k_S, A') \longrightarrow Ш^1(k_S, S \smallsetminus T, A') \longrightarrow \mathrm{coker}^1(k_S, T, A)^\vee \longrightarrow 0.$$

Using the Hasse principles of the last section, we obtain the following result.

(9.2.3) Theorem. *Let k be a global field, and let $T \subseteq S$ be sets of primes of k where T is finite and S contains S_∞ if k is a number field. Let A be a finite $G_S(k)$-module with $\#A \in \mathbb{N}(S)$. Then the canonical homomorphism*

$$res^1(S, T, A) : \; H^1(k_S|k, A) \longrightarrow \bigoplus_{T} H^1(k_\mathfrak{p}, A)$$

is surjective in the following cases:

(i) *A' is a trivial $G_S(k)$-module and $\delta_k(S) > 1/p$, where p is the smallest prime divisor of $\#A$.*

(ii) *$A = \mathbb{Z}/m\mathbb{Z}$ with $m = p_1^{r_1} \cdots p_n^{r_n}$, where p_i are pairwise different prime numbers in $\mathbb{N}(S)$, and*

$$\delta(cs(k(\mu_{p_i^{r_i}})|k) \cap S) > \frac{1}{p_i \cdot [k(\mu_{p_i^{r_i}}) : k]}$$

for $i = 1, \ldots, n$, except we are in the special case $(k, m, S \smallsetminus T)$. In this case the cokernel of $res^1(S, T, \mathbb{Z}/m\mathbb{Z})$ is of order 1 or 2.

(iii) *$cs(k(A')|k) \subseteq S$ and $\#G(k(A')|k) = lcm\{\#G(k(A')_\mathfrak{p}|k_\mathfrak{p}) \mid \mathfrak{p} \in S \smallsetminus T\}$.*

(iv) *A is simple, $G(k(A')|k)$ is solvable and $cs(k(A')|k) \subsetneq S$.*

(v) *A is simple, $G(k(A)|k)$ is solvable and $cs(k(A)|k) \subsetneq S$.*

(vi) *A is simple, $pA = 0$ for a prime number $p \in \mathbb{N}(S)$, and there are finite Galois extensions $K \subseteq \Omega$ of k inside k_S such that $p \nmid [\Omega : K]$,*

$$\mu_p \nsubseteq K \text{ and } k(A) \subseteq K,$$
$$\mu_p \subseteq \Omega \text{ and } cs(\Omega|k) \subsetneq S.$$

(vii) *$\delta_k(S) = 1$ and $G(k(A')_{\mathfrak{p}}|k_{\mathfrak{p}})$ is cyclic for all $\mathfrak{p} \in T$.*

Proof: By (9.2.2), we know that

$$\operatorname{coker}^1(k_S, T, A) \subseteq Ш^1(k_S, S \smallsetminus T, A')^\vee.$$

Thus assertions (i)-(vi) follow from the corresponding statements in (9.1.9) and (9.1.15), respectively. In order to prove (vii), we first note that, since $G'_{\mathfrak{p}} = G(k(A')_{\mathfrak{p}}|k_{\mathfrak{p}})$ is cyclic for all $\mathfrak{p} \in T$ and since $\delta_k(S) = 1$, by Čebotarev's density theorem there exists a prime $\tilde{\mathfrak{p}} \in S \smallsetminus T$ for each $\mathfrak{p} \in T$ such that $G'_{\mathfrak{p}} = G'_{\tilde{\mathfrak{p}}} \subseteq G' = G(k(A')|k)$. If $x \in Ш^1(k(A')|k, S \smallsetminus T, A')$, then $res^1_{\tilde{\mathfrak{p}}}\, x = 0$ by definition, hence $res^1_{\mathfrak{p}}\, x = 0$ and therefore $x \in Ш^1(k(A')|k, S, A')$. Thus the injection $Ш^1(k(A')|k, S, A') \hookrightarrow Ш^1(k(A')|k, S \smallsetminus T, A')$ is an isomorphism. Since $\delta_{k(A')}(S \smallsetminus T) = 1$, we have

$$Ш^1(k_S|k(A'), A') = 0 = Ш^1(k_S|k(A'), S \smallsetminus T, A')$$

and the commutative exact diagram

$$\begin{array}{ccccc} Ш^1(k_S|k, A') & \hookrightarrow & Ш^1(k_S|k, S \smallsetminus T, A') & \twoheadrightarrow & \operatorname{coker}^1(k_S, T, A)^\vee \\ \uparrow\wr & & \uparrow\wr & & \\ Ш^1(k(A')|k, S, A') & \xrightarrow{\sim} & Ш^1(k(A')|k, S \smallsetminus T, A') & & \end{array}$$

implies (vii). This finishes the proof of the theorem. □

(9.2.4) Corollary. *Let k be a global field, and let $T \subseteq S$ be sets of primes of k where T is finite and S contains S_∞ if k is a number field. Let K be a finite Galois extension of k inside k_S with Galois group $G = G(K|k)$ and let $A = \mathbb{F}_p[G]^n$, where $p \in \mathbb{N}(S)$ is a prime number. If $\delta(S) = 1$, then the map*

$$res^1(S, T, A) : \ H^1(k_S|k, A) \longrightarrow \bigoplus_{\mathfrak{p} \in T} H^1(k_{\mathfrak{p}}, A)$$

is surjective.

Proof: By (9.2.3)(ii), the lower restriction map in the commutative diagram

$$\begin{array}{ccc} H^1(k_S|k,A) & \xrightarrow{res} & \bigoplus\limits_{\mathfrak{p}\in T} H^1(k_\mathfrak{p},A) \\ \downarrow\wr & & \downarrow\wr \\ H^1(k_S|K,\mathbb{F}_p^n) & \xrightarrow{res} & \bigoplus\limits_{\mathfrak{P}\in T(K)} H^1(K_\mathfrak{P},\mathbb{F}_p^n) \end{array}$$

is surjective, hence the same is true for the upper restriction map. □

The statement of (9.2.4) also holds if $p = \mathrm{char}(k)$. In fact, we have the following stronger result.

(9.2.5) Theorem. *Let k be a global field of characteristic $p > 0$ and let $T \subseteq S$ be sets of primes of k where T is finite and $S \smallsetminus T \neq \varnothing$. Let A be a p-primary $G_S(k)$-module. Then the canonical homomorphism*

$$res^1(S,T,A):\ H^1(k_S|k,A) \longrightarrow \bigoplus_T H^1(k_\mathfrak{p},A)$$

is surjective.

Proof: We start with the case $A = \mathbb{F}_p$. By (8.3.2), we have

$$H^1(G_S(k),\mathbb{F}_p) = \mathcal{O}_{k,S}/\wp\mathcal{O}_{k,S},$$

where $\wp(x) = x^p - x$. For any prime $\mathfrak{p}$ of k, (6.1.2) yields $H^1(k_\mathfrak{p},\mathbb{F}_p) = k_\mathfrak{p}/\wp k_\mathfrak{p}$. Furthermore, $\wp k_\mathfrak{p}$ is open in $k_\mathfrak{p}$ by (6.1.6). Since $S \smallsetminus T \neq \varnothing$, the strong approximation theorem (cf. [22], th. 15.1) shows that the natural map

$$\mathcal{O}_{k,S} \longrightarrow \bigoplus_{\mathfrak{p}\in T} k_\mathfrak{p}/\wp k_\mathfrak{p}$$

is surjective. This proves the assertion for $A = \mathbb{F}_p$. Using the same argument as in the proof of (9.2.4), we obtain the result for modules of the form $A = \mathbb{F}_p[G]^n$, where G is any finite quotient of $G_S(k)$.

Now let $0 \to A' \to A \to A'' \to 0$ be an exact sequence of p-primary $G_S(k)$-modules. Since $cd_p\, G_S(k) = 1$ by (8.3.3), the exact and commutative diagram

$$\begin{array}{ccccccc} H^1(k_S|k,A') & \longrightarrow & H^1(k_S|k,A) & \longrightarrow & H^1(k_S|k,A'') & \longrightarrow & 0 \\ \downarrow & & \downarrow & & \downarrow & & \\ \bigoplus\limits_T H^1(k_\mathfrak{p},A') & \longrightarrow & \bigoplus\limits_T H^1(k_\mathfrak{p},A) & \longrightarrow & \bigoplus\limits_T H^1(k_\mathfrak{p},A'') & \longrightarrow & 0 \end{array}$$

shows that the surjectivity statement for A implies that for A'', and that surjectivity for A' and A'' shows surjectivity for A. Any finite $G_S(k)$-module A with $pA = 0$ is a quotient of a module of the form $\mathbb{F}_p[G]^n$. Hence the result is true for such modules. Any finite p-primary $G_S(k)$-module has a composition series with graded pieces annihilated by p. Hence the result is true for all finite p-primary $G_S(k)$-modules. Finally, the general case follows by passing to the inductive limit over all finite submodules. □

(9.2.6) Corollary. *Let k be a global field of characteristic $p > 0$ and let T be a finite set of primes of k. If k^T is the maximal extension of k which is completely decomposed at the primes of T, then*

$$cd_p\, G(k^T|k) = 1\,.$$

Proof: Since the Galois group G_{k^T} is generated by the decomposition groups $G_{k_{\mathfrak{P}}}$, $\mathfrak{P}|\mathfrak{p}$ and $\mathfrak{p} \in T$, the map res_2 in the commutative diagram

$$\begin{array}{ccccccc}
H^1(k,A) & \xrightarrow{res_1} & \bigoplus_{\mathfrak{p}\in T} H^1(k_{\mathfrak{p}},A) & & & & \\
\| & & \uparrow{\scriptstyle res_2} & & & & \\
H^1(k,A) & \longrightarrow & H^1(k^T,A)^{G(k^T|k)} & \longrightarrow & H^2(k^T|k,A) & \longrightarrow & 0
\end{array}$$

is injective; here A is a p-primary $G(k^T|k)$-module, also considered as a G_k-module. The zero in the lower exact sequence is a consequence of (6.5.10). Since res_1 is surjective by the theorem above, it follows that res_2 is an isomorphism, and therefore $H^2(k^T|k,A) = 0$. Hence $cd_p G(k^T|k) \le 1$ by (3.3.2). As the p-Sylow subgroups of $G(k^T|k)$ are nontrivial, we obtain the result. □

For the trivial module $A = \mathbb{Z}/p^r\mathbb{Z}$ we are going to prove a more far reaching surjectivity result.

(9.2.7) Theorem. *Let k be a global field, p a prime number and S a set of primes of k with $\delta(S) = 1$. Let T_0 and T be disjoint subsets of S such that T_0 is finite and $\delta(T) = 0$.*

If k is a function field or if k is a number field and we are not in the special case $(k, p^r, S \smallsetminus (T_0 \cup T))$, then the canonical homomorphism

$$H^1(G_S(k),\mathbb{Z}/p^r\mathbb{Z}) \longrightarrow \bigoplus_{\mathfrak{p}\in T_0} H^1(k_{\mathfrak{p}},\mathbb{Z}/p^r\mathbb{Z}) \oplus \bigoplus_{\mathfrak{p}\in T} H^1(T_{\mathfrak{p}},\mathbb{Z}/p^r\mathbb{Z})^{G_{\mathfrak{p}}}$$

is surjective, where $G_{\mathfrak{p}} = G(\bar{k}_{\mathfrak{p}}|k_{\mathfrak{p}})$ and $T_{\mathfrak{p}} \subseteq G_{\mathfrak{p}}$ is the inertia group.

If k is a number field and we are in the special case $(k, p^r, S \smallsetminus (T_0 \cup T))$, then the cokernel of this map is of order 1 or 2.

Proof: By convention, $T_\mathfrak{p} = G_\mathfrak{p}$, hence $H^1(T_\mathfrak{p}, \mathbb{Z}/p^r\mathbb{Z})^{G_\mathfrak{p}} = H^1(k_\mathfrak{p}, \mathbb{Z}/p^r\mathbb{Z})$, if $\mathfrak{p}$ is archimedean. Therefore we may move all archimedean primes of T to T_0 and assume $T \cap S_\infty = \varnothing$. We first assume that $S \supseteq S_p \cup S_\infty$ if k is a number field. In the following we write $H^i(-)$ for $H^i(-, \mathbb{Z}/p^r\mathbb{Z})$.

Let $T_1 \subseteq T$ be a finite subset. Noting that $H^1(k_\mathfrak{p}) \to H^1(T_\mathfrak{p})^{G_\mathfrak{p}}$ is surjective, (9.2.3)(ii) and (9.2.5) imply the result for the natural map

$$H^1(G_{(S \smallsetminus T)\cup T_1}(k)) \longrightarrow \bigoplus_{\mathfrak{p}\in T_0} H^1(k_\mathfrak{p}) \oplus \bigoplus_{\mathfrak{p}\in T_1} H^1(T_\mathfrak{p})^{G_\mathfrak{p}}.$$

For $T_1 \subseteq T_2 \subseteq T$, we obtain the commutative diagram

$$\begin{array}{ccc} H^1(G_{(S \smallsetminus T)\cup T_2}) & \longrightarrow & \bigoplus_{\mathfrak{p}\in T_0} H^1(k_\mathfrak{p}) \oplus \bigoplus_{\mathfrak{p}\in T_2} H^1(T_\mathfrak{p})^{G_\mathfrak{p}} \\ \uparrow & & \uparrow \\ H^1(G_{(S \smallsetminus T)\cup T_1}) & \longrightarrow & \bigoplus_{\mathfrak{p}\in T_0} H^1(k_\mathfrak{p}) \oplus \bigoplus_{\mathfrak{p}\in T_1} H^1(T_\mathfrak{p})^{G_\mathfrak{p}}. \end{array}$$

Passing to the limit over all finite subsets of T, we obtain the desired result.

It remains to consider the case when k is a number field and S does not contain $S_p \cup S_\infty$. Let $V = (S_p \cup S_\infty) \smallsetminus S$. By what we just proved, the map

$$H^1(G_{S\cup V}(k)) \longrightarrow \bigoplus_{\mathfrak{p}\in T_0\cup V} H^1(k_\mathfrak{p}) \oplus \bigoplus_{\mathfrak{p}\in T} H^1(T_\mathfrak{p})^{G_\mathfrak{p}},$$

is surjective, unless we are in the special case $(k, p^r, S \smallsetminus (T_0 \cup T))$ where the cokernel is of order 1 or 2. Since $H^1(k_{S\cup V}|k_S)^{G_S(k)} \hookrightarrow \bigoplus_{\mathfrak{p}\in V} H^1(T_\mathfrak{p})^{G_\mathfrak{p}}$ is injective, the same statement as above holds true also for the map

$$H^1(G_S(k)) \longrightarrow \bigoplus_{\mathfrak{p}\in T_0} H^1(k_\mathfrak{p}) \oplus \bigoplus_{\mathfrak{p}\in V} H^1_{nr}(G_\mathfrak{p}) \oplus \bigoplus_{\mathfrak{p}\in T} H^1(T_\mathfrak{p})^{G_\mathfrak{p}}.$$

This implies the desired result. □

The following assertion is known as the theorem of GRUNWALD-WANG, who, however, only considered the case of a cyclic group A.

(9.2.8) Theorem. *Let S be a finite set of primes of a global field k and let A be a finite abelian group. Let for all $\mathfrak{p} \in S$ finite abelian extensions $K_\mathfrak{p}|k_\mathfrak{p}$ be given such that $G(K_\mathfrak{p}|k_\mathfrak{p})$ may be embedded into A. If k is a number field assume we are not in the special case $(k, \exp(A), \mathcal{P}(k) \smallsetminus S)$, where $\mathcal{P}(k)$ is the set of all places of k and $\exp(A)$ denotes the exponent of A.*

Then there exists a global abelian extension $K|k$ with Galois group A such that K has the given completions $K_\mathfrak{p}$ for all $\mathfrak{p} \in S$.

Proof: We show the following stronger statement: the map

$$\mathrm{Epi}(G_k, A) \longrightarrow \bigoplus_S \mathrm{Hom}(G_{k_\mathfrak{p}}, A)$$

is surjective, where $\mathrm{Epi}(G_k, A)$ denotes the set of surjective homomorphisms from G_k onto A.

Let $\mathfrak{q}_1, \dots, \mathfrak{q}_r$ be primes not in S (and not dividing 2 in the number field case), and let

$$\varphi_{\mathfrak{q}_i} : G_{k_{\mathfrak{q}_i}} \longrightarrow A$$

be homomorphisms such that the images of the $\varphi_{\mathfrak{q}_i}$ generate the group A. For each prime $\mathfrak{p} \in S$, let $\varphi_\mathfrak{p} : G(K_\mathfrak{p}|k_\mathfrak{p}) \hookrightarrow A$ be an embedding of the local group into A. Denote by S' the union $S \cup \{\mathfrak{q}_1, \dots, \mathfrak{q}_r\}$ and let T and T' be the complements of S and S' in $\mathcal{P}(k)$.

Since $\delta(T) = \delta(T') = 1$, (9.1.8) implies that the special cases $(k, 2^r, T)$ and $(k, 2^r, T')$ can only occur if k is a number field and $k(\mu_{2^r})|k$ is not cyclic. Then all primes $\mathfrak{p} \nmid 2$ decompose in $k(\mu_{2^r})|k$ and therefore $(k, 2^r, T)$ and $(k, 2^r, T')$ are equivalent. Furthermore $(k, 2^r, T)$ implies $(k, 2^{r+1}, T)$. Decomposing A into a product of cyclic groups, (9.2.3) (ii) and (9.2.5) imply the surjectivity of the map $H^1(k, A) \longrightarrow \bigoplus_{S'} H^1(k_\mathfrak{p}, A)$ if we are not in the special case $(k, \exp(A), T') = (k, \exp(A), T)$. Now a pre-image $(\varphi : G_k \to A) \in \mathrm{Hom}(G_k, A) = H^1(k, A)$ of

$$(\varphi_{\mathfrak{q}_1}, \dots, \varphi_{\mathfrak{q}_r}, \varphi_\mathfrak{p}, \mathfrak{p} \in S) \in \bigoplus_{S'} \mathrm{Hom}(G_{k_\mathfrak{p}}, A) = \bigoplus_{S'} H^1(k_\mathfrak{p}, A)$$

realizes the local extensions $K_\mathfrak{p}|k_\mathfrak{p}$ and it is surjective by the choice of the homomorphisms $\varphi_{\mathfrak{q}_i}$. □

Remark: The conclusion of (9.2.8) may hold even if we are in the special case $(k, \exp(A), \mathcal{P}(k) \smallsetminus S)$. By (9.2.2) this is exactly the case when

$$\text{Ш}^1(k_S, \mathcal{P}(k), A') \cong \text{Ш}^1(k_S, \mathcal{P}(k) \smallsetminus S, A').$$

Finally, we give an application to embedding problems with induced G-modules as kernel.

(9.2.9) Proposition. *Let $K|k$ be a finite Galois extension of global fields with Galois group $G = G(K|k)$ and let $A = \mathbb{F}_p[G]^n$, where p is a prime number. Then the embedding problem*

$$\begin{array}{ccccccccc} & & & & & & G_k & & \\ & & & & & & \downarrow & & \\ 1 & \longrightarrow & A & \longrightarrow & E & \longrightarrow & G & \longrightarrow & 1 \end{array}$$

is properly solvable.

Proof: Since $H^2(G, A)$ vanishes, the embedding problem has a solution $\psi_0 : G_k \longrightarrow E$, see (3.5.9). Let $\mathfrak{p}_1, \ldots, \mathfrak{p}_r$ be primes of k which split completely in K and let $\varphi_i : G_{k_{\mathfrak{p}_i}} \longrightarrow A$ be homomorphisms such that their images generate A. By (9.2.4) and (9.2.5) the natural map

$$H^1(k, A) \longrightarrow \bigoplus_{i=1}^{r} H^1(k_{\mathfrak{p}_i}, A)$$

is surjective. Let $[x] \in H^1(k, A)$ be a 1-class such that $res_i[x] = \varphi_i - \psi_0|_{G_{k_{\mathfrak{p}_i}}}$ for $i = 1, \ldots, r$. Then $\psi = x \cdot \psi_0 : G_k \longrightarrow E$ is a new solution of the embedding problem; this is proper, since $\psi|_{G_{k_{\mathfrak{p}_i}}} = \varphi_i$ for $i = 1, \ldots, r$, and so $\psi(G_K) = A$. □

Exercise 1. Let k be a global field. Prove that every finite abelian group A occurs as a Galois group of a finite abelian extension of k.

Exercise 2. Prove the following generalization of the theorem of GRUNWALD-WANG.

Let S be a finite set of primes of a global field k and let A be a finite abelian group. Let for all $\mathfrak{p} \in S$ finite abelian extensions $K_\mathfrak{p}|k_\mathfrak{p}$ be given such that $G(K_\mathfrak{p}|k_\mathfrak{p})$ may be embedded into A. Let T be a set of primes of k of density zero which is disjoint to S. Assume we are not in the special case $(k, \exp(A), \mathcal{P}(k) \smallsetminus (S \cup T))$.

Then there exists a global abelian extension $K|k$ with Galois group A such that K has the given completions $K_\mathfrak{p}$ for all $\mathfrak{p} \in S$ and $K|k$ is unramified at all $\mathfrak{p} \in T$.

§3. Construction of Cohomology Classes

The aim of this section is to establish the existence of global cohomology classes with given local behaviour. More precisely, given classes $x_\mathfrak{p} \in H^1(k_\mathfrak{p}, A)$ for all $\mathfrak{p}$ in a finite set of primes T, we are looking for a global class $x \in H^1(k, A)$ which maps to $x_\mathfrak{p}$ for all $\mathfrak{p} \in T$ and satisfies additional conditions at all other primes. We start by introducing some notation.

(9.3.1) Definition. *Let $k_\mathfrak{p}$ be a local field and let A be a $G_{k_\mathfrak{p}}$-module.*

(i) *We call a class $x_\mathfrak{p} \in H^1(k_\mathfrak{p}, A)$* **cyclic** *if it is split by a cyclic extension of $k_\mathfrak{p}$, i.e. if there exists a cyclic extension $K_\mathfrak{p}|k_\mathfrak{p}$ such that $x_\mathfrak{p}$ lies in the kernel of the restriction map $H^1(k_\mathfrak{p}, A) \to H^1(K_\mathfrak{p}, A)$.*

(ii) *We call $x_\mathfrak{p}$* **unramified** *if it is split by an unramified extension, i.e. if it is contained in the unramified part $H^1_{nr}(k_\mathfrak{p}, A)$ of $H^1(k_\mathfrak{p}, A)$. In particular, if $x_\mathfrak{p}$ is unramified, then it is cyclic.*

For $A = \mu_p$ the following theorem is equivalent, via Kummer theory, to Šafarevič's theorem about the existence of certain algebraic numbers ([188]) and we shall make use of the ideas of Šafarevič's proof.

(9.3.2) Theorem. *Let k be a global field and let p be a prime number different to* char(k). *Let $K \subseteq \Omega$ be finite Galois extensions of k and assume that K contains the group μ_p of p-th roots of unity. Let S_0 be a nonempty finite set of primes of k containing $S_p \cup S_\infty$ if k is a number field. Let T be a finite set of primes of k containing* $\mathrm{Ram}(\Omega|k) \cup S_0$ *and let* $S = cs(\Omega|k) \cup T$.

Let A be a finite $\mathbb{F}_p[G(K|k)]$-module and assume that we are given a class y in $H^1(k_S|K, A)$ such that

$y_{\mathfrak{P}}$ *is unramified for* $\mathfrak{P} \in T(K)$ *and*
$y_{\mathfrak{P}} = 0$ *for* $\mathfrak{P} \cap k \in \mathrm{Ram}(K|k) \cup S_0$.

Then there exists an element $x \in H^1(k_S|k, A)$ such that

$$x_{\mathfrak{p}} = (cor_k^K\, y)_{\mathfrak{p}} \ \text{ for } \mathfrak{p} \in T \ \text{ and } \ x_{\mathfrak{p}} \text{ is cyclic for all } \mathfrak{p} \notin T.$$

Proof: Setting $x = cor_k^K\, z$, it suffices to construct $z \in H^1(k_S|K, A)$ with

(a) $z_{\mathfrak{P}} = y_{\mathfrak{P}}$ for $\mathfrak{P} \in T(K)$,

(b) if $\mathfrak{P} \notin T(K)$ and $z_{\mathfrak{P}}$ is ramified (i.e. not contained in $H^1_{nr}(K_{\mathfrak{P}}, A)$), then $z_{\mathfrak{P}}$ is cyclic and $z_{\sigma\mathfrak{P}} = 0$ for every $\sigma \in G(K|k) \setminus \{1\}$.

We first prove the existence of z in the case $A = \mu_p$, when the cyclicity condition for $z_{\mathfrak{P}}$ is trivially satisfied. Assume first that p is odd. We will apply the method of [188] in order to obtain the element z which we are looking for, as a sum of two members of a sequence

$$z_1, z_2, z_3, \ldots \in H^1(k_S|K, \mu_p)$$

which will be constructed having the following properties.

(1) There exists a prime $\mathfrak{P}_i \in S \setminus T(K)$ such that
- $(z_i)_{\mathfrak{P}_i} \equiv \mathrm{Frob}_{\mathfrak{P}_i}$ modulo $H^1_{nr}(K_{\mathfrak{P}_i}, \mu_p)$,
- $(z_i)_{\mathfrak{P}} \in H^1_{nr}(K_{\mathfrak{P}}, \mu_p)$ for all $\mathfrak{P} \neq \mathfrak{P}_i$,

(2) $(z_i)_{\mathfrak{P}} = \frac{1}{2} y_{\mathfrak{P}}$ for $\mathfrak{P} \in T(K)$,

(3) $(z_{n+1})_{\sigma\mathfrak{P}_i} = -(z_i)_{\sigma\mathfrak{P}_i}$ for $i \leq n$ and all $\sigma \in G(K|k) \setminus \{1\}$.

In (1) we view $\mathrm{Frob}_{\mathfrak{P}}$ as an element of $H^1(K_{\mathfrak{P}}, \mu_p)/H^1_{nr}(K_{\mathfrak{P}}, \mu_p)$ via

$$G(K_{\mathfrak{P}}^{nr}|K_{\mathfrak{P}}) \twoheadrightarrow H^1_{nr}(K_{\mathfrak{P}}, \mathbb{Z}/p\mathbb{Z})^\vee \cong H^1(K_{\mathfrak{P}}, \mu_p)/H^1_{nr}(K_{\mathfrak{P}}, \mu_p),$$

where the isomorphism is induced by the local duality theorem; see (7.2.6) and (7.2.15). Assume that we have already constructed $z_1, \ldots, z_n$ $(n \geq 0)$ and set $T_n = T \cup \{(\mathfrak{P}_1 \cap k), \ldots, (\mathfrak{P}_n \cap k)\}$, i.e. $T_n(K)$ consists of $T(K)$ and of all $G(K|k)$-conjugates of $\mathfrak{P}_1, \ldots, \mathfrak{P}_n$. Observe that $T_n \subseteq S$ and consider the commutative diagram

$$\begin{array}{ccccc}
H^1(k_S|K,\mu_p) & \longrightarrow & \prod\limits_{S} H^1(K_{\mathfrak{P}},\mu_p) & \longrightarrow & H^1(k_S|K,\mathbb{Z}/p\mathbb{Z})^\vee . \\
\Vert & & \downarrow & & \downarrow \\
H^1(k_S|K,\mu_p) & \longrightarrow & \prod\limits_{T_n} H^1(K_{\mathfrak{P}},\mu_p) \times \bigoplus\limits_{S\setminus T_n} H^1(K_{\mathfrak{P}},\mu_p)/H^1_{nr} & \longrightarrow & H^1(k_{T_n}|K,\mathbb{Z}/p\mathbb{Z})^\vee \\
 & & \downarrow & & \downarrow \\
 & & \prod\limits_{T} H^1(K_{\mathfrak{P}},\mu_p) & \longrightarrow & H^1(\Omega|K,\mathbb{Z}/p\mathbb{Z})^\vee
\end{array}$$

The upper row is part of the long exact sequence of Poitou-Tate, and the exactness of the middle row follows in a straightforward manner. By definition of S, all $\mathfrak{P} \in S \smallsetminus T(K)$ are completely decomposed in $\Omega|K$. Therefore, for $\mathfrak{P} \in S \smallsetminus T(K)$, the homomorphism

$$H^1(K_{\mathfrak{P}},\mu_p) \xrightarrow{\sim} H^1(K_{\mathfrak{P}},\mathbb{Z}/p\mathbb{Z})^\vee \longrightarrow H^1(\Omega|K,\mathbb{Z}/p\mathbb{Z})^\vee$$

is zero, showing that the lower square in the diagram commutes. Consider the element

$$\xi \in \prod_{T_n} H^1(K_{\mathfrak{P}},\mu_p) \times \bigoplus_{S\setminus T_n} H^1(K_{\mathfrak{P}},\mu_p)/H^1_{nr}$$

given by $\xi_{\mathfrak{P}} = \frac{1}{2}y_{\mathfrak{P}}$ for $\mathfrak{P} \in T(K)$, $\xi_{\sigma\mathfrak{P}_i} = -(z_i)_{\sigma\mathfrak{P}_i}$ for $i \le n$, $\sigma \in G(K|k) \smallsetminus \{1\}$ and $\xi_{\mathfrak{P}} = 0$ for all other $\mathfrak{P}$. Then ξ has the same image in $H^1(\Omega|K,\mathbb{Z}/p\mathbb{Z})^\vee$ as $\frac{1}{2}y \in H^1(k_S|K,\mu_p)$. By the exactness of the rows, this image is trivial. Let $K^{el}_{T_n}$ be the maximal elementary abelian p-extension of K inside k_{T_n}. We consider the diagram of fields

$$\begin{array}{ccccc}
 & & k_{T_n} & & \\
 & & | & & \\
 & & K^{el}_{T_n}\Omega & & \\
 & / & & \backslash & \\
K^{el}_{T_n} & & & & \Omega \\
 & \backslash & & / & \\
 & & K^{el}_{T_n}\cap\Omega & & \\
 & & | & & \\
 & & K & & \\
 & & | & & \\
 & & k\,. & &
\end{array}$$

Let η be the image of ξ in $H^1(k_{T_n}|K,\mathbb{Z}/p\mathbb{Z})^\vee = G(K^{el}_{T_n}|K)$. As the image of ξ in $H^1(\Omega|K,\mathbb{Z}/p\mathbb{Z})^\vee$ is zero, η lies in the subgroup $G(K^{el}_{T_n}|K^{el}_{T_n}\cap\Omega)$. By Čebotarev's density theorem applied to $K^{el}_{T_n}\Omega|k$, we can choose a prime $\mathfrak{P}^\bullet$ in $K^{el}_{T_n}\Omega$ outside $T_n(K^{el}_{T_n}\Omega)$ such that $\mathrm{Frob}_{\mathfrak{P}^\bullet}|_{K^{el}_{T_n}} = -\eta$ and $\mathrm{Frob}_{\mathfrak{P}^\bullet}|_{\Omega} = 1$. Hence $\mathfrak{P}^\bullet \in S \smallsetminus T_n(K^{el}_{T_n}\Omega)$. We put $\mathfrak{P}_{n+1} = \mathfrak{P}^\bullet|_K \in S \smallsetminus T_n(K)$. Then the image of $-\xi$ in $H^1(k_{T_n}|K,\mathbb{Z}/p\mathbb{Z})^\vee$ is equal to $\mathrm{Frob}_{\mathfrak{P}_{n+1}}$. By the exactness of the middle row, we find a class $z_{n+1} \in H^1(k_S|K,\mu_p)$ with properties (1), (2), (3).

Let $G(K|k) \smallsetminus \{1\} = \{\sigma_1, \ldots, \sigma_r\}$ and consider the maps

$$\psi_n \colon \{z_1, \ldots, z_n\} \longrightarrow \prod_1^r \mu_p \,,$$

for $n = 1, 2, \ldots$ given by

$$\psi_n(z_i) = \Big((z_i)_{\sigma_1 \mathfrak{P}_i}(\mathrm{Frob}_{\sigma_1 \mathfrak{P}_i}), \ldots, (z_i)_{\sigma_r \mathfrak{P}_i}(\mathrm{Frob}_{\sigma_r \mathfrak{P}_i}) \Big).$$

Observe that by construction $z_i \in H^1_{nr}(K_{\sigma_j \mathfrak{P}_i}, \mu_p)$ for $j = 1, \ldots, r$. By the pigeonhole principle, there exists an N with $\psi_N(z_N) = \psi_N(z_i)$ for some $i < N$. We claim that

$$z = z_i + z_N \in H^1(k_S|K, \mu_p)$$

satisfies conditions (a) and (b) above. Indeed, (a) is trivial by condition (2). It therefore remains to show that if $z_{\mathfrak{P}}$ is ramified for some $\mathfrak{P}$, then $\mathfrak{P} \in S \smallsetminus T(K)$ and $z_{\sigma\mathfrak{P}} = 0$ for $\sigma \in G(K|k) \smallsetminus \{1\}$. By construction, z is only ramified at $\mathfrak{P}_i$ and $\mathfrak{P}_N$, and by condition (3), we know that $z_{\sigma \mathfrak{P}_i} = 0$ for $\sigma \neq 1$. In order to show the corresponding statement at $\sigma\mathfrak{P}_N$ ($\sigma \neq 1$), recall that for arbitrary classes $a, b \in H^1(k_S|K, \mu_p)$, we have the product formula

$$\prod_{\mathfrak{P} \in S(K)} (a, b)_{\mathfrak{P}} = 1$$

for the Hilbert symbol. (The symbol $(a, b)_{\mathfrak{P}}$ is defined as the image of $a \cup b$ under the trace homomorphism $H^2(K_{\mathfrak{P}}, \mu_p^{\otimes 2}) \xrightarrow{\sim} \mu_p$.) Since z_i and z_N are unramified at $\sigma\mathfrak{P}_N$, it suffices to show that their values on $\mathrm{Frob}_{\sigma\mathfrak{P}_N}$ are mutually inverse in μ_p. We have

$$\begin{aligned}
z_N(\mathrm{Frob}_{\sigma\mathfrak{P}_N}) &= z_i(\mathrm{Frob}_{\sigma\mathfrak{P}_i}) && \text{because } \psi_n(z_i) = \psi_n(z_N),\\
&= (z_i, \sigma z_i)_{\sigma\mathfrak{P}_i} && \text{by condition (1) for } z_i,\\
&= (z_i, \sigma z_i)_{\mathfrak{P}_i}^{-1} && \text{by the product formula and (1), (2),}\\
&= (z_i, \sigma z_N)_{\mathfrak{P}_i} && \text{by condition (3),}\\
&= (z_i, \sigma z_N)_{\sigma\mathfrak{P}_N}^{-1} && \text{by the product formula and (1), (2),}\\
&= z_i(\mathrm{Frob}_{\sigma\mathfrak{P}_N})^{-1} && \text{by condition (1) for } z_N.
\end{aligned}$$

This finishes the case $A = \mu_p$, for p odd.

In the case $p = 2$ we have to modify the construction, and we will obtain z as a sum of three other elements. We use the combinatorial method of [85], chap.5, §3.

Let $\{G_1, G_2, G_3\}$ be a partition of the set $G(K|k) \smallsetminus \{1\}$ such that G_1 consists of all elements of order 2 and $G_2 = G_3^{-1}$.

We construct recursively a sequence $z_1, z_2, \ldots$ of elements in $H^1(k_S|K, \mu_2)$ satisfying the following properties:

(1) There exists a prime $\mathfrak{P}_i \in S \smallsetminus T(K)$ such that

- $(z_i)_{\mathfrak{P}_i} \equiv \mathrm{Frob}_{\mathfrak{P}_i}$ modulo $H^1_{nr}(K_{\mathfrak{P}_i}, \mu_2)$,
- $(z_i)_{\mathfrak{P}} \in H^1_{nr}(K_{\mathfrak{P}}, \mu_2)$ for all $\mathfrak{P} \neq \mathfrak{P}_i$,

(2) $(z_i)_{\mathfrak{P}} = y_{\mathfrak{P}}$ for $\mathfrak{P} \in T(K)$,

(3) $(z_{n+1})_{\sigma\mathfrak{P}_i} = 0$ for $i \le n$ and all $\sigma \in G_1$.

(4) If $\psi_n(z_i) \neq \psi_n(z_j)$ for all j with $n \ge j > i$, then

$$(z_{n+1})_{\sigma\mathfrak{P}_i} = \begin{cases} 0 & \text{if } \sigma \in G_2, \\ (\sigma^2 z_i)_{\sigma\mathfrak{P}_i} & \text{if } \sigma \in G_3, \end{cases}$$

and otherwise $(z_{n+1})_{\sigma\mathfrak{P}_i} = \begin{cases} (z_i)_{\sigma\mathfrak{P}_i} & \text{if } \sigma \in G_2, \\ 0 & \text{if } \sigma \in G_3. \end{cases}$

The existence of the sequence of classes $z_1, z_2, \dots$ is proved similarly to the case of odd p. In addition, we obtain

Claim: $(z_i)_{\sigma\mathfrak{P}_i} = 0$ for $\sigma \in G_1$.

Proof of the claim: Let $\tilde{z}_i \in K^\times$ be a representative of $z_i \in H^1(k_S|K, \mu_2) \subseteq H^1(K, \mu_2) \cong K^\times/K^{\times 2}$. By condition (1), we have $\tilde{z}_i\mathcal{O}_{K,S_0} = \mathfrak{P}_i\mathfrak{A}^2$ for some fractional ideal $\mathfrak{A}$ of the Dedekind ring $\mathcal{O}_{K,S_0}$. By Čebotarev's density theorem, there exists a prime ideal $\mathfrak{Q}$ of K with $\mathfrak{Q} \notin (T(K) \cup \{\mathfrak{P}_i, \sigma\mathfrak{P}_i\})$ and $\mathfrak{Q} \neq \sigma\mathfrak{Q}$ such that $\mathfrak{Q} = \mathfrak{A} \cdot x\mathcal{O}_{K,S_0}$ with $x \in K^\times$. Hence we have $\tilde{z}_i x^2\mathcal{O}_{K,S_0} = \mathfrak{P}_i\mathfrak{Q}^2$ in $\mathcal{O}_{K,S_0}$. Replacing $\tilde{z}_i$ by $\tilde{z}_i x^2$ and using condition (2) we obtain

$$\tilde{z}_i \in \mathcal{O}_{K,S_0},\ \tilde{z}_i\mathcal{O}_{K,S_0} = \mathfrak{P}_i\mathfrak{Q}^2 \text{ and } (\tilde{z}_i)_\mathfrak{P} \in K_\mathfrak{P}^{\times 2} \text{ for } \mathfrak{P} \in S_0\,.$$

In addition, choose $\delta \in \mathcal{O}_{K,S_0}$ such that $K = K^\sigma(\delta)$ and $\delta^2 \in \mathcal{O}_{K^\sigma,S_0}$, where K^σ is the fixed field of K with respect to $\langle\sigma\rangle$. Then there are $a, b \in K^\sigma$ with $\tilde{z}_i = a + b\delta$, $\sigma\tilde{z}_i = a - b\delta$. Since $S_1' \subseteq S_0$, we have $a \in \mathcal{O}_{K^\sigma,S_0}$ and $b\delta \in \mathcal{O}_{K,S_0}$. Note that $a \neq 0$ and $b \neq 0$, because otherwise $\sigma\mathfrak{P}_i$ would divide $\tilde{z}_i = \pm\sigma\tilde{z}_i$.

Let Σ be the set of prime divisors of $b\delta$ in $\mathcal{O}_{K,S_0}$ which are not in $\mathrm{Ram}(K|k)$. Since $v_{\sigma\mathfrak{P}_i}(\tilde{z}_i) = 0$ and $v_{\sigma\mathfrak{P}_i}(\sigma\tilde{z}_i) = 1$, we conclude $v_{\sigma\mathfrak{P}_i}(2b\delta) = 0$, in particular, $\sigma\mathfrak{P}_i \notin \Sigma$. Furthermore,

$$(*) \qquad \tilde{z}_i^{-1}(2b\delta) = 1 - \tilde{z}_i^{-1}\sigma\tilde{z}_i \in U^1_{\sigma\mathfrak{P}_i}\,.$$

For $\mathfrak{P} \notin (\Sigma \cup \mathrm{Ram}(K|k) \cup S_0)$, $\mathfrak{P} \neq \sigma\mathfrak{P}_i$, we have $2b\delta \in U_\mathfrak{P}$ and $\sigma\tilde{z}_i \in U_\mathfrak{P}K_\mathfrak{P}^{\times 2}$. Hence $(2b\delta, \sigma\tilde{z}_i)_\mathfrak{P} = 1$. For $\mathfrak{P} \in \mathrm{Ram}(K|k) \cup S_0$ we have $\sigma\tilde{z}_i \in K_\mathfrak{P}^{\times 2}$, hence $(2b\delta, \sigma\tilde{z}_i)_\mathfrak{P} = 1$ also for these primes. Therefore the product formula for the Hilbert symbol implies the equality

$$(**) \qquad (2b\delta, \sigma\tilde{z}_i)_{\sigma\mathfrak{P}_i} = \prod_{\mathfrak{P}\in\Sigma} (2b\delta, \sigma\tilde{z}_i)_\mathfrak{P}\,.$$

Since $\tilde{z}_i$ and $\sigma\tilde{z}_i$ are coprime in $\mathcal{O}_{K,S_0}$, we have $v_\mathfrak{P}(a) = 0$ for $\mathfrak{P} \in \Sigma$ and therefore

$$(***) \qquad a^{-1}\sigma\tilde{z}_i = 1 - a^{-1}b\delta \in U^1_\mathfrak{P} \text{ for } \mathfrak{P} \in \Sigma\,.$$

We obtain

$$\begin{aligned} z_i(\mathrm{Frob}_{\sigma\mathfrak{P}_i}) &= (\tilde{z}_i, \sigma\tilde{z}_i)_{\sigma\mathfrak{P}_i} && \text{by definition of } \tilde{z}_i, \\ &= (2b\delta, \sigma\tilde{z}_i)_{\sigma\mathfrak{P}_i} && \text{by } (*), \\ &= \textstyle\prod_{\mathfrak{P}\in\Sigma}(2b\delta, \sigma\tilde{z}_i)_\mathfrak{P} && \text{by } (**), \\ &= \textstyle\prod_{\mathfrak{P}\in\Sigma}(2b\delta, a)_\mathfrak{P} && \text{by } (***). \end{aligned}$$

The last product is easily seen to be unity: For $\mathfrak{P} \in \Sigma$, the element $a \in K^\sigma$ is a unit in the completion with respect to $\mathfrak{P} \cap K^\sigma$. If $\mathfrak{P}$ is inert in $K|K^\sigma$, then a is a square in the unramified quadratic extension $K_\mathfrak{P}$ of $K^\sigma_{\mathfrak{P}\cap K^\sigma}$, so that $(2b\delta, a)_\mathfrak{P} = 1$. If $\mathfrak{P} \in \Sigma$ splits in $K|K^\sigma$, then $\sigma\mathfrak{P} \in \Sigma$ and

$$(2b\delta, a)_\mathfrak{P} \cdot (2b\delta, a)_{\sigma\mathfrak{P}} = (2b\delta, a)_\mathfrak{P} \cdot (-2b\delta, a)_\mathfrak{P} = (-1, a)_\mathfrak{P} = 1.$$

This proves the claim.

Now choose N minimal such that

$$\psi_N(z_i) = \psi_N(z_j) = \psi_N(z_N)$$

for numbers $i < j < N$. We claim that $z = z_i + z_j + z_N$ satisfies conditions (a) and (b). Indeed, (a) follows immediately from (2). It therefore remains to show that if $z_\mathfrak{P}$ is ramified for some $\mathfrak{P}$, then $\mathfrak{P} \in S \smallsetminus T(K)$ and $z_{\sigma\mathfrak{P}} = 0$ for $\sigma \in G(K|k) \smallsetminus \{1\}$. By construction, z is only ramified at $\mathfrak{P}_i$, $\mathfrak{P}_j$ and $\mathfrak{P}_N$. For $\sigma \in G_1$ and $N \geq s,\ t \geq 1$, we have

$$z_s(\mathrm{Frob}_{\sigma\mathfrak{P}_t}) = 1,$$

which is seen for $s > t$ by condition (3), for $s = t$ by the claim and follows for $s < t$ by (1)-(3) and the product formula

$$z_s(\mathrm{Frob}_{\sigma\mathfrak{P}_t}) = (z_s, \sigma z_t)_{\sigma\mathfrak{P}_t} = (\sigma z_s, z_t)_{\mathfrak{P}_t} = (\sigma z_s, z_t)_{\sigma\mathfrak{P}_s} = z_t(\mathrm{Frob}_{\sigma\mathfrak{P}_s}) = 1.$$

Summing up, we obtain $z_{\sigma\mathfrak{P}_i} = z_{\sigma\mathfrak{P}_j} = z_{\sigma\mathfrak{P}_N} = 0$ for $\sigma \in G_1$.

If $\sigma \in G_2$, then by condition (4)

$$z_{\sigma\mathfrak{P}_i} = (z_i)_{\sigma\mathfrak{P}_i} + (z_j)_{\sigma\mathfrak{P}_i} + (z_N)_{\sigma\mathfrak{P}_i} = (z_i)_{\sigma\mathfrak{P}_i} + 0 + (z_i)_{\sigma\mathfrak{P}_i} = 0.$$

Furthermore, since $(z_N)_{\sigma\mathfrak{P}_j} = 0$ by condition (4) and

$$\begin{aligned}
z_i(\mathrm{Frob}_{\sigma\mathfrak{P}_j}) &= (z_i, \sigma z_j)_{\sigma\mathfrak{P}_j} && \text{by condition (1),}\\
&= (z_i, \sigma z_j)_{\mathfrak{P}_i} && \text{by the product formula,}\\
&= (\sigma^{-1} z_i, z_j)_{\sigma^{-1}\mathfrak{P}_i} && \text{using Galois invariance,}\\
&= z_j(\mathrm{Frob}_{\sigma^{-1}\mathfrak{P}_i}) && \text{by condition (1),}\\
&= \sigma^{-2} z_i(\mathrm{Frob}_{\sigma^{-1}\mathfrak{P}_i}) && \text{by condition (4),}\\
&= z_i(\mathrm{Frob}_{\sigma\mathfrak{P}_i}) && \text{using Galois invariance,}\\
&= z_j(\mathrm{Frob}_{\sigma\mathfrak{P}_j}) && \text{because } \psi_N(z_i) = \psi_N(z_j),
\end{aligned}$$

we obtain

$$z_{\sigma\mathfrak{P}_j} = (z_i)_{\sigma\mathfrak{P}_j} + (z_j)_{\sigma\mathfrak{P}_j} + (z_N)_{\sigma\mathfrak{P}_j} = (z_i)_{\sigma\mathfrak{P}_j} + (z_j)_{\sigma\mathfrak{P}_j} + 0 = 0.$$

Finally,

$$z_{\sigma\mathfrak{P}_N} = (z_i)_{\sigma\mathfrak{P}_N} + (z_j)_{\sigma\mathfrak{P}_N} + (z_N)_{\sigma\mathfrak{P}_N} = 0,$$

since

$$\begin{aligned}
z_i(\mathrm{Frob}_{\sigma\mathfrak{P}_N}) &= (z_i, \sigma z_N)_{\sigma\mathfrak{P}_N} && \text{by condition (1),}\\
&= (z_i, \sigma z_N)_{\mathfrak{P}_i} && \text{by the product formula,}\\
&= (\sigma^{-1} z_i, z_N)_{\sigma^{-1}\mathfrak{P}_i} && \text{using Galois invariance,}\\
&= z_N(\mathrm{Frob}_{\sigma^{-1}\mathfrak{P}_i}) && \text{by condition (1),}\\
&= 0 && \text{by condition (4), } \sigma^{-1} \in G_3,
\end{aligned}$$

and

$$\begin{aligned} z_j(\mathrm{Frob}_{\sigma\mathfrak{P}_N}) &= z_N(\mathrm{Frob}_{\sigma^{-1}\mathfrak{P}_j}) \\ &= \sigma^{-2} z_j(\mathrm{Frob}_{\sigma^{-1}\mathfrak{P}_j}) && \text{by condition (4), } \sigma^{-1} \in G_3\,, \\ &= z_j(\mathrm{Frob}_{\sigma\mathfrak{P}_j}) && \text{using Galois invariance,} \\ &= z_N(\mathrm{Frob}_{\sigma\mathfrak{P}_N}) && \text{because } \psi_N(z_N) = \psi_N(z_j). \end{aligned}$$

In the same way one verifies that the local classes $z_{\sigma\mathfrak{P}_i}$, $z_{\sigma\mathfrak{P}_j}$ and $z_{\sigma\mathfrak{P}_N}$ also vanish if $\sigma \in G_3$. This finishes the proof for $A = \mu_p$.

The general case will be proven by induction on $\dim_{\mathbb{F}_p} A$. Let $A = A' \oplus \mu_p$ (as an abelian group). For each $z \in H^1(K_S|K, A)$, let $z = z' + z''$ be the decomposition of z into the components $z' \in H^1(K_S|K, A')$ and $z'' \in H^1(K_S|K, \mu_p)$, and similarly $z_\mathfrak{P} = z'_\mathfrak{P} + z''_\mathfrak{P}$ for $z_\mathfrak{P} \in H^1(K_\mathfrak{P}, A)$. By induction we find an element $z' \in H^1(K_S|K, A')$ such that

(a′) $z'_\mathfrak{P} = y'_\mathfrak{P}$ for $\mathfrak{P} \in T(K)$,
(b′) if $\mathfrak{P} \notin T(K)$ and $z'_\mathfrak{P}$ is ramified, then $z'_\mathfrak{P}$ is cyclic and $z'_{\sigma\mathfrak{P}} = 0$ for every $\sigma \in G(K|k) \smallsetminus \{1\}$.

Let $K'|K$ be the extension defined by the homomorphism z', so that

$$z'\colon G(k_S|K) \twoheadrightarrow G(K'|K) \subseteq A',$$

and let $\tilde{K}$ be its Galois closure over k. By construction, $K'|K$, and hence also $\tilde{K}|K$, is unramified at all primes in $T(K)$ and it only ramifies at primes in $cs(\Omega|k)(K)$.

Set $T' = T \cup \mathrm{Ram}(\tilde{K}|k)$, $\Omega' = \tilde{K}\Omega$, $S' = cs(\Omega'|k) \cup T'$ and suppose we have found a class $\tilde{y} \in H^1(K_{S'}|K, \mu_p)$ with

· $\tilde{y}_\mathfrak{P} = y''_\mathfrak{P}$ for $\mathfrak{P} \in T(K)$,
· $\tilde{y}_\mathfrak{P} = 0$ for $\mathfrak{P} \in T' \smallsetminus T(K)$.

Then we can apply the induction hypothesis to the extensions $\Omega'|K|k$, the sets T', S' and the module $A = \mu_p$, in order to find a class z'' in $H^1(K_{S'}|K, \mu_p)$ with

(a″) $z''_\mathfrak{P} = \tilde{y}_\mathfrak{P}$ for $\mathfrak{P} \in T'(K)$,
(b″) if $\mathfrak{P} \notin T'(K)$ and $z''_\mathfrak{P}$ is ramified, then $z''_\mathfrak{P}$ is cyclic and $z''_{\sigma\mathfrak{P}} = 0$ for every $\sigma \in G(K|k) \smallsetminus \{1\}$.

Noting that $T' \smallsetminus T \subseteq cs(\Omega|k)$, so that $S' \subseteq S$, it is now easily verified that the class $z = z' + \mathit{inf}\, z'' \in H^1(K_S|K, A)$ satisfies conditions (a) and (b). Indeed, for $\mathfrak{P} \in T(K)$ we have

$$z_\mathfrak{P} = z'_\mathfrak{P} + z''_\mathfrak{P} = y'_\mathfrak{P} + y''_\mathfrak{P} = y_\mathfrak{P}\,.$$

For $\mathfrak{P} \in T' \smallsetminus T$ we get

$$z_\mathfrak{P} = z'_\mathfrak{P} + z''_\mathfrak{P} = z'_\mathfrak{P}\,.$$

Therefore $z_\mathfrak{P}$ is cyclic and if $z_\mathfrak{P} = z'_\mathfrak{P}$ is ramified, then the underlying prime $\mathfrak{p}$ of $\mathfrak{P}$ splits completely in Ω and $z_{\sigma\mathfrak{P}} = z'_{\sigma\mathfrak{P}} = 0$ for $\sigma \in G(K|k) \smallsetminus \{1\}$.

Finally suppose $\mathfrak{P} \notin T'$. If $z_{\mathfrak{P}}$ is unramified, then it is cyclic. Suppose $z_{\mathfrak{P}}$ is ramified. Since $z_{\mathfrak{P}} = z'_{\mathfrak{P}} + z''_{\mathfrak{P}}$ and $z'_{\mathfrak{P}}$ is unramified, $z''_{\mathfrak{P}}$ must be ramified. Thus $\mathfrak{p}$ splits completely in Ω', and hence in K'. From this we obtain $z'_{\sigma\mathfrak{P}} = 0$ for all $\sigma \in G(K|k)$, since the element z' becomes zero in $H^1(K', A')$ by definition of K' and thus $z'_{\sigma\mathfrak{P}}$ is zero in $H^1(K'_{\mathfrak{Q}}, A') = H^1(K_{\sigma\mathfrak{P}}, A')$, where $\mathfrak{Q}$ is a prime of K' above $\sigma\mathfrak{P}$. Therefore $z_{\mathfrak{P}} = z''_{\mathfrak{P}}$, i.e. $z_{\mathfrak{P}}$ is cyclic and $z_{\sigma\mathfrak{P}} = z''_{\sigma\mathfrak{P}} = 0$ for $\sigma \in G(K|k) \smallsetminus \{1\}$.

It therefore remains to construct a class $\tilde{y}$ with the above properties. Consider the commutative exact diagram

$$\begin{array}{ccccc}
 & & \prod\limits_{S'\setminus T'} H^1(K_{\mathfrak{P}}, \mu_p) & \xrightarrow{\ \sim\ } & \Big(\prod\limits_{S'\setminus T'} H^1(K_{\mathfrak{P}}, \mathbb{Z}/p\mathbb{Z})\Big)^{\vee} \\
 & & \big\downarrow & & \big\downarrow \\
H^1(k_{S'}|K, \mu_p) & \longrightarrow & \prod\limits_{S'} H^1(K_{\mathfrak{P}}, \mu_p) & \longrightarrow & H^1(k_{S'}|K, \mathbb{Z}/p\mathbb{Z})^{\vee} \\
 & & \big\downarrow & & \big\downarrow \\
 & & \prod\limits_{T'} H^1(K_{\mathfrak{P}}, \mu_p) & \xrightarrow{\ \alpha\ } & H^1(\Omega'|K, \mathbb{Z}/p\mathbb{Z})^{\vee}.
\end{array}$$

If we can show that α annihilates the element $\xi = (\xi_{\mathfrak{P}})_{\mathfrak{P} \in T'(K)}$ of the group $\prod_{T'} H^1(K_{\mathfrak{P}}, \mu_p)$ given by $\xi_{\mathfrak{P}} = y''_{\mathfrak{P}}$ for $\mathfrak{P} \in T(K)$ and $\xi_{\mathfrak{P}} = 0$ for $\mathfrak{P} \in T' \smallsetminus T(K)$, then the existence of $\tilde{y}$ follows by diagram chasing. We use the injection

$$H^1(\Omega'|K, \mathbb{Z}/p\mathbb{Z})^{\vee} \hookrightarrow H^1(\Omega|K, \mathbb{Z}/p\mathbb{Z})^{\vee} \oplus H^1(\tilde{K}|K, \mathbb{Z}/p\mathbb{Z})^{\vee}$$

in order to write the image of ξ in the form $\alpha(\xi) = (\alpha_1(\xi), \alpha_2(\xi))$. Since $\tilde{K}|K$ is unramified at all $\mathfrak{P} \in T(K)$, α_2 factors through the quotient

$$\prod_{T'} H^1(K_{\mathfrak{P}}, \mu_p) / \Big(\prod_{T} H^1_{nr}(K_{\mathfrak{P}}, \mu_p) \times \coprod_{T'\setminus T} \{0\}\Big).$$

Hence $\alpha_2(\xi) = 0$. Finally, the diagram

$$\begin{array}{ccccc}
 & & \prod\limits_{T'} H^1(K_{\mathfrak{P}}, \mu_p) & \xrightarrow{\ \alpha\ } & H^1(\Omega'|K, \mathbb{Z}/p\mathbb{Z})^{\vee} \\
 & & \big\downarrow & & \big\downarrow \\
 & & \prod\limits_{T} H^1(K_{\mathfrak{P}}, \mu_p) & \longrightarrow & H^1(\Omega|K, \mathbb{Z}/p\mathbb{Z})^{\vee} \\
 & & \big\uparrow & & \big\uparrow \\
H^1(k_S|K, \mu_p) & \longrightarrow & \prod\limits_{S} H^1(K_{\mathfrak{P}}, \mu_p) & \longrightarrow & H^1(k_S|K, \mathbb{Z}/p\mathbb{Z})^{\vee}
\end{array}$$

shows that $\alpha_1(\xi)$ is equal to the image of $y'' \in H^1(k_S|K, \mu_p)$ in the group $H^1(\Omega|K, \mathbb{Z}/p\mathbb{Z})^{\vee}$, hence is trivial by the exactness of the lower row. This finishes the proof. □

The proof of the next theorem uses ideas of [158].

(9.3.3) Theorem. *Let k be a global field and let A be a simple G_k-module annihilated by the prime number $p \neq \mathrm{char}(k)$. Let S be a set of primes of k containing $S_p \cup S_\infty$ if k is a number field. Assume there exist finite Galois extensions $K \subseteq \Omega$ of k such that $p \nmid [\Omega : K]$,*

(a) $\mu_p \nsubseteq K$ *and* $k(A) \subseteq K$,

(b) $\mu_p \subseteq \Omega$, $\mathrm{Ram}(\Omega|k) \subseteq S$ *and* $cs(\Omega|k) \subsetneqq S$.

Let $T \subseteq S$ be a finite subset and let local classes $y_\mathfrak{p} \in H^1(k_\mathfrak{p}, A)$ for $\mathfrak{p} \in T$ be given. Then there exists an element $x \in H^1(k_S|k, A)$ such that

(i) $x_\mathfrak{p} = y_\mathfrak{p}$ *for* $\mathfrak{p} \in T$,

(ii) *if $\mathfrak{p} \notin T$ then $x_\mathfrak{p}$ is cyclic, and if $x_\mathfrak{p}$ is ramified then $\mathfrak{p}$ splits completely in Ω.*

Proof: By condition (b), S is nonempty. Let $S_0 \subseteq S$ be a finite nonempty subset which contains $S_p \cup S_\infty$ if k is a number field.

If the assertion of the theorem holds for a finite set $T' \subseteq S$ with $T \subseteq T'$, then it also holds for T (impose the local conditions $y_\mathfrak{p} = 0$ for $\mathfrak{p} \in T' \smallsetminus T$). Therefore we may assume $\mathrm{Ram}(\Omega|k) \cup S_0 \subseteq T$; in particular, A is a $G_T(k)$-module. Let $S_1 = (S \cap cs(\Omega|k)) \cup T$. By (9.2.3)(vi), there exists an element

$$z \in H^1(k_{S_1}|k, A) \subseteq H^1(k_S|k, A)$$

such that $z_\mathfrak{p} = y_\mathfrak{p}$ for $\mathfrak{p} \in T$. Let

$$V = T \cup \{\mathfrak{p} \in S_1 \mid z_\mathfrak{p} \text{ is ramified }\}.$$

Note that all $\mathfrak{p} \in V \smallsetminus T$ split completely in Ω. Thus it remains to show that we can choose the element z in such a way that the ramified local classes are cyclic.

We introduce two additional fields. Let z' be the image of z under the restriction map $H^1(k_{S_1}|k, A) \to H^1(k_{S_1}|K, A)^{G(K|k)}$. Then z' is a $G(K|k)$-homomorphism $z'\colon G_{S_1}(K) \to A$ and defines a finite abelian p-extension $K'|K$ which is Galois over k. We set $\Omega' = K'\Omega$. It follows that $\Omega'|k$ is Galois, $\mu_p \nsubseteq K'$ and $p \nmid [\Omega' : K']$. Furthermore, $\mathrm{Ram}(\Omega'|k) \subseteq V$, since the set $\mathrm{Ram}(\Omega|k)$ is contained in T and all primes $\mathfrak{p}$ with $z_\mathfrak{p}$ ramified are in V. Let

$$S_1' = (S \cap cs(\Omega'|k)) \cup V \subseteq S_1.$$

Since $cs(\Omega|k) \subsetneqq S$, we have $cs(\Omega'|k) \subsetneqq S_1'$. Suppose we have constructed a class $\xi \in H^1(k_{S_1'}|k, A) \subseteq H^1(k_S|k, A)$ such that

(1) $\xi_\mathfrak{p} = 0$ for $\mathfrak{p} \in T$,

(2) $z_\mathfrak{p} + \xi_\mathfrak{p}$ is cyclic for $\mathfrak{p} \in V \smallsetminus T$,

(3) $\xi_\mathfrak{p}$ is cyclic for $\mathfrak{p} \notin V$.

We verify that $x = z + \xi \in H^1(k_S|k, A)$ satisfies conditions (i) and (ii) of the theorem. For $\mathfrak{p} \in T$ we have $x_\mathfrak{p} = z_\mathfrak{p} = y_\mathfrak{p}$. If $\mathfrak{p} \in V \smallsetminus T$, then $x_\mathfrak{p} = z_\mathfrak{p} + \xi_\mathfrak{p}$ is cyclic and $\mathfrak{p}$ splits completely in Ω. Let $\mathfrak{p} \notin V$. Then $z_\mathfrak{p}$ is unramified. If $x_\mathfrak{p}$ is unramified, it is cyclic. So assume that $x_\mathfrak{p}$ is ramified. Then also $\xi_\mathfrak{p} = x_\mathfrak{p} - z_\mathfrak{p}$ is ramified, and so $\mathfrak{p} \in S_1' \smallsetminus V$. Hence $\mathfrak{p}$ splits completely in $\Omega'|k$, in particular, it splits completely in $K'|k$. By construction of K', the image of z in $H^1(k_{S_1}|K', A)$ is zero. If $\mathfrak{P}$ is any prime of K' above $\mathfrak{p}$, we obtain that $z_\mathfrak{p}$ is zero in $H^1(K'_\mathfrak{P}, A) = H^1(k_\mathfrak{p}, A)$. Thus we have $x_\mathfrak{p} = \xi_\mathfrak{p}$, and $x_\mathfrak{p}$ is cyclic.

Therefore it remains to construct a class $\xi \in H^1(k_{S_1'}|k, A)$ satisfying conditions (1)–(3) above.

Claim. For $\mathfrak{p} \in V \smallsetminus T$ there exists a local class $w_\mathfrak{p} \in H^1_{nr}(k_\mathfrak{p}, A)$ such that $z_\mathfrak{p} + w_\mathfrak{p}$ is cyclic.

Proof of the claim: The prime $\mathfrak{p}$ splits completely in Ω, hence $\mu_p \subseteq k_\mathfrak{p}$ and A is a trivial $G_{k_\mathfrak{p}}$-module. Since $\mathfrak{p} \notin S_0$, the $\mathbb{F}_p$-vector space $H^1(k_\mathfrak{p}, \mathbb{Z}/p\mathbb{Z})$ is 2-dimensional. Therefore, if $\kappa|k_\mathfrak{p}$ is a cyclic ramified extension of degree p, the natural map $H^1(\kappa|k_\mathfrak{p}, A) \oplus H^1_{nr}(k_\mathfrak{p}, A) \to H^1(k_\mathfrak{p}, A)$ is an isomorphism. This shows the claim.

Consider the commutative diagram

$$\begin{array}{ccc}
H^1(k_{S_1'}|K(\mu_p), A) & & \\
{\scriptstyle res}\uparrow\downarrow{\scriptstyle cor} & & \\
H^1(k_{S_1'}|K, A) & \xrightarrow{\ \varepsilon\ } \!\!\!\rightarrow & \prod\limits_{\mathfrak{P} \in V(K)} H^1(K_\mathfrak{P}, A) \\
\downarrow {\scriptstyle cor} & & \downarrow {\scriptstyle cor} \\
H^1(k_{S_1'}|k, A) & \longrightarrow & \prod\limits_{\mathfrak{p} \in V} H^1(k_\mathfrak{p}, A)\,.
\end{array}$$

Since $cs(\Omega'|K) \supsetneq cs(\Omega'|k)(K)$ and $cs(\Omega'|k) \supsetneq S_1'$, we have $cs(\Omega'|K) \supsetneq S_1'(K)$. As A is a trivial $G_{S_1'}(K)$-module, (9.2.3)(vi), applied to the simple module $\mathbb{F}_p$ and the Galois extensions $K' \subseteq \Omega'$ of k, implies that the natural map ε is surjective.

We consider the element

$$(\eta_\mathfrak{p})_{\mathfrak{p} \in V} \in \prod_{\mathfrak{p} \in V} H^1_{nr}(k_\mathfrak{p}, A), \ \text{ with } \eta_\mathfrak{p} = 0 \text{ for } \mathfrak{p} \in T \text{ and } \eta_\mathfrak{p} = w_\mathfrak{p} \text{ for } \mathfrak{p} \in V \smallsetminus T,$$

where the elements $w_\mathfrak{p} \in H^1_{nr}(k_\mathfrak{p}, A)$ are chosen as in the claim. Since all $\mathfrak{p} \in V \smallsetminus T$ split completely in $K|k$, we have a pre-image

$$(\tilde{\eta}_\mathfrak{P})_{\mathfrak{P} \in V(K)} \in \prod_{\mathfrak{P} \in V(K)} H^1_{nr}(K_\mathfrak{P}, A) \subseteq \prod_{\mathfrak{P} \in V(K)} H^1(K_\mathfrak{P}, A)$$

of $(\eta_\mathfrak{p})_{\mathfrak{p}\in V}$ under the corestriction map. Let $\tilde{\eta} \in H^1(k_{S'_1}|K, A)$ be a pre-image of $(\tilde{\eta}_\mathfrak{P})_{\mathfrak{P}\in V(K)}$ under ε and let $d = [K(\mu_p) : K]$. Note that $(d, p) = 1$ and consider the class

$$y = res\,^K_{K(\mu_p)}(\tilde{\eta}/d) \in H^1(k_{S'_1}|K(\mu_p), A).$$

We have $y_\mathfrak{P} = 0$ for $\mathfrak{P} \in T(K(\mu_p))$ and $y_\mathfrak{P} \in H^1_{nr}(k_{S'_1}|K(\mu_p), A)$ for $\mathfrak{P} \in V \smallsetminus T(K(\mu_p))$. By (9.3.2), we find a class $\xi \in H^1(k_{S'_1}|k, A)$ with

$$\xi_\mathfrak{p} = (cor\,^{K(\mu_p)}_k y)_\mathfrak{p} = (cor\,^K_k \tilde{\eta})_\mathfrak{p} = \eta_\mathfrak{p} \;\text{ for } \mathfrak{p} \in V$$

such that $\xi_\mathfrak{p}$ is cyclic for all $\mathfrak{p} \notin V$. The class ξ satisfies conditions (1)–(3). This finishes the proof. □

§4. Local Galois Groups in a Global Group

With the notation of the preceding sections we now consider the maximal pro-$\mathfrak{c}$-factor group $G_k(\mathfrak{c})$ of G_k, where $\mathfrak{c}$ is any full class of finite groups.

(9.4.1) Definition. *Let k be a field, $\bar{k}$ a separable closure of k and $\mathfrak{c}$ a full class of groups. We denote by $k(\mathfrak{c})$ the composite of all finite Galois extensions K of k inside $\bar{k}$ with $G(K|k) \in \mathfrak{c}$.*

If k is a global field and S is a set of places of k, we denote by $k_S(\mathfrak{c})$ the composite of all finite Galois extensions K of k inside k_S with $G(K|k) \in \mathfrak{c}$.

The Galois group $G(k(\mathfrak{c})|k)$ is nothing else but the maximal pro-$\mathfrak{c}$-factor group of G_k. Similarly, $G(k_S(\mathfrak{c})|k)$ is the maximal pro-$\mathfrak{c}$-factor group of $G_S(k)$, cf. III §5.

(9.4.2) Lemma.

(i) *The field $k(\mathfrak{c})$ admits no finite Galois extension K with $G(K|k(\mathfrak{c})) \in \mathfrak{c}$.*

(ii) *Let $k = k_0 \subseteq k_1 \subseteq \cdots \subseteq k_n$ be finite extensions of k in $\bar{k}$ such that $k_{i+1}|k_i$ is Galois and $G(k_{i+1}|k_i) \in \mathfrak{c}$ for $i = 0, \ldots, n-1$. Then $k_n \subseteq k(\mathfrak{c})$; in particular, $G(\tilde{k}_n|k) \in \mathfrak{c}$, where $\tilde{k}_n$ is the Galois closure of k_n over k.*

Proof: Assertion (i) follows from (3.5.4)(i), and (ii) follows from (i). □

If k is a global field, we are interested in the question of whether for a given prime $\mathfrak{p}$ of k the local Galois group $G_{k_\mathfrak{p}}(\mathfrak{c})$ is a subgroup of $G_k(\mathfrak{c})$. This is equivalent to the question of whether the equality

$$(k(\mathfrak{c}))_\mathfrak{p} = k_\mathfrak{p}(\mathfrak{c})$$

holds. If $\mathfrak{c}$ is the class of all finite groups, i.e. $k(\mathfrak{c})$ is the separable closure $\bar{k}$ of k, then we have seen this in (8.1.5). If $\mathfrak{c}$ is arbitrary, we will prove a slightly more general statement using the theorem of Grunwald-Wang.

(9.4.3) Theorem. *Let k be a global field and let $\mathfrak{c}$ be a full class of finite groups. Let S be a set of primes of k of density $\delta(S) = 1$. Then, for the maximal $\mathfrak{c}$-extension $k_S(\mathfrak{c})$ of k unramified outside S and a prime $\mathfrak{p} \in S$, we have*

$$(k_S(\mathfrak{c}))_\mathfrak{p} = k_\mathfrak{p}(\mathfrak{c}),$$

or equivalently, the canonical map

$$G_{k_\mathfrak{p}}(\mathfrak{c}) \twoheadrightarrow G_\mathfrak{p}(k_S(\mathfrak{c})|k) \subseteq G_S(k)(\mathfrak{c})$$

is injective.

Proof: We have to show that every finite Galois extension $K_\mathfrak{p}|k_\mathfrak{p}$ with $G(K_\mathfrak{p}|k_\mathfrak{p}) \in \mathfrak{c}$ can be realized by a global Galois extension $K|k$ unramified outside S with $G(K|k) \in \mathfrak{c}$. Since $G_{k_\mathfrak{p}}$ is a prosolvable group and by (9.4.2)(ii), we may assume that $G(K_\mathfrak{p}|k_\mathfrak{p}) \cong \mathbb{Z}/p\mathbb{Z}$ for a prime number p with $\mathbb{Z}/p\mathbb{Z} \in \mathfrak{c}$. Choose a class $\alpha_\mathfrak{p} \in H^1(k_\mathfrak{p}, \mathbb{Z}/p\mathbb{Z})$, $\alpha \neq 0$, which is split by $K_\mathfrak{p}$, i.e. $res\, \alpha_\mathfrak{p} \in H^1(K_\mathfrak{p}, \mathbb{Z}/p\mathbb{Z})$ is trivial. By (9.2.7), the homomorphism

$$H^1(G_S(k), \mathbb{Z}/p\mathbb{Z}) \longrightarrow H^1(k_\mathfrak{p}, \mathbb{Z}/p\mathbb{Z})$$

is surjective. Let $\alpha \in H^1(G_S(k), \mathbb{Z}/p\mathbb{Z})$ be a pre-image of $\alpha_\mathfrak{p}$. Then α is split by a cyclic extension $K|k$ of degree p which realizes $K_\mathfrak{p}$. □

In particular, if $\mathfrak{c}$ is the class of p-groups and p a prime number, then $G_{k_\mathfrak{p}}(p)$ injects into $G_k(p)$ by the theorem above. Now we ask the much deeper question for the interaction of finitely many local subgroups $G_{k_\mathfrak{P}}(p)$ in the global group $G_k(p)$ (where we index these groups by primes $\mathfrak{P}$ of $k(p)$).

A first question is whether the intersection of $G_{k_\mathfrak{P}}(p)$ and $G_{k_{\mathfrak{P}'}}(p)$ for different primes $\mathfrak{P}$ and $\mathfrak{P}'$ is trivial. We will prove that even more is true: under a mild assumption, finitely many local groups $G_{k_\mathfrak{P}}(p)$ are as independent as possible, i.e. the subgroup which they generate inside $G_k(p)$ is their free pro-p-product.

Of course, one wants to know this not only for the maximal pro-p-factors, but also for general classes $\mathfrak{c}$; in particular, if $\mathfrak{c}$ is the class of all finite groups. This statement is also true in a "measure theoretical sense": If $\mathfrak{P}_1, \ldots, \mathfrak{P}_r$ are primes of $\bar{k}$, then for "almost all" r-tuples $(\sigma_1, \ldots, \sigma_r) \in (G_k)^r$, the subgroup $\langle G_{k_{\sigma_1 \mathfrak{P}_1}}, \ldots, G_{k_{\sigma_r \mathfrak{P}_r}} \rangle \subseteq G_k$ is the free profinite product of the groups $G_{k_{\sigma_i \mathfrak{P}_i}}$, $i = 1, \ldots, r$. This result is due to *W.-D. Geyer*; see [55] for a precise statement. In view of (4.2.3), this assertion cannot be true for *all* r-tuples $(\sigma_1, \ldots, \sigma_r) \in (G_k)^r$.

For the case of pro-p-groups we now come to the statement mentioned above.

(9.4.4) Theorem. *Let k be a global field, S a set of primes of k of density $\delta(S) = 1$ and $k_S(p)$ the maximal p-extension of k unramified outside S. Let $T \subseteq S(k_S(p))$ be a finite subset,*

$$H := \langle G_{\mathfrak{P}}(k_S(p)|k), \mathfrak{P} \in T \rangle \subseteq G_S(k)(p)$$

the closed subgroup generated by the decomposition groups of the primes in T and $K = k_S(p)^H$ the fixed field of H.

(i) *Let $\mathfrak{Q} \in S(k_S(p))$ be a prime with $\mathfrak{Q}|_K \neq \mathfrak{P}|_K$ for all $\mathfrak{P} \in T$. Then $K_{\mathfrak{Q}} = k_{\mathfrak{Q}}(p)$. In particular, $\mathfrak{Q}$ splits completely in $k_S(p)|K$.*

(ii) *Let $T_0 \subseteq T$ be a maximal subset such that the primes in T_0 have pairwise different restrictions to K. Then the natural map*

$$\mathop{\ast}_{\mathfrak{P} \in T_0} G_{k_{\mathfrak{P}}}(p) \to G_S(k)(p)$$

is injective, or equivalently, the subgroup H is the free pro-p-product

$$H = \mathop{\ast}_{\mathfrak{P} \in T_0} G_{k_{\mathfrak{P}}}(p).$$

Proof: By (9.4.3), for any prime $\mathfrak{P}$ in $k_S(p)$, the decomposition group $G_{\mathfrak{P}}(k_S(p)|k)$ is the full local group $G_{k_{\mathfrak{P}}}(p)$. Let

$$H_0 := \langle G_{\mathfrak{P}}(k_S(p)|k), \mathfrak{P} \in T_0 \rangle \subseteq H\,.$$

Since the conjugates hH_0h^{-1}, $h \in H$, generate H, we obtain $H_0 = H$, as H is a pro-p-group, see (3.9.1). Therefore we may replace T by T_0.

Now let $T' \subseteq S(k_S(p))$ be a finite set of primes having pairwise different restrictions to $K = k_S(p)^H$. For a subfield $k' \subseteq k_S(p)$ we denote by $T'(k')$ the (finite) set of restrictions of the primes in T' to k'. Replacing k by a finite extension inside K if necessary, we may assume that all primes of T' have pairwise different restrictions to k. From (9.2.7) we obtain a surjection

$$H^1(k_S|k, \mathbb{Z}/p\mathbb{Z}) \twoheadrightarrow \prod_{T'(k)} H^1(k_\mathfrak{p}, \mathbb{Z}/p\mathbb{Z}),$$

and passing to the direct limit we get a surjection

$$(*) \qquad H^1(k_S|K, \mathbb{Z}/p\mathbb{Z}) \twoheadrightarrow \prod_{T'(K)} H^1(k_\mathfrak{P}(p)|K_\mathfrak{P}, \mathbb{Z}/p\mathbb{Z}).$$

If $T \subseteq T'$, then the map $(*)$ is injective because the groups $G_{k_\mathfrak{P}}(p)$, $\mathfrak{P} \in T$, generate H. Now let $\mathfrak{Q}$ be a prime of $k_S(p)$ with $\mathfrak{Q}|_K \notin T(K)$. Using the isomorphism $(*)$ for $T' = T$ and $T' = T \cup \{\mathfrak{Q}\}$, we obtain

$$H^1(k_\mathfrak{Q}(p)|K_\mathfrak{Q}, \mathbb{Z}/p\mathbb{Z}) = 0,$$

and so

$$G(k_\mathfrak{Q}(p)|K_\mathfrak{Q}) = 1.$$

This shows (i).

In the extension $k_S(p)|K$, it are exactly the decomposition groups $G_\mathfrak{P}(k_S(p)|K) = G_{k_\mathfrak{P}}(p)$ with $\mathfrak{P} \in T(K)$ which are nontrivial. In the commutative diagram

$$\begin{array}{ccccc} H^2(k_S|K, \mathbb{Z}/p\mathbb{Z}) & \longrightarrow & \bigoplus_{S(K)} H^2(K_\mathfrak{P}, \mathbb{Z}/p\mathbb{Z}) & & \\ \uparrow{\scriptstyle inf} & & \uparrow{\scriptstyle inf} & & \\ H^2(H, \mathbb{Z}/p\mathbb{Z}) & \longrightarrow & \bigoplus_{S(K)} H^2(k_\mathfrak{P}(p)|K_\mathfrak{P}, \mathbb{Z}/p\mathbb{Z}) & = & \bigoplus_{T(K)} H^2(G_{k_\mathfrak{P}}(p), \mathbb{Z}/p\mathbb{Z}) \end{array}$$

the inflation map on the left is injective, since

$$H^1(k_S|k_S(p), \mathbb{Z}/p\mathbb{Z}) = \mathrm{Hom}(G(k_S|k_S(p)), \mathbb{Z}/p\mathbb{Z}) = 0,$$

i.e. $k_S(p)$ has no p-extension inside k_S by definition. (The inflation map on the right is an isomorphism by (7.5.8), but we do not need this.) From (9.1.10)(i) and passing to the inductive limit, we obtain the injectivity of the upper horizontal map in the diagram, hence the lower map is also injective. Now (4.1.5) gives the result. □

Remark: If T consists of two primes, then $T_0 = T$, but if T contains at least three primes, then T_0 may be a proper subset of T.

Let us consider the case $T = \{\mathfrak{P}_1, \mathfrak{P}_2\}$, $\mathfrak{P}_1 \neq \mathfrak{P}_2$, first. Let K_1 and K_2 be the decomposition fields of $\mathfrak{P}_1$ and $\mathfrak{P}_2$ in $k_S(p)$ and $K = K_1 \cap K_2$. If $\mathfrak{P}_1 \cap K \neq \mathfrak{P}_2 \cap K$, then H is the free product of G_1 and G_2 by (9.4.4). Now suppose that the restrictions to K are equal. Then G_1 and G_2 are conjugate in H and therefore (say) G_1 generates H as a normal subgroup. Since H is a pro-p-group, we conclude that $H = G_1 = G_2$ and $K = K_1 = K_2$. But

$\mathfrak{P}_1 \cap K_1 = \mathfrak{P}_2 \cap K_2$ has exactly one extension to $k_S(p)$, which shows $\mathfrak{P}_1 = \mathfrak{P}_2$ in contrast to our assumption.

To construct an example with a set of three primes, choose any element $\sigma \in H = G_1 * G_2$ not contained in G_1 and put $\mathfrak{P}_3 = \sigma\mathfrak{P}_1$. Then $G_3 = \sigma G_1 \sigma^{-1}$ is contained in H. Hence $K = K_1 \cap K_2 \cap K_3$ and $\mathfrak{P}_1|_K = \mathfrak{P}_3|_K$. This gives the counterexample.

Of greater arithmetical importance, but also much deeper lying, is the question of analogous results in the case where we restrict the ramification to a given finite set of primes. Is it true that $(k_S(p))_\mathfrak{p} = k_\mathfrak{p}(p)$, and do finitely many decomposition groups $G_\mathfrak{P}(k_S(p)|k) \cong G_{k_\mathfrak{p}}(p)$ and their conjugates form a free pro-p-product inside $G_S(k)(p)$, where S is a finite set of primes? This would mean on the one hand that a local p-extension with respect to a prime $\mathfrak{p} \in S$ can be realized by a global p-extension which is unramified outside S and on the other hand that finitely many local groups are as independent as possible in the global group $G_S(k)(p)$. Most difficult is the situation in the number field case for primes dividing p. We will consider this problem in X §5-§7.

§5. Solvable Groups as Galois Groups

Class field theory provides us with a complete solution for a great number of the abstract and arithmetic problems concerning abelian extensions of global fields. The next natural step forward to the general case is to consider solvable extensions. The reason for this is that the solvable extensions are built up by abelian extensions and one is necessarily led to the so-called embedding problem in number theory: a given Galois extension $K|k$ has to be embedded in a larger extension $L|k$ in such a way that the Galois groups of $L|k$ and $K|k$ realize a given group extension. For a global field such an embedding problem has an arithmetic structure. Each embedding problem defined over a global field k canonically induces local embedding problems over the completions $k_\mathfrak{p}$. Hence the local-global question naturally arises: if all local problems are solvable, is then the global one also solvable? This question is connected with the Hasse principle studied in §1. A refinement of the above question is: does there exist a global solution of an embedding problem which induces given local solutions? Here the theorem of Grunwald-Wang comes into play.

One main result of this section is a theorem of K. IWASAWA [86] concerning the structure of the Galois group of the maximal prosolvable extension $\tilde{k}$ over some large field (e.g. over the maximal abelian extension of k). But first we need the notion of a powerful global field.

(9.5.1) Definition. *A (not necessarily finite) separable extension K of a global field k is called* **powerful** *if it has the following property:*

Let E be a finite group and let $K'|K$ be a finite separable extension of K. Then there exist primes $\mathfrak{P}_1, \ldots, \mathfrak{P}_s$ of K which decompose completely in K' and group homomorphisms

$$\varphi_i \colon G(\bar{K}_{\mathfrak{P}_i}|K_{\mathfrak{P}_i}) \longrightarrow E, \quad i = 1, \ldots, s,$$

such that the images of φ_i generate E.

Remark: It is easy to find infinite extensions of a global field k which are not powerful. An obvious example is a Sylow-field K of k, i.e. G_K is a p-Sylow subgroup of G_k for some prime number p.

(9.5.2) Proposition. *Let k be a global field and $K|k$ a (not necessarily finite) abelian extension. Then K is powerful. In particular, each global field is powerful.*

Proof: Let $K'|K$ be a finite separable extension, and let ℓ be a prime number. We will show that there are infinitely many primes $\mathfrak{P}$ of K which are completely decomposed in $K'|K$ and having the property that the local Galois groups $G(\bar{K}_{\mathfrak{P}}|K_{\mathfrak{P}})$ have a surjection onto $\mathbb{Z}_\ell$. Then the desired result follows.

Let $K_0|k$ be a finite extension inside $K|k$ and $K_0'|K_0$ a finite separable extension such $K' = KK_0'$. Let ζ be a primitive ℓ-th root of unity if ℓ is different to the characteristic of k and equal to 1 otherwise.

By Čebotarev's density theorem (9.1.3), there are infinitely many nonarchimedean primes of K_0 which are completely decomposed in $K_0'(\zeta)|K_0$. Considering their prolongations to K, we get infinitely many nonarchimedean primes of K which are completely decomposed in $K'|K$, and the completions of K with respect to these primes contain ζ.

Let $\mathfrak{P}$ be one them, and let $\mathfrak{p} = \mathfrak{P} \cap k$ be the underlying prime of k and p the characteristic of the residue field of $k_\mathfrak{p}$. We identify the decomposition group $G_\mathfrak{P}(K)$ with the local group $G_{K_\mathfrak{P}} = G(\bar{K}_\mathfrak{P}|K_\mathfrak{P})$, see (8.1.5). If the maximal unramified ℓ-extension of $k_\mathfrak{p}$ inside $K_\mathfrak{P}$ is finite, then we have a surjection $G_{K_\mathfrak{P}}(\ell) \twoheadrightarrow G_{K_\mathfrak{P}}(\ell)/T_{K_\mathfrak{P}}(\ell) \xrightarrow{\sim} \mathbb{Z}_\ell$.

Assume now that the maximal unramified ℓ-extension inside $K_{\mathfrak{P}}|k_{\mathfrak{p}}$ is infinite, thus $K_{\mathfrak{P}}$ realizes the maximal unramified ℓ-extension $\tilde{k}_{\mathfrak{p}}(\ell)$ of $k_{\mathfrak{p}}$. First we consider the case $\ell \neq p$ (and so $\ell \neq \operatorname{char} k$). As $\zeta \in K_{\mathfrak{P}}$, we have $T_{K_{\mathfrak{P}}}(\ell) \twoheadrightarrow G_{K_{\mathfrak{P}}}(\ell)$. Since $K_{\mathfrak{P}}|k_{\mathfrak{p}}$ is abelian, it follows from (7.5.9) that $T(K_{\mathfrak{P}}|k_{\mathfrak{p}})(\ell)$ is finite, and since $T_{k_{\mathfrak{p}}}(\ell) \cong \mathbb{Z}_\ell$, see (7.5.2), we have $T_{K_{\mathfrak{P}}}(\ell) \cong \mathbb{Z}_\ell$. Thus we get the desired surjection.

Now let $\ell = p$ and let $\tilde{K}$ be the maximal p-extension of k inside $K|k$. Since $\tilde{K}_{\mathfrak{P}}|k_{\mathfrak{p}}$ is an abelian infinite p-extension, the group $G_{\tilde{K}_{\mathfrak{P}}}(p)$ is a nontrivial free pro-p-group (in fact of infinite rank), see (7.5.10) for $p = \operatorname{char} k$, and (7.1.8) and (7.5.11) for $\mathfrak{p}$-adic fields. Therefore the same holds for the group $G_{K_{\mathfrak{P}}}(p)$, and we get a surjection of $G_{K_{\mathfrak{P}}}$ onto $\mathbb{Z}_p$. □

(9.5.3) Theorem (IWASAWA). *Let K be an infinite powerful extension of a global field with cd $G_K \le 1$ and let $\tilde{K}$ be the maximal solvable extension of K. Then*

$$G(\tilde{K}|K) \cong F_\omega(solv),$$

where $(solv)$ denotes the class of all solvable finite groups.

Proof: Since the set of finite separable extensions of K (inside a fixed separable closure) is countable, the group $G = G(\tilde{K}|K)$ has rank at most $\aleph_0$. We will show that every embedding problem for G with finite solvable E has a proper solution. Then, using (3.5.20), the result follows. So let

$$\begin{array}{ccccccccc} & & & & G_K & & \\ & & & & \downarrow{\scriptstyle can} & & \\ & & & & G & & \\ & & & & \downarrow{\scriptstyle \varphi} & & \\ 1 & \longrightarrow & A & \longrightarrow & E & \longrightarrow & \overline{G} & \longrightarrow & 1 \end{array}$$

be an embedding problem for G with E finite and solvable. In order to find a proper solution, we may proceed inductively and assume that A is abelian and even more that A is a simple $\overline{G}$-module. Since $H^2(G_K, A) = 0$ by assumption, it follows from (3.5.9) that we have a solution $\psi\colon G_K \to E$ for the embedding problem for G_K. Let $N = \ker(\varphi \circ can)$. The image of $\psi_0 = \psi|_N$ in A is a $\overline{G}$-submodule of A, hence equal to A or trivial. In the first case we found a proper solution, which factors through G because E is solvable, and we are done. So let us assume that $\operatorname{im}\psi_0 = 0$ (so that the group extension splits). Let $\overline{G} = G(K'|K)$, $K' \subseteq \tilde{K}$, then A is a trivial $G(\tilde{K}|K')$-module. Since $G(\tilde{K}|K) = \varprojlim_k G(\tilde{K}|k)$ where k runs through all finite subextensions of K, we can find a global field $k_0 \subseteq K$ and a Galois extension $k_0'|k_0$ such that

$K' = Kk'_0$ and $G(K'|K) \cong G(k'_0|k_0)$. Obviously this holds for all intermediate fields $k_0 \subseteq k \subseteq K$:

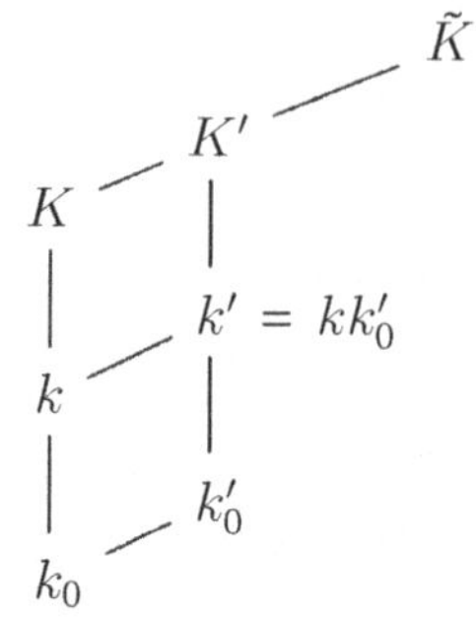

Via the canonical homomorphism $G_k \twoheadrightarrow G(k'|k) \cong G(K'|K)$, we consider A as a G_k-module. Since K is powerful, there exist primes $\mathfrak{P}_1, \dots, \mathfrak{P}_s$ of K such that

(i) $\mathfrak{P}_1, \dots, \mathfrak{P}_s$ split completely in K',
(ii) there are group homomorphisms $\varphi_i\colon G(\bar{K}_{\mathfrak{P}_i}|K_{\mathfrak{P}_i})) \longrightarrow A$ such that the images $\varphi_i(G(\bar{K}_{\mathfrak{P}_i}|K_{\mathfrak{P}_i}))$ generate A.

Enlarging k_0, we may assume that the underlying primes $\mathfrak{p}_i = \mathfrak{P}_i \cap k_0$, $i = 1, \dots, s$, of k_0 split completely in k'_0. Let p be the prime number with $pA = 0$. If $p \neq \mathrm{char}(k)$, (9.2.3)(v) (recalling that $G(k'|k)$ is solvable) implies that for all k between k_0 and K the map

$$H^1(k, A) \longrightarrow \prod_{i=1}^{s} H^1(k_{\mathfrak{P}_i}, A)$$

is surjective. If $p = \mathrm{char}(k)$, then the same assertion follows from (9.2.5). Passing to the direct limit, we see that

$$H^1(K, A) \longrightarrow \prod_{i=1}^{s} H^1(K_{\mathfrak{P}_i}, A)$$

is surjective. Therefore we obtain a class $[x] \in H^1(K, A)$ mapping to $res_i[x] = \varphi_i \in \mathrm{Hom}(G(\bar{K}_{\mathfrak{P}_i}|K_{\mathfrak{P}_i}), A) = H^1(K_{\mathfrak{P}_i}, A)$; observe that A is a trivial $G_{K_{\mathfrak{P}_i}}$-module since $G(K'_{\mathfrak{P}_i}|K_{\mathfrak{P}_i}) = 1$ by condition (i). Now let $\psi' = {}^x\psi \in \mathrm{Hom}(G_K, E)$ be the solution of the given embedding problem for G_K obtained by multiplying the old solution ψ by the 1-cocycle x. Then

$$\psi'_0 = \psi'|_N = x \cdot \psi_0 = x\colon N \to A$$

is surjective, since $G(\bar{K}_{\mathfrak{P}_i}|K_{\mathfrak{P}_i}) \subseteq N$ and the images of $\psi'|_{G(\bar{K}_{\mathfrak{P}_i}|K_{\mathfrak{P}_i})} = \varphi_i$ generate A by condition (ii). Since E is solvable, ψ' factors through G. This finishes the proof of the theorem. □

(9.5.4) Corollary. *Let K be an infinite separable extension of a global field. Then the Galois group of the maximal solvable extension $\tilde{K}$ over K is a free prosolvable group of countable rank,*

$$G(\tilde{K}|K) \cong F_\omega(solv)\,,$$

in the following cases:

(i) *K is a $\hat{\mathbb{Z}}$-extension* *) *of a global field k such that $p^\infty | [K_\mathfrak{p} : k_\mathfrak{p}]$ for all prime numbers p and all nonarchimedean primes $\mathfrak{p}$ of k and k is totally imaginary in the number field case.*

(ii) *K is the* **Kronecker field** *of a global field k, i.e. $K = k(\mu)$ where μ is the group of all roots of unity in $\bar{k}$.*

(iii) *$K = k^{ab}$ is the maximal abelian extension of a global field k.*

Proof: By (8.1.18) and (9.5.2), we know in all these cases that $cd\, G_K \le 1$ and that the field K is powerful. Thus (9.5.3) implies the corollary. □

In §2 we dealt with the problem of whether a given finite family of abelian local Galois extensions of a global field can be simultaneously realized by an abelian global extension. By the theorem of Grunwald-Wang, this is always possible, unless we are in a very special situation.

As a next step of investigation it would be natural to ask whether a similar statement is true for a given finite family of non-abelian local extensions. Since local Galois groups are automatically solvable, it is natural to ask whether there exists a global solvable extension which simultaneously realizes the given local extension.

As the reader may have expected, this question is extremely difficult to answer. It is, however, surprising that even the much weaker question of whether we can find *any* global extension realizing a given solvable group as a Galois group (i.e. without any local conditions at all) is highly nontrivial. A positive answer to this problem has been given by *I. R. Šafarevič*.

Theorem. *Let k be a global field and let G be a finite solvable group. Then there exists a Galois extension $K|k$ with $G(K|k) \cong G$.*

We will prove this theorem in §6. Let us first make some general comments. By definition, a solvable group is built up by successive extensions of abelian groups. Constructing the required global extension recursively by abelian steps, the first step is given by the theorem of Grunwald-Wang. In the second, and every subsequent step, we have to solve embedding problems with abelian

*) i.e. a Galois extension with Galois group isomorphic to $\hat{\mathbb{Z}}$.

kernel. These are not always solvable; in fact we can reach a deadlock very soon as the following example might indicate (for a proof see [216], th. 1.2.4).

Suppose that k is a field of characteristic not equal to 2. Then the quadratic extension $k(\sqrt{a})|k$ can be embedded into a cyclic extension of degree 4 if and only if a is a sum of two squares in k.

We learn from the above example that although we did not impose local conditions, there might be a global obstruction to the existence of a solution of our embedding problem. Therefore it is not very promising to solve the embedding problems of every induction step separately.

The case when all groups occurring are of order prime to $\#\mu(k)$, the order of the group of roots of unity contained in the global field k, can be tackled by a method which had been developed in 1936 independently by *A. Scholz* [206] and *H. Reichardt* [177]. Their idea was to choose the solutions of the inductively given embedding problems in a very special way, in order to avoid deadlocks. Scholz and Reichardt used this method in order to show that every finite nilpotent group of odd order occurs as a Galois group over $\mathbb{Q}$. The reader can find a proof of this statement in the spirit of Scholz and Reichardt in Serre's book [216].

The most far reaching result exploiting the Scholz-Reichardt method, is the following theorem of Neukirch [158], which we are going to explain next.

Let Γ be a fixed profinite group. The (continuous) homomorphisms $G \xrightarrow{f} \Gamma$ of arbitrary profinite groups into Γ are the objects of a category if one defines as morphisms from $G \xrightarrow{f} \Gamma$ to $G' \xrightarrow{f'} \Gamma$ all homomorphisms $G \xrightarrow{\psi} G'$ with $f' \circ \psi = f$. We call two such morphisms ψ and ψ' equivalent if there exists an element $a \in \ker f'$ such that

$$\psi'(\sigma) = a^{-1}\psi(\sigma)a \quad \text{for all } \sigma \in G\,,$$

and we denote the set of all equivalence classes $[\psi]$ by $\mathcal{H}om_\Gamma(G, G')$ and the subset of all $[\psi]$ with surjective $\psi\colon\ G \twoheadrightarrow G'$ by

$$\mathcal{H}om_\Gamma(G, G')_{epi}\,.$$

Let

$$G_k \xrightarrow{\varphi} \Gamma$$

be a homomorphism of the absolute Galois group of a global field k to the profinite group Γ and let

$$G_{k_\mathfrak{p}} \xrightarrow{\varphi_\mathfrak{p}} \Gamma$$

be its restriction to the decomposition group $G_{k_\mathfrak{p}}$ of G_k with respect to a prime $\mathfrak{p}$. Then if $f\colon G \to \Gamma$ is a homomorphism of an arbitrary profinite group G into Γ with kernel H, we obtain diagrams

$$\begin{array}{ccccccc} & & & & & & G_k \\ & & & & & \overset{\psi}{\swarrow} & \downarrow{\scriptstyle\varphi} \\ 1 & \longrightarrow & H & \longrightarrow & G & \xrightarrow{\;f\;} & \Gamma \end{array}$$

and

$$\begin{array}{ccccccc} & & & & & & G_{k_\mathfrak{p}} \\ & & & & & \overset{\psi_\mathfrak{p}}{\swarrow} & \downarrow{\scriptstyle\varphi_\mathfrak{p}} \\ 1 & \longrightarrow & H & \longrightarrow & G & \xrightarrow{\;f\;} & \Gamma . \end{array}$$

Furthermore, we have a canonical restriction map

$$\mathcal{H}om_\Gamma(G_k, G) \longrightarrow \prod_\mathfrak{p} \mathcal{H}om_\Gamma(G_{k_\mathfrak{p}}, G)\,.$$

The following theorem was proved by *J. Neukirch* in [158] (for number fields).

(9.5.5) Theorem. *Let k be a global field and let $\varphi : G_k \twoheadrightarrow \Gamma$ be a surjective homomorphism onto the finite group Γ, i.e. $\Gamma = G(K|k)$ for a finite Galois extension K of k.*

Let $f\colon G \twoheadrightarrow \Gamma$ be a surjective homomorphism whose kernel is separable), prosolvable and of finite exponent which is prime to the order of the group $\mu(K)$ of roots of unity of K. If*

$$\prod_\mathfrak{p} \mathcal{H}om_\Gamma(G_{k_\mathfrak{p}}, G) \neq \varnothing\,,$$

then the map

$$\mathcal{H}om_\Gamma(G_k, G)_{epi} \longrightarrow \prod_{\mathfrak{p}\in S} \mathcal{H}om_\Gamma(G_{k_\mathfrak{p}}, G)$$

is surjective for every finite set S of primes of k.

We shall deduce (9.5.5) from the special case, where the kernel of the homomorphism $f\colon G \twoheadrightarrow \Gamma$ is a finite abelian group A, i.e. we have an embedding

*) A profinite group is separable, if it has a countable basis of neighbourhoods of the identity.

problem $\mathscr{E}(G_k, \varphi, f)$

$$\begin{array}{ccccccccc} & & & & & & G_k & & \\ & & & & & & \big\downarrow{\scriptstyle\varphi} & & \\ 1 & \longrightarrow & A & \longrightarrow & G & \xrightarrow{f} & \Gamma & \longrightarrow & 1 \end{array}$$

with finite abelian kernel A on which Γ acts by conjugation and A becomes a G_k-module via φ. The space $\mathscr{S}_{\mathscr{E}(G_k)}$ of solutions of this embedding problem is nothing else but the set $\mathscr{H}om_\Gamma(G_k, G)$. Therefore $\mathscr{H}om_\Gamma(G_k, G)$ is either empty or a principal homogeneous space over $H^1(G_k, A)$ by (3.5.11).

For a prime $\mathfrak{p}$ we obtain the associated diagram

$$\begin{array}{ccccccccc} & & & & & & G_{k_\mathfrak{p}} & & \\ & & & & & & \big\downarrow{\scriptstyle\varphi_\mathfrak{p}} & & \\ 1 & \longrightarrow & A & \longrightarrow & G & \xrightarrow{f} & \Gamma & \longrightarrow & 1\,. \end{array}$$

Let $\Gamma_\mathfrak{p} \subseteq \Gamma$ be the image of $\varphi_\mathfrak{p}$ and we set $G_\mathfrak{p} = f^{-1}(\Gamma_\mathfrak{p})$. Then

$$\mathscr{H}om_\Gamma(G_{k_\mathfrak{p}}, G) = \mathscr{H}om_{\Gamma_\mathfrak{p}}(G_{k_\mathfrak{p}}, G_\mathfrak{p})$$

and the latter group is the set of solutions $\mathscr{S}_{\mathscr{E}(G_{k_\mathfrak{p}})}$ of the local embedding problem $\mathscr{E}(G_{k_\mathfrak{p}}, \varphi_\mathfrak{p} : G_{k_\mathfrak{p}} \twoheadrightarrow \Gamma_\mathfrak{p}, f_\mathfrak{p} : G_\mathfrak{p} \twoheadrightarrow \Gamma_\mathfrak{p})$. Therefore $\mathscr{H}om_\Gamma(G_{k_\mathfrak{p}}, G)$ is either empty or a principal homogeneous space over $H^1(G_{k_\mathfrak{p}}, A)$.

Thus the canonical map

$$\mathscr{H}om_\Gamma(G_k, G) \longrightarrow \prod_\mathfrak{p} \mathscr{H}om_\Gamma(G_{k_\mathfrak{p}}, G)$$

has to be understood as the map which assigns to a solution of the global embedding problem $\mathscr{E}(G_k, \varphi, f)$ the associated solutions of the local embedding problems $\mathscr{E}(G_{k_\mathfrak{p}}, \varphi_\mathfrak{p}, f_\mathfrak{p})$ for all $\mathfrak{p}$.

An element $[\psi] \in \mathscr{H}om_\Gamma(G_{k_\mathfrak{p}}, G)$ is called *unramified* if the inertia group $T_{k_\mathfrak{p}}$ of $G_{k_\mathfrak{p}}$ lies in the kernel of ψ. As before, we set $T_{k_\mathfrak{p}} = G_{k_\mathfrak{p}}$ if $\mathfrak{p}$ is an infinite prime. By $\mathscr{H}om_\Gamma(G_{k_\mathfrak{p}}, G)_{nr}$ we denote the subset of unramified elements in $\mathscr{H}om_\Gamma(G_{k_\mathfrak{p}}, G)$. If A is an unramified $G_{k_\mathfrak{p}}$-module, then $\mathscr{H}om_\Gamma(G_{k_\mathfrak{p}}, G)_{nr}$ is either empty or a principal homogenous space over $H^1_{nr}(G_{k_\mathfrak{p}}, A)$. Observe that

$$\varnothing \neq \mathscr{H}om_\Gamma(G_{k_\mathfrak{p}}, G)_{nr} \subseteq \mathscr{H}om_\Gamma(G_{k_\mathfrak{p}}, G),$$

if $G_{k_\mathfrak{p}} \overset{\varphi_\mathfrak{p}}{\twoheadrightarrow} \Gamma$ is unramified. Indeed, since $\varphi_\mathfrak{p}$ factors through $G_{k_\mathfrak{p}}/T_{k_\mathfrak{p}} \to \Gamma$, it can be lifted to a homomorphisms $G_{k_\mathfrak{p}} \twoheadrightarrow G_{k_\mathfrak{p}}/T_{k_\mathfrak{p}} \to G$ as $G_{k_\mathfrak{p}}/T_{k_\mathfrak{p}} \cong \hat{\mathbb{Z}}$ is free.

Next we establish a local-global principle.

(9.5.6) Lemma. *If the kernel A of the homomorphism $f\colon G \twoheadrightarrow \Gamma = G(K|k)$ is a simple Γ-module such that $pA = 0$ and $p \nmid \#\mu(K)$, then*

$$\mathcal{H}om_\Gamma(G_k, G) \neq \varnothing \iff \prod_{\mathfrak{p}} \mathcal{H}om_\Gamma(G_{k_\mathfrak{p}}, G) \neq \varnothing\,.$$

Proof: Let us consider the commutative diagram

$$\begin{array}{ccc} H^2(G_k, A) & \hookrightarrow & \bigoplus_{\mathfrak{p}} H^2(G_{k_\mathfrak{p}}, A) \\ {\scriptstyle \varphi^*}\nwarrow & & \nearrow{\scriptstyle (\varphi_\mathfrak{p}^*)_\mathfrak{p}} \\ & H^2(\Gamma, A). & \end{array}$$

If $p \neq \mathrm{char}(k)$, then the horizontal map is injective by (9.1.16)(iii) applied with S being the set of all primes and $\Omega = K(\mu_p)$. If $p = \mathrm{char}(k)$, then the horizontal map is trivially injective as $H^2(G_k, A) = 0$ by (8.3.3). Let $x \in H^2(\Gamma, A)$ be the class belonging to the group extension $1 \to A \to G \to \Gamma \to 1$. Using (3.5.9), we obtain the equivalences

$$\begin{aligned} \mathcal{H}om_\Gamma(G_k, G) \neq \varnothing \quad &\iff \quad \varphi^*(x) = 0 \\ &\iff \quad \varphi_\mathfrak{p}^*(x) = 0 \ \text{ for all } \mathfrak{p} \\ &\iff \quad \prod_{\mathfrak{p}} \mathcal{H}om_\Gamma(G_{k_\mathfrak{p}}, G) \neq \varnothing. \end{aligned}$$ □

Besides the group extension

$$1 \longrightarrow A \longrightarrow G \longrightarrow \Gamma \longrightarrow 1$$

with the finite group Γ, the simple Γ-module A ($pA = 0$) and the homomorphism

$$\varphi\colon G_k \twoheadrightarrow \Gamma = G(K|k)$$

we now consider a further surjective homomorphism

$$E \overset{g}{\twoheadrightarrow} G$$

with a prosolvable kernel of finite exponent e. Let n be a multiple of ep.

(9.5.7) Main Lemma. *Let T be an arbitrary finite set of primes of the global field k and assume that $(n, \#\mu(K)) = 1$. If $\prod_\mathfrak{p} \mathcal{H}om_\Gamma(G_{k_\mathfrak{p}}, G) \neq \varnothing$, then there exists an element $[\psi] \in \mathcal{H}om_\Gamma(G_k, G)_{epi}$ with the following properties:*

(i) *$[\psi]$ induces given elements $[\psi_\mathfrak{p}] \in \mathcal{H}om_\Gamma(G_{k_\mathfrak{p}}, G)$ at the primes $\mathfrak{p} \in T$.*

(ii) *If $\mathfrak{p} \notin T$ is unramified in $K|k$, then $\mathcal{H}om_G(G_{k_\mathfrak{p}}, E) \neq \varnothing$, i.e. $\psi_\mathfrak{p}$ can be lifted to a G-homomorphism $G_{k_\mathfrak{p}} \to E$.*

(iii) *For the field N defined by $\psi\colon G_k \twoheadrightarrow G$ we have $(n, \#\mu(N)) = 1$.*

Proof: If $T' \supseteq T$ is a larger finite set of primes, then the main lemma holds for T if it holds for T'. In order to see this, choose for every $\mathfrak{p} \in T' \smallsetminus T$ an element $[\psi_\mathfrak{p}]$ in the nonempty set $\mathscr{H}om_\Gamma(G_{k_\mathfrak{p}}, G)$. If $\mathfrak{p} \in T' \smallsetminus T$ is unramified, then we choose $[\psi_\mathfrak{p}] \in \mathscr{H}om_\Gamma(G_{k_\mathfrak{p}}, G)_{nr} \neq \varnothing$. If $[\psi] \in \mathscr{H}om_\Gamma(G_k, G)_{epi}$ satisfies conditions (i)–(iii) for T', then so it does for the set T. Indeed, (i) and (iii) are trivially satisfied and (ii) follows from the fact that $\mathscr{H}om_G(G_{k_\mathfrak{p}}, E)_{nr} \neq \varnothing$ for unramified primes $\mathfrak{p} \in T' \smallsetminus T$, since we have chosen unramified elements $[\psi_\mathfrak{p}]$. Therefore we may assume that

$$\mathrm{Ram}(K|k) \cup S_n \cup S_\infty \subseteq T,$$

where S_n is the set of all primes dividing n if k is a number field, and empty if k is a function field.

We next show that it suffices to prove the existence of an element $[\psi]$ in $\mathscr{H}om_\Gamma(G_k, G)$ with (i)–(iii), i.e. we can drop the surjectivity of $[\psi]$. Let $\mathfrak{q} \notin T$ be a prime which splits completely in $K|k$ and let $a \in A$, $a \neq 0$. Let further $\psi_\mathfrak{q}$: $G_{k_\mathfrak{q}} \twoheadrightarrow G_{k_\mathfrak{q}}/T_{k_\mathfrak{q}} \twoheadrightarrow \langle a \rangle$ be an unramified surjective homomorphism onto the cyclic group $\langle a \rangle \subseteq G$. Since $\mathfrak{q}$ splits completely in K, we have $\varphi_\mathfrak{q}(G_{k_\mathfrak{q}}) = \{1\}$, hence $[\psi_\mathfrak{q}] \in \mathscr{H}om_\Gamma(G_{k_\mathfrak{q}}, G)$. Let $T' = T \cup \{\mathfrak{q}\}$ and let $[\psi]$ in $\mathscr{H}om_\Gamma(G_k, G)$ satisfying the conditions (i)–(iii) for T'. Then it satisfies conditions (i)–(iii) also for T, as is seen in the same way as above. Moreover, ψ is surjective. Namely, $\psi(G_{k_\mathfrak{q}}) \subseteq G$ is conjugate to $\psi_\mathfrak{q}(G_{k_\mathfrak{q}}) = \langle a \rangle$ under an element of A, and so $a \in \psi(G_{k_\mathfrak{q}}) \subseteq \psi(G_k)$. Therefore $A \cap \psi(G_k)$ is a nontrivial Γ-submodule of A. Hence $A \subseteq \psi(G_k)$ since A is simple; and ψ is surjective.

A last reduction step is the following: it suffices to prove the existence of an element $[\psi] \in \mathscr{H}om_\Gamma(G_k, G)$ satisfying only conditions (i) and (ii).

Let $n' = n$ if k is a number field and let n' be the maximal divisor of n prime to char(k) if k is a function field. Let $\zeta_{n'}$ be a primitive n'-th root of unity and let $\sigma_1, \ldots, \sigma_r$ be generators of $G(K(\zeta_{n'})|K)$. Let $\mathfrak{q}_1, \ldots, \mathfrak{q}_r \notin T$ be different primes such that $\mathfrak{q}_i$ splits completely in K and

$$\mathrm{Frob}_{\mathfrak{Q}_i} = \sigma_i\,, \quad i = 1, \ldots, r\,,$$

where $\mathfrak{Q}_i$ is a prime of K above $\mathfrak{q}_i$. Then $\varphi_{\mathfrak{q}_i}$ is the trivial homomorphism and the trivial homomorphism $\psi_{\mathfrak{q}_i}$: $G_{k_{\mathfrak{q}_i}} \to G$ yields an element in $\mathscr{H}om_\Gamma(G_{k_{\mathfrak{q}_i}}, G)$, $i = 1, \ldots, r$. Let $T' = T \cup \{\mathfrak{q}_1, \ldots, \mathfrak{q}_r\}$ and let $[\psi]$ in $\mathscr{H}om_\Gamma(G_k, G)$ satisfying the conditions (i) and (ii) for T', and so for T. Moreover it satisfies condition (iii): the restriction of ψ to $G_{k_{\mathfrak{q}_i}}$ is the trivial homomorphism, i.e. $\mathfrak{q}_1, \ldots, \mathfrak{q}_r$ split completely in the field $N|k$, which is defined by ψ: $G_k \to G$. Therefore the images of the automorphisms $\sigma_i = \mathrm{Frob}_{\mathfrak{Q}_i}$ under the surjection $G(K(\zeta_{n'})|K) \twoheadrightarrow G(N \cap K(\zeta_{n'})|K)$ are trivial. It follows that $N \cap K(\zeta_{n'}) = K$. Let $d = (n', \#\mu(N))$ and let ζ_d be a primitive d-th root of unity. Then $\zeta_d \in N \cap K(\zeta_{n'}) = K$, i.e. $d|(n', \#\mu(K))$. Since n/n' is prime to $\#\mu(K)$ if K is a function field, we have $(n', \#\mu(K)) = (n, \#\mu(K)) = 1$, and so $d = 1$.

We now come to the essential step in the proof of the lemma. By the assumption $\prod_{\mathfrak{p}} \mathcal{H}om_\Gamma(G_{k_\mathfrak{p}}, G) \neq \varnothing$ and by (9.5.6), there is an element

$$[\psi_0] \in \mathcal{H}om_\Gamma(G_k, G).$$

We will change this element by a cohomology class $x \in H^1(G_k, A)$ into an element ψ which satisfies the conditions of the lemma. Let $N_0|k$ be the field defined by $\psi_0\colon G_k \to G$. The extension $N_0|K$ is abelian. Let $[\psi_{1\mathfrak{p}}] \in \mathcal{H}om_\Gamma(G_{k_\mathfrak{p}}, G)$ be the elements which are given in advance for $\mathfrak{p} \in T$. Let $\mathfrak{p}_1, \ldots, \mathfrak{p}_r$ be the primes $\mathfrak{p} \notin T$ for which $\psi_{0\mathfrak{p}}$ is ramified. Since $\mathrm{Ram}(K|k) \subseteq T$, the primes $\mathfrak{p}_i$ are unramified in $K|k$, i.e. the homomorphisms $\varphi_{\mathfrak{p}_i}$ are unramified and can therefore be lifted to unramified Γ-homomorphisms $\psi_{1\mathfrak{p}_i}\colon G_{k_{\mathfrak{p}_i}} \to G$. Let $T^* = T \cup \{\mathfrak{p}_1, \ldots, \mathfrak{p}_r\}$ and for each $\mathfrak{p} \in T^*$ let $y_\mathfrak{p} \in H^1(G_{k_\mathfrak{p}}, A)$ be the cohomology class which sends $[\psi_{0\mathfrak{p}}]$ to $[\psi_{1\mathfrak{p}}]$, i.e.

$$[\psi_{0\mathfrak{p}}]^{y_\mathfrak{p}} = [\psi_{1\mathfrak{p}}].$$

We will deal with the case $p = \mathrm{char}(k)$ later and assume $p \neq \mathrm{char}(k)$ for the moment. Let n' be defined as above and let $\Omega = N_0(\zeta_{n'})$, $\zeta_{n'}$ a primitive n'-th root of unity. Let K' be the subfield of Ω which corresponds to the prime-to-p part $G(\Omega|K)(p')$ of the abelian group $G(\Omega|K)$. Then $K'|K$ is a p-extension, hence $\mu_p \not\subseteq K'$ because $\mu_p \not\subseteq K$. Therefore we can apply (9.3.3) with S being the set of all primes to find an element $x \in H^1(G_k, A)$ such that

(a) $x_\mathfrak{p} = y_\mathfrak{p}$ for $\mathfrak{p} \in T^*$,

(b) if $\mathfrak{p} \notin T^*$, then $x_\mathfrak{p}$ is cyclic and if $x_\mathfrak{p}$ is ramified, then $\mathfrak{p}$ splits completely in Ω.

We now prove that the element

$$[\psi] = [\psi_0]^x \in \mathcal{H}om_\Gamma(G_k, G)$$

satisfies conditions (i) and (ii) of the lemma. If $\mathfrak{p} \in T^*$, then

$$[\psi_\mathfrak{p}] = [\psi_{0\mathfrak{p}}]^{x_\mathfrak{p}} = [\psi_{0\mathfrak{p}}]^{y_\mathfrak{p}} = [\psi_{1\mathfrak{p}}],$$

thus condition (i) is satisfied. Let $\mathfrak{p} \notin T$ be unramified in $K|k$. If $\mathfrak{p} \in T^* \smallsetminus T$, then $[\psi_\mathfrak{p}] = [\psi_{1\mathfrak{p}}]$ is unramified, and so $\psi_\mathfrak{p}$ can be lifted to a G-homomorphism $G_{k_\mathfrak{p}} \to E$, i.e. $\mathcal{H}om_G(G_{k_\mathfrak{p}}, E) \neq \varnothing$.

Let $\mathfrak{p} \notin T^*$. If $[\psi_\mathfrak{p}]$ is unramified, then $\mathcal{H}om_G(G_{k_\mathfrak{p}}, E) \neq \varnothing$ for the same reason. Let $[\psi_\mathfrak{p}] = [\psi_{0\mathfrak{p}}]^{x_\mathfrak{p}}$ be ramified. Since $[\psi_{0\mathfrak{p}}]$ is unramified for $\mathfrak{p} \notin T^*$, the cohomology class $x_\mathfrak{p}$ must be ramified. Thus the prime $\mathfrak{p}$ splits completely in $\Omega|k$ by (b), and in particular it splits in $N_0|k$ and in $k(\zeta_{n'})|k$. Therefore $\psi_{0\mathfrak{p}}\colon G_{k_\mathfrak{p}} \to G$ is the trivial homomorphism, i.e. $N_{0\mathfrak{p}} = k_\mathfrak{p}$, and $\zeta_{n'} \in k_\mathfrak{p}$. Because $K_\mathfrak{p} = k_\mathfrak{p}$ the Galois module A is a trivial $G_{k_\mathfrak{p}}$-module, i.e.

$$x_\mathfrak{p}\colon G_{k_\mathfrak{p}} \longrightarrow A \subseteq G$$

is a homomorphism which represents the class $[\psi_\mathfrak{p}] = [\psi_{0\mathfrak{p}}]^{x_\mathfrak{p}}$. It remains to show that $x_\mathfrak{p}$ can be lifted to a G-homomorphism $G_{k_\mathfrak{p}} \to E$.

Since $x_{\mathfrak{p}}$ is cyclic, it defines a cyclic ramified extension $N_{\mathfrak{p}}|k_{\mathfrak{p}}$ of degree p. Since $\mathfrak{p} \nmid p$, this extension is tamely ramified. Thus $N_{\mathfrak{p}} = k_{\mathfrak{p}}(\sqrt[p]{\pi})$ with a prime element π. Let $\bar{\sigma}$ be a generator of

$$G(N_{\mathfrak{p}}|k_{\mathfrak{p}}) = x_{\mathfrak{p}}(G_{k_{\mathfrak{p}}}) \subseteq G$$

and let σ be an element of E which is mapped onto $\bar{\sigma}$ under $g : E \twoheadrightarrow G$. As $\bar{\sigma}$ has order p, we may choose σ in a p-Sylow subgroup E_p of E. Since $\ker(E \twoheadrightarrow G)$ has exponent $e = n/p$, the order m of σ is a p-power which divides n'. Since $\zeta_{n'} \in k_{\mathfrak{p}}$, the m-th roots of unity are in $k_{\mathfrak{p}}$. Therefore the extension $M_{\mathfrak{p}} = k_{\mathfrak{p}}(\sqrt[m]{\pi})|k_{\mathfrak{p}}$ is cyclic of degree m. Mapping a generator τ of $G(M_{\mathfrak{p}}|k_{\mathfrak{p}})$ with image $\bar{\sigma}$ in $G(N_{\mathfrak{p}}|k_{\mathfrak{p}})$ onto σ, we obtain the commutative diagram

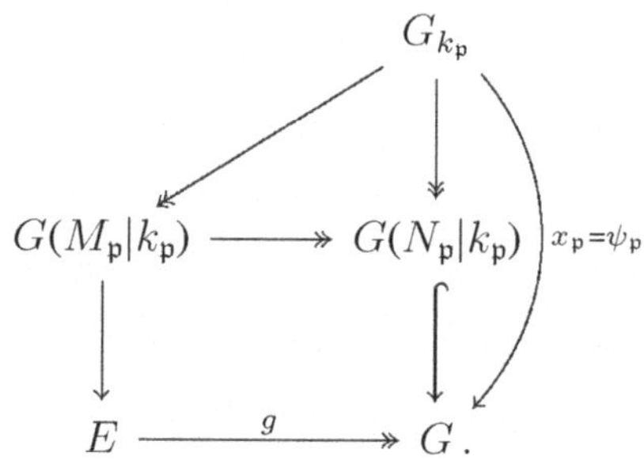

Hence $\psi_{\mathfrak{p}}$ can be lifted to a G-homomorphism $G_{k_{\mathfrak{p}}} \to E$. This finishes the proof of the lemma in the case $p \neq \mathrm{char}(k)$.

Now let $p = \mathrm{char}(k)$. Choose any prime $\mathfrak{q} \notin T^*$ of k which is completely decomposed in $K|k$. By (9.2.5), we find an element $x \in H^1(G_k, A)$ such that

(a) $x_{\mathfrak{p}} = y_{\mathfrak{p}}$ for $\mathfrak{p} \in T^*$,

(b) x is unramified outside $T^* \cup \{\mathfrak{q}\}$.

We now prove that the element

$$[\psi] = [\psi_0]^x \in \mathscr{H}om_\Gamma(G_k, G)$$

satisfies conditions (i) and (ii) of the lemma. If $\mathfrak{p} \in T^*$, then

$$[\psi_{\mathfrak{p}}] = [\psi_{0\mathfrak{p}}]^{x_{\mathfrak{p}}} = [\psi_{0\mathfrak{p}}]^{y_{\mathfrak{p}}} = [\psi_{1\mathfrak{p}}],$$

thus condition (i) is satisfied. Let $\mathfrak{p} \notin T$ be unramified in $K|k$. If $\mathfrak{p} \in T^* \smallsetminus T$, then $[\psi_{\mathfrak{p}}] = [\psi_{1\mathfrak{p}}]$ is unramified, and so $\psi_{\mathfrak{p}}$ can be lifted to a G-homomorphism $G_{k_{\mathfrak{p}}} \to E$, i.e. $\mathscr{H}om_G(G_{k_{\mathfrak{p}}}, E) \neq \varnothing$.

Let $\mathfrak{p} \notin T^*$. If $[\psi_{\mathfrak{p}}]$ is unramified, then $\mathscr{H}om_G(G_{k_{\mathfrak{p}}}, E) \neq \varnothing$ for the same reason. Let $[\psi_{\mathfrak{p}}] = [\psi_{0\mathfrak{p}}]^{x_{\mathfrak{p}}}$ be ramified. Since $[\psi_{0\mathfrak{p}}]$ is unramified for $\mathfrak{p} \notin T^*$, the cohomology class $x_{\mathfrak{p}}$ must be ramified. Thus $\mathfrak{p} = \mathfrak{q}$. By the choice of $\mathfrak{q}$, it splits completely in $\Omega|k$. Therefore $\psi_{0\mathfrak{q}}\colon G_{k_\mathfrak{q}} \to G$ is the trivial homomorphism, i.e. $N_{0\mathfrak{q}} = k_\mathfrak{q}$. Because $K_\mathfrak{q} = k_\mathfrak{q}$ the Galois module A is a trivial $G_{k_\mathfrak{q}}$-module, i.e.

$$x_\mathfrak{q}\colon G_{k_\mathfrak{q}} \longrightarrow A \subseteq G$$

is a homomorphism which represents the class $[\psi_\mathfrak{q}] = [\psi_{0\mathfrak{q}}]^{x_\mathfrak{q}}$. It remains to show that $x_\mathfrak{q}$ can be lifted to a G-homomorphism $G_{k_\mathfrak{q}} \to E$. However, $x_\mathfrak{q}$ factors through the maximal pro-p-factor group $G_{k_\mathfrak{q}}(p)$, which is a free pro-p-group by (6.1.4), and therefore a projective profinite group by (3.5.13). We obtain a commutative diagram

$$\begin{array}{ccc}
 & G_{k_\mathfrak{q}} & \\
 & \downarrow & \searrow{\scriptstyle x_\mathfrak{q}=\psi_\mathfrak{q}} \\
 & G_{k_\mathfrak{q}}(p) & \\
\swarrow & \downarrow & \\
E \overset{g}{\twoheadrightarrow} & G\,. &
\end{array}$$

Hence also in the case $p = \mathrm{char}(k)$ the homomorphism $\psi_\mathfrak{q}$ can be lifted to a G-homomorphism $G_{k_\mathfrak{q}} \to E$. This finishes the proof of the lemma. □

Proof of (9.5.5): We consider the embedding problem

$$\begin{array}{ccccccccc}
 & & & & & & G_k & & \\
 & & & & & & \downarrow{\scriptstyle\varphi} & & \\
1 & \longrightarrow & H & \longrightarrow & G & \xrightarrow{f} & \Gamma & \longrightarrow & 1,
\end{array}$$

where $\Gamma = G(K|k)$ and the kernel H is a separable, prosolvable group of finite exponent n and $(n, \#\mu(K)) = 1$. Therefore we find a chain

$$H = H_0 \supseteq H_1 \supseteq H_2 \supseteq H_3 \supseteq \ldots$$

of open normal subgroups of H such that the quotients $A_i = H_{i-1}/H_i$ are abelian and $\bigcap_i H_i = \{1\}$. Since Γ is finite, we can assume that the H_i are normal in G. Then the A_i are Γ-modules and become G_k-modules via φ. Taking a maximal refinement of such a chain we can assume that each A_i is a simple G_k-module.

For every $\mathfrak{p} \in T$ we prescribe an element $[\psi_\mathfrak{p}] \in \mathscr{H}om_\Gamma(G_{k_\mathfrak{p}}, G)$ and select a Γ-homomorphism $\psi_\mathfrak{p}\colon G_{k_\mathfrak{p}} \to G$ in $[\psi_\mathfrak{p}]$. We will construct an element $[\psi] \in \mathscr{H}om_\Gamma(G_k, G)_{epi}$ such that $[\psi|_{G_{k_\mathfrak{p}}}] = [\psi_\mathfrak{p}]$ for $\mathfrak{p} \in T$. Let $G_i = G/H_i$ and consider the commutative diagrams with exact rows

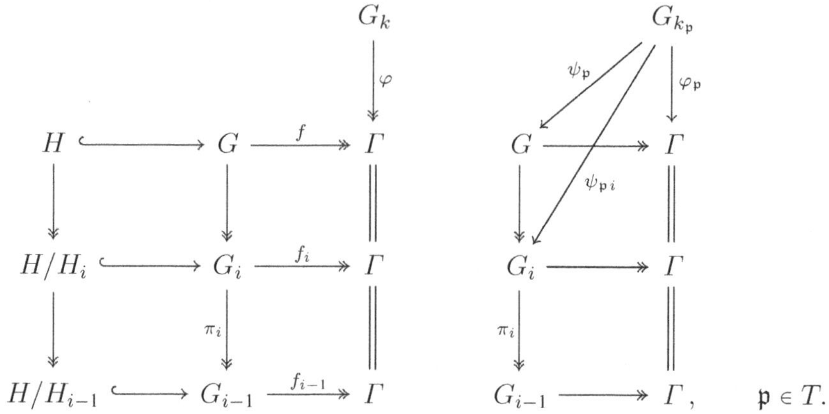

The kernel of $\pi_i\colon G_i \twoheadrightarrow G_{i-1}$ is the simple G_k-module A_i. Let $p_i A_i = 0$ and $n_i = \exp(H_i)$. We denote the composite of $\psi_\mathfrak{p}$ and $G \twoheadrightarrow G_i$ by $\psi_{\mathfrak{p}i}$.

Claim. There exists a sequence of surjective Γ-homomorphisms $\psi_i\colon G_k \twoheadrightarrow G_i$ with the following properties:

1) $\psi_{i-1} = \pi_i \circ \psi_i, \quad i = 1, 2, \ldots .$

2) $[\psi_i|_{G_{k_\mathfrak{p}}}] = [\psi_{\mathfrak{p}i}]$ in $\mathcal{H}om_\Gamma(G_{k_\mathfrak{p}}, G_i)$ for $\mathfrak{p} \in T$.

3) If the field K_i is defined by ψ_i, then we have $(n, \#\mu(K_i)) = 1$.

4) $\prod\limits_{\mathfrak{p}} \mathcal{H}om_{G_i}(G_{k_\mathfrak{p}}, G) \neq \varnothing$.

Proof of the claim: If $i = 0$, then we can identify G_0 with Γ and choose $\psi_0 = \varphi$. The first condition is empty, the second is true since $\varphi|_{G_{k_\mathfrak{p}}} = \varphi_\mathfrak{p}$, and the third and fourth are assumptions of the theorem.

Let us assume that the homomorphism $\psi_{i-1}\colon G_k \twoheadrightarrow G_{i-1}$ is already constructed. Let $T_{i-1} = T \cup \mathrm{Ram}(K_{i-1}|k)$, where the field K_{i-1} is defined by ψ_{i-1}. For each $\mathfrak{p} \in T_{i-1} \smallsetminus T$ we can choose a G_{i-1}-homomorphism $\psi_\mathfrak{p}\colon G_{k_\mathfrak{p}} \to G$ since the set $\mathcal{H}om_{G_{i-1}}(G_{k_\mathfrak{p}}, G)$ is not empty by condition 4. We consider the diagrams

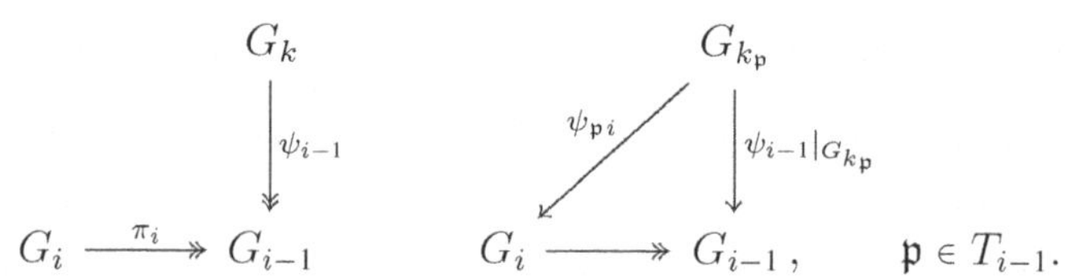

Since $[\psi_{\mathfrak{p}i}] \in \mathcal{H}om_{G_{i-1}}(G_{k_\mathfrak{p}}, G_i)$, the right diagrams are commutative. The kernel of π_i is the simple G_{i-1}-module A_i. Its exponent p_i divides n. Since

the exponent n_i of H_i divides n, it follows that np_i is a multiple of $n_i p_i$ and $(np_i, \#\mu(K_{i-1})) = (n, \#\mu(K_{i-1})) = 1$ by condition 3. Furthermore, by condition 4 we have

$$\prod_{\mathfrak{p}} \mathscr{H}om_{G_{i-1}}(G_{k_\mathfrak{p}}, G_i) \neq \varnothing ,$$

and so we can apply (9.5.7) to the situation replacing T by T_{i-1} and $E \twoheadrightarrow G$ by $G \twoheadrightarrow G_i$ and φ by ψ_{i-1}. We find an element $[\psi_i]$ in $\mathscr{H}om_{G_{i-1}}(G_k, G_i)_{epi}$ with the properties:

(i) $[\psi_i|_{G_{k_\mathfrak{p}}}] = [\psi_{\mathfrak{p}i}]$ in $\mathscr{H}om_{G_{i-1}}(G_{k_\mathfrak{p}}, G_i)$ for $\mathfrak{p} \in T_{i-1}$.

(ii) If $\mathfrak{p} \notin T_{i-1}$, then $\mathscr{H}om_{G_i}(G_{k_\mathfrak{p}}, G) \neq \varnothing$.

(iii) For the field $K_i|k$ defined by ψ_i we have $(n, \#\mu(K_i)) = 1$.

Let ψ_i: $G_k \twoheadrightarrow G_i$ be a representative of $[\psi_i]$. Then ψ_i is a surjective Γ-homomorphism, and we show that it satisfies conditions 1 – 4 of the claim.

Condition 1 is satisfied because of $[\psi_i] \in \mathscr{H}om_{G_{i-1}}(G_k, G_i)$. By (i) the Γ-homomorphisms $\psi_i|_{G_{k_\mathfrak{p}}}$ and $\psi_{\mathfrak{p}i}$, $\mathfrak{p} \in T_{i-1}$, are conjugate by an element of

$$A_i = H_{i-1}/H_i \subseteq H/H_i = \ker(G_i \twoheadrightarrow \Gamma).$$

Therefore $[\psi_i|_{G_{k_\mathfrak{p}}}] = [\psi_{\mathfrak{p}i}]$ in $\mathscr{H}om_\Gamma(G_{k_\mathfrak{p}}, G_i)$, showing condition 2. Condition 3 is satisfied by (iii) and, finally, condition 4 follows from (ii) and from $[\psi_\mathfrak{p}] \in \mathscr{H}om_{G_i}(G_{k_\mathfrak{p}}, G)$ for $\mathfrak{p} \in T_{i-1}$. This finishes the proof of the claim.

We have $G = \varprojlim G_i$, since $\bigcap_i H_i = \{1\}$. Therefore the surjective Γ-homomorphisms ψ_i: $G_k \twoheadrightarrow G_i$ define a surjective Γ-homomorphism ψ: $G_k \twoheadrightarrow G$, i.e. an element in $[\psi] \in \mathscr{H}om_\Gamma(G_k, G)_{epi}$. For each $\mathfrak{p} \in T$ we have $[\psi_i|_{G_{k_\mathfrak{p}}}] = [\psi_{\mathfrak{p}i}]$ in $\mathscr{H}om_\Gamma(G_{k_\mathfrak{p}}, G_i)$, and so there exist elements $\tilde{a}_{\mathfrak{p}i} \in \ker(G_i \twoheadrightarrow \Gamma) = H/H_i$ such that

$$\psi_i|_{G_{k_\mathfrak{p}}} = \tilde{a}_{\mathfrak{p}i}^{-1} \psi_{\mathfrak{p}i} \tilde{a}_{\mathfrak{p}i}, \quad i = 1, 2, \ldots .$$

Using a compactness argument, we find an element $a_\mathfrak{p} \in H$ such that

$$\psi_i|_{G_{k_\mathfrak{p}}} = a_{\mathfrak{p}i}^{-1} \psi_{\mathfrak{p}i} a_{\mathfrak{p}i}, \quad i = 1, 2, \ldots ,$$

where the elements $a_{\mathfrak{p}i}$ are the images of $a_\mathfrak{p}$ in H/H_i. We obtain

$$\psi|_{G_{k_\mathfrak{p}}} = a_\mathfrak{p}^{-1} \psi_\mathfrak{p} a_\mathfrak{p},$$

i.e. $[\psi|_{G_{k_\mathfrak{p}}}] = [\psi_\mathfrak{p}]$ in $\mathscr{H}om_\Gamma(G_{k_\mathfrak{p}}, G)$. This finishes the proof of (9.5.5). □

In the language of embedding problems, (9.5.5) implies the following local-global principle.

(9.5.8) Corollary. *Let $\mathscr{E}(G_k, \varphi, f)$ be an embedding problem of the global field k,*

$$\begin{array}{ccccccccc} & & & & & & G_k & & \\ & & & & & & \downarrow{\scriptstyle\varphi} & & \\ 1 & \longrightarrow & H & \longrightarrow & G & \xrightarrow{f} & \Gamma & \longrightarrow & 1, \end{array}$$

and assume that $\Gamma = G(K|k)$ is finite and H is a separable, prosolvable group of finite exponent prime to $\#\mu(K)$. Then the following holds.

(i) *$\mathscr{E}(G_k, \varphi, f)$ has a proper solution if and only if all local embedding problems $\mathscr{E}(G_{k_\mathfrak{p}}, \varphi_\mathfrak{p}, f_\mathfrak{p})$ are solvable.*

(ii) *$\mathscr{E}(G_k, \varphi, f)$ has a proper solution in the following cases:*

   (a) *$K|k$ is unramified.*

   (b) *$\mathscr{E}(G_k, \varphi, f)$ has a solution.*

   (c) *The group extension splits.*

Remark: Assertion (ii)(b) was proven by Ikeda [83] in the case that k is a number field and H a finite abelian group.

Proof: The solvability of all local embedding problems is equivalent with $\prod_\mathfrak{p} \mathscr{H}om_\Gamma(G_{k_\mathfrak{p}}, G) \neq \varnothing$. Thus (i) follows from (9.5.5).

Since the local embedding problem $\mathscr{E}(G_{k_\mathfrak{p}}, \varphi_\mathfrak{p}, f_\mathfrak{p})$ is always solvable if the extension $K_\mathfrak{p}|k_\mathfrak{p}$ is unramified, we obtain (ii)(a).

The assertion $\mathscr{H}om_\Gamma(G_k, G) \neq \varnothing$ implies $\prod_\mathfrak{p} \mathscr{H}om_\Gamma(G_{k_\mathfrak{p}}, G) \neq \varnothing$ and so $\mathscr{H}om_\Gamma(G_k, G)_{epi} \neq \varnothing$. This proves (ii)(b) and assertion (c) is a trivial consequence of (b). □

Furthermore, we obtain the following existence theorem of Grunwald/Wang type.

(9.5.9) Theorem. *Let k be a global field and let T be a finite set of primes of k. Let G be a separable, prosolvable group of finite exponent prime to $\#\mu(k)$, and for $\mathfrak{p} \in T$ let $K_\mathfrak{p}|k_\mathfrak{p}$ be Galois extensions whose Galois groups $G(K_\mathfrak{p}|k_\mathfrak{p})$ are embedable into G.*

Then there exists a Galois extension $K|k$ with Galois group isomorphic to G, which has the given extensions $K_\mathfrak{p}|k_\mathfrak{p}$ as completions for the primes $\mathfrak{p} \in T$.

Proof: Apply (9.5.5) to the case $\Gamma = \{1\}$ and the homomorphisms

$$\psi_{\mathfrak{p}}\colon G_{k_{\mathfrak{p}}} \twoheadrightarrow G(K_{\mathfrak{p}}|k_{\mathfrak{p}}) \hookrightarrow G, \quad \mathfrak{p} \in T. \qquad \square$$

Finally, we obtain existence results without local conditions.

(9.5.10) Corollary. *Let k be a number field and let G be a separable, prosolvable group of finite odd exponent.*

Then there exists a Galois extension $K|k$ with Galois group isomorphic to the group G.

Remarks: 1. As already mentioned, the theorem of Šafarevič, which we prove in the next section, shows that the assertions above holds for every finite solvable group.

2. The finiteness of the exponent of G is in fact a necessary condition. For example, the group $\mathbb{Z}_p \times \mathbb{Z}_p$ is not the Galois group of an extension of $\mathbb{Q}$.

Proof: We apply (9.5.9) to the field $\mathbb{Q}$, the set $T = \mathrm{Ram}(k|\mathbb{Q})$ and the trivial extensions $K_v = \mathbb{Q}_v$ for $v \in T$. Then trivially the groups $G(K_v|\mathbb{Q}_v)$ are embedable into G and the exponent of G is prime to $\#\mu(\mathbb{Q}) = 2$. Therefore there exists a surjective homomorphism $\psi\colon G_{\mathbb{Q}} \twoheadrightarrow G$ such that the restrictions $\psi_v\colon G_{\mathbb{Q}_v} \to G$ are trivial for $v \in T$. Let K' be the fixed field of $\ker(\psi)$. Since $\psi(G_{\mathbb{Q}_v}) = 1$ for all $v \in T$, the primes in T split completely in K', and so $K' \cap k|\mathbb{Q}$ is an unramified extension. It follows that $K' \cap k = \mathbb{Q}$. Setting $K = K'k$, we obtain $G(K|k) \cong G(K'|\mathbb{Q}) \cong G$. $\square$

The function field analogue of (9.5.10) is the following

(9.5.11) Corollary. *Let k be a global field of characteristic $p > 0$ and let G be a separable, prosolvable group of finite exponent e which is prime to $p - 1$.*

Then there exists a Galois extension $K|k$ with Galois group isomorphic to the group G.

Remark: If $\mathrm{char}(k) = 2$, then the corollary is sharper than the statement of Šafarevič's theorem.

Proof: We choose an embedding $\mathbb{F}_p(t) \subseteq k$, such that $k|\mathbb{F}_p(t)$ is finite and separable. We apply (9.5.9) to $\mathbb{F}_p(t)$, $T = \varnothing$ and the n-fold self-product G^n of G, where n is some natural number. Since e is prime to $\#\mu(\mathbb{F}_p(t)) = p-1$, we obtain a Galois extension $K|\mathbb{F}_p(t)$ with Galois group G^n. The projections $pr_i : G^n \twoheadrightarrow G$, $i = 1, \ldots, n$, yield n linearly disjoint extensions $K_i|\mathbb{F}_p(t)$ with Galois group G. If n was chosen sufficiently large, at least one of the fields K_i is linearly disjoint to k, and $K_ik|k$ is an extension with Galois group G. □

Finally, we would like to mention a result of *G. Henniart* in [78], where only one given local Galois extension at a nonarchimedean prime is considered, but no restriction on the order of the Galois group is made:

Let k be a totally real number field and $\mathfrak{p}$ a prime of k dividing the odd prime number p. Let $K_\mathfrak{p}|k_\mathfrak{p}$ be a finite Galois extension. Then there exists a totally real Galois extension $K|k$ such that $[K : k] = [K_\mathfrak{p} : k_\mathfrak{p}]$ and the completions with respect to the primes $\mathfrak{P}$ above $\mathfrak{p}$ are isomorphic to $K_\mathfrak{p}$.

§6. Šafarevič's Theorem

In this section we will prove the celebrated theorem of *I. R. Šafarevič* concerning the existence of an extension of a global field realizing a given solvable group as a Galois group.

(9.6.1) Theorem (*Šafarevič*). *Let k be a global field and let G be a finite solvable group. Then there exists a Galois extension $K|k$ with $G(K|k) \cong G$.*

Since 1954, when Šafarevič proved this result (see [187], [188], [189], [190]), several mathematicians have made suggestions how to reprove it using the subsequent developments in number theory and, in particular, the duality theorem of Poitou-Tate. Unfortunately, however, there is no such proof of theorem (9.6.1) accessible in the literature. Moreover, the original article [187] contains a mistake relative to the prime 2 *). We have therefore decided to include a complete proof of Šafarevič's theorem in this section. Of course, we use the original ideas of Šafarevič, in particular, the remarkable technique of *shrinking*

*) Šafarevič explains how to correct this in [192].

obstructions, which is highly instructive. The authors do not know any other argument in number theory which utilizes a similar technique.

As already explained in §5 the idea of Scholz and Reichardt was to choose the solutions of the inductively given embedding problems in a very special way, in order to avoid deadlocks when constructing the required global solvable extension inductively by abelian steps. This works if the order of G is prime to the order of $\mu(k)$. Šafarevič uses the same special kind of solutions of the inductively given embedding problems ("Scholz solutions"), but in the general situation one can run into a deadlock. The key idea of Šafarevič's approach is to modify the solutions of the first $i-1$ induction steps already found, in order to leave a deadlock within the i-th step. This happens in a rather complicated way within a *shrinking procedure*. This method works without any restriction on the group, but unfortunately it seems to be impossible to realize given local extensions. Therefore with Šafarevič's approach, one only can say that every solvable group occurs as a Galois group over k such that the associated local extensions are of a particular type.

We now explain Šafarevič's method in detail, and give a proof of theorem (9.6.1). To begin with, let us explain the main ideas. In the first reduction step, one shows that the result follows from the assertion that every split embedding problem of finite groups with nilpotent kernel N

$$\begin{array}{ccccccccc} & & & & & & G_k & & \\ & & & & & & \downarrow & & \\ 1 & \longrightarrow & N & \longrightarrow & E & \rightleftarrows & G & \longrightarrow & 1 \end{array}$$

has a proper solution. In order to solve these split embedding problems, we may assume that N is a p-group and it suffices to consider the generic kernel, i.e. $N = \mathcal{F}(n)/\mathcal{F}(n)^{(\nu)}$, where $\mathcal{F}(n)$ is a free pro-p-G operator group of rank n and $\mathcal{F}(n)^{(\nu)}$ denotes the ν-th term of a filtration of $\mathcal{F}(n)$, which we defined in III §8 and which refines the descending p-central series.*)

We proceed by induction on ν whereas n is arbitrary. The case $\nu = 1$ is trivial. The problems which have to be solved within every induction step are of the form

$$\begin{array}{c} \qquad\qquad\qquad\qquad\qquad\qquad\qquad\qquad G_k \\ \qquad\qquad\qquad\qquad\qquad\qquad\qquad\qquad \downarrow \\ 1 \to \mathcal{F}(n)^{(\nu)}/\mathcal{F}(n)^{(\nu+1)} \to \mathcal{F}(n)/\mathcal{F}(n)^{(\nu+1)} \rtimes G \to \mathcal{F}(n)/\mathcal{F}(n)^{(\nu)} \rtimes G \to 1. \end{array}$$

*) This refinement, which was proposed by Šafarevič in his correction note, is necessary in order to deal with the case $p = 2$.

This induction step is proved in four substeps. In the first step, one shows that this problem is locally solvable everywhere, i.e. for every prime $\mathfrak{p}$ of k the induced local problem

$$\begin{array}{ccccccccc} & & & & & & G_{k_{\mathfrak{p}}} & & \\ & & & & & & \downarrow & & \\ 1 & \longrightarrow & \mathcal{F}(n)^{(\nu)}/\mathcal{F}(n)^{(\nu+1)} & \longrightarrow & E_{\mathfrak{p}} & \longrightarrow & (\mathcal{F}(n)/\mathcal{F}(n)^{(\nu)} \rtimes G)_{\mathfrak{p}} & \longrightarrow & 1 \end{array}$$

has a solution. This can be done if the old solution $N_\nu^n|k$ with

$$G(N_\nu^n|k) = \mathcal{F}(n)/\mathcal{F}(n)^{(\nu)} \rtimes G$$

is locally of certain type, namely a so-called "Scholz solution".

In the second step one uses the local-global principle in order to show that a global solution exists.

In step three and step four we modify the global solution in order to get a proper "Scholz solution", in such a way that the new local problems for the next induction step $\nu + 1$ will be solvable.

Within the induction steps, obstructions to the existence of solutions of the given embedding problems occur. These obstructions really exist and are nontrivial. Šafarevič's idea how to overcome this problem is the following:

We revise the solution found in the $(\nu - 1)$-th induction step. Instead we use the $(\nu - 1)$-th induction step, not for n but for a very large $m > n$. The solution of that problem (which exists by the induction hypothesis) induces a new solution for our original problem via any surjective G-invariant homomorphism

$$\psi : \mathcal{F}(m) \twoheadrightarrow \mathcal{F}(n).$$

If m is sufficiently large, the shrinking process, which we will explain in a moment, gives a homomorphism ψ in such a way that all obstructions for the embedding problem in the ν-th induction step for $\mathcal{F}(n)$ vanish. In this way one proves the induction step from $\nu - 1$ to ν and for every n. Having the result for all n, we can use the shrinking procedure in the next induction step again.

The next proposition is the technical skeleton of the shrinking process. A similar statement was already contained in the original paper [190].

(9.6.2) Proposition. *Let G be a finite group. Suppose that M and N are finitely generated $\mathbb{F}_p[G]$-modules and let $s, t \in \mathbb{N}$. Then for $r \in \mathbb{N}$ sufficiently large*) the following holds:*

---

*) i.e. for all $r \geq r_0$

Given elements

$$z_1, \ldots, z_t \in (\bigoplus_r M)^{\otimes s} \otimes N,$$

there exists $a = (a_i)_{i=1,\ldots,r} \in \mathbb{F}_p^r$ *such that*

$$\varphi_a : \bigoplus_r M \twoheadrightarrow M, \quad (x_i)_{i=1,\ldots,r} \longmapsto \sum_{i=1}^r a_i x_i,$$

is a surjective $\mathbb{F}_p[G]$*-homomorphism (i.e. not all* a_i *are zero) and such that the induced homomorphism*

$$\psi_a = (\varphi_a^{\otimes s}) \otimes id : \quad (\bigoplus_r M)^{\otimes s} \otimes N \longrightarrow M^{\otimes s} \otimes N$$

maps all z_i, $i = 1, \ldots, t$, *to zero.*

Proof: *) Let r be arbitrary. Then φ_a is $\mathbb{F}_p[G]$-invariant and surjective if $a \neq 0$. Let $n = t \cdot \dim_{\mathbb{F}_p}(M^{\otimes s} \otimes N)$ and suppose that $r > s\,n$. Consider the set

$$V = \{a \in \mathbb{F}_p^r \mid \psi_a(z_1) = \ldots = \psi_a(z_t) = 0\}.$$

Then V is the set of common zeros of n polynomials of degree s. It contains the trivial element and by the theorem of Chevalley-Warning (see (6.5.7)), it follows that it must also contain a nontrivial point a. Then φ_a has all desired properties. □

Let us recall some filtrations that we defined in III §8. First, we have the descending p-central series $\{P^i\}_{i\geq 1}$ of a pro-p-group P which is given by

$$P^1 = P, \quad P^{i+1} = (P^i)^p[P^i, P], \ i \geq 1.$$

In addition, we use the descending central series $\{P_i\}_{i\geq 1}$ of P which is defined by

$$P_1 = P, \quad P_{i+1} = [P_i, P], \ i \geq 1.$$

Setting

$$P^{(i,j)} := (P^i \cap P_j)P^{i+1},$$

we have

$$P^{(i,1)} = P^i \text{ and } P^{(i,j)} = P^{i+1} \text{ for } j > i \geq 1.$$

Recall the notational convention:

The letter ν *always stands for a pair* (i, j), $i \geq j \geq 1$, *and we order these pairs lexicographically. We say that* $\nu + 1 = (i, j + 1)$ *if* $i > j$ *and* $\nu + 1 = (i + 1, 1)$ *if* $\nu = (i, i)$.

---

*) This proof is based on an idea of *J. Sonn*. We also want to thank *A. Deitmar* for a further simplification.

The descending chain $\{P^{(\nu)}\}$ of normal characteristic subgroups is a refinement of the descending p-central series. In particular, $P^{(\nu)}/P^{(\nu+1)}$ is an $\mathbb{F}_p$-vector space for all ν, and by (3.8.8) we have surjective $\mathbb{F}_p$-vector space homomorphisms $^{*)}$

$$\begin{array}{rcll} \psi_{(i,j)}: & (P/P^2)^{\otimes j} & \longrightarrow & P^{(i,j)}/P^{((i,j)+1)}, \\ & \bar{x}_1 \otimes \cdots \otimes \bar{x}_j & \longmapsto & ([x_1,[x_2,[\cdots[x_{j-1},x_j]]\cdots])^{p^{i-j}} \bmod P^{((i,j)+1)}. \end{array}$$

Recall the definition of a free pro-p-G operator group: if G is a finite group and F_d a free pro-p-group of rank d, then we set

$$\mathcal{F}(d) = \mathop{\ast}_{G} F_d\,,$$

a free pro-p-G operator group of rank d. The group $\mathcal{F}(d)$ is a free object in the category of pro-p-groups which are endowed with a continuous action of the group G. As a pro-p-group, $\mathcal{F}(d)$ is free of rank $\#G \cdot d$ and one can choose a basis $\{x_{i,g} \mid i = 1,\dots,d,\ g \in G\}$ of $\mathcal{F}(d)$ such that G acts by

$$g' x_{i,g} = x_{i,gg'}\,,\ g' \in G,$$

compare IV §3, example 3.

Now we will apply the shrinking process to cohomology groups with respect to G and $\mathcal{F}(d)$ in order to annihilate given cohomology classes.

(9.6.3) Proposition. *Let G be a finite group and let $\mathcal{F}(d)$ be the free pro-p-G operator group of rank d. Let $n,t \in \mathbb{N}$, $k \in \mathbb{Z}$, a finitely generated $\mathbb{F}_p[G]$-module T and $\nu = (i,j)$ be given. Then for $m \ge n$ large enough$^{**)}$ the following holds:*

Given elements

$$x_1,\dots,x_t \in \hat{H}^k(G,\mathcal{F}(m)^{(\nu)}/\mathcal{F}(m)^{(\nu+1)} \otimes T)\,,$$

there exists a surjective pro-p-G operator homomorphism

$$\psi : \mathcal{F}(m) \twoheadrightarrow \mathcal{F}(n)$$

such that the induced homomorphism

$$\psi_* : \hat{H}^k(G,\mathcal{F}(m)^{(\nu)}/\mathcal{F}(m)^{(\nu+1)} \otimes T) \longrightarrow \hat{H}^k(G,\mathcal{F}(n)^{(\nu)}/\mathcal{F}(n)^{(\nu+1)} \otimes T)$$

maps $x_1,\dots,x_t$ to zero.

$^{*)}$ $\psi_{(i,j)}$ can be lifted to a map $(P/P^2)^{\otimes j} \to P^{(i,j)}/P^{i+1} \subseteq P^i/P^{i+1}$ if either $j > 1$ or if p is odd (hence $p \mid \binom{p}{2}$). This is the reason why we could also work with the descending p-central series if p is odd and why this is not possible in the case $p = 2$.

$^{**)}$ i.e. for all $m \ge m_0 \ge n$

Proof: We set $\mathcal{E}(n,\nu) = \mathcal{F}(n)^{(\nu)}/\mathcal{F}(n)^{(\nu+1)}$. Using dimension shifting (cf. I §3) we have an isomorphism

$$\hat{H}^k(G, \mathcal{E}(n,\nu)\otimes T) \xrightarrow{\sim} \hat{H}^{-1}(G, \mathcal{E}(n,\nu)\otimes T \otimes A_k),$$

where $A_k = J_G^{\otimes(k+1)}$ for $k > -1$, $A_{-1} = \mathbb{F}_p$ and $A_k = I_G^{\otimes -(k+1)}$ for $k < -1$, and I_G and J_G are given by the exact sequences

$$0 \longrightarrow I_G \longrightarrow \mathbb{F}_p[G] \longrightarrow \mathbb{F}_p \longrightarrow 0,$$

$$0 \longrightarrow \mathbb{F}_p \longrightarrow \mathbb{F}_p[G] \longrightarrow J_G \longrightarrow 0.$$

Since T was arbitrary, we may restrict to the case $k = -1$.

Observe that for every $d \geq 1$ the canonical surjective $\mathbb{F}_p$-homomorphism defined in (3.8.8)

$$\theta_\nu(d) : (\mathcal{F}(d)/\mathcal{F}(d)^2)^{\otimes j} \twoheadrightarrow \mathcal{F}(d)^{(\nu)}/\mathcal{F}(d)^{(\nu+1)}$$

is obviously G-invariant, and given a pro-p-G operator homomorphism $\psi : \mathcal{F}(d') \to \mathcal{F}(d)$ we have the compatibility $\theta_\nu(d')\psi_* = \psi_*\theta_\nu(d)$.

Now let $m = rn$, r sufficiently large, so that we can apply (9.6.2) with the G-module T. Given elements

$$z_1, \ldots, z_t \in (\mathcal{F}(m)/\mathcal{F}(m)^2)^{\otimes j} \otimes T = (\mathbb{F}_p[G]^{rn})^{\otimes j} \otimes T,$$

there exists a surjective $\mathbb{F}_p[G]$-homomorphism

$$\bar{\psi} : \mathcal{F}(m)/\mathcal{F}(m)^2 \twoheadrightarrow \mathcal{F}(n)/\mathcal{F}(n)^2,$$

such that $(\bar{\psi}^{\otimes j} \otimes id)(z_\alpha) = 0$, $\alpha = 1, \ldots, t$. By the universal property of free pro-p-G operator groups, $\bar{\psi}$ extends to a pro-p-G operator homomorphism

$$\psi : \mathcal{F}(m) \twoheadrightarrow \mathcal{F}(n),$$

which is necessarily surjective (by the Frattini argument).

Now we consider the commutative diagram

$$\begin{array}{ccccc}
(\mathcal{F}(m)/\mathcal{F}(m)^2)^{\otimes j} \otimes T & \twoheadrightarrow & (\mathcal{E}(m,\nu)\otimes T)_G & \hookleftarrow & \hat{H}^{-1}(G, \mathcal{E}(m,\nu)\otimes T) \\
\downarrow{\scriptstyle \bar{\psi}^{\otimes j}\otimes id} & & \downarrow{\scriptstyle \psi_*\otimes id} & & \downarrow{\scriptstyle \psi_*} \\
(\mathcal{F}(n)/\mathcal{F}(n)^2)^{\otimes j} \otimes T & \twoheadrightarrow & (\mathcal{E}(n,\nu)\otimes T)_G & \hookleftarrow & \hat{H}^{-1}(G, \mathcal{E}(n,\nu)\otimes T)
\end{array}$$

and choose z_α as a pre-image of the image of x_α in the group $(\mathcal{E}(m,\nu)\otimes T)_G$. Choosing an appropriate ψ, the diagram shows that $\psi_*(x_\alpha) = 0$ for $\alpha = 1, \ldots, t$. This proves the proposition. □

We will apply (9.6.3) only for $k = 2$ and $k = -2$, and in the latter dimension we also need the following variant, which goes back to an idea of *V. V. Išhanov*.

(9.6.4) Proposition. *Let G be a finite group and let $\mathcal{F}(d)$ be the free pro-p-G operator group of rank d. Let $n, t \in \mathbb{N}$, a finitely generated $\mathbb{F}_p[G]$-module T and $\nu = (i, j)$ be given. Then for sufficiently large $m \geq n$ the following hold:*

(i) *Given elements*

$$x_1, \ldots, x_t \in H^{-2}(\mathcal{F}(m)/\mathcal{F}(m)^{(\nu)} \rtimes G, \mathcal{F}(m)^{(\nu)}/\mathcal{F}(m)^{(\nu+1)} \otimes T),$$

there exists a surjective pro-p-G operator homomorphism

$$\psi : \mathcal{F}(m) \twoheadrightarrow \mathcal{F}(n),$$

such that the induced homomorphism

$$\psi_* : H^{-2}(\mathcal{F}(m)/\mathcal{F}(m)^{(\nu)} \rtimes G, \mathcal{F}(m)^{(\nu)}/\mathcal{F}(m)^{(\nu+1)} \otimes T) \longrightarrow$$
$$H^{-2}(\mathcal{F}(n)/\mathcal{F}(n)^{(\nu)} \rtimes G, \mathcal{F}(n)^{(\nu)}/\mathcal{F}(n)^{(\nu+1)} \otimes T)$$

maps $x_1, \ldots, x_t$ to zero.

(ii) *Given elements*

$$x_1, \ldots, x_t \in H^{-2}(\mathcal{F}(m)/\mathcal{F}(m)^{(\nu)}, \mathcal{F}(m)^{(\nu)}/\mathcal{F}(m)^{(\nu+1)} \otimes T),$$

there exists a surjective pro-p-G operator homomorphism

$$\psi : \mathcal{F}(m) \twoheadrightarrow \mathcal{F}(n)$$

such that the induced homomorphism

$$\psi_* : H^{-2}(\mathcal{F}(m)/\mathcal{F}(m)^{(\nu)}, \mathcal{F}(m)^{(\nu)}/\mathcal{F}(m)^{(\nu+1)} \otimes T) \longrightarrow$$
$$H^{-2}(\mathcal{F}(n)/\mathcal{F}(n)^{(\nu)}, \mathcal{F}(n)^{(\nu)}/\mathcal{F}(n)^{(\nu+1)} \otimes T)$$

maps $x_1, \ldots, x_t$ to zero.

Proof: We keep the notation

$$\mathcal{F}(n)/\nu = \mathcal{F}(n)/\mathcal{F}(n)^{(\nu)} \text{ and } \mathcal{E}(n, \nu) = \mathcal{F}(n)^{(\nu)}/\mathcal{F}(n)^{(\nu+1)}.$$

If $\nu = 1$, then the statement to prove is just a special case of (9.6.3). So we may assume that $\nu = (i, j) \geq (2, 1)$. Recall that $H^{-2} = H_1$ and consider the exact sequence

$$H_1(\mathcal{F}(n)/\nu, \mathcal{E}(n, \nu) \otimes T) \longrightarrow H_1(\mathcal{F}(n)/\nu \rtimes G, \mathcal{E}(n, \nu) \otimes T) \longrightarrow$$
$$H_1(G, \mathcal{E}(n, \nu) \otimes T) \longrightarrow 0,$$

which is induced by the homological Hochschild-Serre sequence. Since $\mathcal{E}(n, \nu)$ and T are trivial $\mathcal{F}(n)/\nu$-module, we obtain

$$H_1(\mathcal{F}(n)/\nu, \mathcal{E}(n, \nu) \otimes T) \cong \mathcal{F}(n)/\mathcal{F}(n)^2 \otimes \mathcal{E}(n, \nu) \otimes T$$

and (3.8.8) implies the existence of a G-invariant surjection

$$(\mathcal{F}(n)/\mathcal{F}(n)^2)^{\otimes(j+1)} \otimes T \twoheadrightarrow H_1(\mathcal{F}(n)/\nu, \mathcal{E}(n,\nu) \otimes T),$$

where $\nu = (i,j)$. This is obviously true for arbitrary n, and the maps are compatible. Let $r \geq n$ have the property that t arbitrary elements in $(\mathcal{F}(r)/\mathcal{F}(r)^2)^{\otimes(j+1)} \otimes T$ are annihilated by the homomorphism induced by a suitably chosen G-invariant surjection $\mathcal{F}(r) \twoheadrightarrow \mathcal{F}(n)$ (and which exists by (9.6.2)). The above surjection shows that r has the same property with respect to t given arbitrary elements in

$$H_1(\mathcal{F}(r)/\nu, \mathcal{E}(r,\nu) \otimes T).$$

This proves (ii). In order to show (i), let $m \geq r$ have the property that t arbitrary elements in

$$H_1(G, \mathcal{E}(m,\nu) \otimes T)$$

are annihilated by the homomorphism induced by a suitably chosen G-invariant surjection $\mathcal{F}(m) \twoheadrightarrow \mathcal{F}(r)$ (and which exists by (9.6.3)). We obtain the commutative exact diagram

$$\begin{array}{ccccc}
H_1(\mathcal{F}(m)/\nu, \mathcal{E}(m,\nu)\otimes T) & \longrightarrow & H_1(\mathcal{F}(m)/\nu \rtimes G, \mathcal{E}(m,\nu)\otimes T) & \xrightarrow{\alpha} & H_1(G, \mathcal{E}(m,\nu)\otimes T) \\
\downarrow & & \downarrow{\scriptstyle\pi} & & \downarrow \\
H_1(\mathcal{F}(r)/\nu, \mathcal{E}(r,\nu)\otimes T) & \xrightarrow{\beta} & H_1(\mathcal{F}(r)/\nu \rtimes G, \mathcal{E}(r,\nu)\otimes T) & \twoheadrightarrow & H_1(G, \mathcal{E}(r,\nu)\otimes T) \\
\downarrow{\scriptstyle\varepsilon} & & \downarrow & & \downarrow \\
H_1(\mathcal{F}(n)/\nu, \mathcal{E}(n,\nu)\otimes T) & \longrightarrow & H_1(\mathcal{F}(n)/\nu \rtimes G, \mathcal{E}(n,\nu)\otimes T) & \twoheadrightarrow & H_1(G, \mathcal{E}(n,\nu)\otimes T),
\end{array}$$

in which the vertical maps are induced by G-invariant surjections

$$\mathcal{F}(m) \twoheadrightarrow \mathcal{F}(r) \twoheadrightarrow \mathcal{F}(n),$$

which we choose in the following way. Let arbitrary elements $x_1, \ldots, x_t \in H_1(\mathcal{F}(m)/\nu \rtimes G, \mathcal{E}(m,\nu) \otimes T)$ be given. Choose $\mathcal{F}(m) \twoheadrightarrow \mathcal{F}(r)$ such that the induced homomorphism annihilates the elements $\alpha(x_1), \ldots, \alpha(x_t)$. Hence $\pi(x_1), \ldots, \pi(x_t)$ are contained in the image of β and we choose the surjection $\mathcal{F}(r) \twoheadrightarrow \mathcal{F}(n)$ such that ε annihilates arbitrarily chosen β-pre-images of $\pi(x_1), \ldots, \pi(x_t)$. The composite $\mathcal{F}(m) \twoheadrightarrow \mathcal{F}(n)$ has the desired property.

□

For a prime number $p \neq \mathrm{char}(k)$ we denote, as before, the set of primes of k with residue characteristic p by $S_p = S_p(k)$. The set $S_p(k)$ is finite and it is empty if k is a function field. In the number field case we denote the set of archimedean places of k by $S_\infty = S_\infty(k)$. In the function field case we *choose* (only in this section) any finite, nonempty set of primes of k and *call* it $S_\infty = S_\infty(k)$. For every extension field $K|k$, we denote by $S_\infty(K)$ the set of primes of K which lie over $S_\infty(k)$.

In the next technical lemma, which will be needed later, we freely use the notation introduced in §§1,2; in particular, for a G_k-module A the minimal trivializing extension of k is denoted by $k(A)$.

(9.6.5) Lemma. *Let k be a global field, $p \neq \mathrm{char}(k)$ a prime number, and assume that we are given sets of primes of k*

$$S' \supseteq S \supseteq T \supseteq S_p \cup S_\infty ,$$

where T is finite. Let A be a finite G_S-module which is annihilated by p. In addition, suppose that we are given a finite subextension $N \subseteq k_S$, with

a) $k(A) \subseteq N$,

b) $S' \smallsetminus T \subseteq cs(N|k)$,

c) $\mu_p \subseteq N$.

Consider the diagram with solid arrows (the rows are not exact)

$$\begin{array}{ccccc}
\mathrm{coker}(k_S,T,A) & \hookrightarrow & \text{Ш}^1(k_S,S\backslash T,A')^\vee & \overset{\eta}{\twoheadrightarrow} & H^1(N|k,A')^\vee \\
\downarrow & & \downarrow{\scriptstyle \phi} & & \| \\
\mathrm{coker}(k_{S'},T,A) & \hookrightarrow & \text{Ш}^1(k_{S'},S'\backslash T,A')^\vee & \overset{\eta'}{\twoheadrightarrow} & H^1(N|k,A')^\vee
\end{array}$$

in which $A' := A^\vee(1) = \mathrm{Hom}(A,\mu_p)$. The horizontal maps on the left are induced by (9.2.2) and those on the right come from the Hochschild-Serre sequence and from conditions a), b), c).

Then in the above situation a natural dotted arrow ϕ exists which makes the diagram commutative. If in addition

$$cs(N|k) \subsetneqq S' \smallsetminus T,$$

then the surjection η' is an isomorphism.

Proof: First observe that the homomorphism η' is obtained from the following commutative exact diagram

$$\begin{array}{ccccc}
 & & H^1(N_{S'},A') & \overset{\iota}{\longrightarrow} & \prod\limits_{S'\backslash T} H^1(N_{\mathfrak{P}},A') \\
 & & \uparrow & & \uparrow \\
\text{Ш}^1(k_{S'},S'\backslash T,A') & \hookrightarrow & H^1(k_{S'},A') & \longrightarrow & \prod\limits_{S'\backslash T} H^1(k_{\mathfrak{p}},A') \\
 & \nwarrow{\scriptstyle (\eta')^*} & \uparrow & \nearrow{\scriptstyle 0} & \\
 & & H^1(N|k,A') & &
\end{array}$$

and in a similar way we get the homomorphism η.

Now consider the commutative exact diagram with natural homomorphisms

$$\begin{array}{ccccc}
Ш^1(k_S, S\backslash T, A') & \hookrightarrow & H^1(G_S, A') & \longrightarrow & \prod\limits_{S\backslash T} H^1(k_{\mathfrak p}, A') \\
\uparrow{\scriptstyle \kappa} & & \| & & \uparrow \\
\ker(\alpha) & \hookrightarrow & H^1(G_S, A') & \xrightarrow{\alpha} & \prod\limits_{S\backslash T} H^1(k_{\mathfrak p}, A') \times \prod\limits_{S'\backslash S} H^1_{nr}(k_{\mathfrak p}, A') \\
\downarrow{\scriptstyle \varepsilon} & & \downarrow & & \downarrow \\
Ш^1(k_{S'}, S'\backslash T, A') & \hookrightarrow & H^1(G_{S'}, A') & \longrightarrow & \prod\limits_{S\backslash T} H^1(k_{\mathfrak p}, A') \times \prod\limits_{S'\backslash S} H^1(k_{\mathfrak p}, A') \\
 & & \downarrow & & \downarrow \\
 & & H^1(k_{S'}|k_S, A')^{G_S} & \xhookrightarrow{\beta} & \prod\limits_{S'\backslash S} H^1(T_{\mathfrak p}, A')^{G_{k_{\mathfrak p}}} ,
\end{array}$$

in which $T_{\mathfrak p} \subseteq G_{k_{\mathfrak p}}$ denotes the inertia group. Observe that β is injective, since A' is a trivial $G(k_{S'}|k_S)$-module and since k_S has no extensions in $k_{S'}$ which are unramified at all places in $S' \smallsetminus S$ by definition. Diagram chasing then shows that ε is an isomorphism. We can now define ϕ as the dual homomorphism to $\kappa \circ (\varepsilon^{-1})$. Conditions a), b), c) imply that $H^1(N|k, A')$ is canonically contained in the groups $Ш^1(k_S, S\backslash T, A')$, $\ker(\alpha)$ and $Ш^1(k_{S'}, S'\backslash T, A')$. Thus we have constructed ϕ and we see that the right-hand part of the diagram is commutative. But that the left-hand part of the diagram is also commutative can be seen from the diagram preceding lemma (9.2.2) which defines the maps occurring.

Now assume that $cs(N|k) \subsetneq S' \smallsetminus T$. Then $\delta_N(S'\backslash T) = 1$, and by (9.1.9)(i), the homomorphism ι in the exact commutative diagram at the beginning of the proof is injective. Hence the inclusion $(\eta')^*$ is an isomorphism. □

The essential step in the proof of Šafarevič's theorem (9.6.1) is the following

(9.6.6) Theorem. *Let $K|k$ be a finite Galois extension of the global field k and let $\varphi : G_k \twoheadrightarrow G(K|k) = G$. Then every split embedding problem*

$$\begin{array}{ccccccccc}
 & & & & & & G_k & & \\
 & & & & & & \downarrow & & \\
1 & \longrightarrow & H & \longrightarrow & H \rtimes G & \rightleftarrows & G & \longrightarrow & 1
\end{array}$$

with finite nilpotent kernel H has a proper solution.

Since a finite nilpotent group is the direct product of its p-Sylow subgroups and since every finite G-operator p-group is a quotient of $\mathcal{F}(n)/\mathcal{F}(n)^{(\nu)}$ for some n and some ν, it suffices to show the following assertion.

For every prime number p, all $n \in \mathbb{N}$ and all $\nu = (i, j)$, the split embedding problem

$$\begin{array}{ccccccc} & & & & G_k & & \\ & & & & \downarrow & & \\ 1 \longrightarrow & \mathcal{F}(n)/\mathcal{F}(n)^{(\nu)} & \longrightarrow & \mathcal{F}(n)/\mathcal{F}(n)^{(\nu)} \rtimes G & \rightleftarrows & G & \longrightarrow 1 \end{array}$$

has a proper solution $N_\nu^n|k$.

Let us first assume that $p \neq \mathrm{char}(k)$. We will proceed by induction on ν whereas n will be arbitrary. If $\nu = (1, 1)$, there is nothing to show. Now we assume that we have already found a solution $\varphi_{n,\nu} : G_k \twoheadrightarrow \mathcal{F}(n)/\mathcal{F}(n)^{(\nu)} \rtimes G$ and we consider the embedding problem

(∗)

$$\begin{array}{ccccc} & & & & G_k \\ & & & & \downarrow{\scriptstyle \varphi_{n,\nu}} \\ \mathcal{F}(n)^{(\nu)}/\mathcal{F}(n)^{(\nu+1)} & \hookrightarrow & \mathcal{F}(n)/\mathcal{F}(n)^{(\nu+1)} \rtimes G & \twoheadrightarrow & \mathcal{F}(n)/\mathcal{F}(n)^{(\nu)} \rtimes G. \end{array}$$

This embedding problem is in general not solvable, but we are going to solve it after replacing $\varphi_{n,\nu}$ by another solution $\tilde{\varphi}_{n,\nu}$ for induction step ν on level n. This new solution is induced by a solution

$$\varphi_{m,\nu} : G_k \to \mathcal{F}(m)/\mathcal{F}(m)^{(\nu)} \rtimes G$$

for some large $m \geq n$ via a suitably chosen G-invariant surjection $\psi : \mathcal{F}(m) \twoheadrightarrow \mathcal{F}(n)$. Let us consider the associated commutative exact diagram

(∗∗)

$$\begin{array}{ccccc} & & & & G_k \\ & & & & \downarrow{\scriptstyle \varphi_{m,\nu}} \\ \mathcal{F}(m)^{(\nu)}/\mathcal{F}(m)^{(\nu+1)} & \hookrightarrow & \mathcal{F}(m)/\mathcal{F}(m)^{(\nu+1)} \rtimes G & \twoheadrightarrow & \mathcal{F}(m)/\mathcal{F}(m)^{(\nu)} \rtimes G \\ \downarrow{\scriptstyle \bar{\psi}} & & \downarrow{\scriptstyle \psi_{\nu+1}} & & \downarrow{\scriptstyle \psi_{\nu}} \\ \mathcal{F}(n)^{(\nu)}/\mathcal{F}(n)^{(\nu+1)} & \hookrightarrow & \mathcal{F}(n)/\mathcal{F}(n)^{(\nu+1)} \rtimes G & \twoheadrightarrow & \mathcal{F}(n)/\mathcal{F}(n)^{(\nu)} \rtimes G. \end{array}$$

To shorten notation we again set

$$\mathcal{F}(n)/\nu = \mathcal{F}(n)/\mathcal{F}(n)^{(\nu)}, \quad \mathcal{E}(n, \nu) = \mathcal{F}(n)^{(\nu)}/\mathcal{F}(n)^{(\nu+1)}.$$

Since $\mathcal{E}(n, \nu)$ is contained in the center of $\mathcal{F}(n)/\nu+1$, the action of $\mathcal{F}(n)/\nu \rtimes G$ on $\mathcal{E}(n, \nu)$ factors through the canonical projection

$$\mathcal{F}(n)/\nu \rtimes G \twoheadrightarrow G.$$

In particular, $G_K \subseteq G_k$ acts trivially on $\mathcal{E}(n,\nu)$.

Let α_m and α_n denote the 2-classes corresponding to the group extensions in (∗∗) and consider the commutative exact diagram

$$(***)\qquad \begin{array}{ccc} H^2(\mathcal{F}(m)/\nu \rtimes G, \mathcal{E}(m,\nu)) & \xrightarrow{\varphi^*_{m,\nu}} & H^2(k,\mathcal{E}(m,\nu)) \\ \bar{\psi}_*\downarrow & & \downarrow\bar{\psi}_* \\ H^2(\mathcal{F}(m)/\nu \rtimes G, \mathcal{E}(n,\nu)) & \xrightarrow{\varphi^*_{m,\nu}} & H^2(k,\mathcal{E}(n,\nu)) \\ \psi^*_\nu\uparrow & & \| \\ H^2(\mathcal{F}(n)/\nu \rtimes G, \mathcal{E}(n,\nu)) & \xrightarrow{\varphi^*_{n,\nu}} & H^2(k,\mathcal{E}(n,\nu)) . \end{array}$$

By ex.4 in I §5, we have $\psi^*_\nu(\alpha_n) = \bar{\psi}_*(\alpha_m)$, and by (3.5.9), the embedding problem on level n is solvable if and only if $\varphi^*_{n,\nu}(\alpha_n) = 0$. We are searching for an $m \geq n$, a solution $\varphi_{m,\nu}$ on level m and a suitable surjective G-homomorphism $\psi : \mathcal{F}(m) \twoheadrightarrow \mathcal{F}(n)$ such that

$$\bar{\psi}_*(\varphi^*_{m,\nu}(\alpha_m)) = 0 \in H^2(k,\mathcal{E}(n,\nu)).$$

As we will show below, this can be achieved for m large enough if the solution $\varphi_{m,\nu}$ is of a special type. If we can guarantee that the new solution is also of this special type, then the induction process works for a modified, stronger statement. In fact we are going to prove the sharper

(9.6.7) Theorem. *Let $K|k$ be a finite Galois extension of the global field k and let $\varphi : G_k \twoheadrightarrow G(K|k) = G$. In addition, let a natural number e be given. Then for every prime number p, all $n \in \mathbb{N}$ and all $\nu = (i,j)$, the split embedding problem*

$$\begin{array}{ccccccccc} & & & & & & G_k & & \\ & & & & & & \downarrow\varphi & & \\ 1 & \longrightarrow & \mathcal{F}(n)/\mathcal{F}(n)^{(\nu)} & \longrightarrow & \mathcal{F}(n)/\mathcal{F}(n)^{(\nu)} \rtimes G & \rightleftarrows & G & \longrightarrow & 1 \end{array}$$

has a proper solution $N^n_\nu|k$. If $p \neq \mathrm{char}(k)$, we can choose the solution in such a way that the following conditions are satisfied:

(i) *All $\mathfrak{p} \in \mathrm{Ram}(K|k) \cup S_p \cup S_\infty$ are completely decomposed in $N^n_\nu|K$.*

(ii) *If $\mathfrak{p}$ is ramified in $N^n_\nu|K$, then $\mathfrak{p}$ splits completely in $K(\mu_{p^e})|k$ and $N^n_{\nu,\mathfrak{p}}|k_\mathfrak{p}$ is a (cyclic) totally ramified extension of local fields.*

Proof: We prove the theorem by induction on ν, whereas e, n and G are arbitrary. We defer the case $\mathrm{char}(k) = p$ and assume that $\mathrm{char}(k) \neq p$. Then we may restrict to the case $\mu_p \subseteq K$: For this, we lift the embedding problem via

$$G_k \twoheadrightarrow G(K(\mu_p)|k) \twoheadrightarrow G(K|k),$$

thus enlarging G. Note that this does not affect conditions (i) and (ii).

If $\nu = (1,1)$, there is nothing to show. For the induction step $\nu \mapsto \nu + 1$, we may assume that p^e is greater than the exponent of the group $\mathcal{F}(n)/\mathcal{F}(n)^{(\nu)}$ (which does depend on ν but not on n and G). We prove the induction step, i.e. we solve the embedding problem defined by the diagram $(*)$ above in four substeps.

First Step: *After changing $\varphi_{n,\nu}$ the problem $(*)$ induces local split embedding problems at all $\mathfrak{p} \in \mathrm{Ram}(K|k) \cup S_p \cup S_\infty$ and is locally solvable (not necessarily properly) at every prime $\mathfrak{p}$.*

a) If $\mathfrak{p} \in \mathrm{Ram}(K|k) \cup S_p \cup S_\infty$, then $G_\mathfrak{p}(N_\nu^n|k) = (\mathcal{F}(n)/\mathcal{F}(n)^{(\nu)} \rtimes G)_\mathfrak{p} \cong G_\mathfrak{p}(K|k)$ by (i). We show that, after changing $\varphi_{n,\nu}$, the local group extensions corresponding to these primes are split extensions. In particular, the associated local embedding problems are solvable in a trivial way.

Let $\alpha_n(\mathfrak{p})$ be the 2-class in $H^2((\mathcal{F}(n)/\mathcal{F}(n)^{(\nu)} \rtimes G)_\mathfrak{p}, \mathcal{F}(n)^{(\nu)}/\mathcal{F}(n)^{(\nu+1)})$ which corresponds to the group extension given by the upper row of the diagram

$$\begin{array}{ccccc}
\mathcal{F}(n)^{(\nu)}/\mathcal{F}(n)^{(\nu+1)} & \hookrightarrow & E_\mathfrak{p} & \twoheadrightarrow & (\mathcal{F}(n)/\mathcal{F}(n)^{(\nu)} \rtimes G)_\mathfrak{p} \\
\| & & \downarrow & & \downarrow \\
\mathcal{F}(n)^{(\nu)}/\mathcal{F}(n)^{(\nu+1)} & \hookrightarrow & \mathcal{F}(n)/\mathcal{F}(n)^{(\nu+1)} \rtimes G & \twoheadrightarrow & \mathcal{F}(n)/\mathcal{F}(n)^{(\nu)} \rtimes G\,.
\end{array}$$

Apply the induction hypothesis to the corresponding embedding problem on some large level m. The number m and a surjective G-invariant homomorphism $\psi : \mathcal{F}(m) \twoheadrightarrow \mathcal{F}(n)$ will be chosen below. By (i), we know that $\mathfrak{p}$ is completely decomposed in $N_\nu^m|K$ and we have a commutative diagram for the associated local groups (writing $G_\mathfrak{p}$ for $G_\mathfrak{p}(K|k)$)

$$\begin{array}{ccc}
(\mathcal{F}(m)/\nu \rtimes G)_\mathfrak{p} & \xrightarrow{\sim} & G_\mathfrak{p} \\
\downarrow \wr & & \| \\
(\mathcal{F}(n)/\nu \rtimes G)_\mathfrak{p} & \xrightarrow{\sim} & G_\mathfrak{p}\,.
\end{array}$$

Therefore we obtain the diagram

$$\begin{array}{ccc}
H^2((\mathcal{F}(m)/\nu \rtimes G)_\mathfrak{p}, \mathcal{E}(m,\nu)) & \xleftarrow[\sim]{inf} & H^2(G_\mathfrak{p}, \mathcal{E}(m,\nu)) \\
\downarrow \bar{\psi}_* & & \downarrow \bar{\psi}_* \\
H^2((\mathcal{F}(m)/\nu \rtimes G)_\mathfrak{p}, \mathcal{E}(n,\nu)) & \xleftarrow[\sim]{inf} & H^2(G_\mathfrak{p}, \mathcal{E}(n,\nu)) \\
\wr\uparrow \psi_\nu^* & & \| \\
H^2((\mathcal{F}(n)/\nu \rtimes G)_\mathfrak{p}, \mathcal{E}(n,\nu)) & \xleftarrow[\sim]{inf} & H^2(G_\mathfrak{p}, \mathcal{E}(n,\nu))\,.
\end{array}$$

Using (9.6.3) with $G, k = 2$ and $T = \mathrm{Ind}_G^{G_\mathfrak{p}} \mathbb{F}_p$, we find an $m \geq n$ and a surjective pro-p-G homomorphism $\psi : \mathcal{F}(m) \twoheadrightarrow \mathcal{F}(n)$ such that the homomorphism

$$\begin{array}{rcl} H^2(G_\mathfrak{p}, \mathcal{E}(m,\nu)) & = & H^2(G, \mathcal{E}(m,\nu) \otimes T) \\ & & \Big\downarrow \bar{\psi}_* \\ H^2(G_\mathfrak{p}, \mathcal{E}(n,\nu)) & = & H^2(G, \mathcal{E}(n,\nu) \otimes T) \end{array}$$

maps $inf^{-1}(\alpha_m(\mathfrak{p}))$ to 0, so that the above diagram implies that $\alpha_n(\mathfrak{p}) = 0$.

Now we can execute the above procedure for all the finitely many primes $\mathfrak{p} \in \mathrm{Ram}(K|k) \cup S_p \cup S_\infty$, making m bigger each time. Note that we do not destroy the success already achieved for the primes $\mathfrak{p}_1, \ldots, \mathfrak{p}_r$ for which the local embedding problems already split. Indeed, the property of inducing a split embedding problem at a prime $\mathfrak{p}$ survives the shrinking process from m to n if $\mathfrak{p}$ is completely decomposed in $N_\nu^m|K$ (and we suppose this holds for $\mathfrak{p} \in \mathrm{Ram}(K|k) \cup S_p \cup S_\infty$). Therefore we can perform a shrinking process for the prime $\mathfrak{p}_{r+1}$, inducing a solution from $\mathcal{F}(m)$, which already has the desired property for $\mathfrak{p}_1, \ldots, \mathfrak{p}_r$ (and which we could have produced by another shrinking $\mathcal{F}(m') \twoheadrightarrow \mathcal{F}(m)$).

An alternative way to proceed at this point is to replace the module T in the above argument by the direct sum of $\mathrm{Ind}_G^{G_\mathfrak{p}} \mathbb{F}_p$, where $\mathfrak{p}$ runs through $\mathrm{Ram}(K|k) \cup S_p \cup S_\infty$. In this way we deal with all these primes within one shrinking process.

Finally, we see that the embedding problem

$$\begin{array}{ccccc} & & & & G_k \\ & & & & \Big\downarrow \varphi_{m,\nu} \\ & & & & \mathcal{F}(m)/\mathcal{F}(m)^{(\nu)} \rtimes G \\ & & & & \Big\downarrow \psi_\nu \\ \mathcal{F}(n)^{(\nu)}/\mathcal{F}(n)^{(\nu+1)} & \hookrightarrow & \mathcal{F}(n)/\mathcal{F}(n)^{(\nu+1)} \rtimes G & \twoheadrightarrow & \mathcal{F}(n)/\mathcal{F}(n)^{(\nu)} \rtimes G \end{array}$$

induces split group extensions for all $\mathfrak{p} \in \mathrm{Ram}(K|k) \cup S_p \cup S_\infty$.

We have just changed $\varphi_{n,\nu}$ to $\bar{\psi} \circ \varphi_{m,\nu}$, and thus N_ν^n to some $\tilde{N}_\nu^n$. But by assumption conditions (i) *and* (ii) *are also satisfied for N_ν^m, hence also for the new field $\tilde{N}_\nu^n$. We will not change the notation.*

b) If $\mathfrak{p}$ is unramified in $N_\nu^n|k$, then the homomorphism

$$\hat{\mathbb{Z}} \cong G_\mathfrak{p}(k)/T_\mathfrak{p}(k) \xrightarrow{\varphi_\mathfrak{p}} G_\mathfrak{p}(N_\nu^n|k)$$

obviously extends since $\hat{\mathbb{Z}}$ is free.

c) Let $\mathfrak{p} \in \operatorname{Ram}(N_\nu^n|K)$. Then $\mathfrak{p}$ splits completely in the extension $K(\mu_{p^e})|k$ and $G_\mathfrak{p}(N_\nu^n|K) \cong \mathbb{Z}/p^a\mathbb{Z}$ with $a \le e-1$ by the induction assumption (ii) and our condition on e. Since $N_{\nu,\mathfrak{p}}^n|K_\mathfrak{p}$ is totally ramified by assumption, there exists a prime element $\pi_\mathfrak{p}$ of $K_\mathfrak{p}$ such that $N_{\nu,\mathfrak{p}}^n = K_\mathfrak{p}(\sqrt[p^a]{\pi_\mathfrak{p}})$. An arbitrarily chosen pre-image of a generator of the cyclic group $G_\mathfrak{p}(N_\nu^n|K)$ in $\mathcal{F}(n)/\mathcal{F}(n)^{(\nu+1)} \rtimes G$ has order $p^{a+\varepsilon}$, where $0 \le \varepsilon \le 1$ We can solve our embedding problem by taking a $p^{a+\varepsilon}$-th root of $\pi_\mathfrak{p}$, since $\mu_{p^{a+\varepsilon}} \subseteq \mu_{p^e} \subseteq K_\mathfrak{p}$.

Second Step: *The problem* $(*)$ *induces local split embedding problems at all* $\mathfrak{p} \in \operatorname{Ram}(K|k) \cup S_p \cup S_\infty$ *and is globally solvable (not necessarily properly) after changing* $\varphi_{n,\nu}$.

As above we consider the problem for different numbers $m \ge n$:

$$\begin{array}{ccccc} & & & & G_k \\ & & & & \downarrow \varphi_{m,\nu} \\ \mathcal{F}(m)^{(\nu)}/\mathcal{F}(m)^{(\nu+1)} & \hookrightarrow & \mathcal{F}(m)/\mathcal{F}(m)^{(\nu+1)} \rtimes G & \twoheadrightarrow & \mathcal{F}(m)/\mathcal{F}(m)^{(\nu)} \rtimes G \\ \downarrow \bar{\psi} & & \downarrow \psi_{\nu+1} & & \downarrow \psi_\nu \\ \mathcal{F}(n)^{(\nu)}/\mathcal{F}(n)^{(\nu+1)} & \hookrightarrow & \mathcal{F}(n)/\mathcal{F}(n)^{(\nu+1)} \rtimes G & \twoheadrightarrow & \mathcal{F}(n)/\mathcal{F}(n)^{(\nu)} \rtimes G. \end{array}$$

Let α_m and α_n denote the 2-classes corresponding to the above group extensions. A surjective pro-p-G homomorphism $\psi : \mathcal{F}(m) \twoheadrightarrow \mathcal{F}(n)$ (inducing ψ_ν, $\psi_{\nu+1}$ and $\bar{\psi}$) will be defined below. Both problems (for m with $\varphi_{m,\nu}$ and for n with $\varphi_{n,\nu} = \psi_\nu \circ \varphi_{m,\nu}$) are locally solvable by step 1.
In order to show the existence of a solution $\varphi_{n,\nu+1}$, we have to prove that the 2-class α_n maps to zero under the inflation map $\varphi_{n,\nu}^*$, cf. (3.5.9):

$$\begin{array}{ccc} & & \prod_\mathfrak{p} H^2(k_\mathfrak{p}, \mathcal{E}(n,\nu)) \\ & & \uparrow \\ H^2(\mathcal{F}(n)/\mathcal{F}(n)^{(\nu)} \rtimes G,\, \mathcal{E}(n,\nu)) & \xrightarrow{\varphi_{n,\nu}^*} & H^2(k, \mathcal{E}(n,\nu)) \\ & & \uparrow \\ & & Ш^2(k, \mathcal{E}(n,\nu)). \end{array}$$

Here as before we have set $\mathcal{E}(n,\nu) = \mathcal{F}(n)^{(\nu)}/\mathcal{F}(n)^{(\nu+1)}$. By the first step we can assume that

$$\varphi_{n,\nu}^*(\alpha_n) \in Ш^2(k, \mathcal{E}(n,\nu)), \qquad \varphi_{m,\nu}^*(\alpha_m) \in Ш^2(k, \mathcal{E}(m,\nu)).$$

As we already observed at the start,

$$\psi_\nu^*(\alpha_n) = \bar{\psi}_*(\alpha_m),$$

which gives us

$$\begin{aligned} \varphi_{n,\nu}^*(\alpha_n) &= (\psi_\nu \circ \varphi_{m,\nu})^*(\alpha_n) = (\varphi_{m,\nu}^* \circ \psi_\nu^*)(\alpha_n) \\ &= (\varphi_{m,\nu}^* \circ \bar{\psi}_*)(\alpha_m) = (\bar{\psi}_* \circ \varphi_{m,\nu}^*)(\alpha_m) \end{aligned}$$

(see diagram (∗∗∗)). In order to shrink the obstruction, we look for a surjective homomorphism onto $Ш^2(k, \mathcal{E}(n,\nu))$, which has a "shrinkable" source, i.e. to which (9.6.3) applies.

Claim: We have a commutative diagram

$$\begin{array}{ccc} H^{-2}(G, \mathcal{E}(m,\nu)(-1)) & \twoheadrightarrow & Ш^2(k, \mathcal{E}(m,\nu)) \\ \downarrow{\scriptstyle \bar{\psi}_*} & & \downarrow{\scriptstyle \bar{\psi}_*} \\ H^{-2}(G, \mathcal{E}(n,\nu)(-1)) & \twoheadrightarrow & Ш^2(k, \mathcal{E}(n,\nu)) \end{array}$$

with surjective horizontal maps, where (-1) *denotes the* (-1)*-Tate twist.*

Using this claim and (9.6.3) with $k = -2$, the $\mathbb{F}_p[G]$-module $T = \mathrm{Hom}(\mu_p, \mathbb{Z}/p\mathbb{Z})$ and an element x_1 which is a pre-image of $\varphi^*_{m,\nu}(\alpha_m)$ in $H^{-2}(G, \mathcal{E}(m,\nu)(-1))$, we obtain a surjective pro-p-G operator homomorphism $\psi : \mathcal{F}(m) \twoheadrightarrow \mathcal{F}(n)$ such that $\bar{\psi}_* \varphi^*_{m,\nu}(\alpha_m) = 0$, and so $\varphi^*_{n,\nu}(\alpha_n) = 0$. Thus the embedding problem is solvable. Furthermore, as explained in the first step, the local condition at the primes in $\mathrm{Ram}(K|k) \cup S_p \cup S_\infty$ remains unaffected by the shrinking process. In order to finish step 2, it remains to give the

Proof of the claim: Let $\mathcal{E}(n,\nu)' = \mathrm{Hom}(\mathcal{E}(n,\nu), \mu_p)$. By the Poitou-Tate duality theorem, we know that

$$Ш^2(k, \mathcal{E}(n,\nu)) \cong Ш^1(k, \mathcal{E}(n,\nu)')^* .$$

Using the Hasse principle and the fact that $\mathcal{E}(n,\nu)'$ is a trivial G_K-module ($\mu_p \subseteq K$), we see that the homomorphism ι in the commutative exact diagram

$$\begin{array}{ccccccc} & & & & H^1(K, \mathcal{E}(n,\nu)') & \overset{\iota}{\hookrightarrow} & \prod_{\mathfrak{P}} H^1(K_{\mathfrak{P}}, \mathcal{E}(n,\nu)') \\ & & & & \uparrow & & \uparrow \\ 0 & \to & Ш^1(k, \mathcal{E}(n,\nu)') & \longrightarrow & H^1(k, \mathcal{E}(n,\nu)') & \longrightarrow & \prod_{\mathfrak{p}} H^1(k_{\mathfrak{p}}, \mathcal{E}(n,\nu)') \\ & & & & \uparrow & & \\ & & & & H^1(K|k, \mathcal{E}(n,\nu)') & & \\ & & & & \uparrow & & \\ & & & & 0 & & \end{array}$$

is injective. Hence we get an injective homomorphism $Ш^1(k, \mathcal{E}(n,\nu)') \hookrightarrow H^1(K|k, \mathcal{E}(n,\nu)')$, and noting that the dual of cohomology is homology (hence is cohomology in negative dimensions for finite groups), we obtain a canonical surjection

$$\begin{array}{ccc} H^1(G, \mathcal{E}(n,\nu)')^* & \twoheadrightarrow & Ш^1(k, \mathcal{E}(n,\nu)')^* \\ \wr\Vert & & \wr\Vert \\ H^{-2}(G, \mathcal{E}(n,\nu)(-1)) & & Ш^2(k, \mathcal{E}(n,\nu)) . \end{array}$$

This proves the claim.

Third step: *After changing $\varphi_{n,\nu}$, the problem $(*)$ has a proper global solution which satisfies condition* (i) *and all primes $\mathfrak{p} \in \mathrm{Ram}(N^n_{\nu+1}|N^n_\nu) \smallsetminus \mathrm{Ram}(N^n_\nu|K)$ are completely decomposed in $N^n_\nu|k$. Furthermore, the local extension $N^n_{\nu+1,\mathfrak{p}}|K_\mathfrak{p}$ is (cyclic) totally ramified for $\mathfrak{p} \in \mathrm{Ram}(N^n_\nu|K)$.*

We achieve this with the following procedure. Consider a solution $\varphi_{n,\nu+1}$ of the embedding problem $(*)$ which we obtained in step 2. Its equivalence class $[\varphi_{n,\nu+1}]$ is an element of the space $\mathscr{S}_{(*)}$ of solutions of $(*)$ modulo equivalence. Conditions (i), (ii), properness and the other conditions that we want to achieve in this third step, only depend on the equivalence class of a solution. The space $\mathscr{S}_{(*)}$ is a principal homogeneous space over $H^1(G_k, \mathcal{E}(n,\nu))$; see (3.5.11). Recall that the action is defined as follows: choose a representing cocycle $G_k \to \mathcal{E}(n,\nu)$ and multiply a solution $G_k \to \mathcal{F}(n)/\nu{+}1 \rtimes G$ of the embedding problem with the cocycle. This yields a map $G_k \to \mathcal{F}(n)/\nu{+}1 \rtimes G$ which is a homomorphism, and the equivalence class of this new solution is independent of the choices made.

Now we look for a suitable cohomology class $\varepsilon \in H^1(G_k, \mathcal{E}(n,\nu))$ such that the new solution

$$\tilde{\varphi}_{n,\nu+1} = {}^{\varepsilon}\varphi_{n,\nu+1}$$

has the required properties. We assume that $\varphi_{m,\nu+1}$ is obtained from step 2 and that $\varphi_{m,\nu}$ satisfies (i) and (ii). Note that the properness of the solution is only a problem for the first step $(1,1) \to (2,1)$, since in all higher induction steps, the properness follows automatically from the induction hypothesis and from the Frattini argument.

Let us consider how the local behaviour of $\tilde{\varphi}_{n,\nu+1} = {}^{\varepsilon}\varphi_{n,\nu+1}$ is connected to that of $\varphi_{n,\nu+1}$. By this we mean that we want to compare the ramification and decomposition of primes in the associated field extensions $\tilde{N}^n_{\nu+1}|N^n_\nu$ and $N^n_{\nu+1}|N^n_\nu$. (Since we do not know whether the solutions are proper, one or both of these field extensions might be trivial.) Let $\mathfrak{p}$ be a prime in N^n_ν. The behaviour of $\mathfrak{p}$ in $N^n_{\nu+1}|N^n_\nu$ is characterized by the homomorphism

$$\varphi_{n,\nu+1}|_{G_{(N^n_\nu)_\mathfrak{p}}} \in \mathrm{Hom}(G_{(N^n_\nu)_\mathfrak{p}}, \mathcal{E}(n,\nu)).$$

Since $G_{(N^n_\nu)_\mathfrak{p}}$ acts trivially on $\mathcal{E}(n,\nu)$, we can interpret $\varphi_{n,\nu+1}|_{G_{(N^n_\nu)_\mathfrak{p}}}$ as an element in

$$H^1((N^n_\nu)_\mathfrak{p}, \mathcal{E}(n,\nu))^{G_{k_\mathfrak{p}}}.$$

Consider the exact sequence

$$0 \longrightarrow H^1((N^n_\nu)_\mathfrak{p}|k_\mathfrak{p}) \longrightarrow H^1(k_\mathfrak{p}) \xrightarrow{\alpha} H^1((N^n_\nu)_\mathfrak{p})^{G_{k_\mathfrak{p}}} \xrightarrow{\beta} H^2((N^n_\nu)_\mathfrak{p}|k_\mathfrak{p}),$$

which is obtained from the Hochschild-Serre sequence for the tower of fields $\bar{k}_\mathfrak{p} \,|\, (N^n_\nu)_\mathfrak{p} \,|\, k_\mathfrak{p}$ and in which $\mathcal{E}(n,\nu)$ are the coefficients (not written) of the cohomology groups. We see that $\tilde{\varphi}_{n,\nu+1}|_{G_{(N^n_\nu)_\mathfrak{p}}}$ is given by

$$\varphi_{n,\nu+1}|_{G_{(N^n_\nu)_\mathfrak{p}}} + \alpha(\varepsilon) \in H^1((N^n_\nu)_\mathfrak{p}, \mathcal{E}(n,\nu))^{G_{k_\mathfrak{p}}}.$$

Now we choose a finite set T^0 of primes in $cs(N_\nu^n|k)$ and homomorphisms

$$x_\mathfrak{p} : G_{k_\mathfrak{p}}/T_{k_\mathfrak{p}} = G_{(N_\nu^n)_\mathfrak{p}}/T_{(N_\nu^n)_\mathfrak{p}} \longrightarrow \mathcal{E}(n,\nu)$$

for $\mathfrak{p} \in T^0$ such that their images generate $\mathcal{E}(n,\nu)$. (The set T^0 will be responsible for the properness of the new solution.)

Set

$$\begin{aligned} T^1 &= \mathrm{Ram}(K|k) \cup S_p \cup S_\infty\,, \\ T^2 &= \mathrm{Ram}(N_\nu^n|K)\,, \\ T^3 &= \mathrm{Ram}(N_{\nu+1}^n|K) \smallsetminus (\mathrm{Ram}(N_\nu^n|k) \cup S_p \cup S_\infty)\,, \\ T &= T^0 \cup T^1 \cup T^2 \cup T^3 \quad \text{and} \\ S &= cs(N_\nu^n(\mu_{p^e})|k) \cup T\,. \end{aligned}$$

Let $\mathfrak{p} \in T^0$. Then, since $\mathfrak{p}$ splits completely in $N_\nu^n|k$, there exists $\xi_\mathfrak{p} \in H^1(k_\mathfrak{p}, \mathcal{E}(n,\nu))$ with

$$\alpha(\xi_\mathfrak{p}) = x_\mathfrak{p} : G_{(N_\nu^n)_\mathfrak{p}} \longrightarrow \mathcal{E}(n,\nu)\,.$$

Let $\mathfrak{p} \in T^1$. Then, by the induction hypothesis, the extension $(N_\nu^n)_\mathfrak{p}|K_\mathfrak{p}$ is trivial and the group extension in the diagram

$$\begin{array}{ccccccccc} & & & & & & G_{k_\mathfrak{p}} & & \\ & & & \varphi_{n,\nu+1}|G_{k_\mathfrak{p}} \swarrow & & & \downarrow \varphi_{n,\nu}|G_{k_\mathfrak{p}} & & \\ 1 & \longrightarrow & \mathcal{E}(n,\nu) & \longrightarrow & E_\mathfrak{p} & \longrightarrow & G_\mathfrak{p}(N_\nu^n|k) & \longrightarrow & 1 \end{array}$$

splits. Hence $\beta(\varphi_{n,\nu+1}|_{G_{(N_\nu^n)_\mathfrak{p}}}) = 0$ and we can therefore find an element $\xi_\mathfrak{p} \in H^1(k_\mathfrak{p}, \mathcal{E}(n,\nu))$ such that

$$\alpha(\xi_\mathfrak{p}) = -\varphi_{n,\nu+1}|_{G_{(N_\nu^n)_\mathfrak{p}}} : G_{(N_\nu^n)_\mathfrak{p}} \longrightarrow \mathcal{E}(n,\nu)\,.$$

If $\mathfrak{p} \in T^2$, then by the induction hypothesis $\mathfrak{p} \notin T^1$, $K_\mathfrak{p}|k_\mathfrak{p}$ is trivial and $(N_\nu^n)_\mathfrak{p}|K_\mathfrak{p}$ is a (cyclic) totally ramified extension. We consider the commutative exact diagram

$$\begin{array}{ccccc} & & H^1_{nr}(k_\mathfrak{p}) & \dashrightarrow & H^1_{nr}((N_\nu^n)_\mathfrak{p})^{G_{k_\mathfrak{p}}} \\ & & \downarrow & & \downarrow \\ H^1((N_\nu^n)_\mathfrak{p}|k_\mathfrak{p}) & \hookrightarrow & H^1(k_\mathfrak{p}) & \xrightarrow{\alpha} & H^1((N_\nu^n)_\mathfrak{p})^{G_{k_\mathfrak{p}}}\,. \end{array}$$

Since $\mathfrak{p} \in T^2$, the dotted arrow in the diagram above is an isomorphism. Thus there is a $\xi_\mathfrak{p} \in H^1(k_\mathfrak{p}, \mathcal{E}(n,\nu))$ such that

$$\varphi_{n,\nu+1}|_{G_{(N_\nu^n)_\mathfrak{p}}} + \alpha(\xi_\mathfrak{p}) : G_{(N_\nu^n)_\mathfrak{p}} \longrightarrow \mathcal{E}(n,\nu)$$

is either trivial (if $\varphi_{n,\nu+1}|_{G_{(N_\nu^n)_\mathfrak{p}}}$ is unramified) or induces a totally ramified extension of degree p of $(N_\nu^n)_\mathfrak{p}$.

If $\mathfrak{p} \in T^3$, then we have a commutative exact diagram

$$\begin{array}{ccccc} H^1((N_\nu^n)_\mathfrak{p}|k_\mathfrak{p}) & \hookrightarrow & H^1(k_\mathfrak{p}) & \xrightarrow{\alpha} & H^1((N_\nu^n)_\mathfrak{p})^{G_{k_\mathfrak{p}}} \\ & & \downarrow & & \downarrow res \\ & & H^1(T_{k_\mathfrak{p}})^{G_{k_\mathfrak{p}}} & \dashrightarrow & H^1(T_{(N_\nu^n)_\mathfrak{p}})^{G_{k_\mathfrak{p}}} \end{array}$$

where now the lower dotted arrow is an isomorphism, since $(N_\nu^n)_\mathfrak{p}|k_\mathfrak{p}$ is unramified. Let $\xi_\mathfrak{p} \in H^1(k_\mathfrak{p}, \mathcal{E}(n,\nu))$ be such that

$$\varphi_{n,\nu+1}|G_{(N_\nu^n)_\mathfrak{p}} + \alpha(\xi_\mathfrak{p}) : G_{(N_\nu^n)_\mathfrak{p}} \longrightarrow \mathcal{E}(n,\nu)$$

is unramified.

In order to complete step 3, it is therefore sufficient to show the existence of an element $\varepsilon \in H^1(G_S, \mathcal{E}(n,\nu)) \subseteq H^1(G_k, \mathcal{E}(n,\nu))$ with $\varepsilon_\mathfrak{p} = \xi_\mathfrak{p}$ for all $\mathfrak{p} \in T$.

The exact sequence

$$H^1(G_S, \mathcal{E}(n,\nu)) \longrightarrow \prod_T H^1(k_\mathfrak{p}, \mathcal{E}(n,\nu)) \xrightarrow{\pi_n} \text{coker}(k_S, T, \mathcal{E}(n,\nu))$$

shows that the obstruction to the existence of such an ε is the vanishing of $\pi_n(\xi_n)$ with

$$\xi_n = \prod_{\mathfrak{p}\in T_n} \xi_\mathfrak{p} \in \prod_{\mathfrak{p}\in T_n} H^1(k_\mathfrak{p}, \mathcal{E}(n,\nu)).$$

(In the following we denote the sets T^i and T by T_n^i and T_n, respectively, in order to indicate at which level the embedding problem is considered.) By (9.2.2), we have a canonical injection

(1) $$\text{coker}(k_{S_n}, T_n, \mathcal{E}(n,\nu)) \hookrightarrow Ш^1(k_{S_n}, S_n \smallsetminus T_n, \mathcal{E}(n,\nu)')^\vee,$$

where $\mathcal{E}(n,\nu)' = \text{Hom}(\mathcal{E}(n,\nu), \mu_p)$.

Recall that $H^{-2} = H_1$ and that by the induction hypothesis, the solution $\varphi_{n,\nu}$ is proper. Thus we obtain

$$\begin{aligned} H_1(\mathcal{F}(n)/\nu \rtimes G, \mathcal{E}(n,\nu)(-1)) &\cong H^1(\mathcal{F}(n)/\nu \rtimes G, \mathcal{E}(n,\nu)^\vee(1))^\vee \\ &\cong H^1(N_\nu^n|k, \mathcal{E}(n,\nu)')^\vee. \end{aligned}$$

Therefore there exists a canonical isomorphism

(2) $$H^{-2}(\mathcal{F}(n)/\nu \rtimes G, \mathcal{E}(n,\nu)(-1)) \xrightarrow{\sim} H^1(N_\nu^n|k, \mathcal{E}(n,\nu)')^\vee.$$

Now we are going to shrink the obstruction to the existence of a 1-class ε as above. If $\varphi_{n,\nu}$ is induced by a $\varphi_{m,\nu}$ for $m \geq n$ via a G-invariant surjection $\mathcal{F}(m) \twoheadrightarrow \mathcal{F}(n)$, then the inclusion (1) is obviously also true in the form

(1)′ $$\text{coker}(k_{S_m}, T_m, \mathcal{E}(n,\nu)) \hookrightarrow Ш^1(k_{S_m}, S_m \smallsetminus T_m, \mathcal{E}(n,\nu)')^\vee,$$

where S_m and T_m are chosen as above but at the level N_ν^m. Using (9.6.4) we choose $m \geq n$ so that an arbitrarily chosen element in $H^{-2}(\mathcal{F}(m)/\nu \rtimes G, \mathcal{E}(m,\nu)(-1))$ is annihilated by the map which is induced by a suitably chosen surjection $\mathcal{F}(m) \twoheadrightarrow \mathcal{F}(n)$. Then we consider the diagrams, in which we write c for coker, $\mathcal{E}_d$ for $\mathcal{E}(d,\nu)$ $(d = m, n)$ and $\tilde{S}_n = cs(N_\nu^n|k) \cup T_m$:

$$\begin{array}{ccccccc}
H^1(k_{S_m}|k,\mathcal{E}_m) & \longrightarrow & \prod\limits_{T_m} H^1(k_\mathfrak{p},\mathcal{E}_m) & \xrightarrow{\pi} & \mathrm{c}(k_{S_m},T_m,\mathcal{E}_m) & \hookrightarrow & Ш^1(k_{S_m},S_m\backslash T_m,\mathcal{E}'_m)^\vee \\
\downarrow{\scriptstyle\psi} & & \downarrow & & \downarrow & & \downarrow \\
H^1(k_{S_m}|k,\mathcal{E}_n) & \longrightarrow & \prod\limits_{T_m} H^1(k_\mathfrak{p},\mathcal{E}_n) & \xrightarrow{\pi} & \mathrm{c}(k_{S_m},T_m,\mathcal{E}_n) & \hookrightarrow & Ш^1(k_{S_m},S_m\backslash T_m,\mathcal{E}'_n)^\vee \\
\downarrow{\scriptstyle inf} & & \| & & \downarrow & & \downarrow \\
H^1(k_{\tilde{S}_n}|k,\mathcal{E}_n) & \longrightarrow & \prod\limits_{T_m} H^1(k_\mathfrak{p},\mathcal{E}_n) & \xrightarrow{\pi} & \mathrm{c}(k_{\tilde{S}_n},T_m,\mathcal{E}_n) & \hookrightarrow & Ш^1(k_{\tilde{S}_n},\tilde{S}_n\backslash T_m,\mathcal{E}'_n)^\vee,
\end{array}$$

$$\begin{array}{ccccc}
Ш^1(k_{S_m},S_m\backslash T_m,\mathcal{E}'_m)^\vee & \xrightarrow{\sim} & H^1(N_\nu^m|k,\mathcal{E}'_m)^\vee & \xleftarrow{\sim} & H^{-2}(\mathcal{F}(m)/\nu \rtimes G,\mathcal{E}_m(-1)) \\
\downarrow & & \downarrow & & \\
Ш^1(k_{S_m},S_m\backslash T_m,\mathcal{E}'_n)^\vee & \twoheadrightarrow & H^1(N_\nu^n|k,\mathcal{E}'_n)^\vee & & \downarrow \\
\downarrow & & \| & & \\
Ш^1(k_{\tilde{S}_n},\tilde{S}_n\backslash T_m,\mathcal{E}'_n)^\vee & \xrightarrow{\sim} & H^1(N_\nu^n|k,\mathcal{E}'_n)^\vee & \xleftarrow{\sim} & H^{-2}(\mathcal{F}(n)/\nu \rtimes G,\mathcal{E}_n(-1)).
\end{array}$$

The existence of all maps and the fact that the diagrams are commutative follow from the arguments above and from (9.6.5).

Now let

$$\xi_m = \prod_{\mathfrak{p}\in T_m} \xi_\mathfrak{p} \in \prod_{\mathfrak{p}\in T_m} H^1(k_\mathfrak{p},\mathcal{E}(n,\nu))$$

be arbitrary. By theorem (9.6.4)(i), we can choose a G-invariant surjection $\psi : \mathcal{F}(m) \twoheadrightarrow \mathcal{F}(n)$ such that $\pi_n\psi_*(\xi_m) = 0$.

Observe that we have not yet got precisely what we want, because ε has the required property with respect to the sets of primes T_m^i and $\tilde{S}_n$. Nevertheless, one easily verifies that if we modify the solution, which we have obtained after shrinking, by the cocycle ε (which now exists), then we obtain a solution satisfying all required properties. This finishes step 3.

Fourth Step: *After changing $\varphi_{n,\nu}$ again, there exists a proper solution $\varphi_{n,\nu+1}$ of* $(*)$ *which satisfies properties* (i) *and* (ii).

The solution $\varphi_{n,\nu+1}$ which we obtained in step 3, has almost all properties we need, except that for $\mathfrak{p} \in \mathrm{Ram}(N_{\nu+1}^n|K) \smallsetminus \mathrm{Ram}(N_\nu^n|K)$ the local extension $(N_{\nu+1}^n)_\mathfrak{p}|k_\mathfrak{p}$ might not be (cyclic) totally ramified. But we know that for such a prime $\mathfrak{p}$ the extension $(N_\nu^n)_\mathfrak{p}|k_\mathfrak{p}$ is trivial (in fact, $N_\nu^n(\mu_{p^e})_\mathfrak{p}|k_\mathfrak{p}$ is trivial). In order to get a totally ramified cyclic extension, we have to remove the unramified part of the extension $(N^n_{\nu+1})_\mathfrak{p}|(N^n_\nu)_\mathfrak{p}$ and to make sure that at places where new ramification occurs by this procedure, we have cyclic local extensions (these are automatically totally ramified, since the decomposition group is an elementary abelian p-group).

In order to retain the properness of the solution obtained in step 3, we choose a finite set of primes $T^0 \subseteq cs(N^n_\nu|k) \smallsetminus (\mathrm{Ram}(N^n_{\nu+1}|k) \cup S_p \cup S_\infty)$ such that $G_\mathfrak{p}(N^n_{\nu+1}|N^n_\nu)$, $\mathfrak{p} \in T^0$, generate $G(N^n_{\nu+1}|N^n_\nu)$.

We want to alter the solution found in step 3 once again using a class x in $H^1(k_S|K, \mathcal{E}(n,\nu))$, where

- $S = cs(N^n_\nu(\mu_{p^e})|k) \cup T$ with $T = \mathrm{Ram}(N^n_{\nu+1}|k) \cup S_p \cup S_\infty \cup T^0$,
- for $\mathfrak{p} \in \mathrm{Ram}(N^n_\nu|k) \cup S_p \cup S_\infty \cup T^0$, we have $x_\mathfrak{p} = 0$,
- if the prolongations of $\mathfrak{p}$ to K are in $\mathrm{Ram}(N^n_{\nu+1}|K) \smallsetminus \mathrm{Ram}(N^n_\nu|K)$, then $x_\mathfrak{p} \in H^1_{nr}(k_\mathfrak{p}, \mathcal{E}(n,\nu)) = H^1_{nr}((N^n_\nu)_\mathfrak{p}, \mathcal{E}(n,\nu))$ has the property that
$$\varphi_{n,\nu+1}|_{G_{(N^n_\nu)_\mathfrak{p}}} + x_\mathfrak{p} \in H^1((N^n_\nu)_\mathfrak{p}, \mathcal{E}(n,\nu))$$
is cyclic,
- $x_\mathfrak{p}$ is cyclic for all $\mathfrak{p} \notin T$.

For every $\mathfrak{p} \in S(k)$ such that prolongations of $\mathfrak{p}$ to K are in $\mathrm{Ram}(N^n_{\nu+1}|K)$ but not in $\mathrm{Ram}(N^n_\nu|K)$, we fix a prolongation $\mathfrak{p}_0 \in S(K)$ of $\mathfrak{p}$ to K (note that $\mathfrak{p}$ splits completely in $K|k$). Let
$$\eta \in \prod_T H^1(K_\mathfrak{P}, \mathcal{E}(n,\nu))$$
be such that

- $\eta_\mathfrak{P} = 0$ if $\mathfrak{P} \cap k \in \mathrm{Ram}(N^n_\nu|k) \cup S_p \cup S_\infty \cup T^0$,
- if $\mathfrak{P} \in \mathrm{Ram}(N^n_{\nu+1}|K) \smallsetminus \mathrm{Ram}(N^n_\nu|K)$ and $\mathfrak{P} \neq (\mathfrak{P} \cap k)_0$, then $\eta_\mathfrak{P} = 0$,
- if $\mathfrak{P} \in \mathrm{Ram}(N^n_{\nu+1}|K) \smallsetminus \mathrm{Ram}(N^n_\nu|K)$ and $\mathfrak{P} = (\mathfrak{P} \cap k)_0$, then $\eta_\mathfrak{P} \in H^1_{nr}(K_\mathfrak{P}, \mathcal{E}(n,\nu)) = H^1_{nr}((N^n_\nu)_\mathfrak{P}, \mathcal{E}(n,\nu))$ has the property that
$$\varphi_{n,\nu+1}|_{G_{(N^n_\nu)_\mathfrak{P}}} + \eta_\mathfrak{P} \in H^1((N^n_\nu)_\mathfrak{P}, \mathcal{E}(n,\nu))$$
is cyclic.

Applying theorem (9.3.2) in the situation where Ω is the field $N^n_\nu(\mu_{p^e})$ and $A = \mathcal{E}(n,\nu)$, we see that in order to finish the proof, it suffices to construct an element y in $H^1(k_S|K, \mathcal{E}(n,\nu))$ with $y_\mathfrak{P} = \eta_\mathfrak{P}$ for all $\mathfrak{P} \in T$. Indeed, using this procedure, we get new ramification only at places which are completely decomposed in $N^n_\nu(\mu_{p^e})|k$. Hence their decomposition groups are cyclic (by (9.3.2)) and contained in the p-elementary abelian group $G(N^n_{\nu+1}|N^n_\nu) \cong \mathcal{E}(n,\nu)$. Thus the local extensions associated to these new ramification primes are cyclic of order p, and, in particular, are totally ramified. Furthermore, by the choice of T^0, the new solution remains proper.

Similarly to the situation with the class ε in step 3, the exact sequence

$$H^1(K_S|K, \mathcal{E}(n,\nu)) \longrightarrow \prod_T H^1(K_{\mathfrak{p}}, \mathcal{E}(n,\nu)) \xrightarrow{\pi_n} \mathrm{coker}(K_S, T, \mathcal{E}(n,\nu))$$

shows that the obstruction to the existence of such a y is $\pi_n(\eta) = 0$.

Now we apply the shrinking procedure as in step 3, but the commutative diagrams used there have to be modified as follows: replace k by K in the first diagram and instead of the second consider the following diagram

$$\begin{array}{ccccc}
Ш^1(K_{S_m}, S_m\backslash T_m, \mathcal{E}'_m)^\vee & \xrightarrow{\sim} & H^1(N_\nu^m|K, \mathcal{E}'_m)^\vee & \xleftarrow{\sim} & H^{-2}(\mathcal{F}(m)/\nu, \mathcal{E}_m(-1)) \\
\downarrow & & \downarrow & & \\
Ш^1(K_{S_m}, S_m\backslash T_m, \mathcal{E}'_n)^\vee & \twoheadrightarrow & H^1(N_\nu^n|K, \mathcal{E}'_n)^\vee & & \downarrow \\
\downarrow & & \| & & \\
Ш^1(K_{\tilde{S}_n}, \tilde{S}_n\backslash T_m, \mathcal{E}'_n)^\vee & \xrightarrow{\sim} & H^1(N_\nu^n|K, \mathcal{E}'_n)^\vee & \xleftarrow{\sim} & H^{-2}(\mathcal{F}(n)/\nu, \mathcal{E}_n(-1)).
\end{array}$$

Then we use part (ii) of theorem (9.6.4) instead of part (i).

Therefore, after a further shrinking, we get a class y with the properties above, and theorem (9.3.2) then induces the existence of the desired class $x \in H^1(k_S|k, \mathcal{E}(n,\nu))$.

The new solution $\tilde{\varphi}_{n,\nu+1} = {}^x\varphi_{n,\nu+1}$ fulfills conditions (i) and (ii), hence step 4 and the proof in the case $p \neq \mathrm{char}(k)$ of theorem (9.6.7) are complete.

The proof in the case $p = \mathrm{char}(k)$ is comparatively easy. Again we proceed by induction on ν, where n is arbitrary. The case $\nu = (1,1)$ is trivial. In the case $\nu = (2,1)$ we get an embedding problem with abelian kernel isomorphic to $\mathbb{F}_p[G]^n$, which is properly solvable by (9.2.9). In the next induction steps we do not care about the properness of the solutions, because they are automatically proper by the Frattini argument. By (6.1.4), we have $cd_p\, G_k = 1$, so that G_k is p-projective by (3.5.6) and we can solve the embedding problems in all induction steps. Therefore the proof of (9.6.7) and also that of (9.6.6) is complete. □

In order to deduce the theorem of Šafarevič, we use an argument which goes back to *O. Ore* [165]. We need two facts from group theory and we recall the following notation. If G is a finite nontrivial group, then

$\Phi(G)$ is the intersection of all maximal subgroups of G and is called the *Frattini subgroup* of G,

$F(G)$ is the composite of all nilpotent normal subgroups of G and is called the *Fitting subgroup* of G.

The group $\Phi(G)$ is a characteristic subgroup of G and is contained in $F(G)$. The group $F(G)$ is a normal nilpotent subgroup of G. We cite the following two facts, see [81], Kap.III, Satz 3.2 (b) and Satz 4.2 (c).

(9.6.8) Proposition. *Let N be a normal subgroup of the finite group G such that $N \nsubseteq \Phi(G)$. Then there exists a partial complement U of N in G, i.e. $U \neq G$ and $G = N \cdot U$.*

(9.6.9) Proposition. *Let G be a nontrivial finite solvable group. Then $\Phi(G)$ is a proper subgroup of $F(G)$.*

Proof of Šafarevič's theorem: Let $F(G)$ be the Fitting subgroup of $G \neq \{1\}$. By the two propositions above, $F(G)$ has a (solvable) partial complement $U \subsetneqq G$, so there exists a surjection

$$F(G) \rtimes U \twoheadrightarrow G\,.$$

Assuming inductively (on the order of G) that U is the Galois group of a finite normal extension of k, we obtain the result using theorem (9.6.6). □

Finally, we would like to mention the following corollary to (9.6.7).

(9.6.10) Theorem. *Let $K|k$ be a finite Galois extension of the global field k and let $\varphi : G_k \twoheadrightarrow G(K|k) = G$ be the natural projection. Then every embedding problem*

$$\begin{array}{ccccccccc} & & & & & & G_k & & \\ & & & & & & \downarrow{\scriptstyle\varphi} & & \\ 1 & \longrightarrow & N & \longrightarrow & E & \longrightarrow & G & \longrightarrow & 1 \end{array}$$

with finite nilpotent kernel N which has a solution, can also be solved properly.

Proof: Let $\psi : G_k \to E$ be a solution of the embedding problem. Then we obtain a commutative exact diagram

$$\begin{array}{ccccccccc} & & & & & & G_k & & \\ & & & & & & \downarrow & & \\ 1 & \longrightarrow & N & \longrightarrow & N \rtimes \mathrm{im}(\psi) & \longrightarrow & \mathrm{im}(\psi) & \longrightarrow & 1 \\ & & \| & & \downarrow & & \downarrow & & \\ 1 & \longrightarrow & N & \longrightarrow & E & \longrightarrow & G & \longrightarrow & 1\,, \end{array}$$

where the action of $\mathrm{im}(\psi) \subseteq E$ on N is induced by the action of E on N, and the surjection $N \rtimes \mathrm{im}(\psi) \twoheadrightarrow E$ is given by the inclusions of N and $\mathrm{im}(\psi)$ into E. By (9.6.6), the split embedding problem given by the upper row of the diagram above has a proper solution, hence the initial problem is also properly solvable. □

Exercise: (a) Go through the proof of Šafarevič's theorem and show that one can choose K in such a way that finitely many given primes of k split completely in K (if $\mathrm{char}(k) = p > 0$, use (9.2.6)).

(b) If k is a number field, prove that the extension $K|k$ which realizes the solvable group G can be chosen to be tamely ramified.

Chapter X
Restricted Ramification

In this chapter we will investigate Galois groups of global fields with restricted ramification. Let K be a global field and S be any set of primes of K. We are interested in properties of the Galois group

$$G_S = G_S(K) = G(K_S|K),$$

where K_S is the maximal extension of K which is unramified outside the primes in S.

The case $S = \{\text{all places of } K\}$, i.e. $G_S = G_K$, was extensively studied in chapter IX. The case when S is finite and especially the case $S = \varnothing$, where G_S is the Galois group of the maximal unramified extension of K, is of particular arithmetic interest.

One soon observes that the function field case is comparatively easy to deal with. The reason is that if S is finite, then we can view G_S as the *étale fundamental group* of a smooth algebraic curve over a finite field. The closed points of this curve X correspond to the places of K which are not in S. Then we can translate results from algebraic geometry into properties of G_S.

The reason why such a method works is the fact that curves (except proper, rational curves) are $K(\pi, 1)$-spaces in the sense of (étale) homotopy. This means that the *universal covering space* $\tilde{X}$ of a curve X has trivial homotopy groups (and is contractible, in a rather vague sense). In particular, the Hochschild-Serre spectral sequence for the covering $\tilde{X}|X$ is degenerate, i.e. identifies the cohomology of the fundamental group with the cohomology of the space itself.

Having this geometric background in mind, it is not surprising that there are qualitatively different results in the cases $S = \varnothing$ (projective curve) and $S \neq \varnothing$ (affine curve), and that the case of a rational curve causes an additional exception.

We will treat the function field case in §1, referring there to some well-known results of algebraic geometry.

The rest of this chapter will be devoted to the much more complicated number field case, where we are mainly interested in the case when S is finite. The reasons that the number field case is more difficult are at least twofold.

J. Neukirch et al., *Cohomology of Number Fields*,
Grundlehren der mathematischen Wissenschaften 323,
DOI 10.1007/978-3-540-37889-1_10,

In the first instance, the affine scheme $\mathrm{Spec}(\mathcal{O}_{K,S})$ is in general not a $K(\pi,1)$. However, for the finitely many $p \in \mathbb{N}(S) = \mathbb{N} \cap \mathcal{O}_{K,S}^\times$ do the p-parts of the higher homotopy groups $\pi_i^{et}(\mathrm{Spec}(\mathcal{O}_{K,S}))$ vanish.*)

The second reason is the occurrence of wild ramification. In contrast to the function field case, there is not only one but infinitely many different residue characteristics in $\mathcal{O}_{K,S}$. In order to get information about the p-part of the cohomology of G_S we must, by the reasons mentioned before (or just because of the necessary conditions of the Poitou-Tate theorem), enlarge the set S to $S \cup S_p$. But then we are led to consider the rather difficult higher ramification groups at the places dividing p. For example, we will see in §3 that the question of the strict cohomological p-dimension of G_S is independent of the finite set $S \supseteq S_p \cup S_\infty$ and is related to the Leopoldt conjecture.

As a consequence of these difficulties we will restrict to easier classes of Galois groups, i.e. we consider the maximal pro-$\mathfrak{c}$-factor of G_S, where $\mathfrak{c}$ is a full class of finite groups, which will be in most cases the class of p-groups for some prime number p.

In order to do this, our first task is to calculate the cohomology groups of $G_S(\mathfrak{c})$. In §4 we will investigate the group $G_S(L)$ for large number fields, i.e. for number fields of infinite degree over $\mathbb{Q}$. Given a prime number p, we will give several criteria on L and S for the vanishing of $H^i(G_S(L), \mathbb{Z}/p\mathbb{Z})$ for $i \geq 1$. As a corollary we obtain a result due to *O. Neumann*, which asserts that the inflation maps

$$H^i(G_S(p), A)(p) \longrightarrow H^i(G_S, A)(p)$$

are isomorphisms for every discrete $G_S(p)$-module A and all i, provided that $p \in \mathbb{N}(S)$.

The next problem we are confronted with is the question of whether, for a given prime $\mathfrak{p} \in S(K)$, the canonical homomorphism

$$G(K_\mathfrak{p}(\mathfrak{c})|K_\mathfrak{p}) \twoheadrightarrow G_\mathfrak{p}(K_S(\mathfrak{c})|K) \subseteq G(K_S(\mathfrak{c})|K)$$

is injective. This is a local-global problem and equivalent to the question of which local $\mathfrak{c}$-extensions can be globally realized inside $K_S(\mathfrak{c})$. In chapter IX we have given a positive answer for every full class $\mathfrak{c}$ if S is of density 1. The similar question for abelian extensions (the class of abelian groups is not a full class) is answered by the theorem of Grunwald-Wang.

However, if S is finite, then there are arithmetic obstructions to the global realization of local extensions. For instance, the group G_S might be finite or

*) We will neither explain nor use étale homotopy in the following. We refer the reader to [54] for the definition of étale homotopy groups. See [193] for a proof that $\pi_i^{et}(\mathrm{Spec}(\mathcal{O}_{K,S}))(p) = 0$ for $i \geq 2$ and $p \in \mathbb{N}(S)$.

even trivial. Therefore we restrict to the case when $\mathfrak{c}$ is the class of p-groups where p is a prime number such that $S \supseteq S_p$. In §5 we show the number theoretical analogue of Riemann's existence theorem which asserts that the Galois group $G(k(p)|k_S(p))$ decomposes into a free pro-p-product of inertia groups.

The prime number $p = 2$ frequently causes technical problems and requires additional efforts. In Galois cohomology these problems are essentially due to the fact that the decomposition groups of the real places are 2-groups. This was the reason that, so far, we have always assumed k to be a totally imaginary number field if we considered the case $p = 2$. In section §6 we consider the group $G_S(2)$ when not all real places are in S.

In §7 we calculate the dimensions of the cohomology groups $H^i(G_S^T, \mathbb{Z}/p\mathbb{Z})$ for $i = 1, 2$, where $G_S^T = G(K_S^T|K)$ is the Galois group of the maximal extension of K which is unramified outside S and completely decomposed at the primes of the set T.

In §8 we present the following result due to *L. V. Kuz'min*: if $S \supseteq S_p \cup S_\infty$ for a prime number p with $\mu_p \subseteq k$, then in the generic case

$$(K_S(p))_\mathfrak{p} = K_\mathfrak{p}(p)$$

for all primes $\mathfrak{p}$ in K.

In §9 we derive a criterion for the group $G_S(p)$ to be a free pro-p-product of local groups. If $G_S(p)$ does not decompose in this way, then it is a pro-p duality group of dimension 2 in many cases.

The question whether for a given prime number p the group $G_S(p)$ is infinite or not, is trivial if $S_p \subseteq S$. In the case $S = \varnothing$ it became prominent as the problem of "p-class field towers". We will consider that problem in §10; in particular, we derive a sharpened form of the classical result of *E. S. Golod* and *I. R. Šafarevič*, who constructed the first infinite p-class field towers in 1964.

In the final section §11, we investigate the profinite group G_S rather than its pro-p-factors. We will prove a duality theorem for this group and a second form of the Poitou-Tate duality for finitely generated G_S-modules. Furthermore, using analytic results, we will present a theorem of *Y. Ihara* which asserts that for an infinite unramified Galois extension L of a number field K, the set of primes of K which split completely in L cannot be "too big".

§1. The Function Field Case

Assume that K is a global field of char(K) = $p > 0$ and let $k \subseteq K$ be the finite constant field with algebraic closure $\bar{k}$. Let S be a finite set of places of K and let $G_S = G(K_S|K)$. Up to this point we have always assumed S to be nonempty but from now on the case $S = \varnothing$ will be included. We denote the complete curve over k which is associated to K by X. Then the extensions of K inside K_S correspond to the unramified covers of the curve $X \smallsetminus S$, which is obtained by removing the finitely many points from X that correspond to places in S. We denote the (geometric) genus of the curve X by $g = g(X) = g(K)$. Now consider the exact sequence

$$1 \longrightarrow G_S(K\bar{k}) \longrightarrow G_S \longrightarrow G(\bar{k}|k) \longrightarrow 1.$$

The group $G(\bar{k}|k)$ is a free profinite group of rank 1, and the Frobenius automorphism Frob $\in G(\bar{k}|k)$ is a canonical topological generator. We will denote the unique extension of k inside $\bar{k}$ of degree $n \in \mathbb{N}$ by k_n.

The group $G_S(K\bar{k})$ is related to the unramified covers of $X_{\bar{k}} \smallsetminus S$, where $X_{\bar{k}}$ is the base change of X to $\bar{k}$. Recall that the group $Cl_S(K)$, resp. $Cl_S(K\bar{k})$, may be geometrically interpreted as the Picard group Pic($X \smallsetminus S$), resp. Pic($X_{\bar{k}} \smallsetminus S$). When $S = \varnothing$, there exists a degree map

$$\deg : Cl(K) \to \mathbb{Z}$$

which assigns to the class $[\mathfrak{p}]$ of a prime divisor the degree $[k(\mathfrak{p}) : k]$. The homomorphism deg is surjective and we denote its kernel by $Cl^0(K)$. Passing from k to $\bar{k}$, we obtain an exact sequence

$$0 \longrightarrow Cl^0(K\bar{k}) \longrightarrow Cl(K\bar{k}) \xrightarrow{\deg} \mathbb{Z} \longrightarrow 0.$$

The following result is not very deep from the point of view of algebraic geometry, but it will be of crucial importance for our investigation of G_S. A similar result in the number field case would be desirable; compare the discussion of "$\mu = 0$" in chapter XI.

(10.1.1) Proposition. *The group $Cl^0(K\bar{k})$ is divisible. More precisely,*

$$Cl^0(K\bar{k}) \cong \bigoplus_{\ell \neq p} (\mathbb{Q}_\ell/\mathbb{Z}_\ell)^{2g} \oplus (\mathbb{Q}_p/\mathbb{Z}_p)^h,$$

where g is the genus of X and $0 \le h \le g$. If S is nonempty, then the group $Cl_S(K\bar{k})$ is also divisible.

Remark: The number h is the *height* of the *p-divisible group* attached to the *Jacobian variety* of X. It is also called the *p-rank* or the *Hasse-Witt invariant* of X.

Proof: Proposition (10.1.1) is a consequence of the existence of the Jacobian variety of the curve X. In fact, the group $Cl^0(K\bar{k})$ can be canonically identified with the group of $\bar{k}$-rational points of the abelian variety Jac(X) (see [141]). The abelian variety Jac(X) is self-dual and its dimension is equal to the genus g of X (loc.cit.). Therefore the torsion subgroup of $Cl^0(K\bar{k})$ has the desired structure by the general theory of abelian varieties (see [140]). On the other hand, the group $Cl^0(Kk_n)$ is finite for every $n \in \mathbb{N}$, because it is a subgroup of the group of k_n-rational points of Jac(X) (see [141], Remark 1.5)*). But the latter is a finite group. Hence $Cl^0(Kk_n)$ is finite. (This is a classical result, which can be found in various places in the literature as the function field analogue of the finiteness of the ideal class group of number fields.) Therefore $Cl^0(K\bar{k}) = \varinjlim_n Cl^0(Kk_n)$ is the direct limit of finite groups, hence torsion. Finally, if S is nonempty, then $Cl_S(K\bar{k})$ is a quotient of $Cl^0(K\bar{k})$. □

We denote the class of finite groups of order prime to $p = \mathrm{char}(K)$ by (p') and we write $A^{(p')}$ for the (p')-torsion subgroup of an abelian group A, i.e. the subgroup of elements of finite prime-to-p order. The maximal pro-(p')-quotient of a profinite group G will be denoted by $G^{(p')}$. Let n be the number of geometric points in S, i.e. $n = \#S(K\bar{k})$.

(10.1.2) Theorem. (i) *The group $G_S(K\bar{k})$ has the following properties.*

a) *If $S = \varnothing$ and $g = 0$, then $G_S(K\bar{k}) = \{1\}$.*

b) *If $S = \varnothing$ and $g > 0$, then $G_S(K\bar{k})$ is a Poincaré group of dimension 2 at the class (p') and $scd_{(p')}\, G_S(K\bar{k}) = 3$.*

c) *If $S \neq \varnothing$, then $cd_\ell\, G_S(K\bar{k}) = 1$ for every $\ell \in (p')$.*

(ii) *For every prime number $\ell \in (p')$ there exists a presentation of the maximal pro-ℓ-factor group of $G_S(K\bar{k})$ as a pro-ℓ-group by $2g + n$ generators and one relation of the form*

$$G_S(K\bar{k})(\ell) \cong \langle x_1, \dots, x_{2g}, y_1, \dots, y_n \mid (x_1, x_2) \cdots (x_{2g-1}, x_{2g})\, y_1 \cdots y_n = 1 \rangle.$$

In particular, if $S \neq \varnothing$, then $G_S(K\bar{k})(\ell)$ is a free pro-ℓ-group of rank $2g+n-1$. Furthermore, the elements $y_1, \dots, y_n$ may be chosen as generators of the procyclic inertia groups of prolongations of primes in $S(K\bar{k})$ to $K_S(\ell)$.

---

*) In fact, this canonical inclusion is an isomorphism in our situation, since $Br(k_n) = 0$ (see [141], Remark 1.6.).

(10.1.3) Corollary. *The group $G_S = G_S(K)$ has the following properties.*

(i) *If $S = \varnothing$ and $g = 0$, then $G_S = G(\bar{k}|k) \cong \hat{\mathbb{Z}}$.*

(ii) *If $S = \varnothing$ and $g > 0$, then G_S is a Poincaré group at (p') of dimension 3 with dualizing module μ, i.e. the module of all roots of unity is contained in K_S. Furthermore, in this case $scd_{(p')}\, G_S = 3$.*

(iii) *If $S \neq \varnothing$, then G_S is a duality group at (p') of dimension 2. Its (p')-dualizing module $D_2 = D_2(\mathbb{Z}_{(p')})$ is canonically isomorphic to the (p')-torsion part of the S-idèle class group of K_S. In particular, there exists an exact sequence*

$$0 \longrightarrow \mu \longrightarrow \bigoplus_{\mathfrak{p}\in S} \mathrm{Ind}_{G_S}^{G_\mathfrak{p}}\, \mu \longrightarrow D_2 \to 0,$$

where $G_\mathfrak{p}$ is the decomposition group in G_S for a fixed prolongation of the prime $\mathfrak{p} \in S$ to K_S.

Proof of theorem (10.1.2) and corollary (10.1.3): If $g = 0$, we have $X_{\bar{k}} \cong \mathbb{P}^1_{\bar{k}}$ and statement (i) a) follows from the Hurwitz genus formula, which shows that there must be ramification in every separable cover of $\mathbb{P}^1_{\bar{k}}$ (see [77], chap.IV, §2). This also shows assertion (i) of the corollary.

In order to prove c), observe that $(K\bar{k})_\mathfrak{p} = K_\mathfrak{p}^{nr}$, where $K_\mathfrak{p}^{nr}$ denotes the maximal unramified extension of the local field $K_\mathfrak{p}$. Since $cd_{(p')}\, G(\bar{K}_\mathfrak{p}|K_\mathfrak{p}^{nr}) \leq 1$ by (7.1.8), we can calculate for $\ell \in (p')$

$$\begin{aligned} H^2(G_S(K\bar{k}), \mu_\ell) &= \varinjlim_n Ш^2(G_S(Kk_n), \mu_\ell) \\ &= \varinjlim_n Ш^1(G_S(Kk_n), \mathbb{Z}/\ell\mathbb{Z})^\vee \\ &= Cl_S(K\bar{k})/\ell \;=\; 0, \end{aligned}$$

by (10.1.1). Since the cyclotomic character is trivial on $G_S(K\bar{k})$ and since the same arguments also hold for every finite extension of K inside K_S, we conclude that $cd_\ell\, G_S(K\bar{k}) \leq 1$ for every $\ell \in (p')$. Furthermore, the Kummer sequence $0 \to \mu_\ell \to \mathcal{O}_S^\times \to \mathcal{O}_S^\times \to 0$, together with the equality $H^1(G_S(K\bar{k}), \mathcal{O}_S^\times) = Cl_S(K\bar{k})$ (see (8.3.11)), implies that there is an exact sequence

$$0 \to \mathcal{O}_{K\bar{k},S}^\times/\ell \to H^1(G_S(K\bar{k}), \mu_\ell) \to {}_\ell Cl_S(K\bar{k}) \to 0.$$

In particular, $H^1(G_S(K\bar{k}), \mu_\ell)$ is finite and nontrivial unless $g = 0$ and $n = 1$. In the latter case, $X_{\bar{k}} \smallsetminus S \cong \mathbb{A}^1_{\bar{k}} = \mathrm{Spec}(\bar{k}[t])$ and there exist no cyclic unramified covers of degree prime to p. But in this case there exist (many) cyclic covers of degree p which have positive genus (see ex.1). Hence G_S has open subgroups which correspond to curves of positive genus. We conclude that $cd_\ell\, G_S(K\bar{k}) = 1$ for every $\ell \in (p')$ if $S \neq \varnothing$, which shows c). Furthermore, the assumptions of lemma (3.7.5) are fulfilled and therefore $G_S(K)$ is a duality group of dimension 2 at (p') in this case. Finally, from (8.4.3)(i) it follows that $N_{G_S}({}_mC_S) = 0$, thus by the duality theorem (8.4.4) we have a canonical isomorphism

$$ {}_mC_S(L) \cong H^2(G_S(L), \mathbb{Z}/m\mathbb{Z})^\vee $$

for $(m,p) = 1$ and for every finite extension L of K in K_S. Passing to the limit over all L and m, we deduce the statement about the dualizing module. The exact sequence for D_2 then follows from the exact sequence

$$ 0 \longrightarrow \mathcal{O}_S^\times \longrightarrow I_S \longrightarrow C_S \longrightarrow 0, $$

noting that $\mathcal{O}_S^\times$ is ℓ-divisible for all $\ell \in (p')$.

Now assume that $S = \varnothing$ and $g > 0$. In order to prove b), it is sufficient to show that for every $\ell \in (p')$ and every open subgroup $U \subseteq G_\varnothing(K\bar{k})$, the maximal pro-$\ell$-quotient $U(\ell)$ is an (infinite) Demuškin group. Since the assumptions carry over to every finite separable extension of K, we may restrict to the case $U = G_\varnothing(K\bar{k})$. The Kummer sequence implies that

$$ \begin{aligned} G_\varnothing(K\bar{k})^{ab\,(p')} &\cong H^1(G_\varnothing(K\bar{k}), \mathbb{Q}/\mathbb{Z}^{(p')})^\vee \\ &\cong H^1(G_\varnothing(K\bar{k}), \mu)^\vee(1) \\ &\cong (Cl^0(K\bar{k})^{(p')})^\vee(1) \end{aligned} $$

and by (10.1.1) we know the structure of $Cl^0(K\bar{k})$ as an abelian group. In particular, $G_\varnothing(K\bar{k})^{ab}(\ell)$ is nontrivial and torsion-free for every $\ell \in (p')$, and for every open subgroup $V \subseteq G_\varnothing(K\bar{k})(\ell)$, the Hurwitz formula implies the rank equality

$$ d(V) = (G_\varnothing(K\bar{k})(\ell) : V)(d(G_\varnothing(K\bar{k})(\ell)) - 2) + 2, $$

where $d(V) = \dim_{\mathbb{F}_\ell} H^1(V, \mathbb{F}_\ell)$ (cf. III §9). Therefore $G_\varnothing(K\bar{k})(\ell)$ is not free (see (4.2.2)).

Since $H^2(G_\varnothing(K\bar{k})(\ell), \mu_\ell)$ is a subgroup of $H^2(G_\varnothing(K\bar{k}), \mu_\ell)$ (use the Hochschild-Serre spectral sequence for the group extension $1 \to R \to G_\varnothing(K\bar{k}) \to G_\varnothing(K\bar{k})(\ell) \to 1$ and note that $H^1(R, \mu_\ell) = 0$), we conclude that

$$ \dim_{\mathbb{F}_\ell} H^2(G_\varnothing(K\bar{k}), \mu_\ell) \geq 1. $$

Now let T be a finite and nonempty set of places of K, and suppose that the primes in T are of degree 1, i.e. they do not decompose in $K\bar{k}|K$. (We may replace K by Kk_n for a suitable n.) If $H := G(K_T|K_\varnothing)$, we have an exact sequence

$$ (*) \qquad 1 \longrightarrow H \longrightarrow G_T(K\bar{k}) \longrightarrow G_\varnothing(K\bar{k}) \longrightarrow 1. $$

The inertia groups $T_\mathfrak{p}(K_T|K_\varnothing) = G_\mathfrak{p}(K_T|K_\varnothing)$ for $\mathfrak{p} \in T$ (fix any prolongation to K_T) generate H as a normal subgroup in $G_T(K\bar{k})$. Therefore the canonical homomorphism

$$\beta : H^1(H, \mathbb{Z}/\ell\mathbb{Z})^{G_\varnothing(K\bar{k})} \longrightarrow \bigoplus_{\mathfrak{p}\in T} H^1(K_\mathfrak{p}^{nr}, \mathbb{Z}/\ell\mathbb{Z})$$

is injective.

Denote the ℓ-dualizing module of $G_T(K)$ by I. It is also the ℓ-dualizing module for $G_T(Kk_n)$, $n \geq 1$, and by (3.7.5) I is also the ℓ-dualizing module for the group $G_T(K\bar{k})$.

Now consider the following commutative exact diagram

$$\begin{array}{ccccccc}
 & & 0 \to & H^1(G_\varnothing(K\bar{k}), \mu_\ell) & \to & H^1(G_T(K\bar{k}), \mu_\ell) & \to \\
 & & & \uparrow \alpha & & \| & \\
0 \to & H^0(G_T(K\bar{k}), \mu_\ell) & \to & H^0(G_T(K\bar{k}), {}_\ell I) & \to & H^1(G_T(K\bar{k}), \mu_\ell) & \to
\end{array}$$

$$\begin{array}{ccccc}
\to & H^0(G_\varnothing(K\bar{k}), H^1(H, \mu_\ell)) & \to & H^2(G_\varnothing(K\bar{k}), \mu_\ell) & \to 0 \\
 & \beta \downarrow & & \gamma \downarrow & \\
\longrightarrow & \bigoplus_{\mathfrak{p}\in T} H^1(K_\mathfrak{p}^{nr}, \mu_\ell) & \longrightarrow & H^1(G_T(K\bar{k}), {}_\ell I) & \to 0.
\end{array}$$

The upper sequence is the Hochschild-Serre spectral sequence for the group extension $(*)$ and the (trivial) module μ_ℓ. By c) we know that $cd_\ell\, H \leq cd_\ell\, G_T(K\bar{k}) = 1$. The lower exact sequence is obtained by passing to the limit over n and over the Poitou-Tate sequences for $G_T(Kk_n)$ and the module μ_ℓ, where we use in addition the duality group property of $G_T(Kk_n)$.

Since β is injective, α exists and is surjective, and diagram chasing shows that γ is injective. But $H^1(G_T(K\bar{k}), {}_\ell I) \cong \mathbb{Z}/\ell\mathbb{Z}$, hence γ is an isomorphism, because we already know that $H^2(G_\varnothing(K\bar{k}), \mu_\ell)$ is nontrivial. (Thus β is also an isomorphism.)

We conclude that $G_\varnothing(K\bar{k})(\ell)$ is a one-relator pro-ℓ-group and the generator ranks of open subgroups can be calculated by the Hurwitz genus formula. This implies already that $G_\varnothing(K\bar{k})(\ell)$ is a Demuškin group by (3.9.15). However, we will show directly that the pairing

$$H^1(G_\varnothing(K\bar{k}), \mathbb{Z}/\ell\mathbb{Z}) \times H^1(G_\varnothing(K\bar{k}), \mu_\ell) \xrightarrow{\cup} H^2(G_\varnothing(K\bar{k}), \mu_\ell) \cong \mathbb{Z}/\ell\mathbb{Z}$$

is non-degenerate. Since the module μ_ℓ has trivial $G_\varnothing(K\bar{k})$-action and since the associated cohomology groups of $G_\varnothing(K\bar{k})(\ell)$ are canonically the same for μ_ℓ and $\mathbb{Z}/\ell\mathbb{Z}$ (which is trivial for H^1 and was seen above for H^2), it then follows from (3.7.6) that $G_\varnothing(K\bar{k})(\ell)$ is a Demuškin group, hence showing assertion (i) b) of (10.1.2).

It remains to show that the cup-product pairing is non-degenerate. Consider the commutative diagram

$$\begin{array}{ccccc} H^1(G_\varnothing(K\bar{k}),\mathbb{Z}/\ell\mathbb{Z}) & \times & H^1(G_\varnothing(K\bar{k}),\mu_\ell) & \xrightarrow{\;\cup\;} & H^2(G_\varnothing(K\bar{k}),\mu_\ell) \\ {\scriptstyle inf}\downarrow & & \uparrow{\scriptstyle \alpha} & & {\scriptstyle \gamma}\downarrow\wr \\ H^1(G_T(K\bar{k}),\mathbb{Z}/\ell\mathbb{Z}) & \times & H^0(G_T(K\bar{k}),{}_\ell I) & \xrightarrow{\;\cup\;} & H^1(G_T(K\bar{k}),{}_\ell I). \end{array}$$

Since the lower pairing is non-degenerate, it follows that the upper pairing is non-degenerate from the left. But μ_ℓ is a trivial $G_\varnothing(K\bar{k})$-module and $H^1(G_\varnothing(K\bar{k}),\mu_\ell) = H^1(G_\varnothing(K\bar{k}),\mathbb{Z}/\ell\mathbb{Z})(1)$ is finite. Thus the upper pairing is a perfect pairing of finite dimensional $\mathbb{F}_\ell$-vector spaces. As explained above, this shows b), and statement (ii) of the corollary follows from (3.7.4) and from the Serre criterion (3.4.5).

Finally, observe that $G_\varnothing(K\bar{k})$ is either trivial (if $g = 0$) or a Demuškin group of rank $2g$ with torsion-free abelianization. Therefore the remaining statement (ii) of the theorem follows from theorem (3.9.11) and from the group theoretical lemma (3.9.20) applied to the exact sequence $(*)$ above. □

Remark: Using étale cohomology, a straightforward and natural way to prove the above results about G_S in the function field case is the following: if S is empty and $g = 0$, then $G_S(K\bar{k}) = 0$ since there are no connected étale covers of the projective line. In all other cases, show that for all $i > 0$ and every prime number $\ell \neq p = \mathrm{char}(K)$

$$\varinjlim H^i_{et}(Y,\mathbb{Z}/\ell\mathbb{Z}) = 0,$$

where Y runs through the connected étale covers of $X \smallsetminus S$. Conclude that

$$H^i(G_S(K\bar{k}),\mathbb{Z}/\ell\mathbb{Z}) \cong H^i_{et}(X_{\bar{k}} \smallsetminus S,\mathbb{Z}/\ell\mathbb{Z})$$

for all $i \geq 0$ and all $\ell \neq p$. Then extend this isomorphism to arbitrary prime-to-p torsion G_S-modules resp. locally constant sheaves on $X_{\bar{k}} \smallsetminus S$. Finally, calculate the étale cohomology groups on the right and apply the étale Poincaré duality theorem.

(10.1.4) Corollary. *Let S be a nonempty (possibly infinite) set of places of K and let $\mathfrak{c}$ be a full class of finite groups. Then, for every discrete torsion $G_S(K)(\mathfrak{c})$-module A and every $i \geq 0$, the inflation map*

$$H^i(G_S(K)(\mathfrak{c}),A) \longrightarrow H^i(G_S(K),A)(\mathfrak{c})$$

is an isomorphism.

Proof: We may assume that A is ℓ-torsion, where ℓ is a prime number such that $\mathbb{Z}/\ell\mathbb{Z} \in \mathfrak{c}$. By (10.1.2)(c) (if $\ell \neq p$) and by (8.3.3) (if $\ell = p$), we know that $cd_\ell\, G(K_S|K_S(\mathfrak{c})\bar{k}) \leq 1$. As the index of $G(K_S|K_S(\mathfrak{c})\bar{k})$ in $G(K_S|K_S(\mathfrak{c}))$ is prime to ℓ, we obtain $cd_\ell\, G(K_S|K_S(\mathfrak{c})) \leq 1$. Noting that $H^1(G(K_S|K_S(\mathfrak{c})),A) = 0$, the Hochschild-Serre spectral sequence

$$H^i(G_S(K)(\mathfrak{c}),H^j(G(K_S|K_S(\mathfrak{c})),A)) \Rightarrow H^{i+j}(G_S(K),A)$$

implies the result. □

Unfortunately, using the methods above, we can only determine the structure of pro-ℓ-factor groups of G_S. In fact, it seems that algebraic methods are not very effective in determining the structure (in terms of generators and relations) of profinite groups. One can show that a profinite group of countable rank is free by solving embedding problems (see (3.5.20)). Sometimes global Galois groups can be shown to be the free product of local Galois groups. And, as the most far reaching result in this direction, one can determine the structure of the absolute Galois group $G_{K_\mathfrak{p}}$ of a local field $K_\mathfrak{p}$ (cf. VII §5). This is rather deep, and the proof extensively exploits the fact that $G_{K_\mathfrak{p}}$ contains a normal subgroup, the ramification group, which is a pro-p-group. In other words $G_{K_\mathfrak{p}}$ is "not too profinite", which makes it accessible to algebraic methods.

But there seems to be no idea how one should determine the structure of G_S using algebraic methods. In the function field case, however, one can obtain very deep results by exploiting the connections with topology. This is possible by using the *specialization map*, defined by A. GROTHENDIECK. With this method, one can show that the quotient $G_S^{tame}(K\bar{k})$ (see below) of $G_S(K\bar{k})$ is topologically finitely generated and one can determine the maximal prime-to-p-factor group of $G_S(K\bar{k})$ by relating it to the (well-known) topological fundamental group of a Riemann surface. We will briefly explain this beautiful and strong method below; however, to number theorists' sorrow, there seems to be no way to exploit similar techniques in the number field case.

In order to explain Grothendieck's approach, let us change the notation for a moment. Assume that k is an algebraically closed field of char(k) = $p > 0$. Assume that X is a smooth, proper curve of genus g over k and let $\{P_1, \ldots, P_n\}$ be a finite (possibly empty if $n = 0$) set of points in $X(k)$. The points in $X(k)$ correspond to primes (valuations) of the function field $k(X)$, and the étale fundamental group $\pi_1^{et}(X \smallsetminus \{P_1, \ldots, P_n\})$ (we omit the base point) is isomorphic to the Galois group of the maximal extension of $k(X)$ unramified outside the primes which are associated to $P_1, \ldots, P_n$. This group is not finitely generated for $n \geq 1$, because there are many covers of $X \smallsetminus \{P_1, \ldots, P_n\}$ with wild ramification along $P_1, \ldots, P_n$.

Let us consider the **tame fundamental group** $\pi_1^{tame}(X \smallsetminus \{P_1, \ldots, P_n\})$, which classifies étale covers with at most tame ramification along the divisor $P_1 + \cdots + P_n$. It is isomorphic to the quotient of $\pi_1^{et}(X \smallsetminus \{P_1, \ldots, P_n\})$ by the normal subgroup generated by the ramification groups of $P_1, \ldots, P_n$. Galois covers of prime-to-p degree are tamely ramified, so we have an isomorphism

$$\pi_1^{et}(X \smallsetminus \{P_1, \ldots, P_n\})^{(p')} \cong \pi_1^{tame}(X \smallsetminus \{P_1, \ldots, P_n\})^{(p')}$$

of the maximal pro-(p')-factor groups.

Now let A be a complete discrete valuation ring with residue field k and quotient field K of characteristic zero (e.g., the ring of Witt vectors over k). Then one can show that there exists a "lift to characteristic 0" of $(X, \{P_1, \dots, P_n\})$, i.e. a connected, smooth and proper scheme $\mathfrak{X}$ defined over Spec(A), and sections $s_i : \mathrm{Spec}(A) \to \mathfrak{X}$, $i = 1, \dots, n$, such that $X = X_k$ can be identified with the special fibre $\mathfrak{X}_k$ of $\mathfrak{X}$ in such a way that the sections $s_1, \dots, s_n$ specialize to the points $P_1, \dots, P_n$. It may be seen as a consequence of Hensel's lemma that (omitting the suitably chosen base points) the canonical homomorphism

$$\varphi : \pi_1^{tame}(X_k \smallsetminus \{P_1, \dots, P_n\}) \longrightarrow \pi_1^{tame}(\mathfrak{X} \smallsetminus \{s_1, \dots, s_n\}),$$

which is induced by the inclusion of the special fibre, is an isomorphism.

Let X_K be the generic fibre of $\mathfrak{X}$ and denote the specializations of $s_1, \dots, s_n$ to X_K by $P_1, \dots, P_n$ (no confusion should arise by using the same letters as for the points in X_k). Let $\overline{K}$ be an algebraic closure of K. Then the inclusion of the generic fibre composed with the inverse of φ induces a homomorphism of profinite groups

$$sp : \pi_1^{et}(X_{\overline{K}} \smallsetminus \{P_1, \dots, P_n\}) \longrightarrow \pi_1^{tame}(X_k \smallsetminus \{P_1, \dots, P_n\}),$$

which is called the **specialization map**. For this map we have the

(10.1.5) Theorem (Grothendieck). *The specialization map sp is surjective. It defines an isomorphism*

$$sp^{(p')} : \pi_1^{et}(X_{\overline{K}} \smallsetminus \{P_1, \dots, P_n\})^{(p')} \xrightarrow{\sim} \pi_1^{et}(X_k \smallsetminus \{P_1, \dots, P_n\})^{(p')}$$

on the maximal pro-(p')-factor groups.

For a proof of this theorem and of the facts mentioned before we refer the reader to [67], exp. XIII.

Having "lifted" our problem, there remains the easier problem of determining the structure of the algebraic fundamental group of a smooth curve over an algebraically closed field K of characteristic 0. By a general principle we may assume that $K = \mathbb{C}$, so that $Y = X_K \smallsetminus \{P_1, \dots, P_n\}$ is an algebraic Riemann surface. Every topological covering of $Y(\mathbb{C})$ can be uniquely endowed with a holomorphic structure and, by the famous **Riemann existence theorem**, these coverings are in fact algebraic curves. Since topological coverings are classified by the subgroups of the topological (i.e. path-) fundamental group, this implies that there is a canonical isomorphism

$$co : \pi_1^{top}(Y(\mathbb{C}))^\wedge \cong \pi_1^{et}(Y)$$

from the profinite completion of the topological fundamental group to the algebraic fundamental group of Y.

Now the structure of the topological fundamental group is well-known. Denoting the genus of X by g, $\pi_1^{top}(X \smallsetminus \{P_1, \dots, P_n\})$ is the group

$$\Pi_{g,n} = \langle x_1, \dots, x_{2g}, y_1, \dots, y_n \,|\, (x_1, x_2) \cdots (x_{2g-1}, x_{2g})\, y_1 \cdots y_n = 1 \rangle,$$

where $x_1, \dots, x_{2g}$ generate the fundamental group of the complete curve X and $y_1, \dots, y_n$ are loops around the removed points $P_1, \dots, P_n \in X$. In particular, co maps the loop y_i to a generator of the procyclic inertia group of the prime associated to P_i *) for $i = 1, \dots, n$.

Returning to our original notation, we deduce the following theorem, which generalizes the statement of (10.1.2) (ii) from the maximal pro-ℓ-quotients ($\ell \in (p')$) to the full maximal pro-(p')-factor group. Let $G_S^{tame}(K)$ denote the Galois group of the maximal extension of K in K_S with at most tame ramification at the primes in S.

(10.1.6) Theorem. *The group $G_S^{tame}(K\bar{k})$ is topologically finitely generated. The group $G_S(K\bar{k})^{(p')}$ has as a pro-(p')-group a presentation by $2g{+}n$ generators and one relation of the form*

$$\{x_1, \dots, x_{2g}, y_1, \dots, y_n \,|\, (x_1, x_2) \cdots (x_{2g-1}, x_{2g})\, y_1 \cdots y_n = 1\}.$$

In particular, if $S \neq \varnothing$, then $G_S(K\bar{k})^{(p')}$ is a free pro-(p')-group of rank $2g + n - 1$. Furthermore, the elements $y_1, \dots, y_n$ may be chosen as generators of the procyclic inertia groups of prolongations of primes in $S(K\bar{k})$ to $K_S^{(p')}$.

(10.1.7) Corollary. *The group $G_S^{tame}(K)$ is topologically finitely generated.*

In the topological situation, consider the (topological) universal cover $\tilde{X}$ of $X(\mathbb{C})$. The set of points lying over $P_1, \dots, P_n$ is a *discrete* subset in $\tilde{X}$. Therefore the normal subgroup in $\Pi_{g,n}$ generated by $y_1, \dots, y_n$ is the free (discrete) group over the set of loops around the pre-images of $P_1, \dots, P_n$ in $\tilde{X}$. In order to formulate an algebraic version of this result, we have to make some preparations.

Let K be a global (number or function) field and let $M|K$ be a (possibly infinite) Galois extension of K. We denote the one-point compactification of the discrete set of places of K by Sp(K). The compactifying point, which will be denoted by η_K, should be thought as the generic point of Sp(K) in the sense of algebraic geometry or as the trivial valuation from the valuative point of view. We set

$$\mathrm{Sp}(M) := \varprojlim_{L \subseteq M} \mathrm{Sp}(L),$$

*) In fact the inertia group is associated to a prolongation of P_i which is determined by the choice of base points that we omitted from our notation.

where the limit is taken over all finite extensions L of K inside M. In particular, Sp(M) is a profinite space.

As is the case for Sp(K), the set Sp(M) consists of the places of M plus one generic point η_M. Topologically, however, Sp(M) might not be the one-point compactification of Sp$(M) \smallsetminus \{\eta_M\}$. Its topology reflects the fact that M is an inductive limit of global fields.

One can also give an intrinsic definition of the topology of Sp(M). In order to do this, we remind the reader of the definition of the **constructible topology** (see [70], chap.I, §7, (7.2.11)). For a (commutative) ring A, the subsets $f^*(\mathrm{Spec}(B)) \subset \mathrm{Spec}(A)$, where $f : A \to B$ is a ring homomorphism, satisfy the axioms for closed sets in a topological space. The associated topology is the constructible topology. Spec(A) with the constructible topology is a Hausdorff, compact and totally disconnected topological space, i.e. a profinite space (see I §1). This definition extends to arbitrary schemes by gluing.

In the function field case, consider the integral closure X_M of the curve X_K in M. This one-dimensional scheme can be constructed from the field M in the same way as X_K from K; in particular, it depends only on M but not on K. Then one verifies that there is a canonical isomorphism (of sets)

$$\mathrm{Sp}(M) \cong X_M\,,$$

which becomes a homeomorphism if we endow the (not necessarily noetherian) scheme X_M with the constructible topology.

If we exclude the archimedean primes, a similar statement is true in the number field case:

$$\mathrm{Sp}(M) \smallsetminus S_\infty(M) \cong \mathrm{Spec}(\mathcal{O}_M)_{\text{constr.top.}}\,.$$

If S is a finite set of primes in K, then the set $S(M) \subseteq \mathrm{Sp}(M)$ of places in M lying over S is closed in the constructible topology. If S consists of a single prime $\mathfrak{p} \in \mathrm{Sp}(K)$, then the set $S(M)$ is topologically isomorphic to the compact set of coset classes $G(M|K)/G_{\mathfrak{P}}(M|K)$, where $G_{\mathfrak{P}}(M|K)$ is the decomposition group in $M|K$ of an arbitrary chosen prolongation $\mathfrak{P}$ of $\mathfrak{p}$ to M.

Assume that we are given a closed subset $S \subseteq \mathrm{Sp}(M)$. Then the set of inertia groups $(\,T_{\mathfrak{P}}(M|K)\,)_{\mathfrak{P}\in S}$ is a continuous family of subgroups*) of $G(M|K)$. This follows from the fact that only finitely many primes ramify in a finite separable extension of global fields.

(10.1.8) Definition. *The free product over the bundle of profinite groups associated to the continuous family* $(\,T_{\mathfrak{P}}(M|K)\,)_{\mathfrak{P}\in S}$ *by* (4.3.3), *is called the* **free product of the inertia groups** *in* S.

---

*) See IV §3. We set $T_\eta(M|K) = \{1\}$ here, which is justified because separable extensions of K are unramified at η in the sense of algebraic geometry.

There exists a canonical continuous homomorphism

$$\mathop{\ast}_{\mathfrak{P}\in S} T_{\mathfrak{P}}(M|K) \longrightarrow G(M|K).$$

If $\eta_M \notin S$, then the restriction of S to every finite extension of K inside M is finite. In this case the family $(\, G_{\mathfrak{P}}(M|K)\,)_{\mathfrak{P}\in S}$ of decomposition groups is also continuous and we can form the free product of decomposition groups in the same way.

Suppose that we are given an intermediate field $L \subseteq M$ and a closed subset $S \subseteq \mathrm{Sp}(L)$.

(10.1.9) Definition. *We say that $G(M|L)$ is the free product of the inertia groups of the primes in S if there exists a continuous section $s : S \to \mathrm{Sp}(M)$ over S to the projection $\pi : \mathrm{Sp}(M) \to \mathrm{Sp}(L)$ such that the canonical homomorphism*

$$\mathop{\ast}_{\mathfrak{P}\in s(S)} T_{\mathfrak{P}}(M|L) \longrightarrow G(M|L)$$

is an isomorphism. In this case, we write (by abuse of notation)

$$\mathop{\ast}_{\mathfrak{p}\in S} T_{\mathfrak{p}}(M|L) \cong G(M|L)\,,$$

and we will omit the generic point, if it is contained in S, from the notation.

Remark: There always exist (many) sections $s : S \to \mathrm{Sp}(M)$; see [74], lemma 8.1 or [135], lemma 4.7, and also ex.5 below. However, it is clear from the discussion in chapter IV §2 that we cannot expect the homomorphism

$$\mathop{\ast}_{\mathfrak{P}\in s(S)} T_{\mathfrak{P}}(M|L) \to G(M|L)$$

to be an isomorphism for every choice of a section $s : S \to \mathrm{Sp}(M)$, unless we pass to the maximal pro-p-factor groups.

We denote the inertia group of $G(\bar{L}_{\mathfrak{p}}|L_{\mathfrak{p}})$ by $\mathcal{T}_{\mathfrak{p}}$. If, in the above situation, the canonical surjection $\mathcal{T}_{\mathfrak{p}} \twoheadrightarrow T_{\mathfrak{p}}(M|L)$ is an isomorphism for every $\mathfrak{p} \in S$, then we write

$$\mathop{\ast}_{\mathfrak{p}\in S} \mathcal{T}_{\mathfrak{p}} \cong G(M|L)$$

and we say that $G(M|L)$ is the **free product of full local inertia groups**.

If $M|K$ is a $\mathfrak{c}$-extension for a full class of finite groups $\mathfrak{c}$, then the above concepts generalize to pro-$\mathfrak{c}$-products in a straightforward manner.

Now we are in position to formulate the algebraic analogue of Riemann's existence theorem. For finite sets S it follows from the topological results above, and we can pass to infinite S by (4.3.11).

(10.1.10) Riemann's Existence Theorem (Algebraic Form). *Assume that K is a global function field of* char$(K) = p > 0$ *with finite constant field k. Let S be any nonempty set of places of K and assume that $g(K) > 0$. Then there exists a canonical exact sequence of pro-(p')-groups*

$$1 \to \mathop{\ast}_{\mathfrak{p}\in S(K_\varnothing(p'))} \mathcal{T}_\mathfrak{p}^{(p')} \to G_S(K\bar{k})^{(p')} \to G_\varnothing(K\bar{k})^{(p')} \to 1,$$

where $\ast$ denotes the free pro-(p')-product and $\mathcal{T}_\mathfrak{p}^{(p')} \cong \hat{\mathbb{Z}}(1)^{(p')}$ is the full prime-to-p part of the local inertia group at $\mathfrak{p}$. If $g = 0$ (so that $G_\varnothing(K\bar{k}) = \{1\}$), we obtain an isomorphism

$$\mathop{\ast}_{\substack{\mathfrak{p}\in S(K\bar{k})\\ \mathfrak{p}\neq\mathfrak{p}_0}} \mathcal{T}_\mathfrak{p}^{(p')} \xrightarrow{\sim} G_S(K\bar{k})^{(p')},$$

where $\mathfrak{p}_0$ is an arbitrarily chosen prime in $S(K\bar{k})$.

Remark: It is not difficult to deduce a pro-ℓ version of Riemann's existence theorem in a purely algebraic way using similar arguments to the proof of theorem (10.1.2).

This section would be incomplete without mentioning **Frobenius weights**. Choose an $\ell \in (p')$ and consider (if $g > 0$) the $\mathbb{Q}_\ell$-vector spaces

$$\mathbb{H}_i := H_i(G_\varnothing(K\bar{k}), \mathbb{Z}_\ell) \otimes_{\mathbb{Z}_\ell} \mathbb{Q}_\ell$$

for $i = 0, 1, 2$. Trivially $\mathbb{H}_0 = \mathbb{Q}_\ell$ and by (10.1.2) we have $\mathbb{H}_2 \cong \mathbb{Q}_\ell(1)$ and

$$\mathbb{H}_1 \cong (\varprojlim_n H^1(G_\varnothing(K\bar{k}), \mu_{\ell^n})) \otimes_{\mathbb{Z}_\ell} \mathbb{Q}_\ell \cong T_\ell(\mathrm{Jac}X) \otimes_{\mathbb{Z}_\ell} \mathbb{Q}_\ell,$$

where $T_\ell(\mathrm{Jac}X)$ is the ℓ-adic **Tate module** of the Jacobian variety associated to X.

The general theory of abelian varieties over finite fields (see [140]) shows that the characteristic polynomial of the Frobenius automorphism acting on $T_\ell(\mathrm{Jac}X)$ is contained in $\mathbb{Z}[T]$ (rather than in $\mathbb{Z}_\ell[T]$) and is independent of $\ell \in (p')$. The Frobenius eigenvalues are therefore algebraic integers and (loc.cit.) they have absolute value $q^{1/2}$ in every complex embedding, where $q = p^f = \#k$. We conclude that the Frobenius eigenvalues on $\mathbb{H}_i$ have absolute value $q^{i/2}$ for $i = 0, 1, 2$. This result can be reformulated into a statement about zeros and poles of the zeta function $Z(s, X_k)$ of X_k (cf. XI §6). In particular, $\mathrm{Re}(s) = 1/2$ for every zero of $Z(s, X_k)$. This result is therefore called the 1-dimensional Riemann hypothesis in positive characteristic.

Now consider the exact sequence

$$0 \to H :\to H_1(G_S(K\bar{k}), \mathbb{Z}_\ell) \otimes_{\mathbb{Z}_\ell} \mathbb{Q}_\ell \to \mathbb{H}_1 \to 0.$$

The kernel H is generated by the images of $H_1(\mathcal{T}_\mathfrak{p}, \mathbb{Z}_\ell) \cong \mathbb{Z}_\ell(1)$ for the places $\mathfrak{p} \in S$. Therefore the above exact sequence provides a filtration (*weight filtration*) of $H_1(G_S(K\bar{k}), \mathbb{Z}_\ell) \otimes_{\mathbb{Z}_\ell} \mathbb{Q}_\ell$ with respect to the absolute values (the weights) of the Frobenius eigenvalues.

The generalization of the above results to varieties of higher dimensions, the "Weil conjecture", proved by *P. Deligne* [35], [36], is one of the major achievements of modern algebraic geometry. The use of étale cohomology is essential here and in fact étale cohomology was introduced with this application in mind.

We finish this section with the treatment of the cohomological properties of G_S with respect to p = char(K). The following theorem is a somewhat weakened formulation of a result of *M. Raynaud* (see [176]). Recall the definition of the p-rank $h = h(X)$ from the remark after (10.1.1).

(10.1.11) Theorem. *Assume that X is a smooth, projective curve of genus greater than or equal to 2 over an algebraically closed field of characteristic $p > 0$. Then there exist cyclic covers of degree prime to p with arbitrary large p-rank h.*

For a proof see [176], théorème 4.3.1.

(10.1.12) Theorem. *If K is a global function field of* char(K) $= p > 0$, *then $scd_p\, G_S = 2$. Furthermore, the following hold:*

(i) *If $g = 0$, then $G_\varnothing(K) \cong G(\bar{k}|k) \cong \hat{\mathbb{Z}}$.*

(ii) *If $g = 1$, then $G_\varnothing(K\bar{k})$ is abelian and we distinguish the following two cases:*
   - (a) *If $h = 0$, then $G_\varnothing(K\bar{k})(p) = 0$; in particular, $cd_p\, G_\varnothing(K) = 1$.*
   - (b) *If $h = 1$, then $G_\varnothing(K\bar{k})(p) \cong \mathbb{Z}_p$ and the group $G_\varnothing(K)$ is a Poincaré group at p of dimension 2.*

(iii) *If $g \geq 2$, then $cd_p\, G_\varnothing(K\bar{k}) = 1$. The group $G_\varnothing(K)$ is a duality group at p of dimension 2 with dualizing module $I = Cl(K_\varnothing)(p)$.*

(iv) *If $S \neq \varnothing$, then*

$$cd_p\, G_S(K\bar{k}) = cd_p\, G_S(K) = 1.$$

Proof: If S is nonempty, then $scd_p\, G_S = 2$ by (8.3.17). We will slightly modify the proof of (8.3.17) for the case $S = \varnothing$. As in chapter VIII, we use the notation

$$U_K := \prod_{\mathfrak{p}} U_{\mathfrak{p}},$$

where $\mathfrak{p}$ runs through all primes of K and $U_{\mathfrak{p}}$ is the unit group of the completion of K at $\mathfrak{p}$. Then global class field theory gives the exact sequence

$$0 \to k^\times \to U_K \to C(K) \to Cl(K) \to 0.$$

Since finite fields are perfect, $k^\times$ is uniquely p-divisible. Further, recall that U_L is a cohomologically trivial $G(L|K)$-module if $L|K$ is unramified. Thus we obtain

$$\begin{array}{rcl} Cl(L)^{G(L|K)} & = & Cl(K), \\ H^i(G(L|K), C(L)) & \cong & H^i(G(L|K), Cl(L)) \end{array}$$

for $i \ge 1$ and every unramified p-extension $L|K$. This shows that the pair $(G(K_\varnothing(p)|K), Cl(K_\varnothing(p))$ is a class formation. Furthermore, we have an isomorphism

$$G_\varnothing(p)^{ab} \cong Cl(K) \otimes_{\mathbb{Z}} \mathbb{Z}_p$$

by global class field theory. Therefore we obtain that $scd_p\, G_\varnothing(K)(p) \le 2$ by (3.6.4). The same argument applies to every finite extension of K inside $K_\varnothing$. By a limit process, we conclude that a (every) p-Sylow subgroup of $G_\varnothing(K)$ has $scd_p \le 2$, which shows the assertion.

Statement (i) is a repetition of (10.1.3)(i). Now assume that $g = 1$. Fix any point O in $X(\bar{k})$ in order to make it into an elliptic curve $E = (X_{\bar{k}}, O)$. Then it is well-known (see [140]) that every étale covering of E is an isogeny. Hence $\pi_1(E)$ is abelian and (loc.cit.) isomorphic to

$$\varprojlim_n {}_nCl(X).$$

Hence $G_\varnothing(K\bar{k})(p) \cong 0$ or $G_\varnothing(K\bar{k})(p) \cong \mathbb{Z}_p$ depending on the p-rank h. The remaining statement in (b) then follows from (3.7.4).

In order to prove assertion (iii), we set $G = G_\varnothing(K)$. We investigate the terms

$$D_i(\mathbb{Z}) = \varinjlim H^i(U, \mathbb{Z})^*,$$

where U runs through the open subgroups in G, cf. III §4. We will show that $D_0(\mathbb{Z}) = 0 = D_1(\mathbb{Z})$ and that $D_2(\mathbb{Z})$ is p-divisible. By (3.4.8) this shows that G is a duality group of dimension 2 at p with p-dualizing module $D_2(\mathbb{Z})(p)$.

First we observe that $D_0(\mathbb{Z}) = 0$, since $K\bar{k} \subseteq K_\varnothing$ and hence every prime number divides $\#G$ infinitely often. Further, $D_1(\mathbb{Z}) = 0$ for trivial reasons.

Finally, we show that $D_2(\mathbb{Z})$ is p-divisible. Recalling that $H^2(U,\mathbb{Z}) \cong (U^{ab})^*$, global class field theory induces the exact sequence

$$0 \to Cl(K_U) \to H^2(U,\mathbb{Z})^* \to \hat{\mathbb{Z}}/\mathbb{Z} \to 0,$$

where K_U denotes the subextension in $K_\varnothing$ which is associated to U. Passing to the limit over all open subgroups $U \subseteq G$, we obtain

$$D_2(\mathbb{Z})(p) \cong Cl(K_\varnothing)(p) \text{ and } D_2(\mathbb{Z})/p \cong Cl(K_\varnothing)/p.$$

From the observations at the beginning of the proof, we know that

$$Cl(K_U)(p) = Cl(K_\varnothing)(p)^U$$

for every open subgroup $U \subseteq G$. By (10.1.1), we have

$$Cl(K_\varnothing)/p \cong \big(Cl(K_\varnothing)/Cl^0(K_\varnothing)\big)/p.$$

Now let L be a finite extension of K in $K_\varnothing$. Then we have the commutative diagram

$$\begin{array}{ccccccccc} 0 & \longrightarrow & Cl^0(L\bar{k}) & \longrightarrow & Cl(L\bar{k}) & \longrightarrow & \mathbb{Z} & \longrightarrow & 0 \\ & & \uparrow & & \uparrow & & \uparrow{\scriptstyle [L\bar{k}:K\bar{k}]} & & \\ 0 & \longrightarrow & Cl^0(K\bar{k}) & \longrightarrow & Cl(K\bar{k}) & \longrightarrow & \mathbb{Z} & \longrightarrow & 0. \end{array}$$

By (10.1.11), we know that the order of $G_\varnothing(K\bar{k})$ is divisible by p^∞. Hence $D_2(\mathbb{Z})/p = 0$. It remains to show $cd_p\, G_\varnothing(K\bar{k}) \le 1$. We have

$$\begin{aligned} H^2(G_\varnothing(K\bar{k}),\mathbb{Z}/p\mathbb{Z}) &= \varinjlim_n H^2(G_\varnothing(Kk_n),\mathbb{Z}/p\mathbb{Z}) \\ &= (\varprojlim_n {}_pG_\varnothing(K\bar{k})^{ab})^* \\ &= (\varprojlim_n {}_pCl(Kk_n))^*. \end{aligned}$$

But ${}_pCl(Kk_n) = {}_pCl(Kk_m) = {}_pCl(K\bar{k})$ for $m \ge n \gg 0$, so that the last projective limit vanishes (observe that the transition maps are induced by the norm). Since the same argument applies to every finite separable extension of K, this shows (iii).

Now assume $S \neq \varnothing$. Then $H^i(G_S,\mathbb{Z}/p\mathbb{Z}) = 0$ for $i \ge 2$ by (8.3.2) and since the same is true for every open subgroup, we see that

$$cd_p\, G_S(K\bar{k}) \le cd_p\, G_S(K) \le 1.$$

It remains to show that $G_S(K\bar{k})$ has a nontrivial p-Sylow group. This may be achieved by a direct computation of

$$H^1(G_S(K\bar{k}),\mathbb{Z}/p\mathbb{Z}) = \mathcal{O}_S(K\bar{k})/\wp,$$

where $\wp$ is the map $x \mapsto x^p - x$. But we will give a slightly different argument. By the result of ex.1 below, we know that there are many étale covers of degree p of the affine line $\mathbb{A}^1_{\bar{k}}$. Now choose any point $\mathfrak{p}_0 \in S(K\bar{k})$ and construct a morphism $f : X_{\bar k} \to \mathbb{P}^1_{\bar k}$ such that $f^{-1}(\infty) = \mathfrak{p}_0$ (such an f exists by the Riemann-Roch Theorem). Then we obtain many unramified covers of $X_{\bar k} \smallsetminus S$ by base change from $\mathbb{A}^1_{\bar k}$. □

Exercise 1. Let k be an algebraically closed field of characteristic $p > 0$. Consider the coverings Y_m of $\mathbb{A}^1_k = \operatorname{Spec}(k[X])$ which are given by the Artin-Schreier equations $Y^p - Y = X^m$ for $m \in \mathbb{N}$. Show that these coverings are cyclic and unramified (i.e. they are ramified only over the infinite point). Assume that $m = m'p^e$ with $(m', p) = 1$. Show that

$$g(Y_m) = \frac{1}{2}(m' - 1)(p - 1).$$

Exercise 2. Let K be a global field of $\operatorname{char}(K) = p > 0$ and let $\ell \neq p$ be a prime number. Show that

$$G_\varnothing(K)(\ell) \cong \begin{cases} \text{a Poincaré group of dimension 3} & \text{if } Cl(K)(\ell) \neq 0 \text{ and } \mu_\ell \subset K, \\ \text{a duality group of dimension 2 with} & \\ \quad\text{dualizing module } Cl(K_\varnothing(\ell))(\ell) & \text{if } Cl(K)(\ell) \neq 0 \text{ and } \mu_\ell \not\subset K, \\ \mathbb{Z}_\ell & \text{if } Cl(K)(\ell) = 0. \end{cases}$$

Exercise 3. Let K be a global field of $\operatorname{char}(K) = p > 0$. Show that

$$G_\varnothing(K)(p) \cong \begin{cases} \text{a duality group of dimension 2 with} & \\ \quad\text{dualizing module } Cl(K_\varnothing(p))(p) & \text{if } Cl(K)(p) \neq 0, \\ \mathbb{Z}_p & \text{if } Cl(K)(p) = 0. \end{cases}$$

Exercise 4. Let K be a global field of $\operatorname{char}(K) = p > 0$ and let S be a set of places of K. Show that the maximal pro-p-factor group of $G_S(K)$ is finitely generated if and only if $S = \varnothing$.

Exercise 5. Let K be a global field and $M|K$ be a Galois extension. Let $L \subseteq M$ be an intermediate field which is Galois over K. Show that there exists a continuous section

$$s : \operatorname{Sp}(L) \to \operatorname{Sp}(M)$$

to the canonical projection $\pi : \operatorname{Sp}(M) \to \operatorname{Sp}(L)$.

Hint: First construct the section over the sets $S_\mathfrak{p}(L)$ of primes in L which lie over a fixed prime $\mathfrak{p}$ of K. Observe that, fixing a prolongation $\mathfrak{P}$ of $\mathfrak{p}$ to M, there are isomorphisms

$$\begin{aligned} S_\mathfrak{p}(M) &\cong G(M|K)/G_\mathfrak{P}(M|K), \\ S_\mathfrak{p}(L) &\cong G(L|K)/G_{\mathfrak{P}\cap L}(L|K) \cong G(M|K)/G(M|L)G_\mathfrak{P}(M|K). \end{aligned}$$

Then use the result of ex.4 in I §1.

Exercise 6. Let K be a global field of $\operatorname{char}(K) = p > 0$. Let $T \subseteq S$ be sets of primes of K where T is finite and $S \smallsetminus T \neq \varnothing$, and let $K_S^T(p)$ be the maximal p-extension of K inside K_S in which all primes of T are completely decomposed. Show that the canonical homomorphism

$$\mathop{*}_{\mathfrak{P} \in T(K_S^T(p))} G(K_\mathfrak{P}(p)|K_\mathfrak{P}) \longrightarrow G(K_S(p)|K_S^T(p))$$

is an isomorphism.

Hint: Both groups are free pro-p-groups. Use (9.2.5) and (1.6.15).

§2. First Observations on the Number Field Case

In this section we start our investigations into the number field case. Let us fix some notation. Suppose we are given

K	a number field,
S	a set of places of K,
$\mathfrak{c}$	a full class of finite groups,
$K_S(\mathfrak{c})$	the maximal $\mathfrak{c}$-extension of K which is unramified outside the primes in S,
$G_S(K)(\mathfrak{c})$	$= G(K_S(\mathfrak{c})\vert K)$.

We will omit $\mathfrak{c}$ if it is the class of all finite groups and we will write G_S instead of $G_S(K)$ if K is clear from the context.

In contrast to chapter IX, where our main interest was devoted to the case $S = \{\text{all places of } K\}$ (i.e. $G_S = G_K$), in the present chapter we are mainly interested in the case when S is finite.

Let us first assume that $\mathfrak{c}$ is the class of all finite groups and that $S \supseteq S_\infty$. Then, recalling the notation $\mathbb{N}(S) = \mathbb{N} \cap \mathcal{O}_{K,S}^\times$ of VIII §3, we know from (8.3.18), (8.3.20) that

- $cd_p\, G_S \leq 2$ if $p \in \mathbb{N}(S)$, provided that $S_\mathbb{R}(K) = \varnothing$ if $p = 2$,
- if S is finite and if A is a finite G_S-module such that $\#A \in \mathbb{N}(S)$, then $H^i(G_S, A)$ is finite for $i = 0, 1, 2$.

Several questions naturally arise.

(1) How big is the decomposition group of a prime $\mathfrak{p}$ of K in G_S? Do we attain the theoretical maximum $(K_S)_\mathfrak{p} = \overline{(K_\mathfrak{p})}$ (resp. $(K_S)_\mathfrak{p} = K_\mathfrak{p}^{nr}$) for $\mathfrak{p} \in S$ (resp. $\mathfrak{p} \notin S$) ?
(2) Is $cd_p\, G_S$ finite for $p \notin \mathbb{N}(S)$?
(3) Is $cd_p\, G_S$ equal to 1 or equal to 2 for $p \in \mathbb{N}(S)$?
(4) Is $scd_p\, G_S$ equal to 2 for $p \in \mathbb{N}(S)$?
(5) If S is finite, is $H^i(G_S, A)$ finite for $i \geq 0$, if also $\#A \notin \mathbb{N}(S)$?
(6) For which S is the group G_S topologically finitely generated?

1. If S contains all but finitely many primes, the results of IX §4 imply that the maximal possible local extensions are indeed realized by the global field K_S. But if S is smaller, for instance if S is finite, the situation is much more complicated. In general G_S might be finite or even trivial (for instance for $K = \mathbb{Q}$ and $S = S_\infty$), and so we find examples of primes with trivial decomposition group in G_S. But we will prove the following (see §§8,11):

If p is a prime number and if $S \supseteq S_p \cup S_\infty$, then the local field $(K_S)_\mathfrak{p}$ is closed under p-extensions (closed under unramified p-extensions) if $\mathfrak{p}$ is contained (not contained) in S.

2. Question (2) is also rather difficult. Even the much weaker question of determining the set of prime numbers $p \notin \mathbb{N}(S)$ which divide the order of G_S is hard to answer. We will see in §11 that if $S \supseteq S_\ell$ for at least one prime number ℓ, then there exist infinitely many different prime numbers p dividing the order of G_S. But the p-Sylow subgroup for such a prime number could also be finite. There exist examples with $cd\ G_S(p) = 2$ for a prime number $p \notin \mathbb{N}(S)$, see [119], [198].

3. The answer to the third question is that $cd_p\, G_S = 2$ always for $p \in \mathbb{N}(S)$, provided that $S_\mathbb{R}(K) = \varnothing$ if $p = 2$. However, it might happen that the maximal pro-p-factor group $G_S(p)$ is of cohomological p-dimension equal to 1. We will deduce a criterion for this to occur in §9. In §11 we will show that subgroups U with $cd_p\, U(p) = 2$ are cofinal among the open subgroups of G_S, which shows that always $cd_p\, G_S = 2$.

In most cases however, we can read off the cohomological p-dimension directly from the module $I = D_2(\mathbb{Z}_p)$. Assume that K is totally imaginary if $p = 2$ and $S \supseteq S_p \cup S_\infty$. Since $cd_p\, G_S \le 2$, two alternatives exist: either I is nontrivial, then $cd_p\, G_S = 2$ and I is the p-dualizing module of G_S, or I is trivial, in which case $cd_p\, G_S = 1$. Let us denote the subset of finite primes in S by S^f. From the calculation below we can see that I is nontrivial (and hence $cd_p\, G_S = 2$) if there is more than one prime in $S^f(K_S)$. This covers most cases, for example if $\#S^f(K) > 1$. However, if $S = S_p \cup S_\infty$ and if there is exactly one prime dividing p in K, then we cannot easily decide whether there exists an extension of K inside K_S splitting this prime.

(10.2.1) Proposition. *Let K be a number field, let p be a prime number and assume that $S \supseteq S_p \cup S_\infty$ is a set of primes of K. Then the G_S-module $I = D_2(\mathbb{Z}_p)$ is characterized by the exact sequence*

$$0 \longrightarrow \mu_{p^\infty}(K_S) \longrightarrow \varinjlim_{L,n} \prod_{\mathfrak{p}\in S^f(L)} \mu_{p^n}(L_\mathfrak{p}) \longrightarrow I \longrightarrow 0,$$

where L runs through the finite extensions of K inside K_S. In particular, if S is finite, then there exists an exact sequence

$$0 \longrightarrow \mu_{p^\infty} \longrightarrow \bigoplus_{\mathfrak{p}\in S^f(K)} \mathrm{Ind}_{G_S}^{G_\mathfrak{p}} \mu_{p^\infty} \longrightarrow I \longrightarrow 0,$$

where $G_\mathfrak{p}$ is the decomposition group of a prolongation of $\mathfrak{p}$ to K_S.

Proof: Recall (III §4) that

$$I = \varinjlim_{n,U} H^2(U, \mathbb{Z}/p^n\mathbb{Z})^*,$$

where U runs through the open subgroups of G_S and the transition maps with respect to the subgroups are the duals of the corestriction maps. The duality theorem of Poitou-Tate yields

$$\varinjlim_{n,U} Ш^2(U, \mathbb{Z}/p^n\mathbb{Z})^* = \varinjlim_{n,U} Ш^1(U, \mu_{p^n}) = 0,$$

since the transition maps with respect to U on the right-hand side are the restriction maps. Therefore the result follows from the second part of the Poitou-Tate sequence for the module $\mathbb{Z}/p^n\mathbb{Z}$, by passing to the limit over all open subgroups $U \subseteq G_S$. □

We can also interpret the last result (10.2.1) in the following way. The p-dualizing module I of G_S is the quotient of the p-torsion subgroup of C_S, by the subgroup of those classes which are represented by an idèle with support in the archimedean places. Recall that the S-idèle class group $C_S(K)$ was defined as the quotient of C_K by the subgroup $U_{K,S} = \prod_{\mathfrak{P}\in S}\{1\} \times \prod_{\mathfrak{P}\notin S} U_{\mathfrak{P}}$. Using the convention that $U_{\mathfrak{P}} = K_{\mathfrak{P}}^\times$ for an archimedean prime $\mathfrak{P}$, we define

$$U_{K,S^f} = \prod_{\mathfrak{P}\in S^f}\{1\} \times \prod_{\mathfrak{P}\notin S^f} U_{\mathfrak{P}}.$$

If S is strictly larger than S_∞, which we assume here, we may regard U_{K,S^f} as a subgroup of C_K. We set

$$C_{S^f}(K) = C_K/U_{K,S^f}$$

and we call it the **S^f-idèle class group** of K. It is not difficult to see that we have an exact sequence

$$0 \to \mathcal{O}_{K,S}^\times \to \prod_{\mathfrak{P}\in S^f} K_{\mathfrak{P}}^\times \to C_{S^f}(K) \to Cl_0(K) \to 0,$$

where $Cl_0(K)$ is the *ideal class group in the narrow sense* of K, i.e. the quotient of the ideal group by the subgroup of those principal ideals which are generated by a nontrivial totally positive element*). If K is totally imaginary, then $Cl_0(K) = Cl(K)$ and the pair $(G_S, C_{S^f}(K_S))$ is a class formation which is a slight modification of the usual formation $(G_S, C_S(K_S))$. The difference is that it has a smaller group of universal norms, because the infinite idèles have been divided out. However, we will not use this fact in the following. The next corollary is a reformulation of (10.2.1) using the terminology just introduced.

*) We call an element totally positive if its image in every real embedding of K is positive; in particular, if K is totally imaginary, then every element is totally positive.

(10.2.2) Corollary. *Assume that $S \supseteq S_p \cup S_\infty$. Then the p-dualizing module $I = D_2(\mathbb{Z}_p)$ of G_S is canonically isomorphic to the p-torsion subgroup of the S^f-idèle class group of K_S.*

4. We clearly have to assume that K is totally imaginary if $p = 2$. If the set S is finite, then we will see in the next section that for $p \in \mathbb{N}(S)$ the vanishing of $H^2(G_S, \mathbb{Q}_p/\mathbb{Z}_p)$ is equivalent to the Leopoldt conjecture. Although this conjecture is verified only in special cases (see §3), we strongly believe that the answer to question (4) "must" always be yes, but we cannot prove this for any number field at all.

However, if the set S is of density 1, we have the following positive answer to question (4).

(10.2.3) Theorem. *Let K be a number field. Suppose that the set of primes S contains S_∞ and has Dirichlet density 1. Then*

$$scd_p\, G_S = 2$$

for every $p \in \mathbb{N}(S)$, provided that K is totally imaginary if $p = 2$.

Proof: Since $cd_p\, G_S \le 2$ by (8.3.18), it suffices to show that the cohomology group $H^2(U, \mathbb{Q}_p/\mathbb{Z}_p)$ vanishes for every $p \in \mathbb{N}(S)$ and every open subgroup $U \subseteq G_S$ (see (3.3.4)). Since the assumptions carry over to every finite extension of K inside K_S, we can restrict to the case $U = G_S$. The set S has Dirichlet density 1 and therefore we obtain (using the notation of IX §1)

$$cs(k(\mu_{p^r})|k) \subsetneqq S$$

for every prime number p and every $r \in \mathbb{N}$. Using (9.1.10) we obtain injections

$$H^2(G_S, \mathbb{Z}/p^r\mathbb{Z}) \hookrightarrow \bigoplus_{\mathfrak{p}\in S} H^2(K_\mathfrak{p}, \mathbb{Z}/p^r\mathbb{Z})$$

for every r and every $p \in \mathbb{N}(S)$, provided we are not in the special case. Then, passing to the limit over r and observing that local fields have $scd = 2$, cf. (7.2.5), we deduce the result.

If we are in the special case, cf. (9.1.9), then $p = 2$ and $i = \sqrt{-1} \notin K$. In particular, we are not in the special situation if we replace K by $K(i)$, and so we obtain by the above argument that $H^2(G_S(K(i)), \mathbb{Q}_2/\mathbb{Z}_2) = 0$. By (3.3.11) the corestriction map

$$0 = H^2(G_S(K(i)), \mathbb{Q}_2/\mathbb{Z}_2) \longrightarrow H^2(G_S(K), \mathbb{Q}_2/\mathbb{Z}_2)$$

is surjective. This finishes the proof. □

5. The answer to question (5) is "yes" for $i \le 2$, by the more general theorem below. When $\mathfrak{c}$ is the class of p-groups, it is due to *H. Koch* (see [110]), whose original proof did not use the duality theorem of Poitou-Tate.

(10.2.4) Theorem. *Let $\mathfrak{c}$ be a full class of finite groups and let S be a finite set of primes of the number field K. Then the cohomology groups*

$$H^i(G_S(\mathfrak{c}), M)$$

are finite for $i = 0, 1, 2$ and every finite $G_S(\mathfrak{c})$-module M.

Proof: Passing to a trivializing open subgroup of $G_S(\mathfrak{c})$ and using the Hochschild-Serre spectral sequence, we may assume that M is a trivial module and we immediately reduce to the case $M = \mathbb{Z}/p\mathbb{Z}$ for some prime number p.

Next we see that the theorem is true for S if it is true for a finite set of primes $T \supseteq S$. This follows from the Hochschild-Serre spectral sequence associated to the group extension

$$1 \to G(K_T(\mathfrak{c})|K_S(\mathfrak{c})) \to G_T(\mathfrak{c}) \to G_S(\mathfrak{c}) \to 1$$

and the fact that $G(K_T(\mathfrak{c})|K_S(\mathfrak{c}))$, being a normal subgroup in G_T, is generated by the finitely many inertia groups $T_\mathfrak{p}(K_T(\mathfrak{c})|K_S(\mathfrak{c}))$, $\mathfrak{p} \in T \smallsetminus S$ (choose prolongations to $K_T(\mathfrak{c})$).

We denote the full local group $G(\overline{K}_\mathfrak{p}|K_\mathfrak{p})$ by $\mathcal{G}_\mathfrak{p}$ and its inertia group by $\mathcal{T}_\mathfrak{p}$. Then there are surjections

$$\mathcal{G}_\mathfrak{p} \twoheadrightarrow G_\mathfrak{p}(K_T(\mathfrak{c})|K) \quad \text{and} \quad \mathcal{T}_\mathfrak{p} \twoheadrightarrow T_\mathfrak{p}(K_T(\mathfrak{c})|K_S(\mathfrak{c})).$$

We obtain an exact sequence

$$0 \to H^1(G_S(\mathfrak{c}), M) \to H^1(G_T(\mathfrak{c}), M) \to H^1(G(K_T(\mathfrak{c})|K_S(\mathfrak{c})), M)^{G_S(\mathfrak{c})} \to$$
$$\to H^2(G_S(\mathfrak{c}), M) \to H^2(G_T(\mathfrak{c}), M)$$

and canonical injections

$$H^1(G(K_T(\mathfrak{c})|K_S(\mathfrak{c})), M)^{G_S(\mathfrak{c})} \hookrightarrow \bigoplus_{\mathfrak{p}\in T\setminus S} H^1(T_\mathfrak{p}(K_T(\mathfrak{c})|K_S(\mathfrak{c})), M)^{G_\mathfrak{p}(K_T(\mathfrak{c})|K)}$$
$$\hookrightarrow \bigoplus_{\mathfrak{p}\in T\setminus S} H^1(\mathcal{T}_\mathfrak{p}, M)^{\mathcal{G}_\mathfrak{p}}.$$

The last group is finite by the results of chapter VII, which shows that the middle term in the five term exact sequence above is finite. Hence we may enlarge S; in particular, we may assume that $S \supseteq S_p \cup S_\infty$. If $\mathfrak{c}$ is the class of all finite groups, we are done by the result of (8.3.20). If $\mathfrak{c}$ is a class which does not contain $\mathbb{Z}/p\mathbb{Z}$, then the cohomology groups in question are trivial for $i \ge 1$. Finally, if $\mathfrak{c}$ is not the class of all finite groups and $\mathbb{Z}/p\mathbb{Z} \in \mathfrak{c}$, then the result follows from (10.4.8) (and we do not use the result for such $\mathfrak{c}$ before then). □

6. If S is the set of all primes, then G_S is not finitely generated (see chapter IX). Surprisingly, the answer to question (6) is unknown for finite S. It is not difficult to see that G_S^{ab} is finitely generated if S is finite, because, on the one hand, the decomposition groups for the finitely many primes in S are finitely generated, and, on the other hand, $G_\varnothing^{ab}$ is of finite order. This, however, only implies that the maximal pro-p-factor group of G_S is finitely generated for every prime number p.

In contrast to many other questions about G_S, where we "know" what the right answer should be, but are not able to give a real proof, it is even not clear what one should believe to be the right answer to question (6). Many mathematicians (including the authors) tend to think of G_S as being not finitely generated. In order to actually prove such a statement, one should construct many extensions which are unramified outside S. One method of constructing such extensions is the adjunction of points of geometric objects (e.g. moduli spaces) which have good reduction outside S. One of the major achievements of the number theoretical research of the last decades is some insight into the moduli of elliptic curves. (This was crucial for the proof of the Main Conjecture of Iwasawa Theory by *B. Mazur* and *A. Wiles*, and also for the proof of Fermat's Last Theorem by *A. Wiles*.) But it is not clear whether for the purpose explained above it suffices to use only the moduli spaces of elliptic curves. Since our knowledge of other moduli spaces is even smaller, it seems to be very hard to achieve progress in this direction. However, we have the following weaker result.

(10.2.5) Theorem. *Assume that K is a number field and that S is a finite set of primes in K. Then the group $G_S = G_S(K)$ is (topologically) generated by the conjugacy classes of finitely many elements.*

In chapter XII we will deduce theorem (10.2.5) as a corollary of a result of *Y. Ihara* on the decomposition of primes in infinite unramified extensions of number fields. It can also be deduced from the fact that G_S^{ab} is finitely generated and from the following purely group theoretical result.

(10.2.6) Theorem (*Guralnick, Weiss*). *Let G be a profinite group with abelianization of rank r* *). *Then G can be topologically generated by r (or 1 if $r = 0$) conjugacy classes.*

We omit the proof of (10.2.6), referring the reader to the exercises below.

---

*) i.e. G^{ab} can be topologically generated by r elements and r is minimal.

Exercise 1. Let G be a group. We define the **Frattini subgroup** $\Phi(G)$ of G as the intersection of all maximal subgroups of G (see also IX §6). Show that $\Phi(G)$ is a normal subgroup of G and that an element $g \in G$ is contained in $\Phi(G)$ if and only if it is a "non-generator", i.e. if it can be removed from every generating family of elements of G.

Exercise 2. Assume that $N \lhd G$ is a minimal normal subgroup, i.e. generated by the conjugates of one element $n \in G$. Suppose that N is not contained in $\Phi(G)$. Show that the canonical projection

$$G \longrightarrow G/N$$

has a section, i.e. there exists a decomposition of G as a semi-direct product $G = N \rtimes H$ for some subgroup $H \subseteq G$.

Exercise 3. Let G be a finite group and let $N \lhd G$ be a minimal normal subgroup. Suppose that the abelianizations of G and of G/N have the same rank $r \geq 1$ and assume that G/N is generated by the r conjugacy classes $\overline{C}_1, \dots, \overline{C}_r$. Show that these conjugacy classes can be lifted to conjugacy classes $C_1, \dots, C_r$ in G, such that

$$G = \langle C_1, \dots, C_r \rangle.$$

Hint: Use ex.1 and 2 in order to reduce to the case that $G = N \times H$ and N is simple. Choose elements $c_1, \dots, c_r \in H$ with $c_i \bmod N \in \overline{C}_i$ for $i = 1, \dots, r$ and choose any nontrivial element $n \in N$. Then the conjugacy classes of $n \cdot c_1$ and of $c_2, \dots, c_r$ generate G.

Exercise 4. Use induction on $\#G$ and the exercises above in order to prove theorem (10.2.6) in the case that G is finite.

Exercise 5. Extend the result of ex.4 to an arbitrary profinite group G.

§3. Leopoldt's Conjecture

It is one of the fundamental principles in number theory that none of the places of a number field should be privileged above the others, i.e. they all play comparable roles. In particular, this includes the archimedean places, but in practice we are often confronted with the situation that statements which are well-known in the archimedean case turn out to be difficult or even unsolved questions in their p-adic formulations.

However, the truth of these assertions (maybe up to minor modifications) is predicted by the analogy and many conjectures in number theory originate from this source. One prominent example for this phenomenon is the p-adic version of Minkowski theory, which will be described in this section. Several arguments below are taken from *P. Schneider*'s thesis [202].

Let K be a number field and let $r_1 = \#S_{\mathbb{R}}(K)$, resp. $r_2 = \#S_{\mathbb{C}}(K)$, be the number of real, resp. complex, places of K. In this chapter we will also use the notation E_K for the group of units $\mathcal{O}_K^\times$ of K. Recall (cf. [160], chap.I,

§§5,7) that (the multiplicative form of) Minkowski theory investigates the homomorphism

$$j_\infty : K^\times \longrightarrow \prod_{\sigma\in\mathrm{Hom}(K,\mathbb{C})} \mathbb{C}^\times \xrightarrow{\prod_\sigma \log(|\ |)} \prod_{\sigma\in\mathrm{Hom}(K,\mathbb{C})} \mathbb{R}$$

which is given by $j_\infty(x)_\sigma := \log(|\sigma(x)|)$. Dirichlet's unit theorem asserts that $j_\infty(E_K)$ is a lattice of rank r_1+r_2-1 in $\prod_\sigma \mathbb{R}$ (and the $\mathbb{R}$-subspace spanned by $j_\infty(E_K)$ is the space of elements of trace 0 which are invariant under complex conjugation).

Now we are going to investigate the p-adic analogue of the above situation. Let p be a prime number, which will be fixed for the rest of this section. Fix an algebraic closure $\bar{\mathbb{Q}}_p$ of the field $\mathbb{Q}_p$. The field $\bar{\mathbb{Q}}_p$ is not complete with respect to the p-adic absolute value and we denote its completion by $\mathbb{C}_p$.

(10.3.1) Definition. *We call* $\mathbb{C}_p$ *the field of* ***p*-adic complex numbers**.

The p-adic absolute value naturally extends to $\mathbb{C}_p$ and $\bar{\mathbb{Q}}_p$ is dense in $\mathbb{C}_p$.

(10.3.2) Proposition. *The field* $\mathbb{C}_p$ *is algebraically closed.*

Proof: Let $\alpha \in \bar{\mathbb{C}}_p$ and let $f \in \mathbb{C}_p[X]$ be the minimal polynomial of α over $\mathbb{C}_p$. Note that f is separable because $\mathrm{char}(\mathbb{C}_p) = 0$. Since $\bar{\mathbb{Q}}_p$ is dense in $\mathbb{C}_p$, we can choose a polynomial $g \in \bar{\mathbb{Q}}_p[X]$ near to f. Then $|g(\alpha)| = |g(\alpha) - f(\alpha)|$ is small. Writing $g(X) = \prod(X - \beta_j)$, with $\beta_j \in \bar{\mathbb{Q}}_p$, we see that $|\alpha - \beta|$ is small for some root β of $g(X)$. In particular, we can choose $g(X)$ and then β such that $|\beta - \alpha| < |\alpha_i - \alpha|$ for all conjugates $\alpha_i \neq \alpha$ of α over $\mathbb{C}_p$. By Krasner's lemma (8.1.6), we obtain $\alpha \in \mathbb{C}_p(\beta) = \mathbb{C}_p$. □

Remark: The fields $\mathbb{C}$ and $\mathbb{C}_p$ are isomorphic as abstract fields, because they are algebraically closed fields of the same transcendence degree over $\mathbb{Q}$. However, they are not topologically isomorphic.

Recall (cf. [160], chap.II, (5.4)) that for every local field $k|\mathbb{Q}_p$ we have a uniquely defined p-adic logarithm

$$\log_p : k^\times \longrightarrow k.$$

It satisfies $\log_p(p) = 0$ and for a principal unit $(1 + x) \in U_k^1$ it is given by the (convergent) series

$$\log_p(1+x) = x - \frac{x^2}{2} + \frac{x^3}{3} - \cdots .$$

This p-adic logarithm naturally extends to $\bar{\mathbb{Q}}_p$ and, by continuity, also to a function: $\log_p : \mathbb{C}_p^\times \to \mathbb{C}_p$. For a given number field K consider the homomorphism

$$j_p : K^\times \longrightarrow \prod_{\sigma \in \mathrm{Hom}(K,\mathbb{C}_p)} \mathbb{C}_p^\times \xrightarrow{\prod_\sigma \log_p(\,)} \prod_{\sigma \in \mathrm{Hom}(K,\mathbb{C}_p)} \mathbb{C}_p$$

which is given by $j_p(x)_\sigma := \log_p(\sigma(x))$. Let $\varepsilon_1, \dots, \varepsilon_{r_1+r_2-1}$ be a basis of E_K modulo its torsion subgroup. We list the elements of $\mathrm{Hom}(K, \mathbb{C}_p)$ as $\sigma_1, \dots, \sigma_d$ with $d := [K : \mathbb{Q}]$.

(10.3.3) Definition. *We define the* **regulator matrix**

$$\mathcal{R}_p(\varepsilon_1, \dots, \varepsilon_{r_1+r_2-1}) := \begin{pmatrix} \log_p \sigma_1(\varepsilon_1) & \cdots & \log_p \sigma_d(\varepsilon_1) \\ \vdots & \ddots & \vdots \\ \log_p \sigma_1(\varepsilon_{r_1+r_2-1}) & \cdots & \log_p \sigma_d(\varepsilon_{r_1+r_2-1}) \end{pmatrix}$$

and set

$$\begin{aligned} \mathrm{rr}_p(K) &:= \operatorname{rank} \mathcal{R}_p(\varepsilon_1, \dots, \varepsilon_{r_1+r_2-1}) \\ \mathrm{vol}_p(K) &:= \max \{ |\det R|_p \mid R \text{ is a } (r_1+r_2-1) \times (r_1+r_2-1)\text{-} \\ &\qquad \text{minor of } \mathcal{R}_p(\varepsilon_1, \dots, \varepsilon_{r_1+r_2-1}) \}. \end{aligned}$$

We call $\mathrm{rr}_p(K)$ *the* **p-adic regulator rank** *of* K.

Remarks: 1. It is easy to see that $\mathrm{rr}_p(K)$ and $\mathrm{vol}_p(K)$ are independent of the choice of the system $\varepsilon_1, \dots, \varepsilon_{r_1+r_2-1}$ and of the chosen ordering $\sigma_1, \dots, \sigma_d$ of the elements of $\mathrm{Hom}(K, \mathbb{C}_p)$, i.e. they are arithmetic invariants of K.

2. We think of $\mathrm{vol}_p(K)$ as the p-adic covolume of the unit lattice of K, because of the analogy with the archimedean case, where the corresponding value is (up to the factor $\sqrt{r_1+r_2}$ and a power of 2) the volume of a fundamental domain of the unit lattice in the Minkowski space (cf. [160], chap.I, (7.5)).

Assume for a moment that the number field K is totally real, so that $r_1 + r_2 - 1 = d - 1$. Then the regulator matrix is a $(d-1) \times d$-matrix and the sum of all columns equals zero, since

$$\sum_\sigma \log_p \sigma(\varepsilon_i) = \log_p(N_{K|\mathbb{Q}}\, \varepsilon_i) = \log_p(\pm 1) = 0.$$

In this case the determinant of a $(d-1) \times (d-1)$-minor is (up to sign) independent of the choice of the minor. Also, changing the basis of the unit group modulo torsion or changing the ordering $\sigma_1, \dots, \sigma_d$ might introduce a factor (± 1).

(10.3.4) Definition. *If K is totally real, we call*

$$R_p(K) := \det \begin{pmatrix} \log_p \sigma_1(\varepsilon_1) & \cdots & \log_p \sigma_{d-1}(\varepsilon_1) \\ \vdots & \ddots & \vdots \\ \log_p \sigma_1(\varepsilon_{d-1}) & \cdots & \log_p \sigma_{d-1}(\varepsilon_{d-1}) \end{pmatrix}$$

the **p-adic regulator** *of K. It is well-defined up to sign.*

Returning to the case of an arbitrary number field we obviously have the inequality $\mathrm{rr}_p(K) \le r_1 + r_2 - 1$ and the analogy to the archimedean case predicts the following conjecture.

(10.3.5) Leopoldt's Conjecture. *For every number field K and every prime number p, the p-adic regulator rank $\mathrm{rr}_p(K)$ is equal to $r_1 + r_2 - 1$.*

For totally real number fields the Leopoldt conjecture is equivalent to the non-vanishing of the p-adic regulator R_p. It is also equivalent to the non-vanishing of certain p-adic L-functions at $s = 1$ (see [246]) in this case.

If ε is a unit which is not a root of unity, then $\log_p(\varepsilon) \neq 0$. Therefore $\mathrm{rr}_p(K) = r_1 + r_2 - 1$ provided that $r_1 + r_2 \le 2$. In other words, the Leopoldt conjecture is true for K and every prime number p if K is a quadratic number field or if K is an extension of degree 2 of an imaginary quadratic number field or if K is a cubic field which is not totally real. *M. Waldschmidt* has shown that $\mathrm{rr}_p(K) \ge \frac{1}{2}(r_1 + r_2 - 1)$ for every number field K and every prime number p (see [244]). Later on in this section, we will see that the Leopoldt conjecture is true for abelian extensions $K|k$ where $k = \mathbb{Q}$ or k is imaginary quadratic.

There exist a large number of equivalent formulations of the Leopoldt conjecture, some of which we will describe below. Let us fix some notation.

As before we let the prime number p be fixed and we denote the p-adic completion of an abelian group A by $\hat{A}$, i.e.

$$\hat{A} := \varprojlim_n A/p^n A.$$

Observe that $\hat{A}$ is a $\mathbb{Z}_p$-module in a natural way. Further, $\hat{A} = A \otimes_{\mathbb{Z}} \mathbb{Z}_p$ if A is a finitely generated $\mathbb{Z}$-module, and $\hat{A} = A$ if A is a finitely generated $\mathbb{Z}_p$-module. If A is p-torsion-free, then we have an exact sequence

$$0 \longrightarrow \hat{A} \longrightarrow \hat{A} \otimes_{\mathbb{Z}_p} \mathbb{Q}_p \longrightarrow A \otimes_{\mathbb{Z}} \mathbb{Q}_p/\mathbb{Z}_p \longrightarrow 0.$$

For a finite prime $\mathfrak{p}$ of K we denote the group of units of the local field $K_\mathfrak{p}$ by $U_\mathfrak{p}$ and we set $U_\mathfrak{p} := K_\mathfrak{p}^\times$ for an archimedean $\mathfrak{p}$. Then $\hat{U}_\mathfrak{p}$ is finite if $\mathfrak{p} \notin S_p(K)$. If $\mathfrak{p} \in S_p(K)$, then the inclusion of the principal units $U_\mathfrak{p}^1 \subseteq U_\mathfrak{p}$ induces an isomorphism $U_\mathfrak{p}^1 \xrightarrow{\sim} \hat{U}_\mathfrak{p}$.

Assume that $S \supseteq S_p \cup S_\infty$ and $T \subseteq S$ are finite sets of places of K. We denote by S^f the subset of finite (i.e. nonarchimedean) primes in S. Consider the diagonal embedding*)

$$\Delta : E_{K,T} \hookrightarrow \prod_{\mathfrak{p}\in S\setminus T} U_\mathfrak{p} \times \prod_{\mathfrak{p}\in T} K_\mathfrak{p}^\times, \quad e \longmapsto (e,\ldots,e).$$

It induces a homomorphism $\Delta' : E_{K,T} \to \prod_{\mathfrak{p}\in S\setminus T} \hat{U}_\mathfrak{p} \times \prod_{\mathfrak{p}\in T} \hat{K}_\mathfrak{p}^\times$. The kernel of Δ' is the group $\mu_{p'}(K)$ of roots of unity of order prime to p contained in K. Indeed, we clearly have the inclusion $\mu_{p'} \subseteq \ker \Delta'$. On the other hand, assume $e \in \ker \Delta'$ and let $\mathfrak{p} \in S_p$ be arbitrary. Then $e = \zeta e'$ in $K_\mathfrak{p}$, with $\zeta \in \mu_{p'}(K_\mathfrak{p})$ and $e' \in U_\mathfrak{p}^1$. The image of ζ in $\hat{U}_\mathfrak{p}$ is trivial. Since $U_\mathfrak{p}^1$ maps isomorphically onto $\hat{U}_\mathfrak{p}$, we conclude that $e' = 1$. Therefore $e = \zeta$ in $K_\mathfrak{p}$ and hence also in K.

We denote the topological closure of the image of $E_{K,T}$ under Δ' in the group $\prod_{\mathfrak{p}\in S\setminus T} \hat{U}_\mathfrak{p} \times \prod_{\mathfrak{p}\in T} \hat{K}_\mathfrak{p}^\times$ by $\bar{E}_{K,T}^{(S)}$. It is equal to the image of the induced homomorphism

$$\hat{\Delta} : E_{K,T} \otimes_\mathbb{Z} \mathbb{Z}_p \to \prod_{\mathfrak{p}\in S\setminus T} \hat{U}_\mathfrak{p} \times \prod_{\mathfrak{p}\in T} \hat{K}_\mathfrak{p}^\times .$$

In particular, $\bar{E}_{K,T}^{(S)}$ is a finitely generated $\mathbb{Z}_p$-module. If $T = \varnothing$, so that $E_{K,T} = E_K$, we denote $\bar{E}_{K,T}^{(S)}$ by $\bar{E}_K^{(S)}$.

(10.3.6) Theorem. *Let K be a number field, p be a prime number and assume that S is a finite set of places of K containing $S_p \cup S_\infty$. Then the following assertions are equivalent.*

(i) *Leopoldt's conjecture is true for K and p.*

(ii) $\operatorname{rank}_{\mathbb{Z}_p} \bar{E}_K^{(S)} = r_1 + r_2 - 1$.

(iii) *The canonical homomorphism*
$$E_K \otimes_\mathbb{Z} \mathbb{Z}_p \longrightarrow \prod_{\mathfrak{p}\in S} \hat{U}_\mathfrak{p}$$
is injective.

(iv) *The kernel of the canonical homomorphism*
$$E_K \otimes_\mathbb{Z} \mathbb{Q}_p/\mathbb{Z}_p \longrightarrow \prod_{\mathfrak{p}\in S} U_\mathfrak{p} \otimes_\mathbb{Z} \mathbb{Q}_p/\mathbb{Z}_p$$
is finite.

(v) $\operatorname{rank}_{\mathbb{Z}_p} H_1(G_S(K), \mathbb{Z}_p) = r_2 + 1$.

(vi) $H_2(G_S(K), \mathbb{Z}_p) = 0$.

(vii) *The p-torsion subgroup of the universal norm group $D_S(K)$ of the class formation $(G_S(K), C_S(K_S))$ is isomorphic to*
$$\prod_{v\in S_\mathbb{C}(K)} \mu_p(K_v).$$

*) Archimedean primes are ignored in T.

(10.3.7) Corollary. *The following numbers agree and are independent of the finite set $S \supseteq S_p \cup S_\infty$:*

- $\operatorname{rank}_{\mathbb{Z}_p} \ker(E_K \otimes_{\mathbb{Z}} \mathbb{Z}_p \longrightarrow \prod_{\mathfrak{p}\in S} \hat{U}_{\mathfrak{p}})$,
- $\operatorname{corank}_{\mathbb{Z}_p} \ker(E_K \otimes_{\mathbb{Z}} \mathbb{Q}_p/\mathbb{Z}_p \longrightarrow \prod_{\mathfrak{p}\in S} U_{\mathfrak{p}} \otimes_{\mathbb{Z}} \mathbb{Q}_p/\mathbb{Z}_p)$,
- $r_1 + r_2 - 1 - \operatorname{rank}_{\mathbb{Z}_p} \bar{E}_K^{(S)}$,
- $r_1 + r_2 - 1 - \operatorname{rr}_p(K)$,
- $\operatorname{rank}_{\mathbb{Z}_p} H_1(G_S(K), \mathbb{Z}_p) - r_2 - 1$,
- $\operatorname{rank}_{\mathbb{Z}_p} H_2(G_S(K), \mathbb{Z}_p)$.

We denote this number by $\mathfrak{d}_p(K)$ and call it the **Leopoldt defect**. *One has*

$$0 \le \mathfrak{d}_p(K) \le r_1 + r_2 - 1,$$

and the Leopoldt conjecture for K and p is true if and only if $\mathfrak{d}_p(K) = 0$.

(10.3.8) Corollary. *Assume that the Leopoldt conjecture is true for K and p. If p is unramified in $K|\mathbb{Q}$, then*

$$\#\ker\Big(E_K \otimes_{\mathbb{Z}} \mathbb{Q}_p/\mathbb{Z}_p \to \prod_{\mathfrak{p}\in S_p} U_{\mathfrak{p}} \otimes_{\mathbb{Z}} \mathbb{Q}_p/\mathbb{Z}_p\Big) = p^{r_1+r_2-1} \cdot \operatorname{vol}_p(K)^{-1}.$$

(10.3.9) Corollary. *Suppose that $p \neq 2$ or that K is totally imaginary. If $S \supseteq S_p \cup S_\infty$, then $scd_p\, G_S(K) = 2$ if and only if the Leopoldt conjecture holds for p and every finite extension L of K inside K_S.*

We can also formulate the Leopoldt conjecture in terms of continuous cochain cohomology.

(10.3.10) Corollary. *We have*

$$\begin{aligned} \mathfrak{d}_p(K) &= \operatorname{rank}_{\mathbb{Z}_p} H^1_{cts}(G_S(K), \mathbb{Z}_p) - r_2 - 1 \\ &= \operatorname{rank}_{\mathbb{Z}_p} H^2_{cts}(G_S(K), \mathbb{Z}_p)\,. \end{aligned}$$

In particular, the Leopoldt conjecture holds if and only if $H^2_{cts}(G_S(K), \mathbb{Z}_p)$ is finite.

(10.3.11) Corollary. *Let $K|k$ be a finite extension. Then the following is true:*

(i) *If the Leopoldt conjecture holds for K and p, then it holds for k and p.*

(ii) *If K is a CM-field and $k = K^+$ is its maximal totally real subfield, then the Leopoldt conjecture holds for K and p if and only if it holds for K^+ and p.*

For the proof of theorem (10.3.6) and its corollaries we need two lemmas.

(10.3.12) Lemma. *Let $S \supseteq T$ be sets of primes of K. Then there is a canonical exact sequence*

$$0 \to E_{K,T} \to E_{K,S} \to \bigoplus_{\mathfrak{p} \in S \setminus T} K_{\mathfrak{p}}^{\times}/U_{\mathfrak{p}} \xrightarrow{\alpha} Cl_T(K) \to Cl_S(K) \to 0,$$

where α is the map sending $\oplus_{\mathfrak{p}} x_{\mathfrak{p}} \bmod U_{\mathfrak{p}}$ to the class of $\prod_{\mathfrak{p}} \mathfrak{p}^{v_{\mathfrak{p}}(x_{\mathfrak{p}})}$.

Proof: This is clear from the definition of the objects and maps occurring. □

(10.3.13) Lemma. *For $S \supseteq S_p \cup S_\infty$ and $T \subseteq S$ there is a canonical commutative exact diagram*

$$\begin{array}{ccccccccc}
H_2(G_S, \mathbb{Z}_p) & \hookrightarrow & E_{K,T} \otimes \mathbb{Z}_p & \longrightarrow & \prod\limits_{\mathfrak{p} \in S \setminus T} \hat{U}_{\mathfrak{p}} \times \prod\limits_{\mathfrak{p} \in T} \hat{K}_{\mathfrak{p}}^{\times} & \to & G_S^{ab}(p) & \twoheadrightarrow & Cl_T(K)(p) \\
\| & & \downarrow & & \downarrow & & \| & & \downarrow \\
H_2(G_S, \mathbb{Z}_p) & \hookrightarrow & E_{K,S} \otimes \mathbb{Z}_p & \longrightarrow & \prod\limits_{\mathfrak{p} \in S} \hat{K}_{\mathfrak{p}}^{\times} & \longrightarrow & G_S^{ab}(p) & \twoheadrightarrow & Cl_S(K)(p).
\end{array}$$

(Our notational convention implies for an archimedean prime $\mathfrak{p}$ that $\hat{U}_{\mathfrak{p}} = \hat{K}_{\mathfrak{p}}^{\times} = \mu_2$ if $\mathfrak{p}$ is real and $p = 2$, and $\hat{U}_{\mathfrak{p}} = \hat{K}_{\mathfrak{p}}^{\times} = 0$ otherwise.) In particular, there is an exact sequence

$$0 \longrightarrow H_2(G_S, \mathbb{Z}_p) \longrightarrow E_{K,T} \otimes_{\mathbb{Z}} \mathbb{Z}_p \longrightarrow \bar{E}_{K,T}^{(S)} \longrightarrow 0\,.$$

Proof: We first show the exactness of the lower sequence. Dualizing the Poitou-Tate sequence for the module $\mathbb{Z}/p^n\mathbb{Z}$, we obtain an exact sequence

$$0 \to Ш^2(G_S, \mathbb{Z}/p^n\mathbb{Z})^\vee \to H^1(G_S, \mu_{p^n}) \to \bigoplus_{\mathfrak{p} \in S} H^1(K_{\mathfrak{p}}, \mu_{p^n})$$

$$\to H_1(G_S, \mathbb{Z}/p^n\mathbb{Z}) \to Ш^1(G_S, \mathbb{Z}/p^n\mathbb{Z})^\vee \to 0.$$

Local Kummer theory implies

$$\varprojlim_n H^1(K_{\mathfrak{p}}, \mu_{p^n}) = \hat{K}_{\mathfrak{p}}^{\times}\,,$$

and since $H^2(K_{\mathfrak{p}}, \mathbb{Q}_p/\mathbb{Z}_p) = 0$ (see (7.2.5)), we obtain

$$Ш^2(G_S, \mathbb{Q}_p/\mathbb{Z}_p) = H^2(G_S, \mathbb{Q}_p/\mathbb{Z}_p).$$

The global Kummer sequence, together with the finiteness of $Cl_S(K)$, implies that there is an isomorphism

$$E_{K,S} \otimes_{\mathbb{Z}} \mathbb{Z}_p = \varprojlim_n H^1(G_S, \mu_{p^n}).$$

Further, observe that

$$Ш^1(G_S, \mathbb{Z}/p^n\mathbb{Z})^\vee = Cl_S(K)/p^n.$$

Therefore we obtain the lower exact sequence of the lemma by passing to the inverse limit over n in the dualized Poitou-Tate sequence above.

In order to obtain the right-hand part of the upper exact sequence, observe that the cokernel of the homomorphism

$$\prod_{\mathfrak{p} \in S\setminus T} \hat{U}_\mathfrak{p} \times \prod_{\mathfrak{p} \in T} \hat{K}_\mathfrak{p}^\times \to \prod_{\mathfrak{p} \in S} \hat{K}_\mathfrak{p}^\times \to H_1(G_S, \mathbb{Z}_p)$$

is canonically isomorphic to $Cl_T(K)(p)$ by global class field theory. In order to complete the diagram, we use the exact sequence

$$E_{K,T} \otimes \mathbb{Z}_p \hookrightarrow E_{K,S} \otimes \mathbb{Z}_p \to \bigoplus_{\mathfrak{p} \in S\setminus T} \hat{K}_\mathfrak{p}^\times/\hat{U}_\mathfrak{p} \longrightarrow Cl_T(K)(p) \twoheadrightarrow Cl_S(K)(p),$$

which is obtained from (10.3.12) by tensoring by the flat $\mathbb{Z}$-module $\mathbb{Z}_p$. □

Proof of (10.3.6) and its corollaries: We set $t = r_1 + r_2 - 1$ and we omit the suffix K during the proof.

(i) ⇔ (ii): Since $\hat{U}_\mathfrak{p}$ is finite for primes $\mathfrak{p} \notin S_p$, we have

$$\operatorname{rank}_{\mathbb{Z}_p} \bar{E}^{(S)} = \operatorname{rank}_{\mathbb{Z}_p} \bar{E}^{(S_p)}.$$

Thus we may work with $S = S_p$ instead. The $\mathfrak{p}$-adic logarithms induce a continuous homomorphism

$$Log : \prod_{\mathfrak{p} \in S_p} \hat{U}_\mathfrak{p} \xrightarrow{\prod \log_\mathfrak{p}} \prod_{\mathfrak{p} \in S_p} K_\mathfrak{p}$$

with finite kernel. Therefore we can equivalently calculate the rank of the image of $\bar{E}^{(S_p)}$ under Log.

For $\mathfrak{p} \in S_p$, let $d_\mathfrak{p} := [K_\mathfrak{p} : \mathbb{Q}_p]$ and let $a_1^\mathfrak{p}, \dots, a_{d_\mathfrak{p}}^\mathfrak{p}$ be a $\mathbb{Z}_p$-basis of $\mathcal{O}_\mathfrak{p}$. Further, we fix a basis $\varepsilon_1, \dots, \varepsilon_t$ of E modulo torsion, and suppose that

$$\log_\mathfrak{p}(\varepsilon_i) = \sum_{j=1}^{d_\mathfrak{p}} \xi_{ij}^\mathfrak{p} a_j^\mathfrak{p}$$

for $\mathfrak{p} \in S_p$ and $1 \le i \le t$ with $\xi_{ij}^\mathfrak{p} \in \mathbb{Q}_p$. Now let X be the $(t \times d)$-matrix

$$X := \begin{pmatrix} \xi_{11}^{\mathfrak{p}_1} & \cdots & \xi_{1d_{\mathfrak{p}_1}}^{\mathfrak{p}_1} & \xi_{11}^{\mathfrak{p}_2} & \cdots & \xi_{1d_{\mathfrak{p}_2}}^{\mathfrak{p}_2} & \cdots \\ \xi_{21}^{\mathfrak{p}_1} & \ddots & & & & & \\ \vdots & & & & & & \\ \xi_{t1}^{\mathfrak{p}_1} & \cdots & & & & & \end{pmatrix},$$

where $\mathfrak{p}_v$ runs through S_p. Further, let

$$\phi_1^{\mathfrak{p}}, \ldots, \phi_{d_{\mathfrak{p}}}^{\mathfrak{p}} : K_{\mathfrak{p}} \hookrightarrow \mathbb{C}_p$$

be the $d_{\mathfrak{p}}$ embeddings of $K_{\mathfrak{p}}$ into $\mathbb{C}_p$. We order the $d = [K : \mathbb{Q}]$ embeddings $K \hookrightarrow \mathbb{C}_p$ in the following way

$$\phi_1, \ldots, \phi_d = \phi_1^{\mathfrak{p}_1}\big|_K, \ldots, \phi_{d_{\mathfrak{p}_1}}^{\mathfrak{p}_1}\big|_K, \phi_1^{\mathfrak{p}_2}\big|_K, \ldots .$$

Finally, let $D_{\mathfrak{p}}$ (resp. D) be the $(d_{\mathfrak{p}} \times d_{\mathfrak{p}})$ (resp. $(d \times d)$) matrices

$$D_{\mathfrak{p}} := \begin{pmatrix} \phi_1^{\mathfrak{p}}(a_1^{\mathfrak{p}}) & \ldots & \phi_{d_{\mathfrak{p}}}^{\mathfrak{p}}(a_1^{\mathfrak{p}}) \\ \vdots & \ddots & \vdots \\ \phi_1^{\mathfrak{p}}(a_{d_{\mathfrak{p}}}^{\mathfrak{p}}) & \ldots & \phi_{d_{\mathfrak{p}}}^{\mathfrak{p}}(a_{d_{\mathfrak{p}}}^{\mathfrak{p}}) \end{pmatrix}$$

for $\mathfrak{p} \in S_p$ and

$$D := \begin{pmatrix} D_{\mathfrak{p}_1} & & 0 \\ & D_{\mathfrak{p}_2} & \\ & & \ddots \\ 0 & & \end{pmatrix}.$$

A simple calculation yields

$$\mathcal{R}_p(\varepsilon_1, \ldots, \varepsilon_t) = X \cdot D.$$

By the discriminant-product formula (see [160], chap.III, (2.11)), we have

$$|\det D|_p^2 = |\mathrm{disc}_{K|\mathbb{Q}}|_p \neq 0.$$

Summarizing, we obtain

$$\begin{aligned} \mathrm{rank}_{\mathbb{Z}_p} \bar{E}^{(S)} &= \mathrm{rank}_{\mathbb{Z}_p} Log(\bar{E}^{(S_p)}) \\ &= \mathrm{rank}\, X \\ &= \mathrm{rank}\, \mathcal{R}_p(\varepsilon_1, \ldots, \varepsilon_t) \\ &= \mathrm{rr}_p(K). \end{aligned}$$

This shows the equivalence (i) $\Leftrightarrow$ (ii). The equivalence (ii) $\Leftrightarrow$ (iii) $\Leftrightarrow$ (vi) follows from (10.3.13) together with the fact that $H_2(G_S, \mathbb{Z}_p)$ is torsion-free (see (8.7.7). Furthermore, the equivalence (v) $\Leftrightarrow$ (vi) follows similarly by counting $\mathbb{Z}_p$-ranks in (10.3.13) or from the global Euler-Poincaré characteristic formula (8.7.6).

We denote the torsion subgroup of an abelian group A by $\mathrm{tor}(A)$ and the maximal torsion-free quotient by A/tor. Consider the commutative exact diagram

$$\begin{array}{ccccc} E/\mathrm{tor} \otimes_{\mathbb{Z}} \mathbb{Z}_p & \hookrightarrow & E/\mathrm{tor} \otimes_{\mathbb{Z}} \mathbb{Q}_p & \twoheadrightarrow & E \otimes_{\mathbb{Z}} \mathbb{Q}_p/\mathbb{Z}_p \\ \downarrow{\scriptstyle \varphi} & & \downarrow{\scriptstyle \varphi \otimes \mathbb{Q}_p} & & \downarrow{\scriptstyle \psi} \\ \prod_{\mathfrak{p} \in S} \hat{U}_{\mathfrak{p}}/\mathrm{tor} & \hookrightarrow & \prod_{\mathfrak{p} \in S} (\hat{U}_{\mathfrak{p}}/\mathrm{tor}) \otimes_{\mathbb{Z}_p} \mathbb{Q}_p & \twoheadrightarrow & \prod_{\mathfrak{p} \in S} \hat{U}_{\mathfrak{p}} \otimes_{\mathbb{Z}_p} \mathbb{Q}_p/\mathbb{Z}_p. \end{array}$$

This diagram shows

$$\operatorname{rank}_{\mathbb{Z}_p} \ker \varphi = \operatorname{corank}_{\mathbb{Z}_p} \ker \psi.$$

Since E is a finitely generated abelian group this implies the equivalence (iii) $\Leftrightarrow$ (iv).

Now assume that (i)–(vi) are true. Then the snake lemma shows that

$$\ker \psi = \operatorname{tor}((\prod_{\mathfrak{p}\in S} \hat{U}_\mathfrak{p}/\operatorname{tor})/\operatorname{im}\varphi).$$

If $K_\mathfrak{p}|\mathbb{Q}_p$ is unramified at every $\mathfrak{p} \in S_p$, then the p-adic logarithm $\log_\mathfrak{p} : \hat{U}_\mathfrak{p} \to K_\mathfrak{p}$ has image exactly $p\mathcal{O}_\mathfrak{p}$ and its kernel is equal to μ_2 if $p = 2$ and trivial otherwise (see [160], chap.II, §5). Hence

$$\begin{aligned}
&\#\operatorname{tor}((\prod_{\mathfrak{p}\in S_p} \hat{U}_\mathfrak{p}/\operatorname{tor})/\operatorname{im}\varphi)\\
&= p^t \cdot |\text{product of the elementary divisors of } X|_p^{-1}\\
&= p^t \cdot \min\{|\det Y|_p^{-1} \mid Y \text{ is } (t\times t)\text{-minor of } X\}\\
&= p^t \cdot \min\{|\det R|_p^{-1} \mid R \text{ is } (t\times t)\text{-minor of } \mathcal{R}_p(\varepsilon_1,\ldots,\varepsilon_t)\}\\
&= p^t \cdot \operatorname{vol}_p(K)^{-1}.
\end{aligned}$$

This shows corollary (10.3.8).

It remains to show the equivalence of (vii) to the other conditions. Observe that ${}_pG_S^{ab} = {}_pH_1(G_S,\mathbb{Z}_p)$ and consider the diagram

$$\begin{array}{ccccccccc}
 & & & & & & H_2(G_S,\mathbb{Z}_p)/p & & \\
 & & & & & & \downarrow & & \\
0 \longrightarrow & \prod\limits_{v\in S_\mathbb{C}(K)} \mu_p & \longrightarrow & H^0(G_S, C_S(\mathbb{Z}/p\mathbb{Z})) & \longrightarrow & & H_2(G_S,\mathbb{Z}/p\mathbb{Z}) & \longrightarrow & 0\\
 & \downarrow & & \| & & & \downarrow & & \\
0 \longrightarrow & {}_pD_S(K) & \longrightarrow & {}_pC_S(K) & \xrightarrow{rec} & & {}_pG_S^{ab} & \longrightarrow & 0.
\end{array}$$

The exactness of the upper row follows from (8.6.12), and the exactness of the lower row follows from the divisibility of $D_S(K)$. The exactness of the column follows from the short exact sequence $\mathbb{Z}_p \hookrightarrow \mathbb{Z}_p \twoheadrightarrow \mathbb{Z}/p\mathbb{Z}$ and the dotted arrow is the induced one. Hence the snake lemma gives the exact sequence

$$0 \to \prod_{v\in S_\mathbb{C}(K)} \mu_p \to {}_pD_S(K) \to H_2(G_S,\mathbb{Z}_p)/p \to 0.$$

By Nakayama's lemma, the vanishing of $H_2(G_S,\mathbb{Z}_p)/p$ is equivalent to the vanishing of $H_2(G_S,\mathbb{Z}_p)$ itself and therefore we have proved the equivalence (vi) $\Leftrightarrow$ (vii).

Corollary (10.3.7) follows from the above proof, and (10.3.9) follows from the equivalence (i) ⇔ (vi) and from corollary (3.3.4). Using (2.7.12), one obtains statement (10.3.10) for continuous cochain cohomology. Corollary (10.3.11)(i) follows from (10.3.7) and from the commutative diagram

$$\begin{array}{ccc} E_K \otimes_{\mathbb{Z}} \mathbb{Z}_p & \longrightarrow & \prod\limits_{\mathfrak{P}\in S_p(K)} \hat{U}_{\mathfrak{P}}(K) \\ \uparrow & & \uparrow \\ E_k \otimes_{\mathbb{Z}} \mathbb{Z}_p & \longrightarrow & \prod\limits_{\mathfrak{p}\in S_p(k)} \hat{U}_{\mathfrak{p}}(k). \end{array}$$

Finally, (10.3.11)(ii) follows from the same diagram, noting that $\operatorname{rank}_{\mathbb{Z}} E_K = \operatorname{rank}_{\mathbb{Z}} E_k$ if k is the maximal totally real subfield of the CM-field K. □

Remark: From the equivalence (i) ⇔ (ii) in (10.3.6) and from the exact sequence in (10.3.13) it follows that the Leopoldt conjecture is true for K and p if and only if the canonical surjective homomorphism

$$E_{K,T} \otimes_{\mathbb{Z}} \mathbb{Z}_p \longrightarrow \bar{E}_{K,T}^{(S)}$$

is an isomorphism for one (and hence every) pair $S \supseteq T$ of finite sets of primes of K with $S \supseteq S_p \cup S_\infty$.

Having established a number of equivalent formulations of the Leopoldt conjecture, the question arises as to which of them might be the best one to actually prove the conjecture. We will see later in §6 that the cohomological condition (vi) can be verified for pairs (K, p) provided that a certain arithmetic invariant vanishes. In this situation one can calculate $H_2(G_S, \mathbb{Z}_p)$ in terms of local cohomology groups and then one uses (7.2.5). This method, however, only covers cases which one should view as degenerate.

The authors do not know an algebraic proof of the Leopoldt conjecture in any generic case.*)

If the number field K is a finite abelian extension of a number field k with $r_1(k) + r_2(k) = 1$ (i.e. $k = \mathbb{Q}$ or k imaginary quadratic), then the Leopoldt conjecture can be deduced from the following deep result from transcendence theory. It was proved by *A. Brumer* and generalizes an archimedean result of *A. Baker* to the p-adic case.

*) Attempts to give an algebraic proof often run into the problem that projective and inductive limits do not commute.

(10.3.14) Theorem (*Brumer*). *Let $\alpha_1, \dots, \alpha_n \in \mathbb{C}_p^\times$ be algebraic over $\mathbb{Q}$ and assume that*

$$\log_p(\alpha_1), \dots, \log_p(\alpha_n) \in \mathbb{C}_p$$

are linearly independent over $\mathbb{Q}$. Then these logarithms are also linearly independent over the algebraic closure of $\mathbb{Q}$ in $\mathbb{C}_p$.

For the proof we refer the reader to [20]. In the following we will make use of Dedekind's determinant relation:

(10.3.15) Lemma. *Let G be a finite group and let $f : G \to \mathbb{C}_p$ be a map. Then*

$$\operatorname{rank}(f(\sigma^{-1}\tau)_{\sigma,\tau\in G}) = \#\{\chi \in \operatorname{Hom}(G, \mathbb{C}_p^\times) \mid \sum_{\sigma\in G} \chi(\sigma)f(\sigma) \neq 0\}.$$

Proof: The $\mathbb{C}_p$-vector space of all maps of G to $\mathbb{C}_p$ has two natural bases, namely

(1) the characters $\chi \in \operatorname{Hom}(G, \mathbb{C}_p^\times)$ and

(2) the characteristic maps $\mathfrak{d}_\sigma$, $\sigma \in G$, with $\mathfrak{d}_\sigma(\tau) = \begin{cases} 1 & \text{for } \sigma = \tau, \\ 0 & \text{for } \sigma \neq \tau. \end{cases}$

Consider the linear operator T with

$$Tg(\tau) := \sum_{\sigma\in G} f(\sigma)g(\sigma\tau) \quad \text{for} \quad g : G \to \mathbb{C}_p$$

on this vector space. Then one easily calculates that T is represented by the diagonal matrix $\operatorname{diag}(\sum_{\sigma\in G}\chi(\sigma)f(\sigma))_{\chi\in\operatorname{Hom}(G,\mathbb{C}_p^\times)}$ with respect to basis (1) and is represented by the matrix $(f(\sigma^{-1}\tau)_{\sigma,\tau\in G})$ with respect to basis (2). □

(10.3.16) Theorem. *Assume that the number field K is an abelian extension of $\mathbb{Q}$ or of an imaginary quadratic number field. Then the Leopoldt conjecture holds for K and every prime number p.*

Proof: If K is an abelian extension of $\mathbb{Q}$, we can replace K by its maximal real subfield if necessary, because of the equivalence (i) ⇔ (iii) of (10.3.6). Hence we may assume that K is an abelian extension of a subfield k such that E_k is finite and no archimedean prime splits in $K|k$. Then by (8.7.2) there exists an isomorphism

$$E_K \otimes \mathbb{Q} \cong I_G \otimes \mathbb{Q}$$

as $\mathbb{Q}[G]$-modules, where $G = G(K|k)$ and I_G is the augmentation ideal of $\mathbb{Z}[G]$. Therefore there exists a unit $\varepsilon \in E_K$ *) such that $\{\sigma(\varepsilon)\}_{1\neq\sigma\in G}$ is an independent system of units of K.

*) If $K|\mathbb{Q}$ is abelian, ε is called a Minkowski unit.

Now fix any embedding $\phi : K \hookrightarrow \mathbb{C}_p$ and consider the map

$$f: \begin{array}{ccc} G & \to & \mathbb{C}_p, \\ \sigma & \mapsto & \log_p \phi(\sigma\varepsilon). \end{array}$$

Clearly $\{\phi \circ \sigma^{-1}\}_{\sigma \in G}$ are $r_1 + r_2 = \#G$ different embeddings $K \hookrightarrow \mathbb{C}_p$ and

$$\sum_{\sigma \in G} f(\sigma) = \sum_{\sigma \in G} \log_p \phi(\sigma\varepsilon) = \log_p \phi(\prod_{\sigma \in G} \sigma\varepsilon) = \log_p \pm 1 = 0.$$

Now assume that $\sum_{\sigma \in G} \chi(\sigma) \log_p \phi(\sigma\varepsilon) = 0$ for a nontrivial character $\chi \in \operatorname{Hom}(G, \mathbb{C}_p^\times)$. Then

$$\sum_{1 \neq \sigma \in G} (1 - \chi(\sigma)) \cdot \log_p \phi(\sigma\varepsilon) = 0,$$

and the elements $(1 - \chi(\sigma))$ are algebraic over $\mathbb{Q}$ and not all are zero. By (10.3.14), there exist $n_\sigma \in \mathbb{Z}$, not all zero, with

$$\sum_{1 \neq \sigma \in G} n_\sigma \cdot \log_p \phi(\sigma\varepsilon) = 0, \text{ i.e. } \prod_{1 \neq \sigma \in G} (\sigma\varepsilon)^{n_\sigma} \in \mu(K).$$

This, however, contradicts the choice of the unit ε. With the help of (10.3.15), we obtain

$$r_1 + r_2 - 1 = \#G - 1 = \operatorname{rank}(\log_p \phi(\sigma^{-1}\tau\varepsilon)_{\sigma,\tau \in G}) \leq \operatorname{rr}_p(K).$$

This yields the nontrivial inequality, and hence the theorem is proved. □

The Leopoldt conjecture is closely related to the existence of certain infinite Galois extensions of K, the so-called $\mathbb{Z}_p$-extensions.

(10.3.17) Definition. *Let $L|K$ be a Galois extension of fields. We call L a* **$\mathbb{Z}_p$-extension** *of K if $G(L|K)$ is a free pro-p-group of rank 1, i.e. (noncanonically) isomorphic to the additive group $\mathbb{Z}_p$.*

The closed subgroups of $\mathbb{Z}_p$ are exactly the groups $p^n\mathbb{Z}_p$ for $n = 0, 1, 2, \dots, \infty$, where by convention $p^\infty\mathbb{Z}_p = 0$. Hence we can list the extensions of K inside L in the form

$$K = K_0 \subsetneqq K_1 \subsetneqq K_2 \subsetneqq \cdots \subsetneqq K_\infty = L,$$

where $G(K_n|K_m) \cong \mathbb{Z}_p/p^{n-m}\mathbb{Z}_p$ for $n \geq m$. In particular, $L = K_\infty$ is also a $\mathbb{Z}_p$-extension of K_n for every $0 \leq n < \infty$.

Assume that $p \neq \mathrm{char}(K)$ and consider the extension

$$K(\mu_{p^\infty})|K,$$

which is obtained by adjoining all roots of unity of p-power order to K. This is a Galois extension and $G(K(\mu_{p^\infty})|K)$ is canonically isomorphic to the image of the p-part of the cyclotomic character

$$\kappa_p : G_K \to \mathbb{Z}_p^\times,$$

which is given by $g(\zeta) = \zeta^{\kappa_p(g)}$ for $\zeta \in \mu_{p^\infty}$, $g \in G$, cf. (7.3.6). Since $\mathbb{Z}_p^\times \cong \mathbb{Z}/(p-1)\mathbb{Z} \oplus \mathbb{Z}_p$ for $p \neq 2$ and $\mathbb{Z}_2^\times \cong \mathbb{Z}/2\mathbb{Z} \oplus \mathbb{Z}_2$, the image of κ_p is either finite or has a unique quotient isomorphic to $\mathbb{Z}_p$.

(10.3.18) Definition. *If $K(\mu_{p^\infty})|K$ is infinite (which is the case if K is a local or global field of characteristic $\neq p$), then it contains a unique $\mathbb{Z}_p$-extension K_∞, which we call the* **cyclotomic $\mathbb{Z}_p$-extension** *of K.*

If $\zeta_{2p} \in K$, then clearly $K_\infty = K(\mu_{p^\infty})$. A finite field has a unique $\mathbb{Z}_p$-extension for every p, which is the cyclotomic $\mathbb{Z}_p$-extension if $\mathrm{char}(K) \neq p$.

In the following we will always assume that $p \neq \mathrm{char}(K)$ if K is a local or global field.

Let K be a local field with residue field k. Then K has a unique unramified $\mathbb{Z}_p$-extension, which is the cyclotomic one if $p \neq \mathrm{char}(k)$. If $p = \mathrm{char}(k)$, then the cyclotomic $\mathbb{Z}_p$-extension is ramified, so it is different to the unramified $\mathbb{Z}_p$-extension (however, their first n steps may coincide for some $n < \infty$).

The composite of all $\mathbb{Z}_p$-extensions of K is equal to the maximal p-extension of K with a torsion-free abelian Galois group. Hence

$$\mathrm{rank}_{\mathbb{Z}_p} H_1(G_K, \mathbb{Z}_p)$$

is the (possibly infinite) number of independent $\mathbb{Z}_p$-extensions of K. Therefore the following lemma is an easy consequence of (7.5.3) and (7.3.10).

(10.3.19) Lemma. (i) *If K is a local field with $p \neq \mathrm{char}(K) > 0$, then K has exactly one (namely the unramified) $\mathbb{Z}_p$-extension which comes by base change from the unique $\mathbb{Z}_p$-extension of the finite constant field.*

(ii) *If K is a finite extension of $\mathbb{Q}_\ell$, then there exists exactly one (namely the unramified) $\mathbb{Z}_p$-extension of K if $p \neq \ell$, and when $p = \ell$, the number of independent $\mathbb{Z}_p$-extensions of K is equal to $[K : \mathbb{Q}_p] + 1$.*

Now assume that K is a global field (of characteristic $\neq p$). Then by (10.3.19) a $\mathbb{Z}_p$-extension is unramified at all primes of residue characteristic $\neq p$. Since $Cl(K)$ (resp. $Cl^0(K)$ in the function field case) is finite, we observe by (10.3.7) the

(10.3.20) Proposition. (i) *A global function field K of characteristic $\neq p$ has exactly one $\mathbb{Z}_p$-extension which comes by base change from the unique $\mathbb{Z}_p$-extension of the finite constant field; in particular, it is unramified.*

(ii) *If K is a number field, then every $\mathbb{Z}_p$-extension is unramified outside p and is ramified at least at one prime dividing p. The number of independent $\mathbb{Z}_p$-extensions of K is equal to $r_2 + 1 + \mathfrak{d}_p$, where r_2 is the number of complex places of K and $\mathfrak{d}_p$ is the Leopoldt defect* (see (10.3.7))*; in particular,*

$$\operatorname{rank}_{\mathbb{Z}_p} G_S(K)^{ab}(p) = r_2 + 1 + \mathfrak{d}_p \leq [K : \mathbb{Q}].$$

Hence another formulation of the Leopoldt conjecture is that there are at most $r_2 + 1$ independent $\mathbb{Z}_p$-extensions of K. If K is totally real, this means that the cyclotomic $\mathbb{Z}_p$-extension is the unique $\mathbb{Z}_p$-extension of K.

From now on we will restrict to the number field case. Let K be a number field and let $K_\infty|K$ be any $\mathbb{Z}_p$-extension.

(10.3.21) Definition. *We say that the* **weak Leopoldt conjecture** *holds for $K_\infty|K$ if the numbers $\mathfrak{d}_p(K_n)$ are bounded independently from $n \in [0, \infty)$.*

We will see in a moment that the (strong) Leopoldt conjecture for K and p implies the weak Leopoldt conjecture for every $\mathbb{Z}_p$-extension of K, which justifies the name.

In order to give a couple of equivalent formulations of the weak Leopoldt conjecture, let us fix some notation. Let

$K_\infty|K$ be a fixed $\mathbb{Z}_p$-extension,
$\Gamma = G(K_\infty|K) \cong \mathbb{Z}_p$,
$\Gamma_n = G(K_\infty|K_n) \subseteq \Gamma$,
$S \supseteq S_p \cup S_\infty$ be a finite set of primes of K,
$G_S = G(K_S|K)$,
$H_S = G(K_S|K_\infty) \subseteq G_S$,
$X_S = H_1(H_S, \mathbb{Z}_p)$.

Then X_S is a $\Lambda = \mathbb{Z}_p[[\Gamma]]$-module in a natural way.

(10.3.22) Theorem. *The following assertions are equivalent:*

(i) *The weak Leopoldt conjecture is true for* $K_\infty|K$.

(ii) $H_2(H_S, \mathbb{Z}_p) = 0$.

(iii) X_S *has no finite nontrivial submodules and* $\mathrm{rank}_\Lambda X_S = r_2$.

Observe that (i) is independent of the choice of the finite set of primes $S \supseteq S_\infty \cup S_p$. Hence (ii) and (iii) are true for all S if they are true for one S.

Proof: By (8.3.20), we can apply the results at the end of V §6 (p.331 ff) to the triple H_S, G_S, Γ. By (8.7.7), $H_2(G_S, \mathbb{Z}_p)$ is p-torsion-free, and by (8.7.5), $\chi_2(G_S, \mathbb{Z}/p\mathbb{Z}) = -r_2$. Hence the equivalence (ii) $\Leftrightarrow$ (iii) follows from (5.6.17).

Now assume $H_2(H_S, \mathbb{Z}_p) = 0$. Then the Hochschild-Serre spectral sequence implies that

$$H_2(G_S(K_n), \mathbb{Z}_p) \cong H_1(\Gamma_n, H_1(H_S, \mathbb{Z}_p)) = X_S^{\Gamma_n}.$$

Therefore $\mathfrak{d}_p(K_n)$ is globally bounded by $\mathrm{rank}_{\mathbb{Z}_p} X_S^\delta$, where X_S^δ is the maximal discrete submodule of X_S (see (5.3.12)). This shows (ii) $\Rightarrow$ (i).

Suppose that (i) holds. Then by (10.3.7), the group

$$H_2(H_S, \mathbb{Z}_p) = \varprojlim_n H_2(G_S(K_n), \mathbb{Z}_p)$$

is a finitely generated $\mathbb{Z}_p$-module. Assume for a moment that $cd_p\, G_S \le 2$ (i.e. $p \neq 2$ or $S_\mathbb{R}(K) = \varnothing$). Then $H_2(H_S, \mathbb{Z}_p)$ is also a free Λ-module by (5.6.15)(ii). This shows (ii).

If $p = 2$ and $S_\mathbb{R}(K) \neq \varnothing$, we have to modify the above argument. Since $K_\infty|K$ is unramified at all infinite places, it is disjoint from $K(i)|K$ (or any totally imaginary quadratic extension of K). We identify Γ with $G(K_\infty(i)|K(i))$ and we denote the open normal subgroup $G_S(K_\infty(i)) \subseteq H_S$ by H'_S. Then, as above, we conclude that $H_2(H'_S, \mathbb{Z}_2)$ is a free Λ-module. It is therefore sufficient to find a Λ-invariant injection $H_2(H_S, \mathbb{Z}_2) \hookrightarrow H_2(H'_S, \mathbb{Z}_2)$ in order to finish the proof. Recall (see (8.7.7)) that $H_2(G_S(K_n), \mathbb{Z}_2)$ is torsion-free for all n. Therefore $H^2(H_S, \mathbb{Q}_2/\mathbb{Z}_2)$ is divisible, and the usual restriction-corestriction argument (use (1.5.7)) implies that

$$cor : H^2(H'_S, \mathbb{Q}_2/\mathbb{Z}_2) \to H^2(H_S, \mathbb{Q}_2/\mathbb{Z}_2)$$

is surjective. This implies the required injection on the homology. □

(10.3.23) Corollary. *If the (strong) Leopoldt conjecture is true for* K *and* p, *then the weak Leopoldt conjecture is true for every* $\mathbb{Z}_p$*-extension* K_∞ *of* K.

Proof: This follows from the injectivity of the map

$$H_2(H_S,\mathbb{Z}_p)_\Gamma \hookrightarrow H_2(G_S,\mathbb{Z}_p)$$

and from Nakayama's lemma. □

(10.3.24) Corollary. *The weak Leopoldt conjecture is true for a $\mathbb{Z}_p$-extension $K_\infty|K$ if and only if the canonical homomorphism*

$$\varprojlim_n E_{K_n,T}\otimes_{\mathbb{Z}}\mathbb{Z}_p \longrightarrow \varprojlim_n \bar{E}^{(S)}_{K_n,T}$$

is an isomorphism for one (and hence every) pair $S \supseteq T$, $S \supseteq S_p \cup S_\infty$, of finite sets of primes of K.

Proof: By (10.3.13), we have canonical exact sequences of compact abelian groups for all n

$$0 \longrightarrow H_2(G_{K_n,S},\mathbb{Z}_p) \longrightarrow E_{K_n,T}\otimes_{\mathbb{Z}}\mathbb{Z}_p \longrightarrow \bar{E}^{(S)}_{K_n,T} \longrightarrow 0\,.$$

Passing to the projective limit over n, the result follows from (10.3.22). □

In chapter XI we will investigate how the (strong) Leopoldt conjecture is encoded in the Iwasawa module structure of $X_S = H_1(H_S,\mathbb{Z}_p)$. We finish this section with the

(10.3.25) Theorem. *Let K be a number field and p be a prime number. Then the weak Leopoldt conjecture is true for the cyclotomic $\mathbb{Z}_p$-extension of K.*

Proof: We verify condition (ii) of (10.3.22). Since the weak Leopoldt conjecture descends, we may replace K by any finite extension. Therefore we may assume $\mu_{2p} \subseteq K$, and so $\mu_{p^\infty} \subseteq K_\infty$, i.e. the p-part of the cyclotomic character is trivial on H_S. We obtain

$$H_2(H_S,\mathbb{Z}_p)(-1)^\vee = Ш^2(H_S,\mathbb{Q}_p/\mathbb{Z}_p)(1) = Ш^2(H_S,\mu_{p^\infty}),$$

where the first equality follows from (7.2.5). Then Poitou-Tate duality implies

$$\begin{aligned} Ш^2(H_S,\mu_{p^\infty}) &= \varinjlim_{n,m} Ш^2(G_S(K_n),\mu_{p^m}) \\ &= \varinjlim_{n,m} Ш^1(G_S(K_n),\mathbb{Z}/p^m\mathbb{Z})^\vee \\ &= \varinjlim_{n,m} Cl_S(K_n)/p^m \\ &= \varinjlim_{n} Cl_S(K_n)\otimes_{\mathbb{Z}}\mathbb{Q}_p/\mathbb{Z}_p \;=\; 0\,, \end{aligned}$$

since $Cl_S(K_n)$ is finite for every n. □

Remark: Instead of using Poitou-Tate duality, we could have considered the exact sequence

$$0 \longrightarrow \mu_{p^\infty} \longrightarrow E_{K_S,S} \longrightarrow E_{K_S,S}/\mu_{p^\infty} \longrightarrow 0.$$

Since E_S/μ_{p^∞} is uniquely p-divisible, the result then follows from proposition (8.3.11).

(10.3.26) Corollary. *Let p be a prime number and let K be a number field which is totally imaginary if $p = 2$. Let K_∞ be the cyclotomic $\mathbb{Z}_p$-extension of K. Then for every set S of primes of K containing $S_p \cup S_\infty$ we have*

$$scd_p\, G(K_S|K_\infty) \le 2\,.$$

Proof: If U is an open subgroup of $G(K_S|K_\infty)$, then there exists a finite extension $L|K$ inside K_S such that $U = G(K_S|L_\infty)$. Thus by (10.3.25) we have $H^2(U, \mathbb{Q}_p/\mathbb{Z}_p) = 0$ and therefore $H^3(U, \mathbb{Z})(p) = 0$. Since $cd_p\, G(K_S|K_\infty) \le cd_p\, G(K_S|K) \le 2$, the result follows from (3.3.4). □

Remark: We will see in (10.11.3) that under the above assumptions the field K_S is always strictly larger than K_∞, hence $scd_p\, G(K_S|K_\infty) = 2$.

We have seen in this section that the Leopoldt conjecture is equivalent to the vanishing of $H^2(G_S, \mathbb{Q}_p/\mathbb{Z}_p)$, where S is any finite set of primes containing $S_p \cup S_\infty$. It is natural to consider also other Tate twists of $\mathbb{Q}_p/\mathbb{Z}_p$. For instance it is not difficult to show that $Ш^2(G_S, \mu_{p^\infty}) = 0$, and so we have an isomorphism

$$H^2(G_S, \mathbb{Q}_p/\mathbb{Z}_p(1)) \cong (\mathbb{Q}_p/\mathbb{Z}_p)^{\#S^f - 1}.$$

P. Schneider [201] has conjectured that the twist by +1 is the only Tate twist of $\mathbb{Q}_p/\mathbb{Z}_p$ having a nontrivial second cohomology group. For positive twists this has been proved by *C. Soulé* [219] by relating Galois cohomology to the higher K-groups of $\mathcal{O}_k$, which were defined and shown to be finitely generated by *D. Quillen*:

(10.3.27) Theorem. *Let k be a number field, let p be an odd prime number and assume that S is a finite set of primes of k containing $S_p \cup S_\infty$. Then*

$$H^2(G_S, \mathbb{Q}_p/\mathbb{Z}_p(i)) = 0$$

for $i \ge 2$.

Exercise: Show that the Leopoldt conjecture holds for k and p if and only if for all finite sets $T \supseteq S \supseteq S_p \cup S_\infty$, the natural map

$$\bigoplus_{\mathfrak{p}\in T\setminus S} T(k_{\mathfrak{p}}^{ab}(p)|k_{\mathfrak{p}}) \longrightarrow G(k_T^{ab}(p)|k)$$

is injective. Here $T(k_{\mathfrak{p}}^{ab}(p)|k_{\mathfrak{p}})$ denotes the inertia group of the Galois group of the maximal abelian p-extension of the local field $k_{\mathfrak{p}}$ (which is isomorphic to $\mu(k_{\mathfrak{p}})(p)$).

Hint: Use the Hochschild-Serre sequence and (10.5.3) to obtain the exact sequence

$$H_2(G_T(k),\mathbb{Z}_p) \to H_2(G_S(k),\mathbb{Z}_p) \to \bigoplus_{\mathfrak{p}\in(T\setminus S)(k)} T(k_{\mathfrak{p}}^{ab}(p)|k_{\mathfrak{p}}) \to G_T(k)(p)^{ab} \to G_S(k)(p)^{ab} \to 0.$$

Then pass to the limit over all T and use $H^2(G_k,\mathbb{Q}_p/\mathbb{Z}_p) = 0$.

§4. Cohomology of Large Number Fields

In this section we investigate algebraic extensions $K|\mathbb{Q}$ of not necessarily finite degree. We call (by abuse of language, and only in this and the next section) such a field K a number field and we will say that K is finite if it is of finite degree over $\mathbb{Q}$, i.e. if it is a number field in the strict sense.

Suppose that we are given sets of primes S and T of the number field K. Without further mention we will tacitly assume that $S = S'(K)$, respectively $T = T'(K)$, where S' and T' are sets of primes of a finite number field $k \subseteq K$. In other words, we assume that S and T are closed and open in $\mathrm{Sp}(K)$ (see §1). As before, S^f denotes the subset of finite primes in S. Finally, recall the notation

- K^T is the maximal extension of K which is completely decomposed at T,
- K_S^T is the maximal extension of K which is unramified outside S and completely decomposed at T.

(10.4.1) Definition. *Let p be a prime number. The field K is called p-(S,T)-closed if $K_S^T(p) = K$, i.e. if there is no nontrivial Galois p-extension of K inside K_S^T. We call p-$(S,\varnothing)$-closed fields p-S-closed for short.*

The following theorem generalizes a result of O. Neumann [162] about Galois groups of extensions of p-S-closed number fields.

(10.4.2) Theorem. *Let $L|K$ be a Galois extension of number fields and let p and $S \supseteq S_p \cup S_\infty$ be as above. Suppose that*

(i) *L is p-S-closed,*
(ii) *K is p-$(S_p \cup S_\infty)$-closed, or*
(ii)′ *p^∞ divides the absolute degree of $K_\mathfrak{p}$ for all $\mathfrak{p} \in S^f$, $S_\mathbb{R}(K) = \varnothing$ if $p = 2$, K is p-(S_∞, T)-closed for some $T \subseteq S$ and $\mu_p \subseteq K^T_{S_\infty}$.*

Then $H^2(G(L|K), \mathbb{Z}/p\mathbb{Z}) = 0$. (In particular, $cd_p\, G(L|K) \leq 1$ if $L|K$ is a p-extension.) If in addition $L \subseteq K_S$, then

$$H^i(G(L|K), \mathbb{Z}/p\mathbb{Z}) = 0 \quad \text{for all } i \geq 2.$$

Remark: Theorem (10.4.2) has the following function field analogue. Assume that $S \neq \varnothing$ and $p \neq \mathrm{char}(K)$. Then the statement of (10.4.2) remains true with the following modifications of the second assumption:

- replace (ii) by: K is p-$\varnothing$-closed,
- replace (ii)′ by: p^∞ divides the absolute degree of $K_\mathfrak{p}$ for all $\mathfrak{p} \in S$, K is p-$(\varnothing, T)$-closed for some $T \subseteq S$ and $\mu_p \subseteq K^T_\varnothing$.

All corollaries have their obvious function field analogue.

(10.4.3) Corollary (Neumann). *Let $L|K$ be a Galois extension of p-S-closed number fields with $S \supseteq S_p \cup S_\infty$ and $L \subseteq K_S$. Then*

$$H^i(G(L|K), \mathbb{Z}/p\mathbb{Z}) = 0 \quad \text{for all } i \geq 1.$$

(10.4.4) Corollary. *Let k be a finite number field and let $S \supseteq T \supseteq S_p \cup S_\infty$ be sets of primes of k. Then*

$$cd_p\, G(k_S(p)|k_T(p)) \leq 1.$$

(10.4.5) Corollary. *Let k be a totally imaginary finite number field, $S \supseteq S_p \cup S_\infty$ be a set of places of k and assume that $\mu_p \subseteq k$. Let $L' = (k_\infty)^{S_p}_{S_\infty}(p)$ be the maximal unramified p-extension of the cyclotomic $\mathbb{Z}_p$-extension k_∞ of k which is completely decomposed at all primes in S_p. Then*

$$cd_p\, G(k_S(p)|L') \leq 1.$$

For the proof of (10.4.2) we need the

(10.4.6) Lemma. *Let $L|K$ be a Galois extension of number fields and let $S \supseteq S_p \cup S_\infty$ be a set of primes in K. Then the inflation maps*

$$H^i(G(K_S|K), \mathcal{O}_{K_S,S}^\times)(p) \longrightarrow H^i(G(L_S|K), \mathcal{O}_{L_S,S}^\times)(p)$$

are isomorphisms for all $i \geq 0$.

Proof: Consider the exact sequence

$$0 \to \mathcal{O}_{k,S}^\times \to k^\times \to \bigoplus_{\mathfrak{p} \notin S(k)} \mathbb{Z} \to Cl_S(k) \to 0,$$

where k is any finite number field (containing the finite number field over which we assume the set S to be defined). Passing to the limit over all k, and since $Cl(\bar{K}) = 0$, we obtain the exact sequence

$$0 \to \mathcal{O}_{\bar{K},S}^\times \to \bar{K}^\times \to \varinjlim_{k \subseteq \bar{K}} \bigoplus_{\mathfrak{p} \notin S(k)} \mathbb{Z} \to 0.$$

The transition maps in the limit on the right are induced by multiplication with the ramification index. Therefore the limit on the right is a $\mathbb{Q}$-vector space; in particular, it is cohomologically trivial as a $G(\bar{K}|K)$-module. Taking $G(\bar{K}|K_S)$-cohomology and recalling that $Cl_S(K_S) = 0$ (take the direct limit over $Cl_S(k_S) = 0$, $k \subseteq K$ finite), we obtain isomorphisms for all $i \geq 1$

$$H^i(G(\bar{K}|K_S), \mathcal{O}_{\bar{K},S}^\times) \xrightarrow{\ \sim\ } H^i(G(\bar{K}|K_S), \bar{K}^\times).$$

Hence $H^i(G(\bar{K}|K_S), \mathcal{O}_{\bar{K},S}^\times)(p)$ is zero for all $i \geq 1$: for $i = 1$ by Hilbert's Satz 90, for $i = 2$ by (8.1.14) (ii) since the absolute local degree of every $\mathfrak{p} \in S^f$ is divisible by p^∞ (observe that $\mathbb{Q}(\mu_{p^\infty}) \subseteq K_S$), and for $i \geq 3$ since $scd_p\, G(\bar{K}|K_S) \leq 2$ by (10.2.3). Therefore the Hochschild-Serre spectral sequence implies isomorphisms for all i

$$H^i(G(K_S|K), \mathcal{O}_{K_S,S}^\times)(p) \xrightarrow{\ \sim\ } H^i(G(\bar{K}|K), \mathcal{O}_{\bar{K},S}^\times)(p).$$

We also can replace K_S by L_S in the above argument. Then the diagram

$$\begin{array}{ccc} H^i(G(L_S|K), \mathcal{O}_{L_S,S}^\times)(p) & \xrightarrow{\ \sim\ } & H^i(G(\bar{K}|K), \mathcal{O}_{\bar{K},S}^\times)(p) \\ \uparrow & \nearrow{\scriptstyle\sim} & \\ H^i(G(K_S|K), \mathcal{O}_{K_S,S}^\times)(p) & & \end{array}$$

shows the statement of the lemma. □

Proof of (10.4.2): First observe that in case (ii), K contains the cyclotomic $\mathbb{Z}_p$-extension of $\mathbb{Q}$. Hence the local field $K_\mathfrak{p}$ has absolute degree divisible by p^∞ for every prime $\mathfrak{p} \in S^f$. In addition $S_\mathbb{R}(K) = \varnothing$ if $p = 2$ in case (ii). Let us write $H^i(L|K)$ for $H^i(G(L|K), \mathbb{Z}/p\mathbb{Z})$ in the following.

We first assume that $L = L_S$. Let $K' := K(\mu_p)$. From (8.3.11)(iii) and lemma (10.4.6) we obtain a inclusion

$$ {}_pH^2(G(L|K'), \mathcal{O}_{L,S}^\times) \hookrightarrow {}_p\Big(\prod_{\mathfrak{p}\in S} H^2(K'_\mathfrak{p}, \mu_{p^\infty})\Big) $$

showing that

$$ {}_pH^2(G(L|K'), \mathcal{O}_{L,S}^\times) = 0 $$

by (7.1.8)(i). Therefore the Kummer sequence implies the isomorphism

$$ H^1(G(L|K'), \mathcal{O}_{L,S}^\times)/p \xrightarrow{\sim} H^2(G(L|K'), \mu_p). $$

Twisting by (-1) and taking $G = G(K'|K)$-invariants, we obtain isomorphisms (observe that G is of order prime to p)

$$ (*) \qquad H^2(L|K) \cong H^2(L|K')^G \cong (H^1(G(L|K'), \mathcal{O}_{L,S}^\times)/p)(-1)^G. $$

Now let

$$ \mathcal{K} := \begin{cases} K_{S_p\cup S_\infty} & \text{in case (ii)} \\ K^T_{S_\infty} & \text{in case (ii)}'. \end{cases} $$

Then $K' \subseteq \mathcal{K}$ and $\mathcal{K}$ is closed under p-extensions, which are (everywhere) unramified and completely decomposed at the primes in S. Therefore (8.3.11), (10.4.6) and the principal ideal theorem, imply

$$ H^1(G(L|\mathcal{K}), \mathcal{O}_{L,S}^\times)(p) \cong \varinjlim_{k\subseteq\mathcal{K}} Cl_S(k)(p) = 0. $$

The Hochschild-Serre sequence therefore yields an isomorphism

$$ (**) \qquad H^1(G(\mathcal{K}|K'), \mathcal{O}_{\mathcal{K},S}^\times)(p) \cong H^1(G(L|K'), \mathcal{O}_{L,S}^\times)(p). $$

By definition of $\mathcal{K}$ and by our assumptions on K, we have

$$ H^1(G(\mathcal{K}|K'), \mu_p)(-1)^G \cong H^1(\mathcal{K}|K) = 0. $$

Using the Kummer sequence, the last equality implies

$$ \big({}_pH^1(G(\mathcal{K}|K'), \mathcal{O}_{\mathcal{K},S}^\times)(-1)\big)^G = 0 $$

and hence by the isomorphism $(**)$ that

$$ (***) \qquad \big({}_pH^1(G(L|K'), \mathcal{O}_{L,S}^\times)(-1)\big)^G = 0. $$

Since $H^1(G(L|K'), \mathcal{O}_{L,S}^\times)(p) \cong Cl_S(K')(p)$ and $Cl_S(K')$ is a torsion group, we conclude that

$$ (H^1(G(L|K'), \mathcal{O}_{L,S}^\times)/p)(-1)^G = 0. $$

Finally, the isomorphism $(*)$ shows that $H^2(L|K) = 0$.

Now drop the assumption that $L = L_S$. The group $H^1(L_S|L)$ vanishes by condition (i), so that

$$ H^2(L|K) \hookrightarrow H^2(L_S|K), $$

which proves $H^2(L|K) = 0$; in particular, $G(L|K)$ is free if $L|K$ is a p-extension.

If $L \subseteq K_S$, then consider the Hochschild-Serre spectral sequence for the extensions $K_S|L|K$. By the above arguments, we know that $H^2(K_S|L) = 0$, and since $cd_p\, G(K_S|L) \le cd_p\, G_S(K) \le 2$, we obtain isomorphisms

$$H^i(L|K) \cong H^i(K_S|K) = 0$$

for all $i \ge 3$. □

(10.4.7) Corollary (NEUMANN). *Let K be a p-$(S_p \cup S_\infty)$-closed number field. Then for every set of primes $S \supseteq S_p \cup S_\infty$,*

$$Cl_S(K(\mu_p))(p)(j)^{G(K(\mu_p)|K)} = 0$$

for the Tate twists $j = 0, -1$.

Proof: The case $j = -1$ follows from equation $(***)$ in the last proof. The case $j = 0$ is a direct consequence of the principal ideal theorem. □

The following corollary is a result of O. NEUMANN in the case that $\mathfrak{c}$ is the class of p-groups and that M is a finite p-torsion module.

(10.4.8) Corollary. *Let $\mathfrak{c}$ be a full class of finite groups with $\mathbb{Z}/p\mathbb{Z} \in \mathfrak{c}$. Let k be a finite number field and $S \supseteq S_p \cup S_\infty$ be a set of primes of k. Then the inflation maps*

$$H^i(G(k_S(\mathfrak{c})|k), M)(p) \longrightarrow H^i(G(k_S|k), M)(p)$$

are isomorphisms for all i and every $G(k_S(\mathfrak{c})|k)$-module M.

Proof: Using the Hochschild-Serre spectral sequence it suffices to show that

$$H^i(G(k_S|k_S(\mathfrak{c})), M) = 0$$

for all $i \ge 1$ and every trivial $G(k_S|k_S(\mathfrak{c}))$-module M. Since cohomology commutes with direct limits, one easily reduces to the cases $M = \mathbb{Z}, \mathbb{Z}/p\mathbb{Z}$. Using the exact sequence $0 \to \mathbb{Z} \to \mathbb{Q} \to \mathbb{Q}/\mathbb{Z} \to 0$, we finally reduce to the case $M = \mathbb{Z}/p\mathbb{Z}$. Now the corollary follows from (10.4.2), since conditions (i) and (ii) are clearly satisfied. □

Remark: In §6 we will prove results corresponding to (10.4.7) and (10.4.8) in the case where $p = 2$ and $S \supseteq S_2$ does not necessarily contain all infinite primes, see (10.6.6) and (10.6.7).

(10.4.9) Corollary. *Let p be a prime number, k a number field and S a set of primes of k containing $S_p \cup S_\infty$. Then*

(i) $cd_p\, G(k_S(p)|k) \le cd_p\, G(k_S|k)$,

(ii) $scd_p\, G(k_S(p)|k) \le scd_p\, G(k_S|k)$.

Assume, in addition, that k is totally imaginary if $p = 2$ and let k_∞ be the cyclotomic $\mathbb{Z}_p$-extension of k. Then

(iii) $scd_p\, G(k_S(p)|k_\infty) \le 2$.

Proof: This follows from (10.4.8) and (10.3.26). □

§5. Riemann's Existence Theorem

Now we are prepared to prove the number field analogue of Riemann's existence theorem. In the special case $k = \mathbb{Q}$, it is due to *J. Neukirch* (see [155]). A first attempt to generalize Neukirch's theorem to arbitrary number fields was made by *O. Neumann* [163]; however, he was still lacking the notion of generalized free products of profinite groups over a topological base. In the special case $k = \mathbb{Q}$, the product occurring is a product over a discrete base, hence the use of generalized products may be dispensed with. In the general case its use is crucial. In the form presented below the theorem was proved in [251]*). In this section we keep the convention of calling an algebraic extension of k of $\mathbb{Q}$ a number field and to call k a finite number field if $[k:\mathbb{Q}] < \infty$.

(10.5.1) Riemann's Existence Theorem (Number Theoretical Analogue). *Let k be a finite number field, p a prime number and $T \supseteq S \supseteq S_p \cup S_\infty$ sets of primes of k. Then the canonical homomorphism*

$$\phi_{T,S}\colon \mathop{\ast}_{\mathfrak{p}\in T\setminus S(k_S(p))} T(k_\mathfrak{p}(p)|k_\mathfrak{p}) \longrightarrow G(k_T(p)|k_S(p))$$

is an isomorphism. Here $T(k_\mathfrak{p}(p)|k_\mathfrak{p}) \subseteq G(k_\mathfrak{p}(p)|k_\mathfrak{p})$ is the inertia group and $\ast$ denotes the free pro-p-product.

*) In [251] one has to replace the definition of generalized free products by that given in chap. IV §3.

Remarks: 1. The index set $T \setminus S(k_S(p))$, resp. its closure in $\mathrm{Sp}(k_S(p))$, is a profinite index space (cf. the discussion in §1). The inertia groups $T(k_\mathfrak{p}(p)|k_\mathfrak{p})$ form a bundle of profinite groups over this base space. The bundle structure can either be defined directly, or we can view it as the bundle which is associated to the continuous family $T(k_\mathfrak{p}(p)|k_\mathfrak{p}) = T_\mathfrak{p}(k(p)|k)$ of subgroups in $G(k(p)|k)$ (cf. (9.4.3)).

As before, we do not specify the choice of a continuous section of the projection $T \setminus S(k_T(p)) \to T \setminus S(k_S(p))$. Since we work in the category of pro-p-groups, it is clear from the discussion in chapter IV that the particular isomorphism claimed in the theorem depends on the choice of such a section; but *whether the canonical homomorphism is an isomorphism or not* is independent of the chosen section. The straightforward extension of (4.1.5) to the case of generalized free products provides us with a convenient cohomological criterion. Finally, note that $T(k_\mathfrak{p}(p)|k_\mathfrak{p}) = 1$ for primes $\mathfrak{p} \in T \setminus S(k)$ for which $N(\mathfrak{p}) \not\equiv 1 \bmod p$, i.e. if there is no primitive p-th root of unity contained in the local field $k_\mathfrak{p}$, and $T(k_\mathfrak{p}(p)|k_\mathfrak{p}) \cong \mathbb{Z}_p(1)$ otherwise (see 7.5.2).

2. The assumption $S \supseteq S_\infty$ (only interesting if $p = 2$ and k has real places) can be removed from the theorem, see (10.6.1).

Before proving theorem (10.5.1), let us deduce some corollaries.

(10.5.2) Corollary. *Let K be a p-S-closed number field and assume that $T \supseteq S \supseteq S_p \cup S_\infty$. Then the canonical homomorphism*

$$\mathop{\ast}_{\mathfrak{p} \in T \setminus S(K)} G(K_\mathfrak{p}(p)|K_\mathfrak{p}) \longrightarrow G(K_T(p)|K)$$

is an isomorphism.

Proof: Let $k \subseteq K$ be a finite number field over which S and T are defined. Since K is p-S-closed, it contains the cyclotomic $\mathbb{Z}_p$-extension k_∞ of k. Hence $T(K_\mathfrak{p}(p)|K_\mathfrak{p}) = G(K_\mathfrak{p}(p)|K_\mathfrak{p})$ for every prime $\mathfrak{p} \in T \setminus S(K)$. Now the corollary follows from theorem (10.5.1) by passing to the limit over all finite subextensions of k in K. □

Specializing to the case $K = k_S$, where $S_\infty \cup S_p \subseteq S$, and recalling that $\mu_p \subseteq k_S$ in this case, we obtain the

(10.5.3) Corollary. *If $S_\infty \cup S_p \subseteq S \subseteq T$, then there is a canonical isomorphism*

$$\mathop{\ast}_{\mathfrak{p} \in T \setminus S(k_S)} \mathcal{T}_\mathfrak{p}(p) \longrightarrow G(k_T|k_S)(p),$$

where $\mathcal{T}_\mathfrak{p}$ is the inertia group in the full local group $\mathcal{G}_\mathfrak{p} = G_{k_\mathfrak{p}}$.

Proof of (10.5.1): By (10.4.4), the pro-p-group $G(k_T(p)|k_S(p))$ is free and the same is true for the free pro-p-product on the left. Therefore it suffices to show that the induced homomorphism on the abelianizations is an isomorphism. Using (4.3.14) and local class field theory, the abelianization of the free product can be calculated as

$$\Big(\mathop{\ast}_{\mathfrak{p}\in T\setminus S(k_S(p))} T(k_\mathfrak{p}(p)|k_\mathfrak{p})\Big)^{ab} \cong \varprojlim_{k'\subseteq k_S(p)} \prod_{\mathfrak{p}\in T\setminus S(k')} (T(k'_\mathfrak{p}(p)|k'_\mathfrak{p}))^{ab} \cong \varprojlim_{k'\subseteq k_S(p)} \prod_{\mathfrak{p}\in T\setminus S(k')} \hat{U}_\mathfrak{p},$$

where k' runs through the finite subextensions of k in $k_S(p)$ and $\hat{U}_\mathfrak{p}$ is the pro-p completion of the unit group of the local field $k'_\mathfrak{p}$. Comparing two copies of the upper sequence of (10.3.13) (for T and S) we therefore obtain the commutative exact diagram (writing $E_{k'}$ for $\mathcal{O}_{k'}^\times$ and G_T for $G_T(k')$)

$$\begin{array}{ccccccccc}
H_2(G_T,\mathbb{Z}_p) & \hookrightarrow & E_{k'}\otimes\mathbb{Z}_p & \to & \prod_{\mathfrak{p}\in T(k')} \hat{U}_\mathfrak{p} & \to & H_1(G_T,\mathbb{Z}_p) & \twoheadrightarrow & Cl(k')(p) \\
\downarrow & & \| & & \downarrow & & \downarrow & & \| \\
H_2(G_S,\mathbb{Z}_p) & \hookrightarrow & E_{k'}\otimes\mathbb{Z}_p & \to & \prod_{\mathfrak{p}\in S(k')} \hat{U}_\mathfrak{p} & \to & H_1(G_S,\mathbb{Z}_p) & \twoheadrightarrow & Cl(k')(p).
\end{array}$$

Now we pass to the limit over all $k' \subseteq k_S(p)$. Observe that

$$\begin{aligned}
H_2(G_T(k_S(p)),\mathbb{Z}_p) &= 0 \quad \text{by (10.4.2)},\\
H_2(G_S(k_S(p)),\mathbb{Z}_p) &= 0 \quad \text{by (10.4.2)},\\
H_1(G_S(k_S(p)),\mathbb{Z}_p) &= 0 \quad \text{by definition},\\
\varprojlim_{k'\subseteq k_S(p)} Cl(k')(p) &= 0 \quad \text{since } k_S(p) \text{ has no unramified } p\text{-extensions}.
\end{aligned}$$

Therefore we obtain the commutative exact diagram

$$\begin{array}{ccccc}
\varprojlim_{k'\subseteq k_S(p)} E_{k'}\otimes\mathbb{Z}_p & \hookrightarrow & \varprojlim_{k'\subseteq k_S(p)} \prod_{\mathfrak{p}\in T(k')} \hat{U}_\mathfrak{p} & \twoheadrightarrow & G(k_T(p)|k_S(p))^{ab} \\
\| & & \downarrow & & \\
\varprojlim_{k'\subseteq k_S(p)} E_{k'}\otimes\mathbb{Z}_p & \xrightarrow{\sim} & \varprojlim_{k'\subseteq k_S(p)} \prod_{\mathfrak{p}\in S(k')} \hat{U}_\mathfrak{p} & &
\end{array}$$

in the limit. This finishes the proof of theorem (10.5.1). □

The next corollary follows from (10.5.2) and (4.3.14).

(10.5.4) Corollary. *Let k be a finite number field, let p be a prime number and let $T \supseteq S \supseteq S_p \cup S_\infty$ be sets of primes of k. Assume that K is a p-S-closed Galois extension of k. Furthermore, let A be a p-primary abelian group. Then*

$$H^i(G(K_T|K), A) = 0 \quad \text{for } i \geq 2,$$

and there exists an isomorphism of $G(K|k)$-modules

$$H^1(G(K_T|K), A) \cong \bigoplus_{\mathfrak{p}\in T\setminus S(K)}{}' H^1(K_\mathfrak{p}, A).$$

Remark: Recall that

$$\bigoplus_{\mathfrak{p}\in T\setminus S(K)}{}' H^1(K_\mathfrak{p}, A) = \varinjlim_{k'} \bigoplus_{\mathfrak{p}\in T\setminus S(k')} H^1(k'_\mathfrak{p}, A),$$

where k' runs through the finite extensions of k inside K; see the remark after (4.3.13).

(10.5.5) Corollary. *Let A be a finite p-primary, trivial $G(k_S|k_\infty)$-module and let $T \supseteq S \supseteq S_p \cup S_\infty$ be sets of primes of the finite number field k. If k_∞ is the cyclotomic $\mathbb{Z}_p$-extension of k, then the canonical inflation map*

$$H^2(G(k_S|k_\infty), A) \longrightarrow H^2(G(k_T|k_\infty), A)$$

is an isomorphism.

Proof: Consider the Hochschild-Serre spectral sequence for the extensions $k_T|k_S|k_\infty$:

$$E_2^{ij} = H^i(G(k_S|k_\infty), H^j(G(k_T|k_S), A)) \Rightarrow H^{i+j}(G(k_T|k_\infty), A).$$

By (10.5.3), the spectral term $E_2^{1,1}$ vanishes and by (10.4.2) $E_2^{0,2} = 0$ also. Since $E_2^{2,0} = E_\infty^{2,0}$, this gives the statement of the corollary. □

Let us consider the special case $k = \mathbb{Q}$ and assume that p is odd. Then

$$\mathbb{Q}_{S_p \cup S_\infty}(p) = \mathbb{Q}_\infty.$$

Indeed, by the theorem of Kronecker-Weber, the maximal abelian extension of $\mathbb{Q}$ is obtained by adjoining all roots of unity. The ramification behaviour of these extensions is well-known, and

$$\mathbb{Q}_{S_p \cup S_\infty}(p)^{ab} = \mathbb{Q}_\infty.$$

Therefore $\dim_{\mathbb{F}_p} H^1(G_{S_p\cup S_\infty}, \mathbb{F}_p) = 1$, i.e. $G_{S_p\cup S_\infty}(p)$ is procyclic, which shows the desired identity.

Applying Riemann's existence theorem in this special case yields the following result of *Neukirch* [155]:

(10.5.6) Theorem (*Neukirch*). *Let $S \supseteq S_p \cup S_\infty$ be a finite set of primes of $\mathbb{Q}$ and assume that p is odd. Then the canonical homomorphisms define an isomorphism*

$$\mathop{\ast}_{\mathfrak{p}\in S\setminus(S_p\cup S_\infty)(\mathbb{Q}_\infty)} T(\mathbb{Q}_\mathfrak{p}(p)|\mathbb{Q}_\mathfrak{p}) \cong G(\mathbb{Q}_S(p)|\mathbb{Q}_\infty).$$

(10.5.7) Corollary. *Let $S \supseteq S_p \cup S_\infty$ be a finite set of primes of $\mathbb{Q}$. Assume that p is odd and that S contains at least one prime number $\equiv 1 \bmod p$. Then $G(\mathbb{Q}_S(p)|\mathbb{Q})$ is a pro-p duality group of dimension 2. If S does not contain a prime number $\equiv 1 \bmod p$, then*

$$\mathbb{Q}_S(p) = \mathbb{Q}_\infty.$$

Proof: Consider the group extension

$$(*) \qquad 1 \longrightarrow G(\mathbb{Q}_S(p)|\mathbb{Q}_\infty) \longrightarrow G(\mathbb{Q}_S(p)|\mathbb{Q}) \longrightarrow G(\mathbb{Q}_\infty|\mathbb{Q}) \longrightarrow 1.$$

By (10.5.6), the group on the left is a free product of inertia groups. These are trivial if $N(\mathfrak{p}) \not\equiv 1 \bmod p$ and isomorphic to $\mathbb{Z}_p$ otherwise. Therefore the group on the left is either trivial or a free pro-p-group of finite rank. In the latter case it follows from (3.7.5) that $G(\mathbb{Q}_S(p)|\mathbb{Q})$ is a duality group. □

Corollary (10.5.7) provides the first example where the maximal pro-p-factor of G_S is a duality group. We will investigate this phenomenon in more generality in §9.

The above exact sequence $(*)$ provides a canonical filtration of the group $G(\mathbb{Q}_S(p)|\mathbb{Q})$. Since $\Gamma := G(\mathbb{Q}_\infty|\mathbb{Q})$ is a free pro-p-group, the sequence splits. After choosing of a section $s : \Gamma \to G(\mathbb{Q}_S(p)|\mathbb{Q})$, the group Γ acts on the free product. If there is only one nontrivial factor in the free product (so that $G(\mathbb{Q}_S(p)|\mathbb{Q}_\infty)$ is abelian), then this action is easily understood. If there is more than one nontrivial factor in the free product, the question arises, whether it is possible to understand the non-abelian action of Γ (maybe for a specific choice of s). This interesting problem seems to be unsolved. A solution would give us a description of $G(\mathbb{Q}_S(p)|\mathbb{Q})$ as a pro-p-group in terms of generators and relations.

If the set of primes S has Dirichlet density $\delta(S) = 1$, then we have a more general form of Riemann's Existence Theorem, which also involves decomposition groups. It was first proved in [258]. Recall the following notation for sets of primes R and S:

$k^R(p)$ is the maximal p-extension of k which is completely decomposed at R,

$k_S^R(p)$ is the maximal p-extension of k which is unramified outside S and completely decomposed at R.

(10.5.8) Theorem. *Let k be a number field, p a prime number and $T \supseteq S \supseteq R$ sets of primes of k. If R is finite and $\delta(S) = 1$, then the natural homomorphism*

$$\phi_{T,S}^R : \mathop{\ast}_{\mathfrak{p} \in R(k_S^R(p))} G(k_\mathfrak{p}(p)|k_\mathfrak{p}) * \mathop{\ast}_{\mathfrak{p} \in T \backslash S(k_S^R(p))} T(k_\mathfrak{p}(p)|k_\mathfrak{p}) \xrightarrow{\sim} G(k_T(p)|k_S^R(p))$$

is an isomorphism.

Remarks: 1. We have made no assumption on the set $S_p \cup S_\infty$. In particular, the cyclotomic $\mathbb{Z}_p$-extension of k is not necessarily contained in $k_S^R(p)$. Further note that the obvious analogs of (10.5.2) and (10.5.3) hold.
2. The situation for $\mathfrak{c}$-extensions is much harder, where $\mathfrak{c}$ is a class of finite groups which is strictly larger than the class of p-groups for some prime number p. Recall that whether or not the homomorphism

$$\phi_{T,S}^R(\mathfrak{c}) : \mathop{\ast}_{\mathfrak{p} \in R(k_S^R(\mathfrak{c}))} G(k_\mathfrak{p}(\mathfrak{c})|k_\mathfrak{p}) * \mathop{\ast}_{\mathfrak{p} \in T \backslash S(k_S^R(\mathfrak{c}))} T(k_\mathfrak{p}(\mathfrak{c})|k_\mathfrak{p}) \longrightarrow G(k_T(\mathfrak{c})|k_S^R(\mathfrak{c}))$$

is an isomorphism depends on the choice of a continuous section s: $R(k^R(\mathfrak{c})) \to R(k(\mathfrak{c}))$.

If $\mathfrak{c}$ is the class of all finite groups, $T = S$ is the set of all primes and R is finite, then there exists a continuous section $s : R(k^R) \to R(\bar{k})$ such that ϕ^R is an isomorphism of profinite groups. This was proven in the special case that $R = S_\infty$ is the set of archimedean primes by *Fried-Haran-Völklein* in [52]; the general case was proven by *Pop* in [174].

In the proof of (10.5.8) we will use the following result.

(10.5.9) Proposition. *Let $R \subseteq S$ be sets of primes of k. If R is finite and $\delta(S) = 1$, then*

$$(k_S^R(p))_\mathfrak{p} = \begin{cases} k_\mathfrak{p}(p), & \text{if } \mathfrak{p} \in S \smallsetminus R, \\ k_\mathfrak{p}^{nr}(p), & \text{if } \mathfrak{p} \notin S, \\ k_\mathfrak{p}, & \text{if } \mathfrak{p} \in R, \end{cases}$$

where $k_\mathfrak{p}^{nr}(p)$ denotes the maximal unramified p-extension of $k_\mathfrak{p}$.

Proof: The assertion is trivial for primes $\mathfrak{p} \in R$. In order to show the statement for primes $\mathfrak{p} \notin R$, it suffices to show that, for each finite sets of primes T_1, T_2 with $T_1 \subseteq S \setminus R$ and $T_2 \cap S = \varnothing$, and each finite extension k' of k inside $k_S^R(p)$, the restriction homomorphism

$$(1)\quad H^1(k_S^R(p)|k', \mathbb{Z}/p\mathbb{Z}) \longrightarrow \bigoplus_{\mathfrak{p}\in T_1(k')} H^1(k'_\mathfrak{p}, \mathbb{Z}/p\mathbb{Z}) \oplus \bigoplus_{\mathfrak{p}\in T_2(k')} H^1_{nr}(k'_\mathfrak{p}, \mathbb{Z}/p\mathbb{Z})$$

is surjective. Indeed, let $\mathfrak{p} \in S \setminus R$. Putting $T_1 = \{\mathfrak{p}\}$, we see that the field $(k_S^R(p))_\mathfrak{p}$ has no elementary abelian p-extension and is therefore equal to $k_\mathfrak{p}(p)$. The argument for $\mathfrak{p} \notin S$ is similar.

For T_1 and T_2 as above, the restriction homomorphism

$$(2)\qquad H^1(k_{S\cup T_2}|k', \mathbb{Z}/p\mathbb{Z}) \longrightarrow \bigoplus_{\mathfrak{p}\in T_1\cup T_2\cup R(k')} H^1(k'_\mathfrak{p}, \mathbb{Z}/p\mathbb{Z})$$

is surjective by (9.2.7). Therefore, each $\alpha \in \bigoplus_{\mathfrak{p}\in T_1\cup T_2\cup R(k')} H^1(k'_\mathfrak{p}, \mathbb{Z}/p\mathbb{Z})$ with $\alpha_\mathfrak{p} \in H^1_{nr}(k'_\mathfrak{p}, \mathbb{Z}/p\mathbb{Z})$ for $\mathfrak{p} \in T_2$ and $\alpha_\mathfrak{p} = 0$ for $\mathfrak{p} \in R$ has a pre-image in $H^1(k_{S\cup T_2}|k', \mathbb{Z}/p\mathbb{Z})$, which obviously lies in $H^1(k_S^R(p)|k', \mathbb{Z}/p\mathbb{Z})$. Hence the restriction homomorphism in (1) is surjective. This finishes the proof. □

Proof of (10.5.8): It suffices to show the theorem in the case $T = \{\text{all places}\}$, because the statement for general $T \supseteq S$ follows by dividing out the normal subgroup generated by the inertia groups of the primes $\mathfrak{p} \notin T(k_S^R(p))$.

We write $H^*(-)$ for $H^*(-, \mathbb{Z}/p\mathbb{Z})$. By (1.6.15) it suffices to show that the homomorphism

$$\phi_i : H^i\Big(G\big(k(p)|k_S^R(p)\big)\Big) \to H^i\Big(\underset{\mathfrak{p}\in R(k_S^R(p))}{\ast} G(k_\mathfrak{p}(p)|k_\mathfrak{p}) \ast \underset{\mathfrak{p}\notin S(k_S^R(p))}{\ast} T(k_\mathfrak{p}(p)|k_\mathfrak{p})\Big)$$

is an isomorphism for $i = 1$ and injective for $i = 2$ (we omit the coefficients $\mathbb{Z}/p\mathbb{Z}$ from the notation). By (4.3.14), the right hand side is isomorphic to

$$\sideset{}{'}\bigoplus_{\mathfrak{p}\in R(k_S^R(p))} H^i(G(k_\mathfrak{p}(p)|k_\mathfrak{p})) \oplus \sideset{}{'}\bigoplus_{\mathfrak{p}\notin S(k_S^R(p))} H^i(T(k_\mathfrak{p}(p)|k_\mathfrak{p})).$$

The map ϕ_1 is injective since $G(k_T(p)|k_S^R(p))$ is generated by the inertia groups of the primes in $T \setminus S$ and the decomposition groups of the primes in R. Next we show that ϕ_1 is surjective.

Let k' be a finite extension of k inside $k_S^R(p)$. By (10.4.8) and (7.5.8), the maps $H^i(G(k(p)|k') \to H^i(G(\bar{k}|k'))$ and $H^i(G(k_\mathfrak{p}(p)|k'_\mathfrak{p})) \to H^i(G(\bar{k}_\mathfrak{p}|k'_\mathfrak{p}))$ are isomorphisms for all i.

As the complement of S has density zero, the restriction homomorphism

$$(*)\qquad H^1(G(\bar{k}|k')) \longrightarrow \bigoplus_{\mathfrak{p}\in R(k')} H^1(G(\bar{k}_\mathfrak{p}|k'_\mathfrak{p}) \oplus \bigoplus_{\mathfrak{p}\notin S(k')} H^1(T(\bar{k}_\mathfrak{p}|k'_\mathfrak{p}))^{G(\bar{k}_\mathfrak{p}|k'_\mathfrak{p})}$$

is surjective by (9.2.7). For $\mathfrak{p} \notin S$, we have isomorphisms

$$\begin{aligned} H^1\big(T(\bar{k}_\mathfrak{p}|k_\mathfrak{p})\big)^{G(\bar{k}_\mathfrak{p}|k_\mathfrak{p}^{nr}(p))} &\cong H^1\big(G(\bar{k}_\mathfrak{p}|k_\mathfrak{p}^{nr})\big)^{G(\bar{k}_\mathfrak{p}|k_\mathfrak{p}^{nr}(p))} \\ &\cong H^1\big(G(\bar{k}_\mathfrak{p}|k_\mathfrak{p}^{nr}(p))\big) \\ &\cong H^1\big(G(k_\mathfrak{p}(p)|k_\mathfrak{p}^{nr}(p))\big) \\ &\cong H^1\big(T(k_\mathfrak{p}(p)|k_\mathfrak{p})\big). \end{aligned}$$

By (10.5.9), $k'_\mathfrak{p}$ approaches $k_\mathfrak{p}^{nr}(p)$ when k' runs through the finite extensions of k inside $k_S^R(p)$. Therefore, passing to the limit in $(*)$, we obtain the surjectivity of ϕ_1.

It remains to show that ϕ_2 is injective. For any finite extension k' of k in $k_S^R(p)$, the Hasse principle (9.1.10)(i) yields an injection

$$H^2(k(p)|k') \hookrightarrow \bigoplus_\mathfrak{p} H^2(k_\mathfrak{p}(p)|k'_\mathfrak{p}).$$

Passing to the limit over all k', (10.5.9) implies that only the entries for $\mathfrak{p} \in R$ survive on the right hand side. Consequently, ϕ_2 is injective. □

Using (10.5.8), we are now able to show the following theorem on cohomological dimensions.

(10.5.10) Theorem. *Let k be a number field, p a prime number and $S \supseteq R$ sets of primes of k. Assume that k is totally imaginary if $p = 2$. If R is finite and $\delta(S) = 1$, then*

$$cd\, G(k_S^R(p)|k) = scd\, G(k_S^R(p)|k) = 2.$$

Proof: For a nonarchimedean prime $\mathfrak{p}$, the groups $G(k_\mathfrak{p}(p)|k_\mathfrak{p}^{nr}(p))$ and $G(k_\mathfrak{p}^{nr}(p)|k_\mathfrak{p})$ are free. Therefore the Hochschild-Serre spectral sequence provides an isomorphism

$$H^1(k_\mathfrak{p}^{nr}(p)|k_\mathfrak{p}, H^1(k_\mathfrak{p}(p)|k_\mathfrak{p}^{nr}(p))) \xrightarrow{\sim} H^2(k_\mathfrak{p}(p)|k_\mathfrak{p}),$$

where we omit the coefficients $\mathbb{Z}/p\mathbb{Z}$ from the notation. Next we consider the Hochschild-Serre spectral sequence for the tower $k(p)|k_S(p)|k$. By (10.4.9) and (8.3.18), we have $cd\, G(k(p)|k) \le 2$, hence $H^3(k(p)|k) = 0$. By (10.5.8) for $T = \{\text{all places}\}$ and $R = \varnothing$, the group $G(k(p)|k_S(p))$ is free (hence $E_2^{ij} = 0$ for $j \ge 2$) and we have an isomorphism

$$H^1(k(p)|k_S(p)) \cong \bigoplus_{\mathfrak{p}\notin S} \mathrm{Ind}_{G(k_S(p)|k)}^{G_\mathfrak{p}(k_S(p)|k)} H^1(T(k_\mathfrak{p}(p)|k_\mathfrak{p})).$$

Together with (10.5.9) this implies for $E_2^{1,1} = H^1(k_S(p)|k, H^1(k(p)|k_S(p)))$:

$$\begin{aligned}
E_2^{1,1} &\cong \bigoplus_{\mathfrak{p}\notin S} H^1(k_S(p)|k, \operatorname{Ind}_{G(k_S(p)|k)}^{G_{\mathfrak{p}}(k_S(p)|k)} H^1(T(k_{\mathfrak{p}}(p)|k_{\mathfrak{p}}))) \\
&\cong \bigoplus_{\mathfrak{p}\notin S} H^1(k_{\mathfrak{p}}^{nr}(p)|k_{\mathfrak{p}}, H^1(T(k_{\mathfrak{p}}(p)|k_{\mathfrak{p}}))) \\
&\cong \bigoplus_{\mathfrak{p}\notin S} H^2(k_{\mathfrak{p}}(p)|k_{\mathfrak{p}}).
\end{aligned}$$

Therefore the Hochschild-Serre spectral sequence, the long exact sequence of Poitou-Tate, the Hasse principle (9.1.10)(i) and (10.4.8) imply a commutative exact diagram

$$\begin{array}{ccccc}
 & & \displaystyle\bigoplus_{\mathfrak{p}\in S} H^2(k_{\mathfrak{p}}(p)|k_{\mathfrak{p}}) & & \\
 & & \big\downarrow & \searrow\!\!\!\!\searrow & \\
H^2(k(p)|k) & \hookrightarrow & \displaystyle\bigoplus_{\mathfrak{p}} H^2(k_{\mathfrak{p}}(p)|k_{\mathfrak{p}}) & \twoheadrightarrow & H^0(\bar{k}|k, \mu_p)^\vee \\
\big\downarrow & & \big\downarrow\!\!\!\downarrow & & \\
H^1(k_S(p)|k, H^1(k(p)|k_S(p))) & \xrightarrow{\ \sim\ } & \displaystyle\bigoplus_{\mathfrak{p}\notin S} H^2(k_{\mathfrak{p}}(p)|k_{\mathfrak{p}}) & & \\
\big\downarrow\!\!\!\downarrow & & & & \\
H^3(k_S(p)|k), & & & &
\end{array}$$

which shows that $H^3(k_S(p)|k) = 0$. Next we consider the Hochschild-Serre spectral sequence for the tower $k_S(p)|k_S^R(p)|k$. By (10.5.8) for $T = S$, the groups

$$H^j(k_S(p)|k_S^R(p))$$

are induced $G(k_S^R(p)|k)$-modules for $j \geq 1$, hence $E_2^{ij} = 0$ for $i, j \geq 1$. We obtain an injection

$$H^3(k_S^R(p)|k) = E_2^{2,0} \hookrightarrow H^3(k_S(p)|k) = 0,$$

implying that $cd\, G(k_S^R(p)|k) \leq 2$. It remains to show equality. Since $\delta(S) = 1$ and R is finite, we can find a prime $\mathfrak{p} \in S \smallsetminus (S_p \cup S_\infty \cup R)$ with $\zeta_p \in k_{\mathfrak{p}}$. By (10.5.9), the decomposition group of $\mathfrak{p}$ in $G(k_S^R(p)|k)$ is isomorphic to $G(k_{\mathfrak{p}}(p)|k_{\mathfrak{p}})$, which is a group of cohomological dimension 2 by (7.5.9). This implies that $cd\, G(k_S^R(p)|k) = 2$.

In order to show the assertion concerning the strict cohomological dimension, recall that $H^2(G(k(p)|k), \mathbb{Q}_p/\mathbb{Z}_p) = 0$ by (10.2.3) and (10.4.8). Thus we have the exact sequence (we omit the coefficients $\mathbb{Q}_p/\mathbb{Z}_p$ from the notation)

$$H^1(k(p)|k) \xrightarrow{res} H^1(k(p)|k_S^R(p))^{G(k_S^R(p)|k)} \twoheadrightarrow H^2(k_S^R(p)|k).$$

Using (10.5.8), we obtain a canonical isomorphism

$$H^1(k(p)|k_S^R(p))^{G(k_S^R(p)|k)} \xrightarrow{\sim} \bigoplus_{\mathfrak{p}\in R(k)} H^1(\mathcal{G}_\mathfrak{p}) \oplus \bigoplus_{\mathfrak{p}\notin S(k)} H^1(\mathcal{T}_\mathfrak{p})^{\mathcal{G}_\mathfrak{p}}.$$

By (9.2.7), the restriction map

$$H^1(k(p)|k, \mathbb{Z}/p^m\mathbb{Z}) \longrightarrow \bigoplus_{\mathfrak{p}\in R(k)} H^1(\mathcal{G}_\mathfrak{p}, \mathbb{Z}/p^m\mathbb{Z}) \oplus \bigoplus_{\mathfrak{p}\notin S(k)} H^1(\mathcal{T}_\mathfrak{p}, \mathbb{Z}/p^m\mathbb{Z})^{\mathcal{G}_\mathfrak{p}}$$

is surjective for every $m \geq 1$, except we are in the special case, where the cokernel is annihilated by 2. Passing to the limit over all m, we see that $\mathrm{coker}(\mathit{res}) \cong H^2(k_S^R(p)|k, \mathbb{Q}_p/\mathbb{Z}_p)$ is annihilated by 2. Since $\mathit{cd}\, G(k_S^R(p)|k) = 2$, the group $H^2(k_S^R(p)|k, \mathbb{Q}_p/\mathbb{Z}_p)$ is divisible, hence trivial. This remains true for every finite extension k' of k inside $k_S^R(p)$, and therefore $\mathit{scd}\, G(k_S^R(p)|k) = 2$ by (3.3.4). □

By considering p-Sylow subgroups, we obtain the

(10.5.11) Corollary. *Let k be a number field, p a prime number and $S \supseteq R$ sets of primes of k. Assume that k is totally imaginary if $p = 2$. If R is finite and $\delta(S) = 1$, then*

$$\mathit{cd}_p\, G(k_S^R|k) = \mathit{scd}_p\, G(k_S^R|k) = 2.$$

§6. The Relation between 2 and ∞

Let k be a number field, p a prime number, S a set of primes of k containing the set S_p of primes above p, and $G_S(p) = G(k_S(p)|k)$ the Galois group of the maximal p-extension $k_S(p)$ of k which is unramified outside S. If p is odd or if k is totally imaginary, then $G_S(p) = G_{S\cup S_\infty}(p)$, and the latter group has been studied in the previous sections. In this section we consider the exceptional case where $p = 2$ and k has at least one real place. If S contains all real places, then these places become complex in $k_S(2)$ and therefore $G_S(2)$, containing involutions, has infinite cohomological dimension. Furthermore, the virtual cohomological dimension $\mathit{vcd}\, G_S(2)$ is less than or equal to two in this case, i.e. $G_S(2)$ has an open subgroup U with $\mathit{cd}\, U \leq 2$. The case when not all real places are in S will be considered below. The results are taken from [197].

(10.6.1) Theorem. *Let k be a number field and let S be a set of primes of k which contains all primes dividing 2. If no real prime is in S, then $cd\, G_S(2) \le 2$. If S contains real primes, then they become complex in $k_S(2)$ and $cd\, G_S(2) = \infty$, $vcd\, G_S(2) \le 2$.*

If S is finite, then $H^i(G_S(2), \mathbb{Z}/2\mathbb{Z})$ is finite for all i and

$$\chi_2(G_S(2)) = -r_2,$$

where $\chi_2(G_S(2)) = \sum_{i=0}^{2}(-1)^i \dim_{\mathbb{F}_2} H^i(G_S(2), \mathbb{Z}/2\mathbb{Z})$ is the second partial Euler characteristic and r_2 is the number of complex places of k.

The key to the proof of (10.6.1) is the following theorem (10.6.2) in the case $p = 2$ and $T = S \cup S_{\mathbb{R}}$, where $S_{\mathbb{R}}$ is the set of real places of k. Theorem (10.6.2) is the number theoretical analogue of Riemann's existence theorem which we proved in §5 under the assumption that p is odd or that S contains $S_{\mathbb{R}}$. Again we often write $H^*(-)$ for $H^*(-, \mathbb{Z}/2\mathbb{Z})$.

(10.6.2) Riemann's Existence Theorem. *Let k be a number field, p a prime number and $T \supseteq S \supseteq S_p$ sets of primes of k. Then the canonical homomorphism*

$$\mathop{\ast}_{\mathfrak{p} \in T\setminus S(k_S(p))} T(k_{\mathfrak{p}}(p)|k_{\mathfrak{p}}) \longrightarrow G(k_T(p)|k_S(p))$$

is an isomorphism. Here $T(k_{\mathfrak{p}}(p)|k_{\mathfrak{p}})$ is the inertia subgroup of $G(k_{\mathfrak{p}}(p)|k_{\mathfrak{p}})$.

In the proof of (10.6.2) we will use the following

(10.6.3) Lemma. *Let p be a prime number, k a number field, and*

$$S_p \subseteq S \subseteq S' \subseteq S''$$

sets of primes in k. Let K be a p-S-closed extension of k. Then the following assertions are equivalent:

(i) *The natural homomorphism*

$$\phi_{S'',S} : \mathop{\ast}_{\mathfrak{p}\in S''\setminus S(K)} T(K_{\mathfrak{p}}(p)|K_{\mathfrak{p}}) \to G(K_{S''}(p)|K)$$

is an isomorphism.

(ii) *The natural homomorphisms*

$$\phi_{S',S} : \mathop{\ast}_{\mathfrak{p}\in S'\setminus S(K)} T(K_{\mathfrak{p}}(p)|K_{\mathfrak{p}}) \to G(K_{S'}(p)|K)$$

and

$$\phi_{S'',S'} : \mathop{\ast}_{\mathfrak{p}\in S''\setminus S'(K_{S'}(p))} T(K_{\mathfrak{p}}(p)|K_{\mathfrak{p}}) \to G(K_{S''}(p)|K_{S'}(p))$$

are isomorphisms.

Proof: The homomorphisms $\phi_{S'',S}$, $\phi_{S'',S'}$ and $\phi_{S',S}$ are obviously surjective. Furthermore, all primes $\mathfrak{p} \in S'' \smallsetminus S'(K)$ are completely decomposed in $K_{S'}(p)|K$ because these primes are unramified and K contains the cyclotomic $\mathbb{Z}_p$-extension of k, i.e. $K_\mathfrak{p}$ has no nontrivial unramified p-extensions for $\mathfrak{p} \notin S_p$.

Let us assume that $\phi_{S'',S}$ is an isomorphism. By dividing out the inertia groups of the primes in $S'' \smallsetminus S'$, we conclude that $\phi_{S',S}$ is also an isomorphism. Using the subgroup theorem for free products (4.2.1), one verifies that for each finite extension K' of K inside $K_S(p)$ the homomorphism

$$\mathop{\ast}_{\mathfrak{p}\in S''\setminus S'(K')} T(K_\mathfrak{p}(p)|K_\mathfrak{p}) \longrightarrow G(K_{S''}(p)|K')$$

is injective. Passing to the limit over all K', we conclude that $\phi_{S'',S'}$ is injective, hence an isomorphism. This shows that (i) implies (ii).

Now assume that (ii) holds, i.e. that $\phi_{S'',S'}$ and $\phi_{S',S}$ are isomorphisms. By (1.6.15) and (4.3.14), we have to show that the induced homomorphism

$$H^i(\phi_{S'',S}) : H^i(G(K_{S''}(p)|K)) \longrightarrow \bigoplus_{\mathfrak{p}\in S''\setminus S(K)}{}' H^i(T(K_\mathfrak{p}(p)|K_\mathfrak{p}))$$

is bijective for $i = 1$ and injective for $i = 2$. As $\phi_{S',S}$ is an isomorphism and all primes $\mathfrak{p} \in S'' \smallsetminus S'(K)$ are completely decomposed in $K_{S'}(p)|K$, the short exact sequence of pro-p-groups

$$1 \to G(K_{S''}(p)|K_{S'}(p)) \to G(K_{S''}(p)|K) \to G(K_{S'}(p)|K) \to 1$$

splits. Therefore, in the associated Hochschild-Serre spectral sequence, the differential

$$d_2^{0,1} : E_2^{0,1} = H^1(G(K_{S''}(p)|K_{S'}(p)))^{G(K_{S'}(p)|K)} \to E_2^{2,0} = H^2(G(K_{S'}(p)|K))$$

is zero by (2.4.4).

Since $\phi_{S'',S'}$ is an isomorphism and all primes $\mathfrak{p} \in S'' \smallsetminus S'(K)$ are completely decomposed in $K_{S'}(p)|K$, we have

$$H^i(G(K_{S''}(p)|K_{S'}(p))) \cong \mathrm{Ind}_{G(K_{S'}(p)|K)} \bigoplus_{\mathfrak{p}\in S''\setminus S'(K)}{}' H^i(T(K_\mathfrak{p}(p)|K_\mathfrak{p}))$$

for $i \geq 1$. This implies that $E_2^{ij} = 0$ for $i, j \geq 1$. For $i \geq 1$, we obtain the commutative exact diagram

$$\begin{array}{ccc}
H^i(G(K_{S'}(p)|K)) & \xrightarrow[\sim]{H^i(\phi_{S',S})} & \bigoplus'_{\mathfrak{p}\in S'\setminus S(K)} H^i(T(K_\mathfrak{p}(p)|K_\mathfrak{p})) \\
\downarrow{\scriptstyle \alpha_i} & & \downarrow \\
H^i(G(K_{S''}(p)|K)) & \xrightarrow{H^i(\phi_{S'',S})} & \bigoplus'_{\mathfrak{p}\in S''\setminus S(K)} H^i(T(K_\mathfrak{p}(p)|K_\mathfrak{p})) \\
\downarrow{\scriptstyle \beta_i} & & \downarrow \\
H^i(G(K_{S''}(p)|K_{S'}(p)))^{G(K_{S'}(p)|K)} & \xrightarrow[\sim]{H^i(\phi_{S'',S'})} & \bigoplus'_{\mathfrak{p}\in S''\setminus S'(K)} H^i(T(K_\mathfrak{p}(p)|K_\mathfrak{p})).
\end{array}$$

The map α_1 is injective, and since $d_2^{0,1} = 0$, the map β_1 is surjective and α_2 is injective. Hence $H^1(\phi_{S'',S})$ is an isomorphism and $H^2(\phi_{S'',S})$ is injective. Therefore $\phi_{S'',S}$ is an isomorphism. □

Proof of (10.6.2): If p is odd or if $p = 2$ and $S \supseteq S_{\mathbb{R}}$, then (10.6.2) is just theorem (10.5.1). So we assume that $p = 2$ and $S \not\supseteq S_{\mathbb{R}}$.

We show (10.6.2) first in the special case that $T = S_2 \cup S_{\mathbb{R}}$, $S = S_2$. The groups $\ast_{\mathfrak{p}\in S_{\mathbb{R}}(k_{S_2}(2))} T(k_{\mathfrak{p}}(2)|k_{\mathfrak{p}})$ and $G(k_{S_2\cup S_{\mathbb{R}}}(2)|k_{S_2}(2))$ are both generated by involutions and so their maximal abelian factor groups are annihilated by 2. Since $H^2(T(k_{\mathfrak{p}}(2)|k_{\mathfrak{p}}), \mathbb{Q}_2/\mathbb{Z}_2) = 0$ for every $\mathfrak{p} \in S_{\mathbb{R}}(k_{S_2}(2))$, we have

$$H^2(\underset{\mathfrak{p}\in S_{\mathbb{R}}(k_{S_2}(2))}{\ast} T(k_{\mathfrak{p}}(2)|k_{\mathfrak{p}}), \mathbb{Q}_2/\mathbb{Z}_2) = 0.$$

By (10.4.8), the inflation map

$$H^2(G(k_{S_2\cup S_{\mathbb{R}}}(2)|k_{S_2}(2)), \mathbb{Q}_2/\mathbb{Z}_2) \longrightarrow H^2(G(k_{S_2\cup S_{\mathbb{R}}}|k_{S_2}(2)), \mathbb{Q}_2/\mathbb{Z}_2)$$

is an isomorphism and, since $k_{S_2}(2)$ contains the cyclotomic $\mathbb{Z}_2$-extension $k_\infty(2)$ of k, the validity of the weak Leopoldt conjecture for the cyclotomic $\mathbb{Z}_p$-extension (see (10.3.25)) implies (by (10.3.22)) that

$$H^2(G(k_{S_2\cup S_{\mathbb{R}}}(2)|k_{S_2}(2)), \mathbb{Q}_2/\mathbb{Z}_2) = 0.$$

By (1.6.16) and the calculation of the cohomology of free products (see (4.3.14)), it therefore suffices to show that the natural map

$$H^2(\phi) : H^2(G(k_{S_2\cup S_{\mathbb{R}}}(2)|k_{S_2}(2)) \to \bigoplus_{\mathfrak{p}\in S_{\mathbb{R}}(k_{S_2}(2))}' H^2(T(k_{\mathfrak{p}}(2)|k_{\mathfrak{p}}))$$

is an isomorphism.

Now let K be a finite extension of k inside $k_S(2)$. The long exact sequence of Poitou-Tate induces the exact sequence

$$0 \to Ш^2(K_{S_2\cup S_{\mathbb{R}}}, \mathbb{Z}/2\mathbb{Z}) \to H^2(G(k_{S_2\cup S_{\mathbb{R}}}|K)) \to$$

$$\bigoplus_{\mathfrak{p}\in S_2\cup S_{\mathbb{R}}(K)} H^2(G(\bar{k}_{\mathfrak{p}}|K_{\mathfrak{p}})) \to H^0(G(k_{S_2\cup S_{\mathbb{R}}}|K), \mu_2)^\vee \to 0.$$

Furthermore, we have

$$Ш^2(K_{S_2\cup S_{\mathbb{R}}}, \mathbb{Z}/2\mathbb{Z}) \cong Ш^1(K_{S_2\cup S_{\mathbb{R}}}, \mu_2)^\vee = Cl_{S_2}(K)/2.$$

For a finite, nontrivial extension K' of K inside $k_{S_2}(2)$, the corresponding homomorphism

$$H^0(G(k_{S_2\cup S_{\mathbb{R}}}|K), \mu_2)^\vee \to H^0(G(k_{S_2\cup S_{\mathbb{R}}}|K'), \mu_2)^\vee$$

is the dual of the norm map, hence trivial. Furthermore, by (7.1.8)(i),

$$H^2(G(\bar{k}_{\mathfrak{p}}|k_{S_2}(2)_{\mathfrak{p}})) = 0 \quad \text{for} \quad \mathfrak{p} \in S_2(k_{S_2}(2)).$$

Therefore we obtain the following exact sequence in the limit over all finite subextensions $K|k$ inside $k_{S_2}(2)|k$:

$$0 \to Cl_{S_2}(k_{S_2}(2))/2 \to H^2(G(k_{S_2\cup S_\mathbb{R}}|k_{S_2}(2))) \to \bigoplus_{\mathfrak{p}\in S_\mathbb{R}(k_{S_2}(2))}{}' H^2(G(\bar{k}_\mathfrak{p}|k_\mathfrak{p})) \to 0.$$

The principal ideal theorem implies that $Cl_{S_2}(k_{S_2}(2))/2 = 0$. Furthermore, $G(\bar{k}_\mathfrak{p}|k_\mathfrak{p}) = T(k_\mathfrak{p}(2)|k_\mathfrak{p})$ for $\mathfrak{p} \in S_\mathbb{R}(k_{S_2}(2))$, and the inflation map

$$H^2(G(k_{S_2\cup S_\mathbb{R}}(2)|k_{S_2}(2))) \longrightarrow H^2(G(k_{S_2\cup S_\mathbb{R}}|k_{S_2}(2)))$$

is an isomorphism, see (10.4.8). This concludes the proof in the case when $T = S_2 \cup S_\mathbb{R}$, $S = S_2$. Furthermore, the assertion holds for $T = \{\text{all primes}\}$ and $S = S_2 \cup S_\infty$ by (10.5.1). Applying (10.6.3), we obtain the assertion in the 'extremal' case $T = \{\text{all primes}\}$, $S = S_2$. Applying (10.6.3) again, we obtain the case $T = \{\text{all primes}\}$ and S arbitrary, and then the general case. This concludes the proof of (10.6.2). □

A straightforward limit process shows the following variant of (10.6.2).

(10.6.4) Corollary. *Let k be a number field, p a prime number and $T \supseteq S \supseteq S_p$ sets of primes of k. Let K be a p-S-closed extension field of k. Then the canonical homomorphism*

$$\mathop{\ast}_{\mathfrak{p}\in T\setminus S(K)} T(K_\mathfrak{p}(p)|K_\mathfrak{p}) \longrightarrow G(K_T(p)|K)$$

is an isomorphism.

Proof of (10.6.1): We may assume that $S \not\supseteq S_\mathbb{R}$ and we investigate the Hochschild-Serre spectral sequence

$$E_2^{ij} = H^i(G_S(2), H^j(G(k_{S\cup S_\mathbb{R}}(2)|k_S(2))) \Rightarrow H^{i+j}(G_{S\cup S_\mathbb{R}}(2)),$$

where we omit the coefficients $\mathbb{Z}/2\mathbb{Z} = \mu_2$ from the notation. By (10.5.1), we have complete control over the $G_S(2)$-modules $H^j(G(k_{S\cup S_\mathbb{R}}(2)|k_S(2))$, which are isomorphic to

$$\operatorname{Ind}_{G_S(2)} \bigoplus_{\mathfrak{p}\in S_\mathbb{R}\setminus S(k)} H^j(G(\mathbb{C}|\mathbb{R}))$$

for $j \geq 1$. In particular, $E_2^{ij} = 0$ if $i, j \geq 1$. Therefore the spectral sequence induces an exact sequence

$$(*)\quad \begin{aligned} 0 \to H^1(G_S(2)) \to H^1(G_{S\cup S_\mathbb{R}}(2)) \to \bigoplus_{\mathfrak{p}\in S_\mathbb{R}\setminus S(k)} H^1(G(\mathbb{C}|\mathbb{R})) \to \\ H^2(G_S(2)) \to H^2(G_{S\cup S_\mathbb{R}}(2)) \to \bigoplus_{\mathfrak{p}\in S_\mathbb{R}\setminus S(k)} H^2(G(\mathbb{C}|\mathbb{R})) \to 0 \end{aligned}$$

and exact sequences

$$(**) \qquad 0 \to H^i(G_S(2)) \to H^i(G_{S\cup S_{\mathbb{R}}}(2)) \to \bigoplus_{\mathfrak{p}\in S_{\mathbb{R}}\setminus S(k)} H^i(G(\mathbb{C}|\mathbb{R})) \to 0.$$

for $i \geq 3$. If S is finite, this shows the finiteness statement for the cohomology of $G_S(2)$ and that

$$\chi_2(G_S(2)) = \chi_2(G_{S\cup S_{\mathbb{R}}}(2)).$$

But $\chi_2(G_{S\cup S_{\mathbb{R}}}(2)) = \chi_2(G_{S\cup S_{\mathbb{R}}}) = -r_2$ by (8.7.5) and (10.4.8).

For arbitrary S and $i \geq 3$, the restriction map

$$H^i(G_{S\cup S_{\mathbb{R}}}(2)) \to \bigoplus_{\mathfrak{p}\in S_{\mathbb{R}}(k)} H^i(G(\mathbb{C}|\mathbb{R}))$$

is an isomorphism, see (8.6.10)(ii) and (10.4.8). This, together with the exact sequence $(**)$, shows that the natural homomorphism

$$H^i(G_S(2)) \to \bigoplus_{\mathfrak{p}\in S\cap S_{\mathbb{R}}(k)} H^i(G(\mathbb{C}|\mathbb{R}))$$

is an isomorphism for $i \geq 3$. Therefore $cd\ G_S(2) \leq 2$ if $S \cap S_{\mathbb{R}} = \varnothing$.

In order to conclude the proof of (10.6.1), it remains to show that every real prime in S ramifies in $k_S(2)$. Let S^f be the subset of nonarchimedean primes in S. Then (10.5.1) yields an isomorphism

$$\mathop{\ast}_{\mathfrak{p}\in S_{\mathbb{R}}(k_{S^f}(2))} T(k_{\mathfrak{p}}(2)|k_{\mathfrak{p}}) \cong G(k_S(2)|k_{S^f}(2))$$

which shows the required assertion. This finishes the proof of (10.6.1). □

For later use we formulate the result just obtained in the proof of (10.6.1) as a proposition.

(10.6.5) Proposition. *Let k be a number field and let $S \supseteq S_2$ be a set of primes. Then the natural homomorphism*

$$H^i(G_S(2), \mathbb{Z}/2\mathbb{Z}) \to \bigoplus_{\mathfrak{p}\in S\cap S_{\mathbb{R}}(k)} H^i(G(\mathbb{C}|\mathbb{R}), \mathbb{Z}/2\mathbb{Z})$$

is an isomorphism for $i \geq 3$.

To fix conventions, we recall the following notation: For a set S of primes of k, the group $\mathcal{O}_{k,S}^{\times}$ of S-units is defined as the subgroup of $k^{\times}$ whose elements are units at every finite prime not in S and positive at every real prime not in S. The S-ideal class group $Cl_{0,S}(k)$ in the **narrow sense** of k is the quotient of the group of fractional ideals of k by the subgroup generated by the nonarchimedean primes in S and the principal ideals (a) with a positive at every real place of k not contained in S. In particular, $Cl_{0,\varnothing}(k) = Cl_0(k)$ is the ideal class group in the narrow sense and $Cl_{0,S\cup S_{\mathbb{R}}}(k) = Cl_S(k)$ is the usual S-ideal class group. By class field theory, $Cl_{0,S}(k)$ is isomorphic to the Galois group of the maximal abelian extension of k which is unramified outside $S_{\mathbb{R}}$ and in which every prime in S splits completely.

Let S be a set of finite primes of k and $\Sigma = S \cup S_{\mathbb{R}}$. If S is finite, then the image of the group of Σ-units of k under the logarithm map

$$Log : \mathcal{O}_{k,\Sigma}^{\times} \longrightarrow \bigoplus_{v\in\Sigma} \mathbb{R}, \quad a \mapsto (\log |a|_v)_{v\in S}$$

is a lattice of rank equal to $\#S + r_1 + r_2 - 1$ by Dirichlet's unit theorem. Complementary to this map is the **signature map** (which is also defined for infinite S)

$$Sign_{k,S} : \mathcal{O}_{k,\Sigma}^{\times} \longrightarrow \bigoplus_{v\in S_{\mathbb{R}}} \mathbb{R}^{\times}/\mathbb{R}^{\times 2}.$$

More or less by definition, there exists a five-term exact sequence

$$0 \to \mathcal{O}_{k,S}^{\times} \to \mathcal{O}_{k,\Sigma}^{\times} \to \bigoplus_{v\in S_{\mathbb{R}}(k)} \mathbb{R}^{\times}/\mathbb{R}^{\times 2} \to Cl_{0,S}(k) \to Cl_{0,\Sigma}(k) \to 0,$$

and so the cokernel of $Sign_{k,S}$ measures the difference between the usual S-ideal class group $Cl_S(k) = Cl_{0,\Sigma}(k)$ and that in the narrow sense. Of course this discussion is void if k is totally imaginary. If K is an infinite extension of k, we define the signature map

$$Sign_{K,S} : \mathcal{O}_{K,\Sigma}^{\times} \longrightarrow \varinjlim_{k'} \bigoplus_{v\in S_{\mathbb{R}}(k')} \mathbb{R}^{\times}/\mathbb{R}^{\times 2}$$

as the limit over the signature maps $Sign_{k',S}$, where k' runs through all finite subextensions $k'|k$ of $K|k$.

Now we consider p-S-closed extensions. If p is odd and K is p-S-closed, then the group

$$Cl_S(K(\mu_p))(p)(j)^{G(K(\mu_p)|K)}$$

is trivial for $j = 0, -1$ by (10.4.7). The corresponding result for $p = 2$ is the following

(10.6.6) Theorem. *Let k be a number field, $S \supseteq S_2$ a set of primes of k and K a 2-S-closed extension of k. Then the following hold.*

(i) $Cl_S(K(\mu_4))(2) = 0$.

(ii) $Cl_{0,S}(K)(2) = 0$.

Remarks: 1. The triviality of $Cl(K)(2)$, and hence also that of $Cl_S(K)(2)$, follows easily from the principal ideal theorem; assertions (i) and (ii) do not.

2. In (i) one can replace $K(\mu_4)$ by any totally imaginary extension of degree 2 of K in $K_S(2)$.

Proof: If K is 2-S-closed, then $Cl_S(K)(2) = 0$ and so statement (ii) of (10.6.6) is equivalent to the statement that the signature map $Sign_K$ is surjective.

By (10.6.1) all real places in S become complex in K. By the principal ideal theorem, $Cl(K)(2) = 0$ and so statements (i) and (ii) are trivial if K is totally imaginary (note that $K = K(\mu_4)$ in this case). So we may assume that $S_{\mathbb{R}}(K) \neq \varnothing$ and, by (10.6.1), we may suppose that $S \cap S_{\mathbb{R}} = \varnothing$.

Let $K' = K(\mu_4)$. Then K' is totally imaginary and $G = G(K'|K)$ is cyclic of order 2. Let $\Sigma = S \cup S_{\mathbb{R}}$ and let K_Σ be the maximal (not just the pro-2) extension of K which is unramified outside Σ. Inspecting the Hochschild-Serre spectral sequence associated to $K_\Sigma|K_\Sigma(2)|K$ and using (10.4.8), we see that

$$H^1(G(K_\Sigma(2)|K), \mathcal{O}^\times_{K_\Sigma(2),\Sigma}) = H^1(G(K_\Sigma|K), \mathcal{O}^\times_{K_\Sigma,\Sigma})(2) \cong Cl_S(K)(2) = 0,$$

and the same argument shows that

$$H^1(G(K_\Sigma(2)|K'), \mathcal{O}^\times_{K_\Sigma(2),\Sigma}) \cong Cl_S(K')(2).$$

Next we consider the Hochschild-Serre spectral sequence for the extension $K_\Sigma(2)|K'|K$ and the module $\mathcal{O}^\times_{K_\Sigma(2),\Sigma}$. Using the two isomorphisms obtained above, we obtain an exact sequence

$$0 \to Cl_S(K')(2)^G \to H^2(G, \mathcal{O}^\times_{K',\Sigma}) \xrightarrow{\phi} H^2(G(K_\Sigma(2)|K), \mathcal{O}^\times_{K_\Sigma(2),\Sigma}).$$

In order to prove assertion (i), it suffices to show that ϕ is injective. Let c be a generator of the cyclic group $H^2(G, \mathbb{Z})$. For each prime $\mathfrak{p} \in S_{\mathbb{R}}(K)$ (respectively for the chosen prolongation of $\mathfrak{p}$ to $K_\Sigma(2)$), the composition $T_\mathfrak{p}(K_\Sigma(2)|K) \to G(K_\Sigma(2)|K) \to G$ is an isomorphism, and we denote the image of c in $H^2(T_\mathfrak{p}(K_\Sigma(2)|K), \mathbb{Z})$ by $c_\mathfrak{p}$. The cup-product with c induces an isomorphism $\hat{H}^0(G, \mathcal{O}^\times_{K',\Sigma}) \xrightarrow{\sim} H^2(G, \mathcal{O}^\times_{K',\Sigma})$, and the analogous statement holds for each $c_\mathfrak{p}$, $\mathfrak{p} \in S_{\mathbb{R}}(K)$.

The quotient $\mathcal{O}^\times_{K_\Sigma(2),\Sigma}/\mu_{2^\infty}$ is uniquely 2-divisible, and so we obtain a natural isomorphism

$$H^2(G(K_\Sigma(2)|K), \mu_{2^\infty}) \xrightarrow{\sim} H^2(G(K_\Sigma(2)|K), \mathcal{O}^\times_{K_\Sigma(2),\Sigma}).$$

Furthermore, for each $\mathfrak{p} \in S_{\mathbb{R}} \smallsetminus S$, we obtain an isomorphism

$$\begin{aligned} H^2(T_\mathfrak{p}(K_\Sigma(2)|K), \mu_{2^\infty}) &\xrightarrow{\sim} H^2(T_\mathfrak{p}(K_\Sigma(2)|K), \mathcal{O}^\times_{K_\Sigma(2),\Sigma}) \\ &\cong H^2(G(\bar{K}_\mathfrak{p}|K_\mathfrak{p}), \bar{K}^\times_\mathfrak{p}). \end{aligned}$$

Therefore the calculation of the cohomology in dimension $i \geq 2$ of free products with values in torsion modules, see (4.1.4), and (10.5.1) for the pair (Σ, S) show that we have a natural isomorphism

$$H^2(G(K_\Sigma(2)|K), \mathcal{O}^\times_{K_\Sigma(2),\Sigma}) \xrightarrow{\sim} \bigoplus_{\mathfrak{p} \in S_{\mathbb{R}}(K)}' H^2(G(\bar{K}_\mathfrak{p}|K_\mathfrak{p}), \bar{K}^\times_\mathfrak{p}).$$

(Alternatively, we could have obtained this isomorphism from the calculation of the cohomology of the Σ-units, cf. (8.3.11)(iii), by passing to the limit over all finite extensions of k inside K.) We obtain the following commutative diagram

$$\begin{array}{ccc}
\hat{H}^0(G,\mathcal{O}_{K',\Sigma}^\times) & \overset{\psi}{\dashrightarrow} & \bigoplus'_{\mathfrak{p}\in S_{\mathbb{R}}(K)} \hat{H}^0(G(\bar{K}_\mathfrak{p}|K_\mathfrak{p}),\bar{K}_\mathfrak{p}^\times) \\
\wr\downarrow \cup c & & \wr\downarrow \bigoplus' \cup c_\mathfrak{p} \\
H^2(G,\mathcal{O}_{K',\Sigma}^\times) \xrightarrow{\phi} H^2(G(K_\Sigma(2)|K),\mathcal{O}_{K_\Sigma(2),\Sigma}^\times) & \rightrightarrows & \bigoplus'_{\mathfrak{p}\in S_{\mathbb{R}}(K)} H^2(G(\bar{K}_\mathfrak{p}|K_\mathfrak{p}),\bar{K}_\mathfrak{p}^\times).
\end{array}$$

Hence $\ker(\phi)\cong\ker(\psi)$ and $\operatorname{coker}(\phi)\cong\operatorname{coker}(\psi)$. Since $\hat{H}^0(G,\mathcal{O}_{K',\Sigma}^\times)=\mathcal{O}_{K,\Sigma}^\times/N_{K'|K}(\mathcal{O}_{K',\Sigma}^\times)$, each element in $\ker(\psi)$ is represented by an S-unit in K, and we have to show that all these are norms of Σ-units in K'. Let $e\in\mathcal{O}_{K,S}^\times$. Then $K(\sqrt{e})|K$ is a 2-extension which is unramified outside S, hence trivial. Therefore e is a square in K and if $f^2=e$, then $f\in\mathcal{O}_{K,\Sigma}^\times$ and $e=N_{K'|K}(f)$. This concludes the proof of assertion (i).

To show assertion (ii), it remains to show that $\operatorname{coker}(Sign_{K,S})=\operatorname{coker}(\psi)\cong\operatorname{coker}(\phi)$ is trivial. Using the same spectral sequence as before, in order to see that $\operatorname{coker}(\phi)=0$, it suffices to show that the spectral terms

- $E_2^{02}=H^0(G,H^2(G(K_\Sigma(2)|K'),\mathcal{O}_{K_\Sigma(2),\Sigma}^\times))$ and

- $E_2^{11}=H^1(G,Cl_S(K')(2))$

are trivial. The first assertion is easy, because K' is totally imaginary and contains $k_\infty(2)$ and so

$$H^2(G(K_\Sigma(2)|K'),\mathcal{O}_{K_\Sigma(2),\Sigma}^\times)=0.$$

That the second spectral term is trivial follows from (i). This completes the proof of (10.6.6). □

Finally, we consider the full extension k_S, i.e. the maximal extension of k which is unramified outside S, and its Galois group $G_S=G(k_S|k)$.

(10.6.7) Theorem. *Let k be a number field and S a set of primes of k containing all primes dividing 2. Then $vcd_2G_S\le 2$ and $cd_2G_S\le 2$ if and only if S contains no real primes. For every discrete $G_S(2)$-module A, the inflation maps*

$$inf: H^i(G_S(2),A)\longrightarrow H^i(G_S,A)(2)$$

are isomorphisms for all $i\ge 1$.

Remark: If $cd\, G_S(K)(2) = 2$ (e.g. if K contains at least two primes dividing 2, see (10.9.2) for some finite extension K of k inside k_S, then $vcd_2 G_S = 2$. This is always the case if $S \supseteq S_{\mathbb{R}}$ because the class numbers of the cyclotomic fields $\mathbb{Q}(\mu_{2^n})$ are nontrivial for $n \gg 0$. But, for example, we do not know whether $cd_2 G(\mathbb{Q}_{S_2}|\mathbb{Q})$ equals 1 or 2. The answer would be '2' if at least one of the real cyclotomic fields $\mathbb{Q}(\mu_{2^n})^+$, $n = 2, 3, \ldots$, had a nontrivial class number. But this is unknown.

Proof: The statement on $cd_2 G_S$ and $vcd_2 G_S$ follows by choosing a 2-Sylow subgroup $H \subseteq G_S$ and applying theorem (10.6.1) to all finite extensions of k inside $(k_S)^H$. It remains to show the statement about the inflation map. It is equivalent to the statement that

$$inf \otimes \mathbb{Z}_{(2)} : H^i(G_S(2), A) \otimes \mathbb{Z}_{(2)} \longrightarrow H^i(G_S, A) \otimes \mathbb{Z}_{(2)}$$

is an isomorphism for every discrete $G_S(2)$-module A and all $i \geq 0$, where $\mathbb{Z}_{(2)}$ denotes the localization of $\mathbb{Z}$ at the prime ideal (2).

Since cohomology commutes with inductive limits, we may assume that A is finitely generated as a $\mathbb{Z}$-module. Using the exact sequences

$$0 \longrightarrow \mathrm{tor}(A) \longrightarrow A \longrightarrow A/\mathrm{tor}(A) \longrightarrow 0,$$

$$0 \longrightarrow A/\mathrm{tor}(A) \longrightarrow (A/\mathrm{tor}(A)) \otimes \mathbb{Q} \longrightarrow (A/\mathrm{tor}(A)) \otimes \mathbb{Q}/\mathbb{Z} \longrightarrow 0,$$

and using the limit argument for $(A/\mathrm{tor}(A)) \otimes \mathbb{Q}/\mathbb{Z}$ again, we are reduced to the case that A is finite. Every finite $G_S(2)$-module is the direct sum of its 2-part and its prime-to-2-part. The statement is obvious for the prime-to-2-part and every finite 2-primary $G_S(2)$-module has a composition series whose quotients are isomorphic to $\mathbb{Z}/2\mathbb{Z}$. Therefore we are reduced to showing the statement about the inflation map for $A = \mathbb{Z}/2\mathbb{Z}$. But it is more convenient to work with $A = \mathbb{Q}_2/\mathbb{Z}_2$ (with trivial action), which is possible because of the exact sequence

$$0 \longrightarrow \mathbb{Z}/2\mathbb{Z} \longrightarrow \mathbb{Q}_2/\mathbb{Z}_2 \longrightarrow \mathbb{Q}_2/\mathbb{Z}_2 \longrightarrow 0.$$

Using the Hochschild-Serre spectral sequence for the extensions $k_S|k_S(2)|k$, we thus have to show that

$$H^i(G(k_S|k_S(2)), \mathbb{Q}_2/\mathbb{Z}_2) = 0$$

for $i \geq 1$. The case $i = 1$ is obvious by the definition of the field $k_S(2)$. By theorem (10.6.1), every real prime in S becomes complex in $k_S(2)$ and therefore $cd_2\, G(k_S|k_S(2)) \leq 2$. It remains to show that $H^2(G(k_S|k_S(2)), \mathbb{Q}_2/\mathbb{Z}_2) = 0$. Therefore the next proposition implies the remaining statement of (10.6.7).

(10.6.8) Proposition. *Let k be a number field, $S \supseteq S_2$ a set of primes in k and $K \supseteq k_\infty(2)$ an extension of K inside k_S. Then*

$$H^2(G(k_S|K), \mathbb{Q}_2/\mathbb{Z}_2) = 0.$$

Proof: Let H be a 2-Sylow subgroup of $G(k_S|K)$ and $L = (k_S)^H$. Then the restriction map

$$H^2(G(k_S|K), \mathbb{Q}_2/\mathbb{Z}_2) \longrightarrow H^2(G(k_S|L), \mathbb{Q}_2/\mathbb{Z}_2)$$

is injective and so, replacing K by L, we may suppose that $k_S = K_S(2)$. Applying (10.6.4) to the 2-S-closed field $K_S(2)$, we obtain an isomorphism

$$G(K_{S\cup S_\mathbb{R}}(2)|K_S(2)) \cong \mathop{*}_{\mathfrak{p}\in S_\mathbb{R}(K_S(2))} T_\mathfrak{p}(K_\mathfrak{p}(2)|K_\mathfrak{p}).$$

Hence we have complete control over the Hochschild-Serre spectral sequence associated to $K_{S\cup S_\mathbb{R}}(2)|K_S(2)|K$. Furthermore, the weak Leopoldt conjecture holds for the cyclotomic $\mathbb{Z}_2$-extension and $K \supseteq k_\infty(2)$, which implies that $H^2(G(K_{S\cup S_\mathbb{R}}(2)|K), \mathbb{Q}_2/\mathbb{Z}_2) = 0$. The exact sequence $(*)$ in the proof of (10.6.1), applied to all finite subextensions $k'|k$ of $K|k$, yields a surjection

$$\bigoplus_{\mathfrak{p}\in S_\mathbb{R}\setminus S(K)}{}' H^1(T(K_\mathfrak{p}(2)|K_\mathfrak{p}), \mathbb{Q}_2/\mathbb{Z}_2) \twoheadrightarrow H^2(G(K_S(2)|K), \mathbb{Q}_2/\mathbb{Z}_2)$$

and therefore, in order to prove the proposition, it suffices to show that the group $H^2(G(K_S(2)|K), \mathbb{Q}_2/\mathbb{Z}_2)$ is 2-divisible. This is trivial if $S \cap S_\mathbb{R}(K) = \varnothing$ because then $cd\, G(K_S(2)|K) \le 2$. Otherwise, this follows from the commutative diagram

$$\begin{array}{ccc} H^2(G(K_S(2)|K), \mathbb{Q}_2/\mathbb{Z}_2)/2 & \hookrightarrow & H^3(G(K_S(2)|K), \mathbb{Z}/2\mathbb{Z}) \\ \downarrow & & \downarrow \wr \\ \bigoplus_{\mathfrak{p}\in S\cap S_\mathbb{R}(K)}{}' H^2(T(K_\mathfrak{p}(2)|K_\mathfrak{p}), \mathbb{Q}_2/\mathbb{Z}_2)/2 & \hookrightarrow & \bigoplus_{\mathfrak{p}\in S\cap S_\mathbb{R}(K)}{}' H^3(T(K_\mathfrak{p}(2)|K_\mathfrak{p}), \mathbb{Z}/2\mathbb{Z}). \end{array}$$

The right hand vertical arrow is an isomorphism by (10.6.5). But the group $H^2(T(K_\mathfrak{p}(2)|K_\mathfrak{p}), \mathbb{Q}_2/\mathbb{Z}_2)$ is trivial for all $\mathfrak{p} \in S \cap S_\mathbb{R}(K)$ and therefore the object in the lower left corner is zero. □

§7. Dimension of $H^i(G_S^T, \mathbb{Z}/p\mathbb{Z})$

Throughout this section k denotes a global field and $p \neq \operatorname{char} k$ is a fixed prime number. Let S and T be sets of primes of k, which are allowed to be empty. We want to calculate the dimensions

$$h^i(G_S^T) = \dim_{\mathbb{F}_p} H^i(G_S^T, \mathbb{Z}/p\mathbb{Z}) \text{ and } h^i(G_S^T(p)) = \dim_{\mathbb{F}_p} H^i(G_S^T(p), \mathbb{Z}/p\mathbb{Z})$$

for $i = 1, 2$, where $G_S^T = G_S^T(k)$ is the Galois group $G(k_S^T|k)$ and

k_S^T is the maximal extension of k which is unramified outside S and completely decomposed at T.

Let us recall some conventions. If k is a number field, then

$$S_p = S_p(k) = \{\mathfrak{p} \text{ a prime of } k \text{ dividing } p\}.$$

We denote the set of real primes and of complex primes of k by $S_{\mathbb{R}}$ and $S_{\mathbb{C}}$, respectively, and we put $r_1 = \#S_{\mathbb{R}}$ and $r_2 = \#S_{\mathbb{C}}$.

If k is a function field, $Cl(k)$ denotes the group of divisor classes of k and $\mathcal{O}_k = \mathcal{O}_{k,\varnothing}$ is the finite field of constants of k. The sets S_p, S_∞, $S_{\mathbb{R}}$ and $S_{\mathbb{C}}$ are empty and $r_1 = r_2 = 0$.

We introduce the following notation:

(10.7.1) Definition. *We set*

$$\delta = \begin{cases} 1, & \mu_p \subseteq k, \\ 0, & \mu_p \nsubseteq k, \end{cases} \qquad \text{and} \qquad \delta_{\mathfrak{p}} = \begin{cases} 1, & \mu_p \subseteq k_{\mathfrak{p}}, \\ 0, & \mu_p \nsubseteq k_{\mathfrak{p}}, \end{cases}$$

where $\mathfrak{p}$ is a prime of k.

Recall the group

$$V_S(k) = V_S(k,p) = \ker\Big(k^\times/k^{\times p} \longrightarrow \prod_{\mathfrak{p}\in S} k_{\mathfrak{p}}^\times/k_{\mathfrak{p}}^{\times p} \times \prod_{\mathfrak{p}\notin S} k_{\mathfrak{p}}^\times/U_{\mathfrak{p}}k_{\mathfrak{p}}^{\times p}\Big),$$

which was defined in IX §1 and shown to be finite in (9.1.12). For $S = \varnothing$ we have the

(10.7.2) Proposition. *For $p \neq \operatorname{char} k$ there is an exact sequence*

$$0 \longrightarrow \mathcal{O}_k^\times/p \longrightarrow V_\varnothing(k) \longrightarrow {}_pCl(k) \longrightarrow 0.$$

In particular, the dimension of $\text{Б}_\varnothing(k)$ is given by

$$\dim_{\mathbb{F}_p} \text{Б}_\varnothing(k) = \dim_{\mathbb{F}_p} {}_pCl(k) + \dim_{\mathbb{F}_p} \mathcal{O}_k^\times/p.$$

Proof: The homomorphism

$$\{a \in k^\times \mid a \in U_{\mathfrak{p}}k_{\mathfrak{p}}^{\times p} \text{ for all } \mathfrak{p}\} \longrightarrow {}_pCl(k), \quad a \longmapsto \mathfrak{a} \text{ with } (a) = \mathfrak{a}^p,$$

induces a surjection from $V_\varnothing(k)$ onto ${}_pCl(k)$ whose kernel is isomorphic to $\mathcal{O}_k^\times/p$. □

Remark: For arbitrary S, the group $V_S(k)$ is a subgroup of $V_\varnothing(k)$, and we obtain $\dim_{\mathbb{F}_p} \text{Б}_S(k) \le \dim_{\mathbb{F}_p} {}_pCl(k) + \dim_{\mathbb{F}_p} \mathcal{O}_k^\times/p$.

In the following we denote $H^i(-, \mathbb{Z}/p\mathbb{Z})$ by $H^i(-)$ and put $h^i(-) = \dim_{\mathbb{F}_p} H^i(-)$. For a prime $\mathfrak{p}$ of k, we denote by $\mathcal{G}_{\mathfrak{p}}$ the absolute Galois group of field $k_{\mathfrak{p}}$ and by $\mathcal{T}_{\mathfrak{p}}$ the inertia group in $\mathcal{G}_{\mathfrak{p}}$.

Using the Poitou-Tate theorem, we now determine $h^i(G_S)$ for a nonempty set S which contains the archimedean primes and the primes above p if k is a number field.

(10.7.3) Theorem. *Let k be a global field, $p \neq \operatorname{char} k$ a prime number and S a nonempty set of primes of k containing $S_p \cup S_\infty$ if k is a number field. Then*

$$\begin{aligned} h^1(G_S) &= 1 + \sum_{\mathfrak{p}\in S} \delta_\mathfrak{p} - \delta + \dim_{\mathbb{F}_p} Б_S(k)\,, \\ h^2(G_S) &= \sum_{\mathfrak{p}\in S\setminus S_\mathbb{C}} \delta_\mathfrak{p} - \delta + \dim_{\mathbb{F}_p} Б_S(k)\,. \end{aligned}$$

In particular, $h^1(G_S)$ and $h^2(G_S)$ are finite if and only if S contains only a finite number of primes $\mathfrak{p}$ with $\delta_\mathfrak{p} = 1$. If $\mu_p \subseteq k$ we have $\dim_{\mathbb{F}_p} Б_S(k) = \dim_{\mathbb{F}_p} Cl_S(k)/p$.

Proof: The formula for h^2 follows from the exact sequence

$$Ш^2(G_S, \mathbb{Z}/p\mathbb{Z}) \hookrightarrow H^2(G_S, \mathbb{Z}/p\mathbb{Z}) \to \bigoplus_{\mathfrak{p}\in S} H^2(k_\mathfrak{p}, \mathbb{Z}/p\mathbb{Z}) \twoheadrightarrow H^0(G_S, \mu_p)^\vee$$

and the isomorphism

$$Ш^2(G_S, \mathbb{Z}/p\mathbb{Z}) \cong Ш^1(G_S, \mu_p)^\vee = Б_S(k)\,.$$

The formula for h^1 is then a consequence of the Euler-Poincaré characteristic formula (8.7.5). For the last statement, observe that when $\mu_p \subseteq k$

$$\begin{aligned} Б_S(k) &= \ker(H^1(G_S, \mu_p) \to \textstyle\prod_{\mathfrak{p}\in S} H^1(k_\mathfrak{p}, \mu_p))^\vee \\ &= (Cl_S(k)/p)(-1)\,. \end{aligned}$$

□

In order to proceed, we need some exact sequences.

(10.7.4) Lemma. *Let k be a global field, $p \neq \operatorname{char} k$ a prime number and $S \supseteq T$ sets of primes of k. Then there are exact sequences*

(i) $0 \to H^1(G_T) \to H^1(G_S) \to \bigoplus_{\mathfrak{p}\in S\setminus T} H^1(\mathcal{T}_\mathfrak{p})^{\mathcal{G}_\mathfrak{p}} \to Б_T(k) \to Б_S(k) \to 0.$

(ii) $H^1(G_S) \hookrightarrow \prod_{\mathfrak{p}\in S} H^1(\mathcal{G}_\mathfrak{p}) \times \prod_{\mathfrak{p}\notin S} H^1_{nr}(\mathcal{G}_\mathfrak{p}) \to H^1(G_k, \mu_p)^\vee \twoheadrightarrow Б_S(k).$

Proof: We consider the commutative exact diagram

$$\begin{array}{ccccc}
H^1(G_{k_S})^{G_S} & \hookrightarrow & \bigoplus\limits_{\mathfrak{p}\notin S} H^1(\mathcal{T}_{\mathfrak{p}})^{\mathcal{G}_{\mathfrak{p}}} & & \\
\uparrow & & \twoheaduparrow & & \\
H^1(G_k) & \hookrightarrow & \prod\limits_{\mathfrak{p}} H^1(\mathcal{G}_{\mathfrak{p}}) & \twoheadrightarrow & H^1(G_k, \mu_p)^\vee \\
\uparrow & & \uparrow & & \| \\
H^1(G_S) & & \prod\limits_{\mathfrak{p}\in S} H^1(\mathcal{G}_{\mathfrak{p}}) \times \prod\limits_{\mathfrak{p}\notin S} H^1_{nr}(\mathcal{G}_{\mathfrak{p}}) & \longrightarrow & H^1(G_k, \mu_p)^\vee \twoheadrightarrow Б_S(k).
\end{array}$$

The upper horizontal map is injective since $G_{k_S} = G(\bar{k}|k_S)$ is generated by the inertia groups of the primes outside S in G_k. The exactness of the middle row follows from the long exact sequence of Poitou-Tate together with $Ш^1(G_k, \mathbb{Z}/p\mathbb{Z}) = 0 = Ш^1(G_k, \mu_p)$, see (9.1.9). The exactness of the bottom row is just the definition of $Б_S(k) = V_S(k)^\vee$ as $H^1_{nr}(\mathcal{G}_{\mathfrak{p}})^\vee = k_{\mathfrak{p}}^\times / U_{\mathfrak{p}} k_{\mathfrak{p}}^{\times p}$. By diagram chasing, we obtain exact sequence (ii). Thus the rows in the commutative diagram

$$\begin{array}{ccccccc}
 & & \bigoplus\limits_{\mathfrak{p}\in S\setminus T} H^1(\mathcal{T}_{\mathfrak{p}})^{\mathcal{G}_{\mathfrak{p}}} & & & & \\
 & & \twoheaduparrow & & & & \\
H^1(G_S) & \hookrightarrow & \prod\limits_{\mathfrak{p}\in S} H^1(\mathcal{G}_{\mathfrak{p}}) \times \prod\limits_{\mathfrak{p}\notin S} H^1_{nr}(\mathcal{G}_{\mathfrak{p}}) & \longrightarrow & H^1(G_k, \mu_p)^\vee & \twoheadrightarrow & Б_S(k) \\
\uparrow & & \uparrow & & \| & & \twoheaduparrow \\
H^1(G_T) & \hookrightarrow & \prod\limits_{\mathfrak{p}\in T} H^1(\mathcal{G}_{\mathfrak{p}}) \times \prod\limits_{\mathfrak{p}\notin T} H^1_{nr}(\mathcal{G}_{\mathfrak{p}}) & \longrightarrow & H^1(G_k, \mu_p)^\vee & \twoheadrightarrow & Б_T(k)
\end{array}$$

are exact. Using the snake lemma, we obtain exact sequence (i). □

In the function field case, we have the following result in the case $S = \varnothing$.

(10.7.5) Theorem. *Let k be a global field of positive characteristic and let $p \neq \operatorname{char} k$ be a prime number. Let $g = g(k)$ be the genus. Then $h^i(G_\varnothing) = 0$ for $i \geq 4$ and*

$$\begin{aligned}
h^1(G_\varnothing) &= 1 + \dim_{\mathbb{F}_p} {}_pCl(k) \\
h^2(G_\varnothing) &= \begin{cases} 0, & \text{if } g = 0, \\ \dim_{\mathbb{F}_p} {}_pCl(k) + \delta, & \text{if } g \geq 1, \end{cases} \\
h^3(G_\varnothing) &= \begin{cases} 0, & \text{if } g = 0, \\ \delta, & \text{if } g \geq 1. \end{cases}
\end{aligned}$$

Proof: If $g = 0$, then ${}_pCl(k) = 0$ and $G_\varnothing \cong \hat{\mathbb{Z}}$ by (10.1.3)(i), and the assertions made are obvious. From now on we assume $g \geq 1$. Let S be an arbitrary nonempty finite set of primes of k. Using (10.7.3) and the exact sequence (10.7.4)(i), we obtain $h^1(G_\varnothing) = 1 - \delta + \dim_{\mathbb{F}_p} Б_\varnothing(k)$. Therefore (10.7.2) shows the statement for h^1. By (10.1.3)(ii), we obtain $h^i(G_\varnothing) = 0$ for $i \geq 4$ and $h^3(G_\varnothing) = h^0(G_\varnothing, \mu_p) = \delta$. It remains to show the statement about h^2.

Let κ be the finite field of constants of k and $\Gamma = G(\bar{\kappa}|\kappa) \cong \hat{\mathbb{Z}}$. We consider the Hochschild-Serre spectral sequence

$$E_2^{ij} = H^i(\Gamma, H^j(G_\varnothing(k\bar{\kappa}))) \Rightarrow H^{i+j}(G_\varnothing),$$

which degenerates at E_2 since $cd\,\hat{\mathbb{Z}} = 1$. Consider the exact sequence

$$H^j(G_\varnothing(k\bar{\kappa}))^\Gamma \hookrightarrow H^j(G_\varnothing(k\bar{\kappa})) \xrightarrow{\text{Frob}-1} H^j(G_\varnothing(k\bar{\kappa})) \twoheadrightarrow H^j(G_\varnothing(k\bar{\kappa}))_\Gamma .$$

Since the groups $H^j(G_\varnothing(k\bar{\kappa}))$ are finite and

$$H^1(\Gamma, H^j(G_\varnothing(k\bar{\kappa}))) \cong H^j(G_\varnothing(k\bar{\kappa}))_\Gamma$$

by (1.7.7), we obtain

$$\dim_{\mathbb{F}_p} H^0(\Gamma, H^j(G_\varnothing(k\bar{\kappa}))) = \dim_{\mathbb{F}_p} H^1(\Gamma, H^j(G_\varnothing(k\bar{\kappa})))$$

for $j \geq 0$. The spectral sequence implies

$$\sum_{i=0}^{3} h^i(G_\varnothing) = 0,$$

and the result for h^2 follows from the calculation of h^i for $i \neq 2$. □

It remains to consider the group G_S in the number field case with an arbitrary set S. The following results are due to *I. R. Šafarevič* and *H. Koch*, see [191] and [110].

For an arbitrary set S of primes of k we define the Šafarevič-Tate groups $Ш^i(G_S) = Ш^i(G_S, \mathbb{Z}/p\mathbb{Z})$ by the exactness of the sequences

$$0 \longrightarrow Ш^i(G_S) \longrightarrow H^i(G_S) \longrightarrow \prod_{\mathfrak{p}\in S} H^i(\mathcal{G}_\mathfrak{p}).$$

If $p \neq \operatorname{char} k$, and S nonempty containing $S_p \cup S_\infty$ in the number field case, then we have isomorphisms of finite abelian groups

$$Ш^2(G_S) \cong Ш^1(G_S, \mu_p)^\vee \cong Б_S(k),$$

by (8.6.7) and (9.1.12)(iii). Also, for arbitrary sets S we can relate $Ш^2(G_S)$ to the group $Б_S(k)$.

(10.7.6) Theorem (*H. Koch*). *Let k be a global field and let $p \neq \operatorname{char} k$ be a prime number. Then, for any set S of primes of k, there exists a natural injection*

$$Ш^2(G_S) \hookrightarrow Б_S(k).$$

In particular, $Ш^2(G_S)$ is finite.

Proof: Consider the commutative diagram

$$\begin{array}{ccccccccc} 0 \to H^1(G_S) \to H^1(G_k) \to H^1(G(\bar{k}|k_S))^{G_S} & \longrightarrow & H^2(G_S) & \longrightarrow & H^2(G_k) \\ & & \downarrow & & \downarrow \\ & & \bigoplus_{\mathfrak{p}\in S} H^2(\mathcal{G}_\mathfrak{p}) & \hookrightarrow & \bigoplus_{\mathfrak{p}} H^2(\mathcal{G}_\mathfrak{p}). \end{array}$$

The row is the 5-term sequence associated to the group extension

$$1 \longrightarrow G(\bar{k}|k_S) \longrightarrow G_k \longrightarrow G_S \longrightarrow 1,$$

hence exact. The right-hand vertical map is injective by (9.1.10)(i). Thus the sequence

$$0 \to H^1(G_S) \to H^1(G_k) \to H^1(G(\bar{k}|k_S))^{G_S} \to H^2(G_S) \to \bigoplus_{\mathfrak{p}\in S} H^2(\mathcal{G}_\mathfrak{p})$$

is exact. Using (10.7.4)(i) (where $S := \{\text{all primes}\}$ and $T := S$) and the vanishing of $Б_{\{\text{all primes}\}}(k)$ (see (9.1.12)(ii)), we obtain a commutative and exact diagram

$$\begin{array}{ccccccc} H^1(G_k) & \longrightarrow & \bigoplus_{\mathfrak{p}\notin S} H^1(\mathcal{T}_\mathfrak{p}(k))^{\mathcal{G}_\mathfrak{p}(k)} & \longrightarrow & Б_S(k) & \longrightarrow & 0 \\ \| & & \uparrow & & \uparrow & & \\ H^1(G_k) & \longrightarrow & H^1(G(\bar{k}|k_S))^{G_S} & \longrightarrow & Ш^2(G_S) & \longrightarrow & 0. \end{array}$$

Thus the desired result follows. □

Remark: If $S = \varnothing$, then (10.7.6) can be deduced in a convenient way from duality theorems for arithmetic curves: If k is a function field, let X be the regular projective curve with function field k. Étale Poincaré duality shows

$$H^2_{et}(X, \mathbb{Z}/p\mathbb{Z}) \cong H^1_{et}(X, \mu_p)^\vee \cong Б_\varnothing(k).$$

If k is a number field (and $p \neq 2$), let $X = \operatorname{Spec}(\mathcal{O}_k)$. The flat duality theorem of Artin-Mazur shows that

$$H^2(X_{et}, \mathbb{Z}/p\mathbb{Z}) \cong H^2(X_{fl}, \mathbb{Z}/p\mathbb{Z}) \cong H^1(X_{fl}, \mu_p)^\vee \cong Б_\varnothing(k).$$

In both cases, the Hochschild-Serre spectral sequence yields an inclusion

$$Ш^2(G_\varnothing) = H^2(G_\varnothing, \mathbb{Z}/p\mathbb{Z}) \hookrightarrow H^2(X_{et}, \mathbb{Z}/p\mathbb{Z}).$$

(10.7.7) Corollary. *Let S be a set of primes of the global field k and let $p \neq \operatorname{char} k$ be a prime number. Then there is an equality*

$$h^1(G_S) = 1 + \sum_{\mathfrak{p}\in S\setminus S_{\mathbb{C}}} \delta_{\mathfrak{p}} - \delta + \dim_{\mathbb{F}_p} Б_S(k) + \sum_{\mathfrak{p}\in S\cap S_p} n_{\mathfrak{p}} - r\,,$$

where $n_{\mathfrak{p}} = [k_{\mathfrak{p}} : \mathbb{Q}_p]$ is the local degree with respect to $\mathfrak{p} \in S_p$ and $r = r_1 + r_2$ denotes the number of archimedean primes. We have an inequality

$$h^2(G_S) \le \sum_{\mathfrak{p}\in S\setminus S_{\mathbb{C}}} \delta_{\mathfrak{p}} - \delta + \dim_{\mathbb{F}_p} Б_S(k) + \theta\,,$$

where $\theta = \theta(k,S)$ is equal to 1 if $\delta = 1$ and $S \subseteq S_\infty$ if $p \neq 2$ (resp. $S \subseteq S_{\mathbb{C}}$ if $p = 2$), and zero in all other cases.

Proof: Let us first assume that $\delta = 1$. Recall that $Ш^2(G_k) = 0$, see (9.1.10). From the exact sequence

$$0 \longrightarrow H^2(G_k) \longrightarrow \bigoplus_{\mathfrak{p}} H^2(G_{\mathfrak{p}}) \longrightarrow H^0(G_k, \mu_p)^\vee \longrightarrow 0,$$

we see that the map

$$H^2(G_k) \longrightarrow \bigoplus_{\mathfrak{p}\neq\mathfrak{p}_0} H^2(G_{\mathfrak{p}})$$

remains injective, where $\mathfrak{p}_0$ is an arbitrary nonarchimedean prime or a real prime if $p = 2$ (if one exists). Therefore if S is not contained in S_∞ if $p \neq 2$ resp. $S_{\mathbb{C}}$ if $p = 2$, i.e. $S = S' \cup \{\mathfrak{p}_0\}$ for some prime $\mathfrak{p}_0$ as above, the commutative diagram

$$\begin{array}{ccc} H^2(G_S) & \xrightarrow{\ inf\ } & H^2(G_k) \\ \big\downarrow{\scriptstyle \varphi_{S'}} & & \big\downarrow \\ \bigoplus_{\mathfrak{p}\in S'} H^2(G_{\mathfrak{p}}) & \hookrightarrow & \bigoplus_{\mathfrak{p}\neq\mathfrak{p}_0} H^2(G_{\mathfrak{p}}) \end{array}$$

shows that $Ш^2(G_S) \cong \ker inf \cong \ker \varphi_{S'}$.

Now let δ be arbitrary. From the definition of $Ш^2(G_S)$ and the consideration above, it follows that

$$h^2(G_S) \le \sum_{\mathfrak{p}\in S\setminus S_{\mathbb{C}}} \delta_{\mathfrak{p}} - \delta + \dim_{\mathbb{F}_p} Ш^2(G_S) + \theta\,.$$

Using (10.7.6), we obtain the result for h^2. The assertion for h^1 follows from the exact sequence (10.7.4)(i) with $T = \varnothing$ and (10.7.2) together with $h^1(G_\varnothing) = \dim_{\mathbb{F}_p} Cl(k)$, and the local computations (7.5.9) and (7.5.10). □

Next we extend our calculations and consider the group G_S^T, where T and S are sets of primes. First, we introduce further notation.

(10.7.8) Definition. *Let $p \neq$ char k and let $S_0 \subseteq S$ be sets of primes of k. Then*

$$V_{S_0}^S(k) = \{a \in k^\times \mid a \in k_v^{\times p} \text{ for } v \in S_0 \text{ and } a \in U_v k_v^{\times p} \text{ for } v \notin S\}/k^{\times p},$$

where U_v is the unit group of the local field k_v (by convention $U_v = k_v^\times$ if v is archimedean). The dual group is denoted by $\text{Б}_{S_0}^S(k) = V_{S_0}^S(k)^\vee$.

We have canonical inclusions

$$V_S(k) = V_S^S(k) \subseteq V_{S_0}^{S_0}(k) \subseteq V_{S_0}^S(k).$$

If $S \supseteq S_p \cup S_\infty$, then Kummer theory implies

$$V_{S_0}^S(k) = \text{Ш}^1(k_S, S_0, \mu_p) = \ker\Big(H^1(G_S, \mu_p) \longrightarrow \bigoplus_{v \in S_0} H^1(k_v, \mu_p)\Big).$$

(10.7.9) Lemma. *Let k be a global field, $p \neq$ char k a prime number and $S \supseteq T$ sets of primes of k. Then there is an exact sequence*

$$0 \to H^1(G_S^T) \to H^1(G_S) \to \prod_{\mathfrak{p} \in T} H^1(\mathcal{G}_\mathfrak{p}) \to \text{Б}_{S \setminus T}^S(k) \to \text{Б}_S(k) \to 0.$$

Proof: Consider the commutative and exact diagram

$$\begin{array}{ccccccc}
H^1(k_S|k_S^T)^{G_S^T} & \hookrightarrow & \prod_{\mathfrak{p} \in T} H^1(\mathcal{G}_\mathfrak{p}) & & & & \\
\uparrow & & \uparrow\!\!\uparrow & & & & \\
H^1(G_S) & \hookrightarrow & \prod_{\mathfrak{p} \in S} H^1(\mathcal{G}_\mathfrak{p}) \times \prod_{\mathfrak{p} \notin S} H^1_{nr}(\mathcal{G}_\mathfrak{p}) & \longrightarrow & H^1(G_k, \mu_p)^\vee & \twoheadrightarrow & \text{Б}_S(k) \\
\uparrow & & \uparrow & & \| & & \uparrow\!\!\uparrow \\
H^1(G_S^T) & & \prod_{\mathfrak{p} \in S \setminus T} H^1(\mathcal{G}_\mathfrak{p}) \times \prod_{\mathfrak{p} \notin S} H^1_{nr}(\mathcal{G}_\mathfrak{p}) & \longrightarrow & H^1(G_k, \mu_p)^\vee & \twoheadrightarrow & \text{Б}_{S \setminus T}^S(k).
\end{array}$$

The upper horizontal map is injective since $G(k_S|k_S^T)$ is generated by the decomposition groups of the primes in T. The middle row is exact by (10.7.4)(ii). The exactness of the bottom row is just the definition of $\text{Б}_{S\setminus T}^S(k) = V_{S\setminus T}^S(k)^\vee$, as $H^1_{nr}(\mathcal{G}_\mathfrak{p})^\vee = k_\mathfrak{p}^\times / U_\mathfrak{p} k_\mathfrak{p}^{\times p}$ and $H^1(\mathcal{G}_\mathfrak{p})^\vee = k_\mathfrak{p}^\times / k_\mathfrak{p}^{\times p}$. By diagram chasing, we obtain the claimed exact sequence. □

Using (10.7.9), our results on G_S extend to G_S^T. We obtain the

(10.7.10) Theorem. *Let S and T be finite sets of primes of the global field k. Then we have an equality*

$$h^1(G_S^T) = 1 + \sum_{\mathfrak{p}\in S\backslash(T\cup S_{\mathbb{C}})} \delta_{\mathfrak{p}} - \delta + \dim_{\mathbb{F}_p} Б_{S\backslash T}^{S\cup T}(k) + \sum_{\mathfrak{p}\in(S\backslash T)\cap S_p} n_{\mathfrak{p}} - r - \#T\backslash S_\infty .$$

Furthermore, we have an inequality

$$h^2(G_S^T) \le \sum_{\mathfrak{p}\in S\backslash(T\cup S_{\mathbb{C}})} \delta_{\mathfrak{p}} - \delta + \dim_{\mathbb{F}_p} Б_{S\backslash T}^{S\cup T}(k) + \theta ,$$

where $\theta = \theta(k, S)$ is equal to 1 if $\delta = 1$ and $S \subseteq S_\infty$ if $p \neq 2$ (resp. $S \subseteq S_{\mathbb{C}}$ if $p = 2$), and zero in all other cases.

Proof: Since $G_S^T = G_{S\cup T}^T$, we may assume that $T \subseteq S$. Now the formula for $h^1(G_S)$ given in (10.7.7) together with (10.7.9) imply the equality for $h^1(G_S^T)$. Since $G_S^T = G_{S\backslash T}^T$, we may assume that $S \cap T = \varnothing$ when proving the inequality for $h^2(G_S^T)$. We consider the exact sequence

$$0 \to H^1(G_S^T) \to H^1(G_S) \xrightarrow{res} H^1(G(k_S|k_S^T))^{G_S^T} \to H^2(G_S^T) \to H^2(G_S).$$

Using (10.7.7), the result for $h^1(G_S^T)$ and the injectivity of the map

$$H^1(G(k_S|k_S^T))^{G_S^T} \longrightarrow \prod_{\mathfrak{p}\in T} H^1_{nr}(\mathcal{G}_{\mathfrak{p}}),$$

we obtain

$$\begin{aligned} h^2(G_S^T) &\le h^2(G_S) - h^1(G_S) + h^1(G_S^T) + \dim_{\mathbb{F}_p} H^1(G(k_S|k_S^T))^{G_S^T} \\ &\le h^2(G_S) - h^1(G_S) + h^1(G_S^T) + \textstyle\sum_{\mathfrak{p}\in T} \dim_{\mathbb{F}_p} H^1_{nr}(\mathcal{G}_{\mathfrak{p}}) \\ &\le \textstyle\sum_{\mathfrak{p}\in S\backslash(T\cup S_{\mathbb{C}})} \delta_{\mathfrak{p}} - \delta + \dim_{\mathbb{F}_p} Б_{S\backslash T}^{S\cup T}(k) + \theta . \end{aligned}$$

□

For the partial Euler-Poincaré characteristic

$$\chi_2(G_S^T) = \sum_{i=0}^{2} (-1)^i \, h^i(G_S^T) ,$$

we obtain the

(10.7.11) Corollary. *Let S and T be finite sets of primes of the global field k and let $p \neq$ char k. Then the groups $H^i(G_S^T, \mathbb{Z}/p\mathbb{Z})$, $0 \le i \le 2$, are finite and*

$$\chi_2(G_S^T) \le \theta - \sum_{\mathfrak{p}\in(S\backslash T)\cap S_p} n_{\mathfrak{p}} + r + \#T\backslash S_\infty ,$$

where $\theta = \theta(k, S)$ is equal to 1 if $\delta = 1$ and $S \subseteq S_\infty$ if $p \neq 2$ (resp. $S \subseteq S_{\mathbb{C}}$ if $p = 2$), and zero in all other cases.

Next we consider the maximal pro-p-factor group $G_S^T(p)$ of G_S^T. We have the following equality for the generator rank $h^1(G_S^T(p))$ and the inequality for the relation rank $h^2(G_S^T(p))$ of $G_S^T(p)$.

(10.7.12) Theorem. *Let S and T be finite sets of primes of a global field k and let $p \neq \operatorname{char} k$ be a prime number. Then we have an equality*

$$h^1(G_S^T(p)) = 1 + \sum_{\mathfrak{p} \in S \setminus (T \cup S_{\mathbb{C}})} \delta_{\mathfrak{p}} - \delta + \dim_{\mathbb{F}_p} Б_{S\setminus T}^{S\cup T}(k) + \sum_{\mathfrak{p} \in (S\setminus T) \cap S_p} [k_{\mathfrak{p}} : \mathbb{Q}_p] - r - \# T \setminus S_\infty,$$

and an inequality

$$h^2(G_S^T(p)) \le \sum_{\mathfrak{p} \in S \setminus (T \cup S_{\mathbb{C}})} \delta_{\mathfrak{p}} - \delta + \dim_{\mathbb{F}_p} Б_{S\setminus T}^{S\cup T}(k) + \theta\,.$$

Here $r = r_1 + r_2$ is the number of archimedean primes and $\theta = \theta(k, S)$ is equal to 1 if $\delta = 1$ and $S \subseteq S_\infty$ if $p \neq 2$ (resp. $S \subseteq S_{\mathbb{C}}$ if $p = 2$), and zero in all other cases. Furthermore,

$$\chi_2(G_S^T(p)) \le \theta - \sum_{\mathfrak{p} \in (S\setminus T) \cap S_p} n_{\mathfrak{p}} + r + \# T \setminus S_\infty\,.$$

Proof: The homomorphism

$$H^i(G_S^T(p), \mathbb{Z}/p\mathbb{Z}) \longrightarrow H^i(G_S^T, \mathbb{Z}/p\mathbb{Z})$$

is bijective for $i = 1$ and injective for $i = 2$, since $H^1(k_S^T | k_S^T(p), \mathbb{Z}/p\mathbb{Z}) = 0$. Together with (10.7.10) and (10.7.11), this gives the result. □

We get an equality for the relation rank in the following situation.

(10.7.13) Theorem. *Let k be a global field and $p \neq \operatorname{char} k$ a prime number. Let S be a nonempty set of primes of k containing $S_p \cup S_\infty$ if k is a number field. Then*

$$\begin{aligned} h^1(G_S(p)) &= 1 + \sum_{\mathfrak{p} \in S} \delta_{\mathfrak{p}} - \delta + \dim_{\mathbb{F}_p} Б_S(k)\,, \\ h^2(G_S(p)) &= \sum_{\mathfrak{p} \in S \setminus S_{\mathbb{C}}} \delta_{\mathfrak{p}} - \delta + \dim_{\mathbb{F}_p} Б_S(k)\,. \end{aligned}$$

In particular, we have $\chi_2(G_S(p)) = -r_2$.

Proof: By (10.1.4) and (10.4.8), the inflation maps

$$H^i(G_S(p), \mathbb{Z}/p\mathbb{Z}) \to H^i(G_S, \mathbb{Z}/p\mathbb{Z})$$

are isomorphisms for all $i \ge 0$. Therefore we obtain the desired equalities from (10.7.3). □

(10.7.14) Corollary. *Let k be a global field and $p \neq \operatorname{char} k$ a prime number. Let S be a nonempty set of primes of k containing $S_p \cup S_\infty$ if k is a number field. Then the pro-p-group $G_S(p)$ is free if and only if*

$$\sum_{\mathfrak{p} \in S \setminus S_{\mathbb{C}}} \delta_{\mathfrak{p}} = \delta \quad \text{and} \quad \text{Б}_S(k) = 0\,.$$

If in addition $\mu_p \subseteq k$, then $G_S(p)$ is free if and only if

$${}_pCl_S(k) = 0 \;\text{ and }\; \begin{cases} \#S = 1, & \text{if } k \text{ is a function field,} \\ \#(S \setminus S_{\mathbb{C}}) = \#S_p = 1, & \text{if } k \text{ is a number field.} \end{cases}$$

Proof: This follows from (3.9.5), (10.7.13) and that $\text{Б}_S(k) = (Cl_S(k)/p)(-1)$ if $\mu_p \subseteq k$. □

Remarks: 1. If k is a number field and S is a finite set of primes disjoint from S_p, then $G_S(k)(p)^{ab}$ is finite by (10.3.20)(ii). Therefore $G_S(k)(p)$ is either trivial or not free in this case.
2. Let k be a number field, $p = 2$ and $S \supseteq S_2$. If S contains a real place, then $G_S(2)$ has infinite cohomological dimension by (10.6.1). If $S \cap S_{\mathbb{R}} = \varnothing$, then (10.9.2) will give a criterion for the freeness of $G_S(2)$.
3. If k is a function field, then $G_\varnothing(k)(p)$ is free if and only if $Cl(k)(p) = 0$ (cf. ex. 2 and 3 in §1).

Finally, we calculate the numbers $h^i(G_S(p))$ for the number field $k = \mathbb{Q}$.

(10.7.15) Theorem. *Let p be a prime number. Let S be a finite set of places of $\mathbb{Q}$ and $G_S(p) = G(\mathbb{Q}_S(p)|\mathbb{Q})$. If $p \neq 2$ assume that S is minimal in the sense that it does not contain ∞ and contains only prime numbers ℓ with $\ell \equiv 0, 1 \bmod p$. If $S = \varnothing$, then $G_S(p) = 1$, and if $S \neq \varnothing$, then the following holds.*

(i) *If $p \neq 2$, then*

$$\begin{aligned} h^1(G_S(p)) &= \#S, \\ h^2(G_S(p)) &= \begin{cases} \#S - 1, & \text{if } p \in S, \\ \#S, & \text{if } p \notin S, \end{cases} \\ h^i(G_S(p)) &= 0, \qquad \text{if } p \in S, \quad i \geq 3. \end{aligned}$$

(ii) *If $p = 2$, then*

$$\begin{aligned} h^1(G_S(2)) &= \begin{cases} \#S, & \text{if } 2 \in S, \\ \#S + \alpha - 1, & \text{if } 2 \notin S, \end{cases} \\ h^2(G_S(2)) &= \#S + \alpha - 1, \\ h^i(G_S(2)) &= \begin{cases} 1, & \text{if } \infty \in S \text{ and } 2 \in S, \\ 0, & \text{if } \infty \notin S \text{ and } 2 \in S, \end{cases} \qquad i \geq 3, \end{aligned}$$

where $\alpha = 0$ if $2 \in S$, $\infty \in S$ or $\ell \in S$ for some $\ell \equiv 3 \bmod 4$, and $\alpha = 1$ in all other cases.

Remark: The theorem omits the calculation of $h^i(G_S(p))$, $i \geq 3$, if $p \notin S$. This is difficult. There exist (so-called "circular") finite sets S of prime numbers, not containing p, such that $cd\, G_S(p) = 2$. These sets are cofinal among all finite sets of primes S with $p \notin S$, see [119], [198] (see [199] for the case of a general number field).

Proof: If $S = \varnothing$, then $G_S(p)$ is the trivial group as there are no unramified extensions of $\mathbb{Q}$. We assume that $S \neq \varnothing$ and start with the case $p \neq 2$. Then $Б_S(\mathbb{Q}) = 0$ and $\delta = 0 = \delta_p$. Therefore the statement for h^1 follows from (10.7.12). If $p \in S$, then the assertion for h^2 follows from (10.7.13). If $p \notin S$, then (10.7.12) yields the inequality $h^2(G_S(p)) \leq \#S$. From the exact sequence

$$0 \to \mathbb{Z}/p\mathbb{Z} \to \mathbb{Q}/\mathbb{Z} \xrightarrow{p} \mathbb{Q}/\mathbb{Z} \to 0$$

we obtain the inclusion

$$H^1(G_S(p), \mathbb{Q}/\mathbb{Z})/p \hookrightarrow H^2(G_S(p), \mathbb{Z}/p\mathbb{Z}).$$

By class field theory, $G_S(p)^{ab}$ is finite if $p \notin S$. It follows that

$$\begin{aligned} h^1(G_S(p)) &= \dim_{\mathbb{F}_p} {}_p H^1(G_S(p), \mathbb{Q}/\mathbb{Z}) \\ &= \dim_{\mathbb{F}_p} H^1(G_S(p), \mathbb{Q}/\mathbb{Z})/p \\ &\leq h^2(G_S(p)), \end{aligned}$$

hence the above inequality is an equality. Finally, $cd\, G_S(p) \leq 2$ if $p \in S$. Thus the proof of (i) is complete.

Now let $p = 2$. Then $\dim_{\mathbb{F}_2} Б_S(\mathbb{Q}) = \alpha$ (we leave the verification to the reader, see ex.1). Therefore (10.7.12) implies the statement for h^1. If $2 \in S$, then $\chi_2(G_S(2)) = 0$ by (10.6.1), showing the statement for h^2. Furthermore, if $2 \in S$, then (10.6.5) implies the statement for h^i, $i \geq 3$. There remains the case $2 \notin S$. By (10.7.12), we obtain the inequality $h^2(G_S(p)) \leq \#S + \alpha - 1$, and the same argument as in the case $p \neq 2$ shows that this inequality is an equality. This finishes the proof. □

Exercise 1. Let $k = \mathbb{Q}$. Show that $Б_S(\mathbb{Q}, p) = 0$ if $p \neq 2$, and $Б_S(\mathbb{Q}, 2) = 0$ if and only if S contains $2, \infty$ or a prime number $\ell \equiv 3 \bmod 4$.

Exercise 2. Let $\mathbb{F}$ be a finite field, $k = \mathbb{F}(t)$ and $p \neq \operatorname{char} \mathbb{F}$ a prime number. Show that $Б_S(k) = 0$ for every S if μ_p is not contained in k, and if $\#\mu(k)(p) = p^s$, $s \geq 1$, then $Б_S(k) = 0$ if and only if S contains a prime $\mathfrak{p}_0$ such that $N(\mathfrak{p}_0) \not\equiv 1 \bmod p^{s+1}$.

§8. The Theorem of Kuz'min

This section is devoted to the question of which local p-extensions are globally realized. More exactly, assume k is a number field, p is a prime number and let $S \supseteq S_p \cup S_\infty$ be a finite set of primes in k. We will exclusively consider p-extensions in this section, and so we make the following **notational convention**:

Unless the contrary is explicitly stated, every field extension is tacitly assumed to be a p-extension. We write k_S for $k_S(p)$, G_S for $G_S(p)$, G_k for $G_k(p)$ and so on. Furthermore, we write $Cl_S(k)$ for $Cl_S(k)(p)$ and k_∞ for the cyclotomic $\mathbb{Z}_p$-extension of k.

Recall that every proper subgroup of a pro-p-group is contained in a proper, normal subgroup, cf. (1.6.14)(i). Therefore a subgroup whose normal closure is the full group, is the full group itself. We conclude that the vanishing of $Cl_S(k)$ implies that G_S is generated by the decomposition groups of (arbitrary chosen prolongations of) the primes in S.

Suppose that we are given a finite prime $\mathfrak{p} \in S$ and a p-extension $N|k_\mathfrak{p}$. Our central question is:

"Does there exist an extension K of k inside k_S with $K_\mathfrak{p} = N$?"

The answer is in general "no", as the following examples show.

Example 1. $k = \mathbb{Q}$, p odd, $S = S_p \cup S_\infty$. Then

$$G_S = G(\mathbb{Q}_\infty|\mathbb{Q}) = \mathbb{Z}_p,$$

where $\mathbb{Q}_\infty$ denotes the cyclotomic $\mathbb{Z}_p$-extension of $\mathbb{Q}$. (See the discussion before (10.5.6)).

On the other hand, we know by (7.5.11) that $G_{\mathbb{Q}_p}$ is a free pro-p-group of rank 2. Hence not every given local p-extension at the prime p can be globally realized.

Example 2. $k = \mathbb{Q}(\mu_p)$ with a regular, odd prime number p, $S = S_p \cup S_\infty$. In this case there is exactly one prime $\mathfrak{p} \in S_p(k)$ and in the long Poitou-Tate sequence (recalling (7.5.11)) there occurs an isomorphism

$$H^2(k_\mathfrak{p}, \mathbb{Z}/p\mathbb{Z}) = P^2(G_S, \mathbb{Z}/p\mathbb{Z}) \xrightarrow{\sim} H^0(G_S, \mu_p)^\vee.$$

Using (10.4.8), we obtain

$$H^2(G_S, \mathbb{Z}/p\mathbb{Z}) \cong Ш^1(G_S, \mu_p)^\vee \cong (Cl_S(k)/p)\,(-1) = 0.$$

Hence G_S is a free pro-p-group of rank $(p+1)/2$ (use the Euler-Poincaré formula (8.7.5)). The group $G_{k_\mathfrak{p}}$ is a Demuškin group of rank $p+1$ by (7.5.11).

Furthermore, since $Cl_S(k) = 0$, the canonical homomorphism

$$G_{k_\mathfrak{p}} \longrightarrow G_S$$

is surjective (see above). From our calculations this map is not an isomorphism, so also in this example not every local extension is globally realized. As a by-product, we conclude (see (10.3.6)) that the (strong) Leopoldt conjecture holds for p and every subextension $K \subseteq k_S$.

We denote the decomposition group of a prime v in G_S by G_v and the local Galois group G_{k_v} by $\mathcal{G}_v$. We write $\mathcal{T}_v$ for the inertia subgroup in $\mathcal{G}_v$.

(10.8.1) Theorem. *For $v \in S \smallsetminus S_p$, the canonical homomorphism*

$$\mathcal{G}_v \longrightarrow G_S$$

is injective. If k is not totally real, then the image is not open, i.e. of infinite index in G_S.

Remark: The example $k = \mathbb{Q}$, $p = 3$, $S = \{3, 7, \infty\}$ shows that the assumption that k is not totally real is essential. Indeed, by (10.5.6), we see that $G_S = G_7$ in this example.

Proof: If v is archimedean, then the assertion is obvious, therefore we may assume that v is finite. Since the cyclotomic $\mathbb{Z}_p$-extension of k globally realizes the maximal unramified extension of k_v, the kernel of the homomorphism $\mathcal{G}_v \to G_S$ must be contained in $\mathcal{T}_v$. Thus this kernel is trivial since $\mathcal{T}_v$ maps injectively to G_S by Riemann's existence theorem (10.5.1)(i) and (4.3.12).
It therefore remains to show the statement about the infinite index. By the trivial Leopoldt inequality (see (10.3.7)) G_S has a $\mathbb{Z}_p$-free abelian quotient of rank $r_2 + 1$ which is unramified outside p. A decomposition group of a prime not dividing p therefore has an image of rank 1 in this group. Hence such a decomposition group cannot be open in G_S if $r_2 > 0$. □

(10.8.2) Theorem. *Let k be totally imaginary and let $v \in S^f$. Suppose that G_v is open in G_S. Then $v \in S_p$ and either $G_v = G_S$ or $p = 2$, $(G_S : G_v) = 2$, $\#S^f(k) = 1$ and $S^f(k_S) = 2$.*

Proof: By theorem (10.8.1), we know that $v \in S_p$. Let k^v be the decomposition field of v in k_S, i.e. $k^v = k_S^{G_v}$. Assume that k^v is of finite degree over k. The trivial Leopoldt inequality (see (10.3.7)) shows that

$$r_2(k^v) + 1 \le \operatorname{rank}_{\mathbb{Z}_p} G_S(k^v)^{ab}.$$

Since k, and hence k^v, is totally imaginary, we obtain

$$\begin{aligned}[k^v:\mathbb{Q}]/2+1 &\leq \operatorname{rank}_{\mathbb{Z}_p} G_S(k^v)^{ab} \\ &= \operatorname{rank}_{\mathbb{Z}_p} G_v^{ab} \leq \operatorname{rank}_{\mathbb{Z}_p} \mathcal{G}_v^{ab} = [k_v:\mathbb{Q}_v]+1 \\ &\leq [k:\mathbb{Q}]+1.\end{aligned}$$

Hence $[k^v:k] \leq 2$. Thus it remains to consider the case $(G_S:G_v)=2$; in particular, $p=2$. Observe that $\dim_{\mathbb{F}_2} H^2(K_\mathfrak{p},\mathbb{F}_2)=1$ for every number field K and every $\mathfrak{p} \in S^f(K)$. If K is totally imaginary, we therefore obtain by counting dimensions in the long exact Poitou-Tate sequence

$$\dim_{\mathbb{F}_2} H^2(G_S(K),\mathbb{F}_2) \geq \#S^f(K)-1.$$

The Euler-Poincaré characteristic formula (8.7.5) yields

$$(*)\qquad \dim_{\mathbb{F}_2} H^1(G_S(K),\mathbb{F}_2) \geq r_2(K)+\#S^f(K).$$

Applying $(*)$ to every $K \subseteq k_S$ containing k^v, we obtain

$$\begin{aligned}[K_v:\mathbb{Q}_2]+2 &= \dim_{\mathbb{F}_2} H^1(G_{K_v},\mathbb{F}_2) \\ &\geq \dim_{\mathbb{F}_2} H^1(G_v(k_S|K),\mathbb{F}_2) \\ &= \dim_{\mathbb{F}_2} H^1(G_S(K),\mathbb{F}_2) \\ &\geq [K:\mathbb{Q}]/2+\#S^f(K).\end{aligned}$$

By our assumption, v splits in $k^v|k$, so that $[K_v:\mathbb{Q}_2] \leq [K:\mathbb{Q}]/2$ and $\#S^f(K) \geq 2$. Comparing with the above inequality, we deduce that $\#S^f(K)=2$ and finally $\#S^f(k)=1$. □

Next we investigate the decomposition groups of primes dividing p. In order to simplify notation, we write $H^i(G)$ for $H^i(G,\mathbb{F}_p)$ if G is a pro-p-group. We denote the extension of degree p^m in the cyclotomic $\mathbb{Z}_p$-extension of a field k by k_m. If $v \in S^f(k)$, then $G_{v,m} := G((k_S)_v|k_{v,m}) = G(k_S|(k^v)_m)$.

(10.8.3) Proposition. *Assume $S_\mathbb{R}(k)=\varnothing$ if $p=2$ and let $v \in S_p$. Then*

(i) (a) *$cd_p\, G_v \leq 2$,*
(b) *$\dim_{\mathbb{F}_p} H^i(G_v) < \infty$ for all i.*

Furthermore, if $\mu_p \subseteq k$, then the following is true.

(ii) *$\dim_{\mathbb{F}_p} H^2(G_{v,m})$ is bounded for $m \to \infty$.*

(iii) *$scd_p\, G_v = 2$.*

(iv) *If $p^\infty|(G_S:G_v)$, then*

$$H^2(G_v) \xrightarrow[\sim]{res} \bigoplus_{w \in S^f(k^v)} H^2(G_{k^v_w}),$$

where k^v_w is the local field associated to the prime w of the decomposition field k^v of v.

Proof: By (8.3.18) and (10.4.8), $cd_p\, G_S \le 2$, so that $cd_p\, G_v \le 2$, showing (a). Being a quotient of $\mathcal{G}_v$, the group G_v is finitely generated, so in order to show (b) we can restrict to the case $i = 2$. Denote the Galois group of the local cyclotomic $\mathbb{Z}_p$-extension by $\Gamma_v = G(k_{v,\infty}|k_v)$ and consider the exact sequence

$$1 \longrightarrow G_{v,\infty} \longrightarrow G_v \longrightarrow \Gamma_v \longrightarrow 1.$$

By the Hochschild-Serre sequence, the group $H^1(G_{v,\infty})^{\Gamma_v}$ is finite, and so $G_{v,\infty}^{ab}$ is a finitely generated $\mathbb{Z}_p[[\Gamma_v]]$-module by the topological Nakayama lemma (5.2.18). Since $H^2(G_{v,\infty}, \mathbb{Q}_p/\mathbb{Z}_p) = 0$ by (10.3.26), we have

$$H^1(\Gamma_v, H^1(G_{v,\infty}, \mathbb{Q}_p/\mathbb{Z}_p)) \cong H^2(G_v, \mathbb{Q}_p/\mathbb{Z}_p)\,.$$

From the exact sequence

$$0 \longrightarrow ({}_pG_v^{ab})^* \longrightarrow H^2(G_v) \longrightarrow {}_pH^2(G_v, \mathbb{Q}_p/\mathbb{Z}_p) \longrightarrow 0\,,$$

it follows that

$$\begin{aligned}
\dim_{\mathbb{F}_p} H^2(G_v) &= \dim_{\mathbb{F}_p} {}_pG_v^{ab} + \dim_{\mathbb{F}_p} {}_pH^1(\Gamma_v, H^1(G_{v,\infty}, \mathbb{Q}_p/\mathbb{Z}_p)) \\
&= \dim_{\mathbb{F}_p} {}_pG_v^{ab} + \dim_{\mathbb{F}_p} (G_{v,\infty}^{ab})^{\Gamma_v}/p \\
&\le \dim_{\mathbb{F}_p} H^1(G_v) + \dim_{\mathbb{F}_p} (G_{v,\infty}^{ab})^{\delta}/p < \infty\,,
\end{aligned}$$

where X^δ denotes the maximal discrete submodule of an Iwasawa module X (see (5.3.12)). This shows (b).

From now on we assume that $\mu_p \subseteq k$. Since $(k^v)_\infty$ contains the maximal unramified extension of k_∞ which is completely decomposed at all primes in S, we conclude from (10.4.5) that

$$cd_p\, G_{v,\infty} \le 1.$$

This yields

$$H^2(G_{v,m}) = H^1(\Gamma_{v,m}, H^1(G_{v,\infty})),$$

and therefore

$$\dim_{\mathbb{F}_p} H^2(G_{v,m}) \le \dim_{\mathbb{F}_p} (G_{v,\infty}^{ab}/p)^{\delta} < \infty.$$

This shows (ii).

Since $\mu_p \subseteq k$, we have $k^v(\mu_{p^m}) \subseteq k_S$ and these fields are closed under extensions which are unramified outside S and completely decomposed at v. The principal ideal theorem implies

$$Cl_S(k^v(\mu_{p^m})) = 0.$$

(Recall our notational convention $Cl_S = Cl_S(p)$!) Now consider the commutative exact diagrams

$$\begin{array}{ccccc}
H^1(K(\mu_{p^m})|K) & \hookrightarrow & H^1(G_S(K)) & \longrightarrow & H^1(G_S(K(\mu_{p^m}))) \\
\downarrow{\scriptstyle\alpha} & & \downarrow{\scriptstyle\beta} & & \downarrow{\scriptstyle\gamma} \\
\bigoplus_{w\in S(K)} H^1(K_w(\mu_{p^m})|K_w) & \hookrightarrow & \bigoplus_{w\in S(K)} H^1(K_w) & \longrightarrow & \bigoplus_{w\in S(K(\mu_{p^m}))} H^1(K_w(\mu_{p^m})),
\end{array}$$

where K is a finite extension of k inside k^v and the coefficients of the cohomology groups are μ_{p^m} .

For sufficiently large $K \subseteq k^v$, the map α is injective since v occurs as one of the places in $S(K)$ (α is the map $0 \to 0$ in almost all cases, cf. (9.1.6)). We conclude that for sufficiently large $K \subseteq k^v$

$$\ker \beta = Ш^1(G_S(K), \mu_{p^m}) \hookrightarrow \ker \gamma = (Cl_S(K(\mu_{p^m}))/p^m)^\vee(1).$$

Dualizing and passing to the limit over all finite $K \subseteq k^v$, we obtain a surjection

$$0 = (Cl_S(k^v(\mu_{p^m}))/p^m)(-1) \twoheadrightarrow Ш^2(G_S(k^v), \mathbb{Z}/p^m\mathbb{Z}).$$

The Poitou-Tate sequence therefore implies the exactness of

$$H^2(G_v, \mathbb{Z}/p^m\mathbb{Z}) \hookrightarrow \varinjlim_K \bigoplus_{w \in S(K)} H^2(K_w, \mathbb{Z}/p^m\mathbb{Z}) \to \varinjlim_K H^0(G_S(K), \mu_{p^m})^\vee.$$

Passing to the limit over all $K \subseteq k^v$, the limit of the right-hand term is finite and vanishes if $p^\infty | [k^v : k]$.
Furthermore, $H^2(G_v, \mathbb{Z}/p^m\mathbb{Z})$ is finite by (i), and so

$$H^2(k^v_w, \mathbb{Z}/p^m\mathbb{Z}) = 0$$

for all but finitely many $w \in S(k^v)$. This shows (iv) (setting $m = 1$). Finally, passing to the limit over $m \geq 1$ and since the strict cohomological dimension of local fields is 2, we obtain

$$H^2(G_v, \mathbb{Q}_p/\mathbb{Z}_p) = 0.$$

The same argument applies to every open subgroup of G_v, hence showing assertion (iii). □

The following theorem was first proved by *L. Kuz'min* [115]. The proof presented below is a slight modification of that given in [254], Appendix.

(10.8.4) Theorem (*Kuz'min*). *Let k be a totally imaginary number field with $\mu_p \subseteq k$ and let $S \supseteq S_p \cup S_\infty$ be a finite set of primes in k. Suppose that for a prime $v \in S(k)$ the group G_v is not open in G_S. Then the canonical map*

$$\mathcal{G}_v \longrightarrow G_v$$

is an isomorphism, i.e. every p-extension of the local field k_v is realized by a p-extension of the global field k which is unramified outside S.

Remark: When $v \notin S_p(k)$, this result is already contained in (10.8.1).

Let us explain the strategy of the proof, which we will give below. The prime v clearly must be nonarchimedean. By (10.8.3), both groups are of strict cohomological dimension 2, therefore it is sufficient to show that the associated homomorphism of abelianized groups is an isomorphism. This comes down showing that every abelian extension of k_v is realized by a (not necessarily abelian) extension of k inside k_S. By local class field theory, we have to show that the group of universal norms of $(k_S)_v|k_v$ is contained in the subgroup of p-divisible elements in $k_v^\times$ (which is the group of roots of unity of order prime to p). This will be achieved in two steps. In the first step we show that every element of this norm group is a p-th power in $k_v^\times$. We do this by constructing a suitable cyclic extension of degree p of k inside k_S. The construction uses Kummer theory; this is where we need $\mu_p \subseteq k$.

In the second step we go up the cyclotomic tower in order to show that if G_v were not the full local group, then a suitable extension k' of k inside k_S (for which all assumptions remain valid) would produce a contradiction to the result of step 1.

However, the reader will find that the ideas explained above really lie *behind* the proof. In fact, no universal norm groups will explicitly occur. Instead our main technical tool is a careful analysis of the $\mathbb{Z}_p[\![G_v]\!]$-module structure of the abelianized kernel of $\mathcal{G}_v \to G_v$. Here we use the results of chapter V (for the group ring of a non-abelian group!) in an essential way.

Proof of (10.8.4): For $w \in S^f(k)$, consider the exact sequence

$$1 \longrightarrow R_w \longrightarrow \mathcal{G}_w \longrightarrow G_w \longrightarrow 1.$$

Here we tacitly have chosen a prolongation of w to k_S and R_w is defined by the exactness of the sequence. We set

$$N_w := \ker(\mathcal{G}_w^{ab}/p \to G_w^{ab}/p) = R_w[\mathcal{G}_w, \mathcal{G}_w]\mathcal{G}_w^p/[\mathcal{G}_w, \mathcal{G}_w]\mathcal{G}_w^p.$$

Global class field theory implies the exactness of the sequence

$$\mathcal{O}_{k,S}^\times/p \xrightarrow{\ \psi\ } I_S(k)/p \xrightarrow{\ rec\ } G(k_S^{ab}|L')/p,$$

where L' is defined as the maximal extension of k inside k_S^{ab} in which all primes of S are completely decomposed. Local class field theory gives an isomorphism

$$I_S(k) \cong \bigoplus_{w\in S} \mathcal{G}_w^{ab}/p$$

and the map *rec* clearly factors through $\bigoplus_{w\in S} G_w^{ab}/p$. Therefore we conclude that

$$\bigoplus_{w\in S} N_w \subseteq \ker(rec) = \operatorname{im}(\psi).$$

Now assume that there exists an $x_v \in N_v$, $x_v \neq 0$. Then there exists a global S-unit $e \in \mathcal{O}_{k,S}^\times$ with

$$\psi(e) = (0, \dots, x_v, \dots, 0) \in \bigoplus_{w \in S} N_w.$$

Consider the extension

$$k' := k(\sqrt[p]{e}).$$

The field k' is a cyclic extension of k inside k_S, v does not decompose in k' and all $w \in S(k)$ with $w \neq v$ split in $k'|k$. We denote the cyclic Galois group by $G = G(k'|k)$, the decomposition group of v in $k_S|k'$ by H_v, the associated full local group by $\mathcal{H}_v$ and we set

$$N_v' = \ker(\mathcal{H}_v^{ab}/p \to H_v^{ab}/p) \;=\; R_v[\mathcal{H}_v, \mathcal{H}_v]\mathcal{H}_v^p/[\mathcal{H}_v, \mathcal{H}_v]\mathcal{H}_v^p.$$

Then local class field theory induces a commutative diagram

$$\begin{array}{ccc}
N_v & \xrightarrow{N_G} & N_v' \\
\downarrow & & \downarrow \\
\mathcal{G}_v^{ab}/p & \xrightarrow{Ver} & \mathcal{H}_v^{ab}/p \\
\uparrow \wr\, rec & & \uparrow \wr\, rec \\
k_v^\times/p & \xrightarrow{\text{"incl"}} & k_v'^\times/p.
\end{array}$$

The $x_v \in N_v$ chosen initially corresponds to the class of the global S-unit e in $k_v^\times/p$. This shows that

$$N_G(x_v) = 0.$$

So far the prime $v \in S^f$ has been arbitrary. From now on we assume that G_v is not open in G_S. We will show that the map

$$N_G : N_v \longrightarrow N_v'$$

is then injective, hence showing $N_v = 0$.

For this we first observe that by (10.8.3) we have isomorphisms

$$\begin{aligned}
H^2(G_v) &\cong \bigoplus_{w \in S(k^v)} H^2(\mathcal{G}_w), \\
H^2(H_v) &\cong \bigoplus_{w \in S(k'^v)} H^2(\mathcal{H}_w).
\end{aligned}$$

Since all $w \neq v$ split in $k'|k$, we obtain an $\mathbb{F}_p[G]$-module isomorphism

$$H^2(H_v) \cong H^2(\mathcal{H}_v) \oplus \mathbb{F}_p[G]^t$$

for some finite t (recall $\dim_{\mathbb{F}_p} H^2(H_v) < \infty$). Furthermore, the corestriction induces an isomorphism

$$cor : H^2(\mathcal{G}_v) \xrightarrow{\sim} H^2(\mathcal{H}_v).$$

Dualizing, we conclude that $H_2(H_v)/H_2(\mathcal{H}_v)$ is a free $\mathbb{F}_p[G]$-module of finite rank and the norm induces an isomorphism

$$H_2(G_v)/H_2(\mathcal{G}_v) \xrightarrow[\sim]{N_G} (H_2(H_v)/H_2(\mathcal{H}_v))^G.$$

Keeping the assumption that G_v is not open in G_S, we next investigate the $\mathbb{Z}_p[\![G_v]\!]$-module structure of R_v^{ab}. Consider the commutative diagram

$$\begin{array}{ccc} H_2(\mathcal{G}_v) & \hookrightarrow & H_2(G_v) \\ \downarrow \wr & & \downarrow \wr \\ {}_p\mathcal{G}_v^{ab} & \longrightarrow & {}_pG_v^{ab}. \end{array}$$

The vertical maps are isomorphisms because $scd_p\mathcal{G}_v = scd_p\, G_v = 2$ and the upper horizontal arrow is injective by (10.8.3)(iv). Hence the lower horizontal arrow is injective and it therefore follows from (5.6.13) that R_v^{ab} is a free $\mathbb{Z}_p[\![G_v]\!]$-module of finite rank.

We consider the cyclic extension $k'|k$ constructed above and we compare the homological Hochschild-Serre sequences for the group extensions $R_v \hookrightarrow \mathcal{G}_v \twoheadrightarrow G_v$ and $R_v \hookrightarrow \mathcal{H}_v \twoheadrightarrow H_v$. Using the notation above, we obtain an exact diagram

$$\begin{array}{ccccccc} 0 \longrightarrow & (H_2(H_v)/H_2(\mathcal{H}_v))^G & \longrightarrow & ((R_v^{ab}/p)_{H_v})^G & \longrightarrow & (N_v')^G & \\ & \wr\uparrow N_G & & \wr\uparrow N_G & & \uparrow N_G & \\ 0 \longrightarrow & H_2(G_v)/H_2(\mathcal{G}_v) & \longrightarrow & (R_v^{ab}/p)_{G_v} & \longrightarrow & N_v & \longrightarrow 0. \end{array}$$

The left-hand and middle vertical arrows are isomorphisms, by the module structure of R_v^{ab} and of $H_2(H_v)/H_2(\mathcal{H}_v)$. Therefore the right-hand vertical arrow is injective.

Since we have already constructed a nontrivial element in the kernel of $N_G : N_v \to N_v'$, we obtain a contradiction. Hence $N_v = 0$ and, since all assumptions remain valid, the same is true for every finite extension of k inside k_S.

Setting $r = \mathrm{rank}_{\mathbb{Z}_p[\![G_v]\!]} R_v^{ab}$, we obtain that

$$X := (R_v^{ab})_{G_{v,\infty}}$$

is a free $\mathbb{Z}_p[\![\Gamma_v]\!]$-module of rank r, where $\Gamma_v = G(k_{v,\infty}|k_v)$. The vanishing of N_v (on every level) yields surjections

$$H_2(G_{v,m}) \twoheadrightarrow (X/p)_{\Gamma_{v,m}} \cong \mathbb{F}_p[\Gamma_v/\Gamma_{v,m}]^r$$

for every $m \geq 1$. Since the $\mathbb{F}_p$-dimension of $H_2(G_{v,m})$ is bounded as $m \to \infty$ by (10.8.3)(ii), we conclude that $r = 0$, and hence $R_v = 0$. □

It is much harder to show results corresponding to (10.8.4) for the full group G_S instead of its maximal pro-p-factor group, which we considered in this section. In 2007, *G. Chenevier* (cf. [25]) succeeded in showing that the full local group $\mathcal{G}_{\mathfrak{p}}$ is realized by G_S under certain conditions. His proof uses ℓ-adic representations attached to automorphic forms. Afterwards he and *L. Clozel* (see [CC]) refined the method and obtained the following theorem, in which G_S and $G_{k_{\mathfrak{p}}}$ denote the full Galois groups, i.e. the notational convention from the beginning of this section does not apply to (10.8.5).

(10.8.5) Theorem (*Chenevier-Clozel*). *Let k be a totally real number field, $\mathfrak{p}$ a finite prime of k and ℓ a rational prime number such that $\mathfrak{p} \nmid \ell$. Then the map*

$$G_{k_{\mathfrak{p}}} \longrightarrow G_{S_\ell \cup S_\infty \cup \{\mathfrak{p}\}}(k)$$

is injective. In particular, any natural number divides the supernatural order of the group $G_{S_\ell \cup S_\infty \cup \{\mathfrak{p}\}}(k)$.

Exercise (see [148]): Assume that k is a CM-field, i.e. k is a totally imaginary extension of degree 2 of a totally real subfield k^+. Let p be an odd prime number and assume that all primes dividing p split in the extension $k|k^+$. Show that the canonical homomorphism

$$\mathcal{G}_{\mathfrak{p}} \longrightarrow\!\!\!\!\!\rightarrow G_{\mathfrak{p}} \subseteq G_S(k)$$

of pro-p-groups is an isomorphism for every finite set of primes $S \supseteq S_p \cup S_\infty$ and for every prime $\mathfrak{p}$ dividing p. (Note that we did not assume $\mu_p \subseteq k$!)

Hint: In this special situation every local abelian p-extension can be realized by a global *abelian* p-extension. In order to prove this, examine the upper exact sequence of lemma (10.3.13). If $scd_p\, G_{\mathfrak{p}}$ were equal to 2, this would suffice to prove the statement. Now change to the cyclotomic $\mathbb{Z}_p$-extension k_∞ of k and apply (10.4.9)(iii).

§9. Free Product Decomposition of $G_S(p)$

In this section we investigate how the decomposition groups of the primes in S lie inside the group $G_S = G_S(p)$, where p is a fixed prime number with $S_p \subseteq S$.

We derive a criterion for the group G_S to be a free product of local groups and we call this the degenerate case. If we are in the generic (i.e. not in the degenerate) case and if $\mu_p \subseteq k$, then we show that G_S is a pro-p duality group of dimension 2 in which all decomposition groups are of infinite index (the case $p = 2$ requires some modifications).

The degenerate case is the easier one. Owing to the free product decomposition, we obtain complete control over all extensions of k inside k_S. In particular, it can be easily deduced that the (strong) Leopoldt conjecture holds in the degenerate case. Most of the results below are taken from [254].

Let k be a number field and let p be a prime number. We keep the notational convention of the last section, i.e. unless the contrary is explicitly stated, we tacitly assume extensions to be p-extensions and we always assume that the finite set of primes S contains $S_p \cup S_\infty$. At those few places where we need the full Galois group of the maximal extension $\mathcal{K}_S$ of k which is unramified outside S, we will denote this group by $\mathscr{G}_S = G(\mathcal{K}_S|k)$, while $G_S = G(k_S|k)$ will always denote its maximal pro-p-factor. We define the group $\mathscr{I}_S$ by the exact sequence

$$1 \longrightarrow \mathscr{I}_S \longrightarrow \mathscr{G}_S \longrightarrow G_S \longrightarrow 1.$$

In addition we use the following notation:

S^f	the set of finite primes in S,
G_v	the decomposition group of the prime v in G_S,
T_v	the inertia subgroup of v in G_v,
$\mathcal{G}_v$	$= G(k_v(p)\|k_v)$, the full local group,
$\mathcal{T}_v$	the inertia subgroup in $\mathcal{G}_v$,
C_S	the S-idèle class group,
C_{S^f}	the S^f-idèle class group (see §2),
$\mathrm{tor}_p(A)$	the p-torsion subgroup of the abelian group A.

Recall the definitions of δ, δ_v and $\text{Б}^S_{S_0} = (V^S_{S_0})^\vee$ from (10.7.1) and (10.7.8). We often write $H^i(-)$ for $H^i(-, \mathbb{Z}/p\mathbb{Z})$.

(10.9.1) Theorem. *Let S_0 be a subset of S^f. Then the following assertions are equivalent.*

(i) *There exists a finite set of primes $T \supseteq S$ such that the canonical homomorphism*

$$\mathop{\ast}_{v\in S\setminus S_0} \mathcal{G}_v \ast \mathop{\ast}_{v\in T\setminus S} \mathcal{G}_v/\mathcal{T}_v \longrightarrow G_S$$

is an isomorphism.

(ii) $\text{Б}^S_{S_0} = 0$ *and* $\sum\limits_{v\in S_0} \delta_v = \delta$.

If $\mu_p \subseteq k$, *then* (i) *and* (ii) *are equivalent to*

(ii)′ $S_0 = \{v_0\}$ *with* $v_0|p$ *and* $G_{v_0} = G_S$.

Furthermore, if (i) *and* (ii) *hold, then*

$$\#(T \setminus S) = 1 + \sum_{v\in S_0\cap S_p} [k_v : \mathbb{Q}_p] - \#(S \setminus S_0).$$

Remarks: 1. If (i) and (ii) hold, then G_{S_0} is a free pro-p-group of rank

$$\mathrm{rk}(G_{S_0}) = \#(T \smallsetminus S) + \#(S \smallsetminus S_0)^f = \sum_{v \in S_p \cap S_0} [k_v : \mathbb{Q}_p] - r_1 - r_2 + 1.$$

Indeed, dividing out the inertia groups $\mathcal{T}_v$, $v \in S \smallsetminus S_0$, we obtain from (i) the isomorphism

$$\mathop{\ast}_{v \in S \setminus S_0} \mathcal{G}_v / \mathcal{T}_v \ast \mathop{\ast}_{v \in T \setminus S} \mathcal{G}_v / \mathcal{T}_v \xrightarrow{\sim} G_{S_0}.$$

2. The set T in (i) is not unique.

Proof of (10.9.1): Suppose that G_S has a free product decomposition as in (i). Since $H^2(\mathcal{G}_v/\mathcal{T}_v, \mathbb{Q}_p/\mathbb{Z}_p) = 0 = H^2(\mathcal{G}_v, \mathbb{Q}_p/\mathbb{Z}_p)$ by (7.2.5), it follows from (4.1.4) that

$$H^2(G_S, \mathbb{Q}_p/\mathbb{Z}_p) = 0,$$

in other words (see (10.3.6), (i) $\Leftrightarrow$ (vi)), the Leopoldt conjecture is true for k and p. Hence (loc.cit.)

$$\begin{aligned} r_2 + 1 &= \mathrm{rank}_{\mathbb{Z}_p} G_S^{ab} \\ &= \sum_{v \in (S \setminus S_0)^f} \mathrm{rank}_{\mathbb{Z}_p} \mathcal{G}_v^{ab} + \sum_{v \in T \setminus S} \mathrm{rank}_{\mathbb{Z}_p} \mathcal{G}_v / \mathcal{T}_v \\ &= \sum_{v \in S_p \cap (S \setminus S_0)} [k_v : \mathbb{Q}_p] + \#(T \smallsetminus S_0)^f. \end{aligned}$$

By (10.7.3), we have

$$\dim_{\mathbb{F}_p} H^1(G_S) = 1 + \sum_{v \in S} \delta_v - \delta + \dim_{\mathbb{F}_p} Б_S^S$$

and by (i), (4.1.4) and (7.5.11), this must be equal to

$$\begin{aligned} &\sum_{v \in S \setminus S_0} \dim_{\mathbb{F}_p} H^1(\mathcal{G}_v) + \sum_{v \in T \setminus S} \dim_{\mathbb{F}_p} H^1(\mathcal{G}_v / \mathcal{T}_v) \\ &\qquad = \sum_{v \in S_p \cap (S \setminus S_0)} [k_v : \mathbb{Q}_p] + \sum_{v \in S \setminus (S_0 \cup S_{\mathbb{C}})} \delta_v + \#(S \smallsetminus S_0)^f + \#(T \smallsetminus S). \end{aligned}$$

Combining these equalities, we obtain

$$\dim_{\mathbb{F}_p} Б_S^S + \sum_{v \in S_0} \delta_v - \delta = 0,$$

and hence

$$Б_S^S = 0 \quad \text{and} \quad \sum_{v \in S_0} \delta_v = \delta.$$

Consider the exact sequence (9.2.2)

$$0 \longrightarrow \mathrm{coker}(k_S, S \smallsetminus S_0, \mathbb{Z}/p\mathbb{Z}) \longrightarrow Б_{S_0}^S \longrightarrow Б_S^S \longrightarrow 0.$$

We have just seen that the term on the right-hand side vanishes and condition (i), in conjunction with (4.1.4), implies the vanishing of the left-hand term. Hence $Б_{S_0}^S = 0$ and the proof of the implication (i) $\Rightarrow$ (ii) is complete.

Now assume that (ii) holds. Then the exact sequence above implies that

$$\operatorname{coker}(k_S, S \smallsetminus S_0, \mathbb{Z}/p\mathbb{Z}) = 0 = \text{Б}_S^S.$$

By Čebotarev's density theorem, we can find a finite set of primes $T \supseteq S$ such that the canonical restriction homomorphism

$$\text{Ш}^1(k_S, S \smallsetminus S_0, \mathbb{Z}/p\mathbb{Z}) \longrightarrow \prod_{v \in T\backslash S} H^1_{nr}(k_v)$$

is an isomorphism. For such a set T, the middle horizontal arrow α in the diagram

$$\begin{array}{ccc}
\text{Ш}^1(k_S, S \smallsetminus S_0, \mathbb{Z}/p\mathbb{Z}) & \xrightarrow{\ \sim\ } & \prod_{v \in T\backslash S} H^1_{nr}(k_v) \\
\big\downarrow & & \big\downarrow \\
H^1(G_S) & \xrightarrow{\ \alpha\ } & \prod_{v \in S\backslash S_0} H^1(k_v) \times \prod_{v \in T\backslash S} H^1_{nr}(k_v) \\
\| & & \big\downarrow \\
H^1(G_S) & \twoheadrightarrow & \prod_{v \in S\backslash S_0} H^1(k_v)
\end{array}$$

is an isomorphism. In order to show that the canonical homomorphism

$$\mathop{\ast}_{v \in S\backslash S_0} \mathcal{G}_v \ast \mathop{\ast}_{v \in T\backslash S} \mathcal{G}_v/\mathcal{T}_v \longrightarrow G_S$$

is an isomorphism of pro-p-groups, it therefore (see (4.1.5)) remains to show that $\text{Ш}^2(k_S, S \smallsetminus S_0, \mathbb{Z}/p\mathbb{Z}) = 0$.

Consider the following diagram, in which the exact row in the middle is part of the long exact sequence of Poitou-Tate (use also (10.4.8)).

$$\begin{array}{ccccccc}
 & & & & \prod_{v \in S_0} H^2(k_v) & & \\
 & & & & \big\downarrow & & \\
\text{Ш}^2(k_S, S, \mathbb{Z}/p\mathbb{Z}) & \hookrightarrow & H^2(G_S) & \longrightarrow & \prod_{v \in S} H^2(k_v) & \twoheadrightarrow & H^0(\mathcal{G}_S, \mu_p)^* \\
\big\downarrow & & \| & & \big\downarrow & & \\
\text{Ш}^2(k_S, S_0, \mathbb{Z}/p\mathbb{Z}) & \hookrightarrow & H^2(G_S) & \longrightarrow & \prod_{v \in S\backslash S_0} H^2(k_v). & &
\end{array}$$

Since $\text{Ш}^2(k_S, S, \mathbb{Z}/p\mathbb{Z}) \cong \text{Б}_S^S = 0$, the snake lemma implies the exact sequence

$$0 \longrightarrow \text{Ш}^2(k_S, S_0, \mathbb{Z}/p\mathbb{Z}) \longrightarrow \prod_{v \in S_0} H^0(k_v, \mu_p)^* \longrightarrow H^0(\mathcal{G}_S, \mu_p)^* \longrightarrow 0,$$

where we used the local duality isomorphism $H^2(k_v) \cong H^0(k_v, \mu_p)^*$. Hence

$$\dim_{\mathbb{F}_p} Ш^2(k_S, S_0, \mathbb{Z}/p\mathbb{Z}) = \sum_{v \in S_0} \delta_v - \delta = 0.$$

This shows the implication (ii) $\Rightarrow$ (i).

If (ii) holds and $\mu_p \subseteq k$, then the condition $\sum_{v\in S_0} \delta_v = \delta$ implies that S_0 consists of exactly one finite prime, say v_0. By Kummer theory, the group $Б_{S_0}^S$ corresponds to the maximal elementary abelian extension of k inside k_S in which v_0 is completely decomposed. As $Б_{S_0}^S = 0$, we obtain $G_S = G_{v_0}$ by the Frattini argument. On the other hand, if (ii)$'$ holds, then the equation $\sum_{v\in S_0} \delta_v = \delta$ holds for trivial reasons and Kummer theory implies $Б_{S_0}^S = 0$. It remains to show that $v_0|p$. Assume the contrary and choose a prime $v_1 \in S_p \subseteq S$. Using the isomorphism in (i), we obtain the contradiction

$$1 = \mathrm{rk}_{\mathbb{Z}_p} G_{v_0}^{ab} = \mathrm{rk}_{\mathbb{Z}_p} G_S^{ab} \geq \mathrm{rk}_{\mathbb{Z}_p} \mathcal{G}_{v_1}^{ab} = [k_{v_1} : \mathbb{Q}_p] + 1.$$

Hence $v_0|p$. Finally, we show the statement about the rank. We have

$$\begin{aligned}
\#(T \smallsetminus S) &= \dim_{\mathbb{F}_p} H^1(G_S) - \sum_{v\in S\setminus S_0} \dim_{\mathbb{F}_p} H^1(k_v) \\
&= 1 + \sum_{v\in S} \delta_v - \delta + \dim_{\mathbb{F}_p} Б_S^S \\
&\quad - \sum_{v\in S_p\cap(S\setminus S_0)} [k_v : \mathbb{Q}_p] - \#(S \smallsetminus S_0)^f - \sum_{v\in(S\setminus S_0)^f} \delta_v - \sum_{v\in S_{\mathbb{R}}} \delta_v \\
&= 1 + r_1 + 2r_2 - \sum_{v\in S_p\cap(S\setminus S_0)} [k_v : \mathbb{Q}_p] - \#(S \smallsetminus S_0) \\
&= 1 + \sum_{v\in S_p\cap S_0} [k_v : \mathbb{Q}_p] - \#(S \smallsetminus S_0).
\end{aligned}$$

This finishes the proof. □

As a first application of the theorem above, we can calculate the cohomological dimension of the group $G_S(p)$ in the difficult case where $p = 2$, $S \supseteq S_2$ and $S \cap S_{\mathbb{R}} = \varnothing$. By (10.6.1), we know that $cd\, G_S(2) \leq 2$. Since the cyclotomic $\mathbb{Z}_2$-extension $k_\infty(2)$ of k is contained in $k_{S_2}(2)$, the group $G_{S_2}(2)$ is infinite; in particular, it is nontrivial. Hence, for $S \supseteq S_2$ and $S \cap S_{\mathbb{R}} = \varnothing$, the group $G_S(2)$ is of cohomological dimension 1 or 2, and the next theorem gives a criterion for which case occurs. Recall that $Cl_{0,S}(k)$ denotes the S-ideal class group in the narrow sense.

(10.9.2) Theorem. *Assume that $S \supseteq S_2$ and $S \cap S_{\mathbb{R}} = \varnothing$. Then $cd\, G_S(2) = 1$ if and only if the following conditions* (1)–(3) *hold.*

(1) $S_2 = \{\mathfrak{p}_0\}$, *i.e. there exists exactly one prime dividing* 2 *in* k.
(2) $S = \{\mathfrak{p}_0\} \cup \{$complex places$\}$.
(3) $Cl_{0,S}(k)(2) = 0$.

In this case, $G_S(2)$ is a free pro-2-group of rank $r_2 + 1$ and $\mathfrak{p}_0$ does not split in $k_{S\cup S_{\mathbb{R}}}(2)$. In particular, if k is totally real and $G_S(2)$ is free, then $k_S(2) = k_\infty(2)$.

Proof: Assume that $G_S(2)$ is free. By (10.6.2), we have an isomorphism

$$\mathop{\ast}_{\mathfrak{p}\in S\setminus S_2(k_{S_2}(2))} T(k_\mathfrak{p}(2)|k_\mathfrak{p}) \xrightarrow{\sim} G(k_S(2)|k_{S_2}(2)).$$

For nonarchimedean primes $\mathfrak{p} \notin S_2$, the maximal unramified 2-extension of $k_\mathfrak{p}$ is realized by $k_\infty(2) \subseteq k_{S_2}(2)$. Therefore, for $\mathfrak{p} \in S \smallsetminus S_2$, the maximal 2-extension of the local field $k_\mathfrak{p}$ is realized by $k_S(2)$ or, in other words, the natural homomorphism

$$G(k_\mathfrak{p}(2)|k_\mathfrak{p}) \longrightarrow G_S(2)$$

is injective. But for these primes we have $cd\, G(k_\mathfrak{p}(2)|k_\mathfrak{p}) \geq 2$, showing that $S \smallsetminus S_2 = \varnothing$, i.e. $S = S_2$.

By Čebotarev's density theorem, we find a finite set of nonarchimedean primes $T \supseteq S_2$ such that the natural homomorphism

$$H^1(G_{S_2}(2)) \longrightarrow \bigoplus_{\mathfrak{p}\in T\setminus S} H^1(\mathcal{G}_\mathfrak{p}/\mathcal{T}_\mathfrak{p})$$

is an isomorphism. Then, by (1.6.15), the natural homomorphism

$$\mathop{\ast}_{\mathfrak{p}\in T\setminus S_2} \mathcal{G}_\mathfrak{p}/\mathcal{T}_\mathfrak{p} \longrightarrow G_{S_2}(2)$$

is an isomorphism. By (10.6.2) with $T = S_2 \cup S_\mathbb{R}$ and $S = S_2$ and by (10.6.3), the natural homomorphism

$$\mathop{\ast}_{\mathfrak{p}\in T\setminus S_2} \mathcal{G}_\mathfrak{p}/\mathcal{T}_\mathfrak{p} \ast \mathop{\ast}_{\mathfrak{p}\in S_\mathbb{R}} \mathcal{G}_\mathfrak{p} \longrightarrow G_{S_2\cup S_\mathbb{R}}(2)$$

is an isomorphism. Then, by (10.9.1), we obtain the conditions (1)–(3) and that the unique prime $\mathfrak{p}_0$ dividing 2 in k does not split in $k_{S_2\cup S_\mathbb{R}}$.

If, on the other hand, conditions (1)–(3) of (10.9.2) are satisfied, then we obtain the above isomorphism and deduce that $G_{S_2}(2)$ is free. The statement about the rank of $G_{S_2}(2)$ follows as $\chi_2(G_{S_2}(2)) = -r_2$. If k is totally real, then the homomorphism

$$G_{S_2}(2) \longrightarrow G(k_\infty(2)|k)$$

is a surjection of free pro-2-groups of rank 1 and hence an isomorphism. This concludes the proof of (10.9.2). □

Theorem (10.9.1) motivates the following

(10.9.3) Definition. *We say that G_S is* **degenerate** *if* $\text{Б}^S_{S_0} = 0$ *for a subset* $S_0 \subseteq S^f$ *which satisfies the property*

$$(+) \qquad \sum_{v\in S_0} \delta_v = \delta.$$

Remarks: 1. For $S_0' \subseteq S_0$ we have a surjection $\text{Б}_{S_0'}^S \twoheadrightarrow \text{Б}_{S_0}^S$. Therefore, to decide whether G_S is degenerate, it suffices to consider maximal subsets $S_0 \subseteq S$ with the property (+).

2. Assume that G_S is degenerate and $\mu_p \subseteq k$. Then, by (10.9.1), there exists a prime $v_0|p$ such that $G_{v_0} = G_S$. By (10.7.3) and (7.3.9), we obtain the inequality

$$\begin{aligned}\#S^f + r_1 + r_2 &\leq \dim_{\mathbb{F}_p} H^1(G_S) = \dim_{\mathbb{F}_p} H^1(G_{v_0})\\ &\leq \dim_{\mathbb{F}_p} H^1(\mathcal{G}_{v_0}) = 2 + [k_{v_0} : \mathbb{Q}_p]\\ &\leq 2 + 2r_2 + r_1,\end{aligned}$$

and thus $\#S^f \leq r_2 + 2$. Therefore G_S is not degenerate if $\mu_p \subseteq k$ and S contains more than $r_2 + 2$ finite primes.

3. Assume that G_S is degenerate and $\mu_p \nsubseteq k$. By definition, there exists a subset S_0 of S such that $\mu_p \nsubseteq k_\mathfrak{p}$ for all $\mathfrak{p} \in S_0$ and

$$\mathop{\ast}_{v\in S\setminus S_0} \mathcal{G}_v * \mathop{\ast}_{v\in T\setminus S} \mathcal{G}_v/\mathcal{T}_v \xrightarrow{\sim} G_S.$$

Using (10.3.20), we obtain

$$\sum_{\mathfrak{p}\in S\setminus S_\infty} \delta_\mathfrak{p} \leq \#(S \setminus (S_0 \cup S_\infty)) \leq \operatorname{rank}_{\mathbb{Z}_p} G_S^{ab} \leq [k : \mathbb{Q}].$$

Therefore G_S is not degenerate if $\mu_p \nsubseteq k$ and S contains more than $[k : \mathbb{Q}]$ finite primes $\mathfrak{p}$ with $\delta_\mathfrak{p} = 1$.

The following are examples of a degenerate G_S (cf. §7).

- $K = \mathbb{Q}$, p arbitrary and $S = S_p \cup S_\infty$.
- $K = \mathbb{Q}(\mu_p)$, p regular and $S = S_p \cup S_\infty$.

(10.9.4) Corollary. *If G_S is degenerate, then it decomposes into the free product of local groups.*

(10.9.5) Corollary. *If G_S is degenerate, then the Leopoldt conjecture holds for p and for every finite extension of k inside k_S.*

Proof: The free product decomposition together with (4.1.4) and (7.2.5) shows that G_S has an open subgroup of strict cohomological dimension 2 (there might occur factors of the form $G(\mathbb{C}|\mathbb{R})$ if $p = 2$). Now the corollary follows from (10.3.9) and (10.3.11). □

The group G_S has an essentially different behaviour, depending on whether or not there exists a prime $v \in S$ such that G_v is open in G_S. As we have seen in §6, if k is not totally real, then this is possible only for primes v dividing p, and if k is totally imaginary, then $(G_S : G_v) \leq 2$ by (10.8.2). If k is totally real, the decomposition groups of primes not dividing p can also be open.

(10.9.6) Definition. *The group G_S is called of* **local type** *if there exists a prime $v \in S$ such that $G_v = G_S$.*

Moreover, we say that G_S is **potentially of local type** *if an open subgroup of G_S is of local type, and otherwise G_S is called of* **global type**.

Mainly for the case $p = 2$, we introduce the following terminology.

(10.9.7) Definition. *We say that a profinite group has* **virtually** *property (P) if property (P) is satisfied for all sufficiently small open subgroups.*

For example, if $S \supseteq S_2 \cup S_\infty$, then G_S is always virtually of cohomological 2-dimension less than or equal to two, since every subgroup which corresponds to a totally imaginary field has this property. In our applications the word "*virtually*" will only occur if $p = 2$, and then it will always mean that the property holds for all subgroups which correspond to totally imaginary extensions of the base field.

(10.9.8) Theorem. *If $p \neq 2$ and $\mu_p \subseteq k$, then the group G_S has one of the following forms.*

(i) *If $Б^S_{\{v\}} \neq 0$ for all primes $v \in S^f$, then G_S is of global type and it is a duality group of dimension 2.*

(ii) *If $Б^S_{\{v_0\}} = 0$ for a prime $v_0 \in S^f$, then G_S is of local type and there exists a finite set of primes $T \supseteq S$ such that the canonical homomorphism*

$$\mathop{\ast}_{v \in S \setminus \{v_0\}} \mathcal{G}_v \ast \mathop{\ast}_{v \in T \setminus S} \mathcal{G}_v/\mathcal{T}_v \longrightarrow G_S$$

is an isomorphism.

In (ii) the prime v_0 is unique but the set T is not. For $p = 2$ we have the following variant of (10.9.8).

(10.9.9) Theorem. *If $p = 2$, then G_S has one of the following forms.*

(i) *If $Б^S_{\{v\}} \neq 0$ for all $v \in S^f$ and $\#S^f(k_S) > 2$, then G_S is of global type and it is a virtual duality group of dimension 2.*

(ii) *If $Б^S_{\{v\}} \neq 0$ for all $v \in S^f$ and $\#S^f(k_S) = 2$, then G_S is potentially of local type and it is a virtual Poincaré group of dimension 2.*

(iii) *If $Б^S_{\{v_0\}} = 0$ for a prime $v_0 \in S^f$, then G_S is of local type and there exists a finite set of primes $T \supseteq S$ such that the canonical homomorphism*

$$\mathop{\ast}_{v \in S \setminus \{v_0\}} \mathcal{G}_v \ast \mathop{\ast}_{v \in T \setminus S} \mathcal{G}_v/\mathcal{T}_v \longrightarrow G_S$$

is an isomorphism.

Before we prove (10.9.8) and (10.9.9), we first calculate the module $I = D_2(\mathbb{Z}_p)$. It is nontrivial if and only if G_S is virtually of cohomological dimension 2, and if $cd_p\, G_S = 2$, it is the dualizing module.

(10.9.10) Lemma. $I \cong \operatorname{tor}_p(C_{S^f}(k_S))$.

Proof: By (10.2.1), the dualizing module of $\mathscr{G}_S$ is canonically isomorphic to

$$\operatorname{tor}_p(C_{S^f}(\mathcal{K}_S)).$$

Therefore the statement of the lemma follows from (10.4.8). □

(10.9.11) Corollary. *Assume that $\mu_p \subseteq k$. Then we have the following equivalences:*

G_S is virtually free	$\Longleftrightarrow$	$\#S^f(k_S) = 1$,
G_S is virtually a Demuškin group	$\Longleftrightarrow$	$\#S^f(k_S) = 2$.

Proof: If $\mu_p \subseteq k$, then $I = \operatorname{tor}_p(C_{S^f}(k_S))$ fits into the exact sequence

$$0 \longrightarrow \mu_{p^\infty} \longrightarrow \bigoplus_{v \in S^f} \operatorname{Ind}^{G_v}_{G_S} \mu_{p^\infty} \longrightarrow I \longrightarrow 0$$

(cf. (10.2.1)). The basic properties of I show that G_S is virtually free, i.e. virtually of $cd = 1$ if and only if $I = 0$. By (3.7.2), we conclude that G_S is a virtual Demuškin group if and only if $I \cong \mathbb{Q}_p/\mathbb{Z}_p$ as an abelian group. This shows the asserted equivalences. □

In order to decide whether a given pro-p-group is a duality group of dimension 2, one has to calculate the term D_1 (see III §4).

(10.9.12) Lemma. $D_1(G_S, \mathbb{Z}/p\mathbb{Z}) \cong C_S(k_S)/p$.

Proof: By global class field theory (and, in particular, since G_S is a finitely generated pro-p-group), we obtain the following equalities, in which the field K runs through all finite extensions of k inside k_S:

$$\begin{aligned} D_1(G_S, \mathbb{Z}/p\mathbb{Z}) &= \varinjlim H^1(G(k_S|K), \mathbb{Z}/p\mathbb{Z})^* \\ &= \varinjlim H^1(G(k_S|K), \mathbb{Z}/p\mathbb{Z})^\vee \\ &= \varinjlim G(k_S/K)^{ab}/p \\ &= \varinjlim C_S(K)/p = C_S(k_S)/p. \end{aligned}$$

□

(10.9.13) Proposition. *The group G_S is a duality group of dimension 2 with dualizing module $\mathrm{tor}_p(C_{S^f}(k_S))$ if and only if $cd_p\, G_S = 2$ and $C_S(k_S)$ is p-divisible.*

Proof: This follows from (3.4.6), (10.9.12) and (10.9.10). □

(10.9.14) Proposition. *If $\mu_p \subseteq k$ and if G_S is of global type, then G_S is a (virtual, if $p = 2$ and $S_{\mathbb{R}}(k) \neq \varnothing$) duality group of dimension 2.*

Proof: Consider the exact sequence

$$0 \longrightarrow E_{k_S,S} \longrightarrow I_S(k_S) \longrightarrow C_S(k_S) \longrightarrow Cl_S(k_S) \longrightarrow 0.$$

By the principal ideal theorem, we have $Cl_S(k_S) = 0$ *), and $E_{k_S,S}$ is p-divisible as $\mu_p \subseteq k$. Thus we have an isomorphism

$$I_S(k_S)/p \cong C_S(k_S)/p.$$

By definition,

$$I_S(k_S) = \varinjlim_{K \subseteq k_S} \bigoplus_{v \in S(K)} K_v^\times \cong \bigoplus_{v \in S(k)} \mathrm{Ind}_{G_S}^{G_v} k_{S_v}^\times.$$

Therefore $C_S(k_S)/p = 0$ if and only if $(k_S)_v = k_v(p)$ for all $v \in S$. By the theorem of Kuz'min (10.8.4), this is the case if all decomposition groups are of infinite index in G_S, in other words, if G_S is of global type. □

*) Recall our notational conventions!

Proof of (10.9.8): By Kummer theory, $Б^S_{\{v\}}$ is non-zero if and only if there exists an extension of k inside k_S in which v splits. If $Б^S_{\{v\}} = 0$ for one v, then we can apply (10.9.1) in order to see that we are in case (ii). Otherwise G_S is of global type by (10.8.2), and by (10.9.14) we are in case (i). □

Proof of (10.9.9): If $Б^S_{\{v_0\}} = 0$ for a prime $v_0 \in S^f$, then we can apply (10.9.1), which gives (iii). Now assume that $Б^S_{\{v\}} \neq 0$ for every prime $v \in S^f$ which satisfies property (+) of (10.9.3). Then we see, using Kummer theory, that $G_v \subsetneqq G_S$ for all v. If G_S is global, then we are in case (i) by (10.9.14). So let us assume that there exists a prime v such that G_v is open in G_S. Let $H \subseteq G_S$ be the subgroup (of index 1 or 2) of G_S which corresponds to the subextension $k(i) \subseteq k_S$. Then $G_v \cap H$ is open in H and we deduce from (10.8.2) that $v \in S_2$ and one of the following cases occurs:

First case: $H \subseteq G_v$.
Since $G_v \neq G_S$, we have $H = G_v$ and $i \notin k$ in this case. Furthermore, v decomposes in $k(i)|k$, so $v = v_1 v_2$ say. Let us consider the situation at the level of $k(i)$. Since v_1 does not split in $k_S|k(i)$, we obtain $Б^S_{\{v_1\}} = 0$. Applying (10.9.9)(iii) we get a free product decomposition

$$G_{v_1} = H \cong \mathcal{G}_{v_2} * (\text{other terms}).$$

But

$$\text{rk}(\mathcal{G}_{v_2}) = \text{rk}(\mathcal{G}_{v_1}) \geq \text{rk}(G_{v_1}).$$

Hence there are no other terms and $H = \mathcal{G}_{v_1} = \mathcal{G}_{v_2}$. In particular, H is a Demuškin group, by (7.5.11), and of local type. Furthermore, $\#S^f(k_S) = 2$ by (10.9.11). Hence we are in case (ii).

Second case: $(H : G_v \cap H) = 2$.
Let $\tilde{v}$ be a prime of $k(i)$ above v (the extension $k(i)|k$ can be trivial). If $G_v \cap H = G_S(k')$, then $\tilde{v}$ decomposes in $k'|k(i)$, so $\tilde{v} = v_1 v_2$ say. As in the first case, it follows that $G_S(k') = \mathcal{G}_{v_1}(k') = \mathcal{G}_{v_2}(k')$, and so $\#S^f(k_S) = 2$ by (10.8.2) (thus v does not decompose in the extension $k(i)|k$). Furthermore, G_S is potentially of local type and virtually a Demuškin group and we are again in case (ii). □

It would be interesting to obtain more information about G_S in the generic case if $\mu_p \not\subseteq k$. *) We have the following proposition, at least, which generalizes (10.5.7). (See XI for the definition of the Iwasawa μ-invariant.)

---

*) The sequel to [254], which appeared in J. reine u. angew. Math. **416** (1991), contains a mistake.

(10.9.15) Proposition. *Let k be totally real and $p \neq 2$. Assume that G_S is non-degenerate and that the Iwasawa μ-invariant of the cyclotomic $\mathbb{Z}_p$-extension k_∞ of k vanishes. (This is known for abelian number fields.) Then G_S is a pro-p duality group of dimension 2 with dualizing module* $\mathrm{tor}_p(C_{S^f}(k_S))$.

Proof: Since G_S is non-degenerate, it is not free, and so $G(k_S|k_\infty) \neq 1$. Since $\mu = 0$, it follows from (11.3.7) that $G(k_S|k_\infty)$ is a free pro-p-group. Furthermore, it is of finite rank since k is totally real. Hence the result follows from (3.7.5) and (10.9.10). □

Remark: Another situation where we know that G_S is a duality group, is the following. Let k be a CM-field with totally real subfield k^+ and let $S \supseteq S_p \cup S_\infty$ be a finite set of primes of k^+. Suppose that

(1) $\mu_p \subseteq k^+_{\mathfrak{p}}$ for all $\mathfrak{p} \in S_p(k^+)$ and
(2) all primes in $S_p(k^+)$ split in $k|k^+$.

Then the result of the exercise of §8 shows that $(k_S)_{\mathfrak{p}} = k_{\mathfrak{p}}(p)$ for all primes $\mathfrak{p} \in S(k)$. Removing all primes $\mathfrak{p}$ with $N(\mathfrak{p}) \not\equiv 0, 1 \bmod p$ from $S(k)$ (this does not alter G_S), we may assume that $k_{\mathfrak{p}}$ contains a primitive p-th root of unity for all primes $\mathfrak{p} \in S(k)$, so that $I_S(k_S)$ and therefore also $C_S(k_S)$, is p-divisible.

§10. Class Field Towers

We define a sequence of extensions of an algebraic number field k as follows. Let $k_0 = k$ and for $n \geq 0$, let k_{n+1} be the Hilbert class field of k_n:

$$k = k_0 \subseteq k_1 \subseteq k_2 \subseteq \cdots .$$

This sequence of fields is called the **class field tower** of k. Obviously, the field $L^{solv} = \bigcup_n k_n$ is the maximal unramified extension of k with prosolvable Galois group. The class field tower is called finite if the extension $L^{solv}|k$ is finite and infinite otherwise. For a prime number p, let $L = L(p)$ be the maximal p-subextension of $L^{solv}|k$. Hence $L|k$ corresponds to the **p-class field tower** of k, which is a sequence of fields

$$k = k_0 \subseteq k_1^{(p)} \subseteq k_2^{(p)} \subseteq \cdots ,$$

where $k_{n+1}^{(p)}$ is the p-Hilbert class field of $k_n^{(p)}$, i.e. the maximal abelian unramified p-extension of $k_n^{(p)}$.

In 1964 *E. S. Golod* and *I. R. Šafarevič* proved, cf. [57], that there exist algebraic number fields which possess infinite class field towers. The existence of such fields is a consequence of a theorem that they proved about finite p-groups. We presented a proof of its sharpened form, due to *W. Gaschütz* and *E. B. Vinberg*, in III §9. The example given by Golod and Šafarevič is the imaginary quadratic number field

$$k = \mathbb{Q}(\sqrt{-2 \cdot 3 \cdot 5 \cdot 7 \cdot 11 \cdot 13 \cdot 17 \cdot 19}),$$

which has an infinite 2-class field tower.

More generally, in this section we consider p-class field towers in which primes of a given set T are completely decomposed, i.e. we ask whether the maximal unramified p-extension $k_\varnothing^T$ of k which is completely split at T is infinite or not. We will see that we can always find a finite Galois p-extension $K|k$ such that $K_\varnothing^T|K$ is infinite. This happens if the p-rank of the T-ideal class group of k is large enough, and this will be the case if sufficiently many (depending on T) primes ramify in $K|k$.

Another method of obtaining infinite p-class field towers can be applied in the case of CM-fields when p is odd. Using the action of complex conjugation, it is possible to find a smaller bound for the p-rank of the ideal class group.

Let us fix k and p and recall the notation δ and $\delta_\mathfrak{p}$ from (10.7.1). We first consider the case of a nonempty set S. Of course, if $S_p \subseteq S$, then G_S is infinite, since then $k_S(p)$ contains the cyclotomic $\mathbb{Z}_p$-extension.

The following primes cannot ramify in a p-extension, and are therefore redundant in S:

1. Complex primes.
2. Real primes if $p \neq 2$.
3. Primes $\mathfrak{p} \nmid p$ with $N(\mathfrak{p}) \not\equiv 1 \bmod p$ (see (7.5.2)).

Removing all these redundant places from S, we obtain a subset $S_{\min} \subseteq S$ which has the property that

$$G_S(p) = G_{S_{\min}}(p).$$

(10.10.1) Theorem. *Let p be a prime number, k an algebraic number field and S an arbitrary set of primes of k. If*

$$\#S_{\min} \geq 1 + r_1 + r_2 + 2\sqrt{r_1 + r_2} + \delta,$$

then the group $G_S(k)(p)$ is infinite.

Proof: From (10.7.12), we obtain

$$\begin{aligned} h^1(G_S(k)(p)) &\geq 1 + \#S_{\min} - \delta - (r_1 + r_2)\,, \\ h^2(G_S(k)(p)) - h^1(G_S(k)(p)) &\leq r_1 + r_2 - 1\,. \end{aligned}$$

If $G_S(k)(p)$ is finite, it follows from the theorem of Golod and Šafarevič that

$$h^1(G_S(k)(p)) < 2 + 2\sqrt{h^2(G_S(k)(p)) - h^1(G_S(k)(p)) + 1}\,,$$

and hence

$$1 + \#S_{\min} - \delta - (r_1 + r_2) < 2 + 2\sqrt{r_1 + r_2}\,. \qquad \square$$

(10.10.2) Corollary. *Let p be a prime number and let S be a set of places of $\mathbb{Q}$. For odd p, the group $G_S(\mathbb{Q})(p)$ is infinite if*

$$\#S_{\min} \geq 4\,.\ ^{*)}$$

The group $G_S(\mathbb{Q})(2)$ is infinite, provided that $\#S_{\min} \geq 5$.

Remark: If $\#S_{\min} \leq 3 + \delta$, then the group $G_S(\mathbb{Q})(p)$ can be finite or infinite, see [109] and [107].

Now we will consider the case when S is empty. Recalling the notation $\mathrm{Ram}(K|k)$ for the set of primes of k which ramify in $K|k$, we start with the

(10.10.3) Proposition. *Let $K|k$ be a cyclic p-extension and let $T \supseteq S_\infty$ be a finite set of primes of k contained in $S := \mathrm{Ram}(K|k) \cup S_\infty$. Then the inequality*

$$\dim_{\mathbb{F}_p} Cl_T(K)/p \geq \#S \smallsetminus T(k) - r_1(k) - r_2(k) - \delta(k) + r_1'(k)$$

holds, where $r_1'(k)$ is the number of real places of k which become complex in K.

Remark: We see that for fixed T we can find extensions $K|k$ of degree p with arbitrary large $\dim_{\mathbb{F}_p} Cl_T(K)/p$. Indeed, by the above proposition, the extension $K|k$ has to be ramified at all finite primes in T $^{**)}$ and at sufficiently many other primes. Such extensions exist by the theorem of Grunwald-Wang (9.2.8).

$^{*)}$The set $S_{\min}$ depends on p.

$^{**)}$It follows from the proof that it suffices that the places in T^f do not split in $K|k$.

Proof: For a finite abelian group A, we set $d_p(A) = \dim_{\mathbb{F}_p} A/pA = \dim_{\mathbb{F}_p} {}_pA$. Note that $d_p(B) \le d_p(A)$ for every subquotient B of A. Since G is cyclic, we have $\hat{H}^i(G,M) \cong \hat{H}^{i+2}(G,M)$ for all i and every G-module M. Consider the exact sequence

$$0 \longrightarrow \mathcal{O}_{K,T}^\times \longrightarrow \prod_{\mathfrak{P}\in T(K)} K_\mathfrak{P}^\times \times \prod_{\mathfrak{P}\in S\setminus T(K)} U_\mathfrak{P} \longrightarrow C_S(K) \longrightarrow Cl_T(K) \longrightarrow 0.$$

Setting $G = G(K|k)$, $I_T = \prod_{\mathfrak{P}\in T(K)} K_\mathfrak{P}^\times$ and $U_{S\setminus T} = \prod_{\mathfrak{P}\in S\setminus T(K)} U_\mathfrak{P}$, we obtain the inequality

$$d_p(\hat{H}^0(G,(I_T\times U_{S\setminus T})/\mathcal{O}_{K,T}^\times)) \ge d_p(\hat{H}^0(G,I_T\times U_{S\setminus T})) - d_p(\hat{H}^0(G,\mathcal{O}_{K,T}^\times)).$$

Further,

$$d_p(\hat{H}^0(G,C_S(K))) = d_p(G(K|k)) = 1$$

and

$$d_p(\hat{H}^0(G,\mathcal{O}_{K,T}^\times)) \le \#T(k) - 1 + \delta.$$

Therefore

$$\begin{aligned}
\dim_{\mathbb{F}_p} Cl_T(K)/p &\ge d_p(\hat{H}^{-1}(G,Cl_T(K)))\\
&\ge d_p(\hat{H}^0(G,(I_T\times U_{S\setminus T})/\mathcal{O}_{K,T}^\times)) - d_p(\hat{H}^0(G,C_S(K)))\\
&\ge d_p(\hat{H}^0(G,I_T\times U_{S\setminus T})) - d_p(\hat{H}^0(G,\mathcal{O}_{K,T}^\times)) - d_p(G^{ab})\\
&\ge \sum_{\mathfrak{p}\in T(k)} d_p(\hat{H}^0(K_\mathfrak{P}|k_\mathfrak{p},K_\mathfrak{P}^\times)) + \sum_{\mathfrak{p}\in S\setminus T(k)} d_p(\hat{H}^0(K_\mathfrak{P}|k_\mathfrak{p},U_\mathfrak{P}))\\
&\quad -\#T(k) + 1 - \delta - 1.
\end{aligned}$$

By assumption, all primes in S^f are ramified in $K|k$, hence we are summing up over terms which are equal to at least 1 for finite primes and which are 0 or 1 for infinite primes depending on whether or not these primes ramify in $K|k$ (which can happen only if $p = 2$). This finishes the proof. □

(10.10.4) Corollary. *Let $K|\mathbb{Q}$ be a quadratic number field and let S = $\mathrm{Ram}(K|\mathbb{Q}) \cup S_\infty$. Then the following inequalities are true:*

$$\dim_{\mathbb{F}_2} Cl(K)/2 \ge \begin{cases} \#S \setminus S_\infty(\mathbb{Q}) - 1\,, & \text{if } K \text{ is imaginary,}\\ \#S \setminus S_\infty(\mathbb{Q}) - 2\,, & \text{if } K \text{ is real.}\end{cases}$$

Proof: This follows from (10.10.3) with $T = S_\infty$. □

For an arbitrary set T of primes of k we now consider the Galois group $G(k_\varnothing^T(p)|k)$, where $k_\varnothing^T(p)$ is the maximal unramified p-extension of k which is completely decomposed at every prime of T. For brevity we set

$$k_\varnothing^T = k_\varnothing^T(p) \quad \text{and} \quad G_\varnothing^T = G(k_\varnothing^T(p)|k)\,,$$

and we denote the maximal unramified p-extension $k_\varnothing(p)$ of k by L. By class field theory, we have

$$h^1(G_\varnothing^T) = \dim_{\mathbb{F}_p} H^1(G_\varnothing^T, \mathbb{Z}/p\mathbb{Z}) = \dim_{\mathbb{F}_p} Cl_T(k)/p\,.$$

(10.10.5) Theorem. *Let T be a finite set of primes of the number field k. Then the extension $k_\varnothing^T|k$ is infinite if*

$$\dim_{\mathbb{F}_p} Cl_T(k)/p \geq 2 + 2\sqrt{r_1 + r_2 + \delta + \#T \setminus S_\infty}\,.$$

In particular, the maximal unramified p-extension L of k is infinite if

$$\dim_{\mathbb{F}_p} Cl(k)/p \geq 2 + 2\sqrt{r_1 + r_2 + \delta}\,.$$

Proof: Assume that $G_\varnothing^T$ is finite. Then the theorem of Golod and Šafarevič (3.9.7) implies the inequality

$$\tfrac{1}{4}\, h^1(G_\varnothing^T)^2 < h^2(G_\varnothing^T),$$

and hence

$$(h^1(G_\varnothing^T) - 2)^2 < 4\,(h^2(G_\varnothing^T) - h^1(G_\varnothing^T) + 1).$$

Using (10.7.12), we obtain

$$\begin{aligned} \dim_{\mathbb{F}_p} Cl(k)/p = h^1(G_\varnothing^T) \;&<\; 2 + 2\sqrt{\chi_2(G_\varnothing^T)} \\ &\leq\; 2 + \sqrt{r_1 + r_2 + \delta + \#T \setminus S_\infty}. \end{aligned}$$

This contradicts our assumption, and therefore $G_\varnothing^T$ is infinite. □

(10.10.6) Corollary. *Let k be a number field and let p be a prime number. Then given an arbitrary finite set of primes T, there exists a Galois extension $K|k$ of degree p such that $K_\varnothing^T|K$ is infinite.*

Proof: By (10.10.3) and the remark after it, we can find cyclic extensions $K|k$ of degree p with $\dim_{\mathbb{F}_p} Cl_T(K)/p$ arbitrary large. The corollary follows from (10.10.5) applied to these fields K, since the number $r_1 + r_2 + \delta + \#T \setminus S_\infty$ is bounded independently of the choice of K. □

(10.10.7) Corollary. *Let $K|\mathbb{Q}$ be a quadratic number field such that at least 8 (resp. 6) prime numbers are ramified if K is real (resp. imaginary). Then K has an infinite 2-class field tower.*

Proof: Let $S = \mathrm{Ram}(K|\mathbb{Q}) \cup S_\infty$. Then by (10.10.4),

$$\dim_{\mathbb{F}_2} Cl(K)/2 \geq \#S\backslash S_\infty(\mathbb{Q}) - 2 \geq 6 \geq 2 + 2\sqrt{2+1}\,,$$

if K is real, and

$$\dim_{\mathbb{F}_2} Cl(K)/2 \geq \#S\backslash S_\infty(\mathbb{Q}) - 1 \geq 5 \geq 2 + 2\sqrt{1+1}\,,$$

if K is imaginary. Therefore the result follows from (10.10.5). □

(10.10.8) Corollary. *Let $K|\mathbb{Q}$ be a quadratic number field such that*

$$\dim_{\mathbb{F}_p} Cl(K)/p \geq \begin{cases} 4, & \text{if } K \text{ is complex and } p \text{ is odd,}\\ 5, & \text{if } K \text{ is complex and } p = 2,\\ 5, & \text{if } K \text{ is real and } p \text{ is odd,}\\ 6, & \text{if } K \text{ is real and } p = 2. \end{cases}$$

Then K has an infinite p-class field tower.

Proof: Since $\delta = 0$ for p odd except for $p = 3$ and the field $K = \mathbb{Q}(\sqrt{-3})$ (which has class number 1), the result follows directly from (10.10.5). □

In the case where p is odd, it is possible to find a smaller bound for the p-rank of the ideal class group of a quadratic field having an infinite p-class field tower. This will follow from the results below where more general fields are considered which are quadratic extensions of subextensions having certain properties. In particular, this can be applied to fields of CM-type. The method was developed in [112], [104] and [256].

Let $k|k_0$ be a quadratic extension with Galois group $\Delta = G(k|k_0) \cong \mathbb{Z}/2\mathbb{Z}$ and let p be an odd prime number. Let T be an arbitrary finite set of primes of k closed under the action of Δ. Then Δ acts on $H^i(G^T_\varnothing) = H^i(G(k^T_\varnothing|k), \mathbb{Z}/p\mathbb{Z})$ and we have the following theorem, where the (+) and (−) sign denote the eigenspaces with respect to the action of Δ.

(10.10.9) Theorem. *In the above situation assume that $H^1(G^T_\varnothing)^+ = 0 = H^2(G^T_\varnothing)^+$ and suppose we have*

$$\dim_{\mathbb{F}_p} Cl_T(k)/p \geq 3 \cdot \max(1, \sqrt[3]{u^- + 2\#T(k)})\,,$$

with $u^- = \dim_{\mathbb{F}_p}(\mathcal{O}_k^\times/p)^-$. Then $G(k^T_\varnothing|k)$ is infinite.

Remark: If k is a CM-field with maximal totally real subfield $k^+ = k_0$, then we have $u^- = \delta$ in the above theorem.

(10.10.10) Corollary. *Let $k|k_0$ be a quadratic extension, p an odd prime number and L the maximal unramified p-extension of k. Assume that $H^1(G(L|k))^+ = 0 = H^2(G(L|k))^+$. Then the extension $L|k$ is infinite if*

$$\dim_{\mathbb{F}_p} Cl(k)/p \geq 3 \cdot \max(1, \sqrt[3]{u^-}).$$

Proof of (10.10.9): Let

$$1 \longrightarrow R \longrightarrow F \longrightarrow G(k_\varnothing^T|k) \longrightarrow 1$$

be a minimal presentation of $G(k_\varnothing^T|k)$ where F is a free pro-p-group. The cup-product

$$H^1(G_\varnothing^T) \times H^1(G_\varnothing^T) \xrightarrow{\cup} H^2(G_\varnothing^T)$$

is a Δ-equivariant symplectic form. Since

$$H^1(G_\varnothing^T) = H^1(G_\varnothing^T)^-,$$

its image is contained in $H^2(G_\varnothing^T)^+ = 0$, and so the pairing is trivial. It follows from (3.9.13)(ii) that the defining relations of $G(k_\varnothing^T|k)$ are contained in $(F)^p F^3$, i.e.

$$R \subseteq (F)^p F^3\,,$$

where F^3 is the third term of the descending p-central series of F. Thus

$$R \subseteq F_{(3)},$$

where $F_{(n)}$ denotes the Zassenhaus filtration of F, cf. the remark (2) before (3.9.9). From the sharpened form of the Golod-Šafarevič theorem due to Koch (loc.cit.), it follows that

$$h^2(G_\varnothing^T) > \tfrac{4}{27}\, h^1(G_\varnothing^T)^3,$$

provided that $G_\varnothing^T$ is finite. Taking the minus part of the exact sequence

$$H^1(k_\varnothing|k_\varnothing^T)^{G_\varnothing^T} \longrightarrow H^2(G_\varnothing^T) \longrightarrow H^2(G_\varnothing)\,,$$

we get the inequality

$$\dim_{\mathbb{F}_p} H^2(G_\varnothing^T)^- \leq \dim_{\mathbb{F}_p} H^2(G_\varnothing)^- + \#T(k)\,.$$

Furthermore, the exact sequence (10.7.2) induces the exact sequence

$$0 \longrightarrow (\mathcal{O}_k^\times/p)^- \longrightarrow \text{Б}_\varnothing(k)^{*-} \longrightarrow {}_pCl(k)^- \longrightarrow 0\,.$$

Using (10.7.6) we now obtain

$$\begin{aligned}\dim_{\mathbb{F}_p} H^2(G^T_{\varnothing}) &= \dim_{\mathbb{F}_p} H^2(G^T_{\varnothing})^- \\ &\le \dim_{\mathbb{F}_p} Ш^2(G_{\varnothing})^- + \#T(k) \\ &\le \dim_{\mathbb{F}_p} Б_{\varnothing}(k)^{*-} + \#T(k) \\ &= \dim_{\mathbb{F}_p} {}_pCl(k)^- + u^- + \#T(k) \\ &\le \dim_{\mathbb{F}_p} {}_pCl_T(k) + u^- + 2\,\#T(k)\,,\end{aligned}$$

so that

$$h^2(G^T_{\varnothing}) - h^1(G^T_{\varnothing}) \le u^- + 2\,\#T(k)\,,$$

and therefore

$$\tfrac{4}{27} h^1(G^T_{\varnothing})^3 - h^1(G^T_{\varnothing}) \le u^- + 2\,\#T(k)$$

if $G^T_{\varnothing}$ is finite. This gives us the desired result. □

(10.10.11) Corollary. *Let p be an odd prime number and let $K|\mathbb{Q}$ be a quadratic number field such that*

$$\dim_{\mathbb{F}_p} Cl(K)/p \ge 3.$$

Then K has an infinite p-class field tower.

Proof: This follows from (10.10.9), since

$$(Cl(K)/p)^+ = Cl(\mathbb{Q})/p = 0,$$

$u^- \le 1$ and $H^2(G(L|K), \mathbb{Z}/p\mathbb{Z})^+ = 0$, which we see as follows:

By (10.7.6), the group $H^2(G(L|K), \mathbb{Z}/p\mathbb{Z})^+ = Ш^2(K)^+$ is contained in $Б_{\varnothing}(K)^+$. Furthermore, the exact sequence (10.7.2)

$$0 \longrightarrow (\mathcal{O}_K^\times/p)^+ \longrightarrow Б_{\varnothing}(K)^{*+} \longrightarrow {}_pCl(K)^+ \longrightarrow 0$$

shows that the latter group is trivial. □

Examples: One can find the following examples of the corollary above. Let

$$\begin{aligned} k_1 &= \mathbb{Q}(\sqrt{-3321607}) \\ k_2 &= \mathbb{Q}(\sqrt{39345017}) \\ k_3 &= \mathbb{Q}(\sqrt{-222637549223}) \end{aligned}$$

and let $p = 3$ in the first two cases and $p = 5$ in the last one. Then $\dim_{\mathbb{F}_p} Cl(k)/p = 3$, and so for these fields the maximal unramified p-extension is infinite. See [207] for references and other examples of fields with infinite class field towers.

Remark: As we have seen in this section, the class field tower of a number field can be infinite, and moreover this is the typical case. However, there is a conjecture due to *J.-M. Fontaine* and *B. Mazur* which claims that an infinite unramified Galois extension of a number field never has the structure of a p-adic Lie group. This means that every unramified finite dimensional p-adic representation of G_K has a finite image. More generally, the precise conjecture is the following:

(10.10.12) Fontaine-Mazur Conjecture. *Let k be a number field and let S be a finite set of places of k. If $S \cap S_p = \varnothing$ and if n is any natural number, then every continuous representation*

$$\rho : G_S(k) \longrightarrow Gl_n(\mathbb{Q}_p)$$

factors through a finite quotient of $G_S(k)$.

The above conjecture follows from a more general principle conjectured by Fontaine and Mazur, which says that Galois representations which "look geometric" indeed arise from algebraic geometry in a well-defined sense. We refer the reader to the original paper [51]; see also [105].

Regarding the Fontaine-Mazur conjecture from a technical point of view, we make the following observation: the Golod-Šafarevič inequality (3.9.7) holds in a slightly modified way also for p-adic analytic groups (see [41], th. 6.29 or [127], prop. 1.3 for the stronger form using the Zassenhaus filtration). All methods to produce infinite class field towers which we have at hand at the moment, use the Golod-Šafarevič inequality. Therefore we do not have a method which could produce counterexamples to the Fontaine-Mazur conjecture.

Since a pro-p-group whose open subgroups have globally bounded rank is analytic, see [41], cor. 9.35, we obtain the

(10.10.13) Corollary. *Assume that the Fontaine-Mazur conjecture is true. If K is an infinite unramified p-extension of the number field k, then the p-rank*

$$\dim_{\mathbb{F}_p} Cl(k')/p$$

becomes arbitrarily large as k' varies over the finite extensions of k in K.

§11. The Profinite Group G_S

In this section we will prove a duality theorem for the profinite Galois group $G_S = G(k_S|k)$, where k_S is the maximal extension of the number field k which is unramified outside the finite set S of primes of k. We assume that S contains all archimedean places S_∞. Again we use the notation S^f for the set of finite primes in S.

Let p be a prime number. As before we denote the set of primes of k dividing p by $S_p(k)$. The following theorem was first proved in [193].

(10.11.1) Theorem. *Let $S \supseteq S_p \cup S_\infty$ be finite and assume that k is totally imaginary if $p = 2$. Then G_S is a duality group at p of dimension 2 with p-dualizing module $I = \mathrm{tor}_p(C_{S^f}(k_S))$, i.e. for every finite p-primary G_S-module M and all i, the cup-product*

$$H^i(G_S, M) \times H^{2-i}(G_S, \mathrm{Hom}(M, I)) \xrightarrow{\cup} H^2(G_S, I) \xrightarrow{\sim} \mathbb{Q}_p/\mathbb{Z}_p$$

defines a perfect pairing of finite groups.

Before we are going to prove this theorem we will collect some facts about the asymptotic behaviour of the class number of the cyclotomic field $\mathbb{Q}(\zeta_{p^n})$ as n tends to infinity. The proof of the following proposition can be found in [246], th. 4.20.

(10.11.2) Proposition. *Let h_n^- be the minus part of the class number of $\mathbb{Q}(\zeta_{p^n})$. Then*

$$\log h_n^- \sim \frac{1}{4}(p-1)p^{(n-1)} n \log p \quad \text{as } n \to \infty . \quad {}^{*)}$$

As an easy consequence of the proposition above, we obtain the

(10.11.3) Proposition. *Let $S \supseteq S_p \cup S_\infty$ be a set of primes of the number field k and assume that k is totally imaginary if $p = 2$. Then*

$$cd_p\, G_S = 2 .$$

---

*) $a \sim b$ means $a/b \to 1$.

Proof: We know from (8.3.18) that $cd_p\, G_S \le 2$. Now let n be large enough so that the class number h_n of $\mathbb{Q}(\zeta_{p^n})$ is greater than 1. By class field theory, there exists a cyclic unramified extension $F|\mathbb{Q}(\zeta_{p^n})$ of degree ℓ, where ℓ is a prime number dividing h_n. Since the only prime of $\mathbb{Q}(\zeta_{p^n})$ which divides p is principal, it splits completely in F. Thus there are ℓ different primes dividing p in F, and if $K = Fk$, then $\#S_p(K) > 1$. From the exact sequence

$$H^2(k_S|K, \mathbb{Z}/p\mathbb{Z}) \longrightarrow \bigoplus_{\mathfrak{p}\in S(K)} H^2(K_\mathfrak{p}, \mathbb{Z}/p\mathbb{Z}) \longrightarrow H^0(k_S|K, \mu_p)^* \longrightarrow 0$$

obtained from (8.6.10) it follows that

$$\dim_{\mathbb{F}_p} H^2(k_S|K, \mathbb{Z}/p\mathbb{Z}) \ge \#S^f(K) - 1 \ge \#S_p(K) - 1 \ge 1\,.$$

Therefore the cohomological p-dimension of G_S is equal to 2. □

The following result due to *L. C. Washington*, see [245], cor.3, is of great importance for the proof of (10.11.1).

(10.11.4) Theorem. *Let p be a prime number and let k be an imaginary abelian number field containing the group μ_{2p}. Then the set*

$$H = \{\ell \mid \ell \text{ a prime number dividing } h_n^- \text{ for some } n\}$$

is infinite. Here h_n^- denotes the minus part of the class number of the n-th layer of the cyclotomic $\mathbb{Z}_p$-extension $k_\infty|k$.

Proof: We may assume that $k = \mathbb{Q}(\zeta_{2p})$. We use the following result, also due to Washington, see [246], th. 16.12: *)

If ℓ is a prime number different to p, then the ℓ-part of h_n is bounded independently of n.

From Iwasawa theory, we know that the asymptotic behaviour of the p-part of h_n^- is given by the formula (11.1.6)

$$e_n = \lambda^- n + \mu^- p^n + \nu^-\,,$$

where p^{e_n} is the exact power of p dividing h_n^- and λ^-, μ^- and ν^- are constants independent of n.**) Thus, using (10.11.2), we see that $p^{-e_n} h_n^-$ tends to infinity as $n \to \infty$ and therefore, recalling the result on the ℓ-parts of h_n, it follows that there must always be new prime numbers dividing h_n^- if $n \to \infty$. □

*) Actually, in the original proof of (10.11.4) which appeared before this result was proved, Washington used a much weaker statement, which is sufficient for this purpose: the ℓ-part of h_n^- is bounded asymptotically by ℓ^{p^n}. Further, the assumption that μ_{2p} is contained in k is not necessary, see [245].

**) In fact, $\mu^- = 0$ by the theorem of Ferrero and Washington, see [246], th. 7.15, but we will not use this result.

In order to prove theorem (10.11.1), we use (3.4.6). We have to verify the vanishing of the limit

$$D_i(\mathbb{Z}/p\mathbb{Z}) = \varinjlim_{\substack{U\subseteq G_S \\ cor^*}} H^i(U,\mathbb{Z}/p\mathbb{Z})^*$$

for $i = 0, 1$, since we already proved in (10.2.2) that the p-dualizing module of G_S,

$$I = \varinjlim_{m} D_2(\mathbb{Z}/p^m\mathbb{Z}) = \varinjlim_{m} \varinjlim_{U\subseteq G_S} H^2(U,\mathbb{Z}/p^m\mathbb{Z})^*\,,$$

is isomorphic to $\mathrm{tor}_p(C_{S^f}(k_S))$. Here the limits run through the open subgroups U of G_S and the transition maps are the duals of the corestriction maps and the canonical projections from $\mathbb{Z}/p^m\mathbb{Z}$ onto $\mathbb{Z}/p^n\mathbb{Z}$, respectively.

Obviously, $D_0(\mathbb{Z}/p\mathbb{Z}) = 0$ since $p^\infty|\#G_S$. In order to prove the vanishing of $D_1(\mathbb{Z}/p\mathbb{Z})$, the following theorem is crucial.

(10.11.5) Theorem. *If $S \supseteq S_p \cup S_\infty$ is finite, then $C_S(k_S)$ is p-divisible.*

(10.11.6) Corollary. *Let $S \supseteq S_p \cup S_\infty$ be a set of primes of the number field k. Then $(k_S)_\mathfrak{p}$ is a p-closed local field for all $\mathfrak{p} \in S(k_S)$.*

Remark: We already knew that $(k_S)_\mathfrak{p}$ is p-closed in the following cases:

- $\mathfrak{p} \in S \smallsetminus S_p$, by (10.8.1),
- $\delta(S) = 1$ and $\mathfrak{p} \in S$ arbitrary, by (10.5.9).

In §9 we investigated a similar pro-p version of theorem (10.11.5) and we will need this result now. We use the following notation:

$k_S(p)$	the maximal p-extension of k inside k_S,
$G_S(p)$	the Galois group of $k_S(p)\|k$,
$k_\mathfrak{p}(p)$	the maximal p-extension of $k_\mathfrak{p}$, $\mathfrak{p}$ a prime of k,
$\mathcal{G}_\mathfrak{p}(p)$	the Galois group of $k_\mathfrak{p}(p)\|k_\mathfrak{p}$,
$G_\mathfrak{p}(p)$	the decomposition group of $\mathfrak{p}$ in $G_S(p)$.

From the results of §9, we obtain the

(10.11.7) Corollary. *Assume $\mu_p \subseteq k$ and let $S \supseteq S_p \cup S_\infty$ be finite. Suppose that $C_S(k_S(p))$ is not p-divisible. Then there exists a prime $\mathfrak{p} \in S_p(k)$ such that the following inequality holds:*

$$[k_\mathfrak{p} : \mathbb{Q}_p] \geq \sum_{\substack{\mathfrak{p}'\in S_p(k) \\ \mathfrak{p}'\neq\mathfrak{p}}} [k_{\mathfrak{p}'} : \mathbb{Q}_p].$$

Proof: Since $\mu_p \subseteq k$, we have by (7.3.9) the (in)equality

$$\operatorname{rank} G_{\mathfrak{p}'}(p) \le \operatorname{rank} \mathcal{G}_{\mathfrak{p}'}(p) = [k_{\mathfrak{p}'} : \mathbb{Q}_p] + 2$$

for primes $\mathfrak{p}' \in S_p(k)$. Now assume that $\#S_p(k) > 1$ (otherwise the statement is trivial) and assume that $C_S(k_S(p))$ is not p-divisible, so that $G_S(p)$ is not a (virtual) duality group of dimension 2 by (10.9.12). From (10.9.8) resp. (10.9.9) and (10.9.1)(ii)$'$, it follows that there exists a prime $\mathfrak{p} \in S_p(k)$ such that $G_S(p) = G_{\mathfrak{p}}(p)$, so that

$$[k_{\mathfrak{p}} : \mathbb{Q}_p] \ge \operatorname{rank} G_{\mathfrak{p}}(p) - 2 = \operatorname{rank} G_S(p) - 2\,,$$

and the free product decomposition of $G_S(p)$ obtained there yields the inequalities

$$\begin{aligned} [k_{\mathfrak{p}} : \mathbb{Q}_p] &\ge \Big(\sum_{\substack{\mathfrak{p}' \in S_p(k) \\ \mathfrak{p}' \neq \mathfrak{p}}} \operatorname{rank} \mathcal{G}_{\mathfrak{p}'}(p) \Big) - 2 \\ &= \Big(\sum_{\substack{\mathfrak{p}' \in S_p(k) \\ \mathfrak{p}' \neq \mathfrak{p}}} ([k_{\mathfrak{p}'} : \mathbb{Q}_p] + 2) \Big) - 2 \\ &\ge \sum_{\substack{\mathfrak{p}' \in S_p(k) \\ \mathfrak{p}' \neq \mathfrak{p}}} [k_{\mathfrak{p}'} : \mathbb{Q}_p]\,. \end{aligned}$$

□

Proof of (10.11.5) and (10.11.6): It suffices to show that $C_S(K_S(p))$ is p-divisible for a cofinal set of finite extensions K of k in k_S. Since $\mu_{2p} \subseteq k_S$, we assume without loss of generality that $\mu_{2p} \subseteq k$; in particular, k is totally imaginary, containing the imaginary abelian field $\mathbb{Q}(\zeta_{2p})$. Assume that K is a finite extension of k inside k_S and that $C_S(K_S(p))$ is not p-divisible.

Claim: There exists a finite extension $K'|K$ in k_S such that $C_S(K'_S(p))$ is p-divisible.

Proof of the claim: Using (10.11.4), we choose a prime number $\ell > p$ and a number n such that

(i) $\ell \mid h(\mathbb{Q}(\zeta_{p^n}))$,
(ii) $(\ell, [K : \mathbb{Q}]) = 1$.

It follows that there exists an unramified extension $F|\mathbb{Q}(\zeta_{p^n})$ of degree ℓ, in which necessarily the only prime of $\mathbb{Q}(\zeta_{p^n})$ above p splits completely. By condition (ii), we see that F and $K(\zeta_{p^n})$ are linearly disjoint over $\mathbb{Q}(\zeta_{p^n})$. Thus every prime $\mathfrak{p}$ of $K(\zeta_{p^n})$ dividing p splits into ℓ different primes in KF. Therefore the field KF has the property that for every prime dividing p there are at least $\ell - 1$ other primes having the same absolute local degree. It follows that there exists no prime $\mathfrak{p} \in S_p(KF)$ satisfying the inequality of (10.11.7). Thus the group $C_S((KF)_S(p))$ is p-divisible, which proves the claim and therefore the theorem.

It remains to show (10.11.6). A straightforward limit argument shows that we may assume that $S \supseteq S_p \cup S_\infty$ is finite. Hence $C_S(k_S)$ is p-divisible by (10.11.5). Consider the exact sequence

$$0 \longrightarrow \mathcal{O}_S^\times(k_S) \longrightarrow I_S(k_S) \longrightarrow C_S(k_S) \longrightarrow 0.$$

As the group $\mathcal{O}_S^\times(k_S)$ is p-divisible, we conclude that $I_S(k_S)$ is also p-divisible. Hence the multiplicative group $(k_S)_\mathfrak{p}^\times$ is p-divisible for all $\mathfrak{p} \in S(k_S)$. Since $\mu_p \subseteq k_S \subseteq (k_S)_\mathfrak{p}$, Kummer theory shows that the local fields $(k_S)_\mathfrak{p}$ are p-closed. □

Proof of (10.11.1): By (10.11.3), we have $cd_p\, G_S = 2$, and using (10.11.5), we can prove the vanishing of $D_1(\mathbb{Z}/p\mathbb{Z})$:

$$\begin{aligned} D_1(\mathbb{Z}/p\mathbb{Z}) &= \varinjlim_K H^1(k_S|K, \mathbb{Z}/p\mathbb{Z})^* \\ &= \varinjlim_K G(k_S|K)^{ab}/p \\ &= \varinjlim_K C_S(K)/p \\ &= C_S(k_S)/p = 0\,. \end{aligned}$$

Since the p-dualizing module I of G_S is isomorphic to $\mathrm{tor}_p(C_{S^f}(k_S))$, the proof of (10.11.1) is complete. □

Using (10.11.5) we obtain the following general duality theorem which was already mentioned in VIII (see (8.4.4)).

(10.11.8) Theorem. *Let p be a prime number and let S be a finite set of primes of the number field k such that $S \supseteq S_\infty$. Assume either that S_p is contained in S or that $S \cap S_p = \varnothing$ and $p^\infty | \#G_S$. Then the cup product*

$$\hat{H}^i(G_S, \mathrm{Hom}(A, C_S)) \times \hat{H}^{2-i}(G_S, A) \xrightarrow{\cup} H^2(G_S, C_S) \xrightarrow[\sim]{inv} \tfrac{1}{\#G_S}\mathbb{Z}/\mathbb{Z}$$

induces isomorphisms

$$\hat{H}^i(G_S, \mathrm{Hom}(A, C_S))(p) \xrightarrow{\sim} \hat{H}^{2-i}(G_S, A)(p)^\vee$$

for all $i \in \mathbb{Z}$ and every discrete G_S-module A which is finitely generated as a $\mathbb{Z}$-module.

Proof: Let us first assume that $S \supseteq S_p$. Then $p^\infty | \#G_S$, for instance because k_S contains the cyclotomic $\mathbb{Z}_p$-extension of k. By (10.11.5), C_S and hence also C_S^0 are p-divisible. Furthermore C_S^0 is a level-compact formation module with divisible universal norm groups. Since $\mathrm{Hom}(A, C_S^0)$ has the same cohomology as $\mathrm{Hom}(A, C_S)$ by (8.4.2), the theorem follows in this case from (3.1.11)(ii).

By the same argument, it remains to show the p-divisibility of C_S in the case $S \cap S_p = \varnothing$. This follows from the principal ideal theorem. Indeed, under the given assumptions, for every finite extension K of k inside k_S, the group $G_S(K)^{ab}(p)$ is finitely generated (since S is finite) and torsion (a $\mathbb{Z}_p$-extension is ramified at least at one prime dividing p by (10.3.20)), hence finite. By the group theoretic form of the principal ideal theorem, see [160] chap.VI (7.6), we conclude that

$$\varinjlim_{K \subseteq k_S} G_S(K)^{ab}(p) = 0.$$

Since $D_S(K)$ is divisible, the exact sequences

$$D_S(K) \longrightarrow C_S(K) \longrightarrow G_S(K)^{ab} \longrightarrow 0$$

induce an isomorphism

$$C_S/p \xrightarrow{\sim} \varinjlim_{K \subseteq k_S} G_S(K)^{ab}/p = 0$$

in the limit. Hence C_S and thus also C_S^0 are p-divisible. □

If A is a finitely generated G_S-module with $\#\mathrm{tor}(A) \in \mathbb{N}(S)$, then we have a perfect pairing between the finite groups $Ш^1(G_S, A')$ and $Ш^2(G_S, A)$, see (8.6.7). If S is finite, we will prove that this holds true if we interchange the roles of A and A', at least for the ℓ-parts, where $\ell \in \mathbb{N}(S)$.

We start by investigating the Shapiro maps

$$sh^i : H^i(G_S, I_S(A)) \longrightarrow P^i(G_S, A'),$$

which were defined in VIII §5. We described sh^0 and obtained partial results for sh^1 and sh^2 in (8.5.5). Using (10.11.6), we are able to sharpen these results.

(10.11.9) Proposition. *Let $S \supseteq S_\infty$ be a finite set of primes of the number field k and let $A \in Mod_S(G_S)$. Then the following hold.*

(i) *sh^0 is surjective with kernel*

$$\ker sh^0 = \prod_{\mathfrak{p} \in S_\infty(k)} N_{\bar{k}_\mathfrak{p}|k_\mathfrak{p}} \mathrm{Hom}(A, \mathbb{C}^\times).$$

(ii) *sh^1 is injective, and bijective if A is $\mathbb{Z}$-free or if S is finite.*

(iii) *sh^2 is injective and for each $\ell \in \mathbb{N}(S)$, the induced map on ℓ-torsion subgroups*

$$H^2(G_S, I_S(A))(\ell) \longrightarrow P^2(G_S, A')(\ell)$$

is an isomorphism.

Proof: In (8.5.5) we already proved assertion (i), the injectivity of sh^1, and assertions (ii) and (iii) in the case that A is $\mathbb{Z}$-free. We go through the proof of this proposition again to show the additional assertions.

Since $A \in Mod_S(G_S)$, we have an isomorphism $A \cong \mathbb{Z}^r \otimes A_0$ of trivial $G(\bar{k}|k_S)$-modules, where $r \geq 0$ and A_0 is a finite abelian group with $\#A_0 \in \mathbb{N}(S)$. As A_0' is a trivial $G(\bar{k}|k_S)$-module and $(k_S)_\mathfrak{p}$ is ℓ-closed for $\mathfrak{p} \in S$ and $\ell \in \mathbb{N}(S)$ by (10.11.6), we obtain

$$H^1(G(\bar{k}_\mathfrak{p}|(k_S)_\mathfrak{p}), A') \cong H^1(G(\bar{k}_\mathfrak{p}|(k_S)_\mathfrak{p}), \bar{k}_\mathfrak{p}^\times)^r \oplus H^1(G(\bar{k}_\mathfrak{p}|(k_S)_\mathfrak{p}), A_0') = 0$$

for $\mathfrak{p} \in S$ using Hilbert's Satz 90. The Hochschild-Serre spectral sequence induces an isomorphism

$$H^1(G((k_S)_\mathfrak{p}|k_\mathfrak{p}), \mathrm{Hom}(A, (k_S)_\mathfrak{p}^\times)) \xrightarrow{\sim} H^1(G(\bar{k}_\mathfrak{p}|k_\mathfrak{p}), \mathrm{Hom}(A, \bar{k}_\mathfrak{p}^\times))$$

and an injection

$$H^2(G((k_S)_\mathfrak{p}|k_\mathfrak{p}), \mathrm{Hom}(A, \bar{k}_\mathfrak{p}^\times)) \hookrightarrow H^2(G(\bar{k}_\mathfrak{p}|k_\mathfrak{p}), \mathrm{Hom}(A, \bar{k}_\mathfrak{p}^\times)).$$

Using the commutative diagram

$$\begin{array}{ccc}
\bigoplus_{\mathfrak{p}\in S} H^2(G((k_S)_\mathfrak{p}|k_\mathfrak{p}), \mathrm{Hom}(A, (k_S)_\mathfrak{p}^\times)) & \hookrightarrow & \bigoplus_{\mathfrak{p}\in S} H^2(G(\bar{k}_\mathfrak{p}|k_\mathfrak{p}), \mathrm{Hom}(A, \bar{k}_\mathfrak{p}^\times)) \\
\uparrow\wr & & \uparrow\wr \\
H^2(G_S, I_S(A)) & \xrightarrow{sh^2} & P^2(G_S, A')
\end{array}$$

we obtain the first assertion of (iii). Now let $\ell \in \mathbb{N}(S)$. Then $\ell^\infty \mid G((k_S)_\mathfrak{p}|k_\mathfrak{p})$ for all nonarchimedean primes $\mathfrak{p} \in S$. As k_S is totally imaginary, we obtain

$$H^i(\bar{k}_\mathfrak{p}|(k_S)_\mathfrak{p}, A')(\ell) = 0$$

for all $\mathfrak{p} \in S$ and $i \geq 1$. Hence the horizontal injection in the last diagram is an isomorphism on the ℓ-torsion subgroups. This finishes the proof of (iii).

Finally, assume that S is finite. Then the commutative diagram

$$\begin{array}{ccc}
\prod_{\mathfrak{p}\in S} H^1(G((k_S)_\mathfrak{p}|k_\mathfrak{p}), \mathrm{Hom}(A, (k_S)_\mathfrak{p}^\times)) & \xrightarrow{\sim} & \prod_{\mathfrak{p}\in S} H^1(G(\bar{k}_\mathfrak{p}|k_\mathfrak{p}), \mathrm{Hom}(A, \bar{k}_\mathfrak{p}^\times)) \\
\uparrow\wr & & \uparrow\wr \\
H^1(G_S, I_S(A)) & \xhookrightarrow{sh^1} & P^1(G_S, A'),
\end{array}$$

shows the remaining assertion in (ii). □

(10.11.10) Theorem (Poitou-Tate Duality, second form). *Let $S \supseteq S_\infty$ be a finite set of primes of the number field k and let A be a finitely generated G_S-module with #tor(A) $\in \mathbb{N}(S)$. Then for all $\ell \in \mathbb{N}(S)$ there is a perfect pairing*

$$Ш^1(G_S, A)(\ell) \times Ш^2(G_S, A')(\ell) \longrightarrow \mathbb{Q}/\mathbb{Z}$$

of finite groups, which is induced by the cup-product, i.e. the diagram

$$\begin{array}{ccccccc}
H^1(G_S, A)(\ell) & \times & H^1(G_S, C_S(A))(\ell) & \xrightarrow{\cup} & H^2(G_S, C_S) & \xrightarrow{inv} & \mathbb{Q}/\mathbb{Z} \\
\uparrow & & \downarrow & & & \nearrow & \\
Ш^1(G_S, A)(\ell) & \times & Ш^2(G_S, A')(\ell) & & & &
\end{array}$$

commutes.

Proof: Since $H^1(G_S, A)$ is a finite group by (8.3.20)(ii), its subgroup $Ш^1(G_S, A)$ is also finite. By (8.6.6), the diagram

$$\begin{array}{ccccccc}
H^1(G_S, I_S(A)) & \longrightarrow & H^1(G_S, C_S(A)) & \xrightarrow{\delta} & H^2(G_S, A') & \longrightarrow & H^2(G_S, I_S(A)) \\
{\scriptstyle sh^1}\downarrow\wr & & \downarrow{\scriptstyle \Delta^1} & & \| & & \downarrow{\scriptstyle sh^2} \\
P^1(G_S, A') & \xrightarrow{(\xi^1)^\vee} & H^1(G_S, A)^\vee & & H^2(G_S, A') & \xrightarrow{\lambda^2} & P^2(G_S, A')
\end{array}$$

commutes. By (10.11.9)(ii), sh^1 is an isomorphism. For $\ell \in \mathbb{N}(S)$, the maps Δ^1 and sh^2 induce isomorphisms on the ℓ-primary parts by (10.11.8) and (10.11.9)(iii), respectively. As the upper sequence is exact, we get an exact sequence for the ℓ-primary parts of the groups in the lower line. We obtain the commutative and exact diagram

$$\begin{array}{ccccc}
 & & H^1(G_S, C_S(A))(\ell) & \xrightarrow{\delta} & Ш^2(G_S, A')(\ell) \\
 & & {\scriptstyle \Delta^1}\downarrow\wr & & \| \\
P^1(G_S, A)(\ell)^\vee & \xrightarrow{(\xi^1)^\vee} & H^1(G_S, A)(\ell)^\vee & \longrightarrow & Ш^2(G_S, A)(\ell)^\vee .
\end{array}$$

Since coker $(\xi^1)^\vee = Ш^1(G_S, A)^\vee$ by (8.6.5), this proves the theorem. □

Remark: For a finite module $A \in Mod_S(G_S)$, the pairing constructed in (10.11.10) coincides with that constructed in (8.6.7) for A' up to sign. Indeed, denoting these pairings by $\sqcup_1$ and $\sqcup_2$ respectively, the diagram

$$\begin{array}{ccccc}
H^0(G_S,C_S(A')) & \times & H^2(G_S,A') & \xrightarrow{\ \cup\ } & \mathbb{Q}/\mathbb{Z} \\
\big\downarrow{\scriptstyle\delta} & & \big\uparrow & & \big\| \\
Ш^1(G_S,A'') & \times & Ш^2(G_S,A') & \xrightarrow{\ \cup_2\ } & \mathbb{Q}/\mathbb{Z} \\
{\scriptstyle(-1)\psi_*}\big\downarrow\wr & & \big\| & & \big\| \\
Ш^1(G_S,A) & \times & Ш^2(G_S,A') & \xrightarrow{\ \cup_1\ } & \mathbb{Q}/\mathbb{Z} \\
\big\downarrow & & \big\uparrow{\scriptstyle\delta} & & \big\| \\
H^1(G_S,A) & \times & H^1(G_S,C_S(A)) & \xrightarrow{\ \cup\ } & \mathbb{Q}/\mathbb{Z}
\end{array}$$

commutes by (1.4.5); here ψ is the canonical isomorphism $A'' \xrightarrow{\sim} A$ and the maps δ are the connecting homomorphisms with respect to the exact sequences $0 \to M' \to I(M) \to C(M) \to 0$, where $M = A'$ and $M = A$, respectively.

In order to obtain further information about the decomposition behaviour of primes in infinite algebraic extensions, we need some facts from analytic number theory.

We use from now on the following notation:

k is an algebraic number field,
$M|k$ is an *infinite* unramified Galois extension of k,
$\mathfrak{p}$ runs through all finite primes of k,
$f(\mathfrak{p})$ is the residue extension degree of $\mathfrak{p}$ in $M|k$ ($1 \le f(\mathfrak{p}) \le \infty$),
T $= T(M|k) = \{\mathfrak{p} \mid f(\mathfrak{p}) < \infty\}$.

Furthermore, let d_k be the discriminant of k and

$$\beta_{\mathfrak{p}} = \begin{cases} \frac{1}{2}(\log 4\pi + \gamma), & \text{if } \mathfrak{p} \in S_\infty \text{ is real}, \\[2ex] \log 2\pi + \gamma, & \text{if } \mathfrak{p} \in S_\infty \text{ is complex}, \end{cases}$$

where $\gamma = \lim\limits_{n\to\infty}(1 + 1/2 + \cdots + 1/n - \log n) = 0.577\cdots$ is the Euler constant. Finally, we set

$$\alpha(k) = \frac{1}{2}\log|d_k| - \sum_{\mathfrak{p}\in S_\infty} \beta_{\mathfrak{p}}\,.$$

We will prove the following theorem, due to Y. Ihara [82]. It says that only a small proportion of primes is almost completely decomposed in $M|k$.

(10.11.11) Theorem.

$$\sum_{\mathfrak{p}\in T(M|k)} \frac{\log N(\mathfrak{p})}{N(\mathfrak{p})^{f(\mathfrak{p})} - 1} \le \alpha(k)\,.$$

In particular, the expression on the left is convergent.

Remark: If the Riemann hypothesis were to be valid for the Dedekind zeta function $\zeta_K(s)$ for all finite algebraic number fields K with $k \subseteq K \subseteq M$, then a much stronger result would hold:

$$\sum_{\mathfrak{p}\in T(M|k)} \frac{\log N(\mathfrak{p})}{N(\mathfrak{p})^{(1/2)f(\mathfrak{p})}-1} \le \frac{1}{2}\log|d_k| - \sum_{\mathfrak{p}\in S_\infty} \alpha_\mathfrak{p}\,,$$

where

$$\alpha_\mathfrak{p} = \begin{cases} \frac{1}{2}(\log 8\pi + \gamma + \frac{\pi}{2}), & \text{if } \mathfrak{p} \in S_\infty \text{ is real,} \\ \log 8\pi + \gamma, & \text{if } \mathfrak{p} \in S_\infty \text{ is complex.} \end{cases}$$

For this see the original paper of Ihara [82] where the following analogue for function fields is also proved:

Let $\mathbb{F}_q$ be the constant field of k and let g be its genus. We denote the degree over $\mathbb{F}_q$ of a prime $\mathfrak{p}$ of k by deg($\mathfrak{p}$). Then

$$\sum_{\mathfrak{p}\in T(M|k)} \frac{\deg(\mathfrak{p})}{N(\mathfrak{p})^{(1/2)f(\mathfrak{p})}-1} \le \max(g-1, 0)\,.$$

The presence of the factor $1/2$ in the exponent of $N(\mathfrak{p})$ is due to the theorem of Weil for curves.

Let $\zeta_K(s)$ be the Dedekind zeta function of K,

$$\zeta_K(s) = \prod_{\mathfrak{P}} (1 - N(\mathfrak{P})^{-s})^{-1}, \quad \mathrm{Re}(s) > 1,$$

and let

$$Z_K(s) = -\frac{\zeta_K'(s)}{\zeta_K(s)}\,.$$

The Dirichlet series corresponding to $Z_K(s)$ is given by

$$-\frac{\zeta_K'(s)}{\zeta_K(s)} = \sum_{\mathfrak{P}} \sum_{m\ge 1} \frac{\log N(\mathfrak{P})}{N(\mathfrak{P})^{ms}}\,.$$

For a real number $x > 1$, let $\psi_K(x)$ be the Čebyšev function obtained as the partial sum of coefficients of the Dirichlet series of $Z_K(x)$, i.e.

$$\psi_K(x) = \sum_{\substack{m\ge 1 \\ N(\mathfrak{P})^m < x}} \log N(\mathfrak{P})\,.$$

We will need the following well-known identity for $Z_K(s)$:

(10.11.12) Lemma. *For* $\mathrm{Re}(s) > 1$

$$-\frac{\zeta_K'(s)}{\zeta_K(s)} = s^{-1} \int_1^\infty \psi_K(x) x^{-s-1} dx\,.$$

Proof: We will use Abel's summation formula. Let (a_n) be a sequence of complex numbers and let $\varphi(x)$ be a complex-valued function on $(0,\infty)$ having a continuous derivative. Let $A(x)=\sum_{n\le x} a(n)$ and assume that $A(x)\varphi(x)\to 0$ as $x\to\infty$. Then

$$\sum_{n=1}^{\infty} a(n)\varphi(n) = -\int_1^{\infty} A(t)\varphi'(t)dt\,,$$

provided that either side is convergent, cf. [24], chap.VII, th. 6.

Now we set

$$a(n)=\begin{cases}\log N(\mathfrak{P}), & \text{if } n \text{ is a power } N(\mathfrak{P})^m,\ m>0,\\ 0, & \text{otherwise,}\end{cases}$$

and $\varphi(x)=x^{-\sigma}$, where σ is real and $\sigma>1$. Then $A(x)=\psi_K(x)$ and $A(x)\varphi(x)\to 0$ as $x\to\infty$, since $\psi_K(x)<x\log x$, so that

$$A(x)\varphi(x)=O(x^{1-\sigma}\log x)=o(1)\,.$$

Thus we can apply the summation formula and obtain the desired result for real $s>1$, and for $\mathrm{Re}(s)>1$ by analytic continuation. □

For an algebraic number field K we set

$$A(K)=\pi^{-r_1(K)/2}(2\pi)^{-r_2(K)}|d_K|^{1/2}\,,$$

where as usual $r_1(K)$ and $r_2(K)$ denote the number of real and complex primes of K respectively. For the proof of (10.11.11) we also need the following lemma due to *E. Landau*, cf. [120], Satz 180.

(10.11.13) Lemma. *We have the partial fraction decomposition of $Z_K(s)$*

$$Z_K(s)=\log A(K)+\frac{r_1(K)}{2}\frac{\Gamma'}{\Gamma}\Big(\frac{s}{2}\Big)+r_2(K)\frac{\Gamma'}{\Gamma}(s)+\Big(\frac{1}{s}+\frac{1}{s-1}\Big)-\sum_{\rho\in\mathfrak{Z}(K)}{}'\frac{1}{s-\rho}\,,$$

where $\mathfrak{Z}(K)$ is the set of all nontrivial zeros of $\zeta_K(s)$ and the sum $\sum_\rho'$ is taken with multiplicity, where the terms for ρ and $\bar\rho$ should be summed together.

Proof: Let

$$\xi_K(s)=s(s-1)A(K)^s\Gamma\Big(\frac{s}{2}\Big)^{r_1(K)}\Gamma^{r_2(K)}\zeta_K(s)\,.$$

Then $\xi_K(s)$ is an entire function and $\xi_K(1-s)=\xi_K(s)$, cf. [160], chap.VII, (5.10). Since $\xi_K(s)$ is of order 1, Hadamard's factorization theorem yields the expression

$$\xi_K(s)=e^{a+bs}\prod_\rho\Big(1-\frac{s}{\rho}\Big)e^{s/\rho}$$

for some complex numbers a and b, where ρ runs through all the zeros of $\xi_K(s)$, which are exactly the nontrivial zeros of $\zeta_K(s)$. *)
Taking the logarithmic derivative of this product, we obtain

$$\frac{\xi_K'}{\xi_K}(s) = b + \sum_\rho \Big(\frac{1}{s-\rho} + \frac{1}{\rho}\Big),$$

where the sum converges absolutely. From

$$\frac{\xi_K'}{\xi_K}(s) = -\frac{\xi_K'}{\xi_K}(1-s)$$

it follows that

$$b + \sum_\rho \Big(\frac{1}{s-\rho} + \frac{1}{\rho}\Big) = -b - \sum_\rho \Big(\frac{1}{(1-\rho)-s} + \frac{1}{\rho}\Big).$$

Since $1-\rho$ is a zero whenever ρ is, we obtain

$$b = -\sum_\rho{}' \frac{1}{\rho},$$

where we now have to sum the ρ and $\bar\rho$ terms together. Thus the formula above becomes

$$\frac{\xi_K'}{\xi_K}(s) = \sum_\rho{}' \Big(\frac{1}{s-\rho}\Big).$$

□

Proof of (10.11.11): Let K run through the finite Galois extensions of k inside M, so $K|k$ is unramified. Then

$$\begin{aligned} Z_K(s) &= -\frac{\zeta_K'(s)}{\zeta_K(s)} = \sum_{\mathfrak{P}} \sum_{m\ge 1} \frac{\log N(\mathfrak{P})}{N(\mathfrak{P})^{ms}} \\ &= [K:k] \cdot \sum_{\mathfrak{p}} \sum_{m\ge 1} \frac{\log N(\mathfrak{p})}{N(\mathfrak{p})^{f(\mathfrak{P}|\mathfrak{p})ms}} \\ &= [K:k] \cdot \sum_{\mathfrak{p}} \frac{\log N(\mathfrak{p})}{N(\mathfrak{p})^{f(\mathfrak{P}|\mathfrak{p})s} - 1}, \qquad \mathrm{Re}(s) > 1, \end{aligned}$$

where $f(\mathfrak{P}|\mathfrak{p})$ is the residue extension degree of $\mathfrak{P}$ in $K|k$, and for the real number $x > 1$, we have for the Čebyšev function

$$\psi_K(x) = \sum_{\substack{m\ge 1 \\ N(\mathfrak{P})^m < x}} \log N(\mathfrak{P}) = [K:k] \cdot \sum_{\substack{m\ge 1 \\ N(\mathfrak{p})^{f(\mathfrak{P}|\mathfrak{p})m} < x}} \log N(\mathfrak{p}).$$

Now we set

$$Z_M(s) := \sum_{\mathfrak{p}\in T} \sum_{m\ge 1} \frac{\log N(\mathfrak{p})}{N(\mathfrak{p})^{f(\mathfrak{p})ms}} = \sum_{\mathfrak{p}\in T} \frac{\log N(\mathfrak{p})}{N(\mathfrak{p})^{f(\mathfrak{p})s} - 1},$$

*) For the definition of the order of an entire function and for the product theorem, see [166], appendix 5.

which is a convergent Dirichlet series holomorphic on $\mathrm{Re}(s) > 1$, and

$$\psi_M(x) := \sum_{\substack{m\geq 1, \mathfrak{p}\in T\\ N(\mathfrak{p})^{f(\mathfrak{p})m}<x}} \log N(\mathfrak{p})\,.$$

From the definitions above it follows that

$$\psi_M(x) \leq [K:k]^{-1}\psi_K(x)\,,$$

and by (10.11.12), which similarly holds for $Z_M(s)$, we obtain the inequality for $\mathrm{Re}(s) > 1$

$$Z_M(s) \leq [K:k]^{-1} Z_K(s)\,.$$

Now let $s = \sigma > 1$ be real. Using (10.11.13) and the fact that

$$\frac{1}{\sigma-\rho} + \frac{1}{\sigma-\bar{\rho}} > 1 \quad \text{for } \sigma > 1,$$

we obtain by letting $K \to M$

$$Z_M(\sigma) = \sum_{\mathfrak{p}\in T} \frac{\log N(\mathfrak{p})}{N(\mathfrak{p})^{f(\mathfrak{p})\sigma}-1} \leq \log A(k) + \frac{r_1}{2}\frac{\Gamma'}{\Gamma}(\frac{\sigma}{2}) + r_2\frac{\Gamma'}{\Gamma}(\sigma) \quad (\sigma > 1)\,.$$

(Observe that $\log|d_K| = [K:k]\log|d_k|$, since $K|k$ is unramified.) Obviously, for any finite subset T' of T we have the inequality above with T' in place of T. Letting $\sigma \to 1$ then gives

$$\sum_{\mathfrak{p}\in T'} \frac{\log N(\mathfrak{p})}{N(\mathfrak{p})^{f(\mathfrak{p})}-1} \leq \log A(k) + \frac{r_1}{2}\frac{\Gamma'}{\Gamma}(\frac{1}{2}) + r_2\frac{\Gamma'}{\Gamma}(1)\,.$$

Since T' is an arbitrary finite subset of T, this last inequality is valid for T in place of T', and since

$$\frac{\Gamma'}{\Gamma}(1) = -\gamma, \quad \frac{\Gamma'}{\Gamma}(\frac{1}{2}) = -\log 4 - \gamma$$

and

$$\log A(k) = -\frac{r_1}{2}\log\pi - r_2\log 2\pi + \frac{1}{2}\log|d_k|,$$

we have proved the theorem. □

Recall that given a number field k and a set of nonarchimedean primes T of k, we denote the maximal unramified extension of k in which all primes of T are completely decomposed by $k_\varnothing^T$. The following corollary, which immediately follows from (10.11.11), asserts that the set T cannot be too big if the extension $k_\varnothing^T|k$ is infinite.

(10.11.14) Corollary. *Assume that the extension $k_\varnothing^T|k$ is infinite. Then*

$$\sum_{\mathfrak{p}\in T} \frac{\log N(\mathfrak{p})}{N(\mathfrak{p})-1} \leq \alpha(k)\,.$$

Another consequence of the theorem of Ihara is the following result, which was already mentioned in X §2 and which can also be obtained by a pure group-theoretical method; see (10.2.6) and the exercises in X §2.

(10.11.15) Corollary. *Let S be a finite set of primes of the algebraic number field k. Then the Galois group $G(k_S|k)$ can be topologically generated by a finite number of conjugacy classes.*

Proof: The decomposition groups of $G(k_S|k)$ with respect to the primes $\mathfrak{p} \in S$ are finitely generated as homomorphic images of the absolute local Galois groups $G_{k_\mathfrak{p}}$. So we are reduced to the case $S = \varnothing$, i.e. k_S is the maximal unramified extension k^{nr} of k.

If $k^{nr}|k$ is finite, we are done, so let us assume that $k^{nr}|k$ is infinite. Since the sum of $\log N(\mathfrak{p})/(N(\mathfrak{p})-1)$ over all nonarchimedean primes $\mathfrak{p}$ of k is divergent, we can find primes $\mathfrak{p}_1, \dots, \mathfrak{p}_n$ of k such that

$$\sum_{i=1}^{n} \frac{\log N(\mathfrak{p}_i)}{N(\mathfrak{p}_i)-1} > \alpha(k)\,,$$

where $\alpha(k)$ was defined in (10.11.11). Now let $M'|k$ be the maximal unramified extension of k which is completely decomposed at all primes $\mathfrak{p}_i$, $i = 1, \dots, n$. The extension $M'|k$ cannot be infinite by (10.11.11). Thus the normal subgroup $(G_{\mathfrak{p}_1}, \dots, G_{\mathfrak{p}_n})$ generated by the decomposition groups $G_{\mathfrak{p}_i} = \langle \mathrm{Frob}_{\mathfrak{p}_i} \rangle$, $i = 1, \dots, n$, has finite index in $G(k^{nr}|k)$. Therefore the latter group is generated by finitely many conjugacy classes. □

Chapter XI

Iwasawa Theory of Number Fields

As shown in the previous chapters, there is a remarkable analogy between the theory of algebraic number fields and the theory of function fields in one variable over a finite field. This analogy should also extend to the theory of ζ-functions and L-functions of global fields. If, for a function field k, one considers the corresponding smooth and proper curve $C/\mathbb{F}$, where $\mathbb{F}$ is the field of constants of k, then the ζ-function of the curve C is a rational function. More precisely, if φ is the arithmetic Frobenius automorphism, i.e. the natural generator of the Galois group $G(\bar{\mathbb{F}}|\mathbb{F}) \cong \hat{\mathbb{Z}}$, then φ acts on the Tate module of the Jacobian variety J of C, which is defined as the projective limit of the groups of p^n-torsion points of $J(\bar{\mathbb{F}})$, where $p \neq \mathrm{char}(\mathbb{F})$ is a prime number. Tensoring with $\mathbb{Q}_p$, one obtains a $\mathbb{Q}_p$-vector space of dimension $2g$, where g is the genus of C. The characteristic polynomial with respect to the endomorphism φ^{-1} is the essential part of the ζ-function of the curve C.

In order to obtain an analogous result for a number field k, the idea of *K. Iwasawa* [88] was to consider $\mathbb{Z}_p$-extensions of k, and in particular, the cyclotomic $\mathbb{Z}_p$-extension $k_\infty|k$ (obtained by adjoining all roots of unity of p-power order if k contains μ_p). The analogy asserts that over k_∞ one is in a *geometric* situation: the group $\Gamma = G(k_\infty|k) \cong \mathbb{Z}_p$ acts on various Galois groups of abelian extensions of k_∞. The **main conjecture** for number fields states that the characteristic polynomials of these actions are related to p-adic L-functions. This was first proved by *B. Mazur* and *A. Wiles* [134] under the assumption that the base field is abelian over $\mathbb{Q}$, and later by *A. Wiles* [249] for general totally real fields.

In the first section we start by proving the beautiful theorem of Iwasawa which describes the behaviour of the p-part of the class number in a $\mathbb{Z}_p$-extension. Furthermore, we study the $\mathbb{Z}_p[[\Gamma]]$-module structure of the Iwasawa modules $X_{nr} = G(H|k_\infty)$ and $X_{cs} = G(H'|k_\infty)$, where H is the maximal abelian unramified p-extension of k_∞ and H' is its maximal subextension which is completely decomposed over k_∞ at the primes above p.

J. Neukirch et al., *Cohomology of Number Fields*,
Grundlehren der mathematischen Wissenschaften 323,
DOI 10.1007/978-3-540-37889-1_11, © Springer-Verlag Berlin Heidelberg 2008

After establishing the Iwasawa theory over local fields in §2, we consider the Λ-module $X_1 = G(k_\Sigma(p)|k_\infty)^{ab}$ in §3, where $k_\Sigma(p)$ is the maximal p-extension of k unramified outside the set $\Sigma = S_p \cup S_\infty$ of primes above p and ∞. Most of the results will follow from the general homotopy theory of Λ-modules developed in chapter V.

In §4 we consider the case of Iwasawa modules over CM-fields. Using complex conjugation we get further insight into the structure of the relevant Iwasawa modules. In §5 we present the (non-abelian) concept of positively ramified extensions of number fields.

Finally in §6, we give an overview of the main conjecture of Iwasawa theory and its applications.

Although not always mentioned explicitly, we owe a lot to a paper of *U. Jannsen* [97]. Further, we have used the original article of *K. Iwasawa* [88] and the paper [252]. Though we do not refer to the beautiful book of *L. C. Washington* [246] in this chapter, the reader is strongly advised to compare our approach with the presentation there and to look at the many other important aspects which are not treated here.

§1. The Maximal Abelian Unramified p-Extension of k_∞

We recall a definition given in X §3. A **$\mathbb{Z}_p$-extension** of a number field k is a Galois extension $k_\infty|k$ with Galois group $G(k_\infty|k) \cong \mathbb{Z}_p$, the additive group of p-adic integers. One can regard such a $\mathbb{Z}_p$-extension as a tower of fields

$$k = k_0 \subseteq k_1 \subseteq \ldots \subseteq k_\infty = \bigcup_n k_n$$

with $G(k_n|k) \cong \mathbb{Z}/p^n\mathbb{Z}$, since the nontrivial closed subgroups of $\mathbb{Z}_p$ are of the form $p^n\mathbb{Z}_p$ for some n. The fields $k_n \subseteq k_\infty$ are uniquely determined by the property $[k_n : k] = p^n$.

Let $k(\mu_{p^\infty})$ be the extension of k obtained by adjoining all roots of unity of p-power order. Then

$$G(k(\mu_{p^\infty})|k) = \Gamma \times \Delta$$

where Γ is isomorphic to $\mathbb{Z}_p$ and $\Delta \subseteq \mathbb{Z}/(p-1)\mathbb{Z}$ if p is odd or $\Delta \subseteq \mathbb{Z}/2\mathbb{Z}$ if $p = 2$. By Galois theory, there exists precisely one $\mathbb{Z}_p$-extension $k_\infty|k$ inside $k(\mu_{p^\infty})$. This extension is called the **cyclotomic $\mathbb{Z}_p$-extension** of k.

The next proposition shows that every $\mathbb{Z}_p$-extension is unramified outside the primes above p.

(11.1.1) Proposition.

(i) *Let $k_\infty|k$ be a $\mathbb{Z}_p$-extension. Then every (possibly archimedean) prime $\mathfrak{p}$ of k not dividing p is unramified in $k_\infty|k$ and at least one prime $\mathfrak{p}$ (necessarily above p) ramifies in $k_\infty|k$.*

(ii) *The cyclotomic $\mathbb{Z}_p$-extension is ramified at every prime $\mathfrak{p}$ above p.*

(iii) *The cyclotomic $\mathbb{Z}_p$-extension $\mathbb{Q}_\infty$ of $\mathbb{Q}$ is totally ramified at p.*

Proof: (i) Let $T_\mathfrak{p}$ be the inertia group of $G(k_\infty|k)$ with respect to $\mathfrak{p} \nmid p$. Then $T_\mathfrak{p} = 0$ or $T_\mathfrak{p} \cong p^n\mathbb{Z}_p$ for some n. In the first case we are done, and in the second case we see that $\mathfrak{p}$ is nonarchimedean. By class field theory, $T_\mathfrak{p}$ is the image of the homomorphism

$$U_\mathfrak{p}(k)(p) \longrightarrow G(\bar{k}|k)^{ab}(p) \longrightarrow\!\!\!\!\rightarrow G(k_\infty|k).$$

But $U_\mathfrak{p}(k)(p) = \mu(k_\mathfrak{p})(p)$ is finite, and so $T_\mathfrak{p} = 0$. Observing that a $\mathbb{Z}_p$-extension cannot be unramified everywhere because of the finiteness of the ideal class group of k, we have therefore proved (i).

The assertion (iii) is known from the theory of cyclotomic fields, cf. [160], chap.I, (10.1). Now statement (ii) follows easily for an arbitrary number field k because $k_\infty = k\mathbb{Q}_\infty$. □

In X §3 we were dealing with the question of how many independent $\mathbb{Z}_p$-extensions of a number field k exist, and we saw that this problem is related to the Leopoldt conjecture. We reformulate (10.3.20)(ii) as follows:

(11.1.2) Theorem. *Let $\tilde{k}$ be the composite of all $\mathbb{Z}_p$-extensions of k. Then*

$$G(\tilde{k}|k) \cong \mathbb{Z}_p^{r_2+1+\mathfrak{d}_p},$$

where r_2 is the number of complex places of k and $\mathfrak{d}_p$ is the Leopoldt defect, i.e. $\mathfrak{d}_p = \operatorname{rank}_\mathbb{Z} \mathcal{O}_k^\times - \operatorname{rank}_{\mathbb{Z}_p} \overline{\mathcal{O}_k^\times}$.

Now we consider unramified p-extensions of number fields. For $n \geq 0$ let L_n be the maximal unramified p-extension of k_n and let L_n' be the maximal unramified p-extension of k_n which is completely decomposed at all primes above p. The maximal abelian extension $H_n = L_n^{ab}$ inside L_n is the p-Hilbert class field of k_n. Let $H_n' = L_n'^{ab}$ and let

$$L = \varinjlim_n L_n \qquad \text{and} \qquad L' = \varinjlim_n L'_n$$

be the corresponding extensions over k_∞ (in X §4 the field L' was denoted by $(k_\infty)_{S_\infty}^{S_p}(p)$). Setting

$$H = L^{ab} = \varinjlim_n H_n \qquad \text{and} \qquad H' = L'^{ab} = \varinjlim_n H'_n,$$

we have the following diagram of fields:

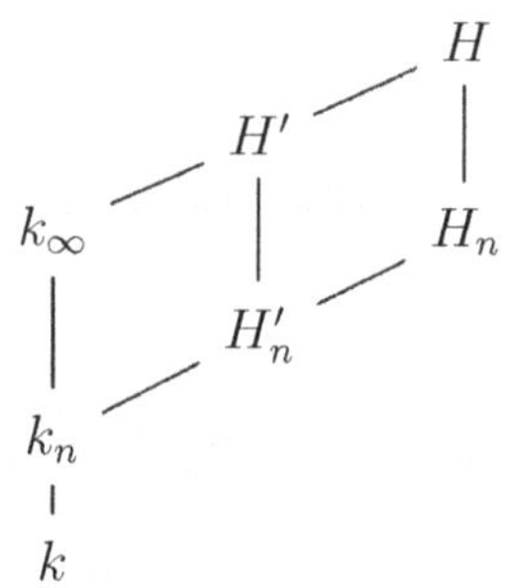

and the Galois groups

$$X_{nr} := G(H|k_\infty) = G(L|k_\infty)^{ab},$$

$$X_{cs} := G(H'|k_\infty) = G(L'|k_\infty)^{ab}$$

are Iwasawa modules for $\Lambda = \mathbb{Z}_p[\![\Gamma]\!]$.

We introduce some further notation.

(11.1.3) Definition. *Let $k_\infty|k$ be a $\mathbb{Z}_p$-extension and let $T_\mathfrak{p} = T_\mathfrak{p}(L|k)$ be the inertia subgroup of $G(L|k)$ with respect to a prime $\mathfrak{p}$. Then $T_\mathfrak{p}$ maps onto an open subgroup $\Gamma_{n_\mathfrak{p}}$ in Γ if $T_\mathfrak{p} \neq \{1\}$. We set*

$n_0 = n_0(k_\infty|k) = \max\{n_\mathfrak{p} \mid \mathfrak{p}$ *ramifies in* $k_\infty|k\}$,
$s_n = s(k_n) = \#\{\mathfrak{p}$ *a prime of* $k_n \mid \mathfrak{p}$ *ramifies in* $k_\infty|k_n\}$,
$s_\infty = s(k_\infty) = \#\{\mathfrak{P}$ *a prime of* $k_\infty \mid \mathfrak{P} \cap k$ *ramifies in* $k_\infty|k\}$.

By (11.1.1)(i), the numbers $s_\infty \geq s_n$ are finite, since ramified primes lie above p and have open decomposition groups in Γ.

(11.1.4) Proposition. *For every $n \geq 0$*

$$\operatorname{rank}_{\mathbb{Z}_p} G(H'|k_\infty)_{\Gamma_n} \leq \operatorname{rank}_{\mathbb{Z}_p} G(H|k_\infty)_{\Gamma_n} \leq s_n - 1 \leq s_\infty - 1.$$

In particular, $X_{nr} = G(H|k_\infty)$ and $X_{cs} = G(H'|k_\infty)$ are finitely generated Λ-torsion modules.

Proof: The last assertion follows from the first and (5.3.10). The exact sequence

$$1 \longrightarrow G(H|k_\infty) \longrightarrow G(H|k_n) \longrightarrow \Gamma_n \longrightarrow 1$$

shows that

$$\operatorname{rank}_{\mathbb{Z}_p} G(H|k_\infty)_{\Gamma_n} = \operatorname{rank}_{\mathbb{Z}_p} G(H|k_n)^{ab} - 1.$$

Let T_n be the normal subgroup of $G(H|k_n)$ generated by the inertia groups $T_{\mathfrak{P}_i}(H|k_n) \cong T_{\mathfrak{p}_i}(k_\infty|k_n) \subseteq \Gamma_n$ with respect to the ramified primes $\mathfrak{p}_i$, $i = 1, \dots, s_n$, in $k_\infty|k_n$. By definition of H_n we obtain the exact sequence

$$T_n/[T_n, G(H|k_n)] \longrightarrow G(H|k_n)^{ab} \longrightarrow G(H_n|k_n) \longrightarrow 1,$$

and so, because $G(H_n|k_n) \cong Cl(k_n)(p)$ is finite,

$$\operatorname{rank}_{\mathbb{Z}_p} G(H|k_n)^{ab} \le \operatorname{rank}_{\mathbb{Z}_p} T_n/[T_n, G(H|k_n)] \le s_n\,.$$

Since $G(H'|k_\infty)$ is a quotient group of $G(H|k_\infty)$, the other inequality follows. □

For $n \ge n_0(k_\infty|k)$, the canonical surjection $G(H|k_n) \twoheadrightarrow G(k_\infty|k_n)$ induces isomorphisms

$$T_{\mathfrak{P}_i}(H|k_n) \xrightarrow{\sim} \Gamma_n \quad \text{for} \quad i = 1, \dots, s_\infty.$$

In particular, for every i the group $G(H|k_n)$ is the semi-direct product of $G(H|k_\infty)$ and $T_{\mathfrak{P}_i}(H|k_n)$ and therefore there exist elements $g_i \in G(H|k_\infty)$ such that

$$\tau_i = g_i\tau_1\,, \qquad i = 2, \dots, s_\infty,$$

where $\tau_i \in T_{\mathfrak{P}_i}(H|k_n)$, $i = 1, \dots, s_\infty$, lifts the generator $\gamma^{p^n} \in \Gamma_n$, where γ is a chosen generator of Γ.

Recall that for $n \ge 0$ the Weierstraß polynomials $\omega_n \in \mathbb{Z}_p[[T]] \cong \Lambda$ are defined by

$$\omega_n = (T+1)^{p^n} - 1 = T^{p^n} + \sum_{i=1}^{p^n-1} \binom{p^n}{i} T^{p^n-i}\,.$$

(11.1.5) Lemma.

(i) *For $n \ge n_0(k_\infty|k)$, the extensions k_∞ and L_n are linearly disjoint. In particular, $G(H_n|k_n) \cong G(H_nk_\infty|k_\infty)$.*

(ii) *For $n \ge n_0(k_\infty|k)$,*

$$\begin{aligned} G(H|H_nk_\infty) &= \langle \omega_n G(H|k_\infty), g_2, \dots, g_{s_\infty}\rangle, \\ G(H|H_{n+1}k_\infty) &= \tfrac{\omega_{n+1}}{\omega_n} G(H|H_nk_\infty), \\ Cl(k_n)(p) &\cong G(H_n|k_n) \\ &\cong G(H|k_\infty)/\tfrac{\omega_n}{\omega_{n_0}} G(H|H_{n_0}k_\infty). \end{aligned}$$

(iii) *Suppose that* $s_\infty = 1$. *Then for* $n \geq n_0(k_\infty|k)$,

$$G(H|k_\infty)_{\Gamma_n} \xrightarrow{\sim} G(H_n|k_n).$$

All statements hold analogously for H' *and* H'_n *if we replace* $Cl(k_n)(p)$ *by* $Cl_{S_p}(k_n)(p)$, $n_0(k_\infty|k)$ *by the finite number*

$$m_0(k_\infty|k) = \max\{m_\mathfrak{p} | \, \mathfrak{p} \in S_p^{fd}(k)\},$$

where $m_\mathfrak{p}$ *is defined by* $G_\mathfrak{p}(k_\infty|k) = \Gamma_{m_\mathfrak{p}}$ *and*

$$S_p^{fd}(k) = \{\mathfrak{p} \in S_p(k) \,|\, \mathfrak{p} \text{ is finitely decomposed in } k_\infty|k\},$$

and s_∞ *by* $t_\infty = \#S_p^{fd}(k_\infty)$, *where*

$$S_p^{fd}(k_\infty) = \{\mathfrak{P} \in S_p(k_\infty) \,|\, \mathfrak{P} \cap k \text{ is finitely decomposed in } k_\infty|k\}.$$

Proof: (i) follows from the fact that for $n \geq n_0$ the extension $k_\infty|k_n$ is totally ramified for some prime $\mathfrak{p}|p$, and (iii) is a direct consequence of (i) and the first assertion of (ii):

$$\begin{aligned} G(H|k_\infty)_{\Gamma_n} &= G(H|k_\infty)/\omega_n G(H|k_\infty) \\ &= G(H|k_\infty)/G(H|H_n k_\infty) \\ &\cong G(H_n|k_n). \end{aligned}$$

In order to prove (ii), observe that by (i) and the fact that H_n is the maximal abelian unramified extension of k_n inside H, we have

$$\begin{aligned} G(H|k_\infty)/G(H|H_n k_\infty) &\cong G(H_n|k_n) \\ &= G(H|k_n)/\langle [G(H|k_n), G(H|k_n)], T_{\mathfrak{P}_1}, \dots, T_{\mathfrak{P}_{s_\infty}} \rangle \\ &= G(H|k_n)/\langle \omega_n G(H|k_\infty), g_2, \dots, g_{s_\infty}, \tau_1 \rangle \\ &\cong G(H|k_\infty)/\langle \omega_n G(H|k_\infty), g_2, \dots, g_{s_\infty} \rangle. \end{aligned}$$

Thus we have proved the first assertion of (ii). Similarly, we obtain

$$G(H|H_{n+1}k_\infty) = \langle \omega_{n+1} G(H|k_\infty), g'_2, \dots, g'_{s_\infty} \rangle,$$

where $g'_i \in G(H|k_\infty)$ such that $\tau_i^p = g'_i \tau_1^p$, $i \geq 2$. It follows that

$$\tau_i^p = (g_i \tau_1)^p = g_i^{1+\tau_1+\cdots+\tau_1^{p-1}} \tau_1^p$$

and therefore

$$\tau_i^p = (\gamma_n g_i) \tau_1^p,$$

where

$$\gamma_n = 1 + \gamma^{p^n} + \gamma^{p^n 2} + \cdots + \gamma^{p^n(p-1)} = \frac{\omega_{n+1}}{\omega_n}.$$

We obtain

$$\begin{aligned} G(H|H_{n+1}k_\infty) &= \langle \omega_{n+1}G(H|k_\infty), \tfrac{\omega_{n+1}}{\omega_n} g_2, \dots, \tfrac{\omega_{n+1}}{\omega_n} g_{s_\infty} \rangle \\ &= \tfrac{\omega_{n+1}}{\omega_n} \langle \omega_n G(H|k_\infty), g_2, \dots, g_{s_\infty} \rangle \\ &= \tfrac{\omega_{n+1}}{\omega_n} G(H|H_n k_\infty). \end{aligned}$$

Finally, by class field theory, we have

$$\begin{aligned} Cl(k_n)(p) &\cong G(H_n|k_n) \\ &\cong G(H_n k_\infty|k_\infty) \\ &\cong G(H|k_\infty)/G(H|H_n k_\infty) \\ &= G(H|k_\infty)/\tfrac{\omega_n}{\omega_{n_0}} G(H|H_{n_0} k_\infty). \end{aligned}$$

The arguments for the extensions H' and H'_n are exactly the same except that one has to consider the decomposition groups

$$G_{\mathfrak{P}_i}(H'|k_n) \cong G_{\mathfrak{p}_i}(k_\infty|k_n)$$

instead of $T_{\mathfrak{P}_i}$ for $\mathfrak{P}_i \in S_p^{fd}(k_\infty)$. □

Now we consider the p-parts of the ideal class groups $Cl(k_n)$ of k_n. The theorem of Iwasawa gives an asymptotic formula for

$$p^{e_n} := \#Cl(k_n)(p).$$

(11.1.6) Theorem (Iwasawa). *Let $k_\infty|k$ be a $\mathbb{Z}_p$-extension. Then there exist integers $\lambda = \lambda(k_\infty|k) \geq 0$, $\mu = \mu(k_\infty|k) \geq 0$ and $\nu = \nu(k_\infty|k)$, all independent of n, such that*

$$e_n = \lambda n + \mu p^n + \nu \qquad \text{for all } n \text{ large enough.}$$

Proof: By (11.1.5), we have

$$Cl(k_n)(p) \cong G(H|k_\infty)/\tfrac{\omega_n}{\omega_{n_0}} G(H|H_{n_0} k_\infty)$$

for $n \geq n_0$, so that

$$\begin{aligned} &\#Cl(k_n)(p) \\ &= \#(G(H|k_\infty)/G(H|H_{n_0}k_\infty)) \cdot \#(G(H|H_{n_0}k_\infty)/\tfrac{\omega_n}{\omega_{n_0}} G(H|H_{n_0}k_\infty)) \\ &= \#G(H_{n_0}|k_{n_0}) \cdot \#(G(H|H_{n_0}k_\infty)/\tfrac{\omega_n}{\omega_{n_0}} G(H|H_{n_0}k_\infty)). \end{aligned}$$

Now for n large enough, (5.3.17) implies the result, since the first factor in the equality above is finite of an order independent of n. □

Remarks: 1. With exactly the same arguments one can also prove such an asymptotic formula for the p-parts of the S_p-ideal class groups $Cl_{S_p}(k_n)$, since $Cl_{S_p}(k_n)(p) \cong G(H'_n|k_n)$.

2. For the invariants λ and μ we obviously have

$$\mu(k_\infty|k_n) = p^n \mu(k_\infty|k),$$
$$\lambda(k_\infty|k_n) = \lambda(k_\infty|k).$$

3. If one considers the analogous situation for a function field and the $\mathbb{Z}_p$-extension given by enlarging the field of constants, then it is known that $\mu = 0$, i.e. $G(L|k_\infty)^{ab}$ is a finitely generated $\mathbb{Z}_p$-module. The same is conjectured for the cyclotomic $\mathbb{Z}_p$-extension of number fields, but is (up to now) only proven for abelian extensions k over $\mathbb{Q}$. This is the famous result of *B. Ferrero* and *L. C. Washington* [49]. Another proof, using p-adic L-functions, was given later by *W. Sinnott* [218]. For an arbitrary $\mathbb{Z}_p$-extension the assertion $\mu = 0$ is not true; it can become arbitrary large [89].

If k is totally real, there is a generalization of the Vandiver conjecture (i.e. p does not divide the class number of $\mathbb{Q}(\zeta_p)^+$) but it is in some sense weaker: for the cyclotomic $\mathbb{Z}_p$-extension k_∞ of k, the Iwasawa invariants are

$$\lambda = 0 \quad \text{and} \quad \mu = 0,$$

which means that the Λ-module $G(H|k_\infty)$ is finite. This conjecture is due to *R. Greenberg*, cf. [62], and is widely believed to be true. However, as far as we know, it is only verified in special cases. For $p = 3$ and small degrees of $k|\mathbb{Q}$, several mathematicians (e.g. [114], [233]) have carried out extensive computations verifying the Greenberg conjecture for many fields.

4. The invariants for the (cyclotomic) $\mathbb{Z}_p$-extension $\mathbb{Q}_\infty$ of $\mathbb{Q}$ are

$$\mu(\mathbb{Q}_\infty|\mathbb{Q}) = 0 = \lambda(\mathbb{Q}_\infty|\mathbb{Q})$$

since $s(\mathbb{Q}_\infty|\mathbb{Q}) = 1$ and $n_0(\mathbb{Q}_\infty|\mathbb{Q}) = 0$, and so by (11.1.5) (iii) we have

$$G(H|\mathbb{Q}_\infty)_\Gamma \cong G(H_0|\mathbb{Q}) = 0\,;$$

thus we even get $H = \mathbb{Q}_\infty$.

The following proposition is a generalization of a classical result due to *H. Weber* and *Ph. Furtwängler* for the field $k = \mathbb{Q}(\mu_p)$.

(11.1.7) Proposition. *Let $k_\infty|k$ be a $\mathbb{Z}_p$-extension in which exactly one prime is ramified. Assume this prime is totally ramified; then*

$$e_0 = 0 \quad \textit{implies} \quad e_n = 0 \ \textit{for all}\ n \ge 0\,.$$

Proof: Using (11.1.5) (iii) and the assumptions $s_\infty = 1$ and $n_0(k_\infty|k) = 0$, we get

$$G(H|k_\infty)_\Gamma \cong G(H_0|k).$$

Therefore $e_0 = 0$ implies that $H = k_\infty$ and consequently

$$G(H_n|k_n) \cong G(H|k_\infty)_{\Gamma_n} = 0. \qquad \square$$

There is a remarkable duality property between the inductive limit and the projective limit of $Cl(k_n)$ (and an analogous result holds for $Cl_{S_p}(k_n)$). In order to state the next theorem, recall that M° denotes the Λ-module M with the inverse action of Γ, cf. (5.5.12).

(11.1.8) Theorem. *Let $k_\infty|k$ be a $\mathbb{Z}_p$-extension. Then there are isomorphisms and pseudo-isomorphisms of Λ-torsion modules*

(i) $\quad \mathrm{Hom}(Cl(k_\infty), \mathbb{Q}_p/\mathbb{Z}_p) \cong E^1(G(H|H_{n_0}k_\infty)) \approx X_{nr}^\circ\,,$

(ii) $\quad \mathrm{Hom}(Cl_{S_p}(k_\infty), \mathbb{Q}_p/\mathbb{Z}_p) \cong E^1(G(H'|H'_{m_0}k_\infty)) \approx X_{cs}^\circ\,.$

Proof: We will only prove the first statement since (ii) follows in exactly the same way. By (11.1.5)(ii), we have

$$Cl(k_n)(p) \cong G(H_n|k_n) \cong G(H|k_\infty)/\nu_{n,n_0}G(H|H_{n_0}k_\infty)$$

where $\nu_{n,n_0} = \frac{\omega_n}{\omega_{n_0}}$. Since $G(H|k_\infty)/G(H|H_{n_0}k_\infty) \cong G(H_{n_0}|k_{n_0})$ is finite, it follows that

$$\begin{aligned} Cl(k_\infty)(p) = \varinjlim_n Cl(k_n)(p) &\cong \varinjlim_n G(H|k_\infty)/\nu_{n,n_0}G(H|H_{n_0}k_\infty) \\ &\cong \varinjlim_n G(H|H_{n_0}k_\infty)/\nu_{n,n_0}G(H|H_{n_0}k_\infty)\,. \end{aligned}$$

Since $G(H|H_{n_0}k_\infty)/\nu_{n,n_0}G(H|H_{n_0}k_\infty)$ is finite, the principal ideals (ν_{n,n_0}), $n \ge n_0$, are disjoint to the prime ideals of height 1 in $\mathrm{supp}(G(H|H_{n_0}k_\infty))$, and so we obtain a canonical isomorphism

$$\mathrm{Hom}(Cl(k_\infty), \mathbb{Q}_p/\mathbb{Z}_p) \cong \alpha(G(H|H_{n_0}k_\infty))\,,$$

where α denotes the Iwasawa-adjoint. Using (5.5.6), it follows that

$$\mathrm{Hom}(Cl(k_\infty), \mathbb{Q}_p/\mathbb{Z}_p) \cong E^1(G(H|H_{n_0}k_\infty))\,,$$

and by (5.5.13) we get a pseudo-isomorphism

$$E^1(G(H|H_{n_0}k_\infty)) \approx G(H|H_{n_0}k_\infty)^\circ \approx X_{nr}^\circ. \qquad \square$$

From the Hochschild-Serre spectral sequence for $n \le m \le \infty$,

$$H^i(\Gamma_n/\Gamma_m, H^j(G(k_S|k_m), \mathcal{O}_S^\times)) \Rightarrow H^{i+j}(G(k_S|k_n), \mathcal{O}_S^\times)\,,$$

we obtain the exact sequence

$$0 \longrightarrow H^1(\Gamma_n/\Gamma_m, \mathcal{O}_{k_m,S}^\times) \longrightarrow Cl_S(k_n) \longrightarrow Cl_S(k_m)^{\Gamma_n}.$$

Here S denotes a finite set of primes of k containing $\Sigma = S_p \cup S_\infty$. We are interested in the behaviour of the kernel $H^1(\Gamma_n/\Gamma_m, \mathcal{O}_{k_m,S}^\times)$ in the tower $k_\infty|k$. This kernel is the obstruction to the **capitulation** of prime ideals of $\mathcal{O}_{k_n,S}$ in the ring $\mathcal{O}_{k_m,S}$.

(11.1.9) Proposition. *Let $k_\infty|k$ be a $\mathbb{Z}_p$-extension and let $S \supseteq \Sigma$ be a finite set of primes of k. Then*

(i) *the order of* $\ker(Cl_S(k_n)(p) \longrightarrow Cl_S(k_m)(p)) = H^1(\Gamma_n/\Gamma_m, \mathcal{O}_{k_m,S}^\times)$ *is bounded independently of $n \le m \le \infty$,*

(ii) *the order of* $\ker(Cl(k_n)(p) \longrightarrow Cl(k_m)(p))$ *is bounded independently of $n \le m \le \infty$.*

Proof: We only show (ii) since the proof of (i) is similar. We may assume that $m \ge n \ge n_1$ for some fixed $n_1 \ge n_0(k_\infty|k)$. Let $Y = G(H|H_{n_0}k_\infty)$. By (11.1.5)(ii), we have a commutative diagram

$$\begin{array}{ccc} X_{nr}/\nu_{n,n_0}Y & \xrightarrow{\nu_{m,n}} & X_{nr}/\nu_{m,n_0}Y \\ \downarrow\wr & & \downarrow\wr \\ Cl(k_n)(p) & \longrightarrow & Cl(k_m)(p) \end{array}$$

where $\nu_{m,n} = \frac{\omega_m}{\omega_n}$ and $m < \infty$. Since $X_{nr} \approx Y \approx Y/T_0(Y) =: \bar{Y}$, where $T_0(Y)$ is the maximal finite Λ-submodule of Y, it is enough to show that the map

$$\bar{Y}/\nu_{n,n_1}\bar{Y} \xrightarrow{\nu_{m,n}} \bar{Y}/\nu_{m,n_1}\bar{Y}$$

has kernel whose order is bounded independently of $m \ge n \ge n_1$. But $\nu_{m,n}$ is even injective if n_1 is large enough. Indeed, as we saw in the claim contained in the proof of (5.3.17),

$$\bar{Y} \xrightarrow{\nu_{m,n}} \bar{Y}$$

is injective for $\infty > m \ge n \ge n_1$. So the commutative and exact diagram

$$\begin{array}{ccccccccc} 0 & \longrightarrow & \bar{Y} & \xrightarrow{\nu_{n,n_1}} & \bar{Y} & \longrightarrow & \bar{Y}/\nu_{n,n_1}\bar{Y} & \longrightarrow & 0 \\ & & \| & & \downarrow \nu_{m,n} & & \downarrow \nu_{m,n} & & \\ 0 & \longrightarrow & \bar{Y} & \xrightarrow{\nu_{m,n_1}} & \bar{Y} & \longrightarrow & \bar{Y}/\nu_{m,n_1}\bar{Y} & \longrightarrow & 0 \end{array}$$

implies the result for finite m. Passing to the limit, we obtain the assertion for $m = \infty$. □

§2. Iwasawa Theory for p-adic Local Fields

Let k be a finite extension of $\mathbb{Q}_\ell$ and let $k_\infty|k$ be a $\mathbb{Z}_p$-extension with Galois group $\Gamma = G(k_\infty|k)$, where p and ℓ may be equal. In this section we want to determine the Λ-module structure of the projective limit $A(k_\infty)$ of the p-completion $A(k_n)$ of the multiplicative group of k_n:

$$A(k_\infty) = \varprojlim_n A(k_n) = \varprojlim_{n,m} k_n^\times / k_n^{\times p^m}$$

where the limit $\varprojlim_n$ is taken with respect to the norm maps. By local class field theory, we have an isomorphism

$$A(k_\infty) \xrightarrow{\sim} G(k(p)|k_\infty)^{ab}.$$

In particular, $A(k_\infty)$ is a finitely generated Λ-module, since the $\mathbb{Z}_p$-module $G(k(p)|k_\infty)^{ab}_\Gamma \subseteq G(k(p)|k)^{ab}$ is finitely generated, cf. (5.3.10).

We want to consider a slightly more general situation. Let $K|k$ be a finite Galois extension of degree prime to p and let $K_\infty = Kk_\infty$. Let

$$G = G(K_\infty|k) = \Gamma \times \Delta\,,$$

where

$$\Gamma = G(K_\infty|K) \qquad \text{and} \qquad \Delta = G(K_\infty|k_\infty)\,.$$

Then $A(K_\infty) = \varprojlim_{L,m} L^\times / L^{\times p^m}$, where L runs through all finite subextensions of $K_\infty|k$, is a finitely generated $\mathbb{Z}_p[[G]]$-module.

Let $\mathscr{G} = G(\bar{k}|k)$ and $\mathscr{H} = G(\bar{k}|K_\infty)$, so that

$$1 \longrightarrow \mathscr{H} \longrightarrow \mathscr{G} \longrightarrow \Gamma \times \Delta \longrightarrow 1$$

is exact and

$$X = \mathscr{H}^{ab}(p) \cong A(K_\infty).$$

(11.2.1) Proposition. $\quad pd_{\mathbb{Z}_p[[G]]} A(K_\infty) \le 1.$

Proof: From (7.2.5) we know that $scd_p\, \mathscr{G} = 2$ and therefore the result follows from (5.6.13). □

(11.2.2) Lemma. *There exists a canonical $\mathbb{Z}_p[[G]]$-isomorphism*

$$E^1(A(K_\infty)) \cong (D_2^{(p)}(\mathscr{G})^{\mathscr{H}})^\vee = \mu(K_\infty)(p)^\vee\,.$$

Proof: Since I_G is projective by (5.2.13), we have (using the notation of V §6) an isomorphism $Y \cong X \oplus I_G$ and, in particular,

$$X \simeq D((D_2^{(p)}(\mathscr{G})^{\mathscr{H}})^\vee) = D(\mu(K_\infty)(p)^\vee)$$

by (7.2.4) and (5.6.9) (observe that $cd_p \mathscr{G} = 2$ by (7.1.8)(i) and $N^{ab}(p)$ is a finitely generated $\mathbb{Z}_p[\![\mathscr{G}]\!]$-module by (7.4.1)). Applying the functor E^1 gives us $E^1(X) \cong E^1(D(\mu(K_\infty)(p)^\vee)) = \mu(K_\infty)(p)^\vee$. □

(11.2.3) Theorem. *Let k be a finite extension of $\mathbb{Q}_\ell$ of degree $n = [k : \mathbb{Q}_\ell]$ and let $k_\infty|k$ be a $\mathbb{Z}_p$-extension. Let $K|k$ be a finite Galois extension of degree prime to p, $K_\infty = Kk_\infty$ and $G = \Gamma \times \Delta$, where $\Gamma = G(K_\infty|K) \cong \mathbb{Z}_p$ and $\Delta = G(K_\infty|k_\infty)$. Let $\mu(K_\infty)(p)$ be the group of roots of unity of p-power order in K_∞.*

(i) *Let $\ell = p$.*

(1) *If $\mu(K_\infty)(p)$ is infinite, so that $K_\infty|K$ is the cyclotomic $\mathbb{Z}_p$-extension, then*

$$A(K_\infty) \cong \mathbb{Z}_p[\![G]\!]^n \oplus \mathbb{Z}_p(1)\,.$$

(2) *If $\mu(K_\infty)(p)$ is finite, then there exists an exact sequence of $\mathbb{Z}_p[\![G]\!]$-modules*

$$0 \longrightarrow A(K_\infty) \longrightarrow \mathbb{Z}_p[\![G]\!]^n \longrightarrow \mu(K_\infty)(p) \longrightarrow 0.$$

(ii) *Let $\ell \neq p$. Then*

$$A(K_\infty) \cong \begin{cases} \mathbb{Z}_p(1)\,, & \text{if } \mu_p \subset K\,, \\ 0\,, & \text{otherwise}\,. \end{cases}$$

Proof: As we saw in the proof of (11.2.2), we have $X \simeq D(\mu(K_\infty)(p)^\vee)$. Let us assume that $\ell = p$. It follows from (5.4.9)(ii), (5.4.15) and the fact that the dualizing module at p of the group G is $D_1^{(p)} = \mathbb{Q}_p/\mathbb{Z}_p$ (since it is equal to the dualizing module of its open subgroup Γ), that

$$E^1(DX) = E^1(\mu(K_\infty)(p)^\vee) \cong \begin{cases} \mathbb{Z}_p(1)\,, & \text{if } \mu(K_\infty)(p) \text{ is infinite,} \\ 0\,, & \text{otherwise,} \end{cases}$$

$$E^2(DX) = E^2(\mu(K_\infty)(p)^\vee) \cong \begin{cases} 0\,, & \text{if } \mu(K_\infty)(p) \text{ is infinite,} \\ \mu(K_\infty)(p)\,, & \text{otherwise.} \end{cases}$$

The $\mathbb{Z}_p[\![G]\!]$-module X^{++} is projective. Indeed, this follows from (5.4.16) and the fact that X^{++} is free (hence projective) as a $\mathbb{Z}_p[\![\Gamma]\!]$-module by (5.1.9). From the exact sequence (5.4.9)(iii)

$$0 \longrightarrow E^1(DX) \longrightarrow X \longrightarrow X^{++} \longrightarrow E^2(DX) \longrightarrow 0\,,$$

we obtain

$$X_U^{++} \otimes \mathbb{Q}_p \cong X_U \otimes \mathbb{Q}_p$$

for an open normal subgroup U of G (observe that $E^1(DX)_U$ is finite). If $G/U = G(k'|k)$, then the exact sequence

$$0 \longrightarrow G(\bar{k}|K_\infty)_U^{ab}(p) \longrightarrow G(\bar{k}|k')^{ab}(p) \longrightarrow U^{ab}(p) \longrightarrow 0$$

implies that

$$\mathbb{Q}_p \oplus X_U \otimes \mathbb{Q}_p \cong \mathbb{Q}_p[G/U]^n \oplus \mathbb{Q}_p\,,$$

where we use (7.4.4)(i). It follows that

$$X_U \otimes \mathbb{Q}_p \cong \mathbb{Q}_p[G/U]^n\,,$$

and so

$$X^{++} \cong \mathbb{Z}_p[\![G]\!]^n$$

by (5.6.11). This already gives us the assertion if $\mu(K_\infty)(p)$ is finite, and if not it suffices to observe that X^{++} is projective.

The case $\ell \neq p$ is trivial, since then K_∞ is the maximal unramified p-extension of K and $G(\bar{K}|K_\infty)(p) = G(K_{tr}|K_\infty)(p)$ where K_{tr} is the maximal tamely ramified extension of K. The latter Galois group is isomorphic to $\mathbb{Z}_p(1)$ if $\mu_p \subseteq K$ and is zero otherwise; see VII §5. □

When $\ell = p$, we are also interested in the $\mathbb{Z}_p[\![G]\!]$-structure of the projective limit of the principal units

$$U^1(K_\infty) = \varprojlim_{L} U^1(L)\,,$$

where L runs through all finite subextensions of $K_\infty|k$.

(11.2.4) Theorem. *With the notation and assumptions of* (11.2.3), *suppose that* $\ell = p$. *Then the following holds.*

(i) *If* $\mu(K_\infty)(p)$ *is infinite, then*

$$U^1(K_\infty) \cong \mathbb{Z}_p[\![G]\!]^n \oplus \mathbb{Z}_p(1).$$

(ii) *If* $\mu(K_\infty)(p)$ *is finite, then* $U^1(K_\infty)$ *is a submodule of* $\mathbb{Z}_p[\![G]\!]^n$ *of finite index equal to or less than* $\#\mu(K_\infty)(p)$.

(iii) *Let* $K_\infty|K$ *be unramified, then*

$$U^1(K_\infty) \cong A(K_\infty).$$

Proof: Consider the inverse system of exact sequences over the finite extensions $K_n|K$

$$0 \longrightarrow U^1(K_n)/p^m \longrightarrow A(K_n)/p^m \longrightarrow \mathbb{Z}/p^m \longrightarrow 0\,,$$

where the transition maps in the middle and on the left are the norm maps and are multiplication by the residue degree on the value groups. Therefore

$$\varprojlim_{m,n} \mathbb{Z}/p^m = \varprojlim_{m} (\varprojlim_{n} \mathbb{Z}/p^m) = 0$$

if $K_\infty|K$ is unramified, and we have proved assertion (iii).

Assume now that $K_\infty|K$ is ramified. Then we get an exact sequence

$$0 \longrightarrow U^1(K_\infty) \longrightarrow A(K_\infty) \xrightarrow{v} \mathbb{Z}_p \longrightarrow 0$$

where v denotes the valuation. Since $pd_{\mathbb{Z}_p[\![G]\!]} A(K_\infty) \leq 1$ and $pd_{\mathbb{Z}_p[\![G]\!]} \mathbb{Z}_p \leq 1$, we see that $pd_{\mathbb{Z}_p[\![G]\!]} U^1(K_\infty) \leq 1$.

If $\mu(K_\infty)(p)$ is infinite, then by (11.2.3) $A(K_\infty) \cong \mathbb{Z}_p[\![G]\!]^n \oplus \mathbb{Z}_p(1)$. Since $\mathbb{Z}_p$ is a trivial $\mathbb{Z}_p[\![G]\!]$-module, the factor $\mathbb{Z}_p(1)$ of $A(K_\infty)$ has to be in the kernel $U^1(K_\infty)$ of v. Changing the basis of $\mathbb{Z}_p[\![G]\!]^n$, we obtain a commutative exact diagram

$$\begin{array}{ccccccccc} 0 & \longrightarrow & U^1(K_\infty) & \longrightarrow & A(K_\infty) & \longrightarrow & \mathbb{Z}_p & \longrightarrow & 0 \\ & & \downarrow & & \downarrow\wr & & \| & & \\ 0 & \longrightarrow & I_G \oplus \mathbb{Z}_p[\![G]\!]^{n-1} \oplus \mathbb{Z}_p(1) & \longrightarrow & \mathbb{Z}_p[\![G]\!]^n \oplus \mathbb{Z}_p(1) & \longrightarrow & \mathbb{Z}_p & \longrightarrow & 0, \end{array}$$

and hence an isomorphism

$$U^1(K_\infty) \cong I_G \oplus \mathbb{Z}_p[\![G]\!]^{n-1} \oplus \mathbb{Z}_p(1) \cong \mathbb{Z}_p[\![G]\!]^n \oplus \mathbb{Z}_p(1),$$

where we use (5.6.12). This proves (i).

Now we assume that $\mu(K_\infty)(p)$ is finite. Since $pd_{\mathbb{Z}_p[\![G]\!]} U^1(K_\infty) \leq 1$, we have

$$D(U^1(K_\infty)) \simeq E^1(U^1(K_\infty))$$

by (5.4.11). It follows from (11.2.3), using (5.5.8)(iv) and (5.4.17), that

$$E^1(D(U^1(K_\infty))) = E^1(E^1(U^1(K_\infty))) = T_1(U^1(K_\infty)) \subseteq T_1(A(K_\infty)) = 0\,.$$

Further, $E^1(A(K_\infty))$, which is isomorphic to $\mu(K_\infty)(p)^\vee$ by (11.2.2), surjects onto $E^1(U^1(K_\infty))$. Thus by (5.4.15)(ii) we get

$$E^2(D(U^1(K_\infty))) = E^2(E^1(U^1(K_\infty))) = E^1(U^1(K_\infty))^\vee \subseteq \mu(K_\infty)(p)\,.$$

As in the proof of theorem (11.2.3), $U^1(K_\infty)^{++}$ is $\mathbb{Z}_p[\![G]\!]$-projective, $U^1(K_\infty)^{++}_U \otimes \mathbb{Q}_p \cong U^1(K_\infty)_U \otimes \mathbb{Q}_p$ for $U \trianglelefteq G$ open, and from the exact sequence

$$0 \longrightarrow U^{ab}(p) \longrightarrow U^1(K_\infty)_U \longrightarrow A(K_\infty)_U \longrightarrow \mathbb{Z}_p \longrightarrow 0$$

it follows that

$$\begin{aligned} \mathbb{Q}_p \oplus U^1(K_\infty)_U \otimes \mathbb{Q}_p &\cong A(K_\infty)_U \otimes \mathbb{Q}_p \oplus \mathbb{Q}_p \\ &\cong \mathbb{Q}_p[G/U]^n \oplus \mathbb{Q}_p\,. \end{aligned}$$

Thus

$$U^1(K_\infty)_U \otimes \mathbb{Q}_p \cong \mathbb{Q}_p[G/U]^n\,,$$

and it follows by (5.6.11) that

$$U^1(K_\infty)^{++} \cong \mathbb{Z}_p[\![G]\!]^n.$$

Putting everything together, the exact sequence

$$0 \longrightarrow E^1(D(U^1(K_\infty))) \longrightarrow U^1(K_\infty) \longrightarrow U^1(K_\infty)^{++} \longrightarrow E^2(D(U^1(K_\infty)))$$

implies (ii). This finishes the proof of the theorem. □

Exercise: Calculate the quotient $\mathbb{Z}_p[\![G]\!]^n/U^1(K_\infty)$ in theorem (11.2.4)(ii).

§3. The Maximal Abelian p-Extension of k_∞ Unramified Outside S

Let k_∞ be a $\mathbb{Z}_p$-extension of the number field k and let S be a finite set of primes of k containing $\Sigma = S_p \cup S_\infty$. We assume throughout this section that k is totally imaginary if $p = 2$. In §1 we considered the Λ-modules

$$X_{nr} = G(L|k_\infty)^{ab},$$
$$X_{cs} = G(L'|k_\infty)^{ab},$$

where L is the maximal unramified p-extension of k_∞ and $L'|k_\infty$ its maximal subextension which is completely decomposed at p. Now we are interested in the Λ-module

$$X_S = G(k_S(p)|k_\infty)^{ab}.$$

For $S = \Sigma$ we set

$$X = G(k_\Sigma(p)|k_\infty)^{ab}.$$

(11.3.1) Proposition. *Let $S \supseteq \Sigma$ be a finite set of primes of k. Then the Λ-module X_S is finitely generated.*

Proof: This follows from (5.3.10) and (8.3.20) since

$$\dim_{\mathbb{F}_p}(X_S/p)_\Gamma = \dim_{\mathbb{F}_p} H^1(G(k_S|k), \mathbb{Z}/p\mathbb{Z}) - 1.$$ □

The Iwasawa invariants of X_S are denoted by

$$\mu_S = \mu(X_S) \qquad \text{and} \qquad \lambda_S = \lambda(X_S)$$

and we set

$$\mu = \mu(X_\Sigma) \qquad \text{and} \qquad \lambda = \lambda(X_\Sigma)\,.$$

Since we have already denoted the Iwasawa invariants of X_{nr} by μ and λ in §1, we will use the notation

$$\mu_{nr} = \mu(X_{nr})\,, \qquad \lambda_{nr} = \lambda(X_{nr})\,,$$

$$\mu_{cs} = \mu(X_{cs})\,, \qquad \lambda_{cs} = \lambda(X_{cs})\,,$$

for the rest of this chapter.

In the following, the weak Leopoldt conjecture, i.e. the vanishing of the group $H^2(G(k_\Sigma|k_\infty), \mathbb{Q}_p/\mathbb{Z}_p)$, will play an important role. The next theorem collects the results on this conjecture which we already know from (10.3.22), (10.3.25) and (10.3.23).

(11.3.2) Theorem.

(i) *$H^2(G(k_S|k_\infty), \mathbb{Q}_p/\mathbb{Z}_p) = 0$ if and only if $H^2(G(k_\Sigma|k_\infty), \mathbb{Q}_p/\mathbb{Z}_p) = 0$.*

(ii) *$H^2(G(k_\Sigma|k_\infty), \mathbb{Q}_p/\mathbb{Z}_p) = 0$ if $k_\infty|k$ is the cyclotomic $\mathbb{Z}_p$-extension.*

(iii) *If the (strong) Leopoldt conjecture holds, i.e. $H^2(G(k_\Sigma|k), \mathbb{Q}_p/\mathbb{Z}_p) = 0$, then the weak conjecture is also true.*

(iv) *The weak Leopoldt conjecture is true if and only if $pd_\Lambda X_S \leq 1$ and $\operatorname{rank}_\Lambda X_S = r_2$, where r_2 denotes the number of complex places of k. In particular, in this case X_S is a Λ-torsion module if k is totally real.*

The dependence between the weak and strong form of this conjecture is contained in the following

(11.3.3) Proposition. *Let $k_\infty = \bigcup k_n$ be a $\mathbb{Z}_p$-extension of k.*

(i) *The (strong) Leopoldt conjecture holds for k if and only if the weak form is satisfied and $X^\Gamma = 0$.*

(ii) *The following assertions are equivalent:*

(1) *The Leopoldt conjecture is true for all layers k_n.*

(2) *There exists some $n_1 \geq \lambda$ such that the Leopoldt conjecture holds for k_{n_1}.*

Proof: (i) follows from the exact sequence (which is obtained from the Hochschild-Serre spectral sequence)

$$0 \to H^1(\Gamma_n, X^\vee) \to H^2(G(k_\Sigma|k_n), \mathbb{Q}_p/\mathbb{Z}_p) \to H^2(G(k_\Sigma|k_\infty), \mathbb{Q}_p/\mathbb{Z}_p)^{\Gamma_n} \to 0$$

for $n = 0$. In order to prove the nontrivial implication (2) $\Rightarrow$ (1) of (ii), we first observe that the weak Leopoldt conjecture is true and $X^{\Gamma_{n_1}} = 0$ by (i). Thus the set of prime divisors of $\omega_{n_1}\Lambda$ is disjoint from the set of prime ideals of height 1 in $\mathrm{supp}(\mathrm{tor}_\Lambda X)$. Since

$$\omega_n = \xi_0 \cdot \xi_1 \cdots \xi_n$$

where

$$\xi_0 = \omega_0 = T\,, \qquad \xi_k = \sum_{i=0}^{p-1} (1+T)^{ip^{k-1}} \quad \text{for } k \geq 1,$$

are the irreducible cyclotomic polynomials, it follows that $(\xi_n) \notin P(X)$ if $n \leq n_1$. But the same holds for $n > n_1$ because

$$\deg(\xi_n) = (p-1)p^{n-1} \geq n > n_1 \geq \lambda = \deg(F_X)\,,$$

where F_X is the characteristic polynomial of $T_\Lambda(X)$. Thus X^{Γ_n} is finite, hence zero, for all n (observe that X^{Γ_n} is $\mathbb{Z}_p$-free by (5.3.19)(i)). Since the weak Leopoldt conjecture holds, the result follows. □

Recall the notation S^f for the subset of finite primes in S. In order to consider the relation between X_S and X we introduce the following

(11.3.4) Definition. *Let $k_\infty|k$ be a $\mathbb{Z}_p$-extension and let S be a set of primes of k. Then*

$S^{fd}(k) = \{\mathfrak{p} \in S^f(k) \mid \mathfrak{p}$ *is finitely decomposed in* $k_\infty|k\}$,

$S^{cd}(k) = \{\mathfrak{p} \in S^f(k) \mid \mathfrak{p}$ *is completely decomposed in* $k_\infty|k\}$.

Furthermore, in this section we will use the following

Notation: If G is a profinite group, H a closed subgroup of G and M a compact $\mathbb{Z}_p[\![H]\!]$-module, then

$$\mathrm{Ind}_G^H M := M \hat{\otimes}_{\mathbb{Z}_p[\![H]\!]} \mathbb{Z}_p[\![G]\!]$$

denotes the **compact induction** of M from H to G. The adjunction between $\hat{\otimes}$ and Hom implies the following compatibility formula, relating the compact induction to the discrete (co)induction defined in I §6

$$(\mathrm{Ind}_G^H M)^\vee = \mathrm{Ind}_G^H (M^\vee).$$

This notation may lead to confusion if M is finite, i.e. compact *and* discrete, at least if H is of infinite index in G. However, it should always be clear from the context which induction is meant. In this section we will only use the compact induction which, unraveling the definition of the complete tensor product, satisfies the formula

$$\mathrm{Ind}_G^H M = \varprojlim_N \mathrm{Ind}_{G/N}^{HN/N} M/I_{N\cap H}M\,,$$

where N runs through all open normal subgroups of G.

(11.3.5) Theorem. *Assume that the weak Leopoldt conjecture holds for the $\mathbb{Z}_p$-extension $k_\infty|k$ and let $S \supseteq \Sigma$ be finite. Then there exists a canonical exact sequence of Λ-modules*

$$0 \longrightarrow \bigoplus_{\mathfrak{p}\in S\backslash\Sigma(k)} \mathrm{Ind}_\Gamma^{\Gamma_\mathfrak{p}}(T(k_\mathfrak{p}(p)|k_\mathfrak{p})_{G_{k_{\infty,\mathfrak{p}}}}) \longrightarrow X_S \longrightarrow X \longrightarrow 0.$$

In particular, there is an exact sequence of Λ-torsion modules

$$0 \longrightarrow \bigoplus_{\mathfrak{p}\in S\backslash\Sigma(k)} \mathrm{Ind}_\Gamma^{\Gamma_\mathfrak{p}}(T(k_\mathfrak{p}(p)|k_\mathfrak{p})_{G_{k_{\infty,\mathfrak{p}}}}) \longrightarrow T_1(X_S) \longrightarrow T_1(X) \longrightarrow 0.$$

Proof: Since $H^2(G(k_\Sigma(p)|k_\infty), \mathbb{Q}_p/\mathbb{Z}_p) = 0$, we have an exact sequence

$$0 \longrightarrow H^1(k_\Sigma(p)|k_\infty) \longrightarrow H^1(k_S(p)|k_\infty) \longrightarrow H^1(k_S(p)|k_\Sigma(p))^{G_\Sigma(k_\infty)} \longrightarrow 0$$

with coefficients in $\mathbb{Q}_p/\mathbb{Z}_p$ or, dually, using (10.5.4)

$$0 \longrightarrow \varprojlim_n \prod_{\mathfrak{p}\in S\backslash\Sigma(k_n)} T(k_\mathfrak{p}(p)|k_\mathfrak{p})_{G_{k_{n,\mathfrak{p}}}} \longrightarrow X_S \longrightarrow X \longrightarrow 0.$$

Suppose $\mu_p \subseteq k_\mathfrak{p}$ for $\mathfrak{p} \in S\backslash\Sigma$ (otherwise $T(k_\mathfrak{p}(p)|k_\mathfrak{p}) = 0$). Then $T(k_\mathfrak{p}(p)|k_\mathfrak{p}) = T(k_\mathfrak{p}(p)|k_\mathfrak{p})_{G_{k_{\infty,\mathfrak{p}}}}$ is isomorphic to $\mathbb{Z}_p(1)$ if $\mathfrak{p}$ is finitely decomposed in $k_\infty|k$, and otherwise $T(k_\mathfrak{p}(p)|k_\mathfrak{p})_{G_{k_{\infty,\mathfrak{p}}}} = \mu(k_{\infty,\mathfrak{p}})(p) = \mu(k_\mathfrak{p})(p)$ is finite. Therefore

$$\varprojlim_n \prod_{\mathfrak{p}\in S\backslash\Sigma(k_n)} T(k_\mathfrak{p}(p)|k_\mathfrak{p})_{G_{k_{n,\mathfrak{p}}}} \cong \bigoplus_{\substack{\mathfrak{p}\in(S\backslash\Sigma)^{fd}(k)\\ \mu_p\subseteq k_\mathfrak{p}}} \Lambda/\omega_{r_\mathfrak{p}}(1) \oplus \bigoplus_{\mathfrak{p}\in(S\backslash\Sigma)^{cd}(k)} \Lambda/p^{t_\mathfrak{p}}$$

where $p^{r_\mathfrak{p}} = [\Gamma : \Gamma_\mathfrak{p}]$ and $p^{t_\mathfrak{p}} = \#\mu(k_\mathfrak{p})(p)$ (cf. (11.3.4) for the definition of $(S\backslash\Sigma)^{fd}$ and $(S\backslash\Sigma)^{cd}$). In particular, this module is Λ-torsion and therefore the second statement follows from the first. □

(11.3.6) Corollary. *Let $k_\infty|k$ be a $\mathbb{Z}_p$-extension for which the weak Leopoldt conjecture is true and let $S \supseteq \Sigma$ be finite. Then*

(i) $\mu_S = \mu + \sum_{\mathfrak{p}\in(S\backslash\Sigma)^{cd}(k)} t_\mathfrak{p}$, *where* $p^{t_\mathfrak{p}} = \#\mu(k_\mathfrak{p})(p)$,

(ii) $\lambda_S = \lambda + \#\{\mathfrak{p} \in (S\backslash\Sigma)^{fd}(k_\infty) \mid \mu_p \subseteq k_\mathfrak{p}\}$.

The μ-invariant not only influences the abelian pro-p-group X_S but its vanishing implies an important property of the whole Galois group $G(k_S(p)|k_\infty)$.

(11.3.7) Theorem. *Assume the weak Leopoldt conjecture holds for the $\mathbb{Z}_p$-extension $k_\infty|k$. Then $G(k_S(p)|k_\infty)$ is a free pro-p-group if and only if $\mu_S = 0$.*

In particular, if k_∞ is the cyclotomic $\mathbb{Z}_p$-extension of k, then $G(k_S(p)|k_\infty)$ is free if and only if $\mu_S = 0$.

Proof: Using the assumption and (10.4.8), we have

$$H^2(k_S(p)|k_\infty, \mathbb{Q}_p/\mathbb{Z}_p) = H^2(k_S|k_\infty, \mathbb{Q}_p/\mathbb{Z}_p) = 0\,.$$

Furthermore, since $cd_p\, G(k_S(p)|k) \leq 2$ (recall that k is totally imaginary if $p = 2$), the group $H^2(k_S(p)|k, \mathbb{Q}_p/\mathbb{Z}_p)$ is p-divisible. Now the desired result follows from (5.6.18). □

We are interested in how properties of X_S are affected by change of base field. This is no longer an abelian question. The next theorem shows that the validity of the weak Leopoldt conjecture and the vanishing of the μ-invariant are properties which have to be considered simultaneously. From (5.6.19) follows the

(11.3.8) Theorem. *Let $K|k$ be a finite Galois p-extension inside k_S, $k_\infty|k$ be a $\mathbb{Z}_p$-extension and $K_\infty = Kk_\infty$. Then*

$$\left\{ \begin{array}{c} \mu_S(K_\infty|K) = 0 \;\; \textit{and} \\ H^2(G(K_S|K_\infty), \mathbb{Q}_p/\mathbb{Z}_p) = 0 \end{array} \right\} \iff \left\{ \begin{array}{c} \mu_S(k_\infty|k) = 0 \;\; \textit{and} \\ H^2(G(k_S|k_\infty), \mathbb{Q}_p/\mathbb{Z}_p) = 0 \end{array} \right\}.$$

In particular, if $k_\infty|k$ is the cyclotomic $\mathbb{Z}_p$-extension, then

$$\mu(K_\infty|K) = 0 \iff \mu(k_\infty|k) = 0.$$

Proof: The last assertion follows from the first, since the weak Leopoldt conjecture holds true for the cyclotomic $\mathbb{Z}_p$-extension and $\mu_S = \mu$ by (11.3.6)(i). □

Now we combine the results of the local and global theory. Let us recall some notation: let $K|k$ be a finite Galois extension of degree prime to p, $k_\infty|k$ a $\mathbb{Z}_p$-extension and $K_\infty = Kk_\infty$. Let $G = G(K_\infty|k)$ and let $S \supseteq \Sigma$ be a finite set of primes of k large enough so that $K \subseteq k_S$. Let $G_S = G(K_S(p)|k)$ and $H_S = G(K_S(p)|K_\infty)$, so that there is an exact sequence

$$1 \longrightarrow H_S \longrightarrow G_S \longrightarrow G \longrightarrow 1.$$

Furthermore, let

$$1 \longrightarrow N \longrightarrow F_d \longrightarrow G_S \longrightarrow 1$$

be a presentation of G_S with a free profinite group F_d of finite rank d. We obtain a commutative exact diagram

$$(*) \qquad \begin{array}{ccccccccc} & & N & = & N & & & & \\ & & \downarrow & & \downarrow & & & & \\ 1 & \longrightarrow & R & \longrightarrow & F_d & \longrightarrow & G & \longrightarrow & 1 \\ & & \downarrow & & \downarrow & & \| & & \\ 1 & \longrightarrow & H_S & \longrightarrow & G_S & \longrightarrow & G & \longrightarrow & 1. \end{array}$$

Finally, let

$$A_S = A_S(K_\infty) = \varprojlim_n \prod_{\mathfrak{p} \in S^f(K_n)} A_\mathfrak{p}(K_n), \qquad A_\mathfrak{p}(K_n) = \varprojlim_m K_{n,\mathfrak{p}}^\times / K_{n,\mathfrak{p}}^{\times\, p^m},$$

$$U_S = U_S(K_\infty) = \varprojlim_n \prod_{\mathfrak{p} \in S^f(K_n)} U_\mathfrak{p}(K_n), \qquad U_\mathfrak{p}(K_n) = \varprojlim_m U_{K_{n,\mathfrak{p}}} / U_{K_{n,\mathfrak{p}}}^{p^m},$$

$$E_S = E_{K_\infty,S} = \varprojlim_n (\mathcal{O}_{K_n,S}^\times \otimes \mathbb{Z}_p),$$

$$E = E_{K_\infty} = \varprojlim_n (\mathcal{O}_{K_n}^\times \otimes \mathbb{Z}_p),$$

where the projective limits are taken with respect to the norm maps and the canonical projections. Analogously to X_{cs} we define

$$X_{cs}^S = G(L^S|K_\infty)^{ab},$$

where L^S is the maximal unramified p-extension of K_∞ which is completely decomposed at every prime above S (so $L^S = (K_\infty)_{S_\infty,S}(p)$ with the notation of (10.4.1)).

(11.3.9) Lemma. $\quad pd_{\mathbb{Z}_p[\![G]\!]} A_S \leq 1$ *and* $pd_{\mathbb{Z}_p[\![G]\!]} U_S \leq 1$.

Proof: Since A_S is the direct product of $A_{S^{fd}}$ and $A_{S^{cd}}$, where S^{fd} and S^{cd} are the subsets of finite primes of S which are finitely and completely decomposed in $k_\infty|k$ respectively, we consider these factors separately. The assertion for $A_{S^{fd}}$ follows from (11.2.1) and (5.4.17). The second factor

$$A_{S^{cd}} = \prod_{\mathfrak{p} \in S^{cd}} \mathrm{Ind}_G^{G_\mathfrak{P}} A_\mathfrak{P}$$

decomposes as a Λ-module into the direct sum of a free Λ-module and a "μ-part", i.e. an elementary Λ-module which is $\mathbb{Z}_p$-torsion. In particular, $A_{S^{cd}}$ has projective Λ-dimension less than or equal to 1 by (5.3.19)(i). Hence the same holds as a $\mathbb{Z}_p[\![G]\!]$-module by (5.4.17). The result for U_S follows analogously using (11.2.4). □

The following theorem generalizes results from [97].

(11.3.10) Theorem. *With the notation as above the following holds:*

(i) *There is a commutative exact diagram of $\mathbb{Z}_p[\![G]\!]$-modules*

$$\begin{array}{ccccccccccccc} 0 & \longrightarrow & H_2(H_S,\mathbb{Z}_p) & \longrightarrow & E & \longrightarrow & U_S & \longrightarrow & X_S & \longrightarrow & X_{nr} & \longrightarrow & 0 \\ & & \| & & \downarrow & & \downarrow & & \| & & \downarrow & & \\ 0 & \longrightarrow & H_2(H_S,\mathbb{Z}_p) & \longrightarrow & E_S & \longrightarrow & A_S & \longrightarrow & X_S & \longrightarrow & X_{cs}^S & \longrightarrow & 0. \end{array}$$

(ii) *There is a canonical exact sequence*

$$0 \longrightarrow E \longrightarrow E_S \longrightarrow \bigoplus_{\mathfrak{p}\in S^{cd}\cup S^r} \mathrm{Ind}_G^{G_\mathfrak{p}} \mathbb{Z}_p \longrightarrow X_{nr} \longrightarrow X_{cs}^S \longrightarrow 0\,,$$

where $S^{cd} = \{\mathfrak{p} \in S^f(k) \mid \mathfrak{p}$ is completely decomposed in $k_\infty|k\}$ and $S^r = \{\mathfrak{p} \in S^f(k) \mid \mathfrak{p}$ is ramified in $k_\infty|k\}$.

(iii) *$N^{ab}(p)$ is a finitely generated projective $\mathbb{Z}_p[\![G_S]\!]$-module and*

$$N_{H_S}^{ab}(p) \cong \bigoplus_{\mathfrak{p}\in S'_\infty} \mathrm{Ind}_G^{G_\mathfrak{p}} \mathbb{Z}_p \oplus \mathbb{Z}_p[\![G]\!]^{d-r_2-r'_1-1}\,,$$

where S'_∞ is the set of all real places of k becoming complex in K_∞, $r'_1 = \#S'_\infty$, and d is chosen greater than or equal to $r_2 + r'_1 + 1$.

(iv) *Let $Z_S := (D_2^{(p)}(G_S)^{H_S})^\vee$. Then*

$$X_S \simeq DZ_S\,.$$

If $H_2(H_S,\mathbb{Z}_p) = 0$, then there exists a canonical $\mathbb{Z}_p[\![G]\!]$-isomorphism

$$E^1(X_S) \cong Z_S\,.$$

Proof: (i) follows from (10.3.13) with $T = \varnothing$ by passing to the projective limit, and (ii) then follows by the snake lemma.

The $\mathbb{Z}_p[\![G_S]\!]$-module $N^{ab}(p)$ is projective by (5.6.7). Considering $N^{ab}(p)$ as a $\mathbb{Z}_p[\![G(K_S(p)|K)]\!]$-module, we see by (5.6.13) that this module is finitely generated, since $H^2(G(K_S(p)|K), \mathbb{Z}/p\mathbb{Z})$ is finite, cf. (8.3.20) and (10.4.8). Thus $N^{ab}(p)$ is also finitely generated as a $\mathbb{Z}_p[\![G_S]\!]$-module. The structure of $N^{ab}_{H_S}(p)$ as a $\mathbb{Z}_p[\![G]\!]$-module can be seen from (5.6.11): First observe that this module is homotopic to the right-hand side of the isomorphism stated in (iii) (both modules are projective). In order to prove assumption (ii)$'$ of (5.6.11), let $\bar{G} = G/\Gamma_n = G(K_n|k)$. Then from the diagram $(*)$ we obtain the commutative exact diagram

$$\begin{array}{ccccccccc}
 & & N & = & N & & & & \\
 & & \big\downarrow & & \big\downarrow & & & & \\
1 & \longrightarrow & M(n) & \longrightarrow & F_d & \longrightarrow & \overline{G} & \longrightarrow & 1 \\
 & & \big\downarrow & & \big\downarrow & & \| & & \\
1 & \longrightarrow & G_S(K_n) & \longrightarrow & G_S & \longrightarrow & \overline{G} & \longrightarrow & 1
\end{array}$$

where $M(n)$ is a profinite group defined by the exactness of the upper row. In the Grothendieck group $K_0'(\mathbb{Q}_p[\overline{G}])$, which we denote here by $K_0(\mathbb{Q}_p[\overline{G}])$ since every finitely generated $\mathbb{Q}_p[\overline{G}]$-module is projective, we obtain using (5.6.6),

$$\begin{aligned}
[N^{ab}_{G_S(K_n)}(p) \otimes \mathbb{Q}_p] &= [M(n)^{ab}(p) \otimes \mathbb{Q}_p] + [H_2(G_S(K_n), \mathbb{Z}_p) \otimes \mathbb{Q}_p] \\
&\quad -[G_S(K_n)^{ab}(p) \otimes \mathbb{Q}_p] \\
&= [\mathbb{Q}_p[\overline{G}]^{d-1}] + [\mathbb{Q}_p] + [H_2(G_S(K_n), \mathbb{Z}_p) \otimes \mathbb{Q}_p] \\
&\quad -[G_S(K_n)^{ab}(p) \otimes \mathbb{Q}_p] .
\end{aligned}$$

From (10.3.13), Dirichlet's unit theorem (8.7.3) and (7.4.4)(i), we then get

$$\begin{aligned}
[N^{ab}_{G_S(K_n)}(p) \otimes \mathbb{Q}_p] &= [\mathbb{Q}_p[\overline{G}]^{d-1}] + [\mathbb{Q}_p] - [U_{K_n,S} \otimes \mathbb{Q}_p] + [E_{K_n} \otimes \mathbb{Q}_p] \\
&= [\mathbb{Q}_p[\overline{G}]^{d-1}] + [\mathbb{Q}_p] - [\mathbb{Q}_p[\overline{G}]^{2r_2+r_1}] \\
&\quad +[\bigoplus_{\mathfrak{p}\in S'_\infty} \mathrm{Ind}_{\overline{G}}^{\overline{G}_\mathfrak{p}} \mathbb{Q}_p] + [\mathbb{Q}_p[\overline{G}]^{r_2+r_1-r_1'}] - [\mathbb{Q}_p] \\
&= [\mathbb{Q}_p[\overline{G}]^{d-r_2-r_1'-1}] + [\bigoplus_{\mathfrak{p}\in S'_\infty} \mathrm{Ind}_{\overline{G}}^{\overline{G}_\mathfrak{p}} \mathbb{Q}_p].
\end{aligned}$$

The proof of (iv) is analogous to (11.2.2). Since $cd_p\, G \le 1$, the augmentation ideal I_G is projective and therefore $Y \cong X_S \oplus I_G$, so that

$$X_S \simeq DZ_S$$

by (5.6.9) (noting (iii)). Since $H_2(H_S, \mathbb{Z}_p) = 0$, it follows that $E^1(X_S) \cong Z_S$, again by (5.6.9). $\square$

(11.3.11) Theorem. *With the notation as above, suppose $H^2(H_S, \mathbb{Q}_p/\mathbb{Z}_p) = 0$. Then the following are true:*

(i) $pd_{\mathbb{Z}_p[\![G]\!]} X_S \le 1$ *and*

$$E^0(X_S) \cong \mathbb{Z}_p[\![G]\!]^{r_2} \oplus \bigoplus_{\mathfrak{p} \in S'_\infty} \mathrm{Ind}_G^{G_\mathfrak{p}} \mathbb{Z}_p^- ,$$

where $\mathbb{Z}_p^-$ is the $G_\mathfrak{p}$-module $\mathbb{Z}_p$ on which the generator of $G_\mathfrak{p} \cong \mathbb{Z}/2\mathbb{Z}$ acts as multiplication by -1.

(ii) $pd_{\mathbb{Z}_p[\![G]\!]} E_S \le 1$ *and there is an exact sequence of $\mathbb{Z}_p[\![G]\!]$-modules*

$$0 \to E_S \to \bigoplus_{\mathfrak{p} \in S^{cd} \cup S'_\infty} \mathrm{Ind}_G^{G_\mathfrak{p}} \mathbb{Z}_p \oplus \mathbb{Z}_p[\![G]\!]^{r_2 + r_1 - r'_1} \to \mu_{p^\infty}(K_\infty) \to 0$$

if $\mu_{p^\infty}(K_\infty)$ is finite, and an isomorphism

$$E_S \cong \bigoplus_{\mathfrak{p} \in S'_\infty} \mathrm{Ind}_G^{G_\mathfrak{p}} \mathbb{Z}_p \oplus \mathbb{Z}_p[\![G]\!]^{r_2 + r_1 - r'_1} \oplus \mathbb{Z}_p(1)$$

if $\mu_{p^\infty}(K_\infty)$ is infinite.

(iii) $pd_{\mathbb{Z}_p[\![G]\!]} E \le 1$ *and there is an exact sequence of $\mathbb{Z}_p[\![G]\!]$-modules*

$$0 \longrightarrow E \longrightarrow \bigoplus_{\mathfrak{p} \in S'_\infty} \mathrm{Ind}_G^{G_\mathfrak{p}} \mathbb{Z}_p \oplus \mathbb{Z}_p[\![G]\!]^{r_2 + r_1 - r'_1} \longrightarrow \mu_{p^\infty}(K_\infty)$$

if $\mu_{p^\infty}(K_\infty)$ is finite, and an isomorphism

$$E \cong \bigoplus_{\mathfrak{p} \in S'_\infty} \mathrm{Ind}_G^{G_\mathfrak{p}} \mathbb{Z}_p \oplus \mathbb{Z}_p[\![G]\!]^{r_2 + r_1 - r'_1} \oplus \mathbb{Z}_p(1)$$

if $\mu_{p^\infty}(K_\infty)$ is infinite.

Proof: (i) By (5.6.12), the augmentation ideal I_G is a free $\mathbb{Z}_p[\![G]\!]$-module of rank 1. Therefore, using (5.6.7), we obtain an exact sequence

$$0 \longrightarrow N_{H_S}^{ab}(p) \longrightarrow \mathbb{Z}_p[\![G]\!]^d \longrightarrow X_S \oplus I_G \longrightarrow 0 ,$$

which shows that $pd_{\mathbb{Z}_p[\![G]\!]} X_S \le 1$ since $N_{H_S}^{ab}(p)$ is projective. In order to prove the assertion concerning $E^0(X_S)$, we proceed as in the proof of part (iii) of theorem (11.3.10). First observe that $E^0(X_S)$ is $\mathbb{Z}_p[\![G]\!]$-projective by (5.4.16) and the fact that it is $\mathbb{Z}_p[\![\Gamma]\!]$-free by (5.5.10)(i). From the exact sequence above, we obtain the exact sequence

$$0 \longrightarrow E^0(X_S) \oplus (I_G)^+ \longrightarrow (\mathbb{Z}_p[\![G]\!]^d)^+ \longrightarrow (N_{H_S}^{ab}(p))^+ \longrightarrow E^1(X_S) \longrightarrow 0.$$

For $\Gamma_n \trianglelefteq G$ let $\bar{G} = G/\Gamma_n = G(K_n|k)$. Since $E^1(X_S)$ is a Γ-torsion module and $I_G \cong \mathbb{Z}_p[\![G]\!]$,

$$\begin{aligned} [E^0(X_S)_{\Gamma_n} \otimes \mathbb{Q}_p] &= [\mathbb{Q}_p[\bar{G}]^{d-1}] - [(N_{HS}^{ab}(p))^+_{\Gamma_n} \otimes \mathbb{Q}_p] \\ &= [\mathbb{Q}_p[\bar{G}]^{r_2 + r'_1}] - [\bigoplus_{\mathfrak{p} \in S'_\infty} \mathrm{Ind}_{\bar{G}}^{\bar{G}_\mathfrak{p}} \mathbb{Q}_p] \\ &= [\mathbb{Q}_p[\bar{G}]^{r_2}] + [\bigoplus_{\mathfrak{p} \in S'_\infty} \mathrm{Ind}_{\bar{G}}^{\bar{G}_\mathfrak{p}} \mathbb{Q}_p^-] \end{aligned}$$

in the Grothendieck group $K_0(\mathbb{Q}_p[\overline{G}])$, where we have used (11.3.10)(iii) and $\mathbb{Z}_p[\![G]\!] = \mathrm{Ind}_G^{G_\mathfrak{p}} \mathbb{Z}_p[G_\mathfrak{p}] = \mathrm{Ind}_G^{G_\mathfrak{p}} \mathbb{Z}_p \oplus \mathrm{Ind}_G^{G_\mathfrak{p}} \mathbb{Z}_p^-$ for $\mathfrak{p} \in S'_\infty$. Now we get the result from (5.6.11).

(ii) Since $pd_{\mathbb{Z}_p[\![G]\!]} A_S \leq 1$ and $pd_{\mathbb{Z}_p[\![G]\!]} U_S \leq 1$ by (11.3.9), this also holds for E_S and E (which is easily seen using (5.3.19)(i) and (5.4.17)).

Now we split the 4-term exact sequence (11.3.10)(i)

$$\begin{array}{ccccccccc} & & & & & B & & & \\ & & & & \nearrow & & \searrow & & \\ 0 \longrightarrow & E_S & \longrightarrow & A_S & & \longrightarrow & & X_S \longrightarrow X_{cs}^S \longrightarrow 0 \end{array}$$

into two short exact sequences and obtain the commutative exact diagram

$$\begin{array}{ccccccc} E^1(X_{cs}^S) & \longrightarrow & E^1(X_S) & \longrightarrow & E^1(B) & \longrightarrow & E^2(X_{cs}^S) \\ & & \| & & \downarrow & & \\ & & E^1(X_S) & \xrightarrow{\varphi} & E^1(A_S) & & \\ & & & & \downarrow & & \\ & & & & E^1(E_S) & & \end{array}$$

using the fact that $E^2(B)$ vanishes, being a quotient of $E^2(X_S)$. We obtain the commutative diagram with exact rows

$$\begin{array}{ccccc} & & & \nearrow\!\!\!\!\nearrow & \mathrm{coker}(\varphi) \\ E^1(X_S) & \longrightarrow & E^1(A_S) & & \downarrow \\ & & & \searrow\!\!\!\!\searrow & E^1(E_S) \\ \downarrow \wr & & \downarrow \wr & & \downarrow \\ (D_2^{(p)}(G_S)^{H_S})^\vee & \longrightarrow & \bigoplus\limits_{\mathfrak{p} \in S(k)} \mathrm{Ind}_G^{G_\mathfrak{P}} (D_2^{(p)}(G_{k_\mathfrak{p}})^{G_{K_{\infty,\mathfrak{P}}}})^\vee & \twoheadrightarrow & \mu_{p^\infty}(K_\infty)^\vee \\ \| & & \| & & \| \\ (\varprojlim\limits_{K',m} H^2(K_S|K', \mathbb{Z}/p^m))_{H_S} & \longrightarrow & (\varprojlim\limits_{K',m} \bigoplus\limits_{S(K')} H^2(K'_\mathfrak{p}, \mathbb{Z}/p^m))_{H_S} & \twoheadrightarrow & \mu_{p^\infty}(K_\infty)^\vee, \end{array}$$

where K' runs through all finite subextensions of $K_S(p)|k$. The lower exact sequence is obtained by taking the limit and taking H_S-coinvariants of the right-hand part of the Poitou-Tate long exact sequence (8.6.10). The left-hand vertical isomorphism is assertion (11.3.10)(iv) and the vertical isomorphism in the middle is obtained as follows. Let $\mathfrak{p} \in S$ be a prime such that $\mu(K_{\infty,\mathfrak{P}})$ is infinite for $\mathfrak{P}|\mathfrak{p}$. Then, by (11.2.3), (5.4.15)(i) and observing that

$$D_1^{(p)}(G) = \mathbb{Q}_p/\mathbb{Z}_p,$$

we obtain

$$E^1(\varprojlim_n \prod_{\mathfrak{P}_n|\mathfrak{p}} A_{\mathfrak{P}_n}(K_n)) = E^1(\mathrm{Ind}_G^{G_\mathfrak{P}} \mathbb{Z}_p(1)) \cong \mathrm{Ind}_G^{G_\mathfrak{P}} (\mu(K_{\infty,\mathfrak{P}})(p))^\vee .$$

If $\mathfrak{p} \in S$ is finitely decomposed in $k_\infty|k$ and $\mu(K_{\infty,\mathfrak{P}})$ is finite for $\mathfrak{P}|\mathfrak{p}$, then by (11.2.3) and (5.4.15)(ii) we obtain

$$E^1(\varprojlim_n \prod_{\mathfrak{P}_n|\mathfrak{p}} A_{\mathfrak{P}_n}(K_n)) = E^2(\mathrm{Ind}_G^{G_\mathfrak{P}} (\mu(K_{\infty,\mathfrak{P}})(p))) \cong \mathrm{Ind}_G^{G_\mathfrak{P}} (\mu(K_{\infty,\mathfrak{P}})(p))^\vee .$$

For a prime $\mathfrak{p} \in S$ which splits completely in $k_\infty|k$, we have by (5.4.13)(iii)

$$\begin{aligned} E^1(\varprojlim_n \prod_{\mathfrak{P}_n|\mathfrak{p}} A_{\mathfrak{P}_n}(K_n)) &= E^1(\mathrm{Ind}_G^{G_\mathfrak{P}} (\mu(K_\mathfrak{P})(p))) \\ &\cong D_0((\mathrm{Ind}_G^{G_\mathfrak{P}} (\mu(K_\mathfrak{P})(p)))^\vee)^\vee \\ &= \mathrm{Ind}_G^{G_\mathfrak{P}} (\mu(K_\mathfrak{P})(p))^\vee \end{aligned}$$

(observe that in this case $\mu(K_\mathfrak{P})(p) = \mu(K_{\infty,\mathfrak{P}})(p)$). Finally, we recall that $\mu(K_{\infty,\mathfrak{P}})(p) = D_2^{(p)}(G_{k_\mathfrak{p}})^{G_{K_{\infty,\mathfrak{P}}}}$ by (7.2.4).

From the diagram above, it now follows that

$$E^1(E_S) \cong \mu_{p^\infty}(K_\infty)^\vee .$$

Because $pd_{\mathbb{Z}_p[\![G]\!]} E_S \le 1$, the last isomorphism implies that $DE_S \simeq \mu_{p^\infty}(K_\infty)^\vee$ by (5.4.11), and therefore

$$T_1(E_S) = E^1(DE_S) \cong \begin{cases} \mathbb{Z}_p(1), & \text{if } \mu_{p^\infty}(K_\infty) \text{ is infinite,} \\ 0, & \text{otherwise,} \end{cases}$$

$$T_2(E_S) = E^2(DE_S) \cong \begin{cases} 0, & \text{if } \mu_{p^\infty}(K_\infty) \text{ is infinite,} \\ \mu_{p^\infty}(K_\infty), & \text{otherwise.} \end{cases}$$

In order to calculate E_S^{++}, we proceed as in (iii) of (11.3.10): E_S^{++} is $\mathbb{Z}_p[\![G]\!]$-projective (because E_S^{++} is Λ-projective) and

$$[(E_S^{++})_{\Gamma_n} \otimes \mathbb{Q}_p] = [(E_S)_{\Gamma_n} \otimes \mathbb{Q}_p]$$

in $K_0(\mathbb{Q}_p[\overline{G}])$ for $\overline{G} = G/\Gamma_n$. We have exact sequences

$$0 \longrightarrow B^{\Gamma_n} \longrightarrow (E_S)_{\Gamma_n} \longrightarrow (A_S)_{\Gamma_n} \longrightarrow B_{\Gamma_n} \longrightarrow 0$$

$$0 \to B^{\Gamma_n} \to X_S{}^{\Gamma_n} \to (X_{cs}^S)^{\Gamma_n} \longrightarrow B_{\Gamma_n} \longrightarrow (X_S)_{\Gamma_n} \to (X_{cs}^S)_{\Gamma_n} \to 0$$

(observe that $(A_S)^{\Gamma_n} = 0$ by (11.2.3)), and from the exact sequence in the proof of (i), we get an exact sequence

$$0 \longrightarrow X_S^{\Gamma_n} \longrightarrow (N_{H_S}^{ab}(p))_{\Gamma_n} \longrightarrow \mathbb{Z}_p[\overline{G}]^d \longrightarrow (X_S)_{\Gamma_n} \oplus (I_G)_{\Gamma_n} \longrightarrow 0 .$$

Thus the facts that X_{cs}^S is Λ-torsion, $I_G \cong \mathbb{Z}_p[[G]]$, (11.3.10)(iii) and (11.2.3), imply that

$$\begin{aligned}
[(E_S)_{\Gamma_n} \otimes \mathbb{Q}_p] &= [(A_S)_{\Gamma_n} \otimes \mathbb{Q}_p] + [B^{\Gamma_n} \otimes \mathbb{Q}_p] - [B_{\Gamma_n} \otimes \mathbb{Q}_p] \\
&= [(A_S)_{\Gamma_n} \otimes \mathbb{Q}_p] + [X_S^{\Gamma_n} \otimes \mathbb{Q}_p] - [(X_S)_{\Gamma_n} \otimes \mathbb{Q}_p] \\
&= [(A_S)_{\Gamma_n} \otimes \mathbb{Q}_p] + [(N_{H_S}^{ab}(p))_{\Gamma_n} \otimes \mathbb{Q}_p] - [\mathbb{Q}_p[\bar{G}]^d] \\
&\quad + [(I_G)_{\Gamma_n} \otimes \mathbb{Q}_p] \\
&= [(A_S)_{\Gamma_n} \otimes \mathbb{Q}_p] + [\bigoplus_{\mathfrak{p} \in S'_\infty} \operatorname{Ind}_{\bar{G}}^{\bar{G}_\mathfrak{p}} \mathbb{Q}_p] - [\mathbb{Q}_p[\bar{G}]^{r_2+r'_1}] \\
&= [\mathbb{Q}_p[\bar{G}]^{r_2+r_1-r'_1}] + [\bigoplus_{\mathfrak{p} \in S^{cd}} \operatorname{Ind}_{\bar{G}}^{\bar{G}_\mathfrak{p}} \mathbb{Q}_p] + [\bigoplus_{\mathfrak{p} \in S'_\infty} \operatorname{Ind}_{\bar{G}}^{\bar{G}_\mathfrak{p}} \mathbb{Q}_p] .
\end{aligned}$$

It follows that

$$E_S^{++} \cong \bigoplus_{\mathfrak{p} \in S^{cd} \cup S'_\infty} \operatorname{Ind}_G^{G_\mathfrak{p}} \mathbb{Z}_p \oplus \mathbb{Z}_p[[G]]^{r_2+r_1-r'_1}$$

and from (5.4.9) (iii), we get the result for E_S.

Assertion (iii) for E follows the same lines as that for E_S by considering U_S instead of A_S and noting that

$$[(U_S)_{\Gamma_n} \otimes \mathbb{Q}_p] = [\mathbb{Q}_p[\bar{G}]^{2r_2+r_1}]$$

and

$$T_1(E) = T_1(E_S), \qquad E^1(E_S) \twoheadrightarrow E^1(E)$$

(which follows from (11.3.10)(ii)). □

Remark: Let $S \supseteq \Sigma$ be a finite set of primes and

$$\bar{E}_S(K_\infty) = \varprojlim_n \bar{E}_{K_n,S}^{(S)} \quad \text{and} \quad \bar{E}(K_\infty) = \varprojlim_n \bar{E}_{K_n}^{(S)},$$

where $\bar{E}_{K_n,S}^{(S)}$ (resp. $\bar{E}_{K_n}^{(S)}$) is the closure of the image of $\mathcal{O}_{K_n,S}^\times$ in $A_S(K_n)$ (resp. of $\mathcal{O}_{K_n}^\times$ in $U_S(K_n)$) with respect to the idèle topology, cf. p.628. We have proved that $\bar{E}_S(K_\infty) = E_{K_\infty,S}$ (resp. $\bar{E}(K_\infty) = E_{K_\infty}$) if and only if the weak Leopoldt conjecture holds, i.e. $H^2(H_S, \mathbb{Q}_p/\mathbb{Z}_p) = 0$, cf. (10.3.24). Therefore we may replace the $\mathbb{Z}_p[[G]]$-modules E_S and E by $\bar{E}_S$ and $\bar{E}$ in (11.3.11).

(11.3.12) Corollary. *Let $K_\infty|K$ be a $\mathbb{Z}_p$-extension for which the weak Leopoldt conjecture holds and let $S \supseteq \Sigma$ be a finite set of primes of K. Then*

$$\operatorname{rank}_\Lambda \bar{E}_S = r_2 + r_1 + d_S \quad \text{and} \quad \operatorname{rank}_\Lambda \bar{E} = r_2 + r_1,$$

where d_S is the cardinality of the set S^{cd} of finite primes in S which completely decompose in $K_\infty|K$.

(11.3.13) Corollary. *Let $K = k(\mu_p)$ and $\Delta = G(K|k)$, $k_\infty|k$ the cyclotomic $\mathbb{Z}_p$-extension and $K_\infty = Kk_\infty = k(\mu_{p^\infty})$ with Galois group $G(K_\infty|k) = \Gamma \times \Delta$. Then there is an exact sequence of $\Lambda[\Delta]$-modules*

$$\begin{array}{ccccccccc} 0 \longrightarrow & X_{cs}(-1) & \longrightarrow & E^1(X_S) & \longrightarrow & E^1(A_S) & \longrightarrow & E^1(E_S) & \longrightarrow 0 \\ & \| & & \| \wr & & \| \wr & & \| \wr & \\ 0 \longrightarrow & X_{cs}(-1) & \longrightarrow & Z_S & \longrightarrow & \bigoplus_{\mathfrak{p}\in S^f} \mathrm{Ind}_{\Gamma\times\Delta}^{(\Gamma\times\Delta)_\mathfrak{p}} \mathbb{Z}_p(-1) & \longrightarrow & \mathbb{Z}_p(-1) & \longrightarrow 0. \end{array}$$

Proof: We merely have to calculate the kernel of $E^1(X_S) \to E^1(A_S)$. From the diagram in the proof of (11.3.11), we see that this is

$$\begin{aligned} \varprojlim_{n,m} Ш^2(K_n, \mathbb{Z}/p^m) &\cong \varprojlim_{n,m} Ш^1(K_n, \mu_{p^m})^* \\ &= \varprojlim_m \varprojlim_n Cl_S(K_n)/p^m(-1) \\ &= \varprojlim_m \varprojlim_n Cl_\Sigma(K_n)/p^m(-1) \\ &= X_{cs}(-1)\,. \end{aligned}$$

□

(11.3.14) Corollary. *With the assumptions of* (11.3.13)*, there is an exact sequence of $\Lambda[\Delta]$-modules*

$$0 \longrightarrow T_1(E_S) \longrightarrow T_1(A_S) \longrightarrow T_1(X_S) \longrightarrow E^1(X_{cs}(-1)) \longrightarrow 0.$$

Proof: Splitting the exact sequence in (11.3.13) into two short exact sequences

$$0 \longrightarrow X_{cs}(-1) \longrightarrow E^1(X_S) \longrightarrow C \longrightarrow 0,$$

$$0 \longrightarrow C \longrightarrow E^1(A_S) \longrightarrow E^1(E_S) \longrightarrow 0,$$

and applying the functor E^1, we obtain the exact sequence

$$0 \to E^1E^1(E_S) \to E^1E^1(A_S) \to E^1E^1(X_S) \to E^1(X_{cs}(-1)) \to E^2(C)$$

because $E^2E^1(E_S) = 0$ by (11.3.11)(ii), (5.5.3)(iv) and (5.4.17). But the projective dimensions of E_S, A_S and X_S are less than or equal to 1, so we can replace E^1E^1 by $E^1D = T_1$ (5.4.11). Furthermore, $E^2(C) = 0$, since

$$pd_{\Lambda[\Delta]}C \le pd_{\Lambda[\Delta]}E^1(A_S) \le 1.$$

□

(11.3.15) Corollary. *With the assumptions of* (11.3.13) *the following holds:*

(i) $T_0(X_S) = 0,\ T_1(X_S) = E^1(Z_S),\ \ T_2(X_S) = T_0(X_{cs}(-1))^\vee$.

(ii) $T_0(X_{cs}) \cong \varprojlim_n H^1(G(K_\infty|K_n), \mathcal{O}^\times_{K_\infty,S})$,
where the limit is taken with respect to the corestriction maps.

(iii) *There is an exact sequence of* $\Lambda[\Delta]$*-modules*

$$0 \to T_1(X_S) \to X_S \to \Lambda[\Delta]^{r_2} \oplus \bigoplus_{\mathfrak{p}\in S'_\infty} \operatorname{Ind}_\Delta^{\Delta_\mathfrak{p}} \Lambda^- \to T_0(X_{cs}(-1))^\vee \to 0,$$

where Λ^- *is the* $\Lambda[\Delta_\mathfrak{p}]$*-module* Λ *with* $\Delta_\mathfrak{p} \cong \mathbb{Z}/2\mathbb{Z}$ *acting by* -1. *In particular,* $F_\Lambda(X_S) = X_S/T_1(X_S)$ *is a free* Λ*-module if and only if* X_{cs} *contains no finite nontrivial* Λ*-submodule.*

Proof: (i) It remains only to show the last assertion, which is obtained as follows:

$$\begin{aligned} T_2(X_S) &= E^2(DX_S) = E^2(E^1(X_S)) = E^2(T_0(E^1(X_S))) \\ &= T_0(E^1(X_S))^\vee = T_0(X_{cs}(-1))^\vee \end{aligned}$$

where we have used (5.4.11), (5.5.3)(iv) (together with (5.4.17) and the remark that for a finitely generated $\mathbb{Z}_p[\![G]\!]$-module M, the $\mathbb{Z}_p[\![\Gamma]\!]$-module $T_0(M)$ is also a $\mathbb{Z}_p[\![G]\!]$-module), (5.4.15)(ii) and (11.3.13).

(ii) follows from the commutative exact diagram for $m \geq n$

$$\begin{array}{ccccccc} 0 \longrightarrow & H^1(K_\infty|K_m, \mathcal{O}^\times_{K_\infty,S}) & \longrightarrow & Cl_S(K_m) & \longrightarrow & Cl_S(K_\infty)^{\Gamma_m} \\ & \downarrow{\scriptstyle cor} & & \downarrow{\scriptstyle N} & & \downarrow{\scriptstyle N} \\ 0 \longrightarrow & H^1(K_\infty|K_n, \mathcal{O}^\times_{K_\infty,S}) & \longrightarrow & Cl_S(K_n) & \longrightarrow & Cl_S(K_\infty)^{\Gamma_n} \end{array}$$

which induces the exact sequence

$$0 \longrightarrow \varprojlim_n H^1(K_\infty|K_n, \mathcal{O}^\times_{K_\infty,S}) \longrightarrow X_{cs} \longrightarrow \varprojlim_n Cl_S(K_\infty)(p)^{\Gamma_n}.$$

Since

$$\Big(\varprojlim_n Cl_S(K_\infty)(p)^{\Gamma_n}\Big)^\Gamma = \varprojlim_n Cl_S(K_\infty)(p)^\Gamma$$

is uniquely p-divisible (the inverse limit on the right-hand side is taken via multiplication by p), we see that $\varprojlim_n Cl_S(K_\infty)(p)^\Gamma$ has no nontrivial finite Γ-submodule. Thus the same is true for $\varprojlim_n Cl_S(K_\infty)(p)^{\Gamma_n}$. Since the order of $H^1(K_\infty|K_n, \mathcal{O}^\times_{K_\infty,S})$ is bounded independently of n by (11.1.9)(i), $\varprojlim_n H^1(K_\infty|K_n, \mathcal{O}^\times_{K_\infty,S})$ is finite, and so $\varprojlim_n H^1(K_\infty|K_n, \mathcal{O}^\times_{K_\infty,S}) = T_0(X_{cs})$.

Now (iii) follows from (5.4.9)(iii) and (11.3.11)(i) since $X_S^{++} = E^0(X_S)^+$ and $\operatorname{Hom}_{\Lambda[\Delta]}(\operatorname{Ind}_\Delta^{\Delta_\mathfrak{p}} \Lambda^-, \Lambda[\Delta]) \cong \operatorname{Ind}_\Delta^{\Delta_\mathfrak{p}} \Lambda^-$. □

We easily obtain from the preceding results the following corollaries for the Iwasawa invariants of X, X_{nr} and X_{cs}.

(11.3.16) Corollary. *Let $k_\infty|k$ be a $\mathbb{Z}_p$-extension such that every prime above p is finitely decomposed in k_∞. Then*

$$\mu_{nr} = \mu_{cs}\,,$$
$$\lambda_{cs} \le \lambda_{nr} \le \lambda_{cs} + s(k_\infty|k)\,,$$

where $s(k_\infty|k)$ is the number of primes in k_∞ ramifying in $k_\infty|k$. Furthermore, there is a pseudo-isomorphism

$$G(H|H') \approx \bigoplus_i \Lambda/\xi_{n_i}$$

with some cyclotomic polynomials ξ_{n_i} satisfying $\sum_i \deg(\xi_{n_i}) \le s(k_\infty|k)$.

Proof: Recall that $\xi_n = \dfrac{\omega_n}{\omega_{n-1}} = \sum_{k=0}^{p-1}(1+T)^{k\,p^{n-1}}$ is the (irreducible) p^n-th cyclotomic polynomial in the variable $1+T$. By (11.3.10)(ii) with $S = \Sigma = S_p \cup S_\infty$, we get the exact sequence

$$\bigoplus_{\mathfrak{p}\in S^r(k)} \Lambda/\omega_{n_\mathfrak{p}} \longrightarrow X_{nr} \longrightarrow X_{cs} \longrightarrow 0,$$

where $p^{n_\mathfrak{p}} = [\Gamma : \Gamma_{n_\mathfrak{p}}]$ and $\sum_{\mathfrak{p}\in S^r(k)} \deg(\omega_{n_\mathfrak{p}}) = \#S^r(k_\infty) = s(k_\infty|k)$, which implies the result. □

(11.3.17) Corollary. *Assume that $\mu_p \subseteq k$ and let $k_\infty|k$ be the cyclotomic $\mathbb{Z}_p$-extension. Then*

$$\mu = \mu_{cs} \quad \text{and} \quad \lambda = \lambda_{cs} + \#S_p(k_\infty) - 1\,.$$

Proof: This follows from (11.3.14) since

$$\mu(T_1(E_\Sigma)) = 0\,, \quad \mu(T_1(A_\Sigma)) = 0\,,$$
$$\lambda(T_1(E_\Sigma)) = 1\,, \quad \lambda(T_1(A_\Sigma)) = \#S_p(k_\infty)\,,$$

by (11.3.11)(ii) and (11.2.3), and the λ- and μ-invariants of $E^1(X_{cs}(-1))$ and X_{cs} coincide by (5.5.13). □

We finish this section by listing all results of this section for the cyclotomic $\mathbb{Z}_p$-extension. The set $S_p \cup S_\infty$ is again denoted by Σ.

Let $K = k(\mu_p)$, $K_\infty = k(\mu_{p^\infty})$ the cyclotomic $\mathbb{Z}_p$-extension of K and $\Delta = G(K_\infty|k_\infty) \cong G(K|k)$. We denote the **$p$-part of the cyclotomic character** by

$$\kappa : G(K_\infty|k) \longrightarrow \mathrm{Aut}(\mu_{p^\infty}) \cong \mathbb{Z}_p^\times ,$$

i.e. $\zeta^\sigma = \zeta^{\kappa(\sigma)}$ for all $\zeta \in \mu_{p^\infty}$ and $\sigma \in G(K_\infty|k)$. For $j \in \mathbb{Z}$ and $\mathfrak{p}|p$ we define

$$\delta_{j,\mathfrak{p}} = \begin{cases} 1, & \kappa^j|_{\Delta_\mathfrak{p}} = 1, \\ 0, & \text{otherwise}, \end{cases} \qquad \delta_j = \begin{cases} 1, & \kappa^j|_{\Delta} = 1, \\ 0, & \text{otherwise}. \end{cases}$$

The $\Lambda[\Delta]$-module X (and analogously X_{nr}, X_{cs}) has a decomposition coming from the action of Δ:

$$X = \bigoplus_{i \bmod d} e_i X ,$$

where $d = \#\Delta$ and $\{e_i = \frac{1}{d}\sum_{\sigma\in\Delta}\kappa^{-i}(\sigma)\sigma\}_{i \bmod d}$ are the idempotents of $\mathbb{Z}_p[\Delta]$; thus e_iX is the i-th eigenspace of the Δ-module X. We set

$$\mu^{(i)} = \mu(e_iX) \quad \text{and} \quad \lambda^{(i)} = \lambda(e_iX),$$

and denote the invariants of e_iX_{nr} and e_iX_{cs} analogously.

(11.3.18) Theorem. *Let $K|k$ be a finite Galois extension of degree prime to p with Galois group Δ and assume that k is totally imaginary if $p = 2$. Let K_∞ be the cyclotomic $\mathbb{Z}_p$-extension of K. Then*

- *the weak Leopoldt conjecture holds for $K_\infty|K$,*
- $E_\Sigma \cong \Lambda[\Delta]^{r_2+r_1-r_1'} \oplus (\bigoplus_{\mathfrak{p}\in S'_\infty} \mathrm{Ind}_\Delta^{\Delta_\mathfrak{p}} \Lambda) \oplus \mathbb{Z}_p(1)$,
- $E \cong \Lambda[\Delta]^{r_2+r_1-r_1'} \oplus (\bigoplus_{\mathfrak{p}\in S'_\infty} \mathrm{Ind}_\Delta^{\Delta_\mathfrak{p}} \Lambda) \oplus \mathbb{Z}_p(1)$,
- $A_\Sigma \cong \Lambda[\Delta]^{r_1+2r_2} \oplus \bigoplus_{\mathfrak{p}\in S_p(k_\infty)} \mathbb{Z}_p(1)$,
- $U_\Sigma \cong \Lambda[\Delta]^{r_1+2r_2} \oplus \bigoplus_{\mathfrak{p}\in S_p(k_\infty)} \mathbb{Z}_p(1)$.

Assume in addition that $K = k(\mu_p)$. Then

- *X_Σ contains no finite nontrivial Λ-submodules, and there is an exact sequence of $\Lambda[\Delta]$-modules*
$$0 \longrightarrow X_\Sigma/T_1(X_\Sigma) \longrightarrow \Lambda[\Delta]^{r_2} \oplus \bigoplus_{\mathfrak{p}\in S'_\infty} \mathrm{Ind}_\Delta^{\Delta_\mathfrak{p}} \Lambda^- \longrightarrow T_0(X_{cs}(-1))^\vee \longrightarrow 0,$$
- $\mu^{(i)} = \mu_{nr}^{(1-i)} = \mu_{cs}^{(1-i)}$ *for all $i \in \mathbb{Z}$; in particular,* $\mu = \mu_{nr} = \mu_{cs}$,
- $\lambda^{(i)} = \lambda_{cs}^{(1-i)} + \sum_{\mathfrak{p}\in S_p(K_\infty)} \delta_{1-i,\mathfrak{p}} - \delta_{1-i}$ *for all $i \in \mathbb{Z}$; in particular,*
$$\lambda = \lambda_{cs} + \#S_p(K_\infty) - 1.$$

Recall that S'_∞ is the set of all real primes of k which become complex in K, $r'_1 = \#S'_\infty$ and $\Lambda[\Delta_\mathfrak{p}] = \Lambda \oplus \Lambda^-$ for $\mathfrak{p} \in S'_\infty$, where $\Delta_\mathfrak{p}$ acts on Λ^- by -1 and trivially on Λ. The assertions concerning the μ- and λ-invariants follow from (11.3.16) and (11.3.13), noting that $e_i E^1(M) = E^1(e_{-i}M)$ if M denotes a $\Lambda[\Delta]$-module.

§4. Iwasawa Theory for Totally Real Fields and CM-Fields

If k is a number field of CM-type, i.e. k is a totally imaginary quadratic extension of its maximal totally real subfield k^+, then we obtain more information on the Iwasawa modules considered in the previous sections by using the involution ρ which generates the Galois group

$$G(k|k^+) \cong \mathbb{Z}/2\mathbb{Z}.$$

We assume in this section that p is odd. Thus, if k contains the group μ_p of p-th roots of unity, then $k = k^+(\mu_p)$. If A is a $G(k|k^+)$-module, then we denote the (+) resp. (−)-eigenspace of A with respect to the action of $G(k|k^+)$ by

$$A^\pm = (1 \pm \rho)A.$$

Obviously, $A = A^+ \oplus A^-$, if multiplication by 2 is an isomorphism on A.

In this section we first consider the following situation:

p	is an odd prime number,
k_0	is a totally real number field,
k	is the CM-field $k_0(\mu_p)$ with Galois group $\Delta = G(k\|k_0)$,
$k_\infty\|k$	is the cyclotomic $\mathbb{Z}_p$-extension, $\Gamma = G(k_\infty\|k)$ and $\Lambda = \mathbb{Z}_p[\![\Gamma]\!]$,
k_Σ	is the maximal extension of k which is unramified outside Σ, where $\Sigma = S_p \cup S_\infty$,
L	is the maximal unramified p-extension of k_∞,
L'	is the maximal unramified p-extension of k_∞ which is completely decomposed everywhere.

We consider the Kummer pairing

$$\langle\ ,\ \rangle : G^{ab}_{k_\infty}(p) \times (k^\times_\infty \otimes \mathbb{Q}_p/\mathbb{Z}_p) \longrightarrow \mu_{p^\infty}$$

which is defined by

$$\langle \sigma, x \rangle = \sigma(\sqrt[p^n]{\alpha})/\sqrt[p^n]{\alpha}$$

if $\sigma \in G^{ab}_{k_\infty}(p)$ and $x = \alpha \otimes p^{-n} \bmod \mathbb{Z}_p$ with $\alpha \in k_\infty^\times$ (the definition obviously does not depend on the chosen p^n-th root of α). It is a non-degenerate, $(\Gamma \times \Delta)$-invariant pairing of the compact abelian group $G^{ab}_{k_\infty}(p) = G(k(p)^{ab}|k_\infty)$ with the discrete abelian group $k_\infty^\times \otimes \mathbb{Q}_p/\mathbb{Z}_p$. If H is a closed subgroup of $G = G^{ab}_{k_\infty}(p)$, then $H^\perp$ denotes the annihilator of H in $k_\infty^\times \otimes \mathbb{Q}_p/\mathbb{Z}_p$. From the theory of Pontryagin duality we get non-degenerate pairings

$$\begin{aligned} H \times (k_\infty^\times \otimes \mathbb{Q}_p/\mathbb{Z}_p)/H^\perp &\longrightarrow \mu_{p^\infty}, \\ G/H \quad \times \quad H^\perp \qquad &\longrightarrow \mu_{p^\infty}. \end{aligned}$$

(11.4.1) Proposition. *With the notation as above, consider the following submodules of* $k_\infty^\times \otimes \mathbb{Q}_p/\mathbb{Z}_p$

$$\begin{aligned} \mathfrak{M} &= \{\alpha \otimes p^{-n} \mid n \geq 0, \alpha \in U_{\infty,\mathfrak{p}} \cdot (k_{\infty,\mathfrak{p}}^\times)^{p^n} \text{ if } \mathfrak{p} \notin \Sigma\}, \\ \mathfrak{L}' &= \{\alpha \otimes p^{-n} \mid n \geq 0, \alpha \in U_{\infty,\mathfrak{p}} \cdot (k_{\infty,\mathfrak{p}}^\times)^{p^n} \text{ if } \mathfrak{p} \notin \Sigma,\ \alpha \in (k_{\infty,\mathfrak{p}}^\times)^{p^n} \text{ if } \mathfrak{p} \in \Sigma\}, \end{aligned}$$

where $U_{\infty,\mathfrak{p}}$ *is the inductive limit over the group of local units* $U_{k_{n,\mathfrak{p}}}$ *of* $k_{n,\mathfrak{p}}$. *Then we have the equalities*

$$G(k(p)^{ab}|k(p)^{ab} \cap k_\Sigma)^\perp = \mathfrak{M} \quad \text{and} \quad G(k(p)^{ab}|k(p)^{ab} \cap L')^\perp = \mathfrak{L}'.$$

Thus there are non-degenerate pairings

$$\begin{aligned} X \times \mathfrak{M} &\longrightarrow \mu_{p^\infty}, \\ X_{cs} \times \mathfrak{L}' &\longrightarrow \mu_{p^\infty} \end{aligned}$$

involving the Iwasawa modules $X = G(k_\Sigma|k_\infty)^{ab}(p)$ *and* $X_{cs} = G(L'|k_\infty)^{ab}$.

Proof: This follows dircctly from the definition of k_Σ and L'. □

(11.4.2) Proposition. *With the notation as above, there are canonical* $(\Gamma \times \Delta)$-*invariant exact sequences*

(i) $\quad 0 \longrightarrow \mathcal{O}_{k_\infty}^\times \otimes \mathbb{Q}_p/\mathbb{Z}_p \longrightarrow \mathfrak{M} \xrightarrow{\varphi_1} Cl(k_\infty)(p) \longrightarrow 0,$

(ii) $\quad 0 \longrightarrow \mathcal{O}_{k_\infty,\Sigma}^\times \otimes \mathbb{Q}_p/\mathbb{Z}_p \longrightarrow \mathfrak{M} \xrightarrow{\varphi_2} Cl_\Sigma(k_\infty)(p) \longrightarrow 0.$

Proof: The second sequence follows from the exact sequence (8.3.4)

$$0 \longrightarrow \mu_{p^m} \longrightarrow \mathcal{O}_\Sigma^\times \xrightarrow{p^m} \mathcal{O}_\Sigma^\times \longrightarrow 0$$

by applying cohomology:

$$0 \longrightarrow \mathcal{O}_{k_\infty,\Sigma}^\times/p^m \longrightarrow H^1(k_\Sigma|k_\infty, \mu_{p^m}) \longrightarrow {}_{p^m}H^1(k_\Sigma|k_\infty, \mathcal{O}_\Sigma^\times) \longrightarrow 0,$$

passing to the direct limit and noting that

$$H^1(k_\Sigma|k_\infty, \mathcal{O}_\Sigma^\times)(p) = Cl_\Sigma(k_\infty)(p),$$

$$H^1(k_\Sigma|k_\infty, \mu_{p^\infty}) \cong G(k_\Sigma|k_\infty)^{ab}(p)(-1)^\vee = \mathfrak{M}\,.$$

It is easy to see that the map φ_2 is given explicitly by

$$\varphi_2 : \mathfrak{M} \longrightarrow Cl_\Sigma(k_\infty)(p),\ \alpha \otimes p^{-n} \longmapsto [\mathfrak{a}_1] \in Cl_\Sigma(k_m)(p),$$

where $\alpha \in k_m^\times$ and $\alpha\mathcal{O}_{k_m} = \mathfrak{a}_1^{p^n}\mathfrak{a}_2$ with an ideal $\mathfrak{a}_2$ having only prime divisors in $S_p(k_m)$. Since all primes dividing p have an infinite ramification index in $k_\infty|k$, we see that, for sufficiently large s, $\alpha\mathcal{O}_{k_{m+s}} = \mathfrak{a}^{p^n}$ for some ideal $\mathfrak{a}$ in k_{m+s}. We define φ_1 by sending $\alpha \otimes p^{-n}$ to the class $[\,\mathfrak{a}\,] \in Cl(k_\infty)(p)$. The map φ_2 is the composition of φ_1 with the natural projection $Cl(k_\infty)(p)$ to $Cl_\Sigma(k_\infty)(p)$.

Restricting φ_1 to $\mathcal{O}_{k_\infty,\Sigma}^\times \otimes \mathbb{Q}_p/\mathbb{Z}_p$ and denoting this map by $\tilde{\varphi}_1$, we get an exact sequence

$$\mathcal{O}_{k_\infty}^\times \otimes \mathbb{Q}_p/\mathbb{Z}_p \hookrightarrow \mathcal{O}_{k_\infty,\Sigma}^\times \otimes \mathbb{Q}_p/\mathbb{Z}_p \xrightarrow{\tilde{\varphi}_1} Cl(k_\infty)(p) \twoheadrightarrow Cl_\Sigma(k_\infty)(p)\,.$$

Now the commutative exact diagram

$$\begin{array}{ccccc} \mathcal{O}_{k_\infty,\Sigma}^\times \otimes \mathbb{Q}_p/\mathbb{Z}_p & \hookrightarrow & \mathfrak{M} & \twoheadrightarrow & \mathfrak{M}/\mathcal{O}_{k_\infty,\Sigma}^\times \otimes \mathbb{Q}_p/\mathbb{Z}_p \\ \downarrow & & \downarrow{\scriptstyle \varphi_1} & & \wr\downarrow{\scriptstyle \bar{\varphi}_2} \\ \mathrm{im}(\tilde{\varphi}_1) & \hookrightarrow & Cl(k_\infty)(p) & \twoheadrightarrow & Cl_\Sigma(k_\infty)(p) \end{array}$$

shows that φ_1 is surjective with kernel $\mathcal{O}_{k_\infty}^\times \otimes \mathbb{Q}_p/\mathbb{Z}_p$. □

Let

$$\kappa : G(k_\infty|k_0) \longrightarrow \mathrm{Aut}(\mu_{p^\infty}) = \mathbb{Z}_p^\times$$

be the p-part of the cyclotomic character, $d = \#\Delta$, and let

$$e_i = \frac{1}{d}\sum_{\sigma\in\Delta} \kappa^{-i}(\sigma)\sigma \in \mathbb{Z}_p[\Delta]\,, \quad i \in \mathbb{Z}/d\mathbb{Z}\,,$$

be the corresponding idempotents.

(11.4.3) Theorem. *The Kummer pairing induces a non-degenerate pairing*

$$e_i\,X \times e_{1-i}\,Cl(k_\infty)(p) \longrightarrow \mu_{p^\infty}$$

for all even i, hence

$$e_i\,X \cong \mathrm{Hom}(e_{1-i}\,Cl(k_\infty), \mu_{p^\infty}).$$

Proof: The subgroup $\langle \mathcal{O}_{k_n^+}^\times, \mu(k_n)\rangle$ of $\mathcal{O}_{k_n}^\times$ has index 1 or 2 (see [246], th. 4.12), and hence, since p is odd,

$$e_i(\mathcal{O}_{k_\infty}^\times \otimes \mathbb{Q}_p/\mathbb{Z}_p) = 0 \qquad \text{for } i \text{ odd.}$$

For i even it follows from (11.4.2)(i) that

$$e_{1-i}\,\mathfrak{M} \cong e_{1-i}\,Cl(k_\infty)(p)$$

and from (11.4.1) that

$$\mathrm{Hom}(e_{1-i}\,\mathfrak{M}, \mu_{p^\infty}) \cong e_i\,X.$$

□

As in §3 we set

$$\mu_\bullet^{(i)} = \mu(e_i X_\bullet), \qquad \lambda_\bullet^{(i)} = \lambda(e_i X_\bullet)$$

for $i \in \mathbb{Z}$, and define

$$\mu_\bullet^\pm := \mu_\bullet^\pm(k) := \mu(X_\bullet^\pm) = \sum_{i \substack{\text{even} \\ \text{odd}}} \mu_\bullet^{(i)},$$

$$\lambda_\bullet^\pm := \lambda_\bullet^\pm(k) := \lambda(X_\bullet^\pm) = \sum_{i \substack{\text{even} \\ \text{odd}}} \lambda_\bullet^{(i)},$$

where the index $\bullet$ is nothing, "*nr*", "*cs*" or a set $S \supseteq \Sigma$. Furthermore, recall that M° denotes the Λ-module M with the inverse action of Γ, cf. (5.5.12).

(11.4.4) Corollary. *There is an isomorphism and a pseudo-isomorphism of Λ-modules for each even i*

$$(e_i X)(-1) \cong E^1(e_{1-i} G(H|H_{n_0}k_\infty)) \approx (e_{1-i} X_{nr})^\circ.$$

In particular,

$$\mu^{(i)} = \mu_{nr}^{(1-i)}, \qquad \lambda^{(i)} = \lambda_{nr}^{(1-i)},$$

and

$$\mu^\pm = \mu_{nr}^\mp, \qquad \lambda^\pm = \lambda_{nr}^\mp.$$

Proof: This follows from the last theorem and (11.1.8). □

Regarding the Λ-ranks of the different eigenspaces of $E_\Sigma = \varprojlim_n (\mathcal{O}_{k_n,\Sigma}^\times \otimes \mathbb{Z}_p)$, $E = \varprojlim_n (\mathcal{O}_{k_n}^\times \otimes \mathbb{Z}_p)$ and X and the λ-invariants of $X^\pm$ and $X_{cs}^\pm$, we obtain the

(11.4.5) Proposition. *For each $i \in \mathbb{Z}$ we have*

(i)
$$\operatorname{rank}_\Lambda e_i X = \begin{cases} [k_0 : \mathbb{Q}], & i \text{ odd}, \\ 0, & i \text{ even}, \end{cases}$$

(ii)
$$e_i E_\Sigma \cong \begin{cases} e_i \Lambda^{[k_0:\mathbb{Q}]}, & i \text{ even}, \\ \mathbb{Z}_p(1), & i \equiv 1 \bmod \#\Delta, \\ 0, & \text{otherwise}, \end{cases}$$

and the same for $e_i E$,

(iii)
$$\lambda^+ = \lambda_{cs}^- + \#S_p(k_\infty) - \#S_p(k_\infty^+),$$
$$\lambda^- = \lambda_{cs}^+ + \#S_p(k_\infty^+) - 1.$$

Proof: All assertions follow from (11.3.18) (observe that $S'_\infty = S_\infty$, so that $r_2 = 0$ and $r'_1 = r_1 = [k_0 : \mathbb{Q}]$). □

Now we want to consider the "difference" between X_{nr} and X_{cs}. Recall the notion of the Weierstraß polynomials ω_n (5.3.13), and the numbers $\delta_{i,\mathfrak{p}}$ defined by

$$(\mathbb{Z}/p\mathbb{Z}(i))^{\Delta_\mathfrak{p}} \cong (\mathbb{Z}/p\mathbb{Z})^{\delta_{i,\mathfrak{p}}}\,,$$

where $\Delta_\mathfrak{p}$ is the decomposition group of $\Delta = G(k|k_0)$ with respect to the prime $\mathfrak{p}$.

(11.4.6) Proposition. *For odd $i \in \mathbb{Z}$, there is an exact sequence of $\Lambda[\Delta]$-modules*

$$0 \longrightarrow \bigoplus_{\mathfrak{p}\in S_p(k)} \Lambda/\omega_{n_\mathfrak{p}}^{\delta_{i,\mathfrak{p}}} \longrightarrow e_i X_{nr} \longrightarrow e_i X_{cs} \longrightarrow 0;$$

consequently,

$$\lambda_{nr}^{(i)} = \lambda_{cs}^{(i)} + \sum_{\mathfrak{p}\in S_p(k_\infty)} \delta_{i,\mathfrak{p}}\,, \qquad \mu_{nr}^{(i)} = \mu_{cs}^{(i)},$$

and

$$\lambda_{nr}^{-} - \lambda_{cs}^{-} = \#S_p(k_\infty) - \#S_p(k_\infty^+).$$

Proof: The exact sequence follows from (11.3.10)(ii) and (11.4.5)(ii). The assertions concerning the Iwasawa invariants may also be obtained from (11.4.4) and (11.3.18). □

It is more difficult to obtain results for the even eigenspaces. According to the conjecture of Greenberg, X_{nr}^+ and X_{cs}^+ should be finite (nevertheless they can be non-zero; see [62] for examples).

(11.4.7) Proposition. *Assume that the (strong) Leopoldt conjecture holds for some layer k_n^+ of $k_\infty^+|k^+$ with $n \geq \lambda^+$. Then for i even*

$$\lambda_{nr}^{(i)} = \lambda_{cs}^{(i)}$$

and, in particular,

$$\lambda_{nr}^{+} = \lambda_{cs}^{+}\,.$$

Proof: From (11.3.3) we know that under our assumption, the Leopoldt conjecture holds for all finite layers of the tower $k_\infty^+|k^+$, and so

$$(e_i X)^{\Gamma_n} = 0 \qquad \text{for } i \text{ even and all } n.$$

Consequently, the characteristic polynomial of the eigenspace $e_i X$, i even, is prime to ω_n for all n and this also holds for its quotient $e_i X_{nr}$, since $e_i X$ is a Λ-torsion module by fact (4) of the beginning of §3. Now the exact sequence (11.3.10)(ii) shows that the kernel of $e_i X_{nr} \twoheadrightarrow e_i X_{cs}$ is finite. □

In the following, we do not assume that μ_p is contained in our CM-field. So let $k|k^+$ be a CM-field with Galois group $\Delta = G(k|k^+) \cong \mathbb{Z}/2\mathbb{Z}$ and let k_∞ be the cyclotomic $\mathbb{Z}_p$-extension of k, for $p > 2$.

(11.4.8) Theorem. *We obtain for the projective dimensions of the Λ-modules X_{nr}^- and X_{cs}^- that*

$$pd_\Lambda X_{nr}^- \le 1 \quad \textit{and} \quad pd_\Lambda X_{cs}^- \le 1.$$

Consequently, X_{nr}^- and X_{cs}^- do not contain a finite nontrivial Λ-submodule.

Proof: First let us assume that $\mu_p \subseteq k$. Then, by (11.3.15)(iii), there is an exact sequence

$$0 \longrightarrow X/T_1(X) \longrightarrow \bigoplus_{\mathfrak{p}\in S_\infty(k^+)} \Lambda^- \longrightarrow T_0(X_{cs}(-1))^\vee \longrightarrow 0\,.$$

Thus $(T_0(X_{cs}(-1))^\vee)^+ = 0$, and so $T_0(X_{cs}^-) = 0$, which gives the result for X_{cs}^-.

If $\mu_p \nsubseteq k$, then let $K = k(\mu_p)$, which is again a CM-field with maximal totally real subfield K^+ such that $K^+ \cap k = k^+$. It follows from the above that $X(K)_{cs}^-$ contains no finite nontrivial Λ-submodule. Since $X(k)_{cs}^-$ is a direct summand of $X(K)_{cs}^-$, the same holds for this module, and the result follows.

The assertion for X_{nr}^- is a consequence of the next lemma, which holds for arbitrary CM-fields $k|k^+$ (without the assumption that μ_p is contained in k). □

(11.4.9) Lemma. *Let $k|k^+$ be a CM-field. Then there is a canonical exact sequence of $\Lambda[G(k|k^+)]$-modules*

$$0 \longrightarrow \Big(\bigoplus_{S_p(k)} \Lambda/\omega_{n_\mathfrak{p}}\Big)^- \longrightarrow X_{nr}^- \longrightarrow X_{cs}^- \longrightarrow 0\,.$$

Proof: This follows from the exact sequence (11.3.10)(ii) and the isomorphism $E^- \xrightarrow{\sim} E_\Sigma^-$ ($\cong \mathbb{Z}_p(1)^\delta$, where $\delta = 1$ if $\mu_p \subseteq k$, and zero otherwise). □

We now come to the question of how the λ-invariants behave under a change of the ground field. Although nothing is known if the base change is given by an extension $K|k$ where $[K:k]$ is not a power of p, there is a remarkable analogy to the Riemann-Hurwitz formula in algebraic geometry if $K|k$ is a Galois p-extension of CM-fields. This was first observed by Y. KIDA [103] and later other authors proved this by different methods. We will proceed as in [90].

Let $k|k^+$ be a CM-field, $K^+|k^+$ be a finite Galois p-extension and $K = K^+k$. We denote the Galois group of the extension $K_\infty|k_\infty$ by G,

$$G = G(K_\infty|k_\infty) \xrightarrow{\sim} G(K_\infty^+|k_\infty^+)\,,$$

where k_∞ is the cyclotomic $\mathbb{Z}_p$-extension and $K_\infty = Kk_\infty$. Since $X(K^+)$ is a Λ-torsion module by (11.3.2)(iv) and (ii), the $\mathbb{Q}_p[G]$-module $X(K^+) \otimes \mathbb{Q}_p$ is finitely generated. Regarding its structure, we obtain the

(11.4.10) Theorem. *Suppose that the Iwasawa invariant $\mu^+ = \mu(X(k^+))$ is zero. Then there is an isomorphism of $\mathbb{Q}_p[G]$-modules*

$$X(K^+) \otimes \mathbb{Q}_p \cong \mathbb{Q}_p[G]^{\lambda^+(k)-1} \oplus \mathbb{Q}_p \oplus \bigoplus_{\substack{\mathfrak{p} \text{ prime of } k_\infty^+ \\ \mathfrak{p} \nmid p}} \mathrm{Ind}_G^{G_{\mathfrak{P}}} I_{G_{\mathfrak{P}}}$$

where $I_{G_{\mathfrak{P}}}$ is the augmentation ideal in $\mathbb{Q}_p[G_{\mathfrak{P}}]$ and $\mathfrak{P}$ denotes an arbitrary extension of the prime $\mathfrak{p}$ of k_∞^+ to K_∞^+.

Proof: First observe that the sum on the right is finite because $G_{\mathfrak{P}} \neq 1$, and so $I_{G_{\mathfrak{P}}} \neq 0$ only for primes $\mathfrak{P}$ which are ramified in $K_\infty^+|k_\infty^+$. Since the weak Leopoldt conjecture holds for the cyclotomic $\mathbb{Z}_p$-extension and since $\mu^+ = 0$, it follows from (11.3.7) and (11.3.6)(i) that $G(k_S^+|k_\infty^+)(p)$ is a finitely generated free pro-p-group for every finite $S \supseteq \Sigma$. Let S be finite and large enough so that $K^+ \subseteq k_S^+(p)$. Then, by (5.6.6) applied to the presentation

$$1 \longrightarrow G(k_S^+(p)|K_\infty^+) \longrightarrow G(k_S^+(p)|k_\infty^+) \longrightarrow G \longrightarrow 1\,,$$

we obtain (using (11.3.6)(ii))

$$\begin{aligned} X_S(K^+) \otimes \mathbb{Q}_p &\cong \mathbb{Q}_p[G]^{\lambda_S^+(k)-1} \oplus \mathbb{Q}_p \\ &\cong \mathbb{Q}_p[G]^{\lambda^+(k)-1} \oplus \mathbb{Q}_p \oplus \mathbb{Q}_p[G]^d\,, \quad d = \sum_{\mathfrak{p} \in S\setminus\Sigma(k_\infty^+)} \delta_{\mathfrak{p}} \\ &\cong \mathbb{Q}_p[G]^{\lambda^+(k)-1} \oplus \mathbb{Q}_p \oplus \bigoplus_{\mathfrak{p} \in S\setminus\Sigma(k_\infty^+)} \mathrm{Ind}_G^{G_{\mathfrak{P}}} \mathbb{Q}_p[G_{\mathfrak{P}}]^{\delta_{\mathfrak{p}}}\,. \end{aligned}$$

From the G-invariant sequence (11.3.5) (for the $\mathbb{Z}_p$-extension $K_\infty^+|K^+$), we see

$$X_S(K^+) \otimes \mathbb{Q}_p \cong X(K^+) \otimes \mathbb{Q}_p \oplus \bigoplus_{\mathfrak{p} \in S \setminus \Sigma(k_\infty^+)} \mathrm{Ind}_G^{G_\mathfrak{P}} \mathbb{Q}_p^{\delta_\mathfrak{p}} .$$

Putting both formulae together and recalling (5.6.10)(i), we obtain the result. □

(11.4.11) Corollary (Riemann-Hurwitz Formula). *Let $K^+|k^+$ be a finite Galois p-extension of totally real fields and assume that $\mu(k^+) = 0$ for the cyclotomic $\mathbb{Z}_p$-extension of k^+, $p > 2$. Then*

$$\lambda^+(K) - 1 = [K_\infty^+ : k_\infty^+](\lambda^+(k) - 1) + \sum_{\mathfrak{P} \nmid p}(e_\mathfrak{P} - 1),$$

where $e_\mathfrak{P}$ denotes the ramification index of $K_\infty^+|k_\infty^+$ with respect to a prime $\mathfrak{P}$ of K_∞^+.

Proof: Take the $\mathbb{Q}_p$-dimensions of both sides of the isomorphism (11.4.10) and recall that $\dim_{\mathbb{Q}_p} I_{G_\mathfrak{P}} = \#G_\mathfrak{P} - 1 = e_\mathfrak{P} - 1$. The formula follows. □

Using duality, we obtain similar results for the modules X_{nr}^- and X_{cs}^- in the case of CM-fields.

(11.4.12) Theorem. *Let $p > 2$ and let $K|k$ be a finite Galois p-extension of CM-fields. Assume that $\mu(k) = 0$ for the cyclotomic $\mathbb{Z}_p$-extension $k_\infty|k$ and let $G = G(K_\infty|k_\infty)$. Then there are isomorphisms of $\mathbb{Q}_p[G]$-modules*

$$X_{nr}(K)^- \otimes \mathbb{Q}_p \cong \mathbb{Q}_p[G]^{\lambda_{nr}^-(k)-\delta} \oplus \mathbb{Q}_p^\delta \oplus \bigoplus_{\substack{\mathfrak{p} \text{ prime of } k_\infty^+ \\ \mathfrak{p} \nmid p,\ \mathfrak{p} \text{ splits in } k_\infty}} \mathrm{Ind}_G^{G_\mathfrak{P}} I_{G_\mathfrak{P}},$$

$$X_{cs}(K)^- \otimes \mathbb{Q}_p \cong \mathbb{Q}_p[G]^{\lambda_{cs}^-(k)-\delta} \oplus \mathbb{Q}_p^\delta \oplus \bigoplus_{\substack{\mathfrak{p} \in S(k_\infty^+) \\ \mathfrak{p} \text{ splits in } k_\infty}} \mathrm{Ind}_G^{G_\mathfrak{P}} I_{G_\mathfrak{P}},$$

where $\delta = 1$ if $\mu_p \subseteq k$ and zero otherwise, and $S(k_\infty^+)$ is the union of $S_p(k_\infty^+)$ and all ramified primes of the extension $K_\infty^+|k_\infty^+$.

Proof: If $\mu_p \subseteq k$, then

$$X_{nr}(K)^- \otimes \mathbb{Q}_p \cong X(K^+) \otimes \mathbb{Q}_p$$

as $\mathbb{Q}_p[G]$-modules by (11.4.4) and $\lambda^+ = \lambda_{nr}^-$. If $I_{G_\mathfrak{P}} \neq 0$ for a prime $\mathfrak{p} \nmid p$ of k_∞^+, then $\mathfrak{p}$ ramifies in K_∞^+. Hence $\mu_p \subseteq k_{\infty,\mathfrak{p}}^+$ and $\mathfrak{p}$ splits in k_∞. Therefore, if $\mu_p \subseteq k$, the first formula follows from (11.4.10).

Now suppose $\mu_p \nsubseteq k$. Using the exact sequence (11.4.2)(i) for the field $K' = K(\mu_p)$ and taking $G(K'_\infty|K_\infty)$-invariants, we obtain the exact sequence

$$0 \longrightarrow \mathcal{O}_{K_\infty}^\times \otimes \mathbb{Q}_p/\mathbb{Z}_p \longrightarrow H^1(K_\Sigma|K_\infty, \mu_{p^\infty}) \longrightarrow Cl(K_\infty)(p) \longrightarrow 0,$$

so that

$$H^1(K_\Sigma|K_\infty, \mu_{p^\infty})^- \cong Cl(K_\infty)(p)^-.$$

Again let $S \supseteq \Sigma$ be finite and sufficiently large that $K \subseteq k_S(p)$. The Hochschild-Serre spectral sequence (2.4.1)

$$E_2^{ij} = H^i(G, H^j(k_S|K_\infty, \mu_{p^\infty})) \Rightarrow H^{i+j}(k_S|k_\infty, \mu_{p^\infty})$$

implies (since $\mu_p \nsubseteq k_\infty$ and $G(k_S|k_\infty)$ is free) that

$$(*) \qquad \begin{aligned} H^0(G, H^1(k_S|K_\infty, \mu_{p^\infty})) &= H^1(k_S|k_\infty, \mu_{p^\infty}), \\ H^i(G, H^1(k_S|K_\infty, \mu_{p^\infty})) &= 0 \qquad \text{for } i \geq 1. \end{aligned}$$

Thus $H^1(k_S|K_\infty, \mu_{p^\infty})^\vee$ is a cohomologically trivial $\mathbb{Z}_p[G]$-module and the same is true for the minus-part. Since

$$\begin{aligned} (H^1(K_\Sigma|K_\infty, \mu_{p^\infty})^-)^\vee &\cong \mathrm{Hom}(Cl(K_\infty), \mathbb{Q}_p/\mathbb{Z}_p)^- \\ &\cong E^1(G(H|H_{n_0}K_\infty)^-) \\ &\approx (X_{nr}(K_\infty)^-)^\circ \end{aligned}$$

by (11.1.8) and $\mu_{nr}(K_\infty|K) = \mu(K_\infty|K) = 0$ by (11.3.8), we see that $(H^1(K_\Sigma|K_\infty, \mu_{p^\infty})^-)^\vee = G(K_\Sigma(p)|K_\infty)^{ab+}(-1)$ is $\mathbb{Z}_p$-free and finitely generated as a $\mathbb{Z}_p[G]$-module. Using (11.3.5), it follows that the same holds for $(H^1(k_S|K_\infty, \mu_{p^\infty})^-)^\vee$. Thus by (5.2.21) and $(*)$,

$$(H^1(k_S|K_\infty, \mu_{p^\infty})^-)^\vee \cong \mathbb{Z}_p[G]^r$$

is a free $\mathbb{Z}_p[G]$-module.

By (11.3.6)(ii) and (11.4.4) its rank is equal to $\lambda_{nr}^-(k) + d$, where d is the sum of the numbers $\delta_\mathfrak{p}$ for the primes $\mathfrak{p} \in S \smallsetminus \Sigma(k_\infty^+)$ which split in k_∞. It follows that

$$X_{nr}(K)^- \otimes \mathbb{Q}_p \cong \mathbb{Q}_p[G]^{\lambda_{nr}^-(k)} \oplus \bigoplus_{\substack{\mathfrak{p} \text{ prime of } k_\infty^+ \\ \mathfrak{p} \nmid p,\ \mathfrak{p} \text{ splits in } k_\infty}} I_{G_\mathfrak{P}}.$$

The assertion for $X_{cs}(K)^-$ is now an easy consequence of the above, using the isomorphism

$$X_{nr}(K)^- \otimes \mathbb{Q}_p \cong X_{cs}(K)^- \otimes \mathbb{Q}_p \oplus \bigoplus_{\substack{\mathfrak{p} \in S_p(k_\infty^+) \\ \mathfrak{p} \text{ splits in } k_\infty}} \bigoplus_{\mathfrak{P}|\mathfrak{p}} \mathbb{Q}_p$$

and the equality $\lambda_{nr}^-(k) = \lambda_{cs}^-(k) + \#\{\mathfrak{p} \in S_p(k_\infty^+) \mid \mathfrak{p}$ splits in $k_\infty\}$, which follows from (11.4.9). □

Considering $\mathbb{Q}_p$-dimensions on both sides of the isomorphisms (11.4.12), we obtain the

(11.4.13) Corollary. *Let $p > 2$ and let $K|k$ be a finite Galois p-extension of CM-fields. Assume that $\mu(k) = 0$ for the cyclotomic $\mathbb{Z}_p$-extension $k_\infty|k$. Then*

$$\lambda_{nr}^-(K) - \delta = [K_\infty : k_\infty](\lambda_{nr}^-(k) - \delta) + \sum_{\substack{\mathfrak{p} \text{ prime of } k_\infty^+ \\ \mathfrak{p}\nmid p,\ \mathfrak{p} \text{ splits in } k_\infty}} \sum_{\mathfrak{P}|\mathfrak{p}} (e_{\mathfrak{P}} - 1)$$

and

$$\lambda_{cs}^-(K) - \delta = [K_\infty : k_\infty](\lambda_{cs}^-(k) - \delta) + \sum_{\substack{\mathfrak{p} \in S(k_\infty^+) \\ \mathfrak{p} \text{ splits in } k_\infty}} \sum_{\mathfrak{P}|\mathfrak{p}} (n_{\mathfrak{P}} - 1),$$

where $\delta = 1$ if $\mu_p \subseteq k$ and zero otherwise, $e_{\mathfrak{P}}$ denotes the ramification index and $n_{\mathfrak{P}}$ the local degree of $K_\infty^+|k_\infty^+$ with respect to a prime $\mathfrak{P}$ of K_∞^+. The set $S(k_\infty^+)$ is the union of $S_p(k_\infty^+)$ and the set of all primes ramifying in $K_\infty^+|k_\infty^+$.

(11.4.14) Proposition. *Let $p \neq 2$ and let $k = k^+(\mu_p)$ be a CM-field such that $\mu_p \nsubseteq k_{\infty,\mathfrak{p}}^+$ for every prime $\mathfrak{p}$ above p. Then there exists a canonical isomorphism*

$$X_{cs}^-(-1) \xrightarrow{\sim} E^1(X^+)$$

for the cyclotomic $\mathbb{Z}_p$-extension.

Proof: This follows directly from (11.3.13). □

We conclude this section by showing that an isomorphism as above can also be obtained on the level of finite coefficients, provided that $\mu = 0$. The isomorphism of (11.4.14) can be recovered from this result by passage to the limit over n, but we will neither prove nor use this fact. We introduce the following notation. In the situation where

p is an odd prime number,
$k = k^+(\mu_p)$ is a CM-field with Galois group $\Delta = G(k|k^+) \cong \mathbb{Z}/2\mathbb{Z}$,

let

$$k_\Sigma^{CM} = k_\Sigma^+(p)(\mu_p)$$

be the **maximal CM-field** of k inside $k_\Sigma(p)$. We denote the maximal p-extension of k_Σ^{CM} which is completely decomposed everywhere by $\tilde{L}'$.

(11.4.15) Proposition. *In the above situation assume that $\mu_p \notin k_\mathfrak{p}^+$ for all primes above p and that the Iwasawa μ-invariant of the cyclotomic $\mathbb{Z}_p$-extension $k_\infty|k$ is zero. Then the cup-product induces a perfect pairing of finite groups*

$$\begin{array}{ccccc}
H^0(k_\Sigma^{CM}|k_\infty, H^1(\tilde{L}'|k_\Sigma^{CM})(1))^- & \times & H^1(k_\Sigma^{CM}|k_\infty) & \xrightarrow{\cup} & H^1(k_\Sigma^{CM}|k_\infty, H^1(\tilde{L}'|k_\Sigma^{CM})(1))_\Delta \\
\| \wr & & \| \wr & & \| \wr \\
(X_{cs}^-/p^m)(-1)^* & & (X^+/p^m)^* & & \mathbb{Z}/p^m\mathbb{Z},
\end{array}$$

where the coefficients of the cohomology groups are $\mathbb{Z}/p^m\mathbb{Z}$.

Proof: We may assume that $\lambda^+ = \lambda_{cs}^-$ is non-zero since otherwise all groups under consideration are trivial. From the exact sequence

$$0 \longrightarrow \mathcal{O}_\Sigma^\times \longrightarrow I_\Sigma \longrightarrow C_\Sigma \longrightarrow 0$$

we obtain the exact cohomology sequence

$$\begin{aligned}
\mathrm{Hom}(\mathbb{Z}/p^m, I_\Sigma)^{G(k_\Sigma|k_\Sigma^{CM})} \to \mathrm{Hom}(\mathbb{Z}/p^m, C_\Sigma)^{G(k_\Sigma|k_\Sigma^{CM})} &\xrightarrow{\delta} H^1(k_\Sigma|k_\Sigma^{CM}, \mu_{p^m}) \\
&\to \prod_{\mathfrak{p}\in\Sigma(k_\Sigma^{CM})} H^1((k_\Sigma^{CM})_\mathfrak{p}, \mu_{p^m}),
\end{aligned}$$

recalling that $\mathcal{O}_\Sigma^\times$ is p-divisible. We get an isomorphism

$$(\mathrm{Hom}(\mathbb{Z}/p^m\mathbb{Z}, C_\Sigma)^{G(k_\Sigma|k_\Sigma^{CM})})_\Delta \xrightarrow[\sim]{\delta} H^1(\tilde{L}'|k_\Sigma^{CM}, \mathbb{Z}/p^m\mathbb{Z})^-(1)$$

since $\mu_p \notin k_\mathfrak{p}^\times$ for all $\mathfrak{p}|p$. Furthermore, the following diagram is commutative (assuming n large enough):

$$\begin{array}{ccccc}
H^0(k_\Sigma|k_n, \mathrm{Hom}(\mathbb{Z}/p^m\mathbb{Z}, C_\Sigma)) & \times & H^2(k_\Sigma|k_n) & \xrightarrow{\cup} & H^2(k_\Sigma|k_n, {}_{p^m}C_\Sigma) \cong \mathbb{Z}/p^m\mathbb{Z} \\
\downarrow & & \uparrow & & \uparrow \wr \\
H^0(k_\Sigma|k_\infty, {}_{p^m}C_\Sigma)_\Delta^{\Gamma_n} & \times & H^1(k_\Sigma|k_\infty)_{\Gamma_n}^\Delta & \xrightarrow{\cup} & H^1(k_\Sigma|k_\infty, {}_{p^m}C_\Sigma)_{\Gamma_n\times\Delta} \\
\downarrow & & \uparrow \wr & & \uparrow inf \\
H^0(k_\Sigma^{CM}|k_\infty, {}_{p^m}C_\Sigma(k_\Sigma^{CM}))_\Delta & \times & H^1(k_\Sigma^{CM}|k_\infty) & \xrightarrow{\cup} & H^1(k_\Sigma^{CM}|k_\infty, {}_{p^m}C_\Sigma(k_\Sigma^{CM}))_\Delta \\
\downarrow \wr\, H^0(\delta) & & \| & & \downarrow \wr\, H^1(\delta) \\
H^0(k_\Sigma^{CM}|k_\infty, H^1(\tilde{L}'|k_\Sigma^{CM})^-(1)) & \times & H^1(k_\Sigma^{CM}|k_\infty) & \xrightarrow{\cup} & H^1(k_\Sigma^{CM}|k_\infty, H^1(\tilde{L}'|k_\Sigma^{CM})^-(1)),
\end{array}$$

where the missing coefficients of the cohomology groups are $\mathbb{Z}/p^m\mathbb{Z}$ (for the compatibility of the upper two pairings observe that $H^1(-)_{\Gamma_n} = H^1(\Gamma, H^1(-))$ and that the Hochschild-Serre spectral sequence is functorial with respect to the cup-product). The isomorphism $H^2(k_\Sigma|k_n, {}_{p^m}C_\Sigma) \cong \mathbb{Z}/p^m\mathbb{Z}$ follows from (10.11.5) and (8.3.9)(ii),(iii). The map *inf* on the right is an isomorphism as one sees as follows:

If $r \geq n$, then the diagram

$$\begin{array}{ccccc} H^1(k_\Sigma|k_\infty,\, {}_{p^m}C_\Sigma)_{\Gamma_n} & \xrightarrow{\sim} & H^2(k_\Sigma|k_n,\, {}_{p^m}C_\Sigma) & \xrightarrow{\sim} & \mathbb{Z}/p^m\mathbb{Z} \\ \uparrow & & \uparrow{\scriptstyle cor}\wr & & \| \\ H^1(k_\Sigma|k_\infty,\, {}_{p^m}C_\Sigma)_{\Gamma_r} & \xrightarrow{\sim} & H^2(k_\Sigma|k_r,\, {}_{p^m}C_\Sigma) & \xrightarrow{\sim} & \mathbb{Z}/p^m\mathbb{Z} \end{array}$$

commutes, showing that $H^1(k_\Sigma|k_\infty, {}_{p^m}C_\Sigma)$ is fixed by Γ_n for some n. Thus *inf* is injective. Now let $K|k_\infty$ be a finite Galois extension inside $k_\Sigma(p)$. Then

$$H^1(k_\Sigma|K, {}_{p^m}C_\Sigma) = \varprojlim_n H^1(k_\Sigma|K, {}_{p^m}C_\Sigma)_{\Gamma_n} \cong \mu_{p^m}(K)(-1)^\vee \cong \mathbb{Z}/p^m\mathbb{Z}$$

and therefore

$$H^1(k_\Sigma|k_\Sigma^{CM}, {}_{p^m}C_\Sigma) = \varinjlim_{K\subseteq k_\Sigma^{CM}} H^1(k_\Sigma|K, {}_{p^m}C_\Sigma) \cong (\varprojlim_{K\subseteq k_\Sigma^{CM}} \mu_{p^m}(K)(-1))^\vee = 0,$$

since the degree of $[k_\Sigma^{CM} : k_\infty]$ is divisible by p^∞ and the projective limit is taken via p-multiplication. This shows that *inf* is surjective.

Since the upper pairing in the diagram above is non-degenerate on the right by the global duality theorem (8.4.4), the same holds for the lower pairing. In order to prove that the pairing is perfect, we will show that the orders of both groups coincide. Obviously,

$$\#H^1(k_\Sigma^{CM}|k_\infty, \mathbb{Z}/p^m\mathbb{Z}) = \#(X^+/p^m) = p^{m\cdot\lambda^+}$$

and

$$\#(X_{cs}^-/p^m) = p^{m\lambda_{cs}^-} = p^{m\lambda^+},$$

by (11.4.8) and the assumption that $\mu = 0$. It remains to show that there is an isomorphism

$$\begin{aligned} (H^1(\tilde{L}'|k_\Sigma^{CM}, \mathbb{Z}/p^m\mathbb{Z})^{G(k_\Sigma^{CM}|k_\infty)})^-(1) &\cong H^1(L'|k_\infty, \mathbb{Z}/p^m\mathbb{Z})^-(1) \\ &\cong (X_{cs}^-/p^m)(-1)^\vee . \end{aligned}$$

But this follows from the commutative exact diagram with coefficients $\mathbb{Z}/p^m\mathbb{Z}$

$$\begin{array}{ccccc} H^1(\tilde{L}'|k_\Sigma^{CM})^{G(k_\Sigma^{CM}|k_\infty)} & \hookrightarrow & H^1(k_\Sigma|k_\Sigma^{CM})^{G(k_\Sigma^{CM}|k_\infty)} & \longrightarrow & \prod\limits_{\mathfrak{p}\in\Sigma(k_\infty)} H^1(k_{\Sigma,\mathfrak{p}}^{CM})^{G_\mathfrak{p}(k_\Sigma^{CM}|k_\infty)} \\ \uparrow & & \uparrow & & \uparrow \\ H^1(L'|k_\infty) & \hookrightarrow & H^1(k_\Sigma|k_\infty) & \longrightarrow & \prod\limits_{\mathfrak{p}\in\Sigma(k_\infty)} H^1(k_{\infty,\mathfrak{p}}) \end{array}$$

if one takes the minus-parts, and recalling again that $\mu_p \notin k_{\infty,\mathfrak{p}}^+$ for $\mathfrak{p}|p$. □

Exercise: (*Leopoldt's Spiegelungssatz*)
(a) Let k be a number field, p an odd prime number and $K|k$ a finite abelian extension containing μ_p. Assume that the Galois group $\Delta = G(K|k)$ has exponent $p-1$. For each character $\chi : \Delta \to \mathbb{Z}_p^\times$ and let $M^{(\chi)} = \{x \in M \mid gx = \chi(g)x \text{ for all } g \in \Delta\}$ be the χ-eigenspace of a $\mathbb{Z}_p[\Delta]$-module M. Prove the following inequality

$$\dim_{\mathbb{F}_p}(Cl_{S_p}(K)/p(-1))^{(\chi^{-1})} \le \dim_{\mathbb{F}_p}(Cl_{S_p}(K)/p)^{(\chi)} + \dim_{\mathbb{F}_p}(E_{S_p}(K)/p)^{(\chi)}.$$

Hint: Consider the exact sequence

$$0 \longrightarrow E_{S_p}(K)/p \longrightarrow H^1(G(K_{S_p}|K), \mu_p) \longrightarrow {}_pCl_{S_p}(K) \longrightarrow 0$$

and the inclusion $\mathrm{Hom}(Cl_{S_p}(K), \mu_p) \hookrightarrow H^1(G(K_{S_p}|K), \mu_p)$.

(b) Let p be an odd prime number, $K = \mathbb{Q}(\zeta_p)$ and $\omega : G(K|\mathbb{Q}) \cong (\mathbb{Z}/p\mathbb{Z})^\times \to \mathbb{Z}_p^\times$ the Teichmüller character, i.e. the character with $\omega(\tilde{a}) \equiv a \bmod p$, where $\tilde{a}$ the automorphism $\tilde{a}(\zeta_p) = \zeta_p^a$. Let i be an even and j an odd integer such that $i + j \equiv 1 \bmod p-1$. Prove the following inequalities:

$$\dim_{\mathbb{F}_p}(Cl_{S_p}(K)/p)^{(\omega^i)} \le \dim_{\mathbb{F}_p}(Cl_{S_p}(K)/p)^{(\omega^j)} \le \dim_{\mathbb{F}_p}(Cl_{S_p}(K)/p)^{(\omega^i)} + 1\,.$$

Hint: Use (a) and

$$\dim_{\mathbb{F}_p}(E_{S_p}(K)/p)^{(\omega^r)} = \begin{cases} 1, & \text{if } r \text{ is even,} \\ 0, & \text{if } r \text{ is odd,} \end{cases} \qquad \text{for } 1 < r < p-1.$$

§5. Positively Ramified Extensions

In 1969, in his foreword to *H. Koch*'s book [110], *I. R. Šafarevič* recalled that already *Hilbert*, when laying the foundations of class field theory, used the analogy between algebraic number fields and function fields as his starting point. He concluded:

> "From this point of view, a noncommutative generalization of class field theory must correspond to the investigation of the fundamental group of a Riemann surface, which, as is well-known, is noncommutative."

The present book confirms this analogy with function fields to a large extent. As explained in X §1, the analogy to Riemann surfaces comes about after a base change to the algebraic closure of the finite ground field.

The natural analogue to this base change for number fields is the passage to the cyclotomic, say, $\mathbb{Z}_p$-extension for a prime number p. But $G(k_S(p)|k_\infty)$ is a free pro-p-group (provided that $\mu = 0$), and thus is analogous to the fundamental group of an open, i.e. noncompact, Riemann surface. This is not surprising since G_S is the fundamental group of the affine curve $\mathrm{Spec}(\mathcal{O}_S)$.

It is a remarkable observation that we can overcome this problem by introducing further restrictions on the wild ramification (which doesn't occur in complex geometry). We will see in this section that dualities between certain Iwasawa modules which were shown in the last section, can be viewed as a perfect cup-product pairing of a (non-abelian) Demuškin group. This group, the Galois group of the maximal *positively ramified* extension, is analogous to the fundamental group of a compact Riemann surface. However, despite the clear analogy, we are far from a deeper understanding of this phenomenon.

Let p be an odd prime number and let $k|k^+$ be a CM-field containing the group μ_p of the p-th roots of unity. Thus $k = k^+(\mu_p)$. Let k_∞ be the cyclotomic $\mathbb{Z}_p$-extension of k, $\Gamma = G(k_\infty|k)$ and $\Lambda = \mathbb{Z}_p[\![\Gamma]\!]$ be the Iwasawa algebra. We assume that

(1) the Iwasawa μ-invariant of $k_\infty|k$ is zero,

(2) $\mu_p \notin k_\mathfrak{p}^+$ for all primes $\mathfrak{p}$ of k above p.

We define a natural Galois p-extension $\tilde{k}_\Sigma$ inside $k_S(p)$ as the maximal p-extension of k unramified outside p and *positively ramified* at p.

(11.5.1) Definition. *Let $k = k^+(\mu_p)$ be a CM-field satisfying condition* (2). *A finite Galois p-extension $K|k$ is called* **positively ramified at $\mathfrak{p}$** *if*

$$K_\mathfrak{p} \subseteq k_\mathfrak{p}^+(p)(\mu_p)$$

for all primes $\mathfrak{p}$ dividing p.

Remarks: 1. The composite of positively ramified p-extensions is again positively ramified. Hence the **maximal positively ramified p-extension** $\tilde{k}$ of k exists. We set

$$\tilde{k}_\Sigma = \tilde{k} \cap k_\Sigma(p)\,,$$

where $\Sigma = S_p \cup S_\infty$. Obviously, $k_\infty \subseteq \tilde{k}_\Sigma$ and

$$k_\Sigma^{CM} = k_\Sigma^+(p)(\mu_p) \subseteq \tilde{k}_\Sigma,$$

where k_Σ^{CM} is the **maximal CM-field** of k inside k_Σ.

2. We also can give a description of $\tilde{k}$ by its Galois group. Setting $\Delta = G(k|k^+) \cong \mathbb{Z}/2\mathbb{Z}$, we have $\Delta = \Delta_\mathfrak{p}$ for all $\mathfrak{p}|p$ by assumption (2). Let $T_\mathfrak{p} = T(k_\mathfrak{p}(p)|k_\mathfrak{p})$ be the inertia subgroup of the local Galois group $G(k_\mathfrak{p}(p)|k_\mathfrak{p})$ for a nonarchimedean prime $\mathfrak{p}$ of k. The sequence

$$1 \longrightarrow G(k_\mathfrak{p}(p)|k_\mathfrak{p}^+(p)(\mu_p)) \longrightarrow G(k_\mathfrak{p}(p)|k_\mathfrak{p}) \longrightarrow G(k_\mathfrak{p}^+(p)|k_\mathfrak{p}^+) \longrightarrow 1$$

is exact and we set

$$T_{\mathfrak{p}}^{-} := G(k_{\mathfrak{p}}(p)|k_{\mathfrak{p}}^{+}(p)(\mu_p)).$$

By definition of $\tilde{k}$, we obtain

$$G(\tilde{k}|k) = G(k(p)|k)/(T_{\mathfrak{p}}^{-},\ \mathfrak{p}|p)$$

and

$$\mathcal{G} := G(\tilde{k}_\Sigma|k_\infty) = G(k(p)|k_\infty)/(T_{\mathfrak{p}}^{-},\ \mathfrak{p}|p\,;\ T_{\mathfrak{p}},\ \mathfrak{p}\nmid p)\,,$$

(recall that for closed subgroups H_i, $i \in I$, of a profinite group H, the subgroup $(H_i\,, i \in I)$ denotes the normal closure of the subgroup which is generated by the groups H_i, $i \in I$). Observe that by (9.4.3) the groups $G(k_{\mathfrak{p}}(p)|k_{\mathfrak{p}})$ can be considered as subgroups of $G(k(p)|k)$. If

$$\mathcal{H} := G(\tilde{k}_\Sigma|k_\Sigma^{CM})\,,$$

then

$$\mathcal{G}/\mathcal{H} = G(k_\Sigma^{CM}|k_\infty) \cong G(k_\Sigma^{+}(p)|k_\infty^{+})\,.$$

3. We have the following diagram of fields

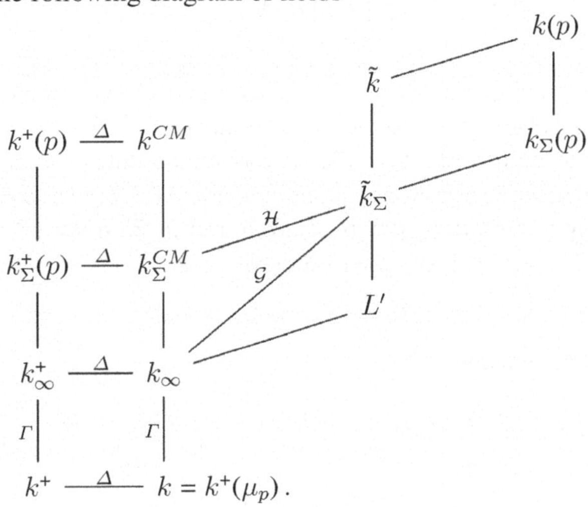

(11.5.2) Proposition. *Under the above assumptions* (1) *and* (2), *the following are true:*

(i) *There is a canonical isomorphism of* $\Lambda[\Delta]$*-modules*

$$G(\tilde{k}_\Sigma|k_\infty)^{ab} \xrightarrow{\ \sim\ } X^{+} \oplus X_{cs}^{-}\,,$$

where $X = G(k_\Sigma(p)|k_\infty)^{ab}$ *and* $X_{cs} = G(L'|k_\infty)^{ab}$.

(ii) *If* $\lambda^{+} \neq 0$, *then*

$$H^2(G(\tilde{k}_\Sigma|k_\infty), \mu_p) \cong \mathbb{Z}/p\mathbb{Z}\,.$$

Proof: As above, let $\mathcal{G} = G(\tilde{k}_\Sigma|k_\infty)$ and $\mathcal{H} = G(\tilde{k}_\Sigma|k_\Sigma^{CM})$. Then $\mathcal{G}/\mathcal{H} \cong G(k_\Sigma^{CM}|k_\infty) \cong G(k_\Sigma^+(p)|k_\infty^+)$ is a free pro-p-group of rank λ^+ by (11.3.7). Considering the Hochschild-Serre spectral sequence for the group extension

$$1 \longrightarrow G(k_\Sigma(p)|\tilde{k}_\Sigma) \longrightarrow G(k_\Sigma(p)|k_\infty) \longrightarrow \mathcal{G} \longrightarrow 1$$

and noting that $G(k_\Sigma(p)|k_\infty)$ is free (since $\mu = 0$, cf. (11.3.7)), we obtain a commutative exact diagram

$$\begin{array}{ccccccccc}
 & & & & & & X^+ & = & X^+ \\
 & & & & & & \uparrow\!\!\uparrow & & \uparrow\!\!\uparrow \\
0 & \longrightarrow & H_2(\mathcal{G},\mathbb{Z}_p) & \longrightarrow & G(k_\Sigma(p)|\tilde{k}_\Sigma)^{ab}_{\mathcal{G}} & \longrightarrow & X & \longrightarrow & \mathcal{G}^{ab} & \longrightarrow & 0 \\
 & & {\scriptstyle\varphi_1}\uparrow\!\!\uparrow & & {\scriptstyle\varphi_2}\uparrow\!\!\uparrow & & \uparrow & & {\scriptstyle\varphi_3}\uparrow \\
0 & \longrightarrow & E_\Sigma^- & \longrightarrow & A_\Sigma^- & \longrightarrow & X^- & \longrightarrow & X_{cs}^- & \longrightarrow & 0,
\end{array}$$

where the canonical map

$$\varphi_2 : A_\Sigma^- \cong U_\Sigma^- \xrightarrow{rec} \prod_{\mathfrak{p}\in S_p(k_\infty)} (T_\mathfrak{p}^-)^{ab}_{G_{k_\infty,\mathfrak{p}}} \longrightarrow G(k_\Sigma(p)|\tilde{k}_\Sigma)^{ab}_{\mathcal{G}}$$

is surjective by definition of $\tilde{k}_\Sigma$. It follows that φ_3 is injective and φ_1 is surjective. This proves (i) since the exact sequence $0 \to X_{cs}^- \to \mathcal{G}^{ab} \to X^+ \to 0$ splits (using the action of Δ). The Λ-torsion modules X^+ and X_{cs}^- have projective dimension less than or equal to one by (11.3.2) (ii),(iv) and (11.4.8). Since $\mu = 0$ by assumption, $\mathcal{G}^{ab}$ is $\mathbb{Z}_p$-free and of $\mathbb{Z}_p$-rank $\lambda^+ + \lambda_{cs}^-$, which equals $2\lambda^+$ by (11.4.14). From the exact sequence

$$0 \longrightarrow \mathcal{H} \longrightarrow \mathcal{G} \longrightarrow \mathcal{G}/\mathcal{H} \longrightarrow 0,$$

we obtain the exact sequence

$$0 \longrightarrow \mathcal{H}/[\mathcal{H},\mathcal{G}] \longrightarrow \mathcal{G}^{ab} \longrightarrow (\mathcal{G}/\mathcal{H})^{ab} \longrightarrow 0\,,$$

since $\mathcal{G}/\mathcal{H}$ is free, and so

$$\mathcal{H}/[\mathcal{H},\mathcal{G}] = X_{cs}^-\,.$$

Furthermore, we get from the diagram above a surjection

$$(\mathbb{Z}/p\mathbb{Z})(1) \cong E_\Sigma^-/p \twoheadrightarrow ({}_pH^2(\mathcal{G},\mathbb{Q}_p/\mathbb{Z}_p))^\vee = H^2(\mathcal{G},\mathbb{Z}/p\mathbb{Z})^\vee\,.$$

It remains to show that $H^2(\mathcal{G},\mathbb{Z}/p\mathbb{Z}) \neq 0$. First observe that

$$G(\tilde{k}_\Sigma|k_\Sigma^{CM})^{ab} \cong G(\tilde{L}'|k_\Sigma^{CM})^{ab-}\,,$$

where $\tilde{L}'$ is the maximal p-extension of k_Σ^{CM} which is completely decomposed everywhere. Indeed, for the $(-)$-part this follows from the commutative exact diagram

$$\begin{array}{ccccccc}
0 \longrightarrow & H^1(\tilde{L}'|k_\Sigma^{CM})^- & \longrightarrow & H^1(k_\Sigma|k_\Sigma^{CM})^- & \longrightarrow & \prod\limits_{\mathfrak{p}\in\Sigma(k_\Sigma^{CM})} H^1(k_{\Sigma,\mathfrak{p}}^{CM})^- \\
& \downarrow\wr & & \| & & \| \\
0 \longrightarrow & H^1(\tilde{k}_\Sigma|k_\Sigma^{CM})^- & \longrightarrow & H^1(k_\Sigma|k_\Sigma^{CM})^- & \longrightarrow & \prod\limits_{\mathfrak{p}\in\Sigma(k_\Sigma^{CM})} H^1(k_{\Sigma,\mathfrak{p}}^{CM})^- ,
\end{array}$$

where the coefficients are $\mathbb{Q}_p/\mathbb{Z}_p$, and the (+)-part of $\mathcal{H}^{ab} = G(\tilde{k}_\Sigma|k_\Sigma^{CM})^{ab}$ is trivial, since this holds for the group $\mathcal{H}_\mathcal{G}^{ab}$ as

$$H^1(\tilde{k}_\Sigma|k_\Sigma^{CM}, \mathbb{Q}_p/\mathbb{Z}_p)^{G(k_\Sigma^{CM}|k_\infty)} = (\mathcal{H}/[\mathcal{H},\mathcal{G}])^\vee = (X_{cs}^-)^\vee .$$

Now we consider the diagram of pairings

$$\begin{array}{ccccc}
H^1(\tilde{k}_\Sigma|k_\Sigma^{CM},\mu_p)^{G(k_\Sigma^{CM}|k_\infty)} & \times & H^1(k_\Sigma^{CM}|k_\infty,\mathbb{Z}/p\mathbb{Z}) & \xrightarrow{\cup} & H^1(k_\Sigma^{CM}|k_\infty, H^1(\tilde{k}_\Sigma|k_\Sigma^{CM},\mu_p))_\Delta \\
\uparrow\wr & & \downarrow\wr & & \downarrow \\
H^1(\mathcal{G},\mu_p)_\Delta & \times & H^1(\mathcal{G},\mathbb{Z}/p\mathbb{Z})^\Delta & \xrightarrow{\cup} & H^2(\mathcal{G},\mu_p)_\Delta \\
\uparrow & & \downarrow & & \uparrow \\
H^1(\mathcal{G},\mu_p) & \times & H^1(\mathcal{G},\mathbb{Z}/p\mathbb{Z}) & \xrightarrow{\cup} & H^2(\mathcal{G},\mu_p),
\end{array}$$

which commutes since the Hochschild-Serre spectral sequence is functorial with respect to the cup-product. We know from (11.4.15) that

$$H^1(k_\Sigma^{CM}|k_\infty, H^1(\tilde{k}_\Sigma|k_\Sigma^{CM},\mu_p))_\Delta \cong H^1(k_\Sigma^{CM}|k_\infty, H^1(\tilde{L}'|k_\Sigma^{CM},\mu_p))_\Delta \cong \mathbb{Z}/p\mathbb{Z}$$

and the upper row is a perfect pairing of finite groups. It follows that

$$H^2(\mathcal{G},\mu_p) \cong \mathbb{Z}/p\mathbb{Z}.$$

□

Now we determine the structure of the group $G(\tilde{k}_\Sigma|k_\infty)$, which is exactly analogous to the structure of the fundamental group of a Riemann surface (cf. p.610 and (10.1.2)(ii)).

(11.5.3) Theorem. *Let $k = k^+(\mu_p)$, $p \neq 2$, be a CM-field such that $\mu_p \not\subseteq k_\mathfrak{p}^+$ for all primes $\mathfrak{p}$ above p. Let $k_\infty|k$ be the cyclotomic $\mathbb{Z}_p$-extension and assume that the Iwasawa μ-invariant of $k_\infty|k$ is zero.*

Then $G(\tilde{k}_\Sigma|k_\infty)$ is trivial or a Demuškin group of rank $2g$ where $g = \lambda^+ = \lambda_{cs}^-$. Moreover, there are $2g$ generators $x_1, y_1, \ldots, x_g, y_g$ of $\mathcal{G} = G(\tilde{k}_\Sigma|k_\infty)$ and one defining relation

$$\prod_{i=1}^{g}[x_i, y_i] = 1 .$$

Proof: We know that $\mathcal{G}^{ab} \cong \mathbb{Z}_p^{2\lambda^+}$ is $\mathbb{Z}_p$-torsion-free. If $\mathcal{G}$ is nontrivial, i.e. $\lambda^+ \neq 0$, then $H^2(\mathcal{G}, \mathbb{Z}/p\mathbb{Z}) \cong \mathbb{Z}/p\mathbb{Z}$ by (11.5.2)(ii). Furthermore, the cup-product pairing

$$H^1(\mathcal{G}, \mathbb{Z}/p\mathbb{Z}) \times H^1(\mathcal{G}, \mathbb{Z}/p\mathbb{Z}) \xrightarrow{\cup} H^2(\mathcal{G}, \mathbb{Z}/p\mathbb{Z})$$

is non-degenerate, as shown in the proof of the previous proposition. The result follows by the theory of Demuškin groups (3.9.11). □

(11.5.4) Corollary. *With the assumptions as above, the Galois group $G(\tilde{k}_\Sigma|k)$ is isomorphic to $\mathbb{Z}_p$ or is a Poincaré group of dimension 3 of rank $2g+1$.*

Proof: This follows from (11.5.3) and (3.7.4). □

Having the function field case (cf. X §1) and the corollary above in mind, one is led to ask whether for an arbitrary number field k the absolute Galois group G_k possesses a pro-p-factor group for every prime number p which is either a Poincaré group of dimension 3 or a duality group of dimension 2 according to whether $\zeta_p \in k$ or not (supposing we are not in the "genus 0" case). And indeed this is the case, at least up to a finite number of "bad" primes, for every number field. One can define an arithmetic site over the rings of integers of algebraic number fields such that the desired factor groups occur as fundamental groups. To be precise, it is very likely that the groups constructed have the expected properties, but we can show this only in special situations. The appendix below gives more details.

Appendix: An Arithmetic Site

In the following we try to give a survey of the generalized theory of positive ramification, and we will freely use the notion of Grothendieck topologies and their associated cohomology theories. The definitions and theorems below are taken from [195].

We start with the definition of *orientable* p-adic local fields. First we define, for an odd prime number p, a Galois extension $\mathbb{Q}_p^{pre}$ which is called the maximal pre-orientable extension of $\mathbb{Q}_p$. If $L(odd)$ denotes the maximal Galois extension of a number field L whose (supernatural) degree is odd, and if L^{nr} denotes the maximal unramified extension of L, then we set

$$\begin{aligned} \mathbb{Q}_p^{pre+} &:= \mathbb{Q}_p(\zeta_p + \zeta_p^{-1})^{nr}(odd)\,, \\ \mathbb{Q}_p^{pre} &:= \mathbb{Q}_p^{pre+}(\zeta_p)\,, \end{aligned}$$

where ζ_p is a primitive p-th root of unity.

If p is odd, then the subgroup $\langle\rho\rangle = G(\mathbb{Q}_p^{pre}|\mathbb{Q}_p^{pre+}) \cong \mathbb{Z}/2\mathbb{Z}$ is a normal subgroup of $G(\mathbb{Q}_p^{pre}|\mathbb{Q}_p)$. Therefore the automorphism ρ acts on every local field contained in $\mathbb{Q}_p^{pre}$. The involution ρ extends the automorphism $\mathbb{Q}_p(\zeta_p) \to \mathbb{Q}_p(\zeta_p)$, $\zeta_p \mapsto \zeta_p^{-1}$, and we think of it as a local analogue of complex conjugation. The field $\mathbb{Q}_p^{pre}$ is the maximal extension of $\mathbb{Q}_p$ naturally having such an involution. If a p-adic local field k is contained in $\mathbb{Q}_p^{pre}$, we denote its subfield of ρ-invariant elements by k^+. The extension $k|k^+$ is of degree 1 or 2.

(11.5.5) Definition.

(i) *Let p be an odd prime number. A finite extension field k of $\mathbb{Q}_p$ is called* **orientable** *if*

- $k \subseteq \mathbb{Q}_p^{pre}$ *and*
- *either $k = k^+$ or ζ_p is contained in k^{nr}.*

(ii) *(Ad hoc definition) No 2-adic local field is orientable.*

Note that if p is odd, every abelian extension of $\mathbb{Q}_p$ which contains μ_p is orientable. Unfortunately, there is still no reasonable definition of an "orientation" on a 2-adic local field.

(11.5.6) Definition. *An extension $L|K$ of number fields is called* **positively ramified** *at a prime $\mathfrak{p}|p$ of L if*

(i) $L_\mathfrak{p} \subseteq \mathbb{Q}_p^{pre} K_\mathfrak{p}$ $\quad(\subseteq \bar{\mathbb{Q}}_2$ *if* $p = 2)$, *and if*

(ii) *the Galois closure $\hat{L}_\mathfrak{p}|K_\mathfrak{p}$ has at most pure wild ramification.*

We remark that

- the cyclotomic $\hat{\mathbb{Z}}$-extension of a number field,
- the maximal p-extension of $\mathbb{Q}(\zeta_p + \zeta_p^{-1})$ unramified outside p, and
- unramified extensions

are everywhere positively ramified.

This leads to a covering type (the positive coverings) over the rings of integers of algebraic number fields and defines a Grothendieck topology which is finer than the étale topology, i.e. étale coverings are positively ramified.

If $K|\mathbb{Q}$ is a number field, then there are only finitely many primes $\mathfrak{p}$ of K such that the completion $K_\mathfrak{p}$ is not orientable. These can be thought as primes where $\mathrm{Spec}(\mathcal{O}_K)$ has "bad reduction", since the local duality pairing associated to $\mathrm{Spec}(\mathcal{O}_{K_\mathfrak{p}})_{pos}$ is degenerate at these primes. Suppose $d(K)$ is the smallest positive integer such that $K_\mathfrak{p}$ is orientable at all primes $\mathfrak{p}$ not dividing $d(K)$. We always have $d(K) \mid 2d_K$, where d_K is the absolute discriminant of K. But usually $d(K)$ is much smaller; for example, $d(K) = 2$ if K is a cyclotomic field.

The cohomology groups of sheaves of abelian groups over the positive topology are denoted by H^*_{pos}. They satisfy the following global duality theorem, which is the exact analogue to the étale Poincaré duality theorem for complete curves over finite fields.

(11.5.7) Theorem. *Let K be a number field and let $X = \mathrm{Spec}(\mathcal{O}_K)$. For every integer n with $(n, d(K)) = 1$, there is a canonical trace map*

$$tr : H^3_{pos}(X, \mu_n) \xrightarrow{\sim} \mathbb{Z}/n\mathbb{Z}$$

and for every locally constant constructible sheaf F of $\mathbb{Z}/n\mathbb{Z}$-modules on X_{pos}, the cup-product

$$H^i_{pos}(X, F) \times H^{3-i}_{pos}(X, \mathrm{Hom}(F, \mu_n)) \xrightarrow{\cup} H^3_{pos}(X, \mu_n) \xrightarrow{\sim} \mathbb{Z}/n\mathbb{Z}$$

induces a perfect pairing of finite groups for all i.

The fundamental group $\pi_1^{pos}(X)$ of $X = \mathrm{Spec}(\mathcal{O}_K)$ exists with respect to the positive topology, since the necessary axioms are fulfilled. Let p be a prime number. We consider the maximal pro-p-factor group $\pi_1^{pos}(X)(p)$ of $\pi_1^{pos}(X)$. From the Hochschild-Serre spectral sequence, we get canonical homomorphisms

$$\phi_{M,i} : H^i(\pi_1^{pos}(X)(p), M) \longrightarrow H^i_{pos}(X, M)$$

for all i and every discrete $\pi_1^{pos}(X)(p)$-module M (also considered as a locally constant sheaf on X_{pos}). We say that

$$X \text{ is a } K(\pi, 1) \text{ for } p$$

if $\phi_{M,i}$ is an isomorphism for all i and every finite p-primary $\pi_1^{pos}(X)(p)$-module M. The analogy to function fields predicts that this should be "generically" true. One can verify this condition in some cases and then one can obtain information about $\pi_1^{pos}(X)(p)$ from the duality theorem above. In particular, we naturally obtain corollary (11.5.4) in the following situation.

(11.5.8) Theorem. *Let p be an odd prime number and let K be an abelian extension of $\mathbb{Q}$. Assume that all primes above p ramify in $K|K^+$. Then the following is true:*

(i) *If $K = K^+$, then*

$$\pi_1^{pos}(X)(p) = \begin{cases} \text{free pro-}p\text{-group of finite rank,} & \text{if } \text{Б}_{S_p}(K) = 0, \\ \text{duality group of dimension 2,} & \text{otherwise.} \end{cases}$$

(ii) *If $[K : K^+] = 2$, then either $\pi_1^{pos}(X)(p) \cong \mathbb{Z}_p$ (the genus 0 case) or*

$$\pi_1^{pos}(X)(p) = \begin{cases} \text{Poincaré group of dimension 3,} & \text{if } \zeta_p \in K, \\ \text{duality group of dimension 2,} & \text{if } \zeta_p \notin K. \end{cases}$$

§6. The Main Conjecture

In this section we want to explain how certain Iwasawa modules are connected to other arithmetic objects such as p-adic L-functions, Euler characteristics and K-groups. A full presentation of the subject is beyond the scope of this book, so we will explain the situation, present the main results and provide the most important references. Again the function field case is the easier one and was a guide for the conjectures (now theorems) in the number field case, which are known under the name "Main Conjecture" of Iwasawa theory.

1. *Function Fields*

Assume that k is a function field in one variable over a finite field $\mathbb{F} = \mathbb{F}_q$ of characteristic p. *) Let us consider the function field analogue of the Dedekind zeta function

$$\zeta_k(s) = \sum_{\mathfrak{a}} \frac{1}{\mathfrak{N}(\mathfrak{a})^s} = \prod_{\mathfrak{p}} \frac{1}{1 - \mathfrak{N}(\mathfrak{p})^{-s}} .$$

If r is a natural number, we denote the field $k\mathbb{F}_{q^r}$ by k_r and we set

$$b_r := \text{number of primes in } k \text{ of degree } r,$$

$$N_r := \text{number of primes in } k_r \text{ of degree } 1.$$

Geometrically speaking, N_r is the number of $\mathbb{F}_{q^r}$-rational points of the projective curve attached to k. In particular, N_r and b_r are finite and linked by the formula $N_n = \sum_{r|n} rb_r$. We can therefore transform the product expansion into

$$\zeta_k(s) = \prod_r \frac{1}{(1 - q^{-rs})^{b_r}} = \exp\Big(\sum_r N_r \frac{q^{-rs}}{r}\Big) \overset{\text{def}}{=} Z(q^{-s}) .$$

The following theorem was proven by *H. Hasse* for elliptic curves and by *A. Weil* in the general case.

(11.6.1) Theorem.

(i) *The function $Z(t)$ is a rational function in t. In particular, $\zeta_k(s) = Z(q^{-s})$ can be uniquely defined as a meromorphic function on the complex plane.*

(ii) *We have the functional equation*

$$\zeta_k(1 - s) = (q^{2-g})^{\frac{1}{2}-s} \zeta_k(s) ,$$

where g is the genus of k.

(iii) *$\zeta_k(s)$ has simple poles at $s = 0$ and $s = 1$ and all zeros lie on the line* $\text{Re}(s) = \frac{1}{2}$.

---

*) We assume $\mathbb{F}$ to be algebraically closed in k.

The crucial step in the proof of theorem (11.6.1) is the following

(11.6.2) Theorem. *Let X be the smooth projective curve associated to k and let $\alpha_1, \cdots, \alpha_{2g}$ be the zeros of the characteristic polynomial of the geometric Frobenius automorphism F acting on the Jacobian variety* Jac X. *Then*

$$Z(t) = \frac{(1-\alpha_1 t)\cdots(1-\alpha_{2g}t)}{(1-t)(1-qt)}.$$

Choose a prime number ℓ different to the characteristic p of k. By [140], 12.1, the characteristic polynomial of F acting on Jac X coincides with the characteristic polynomial of F acting on the $2g$-dimensional $\mathbb{Q}_\ell$-vector space

$$T_\ell(\text{Jac } X) \otimes_{\mathbb{Z}_\ell} \mathbb{Q}_\ell \cong \mathbb{H}_i := H_i(G_\varnothing(k\bar{\mathbb{F}}), \mathbb{Z}_\ell) \otimes \mathbb{Q}_\ell$$

(cf. X §1, p.613). These groups are naturally locally compact $G(\bar{\mathbb{F}}|\mathbb{F})$-modules. Let $\gamma \in G(\bar{\mathbb{F}}|\mathbb{F})$ be the arithmetic Frobenius automorphism. It can be shown that the actions of F and γ are inverse to each other. Hence we can reformulate the result in the form

$$Z(t) = \prod_{i=1}^{2} \det(1-\gamma t \mid \mathbb{H}_i)^{(-1)^{i+1}}\,. \,^{*)}$$

In particular, we see that the above expression is independent of the prime number $\ell \neq p$.

Result: In the function field case, the zeta function is a rational function in q^{-s}. It therefore can also be viewed as a function on $\mathbb{Q}_\ell$ (or $\mathbb{C}_\ell$) and then it is essentially the characteristic polynomial of the Frobenius automorphism acting on an associated finite dimensional $\mathbb{Q}_\ell$-vector space.

The above result has vast generalizations. Instead of smooth projective curves, one can consider arbitrary varieties over finite fields and L-functions associated to constructible sheaves of $\mathbb{Q}_\ell$-vector spaces.

2. *p-adic Zeta Functions*

Now assume that k is a number field. The Dedekind zeta function $\zeta_k(s)$ is a meromorphic function on the complex plane having a simple pole at $s=1$. It is, however, not a rational function. Adding a certain Euler factor $L_\infty(s)$ for the infinite places (see [160], chap.VII, §§4,5), we obtain a modified zeta function with simple poles at $s=0,1$, which satisfies a functional equation like (11.6.1)(ii) with the term (q^{2-g}) replaced by $|d_k|^{-1}$ (loc.cit.). The question

*) Working with ℓ-adic cohomology $\mathbb{H}^i$, one has to replace γ by γ^{-1}.

whether all zeros are on the line $\mathrm{Re}(s) = \frac{1}{2}$ is the famous (and unproven) Riemann hypothesis. We have the

(11.6.3) Theorem (*Siegel-Klingen*). *Let $n \geq 1$ be an integer. Then*

$$\zeta_k(1-n)$$

is a rational number. These values vanish for $n > 1$ either if n is odd or if k has a complex place. If k is totally real and n is even, then these values are non-zero.

These special zeta-values satisfy several congruence relations *) which can be reformulated as the existence of a continuous function on the p-adic numbers. More precisely, let k be a totally real number field, p a prime number and $d = [k(\mu_{2p}) : k]$. Then we have the following result, cf. [37]:

(11.6.4) Theorem. *There exists a unique continuous function*

$$\zeta_{k,p} : \mathbb{Z}_p \smallsetminus \{1\} \longrightarrow \mathbb{Q}_p$$

satisfying

$$\zeta_{k,p}(1-n) = \zeta_k(1-n) \prod_{\mathfrak{p}|p} \left(1 - \mathfrak{N}(\mathfrak{p})^{n-1}\right)$$

for all $n > 1$ with $d \mid n$. The function $\zeta_{k,p}(s)$ is p-adic analytic, having at most a simple pole at $s = 1$.

Observe that the second factor on the right-hand side of the defining equation is just the Euler factor at p which has to be removed before finding a p-adic interpolating function.

The existence of such a p-adic interpolating function has been verified for abelian number fields by *T. Kubota* and *H. W. Leopoldt*. More generally, they showed the existence of p-adic L-functions attached to Dirichlet characters (see below). Many mathematicians made contributions to extend the result of Kubota-Leopoldt to arbitrary totally real number fields. Amongst others, we mention the names *J.-P. Serre*, *N. Katz*, *D. Barsky*, *P. Cassou-Noguès*, *P. Deligne* and *K. Ribet*. The general idea of the construction is to interpret the elements of Λ as p-adic measures and then to obtain the required L-function as a Mellin transform with respect to p-adic integration. The interested reader should consult [122], chap. 4, or [246], chap. 12 and the references given there.

---

*) They are called Bernoulli congruences if $k = \mathbb{Q}$.

There exists the following p-adic analogue of the analytic class number formula in order to compute the residue of $\zeta_{k,p}$ at $s = 1$ (see [246], th. 5.24 for abelian number fields and [30] for the general case).

(11.6.5) Theorem (*p-adic Analytic Class Number Formula*).

$$\lim_{s\to 1}(s-1)\zeta_{k,p}(s) = \frac{2^n\, R_p\, h}{w\,\sqrt{|d_k|}} \prod_{\mathfrak{p}|p}\left(1 - \mathfrak{N}(\mathfrak{p})^{-1}\right),$$

where R_p is the p-adic regulator (see X §3), h is the class number and d_k is the discriminant of k; $w\,(=2)$ is the number of roots of unity contained in the (totally real) number field k and $n = [k:\mathbb{Q}]$.

Remark: In the above formula both R_p and $\sqrt{|d_k|}$ are determined only up to sign but it is possible to give their ratio a well-defined sign, see [2].

(11.6.6) Corollary. *$\zeta_{k,p}$ has a (simple) pole at $s = 1$ if and only if the Leopoldt conjecture is true for k and p.*

In particular, this is the case if k is an abelian number field, by (10.3.16). This connection between the residue at $s = 1$ of the p-adic zeta function and the Leopoldt conjecture can be easily deduced from the main conjecture (see below); however, it is used in the proof.

Let us now formulate the main conjecture for the p-adic zeta function. Since the case $p = 2$ introduces additional, rather subtle difficulties, we will assume that p is odd in what follows.

We denote the cyclotomic $\mathbb{Z}_p$-extension of k by k_∞ and we fix a topological generator γ of $\Gamma = G(k_\infty|k)$. Note that we do not have a canonical generator like the Frobenius automorphism in the function field case. Therefore we choose any generator and all statements and results should be independent of the choice made. Let

$$\kappa : \Gamma = G(k_\infty|k) \cong G(k(\mu_{p^\infty})|k(\mu_p)) \longrightarrow \mathbb{Z}_p^\times$$

be the p-part of the cyclotomic character. We set

$$q := \kappa(\gamma)\,,$$

and we think of q as a substitute for the order of the ground field in the function field case. Since q is a principal unit, we have a well-defined q-exponentiation map

$$\mathbb{Z}_p \longrightarrow \mathbb{Z}_p^\times, \quad s \longmapsto q^s .$$

In contrast to the function field case, the p-adic zeta function has no pole at $s = 0$ and it does not satisfy a functional equation. It is therefore natural to look for an expansion around the pole at $s = 1$. The following result is proved together with the existence of the p-adic zeta function.

(11.6.7) Theorem. *There exists a unique power series $G_{k,p} \in \mathbb{Z}_p[[T]]$ such that*

$$\zeta_{k,p}(s) = G_{k,p}(q^{1-s} - 1)/(q^{1-s} - 1) .$$

The main conjecture of Iwasawa theory claims that $G_{k,p}$ is essentially the characteristic polynomial of the Iwasawa module $X = X_\Sigma$, i.e. of the Galois group of the maximal abelian extension of k_∞ which is unramified outside $\Sigma = S_p \cup S_\infty$. This was first proved by *B. Mazur* and *A. Wiles* [134] under the assumption that the base field is abelian over $\mathbb{Q}$, and later by *A. Wiles* [249] for general totally real fields.

(11.6.8) Theorem (*Main Conjecture*). *Let $F_X(t)$ and $\mu(X)$ be the characteristic polynomial and the Iwasawa μ-invariant of $X = X_\Sigma$. Then we have the following equality of ideals in $\mathbb{Z}_p[[T]]$:*

$$\Big(G_{k,p}(T)\Big) = \Big(p^{\mu(X)} \cdot F_X(T)\Big).$$

In other words, the functions on both sides differ by an invertible power series.

It is not difficult to see that the above result is invariant under a change of the generator $\gamma \in \Gamma$ (changing γ also alters $q = \kappa(\gamma)$). Conjecturally we have $\mu(X) = 0$ and this is proven for abelian number fields. In general, the theorem says that the analytic and algebraic μ-invariants coincide and that $F_X(T)$ is the Weierstraß polynomial associated to $p^{-\mu(X)}G_{k,p}$.

Remark: Consider the field $k(\mu_p)$. By (11.4.4), there is a pseudo-isomorphism of Λ-modules

$$X = e_0 X(k(\mu_p)) \approx (e_1 X_{nr}(k(\mu_p)))^\circ,$$

where e_i, $i = 0, 1$, are certain idempotents in $\mathbb{Z}_p[G(k(\mu_p)|k)]$ (see §4). Therefore the main conjecture can be reformulated in terms of the Iwasawa module $e_1 X_{nr}(k(\mu_p))$.

3. *Applications*

The main conjecture has several important applications. Most of the theorems below were first proved assuming the main conjecture before it was itself proved.

As a first application we will consider the group K_2 of rings of integers in totally real number fields. Consider the **Steinberg group** $St(R)$ of a ring R. It is generated by symbols $s_{ij}(\alpha)$, $i, j \in \mathbb{N}, i \neq j$, $\alpha \in R$, modulo the relations

$$s_{ij}(\alpha)s_{ij}(\beta) = s_{ij}(\alpha+\beta), \quad [s_{ij}(\alpha), s_{kl}(\beta)] = \begin{cases} 1 & \text{for } \; i \neq l, j \neq k, \\ s_{il}(\alpha\beta) & \text{for } \; j = k, i \neq l. \end{cases}$$

The group $\boldsymbol{K_2(R)}$ is defined as the center of the Steinberg group $St(R)$. If R is a field, then by Matsumoto's theorem, $K_2(R)$ coincides with the Milnor K-group defined in VI §4. Its arithmetic importance lies in the following exact sequence, which is part of the long localization sequence for the higher K-theory of Quillen:

$$0 \longrightarrow K_2(\mathcal{O}_k) \longrightarrow K_2(k) \stackrel{d}{\longrightarrow} \bigoplus_{v \text{ finite}} k(v)^\times \longrightarrow 0.$$

Here k is a number field and d is defined by the *tame symbols*

$$d_v : K_2(k) \to k(v)^\times$$

for all finite primes v (see VI §4, ex.1). It can be shown that $K_2(\mathcal{O}_k)$ is finite, and using Tate's result (6.4.4), *J. Coates* [26] showed the existence of an isomorphism

$$K_2(\mathcal{O}_k)(p) \cong \Big(Cl(k(\mu_{p^\infty}))(p) \otimes \mathbb{Z}_p(1)\Big)^{G(k(\mu_{p^\infty})|k)}.$$

This allows us to compute the odd part of $\#K_2(\mathcal{O}_k)$ via the main conjecture. Additional results of *M. Kolster* and *A. Wiles* for the 2-primary part (see [113], [249]) then imply the

(11.6.9) Theorem (*Birch-Tate Conjecture*). *Let k be a totally real number field and let $w_2(k)$ be the largest positive integer N such that $G(k(\mu_N)|k)$ has exponent 2. Then*

$$\#K_2(\mathcal{O}_k) = w_2(k) \cdot |\zeta_k(-1)|.$$

Another application of the main conjecture (see [10]) is the calculation of special values of zeta functions as Euler characteristics, as was conjectured by *S. Lichtenbaum*.

(11.6.10) Theorem. *Let k be a totally real number field, p an odd prime number and n be an even positive integer. Then*

$$\begin{aligned} |\zeta_k(1-n)|_p &= \prod_{i=0,1} \left(\#H^i(G_{S_p}, \mathbb{Q}_p/\mathbb{Z}_p(n))\right)^{(-1)^{i+1}} \\ &= \prod_{i=1,2} \left(\#H^i_{cts}(G_{S_p}, \mathbb{Z}_p(n))\right)^{(-1)^{i}}. \end{aligned}$$

The cohomology groups on the right-hand side have finite order by Soulé 's theorem (10.3.27) in conjunction with (8.7.6).

As another application, let us consider the group of **cyclotomic units**. It is defined for abelian number fields in the following way (see [217]): if k is an abelian number field, we denote its unit group by E_k. Let n be an integer > 1, and let a be any integer not divisible by n. The number

$$N_{\mathbb{Q}(\mu_n)|\,k\cap\mathbb{Q}(\mu_n)}(1-\zeta_n^a)$$

lies in $k^\times$; we define the cyclotomic numbers D_k of k to be the group generated in $k^\times$ by -1 and all such elements $N_{\mathbb{Q}(\mu_n)|k\cap\mathbb{Q}(\mu_n)}(1-\zeta_n^a)$. The cyclotomic units are then defined by

$$C_k = E_k \cap D_k.$$

We have (loc.cit.) the following

(11.6.11) Theorem. *If k is a real abelian number field, then the group of cyclotomic units is of finite index in the full unit group and*

$$[E_k : C_k] = h_k \cdot c_k,$$

where h_k is the class number and c_k is a rational number which has a nontrivial p-part only for primes p dividing $2\cdot[k:\mathbb{Q}]$.

We conclude that the $\mathbb{Z}_p[G(k|\mathbb{Q})]$-modules

$$Cl(k)\otimes\mathbb{Z}_p \quad \text{and} \quad (E_k/C_k)\otimes\mathbb{Z}_p$$

have the same order for $p \nmid 2[k:\mathbb{Q}]$. In general, however, they are not isomorphic. For example (see the remark after th. 8.2. in [246]), for the real quadratic field $k = \mathbb{Q}(\sqrt{62501})$ the 3-part of E_k/C_k is cyclic of order 9 but $Cl(k)(3) \cong \mathbb{Z}/3\mathbb{Z}\times\mathbb{Z}/3\mathbb{Z}$. Nevertheless, *G. Gras* conjectured that, if $p \nmid 2[k:\mathbb{Q}]$, both modules have the same image in $K_0'(\mathbb{Z}_p[G(k|\mathbb{Q})])$. *R. Greenberg* [63] showed that this conjecture is indeed a consequence of the main conjecture.

(11.6.12) Theorem (*Conjecture of Gras*). *Let k be a real abelian number field and assume that the odd prime number p does not divide $[k:\mathbb{Q}]$. Then the $\mathbb{Z}_p[G(k|\mathbb{Q})]$-modules $(E_k/C_k)\otimes\mathbb{Z}_p$ and $Cl(k)\otimes\mathbb{Z}_p$ have isomorphic Jordan-Hölder series.*

The conjecture of Gras is an analogue of an earlier conjecture of Iwasawa and Leopoldt concerning the odd eigenspace of ideal class groups of cyclotomic fields, relating it to the quotient of $\mathbb{Z}_p[G(k|\mathbb{Q})]$ by the Stickelberger ideal. This conjecture was proven by Mazur and Wiles [134], thereby extending earlier results of Greenberg [61].

4. *Characters*

All ideas, conjectures and theorems stated so far can be refined to statements about characters. Moreover, the main conjecture was formulated for characters from the beginning, and was also proved in this greater generality.

We keep the assumption that k is a totally real number field and that p is an odd prime number. Let χ be a one-dimensional even character of G_k, i.e. we are given a continuous homomorphism

$$\chi : G_k \longrightarrow \bar{\mathbb{Q}}^\times$$

which sends every complex conjugation to 1. The image of χ is necessarily finite and let k_χ be the extension attached to χ, i.e. χ defines a faithful representation of $G(k_\chi|k)$. Since χ is even, the field k_χ is again totally real. Choosing embeddings $\bar{\mathbb{Q}} \subseteq \mathbb{C}$ and $\bar{\mathbb{Q}} \subseteq \mathbb{C}_p$, we will likewise view χ as a complex or as a p-adic character. We denote the ring of integers in the field $\mathbb{Q}_p(\chi)$, obtained from $\mathbb{Q}_p$ by adjoining all values of χ, by $\mathcal{O}_\chi = \mathbb{Z}_p[\chi]$. Furthermore, let

$$\omega : G(k(\mu_p)|k) \longrightarrow \mu_{p-1} \subseteq \mathbb{Z}_p^\times$$

be the Teichmüller character, i.e. the tame part of the cyclotomic character. Deligne and Ribet have shown that there exists a continuous p-adic L-function $L_p(s,\chi)$ on $\mathbb{Z}_p \smallsetminus \{1\}$, and even at $s=1$ if χ is not trivial, which satisfies the interpolation property

$$(1) \qquad L_p(1-n,\chi) = L(1-n,\chi\omega^{-n}) \prod_{\mathfrak{p}|p} \left(1 - \chi\omega^{-n}(\mathfrak{p})\mathfrak{N}(\mathfrak{p})^{n-1}\right)$$

for every integer $n \geq 1$. Here, $L(1-n,\chi\omega^{-n})$ is the value of the classical complex L-function. To make sense of this, we recall the formula

$$(2) \qquad L(1-n,\psi) = \sum_{\sigma\in G(k_\psi|k)} \psi(\sigma)\zeta_k(\sigma,1-n)$$

for a complex character ψ. The value $\zeta_k(\sigma,1-n)$ of the partial zeta function is rational by the theorem of Siegel-Klingen (cf. [160], chap.VII, (9.9)).

We therefore can use (2) in order to interpret $L(1-n, \chi\omega^{-n})$ as a p-adic number in (1). If χ is the trivial character, then obviously $L_p(s,\chi) = \zeta_{p,k}(s)$.

Following Greenberg, we say that χ is of **type S** if $k_\chi \cap k_\infty = k$ and of **type W** if $k_\chi \subseteq k_\infty$. Let $H_\chi(T)$ be defined as $(\chi(\gamma)(1+T)-1)$ if χ is of type W and 1 otherwise. Then there exists a power series $G_\chi(T) \in \mathcal{O}_\chi[[T]]$ such that

$$L_p(s,\chi) = G_\chi(q^{1-s}-1)/H_\chi(q^{1-s}-1).$$

If ψ is of type W, then

$$G_{\chi\psi}(T) = G_\chi(\psi(\gamma)(1+T)-1).$$

The main conjecture compares G_χ with an algebraically defined polynomial f_χ for characters χ of type S. The polynomial f_χ will be of the form

$$f_\chi = \pi^{\mu_\chi} \cdot f_\chi^*,$$

where π is a uniformizer of $\mathcal{O}_\chi$ and $f_\chi^* \in \mathcal{O}_\chi[[T]]$ is a Weierstraß polynomial. Let us define f_χ^* first. Consider the $\mathbb{Z}_p[[\Gamma]]$-module

$$X = X_{S_p}(k_\chi).$$

It carries a canonical $G(k_\chi|k)$-module structure. After tensoring with $\bar{\mathbb{Q}}_p$, we can decompose it into eigenspaces with respect to idempotents. We set

$$V = X \otimes_{\mathbb{Z}_p} \bar{\mathbb{Q}}_p,$$

and we denote the χ-eigenspace of V by $V^{(\chi)}$ *).

(11.6.13) Definition. *We define f_χ^* as the characteristic polynomial for the action of $\gamma - 1$ on $V^{(\chi)}$.*

If the order of χ is prime to p, then the definition of μ_χ is straightforward.

(11.6.14) Definition. *If the order of χ is prime to p, then we define μ_χ as the μ-invariant of the $\mathcal{O}_\chi[[T]]$-module*

$$(X \otimes \mathcal{O}_\chi)^{(\chi)} = \{x \in X \otimes \mathcal{O}_\chi \mid gx = \chi(g)x,\ \forall g \in G(k_\chi|k)\}.$$

One can also define a μ-invariant for characters of order divisible by p. By the theorem of Ferrero-Washington, these μ-invariants are equal to zero for Dirichlet characters and in view of the conjecture that $\mu = 0$ the following discussion is presumably empty for all characters. Therefore the reader who

*) i.e. $V^{(\chi)} = \{v \in V \mid gv = \chi(g)v \text{ for all } g \in G(k_\chi|k)\}$.

is only interested in the zeros of the p-adic L-functions may skip the following considerations.

(We wish to thank R. Greenberg for his help with this point.) Assume that we are given a character whose order is divisible by p. We can write it uniquely in the form $\chi = \varphi \cdot \psi$, where φ is of order prime to p and ψ is of p-power order, say of order p^n. We have inclusions $k \subseteq k_\psi \subseteq k_\chi = k_\varphi k_\psi$. The idea of the definition of μ_χ is the following observation from the analytic side:

Consider the p-adic L-function of φ as a character of the field k_ψ and as a character of (the subfield of k_ψ of degree p of) k_{ψ^p}. Their ratio is the product of the p-adic L-functions over k of the characters $\varphi \cdot \psi'$, where ψ' varies over all $\mathbb{Q}$-conjugates of ψ. All these L-functions have the same analytic μ-invariant. This motivates the following definition.

Let $X^{(\varphi)}$ be defined as

$$X^{(\varphi)} = (X_{S_p}(k_\chi) \otimes_{\mathbb{Z}_p} \mathcal{O}_\varphi)^{(\varphi)}.$$

Then $X^{(\varphi)}$ is a $\mathcal{O}_\psi[H][[T]]$-module, where $H := G(k_\psi|k)$. The decomposition of $(t^{p^n} - 1)$ into irreducible cyclotomic polynomials, together with the evaluation of ψ, induces an isomorphism

$$\mathcal{O}_\varphi[H] \otimes_{\mathbb{Z}} \mathbb{Q} \cong \mathbb{Q}_p(\varphi) \oplus \mathbb{Q}_p(\varphi)(\mu_p) \oplus \cdots \oplus \mathbb{Q}_p(\varphi)(\mu_{p^n}).$$

The projection to the highest component therefore induces a homomorphism

$$\alpha : \mathcal{O}_\varphi[H] \longrightarrow \mathcal{O}_\chi \subseteq \mathbb{Q}_p(\varphi)(\mu_{p^n}) = \mathbb{Q}_p(\chi).$$

Let h be a generator of the cyclic p-group H. Then $\ker \alpha$ is generated by $\Phi_{p^n}(h)$, where Φ_{p^n} is the p^n-th cyclotomic polynomial. The action of $\mathcal{O}_\varphi[H]$ on the submodule

$$(h^{p^{n-1}} - 1)X^{(\varphi)} \subseteq X^{(\varphi)}$$

factors through α.

(11.6.15) Definition. *For a general character χ, we define μ_χ as the μ-invariant of $(h^{p^{n-1}} - 1)X^{(\varphi)}$, viewed as an $\mathcal{O}_\chi[[T]]$-module via α.*

Remark: The Weierstraß polynomial f_χ^* defined above coincides with the characteristic polynomial of the Iwasawa module $(h^{p^{n-1}} - 1)X^{(\varphi)}$.

Now we are able to state the main conjecture in its general form (but only for odd p).

(11.6.16) Main Conjecture of Iwasawa Theory. *Assume that χ is even and of type S. Then with the above notation we have an equality of ideals in $\mathcal{O}_\chi[[T]]$:*

$$\big(G_\chi(T)\big) = \big(\pi^{\mu_\chi} \cdot f^*_\chi(T)\big).$$

Wiles [249] only defined the μ-invariant for characters of order prime to p and therefore he stated his main theorem concerning the μ-invariants only for those characters. Using the above definition of μ-invariants for characters of arbitrary order, (11.6.16) is easily deduced from the main results th. 1.3 and 1.4 of [249].

There exists yet another generalization; namely, the existence of p-adic L-functions can be extended to characters of arbitrary degree (provided that k_χ is totally real). This has been proved by *R. Greenberg* [61], who also showed that there exists an analogous power series expansion

$$L_p(s,\chi) = G_\chi(q^{1-s} - 1)/H_\chi(q^{1-s} - 1).$$

However, G_χ is a priori only in the quotient field of $\mathbb{Z}_p[\chi][[T]]$. The following is the p-adic version of the famous Artin conjecture (cf. [160], chap.VII, §10).

(11.6.17) Theorem. *If p is odd, then $G_\chi \in \mathbb{Z}_p[\chi][[T]] \otimes \mathbb{Q}_p$.*

Greenberg (loc.cit.) showed that (11.6.17) follows from the main conjecture.

5. *Motivation*

Let us finally say a few words about the history, the motivation and the proof of the main conjecture. To make things easier, we restrict to the maximal real abelian subfield of $\mathbb{Q}(\zeta_p + \zeta_p^{-1})$ and we assume that p is odd. The analytic part of the arguments sketched below can be found in Washington's book [246].

Recall that the prime number p is called regular if the class number of $k = \mathbb{Q}(\mu_p)$ is not divisible by p. It was already known to *E. Kummer* that this is equivalent to the fact that none of the (numerators of the) Bernoulli numbers $B_2, B_4, \dots, B_{p-3}$ is divisible by p.

This arises as follows. Using (11.4.3) and (11.1.7), it is not difficult to see that for even i we have the equivalence

$$e_i\, G_\Sigma(k)^{ab}/p = 0 \iff e_{1-i}\, Cl(k)(p) = 0,$$

with an obvious modification in the case $i = 0$. This equivalence is known under the name *Leopoldt's Spiegelungssatz*, see §4 ex. In particular, p is regular if and only if p does not divide the order h_k^- of the $(-)$-part of $Cl(k)(p)$ with respect to the action of the complex conjugation. It follows from the analytic class number formula that $p \nmid h_k^-$ if and only if p does not divide the L-values $L(0, \omega^i)$ for $i = 1, 3, \dots, p-4$. Finally, the congruence

$$L(0, \omega^i) \equiv -\frac{B_{i+1}}{i+1} \bmod p$$

implies the statement.

It is then natural to look for a finer correspondence between the non-triviality of $e_i\, Cl(k)$ and the p-valuation of $L(1, \omega^i)$ for every odd i separately. An investigation of the Stickelberger ideal shows that $L(1, \omega^i)$ indeed annihilates the group $e_i\, Cl(k)$ for $i = 3, 5, \dots, p-2$. We deduce (the *Theorem of Herbrand*) that for these i,

$$e_i\, Cl(k)(p) \neq 0 \;\Rightarrow\; p | B_{p-i}.$$

Iwasawa showed that the p-adic L-function can alternatively be constructed using Stickelberger elements in the group ring $\mathbb{Z}_p[G(\mathbb{Q}(\mu_{p^n})|\mathbb{Q})]$. This approach made it possible for him to derive a connection between the p-adic L-functions and the quotient of the local units modulo the cyclotomic units. More precisely, using the notation of §3, we consider the Iwasawa module of local units $U = U_{S_p}(k_\infty)$ and its submodule

$$C = \varprojlim_n (C_{k_n} \otimes \mathbb{Z}_p),$$

where C_{k_n} is the group of cyclotomic units in the field $k_n = \mathbb{Q}(\mu_{p^n})$. Since the weak Leopoldt conjecture is true for the cyclotomic $\mathbb{Z}_p$-extension, cf. (10.3.25), C is a subgroup of U.

(11.6.18) Theorem (IWASAWA). *Let $i \not\equiv 0 \bmod p-1$ be even. Then*

$$e_i\,(U/C) \approx \Lambda/f_i(T), \text{ where } f_i(q^{1-s}-1) = L_p(s, \omega^i).$$

This important theorem relates an analytic object, namely the p-adic L-function, with a purely algebraic defined object, the Iwasawa module $e_i\,(U/C)$.

Now recall that the order of the p-part of E_{k_n}/C_{k_n} is essentially equal to

$$h_{k_n}^+ = \#Cl(\mathbb{Q}(\zeta_{p^n} + \zeta_{p^n}^{-1}))(p)\,,$$

and consider the exact sequence (11.3.10)

$$0 \longrightarrow E \longrightarrow U \longrightarrow X_\Sigma \longrightarrow X_{nr} \longrightarrow 0.$$

We see that if $h_{k_n}^+$ is bounded independently of n, i.e. if the Greenberg conjecture (see p. 728) holds, then we have a pseudo-isomorphism for even $i \not\equiv 0 \bmod p-1$,

$$e_i\,(U/C) \approx e_i\,(U/E) \approx e_i\, X_\Sigma.$$

Therefore the Greenberg conjecture implies the main conjecture. In particular (see (11.1.7)), this proves the main conjecture for the powers of the (mod p)-Teichmüller character if p is an odd prime number such that $p \nmid h(\mathbb{Q}(\zeta_p + \zeta_p^{-1}))$ (i.e. if Vandiver's conjecture holds for p).

The above results are at least an indication that one possible method to prove the main conjecture is to compare the asymptotic orders of $e_i\,(E_{k_n}/C_{k_n})$ and of $e_i\,Cl(k_n)(p)$. Since the product over all eigenspaces is the same for both terms by the analytic class number formula, it suffices to show an inequality in one direction.

Indeed, there now exists a proof along these lines (cf. [184] or the new edition of [246]) making use of the technique of Euler systems introduced by *V. A. Kolyvagin* and *F. Thaine*. However, this technique can only be applied to abelian number fields because we are lacking cyclotomic units in the more general situation.

Mazur and Wiles applied another more geometric method. A similar class number formula argument as above shows that it suffices to prove that one of the ideals in the main conjecture divides the other one. Using the variant with the Iwasawa module X_{nr}, this comes down to constructing many unramified abelian extensions in the $(-)$-eigenspaces. Here, "many" depends on the p-valuation of certain values of L-functions. A first important step in this direction was achieved by *K. Ribet*, who proved the converse to Herbrand's theorem.

(11.6.19) Theorem (*Ribet*). *For* $i = 3, 5, \ldots, p-2$,

$$p | B_{p-i} \;\Rightarrow\; e_i\,Cl(k)(p) \neq 0.$$

Ribet's method of constructing the required unramified extension of $\mathbb{Q}(\mu_p)$ uses arithmetic geometry. The extension comes from a 2-dimensional representation of $G_{\mathbb{Q}}$ which is attached to a certain quotient of the modular variety $J_1(p)$. The quotient is associated to a normalized weight 2 cusp form whose coefficients satisfy certain congruences and which exists if the corresponding Bernoulli number is divisible by p (see [180]).

The proof of the main conjecture given by Mazur and Wiles for abelian number fields, and by Wiles for arbitrary totally real number fields, strongly refines this technique. We refer the reader to the original papers [134], [249].

The main conjecture as presented above is only the starting point of a much more general theory of "Iwasawa theory of motives", which is, however, to a large extent conjectural on both the analytic and the algebraic side. As a first step in this direction, *J. Coates* and *A. Wiles* investigated p-adic L-functions associated to elliptic curves over $\mathbb{Q}$ with complex multiplication by the ring of integers of an imaginary quadratic number field. They established results in the direction of a main conjecture relating these p-adic L-functions to the Iwasawa module structure of the Tate module of the elliptic curve (see [29], [33]). Using techniques developed by *V. A. Kolyvagin*, *K. Rubin* proved this main conjecture, see [185]. For related work on abelian varieties see [132], [133], [203].

For a survey of the general ideas of motivic Iwasawa theory, we refer the reader to the articles [204], [27], [65], [100].

Exercise 1. (Invariance of the main conjecture under liftings.) Show that it is not necessary to work with the minimal field k_χ, i.e. suppose that we are given a finite totally real extension $k'|k_\chi$. Then we may consider the (not faithful) character χ' of $G(k'|k)$ which is the composite of χ with the restriction of $G(k'|k)$ to $G(k_\chi|k)$. Define $f_{\chi'}$ in an analogous manner from $X(k')$ and prove that $f_{\chi'}(T) = f_\chi(T)$ (see [61], prop.1).

Exercise 2. Define the algebraic p-adic L-function associated to a character χ by

$$L_p^{alg}(s,\chi) = f_\chi(q^{1-s} - 1)/H_\chi(q^{1-s} - 1).$$

Show that this definition is independent of the choice of the generator $\gamma \in \Gamma$. Furthermore, assume that we are given a finite abelian extension $K|k$ of totally real number fields with $K \cap k_\infty = k$. Show that

$$\zeta_{p,K}^{alg}(s) = \zeta_{p,k}^{alg}(s) \prod_{\chi \neq 1} L_p^{alg}(s,\chi),$$

where χ runs through all nontrivial characters of $G(K|k)$.

Exercise 3. Deduce the following generalization of the main conjecture: let $S \supseteq S_p$ be a finite set of finite primes of k, and let χ be an even character of G_k as before. Then there exists a continuous p-adic L-function $L_{p,S}(s,\chi)$ satisfying the interpolation property

$$L_{p,S}(1-n,\chi) = L(1-n,\chi\omega^{-n}) \prod_{\mathfrak{p} \in S} \left(1 - \chi\omega^{-n}(\mathfrak{p})\mathfrak{N}(\mathfrak{p})^{n-1}\right)$$

for every integer $n \geq 1$. Show that a variant of the main conjecture holds with the algebraic p-adic L-function which is defined in the same way as before but with the Iwasawa module X_S instead of $X = X_{S_p}$.

Chapter XII
Anabelian Geometry

In this last chapter we want to give some idea of the "anabelian" program of *A. Grothendieck*. The term anabelian should be read as "far from being abelian" and as we understand the matter, a group is far enough away from being abelian if all of its subgroups of finite index have a trivial center. The principal idea is the following: in topology, a space X of type $K(\pi, 1)$ is determined by its fundamental group π up to weak homotopy equivalence. If we require that X is a CW-complex, then X is already determined up to strong homotopy equivalence. The "anabelian" idea is that something similar should also be true for schemes, i.e. a scheme X which is an étale $K(\pi, 1)$ should be essentially reconstructible from its étale fundamental group. This is obviously not correct in general, but it should be true under certain conditions; for example, X should be absolutely finitely generated and $\pi_1^{et}(X)$ is supposed to be "anabelian". The smallest constituents of this anabelian world are points, i.e. spectra of fields which are finitely generated over their prime fields. Here the étale fundamental group is just the absolute Galois group. Finite fields have an abelian absolute Galois group (in fact all these fields have the *same* Galois group $\hat{\mathbb{Z}}$), and so the first objects of interest are global fields. In §1,2 we will present some results on "anabelian properties" of global fields, which already existed before Grothendieck formulated his program. We will explain the general conjectures in §3.

§1. Subgroups of G_k

In the previous chapters we were mainly interested in arithmetically relevant quotients of the absolute Galois group G_k of a global field k; for example, we studied the group G_S of the maximal extension of k unramified outside a set of primes S. This gave the insight that a global field has more structure than just the structure of an abstract field. It comes equipped with a set of valuations which provide a family of local structures fitting together globally in an arithmetically relevant way. On the level of Galois groups we therefore should consider not only G_k, but G_k together with the distinguished family of

J. Neukirch et al., *Cohomology of Number Fields*,
Grundlehren der mathematischen Wissenschaften 323,
DOI 10.1007/978-3-540-37889-1_12, © Springer-Verlag Berlin Heidelberg 2008

closed subgroups $G_{\mathfrak{p}}$, $\mathfrak{p}$ prime of k. It is a remarkable and surprising fact that this distinguished family is already encoded in G_k, i.e. we can see whether a closed subgroup in G_k is the decomposition group for a prime in a purely group theoretical way. In other words, we can reconstruct the arithmetic structure of G_k from its algebraic structure.

We need the following application of Krasner's lemma (8.1.6).

(12.1.1) Lemma. *Let k be a field, complete with respect to a valuation $|\ |$*[*)] *and let $f_1 = a_{0,1} + a_{1,1}X + \cdots + a_{d,1}X^d \in k[X]$ be a separable polynomial. Then every polynomial $f_2 = a_{0,2} + a_{1,2}X + \cdots + a_{d,2}X^d \in k[X]$ with sufficiently small distance*

$$|f_1 - f_2| \stackrel{\mathrm{def}}{=} \max_{j=0,\ldots,d} |a_{j,1} - a_{j,2}|$$

has the same splitting field as f_1.

Proof: If $k = \mathbb{C}$, there is nothing to prove. Let $k = \mathbb{R}$. If f_1 has splitting field $\mathbb{R}$, i.e. if f_1 has d different real zeros, then the same is obviously true for every f_2 sufficiently near to f_1. If f_1 has a zero in $\mathbb{C} \smallsetminus \mathbb{R}$, then every f_2 near to f_1 also has such a zero. Thus by the theorem of Ostrowski, [160], chap.II, (4.2), we may assume that the valuation is nonarchimedean. Assume that f_2 is near to f_1 and that α is a root of f_1 in the separable closure $\bar{k}$ of k. Then $|f_2(\alpha)| = |(f_2 - f_1)(\alpha)|$ is small. Writing $f_2 = c \cdot \prod_j (X - \beta_j)$, we see that $|\alpha - \beta|$ is small for some root β of f_2. In particular, for f_2 sufficiently near to f_1, we can choose a root β of f_2 such that $|\beta - \alpha| < |\alpha_i - \alpha|$ for all other roots $\alpha_i \neq \alpha$ of f_1. This set contains all conjugates of α over k, hence, by (8.1.6), we obtain that $\alpha \in k(\beta)$. Therefore the splitting field of f_1 is contained in that of f_2 if f_2 is sufficiently near to f_1. Now perform the above procedure for all the zeros $\alpha_1, \ldots, \alpha_d$ of f_1, finding for every i a root $\beta_{j(i)}$ of f_2 which is near to α_i. Since $\alpha_1, \ldots, \alpha_d$ are pairwise different, we obtain for f_2 sufficiently near to f_1 a bijection $\alpha_i \leftrightarrow \beta_{j(i)}$ between the zero-sets of both polynomials, in such a way that

$$|\beta_{j(i)} - \alpha_i| < |\beta_{j(i)} - \beta_j|$$

for all i and all $j \neq j(i)$. Another application of Krasner's lemma shows that the splitting field of f_2 is contained in that of f_1, provided that f_2 is sufficiently near to f_1. □

The following proposition is due to *F. K. Schmidt* [200].

[*)] We only consider rank 1 valuations here, i.e. the value group is contained in $\mathbb{R}^{\times}_{>0}$.

(12.1.2) Proposition. *A proper subfield K of the separable closure $\bar{k}$ of a global field k possesses at most one prime which is indecomposable in $\bar{k}$.*

Proof: Suppose $\mathfrak{p}_1$ and $\mathfrak{p}_2$ are two different primes of K indecomposable in $\bar{k}$. Let f_1, f_2 be any separable polynomials of the same degree d over K. By the approximation theorem, for every $\varepsilon > 0$ there exists a polynomial $f \in K[X]$ such that $|f - f_1|_{\mathfrak{p}_1} < \varepsilon$ and $|f - f_2|_{\mathfrak{p}_2} < \varepsilon$. For ε sufficiently small, (12.1.1) implies that the splitting fields of f and f_1 over $K_{\mathfrak{p}_1}$ coincide and the same holds for f and f_2 with respect to $\mathfrak{p}_2$. But by assumption, $\mathfrak{p}_1$ and $\mathfrak{p}_2$ do not split in $\bar{k}|K$, and so the splitting fields of f_1 and f_2 over K are the same. In particular, we can apply this in the case where f_1 is any separable, irreducible polynomial and $f_2 = \prod_{i=1}^{d}(X - x_i)$, where x_i are pairwise different elements in K. Hence $K = \bar{k}$. □

(12.1.3) Corollary. *Let $\mathfrak{P}_1$ and $\mathfrak{P}_2$ be two distinct primes of the separable closure $\bar{k}$ of a global field k. Then*

$$G_{\mathfrak{P}_1} \cap G_{\mathfrak{P}_2} = 1 .$$

(12.1.4) Corollary. *Let $\mathfrak{P}$ be a prime of $\bar{k}$. Then the decomposition group $G_{\mathfrak{P}}$ is its own normalizer in G_k.*

Proof: Let $g \in G_k$ such that $G_{\mathfrak{P}}^g = G_{\mathfrak{P}}$. Then $G_{g\mathfrak{P}} = G_{\mathfrak{P}}$, hence by the previous corollary, $g\mathfrak{P} = \mathfrak{P}$ and therefore $g \in G_{\mathfrak{P}}$. □

(12.1.5) Proposition. *Let $K|k$ be a finite Galois extension of global fields. Then the canonical homomorphism*

$$G_k \longrightarrow \mathrm{Aut}(G_K),$$

which sends $\sigma \in G_k$ to the automorphism $g \mapsto \sigma g \sigma^{-1}$ of G_K, is injective.

Proof: Assume σ lies in the kernel. Then σ fixes the open subgroup $G_{\mathfrak{P}} \cap G_K$ of every decomposition group $G_{\mathfrak{P}}$ of G_k. Hence $G_{\sigma\mathfrak{P}} \cap G_{\mathfrak{P}} \neq \{1\}$, which implies that $\mathfrak{P} = \sigma\mathfrak{P}$, i.e. $\sigma \in G_{\mathfrak{P}}$ for every prime $\mathfrak{P}$. But two different decomposition groups have a trivial intersection. □

In the special case when $K = k$, this implies the

(12.1.6) Corollary. *The absolute Galois group of a global field has a trivial center.*

Since G_k has finite cohomological dimension if k is a function field, it cannot contain finite subgroups. The same is true in the number field case if k is totally imaginary. If k has real places, their decomposition groups are isomorphic to $\mathbb{Z}/2\mathbb{Z}$, and it follows from the above results that all these subgroups are different, and further, no two of them commute.

The following famous result, due to *E. Artin* [4], provides the converse statement.

(12.1.7) Theorem. *Let K be a proper subfield of the algebraic closure $\bar{k}$ of a number field k with $[\bar{k} : K] < \infty$. Then $[\bar{k} : K] = 2$, $\bar{k} = K(\sqrt{-1})$ and K has a real place. In other words, every finite subgroup in G_k is of order 2 and is the decomposition group for a real prime of k.*

Proof: The field $K(\sqrt{-1})$ is a union of totally imaginary number fields, so $cd\,(G_{K(\sqrt{-1})}) \leq 2$ by (8.3.18). On the other hand, the group $G_{K(\sqrt{-1})}$ is finite, hence it is trivial, i.e. $\bar{k} = K(\sqrt{-1})$. Since K is a proper subfield of $\bar{k}$, we conclude that $G_K \cong \mathbb{Z}/2\mathbb{Z}$ and K has a real place by (8.3.19). (K has exactly one real place by (12.1.2).) □

Such a converse statement is also true for finite primes. This was first observed by *J. Neukirch* [152]. We say that a local field κ is of type (p_0, p_1) if it is of characteristic p_0 and has residue characteristic p_1. By convention, $\mathbb{R}$ and $\mathbb{C}$ are of type $(0,0)$. Hence the pair (p_0, p_1) is of one of the following forms: $(0,0)$, $(0,p)$, (p,p), where p is a prime number. Note that we can read off the type of κ from G_κ. Indeed,

- If G_κ is finite, then κ is of type $(0,0)$.
- If $\dim_{\mathbb{F}_p} H^1(G_\kappa, \mathbb{F}_p) = \infty$ for a prime number p, then κ is of type (p,p). This follows from (6.1.2) and (7.1.8)(iii).
- If $H^1(G_\kappa, \mathbb{F}_\ell)$ is finite for all prime numbers ℓ, then there exists exactly one prime number p such that G_κ has a closed subgroup which is a pro-p-group of rank > 2 and then κ is of type $(0,p)$.

It might happen, however, that G_κ contains $G_{\kappa'}$ as a closed subgroup of *infinite* index and κ is of type $(0,p)$, while κ' is of type (p,p). Moreover, this happens quite often as *J.-P. Wintenberger* has shown:

> *Let $\kappa|\mathbb{Q}_p$ be a finite extension and let κ_∞ be a ramified $\mathbb{Z}_p$-extension of κ. Then there exists a local field κ' of characteristic p such that $G_{\kappa_\infty} \cong G_{\kappa'}$.*

For a proof we refer the reader to the original article [259]. The field κ' is called the **field of norms**, since we have an isomorphism $\varprojlim \kappa_n^\times \cong (\kappa')^\times$. Moreover, such a field κ' exists for a larger class of infinite extensions of $\mathbb{Q}_p$: it suffices that $\kappa_\infty|\kappa$ is a so-called **arithmetically profinite** extension (loc.cit.).

Let κ be a finite extension of $\mathbb{Q}_p$. Then, in addition to the prime number p, we can also reconstruct the ramification group V_κ and the inertia group T_κ from G_κ as follows:

- V_κ is the uniquely determined maximal element among the normal, closed subgroups in G_κ which are pro-p-groups, by (7.5.7)(i).
- T_κ/V_κ is the uniquely determined maximal element among the abelian, normal, closed subgroups in G_κ/V_κ, by (7.5.7)(ii). Hence T_κ is also uniquely determined by G_κ.

The order q of the residue field of κ can also be reconstructed from G_κ. Indeed, by (7.3.10), we obtain for $\ell \neq p$ the number $\#\mu_{\ell^\infty}(\kappa) = w_\ell^1$ as the order of the quotient of $H^1(G_\kappa, \mathbb{Q}_\ell/\mathbb{Z}_\ell)$ by its maximal divisible subgroup and $q = \prod_{\ell \neq p} w_\ell^1$.

As usual, we call an element $F \in G_\kappa$ a **Frobenius lift** if its image in $G_\kappa/T_\kappa \cong G(\kappa^{nr}|\kappa)$ is equal to the arithmetic Frobenius automorphism.

(12.1.8) Lemma. *An element $F \in G_\kappa$ is a Frobenius lift if and only if for every $t \in T_\kappa$*

$$FtF^{-1} \equiv t^q \bmod V_\kappa.$$

In particular, the set of Frobenius lifts is determined by G_κ.

Proof: By (7.5.4)(ii), the map $G(\kappa^{nr}|\kappa) \longrightarrow \mathrm{Aut}(T_\kappa/V_\kappa)$ is injective. Therefore F is a Frobenius lift if and only if conjugation by F defines the same automorphism of T_κ/V_κ as the conjugation with a Frobenius lift. □

Returning to our problem of detecting local groups in a global Galois group, we have the following

(12.1.9) Theorem (NEUKIRCH). *Let k be a global field, κ a nonarchimedean local field, and assume that G_k has a closed subgroup $H \cong G_\kappa$. Then there exists a unique prime $\mathfrak{p}$ in k and a unique extension $\mathfrak{P}$ of $\mathfrak{p}$ to $\bar{k}$ such that $H \subseteq G_{\mathfrak{P}}$.*

If κ is a finite extension of $\mathbb{Q}_p$ for some prime number p, then k is a number field and $(G_{\mathfrak{P}} : H) < \infty$. Furthermore, in this case $\mathfrak{p}|p$ and $[\kappa : \mathbb{Q}_p] \geq [k_{\mathfrak{p}} : \mathbb{Q}_p]$.

For the proof we need the

(12.1.10) Lemma. *Let k be a global field, $\mathfrak{P}$ a prime of $\bar{k}$ and H an infinite closed subgroup in G_k such that*

$$(H : H \cap G_{\mathfrak{P}}) < \infty.$$

Then $H \subseteq G_{\mathfrak{P}}$.

Proof: Let $K = \bar{k}^H$ and let $L = \bar{k}^U$ for some open subgroup $U \subseteq H \cap G_{\mathfrak{P}}$ which is normal in H. Then $[L : K] < \infty$ and L is henselian with respect to $\mathfrak{P} \cap L$, i.e. $\mathfrak{P} \cap L$ is indecomposable in $\bar{k}|L$. All extensions of $\mathfrak{P} \cap K$ to L are conjugate, and so L is also henselian with respect to these other primes. Since H was infinite, L is not separably closed, hence by (12.1.2), $\mathfrak{P} \cap L$ must be the only extension of $\mathfrak{P} \cap K$ to L. Therefore $H \subseteq G_{\mathfrak{P}}$. □

Proof of (12.1.9): The uniqueness of $\mathfrak{p}$ and $\mathfrak{P}$ follows from (12.1.3). In order to prove their existence, we can use (12.1.10) to replace H and G_k by suitable open subgroups. Thus, fixing an arbitrary odd prime number ℓ different to char(κ) and char(k), we may assume that $\mu_\ell \subseteq k$ and $\mu_\ell \subseteq \kappa$. Then $H^2(U, \mathbb{F}_\ell) \cong \mathbb{F}_\ell$ for every open subgroup of H by (7.1.8) (ii). Setting $K = \bar{k}^H$, consider the injective map

$$H^2(G_K, \mathbb{F}_\ell) \hookrightarrow \prod_{\mathfrak{P}} H^2(G_{K_{\mathfrak{P}}}, \mathbb{F}_\ell)$$

which is obtained from (9.1.10) by passing to the limit. It follows that

$$H^2(G_{K_{\mathfrak{P}}}, \mathbb{F}_\ell) \neq 0$$

for at least one prime $\mathfrak{P}$ of K, which is nonarchimedean because $\ell \neq 2$. We claim that $\mathfrak{P}$ does not decompose in $\bar{k}|K$. Indeed, every finite separable extension L of K corresponds to an open subgroup of H and (9.2.1) implies a surjection

$$\mathbb{F}_\ell \cong H^2(G_L, \mathbb{F}_\ell) \twoheadrightarrow \prod_{\mathfrak{P}'|\mathfrak{P}} H^2(G_{L_{\mathfrak{P}'}}, \mathbb{F}_\ell).$$

Recall that $H^2(G_{K_{\mathfrak{P}}}, \mathbb{F}_\ell) \neq 0$ implies that $H^2(V, \mathbb{F}_\ell) \neq 0$ for every open subgroup V of $G_{K_{\mathfrak{P}}}$, by (7.1.8) (i),(ii). Hence $\mathfrak{P}$ does not decompose in $L|K$ and L was arbitrary, so that $H = G_{K_{\mathfrak{P}}}$.

We denote the unique extension of $\mathfrak{P}$ to $\bar{k}$ also by $\mathfrak{P}$ and put $\mathfrak{p} = \mathfrak{P} \cap k$. Then we have the inclusion $G_{K_{\mathfrak{P}}} \subseteq G_{\mathfrak{P}}$, where $G_{\mathfrak{P}}$ is the decomposition group of $\mathfrak{P}$ in G_k.

Now assume that κ is a finite extension of $\mathbb{Q}_p$. Then $cd_\ell H = 2$ for all prime numbers ℓ. In particular, k must be a number field, cf. (6.1.3). Furthermore,

$G_{\mathfrak{P}} \supseteq H \cong G_\kappa$ contains closed subgroups which are pro-p-groups of rank greater than 2, so $\mathfrak{p}|p$. An inspection of $H^2(-, \mathbb{F}_p)$ implies that $\mu_p \subseteq \kappa$ if and only if $\mu_p \subseteq K_{\mathfrak{P}}$ and $p^\infty \nmid [K_{\mathfrak{P}} : \mathbb{Q}_p]$. Thus the p-part of $(G_{\mathfrak{P}} : H)$ is finite. Suppose $(G_{\mathfrak{P}} : H)$ is infinite. For open subgroups V in $G_{\mathfrak{P}}$ containing H with $p \nmid (V : H)$, the map $H^1(V, \mathbb{F}_p) \xrightarrow{res} H^1(H, \mathbb{F}_p)$ is injective. But $H^1(H, \mathbb{F}_p)$ is finite and $\#H^1(V, \mathbb{F}_p)$ becomes arbitrarily large as $(G_{\mathfrak{P}} : V)$ tends to infinity. Hence $(G_{\mathfrak{P}} : H) < \infty$, and (7.3.9) shows that $[\kappa : \mathbb{Q}_p] = [K_{\mathfrak{P}} : \mathbb{Q}_p] \geq [k_{\mathfrak{p}} : \mathbb{Q}_p]$. □

(12.1.11) Corollary. *Let k be a global field. A closed subgroup $H \subseteq G_k$ is the decomposition group of a prime if and only if H is maximal among the closed subgroups which are isomorphic to the absolute Galois group of a local field.*

Exercise 1. Let k be a global field and let H be a pro-abelian subgroup of G_k. Then either $H \cong \mathbb{Z}/2\mathbb{Z}$ or

$$H \cong \prod_\ell \mathbb{Z}_\ell^{a_\ell}$$

where $a_\ell \in \{0, 1\}$. The first case can only occur if k is a number field which is not totally imaginary.

Hint: Use $scd_\ell(\mathbb{Z}_\ell \times \mathbb{Z}_\ell) = 3$.

Exercise 2. Let k be a global field. Assume that the closed subgroup $H \subseteq G_k$ is a pro-ℓ Demuškin group. Show that H is contained in the decomposition group $G_{\mathfrak{P}}$ of a uniquely determined prime $\mathfrak{P}$. If H is infinite (i.e. not cyclic of order 2), then $\mathfrak{P}$ is finite and of residue characteristic different to ℓ; in particular, char$(k) \neq \ell$. Furthermore, $H \cong \mathbb{Z}_\ell \rtimes \mathbb{Z}_\ell$ in this case.

§2. The Neukirch-Uchida Theorem

Now we come to the question of what extent a global field is characterized by its absolute Galois group. Recall that the absolute Galois group G_k is not an invariant of k: it depends on the choice of a separable closure $\bar{k}$ of k, and therefore G_k is determined by k only up to inner automorphisms. This detail is unimportant as long as one considers abelian class field theory, and by (1.6.3) it has also no effect on cohomological considerations. In the language of étale fundamental groups (and by analogy to topology) the choice of $\bar{k}$ is the choice of a base point on which the fundamental group depends.

We will restrict to the number field case in this section. The main result is theorem (12.2.1), which is due to *J. Neukirch* and *K. Uchida* [152], [153], [237]. Based on results of Neukirch, this theorem was also proven by *M. Ikeda* [84] and *K. Iwasawa* (unpublished).

Let k_1 and k_2 be two number fields and let $\bar{k}_1$ and $\bar{k}_2$ be fixed separable closures. Let $\mathrm{Iso}(k_2, k_1)$ be the set of field isomorphisms $k_2 \xrightarrow{\sim} k_1$ and let $\mathrm{Iso}(\bar{k}_2|k_2, \bar{k}_1|k_1)$ denote the set

$$\mathrm{Iso}(\bar{k}_2|k_2, \bar{k}_1|k_1) = \left\{\alpha : \bar{k}_2 \xrightarrow{\sim} \bar{k}_1 \mid \alpha(k_2) = k_1\right\}.$$

We consider the Galois groups $G(\bar{k}_1|k_1)$ and $G(\bar{k}_2|k_2)$ *). $G(\bar{k}_2|k_2)$ acts on $\mathrm{Iso}(\bar{k}_2|k_2, \bar{k}_1|k_1)$ by the rule $\sigma(\phi) = \phi \circ \sigma^{-1}$ and we have an isomorphism

$$\mathrm{Iso}(\bar{k}_2|k_2, \bar{k}_1|k_1)/G(\bar{k}_2|k_2) \cong \mathrm{Iso}(k_2, k_1).$$

An element $\alpha \in \mathrm{Iso}(\bar{k}_2|k_2, \bar{k}_1|k_1)$ induces an isomorphism

$$\alpha^* : G(\bar{k}_1|k_1) \xrightarrow{\sim} G(\bar{k}_2|k_2)$$

by $\alpha^*(g_1)(x_2) = \alpha^{-1}(g_1(\alpha(x_2)))$, $g_1 \in G(\bar{k}_1|k_1)$, $x_2 \in \bar{k}_2$.

(12.2.1) Theorem. *Let k_1 and k_2 be number fields and let*

$$\sigma : G(\bar{k}_1|k_1) \xrightarrow{\sim} G(\bar{k}_2|k_2)$$

be an isomorphism of profinite groups. Then there exists a unique element $\alpha \in \mathrm{Iso}(\bar{k}_2|k_2, \bar{k}_1|k_1)$ inducing σ, i.e. $\sigma = \alpha^*$.

In order to reformulate this theorem in terms of absolute Galois groups without fixing a separable closure, we introduce the following notation. Let G_1, G_2 be profinite groups. We denote the set of isomorphisms of profinite groups $G_1 \xrightarrow{\sim} G_2$ by $\mathrm{Iso}(G_1, G_2)$ and the group of inner automorphisms of G_i by $\mathrm{Inn}(G_i)$, $i = 1, 2$.

The group $\mathrm{Inn}(G_2)$ acts on $\mathrm{Iso}(G_1, G_2)$ by the rule $\sigma(\phi) = \sigma \circ \phi$ and we call

$$\mathrm{OutIso}(G_1, G_2) := \mathrm{Iso}(G_1, G_2)/\mathrm{Inn}(G_2)$$

the set of **outer isomorphisms** from G_1 to G_2. This notation has the advantage that for fields k_1, k_2 the set

$$\mathrm{OutIso}(G_{k_1}, G_{k_2})$$

*) In geometric language, these are the groups $\pi_1^{et}(\mathrm{Spec}\, k_i, \mathrm{Spec}\, \bar{k}_i)$.

does not depend on the choice of separable closures. An isomorphism $k_2 \xrightarrow{\sim} k_1$ can be extended to an isomorphism $\bar{k}_2 \xrightarrow{\sim} \bar{k}_1$ of arbitrarily chosen separable closures and therefore induces a well-defined element in $\mathrm{OutIso}(G_{k_1}, G_{k_2})$. Theorem (12.2.1) can now be reformulated in the following way.

(12.2.2) Corollary. *Let k_1 and k_2 be number fields. Then the natural map*

$$\mathrm{Iso}(k_2, k_1) \longrightarrow \mathrm{OutIso}(G_{k_1}, G_{k_2})$$

is an isomorphism.
In particular, for a number field k there is a canonical isomorphism

$$\mathrm{Aut}(k) \xrightarrow{\sim} \mathrm{Aut}(G_k)/\mathrm{Inn}(G_k) =: \mathrm{Out}(G_k).$$

(12.2.3) Corollary. *All automorphisms of $G_{\mathbb{Q}}$ are inner. Moreover, the canonical homomorphism*

$$G_{\mathbb{Q}} \longrightarrow \mathrm{Aut}(G_{\mathbb{Q}})$$

which sends $g \in G_{\mathbb{Q}}$ to the automorphism $h \mapsto ghg^{-1}$ is an isomorphism.

Let us deduce the corollaries first.

Proof of (12.2.2) and (12.2.3): By theorem (12.2.1), we obtain an isomorphism of sets

$$\begin{array}{rcl}\mathrm{Iso}(\bar{k}_2|k_2, \bar{k}_1|k_1) & \longrightarrow & \mathrm{Iso}(G(\bar{k}_1|k_1), G(\bar{k}_2|k_2)) \\ \alpha & \longmapsto & \alpha^*\end{array}$$

which is easily seen to be $G(\bar{k}_2|k_2)$-invariant if we let $G(\bar{k}_2|k_2)$ act by inner automorphisms on the right-hand side. Factoring out by the $G(\bar{k}_2|k_2)$-action, we obtain the required isomorphism

$$\mathrm{Iso}(k_2, k_1) \xrightarrow{\sim} \mathrm{OutIso}(G_{k_1}, G_{k_2}).$$

This shows (12.2.2). Applying this result to the case $k = \mathbb{Q}$, we obtain $\mathrm{Aut}(G_{\mathbb{Q}}) = \mathrm{Inn}(G_{\mathbb{Q}})$, because $\mathrm{Aut}(\mathbb{Q}) = 1$. Finally, the homomorphism $G_{\mathbb{Q}} \to \mathrm{Inn}(G_{\mathbb{Q}})$ is injective and hence an isomorphism, because $G_{\mathbb{Q}}$ has a trivial center by (12.1.6). □

The first step towards a proof of (12.2.1) is to establish a local correspondence. Recall the definition of $\mathrm{Sp}(K)$ from X §1. $\mathrm{Sp}(K)$ is a totally disconnected Hausdorff topological space whose underlying set is the set of primes of K plus one generic point. Suppose that we are given an isomorphism

$$\sigma : G(\bar{k}_1|k_1) \xrightarrow{\sim} G(\bar{k}_2|k_2).$$

Then (12.1.9) shows that for every prime $\mathfrak{P}_1$ of $\bar{k}_1$, the group $\sigma(G_{\mathfrak{P}_1})$ is the decomposition group of a uniquely determined prime $\mathfrak{P}_2$ of $\bar{k}_2$. Sending the generic point of $\mathrm{Sp}(\bar{k}_1)$ to that of $\mathrm{Sp}(\bar{k}_2)$, this induces a bijection $\sigma_* : \mathrm{Sp}(\bar{k}_1) \xrightarrow{\sim} \mathrm{Sp}(\bar{k}_2)$.

(12.2.4) Local Correspondence. *The bijection*

$$\sigma_* : \mathrm{Sp}(\bar{k}_1) \xrightarrow{\sim} \mathrm{Sp}(\bar{k}_2)$$

is a homeomorphism. If the extension field K_1, $k_1 \subseteq K_1 \subseteq \bar{k}_1$, corresponds via σ to the field K_2, $k_2 \subseteq K_2 \subseteq \bar{k}_2$, then σ_* induces a homeomorphism

$$\begin{array}{ccc} \sigma_{*,K_1,K_2} : \mathrm{Sp}(K_1) & \xrightarrow{\quad\sim\quad} & \mathrm{Sp}(K_2) \\ & \searrow \quad \swarrow & \\ & \mathrm{Sp}(\mathbb{Q}) & \end{array}$$

which commutes with the canonical projection to $\mathrm{Sp}(\mathbb{Q})$. *If $\mathfrak{p} \in S_p(K_1)$ for a prime number p, then $[K_{1,\mathfrak{p}} : \mathbb{Q}_p] = [K_{2,\sigma_*(\mathfrak{p})} : \mathbb{Q}_p]$.*

Proof: We have to show that if two primes $\mathfrak{P}$ and $\mathfrak{Q}$ in $\mathrm{Sp}(\bar{k}_1)$ restrict to the same prime in K_1, then $\sigma_*(\mathfrak{P}), \sigma_*(\mathfrak{Q}) \in \mathrm{Sp}(\bar{k}_2)$ restrict to the same prime in K_2. But $\mathfrak{P}$ and $\mathfrak{Q}$ restrict to the same prime in K_1 if and only if $G_{\mathfrak{P}} \cap G(\bar{k}_1|K_1)$ and $G_{\mathfrak{Q}} \cap G(\bar{k}_1|K_1)$ are conjugate subgroups in $G(\bar{k}_1|K_1)$ and this easily translates to the other side of the correspondence. Hence

$$\sigma_{*,K_1,K_2} : \mathrm{Sp}(K_1) \xrightarrow{\sim} \mathrm{Sp}(K_2)$$

is a well-defined bijection which commutes with the natural projection to $\mathrm{Sp}(\mathbb{Q})$, since the type of $K_{\mathfrak{P}}$ is encoded in $G_{\mathfrak{P}}$; see the discussion before (12.1.8). If K_1 is finite over k_1 (and hence also $[K_2 : k_2] < \infty$), then σ_{*,K_1,K_2} is automatically a homeomorphism because $\mathrm{Sp}(K_i)$ is the one-point-compactification of the discrete set of primes of K_i, $i = 1, 2$. If K_1, K_2 are of infinite degree, then σ_{*,K_1,K_2} can be identified with the inverse limit of the corresponding maps on the finite levels. Further, the topology on $\mathrm{Sp}(K_i)$ is the inverse limit topology obtained from the finite levels. Hence σ_{*,K_1,K_2} is also continuous in the general case. In particular, this applies to the case $K_1 = \bar{k}_1$, $K_2 = \bar{k}_2$. Finally, the remaining equality of degrees over $\mathbb{Q}_p$ follows from (12.1.9). □

In order to deduce (12.2.1), we recall the following application of Čebotarev's density theorem, cf. [160], chap.VII, (13.9).

(12.2.5) Theorem. *Let K be a finite normal extension of $\mathbb{Q}$ and let $L|\mathbb{Q}$ be finite. If every prime number p which has a prime factor of degree 1 in L splits completely in $K|\mathbb{Q}$, then $K \subseteq L$.*

Proof: Consider the finite normal extension $KL|L$. Every prime of degree 1 in L splits completely in KL by assumption. But this set of primes has Dirichlet density 1, so that $KL = L$ by Čebotarev's density theorem. □

(12.2.6) Corollary. *Let k be a number field and assume that every prime number p which has a prime factor of degree 1 in k splits completely in $k|\mathbb{Q}$. Then $k|\mathbb{Q}$ is normal.*

Proof: Apply (12.2.5) in the situation when $L = k$ and K is the normal closure of k over $\mathbb{Q}$. □

Proof of theorem (12.2.1): (We follow [152] and [237] with a simplification due to *H. W. Lenstra* at the end.) Let $\mathbb{A}$ be the algebraic closure of $\mathbb{Q}$ in $\mathbb{C}$ (i.e. $\mathbb{A}$ is a fixed model for $\bar{\mathbb{Q}}$). Identifying $\bar{k}_1 \cong \mathbb{A}$ and $\bar{k}_2 \cong \mathbb{A}$ via any isomorphisms, we consider k_1 and k_2 as subfields of $\mathbb{A}$ and we have to show:

Every isomorphism $\sigma : G(\mathbb{A}|k_1) \xrightarrow{\sim} G(\mathbb{A}|k_2)$ is induced by a uniquely determined automorphism $\alpha \in \mathrm{Aut}(\mathbb{A})$ such that $\alpha(k_2) = k_1$ and $\sigma = \alpha^*$.

In order to simplify notation, we make the following convention: if K_1 is an extension of k_1 in $\mathbb{A}$, then we denote by K_2 the extension of k_2 in $\mathbb{A}$ which corresponds to K_1 via the isomorphism σ, and vice versa.

The local correspondence (12.2.4) now shows that the set of prime numbers which have a prime factor of degree 1 in K_1 coincides with the set of prime numbers which have a prime factor of degree 1 in K_2. The same is true for the set of prime numbers which split completely in K_1 resp. K_2. Hence if K_1 is finite and *normal* over $\mathbb{Q}$, then $K_2|\mathbb{Q}$ is normal by (12.2.6), and (12.2.5) implies that $K_1 = K_2$.

From now on the letter N will always denote a subfield of $\mathbb{A}$ which contains the composite k_1k_2 and which is finite and normal over $\mathbb{Q}$. The above observations show that $\sigma(N) = N$, and we obtain an induced homomorphism $\sigma_N : G(N|k_1) \to G(N|k_2)$. Furthermore, for every $\alpha \in \mathrm{Aut}(\mathbb{A})$, we have $\alpha^*(N) = N$ and we obtain an induced homomorphism $\alpha_N^* : G(N|\mathbb{Q}) \to G(N|\mathbb{Q})$.

We first show the uniqueness of α. Assume that we have $\alpha_1, \alpha_2 \in \mathrm{Aut}(\mathbb{A})$ such that $\alpha_i(k_2) = k_1$ for $i = 1, 2$ and $\alpha_1^* = \alpha_2^* : G(\mathbb{A}|k_1) \xrightarrow{\sim} G(\mathbb{A}|k_2)$. Then for every field N which satisfies the above convention, we have $\alpha_i(N) = N$

and $\alpha_1^*|_{G(\mathbb{A}|N)} = \alpha_2^*|_{G(\mathbb{A}|N)}$. Therefore α_1 and α_2 have the same image under the canonical map $\mathrm{Aut}(\mathbb{A}) = G(\mathbb{A}|\mathbb{Q}) \to \mathrm{Aut}(G(\mathbb{A}|N))$. Hence $\alpha_1 = \alpha_2$ by (12.1.5).

In order to prove the existence of α, it suffices to show that for every N, there exists an $\alpha^N \in \mathrm{Aut}(\mathbb{A})$ with $\alpha^N(k_2) = k_1$ and such that the induced isomorphism

$$(\alpha^N)_N^* : G(N|k_1) \xrightarrow{\sim} G(N|k_2)$$

is equal to σ_N, the isomorphism $G(N|k_1) \xrightarrow{\sim} G(N|k_2)$ induced by σ. Indeed, having shown this, the compact sets

$$\mathfrak{A}_N = \{\alpha^N \in \mathrm{Aut}(\mathbb{A}) \mid (\alpha^N)_N^* = \sigma_N\}$$

are nonempty and define a projective system for which $\mathfrak{A} = \varprojlim_N \mathfrak{A}_N$ is nonempty.

Now we fix N and we write $G = G(N|\mathbb{Q})$, $G_i = G(N|k_i)$, $i = 1, 2$. Let us first assume that $N|k_2$ is cyclic and let F be a generator of $G(N|k_2)$. By Čebotarev's density theorem, there exists a prime $\mathfrak{P}$ such that $G_{\mathfrak{P}} \subseteq G(\mathbb{A}|k_2)$, (i.e. p splits completely in $k_2|\mathbb{Q}$), $\mathfrak{P} \cap k_2$ is unramified in the extension $N|k_2$ and $F \equiv \mathrm{Frob}_{\mathfrak{P}} \bmod G(\mathbb{A}|N)$.

Then $\sigma^{-1}G_{\mathfrak{P}} = G_{\tilde{\mathfrak{P}}}$ for some prime $\tilde{\mathfrak{P}}|p$, so there exists an $\alpha^N \in \mathrm{Aut}(\mathbb{A})$ such that $\tilde{\mathfrak{P}} = \alpha^N\mathfrak{P}$. It follows that $\alpha^N(k_2) = k_1$ because

$$\begin{aligned} \alpha^N(k_2) = k_1 \quad &\Longleftrightarrow \quad G(\mathbb{A}|k_2) = (\alpha^N)^* G(\mathbb{A}|k_1) \\ &\Longleftrightarrow \quad G(\mathbb{A}|N)G_{\mathfrak{P}} = (\alpha^N)^*(G(\mathbb{A}|N)G_{\tilde{\mathfrak{P}}}) \\ &\Longleftrightarrow \quad G_{\mathfrak{P}} = (\alpha^N)^{-1}\, G_{\tilde{\mathfrak{P}}}\, \alpha^N \end{aligned}$$

and the last equality is true since $\tilde{\mathfrak{P}} = \alpha^N\mathfrak{P}$.

Fixing any Frobenius lift $\mathrm{Frob}_{\mathfrak{P}} \in G_{\mathfrak{P}}$, we know from lemma (12.1.8) that $\sigma^{-1}(\mathrm{Frob}_{\mathfrak{P}}) \in G_{\alpha^N\mathfrak{P}}$ is a Frobenius lift and the same is trivially true for $\alpha^N \mathrm{Frob}_{\mathfrak{P}}(\alpha^N)^{-1}$. Since $\alpha^N\mathfrak{P}$ is unramified and does not split in the cyclic extension $N|k_1$, we observe that the two homomorphisms

$$(\alpha^N)_N^*, \sigma_N : G(N|k_1) \to G(N|k_2)$$

coincide on a generator of $G(N|k_1)$ and hence are equal.

In order to deal with the general situation, let $n = \#G$ and let p be any prime number greater than n. The split embedding problem

$$\begin{array}{ccccccccc} & & & & & & G(\bar{k}_1|\mathbb{Q}) & & \\ & & & & & & \downarrow & & \\ 1 & \longrightarrow & \mathbb{F}_p[G] & \longrightarrow & E & \longrightarrow & G(N|\mathbb{Q}) & \longrightarrow & 1 \end{array}$$

has a proper solution $M|\mathbb{Q}$ by (9.2.9), i.e. $E = G(M|\mathbb{Q})$ and $\mathbb{F}_p[G] = G(M|N)$. Then σ induces isomorphisms

$$\sigma_M : G(M|k_1) \xrightarrow{\sim} G(M|k_2)$$

and

$$\sigma_M^N := \sigma_M|_{G(M|N)} : \mathbb{F}_p[G] \xrightarrow{\sim} \mathbb{F}_p[G].$$

We are going to describe σ_M^N more explicitly. Let $\lambda \in \mathbb{F}_p[G]$ be any non-zero element of the group ring and let L_1 and L_2 be the fixed field of the subgroup $\langle\lambda\rangle \subseteq \mathbb{F}_p[G] \subseteq E = G(M|\mathbb{Q})$ and of $\sigma_M(\langle\lambda\rangle) \subseteq \mathbb{F}_p[G] \subseteq G(M|\mathbb{Q})$, so that

$$\sigma : G(\mathbb{A}|L_1) \xrightarrow{\sim} G(\mathbb{A}|L_2).$$

We have the following diagram of fields

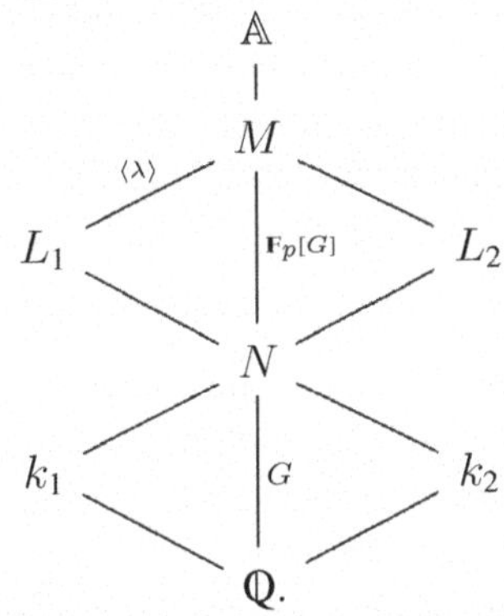

By the above considerations in the cyclic case, we see the existence of an $\alpha = \alpha^M \in \mathrm{Aut}(\mathbb{A})$ such that

$$\sigma_M = \alpha_M^* : G(M|L_1) \to G(M|L_2).$$

The automorphism $\alpha^* : G(\mathbb{A}|\mathbb{Q}) \to G(\mathbb{A}|\mathbb{Q})$ induces an automorphism $\alpha^* : G \to G$, which is just conjugation with $h = \alpha^{-1} \bmod G(\mathbb{A}|N) \in G$. Then

$$\alpha_M^* : G(M|N) \to G(M|N)$$

is left multiplication by h on the G-module $G(M|N) \cong \mathbb{F}_p[G]$. In particular,

$$\sigma_M^N(\lambda) = \alpha_M^*(\lambda) = h\lambda \in \mathbb{F}_p[G].$$

For each $h \in G$ consider

$$U_h = \{\lambda \in \mathbb{F}_p[G] \mid \sigma_M^N(\lambda) = h\lambda\} \subseteq \mathbb{F}_p[G].$$

The union of the subgroups U_h, $h \in G$, is all of $\mathbb{F}_p[G]$. Since $\mathbb{F}_p[G]$ is a p-group and $p > n = \#G$, a counting argument shows that one of the U_h must be equal to $\mathbb{F}_p[G]$. Therefore there exists an $h_0 \in G$ with

$$(*) \qquad \sigma_M^N(\lambda) = h_0\lambda \text{ for all} \lambda \in \mathbb{F}_p[G].$$

Now let $g_1 \in G(N|k_1)$ be arbitrary. Then we have

$$\sigma_M^N(g_1) = \sigma_N(g_1)\sigma_M^N(1)$$

by the definition of the occurring maps. Applying $(*)$ to $\lambda = g_1$, we obtain

$$h_0 g_1 = \sigma_M^N(g_1) = \sigma_N(g_1)\sigma_M^N(1) = \sigma_N(g_1)h_0.$$

Hence $\sigma_N(g_1) = h_0 g_1 h_0^{-1} = \alpha^*(g_1)$, which shows that $\alpha^N := \alpha$ satisfies the required condition. □

Closing remark: *K. Uchida* [238] proved theorem (12.2.1) (with a small modification necessary in positive characteristic) also in the function field case. Furthermore, the result remains true after replacing the absolute Galois groups by their maximal prosolvable quotients. *F. Pop* [172], [173], [175] showed that (12.2.1) is true in the much more general situation that k_1, k_2 are infinite fields which are finitely generated over their prime fields; see also [224].

Finite fields are obvious examples of fields which are not anabelian. Another example is p-adic local fields, which can be seen as follows: as is well-known, every automorphism of a finite extension $k|\mathbb{Q}_p$ is continuous*⁾; in particular, $\mathrm{Aut}(\mathbb{Q}_p) = 1$. If (12.2.1) (and hence also (12.2.2)) were true for p-adic local fields, then taking $k_1 = k_2 = \mathbb{Q}_p$ we would obtain an isomorphism

$$1 = \mathrm{Aut}(\mathbb{Q}_p) \xrightarrow{\sim} \mathrm{Out}(G_{\mathbb{Q}_p}) = \mathrm{Aut}(G_{\mathbb{Q}_p})/\mathrm{Inn}(G_{\mathbb{Q}_p}).$$

But we constructed a nontrivial outer automorphism of $G_{\mathbb{Q}_p}$ in VII §5, so p-adic local fields do not deserve to be called anabelian. This can also be seen using (7.5.3): it is not difficult to construct local fields which are not isomorphic but have the same absolute Galois group.

However, there is the following variant due to *S. Mochizuki* [144].

(12.2.7) Theorem. *Let k_1 and k_2 be finite extensions of $\mathbb{Q}_p$. Let $\mathrm{Iso}_{\mathbb{Q}_p}(k_2, k_1)$ denote the set of $\mathbb{Q}_p$-isomorphisms from k_2 to k_1 and let $\mathrm{OutIso}_{Filt}(G_{k_1}, G_{k_2})$ denote the set of outer isomorphisms of filtered groups between the absolute Galois groups of k_1 and k_2 equipped with the filtrations defined by the higher ramification groups in the upper numbering. Then the natural map*

$$\mathrm{Iso}_{\mathbb{Q}_p}(k_2, k_1) \longrightarrow \mathrm{OutIso}_{Filt}(G_{k_1}, G_{k_2})$$

is an isomorphism.

§3. Anabelian Conjectures

In this last section we present the anabelian conjecture(s) in detail, thereby making free use of the language of schemes and their étale fundamental groups.

---

*⁾Indeed, choose a prime number $\ell \neq p$ such that $\mu_\ell(k) = 1$. Then the unit group U_k can particularly be characterized as the subgroup of ℓ-divisible elements in $k^\times$. Thus every automorphism $\phi \in \mathrm{Aut}(k)$ fixes U_k. An element $a \in \mathcal{O}_k$ is either a unit or of positive valuation and the latter is characterized by the property that $a + u \in U_k$ for every $u \in U_k$. Hence $\phi(\mathcal{O}_k) = \mathcal{O}_k$, which implies that ϕ is continuous.

Let k be a field with separable closure $\bar{k}$ and let X be a scheme of finite type over k which is geometrically connected. Fixing a geometric point $\bar{x}$: Spec $\bar{k} \to X$, we have the following exact sequence of profinite groups

$$1 \longrightarrow \pi_1(\bar{X},\bar{x}) \longrightarrow \pi_1(X,\bar{x}) \xrightarrow{p_X} G(\bar{k}|k) \longrightarrow 1,$$

where $\bar{X}$ is the base change from X to $\bar{k}$ and π_1 denotes the étale fundamental group.

When k is finitely generated over $\mathbb{Q}$, *A. Grothendieck* posed the conjecture that

if X is "anabelian", then it is functorially determined by p_X.

In order to make this more precise, we introduce some notation. If G_1, G_2 denote profinite groups, let

$$\mathrm{Hom}^{op}(G_1, G_2)$$

be the set of continuous homomorphisms $\phi : G_1 \to G_2$ with open image. If G_1, G_2 are augmented, i.e. if we are given homomorphisms $p_i : G_i \to G$, $i = 1, 2$, to another profinite group G, then $\mathrm{Hom}^{op}_G(G_1, G_2)$ is the subset of $\mathrm{Hom}^{op}(G_1, G_2)$ consisting of homomorphisms ϕ with $\phi \circ p_2 = p_1$. If $H_2 \subseteq G_2$ is a closed subgroup with trivial center, then $\mathrm{Inn}(H_2)$ acts on $\mathrm{Hom}^{op}(G_1, G_2)$ and it acts on $\mathrm{Hom}^{op}_G(G_1, G_2)$ if H_2 is contained in the kernel of p_2.

For schemes X_1, X_2, we denote the set of dominant morphisms from X_1 to X_2 by

$$\mathrm{Mor}_{dom}(X_1, X_2).$$

Following Grothendieck's philosophy, there should exist a full subcategory An_k of the category of schemes of finite type over k such that (in a base point free version as in §2) the following holds:

(12.3.1) Anabelian Bijections.

(i) *(For isomorphisms): If $X_1, X_2 \in \mathrm{An}_k$, then the canonical map*

$$\mathrm{Iso}_{\mathrm{Sch}|k}(X_1, X_2) \longrightarrow \mathrm{Iso}_{G_k}(\pi_1(X_1), \pi_1(X_2))/\mathrm{Inn}(\pi_1(\bar{X}_2))$$

is a bijection.

(ii) *(For dominant morphisms): If $X_1, X_2 \in \mathrm{An}_k$, then the canonical map*

$$\mathrm{Mor}_{\mathrm{Sch}|k,dom}(X_1, X_2) \longrightarrow \mathrm{Hom}^{op}_{G_k}(\pi_1(X_1), \pi_1(X_2))/\mathrm{Inn}(\pi_1(\bar{X}_2))$$

is a bijection.

(iii) *(For sections): For every proper curve $X \in \mathrm{An}_k$ and for every finite separable extension $K|k$, the canonical map*

$$\mathrm{Mor}_{\mathrm{Sch}|k}(\mathrm{Spec}\, K, X) \longrightarrow \mathrm{Hom}_{G_k}(G_K, \pi_1(X))/\mathrm{Inn}(\pi_1(\bar{X}))$$

is a bijection.

Clearly (i) follows from (ii). Note that the triviality of the center of $\pi_1(\bar{X}_2)$ is an implicit assumption. It is part of the philosophy that the property of being anabelian should be geometric, i.e. whether X is in An_k should only depend on $\bar{X}$. But which schemes are anabelian?

Following Grothendieck, the category An_k should contain all hyperbolic curves and successive fibrations of such curves over each other. Here a smooth, geometrically connected curve X over k is called hyperbolic if it satisfies: $\chi(X) \stackrel{\text{def}}{=} 2 - 2g - n < 0$, where g is the genus of the smooth compactification C of X and n is the cardinality of $C(\bar{k}) \smallsetminus X(\bar{k})$. In particular, all curves of genus ≥ 2 are hyperbolic and (using "Artin neighbourhoods") every point on a smooth variety admits a fundamental system of anabelian neighbourhoods. Furthermore, Grothendieck conjectured that the moduli spaces $\mathfrak{M}_{g,k}$ of curves of a given genus g over k are also anabelian.

Consider the following variant of (12.3.1)(i).

(12.3.2) Absolute Anabelian Bijection of Isomorphisms.

If $X_1, X_2 \in \mathrm{An}_k$, then the canonical map

$$\mathrm{Iso}_{\mathrm{Schemes}}(X_1, X_2) \longrightarrow \mathrm{Iso}(\pi_1(X_1), \pi_1(X_2))/\mathrm{Inn}(\pi_1(X_2))$$

is a bijection.

The generalization of theorem (12.2.1) to fields which are finitely generated over $\mathbb{Q}$, proved by *F. Pop* (cf. §2), should be seen as a birational version of (12.3.2). A birational version of (12.3.1)(ii) was proved by *S. Mochizuki* [145]. One can ask whether an *absolute* birational version of (12.3.1)(ii) is also true. Since homomorphisms of fields are always injective, the following question naturally arises:

(12.3.3) Question. *Assume that k_1, k_2 are finitely generated over $\mathbb{Q}$. Is every open homomorphism $G_{k_1} \to G_{k_2}$ injective?*

We do not know the answer to question (12.3.3), even in the case when k_1 and k_2 are number fields.

If $k|\mathbb{Q}$ is finitely generated, then the subgroup $\pi_1(\bar{X}, \bar{x}) \subseteq \pi_1(X, \bar{x})$ can be detected in a purely group theoretical way as follows: since X is of finite type, $\pi_1(\bar{X}, \bar{x})$ is finitely generated. On the other hand, the field k is Hilbertian, so $G(\bar{k}|k)$ does not contain a nontrivial finitely generated normal subgroup, by

[53], th. 15.10. Hence $\pi_1(X, \bar{x})$ possesses a unique maximal finitely generated normal subgroup which is just its geometric part $\pi_1(\bar{X}, \bar{x})$. Using Pop's results about absolute Galois groups, we see that (12.3.1)(ii) is equivalent to its absolute version (12.3.2) if $k|\mathbb{Q}$ is finitely generated.

Suppose we are given a proper curve of genus greater than or equal to 2 over a number field. If we are interested in the isomorphism class of the curve, then, just by knowing that it is defined over a number field, we know that there are at most a countable number of possibilities. If we give ourselves, in addition, the Tate module of the curve (and this can be easily determined from the étale fundamental group), then we know the primes where the Jacobian of the curve has bad reduction. Hence by the Šafarevič conjecture, proved by *G. Faltings* [47], and even before we discuss (12.3.1)(i), the curves are already determined up to a finite number of possibilities. It is therefore surprising how difficult the problem is, even in the case of curves. For a period of more than ten years (Grothendieck formulated his conjecture in a letter to Faltings in 1983 [68]) only a few partial results (e.g. [149]) had been known.

This rapidly changed in 1995, when *A. Tamagawa* made the observation that (12.3.2) is true for *affine hyperbolic curves over finite fields* [227]. This came rather unexpectedly since finite fields themselves are not anabelian.*) Let us briefly explain the ideas of Tamagawa's proof:

In the first step a local correspondence is established. Recall that in the case of global fields (see §1), the places could be characterized by their decomposition groups and the latter could be detected via their second cohomology. Let us denote the function field of the curve X by K and let S be the finite set of places of K which do not lie on X. The field K is a global field of positive characteristic and $\pi_1(X) = G_S(K)$. Every prime of K which lies on X has a decomposition group isomorphic to $\hat{\mathbb{Z}}$ in $G_S(K)$. Thus one is confronted with the (seemingly impenetrable) task of characterizing the decomposition groups among the huge set of subgroups of $G_S(K)$ which are isomorphic to $\hat{\mathbb{Z}}$. Tamagawa solved this problem in an elegant way: assume for simplicity that $x \in X(k)$ is a prime of K that corresponds to a k-rational point of X and let $\tilde{x}$ be a prolongation of x to K_S. Let L be a finite extension of K in K_S and let X_L be the normalization of X in L (which is étale over X). Then the restriction of $\tilde{x}$ to L defines a k-rational point on X_L if $G_S(L)$ contains the decomposition group $G_{\tilde{x}}(K_S|K)$. In particular, we have $\#X_L(k) > 0$ for those L. Via the Lefschetz trace formula, the last equality can be reformulated into a cohomological statement which can (for large enough L) be solely expressed in terms of $G_S(L)$. This makes it possible to characterize the decomposition groups by the way in which they lie inside $G_S(K)$.

---

*) Note that (12.3.1)(i) is false over finite fields.

The global part of the proof generalizes that of Uchida [238] for the function field analogue of (12.2.1) to a large extent (at this point, however, the assumption of X being affine enters).

Then (12.3.1)(i) for fields which are finitely generated over $\mathbb{Q}$ can be derived from (12.3.2) for finite fields [227]. In this process the characterization of good reduction of curves via outer pro-ℓ Galois representations (due to *T. Oda*, cf. [164], [227], th. 5.3) is applied. This showed (12.3.1)(i) for affine hyperbolic curves over fields which are finitely generated over $\mathbb{Q}$. Finally, the restriction to affine curves was removed by *S. Mochizuki* using methods of *logarithmic algebraic geometry* [143].**)

Only a short time afterwards, Mochizuki [145] (see also [48]) achieved further progress by changing the point of view. He claimed that the Grothendieck Conjecture for hyperbolic curves is best understood not as a global, number theoretical result, but rather as a p-adic result. He succeeded in proving (12.3.1)(ii) for hyperbolic curves over *sub-p-adic local fields*, i.e. subfields of finitely generated field extensions of $\mathbb{Q}_p$. This particularly includes all fields which are finitely generated over $\mathbb{Q}$. The proof is completely independent of Tamagawa's approach and uses *p-adic Hodge theory*, a theory which investigates properties of p-adic Galois representations arising from arithmetic geometry. Mochizuki obtains even stronger results. Among other generalizations, he also shows, cf. [145], th. D, that (12.3.1)(i) holds for hyperbolically fibred surfaces over a sub-p-adic field.

Concerning (12.3.1)(iii), which is also referred to as the anabelian section conjecture, only the injectivity of the canonical map is known so far ([145], th. C). A positive answer to the section conjecture would be of special interest because of the following vague hope: let X be a proper curve of genus ≥ 2 over a number field. The set $\mathrm{Hom}_{G_k}(G_k, \pi_1(X))/\mathrm{Inn}(\pi_1(\bar{X}))$ should belong in some sense to a "compact" world, while $X(k)$ is "discrete". If the section conjecture holds, this fact could possibly be used in order to show the finiteness of $\#X(k)$. This would yield a new proof of the famous Mordell conjecture, proved by *G. Faltings*.

The analogy between number fields and function fields, together with Tamagawa's results, raises the following question.

(12.3.4) Question. *Are a number field k and a set S of places of k functorially determined by $G_S(k)$ if S is large enough?*

**) *J. Stix* has shown variants of these results for curves over absolutely finitely generated fields of positive characteristic, see [221], [222].

This might be a little bit too optimistic because, in contrast to the function field case, we cannot go up the cyclotomic $\hat{\mathbb{Z}}$-extension of k inside k_S. Writing $k(\mu) = \bigcup_n k(\mu_n)$, we conclude this chapter with the following

(12.3.5) Conjecture. *Let k_1, k_2 be number fields and let S_i be sufficiently large finite sets of primes of k_i, $i = 1, 2$. Then the canonical map*

$$\mathrm{Iso}(\mathcal{O}_{k_2,S_2}, \mathcal{O}_{k_1,S_1}) \longrightarrow \mathrm{OutIso}(G(k_1(\mu)_{S_1}|k_1), G(k_2(\mu)_{S_2}|k_2))$$

is a bijection.

Finally, we should mention that these anabelian ideas are only part of a larger program, initiated by *A. Grothendieck* [69], which aims at a description of the absolute Galois group of $\mathbb{Q}$ in geometric terms.

Literature

Albert, A.A.

[1] Structure of algebras. Am. Math. Soc. Providence 1964

Amice, Y., Fresnel, J.

[2] Fonctions zêta p-adiques des corps des nombres algébriques abéliens reéls. Acta Arith. Warzawa **20** (1972) 353–384

Andožskii, I.V.

[3] Demuškin groups (in Russian). Mat. Zametki **14** (1973) No.1 121–126, English translation in Math. Notes **14** (1974) 626–628

Artin, E.

[4] Kennzeichnung des Körpers der reellen algebraischen Zahlen. Hamb. Abh. **3** (1924) 319–323

Artin, M., Grothendieck, A., Verdier, J.L.

[5] Théorie des Topos et Cohomologie Etale des Schémas (SGA 4), tome 1,2,3. Lecture Notes Math. **269**, **270**, **305**, Springer 1972–1973

Artin, E., Tate, J.

[6] Class Field Theory. Benjamin New York, Amsterdam 1967

Atiyah, M.F., Wall, C.T.C.

[7] Cohomology of groups. Chapter IV in [23], 94–115

Ax, J.

[8] Proof of some conjectures on cohomological dimension. Proc. Amer. Math. Soc. **16** (1965) 1214–1221

Ax, J., Kochen, S.

[9] Diophantine problems over local fields I + II. Am. J. of Math. **87** (1965), III, Ann. of Math. **83** (1966) 437–456

Bayer, P., Neukirch, J.

[10] On values of zeta functions and ℓ-adic Euler characteristics. Invent. Math. **50** (1978) 35–64

Bieri, B., Eckmann, B.

[11] Groups with homological duality generalizing Poincaré duality. Invent. Math. **20** (1973) 103–124

Binz, E., Neukirch, J., Wenzel, G.

[12] A subgroup theorem for free products of profinite groups. J. of Algebra **19** (1971) 104–109

J. Neukirch et al., *Cohomology of Number Fields*,
Grundlehren der mathematischen Wissenschaften 323,
DOI 10.1007/978-3-540-37889-1, © Springer-Verlag Berlin Heidelberg 2008

Bloch, S., Kato, K.

[13] p-adic étale cohomology. Publ. Math. IHES **63** (1986) 107–152

Bourbaki, N.

[14] Algèbre Chap.8. Hermann, Paris 1958
[15] General Topology Part 1,2. Hermann, Paris 1966
[16] Commutative Algebra. Hermann, Paris 1972
[17] Algebra I. Hermann, Paris 1974

Brown, K.

[18] Cohomology of Groups. Springer 1982

Brumer, A.

[19] Pseudocompact algebras, profinite groups and class formations. J. of Algebra **4** (1966) 442–470
[20] On the units of algebraic number fields. Mathematika **14** (1967) 121–124

Cartan, E., Eilenberg, S.

[21] Homological Algebra. Princeton Math. Ser. **19**, Princeton 1956

Cassels, J.W.S.

[22] Global fields. Chapter II in [23], 42–84

Cassels, J.W.S., Fröhlich, A. (ed.)

[23] Algebraic Number Theory. Academic Press, London New York 1967

Chandrasekharan, K.

[24] Introduction to Analytic Number Theory. Springer 1968

Chenevier, G.

[25] On Number Fields with given Ramification. Comp. Math. **143** (2007) 1359–1373

Coates, J.

[26] K-theory and Iwasawa's analogue of the Jacobian. In Algebraic K-theory II, Lecture Notes in Math. **342** Springer 1973, pp. 502–520
[27] p-adic L-functions for motives. In [28], pp. 141–172

Coates, J., Taylor, M.J. (ed.)

[28] L-Functions and Arithmetic. Proceedings of the Durham Symposium, July 1989, Cambridge University Press 1991

Coates, J., Wiles, A.

[29] On the conjecture of Birch and Swinnerton-Dyer. Invent. Math. **39** (1977) 223–251

Colmez, P.

[30] Résidu en $s = 1$ des fonctions zêta p-adiques. Invent. Math. **91** (1988) 371–389

Cornell, G., Silverman, J.H. (ed.)

[31] Arithmetic Geometry. Springer 1986

Curtis, C.W., Reiner, I.

[32] Methods of Representation Theory Vol I. Wiley-Interscience Publication New York, Chichester, Brisbane, Toronto 1981

de Shalit, Ehud

[33] Iwasawa Theory of Elliptic Curves with Complex Multiplication. Perspectives in Math. Vol.3 Academic Press Boston 1987

Deligne, P.

[34] Théorie de Hodge II. Publ. Math. IHES **40** (1971) 5–57

[35] La conjecture de Weil I. Publ. Math. IHES **43** (1974) 273–308

[36] La conjecture de Weil II. Publ. Math. IHES **52** (1981) 313–428

Deligne, P., Ribet, K.

[37] Values of abelian L-functions at negative integers over totally real fields. Invent. Math. **59** (1980) 227–286

Demuškin, S.P.

[38] The group of the maximal p-extension of a local field (in Russian). Izv. Akad. Nauk. USSR Math. Ser. **25** (1961) 329–346

Diekert, V.

[39] Über die absolute Galoisgruppe dyadischer Zahlkörper. J. reine u. angew. Math. **350** (1984) 152–172

[40] Eine Bemerkung zu freien Moduln über regulären lokalen Ringen. J. of Algebra **101** (1986) 188–189

Dixon, J.D., du Sautoy, M. P. F., Mann, A., Segal, D.

[41] Analytic pro-p Groups. London Math. Soc. Series **157** Cambridge University Press 1991

Draxl, P.K.

[42] Skew fields, London Math. Soc. Series **81**, Cambridge University Press 1983

Dress, A.

[43] Contributions to the theory of induced representations. Lect. Notes Math. **342**, Springer 1973

Dummit, J., Labute, J.P.

[44] On a new characterization of Demuškin groups. Invent. Math. **73** (1983) 413–418

Engler, A., Haran, D., Kochloukova, D., Zalesskii, P.A.

[45] Normal subgroups of profinite groups of finite cohomological dimension. J. Lond. Math. Soc. **69** (2004) 317–332

Evens, L.

[46] The Cohomology of Groups. Clarendon Press, Oxford 1991

Faltings, G.

[47] Endlichkeitssätze für abelsche Varietäten über Zahlkörpern. Invent. Math. **73** (1983) 349–366, Erratum **75** (1984) 381. English translation in [31], 9–27

[48] Curves and their fundamental groups (following Grothendieck, Tamagawa and Mochizuki). Séminaire Bourbaki 50ème année, 1997–98, No. 840

Ferrero, B., Washington, L.

[49] The Iwasawa invariant μ_p vanishes for abelian number fields. Ann. of Math. **109** (1979) 377–395

Fesenko, I.B., Vostokov, S.V.

[50] Local Fields and their Extensions. A Constructive Approach. AMS Providence, Rhode Island 1993

Fontaine, J-M., Mazur, B.

[51] Geometric Galois representations. In: Elliptic Curves, Modular Forms, & Fermat's Last Theorem, edited by J. Coates, and S.T. Yau. International Press, Boston 1995

Fried, M.D., Haran,D.,Völklein,H.

[52] Absolute Galois group of the totally real numbers. C.R. Acad. Sci., Paris **317** (1993) 995–999

Fried, M.D., Jarden, M.

[53] Field Arithmetic. Springer 1996

Friedlander, E.M.

[54] Etale Homotopy of Simplicial Schemes. Princeton University Press 1982

Geyer, W.-D.

[55] Galois groups of intersections of local fields. Israel J. of Math. **30** (1978) 382–396

Gildenhuys, D., Ribes, L.

[56] A Kurosh subgroup theorem for free pro-$\mathfrak{c}$-products of pro-$\mathfrak{c}$-groups. Trans. Amer. Math. Soc. **186** (1973) 309–329

Golod, E.S., Šafarevič, I.R.

[57] On class field towers (in Russian). Izv. Akad. Nauk SSSR **28** (1964) 261–272. English translation in Am. Math. Soc. Transl. (2) **48** (1965) 91–102

Gordeev, N.L.

[58] Infinitude of the number of relations in the Galois group of the maximal p-extension of a local field with restricted ramification. Izv. Akad. Nauk SSSR **45** (1981) 592–607 English translation in Math. USSR, Izv. **18** (1982) 513–524

Green, J.A.

[59] Axiomatic representation theory of finite groups. J. of Pure and Applied Algebra **1** (1971) 41–77

Greenberg, M.J.

[60] Lectures on Forms in Many Variables. Benjamin 1969

Greenberg, R.

[61] On a certain ℓ-adic representation. Invent. Math. **21** (1973) 117–124
[62] On the Iwasawa invariants of totally real number fields. Amer. J. Math. **93** (1976) 263–284
[63] On p-adic L-functions and cyclotomic fields II. Nagoya Math. Journ. **67** (1977) 139–158
[64] On p-adic Artin L-functions. Nagoya Math. Journ. **89** (1983) 77–87
[65] Iwasawa theory for motives. In [28], pp. 211–233

Grothendieck, A.

[66] Sur quelques points d'algèbre homologique. Tôhoku Math. J. **9** (1957) 119–221
[67] Revêtements Étales et Groupe de Fondamental (SGA1). Lecture Notes in Math. **224**, Springer 1971
[68] Brief an G. Faltings (1983). Reprinted in [205] 49–58. English translation: loc. cit. 285–293
[69] Esquisse d'un Programme. In [205], 5–48. English translation: loc. cit. 243–284

Grothendieck, A., Dieudonné, J.A.

[70] Eléments de Géométrie Algébrique I (EGA 1). Springer Grundlehren Bd. 166, 1971

Gruenberg, K.W.

[71] Projective profinite groups. J. London Math. Soc. **42** (1967) 155–165

Haberland, K.

[72] Galois Cohomology of Algebraic Number Fields. Deutscher Verlag der Wiss., Berlin 1978

Hall, M.

[73] The Theory of Groups. Macmillan Company, New York 1968

Haran, D.

[74] On closed subgroups of free products of profinite groups. Proc. London Math. Soc. (3) **55** (1987) 266–298
[75] A proof of Serre's theorem. J. of Indian Math. Soc. **55** (1990) 213–234

Harari, D., Szamuely, T.

[76] Arithmetic duality theorems for 1-motives. J. reine u. angew. Math. **578** (2005) 93–128

Hartshorne, R.

[77] Algebraic Geometry. Springer 1977

Henniart, G.

[78] Relèvement global d'extensions locales: quelques problèmes de plongement. Math. Ann. **319** (2001) 75–87

Hilton, P.J., Stammbach, U.

[79] A Course in Homological Algebra. Springer 1971

Hoechsmann, K.

[80] Zum Einbettungsproblem. J. reine u. angew. Math. **229** (1968) 81–106

Huppert, B.

[81] Endliche Gruppen I. Springer 1967

Ihara, Y.

[82] How many primes decompose completely in an infinite unramified Galois extension of a global field? J. Math. Soc. Japan **35** (1983) 693–709

Ikeda, M.

[83] Zur Existenz eigentlicher galoisscher Körper beim Einbettungsproblem. Hamb. Abh. **24** (1960) 126–131

[84] Completeness of the absolute Galois group of the rational number field. J. reine u. angew. Math. **291** (1977) 1–22

Išhanov, V.V., Lur'e, B.B., Faddeev, D.K.

[85] The Embedding Problem in Galois Theory. Trans. of Math. Monographs **165** AMS Providence 1997

Iwasawa, K.

[86] On solvable extensions of algebraic number fields. Ann. of Math. **58** (1953) 548–572

[87] On Galois groups of local fields. Trans. AMS **80** (1955) 448–469

[88] On $\mathbb{Z}_\ell$-extensions of algebraic number fields. Ann. of Math. **98** (1973) 246–326

[89] On the μ-invariant of $\mathbb{Z}_\ell$-extensions. Conf. on number theory, algebraic geometry, and commutative algebra in honor of Y. Akizuki, Tokyo 1977, 1–11

[90] Riemann-Hurwitz formula and p-adic Galois representations for number fields. Tohoku Math. J. **33** (1981) 263–288

Jacobson, N.

[91] Finite-dimensional Division Algebras over Fields. Springer 1996

Jakovlev, A.S.

[92] The Galois group of the algebraic closure of a local field. Math. USSR-Izv. **2** (1968) 1231–1269

[93] Remarks on my paper "The Galois group of the algebraic closure of a local field". Math. USSR-Izv. **12** (1978) 205–206

Jannsen, U.

[94] Über Galoisgruppen lokaler Körper. Invent. Math. **70** (1982) 53–69

[95] Galoismoduln mit Hasse-Prinzip. J. reine u. angew. Math. **337** (1982) 154–158

[96] Continuous étale cohomology. Math. Ann. **280** (1988) 207–245

[97] Iwasawa modules up to isomorphism. Advanced Studies in Pure Mathematics **17** (1989) 171–207

[98] The splitting of the Hochschild-Serre spectral sequence for a product of groups. Canad. Math. Bull. **33** (1990) 181–183

Jannsen, U., Wingberg, K.

[99] Die Struktur der absoluten Galoisgruppe $\mathfrak{p}$-adischer Zahlkörper. Invent. Math. **70** (1982) 71–98

Kato, K.

[100] Iwasawa theory and p-adic Hodge theory. Kodai Math. J. **16** No.1 (1993) 1–31

Kawada, Y.

[101] Class formations. Proc. Symp. Pure Math. Vol. XX, AMS, Providence (1969) 96–114

Kersten, I.

[102] Brauergruppen von Körpern. Vieweg 1990

Kida, Y.

[103] ℓ-extensions of CM-fields and cyclotomic invariants. J. of Number Theory **12** (1980) 519–528

Kisilevsky, H., Labute, J.P.

[104] On a sufficient condition for the p-class tower of a CM-field to be infinite. In: Proceedings of the Int. Number Conf. Laval 1987, Berlin 1989

Kisin, M., Wortmann, S.

[105] A note on Artin motives. Math. Res. Lett. **10** (2003) 375–389

Knus, M.-A.

[106] Quadratic and Hermitian Forms over Rings. Springer 1980

Koch, H.

[107] ℓ-Erweiterungen mit vorgegebenen Verzweigungsstellen. J. reine und angew. Math. **219** (1965) 30–61

[108] Über die Galoissche Gruppe der algebraischen Abschließung eines Potenzreihenkörpers mit endlichem Konstantenkörper. Math. Nachr. **35** (1967) 323–327

[109] Zum Satz von Golod-Schafarewitsch. Math. Nachr. **42** (1969) 321–333

[110] Galoissche Theorie der p-Erweiterungen. Deutscher Verlag der Wiss., 1970, Springer 1970 (Russian translation Moscow 1973)

[111] The Galois group of a p-closed extension of a local field. Soviet. Math. Dokl. **19** (1978) 10–13

Koch, H., Venkov, B.B.

[112] Über den p-Klassenkörperturm eines imaginär-quadratischen Zahlkörpers. Soc. Math. France, Astèrisque **24–25** (1975) 57–67

Kolster, M.

[113] A relation between the 2-primary parts of the main conjecture and the Birch-Tate conjecture. Can. Math. Bull. **32** No.2 (1989) 248–251

Kraft, J.S., Schoof, R.

[114] Computing Iwasawa modules of real quadratic number fields. Comp. Math. **97** (1995) 135–155

Kuz'min, L.V.

[115] Local extensions associated with ℓ-extensions with given ramification. Izv. Akad. Nauk SSSR **39** (1975) No. 4. English translation in Math. USSR Izv. **9** (1975) No. 4 693–726

Labute, J.P.

[116] Classification des groupes de Demuškin. C.R. Acad. Sci. Paris **260** (1965) 1043–1046
[117] Classification of Demuškin groups. Can. J. Math. **19** (1967) 106–132
[118] Algèbres de Lie et pro-p-groupes definis par une seule relation. Invent. Math. **4** (1967) 142–158
[119] Mild pro-p-groups and Galois groups of p-extensions of $\mathbb{Q}$. J. reine u. angew. Math. **596** (2006) 155–182

Landau, E.

[120] Einführung in die elementare und analytische Theorie der algebraischen Zahlen und der Ideale. Chelsea Publ. New York 1949

Lang, S.

[121] On quasi-algebraic closure. Ann. of Math. **55** (1952) 373–390
[122] Cyclotomic Fields I and II. Springer 1990
[123] Algebraic Number Theory. Springer 1994
[124] Rapport sur la cohomologie des groupes. Benjamin, New York-Amsterdam 1966, English translation: Topics in Cohomology of Groups. Springer LNM 1625, 1996

Lazard, M.

[125] Sur les groupes nilpotents et les anneaux de Lie. Ann. Ec. Norm. Sup. **71** (1954) 101–190
[126] Groupes analytiques p-adiques. Publ. Math. I.H.E.S. **26** (1965) 389–603

Lubotzky, A.

[127] Group representations, p-adic analytic groups and lattices in $SL_2\mathbb{C}$. Ann. of Math. **118** (1983) 115–130

MacLane, S.

[128] Homology. Springer 1967

Matsumura, H.

[129] Commutative Ring Theory. Cambridge University Press 1986

Maus, E.

[130] Über die Verteilung der Grundverzweigungszahlen von wild verzweigten Erweiterungen p-adischer Zahlkörper. J. reine u. angew. Math. **257** (1972) 47–79
[131] Relationen in Verzweigungsgruppen. J. reine u. angew. Math. **258** (1973) 23–50

Mazur, B.

[132] Rational points of abelian varieties with values in towers of number fields. Invent. Math. **18** (1972) 183–266

Mazur, B., Tate, J., Teitelbaum, J.

[133] On p-adic analogues of the conjectures of Birch and Swinnerton-Dyer. Invent. Math. **84** (1986) 1–48

Mazur, B., Wiles, A.

[134] Class fields of abelian extensions of $\mathbb{Q}$. Invent. Math. **76** (1984) 179–330

Melnikov, O.V.

[135] Subgroups and homology of free products of profinite groups (in Russian). Izv. Akad. Nauk SSSR **53** (1989) No. 1. English translation in Math. USSR Izv. **34** (1990) No. 1 97–119

Melnikov, O.V., Sharomet, A.A.

[136] The Galois groups of a multidimensional local field of positive characteristic (in Russian). Math. Sborn. **180** (1989), No. 8 English translation in Math. USSR Sborn. **67** (1990), No. 2 595–610

Merkur'ev, A.S.

[137] K_2 of fields and the Brauer group. Contemporary Math. **55** Part 2, AMS 1986

Merkur'ev, A.S., Suslin, A.A.

[138] K-cohomology of Severi-Brauer varieties and the norm residue homomorphism. Izv. Akad. Nauk. SSSR **46** (1982) 1011–1046. English translation in Math. USSR Izv. **21** (1983) 307–340

Milne, J.S.

[139] Etale Cohomology. Princeton Univ. Press 1980
[140] Abelian varieties. Chapter V in [31], 103–150
[141] Jacobian varieties. Chapter VII in [31], 167–212

Milnor, J.

[142] Algebraic K-theory and quadratic forms. Invent. Math. **9** (1969/70) 318–344

Mochizuki, S.

[143] The profinite Grothendieck conjecture for closed hyperbolic curves over number fields. J. Math. Sci. Tokyo **3** (1996) 571–627
[144] A version of the Grothendieck conjecture for p-adic local fields. Int. J. Math. **8** (1997) 499–506
[145] The Local Pro-p Anabelian Geometry of Curves. Invent. Math. **138** (1999) 319–423

Morris, S.A.

[146] Pontryagin Duality and the Structure of Locally Compact Abelian Groups. LMS Lecture Notes Series **29**, Cambridge University Press 1977

Moser, Th.

[147] G-Modulationen und Reziprozität. Diplomarbeit, Regensburg 1992

Mukhamedov, V.G.

[148] Local extensions associated with the ℓ-extensions of number fields with restricted ramification (in Russian). Mat. Zametki **35** (1984) No.4 481–490, English translation in Math. Notes **35** No.3–4 253–258

Nakamura, H.

[149] Rigidity of arithmetic fundamental group of a punctured projective line. J. reine u. angew. Math. **405** (1990) 117–130

Neeman, A.

[150] A counterexample to a 1961 "theorem" in homological algebra. Invent. Math. **148** (2002) 397–420

Neukirch, J.

[151] Klassenkörpertheorie. Bibliographisches Institut, Mannheim 1969
[152] Kennzeichnung der p-adischen und der algebraischen Zahlkörper. Invent. Math. **6** (1969) 296–314
[153] Kennzeichnung der endlich-algebraischen Zahlkörper durch die Galoisgruppe der maximalen auflösbaren Erweiterungen. J. reine u. angew. Math. **238** (1970) 135–147
[154] Freie Produkte pro-endlicher Gruppen und ihre Kohomologie. Archiv der Math. **22** (1971) 337–357
[155] Einbettungsprobleme mit lokaler Vorgabe und freie Produkte lokaler Galoisgruppen. J. reine u. angew. Math. **259** (1973) 1–47
[156] Über das Einbettungsproblem in der algebraischen Zahlentheorie. Invent. Math. **21** (1973) 59–116
[157] Über die absoluten Galoisgruppen algebraischer Zahlkörper. Soc.Math. de France, Astérisque **41–42** (1977) 67–79
[158] On solvable number fields. Invent. Math. **53** (1979) 135–164
[159] Class Field Theory. Springer 1986
[160] Algebraische Zahlentheorie. Springer 1992, English translation: Algebraic Number Theory. Springer 1999
[161] Micro primes. Math. Ann. **298** (1994) 629–666

Neumann, O.

[162] On p-closed algebraic number fields with restricted ramification. Math. USSR Izv. **9** (1975) No.2 243–254
[163] On p-closed number fields and an analogue of Riemann's existence theorem. In: A.Fröhlich (ed.): Algebraic Number Fields. Academic Press London 1977 625–647

Oda, T.

[164] A note on ramification of the Galois representation on the fundamental group of an algebraic curve II. J. Number Theory **53** (1995) 342–355

Ore, O.

[165] Contributions to the theory of groups of finite order. Duke Math. J. **5** (1939) 431–460

Patterson, S.J.

[166] An introduction to the theory of the Riemann Zeta-Function. Cambridge University Press 1989

Pletch, A.

[167] Profinite duality groups I and II. J. Pure Applied Algebra **16** (1980) 55–74 and 285–297

Poitou, G.

[168] Remarques sur l'homologie des groupes profinis. Colloques Int. CNRS **143** (1966) 201–213
[169] Cohomologie galoisienne des modules finis. Dunod, Paris 1967

Pontryagin, L.S.

[170] Topological Groups. Gordon and Breach, New York, London, Paris 1966

Pop, F.

[171] Étale Galois covers of affine smooth curves. Invent. Math. **120** (1995) 555–578
[172] On Grothendieck's conjecture of birational anabelian geometry. Ann. of Math. **138** (1994) 145–182
[173] On Grothendieck's conjecture of birational anabelian geometry II. Preprint 1994
[174] Embedding problems over large fields. Ann. of Math. **144** (1996) 1–34
[175] Alterations and birational anabelian geometry. Resolution of singularities (Obergurgl, 1997), Progr. Math. 181, Birkhäuser Basel 2000 519–532

Raynaud, M.

[176] Section des fibrés vectoriels sur une courbe. Bull. Soc. Math. France **110** (1982) 103–125

Reichardt, H.

[177] Konstruktion von Zahlkörpern mit gegebener Galoisgruppe von Primzahlpotenzordnung. J. reine u. angew. Math. **177** (1937) 1–5

Ribes, L.

[178] On a cohomology theory of pairs of groups. Proc. AMS **21** (1969) 230–234
[179] Amalgamated products of profinite groups. Math. Zeitschrift **123** (1971) 357–364

Ribet, K.

[180] A modular construction of unramified p-extensions of $\mathbb{Q}(\mu_p)$. Invent. Math. **34** (1976) 151–162

Riehm, C.

[181] The construction of algebraic structures. Invent. Math. **11** (1970) 73–198

Roos, J.-E.

[182] Sur les foncteurs dérivés de $\varprojlim$, Applications. C. R. Acad. Sci. Ser. I **252** (1961) 3702–3704

Roquette, P.

[183] On class field towers. Chapter IX in [23], 231–249

Rubin, K.

[184] The main conjecture. Appendix to [122]
[185] The "main conjectures" of Iwasawa theory for imaginary quadratic fields. Invent. Math. **103** (1991) 25–68

Rudin, W.

[186] Fourier Analysis on Groups. Interscience Publishers 1967

Šafarevič, I.R.

[187] On the construction of fields with a given Galois group of order ℓ^α (in Russian). Izv. Akad. Nauk SSSR **18** (1954) 261–296, English translation in Amer. Math. Soc. Transl. **4** (1956) 107–142

[188] On an existence theorem in the theory of algebraic numbers (in Russian). Izv. Akad. Nauk. SSSR **18** (1954) 327–334, English translation in Amer. Math. Soc. Transl. **4** (1956) 143–150

[189] On the problem of imbedding fields (in Russian). Izv. Akad. Nauk SSSR **18** (1954) 389–418, English translation in Amer. Math. Soc. Transl. **4** (1956) 151–183

[190] Construction of fields of algebraic numbers with given solvable groups (in Russian). Izv. Akad. Nauk. SSSR **18** (1954) 525–578, English translation in Amer. Math. Soc. Transl. **4** (1956) 185–237

[191] Extensions with given ramification points (in Russian). Publ. Math. IHES **18** (1963) 295–319, English translation in Amer. Math. Soc. Transl. **59** (1966) 128–149

[192] Factors of a decreasing central series (in Russian). Mat. Zametki **45** (1989) 114–117, English translation in Math. Notes **45** (1989) 262–264

Schmidt, A.

[193] Extensions with restricted ramification and duality for arithmetic schemes. Comp. Math. **100** (1996) 233–245

[194] Bounded defect in partial Euler characteristics. Bull. London Math. Soc. **28** (1996) 463–464

[195] An arithmetic site for the ring of integers of algebraic number fields. Invent. Math. **123** (1996) 575–610

[196] On Poitou's duality theorem. J. reine u. angew. Math. **517** (1999) 145–160

[197] On the relation between 2 and ∞ in Galois cohomology of number fields. Comp. Math. **133** (2002) 267–288

[198] Circular sets of prime numbers and p-extensions of the rationals. J. reine u. angew. Math. **596** (2006) 115–130

[199] Rings of integers of type $K(\pi, 1)$. Doc. Math. **12** (2007) 441–471

Schmidt, F.K.

[200] Mehrfach perfekte Körper. Math. Ann. **108** (1933) 1–25

Schneider, P.

[201] Über gewisse Galoiscohomologiegruppen. Math. Z. **168** (1979) 181–205

[202] Die Galoiscohomologie p-adischer Darstellungen über Zahlkörpern. Dissertation Regensburg 1980

[203] p-adic height pairings I and II. Invent. Math. **69** (1982) 401–409 and Invent. Math. **79** (1985) 329–374

[204] Motivic Iwasawa theory. Advanced Studies in Pure Mathematics **17** (1989) 421–456

Schneps, L., Lochak, P. (ed.)

[205] Geometric Galois Actions I. London Math. Soc. Lect. Notes **242** Cambridge 1997

Scholz, A.

[206] Konstruktion algebraischer Zahlkörper mit beliebiger Gruppe von Primzahlpotenzordnung. Math. Z. **42** (1936) 161–188

Schoof, R.

[207] Infinite Class Field Towers of Quadratic Fields. J. reine u. angew. Math. **372** (1986) 209–220

Serre, J.-P.

[208] Groupes algébriques et corps de classes. Hermann, Paris 1959

[209] Structure de certains pro-p-groupes. Séminaire Bourbaki (1962/63) Exp. 252

[210] Cohomologie Galoisienne. Lecture Notes in Mathematics **5**, Springer 1964 (Cinquième édition 1994)

[211] Sur la dimension cohomologique des groupes profinis. Topology **3** (1965) 413–420

[212] Corps Locaux. Hermann, Paris 1968. English translation: Local Fields. Springer 1995

[213] A Course in Arithmetic. Springer 1973

[214] Linear Representations of Finite Groups. Springer 1977

[215] Trees, Springer 1980

[216] Topics in Galois Theory. Jones and Bartlett Publ. Boston 1992

Sinnott, W.

[217] On the Stickelberger ideal and the circular units of abelian number fields. Invent. Math. **62** (1980) 181–234

[218] On the μ-invariant of the Γ-transform of a rational function. Invent. Math. **75** (1984) 273–282

Soulé, C.

[219] K-théorie des anneaux d'entiers de corps de nombres et cohomologie étale. Invent. Math. **55** (1979) 251–295

Stammbach, U.

[220] Cohomological characterisations of finite solvable and nilpotent groups. J. Pure and Appl. Alg. **11** (1977) 293-301

Stix, J.

[221] Affine anabelian curves in positive characteristic. Compositio Math. **134** (2002) 75–85

[222] Projective anabelian curves in positive characteristic and descent theory for log-étale covers. Dissertation, Bonner Mathematische Schriften **354** (2002)

Suslin, A.A.

[223] Algebraic K-theory and the norm residue homomorphism. J. Soviet. Math. **30** (1985) 2556–2611

Szamuely, T.

[224] Groupes de Galois de corps de type fini (d'après Pop). Astérisque No. 294 (2004), ix, 403–431

Takahashi, T.

[225] Galois cohomology of unramified extensions of algebraic function fields. Tôhoku Math. J. **24** (1972) 33–39

[226] Galois cohomology of finitely generated modules. Tôhoku Math. J. **21** (1969) 102–111

Tamagawa, A.

[227] The Grothendieck conjecture for affine curves. Comp. Math. **109** (1997) 135–194

Tamme, G.

[228] Introduction to Étale Cohomology. Springer 1994

Tate, J.

[229] Duality theorems in Galois cohomology over number fields. Proc. Int. Congress Stockholm 1962
[230] Letter to Serre. Annexe to Chap.I in [210]
[231] Symbols in arithmetic. Proc. Int. Congress Nice 1970, Vol.1, 201–211
[232] Relations between K_2 and Galois cohomology. Invent. Math. **36** (1976) 257–274

Taya, H.

[233] Computation of $\mathbb{Z}_3$-invariants of real quadratic fields. Math. Comp. **65** (1996) 779–784

Terjanian, G.

[234] Un contre-exemple à une conjecture d'Artin. C. R. Acad. Sci. Paris **262** (1966) p. 612

Thévenaz, J. and *Webb, P.*

[235] The structure of Mackey functors. Trans. AMS. **347** (1995) 1865–1961

Uchida, K.

[236] On Tate's duality theorems in Galois cohomology. Tôhoku Math. J. **21**, No. 4 (1969) 92–101
[237] Isomorphisms of Galois groups. J. Math. Soc. Japan **28** (1976) 617–620
[238] Isomorphisms of Galois groups of algebraic function fields. Ann. of Math. **106** (1977) 589–598

van der Kallen, W.

[239] The Merkurjev-Suslin theorem. In I. Reiner, K. W. Roggenkamp (ed.): Orders and their Applications. Lecture Notes in Math. **1142**, Springer 1985, pp. 157–168

Verdier, J.L.

[240] Dualité dans la cohomologie des groupes profinis. In [210].

Vinberg, E.B.

[241] On the dimension theorem for associative algebras (in Russian) Izv. Akad. Nauk. SSSR **29** (1965) 209–214

Voevodsky, V.

[242] Motivic cohomology with $\mathbb{Z}/2$-coefficients. Publ. Math. Inst. Hautes Études Sci. **98** (2003) 59–104
[243] Reduced power operations in motivic cohomology. Publ. Math. Inst. Hautes Études Sci. **98** (2003) 1–57

Waldschmidt, M.

[244] Transcendance et exponentielles en plusieurs variables. Invent. Math. **63** (1981) 97–127

Washington, L.C.

[245] Class numbers and $\mathbb{Z}_p$-extensions. Math. Ann. **214** (1975) 177–193
[246] Introduction to Cyclotomic Fields. Springer 1982 (Second ed. 1997)

Weigel, Th., Zalesskii, P.A.

[247] Profinite groups of finite cohomological dimension. C. R. Acad. Sci. Paris **338** (2004) 353–358

Wiese, G.

[248] Zur Fontaine-Mazur-Vermutung für Erweiterungen mit beschränkter Verzweigung von CM-Körpern. Diplomarbeit, Heidelberg 2002

Wiles, A.

[249] The Iwasawa conjecture for totally real fields. Ann. of Math. **131** (1990) 493–540

Wingberg, K.

[250] Der Eindeutigkeitssatz für Demuškin-Formationen. Invent. Math. **70** (1982) 19–113

[251] Ein Analogon zur Fundamentalgruppe einer Riemannschen Fläche im Zahlkörperfall. Invent. Math. **77** (1984) 557–584

[252] Duality theorems for Γ-extensions of algebraic number fields. Comp. Math. **55** (1985) 333–381

[253] On Poincaré groups. J. London Math. Soc. **33** (1986) 271–278

[254] On Galois groups of p-closed algebraic number fields with restricted ramification. J. reine u. angew. Math. **400** (1989) 185–202

[255] On Demuškin groups with involution. Ann. Sci. Éc. Norm. Sup. **22** (1989) 555–567

[256] On the maximal unramified extension of an algebraic number field. J. reine u. angew. Math. **440** (1993) 129–156

[257] On the Fontaine-Mazur conjecture for CM-fields. Comp. Math. **131** (2002) 341–354

[258] Free Product Decomposition of Galois Groups of Number Fields. Publ. RIMS, Kyoto Univ. **41** (2005) 295–306

Wintenberger, J.-P.

[259] Le corps des normes de certaines extensions infinies de corps locaux; applications. Ann. Sci. Éc. Norm. Sup. **16** (1983) 59–89

Witt, E.

[260] Treue Darstellung Liescher Ringe. J. reine u. angew. Math. **177** (1937) 152–160

Zalesskii, P.A.

[261] Open subgroups of free profinite products. Contemp. Math. **131** (1992) 473–492

Additional References

Arkhipov, G.I., Karatsuba, A.A.

[AK] Local representation of zero by a form. (Russian) Izv. Akad. Nauk SSSR Ser. Mat. **45** (1981) 948–961. English translation in Math. USSR, Izv. **19** (1982) 231–240

Chenevier, G., Clozel, L.

[CC] Corps de nombres peu ramifiés et formes automorphes autoduales. J. Amer. Math. Soc. **22** (2009), no. 2, 467–519

Colliot-Thélène, J.-L., Sansuc, J.-J.

[CS] Principal homogeneous spaces under flasque tori: applications. J. Algebra **106** (1987), no. 1, 148–205

Merkur'ev, A.S., Suslin, A.A.

[MS] Norm residue homomorphism of degree three. Izv. Akad. Nauk SSSR Ser. Mat. 54 (1990), no. 2, 339–356; translation in Math. USSR Izv. 36 (1991), no. 2, 349–367

Rost, M.

[Ro] Hilbert 90 for K_3 for degree-two extensions. Preprint 1986 (available on M. Rost's homepage)

Schmidt, A., Wingberg, K.

[SW] Extensions of profinite duality groups. arXiv:0809.4371 (2008)

Voevodsky, V.

[Vo] On motivic cohomology with Z/l-coeffcients. Ann. of Math. **174** (2011), 401–438

Index

J. Neukirch et al., *Cohomology of Number Fields*,
Grundlehren der mathematischen Wissenschaften 323,
DOI 10.1007/978-3-540-37889-1, © Springer-Verlag Berlin Heidelberg 2008

The manufacturer's authorised representative in the EU is Springer Nature Customer Service Centre GmbH, Europaplatz 3, 69115 Heidelberg, Germany. If you have any concerns regarding our products, please contact ProductSafety@springernature.com

Printed and bound by CPI Group (UK) Ltd, Croydon, CR0 4YY
25/09/2026
02222831-0014